Spoken Language Processing

Spoken Language Processing

A Guide to Theory, Algorithm, and System Development

Xuedong Huang

Alex Acero

Hsiao-Wuen Hon

Microsoft Research

Prentice Hall PTR
Upper Saddle River, New Jersey 07458
www.phptr.com

Library of Congress Cataloging-in-Publication Data

Huang, Xuedong.
Spoken language processing: a guide to theory, algorithm, and system development/ Xuedong Huang, Alex Acero, Hsiao-Wuen Hon.
p. cm.
Includes bibliographical references and index.
ISBN 0-13-022616-5
1. Natural language processing (Computer science) I. Acero, Alex. II. Hon, Hsiao-Wuen. III. Title.

QA76.9.N38 H83 2001
006.3'5—dc21 00-050196

Editorial/production supervision: *Jane Bonnell*
Cover design director: *Jerry Votta*
Cover design: *Anthony Gemmellaro*
Manufacturing buyer: *Maura Zaldivar*
Development editor: *Russ Hall*
Acquisitions editor: *Tim Moore*
Editorial assistant: *Allyson Kloss*
Marketing manager: *Debby van Dijk*

Prentice-Hall, Inc.
Upper Saddle River, New Jersey 07458

Printed in the United States of America
10 9 8 7 6 5 4 3 2 1

ISBN 0-13-022616-5

Prentice-Hall International (UK) Limited, *London*
Prentice-Hall of Australia Pty. Limited, *Sydney*
Prentice-Hall Canada Inc., *Toronto*
Prentice-Hall Hispanoamericana, S.A., *Mexico*
Prentice-Hall of India Private Limited, *New Delhi*
Prentice-Hall of Japan, Inc., *Tokyo*
Pearson Education Asia Pte. Ltd.
Editora Prentice-Hall do Brasil, Ltda., *Rio de Janeiro*

To Yingzhi, Angela, Christina, and Derek

To Donna and Nicolas

To Phen, Stephanie, and Jacqueline

Contents

FOREWORD xxi

PREFACE xxv

1. INTRODUCTION 1

1.1. MOTIVATIONS 2
1.1.1. Spoken Language Interface 2
1.1.2. Speech-to-Speech Translation 3
1.1.3. Knowledge Partners 3
1.2. SPOKEN LANGUAGE SYSTEM ARCHITECTURE 4
1.2.1. Automatic Speech Recognition 4
1.2.2. Text-to-Speech Conversion 6
1.2.3. Spoken Language Understanding 7
1.3. BOOK ORGANIZATION 8
1.3.1. Part I: Fundamental Theory 9
1.3.2. Part II: Speech Processing 9
1.3.3. Part III: Speech Recognition 9
1.3.4. Part IV: Text-to-Speech Systems 10
1.3.5. Part V: Spoken Language Systems 10
1.4. TARGET AUDIENCES 10
1.5. HISTORICAL PERSPECTIVE AND FURTHER READING 11

PART I: FUNDAMENTAL THEORY

2. SPOKEN LANGUAGE STRUCTURE 19

2.1. SOUND AND HUMAN SPEECH SYSTEMS 21
2.1.1. Sound 21
2.1.2. Speech Production 24
2.1.3. Speech Perception 29

2.2. PHONETICS AND PHONOLOGY 36
2.2.1. *Phonemes* 36
2.2.2. *The Allophone: Sound and Context* 47
2.2.3. *Speech Rate and Coarticulation* 49
2.3. SYLLABLES AND WORDS 51
2.3.1. *Syllables* 51
2.3.2. *Words* 53
2.4. SYNTAX AND SEMANTICS 58
2.4.1. *Syntactic Constituents* 58
2.4.2. *Semantic Roles* 63
2.4.3. *Lexical Semantics* 64
2.4.4. *Logical Form* 67
2.5. HISTORICAL PERSPECTIVE AND FURTHER READING 68

3. PROBABILITY, STATISTICS, AND INFORMATION THEORY 73

3.1. PROBABILITY THEORY 74
3.1.1. *Conditional Probability and Bayes' Rule* 75
3.1.2. *Random Variables* 77
3.1.3. *Mean and Variance* 79
3.1.4. *Covariance and Correlation* 82
3.1.5. *Random Vectors and Multivariate Distributions* 83
3.1.6. *Some Useful Distributions* 85
3.1.7. *Gaussian Distributions* 92
3.2. ESTIMATION THEORY 98
3.2.1. *Minimum/Least Mean Squared Error Estimation* 99
3.2.2. *Maximum Likelihood Estimation* 104
3.2.3. *Bayesian Estimation and MAP Estimation* 107
3.3. SIGNIFICANCE TESTING 113
3.3.1. *Level of Significance* 114
3.3.2. *Normal Test (Z-Test)* 115
3.3.3. χ^2 *Goodness-of-Fit Test* 116
3.3.4. *Matched-Pairs Test* 118
3.4. INFORMATION THEORY 120
3.4.1. *Entropy* 120
3.4.2. *Conditional Entropy* 123
3.4.3. *The Source Coding Theorem* 124
3.4.4. *Mutual Information and Channel Coding* 126
3.5. HISTORICAL PERSPECTIVE AND FURTHER READING 128

4. PATTERN RECOGNITION 133

4.1. BAYES' DECISION THEORY 134
4.1.1. *Minimum-Error-Rate Decision Rules* 135

4.1.2. *Discriminant Functions* 138
4.2. HOW TO CONSTRUCT CLASSIFIERS 140
4.2.1. *Gaussian Classifiers* 142
4.2.2. *The Curse of Dimensionality* 144
4.2.3. *Estimating the Error Rate* 146
4.2.4. *Comparing Classifiers* 148
4.3. DISCRIMINATIVE TRAINING 150
4.3.1. *Maximum Mutual Information Estimation* 150
4.3.2. *Minimum-Error-Rate Estimation* 156
4.3.3. *Neural Networks* 158
4.4. UNSUPERVISED ESTIMATION METHODS 163
4.4.1. *Vector Quantization* 163
4.4.2. *The EM Algorithm* 170
4.4.3. *Multivariate Gaussian Mixture Density Estimation* 172
4.5. CLASSIFICATION AND REGRESSION TREES 175
4.5.1. *Choice of Question Set* 177
4.5.2. *Splitting Criteria* 178
4.5.3. *Growing the Tree* 181
4.5.4. *Missing Values and Conflict Resolution* 182
4.5.5. *Complex Questions* 182
4.5.6. *The Right-Sized Tree* 184
4.6. HISTORICAL PERSPECTIVE AND FURTHER READING 190

PART II: SPEECH PROCESSING

5. DIGITAL SIGNAL PROCESSING 201
5.1. DIGITAL SIGNALS AND SYSTEMS 202
5.1.1. *Sinusoidal Signals* 203
5.1.2. *Other Digital Signals* 206
5.1.3. *Digital Systems* 206
5.2. CONTINUOUS-FREQUENCY TRANSFORMS 208
5.2.1. *The Fourier Transform* 208
5.2.2. *Z-Transform* 211
5.2.3. *Z-Transforms of Elementary Functions* 212
5.2.4. *Properties of the Z- and Fourier Transforms* 215
5.3. DISCRETE-FREQUENCY TRANSFORMS 216
5.3.1. *The Discrete Fourier Transform (DFT)* 218
5.3.2. *Fourier Transforms of Periodic Signals* 219
5.3.3. *The Fast Fourier Transform (FFT)* 222
5.3.4. *Circular Convolution* 227
5.3.5. *The Discrete Cosine Transform (DCT)* 228

5.4. DIGITAL FILTERS AND WINDOWS ... 229
5.4.1. *The Ideal Low-Pass Filter* ... 229
5.4.2. *Window Functions* ... 230
5.4.3. *FIR Filters* ... 232
5.4.4. *IIR Filters* ... 238
5.5. DIGITAL PROCESSING OF ANALOG SIGNALS ... 242
5.5.1. *Fourier Transform of Analog Signals* ... 243
5.5.2. *The Sampling Theorem* ... 243
5.5.3. *Analog-to-Digital Conversion* ... 245
5.5.4. *Digital-to-Analog Conversion* ... 246
5.6. MULTIRATE SIGNAL PROCESSING ... 248
5.6.1. *Decimation* ... 248
5.6.2. *Interpolation* ... 249
5.6.3. *Resampling* ... 250
5.7. FILTERBANKS ... 251
5.7.1. *Two-Band Conjugate Quadrature Filters* ... 251
5.7.2. *Multiresolution Filterbanks* ... 254
5.7.3. *The DFT as a Filterbank* ... 255
5.7.4. *Modulated Lapped Transforms* ... 258
5.8. STOCHASTIC PROCESSES ... 260
5.8.1. *Statistics of Stochastic Processes* ... 261
5.8.2. *Stationary Processes* ... 264
5.8.3. *LTI Systems with Stochastic Inputs* ... 267
5.8.4. *Power Spectral Density* ... 268
5.8.5. *Noise* ... 269
5.9. HISTORICAL PERSPECTIVE AND FURTHER READING ... 270

6. SPEECH SIGNAL REPRESENTATIONS ... 275

6.1. SHORT-TIME FOURIER ANALYSIS ... 276
6.1.1. *Spectrograms* ... 281
6.1.2. *Pitch-Synchronous Analysis* ... 283
6.2. ACOUSTICAL MODEL OF SPEECH PRODUCTION ... 283
6.2.1. *Glottal Excitation* ... 284
6.2.2. *Lossless Tube Concatenation* ... 284
6.2.3. *Source-Filter Models of Speech Production* ... 288
6.3. LINEAR PREDICTIVE CODING ... 290
6.3.1. *The Orthogonality Principle* ... 291
6.3.2. *Solution of the LPC Equations* ... 292
6.3.3. *Spectral Analysis via LPC* ... 300
6.3.4. *The Prediction Error* ... 301
6.3.5. *Equivalent Representations* ... 303

6.4. CEPSTRAL PROCESSING 306
6.4.1. The Real and Complex Cepstrum 307
6.4.2. Cepstrum of Pole-Zero Filters 308
6.4.3. Cepstrum of Periodic Signals 311
6.4.4. Cepstrum of Speech Signals 312
6.4.5. Source-Filter Separation via the Cepstrum 314
6.5. PERCEPTUALLY MOTIVATED REPRESENTATIONS 315
6.5.1. The Bilinear Transform 315
6.5.2. Mel-Frequency Cepstrum 316
6.5.3. Perceptual Linear Prediction (PLP) 318
6.6. FORMANT FREQUENCIES 319
6.6.1. Statistical Formant Tracking 320
6.7. THE ROLE OF PITCH 324
6.7.1. Autocorrelation Method 324
6.7.2. Normalized Cross-Correlation Method 327
6.7.3. Signal Conditioning 329
6.7.4. Pitch Tracking 330
6.8. HISTORICAL PERSPECTIVE AND FURTHER READING 332

7. SPEECH CODING 337

7.1. SPEECH CODERS ATTRIBUTES 338
7.2. SCALAR WAVEFORM CODERS 340
7.2.1. Linear Pulse Code Modulation (PCM) 340
7.2.2. μ-law and A-law PCM 342
7.2.3. Adaptive PCM 344
7.2.4. Differential Quantization 345
7.3. SCALAR FREQUENCY DOMAIN CODERS 348
7.3.1. Benefits of Masking 349
7.3.2. Transform Coders 350
7.3.3. Consumer Audio 351
7.3.4. Digital Audio Broadcasting (DAB) 352
7.4. CODE EXCITED LINEAR PREDICTION (CELP) 353
7.4.1. LPC Vocoder 353
7.4.2. Analysis by Synthesis 353
7.4.3. Pitch Prediction: Adaptive Codebook 356
7.4.4. Perceptual Weighting and Postfiltering 357
7.4.5. Parameter Quantization 358
7.4.6. CELP Standards 359
7.5. LOW-BIT RATE SPEECH CODERS 361
7.5.1. Mixed-Excitation LPC Vocoder 362
7.5.2. Harmonic Coding 363
7.5.3. Waveform Interpolation 367
7.6. HISTORICAL PERSPECTIVE AND FURTHER READING 371

PART III: SPEECH RECOGNITION

8. HIDDEN MARKOV MODELS 377

8.1. THE MARKOV CHAIN 378
8.2. DEFINITION OF THE HIDDEN MARKOV MODEL 380
8.2.1. Dynamic Programming and DTW 383
8.2.2. How to Evaluate an HMM—The Forward Algorithm 385
8.2.3. How to Decode an HMM—The Viterbi Algorithm 387
8.2.4. How to Estimate HMM Parameters—Baum-Welch Algorithm 389
8.3. CONTINUOUS AND SEMICONTINUOUS HMMS 394
8.3.1. Continuous Mixture Density HMMs 394
8.3.2. Semicontinuous HMMs 396
8.4. PRACTICAL ISSUES IN USING HMMS 398
8.4.1. Initial Estimates 398
8.4.2. Model Topology 399
8.4.3. Training Criteria 401
8.4.4. Deleted Interpolation 401
8.4.5. Parameter Smoothing 403
8.4.6. Probability Representations 404
8.5. HMM LIMITATIONS 405
8.5.1. Duration Modeling 406
8.5.2. First-Order Assumption 408
8.5.3. Conditional Independence Assumption 409
8.6. HISTORICAL PERSPECTIVE AND FURTHER READING 409

9. ACOUSTIC MODELING 415

9.1. VARIABILITY IN THE SPEECH SIGNAL 416
9.1.1. Context Variability 417
9.1.2. Style Variability 418
9.1.3. Speaker Variability 418
9.1.4. Environment Variability 419
9.2. HOW TO MEASURE SPEECH RECOGNITION ERRORS 419
9.3. SIGNAL PROCESSING—EXTRACTING FEATURES 421
9.3.1. Signal Acquisition 422
9.3.2. End-Point Detection 422
9.3.3. MFCC and Its Dynamic Features 424
9.3.4. Feature Transformation 426
9.4. PHONETIC MODELING—SELECTING APPROPRIATE UNITS 428
9.4.1. Comparison of Different Units 429
9.4.2. Context Dependency 430
9.4.3. Clustered Acoustic-Phonetic Units 432
9.4.4. Lexical Baseforms 436

9.5. ACOUSTIC MODELING—SCORING ACOUSTIC FEATURES 439
9.5.1. Choice of HMM Output Distributions 439
9.5.2. Isolated vs. Continuous Speech Training 441
9.6. ADAPTIVE TECHNIQUES—MINIMIZING MISMATCHES 444
9.6.1. Maximum a Posteriori (MAP) 445
9.6.2. Maximum Likelihood Linear Regression (MLLR) 447
9.6.3. MLLR and MAP Comparison 450
9.6.4. Clustered Models 452
9.7. CONFIDENCE MEASURES: MEASURING THE RELIABILITY 453
9.7.1. Filler Models 453
9.7.2. Transformation Models 454
9.7.3. Combination Models 456
9.8. OTHER TECHNIQUES 457
9.8.1. Neural Networks 457
9.8.2. Segment Models 459
9.9. CASE STUDY: WHISPER 464
9.10. HISTORICAL PERSPECTIVE AND FURTHER READING 465

10. ENVIRONMENTAL ROBUSTNESS 477

10.1. THE ACOUSTICAL ENVIRONMENT 478
10.1.1. Additive Noise 478
10.1.2. Reverberation 480
10.1.3. A Model of the Environment 482
10.2. ACOUSTICAL TRANSDUCERS 486
10.2.1. The Condenser Microphone 486
10.2.2. Directionality Patterns 489
10.2.3. Other Transduction Categories 496
10.3. ADAPTIVE ECHO CANCELLATION (AEC) 497
10.3.1. The LMS Algorithm 499
10.3.2. Convergence Properties of the LMS Algorithm 500
10.3.3. Normalized LMS Algorithm 501
10.3.4. Transform-Domain LMS Algorithm 502
10.3.5. The RLS Algorithm 503
10.4. MULTIMICROPHONE SPEECH ENHANCEMENT 504
10.4.1. Microphone Arrays 505
10.4.2. Blind Source Separation 510
10.5. ENVIRONMENT COMPENSATION PREPROCESSING 515
10.5.1. Spectral Subtraction 516
10.5.2. Frequency-Domain MMSE from Stereo Data 519
10.5.3. Wiener Filtering 520
10.5.4. Cepstral Mean Normalization (CMN) 522
10.5.5. Real-Time Cepstral Normalization 525
10.5.6. The Use of Gaussian Mixture Models 525

10.6. ENVIRONMENTAL MODEL ADAPTATION 528
10.6.1. Retraining on Corrupted Speech 528
10.6.2. Model Adaptation 530
10.6.3. Parallel Model Combination 531
10.6.4. Vector Taylor Series 535
10.6.5. Retraining on Compensated Features 537
10.7. MODELING NONSTATIONARY NOISE 538
10.8. HISTORICAL PERSPECTIVE AND FURTHER READING 540

11. LANGUAGE MODELING 545

11.1. FORMAL LANGUAGE THEORY 546
11.1.1. Chomsky Hierarchy 547
11.1.2. Chart Parsing for Context-Free Grammars 549
11.2. STOCHASTIC LANGUAGE MODELS 554
11.2.1. Probabilistic Context-Free Grammars 554
11.2.2. N-gram Language Models 558
11.3. COMPLEXITY MEASURE OF LANGUAGE MODELS 560
11.4. N-GRAM SMOOTHING 562
11.4.1. Deleted Interpolation Smoothing 564
11.4.2. Backoff Smoothing 565
11.4.3. Class N-grams 570
11.4.4. Performance of N-gram Smoothing 573
11.5. ADAPTIVE LANGUAGE MODELS 574
11.5.1. Cache Language Models 574
11.5.2. Topic-Adaptive Models 575
11.5.3. Maximum Entropy Models 576
11.6. PRACTICAL ISSUES 578
11.6.1. Vocabulary Selection 578
11.6.2. N-gram Pruning 580
11.6.3. CFG vs. N-gram Models 581
11.7. HISTORICAL PERSPECTIVE AND FURTHER READING 584

12. BASIC SEARCH ALGORITHMS 591

12.1. BASIC SEARCH ALGORITHMS 592
12.1.1. General Graph Searching Procedures 593
12.1.2. Blind Graph Search Algorithms 597
12.1.3. Heuristic Graph Search 601
12.2. SEARCH ALGORITHMS FOR SPEECH RECOGNITION 608
12.2.1. Decoder Basics 609
12.2.2. Combining Acoustic and Language Models 610
12.2.3. Isolated Word Recognition 610
12.2.4. Continuous Speech Recognition 611

12.3. LANGUAGE MODEL STATES 613
12.3.1. Search Space with FSM and CFG 613
12.3.2. Search Space with the Unigram 616
12.3.3. Search Space with Bigrams 617
12.3.4. Search Space with Trigrams 619
12.3.5. How to Handle Silences Between Words 621
12.4. TIME-SYNCHRONOUS VITERBI BEAM SEARCH 622
12.4.1. The Use of Beam 624
12.4.2. Viterbi Beam Search 625
12.5. STACK DECODING (A^* SEARCH) 626
12.5.1. Admissible Heuristics for Remaining Path 630
12.5.2. When to Extend New Words 631
12.5.3. Fast Match 634
12.5.4. Stack Pruning 638
12.5.5. Multistack Search 639
12.6. HISTORICAL PERSPECTIVE AND FURTHER READING 640

13. LARGE-VOCABULARY SEARCH ALGORITHMS 645

13.1. EFFICIENT MANIPULATION OF A TREE LEXICON 646
13.1.1. Lexical Tree 646
13.1.2. Multiple Copies of Pronunciation Trees 648
13.1.3. Factored Language Probabilities 650
13.1.4. Optimization of Lexical Trees 653
13.1.5. Exploiting Subtree Polymorphism 656
13.1.6. Context-Dependent Units and Inter-Word Triphones 658
13.2. OTHER EFFICIENT SEARCH TECHNIQUES 659
13.2.1. Using Entire HMM as a State in Search 659
13.2.2. Different Layers of Beams 660
13.2.3. Fast Match 661
13.3. *N*-BEST AND MULTIPASS SEARCH STRATEGIES 663
13.3.1. N-best Lists and Word Lattices 664
13.3.2. The Exact N-best Algorithm 666
13.3.3. Word-Dependent N-best and Word-Lattice Algorithm 667
13.3.4. The Forward-Backward Search Algorithm 670
13.3.5. One-Pass vs. Multipass Search 673
13.4. SEARCH-ALGORITHM EVALUATION 674
13.5. CASE STUDY—MICROSOFT WHISPER 676
13.5.1. The CFG Search Architecture 676
13.5.2. The N-gram Search Architecture 677
13.6. HISTORICAL PERSPECTIVE AND FURTHER READING 681

PART IV: TEXT-TO-SPEECH SYSTEMS

14. TEXT AND PHONETIC ANALYSIS 689

14.1. MODULES AND DATA FLOW 690
14.1.1. Modules 692
14.1.2. Data Flows 694
14.1.3. Localization Issues 696
14.2. LEXICON 697
14.3. DOCUMENT STRUCTURE DETECTION 699
14.3.1. Chapter and Section Headers 700
14.3.2. Lists 701
14.3.3. Paragraphs 702
14.3.4. Sentences 702
14.3.5. Email 704
14.3.6. Web Pages 705
14.3.7. Dialog Turns and Speech Acts 705
14.4. TEXT NORMALIZATION 706
14.4.1. Abbreviations and Acronyms 709
14.4.2. Number Formats 712
14.4.3. Domain-Specific Tags 718
14.4.4. Miscellaneous Formats 719
14.5. LINGUISTIC ANALYSIS 720
14.6. HOMOGRAPH DISAMBIGUATION 724
14.7. MORPHOLOGICAL ANALYSIS 725
14.8. LETTER-TO-SOUND CONVERSION 728
14.9. EVALUATION 730
14.10. CASE STUDY: FESTIVAL 732
14.10.1. Lexicon 733
14.10.2. Text Analysis 733
14.10.3. Phonetic Analysis 735
14.11. HISTORICAL PERSPECTIVE AND FURTHER READING 735

15. PROSODY 739

15.1. THE ROLE OF UNDERSTANDING 740
15.2. PROSODY GENERATION SCHEMATIC 743
15.3. SPEAKING STYLE 744
15.3.1. Character 744
15.3.2. Emotion 744
15.4. SYMBOLIC PROSODY 745
15.4.1. Pauses 747
15.4.2. Prosodic Phrases 749

15.4.3. *Accent* 751
15.4.4. *Tone* 753
15.4.5. *Tune* 757
15.4.6. *Prosodic Transcription Systems* 759
15.5. DURATION ASSIGNMENT 761
15.5.1. *Rule-Based Methods* 762
15.5.2. *CART-Based Durations* 763
15.6. PITCH GENERATION 763
15.6.1. *Attributes of Pitch Contours* 764
15.6.2. *Baseline F0 Contour Generation* 768
15.6.3. *Parametric F0 Generation* 774
15.6.4. *Corpus-Based F0 Generation* 778
15.7. PROSODY MARKUP LANGUAGES 783
15.8. PROSODY EVALUATION 784
15.9. HISTORICAL PERSPECTIVE AND FURTHER READING 785

16. SPEECH SYNTHESIS 793

16.1. ATTRIBUTES OF SPEECH SYNTHESIS 794
16.2. FORMANT SPEECH SYNTHESIS 796
16.2.1. *Waveform Generation from Formant Values* 797
16.2.2. *Formant Generation by Rule* 800
16.2.3. *Data-Driven Formant Generation* 803
16.2.4. *Articulatory Synthesis* 803
16.3. CONCATENATIVE SPEECH SYNTHESIS 804
16.3.1. *Choice of Unit* 805
16.3.2. *Optimal Unit String: The Decoding Process* 810
16.3.3. *Unit Inventory Design* 817
16.4. PROSODIC MODIFICATION OF SPEECH 818
16.4.1. *Synchronous Overlap and Add (SOLA)* 818
16.4.2. *Pitch Synchronous Overlap and Add (PSOLA)* 820
16.4.3. *Spectral Behavior of PSOLA* 822
16.4.4. *Synthesis Epoch Calculation* 823
16.4.5. *Pitch-Scale Modification Epoch Calculation* 825
16.4.6. *Time-Scale Modification Epoch Calculation* 826
16.4.7. *Pitch-Scale Time-Scale Epoch Calculation* 827
16.4.8. *Waveform Mapping* 827
16.4.9. *Epoch Detection* 828
16.4.10. *Problems with PSOLA* 829
16.5. SOURCE-FILTER MODELS FOR PROSODY MODIFICATION 831
16.5.1. *Prosody Modification of the LPC Residual* 832
16.5.2. *Mixed Excitation Models* 832
16.5.3. *Voice Effects* 834

16.6. EVALUATION OF TTS SYSTEMS 834
16.6.1. Intelligibility Tests 837
16.6.2. Overall Quality Tests 840
16.6.3. Preference Tests 842
16.6.4. Functional Tests 842
16.6.5. Automated Tests 843
16.7. HISTORICAL PERSPECTIVE AND FURTHER READING 844

PART V: SPOKEN LANGUAGE SYSTEMS

17. SPOKEN LANGUAGE UNDERSTANDING 853
17.1. WRITTEN VS. SPOKEN LANGUAGES 855
17.1.1. Style 856
17.1.2. Disfluency 857
17.1.3. Communicative Prosody 858
17.2. DIALOG STRUCTURE 859
17.2.1. Units of Dialog 860
17.2.2. Dialog (Speech) Acts 861
17.2.3. Dialog Control 866
17.3. SEMANTIC REPRESENTATION 867
17.3.1. Semantic Frames 867
17.3.2. Conceptual Graphs 872
17.4. SENTENCE INTERPRETATION 873
17.4.1. Robust Parsing 874
17.4.2. Statistical Pattern Matching 878
17.5. DISCOURSE ANALYSIS 881
17.5.1. Resolution of Relative Expression 882
17.5.2. Automatic Inference and Inconsistency Detection 885
17.6. DIALOG MANAGEMENT 886
17.6.1. Dialog Grammars 887
17.6.2. Plan-Based Systems 888
17.6.3. Dialog Behavior 892
17.7. RESPONSE GENERATION AND RENDITION 894
17.7.1. Response Content Generation 895
17.7.2. Concept-to-Speech Rendition 899
17.7.3. Other Renditions 901
17.8. EVALUATION 901
17.8.1. Evaluation in the ATIS Task 901
17.8.2. PARADISE Framework 903
17.9. CASE STUDY—DR. WHO 906
17.9.1. Semantic Representation 906
17.9.2. Semantic Parser (Sentence Interpretation) 908

17.9.3. Discourse Analysis ... 909
17.9.4. Dialog Manager ... 910
17.10. HISTORICAL PERSPECTIVE AND FURTHER READING ... 913

18. APPLICATIONS AND USER INTERFACES ... 919

18.1. APPLICATION ARCHITECTURE ... 920
18.2. TYPICAL APPLICATIONS ... 921
18.2.1. Computer Command and Control ... 921
18.2.2. Telephony Applications ... 924
18.2.3. Dictation ... 926
18.2.4. Accessibility ... 929
18.2.5. Handheld Devices ... 930
18.2.6. Automobile Applications ... 930
18.2.7. Speaker Recognition ... 931
18.3. SPEECH INTERFACE DESIGN ... 931
18.3.1. General Principles ... 931
18.3.2. Handling Errors ... 937
18.3.3. Other Considerations ... 941
18.3.4. Dialog Flow ... 942
18.4. INTERNATIONALIZATION ... 943
18.5. CASE STUDY—MIPAD ... 945
18.5.1. Specifying the Application ... 946
18.5.2. Rapid Prototyping ... 948
18.5.3. Evaluation ... 949
18.5.4. Iterations ... 951
18.6. HISTORICAL PERSPECTIVE AND FURTHER READING ... 952

INDEX ... 957

Foreword

Recognition and understanding of spontaneous unrehearsed speech remains an elusive goal. To understand speech, a human considers not only the specific information conveyed to the ear, but also the context in which the information is being discussed. For this reason, people can understand spoken language even when the speech signal is corrupted by noise. However, understanding the context of speech is, in turn, based on a broad knowledge of the world. And this has been the source of the difficulty and over forty years of research.

It is difficult to develop computer programs that are sufficiently sophisticated to understand continuous speech by a random speaker. Only when programmers simplify the problem—by isolating words, limiting the vocabulary or number of speakers, or constraining the way in which sentences may be formed—is speech recognition by computer possible.

Since the early 1970s, researchers at AT&T, BBN, CMU, IBM, Lincoln Labs, MIT, and SRI have made major contributions in Spoken Language Understanding Research. In 1971, the Defense Advanced Research Projects Agency (DARPA) initiated an ambitious five-year, $15 million, multisite effort to develop speech understanding systems. The goals were to develop systems that would accept continuous speech from many speakers, with minimal speaker adaptation, and operate on a 1000-word vocabulary, artificial syntax, and a constrained task domain. Two of the systems, Harpy and Hearsay-II, both developed at Carnegie Mellon University, achieved the original goals and in some instances surpassed them.

During the last three decades I have been at Carnegie Mellon, I have been very fortunate to be able to work with many brilliant students and researchers. Xuedong Huang, Alex Acero, and Hsiao-Wuen Hon were arguably among the outstanding researchers in the speech group at CMU. Since then, they have moved to Microsoft and have put together a world-class team at Microsoft Research. Over the years, they have contributed standards for building spoken language understanding systems with Microsoft's SAPI/SDK family of products and pushed the technologies forward with the rest of the community. Today, they continue to play a premier leadership role in both the research community and in industry.

This new book, *Spoken Language Processing*, represents a welcome addition to the technical literature on this increasingly important emerging area of Information Technology. As we move from desktop PCs to personal digital assistants (PDAs), wearable computers, and Internet cell phones, speech becomes a central, if not the only, means of communication between the human and machine! Huang, Acero, and Hon have undertaken a commendable task of creating a comprehensive reference that covers theoretical, algorithmic, and systems aspects of the spoken language tasks of recognition, synthesis, and understanding.

The task of spoken language communication requires a system to recognize, interpret, execute, and respond to a spoken query. This task is complicated by the fact that the speech signal is corrupted by many sources: noise in the background, characteristics of the microphone, vocal tract characteristics of the speakers, and differences in pronunciation. In addition, the system has to cope with non-grammaticality of spoken communication and ambiguity of language. An effective system must strive to utilize all the available sources of knowledge—acoustics, phonetics and phonology, lexical, syntactic, and semantic structure of language, and task-specific context-dependent information.

Speech is based on a sequence of discrete sound segments that are linked in time. These segments, called phonemes, are assumed to have unique articulatory and acoustic characteristics. While the human vocal apparatus can produce an almost infinite number of articulatory gestures, the number of phonemes is limited. English as spoken in the United States, for example, contains 16 vowel and 24 consonant sounds. Each phoneme has distinguishable acoustic characteristics and, in combination with other phonemes, forms larger units such as syllables and words. Knowledge about the acoustic differences among these sound units is essential to distinguish one word from another, say, *bit* from *pit*.

When speech sounds are connected to form larger linguistic units, the acoustic characteristics of a given phoneme will change as a function of its immediate phonetic environment because of the interaction among various anatomical structures (such as the tongue, lips, and vocal chords) and their different degrees of sluggishness. The result is an overlap of phonemic information in the acoustic signal from one segment to the other. For example, the same underlying phoneme *t* can have drastically different acoustic characteristics in different words, say, in *tea*, *tree*, *city*, *beaten*, and *steep*. This effect, known as coarticulation, can occur within a given word or across a word boundary. Thus, the word *this* will have very different acoustic properties in phrases such as *this car* and *this ship*.

This book is self-contained for those who wish to familiarize themselves with the current state of spoken language systems technology. However, a researcher or a professional in the field will benefit from a thorough grounding in a number of disciplines, including:

- *Signal processing*: Fourier Transforms, DFT, and FFT
- *Acoustics*: physics of sounds and speech, models of vocal tract
- *Pattern recognition*: clustering and pattern matching techniques
- *Artificial intelligence*: knowledge representation and search, natural language processing
- *Computer science*: hardware, parallel systems, algorithm optimization
- *Statistics*: probability theory, hidden Markov models, dynamic programming
- *Linguistics*: acoustic phonetics, lexical representation, syntax, and semantics

A newcomer to this field, easily overwhelmed by the vast number of different algorithms scattered across many conference proceedings, can find in this book a set of techniques that Huang, Acero, and Hon have found to work well in practice. This book is unique in that it includes both the theory and implementation details necessary to build spoken language systems. If you were able to assemble all the individual material that is covered in the book and put it on a shelf, it would be several times larger than this volume and yet you would be missing vital information. You would not have the material that is in this book that threads it all into one story, one context. If you need additional resources, the authors include extensive references to get that additional detail. *Spoken Language Processing* is very appealing both as a textbook and as a reference book for practicing engineers. Some readers familiar with a specific topic may decide to skip a few chapters; others may want to focus in other chapters. This is not a book that you will pick up and read once from cover to cover, but one you will keep near you for reference as long as you work in this field.

Raj Reddy
Dean, School of Computer Science
Carnegie Mellon University

Preface

Our primary motivation in writing this book is to share our working experience to bridge the gap between the knowledge of industry gurus and newcomers to the spoken language processing community. Many powerful techniques hide in conference proceedings and academic papers for years before becoming widely recognized by the research community or the industry. We spent many years pursuing spoken language technology research at Carnegie Mellon University before we started spoken language R&D at Microsoft. We fully understand that it is by no means a small undertaking to transfer a state-of-the-art spoken language research system into a commercially viable product that can truly help people improve their productivity. Our experience in both industry and academia is reflected in the context of this book, which presents a contemporary and comprehensive description of both theoretic and practical issues in spoken language processing. This book is intended for people of diverse academic and practical backgrounds. Speech scientists, computer scientists, linguists, engineers, physicists, and psychologists all have a unique perspective on spoken language processing. This book will be useful to all of these special interest groups.

Spoken language processing is a diverse subject that relies on knowledge of many levels, including acoustics, phonology, phonetics, linguistics, semantics, pragmatics, and discourse. The diverse nature of spoken language processing requires knowledge in computer science, electrical engineering, mathematics, syntax, and psychology. There are a number of excellent books on the subfields of spoken language processing, including speech recognition, text-to-speech conversion, and spoken language understanding, but there is no single book that covers both theoretical and practical aspects of these subfields and spoken language interface design. We devote many chapters systematically introducing fundamental theories needed to understand how speech recognition, text-to-speech synthesis, and spoken language understanding work. Even more important is the fact that the book highlights what works well in practice, which is invaluable if you want to build a practical speech recognizer, a practical text-to-speech synthesizer, or a practical spoken language system. Using numerous real examples in developing Microsoft's spoken language systems, we concentrate on showing how the fundamental theories can be applied to solve real problems in spoken language processing.

We would like to thank many people who helped us during our spoken language processing R&D careers. We are particularly indebted to Professor Raj Reddy at the School of Computer Science, Carnegie Mellon University. Under his leadership, Carnegie Mellon University has become a center of research excellence on spoken language processing. Today's computer industry and academia benefit tremendously from his leadership and contributions.

Special thanks are due to Microsoft for its encouragement of spoken language R&D. The management team at Microsoft has been extremely generous to the speech technology group. We are particularly grateful to Bill Gates, Nathan Myhrvold, Rick Rashid, Dan Ling, and Jack Breese for the great environment they have created for us at Microsoft Research. We would also like to thank Bob Muglia and Kai-Fu Lee for their leadership role in Microsoft's speech product development.

Scott Meredith helped us write a number of chapters in this book and deserves to be a co-author. His insight and experience in text-to-speech synthesis enriched this book a great deal. We also owe gratitude to many colleagues we worked with in the speech technology group of Microsoft Research. In alphabetic order, Jim Adcock, Bruno Alabiso, Fil Alleva, Eric Bidstrup, Antonio Bigazzi, Ciprian Chelba, Li Deng, James Droppo, Doug Duchene, Joshua Goodman, Mei-Yuh Hwang, Larry Israel, Derek Jacoby, Li Jiang, Yun-Cheng Ju, David Larson, Kevin Larson, Jingsong Liu, Ricky Loynd, Milind Mahajan, Peter Mau, John Merrill, Yunus Mohammed, Salman Mughal, Mike Plumpe, Scott Quinn, Bill Rockenbeck, Mike Rozak, Kevin Schofield, Roxana Teodorescu, Gina Venolia, Kuansan Wang, Ye-Yi Wang, and Shenzhi Zhang.

In addition, we want to thank Les Atlas, Jeff Bilmes, Alan Black, David Caulton, Eric Chang, Phil Chou, Dinei Florencio, Allen Gersho, Francisco Gimenez-Galanes, Hynek Hermansky, Henrique Malvar, Julian Odell, Mari Ostendorf, Joseph Pentheroudakis, Tandy Trower, and Charles Wayne. They provided us with many wonderful comments to refine this book. Tim Moore, Russ Hall, and Jane Bonnell at Prentice Hall helped us finish this book in a finite amount of time.

Finally, writing this book was a marathon that could not have been finished without the support of our spouses, Yingzhi, Donna, and Phen, during the many evenings and weekends we spent on this project.

Xuedong Huang
Alex Acero
Hsiao-Wuen Hon
Redmond, WA

CHAPTER 1

Introduction

From human prehistory to the new media of the future, speech communication has been and will be the dominant mode of human social bonding and information exchange. The spoken word is now extended, through technological mediation such as telephony, movies, radio, television, and the Internet. This trend reflects the primacy of spoken communication in human psychology.

In addition to human-human interaction, this human preference for spoken language communication finds a reflection in human-machine interaction as well. Most computers currently utilize a *graphical user interface* (GUI), based on graphically represented interface objects and functions such as windows, icons, menus, and pointers. Most computer operating systems and applications also depend on a user's keyboard strokes and mouse clicks, with a display monitor for feedback. Today's computers lack the fundamental human abilities to speak, listen, understand, and learn. Speech, supported by other natural modalities, will be one of the primary means of interfacing with computers. And, even before speech-based interaction reaches full maturity, applications in home, mobile, and office segments are incorporating spoken language technology to change the way we live and work.

A spoken language system needs to have both speech recognition and speech synthesis capabilities. However, those two components by themselves are not sufficient to build a useful spoken language system. An understanding and dialog component is required to manage interactions with the user; and domain knowledge must be provided to guide the system's interpretation of speech and allow it to determine the appropriate action. For all these components, significant challenges exist, including robustness, flexibility, ease of integration, and engineering efficiency. The goal of building commercially viable spoken language systems has long attracted the attention of scientists and engineers all over the world. The purpose of this book is to share our working experience in developing advanced spoken language processing systems with both our colleagues and newcomers. We devote many chapters to systematically introducing fundamental theories and to highlighting what works well based on numerous lessons we learned in developing Microsoft's spoken language systems.

1.1. MOTIVATIONS

What motivates the integration of spoken language as the primary interface modality? We present a number of scenarios, roughly in order of expected degree of technical challenges and expected time to full deployment.

1.1.1. Spoken Language Interface

There are generally two categories of users who can benefit from adoption of speech as a control modality in parallel with others, such as the mouse, keyboard, touch-screen, and joystick. For novice users, functions that are conceptually simple should be directly accessible. For example, raising the voice output volume under software control on the desktop speakers, a conceptually simple operation, in some GUI systems of today requires opening one or more windows or menus, and manipulating sliders, check-boxes, or other graphical elements. This requires some knowledge of the system's interface conventions and structures. For the novice user, to be able to say *raise the volume* would be more direct and natural. For expert users, the GUI paradigm is sometimes perceived as an obstacle or nuisance and shortcuts are sought. Frequently these shortcuts allow the power user's hands to remain on the keyboard or mouse while mixing content creation with system commands. For example, an operator of a graphic design system for CAD/CAM might wish to specify a text formatting command while keeping the pointer device in position over a selected screen element.

Speech has the potential to accomplish these functions more powerfully than keyboard and mouse clicks. Speech becomes more powerful when supplemented by information streams encoding other dynamic aspects of user and system status, which can be resolved by the semantic component of a complete multimodal interface. We expect such multimodal interactions to proceed based on more complete user modeling, including speech, visual orientation, natural and device-based gestures, and facial expression, and these will be coordinated with detailed system profiles of typical user tasks and activity patterns.

In some situations you must rely on speech as an input or output medium. For example, with wearable computers, it may be impossible to incorporate a large keyboard. When driving, safety is compromised by any visual distraction, and hands are required for controlling the vehicle. The ultimate speech-only device, the telephone, is far more widespread than the PC. Certain manual tasks may also require full visual attention to the focus of the work. Finally, spoken language interfaces offer obvious benefits for individuals challenged with a variety of physical disabilities, such as loss of sight or limitations in physical motion and motor skills. Chapter 18 contains a detailed discussion on spoken language applications.

1.1.2. Speech-to-Speech Translation

Speech-to-speech translation has been depicted for decades in science fiction stories. Imagine questioning a Chinese-speaking conversational partner by speaking English into an unobtrusive device, and hearing real-time replies you can understand. This scenario, like the spoken language interface, requires both speech recognition and speech synthesis technology. In addition, sophisticated multilingual spoken language understanding is needed. This highlights the need for tightly coupled advances in speech recognition, synthesis, and understanding systems, a point emphasized throughout this book.

1.1.3. Knowledge Partners

The ability of computers to process spoken language as proficient as humans will be a landmark to signal the arrival of truly intelligent machines. Alan Turing [29] introduced his famous *Turing test*. He suggested a game, in which a computer's use of language would form the criterion for intelligence. If the machine could win the game, it would be judged intelligent. In Turing's game, you play the role of an interrogator. By asking a series of questions via a teletype, you must determine the identity of the other two participants: a machine and a person. The task of the machine is to fool you into believing it is a person by responding as a person to your questions. The task of the other person is to convince you the other participant is the machine. The critical issue for Turing was that using language as humans do is sufficient as an operational test for intelligence.

The ultimate use of spoken language is to pass the Turing test in allowing future extremely intelligent systems to interact with human beings as knowledge partners in all aspects of life. This has been a staple of science fiction, but its day will come. Such systems require reasoning capabilities and extensive world knowledge embedded in sophisticated search, communication, and inference tools that are beyond the scope of this book. We expect that spoken language technologies described in this book will form the essential enabling mechanism to pass the Turing test.

1.2. SPOKEN LANGUAGE SYSTEM ARCHITECTURE

Spoken language processing refers to technologies related to speech recognition, text-to-speech, and spoken language understanding. A spoken language system has at least one of the following three subsystems: a speech recognition system that converts speech into words, a text-to-speech system that conveys spoken information, and a spoken language understanding system that maps words into actions and that plans system-initiated actions.

There is considerable overlap in the fundamental technologies for these three subareas. Manually created rules have been developed for spoken language systems with limited success. But, in recent decades, data-driven statistical approaches have achieved encouraging results, which are usually based on modeling the speech signal using well-defined statistical algorithms that can automatically extract knowledge from the data. The data-driven approach can be viewed fundamentally as a pattern recognition problem. In fact, speech recognition, text-to-speech conversion, and spoken language understanding can all be regarded as pattern recognition problems. The patterns are either recognized during the runtime operation of the system or identified during system construction to form the basis of runtime generative models such as prosodic templates needed for text-to-speech synthesis. While we use and advocate the statistical approach, we by no means exclude the knowledge engineering approach from consideration. If we have a good set of rules in a given problem area, there is no need to use the statistical approach at all. The problem is that, at time of this writing, we do not have enough knowledge to produce a complete set of high-quality rules. As scientific and theoretical generalizations are made from data collected to construct data-driven systems, better rules may be constructed. Therefore, the rule-based and statistical approaches are best viewed as complementary.

1.2.1. Automatic Speech Recognition

A source-channel mathematical model described in Chapter 3 is often used to formulate speech recognition problems. As illustrated in Figure 1.1, the speaker's mind decides the source word sequence $\mathbf{W}$ that is delivered through his/her text generator. The source is passed through a noisy communication channel that consists of the speaker's vocal apparatus to produce the speech waveform and the speech signal processing component of the speech recognizer. Finally, the speech decoder aims to decode the acoustic signal $\mathbf{X}$ into a word sequence $\hat{\mathbf{W}}$, which is hopefully close to the original word sequence $\mathbf{W}$.

A typical practical speech recognition system consists of basic components shown in the dotted box of Figure 1.2. Applications interface with the decoder to get recognition results that may be used to adapt other components in the system. *Acoustic models* include the representation of knowledge about acoustics, phonetics, microphone and environment variability, gender and dialect differences among speakers, etc. *Language models* refer to a system's knowledge of what constitutes a possible word, what words are likely to co-occur, and in what sequence. The semantics and functions related to an operation a user may wish to perform may also be necessary for the language model. Many uncertainties exist in these areas, associated with speaker characteristics, speech style and rate, recognition of basic speech segments, possible words, likely words, unknown words, grammatical variation, noise interference, nonnative accents, and confidence scoring of results. A successful speech recognition system must contend with all of these uncertainties. But that is only the beginning. The acoustic uncertainties of the different accents and speaking styles of individual speakers are compounded by the lexical and grammatical complexity and variations of spoken language, which are all represented in the language model.

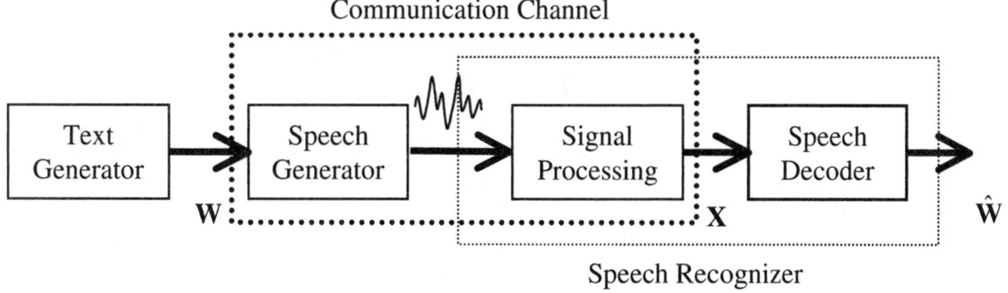

Figure 1.1 A source-channel model for a speech recognition system [15].

The speech signal is processed in the signal processing module that extracts salient feature vectors for the decoder. The decoder uses both acoustic and language models to generate the word sequence that has the maximum posterior probability for the input feature vectors. It can also provide information needed for the adaptation component to modify either the acoustic or language models so that improved performance can be obtained.

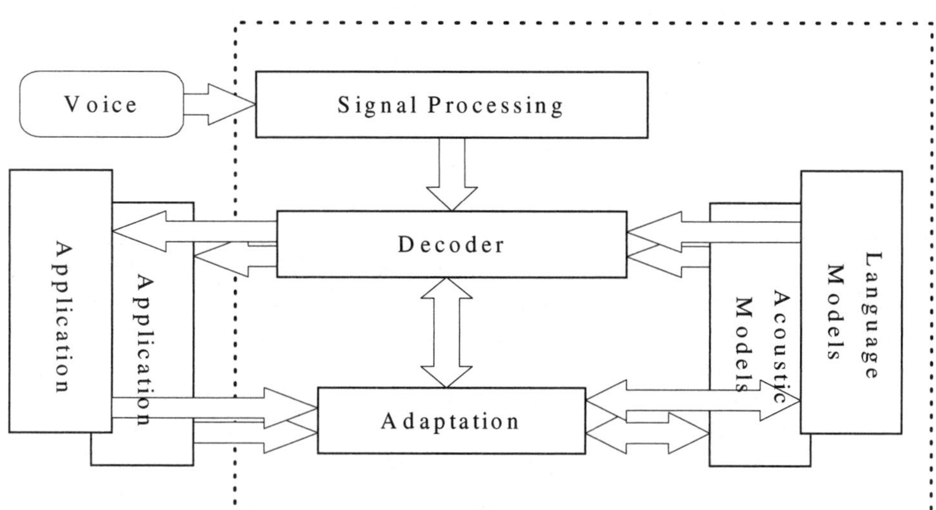

Figure 1.2 Basic system architecture of a speech recognition system [12].

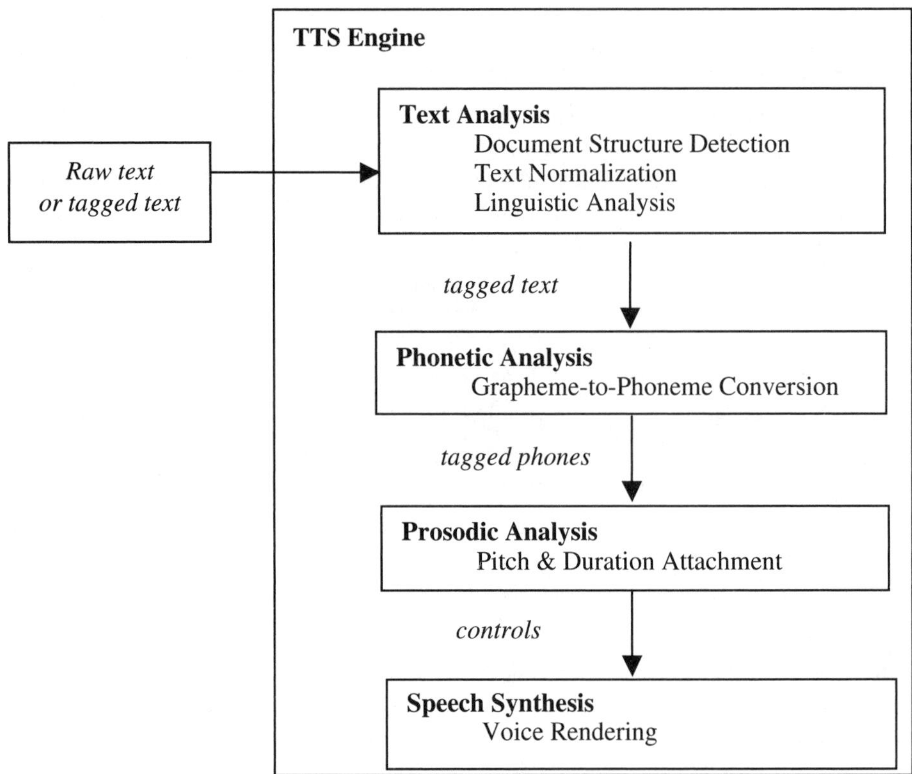

Figure 1.3 Basic system architecture of a TTS system.

1.2.2. Text-to-Speech Conversion

The term *text-to-speech*, often abbreviated as TTS, is easily understood. The task of a text-to-speech system can be viewed as speech recognition in reverse – a process of building a machinery system that can generate human-like speech from any text input to mimic human speakers. TTS is sometimes called *speech synthesis*, particularly in the engineering community.

The conversion of words in written form into speech is nontrivial. Even if we can store a huge dictionary for most common words in English; the TTS system still needs to deal with millions of names and acronyms. Moreover, in order to sound natural, the intonation of the sentences must be appropriately generated.

The development of TTS synthesis can be traced back to the 1930s when Dudley's *Voder*, developed by Bell Laboratories, was demonstrated at the World's Fair [18]. Taking advantage of increasing computation power and storage technology, TTS researchers have been able to generate high-quality commercial multilingual text-to-speech systems, although the quality is inferior to human speech for general-purpose applications.

The basic TTS components are shown in Figure 1.3. The text analysis component normalizes the text to the appropriate form so that it becomes speakable. The input can be either raw text or tagged. These tags can be used to assist text, phonetic, and prosodic analysis. The phonetic analysis component converts the processed text into the corresponding phonetic sequence, which is followed by prosodic analysis to attach appropriate pitch and duration information to the phonetic sequence. Finally, the speech synthesis component takes the parameters from the fully tagged phonetic sequence to generate the corresponding speech waveform.

Various applications have different degrees of knowledge about the structure and content of the text that they wish to speak so some of the basic components shown in Figure 1.3 can be skipped. For example, some applications may have certain broad requirements such as rate and pitch. These requirements can be indicated with simple command tags appropriately located in the text. Many TTS systems provide a set of markups (tags), so the text producer can better express their semantic intention. An application may know a lot about the structure and content of the text to be spoken to greatly improve speech output quality. For engines providing such support, the *text analysis* phase can be skipped, in whole or in part. If the system developer knows the phonetic form, the phonetic analysis module can be skipped as well. The prosodic analysis module assigns a numeric duration to every phonetic symbol and calculates an appropriate pitch contour for the utterance or paragraph. In some cases, an application may have prosodic contours precalculated by some other process. This situation might arise when TTS is being used primarily for compression, or the prosody is *transplanted* from a real speaker's utterance. In these cases, the quantitative prosodic controls can be treated as special tagged field and sent directly along with the phonetic stream to speech synthesis for voice rendition.

1.2.3. Spoken Language Understanding

Whether a speaker is inquiring about flights to Seattle, reserving a table at a Pittsburgh restaurant, dictating an article in Chinese, or making a stock trade, a spoken language understanding system is needed to interpret utterances in context and carry out appropriate actions. Lexical, syntactic, and semantic knowledge must be applied in a manner that permits cooperative interaction among the various levels of acoustic, phonetic, linguistic, and application knowledge in minimizing uncertainty. Knowledge of the characteristic vocabulary, typical syntactic patterns, and possible actions in any given application context for both interpretation of user utterances and planning system activity are the heart and soul of any spoken language understanding system.

A schematic of a typical spoken language understanding system is shown in Figure 1.4. Such a system typically has a speech recognizer and a speech synthesizer for basic speech input and output, and a *sentence interpretation* component to parse the speech recognition results into semantic forms, which often need *discourse analysis* to track context and resolve ambiguities. The *Dialog Manager* is the central component that communicates with applications and the spoken language understanding modules such as discourse analysis, sentence interpretation, and response generation.

While most components of the system may be partly or wholly generic, the dialog manager controls the flow of conversation tied to the action. The dialog manager is responsible for providing status needed for formulating responses, and maintaining the system's idea of the state of the discourse. The discourse state records the current transaction, dialog goals that motivated the current transaction, current objects in focus (temporary center of attention), the object history list for resolving dependent references, and other status information. The discourse information is crucial for sentence interpretation to interpret utterances in context. Various systems may alter the flow of information implied in Figure 1.4. For example, the dialog manager may be able to supply contextual discourse information or pragmatic inferences, as feedback to guide the recognizer's evaluation of hypotheses at the earliest level of search.

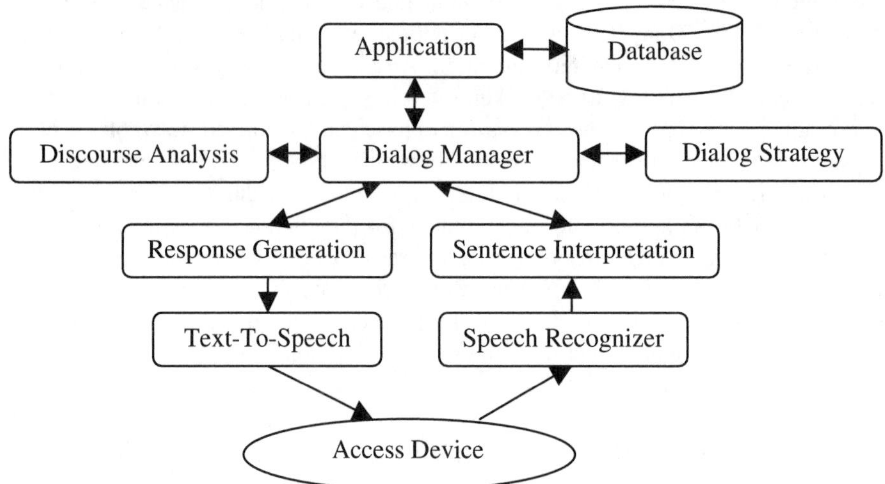

Figure 1.4 Basic system architecture of a spoken language understanding system.

1.3. BOOK ORGANIZATION

We attempt to present a comprehensive introduction to spoken language processing, which includes not only fundamentals but also a practical guide to build a working system that requires knowledge in speech signal processing, recognition, text-to-speech, spoken language understanding, and application integration. Since there is considerable overlap in the fundamental spoken language processing technologies, we have devoted Part I to the foundations needed. Part I contains background on speech production and perception, probability and information theory, and pattern recognition. Parts II, III, IV, and V include chapters on speech processing, speech recognition, speech synthesis, and spoken language systems, respectively. A reader with sufficient background can skip Part I, referring back to it later as needed. For example, the discussion of speech recognition in Part III relies on the pattern recognition algorithms presented in Part I. Algorithms that are used in several chapters within Part III are also included in Parts I and II. Since the field is still evolving, at the end of each chapter we provide a historical perspective and list further readings to facilitate future research.

1.3.1. Part I: Fundamental Theory

Chapters 2 to 4 provide you with a basic theoretic foundation to better understand techniques that are widely used in modern spoken language systems. These theories include the essence of linguistics, phonetics, probability theory, information theory, and pattern recognition. These chapters prepare you fully to understand the rest of the book.

Chapter 2 discusses the basic structure of spoken language including speech science, phonetics, and linguistics. Chapter 3 covers probability theory and information theory, which form the foundation of modern pattern recognition. Many important algorithms and principles in pattern recognition and speech coding are derived based on these theories. Chapter 4 introduces basic pattern recognition, including decision theory, estimation theory, and a number of algorithms widely used in speech recognition. Pattern recognition forms the core of most of the algorithms used in spoken language processing.

1.3.2. Part II: Speech Processing

Part II provides you with necessary speech signal processing knowledge that is critical to spoken language processing. Most of what discuss here is traditionally the subject of electrical engineering.

Chapters 5 and 6 focus on how to extract useful information from the speech signal. The basic principles of digital signal processing are reviewed and a number of useful representations for the speech signal are discussed. Chapter 7 covers how to compress these representations for efficient transmission and storage.

1.3.3. Part III: Speech Recognition

Chapters 8 to 13 provide you with an in-depth look at modern speech recognition systems. We highlight techniques that have been proven to work well in real systems and explain in detail how and why these techniques work from both theoretic and practical perspectives.

Chapter 8 introduces hidden Markov models, the most prominent technique used in modern speech recognition systems. Chapters 9 and 11 deal with acoustic modeling and language modeling respectively. Because environment robustness is critical to the success of practical systems, we devote Chapter 10 to discussing how to make systems less affected by environment noises. Chapters 12 and 13 deal in detail with how to efficiently implement the decoder for speech recognition. Chapter 12 discusses a number of basic search algorithms, and Chapter 13 covers large vocabulary speech recognition. Throughout our discussion, Microsoft's Whisper speech recognizer is used as a case study to illustrate the methods introduced in these chapters.

1.3.4. Part IV: Text-to-Speech Systems

In Chapters 14 through 16, we discuss proven techniques in building text-to-speech systems. The synthesis system consists of major components found in speech recognition systems, except that they are in the reverse order.

Chapter 14 covers the analysis of written documents and the text needed to support spoken rendition, including the interpretation of audio markup commands, interpretation of numbers and other symbols, and conversion from orthographic to phonetic symbols. Chapter 15 focuses on the generation of pitch and duration controls for linguistic and emotional effect. Chapter 16 discusses the implementation of the synthetic voice, and presents algorithms to manipulate a limited voice data set to support a wide variety of pitch and duration controls required by the text analysis. We highlight the importance of trainable synthesis, with Microsoft's Whistler TTS system as an example.

1.3.5. Part V: Spoken Language Systems

As discussed in Section 1.1, spoken language applications motivate spoken language R&D. The central component is the spoken language understanding system. Since it is closely related to applications, we group it together with application and interface design.

Chapter 17 covers spoken language understanding. The output of the recognizer requires interpretation and action in a particular application context. This chapter details useful strategies for dialog management, and the coordination of all the speech and system resources to accomplish a task for a user. Chapter 18 concludes the book with a discussion of important principles for building spoken language interfaces and applications, including general human interface design goals, and interaction with other modalities in specific application contexts. Microsoft's MiPad is used as a case study to illustrate a number of issues in developing spoken language and multimodal applications.

1.4. Target Audiences

This book can serve a variety of audiences:

Integration engineers: Software engineers who want to build spoken language systems, but who do not want to learn detailed speech technology internals, will find plentiful relevant material, including application design and software interfaces. Anyone with a professional interest in aspects of speech applications, integration, and interfaces can also achieve enough understanding of how the core technologies work, to allow them to take full advantage of state-of-the-art capabilities.

Speech technology engineers: Engineers and researchers working on various subspecialties within the speech field will find this book a useful guide to understanding related technologies in sufficient depth to help them gain insight on where their own approaches overlap with, or diverge from, their neighbors' common practice.

Graduate students: This book can serve as a primary textbook in a graduate or advanced undergraduate speech analysis or language engineering course. It can serve as a supplementary textbook in some applied linguistics, digital signal processing, computer science, artificial intelligence, and possibly psycholinguistics course.

Linguists: As the practice of linguistics increasingly shifts to empirical analysis of real-world data, students and professional practitioners alike should find a comprehensive introduction to the technical foundations of computer processing of spoken language helpful. The book can be read at different levels and through different paths, for readers with differing technical skills and background knowledge.

Speech scientists: Researchers engaged in professional work on issues related to normal or pathological speech may find this complete exposition of the state-of-the-art in computer modeling of generation and perception of speech interesting.

Business planners: Increasingly, business and management functions require some level of insight into the vocabulary and common practices of technology development. While not the primary audience, managers, marketers, and others with planning responsibilities and sufficient technical background will find portions of this book useful in evaluating competing proposals, and in making business decisions related to the speech technology components.

1.5. HISTORICAL PERSPECTIVE AND FURTHER READING

Spoken language processing is a diverse field that relies on knowledge of language at the levels of signal processing, acoustics, phonology, phonetics, syntax, semantics, pragmatics, and discourse. The foundations of spoken language processing lie in computer science, electrical engineering, linguistics, and psychology. In the 1970s an ambitious speech understanding project was funded by DARPA, which led to many seminal systems and technologies [17]. A number of human language technology projects funded by DARPA in the 1980s and 1990s further accelerated the progress, as evidenced by many papers published in *The Proceedings of the DARPA Speech and Natural Language/Human Language Workshop*. The field is still rapidly progressing and there are a number of excellent review articles and introductory books. We provide a brief list here. More detailed references can be found within each chapter of this book. Gold and Morgan's *Speech and Audio Signal Processing* [10] also has a strong historical perspective on spoken language processing.

Hyde [14] and Reddy [24] provided an excellent review of early speech recognition work in the 1970s. Some of the principles are still applicable to today's speech recognition research. Waibel and Lee assembled many seminal papers in *Readings in Speech Recognition Speech Recognition* [31]. There are a number of excellent books on modern speech recognition [1, 13, 15, 22, 23].

Where does the state of the art speech recognition system stand today? A number of different recognition tasks can be used to compare the recognition error rate of people vs. machines. Table 1.1 shows five typical recognition tasks with vocabularies ranging from 10 to 5000 words speaker-independent continuous speech recognition. The Wall Street Journal Dictation (WSJ) Task has a 5000-word vocabulary as a continuous dictation application for the WSJ articles. In Table 1.1, the error rate for machines is based on state of the art speech recognizers such as systems described in Chapter 9, and the error rate of humans is based on a range of subjects tested on the similar task. We can see the error rate of humans is at least 5 times smaller than machines except for the sentences that are generated from a trigram language model, where the sentences have the perfect match between humans and machines so humans cannot use high-level knowledge that is not used in machines.[1]

Table 1.1 Word error rate comparisons between human and machines on similar tasks.

Tasks	Vocabulary	Humans	Machines
Connected digits	10	0.009%	0.72%
Alphabet letters	26	1%	5%
Spontaneous telephone speech	2000	3.8%	36.7%
WSJ with clean speech	5000	0.9%	4.5%
WSJ with noisy speech (10-db SNR)	5000	1.1%	8.6%
Clean speech based on trigram sentences	20,000	7.6%	4.4%

We can see that humans are far more robust than machines for normal tasks. The error rate for machine spontaneous conversational telephone speech recognition is above 35%, more than a factor 10 higher than humans on the similar task. In addition, the error rate of humans does not increase as dramatically as machines when the environment becomes noisy (from quiet to 10-db SNR environments on the WSJ task). The relative error rate of humans increases from 0.9% to 1.1% (1.2 times), while the error rate of CSR systems increases from 4.5% to 8.6% (1.9 times). One interesting experiment is that when we generated sentences using the WSJ trigram language model (cf. Chapter 11), the difference between humans and machines disappears (the last row in Table 1.1). In fact, the error rate of humans is even higher than machines. This is because both humans and machines have the same high-level syntactic and semantic models. The test sentences are somewhat random to humans but perfect to machines that used the same trigram model for decoding. This experiment indicates humans make more effective use of semantic and syntactic constraints for improved speech recognition in meaningful conversation. In addition, machines don't have attention problems as humans do on random sentences.

Fant [7] gave an excellent introduction to speech production. Early reviews of text-to-speech synthesis can be found in [3, 8, 9]. Sagisaka [26] and Carlson [6] provide more recent reviews of progress in speech synthesis. A more detailed treatment can be found in [19, 30].

Where does the state of the art text to speech system stand today? Unfortunately, like speech recognition, this is not a solved problem either. Although machine storage capabilities are improving, the quality remains a challenge for many researchers if we want to pass the Turing test.

[1] Some of these experiments were conducted at Microsoft with only a small number of human subjects (3-5 people), which is not statistically significant. Nevertheless, the experiments give some interesting insight on the performance of humans and machines.

Spoken language understanding is deeply rooted in speech recognition research. There are a number of good books on spoken language understanding [2, 5, 16]. Manning and Schutze [20] focuses on statistical methods for language understanding. Like Waibel and Lee, Grosz et al. assembled many foundational papers in *Readings in Natural Language Processing* [11]. More recent reviews of progress in spoken language understanding can be found in [25, 28]. Related spoken language interface design issues can be found in [4, 21, 27, 32].

In comparison to speech recognition and text to speech, spoken language understanding is further away from approaching the level of humans, especially for general-purpose spoken language applications.

A number of good conference proceedings and journals report the latest progress in the field. Major results on spoken language processing are presented at the *International Conference on Acoustics, Speech and Signal Processing (ICASSP), International Conference on Spoken Language Processing (ICSLP), Eurospeech Conference,* the *DARPA Speech and Human Language Technology Workshops,* and many workshops organized by the *European Speech Communications Associations (ESCA)* and *IEEE Signal Processing Society.* Journals include *IEEE Transactions on Speech and Audio Processing, IEEE Transactions on Pattern Analysis and Machine Intelligence (PAMI), Computer Speech and Language, Speech Communication,* and *Journal of Acoustical Society of America (JASA).* Research results can also be found at computational linguistics conferences such as the *Association for Computational Linguistics (ACL), International Conference on Computational Linguistics (COLING),* and *Applied Natural Language Processing (ANLP).* The journals *Computational Linguistics* and *Natural Language Engineering* cover both theoretical and practical applications of language research. *Speech Recognition Update* published by TMA Associates is an excellent industry newsletter on spoken language applications.

REFERENCES

[1] Acero, A., *Acoustical and Environmental Robustness in Automatic Speech Recognition*, 1993, Boston, MA, Kluwer Academic Publishers.

[2] Allen, J., *Natural Language Understanding*, 2nd ed., 1995, Menlo Park, CA, The Benjamin/Cummings Publishing Company.

[3] Allen, J., M.S. Hunnicutt, and D.H. Klatt, *From Text to Speech: The MITalk System*, 1987, Cambridge, UK, University Press.

[4] Balentine, B., and D. Morgan, *How to Build a Speech Recognition Application*, 1999, Enterprise Integration Group.

[5] Bernsen, N., H. Dybkjar, and L. Dybkjar, *Designing Interactive Speech Systems*, 1998, Springer.

[6] Carlson, R., "Models of Speech Synthesis" in *Voice Communications Between Humans and Machines. National Academy of Sciences*, D.B. Roe and J.G. Wilpon, eds., 1994, Washington, D.C., National Academy of Sciences.

[7] Fant, G., *Acoustic Theory of Speech Production*, 1970, The Hague, NL, Mouton.

[8] Flanagan, J., *Speech Analysis Synthesis and Perception*, 1972, New York, Springer-Verlag.

[9] Flanagan, J., "Voices Of Men And Machines," *Journal of Acoustical Society of America*, 1972, **51**, p. 1375.

[10] Gold, B. and N. Morgan, *Speech and Audio Signal Processing: Processing and Perception of Speech and Music*, 2000, John Wiley and Sons.

[11] Grosz, B., F.S. Jones, and B.L. Webber, *Readings in Natural Language Processing*, 1986, Los Altos, CA, Morgan Kaufmann.

[12] Huang, X., *et al.*, "From Sphinx-II to Whisper – Make Speech Recognition Usable" in *Automatic Speech and Speaker Recognition*, C.H. Lee, F.K. Soong, and K.K. Paliwal, eds. 1996, Norwell, MA, Kluwer Academic Publishers.

[13] Huang, X.D., Y. Ariki, and M.A. Jack, *Hidden Markov Models for Speech Recognition*, 1990, Edinburgh, U.K., Edinburgh University Press.

[14] Hyde, S.R., "Automatic Speech Recognition: Literature, Survey, and Discussion" in *Human Communication, A Unified Approach*, E.E. David and P.B. Denes, eds. 1972, New York, McGraw Hill.

[15] Jelinek, F., *Statistical Methods for Speech Recognition*, Language, Speech, and Communication, 1998, Cambridge, MA, MIT Press.

[16] Jurafsky, D. and J. Martin, *Speech and Language Processing: An Introduction to Natural Language Processing, Computational Linguistics, and Speech Recognition*, 2000, Upper Saddle River, NJ, Prentice Hall.

[17] Klatt, D., "Review of the ARPA Speech Understanding Project," *Journal of Acoustical Society of America*, 1977, **62**(6), pp. 1324-1366.

[18] Klatt, D., "Review of Text-to-Speech Conversion for English," *Journal of Acoustical Society of America*, 1987, **82**, pp. 737-793.

[19] Kleijn, W.B. and K.K. Paliwal, *Speech Coding and Synthesis*, 1995, Amsterdam, Netherlands, Elsevier.

[20] Manning, C. and H. Schutze, *Foundations of Statistical Natural Language Processing*, 1999, MIT Press, Cambridge, USA.

[21] Markowitz, J., *Using Speech Recognition*, 1996, Prentice Hall.

[22] Mori, R.D., *Spoken Dialogues with Computers*, 1998, London, UK, Academic Press.

[23] Rabiner, L.R. and B.H. Juang, *Fundamentals of Speech Recognition*, May, 1993, Prentice-Hall.

[24] Reddy, D.R., "Speech Recognition by Machine: A Review," *IEEE Proc.*, 1976, **64**(4), pp. 502-531.

[25] Sadek, D. and R.D. Mori, "Dialogue Systems" in *Spoken Dialogues with Computers*, R.D. Mori, Editor 1998, London, UK, pp. 523-561, Academic Press.

[26] Sagisaka, Y., "Speech Synthesis from Text," *IEEE Communication Magazine*, 1990(1).

[27] Schmandt, C., *Voice Communication with Computers*, 1994, New York, NY, Van Nostrand Reinhold.

[28] Seneff, S., "The Use of Linguistic Hierarchies in Speech Understanding," *Int. Conf. on Spoken Language Processing*, 1998, Sydney, Australia.

[29] Turing, A.M., "Computing Machinery and Intelligence," *Mind*, 1950, **LIX**(236), pp. 433-460.

[30] van Santen, J., *et al.*, *Progress in Speech Synthesis*, 1997, New York, Springer-Verlag.

[31] Waibel, A.H. and K.F. Lee, *Readings in Speech Recognition*, 1990, San Mateo, CA, Morgan Kaufman Publishers.

[32] Weinschenk, S. and D. Barker, *Designing Effective Speech Interfaces*, 2000, John Wiley & Sons, Inc.

PART I

FUNDAMENTAL THEORY

CHAPTER 2

Spoken Language Structure

Spoken language is used to communicate information from a speaker to a listener. Speech production and perception are both important components of the speech chain. Speech begins with a thought and intent to communicate in the brain, which activates muscular movements to produce speech sounds. A listener receives it in the auditory system, processing it for conversion to neurological signals the brain can understand. The speaker continuously monitors and controls the vocal organs by receiving his or her own speech as feedback.

Considering the universal components of speech communication as shown in Figure 2.1, the fabric of spoken interaction is woven from many distinct elements. The speech production process starts with the semantic message in a person's mind to be transmitted to the listener via speech. The computer counterpart to the process of message formulation is the application semantics that creates the concept to be expressed. After the message is created, the next step is to convert the message into a sequence of words. Each word consists of a sequence of phonemes that corresponds to the pronunciation of the words. Each sentence also contains a prosodic pattern that denotes the duration of each phoneme, intonation of the sentence, and loudness of the sounds. Once the language system finishes the mapping, the talker executes a series of neuromuscular signals. The neuromuscular commands perform articulatory mapping to control the vocal cords, lips, jaw, tongue, and velum, thereby producing the sound sequence as the final output. The speech understanding process works in reverse order. First the signal is passed to the cochlea in the inner ear, which performs frequency analysis as a filter bank. A neural transduction process follows and converts the spectral signal into activity signals on the auditory nerve, corresponding roughly to a feature extraction component. Currently, it is unclear how neural activity is mapped into the language system and how message comprehension is achieved in the brain.

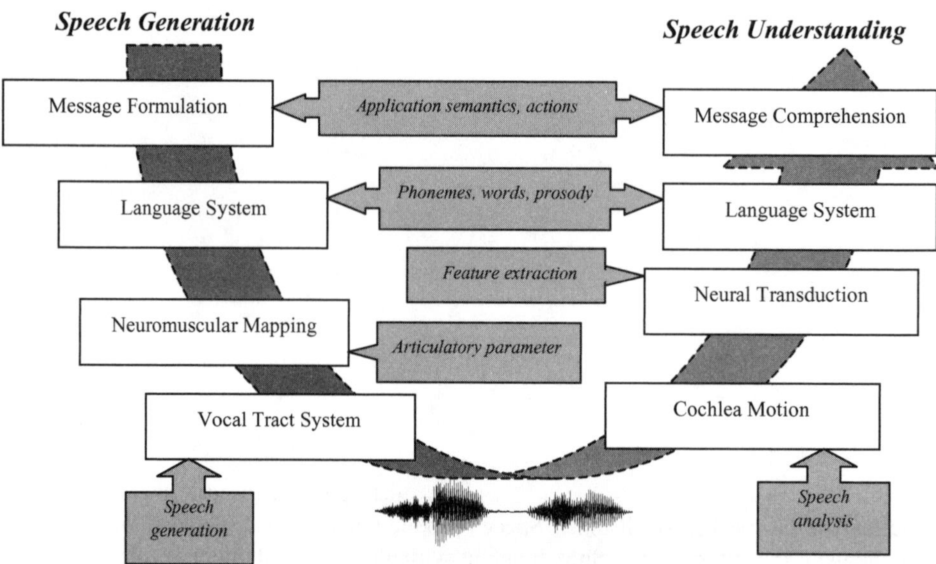

Figure 2.1 The underlying determinants of speech generation and understanding. The gray boxes indicate the corresponding computer system components for spoken language processing.

Speech signals are composed of analog sound patterns that serve as the basis for a discrete, symbolic representation of the spoken language – phonemes, syllables, and words. The production and interpretation of these sounds are governed by the syntax and semantics of the language spoken. In this chapter, we take a bottom up approach to introduce the basic concepts from sound to phonetics and phonology. Syllables and words are followed by syntax and semantics, which form the structure of spoken language processing. The examples in this book are drawn primarily from English, though they are relevant to other languages.

2.1. SOUND AND HUMAN SPEECH SYSTEMS

In this section, we briefly review human speech production and perception systems. We hope spoken language research will enable us to build a computer system that is as good as or better than our own speech production and understanding system.

2.1.1. Sound

Sound is a longitudinal pressure wave formed of compressions and rarefactions of air molecules, in a direction parallel to that of the application of energy. Compressions are zones where air molecules have been forced by the application of energy into a tighter-than-usual configuration, and rarefactions are zones where air molecules are less tightly packed. The alternating configurations of compression and rarefaction of air molecules along the path of an energy source are sometimes described by the graph of a sine wave as shown in Figure 2.2. In this representation, crests of the sine curve correspond to moments of maximal compression and troughs to moments of maximal rarefaction.

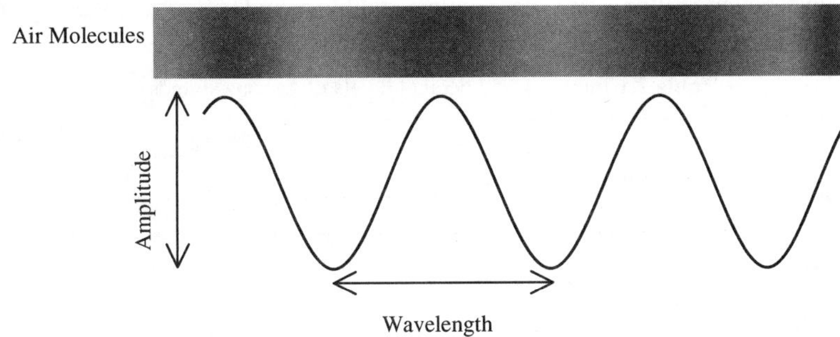

Figure 2.2 Application of sound energy causes alternating compression/rarefaction of air molecules, described by a sine wave. There are two important parameters, amplitude and wavelength, to describe a sine wave. Frequency [cycles/second measured in Hertz (Hz)] is also used to measure of the waveform.

The use of the sine graph in Figure 2.2 is only a notational convenience for charting local pressure variations over time, since sound does not form a transverse wave, and the air particles are just *oscillating in place* along the line of application of energy. The speed of a sound pressure wave in air is approximately $331.5 + 0.6T_c m/s$, where T_c is the Celsius temperature.

The amount of work done to generate the energy that sets the air molecules in motion is reflected in the amount of displacement of the molecules from their resting position. This *degree of displacement* is measured as the amplitude of a sound as shown in Figure 2.2. Because of the wide range, it is convenient to measure sound amplitude on a logarithmic scale in *decibels* (dB). A decibel scale is a means for comparing the intensity of two sounds:

$$10\log_{10}\ (I/I_0) \tag{2.1}$$

where I and I_0 are the two intensity levels, with intensity being proportional to the square of the sound pressure P.

Sound pressure level (SPL) is a measure of absolute sound pressure P in dB:

$$SPL(dB) = 20\log_{10}\left(\frac{P}{P_0}\right) \tag{2.2}$$

where the reference 0 dB corresponds to the threshold of hearing, which is $P_0 = 0.0002\mu bar$ for a tone of 1kHz. The speech conversation level at 3 feet is about 60 dB SPL, and a jackhammer's level is about 120 dB SPL. Alternatively, watts/meter^2 units are often used to indicate intensity. We can bracket the limits of human hearing as shown in Table 2.1. On the low end, the human ear is quite sensitive. A typical person can detect sound waves having an intensity of 10^{-12} W/m^2 (the *threshold of hearing* or TOH). This intensity corresponds to a pressure wave affecting a given region by only one-billionth of a centimeter of molecular motion. On the other end, the most intense sound that can be safely detected without suffering physical damage is one trillion times more intense than the TOH. 0 dB begins with the TOH and advances logarithmically. The faintest audible sound is arbitrarily assigned a value of 0 dB, and the loudest sounds that the human ear can tolerate are about 120 dB.

Table 2.1 Intensity and decibel levels of various sounds.

Sound	**dB Level**	**Times > TOH**
Threshold of hearing (TOH: $10^{-12} W/m^2$)	0	10^0
Light whisper	10	10^1
Quiet living room	20	10^2
Quiet conversation	40	10^4
Average office	50	10^5
Normal conversation	60	10^6
Busy city street	70	10^7
Acoustic guitar – 1 ft. away	80	10^8
Heavy truck traffic	90	10^9
Subway from platform	100	10^{10}
Power tools	110	10^{11}
Pain threshold of ear	120	10^{12}
Airport runway	130	10^{13}
Sonic boom	140	10^{14}
Permanent damage to hearing	150	10^{15}
Jet engine, close up	160	10^{16}
Rocket engine	180	10^{18}
Twelve ft. from artillery cannon muzzle ($10^{10} W/m^2$)	220	10^{22}

The absolute threshold of hearing is the maximum amount of energy of a pure tone that cannot be detected by a listener in a noise free environment. The absolute threshold of hearing is a function of frequency that can be approximated by

$$T_q(f) = 3.64(f/1000)^{-0.8} - 6.5e^{-0.6(f/1000-3.3)^2} + 10^{-3}(f/1000)^4 \quad (dB\ SPL) \tag{2.3}$$

and is plotted in Figure 2.3.

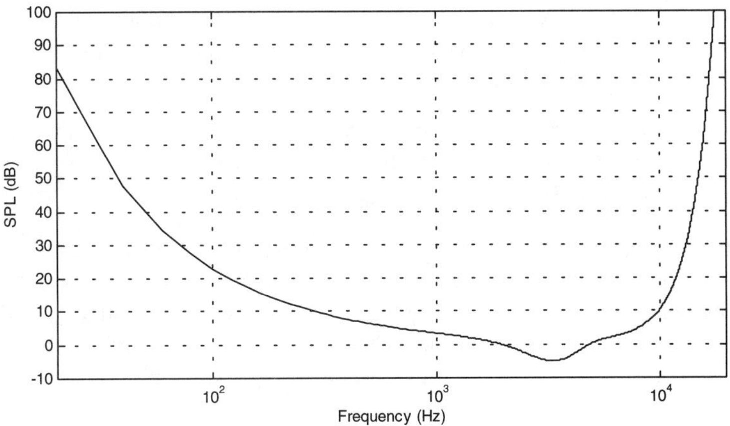

Figure 2.3 The sound pressure level (SPL) level in dB of the absolute threshold of hearing as a function of frequency. Sounds below this level are inaudible. Note that below 100 Hz and above 10 kHz this level rises very rapidly. Frequency goes from 20 Hz to 20 kHz and is plotted in a logarithmic scale from Eq. (2.3).

Let's compute how the pressure level varies with distance for a sound wave emitted by a point source located a distance r away. Assuming no energy absorption or reflection, the sound wave of a point source is propagated in a spherical front, such that the energy is the same for the sphere's surface at all radius r. Since the surface of a sphere of radius r is $4\pi r^2$, the sound's energy is inversely proportional to r^2, so that every time the distance is doubled, the sound pressure level decreases by 6 dB. For the point sound source, the energy (E) transported by a wave is proportional to the square of the amplitude (A) of the wave and the distance (r) between the sound source and the listener:

$$E \propto \frac{A^2}{r^2} \tag{2.4}$$

The typical sound intensity of a speech signal one inch away (close-talking microphone) from the talker is 1 Pascal = 10μbar, which corresponds to 94 dB SPL. The typical sound intensity 10 inches away from a talker is 0.1 Pascal = 1μbar, which corresponds to 74 dB SPL.

2.1.2. Speech Production

We review here basic human speech production systems, which have influenced research on speech coding, synthesis, and recognition.

2.1.2.1. Articulators

Speech is produced by air-pressure waves emanating from the mouth and the nostrils of a speaker. In most of the world's languages, the inventory of *phonemes,* as discussed in Section 2.2.1, can be split into two basic classes:

- consonants – articulated in the presence of constrictions in the throat or obstructions in the mouth (tongue, teeth, lips) as we speak.
- vowels – articulated without major constrictions and obstructions.

The sounds can be further partitioned into subgroups based on certain articulatory properties. These properties derive from the anatomy of a handful of important articulators and the places where they touch the boundaries of the human vocal tract. Additionally, a large number of muscles contribute to articulator positioning and motion. We restrict ourselves to a schematic view of only the major articulators, as diagrammed in Figure 2.4. The

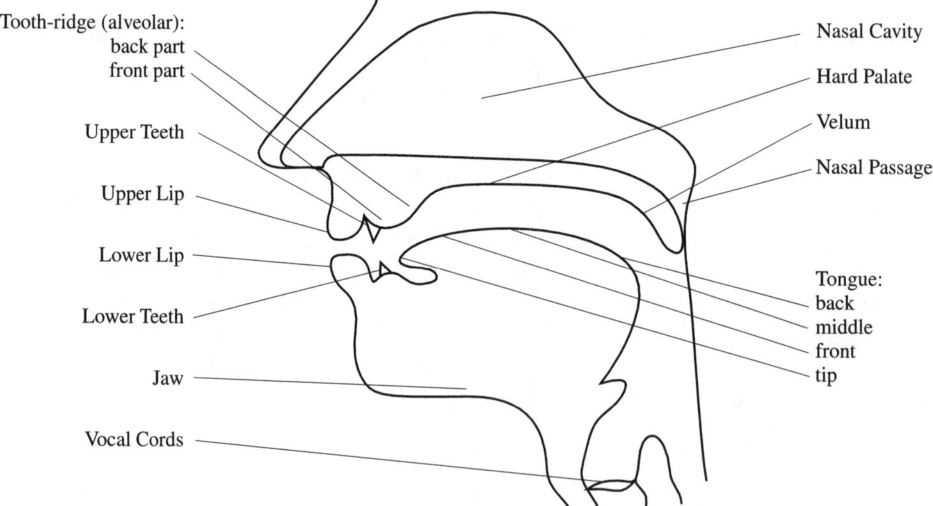

Figure 2.4 A schematic diagram of the human speech production apparatus.

gross components of the speech production apparatus are the lungs, trachea, larynx (organ of voice production), pharyngeal cavity (throat), oral and nasal cavity. The pharyngeal and oral cavities are typically referred to as the vocal tract, and the nasal cavity as the nasal tract. As illustrated in Figure 2.4, the human speech production apparatus consists of:

- *Lungs*: source of air during speech.
- *Vocal cords (larynx):* when the vocal folds are held close together and oscillate against one another during a speech sound, the sound is said to be *voiced*. When the folds are too slack or tense to vibrate periodically, the sound is said to be *unvoiced*. The place where the vocal folds come together is called the *glottis*.
- *Velum (soft palate)*: operates as a *valve*, opening to allow passage of air (and thus resonance) through the nasal cavity. Sounds produced with the flap open include *m* and *n*.
- *Hard palate*: a long relatively hard surface at the roof inside the mouth, which, when the tongue is placed against it, enables consonant articulation.
- *Tongue*: flexible articulator, shaped away from the palate for vowels, placed close to or on the palate or other hard surfaces for consonant articulation.
- *Teeth*: another place of articulation used to brace the tongue for certain consonants.
- *Lips*: can be rounded or spread to affect vowel quality, and closed completely to stop the oral air flow in certain consonants (*p, b, m*).

2.1.2.2. The Voicing Mechanism

The most fundamental distinction between sound types in speech is the voiced/voiceless distinction. Voiced sounds, including vowels, have in their time and frequency structure a roughly regular pattern that voiceless sounds, such as consonants like *s*, lack. Voiced sounds typically have more energy as shown in Figure 2.5. We see here the waveform of the word *sees*, which consists of three phonemes: an unvoiced consonant */s/,* a vowel */iy/*, and a voiced consonant */z/*.

What in the speech production mechanism creates this fundamental distinction? When the vocal folds vibrate during phoneme articulation, the phoneme is considered voiced; otherwise it is unvoiced. Vowels are voiced throughout their duration. The distinct vowel *timbres* are created by using the tongue and lips to shape the main oral resonance cavity in different ways. The vocal folds vibrate at slower or faster rates, from as low as 60 cycles per second (Hz) for a large man, to as high as 300 Hz or higher for a small woman or child. The rate of cycling (opening and closing) of the vocal folds in the larynx during phonation of voiced sounds is called the *fundamental frequency*. This is because it sets the periodic baseline for all higher-frequency harmonics contributed by the pharyngeal and oral resonance cavities above. The fundamental frequency also contributes more than any other single factor to the perception of *pitch* (the semi-musical rising and falling of voice tones) in speech.

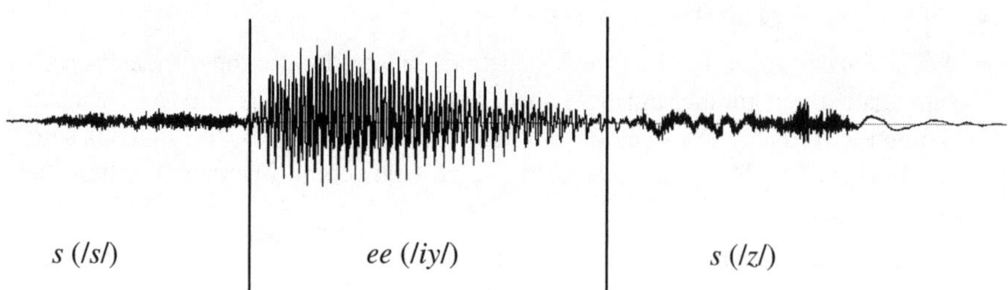

Figure 2.5 Waveform of *sees*, showing a voiceless phoneme /*s*/, followed by a voiced sound, the vowel /*iy*/. The final sound, /*z*/, is a type of voiced consonant.

The glottal cycle is illustrated in Figure 2.6. At stage (a), the vocal folds are closed and the air stream from the lungs is indicated by the arrow. At some point, the air pressure on the underside of the barrier formed by the vocal folds increases until it overcomes the resistance of the vocal fold closure and the higher air pressure below blows them apart (b). However, the tissues and muscles of the larynx and the vocal folds have a natural elasticity which tends to make them fall back into place rapidly, once air pressure is temporarily equalized (c). The successive airbursts resulting from this process are the source of energy for all voiced sounds. The time for a single open-close cycle depends on the stiffness and size of the vocal folds and the amount of subglottal air pressure. These factors can be controlled by a speaker to raise and lower the perceived frequency or pitch of a voiced sound.

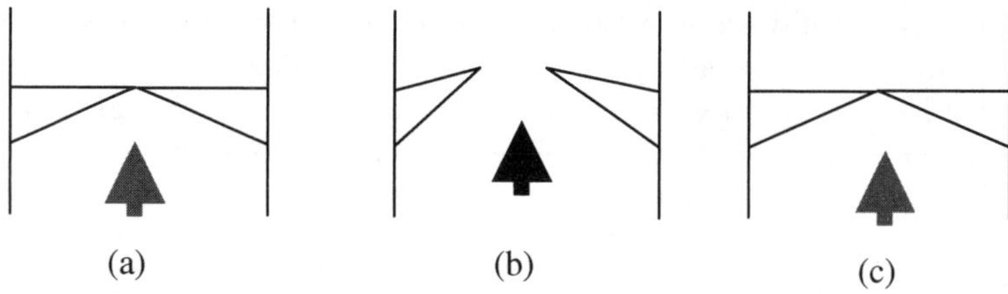

Figure 2.6 Vocal fold cycling at the larynx. (a) Closed with sub-glottal pressure buildup; (b) trans-glottal pressure differential causing folds to blow apart; (c) pressure equalization and tissue elasticity forcing temporary reclosure of vocal folds, ready to begin next cycle.

The waveform of air pressure variations created by this process can be described as a periodic flow, in cubic centimeters per second (after [15]). As shown in Figure 2.7, during the time bracketed as *one cycle*, there is no air flow during the initial closed portion. Then as the glottis opens (open phase), the volume of air flow becomes greater. After a short peak, the folds begin to resume their original position and the air flow declines until complete closure is attained, beginning the next cycle. A common measure is the number of such cycles per second (Hz), or the fundamental frequency (*F0*). Thus the fundamental frequency for the waveform in Figure 2.7 is about 120 Hz.

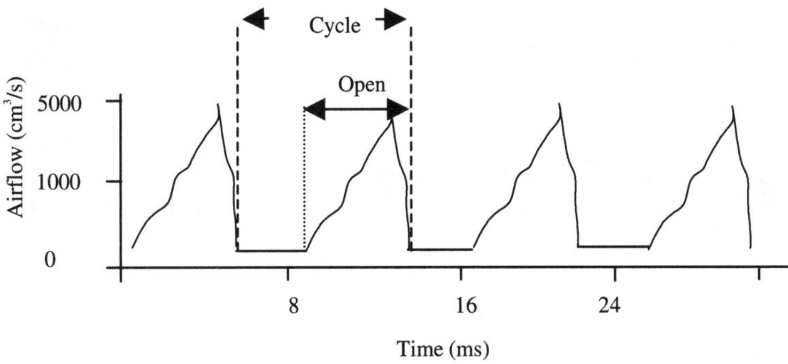

Figure 2.7 Waveform showing air flow during laryngeal cycle.

2.1.2.3. Spectrograms and Formants

Since the glottal wave is periodic, consisting of fundamental frequency (*F0*) and a number of harmonics (integral multiples of *F0*), it can be analyzed as a sum of sine waves as discussed in Chapter 5. The resonances of the vocal tract (above the glottis) are excited by the glottal energy. Suppose, for simplicity, we regard the vocal tract as a straight tube of uniform cross-sectional area, closed at the glottal end, open at the lips. When the shape of the vocal tract changes, the resonances change also. Harmonics near the resonances are emphasized, and, in speech, the resonances of the cavities that are typical of particular articulator configurations (e.g., the different vowel timbres) are called *formants*. The vowels in an actual speech waveform can be viewed from a number of different perspectives, emphasizing either a *cross-sectional* view of the harmonic responses at a single moment, or a longer-term view of the formant track evolution over time. The actual spectral analysis of a vowel at a single time-point, as shown in Figure 2.8, gives an idea of the uneven distribution of energy in resonances for the vowel /*iy*/ in the waveform for *see*, which is shown in Figure 2.5.

Another view of *sees* of Figure 2.5, called a spectrogram, is displayed in the lower part of Figure 2.9. It shows a long-term frequency analysis, comparable to a complete series of single time-point *cross sections* (such as that in Figure 2.8) ranged alongside one another in time and viewed from *above*.

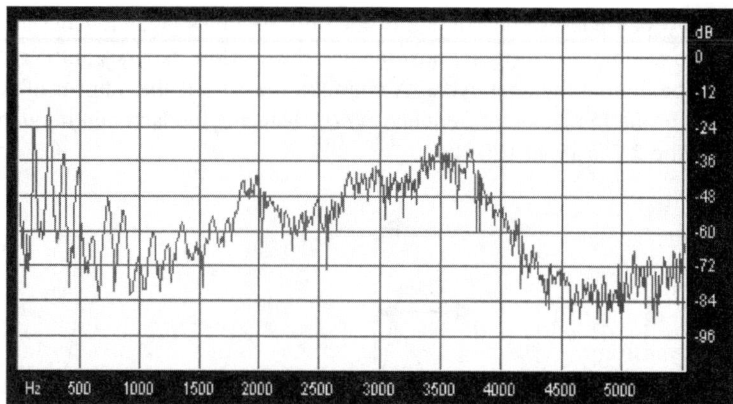

Figure 2.8 A spectral analysis of the vowel /iy/, showing characteristically uneven distribution of energy at different frequencies.

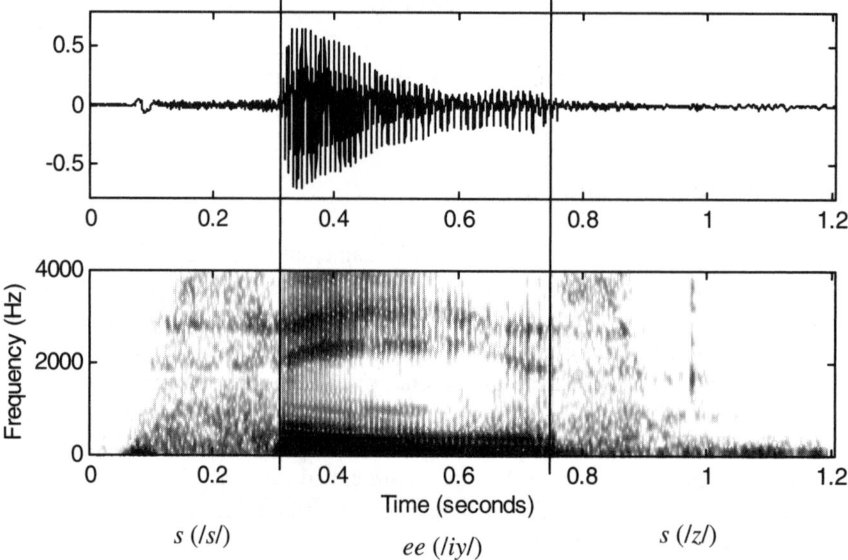

Figure 2.9 The spectrogram representation of the speech waveform *sees* (approximate phone boundaries are indicated with heavy vertical lines).

In the spectrogram of Figure 2.9, the darkness or lightness of a band indicates the relative amplitude or energy present at a given frequency. The dark horizontal bands show the formants, which are harmonics of the fundamental at natural resonances of the vocal tract cavity position for the vowel /iy/ in *see*. The mathematical methods for deriving analyses and representations such as those illustrated above are covered in Chapters 5 and 6.

2.1.3. Speech Perception

There are two major components in the auditory perception system: the peripheral auditory organs (ears) and the auditory nervous system (brain). The ear processes an acoustic pressure signal by first transforming it into a mechanical vibration pattern on the basilar membrane, and then representing the pattern by a series of pulses to be transmitted by the auditory nerve. Perceptual information is extracted at various stages of the auditory nervous system. In this section we focus mainly on the auditory organs.

2.1.3.1. Physiology of the Ear

The human ear, as shown in Figure 2.10, has three sections: the outer ear, the middle ear, and the inner ear. The outer ear consists of the external visible part and the external auditory canal that forms a tube along which sound travels. This tube is about 2.5 cm long and is covered by the eardrum at the far end. When air pressure variations reach the eardrum from the outside, it vibrates, and transmits the vibrations to bones adjacent to its opposite side. The vibration of the eardrum is at the same frequency (alternating compression and rarefaction) as the incoming sound pressure wave. The middle ear is an air-filled space or cavity about 1.3 cm across, and about 6 cm^3 volume. The air travels to the middle ear cavity along the tube (when opened) that connects the cavity with the nose and throat. The oval window shown in Figure 2.10 is a small membrane at the bony interface to the inner ear (cochlea). Since the cochlear walls are bony, the energy is transferred by mechanical action of the stapes into an impression on the membrane stretching over the oval window.

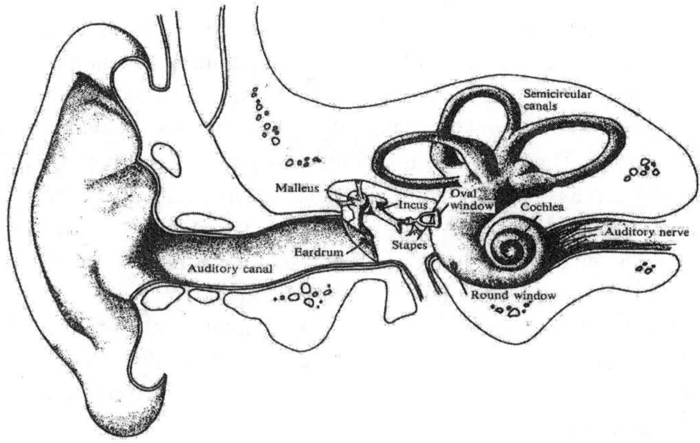

Figure 2.10 The structure of the peripheral auditory system with the outer, middle, and inner ear.

The relevant structure of the inner ear for sound perception is the cochlea, which communicates directly with the auditory nerve, conducting a representation of sound to the brain. The cochlea is a spiral tube about 3.5 cm long, which coils about 2.6 times. The spiral is divided, primarily by the basilar membrane running lengthwise, into two fluid-filled chambers. The cochlea can be roughly regarded as a filter bank, whose outputs are ordered by location, so that a frequency-to-place transformation is accomplished. The filters closest to the cochlear base respond to the higher frequencies, and those closest to its apex respond to the lower.

2.1.3.2. Physical vs. Perceptual Attributes

In psychoacoustics, a basic distinction is made between the perceptual attributes of a sound, especially a speech sound, and the measurable physical properties that characterize it. Each of the perceptual attributes, as listed in Table 2.2, seems to have a strong correlation with one main physical property, but the connection is complex, because other physical properties of the sound may affect perception in complex ways.

Table 2.2 Relation between perceptual and physical attributes of sound.

Physical Quantity	Perceptual Quality
Intensity	Loudness
Fundamental frequency	Pitch
Spectral shape	Timbre
Onset/offset time	Timing
Phase difference in binaural hearing	Location

Although sounds with a greater intensity level usually sound louder, the sensitivity of the ear varies with the frequency and the quality of the sound. One fundamental divergence between physical and perceptual qualities is the phenomenon of non-uniform *equal loudness* perception of tones of varying frequencies. In general, tones of differing pitch have different inherent *perceived loudness*. The sensitivity of the ear varies with the frequency and the quality of the sound. The graph of equal loudness contours adopted by ISO is shown in Figure 2.11. These curves demonstrate the relative insensitivity of the ear to sounds of low frequency at moderate to low intensity levels. Hearing sensitivity reaches a maximum around 4000 Hz, which is near the first resonance frequency of the outer ear canal, and peaks again around 13 kHz, the frequency of the second resonance [38].

Pitch is indeed most closely related to the fundamental frequency. The higher the fundamental frequency, the higher the pitch we perceive. However, discrimination between two pitches depends on the frequency of the lower pitch. Perceived pitch will change as intensity is increased and frequency is kept constant.

In another example of the non-identity of acoustic and perceptual effects, it has been observed experimentally that when the ear is exposed to two or more different tones, it is a common experience that one tone may *mask* the others. Masking is probably best explained as an upward shift in the hearing threshold of the weaker tone by the louder tone. Pure tones, complex sounds, narrow and broad bands of noise all show differences in their ability to mask other sounds. In general, pure tones close together in frequency mask each other more than tones widely separated in frequency. A pure tone masks tones of higher frequency more effectively than tones of lower frequency. The greater the intensity of the masking tone, the broader the range of the frequencies it can mask [18, 31].

Binaural listening greatly enhances our ability to sense the direction of the sound source. The sense of localization attention is mostly focused on side-to-side discrimination or *lateralization*. Time and intensity cues have different impacts for low frequency and high frequency, respectively. Low-frequency sounds are lateralized mainly on the basis of interaural time difference, whereas high-frequency sounds are localized mainly on the basis of interaural intensity differences [5].

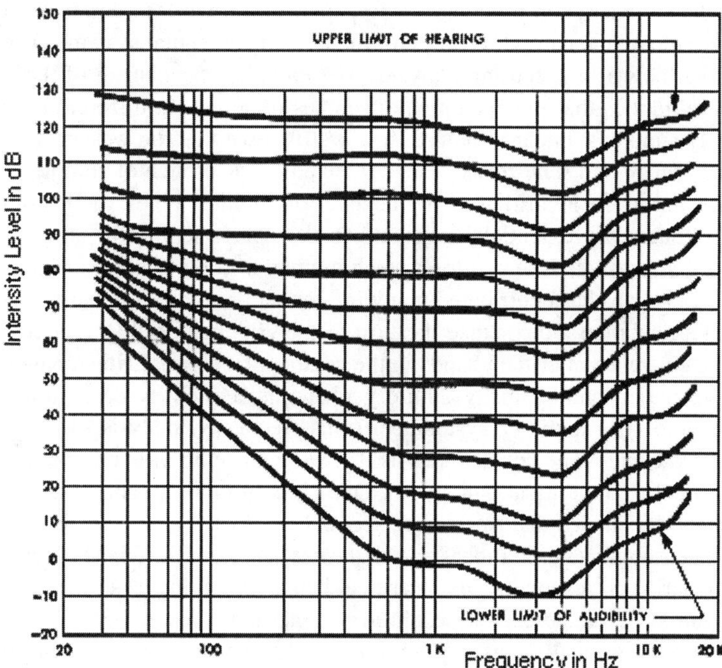

Figure 2.11 Equal-loudness curves indicate that the response of the human hearing mechanism is a function of frequency and loudness levels. This relationship again illustrates the difference between physical dimensions and psychological experience (after ISO 226).

Finally, an interesting perceptual issue is the question of distinctive voice quality. Speech from different people sounds different. Partially this is due to obvious factors, such as differences in characteristic fundamental frequency caused by, for example, the greater mass and length of adult male vocal folds as opposed to female. But there are more subtle effects as well. In psychoacoustics, the concept of *timbre* (of a sound or instrument) is defined as that attribute of auditory sensation by which a subject can judge that two sounds similarly presented and having the same loudness and pitch are dissimilar. In other words, when all the easily measured differences are controlled, the remaining perception of difference is ascribed to timbre. This is heard most easily in music, where the same note in the same octave played for the same duration on a violin sounds different from a flute. The timbre of a sound depends on many physical variables including a sound's spectral power distribution, its temporal envelope, rate and depth of amplitude or frequency modulation, and the degree of inharmonicity of its harmonics.

2.1.3.3. Frequency Analysis

Researchers have undertaken psychoacoustic experimental work to derive frequency scales that attempt to model the natural response of the human perceptual system, since the cochlea of the inner ear acts as a spectrum analyzer. The complex mechanism of the inner ear and auditory nerve implies that the perceptual attributes of sounds at different frequencies may not be entirely simple or linear in nature. It is well known that the western musical pitch is described in *octaves*[1] and *semi-tones.*[2] The perceived musical pitch of complex tones is basically proportional to the logarithm of frequency. For complex tones, the just noticeable difference for frequency is essentially constant on the octave/semi-tone scale. Musical pitch scales are used in prosodic research (on speech intonation contour generation).

AT&T Bell Labs has contributed many influential discoveries in hearing, such as critical band and articulation index, since the turn of the 20th century [3]. Fletcher's work [14] pointed to the existence of critical bands in the cochlear response. Critical bands are of great importance in understanding many auditory phenomena such as perception of loudness, pitch, and timbre. The auditory system performs frequency analysis of sounds into their component frequencies. The cochlea acts as if it were made up of overlapping filters having bandwidths equal to the critical bandwidth. One class of critical band scales is called *Bark frequency scale*. It is hoped that by treating spectral energy over the Bark scale, a more natural fit with spectral information processing in the ear can be achieved. The Bark scale ranges from 1 to 24 Barks, corresponding to 24 critical bands of hearing as shown in Table 2.3. As shown in Figure 2.12, the perceptual resolution is finer in the lower frequencies. It should be noted that the ear's critical bands are continuous, and a tone of any audible frequency always finds a critical band centered on it. The Bark frequency b can be expressed in terms of the linear frequency (in Hz) by

$$b(f) = 13\arctan(0.00076f) + 3.5 * \arctan\left((f/7500)^2\right) \quad (Bark) \tag{2.5}$$

---

[1] A tone of frequency f_1 is said to be an octave above a tone with frequency f_2 if and only if $f_1 = 2f_2$.

[2] There are 12 semitones in one octave, so a tone of frequency f_1 is said to be a semitone above a tone with frequency f_2 if and only if $f_1 = 2^{1/12} f_2 = 1.05946 f_2$.

Table 2.3 The Bark frequency scale.

Bark Band #	Edge (Hz)	Center (Hz)
1	100	50
2	200	150
3	300	250
4	400	350
5	510	450
6	630	570
7	770	700
8	920	840
9	1080	1000
10	1270	1170
11	1480	1370
12	1720	1600
13	2000	1850
14	2320	2150
15	2700	2500
16	3150	2900
17	3700	3400
18	4400	4000
19	5300	4800
20	6400	5800
21	7700	7000
22	9500	8500
23	12000	10500
24	15500	13500

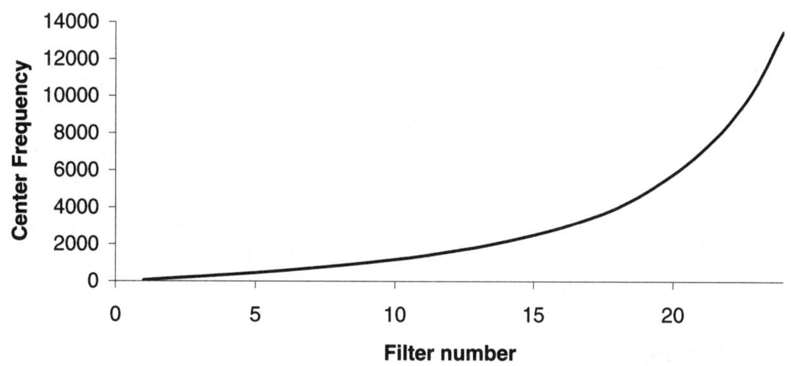

Figure 2.12 The center frequency of 24 Bark frequency filters as illustrated in Table 2.3.

Another such perceptually motivated scale is the mel frequency scale [41], which is linear below 1 kHz, and logarithmic above, with equal numbers of samples taken below and above 1 kHz. The mel scale is based on experiments with simple tones (sinusoids) in which subjects were required to divide given frequency ranges into four perceptually equal intervals or to adjust the frequency of a stimulus tone to be half as high as that of a comparison tone. One mel is defined as one thousandth of the pitch of a 1 kHz tone. As with all such attempts, it is hoped that the mel scale more closely models the sensitivity of the human ear than a purely linear scale and provides for greater discriminatory capability between speech segments. Mel-scale frequency analysis has been widely used in modern speech recognition systems. It can be approximated by:

$$B(f) = 1125 \ln(1 + f / 700) \tag{2.6}$$

The mel scale is plotted in Figure 2.13 together with the Bark scale and the bilinear transform (see Chapter 6).

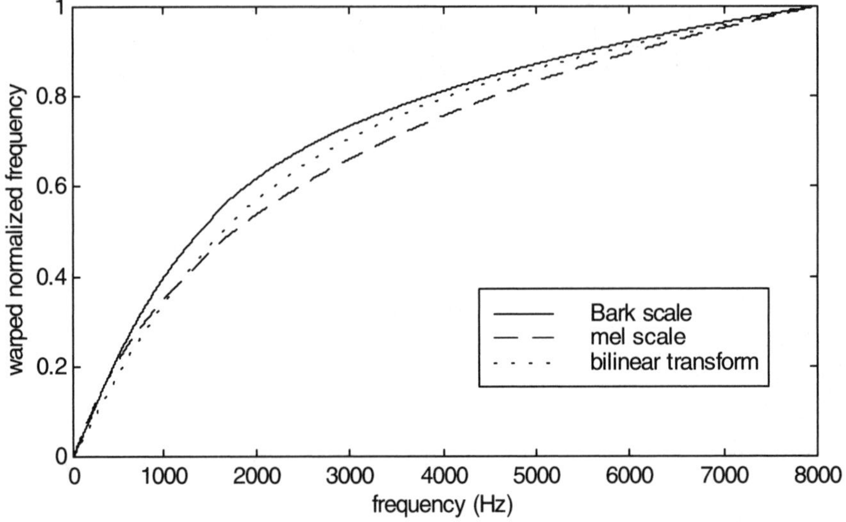

Figure 2.13 Frequency warping according to the Bark scale, ERB scale, mel-scale, and bilinear transform for $\alpha = 0.6$: linear frequency in the x-axis and normalized frequency in the y-axis.

A number of techniques in the modern spoken language system, such as cepstral analysis, and dynamic feature, have benefited tremendously from perceptual research as discussed throughout this book.

2.1.3.4. Masking

Frequency masking is a phenomenon under which one sound cannot be perceived if another sound close in frequency has a high enough level. The first sound *masks* the other one. Frequency-masking levels have been determined empirically, with complicated models that take into account whether the masker is a tone or noise, the masker's level, and other considerations.

We now describe a phenomenon known as *tone-masking noise.* It has been determined empirically that noise with energy E_N (dB) at Bark frequency g masks a tone at Bark frequency b if the tone's energy is below the threshold

$$T_T(b) = E_N - 6.025 - 0.275g + S_m(b-g) \quad (dB\ SPL) \tag{2.7}$$

where the *spread-of-masking* function $S_m(b)$ is given by

$$S_m(b) = 15.81 + 7.5(b+0.474) - 17.5\sqrt{1+(b+0.474)^2} \quad (dB) \tag{2.8}$$

We now describe a phenomenon known as *noise-masking tone.* It has been determined empirically that a tone at Bark frequency g with energy E_T (dB) masks noise at Bark frequency b if the noise energy is below the threshold

$$T_N(b) = E_T - 2.025 - 0.175g + S_m(b-g) \quad (dB\ SPL) \tag{2.9}$$

Masking thresholds are commonly referred to in the literature as Bark scale functions of *just noticeable distortion* (JND). Equation (2.8) can be approximated by a triangular spreading function that has slopes of +25 and –10 dB per Bark, as shown in Figure 2.14.

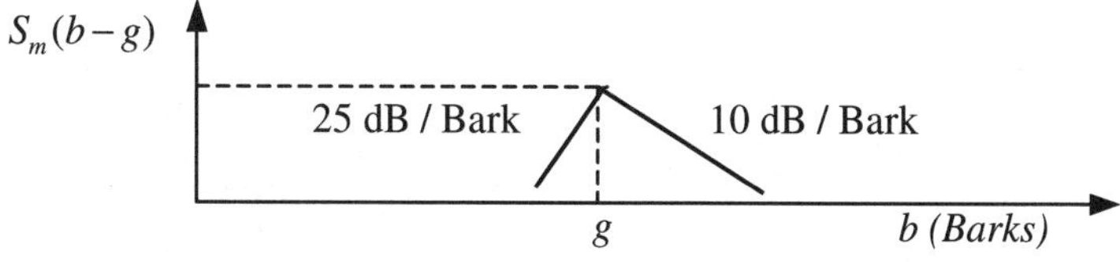

Figure 2.14 Contribution of Bark frequency g to the masked threshold $S_m(b)$.

In Figure 2.15 we show both the threshold of hearing and the masked threshold of a tone at 1 kHz with a 69 dB SPL. The combined masked threshold is the sum of the two in the linear domain

$$T(f) = 10\log_{10}\left(10^{0.1T_h(f)} + 10^{0.1T_T(f)}\right) \tag{2.10}$$

which is approximately the largest of the two.

In addition to frequency masking, there is a phenomenon called temporal masking by which a sound too close in time to another sound cannot be perceived. Whereas premasking tends to last about 5 ms, postmasking can last from 50 to 300 ms. Temporal masking level of a masker with a uniform level starting at 0 ms and lasting 200 ms is shown in Figure 2.16.

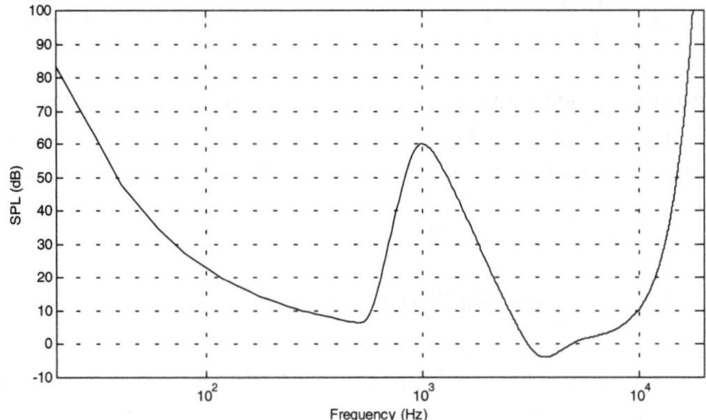

Figure 2.15 Absolute threshold of hearing and spread of masking threshold for a 1 kHz sinewave masker with a 69 dB SPL. The overall masked threshold is approximately the largest of the two thresholds.

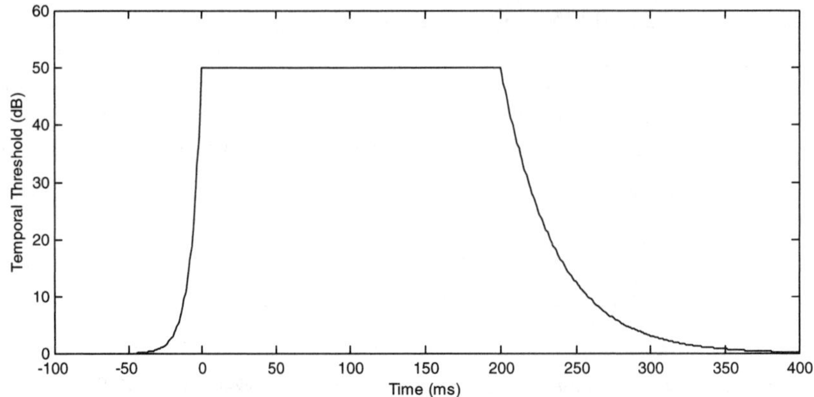

Figure 2.16 Temporal masking level of a masker with a uniform level starting at 0 ms and lasting 200 ms.

2.2. PHONETICS AND PHONOLOGY

We now discuss basic phonetics and phonology needed for spoken language processing. *Phonetics* refers to the study of speech sounds and their production, classification, and transcription. *Phonology* is the study of the distribution and patterning of speech sounds in a language and of the tacit rules governing pronunciation.

2.2.1. Phonemes

Linguist Ferdinand de Saussere (1857-1913) is credited with the observation that the relation between a sign and the object signified by it is arbitrary. The same concept, a certain yellow and black flying social insect, has the sign *honeybee* in English and *mitsubachi* in Japanese.

There is no particular relation between the various pronunciations and the meaning, nor do these pronunciations per se describe the bee's characteristics in any detail. For phonetics, this means that the speech sounds described in this chapter have no inherent meaning, and should be randomly distributed across the lexicon, except as affected by extraneous historical or etymological considerations. The sounds are just a set of arbitrary effects made available by human vocal anatomy. You might wonder about this theory when you observe, for example, the number of words beginning with *sn* that have to do with nasal functions in English: *sneeze, snort, sniff, snot, snore, snuffle*, etc. But Saussere's observation is generally true, except for obvious onomatopoetic (sound) words like *buzz.*

Like fingerprints, every speaker's vocal anatomy is unique, and this makes for unique vocalizations of speech sounds. Yet language communication is based on commonality of form at the perceptual level. To allow discussion of the commonalities, researchers have identified certain gross characteristics of speech sounds that are adequate for description and classification of words in dictionaries. They have also adopted various systems of notation to represent the subset of phonetic phenomena that are crucial for meaning.

As an analogy, consider the system of computer coding of text characters. In such systems, the *character* is an abstraction, e.g. the Unicode character U+0041. The identifying property of this character is its Unicode name *LATIN CAPITAL LETTER A*. This is a genuine abstraction; no particular realization is necessarily specified. As the Unicode 2.1 standard [1] states:

The Unicode Standard does not define glyph images. The standard defines how characters are interpreted, not how glyphs are rendered. The software or hardware-rendering engine of a computer is responsible for the appearance of the characters on the screen. The Unicode Standard does not specify the size, shape, nor orientation of on-screen characters.

Thus, the U+0041 character can be realized differently for different purposes, and in different sizes with different fonts:

U+0041➔ A, ***A***, A, A, A, …

The realizations of the character U+0041 are called glyphs, and there is no distinguished uniquely correct glyph for U+0041. In speech science, the term *phoneme* is used to denote any of the minimal units of speech sound in a language that can serve to distinguish one word from another. We conventionally use the term *phone* to denote a phoneme's acoustic realization. In the example given above, U+0041 corresponds to a phoneme and the various fonts correspond to the phone. For example, English phoneme */t/* have two very different acoustic realizations in the words *sat* and *meter*. You had better treat them as two different phones if you want to build a spoken language system. We will use the terms *phone* or *phoneme* interchangeably to refer to the speaker-independent and context-independent units of meaningful sound contrast. Table 2.4 shows a complete list of phonemes used in American English. The set of phonemes will differ in realization across individual speakers. But phonemes will always function systematically to differentiate meaning in words, just as the phoneme */p/* signals the word *pat* as opposed to the similar-sounding but distinct *bat*. The important contrast distinguishing this pair of words is */p/* vs. */b/*.

In this section we concentrate on the basic qualities that *define and differentiate abstract phonemes.* In Section 2.2.1.3 below we consider *why and how phonemes vary* in their actual realizations by different speakers and in different contexts.

Table 2.4 English phonemes used for typical spoken language systems.

Phonemes	Word Examples	Description
iy	*feel, eve, me*	front close unrounded
ih	*fill, hit, lid*	front close unrounded (lax)
ae	*at, carry, gas*	front open unrounded (tense)
aa	*father, ah, car*	back open unrounded
ah	*cut, bud, up*	open-mid back unrounded
ao	*dog, lawn, caught*	open-mid back round
ay	*tie, ice, bite*	diphthong with quality: aa + ih
ax	*ago, comply*	central close mid (schwa)
ey	*ate, day, tape*	front close-mid unrounded (tense)
eh	*pet, berry, ten*	front open-mid unrounded
er	*turn, fur, meter*	central open-mid unrounded rhoti-
ow	*go, own, tone*	back close-mid rounded
aw	*foul, how, our*	diphthong with quality: aa + uh
oy	*toy, coin, oil*	diphthong with quality: ao + ih
uh	*book, pull, good*	back close-mid unrounded (lax)
uw	*tool, crew, moo*	back close round
b	*big, able, tab*	voiced bilabial plosive
p	*put, open, tap*	voiceless bilabial plosive
d	*dig, idea, wad*	voiced alveolar plosive
t	*talk, sat*	voiceless alveolar plosive &
t	*meter*	alveolar flap
g	*gut, angle, tag*	voiced velar plosive
k	*cut, ken, take*	voiceless velar plosive
f	*fork, after, if*	voiceless labiodental fricative
v	*vat, over, have*	voiced labiodental fricative
s	*sit, cast, toss*	voiceless alveolar fricative
z	*zap, lazy, haze*	voiced alveolar fricative
th	*thin, nothing, truth*	voiceless dental fricative
dh	*then, father, scythe*	voiced dental fricative
sh	*she, cushion, wash*	voiceless postalveolar fricative
zh	*genre, azure*	voiced postalveolar fricative
l	*lid*	alveolar lateral approximant
l	*elbow, sail*	velar lateral approximant
r	*red, part, far*	retroflex approximant
y	*yacht, yard*	palatal sonorant glide
w	*with, away*	labiovelar sonorant glide
hh	*help, ahead, hotel*	voiceless glottal fricative
m	*mat, amid, aim*	bilabial nasal
n	*no, end, pan*	alveolar nasal
ng	*sing, anger*	velar nasal
ch	*chin, archer, march*	voiceless alveolar affricate: t + sh
jh	*joy, agile, edge*	voiced alveolar affricate: d + zh

2.2.1.1. Vowels

The tongue shape and positioning in the oral cavity do not form a major constriction of air flow during vowel articulation. However, variations of tongue placement give each vowel its distinct character by changing the resonance, just as different sizes and shapes of bottles give rise to different acoustic effects when struck. The primary energy entering the pharyngeal and oral cavities in vowel production vibrates at the fundamental frequency. The major resonances of the oral and pharyngeal cavities for vowels are called F1 and F2 – the first and second formants, respectively. They are determined by tongue placement and oral tract shape in vowels, and they determine the characteristic timbre or quality of the vowel.

The relationship of F1 and F2 to one another can be used to describe the English vowels. While the shape of the complete vocal tract determines the spectral outcome in a complex, nonlinear fashion, generally F1 corresponds to the back or pharyngeal portion of the cavity, while F2 is determined more by the size and shape of the oral portion, forward of the major tongue extrusion. This makes intuitive sense – the cavity from the glottis to the tongue extrusion is longer than the forward part of the oral cavity, thus we would expect its resonance to be lower. In the vowel of *see*, for example, the tongue extrusion is far forward in the mouth, creating an exceptionally long rear cavity, and correspondingly low F1. The forward part of the oral cavity, at the same time, is extremely short, contributing to higher F2. This accounts for the wide separation of the two lowest dark horizontal bands in Figure 2.9, corresponding to F1 and F2, respectively. Rounding the lips has the effect of extending the front-of-tongue cavity, thus lowering F2. Typical values of F1 and F2 of American English vowels are listed in Table 2.5.

Table 2.5 Phoneme labels and typical formant values for vowels of English.

Vowel Labels	Mean F1 (Hz)	Mean F2 (Hz)
iy (feel)	300	2300
ih (fill)	360	2100
ae (gas)	750	1750
aa (father)	680	1100
ah (cut)	720	1240
ao (dog)	600	900
ax (comply)	720	1240
eh (pet)	570	1970
er (turn)	580	1380
ow (tone)	600	900
uh (good)	380	950
uw (tool)	300	940

The characteristic F1 and F2 values for vowels are sometimes called formant targets, which are ideal locations for perception. Sometimes, due to fast speaking or other limitations on performance, the speaker cannot quite attain an ideal target before the articulators begin shifting to targets for the following phoneme, which is phonetic context dependent. Additionally, there is a special class of vowels that combine two distinct sets of F1/F2 targets.

These are called *diphthongs.* As the articulators move, the initial vowel targets glide smoothly to the final configuration. Since the articulators are working faster in production of a diphthong, sometimes the *ideal* formant target values of the component values are not fully attained. Typical diphthongs of American English are listed in Table 2.6.

Table 2.6 The diphthongs of English.

Diphthong Labels	Components
ay (**tie**)	*/aa/* ➔ */iy/*
ey (**a**te)	*/eh/* ➔ */iy/*
oy (c**oi**n)	*/ao/* ➔ */iy/*
aw (f**ou**l)	*/aa/* ➔ */uw/*

Figure 2.17 shows the first two formants for a number of typical vowels.

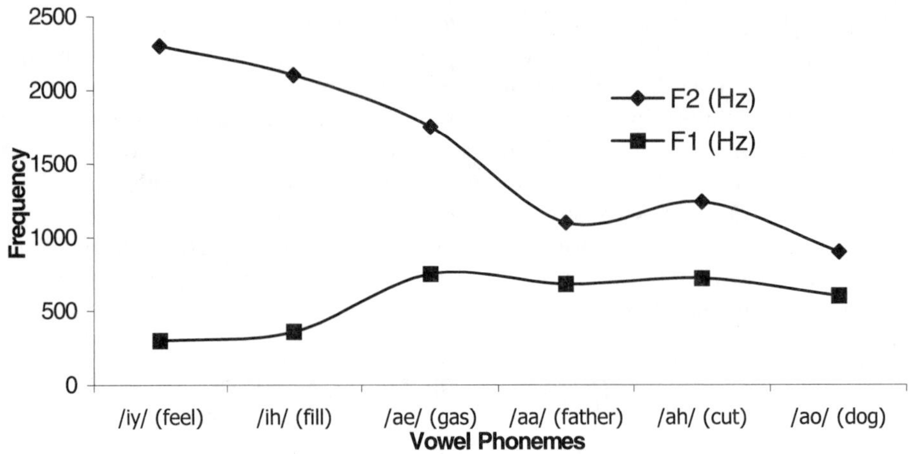

Figure 2.17 F1 and F2 values for articulations of some English vowels.

The major articulator for English vowels is the middle to rear portion of the tongue. The position of the tongue's surface is manipulated by large and powerful muscles in its root, which move it as a whole within the mouth. The linguistically important dimensions of movement are generally the ranges [front ⇔ back] and [high ⇔ low]. You can feel this movement easily. Say mentally, or whisper, the sound */iy/* (as in *see*) and then */aa/* (as in *father*). Do it repeatedly, and you will get a clear perception of the tongue movement from high to low. Now try */iy/* and then */uw/* (as in *blue*), repeating a few times. You will get a clear perception of place of articulation from front */iy/* to back */uw/*. Figure 2.18 shows a schematic characterization of English vowels in terms of relative tongue positions. There are two kinds of vowels: those in which tongue height is represented as a point and those in which it is represented as a vector.

Though the tongue hump is the major actor in vowel articulation, other articulators come into play as well. The most important secondary vowel mechanism for English and many other languages is lip rounding. Repeat the exercise above, moving from the */iy/* (*see*) to the */uw/* (*blue*) position. Now rather than noticing the tongue movement, pay attention to your lip shape. When you say */iy/*, your lips will be flat, slightly open, and somewhat spread. As you move to */uw/*, they begin to *round out*, ending in a more puckered position. This lengthens the oral cavity during */uw/*, and affects the spectrum in other ways.

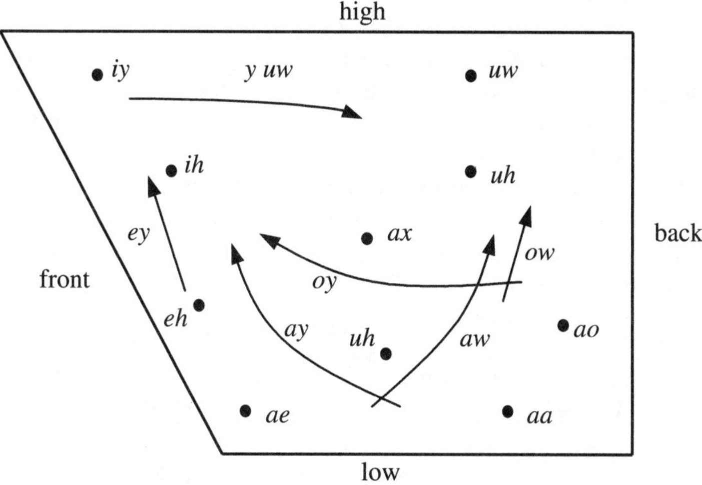

Figure 2.18 Relative tongue positions of English vowels [24].

Though there is always some controversy, linguistic study of phonetic abstractions, called *phonology*, has largely converged on the five binary features: +/- high, +/- low, +/- front, +/- back, and +/- round, plus the phonetically ambiguous but phonologically useful feature +/- tense, as adequate to uniquely characterize the major vowel distinctions of Standard English (and many other languages). Obviously, such a system is a little bit too free with logically contradictory specifications, such as [+high, +low], but these are excluded from real-world use. These features can be seen in Table 2.7.

Table 2.7 Phonological (abstract) feature decomposition of basic English vowels.

Vowel	high	low	front	back	round	tense
iy	+	-	+	-	-	+
ih	+	-	+	-	-	-
ae	-	+	+	-	-	+
aa	-	+	-	-	-	+
ah	-	-	-	-	-	+
ao	-	+	-	+	+	+
ax	-	-	-	-	-	-
eh	-	-	+	-	-	-
ow	-	-	-	+	+	+
uh	+	-	-	+	-	-
uw	+	-	-	+	-	+

This kind of abstract analysis allows researchers to make convenient statements about classes of vowels that behave similarly under certain conditions. For example, one may speak simply of the high vowels to indicate the set */iy, ih, uh, uw/*.

2.2.1.2. Consonants

Consonants, as opposed to vowels, are characterized by significant constriction or obstruction in the pharyngeal and/or oral cavities. Some consonants are voiced; others are not. Many consonants occur in pairs, that is, they share the same configuration of articulators, and one member of the pair additionally has voicing which the other lacks. One such pair is */s, z/*, and the voicing property that distinguishes them shows up in the non-periodic noise of the initial segment */s/* in Figure 2.5 as opposed to the voiced consonant end-phone, */z/*. Manner of articulation refers to the articulation mechanism of a consonant. The major distinctions in manner of articulation are listed in Table 2.8.

Table 2.8 Consonant manner of articulation.

Manner	Sample Phone	Example Words	Mechanism
Plosive	*/p/*	tat, tap	Closure in oral cavity
Nasal	*/m/*	team, meet	Closure of nasal cavity
Fricative	*/s/*	sick, kiss	Turbulent airstream noise
Retroflex liquid	*/r/*	rat, tar	Vowel-like, tongue high and curled back
Lateral liquid	*/l/*	lean, kneel	Vowel-like, tongue central, side airstream
Glide	*/y/,/w/*	yes, well	Vowel-like

The English phones that typically have voicing without complete obstruction or narrowing of the vocal tract are called *semivowels* and include */l, r/*, the *liquid* group, and */y, w/*, the *glide* group. Liquids, glides, and vowels are all *sonorant*, meaning they have continuous voicing. Liquids */l/* and */r/* are quite vowel-like and in fact may become *syllabic* or act entirely as vowels in certain positions, such as the *l* at the end of *edible*. In */l/*, the airstream flows around the sides of the tongue, leading to the descriptive term *lateral*. In */r/*, the tip of the tongue is curled back slightly, leading to the descriptive term *retroflex*. Figure 2.19 shows some semivowels.

Glides */y, w/* are basically vowels */iy, uw/* whose initial position within the syllable require them to be a little shorter and to lack the ability to be stressed, rendering them just different enough from true vowels that they are classed as a special category of consonant. Pre-vocalic glides that share the syllable-initial position with another consonant, such as the */y/* in the second syllable of *computer /k uh m . p* **y** *uw . t er/*, or the */w/* in *quick /k* **w** *ih k/*, are sometimes called *on-glides*. The semivowels, as a class, are sometimes called *approximants,* meaning that the tongue approaches the top of the oral cavity, but does not completely contact so as to obstruct the air flow.

Even the non-sonorant consonants that require complete or close-to-complete obstruction may still maintain some voicing before or during the obstruction, until the pressure differential across the glottis starts to disappear, due to the closure. Such voiced consonants include */b,d,g, z, zh, v/*. They have a set of counterparts that differ only in their characteristic lack of voicing: */p,t,k, s, sh, f/*.

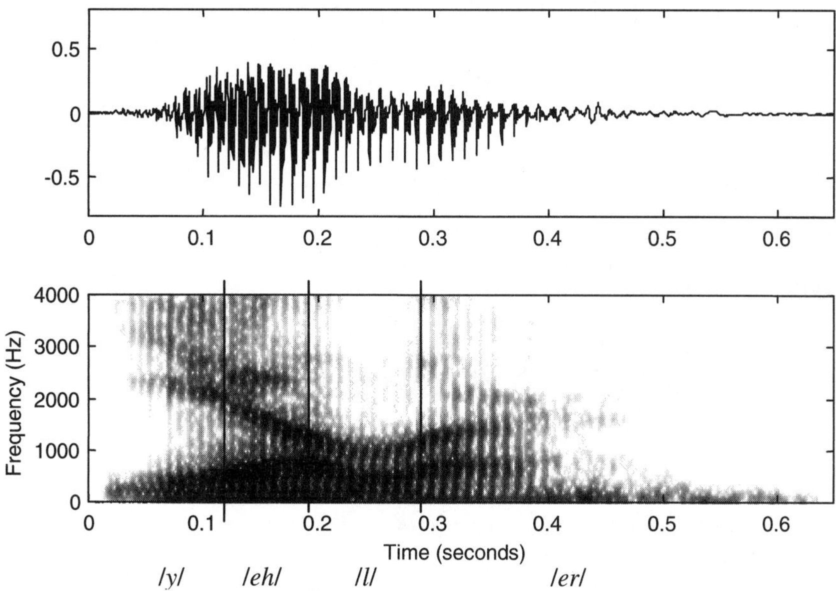

Figure 2.19 Spectrogram for the word *yeller*, showing semivowels */y/*, */l/*, */er/* (approximate phone boundaries shown with vertical lines).

Nasal consonants */m,n/* are a mixed bag: the oral cavity has significant constriction (by the tongue or lips), yet the voicing is continuous, like that of the sonorants, because, with the velar flap open, air passes freely through the nasal cavity, maintaining a pressure differential across the glottis.

A consonant that involves complete blockage of the oral cavity is called an obstruent stop, or plosive consonant. These may be voiced throughout if the trans-glottal pressure drop can be maintained long enough, perhaps through expansion of the wall of the oral cavity. In any case, there can be voicing for the early sections of stops. Voiced, unvoiced pairs of stops include: */b,p/*, */d,t/*, and */g,k/*. In viewing the waveform of a stop, a period of silence corresponding to the oral closure can generally be observed. When the closure is removed (by opening the constrictor, which may be lips or tongue), the trapped air rushes out in a more or less sudden manner. When the upper oral cavity is unimpeded, the closure of the vocal folds themselves can act as the initial blocking mechanism for a type of stop heard at the very beginning of vowel articulation in vowel-initial words like *atrophy*. This is called a *glottal stop*. Voiceless plosive consonants in particular exhibit a characteristic aperiodic *burst* of energy at the (articulatory) point of closure as shown in Figure 2.20 just prior to */i/*. By com-

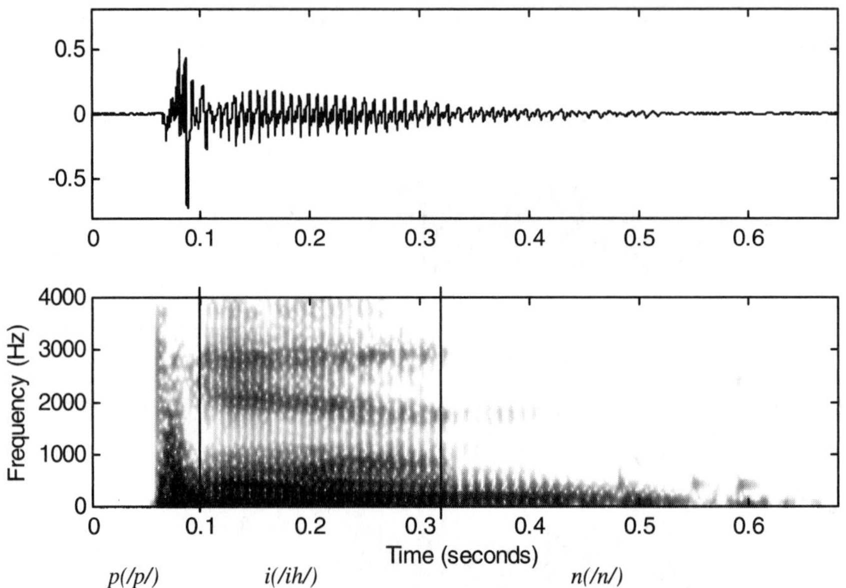

Figure 2.20 Spectrogram: stop release *burst* of /p/ in the word *pin*.

parison, the voicing of voiced plosive consonants may not always be obvious in a spectrogram.

A consonant that involves nearly complete blockage of some position in the oral cavity creates a narrow stream of turbulent air. The friction of this air stream creates a non-periodic hiss-like effect. Sounds with this property are called fricatives and include /*s*, *z*/. There is no voicing during the production of *s*, while there can be voicing (in addition to the frication noise), during the production of *z*, as discussed above. /*s*, *z*/ have a common place of articulation, as explained below, and thus form a natural similarity class. Though controversial, /*h*/ can also be thought of as a (glottal) fricative. /*s*/ in word-initial position and /*z*/ in word-final position are exemplified in Figure 2.5.

Some sounds are complex combinations of manners of articulation. For example, the *affricates* consist of a stop (e.g., /*t*/), followed by a fricative [e.g., /*sh*/) combining to make a unified sound with rapid phases of closure and continuancy (e.g., {*t* + *sh*} = *ch* as in *church*). The affricates in English are the voiced/unvoiced pairs: /*j*/ (*d* + *zh*) and /*ch*/ (*t* + *sh*). The complete consonant inventory of English is shown in Table 2.9.

Consider the set /*m*/, /*n*/, /*ng*/ from Table 2.9. They are all voiced nasal consonants, yet they sound distinct to us. The difference lies in the location of the major constriction along the top of the oral cavity (from lips to velar area) that gives each consonant its unique quality. The articulator used to touch or approximate the given location is usually some spot along the length of the tongue. As shown in Figure 2.21, the combination of articulator and place of articulation gives each consonant its characteristic sound:

Table 2.9 Manner of articulation of English consonants.

Consonant Labels	Consonant Examples	Voiced?	Manner
b	**b**ig, a**b**le, ta**b**	+	plosive
p	**p**ut, o**p**en, ta**p**	-	plosive
d	**d**ig, i**d**ea, wa**d**	+	plosive
t	**t**alk, sa**t**	-	plosive
g	**g**ut, an**g**le, ta**g**	+	plosive
k	**c**ut, oa**k**en, ta**k**e	-	plosive
v	**v**at, o**v**er, ha**v**e	+	fricative
f	**f**ork, a**f**ter, i**f**	-	fricative
z	**z**ap, la**z**y, ha**z**e	+	fricative
s	**s**it, ca**s**t, to**ss**	-	fricative
dh	**th**en, fa**th**er, scy**th**e	+	fricative
th	**th**in, no**th**ing, tru**th**	-	fricative
zh	**g**enre, a**z**ure, bei**g**e	+	fricative
sh	**sh**e, cu**sh**ion, wa**sh**	-	fricative
jh	**j**oy, a**g**ile, e**dg**e	+	affricate
ch	**ch**in, ar**ch**er, mar**ch**	-	affricate
l	**l**id, e**l**bow, sai**l**	+	lateral
r	**r**ed, pa**r**t, fa**r**	+	retroflex
y	**y**acht, on**i**on, **y**ard	+	glide
w	**w**ith, a**w**ay	+	glide
hh	**h**elp, a**h**ead, **h**otel	+	fricative
m	**m**at, a**m**id, ai**m**	+	nasal
n	**n**o, e**n**d, pa**n**	+	nasal
ng	si**ng**, a**ng**er, dri**n**k	+	nasal

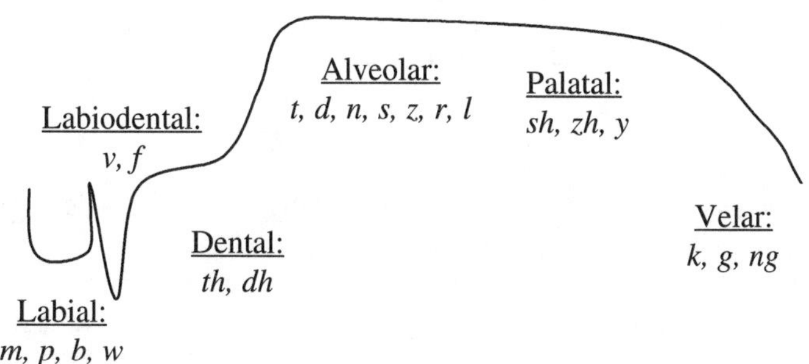

Figure 2.21 The major places of consonant articulation with respect to the human mouth.

- The *labial* consonants have their major constriction at the lips. This includes */p/, /b/* (these two differ only by manner of articulation) and */m/* and */w/*.
- The class of *dental or labio-dental* consonants includes */f, v/* and */th, dh/* (the members of these groups differ in manner, not place).
- *Alveolar* consonants bring the front part of the tongue, called the tip or the part behind the tip called the blade, into contact or approximation to the alveolar ridge, rising semi-vertically above and behind the teeth. These include */t, d, n, s, z, r, l/*. The members of this set again differ in manner of articulation (voicing, continuity, nasality), rather than place.
- *Palatal* consonants have approximation or constriction on or near the roof of the mouth, called the palate. The members include */sh, zh, y/*.
- *Velar* consonants bring the articulator (generally the back of the tongue), up to the rearmost top area of the oral cavity, near the velar flap. Velar consonants in English include */k, g/* (differing by voicing) and the nasal continuant */ng/*.

With the place terminology, we can complete the descriptive inventory of English consonants, arranged by manner (rows), place (columns), and voiceless/voiced (pairs in cells) as illustrated in Table 2.10.

Table 2.10 The consonants of English arranged by place (columns) and manner (rows).

	Labial	**Labio-dental**	**Dental**	**Alveolar**	**Palatal**	**Velar**	**Glottal**
Plosive	*p b*			*t d*		*k g*	*?*
Nasal	*m*			*n*		*ng*	
Fricative		*f v*	*th dh*	*s z*	*sh zh*		*h*
Retroflex sonorant				*r*			
Lateral sonorant				*l*			
Glide	*w*				*y*		

2.2.1.3. Phonetic Typology

The oral, nasal, pharyngeal, and glottal mechanisms actually make available a much wider range of effects than English happens to use. So, it is expected that other languages would utilize other vocal mechanisms, in an internally consistent but essentially arbitrary fashion, to represent their lexicons. In addition, often a vocal effect that is part of the systematic linguistic phonetics of one language is present in others in a less codified, but still perceptible, form. For example, Japanese vowels have a characteristic distinction of length that can be hard for non-natives to perceive and to use when learning the language. The words *kado* (*corner*) and *kaado* (*card*) are spectrally identical, differing only in that *kado* is much shorter in all contexts. The existence of such minimally-contrasting pairs is taken as conclusive evidence that *length* is phonemically distinctive for Japanese. As noted above, what is linguistically distinctive in any one language is generally present as a less *meaningful* signaling dimension in other languages. Thus, vowel length can be manipulated in any English word as well, but this occurs either consciously for emphasis or humorous effect, or unconsciously and very predictably at clause and sentence end positions, rather than to signal lexical identity in all contexts, as in Japanese.

Other interesting sounds that the English language makes no linguistic use of include the trilled *r* sound and the implosive. The trilled *r* sound is found in Spanish, distinguishing (for example) the words *pero* (*but*) and *perro* (*dog*). This trill could be found in times past as a non-lexical sound used for emphasis and interest by American circus ringmasters and other showpersons.

While the world's languages have all the variety of manner of articulation exemplified above and a great deal more, the primary dimension lacking in English that is exploited by a large subset of the world's languages is pitch variation. Many of the huge language families of Asia and Africa are tonal, including all varieties of Chinese. A large number of other languages are not considered strictly tonal by linguistics, yet they make systematic use of pitch contrasts. These include Japanese and Swedish. To be considered tonal, a language should have lexical meaning contrasts cued by pitch, just as the lexical meaning contrast between *pig* and *big* is cued by a voicing distinction in English. For example, Mandarin Chinese has four primary tones (tones can have minor context-dependent variants just like ordinary phones, as well) as shown in Table 2.11.

Table 2.11 The contrastive tones of Mandarin Chinese.

Tone	Shape	Example	Chinese	Meaning
1	High level	*ma*	妈	mother
2	High rising	*ma*	麻	numb
3	Low rising	*ma*	马	horse
4	High falling	*ma*	骂	to scold

Though English does not make systematic use of pitch in its inventory of word contrasts, nevertheless, as we always see with any possible phonetic effect, pitch is systematically varied in English to signal a speaker's emotions, intentions, and attitudes, and it has some linguistic function in signaling grammatical structure as well. Pitch variation in English will be considered in more detail in Chapter 15.

2.2.2. The Allophone: Sound and Context

The vowel and consonant charts provide abstract symbols for the phonemes – major sound distinctions. Phonemic units should be correlated with potential meaning distinctions. For example, the change created by holding the tongue high and front (*/iy/*) vs. directly down from the (frontal) position for /eh/, in the consonant context /m _ n/, corresponds to an important meaning distinction in the lexicon of English: *mean /m iy n/* vs. *men /m eh n/*. This meaning contrast, conditioned by a pair of rather similar sounds, in an identical context, justifies the inclusion of /iy/ and /eh/ as logically separate distinctions.

However, one of the fundamental, meaning-distinguishing sounds is often modified in some systematic way by its phonetic neighbors. The process by which neighboring sounds influence one another is called *coarticulation*. Sometimes, when the variations resulting from coarticulatory processes can be consciously perceived, the modified phonemes are called *allophones*. Allophonic differences are always *categorical*, that is, they can be understood and denoted by means of a small, bounded number of symbols or diacritics on the basic phoneme symbols.

As an experiment, say the word *like* to yourself. Feel the front of the tongue touching the alveolar ridge (cf. Figure 2.21) when realizing the initial phoneme /l/. This is one allophone of /l/, the so-called *light* or *clear* /l/. Now say *kill*. In this word, most English speakers will no longer feel the front part of the tongue touch the alveolar ridge. Rather, the /l/ is realized by stiffening the broad midsection of the tongue in the rear part of the mouth while the continuant airstream escapes laterally. This is another allophone of /l/, conditioned by its syllable-final position, called the *dark* /l/. Predictable contextual effects on the realization of phones can be viewed as a nuisance for speech recognition, as will be discussed in Chapter 9. On the other hand, such variation, because it is systematic, could also serve as a cue to the syllable, word, and prosodic structure of speech.

Now experiment with the sound /p/ by holding a piece of tissue in front of your mouth while saying the word *pin* in a normal voice. Now repeat this experiment with *spin*. For most English speakers, the word *pin* produces a noticeable puff of air, called aspiration. But the same phoneme, /p/, embedded in the consonant cluster /sp/ loses its aspiration (burst, see the lines bracketing the /p/ release in *pin* and *spin* in Figure 2.22), and because these two types of /p/ are in complementary distribution (completely determined by phonetic and syllabic context), the difference is considered allophonic.

Try to speak the word *bat* in a framing phrase *say bat again*. Now speak *say bad again*. Can you feel the length difference in the vowel /ae/? A vowel before a voiced consonant, e.g., /d/, seems typically longer than the same vowel before the unvoiced counterpart, in this case /t/.

A sound phonemicized as /t/ or /d/, that is, a stop made with the front part of the tongue, may be reduced to a quick tongue tap that has a different sound than either /t/ or /d/ in fuller contexts. This process is called flapping. It occurs when /t/ or /d/ closes a stressed vowel (coda position) followed by an unstressed vowel, as in: *bitter, batter, murder, quarter, humidity,* and can even occur across words as long as the preconditions are met, as in *you can say that again*. Sometimes the velar flap opens too soon (anticipation), giving a characteristically nasal quality to some pre-nasal vowels such as /ae/ in *ham* vs. *had*. We have a more detailed discussion on allophones in Chapter 9.

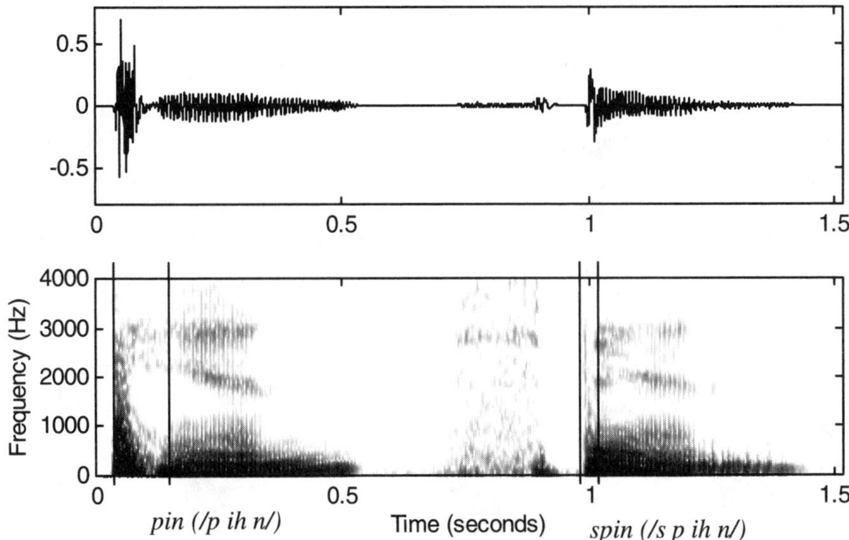

Figure 2.22 Spectrogram: bursts of *pin* and *spin*. The relative duration of a *p*-burst in different phonetic contexts is shown by the differing width of the area between the vertical lines.

2.2.3. Speech Rate and Coarticulation

In addition to allophones, there are other variations in speech for which no small set of established categories of variation can be established. These are *gradient*, existing along a scale for each relevant dimension, with speakers scattered widely. In general, it is harder to become consciously aware of coarticulation effects than of allophonic alternatives.

Individual speakers may vary their rates according to the content and setting of their speech, and there may be great inter-speaker differences as well. Some speakers may pause between every word, while others may speak hundreds of words per minute with barely a pause between sentences. At the faster rates, formant targets are less likely to be fully achieved. In addition, individual allophones may merge.

For example [20], consider the utterance *Did you hit it to Tom*? The pronunciation of this utterance is */d ih d y uw h ih t ih t t uw t aa m/*. However, a realistic, casual rendition of this sentence would appear as */d ih jh ax hh ih dx ih t ix t aa m/,* where */ix/* is a reduced schwa */ax/* that is short and often unvoiced, and */dx/* is a kind of shortened, indistinct stop, intermediate between */d/* and */t/*. The following five phonologic rules have operated on altering the pronunciation in the example:

- Palatalization of /*d*/ before /*y*/ in *di<u>d</u> you*
- Reduction of unstressed /*u*/ to schwa in *y<u>o</u>u*

- Flapping of intervocalic /t/ in *hi<u>t</u> it*
- Reduction of schwa and devoicing of /u/ in *t<u>o</u>*
- Reduction of geminate (double consonant) /t/ in *i<u>t t</u>o*

There are also coarticulatory influences in the spectral appearance of speech sounds, which can only be understood at the level of spectral analysis. For example, in vowels, consonant neighbors can have a big effect on formant trajectories near the boundary. Consider the differences in *F1* and *F2* in the vowel /eh/ as realized in words with different initial consonants *bet*, *debt*, and *get*, corresponding to the three major places of articulation (labial, alveolar, and velar), illustrated in Figure 2.23. You can see the different relative spreads of *F1* and *F2* following the initial stop consonants.

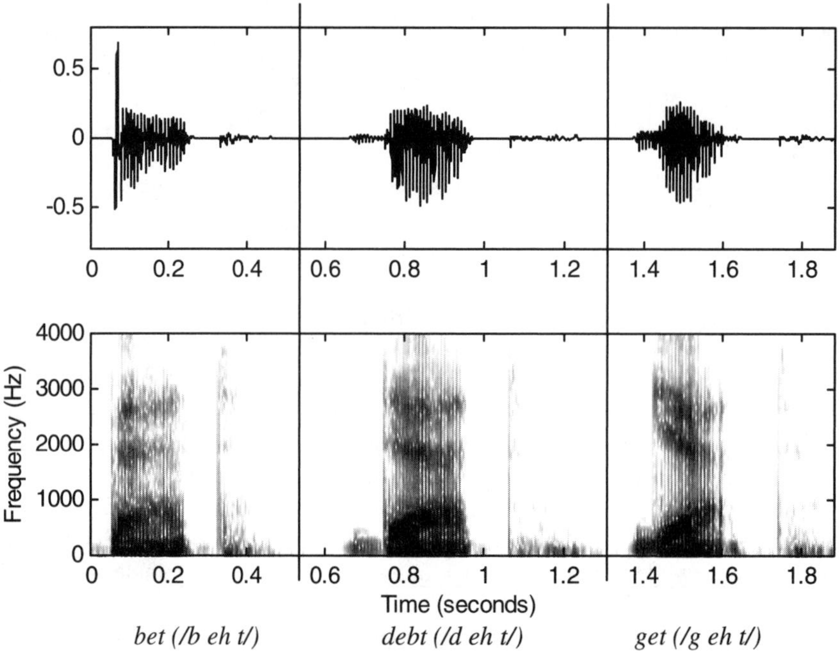

Figure 2.23 Spectrogram: *bet, debt,* and *get* (separated by vertical lines). Note different relative spreads of *F1* and *F2* following the initial stop consonants in each word.

Now let's see different consonants following the same vowel, *ebb*, *head*, and *egg*. In Figure 2.23, the coarticulatory effect is *perseverance*; i.e., in the early part of the vowel the articulators are still somewhat set from realization of the initial consonant. In the *ebb*, *head*, and *egg* examples shown in Figure 2.24, the coarticulatory effect is *anticipation*; i.e., in the latter part of the vowel the articulators are moving to prepare for the upcoming consonant articulation. You can see the increasing relative spread of *F1* and *F2* at the final vowel-consonant transition in each word.

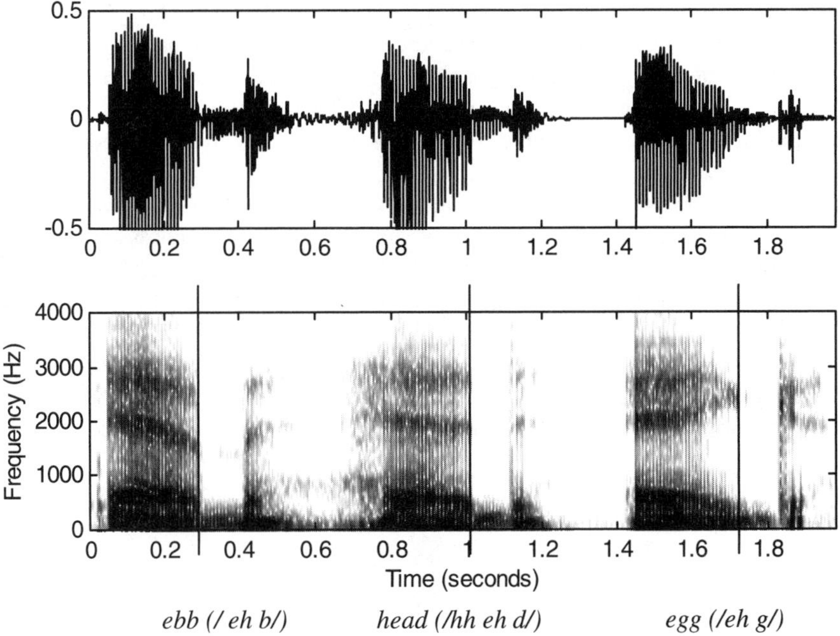

Figure 2.24 Spectrogram: *ebb*, *head*, and *egg*. Note the increasing relative spread of *F1* and *F2* at the final vowel-consonant transition in each word.

2.3. Syllables and Words

Phonemes are small building blocks. To contribute to language meaning, they must be organized into longer cohesive spans, and the units so formed must be combined in characteristic patterns to be meaningful, such as syllables and words in the English language.

2.3.1. Syllables

An intermediate unit, the *syllable*, is sometimes thought to interpose between the phones and the word level. The syllable is a slippery concept, with implications for both production and perception. Here we will treat it as a perceptual unit. Syllables are generally centered around vowels in English, giving two perceived syllables in a word like *tomcat*: */tOm-cAt/*. To completely parse a word into syllables requires making judgments of consonant affiliation (with the syllable peak vowels). The question of whether such judgments should be based on articulatory or perceptual criteria, and how they can be rigorously applied, remains unresolved.

Syllable centers can be thought of as *peaks* in sonority (high-amplitude, periodic sections of the speech waveform). These sonority peaks have affiliated *shoulders* of strictly non-increasing sonority. A scale of sonority can be used, ranking consonants along a continuum of stops, affricates, fricatives, and approximants. So, in a word like *verbal*, the syllabification would be *ver-bal*, or *verb-al*, but not *ve-rbal*, because putting the approximant /r/ before the stop /b/ in the second syllable would violate the non-decreasing sonority requirement heading into the syllable.

As long as the sonority conditions are met, the exact affiliation of a given consonant that could theoretically affiliate on either side can be ambiguous, unless determined by higher-order considerations of word structure, which may block affiliation. For example, in a word like *beekeeper*, an abstract boundary in the compound between the component words *bee* and *keeper* keeps us from accepting the syllable parse: *beek-eeper*, based on lexical interpretation. However, the same phonetic sequence in *beaker* could, depending on one's theory of syllabicity, permit affiliation of the *k: beak-er*. In general, the syllable is a unit that has intuitive plausibility but remains difficult to pin down precisely.

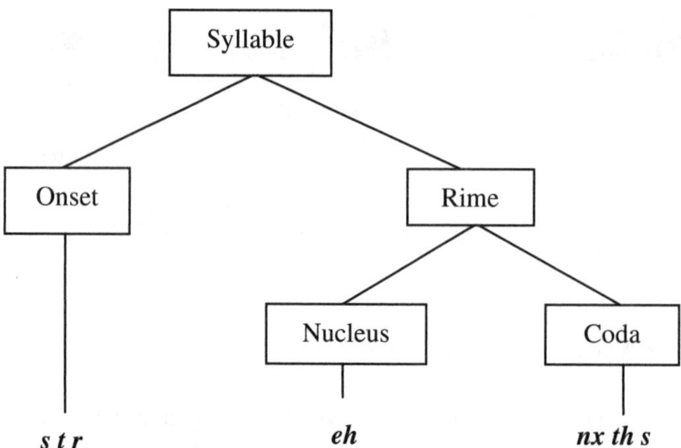

Figure 2.25 The word/syllable *strengths* (*/s t r eh nx th s/*) is the longest syllable of English.

Syllables are thought (by linguistic theorists) to have internal structure, and the terms used are worth knowing. Consider a big syllable such as *strengths /s t r eh nx th s/*. This consists of a vowel peak, called the *nucleus*, surrounded by the other sounds in characteristic positions. The *onset* consists of initial consonants if any, and the rime is the nucleus with trailing consonants (the part of the syllable that matters in determining poetic rhyme). The coda consists of consonants in the rime following the nucleus (in some treatments, the last consonant in a final cluster would belong to an *appendix*). This can be diagrammed as a syllable parse tree as shown in Figure 2.25. The syllable is sometimes thought to be the primary domain of coarticulation, that is, sounds within a syllable influence one another's realization more than the same sounds separated by a syllable boundary.

2.3.2. Words

The concept of words seems intuitively obvious to most speakers of Indo-European languages. It can be loosely defined as a lexical item, with an agreed-upon meaning in a given speech community, that has the freedom of syntactic combination allowed by its type (noun, verb, etc.).

In spoken language, there is a segmentation problem: words *run together* unless affected by a disfluency (unintended speech production problem) or by the deliberate placement of a pause (silence) for some structural or communicative reason. This is surprising to many people, because literacy has conditioned speakers/readers of Indo-European languages to expect a *blank space* between words on the printed page. But in speech, only a few true pauses (the aural equivalent of a blank space) may be present. So, what appears to the reading eye as *never give all the heart, for love* would appear to the ear, if we simply use letters to stand for their corresponding English speech sounds, as *nevergivealltheheart forlove* or, in phonemes, as *n eh v er g ih v ah l dh ax h aa r t* \\ *f ao r l ah v*. The \\ symbol marks a linguistically motivated pause, and the units so formed are sometimes called *intonation phrases*, as explained in Chapter 15.

Certain facts about word structure and combinatorial possibilities are evident to most native speakers and have been confirmed by decades of linguistic research. Some of these facts describe relations among words when considered in isolation, or concern groups of related words that seem intuitively similar along some dimension of form or meaning – these properties are *paradigmatic*. Paradigmatic properties of words include part-of-speech, inflectional and derivational morphology, and compound structure. Other properties of words concern their behavior and distribution when combined for communicative purposes in fully functioning utterances – these properties are *syntagmatic*.

2.3.2.1. Lexical Part-of-Speech

Lexical part-of-speech (POS) is a primitive form of linguistic theory that posits a restricted inventory of word-type categories, which capture generalizations of word forms and distributions. Assignment of a given POS specification to a word is a way of summarizing certain facts about its potential for syntagmatic combination. Additionally, paradigms of word formation processes are often similar within POS types and subtypes as well. The word properties upon which POS category assignments are based may include affixation behavior, very abstract semantic typologies, distributional patterns, compounding behavior, historical development, productivity and generalizabilty, and others.

A typical set of POS categories would include *noun*, *verb*, *adjective*, *adverb, interjection*, *conjunction*, *determiner*, *preposition*, and *pronoun*. Of these, we can observe that certain classes of words consist of infinitely large membership. This means new members can be added at any time. For example, the category of noun is constantly expanded to accommodate new inventions, such as Velcro or Spandex. New individuals are constantly being born, and their names are a type of noun called *proper noun*. The proliferation of words using the descriptive prefix *cyber* is another recent set of examples: *cyberscofflaw, cybersex,* and even *cyberia* illustrate the infinite creativity of humans in manipulating word structure to express new shades of meaning, frequently by analogy with, and using fragments of, existing vocabulary. Another example is the neologism *sheeple*, a noun combining the forms and meanings of *sheep* and *people* to refer to large masses of people who lack the capacity or willingness to take independent action. We can create new words whenever we like, but they had best fall within the predictable paradigmatic and syntagmatic patterns of use summarized by the existing POS generalizations, or there will be little hope of their adoption by any other speaker. These open POS categories are listed in Table 2.12. Nouns are inherently referential. They refer to persons, places, and things. Verbs are predicative; they indicate relations between entities and properties of entities, including participation in events. Adjectives typically describe and more completely specify noun reference, while adverbs describe, intensify, and more completely specify verbal relations. Open-class words are sometimes called *content* words, for their referential properties.

Table 2.12 Open POS categories.

Tag	Description	Function	Example
N	Noun	Names entity	*cat*
V	Verb	Names event or condition	*forget*
Adj	Adjective	Descriptive	*yellow*
Adv	Adverb	Manner of action	*quickly*
Interj	Interjection	Reaction	*oh!*

In contrast to the open-class categories, certain other categories of words only rarely and very slowly admit new members over the history of English development. These closed POS categories are shown in Table 2.13. The closed-category words are fairly stable over time. Conjunctions are used to join larger syntactically complete phrases. Determiners help to narrow noun reference possibilities. Prepositions denote common spatial and temporal relations of objects and actions to one another. Pronouns provide a convenient substitute for noun phrases that are fully understood from context. These words denote grammatical relations of other words to one another and fundamental properties of the world and how humans understand it. They can, of course, change slowly; for example, the Middle English pronoun *thee* is no longer in common use. The closed-class words are sometimes called *function* words.

Table 2.13 Closed POS categories.

Tag	Description	Function	Example
Conj	Conjunction	Coordinates phrases	*and*
Det	Determiner	Indicates definiteness	*the*
Prep	Preposition	Relations of time, space, direction	*from*
Pron	Pronoun	Simplified reference	*she*

The set of POS categories can be extended indefinitely. Examples can be drawn from the Penn Treebank project (http://www.cis.upenn.edu/ldc) as shown in Table 2.14, where you can find the proliferation of sub-categories, such as *Verb, base form* and *Verb, past tense*. These categories incorporate *morphological* attributes of words into the POS label system discussed in Section 2.3.2.2.

Table 2.14 Treebank POS categories – an expanded inventory.

String	Description	Example
CC	Coordinating conjunction	and
CD	Cardinal number	two
DT	Determiner	the
EX	Existential *there*	there (*There was an old lady*)
FW	Foreign word	omerta
IN	Preposition, subord. conjunction	over, but
JJ	Adjective	yellow
JJR	Adjective, comparative	better
JJS	Adjective, superlative	best
LS	List item marker	
MD	Modal	might
NN	Noun, singular or mass	rock, water
NNS	Noun, plural	rocks
NNP	Proper noun, singular	Joe
NNPS	Proper noun, plural	Red Guards
PDT	Predeterminer	all (*all the girls*)
POS	Possessive ending	's
PRP	Personal pronoun	I
PRP$	Possessive pronoun	mine
RB	Adverb	quickly
RBR	Adverb, comparative	higher (*shares closed higher.*)
RBS	Adverb, superlative	highest (*he jumped highest of all.*)
RP	Particle	up (*take up the cause*)
TO	*to*	to
UH	Interjection	hey!
VB	Verb, base form	choose
VBD	Verb, past tense	chose
VBG	Verb, gerund, or present participle	choosing
VBN	Verb, past participle	chosen
VBP	Verb, non-third person sing. present	jump
VBZ	Verb, third person singular present	jumps
WDT	Wh-determiner	which
WP	Wh-pronoun	who
WP$	Possessive wh-pronoun	whose
WRB	Wh-adverb	when (*When he came, it was late.*)

POS tagging is the process of assigning a part-of-speech or other lexical class marker to each word in a corpus. There are many algorithms to automatically tag input sentences into a set of tags. Rule-based methods [45], hidden Markov models (see Chapter 8) [23, 29, 46], and machine-learning methods [6] are used for this purpose.

2.3.2.2. Morphology

Morphology is about the subparts of words, i.e., the patterns of word formation including inflection, derivation, and the formation of compounds. English mainly uses prefixes and suffixes to express *inflection* and *derivational* morphology.

Inflectional morphology deals with variations in word form that reflect the contextual situation of a word in phrase or sentence syntax, and that rarely have direct effect on interpretation of the fundamental meaning expressed by the word. English inflectional morphology is relatively simple and includes person and number agreement and tense markings only. The variation in *cats* (vs. *cat*) is an example. The plural form is used to refer to an indefinite number of cats greater than one, depending on a particular situation. But the basic POS category (*noun*) and the basic meaning (*felis domesticus*) are not substantially affected. Words related to a common lemma via inflectional morphology are said to belong to a common paradigm, with a single POS category assignment. In English, common paradigm types include the verbal set of affixes (pieces of words): *-s, -ed, -ing*; the noun set: *-s*; and the adjectival *-er, -est*. Note that sometimes the base form may change spelling under affixation, complicating the job of automatic textual analysis methods. For historical reasons, certain paradigms may consist of highly idiosyncratic irregular variation as well, e.g., *go, going, went, gone* or *child, children*. Furthermore, some words may belong to defective paradigms, where only the singular (noun: *equipment*) or the plural (noun: *scissors*) is provided for.

In *derivational morphology*, a given root word may serve as the source for wholly new words, often with POS changes as illustrated in Table 2.15. For example, the terms *racial* and *racist*, though presumably based on a single root word *race,* have different POS possibilities (*adjective* vs. *noun-adjective*) and meanings. Derivational processes may induce pronunciation change or stress shift (e.g., *el**e**ctric* vs. *electr**i**city*). In English, typical derivational affixes (pieces of words) that are highly productive include prefixes and suffixes: *re-, pre-, -ial, -ism, -ish, -ity, -tion, -ness, -ment, -ious, -ify, -ize*, and others. In many cases, these can be added successively to create a complex layered form.

Table 2.15 Examples of stems and their related forms across POS categories.

Noun	Verb	Adjective	Adverb
criticism	*criticize*	*critical*	*critically*
fool	*fool*	*foolish*	*foolishly*
industry, industrialization	*industrialize*	*industrial,industrious*	*industriously*
employ, employee, employer	*employ*	*employable*	*employably*
certification	*certify*	*certifiable*	*certifiably*

Generally, word formation operates in layers, according to a kind of word syntax: *(deriv-prefix)* **root** *(root)* (deriv-suffix)* (infl-suffix).* This means that one or more *roots* can be compounded in the inner layer, with one or more optional *derivational prefixes*, followed by any number of optional *derivational suffixes*, capped off with no more than one *inflectional suffix*. There are, of course, limits on word formation, deriving both from semantics of the component words and simple lack of imagination. An example of a nearly maximal word in English might be *autocyberconceptualizations*, meaning (perhaps!) multiple instances of automatically creating computer-related concepts. This word lacks only compounding to be truly maximal. This word has a derivational prefix *auto-*, two root forms compounded (*cyber* and *concept*, though some may prefer to analyze *cyber-* as a prefix), three derivational suffixes (*-ual, -ize, -ation*), and is capped off with the plural inflectional suffix for nouns, *-s*.

2.3.2.3. Word Classes

POS classes are based on traditional grammatical and lexical analysis. With improved computational resources, it has become possible to examine words in context and assign words to groups according to their actual behavior in real text and speech from a statistical point of view. These kinds of classifications can be used in language modeling experiments for speech recognition, text analysis for text-to-speech synthesis, and other purposes.

One of the main advantages of word classification is its potential to derive more refined classes than traditional POS, while only rarely actually crossing traditional POS group boundaries. Such a system may group words automatically according to the similarity of usage with respect to their word neighbors. Consider classes automatically found by the classification algorithms of Brown *et al.* [7]:

{Friday Monday Thursday Wednesday Tuesday Saturday Sunday weekends}
{great big vast sudden mere sheer gigantic lifelong scant colossal}
{down backwards ashore sideways southward northward overboard aloft adrift}
{mother wife father son husband brother daughter sister boss uncle}
{John George James Bob Robert Paul William Jim David Mike}
{feet miles pounds degrees inches barrels tons acres meters bytes}

You can see that words are grouped together based on the semantic meaning, which is different from word classes created purely from syntactic point of view. Other types of classification are also possible, some of which can identify semantic relatedness across traditional POS categories. Some of the groups derived from this approach may include follows:

{problems problem solution solve analyzed solved solving}
{write writes writing written wrote pen}
{question questions asking answer answers answering}
{published publication author publish writer titled}

2.4. SYNTAX AND SEMANTICS

Syntax is the study of the patterns of formation of sentences and phrases from words and the rules for the formation of grammatical sentences. Semantics is another branch of linguistics dealing with the study of meaning, including the ways meaning is structured in language and changes in meaning and form over time.

2.4.1. Syntactic Constituents

Constituents represent the way a sentence can be divided into its grammatical subparts as constrained by common grammatical patterns (which implicitly incorporate normative judgments on acceptability). Syntactic constituents at least respect, and at best explain, the linear order of words in utterances and text. In this discussion, we will not strictly follow any of the many theories of syntax but will instead bring out a few basic ideas common to many approaches. We will not attempt anything like a complete presentation of the grammar of English but instead focus on a few simple phenomena.

Most work in syntactic theory has adopted machinery from traditional grammatical work on written language. Rather than analyze toy sentences, let's consider what kinds of superficial syntactic patterns are lurking in a random chunk of serious English text, excerpted from David Thoreau's essay *Civil Disobedience* [43]:

The authority of government, even such as I am willing to submit to – for I will cheerfully obey those who know and can do better than I, and in many things even those who neither know nor can do so well – is still an impure one: to be strictly just, it must have the sanction and consent of the governed. It can have no pure right over my person and property but what I concede to it. The progress from an absolute to a limited monarchy, from a limited monarchy to a democracy, is a progress toward a true respect for the individual.

2.4.1.1. Phrase Schemata

Words may be combined to form phrases that have internal structure and unity. We use generalized schemata to describe the phrase structure. The goal is to create a simple, uniform template that is independent of POS category.

Let's first consider nouns, a fundamental category referring to persons, places, and things in the world. The noun and its immediate modifiers form a constituent called the noun phrase (*NP*). To generalize this, we consider a word of arbitrary category, say category *X* (which could be a noun *N* or a verb *V*). The generalized rule for a phrase *XP* is *XP* $\Rightarrow$ *(modifiers) X-head (post-modifiers),* where *X* is the head, since it dominates the configuration and names the phrase. Elements preceding the head in its phrase are *premodifiers* and elements following the head are *postmodifiers*. *XP*, the culminating phrase node, is called a *maximal projection* of category *X*. We call the whole structure an *x-template*. Maximal projections, *XP*, are the primary currency of basic syntactic processes. The post-modifiers are usually maximal projections (another head, with its own post-modifiers forming an *XP* on its own) and are sometimes termed *complements*, because they are often required by the lexical properties of the head for a complete meaning to be expressed (e.g., when *X* is a preposition or verb). Complements are typically noun phrases (*NP*), prepositional phrases (*PP*), verb phrases (*VP*), or sentence/clause (*S*), which make an essential contribution to the head's reference or meaning, and which the head requires for semantic completeness. Premodifiers are likely to be adverbs, adjectives, quantifiers, and determiners, i.e., words that help to specify the meaning of the head but may not be essential for completing the meaning. With minor variations, the *XP* template serves for most phrasal types, based on the POS of the head (*N, V, ADJ*, etc.).

For *NP*, we thus have *NP* $\Rightarrow$ *(det) (modifier)* ***head-noun*** *(post-modifier).* This rule describes an *NP* (noun phrase – left side of arrow) in terms of its optional and required internal contents (right side of the arrow). *Det* is a word like *the* or *a* that helps to resolve the reference to a specific or an unknown instance of the noun. The *modifier* gives further information about the noun. The *head* of the phrase, and the only mandatory element, is the noun itself. *Post-modifiers* also give further information, usually in a more elaborate syntactic form than the simpler pre-modifiers, such as a relative clause or a prepositional phrase (covered below). The noun phrases of the passage above can be parsed as shown in Table 2.16. The head nouns may be personal pronouns (*I, it*), demonstrative and relative pronouns (*those*), coordinated nouns (*sanction and consent*), or common nouns (*individual*). The modifiers are mostly adjectives (*impure*, *pure*) or verbal forms functioning as adjectives (*limited*). The post-modifiers are interesting, in that, unlike the (pre-)modifiers, they are typically full phrases themselves, rather than isolated words. They include relative clauses (which are a kind of dependent sentence, e.g., *[those] who know and can do better than I*), as well as prepositional phrases (*of the governed*).

Table 2.16 *NP*s of the sample passage.

NP	Det	Mod	Head Noun	Post-Mod
1	the		**authority**	of government
2		even	**such**	as I am willing to submit to
3			**I**	
4			**those**	who know and can do better than I
5		many	**things**	
6		even	**those**	who neither know nor can do so well
7	an	impure	**one**	
8			**it**	
9	the		**sanction and consent**	of the governed
10	no	pure	**right**	over my person … concede to it.
11	the		**progress**	from an absolute to a limited monarchy
12	an	absolute	**[monarchy]**	
13	a	limited	**monarchy**	
14	a		**democracy**	
15	a		**progress**	
16	a	true	**respect**	for the individual
17	the		**individual**	

Table 2.17 *PP*s of the sample passage.

Head Prep	Complement (Postmodifier)
of	Government
as	I am willing to submit to
than	I
in	many things
of	the governed
over	my person and property
to	it
from	an absolute [monarchy]
to	a limited monarchy
to	a democracy
toward	a true respect [for the individual]
for	the individual

Prepositions express spatial and temporal relations, among others. These are also said to project according to the *X*-template, but usually lack a pre-modifier. Some examples from the sample passage are listed in Table 2.17. The complements of *PP* are generally *NP*s, which may be simple head nouns like *government*. However, other complement types, such as the verb phrase in *after discussing it with Jo*, are also possible.

For verb phrases, the postmodifier (or complement) of a head verb would typically be one or more *NP* (noun phrase) maximal projections, which might, for example, function as a direct object in a *VP* like *pet the cat*. The complement may or may not be optional, depending on characteristics of the head. We can now make some language-specific generalizations about English. Some verbs, such as *give*, may take more than one kind of complement. So an appropriate template for a *VP* maximal projection in English would appear abstractly as *VP* $\Rightarrow$ *(modifier) verb (modifier) (Complement1, Complement2 ComplementN)*. Complements are usually regarded as maximal projections, such as *NP*, *ADJP*, etc., and are enumerated in the template above, to cover possible multi-object verbs, such as *give*, which take both direct and indirect objects. Certain types of adverbs (*really*, *quickly*, *smoothly*, etc.) could be considered fillers for the *VP* modifier slots (before and after the head). In the sample passage, we find the following verb phrases as shown in Table 2.18.

VP presents some interesting issues. First, notice the multi-word verb *submit to*. Multi-word verbs such as *look after* and *put up with* are common. We also observe a number of auxiliary elements clustering before the verb in sentences of the sample passage: *am willing to submit to*, *will cheerfully obey*, and *can do better*. Rather than considering these as simple modifiers of the verbal head, they can be taken to have *scope* over the *VP* as a whole, which implies they are outside the *VP*. Since they are outside the *VP*, we can assume them to be heads in their own right, of phrases which require a *VP* as their complement. These elements mainly express tense (time or duration of verbal action) and modality (likelihood or probability of verbal action). In a full sentence, the *VP* has explicit or implicit inflection (projected from its verbal head) and indicates the person, number, and other context-dependent features of the verb in relation to its arguments. In English, the person (first, second, third) and number (singular, plural) attributes, collectively called agreement features, of subject and verb must match. For simplicity, we will lump all these considerations together as inflectional elements, and posit yet another phrase type, the Inflectional Phrase (*IP*): *IP* ⇒ *premodifier head VP-complement.*

Table 2.18 *VP*s of the sample passage.

Pre-mod	Verb Head	Post-mod	Complement
	submit to		[the authority of government]
cheerfully	**obey**		those who know and can do better than I
	is	still	an impure one
	be		strictly just
	have		the sanction
	have		no pure right
	concede		to it
	is		a progress

The premodifier slot (sometimes called the *specifier* position in linguistic theory) of an *IP* is often filled by the subject of the sentence (typically a noun or *NP*). Since the *IP* unites the subject of a sentence with a *VP*, *IP* can also be considered simply as the sentence category, often written as *S* in speech grammars.

2.4.1.2. Clauses and Sentences

The *subject* of a sentence is what the sentence is mainly about. A *clause* is any phrase with both a subject and a *VP* (*predicate* in traditional grammars) that has potentially independent interpretation – thus, for us, a clause is an *IP*, a kind of sentence. A phrase is a constituent lacking either subject, predicate, or both. We have reviewed a number of phrase types above. There are also various types of clauses and sentences.

Even though clauses are sentences from an internal point of view (having subject and predicate), they often function as simpler phrases or words would, e.g., as modifiers (adjective and adverbs) or nouns and noun phrases. Clauses may appear as post-modifiers for nouns (so-called *relative clauses*), basically a kind of adjective clause, sharing their subjects with the containing sentence. Some clauses function as *NP*s in their own right. One common clause type substitutes a *wh-word* like *who* or *what* for a direct object of a verb in the embedded clause, to create a *questioned noun phrase* or indirect question: (*I don't know who Jo saw.*). In these clauses, it appears to syntacticians that the *questioned* object of the verb [*VP* saw **who**] has been extracted or moved to a new surface position (following the main clause verb *know*). This is sometimes shown in the phrase-structure diagram by co-indexing an empty *ghost* or trace constituent at the original position of the question pronoun with the question-*NP* appearing at the surface site:

I don't know $[_{NPobj}\ [_{IP}\ [_{NPi}$ *who] Jo saw* $[_{NPi}\ \_\ ]]]$

$[_{NPsubj}\ [_{IP}$ *Whoever wins the game]] is our hero.*

There are various characteristic types of sentences. Some typical types include:

- Declarative: *I gave her a book.*
- Yes-no question: *Did you give her a book?*
- Wh-question: *What did you give her?*
- Alternatives question: *Did you give her a book, a scarf, or a knife?*
- Tag question: *You gave it to her, didn't you?*
- Passive: *She was given a book.*
- Cleft: *It must have been a book that she got.*
- Exclamative: *Hasn't this been a great birthday!*
- Imperative: *Give me the book.*

2.4.1.3. Parse Tree Representations

Sentences can be diagrammed in parse trees to indicate phrase-internal structure and linear precedence and immediate dominance among phrases. A typical phrase-structure tree for part of an embedded sentence is illustrated in Figure 2.26.

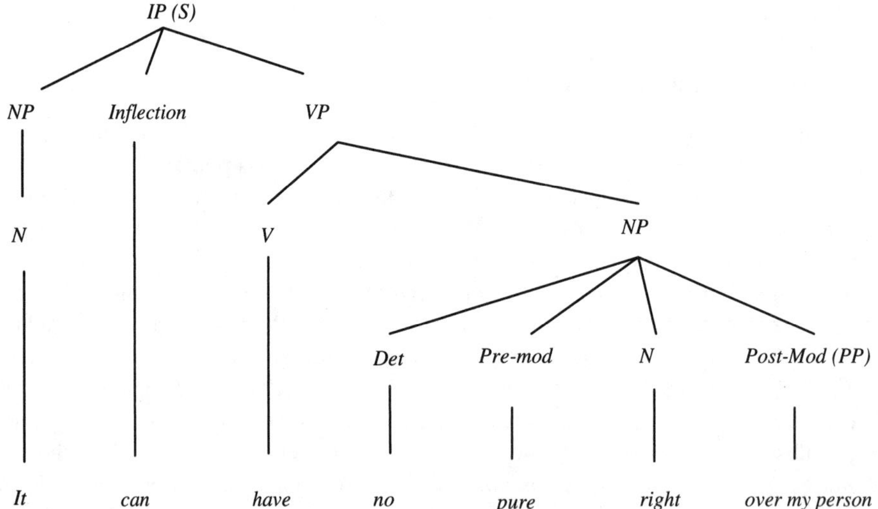

Figure 2.26 A simplified phrase-structure diagram.

For brevity, the same information illustrated in the tree can be represented as a bracketed string as follows:

$$[_{IP}\ [_{NP}\ [_{N}\ \textit{It}\ ]_{N}]_{NP}\ [_{I}\ \textit{can}\ ]_{I}\ [_{VP}\ [_{V}\ \textit{have}\ ]_{V}\ [_{NP}\ \textit{no pure right}\ [_{PP}\ \textit{over my person}\ ]_{PP}]_{NP}]_{VP}]_{IP}$$

With such a bracketed representation, almost every type of syntactic constituent can be coordinated or joined with another of its type, and usually a new phrase node of the common type is added to subsume the constituents such as *NP: We have* $[_{NP}$ $[_{NP}$ *tasty berries] and* $[_{NP}$ *tart juices]], IP/S:* $[_{IP}$ $[_{IP}$ *Many have come] and* $[_{IP}$ *most have remained]], PP: We went* $[_{PP}$ $[_{PP}$ *over the river] and* $[_{PP}$ *into the trees]],* and *VP: We want to* $[_{VP}$ $[_{VP}$ *climb the mountains] and* $[_{VP}$ *sail the seas]].*

2.4.2. Semantic Roles

In traditional syntax, grammatical roles are used to describe the direction or control of action relative to the verb in a sentence. Examples include the ideas of *subject, object, indirect object,* etc. Semantic roles, sometimes called case relations, seem similar but dig deeper. They are used to make sense of the participants in an event, and they provide a vocabulary for us to answer the basic question *who did what to whom.* As developed by [13] and others, the theory of semantic roles posits a limited number of universal roles. Each basic meaning of each verb in our mental dictionary is tagged for the obligatory and optional semantic roles used to convey the particular meaning. A typical inventory of case roles is given below:

Agent	*cause or initiator of action, often intentional*
Patient/Theme	*undergoer of the action*
Instrument	*how action is accomplished*
Goal	*to whom action is directed*
Result	*result of action*
Location	*location of action*

These can be realized under various syntactic identities, and can be assigned to both required complement and optional adjuncts. A noun phrase in the Agentive role might be the surface subject of a sentence, or the object of the preposition *by* in a passive. For example, the verb *put* can be considered a process that has, in one of its senses, the case role specifications shown in Table 2.19.

Table 2.19 Analysis of a sentence with *put.*

Analysis	**Example**			
	Kim	*put*	*the book*	*on the table.*
Grammatical functions	*Subject (NP)*	*Predicate (VP)*	*Object (NP)*	*Adverbial (ADVP)*
Semantic roles	*Agent*	*Instrument*	*Theme*	*Location*

Now consider this passive-tense example, where the semantic roles align with different grammatical roles shown in Table 2.20. Words that look and sound identical can have different meaning or different *senses* as shown in Table 2.21. The sporting sense of *put* (as in the sport of shot-put) illustrates the meaning/sense-dependent nature of the role patterns, because in this sense the Locative case is no longer obligatory, as it is in the original sense illustrated in Table 2.19 and Table 2.20.

Table 2.20 Analysis of passive sentence with *put.*

Analysis	**Example**		
	The book	*was put*	*on the table.*
Grammatical functions	*Subject (NP)*	*Predicate (VP)*	*Adverbial (ADVP)*
Semantic roles	*Agent*	*Instrument*	*Location*

Table 2.21 Analysis of a different pattern of *put.*

Analysis	**Example**		
	Kim	*put*	*the shot.*
Grammatical functions	*Subject (NP)*	*Predicate (VP)*	*Object (NP)*
Semantic roles	*Agent*	*Instrument*	*Theme*

The lexical meaning of a verb can be further decomposed into primitive semantic relations such as CAUSE, CHANGE, and BE. The verb *open* might appear as *CAUSE(NP1,PHYSICAL-CHANGE(NP2,NOT-OPEN,OPEN)).* This says that for an agent (*NP1*) to *open* a theme (*NP2*) is to cause the patient to change from a not-opened state to an opened state. Such systems can be arbitrarily detailed and exhaustive, as the application requires.

2.4.3. Lexical Semantics

The specification of particular meaning templates for individual senses of particular words is called *lexical semantics*. When words combine, they may take on propositional meanings resulting from the composition of their meanings in isolation. We could imagine that a speaker starts with a proposition in mind (logical form as will be discussed in the next section), creating a need for particular words to express the idea (lexical semantics); the proposition is then linearized (syntactic form) and spoken (phonological/phonetic form). Lexical semantics is the level of meaning before words are composed into phrases and sentences, and it may heavily influence the possibilities for combination.

Words can be defined in a large number of ways including by relations to other words, in terms of decomposition semantic primitives, and in terms of non-linguistic cognitive constructs, such as perception, action, and emotion. There are hierarchical and non-hierarchical relations. The main hierarchical relations would be familiar to most object-oriented programmers. One is *is-a* taxonomies (a *crow* is-a *bird*), which have transitivity of properties from type to subtype (inheritance). Another is *has-a* relations (a *car* has-a *windshield*), which are of several differing qualities, including process/subprocess (*teaching* has-a subprocess *giving exams*), and arbitrary or natural subdivisions of part-whole relations (*bread* has-a division into *slices, meter* has-a division into *centimeters*). Then there are non-branching hierarchies (no fancy name) that essentially form scales of degree, such as *frozen* ⇒ *cold* ⇒ *lukewarm* ⇒ *hot* ⇒ *burning*. Non-hierarchical relations include synonyms, such as *big/large*, and antonyms such as *good/bad*.

Words seem to have natural affinities and disaffinities in the semantic relations among the concepts they express. Because these affinities could potentially be exploited by future language understanding systems, researchers have used the generalizations above in an attempt to tease out a parsimonious and specific set of basic relations under which to group entire lexicons of words. A comprehensive listing of the families and subtypes of possible semantic relations has been presented in [10]. In Table 2.22, the leftmost column shows names for families of proposed relations, the middle column differentiates subtypes within each family, and the rightmost column provides examples of word pairs that participate in the proposed relation. Note that case roles have been modified for inclusion as a type of semantic relation within the lexicon.

We can see from Table 2.22 that a single word could participate in multiple relations of different kinds. For example, *knife* appears in the examples for *Similars: invited attribute* (i.e., a desired and expected property) as: *knife-sharp*, and also under *Case Relations*: *action-instrument*, which would label the relation of *knife* to the action *cut* in *He cut the bread with a knife.* This suggests that an entire lexicon could be viewed as a graph of semantic relations, with words or idioms as nodes and connecting edges between them representing semantic relations as listed above. There is a rich tradition of research in this vein.

The biggest practical problem of lexical semantics is the context-dependent resolution of senses of words – so-called polysemy. A classic example is *bank* – *bank of the stream* as opposed to *money in the bank.* While lexicographers try to identify distinct senses when they write dictionary entries, it has been generally difficult to rigorously quantify exactly what counts as a discrete sense of a word and to disambiguate the senses in practical contexts. Therefore, designers of practical speech understanding systems generally avoid the problem by limiting the domain of discourse. For example, in a financial application, generally only the sense of *bank* as a fiduciary institution is accessible, and others are assumed not to exist. It is sometimes difficult to make a principled argument as to how many distinct senses a word has, because at some level of depth and abstraction, what might appears as separate senses seem to be similar or related, as *face* could be *face of a clock* or *face of person*.

Senses are usually distinguished within a given part-of-speech (POS) category. Thus, when an occurrence of *bank* has been identified as a verb, the *shore* sense might be automatically eliminated, though depending on the sophistication of the system's lexicon and goals, there can be sense differences for many English verbs as well. Within a POS category, often the words that occur near a given ambiguous form in the utterance or discourse are clues to interpretation, where links can be established using semantic relations as described above. Mutual information measures as discussed in Chapter 3 can sometimes provide hints. In a context of dialog where other, less ambiguous financial terms come up frequently, the sense of *bank* as fiduciary institution is more likely. Finally, when all else fails, often senses can be ranked in terms of their a priori likelihood of occurrence. It should always be borne in mind that language is not static; it can change form under a given analysis at any time. For example, the stable English form *spinster*, a somewhat pejorative term for an older, never-married female, has recently taken on a new morphologically complex form, with the new sense of a high political official, or media spokesperson, employed to provide bland disinformation (*spin*) on a given topic.

Table 2.22 Semantic relations.

Family	Subtype	Example
Contrasts	Contrary	*old-young*
	Contradictory	*alive-dead*
	Reverse	*buy-sell*
	Directional	*front-back*
	Incompatible	*happy-morbid*
	Asymmetric contrary	*hot-cool*
	Attribute similar	*rake-fork*
Similars	Synonymity	*car-auto*
	Dimensional similar	*smile-laugh*
	Necessary attribute	*bachelor-unmarried*
	Invited attribute	*knife-sharp*
	Action subordinate	*talk-lecture*
Class Inclusion	Perceptual subord.	*animal-horse*
	Functional subord.	*furniture-chair*
	State subord.	*disease-polio*
	Activity subord.	*game-chess*
	Geographic subord.	*country-Russia*
	Place	*Germany-Hamburg*
Case Relations	Agent-action	*artist-paint*
	Agent-instrument	*farmer-tractor*
	Agent-object	*baker-bread*
	Action-recipient	*sit-chair*
	Action-instrument	*cut-knife*
Part-Whole	Functional object	*engine-car*
	Collection	*forest-tree*
	Group	*choir-singer*
	Ingredient	*table-wood*
	Functional location	*kitchen-stove*
	Organization	*college-admissions*
	Measure	*mile-yard*

2.4.4. Logical Form

Because of all the lexical, syntactic, and semantic ambiguity in language, some of which requires external context for resolution, it is desirable to have a metalanguage in which to concretely and succinctly express all linguistically possible meanings of an utterance before discourse and world knowledge are applied to choose the most likely interpretation. The favored metalanguage for this purpose is called the predicate logic, used to represent the logical form, or context-independent meaning, of an utterance. The semantic component of many SLU architectures builds on a substrate of two-valued, first-order, logic. To distinguish *shades of meaning* beyond truth and falsity requires more powerful formalisms for knowledge representation.

In a typical first-order system, predicates correspond to events or conditions denoted by verbs (such as *Believe* or *Like*), states of identity (such as being a *Dog* or *Cat*), and properties of varying degrees of permanence (*Happy*). In this form of logical notation, predicates have open places, filled by arguments, as in a programming language subroutine definition. Since individuals may have identical names, subscripting can be used to preserve unique reference. In the simplest systems, predication ranges over individuals rather than higher-order entities such as properties and relations.

Predicates with filled argument slots map onto sets of individuals (constants) in the universe of discourse, in particular those individuals possessing the properties, or participating in the relation, named by the predicate. One-place predicates like *Soldier*, *Happy*, or *Sleeps* range over sets of individuals from the universe of discourse. Two-place predicates, like transitive verbs such as *loves*, range over a set consisting of ordered pairs of individual members (constants) of the universe of discourse. For example, we can consider the universe of discourse to be *U* = {*Romeo, Juliet, Paris, Rosaline, Tybalt}*, people as characters in a play. They do things with and to one another, such as loving and killing. Then we could imagine the relation *Loves* interpreted as the set of ordered pairs: {<*Romeo, Juliet*>, <*Juliet, Romeo*>, <*Tybalt, Tybalt*>, <*Paris, Juliet*>}, a subset of the Cartesian product of theoretically possible love matches $U \times U$. So, for any ordered pair *x, y* in *U*, *Loves*(*x, y*) is true if the ordered pair <*x,y*> is a member of the extension of the *Loves* predicate as defined, e.g., *Romeo loves Juliet*, *Juliet loves Romeo*, etc.. Typical formal properties of relations are sometimes specially marked by grammar, such as the reflexive relation *Loves*(*Tybalt*, *Tybalt*), which can rendered in natural language as *Tybalt loves himself.* Not every possibility is present; for instance in our example, the individual Rosaline does not happen to participate at all in this extensional definition of *Loves* over *U*, as her omission from the pairs list indicates. Notice that the subset of *Loves*(*x, y*) of ordered pairs involving both *Romeo* and *Juliet* is symmetric, also marked by grammar, as in *Romeo and Juliet love each other.* This general approach extends to predicates with any arbitrary number of arguments, such as intransitive verbs like *give*.

Just as in ordinary propositional logic, connectives such as negation, conjunction, disjunction, and entailment are admitted, and can be used with predicates to denote common natural language meanings:

Romeo isn't happy = ¬Happy(Romeo)
Romeo isn't happy, but Tybalt is (happy) = ¬Happy(Romeo) ∧ Happy(Tybalt)
Either Romeo or Tybalt is happy = Happy(Romeo) ∨ Happy(Tybalt)
If Romeo is happy, Juliet is happy = Happy(Romeo) → Happy(Juliet)

Formulae, such as those above, are also said to bear a binary truth value, true or false, with respect to a world of individuals and relations. The determination of the truth value is compositional, in the sense that the truth value of the whole depends on the truth value of the parts. This is a simplistic but formally tractable view of the relation between language and meaning.

Predicate logic can also be used to denote quantified noun phrases. Consider a simple case such as *Someone killed Tybalt*, predicated over our same U = {*Romeo, Juliet, Paris, Rosaline, Tybalt*}. We can now add an *existential* quantifier, ∃, standing for *there exists* or *there is at least one*. This quantifier will bind a variable over individuals in U, and will attach to a proposition to create a new, quantified proposition in logical form. The use of variables in propositions such as *killed(x, y)* creates open propositions. Binding the variables with a quantifier over them closes the proposition. The quantifier is prefixed to the original proposition: ∃*x Killed(x, Tybalt)*.

To establish a truth (semantic) value for the quantified proposition, we have to satisfy the disjunction of propositions in U: *Killed*(*Romeo, Tybalt*) ∨ *Killed*(*Juliet, Tybalt*) ∨ *Killed*(*Paris, Tybalt*) ∨ *Killed*(*Rosaline, Tybalt*) ∨ *Killed*(*Tybalt, Tybalt*). The set of all such bindings of the variable x is the space that determines the truth or falsity of the proposition. In this case, the binding of x = *Romeo* is sufficient to assign a value true to the existential proposition.

2.5. HISTORICAL PERSPECTIVE AND FURTHER READING

Motivated to improve speech quality over the telephone, AT&T Bell Labs has contributed many influential discoveries in speech hearing, including the critical band and articulation index [2, 3]. The *Auditory Demonstration* CD prepared by Houtsma, Rossing, and Wagenaars [18] has a number of very interesting examples on psychoacoustics and its explanations. *Speech, Language, and Communication* [30] and *Speech Communication – Human and Machine* [32] are two good books that provide modern introductions to the structure of spoken language. Many speech perception experiments were conducted by exploring how phonetic information is distributed in the time or frequency domain. In addition to the formant structures for vowels, frequency importance function [12] has been developed to study how features related to phonetic categories are stored at various frequencies. In the time domain, it has been observed [16, 19, 42] that salient perceptual cues may not be evenly distributed over the speech segments and that certain perceptual critical points exist.

As intimate as speech and acoustic perception may be, there are also strong evidences that lexical and linguistic effects on speech perception are not always consistent with acoustic ones. For instance, it has long been observed that humans exhibit difficulties in distinguishing non-native phonemes. Human subjects also carry out categorical goodness difference assimilation based on their mother tongue [34], and such perceptual mechanism can be observed as early as in six-month-old infants [22]. On the other hand, hearing-impaired listeners are able to effortlessly overcome their acoustical disabilities for speech perception [8]. Speech perception is not simply an auditory matter. McGurk and MacDonald (1976) [27, 28] dramatically demonstrated this when they created a videotape on which the auditory information (phonemes) did not match the visual speech information. The effect of this mismatch between the auditory signal and the visual signal was to create a third phoneme different from both the original auditory and visual speech signals. An example is dubbing the phoneme */ba/* to the visual speech movements */ga/*. This mismatch results in hearing the phoneme */da/*. Even when subjects know of the effect, they report the McGurk effect percept. The McGurk effect has been demonstrated for consonants, vowels, words, and sentences.

The earliest scientific work on phonology and grammars goes back to Panini, a Sanskrit grammarian of the fifth century B.C. (estimated), who created a comprehensive and scientific theory of phonetics, phonology, and morphology, based on data from Sanskrit (the classical literary language of the ancient Hindus). Panini created formal production rules and definitions to describe Sanskrit grammar, including phenomena such as construction of sentences, compound nouns, etc. Panini's formalisms function as ordered rules operating on underlying structures in a manner analogous to modern linguistic theory. Panini's phonological rules are equivalent in formal power to Backus-Nauer form (BNF). A general introduction to this pioneering scientist is Cardona [9].

An excellent introduction to all aspects of phonetics is *A Course in Phonetics* [24]. A good treatment of the acoustic structure of English speech sounds and a through introduction and comparison of theories of speech perception is to be found in [33]. The basics of phonology as part of linguistic theory are treated in *Understanding Phonology* [17]. An interesting treatment of word structure (morphology) from a computational point of view can be found in *Morphology and Computation* [40]. A comprehensive yet readable treatment of English syntax and grammar can be found in *English Syntax* [4] and *A Comprehensive Grammar of the English Language* [36]. Syntactic theory has traditionally been the heart of linguistics, and has been an exciting and controversial area of research since the 1950s. Be aware that almost any work in this area will adopt and promote a particular viewpoint, often to the exclusion or minimization of others. A reasonable place to begin with syntactic theory is *Syntax: A Minimalist Introduction* [37]. An introductory textbook on syntactic and semantic theory that smoothly introduces computational issues is *Syntactic Theory: A Formal Introduction* [39]. For a philosophical and entertaining overview of various aspects of linguistic theory, see *Rhyme and Reason: An Introduction to Minimalist Syntax* [44]. A good and fairly concise treatment of basic semantics is *Introduction to Natural Language Semantics* [11]. Deeper issues are covered in greater detail and at a more advanced level in *The Handbook of Contemporary Semantic Theory* [25]. The intriguing area of lexical semantics (theory of word meanings) is comprehensively presented in *The Generative Lexicon* [35]. *Concise History of the Language Sciences* [21] is a good edited book if you are interested in the history of linguistics.

REFERENCES

[1] Aliprand, J., *et al.*, *The Unicode Standard, Version 2.0*, 1996, Addison Wesley.

[2] Allen, J.B., "How Do Humans Process and Recognize Speech?," *IEEE Trans. on Speech and Audio Processing*, 1994, **2**(4), pp. 567-577.

[3] Allen, J.B., "Harvey Fletcher 1884–1981" in *The ASA Edition of Speech and Hearing Communication* 1995, Woodbury, New York, pp. A1-A34, Acoustical Society of America.

[4] Baker, C.L., *English Syntax*, 1995, Cambridge, MA, MIT Press.

[5] Blauert, J., *Spatial Hearing*, 1983, MIT Press.

[6] Brill, E., "Transformation-Based Error-Driven Learning and Natural Language Processing: A Case Study in Part-of-Speech Tagging," *Computational Linguistics*, 1995, **21**(4), pp. 543-566.

[7] Brown, P., *et al.*, "Class-Based N-gram Models of Natural Language," *Computational Linguistics*, 1992, **18**(4).

[8] Caplan, D. and J. Utman, "Selective Acoustic Phonetic Impairment and Lexical Access in an Aphasic Patient," *Journal of the Acoustical Society of America*, 1994, **95**(1), pp. 512-517.

[9] Cardona, G., *Panini: His Work and Its Traditions: Background and Introduction*, 1988, Motilal Banarsidass.

[10] Chaffin, R., and Herrmann, D., "The Nature of Semantic Relations: A Comparison of Two Approaches" in *Representing Knowledge in Semantic Networks*, M. Evens, ed., 1988, Cambridge, UK, Cambridge University Press.

[11] de Swart, H., *Introduction to Natural Language Semantics*, 1998, Stanford, CA, Center for the Study of Language and Information Publications.

[12] Duggirala, V., *et al.*, "Frequency Importance Function for a Feature Recognition Test Material," *Journal of the Acoustical Society of America*, 1988, **83**(9), pp. 2372-2382.

[13] Fillmore, C.J., "The Case for Case" in *Universals in Linguistic Theory*, E. Bach and R. Harms, eds. 1968, New York, NY, Holt, Rinehart and Winston.

[14] Fletcher, H., "Auditory Patterns," *Rev. Mod. Phys.*, 1940, **12**, pp. 47-65.

[15] Fry, D.B., *The Physics of Speech*, Cambridge Textbooks in Linguistics, 1979, Cambridge, U.K., Cambridge University Press.

[16] Furui, S., "On The Role of Spectral Transition for Speech Perception," *Journal of the Acoustical Society of America*, 1986, **80**(4), pp. 1016-1025.

[17] Gussenhoven, C., and Jacobs, H., *Understanding Phonology*, Understanding Language Series, 1998, Edward Arnold.

[18] Houtsma, A., T. Rossing, and W. Wagenaars, *Auditory Demonstrations*, 1987, Institute for Perception Research, Eindhovern, The Netherlands, Acoustic Society of America.

[19] Jenkins, J., W. Strange, and S. Miranda, "Vowel Identification in Mixed-Speaker Silent-Center Syllables," *Journal of the Acoustical Society of America*, 1994, **95**(2), pp. 1030-1041.

[20] Klatt, D., "Review of the ARPA Speech Understanding Project," *Journal of Acoustical Society of America*, 1977, **62**(6), pp. 1324-1366.

[21] Koerner, E. and E. Asher, eds. *Concise History of the Language Sciences*, 1995, Oxford, Elsevier Science.

[22] Kuhl, P., "Infant's Perception and Representation of Speech: Development of a New Theory," *Int. Conf. on Spoken Language Processing*, 1992, Alberta, Canada, pp. 449-452.

[23] Kupeic, J., "Robust Part-of-Speech Tagging Using a Hidden Markov Model," *Computer Speech and Language*, 1992, **6**, pp. 225-242.

[24] Ladefoged, P., *A Course in Phonetics*, 1993, Harcourt Brace Johanovich.

[25] Lappin, S., *The Handbook of Contemporary Semantic Theory*, Blackwell Handbooks in Linguistics, 1997, Oxford, UK, Blackwell Publishsers Inc.

[26] Lindsey, P. and D. Norman, *Human Information Processing*, 1972, New York and London, Academic Press.

[27] MacDonald, J. and H. McGurk, "Visual Influence on Speech Perception Process," *Perception and Psychophysics*, 1978, **24**(3), pp. 253-257.

[28] McGurk, H. and J. MacDonald, "Hearing Lips and Seeing Voices," *Nature*, 1976, **264**, pp. 746-748.

[29] Merialdo, B., "Tagging English Text with a Probabilistic Model," *Computational Linguistics*, 1994, **20**(2), pp. 155-172.

[30] Miller, J. and P. Eimas, *Speech, Language and Communication*, Handbook of Perception and Cognition, eds. E. Carterette and M. Friedman, 1995, Academic Press.

[31] Moore, B.C., *An Introduction to the Psychology of Hearing*, 1982, London, Academic Press.

[32] O'Shaughnessy, D., *Speech Communication – Human and Machine*, 1987, Addison-Wesley.

[33] Pickett, J.M., *The Acoustics of Speech Communication*, 1999, Needham Heights, MA, Allyn & Bacon.

[34] Polka, L., "Linguistic Influences in Adult Perception of Non-native Vowel Contrast," *Journal of the Acoustical Society of America*, 1995, **97**(2), pp. 1286-1296.

[35] Pustejovsky, J., *The Generative Lexicon*, 1998, Bradford Books.

[36] Quirk, R., Svartvik, J., Leech, G., *A Comprehensive Grammar of the English Language*, 1985, Addison-Wesley Pub. Co.

[37] Radford, A., *Syntax: A Minimalist Introduction*, 1997, Cambridge, U.K., Cambridge Univ. Press.

[38] Rossing, T.D., *The Science of Sound*, 1982, Reading, MA, Addison-Wesley.

[39] Sag, I., Wasow, T., *Syntactic Theory: A Formal Introduction*, 1999, Cambridge, UK, Cambridge University Press.

[40] Sproat, R., *Morphology and Computation*, ACL-MIT Press Series in Natural Language Processing, 1992, Cambridge, MA, MIT Press.

[41] Stevens, S.S. and J. Volkman, "The Relation of Pitch to Frequency," *Journal of Psychology*, 1940, **53**, pp. 329.

[42] Strange, W., J. Jenkins, and T. Johnson, "Dynamic Specification of Coarticulated Vowels," *Journal of the Acoustical Society of America*, 1983, **74**(3), pp. 695-705.

[43] Thoreau, H.D., *Civil Disobedience, Solitude and Life Without Principle*, 1998, Prometheus Books.

[44] Uriagereka, J., *Rhyme and Reason: An Introduction to Minimalist Syntax*, 1998, Cambridge, MA, MIT Press.

[45] Voutilainen, A., "Morphological Disambiguation" in *Constraint Grammar: A Language-Independent System for Parsing Unrestricted Text* 1995, Berlin, Mouton de Gruyter.

[46] Weischedel, R., "BBN: Description of the PLUM System as Used for MUC-6," *The 6th Message Understanding Conferences (MUC-6)*, 1995, San Francisco, Morgan Kaufmann, pp. 55-70.

CHAPTER 3

Probability, Statistics, and Information Theory

Randomness and uncertainty play an important role in science and engineering. Most spoken language processing problems can be characterized in a probabilistic framework. Probability theory and statistics provide the mathematical language to describe and analyze such systems.

The criteria and methods used to estimate the unknown probabilities and probability densities form the basis for estimation theory. Estimation theory is critical to parameter learning in pattern recognition. In this chapter, three widely used estimation methods are discussed. They are *minimum mean squared error estimation* (MMSE), *maximum likelihood estimation* (MLE), and *maximum posterior probability estimation* (MAP).

Significance testing deals with the confidence of statistical inference, such as knowing whether the estimation of some parameter can be accepted with confidence. In pattern recognition, significance testing is important for determining whether the observed difference between two different classifiers is real. In our coverage of significance testing, we describe various methods that are used in pattern recognition, discussed in Chapter 4.

Information theory was originally developed for efficient and reliable communication systems. It has evolved into a mathematical theory concerned with the very essence of the communication process. It provides a framework for the study of fundamental issues, such as the efficiency of information representation and the limitations in reliable transmission of information over a communication channel. Many of these problems are fundamental to spoken language processing.

3.1. PROBABILITY THEORY

Probability theory deals with the averages of mass phenomena occurring sequentially or simultaneously. We often use probabilistic expressions in our day-to-day lives, such as when saying, *It is very likely that the Dow (Dow Jones Industrial index) will hit 12,000 points next month*, or, *The chance of scattered showers in Seattle this weekend is high.* Each of these expressions is based upon the concept of the probability, or the likelihood, that some specific event will occur.

Probability can be used to represent the degree of confidence in the outcome of some actions (observations), which are not definite. In probability theory, the term *sample space, S,* is used to refer to the collection (set) of all possible outcomes. An *event* refers to a subset of the sample space or a collection of outcomes. The *probability of event A,* denoted as $P(A)$, can be interpreted as the *relative frequency* with which event A would occur if the process were repeated a large number of times under similar conditions. Based on this interpretation, $P(A)$ can be computed simply by counting the total number, N_S, of all observations and the number of observations N_A whose outcome belongs to event A. That is,

$$P(A) = \frac{N_A}{N_S} \tag{3.1}$$

$P(A)$ is bounded between zero and one, i.e.,

$$0 \le P(A) \le 1 \quad \text{for all } A \tag{3.2}$$

The lower bound of probability $P(A)$ is zero when the event set A is an empty set. On the other hand, the upper bound of probability $P(A)$ is one when the event set A happens to be S.

If there are n events $A_1, A_2, \cdots, A_n$ in S such that $A_1, A_2, \cdots, A_n$ are disjoint and $\bigcup_{i=1}^{n} A_i = S$, events $A_1, A_2, \cdots, A_n$ are said to form a *partition* of S. The following obvious equation forms a fundamental axiom for probability theory.

$$P(A_1 \cup A_2 \cup \ldots A_n) = \sum_{i=1}^{n} P(A_i) = 1 \tag{3.3}$$

Based on the definition in Eq. (3.1), the *joint probability* of event A and event B occurring concurrently is denoted as $P(AB)$ and can be calculated as:

$$P(AB) = \frac{N_{AB}}{N_S} \tag{3.4}$$

3.1.1. Conditional Probability and Bayes' Rule

It is useful to study the way in which the probability of an event A changes after it has been learned that some other event B has occurred. This new probability denoted as $P(A \mid B)$ is called the *conditional probability* of event A given that event B has occurred. Since the set of those outcomes in B that also result in the occurrence of A is exactly the set AB as illustrated in Figure 3.1, it is natural to define the conditional probability as the proportion of the total probability $P(B)$ that is represented by the joint probability $P(AB)$. This leads to the following definition:

$$P(A \mid B) = \frac{P(AB)}{P(B)} = \frac{N_{AB}/N_S}{N_B/N_S} \tag{3.5}$$

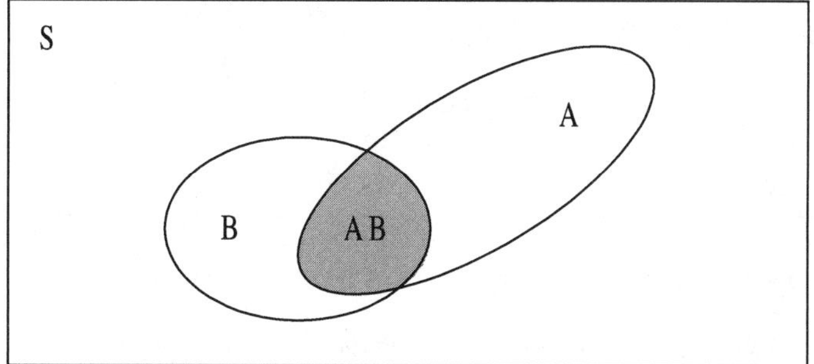

Figure 3.1 The intersection AB represents where the joint event A and B occurs concurrently.

Based on the definition of conditional probability, the following expressions can be easily derived.

$$P(AB) = P(A \mid B)P(B) = P(B \mid A)P(A) \tag{3.6}$$

Equation (3.6) is the simple version of the *chain rule*. The chain rule, which can specify a joint probability in terms of multiplication of several cascaded conditional probabilities, is often used to decompose a complicated joint probabilistic problem into a sequence of stepwise conditional probabilistic problems. Equation (3.6) can be converted to such a general chain:

$$P(A_1 A_2 \cdots A_n) = P(A_n \mid A_1 \cdots A_{n-1}) \cdots P(A_2 \mid A_1)P(A_1) \tag{3.7}$$

When two events, A and B, are independent of each other, in the sense that the occurrence or of either of them has no relation to and no influence on the occurrence of the other, it is obvious that the conditional probability $P(B \mid A)$ equals to the unconditional probability

$P(B)$. It follows that the joint probability $P(AB)$ is simply the product of $P(A)$ and $P(B)$ if A and B, are independent.

If the n events $A_1, A_2, \cdots, A_n$ form a partition of S and B is any event in S as illustrated in Figure 3.2, the events $A_1B, A_2B, \cdots, A_nB$ form a partition of B. Thus, we can rewrite:

$$B = A_1B \cup A_2B \cup \cdots \cup A_nB \tag{3.8}$$

Since $A_1B, A_2B, \cdots, A_nB$ are disjoint,

$$P(B) = \sum_{k=1}^{n} P(A_kB) \tag{3.9}$$

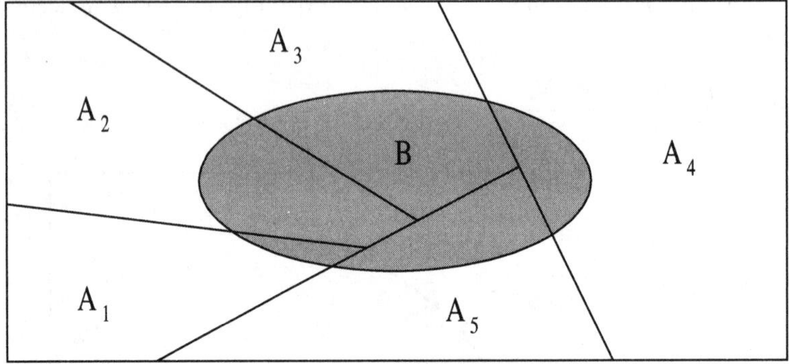

Figure 3.2 The intersections of B with partition events $A_1, A_2, \cdots, A_n$.

Equation (3.9) is called the *marginal probability* of event B, where the probability of event B is computed from the sum of joint probabilities.

According to the chain rule, Eq. (3.6), $P(A_iB) = P(A_i)P(B \mid A_i)$, it follows that

$$P(B) = \sum_{k=1}^{n} P(A_k)P(B \mid A_k) \tag{3.10}$$

Combining Eqs. (3.5) and (3.10), we get the well-known *Bayes' rule*:

$$P(A_i \mid B) = \frac{P(A_iB)}{P(B)} = \frac{P(B \mid A_i)P(A_i)}{\sum_{k=1}^{n} P(B \mid A_k)P(A_k)} \tag{3.11}$$

Bayes' rule is the basis for pattern recognition that is described in Chapter 4.

3.1.2. Random Variables

Elements in a sample space may be numbered and referred to by the numbers given. A variable X that specifies the numerical quantity in a sample space is called a *random variable*. Therefore, a random variable X is a function that maps each possible outcome s in the sample space S onto real numbers $X(s)$. Since each event is a subset of the sample space, an event is represented as a set of $\{s\}$ which satisfies $\{s \mid X(s)=x\}$. We use capital letters to denote random variables and lower-case letters to denote fixed values of the random variable. Thus, the probability that $X = x$ is denoted as:

$$P(X = x) = P(s \mid X(s) = x) \tag{3.12}$$

A random variable X is a *discrete* random variable, or X has a *discrete distribution*, if X can take only a finite number n of different values $x_1, x_2, \cdots, x_n$, or at most, an infinite sequence of different values $x_1, x_2, \cdots$. If the random variable X is a discrete random variable, the *probability function* (pf) or *probability mass function* (pmf) of X is defined to be the function p such that for any real number x,

$$p_X(x) = P(X = x) \tag{3.13}$$

For the cases in which there is no confusion, we drop the subscription X for $p_X(x)$. The sum of probability mass over all values of the random variable is equal to unity.

$$\sum_{i=1}^{n} p(x_i) = \sum_{i=1}^{n} P(X = x_i) = 1 \tag{3.14}$$

The marginal probability, chain rule and Bayes' rule can also be rewritten with respect to random variables:

$$p_X(x_i) = P(X = x_i) = \sum_{k=1}^{m} P(X = x_i, Y = y_k) = \sum_{k=1}^{m} P(X = x_i \mid Y = y_k) P(Y = y_k) \tag{3.15}$$

$$\begin{aligned} &P(X_1 = x_1, \cdots, X_n = x_n) = \\ &P(X_n = x_n \mid X_1 = x_1, \cdots, X_{n-1} = x_{n-1}) \cdots P(X_2 = x_2 \mid X_1 = x_1) P(X_1 = x_1) \end{aligned} \tag{3.16}$$

$$P(X = x_i \mid Y = y) = \frac{P(X = x_i, Y = y)}{P(Y = y)} = \frac{P(Y = y \mid X = x_i) P(X = x_i)}{\sum_{k=1}^{n} P(Y = y \mid X = x_k) P(X = x_k)} \tag{3.17}$$

In a similar manner, if the random variables X and Y are statistically independent, they can be represented as:

$$P(X = x_i, Y = y_j) = P(X = x_i) P(Y = y_j) = p_X(x_i) p_Y(y_j) \quad \forall \text{ all } i \text{ and } j \tag{3.18}$$

A random variable X is a *continuous* random variable, or X has a *continuous distribution,* if there exists a nonnegative function f, defined on the real line, such that for an interval A,

$$P(X \in A) = \int_A f_X(x)dx \tag{3.19}$$

The function f_X is called the *probability density function* (abbreviated pdf) of X. We drop the subscript X for f_X if there is no ambiguity. As illustrated in Figure 3.3, the area of shaded region is equal to the value of $P(a \le X \le b)$.

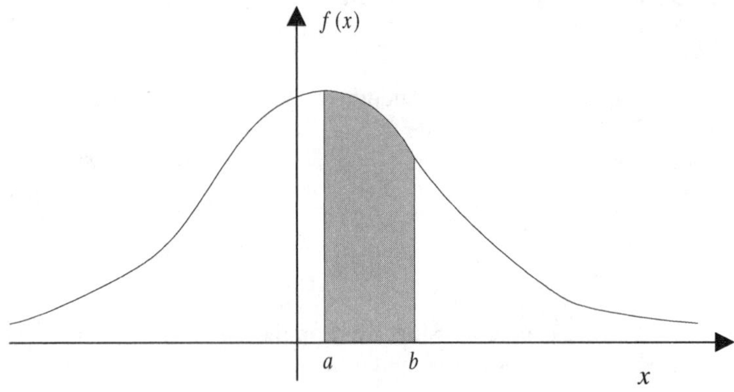

Figure 3.3 An example of pdf. The area of the shaded region is equal to the value of $P(a \le X \le b)$.

Every pdf must satisfy the following two requirements:

$$\begin{aligned} &f(x) \ge 0 \text{ for } -\infty \le x \le \infty \text{ and} \\ &\int_{-\infty}^{\infty} f(x)dx = 1 \end{aligned} \tag{3.20}$$

The marginal probability, chain rule, and Bayes' rule can also be rewritten with respect to continuous random variables:

$$f_X(x) = \int_{-\infty}^{\infty} f_{X,Y}(x,y)dy = \int_{-\infty}^{\infty} f_{X|Y}(x \mid y) f_Y(y)dy \tag{3.21}$$

$$f_{X_1,\cdots,X_n}(x_1,\cdots,x_n) = f_{X_n|X_1,\cdots,X_{n-1}}(x_n \mid x_1,\cdots,x_{n-1}) \cdots f_{X_2|X_1}(x_2 \mid x_1) f_{X_1}(x_1) \tag{3.22}$$

$$f_{X|Y}(x \mid y) = \frac{f_{X,Y}(x,y)}{f_Y(y)} = \frac{f_{Y|X}(y \mid x) f_X(x)}{\int_{-\infty}^{\infty} f_{Y|X}(y \mid x) f_X(x)dx} \tag{3.23}$$

The *distribution function* or *cumulative distribution function* F of a discrete or continuous random variable X is a function defined for every real number x as follows:

$$F(x) = P(X \le x) \quad \text{for } -\infty \le x \le \infty \tag{3.24}$$

For continuous random variables, it follows that:

$$F(x) = \int_{-\infty}^{x} f_X(x)dx \tag{3.25}$$

$$f_X(x) = \frac{dF(x)}{dx} \tag{3.26}$$

3.1.3. Mean and Variance

Suppose that a discrete random variable X has a pf $f(x)$; the *expectation* or *mean* of X is defined as follows:

$$E(X) = \sum_x xf(x) \tag{3.27}$$

Similarly, if a continuous random variable X has a pdf f, the *expectation* or *mean* of X is defined as follows:

$$E(X) = \int_{-\infty}^{\infty} xf(x)dx \tag{3.28}$$

In physics, the mean is regarded as the center of mass of the probability distribution. The expectation can also be defined for any function of the random variable X. If X is a continuous random variable with pdf f, then the expectation of any function $g(X)$ can be defined as follows:

$$E\left[g(X)\right] = \int_{-\infty}^{\infty} g(x)f(x)dx \tag{3.29}$$

The expectation of a random variable is a linear operator. That is, it satisfies both additivity and homogeneity properties:

$$E(a_1X_1 + \cdots + a_nX_n + b) = a_1E(X_1) + \cdots + a_nE(X_n) + b \tag{3.30}$$

where $a_1, \cdots, a_n, b$ are constants.

Equation (3.30) is valid regardless of whether or not the random variables $X_1, \cdots, X_n$ are independent.

Suppose that X is a random variable with mean $\mu = E(X)$. The *variance* of X denoted as $Var(X)$ is defined as follows:

$$Var(X) = \sigma^2 = E\left[(X - \mu)^2\right] \tag{3.31}$$

where σ, the nonnegative square root of the variance is known as the *standard deviation* of random variable *X*. Therefore, the variance is also often denoted as σ^2.

The variance of a distribution provides a measure of the spread or dispersion of the distribution around its mean μ. A small value of the variance indicates that the probability distribution is tightly concentrated around μ, and a large value of the variance typically indicates the probability distribution has a wide spread around μ. Figure 3.4 illustrates three different Gaussian distributions[1] with the same mean, but different variances.

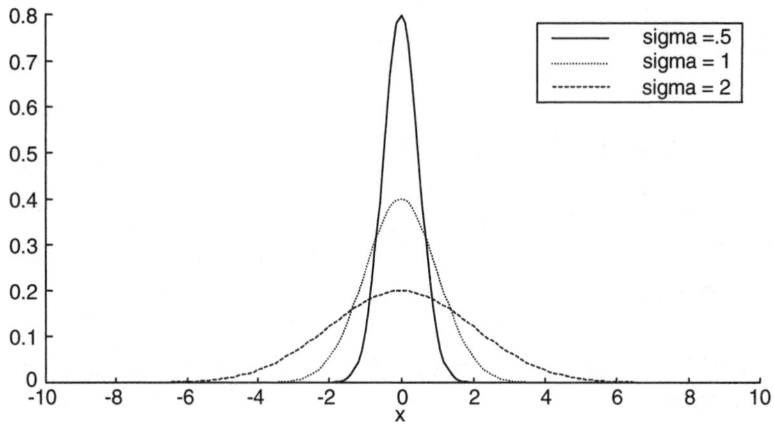

Figure 3.4 Three Gaussian distributions with same mean μ, but different variances, 0.5, 1.0, and 2.0, respectively. The distribution with a large value of the variance has a wide spread around the mean μ.

The variance of random variable *X* can be computed in the following way:

$$Var(X) = E(X^2) - [E(X)]^2 \tag{3.32}$$

In physics, the expectation $E(X^k)$ is called the kth moment of *X* for any random variable *X* and any positive integer *k*. Therefore, the variance is simply the difference between the second moment and the square of the first moment.

The variance satisfies the following additivity property, if random variables *X* and *Y* are independent:

$$Var(X+Y) = Var(X) + Var(Y) \tag{3.33}$$

However, it does not satisfy the homogeneity property. Instead for constant *a*,

$$Var(aX) = a^2 Var(X) \tag{3.34}$$

[1] We describe Gaussian distributions in Section 3.1.7.

Since it is clear that $Var(b) = 0$ for any constant b, we have an equation similar to Eq. (3.30) if random variables $X_1, \cdots, X_n$ are independent.

$$Var(a_1X_1 + \cdots + a_nX_n + b) = a_1^2 Var(X_1) + \cdots + a_n^2 Var(X_n) \tag{3.35}$$

Conditional expectation can also be defined in a similar way. Suppose that *X* and *Y* are discrete random variables and let $f(y \mid x)$ denote the conditional pf of *Y* given $X = x$, then the conditional expectation $E(Y \mid X)$ is defined to be the function of *X* whose value $E(Y \mid x)$ when $X = x$ is

$$E_{Y|X}(Y \mid X = x) = \sum_y y f_{Y|X}(y \mid x) \tag{3.36}$$

For continuous random variables *X* and *Y* with $f_{Y|X}(y \mid x)$ as the conditional pdf. of *Y* given $X = x$, the conditional expectation $E(Y \mid X)$ is defined to be the function of *X* whose value $E(Y \mid x)$ when $X = x$ is

$$E_{Y|X}(Y \mid X = x) = \int_{-\infty}^{\infty} y f_{Y|X}(y \mid x) dy \tag{3.37}$$

Since $E(Y \mid X)$ is a function of random variable *X*, it itself is a random variable whose probability distribution can be derived from the distribution of *X*. It can be shown that

$$E_X\left[E_{Y|X}(Y \mid X)\right] = E_{X,Y}(Y) \tag{3.38}$$

More generally, suppose that *X* and *Y* have a continuous joint distribution and that $g(x, y)$ is any arbitrary function of *X* and *Y*. The conditional expectation $E\left[g(X,Y) \mid X\right]$ is defined to be the function of *X* whose value $E\left[g(X,Y) \mid x\right]$ when $X = x$ is

$$E_{Y|X}\left[g(X,Y) \mid X = x\right] = \int_{-\infty}^{\infty} g(x, y) f_{Y|X}(y \mid x) dy \tag{3.39}$$

Equation (3.38) can also be generalized into the following equation:

$$E_X\left\{E_{Y|X}\left[g(X,Y) \mid X\right]\right\} = E_{X,Y}\left[g(X,Y)\right] \tag{3.40}$$

Finally, it is worthwhile to introduce *median* and *mode*. The median of a distribution of *X* is defined to be a point *m*, such that $P(X \le m) \ge 1/2$ and $P(X \ge m) \ge 1/2$. Thus, the median *m* divides the total probability into two equal parts, i.e., the probability to the left of *m* and the probability to the right of *m* are exactly $1/2$.

Suppose a random variable *X* has either a discrete distribution with pf $p(x)$ or continuous pdf $f(x)$; a point ϖ is called the mode of the distribution if $p(x)$ or $f(x)$ attains the maximum value at the point ϖ. A distribution can have more than one mode.

3.1.3.1. The Law of Large Numbers

The concept of *sample mean* and *sample variance* is important in statistics because most statistical experiments involve sampling. Suppose that the random variables $X_1,\cdots,X_n$ form a random sample of size *n* from some distribution for which the mean is μ and the variance is σ^2. In other words, the random variables $X_1,\cdots,X_n$ are *independent identically distributed* (often abbreviated by iid) and each has mean μ and variance σ^2. Now if we denote $\overline{X}_n$ as the arithmetic average of the *n* observations in the sample, then

$$\overline{X}_n = \frac{1}{n}(X_1 + \cdots + X_n) \tag{3.41}$$

$\overline{X}_n$ is a random variable and is referred to as *sample mean*. The mean and variance of $\overline{X}_n$ can be easily derived based on the definition.

$$E(\overline{X}_n) = \mu \quad \text{and} \quad \text{Var}(\overline{X}_n) = \frac{\sigma^2}{n} \tag{3.42}$$

Equation (3.42) states that the mean of sample mean is equal to mean of the distribution, while the variance of sample mean is only $1/n$ times the variance of the distribution. In other words, the distribution of $\overline{X}_n$ will be more concentrated around the mean μ than was the original distribution. Thus, the sample mean is closer to μ than is the value of just a single observation X_i from the given distribution.

The *law of large numbers* is one of most important theorems in probability theory. Formally, it states that the sample mean $\overline{X}_n$ converges to the mean μ in probability, that is,

$$\lim_{n\to\infty} P\left(|\overline{X}_n - \mu| < \varepsilon\right) = 1 \text{ for any given number } \varepsilon > 0 \tag{3.43}$$

The law of large numbers basically implies that the sample mean is an excellent estimate of the unknown mean of the distribution when the sample size *n* is large.

3.1.4. Covariance and Correlation

Let *X* and *Y* be random variables having a specific joint distribution, and $E(X) = \mu_X$, $E(Y) = \mu_Y$, $Var(X) = \sigma_X^2$, and $Var(Y) = \sigma_Y^2$. The *covariance* of *X* and *Y*, denoted as $Cov(X,Y)$, is defined as follows:

$$Cov(X,Y) = E[(X-\mu_X)(Y-\mu_Y)] = Cov(Y,X) \tag{3.44}$$

In addition, the *correlation coefficient* of *X* and *Y*, denoted as ρ_{XY}, is defined as follows:

$$\rho_{XY} = \frac{Cov(X,Y)}{\sigma_X \sigma_Y} \tag{3.45}$$

It can be shown that $\rho(X,Y)$ should be bound within $[-1\ldots1]$, that is,

$$-1 \le \rho(X,Y) \le 1 \tag{3.46}$$

X and *Y* are said to be *positively correlated* if $\rho_{XY} > 0$, *negatively correlated* if $\rho_{XY} < 0$, and *uncorrelated* if $\rho_{XY} = 0$. It can also be shown that $Cov(X,Y)$ and ρ_{XY} must have the same sign; that is, both are positive, negative, or zero at the same time. When $E(XY) = 0$, the two random variables are called *orthogonal.*

There are several theorems pertaining to the basic properties of covariance and correlation. We list here the most important ones:

Theorem 1 For any random variables *X* and *Y*

$$Cov(X,Y) = E(XY) - E(X)E(Y) \tag{3.47}$$

Theorem 2 If *X* and *Y* are independent random variables, then

$$Cov(X,Y) = \rho_{XY} = 0$$

Theorem 3 Suppose *X* is a random variable and *Y* is a linear function of *X* in the form of $Y = aX + b$ for some constant *a* and *b*, where $a \ne 0$. If $a > 0$, then $\rho_{XY} = 1$. If $a < 0$, then $\rho_{XY} = -1$. Sometimes, ρ_{XY} is referred to as the amount of linear dependency between random variables *X* and *Y*.

Theorem 4 For any random variables *X* and *Y*,

$$Var(X+Y) = Var(X) + Var(Y) + 2Cov(X,Y) \tag{3.48}$$

Theorem 5 If $X_1, \cdots, X_n$ are random variables, then

$$Var(\sum_{i=1}^{n} X_i) = \sum_{i=1}^{n} Var(X_i) + 2\sum_{i=1}^{n}\sum_{j=1}^{i-1} Cov(X_i, X_j) \tag{3.49}$$

3.1.5. Random Vectors and Multivariate Distributions

When a random variable is a vector rather than a scalar, it is called a random vector and we often use boldface variable like $\mathbf{X} = (X_1, \cdots, X_n)$ to indicate that it is a random vector. It is said that *n* random variables $X_1, \cdots, X_n$ have a *discrete joint distribution* if the random vector $\mathbf{X} = (X_1, \cdots, X_n)$ can have only a finite number or an infinite sequence of different values $(x_1,\cdots,x_n)$ in R^n. The joint pf of $X_1,\cdots,X_n$ is defined to be the function $f_{\mathbf{X}}$ such that for any point $(x_1,\cdots,x_n)\in R^n$,

$$f_{\mathbf{X}}(x_1,\cdots,x_n)=P(X_1=x_1,\cdots,X_n=x_n) \tag{3.50}$$

Similarly, it is said that *n* random variables $X_1,\cdots,X_n$ have a *continuous joint distribution* if there is a nonnegative function *f* defined on R^n such that for any subset $A\subset R^n$,

$$P\left[(X_1,\cdots,X_n)\in A\right]=\int_A\cdots\int f_{\mathbf{X}}(x_1,\cdots,x_n)dx_1\cdots dx_n \tag{3.51}$$

The *joint distribution function* can also be defined similarly for *n* random variables $X_1,\cdots,X_n$ as follows:

$$F_{\mathbf{X}}(x_1,\cdots,x_n)=P(X_1\le x_1,\cdots,X_n\le x_n) \tag{3.52}$$

The concept of mean and variance for a random vector can be generalized into *mean vector* and *covariance matrix*. Supposed that **X** is an *n*-dimensional random vector with components $X_1,\cdots,X_n$, under matrix representation, we have

$$\mathbf{X}=\begin{bmatrix}X_1\\ \vdots\\ X_n\end{bmatrix} \tag{3.53}$$

The *expectation (mean) vector* $E(\mathbf{X})$ of random vector **X** is an *n*-dimensional vector whose components are the expectations of the individual components of **X**, that is,

$$E(\mathbf{X})=\begin{bmatrix}E(X_1)\\ \vdots\\ E(X_n)\end{bmatrix} \tag{3.54}$$

The *covariance matrix* $Cov(\mathbf{X})$ of random vector **X** is defined to be an $n\times n$ matrix such that the element in the ith row and j^{th} column is $Cov(X_i,Y_j)$, that is,

$$Cov(\mathbf{X})=\begin{bmatrix}Cov(X_1,X_1) & \cdots & Cov(X_1,X_n)\\ \vdots & & \vdots\\ Cov(X_n,X_1) & \cdots & Cov(X_n,X_n)\end{bmatrix}=E\left[\left[X-E(X)\right]\left[X-E(X)\right]^t\right] \tag{3.55}$$

It should be emphasized that the *n* diagonal elements of the covariance matrix $Cov(\mathbf{X})$ are actually the variances of $X_1,\cdots,X_n$. Furthermore, since the covariance is symmetric, i.e., $Cov(X_i,X_j)=Cov(X_j,X_i)$, the covariance matrix $Cov(\mathbf{X})$ must be a symmetric matrix.

There is an important theorem regarding the mean vector and covariance matrix for a linear transformation of random vector **X**. Suppose **X** is an *n*-dimensional vector as specified by Eq. (3.53), with mean vector $E(\mathbf{X})$ and covariance matrix $Cov(\mathbf{X})$. Now, assume **Y** is a *m*-dimensional random vector which is a linear transform of random vector **X** by the relation: $\mathbf{Y} = \mathbf{AX} + \mathbf{B}$, where $\mathbf{A}$ is a $m \times n$ transformation matrix whose elements are constants, and $\mathbf{B}$ is a m-dimensional constant vector. Then we have the following two equations:

$$E(\mathbf{Y}) = \mathbf{A}E(\mathbf{X}) + \mathbf{B} \tag{3.56}$$

$$Cov(\mathbf{Y}) = \mathbf{A}Cov(\mathbf{X})\mathbf{A}^t \tag{3.57}$$

3.1.6. Some Useful Distributions

In the following two sections, we will introduce several useful distributions that are widely used in applications of probability and statistics, particularly in spoken language systems.

3.1.6.1. Uniform Distributions

The simplest distribution is uniform distribution where the pf or pdf is a constant function. For uniform discrete random variable *X*, which only takes possible values from $\{x_i \mid 1 \le i \le n\}$, the pf for *X* is

$$P(X = x_i) = \frac{1}{n} \qquad 1 \le i \le n \tag{3.58}$$

For uniform continuous random variable *X*, which only takes possible values in the real interval $[a,b]$, as shown in Figure 3.5, the pdf for *X* is

$$f(x) = \frac{1}{b-a} \qquad a \le x \le b \tag{3.59}$$

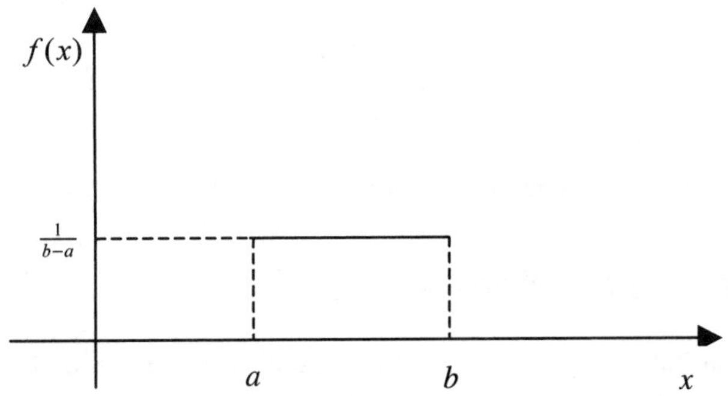

Figure 3.5 A uniform distribution for pdf in Eq. (3.59).

3.1.6.2. Binomial Distributions

The *binomial distribution* is used to describe binary-decision events. For example, suppose that a single coin toss will produce heads with probability p and produce tails with probability $1-p$. Now, if we toss the same coin n times and let X denote the number of heads observed, then the random variable X has the following *binomial* pf:

$$P(X = x) = f(x \mid n, p) = \binom{n}{x} p^x (1-p)^{n-x} \tag{3.60}$$

It can be shown that the mean and variance of a binomial distribution are:

$$E(X) = np \tag{3.61}$$

$$Var(X) = np(1-p) \tag{3.62}$$

Figure 3.6 illustrates three binomial distributions with $p = 0.2$, 0.3, and 0.4, and $n = 10$.

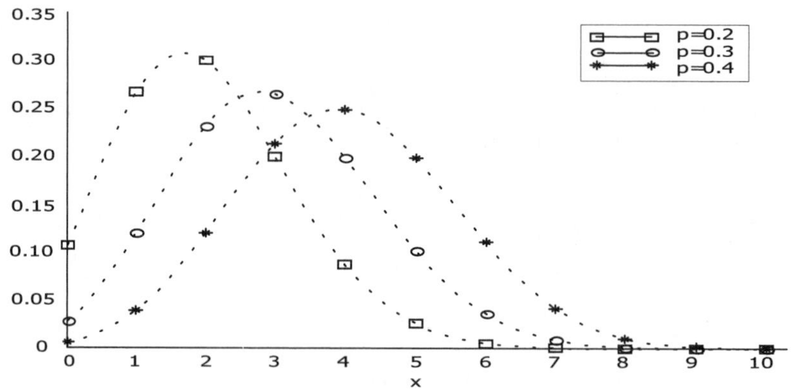

Figure 3.6 Three binomial distributions with $p = 0.2$, 0.3, and 0.4, and $n = 10$.

3.1.6.3. Geometric Distributions

The geometric distribution is related to the binomial distribution. As in the independent coin toss example, heads has a probability p and tails has a probability $1-p$. The geometric distribution is to model the time until tails appears. Let the random variable X

be the time (the number of tosses) until the first tail-up is shown. The pdf of X is in the following form:

$$P(X = x) = f(x \mid p) = p^{x-1}(1-p) \quad x = 1,2,\ldots \text{ and } 0 < p < 1 \tag{3.63}$$

The mean and variance of a geometric distribution are given by:

$$E(X) = \frac{1}{1-p} \tag{3.64}$$

$$Var(X) = \frac{1}{(1-p)^2} \tag{3.65}$$

One example for the geometric distribution is the distribution of the state duration for a hidden Markov model, as described in Chapter 8. Figure 3.7 illustrates three geometric distributions with $p = 0.1$, 0.4, and 0.7.

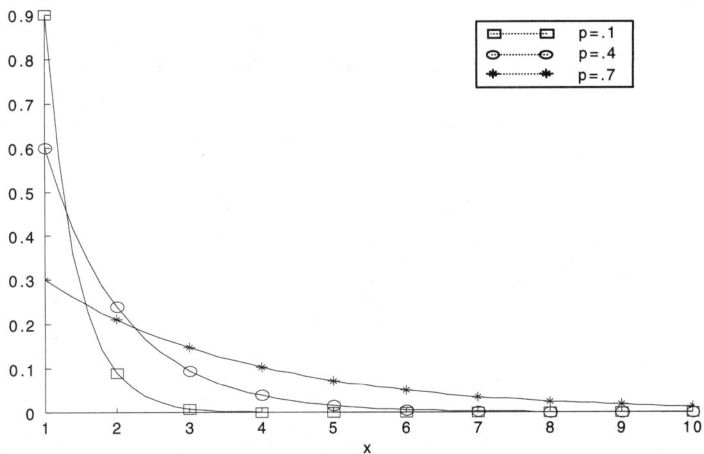

Figure 3.7 Three geometric distributions with different parameter p.

3.1.6.4. Multinomial Distributions

Suppose that a bag contains balls of k different colors, where the proportion of the balls of color i is p_i. Thus, $p_i > 0$ for $i = 1,\ldots,k$ and $\sum_{i=1}^{k} p_i = 1$. Now suppose that n balls are randomly selected from the bag and there are enough balls ($> n$) of each color. Let X_i denote the number of selected balls that are of color *i*. The random vector $\mathbf{X} = (X_1, \ldots, X_k)$ is said to have a *multinomial distribution* with parameters n and $\mathbf{p} = (p_1, \ldots, p_k)$. For a vector $\mathbf{x} = (x_1, \ldots, x_k)$, the pf of $\mathbf{X}$ has the following form:

$$P(\mathbf{X} = \mathbf{x}) = f(\mathbf{x} \mid n, \mathbf{p}) = \begin{cases} \dfrac{n!}{x_1! \ldots x_k!} p_1^{x_1} \ldots p_k^{x_k} & \text{where } x_i \geq 0 \ \forall i = 1 \ .., k \\ & \text{and } x_1 + \cdots + x_k = n \\ 0 & \text{otherwise} \end{cases} \tag{3.66}$$

It can be shown that the mean, variance and covariance of the multinomial distribution are:

$$E(X_i) = np_i \ \text{ and } \ Var(X_i) = np_i(1 - p_i) \ \ \forall i = 1, \ldots, k \tag{3.67}$$

$$Cov(X_i, X_j) = -np_i p_j \tag{3.68}$$

Figure 3.8 shows a multinomial distribution with n = 10, $p_1 = 0.2$, and $p_2 = 0.3$. Since there are only two free parameters x_1 and x_2, the graph is illustrated only using x_1

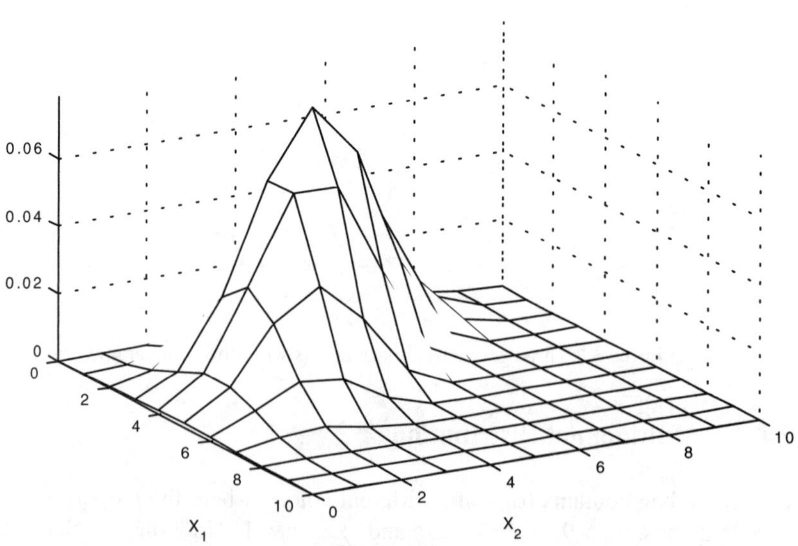

Figure 3.8 A multinomial distribution with n=10, $p_1 = 0.2$, and $p_2 = 0.3$.

and x_2 as axis. Multinomial distributions are typically used with the χ^2 test that is one of the most widely used goodness-of-fit hypotheses testing procedures described in Section 3.3.3.

3.1.6.5. Poisson Distributions

Another popular discrete distribution is the *Poisson distribution*. The random variable *X* has a Poisson distribution with mean λ $(\lambda > 0)$ if the pf of *X* has the following form:

$$P(X = x) = f(x \mid \lambda) = \begin{cases} \dfrac{e^{-\lambda}\lambda^x}{x!} & \text{for } x = 0,1,2,\ldots \\ 0 & \text{otherwise} \end{cases} \tag{3.69}$$

The mean and variance of a Poisson distribution are the same and equal λ:

$$E(X) = Var(X) = \lambda \tag{3.70}$$

Figure 3.9 illustrates three Poisson distributions with $\lambda = 1$, 2, and 4. The Poisson distribution is typically used in queuing theory, where *x* is the total number of occurrences of some phenomenon during a fixed period of time or within a fixed region of space. Examples include the number of telephone calls received at a switchboard during a fixed period of time. In speech recognition, the Poisson distribution is used to model the duration for a phoneme.

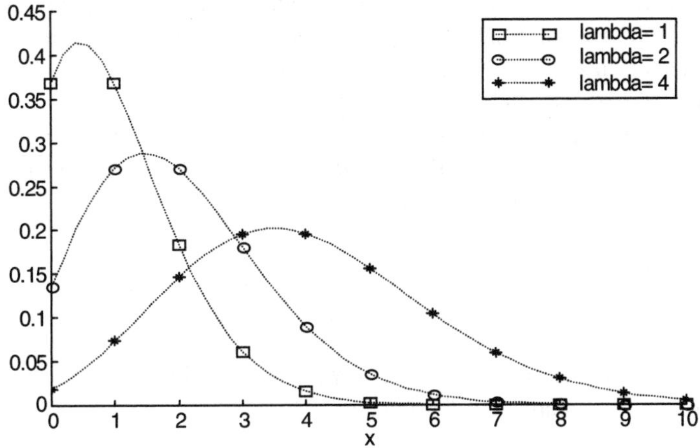

Figure 3.9 Three Poisson distributions with $\lambda = 1$, 2, and 4.

3.1.6.6. Gamma Distributions

A continuous random variable X is said to have a *gamma distribution* with parameters α and β ($\alpha > 0$ and $\beta > 0$) if X has a continuous pdf of the following form:

$$f(x \mid \alpha, \beta) = \begin{cases} \dfrac{\beta^{\alpha}}{\Gamma(\alpha)} x^{\alpha-1} e^{-\beta x} & x > 0 \\ 0 & x \le 0 \end{cases} \tag{3.71}$$

where

$$\Gamma(\alpha) = \int_0^{\infty} x^{\alpha-1} e^{-x} dx \tag{3.72}$$

It can be shown that the function Γ is a factorial function when α is a positive integer.

$$\Gamma(n) = \begin{cases} (n-1)! & n = 2, 3, \ldots \\ 1 & n = 1 \end{cases} \tag{3.73}$$

The mean and variance of a gamma distribution are:

$$E(X) = \frac{\alpha}{\beta} \text{ and } Var(X) = \frac{\alpha}{\beta^2} \tag{3.74}$$

Figure 3.10 illustrates three gamma distributions with $\beta = 1.0$ and $\alpha = 2.0$, 3.0, and 4.0. There is an interesting theorem associated with gamma distributions. If the random variables $X_1, \ldots, X_k$ are independent and each random variable X_i has a gamma distribution with parameters α_i and β, then the sum $X_1 + \cdots + X_k$ also has a gamma distribution with parameters $\alpha_1 + \cdots + \alpha_k$ and β.

A special case of gamma distribution is called *exponential distribution*. A continuous random variable X is said to have an *exponential distribution* with parameters β ($\beta > 0$) if X has a continuous pdf of the following form:

$$f(x \mid \beta) = \begin{cases} \beta e^{-\beta x} & x > 0 \\ 0 & x \le 0 \end{cases} \tag{3.75}$$

It is clear that the exponential distribution is a gamma distribution with $\alpha = 1$. The mean and variance of the exponential distribution are:

$$E(X) = \frac{1}{\beta} \text{ and } Var(X) = \frac{1}{\beta^2} \tag{3.76}$$

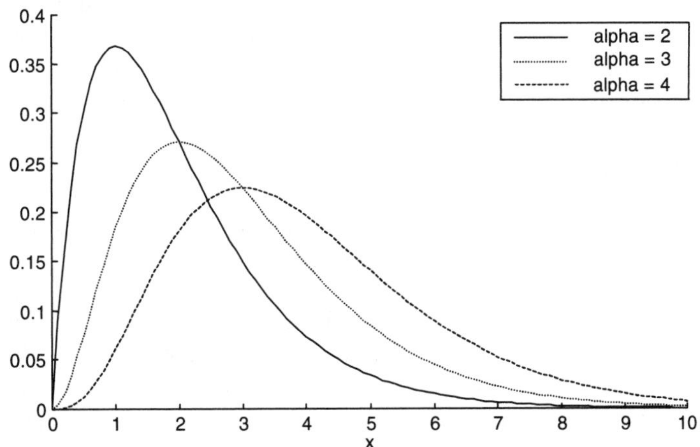

Figure 3.10 Three Gamma distributions with $\beta = 1.0$ and $\alpha = 2.0$, 3.0, and 4.0.

Figure 3.11 shows three exponential distributions with $\beta = 1.0$, 0.6, and 0.3. The exponential distribution is often used in queuing theory for the distributions of the duration of a service or the inter-arrival time of customers. It is also used to approximate the distribution of the life of a mechanical component.

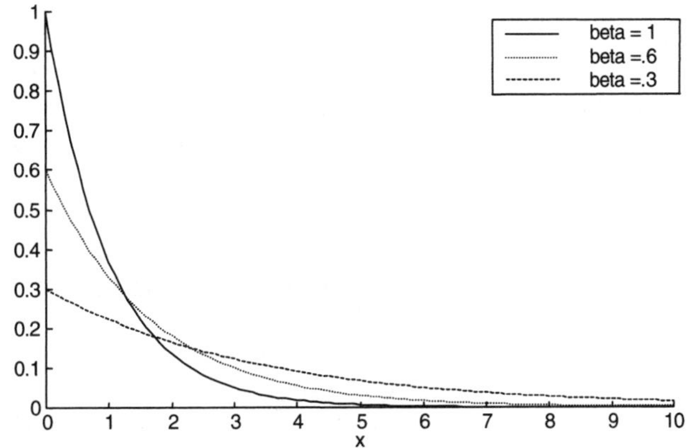

Figure 3.11 Three exponential distributions with $\beta = 1.0$, 0.6, and 0.3.

3.1.7. Gaussian Distributions

Gaussian distribution is by far the most important probability distribution mainly because many scientists have observed that the random variables studied in various physical experiments (including speech signals), often have distributions that are approximately Gaussian. The Gaussian distribution is also referred to as normal distribution. A continuous random variable X is said to have a *Gaussian distribution* with mean μ and variance σ^2 ($\sigma > 0$) if X has a continuous pdf in the following form:

$$f(x \mid \mu, \sigma^2) = N(\mu, \sigma^2) = \frac{1}{\sqrt{2\pi}\sigma} \exp\left[-\frac{(x-\mu)^2}{2\sigma^2} \right] \tag{3.77}$$

It can be shown that μ and σ^2 are indeed the mean and the variance for the Gaussian distribution. Some examples of Gaussians can be found in Figure 3.4.

The use of Gaussian distributions is justified by the *Central Limit Theorem*, which states that observable events considered to be a consequence of many unrelated causes with no single cause predominating over the others, tend to follow the Gaussian distribution [6].

It can be shown from Eq. (3.77) that the Gaussian $f(x \mid \mu, \sigma^2)$ is symmetric with respect to $x = \mu$. Therefore, μ is both the mean and the median of the distribution. Moreover, μ is also the mode of the distribution, i.e., the pdf $f(x \mid \mu, \sigma^2)$ attains its maximum at the mean point $x = \mu$.

Several Gaussian pdfs with the same mean μ, but different variances are illustrated in Figure 3.4. Readers can see that the curve has a *bell* shape. The Gaussian pdf with a small variance has a high peak and is very concentrated around the mean μ, whereas the Gaussian pdf with a large variance is relatively flat and is spread out more widely over the x-axis.

If the random variable X is a *Gaussian distribution* with mean μ and variance σ^2, then any linear function of X also has a Gaussian distribution. That is, if $Y = aX + b$, where a and b are constants and $a \neq 0$, Y has a Gaussian distribution with mean $a\mu + b$ and variance $a^2\sigma^2$. Similarly, the sum $X_1 + \cdots + X_k$ of random variables $X_1, \ldots, X_k$, where each random variable X_i has a Gaussian distribution, is also a Gaussian distribution.

3.1.7.1. Standard Gaussian Distributions

The Gaussian distribution with mean 0 and variance 1, denoted as $N(0,1)$, is called the *standard Gaussian distribution* or *unit Gaussian distribution*. Since the linear transformation of a Gaussian distribution is still a Gaussian distribution, the behavior of a Gaussian distribution can be solely described using a standard Gaussian distribution. If the random variable X is a Gaussian distribution with mean μ and variance σ^2, that is, $X \sim N(\mu, \sigma^2)$, it can be shown that

$$Z = \frac{X - \mu}{\sigma} \sim N(0,1) \tag{3.78}$$

Based on Eq. (3.78), the following property can be shown:

$$P(|X-\mu| \le k\sigma) = P(|Z| \le k) \tag{3.79}$$

Equation (3.79) demonstrates that every Gaussian distribution contains the same total amount of probability within any fixed number of standard deviations of its mean.

3.1.7.2. The Central Limit Theorem

If random variables $X_1,\ldots,X_n$ are i.i.d. according to a common distribution function with mean μ and variance σ^2, then as the random sample size n approaches ∞, the following random variable has a distribution converging to the standard Gaussian distribution:

$$Y_n = \frac{n(\overline{X}_n - \mu)}{\sqrt{n\sigma^2}} \sim N(0,1) \tag{3.80}$$

where $\overline{X}_n$ is the sample mean of random variables $X_1,\ldots,X_n$ as defined in Eq. (3.41).

Based on Eq. (3.80), the sample mean random variable $\overline{X}_n$ can be approximated by a Gaussian distribution with mean μ and variance σ^2/n.

The central limit theorem above is applied to i.i.d. random variables $X_1,\ldots,X_n$. A. Liapounov in 1901 derived another central limit theorem for independent but not necessarily identically distributed random variables $X_1,\ldots,X_n$. Suppose $X_1,\ldots,X_n$ are independent random variables and $E(|X_i-\mu_i|^3)<\infty$ for $1\le i\le n$; the following random variable will converge to standard Gaussian distribution when $n\to\infty$.

$$Y_n = (\sum_{i=1}^{n} X_i - \sum_{i=1}^{n} \mu_i) / \left(\sum_{i=1}^{n} \sigma_i^2\right)^{1/2} \tag{3.81}$$

In other words, the sum of random variables $X_1,\ldots,X_n$ can be approximated by a Gaussian distribution with mean $\sum_{i=1}^{n} \mu_i$ and variance $\left(\sum_{i=1}^{n} \sigma_i^2\right)^{1/2}$.

Both central limit theorems essentially state that regardless of their original individual distributions, the sum of many independent random variables (effects) tends to be distributed like a Gaussian distribution as the number of random variables (effects) becomes large.

3.1.7.3. Multivariate Mixture Gaussian Distributions

When $\mathbf{X}=(X_1,\ldots,X_n)$ is an n-dimensional continuous random vector, the multivariate Gaussian pdf has the following form:

$$f(\mathbf{X}=\mathbf{x}\,|\,\boldsymbol{\mu},\boldsymbol{\Sigma}) = N(\mathbf{x};\boldsymbol{\mu},\boldsymbol{\Sigma}) = \frac{1}{(2\pi)^{n/2}|\boldsymbol{\Sigma}|^{1/2}} \exp\left[-\frac{1}{2}(\mathbf{x}-\boldsymbol{\mu})^t \boldsymbol{\Sigma}^{-1}(\mathbf{x}-\boldsymbol{\mu})\right] \tag{3.82}$$

where $\boldsymbol{\mu}$ is the n-dimensional mean vector, $\boldsymbol{\Sigma}$ is the $n \times n$ covariance matrix, and $|\boldsymbol{\Sigma}|$ is the determinant of the covariance matrix $\boldsymbol{\Sigma}$.

$$\boldsymbol{\mu} = E(\mathbf{x}) \tag{3.83}$$

$$\boldsymbol{\Sigma} = E\left[(\mathbf{x}-\boldsymbol{\mu})(\mathbf{x}-\boldsymbol{\mu})^t\right] \tag{3.84}$$

More specifically, the i-j^{th} element σ_{ij}^2 of covariance matrix $\boldsymbol{\Sigma}$ can be specified as follows:

$$\sigma_{ij}^2 = E\left[(x_i-\mu_i)(x_j-\mu_j)\right] \tag{3.85}$$

If $X_1, \ldots, X_n$ are independent random variables, the covariance matrix $\boldsymbol{\Sigma}$ is reduced to diagonal covariance where all the off-diagonal entries are zero. The distribution can be regarded as n independent scalar Gaussian distributions. The joint pdf is the product of all the individual scalar Gaussian pdfs. Figure 3.12 shows a two-dimensional multivariate Gaussian distribution with independent random variables x_1 and x_2 with the same variance. Figure 3.13 shows another two-dimensional multivariate Gaussian distribution with independent random variables x_1 and x_2 that have different variances.

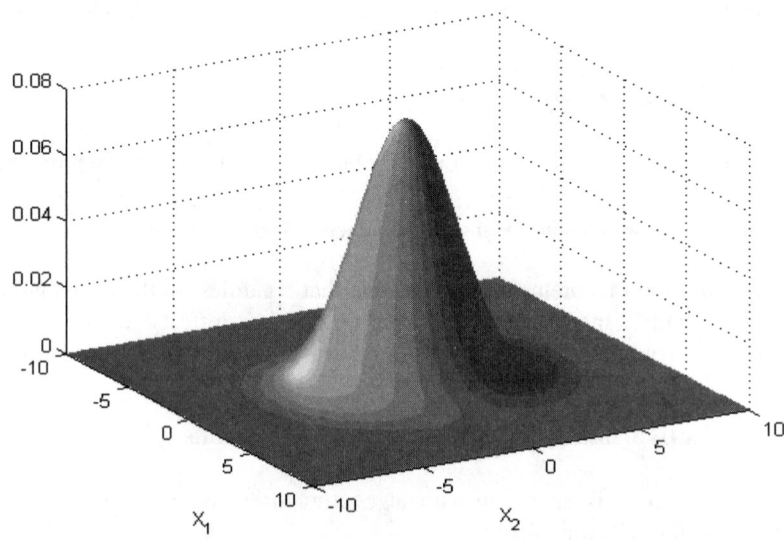

Figure 3.12 A two-dimensional multivariate Gaussian distribution with independent random variables x_1 and x_2 that have the same variance.

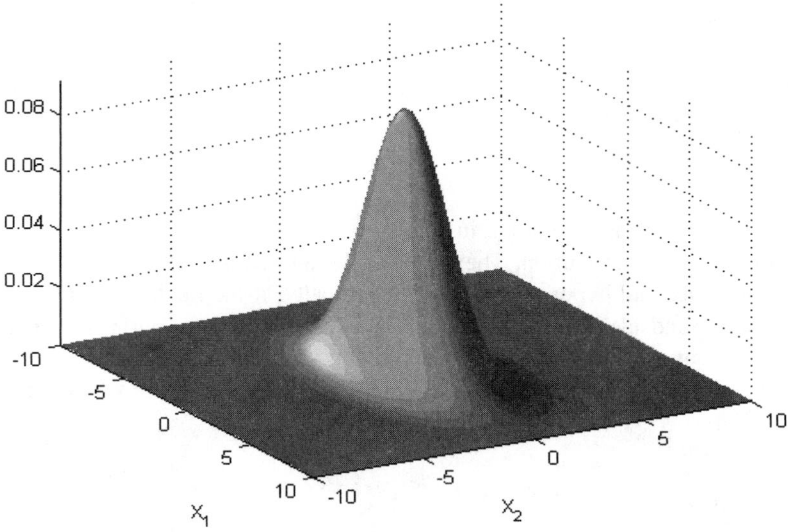

Figure 3.13 Another two-dimensional multivariate Gaussian distribution with independent random variable x_1 and x_2 which have different variances.

Although Gaussian distributions are unimodal,[2] more complex distributions with multiple local maxima can be approximated by Gaussian mixtures:

$$f(\mathbf{x}) = \sum_{k=1}^{K} c_k N_k(\mathbf{x}; \boldsymbol{\mu}_k, \boldsymbol{\Sigma}_k) \tag{3.86}$$

where c_k, the mixture weight associated with kth Gaussian component, is subject to the following constraint:

$$c_k \geq 0 \text{ and } \sum_{k=1}^{K} c_k = 1$$

Gaussian mixtures with enough mixture components can approximate any distribution. Throughout this book, most continuous probability density functions are modeled with Gaussian mixtures.

3.1.7.4. χ^2 Distributions

The gamma distribution with parameters α and β is defined in Eq. (3.71). For any given positive integer n, the gamma distribution for which $\alpha = n/2$ and $\beta = 1/2$ is called the χ^2

[2] A unimodal distribution has a single maximum (bump) for the distribution. For a Gaussian distribution, the maximum occurs at the mean.

distribution with n degrees of freedom. It follows from Eq. (3.71) that the pdf for the χ^2 distribution is

$$f(x \mid n) = \begin{cases} \dfrac{1}{2^{n/2}\Gamma(n/2)} x^{(n/2)-1} e^{-x/2} & x > 0 \\ 0 & x \leq 0 \end{cases} \tag{3.87}$$

χ^2 distributions are important in statistics because they are closely related to random samples of Gaussian distribution. They are widely applied in many important problems of statistical inference and hypothesis testing. Specifically, if the random variables $X_1, \ldots, X_n$ are independent and identically distributed, and if each of these variables has a standard Gaussian distribution, then the sum of square $X_1^2 + \ldots + X_n^2$ can be proved to have a χ^2 distribution with n degrees of freedom. Figure 3.14 illustrates three χ^2 distributions with $n = 2$, 3, and 4.

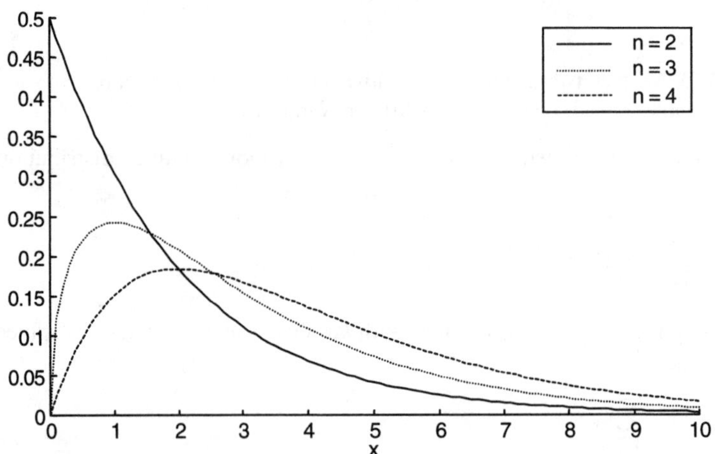

Figure 3.14 Three χ^2 distributions with $n = 2$, 3, and 4.

The mean and variance for the χ^2 distribution are

$$E(X) = n \text{ and } Var(X) = 2n \tag{3.88}$$

Following the additivity property of the gamma distribution, the χ^2 distribution also has the additivity property. That is, if the random variables $X_1, \ldots, X_n$ are independent and if X_i has a χ^2 distribution with k_i degrees of freedom, the sum $X_1 + \ldots + X_n$ has a χ^2 distribution with $k_1 + \ldots + k_n$ degrees of freedom.

3.1.7.5. Log-Normal Distribution

Let x be a Gaussian random variable with mean μ_x and standard deviation σ_x, then

$$y = e^x \tag{3.89}$$

follows the *lognormal* distribution

$$f(y \mid \mu_x, \sigma_x) = \frac{1}{y\sigma_x\sqrt{2\pi}} \exp\left\{-\frac{(\ln y - \mu_x)^2}{2\sigma_x^2}\right\} \tag{3.90}$$

shown in Figure 3.15, and whose mean is given by

$$\begin{aligned}\mu_y &= E\{y\} = E\{e^x\} = \int_{-\infty}^{\infty} \exp\{x\} \frac{1}{\sqrt{2\pi}\sigma_x} \exp\left\{-\frac{(x-\mu_x)^2}{2\sigma_x^2}\right\} dx \\ &= \int_{-\infty}^{\infty} \exp\{\mu_x + \sigma_x^2/2\} \frac{1}{\sqrt{2\pi}\sigma_x} \exp\left\{-\frac{(x-(\mu_x+\sigma_x^2)^2}{2\sigma_x^2}\right\} dx = \exp\{\mu_x + \sigma_x^2/2\}\end{aligned} \tag{3.91}$$

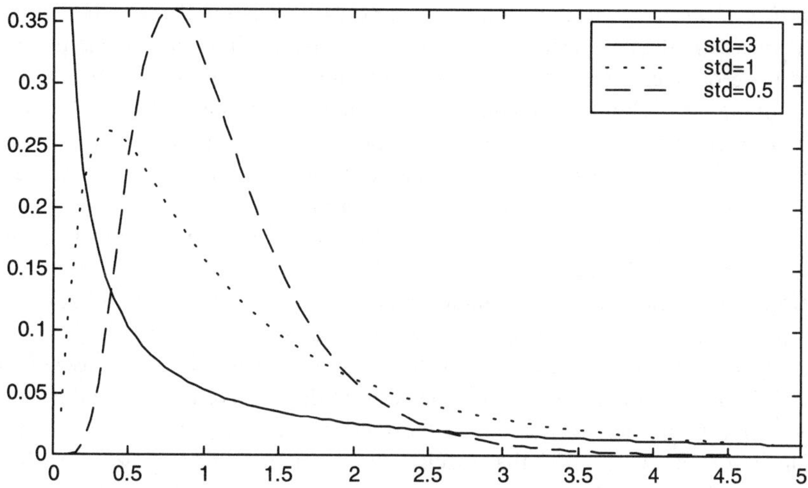

Figure 3.15 Lognormal distribution for $\mu_x = 0$ and $\sigma_x = 3$, 1, and 0.5, according to Eq. (3.90).

where we have rearranged the quadratic form of x and made use of the fact that the total probability mass of a Gaussian is 1. Similarly, the second order moment of y is given by

$$E\{y^2\} = \int_{-\infty}^{\infty} \exp\{2x\} \frac{1}{\sqrt{2\pi}\sigma_x} \exp\left\{-\frac{(x-\mu_x)^2}{2\sigma_x^2}\right\} dx$$
$$= \int_{-\infty}^{\infty} \exp\{2\mu_x + 2\sigma_x^2\} \frac{1}{\sqrt{2\pi}\sigma_x} \exp\left\{-\frac{(x-(\mu_x+2\sigma_x^2))^2}{2\sigma_x^2}\right\} dx = \exp\{2\mu_x + 2\sigma_x^2\} \quad (3.92)$$

and thus the variance of y is given by

$$\sigma_y^2 = E\{y^2\} - (E\{y\})^2 = \mu_y^2\left(\exp\{\sigma_x^2\} - 1\right) \quad (3.93)$$

Similarly, if $\mathbf{x}$ is a Gaussian random vector with mean $\boldsymbol{\mu}_\mathbf{x}$ and covariance matrix $\boldsymbol{\Sigma}_\mathbf{x}$, then random vector $\mathbf{y} = e^{\mathbf{x}}$ is log-normal with mean and covariance matrix [8] given by

$$\boldsymbol{\mu}_\mathbf{y}[i] = \exp\{\boldsymbol{\mu}_\mathbf{x}[i] + \boldsymbol{\Sigma}_\mathbf{x}[i,i]/2\}$$
$$\boldsymbol{\Sigma}_\mathbf{y}[i,j] = \boldsymbol{\mu}_\mathbf{y}[i]\boldsymbol{\mu}_\mathbf{y}[j]\left(\exp\{\boldsymbol{\Sigma}_\mathbf{x}[i,j]\} - 1\right) \quad (3.94)$$

using a similar derivation as in Eqs. (3.91) to (3.93).

3.2. Estimation Theory

Estimation theory and *significance testing* are two of the most important theories and methods of *statistical inference*. In this section, we describe estimation theory while significance testing is covered in the next section. A problem of statistical inference is one in which data generated in accordance with some unknown probability distribution must be analyzed, and some type of inference about the unknown distribution must be made. In a problem of statistical inference, any characteristic of the distribution generating the experimental data, such as the mean μ and variance σ^2 of a Gaussian distribution, is called a parameter of the distribution. The set Ω of all possible values of a *parameter* Φ or a group of parameters $\Phi_1, \Phi_2, \ldots, \Phi_n$ is called the *parameter space*. In this section we focus on how to estimate the parameter Φ from sample data.

Before we describe various estimation methods, we introduce the concept and nature of the estimation problems. Suppose that a set of random variables $\mathbf{X} = \{X_1, X_2, \ldots, X_n\}$ is iid according to a pdf $p(x \mid \Phi)$ where the value of the parameter Φ is unknown. Now, suppose also that the value of Φ must be estimated from the observed values in the sample. An *estimator* of the parameter Φ, based on the random variables $X_1, X_2, \ldots, X_n$, is a real-valued function $\theta(X_1, X_2, \ldots, X_n)$ that specifies the estimated value of Φ for each possible set of values of $X_1, X_2, \ldots, X_n$. That is, if the sample values of $X_1, X_2, \ldots, X_n$ turn out to be $x_1, x_2, \ldots, x_n$, then the estimated value of Φ will be $\theta(x_1, x_2, \ldots, x_n)$.

We need to distinguish between *estimator*, *estimate*, and *estimation*. An estimator $\theta(X_1, X_2, \ldots, X_n)$ is a function of the random variables, whose probability distribution can be derived from the joint distribution of $X_1, X_2, \ldots, X_n$. On the other hand, an estimate is a specific value $\theta(x_1, x_2, \ldots, x_n)$ of the estimator that is determined by using some specific sample values $x_1, x_2, \ldots, x_n$. Estimation is usually used to indicate the process of obtaining such an estimator for the set of random variables or an estimate for the set of specific sample values. If we use the notation $\mathbf{X} = \{X_1, X_2, \ldots, X_n\}$ to represent the vector of random variables and $\mathbf{x} = \{x_1, x_2, \ldots, x_n\}$ to represent the vector of sample values, an estimator can be denoted as $\theta(\mathbf{X})$ and an estimate $\theta(\mathbf{x})$. Sometimes we abbreviate an estimator $\theta(\mathbf{X})$ by just the symbol θ.

In the following four sections we describe and compare three different estimators (estimation methods). They are *minimum mean square estimator, maximum likelihood estimator*, and *Bayes' estimator*. The first one is often used to estimate the random variable itself, while the latter two are used to estimate the parameters of the distribution of the random variables.

3.2.1. Minimum/Least Mean Squared Error Estimation

Minimum mean squared error (MMSE) estimation and *least squared error* (LSE) estimation are important methods for random variables since the goal (minimize the squared error) is an intuitive one. In general, two random variables *X* and *Y* are i.i.d. according to some pdf $f_{X,Y}(x,y)$. Suppose that we perform a series of experiments and observe the value of *X*. We want to find a transformation $\hat{Y} = g(X)$ such that we can predict the value of the random variable *Y*. The following quantity can measure the goodness of such a transformation:

$$E(Y-\hat{Y})^2 = E(Y-g(X))^2 \tag{3.95}$$

This quantity is called *mean squared error* (MSE) because it is the mean of the squared error of the predictor $g(X)$. The criterion of minimizing the mean squared error is a good one for picking the predictor $g(X)$. Of course, we usually specify the class of function *G*, from which $g(X)$ may be selected. In general, there is a parameter vector $\mathbf{\Phi}$ associated with the function $g(X)$, so the function can be expressed as $g(X, \mathbf{\Phi})$. The process to find the parameter vector $\hat{\mathbf{\Phi}}_{MMSE}$ that minimizes the mean of the squared error is called *minimum mean squared error estimation* and $\hat{\mathbf{\Phi}}_{MMSE}$ is called the *minimum mean squared error estimator.* That is,

$$\hat{\mathbf{\Phi}}_{MMSE} = \arg\min_{\Phi} \left[E\left[(Y - g(X, \mathbf{\Phi}))^2 \right] \right] \tag{3.96}$$

Sometimes, the joint distribution of random variables *X* and *Y* is not known. Instead, samples of (*x*,*y*) pairs may be observable. In this case, the following criterion can be used instead,

$$\hat{\mathbf{\Phi}}_{LSE} = \arg\min_{\Phi} \sum_{i=1}^{n} [y_i - g(x_i, \mathbf{\Phi})]^2 \tag{3.97}$$

The argument of the minimization in Eq. (3.97) is called *sum-of-squared-error* (SSE) and the process of finding the parameter vector $\hat{\mathbf{\Phi}}_{LSE}$, which satisfies the criterion is called *least squared error estimation* or *minimum squared error estimation*. LSE is a powerful mechanism for curve fitting, where the function $g(x,\mathbf{\Phi})$ describes the observation pairs (x_i, y_i). In general, there are more points (n) than the number of free parameters in function $g(x,\mathbf{\Phi})$, so the fitting is over-determined. Therefore, no exact solution exists, and LSE fitting becomes necessary.

It should be emphasized that MMSE and LSE are actually very similar and share similar properties. The quantity in Eq. (3.97) is actually n times the sample mean of the squared error. Based on the law of large numbers, when the joint probability $f_{X,Y}(x,y)$ is uniform or the number of samples approaches to infinity, MMSE and LSE are equivalent.

For the class of functions, we consider the following three cases:

- Constant functions, i.e.,

$$G_c = \{g(x) = c, c \in R\} \tag{3.98}$$

- Linear functions, i.e.,

$$G_l = \{g(x) = ax + b \quad a, b \in R\} \tag{3.99}$$

- Other non-linear functions G_{nl}

3.2.1.1. MMSE/LSE for Constant Functions

When $\hat{Y} = g(x) = c$, Eq. (3.95) becomes

$$E(Y - \hat{Y})^2 = E(Y - c)^2 \tag{3.100}$$

To find the MMSE estimate for c, we take the derivatives of both sides in Eq. (3.100) with respect to c and equate it to 0. The MMSE estimate c_{MMSE} is given as

$$c_{MMSE} = E(Y) \tag{3.101}$$

and the minimum mean squared error is exactly the variance of Y, $Var(Y)$.

For the LSE estimate of c, the quantity in Eq. (3.97) becomes

$$\min \sum_{i=1}^{n} [y_i - c]^2 \tag{3.102}$$

Similarly, the LSE estimate c_{LSE} can be obtained as follows:

$$c_{LSE} = \frac{1}{n} \sum_{i=1}^{n} y_i \tag{3.103}$$

The quantity in Eq. (3.103) is the sample mean.

3.2.1.2. MMSE and LSE for Linear Functions

When $\hat{Y} = g(x) = ax + b$, Eq. (3.95) becomes

$$e(a,b) = E(Y - \hat{Y})^2 = E(Y - ax - b)^2 \tag{3.104}$$

To find the MMSE estimate of a and b, we can first set

$$\frac{\partial e}{\partial a} = 0, \text{ and } \frac{\partial e}{\partial b} = 0 \tag{3.105}$$

and solve the two linear equations. Thus, we can obtain

$$a = \frac{\text{cov}(X,Y)}{Var(X)} = \rho_{XY} \frac{\sigma_Y}{\sigma_X} \tag{3.106}$$

$$b = E(Y) - \rho_{XY} \frac{\sigma_Y}{\sigma_X} E(X) \tag{3.107}$$

For LSE estimation, we assume that the sample $\mathbf{x}$ is a d-dimensional vector for generality. Assuming we have n sample-vectors $(\mathbf{x}_i, y_i) = (x_i^1, x_i^2, \cdots, x_i^d, y_i),\ i = 1 \ldots n$, a linear function can be represented as

$$\hat{\mathbf{Y}} = \mathbf{XA} \text{ or } \begin{pmatrix} y_1 \\ y_2 \\ \vdots \\ y_n \end{pmatrix} = \begin{pmatrix} 1 & x_1^1 & \cdots & x_1^d \\ 1 & x_2^1 & \cdots & x_2^d \\ \vdots & \vdots & & \vdots \\ 1 & x_n^1 & \cdots & x_n^d \end{pmatrix} \begin{pmatrix} a_0 \\ a_1 \\ \vdots \\ a_d \end{pmatrix} \tag{3.108}$$

The sum of squared error can then be represented as

$$e(\mathbf{A}) = \| \hat{\mathbf{Y}} - \mathbf{Y} \|^2 = \sum_{i=1}^{n} \left(\mathbf{A}^t \mathbf{x}_i - y_i \right)^2 \tag{3.109}$$

A closed-form solution of the LSE estimate of $\mathbf{A}$ can be obtained by taking the gradient of $e(\mathbf{A})$,

$$\nabla e(\mathbf{A}) = \sum_{i=1}^{n} 2(\mathbf{A}^t \mathbf{x}_i - y_i)\mathbf{x}_i = 2\mathbf{X}^t(\mathbf{XA} - \mathbf{Y}) \tag{3.110}$$

and equating it to zero. This yields the following equation:

$$\mathbf{X}^t \mathbf{XA} = \mathbf{X}^t \mathbf{Y} \tag{3.111}$$

Thus the LSE estimate $\mathbf{A}_{LSE}$ will be of the following form:

$$\mathbf{A}_{LSE} = (\mathbf{X}^t\mathbf{X})^{-1}\mathbf{X}^t\mathbf{Y} \tag{3.112}$$

$(\mathbf{X}^t\mathbf{X})^{-1}\mathbf{X}^t$ in Eq. (3.112) is also refereed to as the *pseudo-inverse* of $\mathbf{X}$ and is sometimes denoted as $\mathbf{X}^{\perp}$.

When $\mathbf{X}^t\mathbf{X}$ is singular or some boundary conditions cause the LSE estimation in Eq. (3.112) to be unattainable, some numeric methods can be used to find an approximate solution. Instead of minimizing the quantity in Eq. (3.109), one can minimize the following quantity:

$$e(\mathbf{A}) = \| \mathbf{XA} - \mathbf{Y} \|^2 + \alpha \| \mathbf{A} \|^2 \tag{3.113}$$

Following a similar procedure, one can obtain the LSE estimate to minimize the quantity above in the following form.

$$\mathbf{A}^*_{LSE} = (\mathbf{X}^t\mathbf{X} + \alpha\mathbf{I})^{-1}\mathbf{X}^t\mathbf{Y} \tag{3.114}$$

The LSE solution in Eq. (3.112) can be used for polynomial functions too. In the problem of polynomial curve fitting using the least square criterion, we are aiming to find the coefficients $\mathbf{A} = (a_0, a_1, a_2, \cdots, a_d)^t$ that minimize the following quantity:

$$\min_{a_0, a_1, a_2, \cdots, a_d} E(Y - \hat{Y})^2 \tag{3.115}$$

where $\hat{Y} = a_0 + a_1 x + a_2 x^2 + \cdots + a_d x^d$

To obtain the LSE estimate of coefficients $\mathbf{A} = (a_0, a_1, a_2, \cdots, a_d)^t$, simply change the formation of matrix $\mathbf{X}$ in Eq. (3.108) to the following:

$$\mathbf{X} = \begin{pmatrix} 1 & x_1 & \cdots & x_1^d \\ 1 & x_2 & \cdots & x_2^d \\ \vdots & \vdots & & \vdots \\ 1 & x_n & \cdots & x_n^d \end{pmatrix} \tag{3.116}$$

Note that x_i^j in Eq. (3.108) means the *j*-th dimension of sample $\mathbf{x}_i$, while x_i^j in Eq. (3.116) means *j*-th order of value x_i. Therefore, the LSE estimate of polynomial coefficients $\mathbf{A}_{LSE} = (a_0, a_1, a_2, \cdots, a_d)^t$ has the same form as Eq. (3.112).

3.2.1.3. MMSE/LSE for Nonlinear Functions

As the most general case, consider solving the following minimization problem:

$$\min_{g(\bullet) \in G_{nl}} E[Y - g(X)]^2 \tag{3.117}$$

Since we need to deal with all possible nonlinear functions, taking a derivative does not work here. Instead, we use the property of conditional expectation to solve this minimization problem. By applying Eq. (3.38) to (3.117), we get

$$\begin{aligned} E_{X,Y}\left[Y-g(X)\right]^2 &= E_X\left\{E_{Y|X}\left[\left[Y-g(X)\right]^2 \mid X=x\right]\right\} \\ &= \int_{-\infty}^{\infty} E_{Y|X}\left[\left[Y-g(X)\right]^2 \mid X=x\right] f_X(x)dx \\ &= \int_{-\infty}^{\infty} E_{Y|X}\left[\left[Y-g(x)\right]^2 \mid X=x\right] f_X(x)dx \end{aligned} \tag{3.118}$$

Since the integrand is nonnegative in Eq. (3.118), the quantity in Eq. (3.117) will be minimized at the same time the following equation is minimized.

$$\min_{g(x)\in R} E_{Y|X}\left[\left[Y-g(x)\right]^2 \mid X=x\right] \tag{3.119}$$

Since $g(x)$ is a constant in the calculation of the conditional expectation above, the MMSE estimate can be obtained in the same way as the constant functions in Section 3.2.1.1. Thus, the MMSE estimate should take the following form:

$$\hat{Y} = g_{MMSE}(X) = E_{Y|X}(Y \mid X) \tag{3.120}$$

If the value $X = x$ is observed and the value $E(Y \mid X = x)$ is used to predict *Y*, the mean squared error (MSE) is minimized and specified as follows:

$$E_{Y|X}\left[\left[Y - E_{Y|X}(Y \mid X = x)\right]^2 \mid X = x\right] = Var_{Y|X}(Y \mid X = x) \tag{3.121}$$

The overall MSE, averaged over all the possible values of *X*, is:

$$E_X\left[Y - E_{Y|X}(Y \mid X)\right]^2 = E_X\left\{E_{Y|X}\left[\left[Y - E_{Y|X}(Y \mid X)\right]^2 \mid X\right]\right\} = E_X\left[{}_{Y|X}Var(Y \mid X = x)\right] \tag{3.122}$$

It is important to distinguish between the overall MSE $E_X\left[Var_{Y|X}(Y \mid X)\right]$ and the MSE of the particular estimate when $X = x$, which is $Var_{Y|X}(Y \mid X = x)$. Before the value of *X* is observed, the expected MSE for the process of observing *X* and predicting *Y* is $E_X\left[Var_{Y|X}(Y \mid X)\right]$. On the other hand, after a particular value *x* of *X* has been observed and the prediction $E_{Y|X}(Y \mid X = x)$ has been made, the appropriate measure of MSE of the prediction is $Var_{Y|X}(Y \mid X = x)$.

In general, the form of the MMSE estimator for nonlinear functions depends on the form of the joint distribution of *X* and *Y*. There is no mathematical closed-form solution. To get the conditional expectation in Eq. (3.120), we have to perform the following integral:

$$\hat{Y}(x) = \int_{-\infty}^{\infty} y f_Y(y \mid X = x)dy \tag{3.123}$$

It is difficult to solve this integral calculation. First, different measures of x could determine different conditional pdf for the integral. Exact information about the pdf is often impossible to obtain. Second, there could be no analytic solution for the integral. Those difficulties reduce the interest of the MMSE estimation of nonlinear functions to theoretical aspects only. The same difficulties also exist for LSE estimation for nonlinear functions. Some certain classes of well-behaved nonlinear functions are typically assumed for LSE problems and numeric methods are used to obtain LSE estimate from sample data.

3.2.2. Maximum Likelihood Estimation

Maximum likelihood estimation (MLE) is the most widely used parametric estimation method, largely because of its efficiency. Suppose that a set of random samples $\mathbf{X} = \{X_1, X_2, \ldots, X_n\}$ is to be drawn independently according to a discrete or continuous distribution with the pf or the pdf $p(x \mid \mathbf{\Phi})$, where the parameter vector $\mathbf{\Phi}$ belongs to some parameter space Ω. Given an observed vector $\mathbf{x} = (x_1, \cdots, x_n)$, the *likelihood* of the set of sample data vectors $\mathbf{x}$ with respect to $\mathbf{\Phi}$ is defined as the joint pf or joint pdf $p_n(\mathbf{x} \mid \mathbf{\Phi})$; $p_n(\mathbf{x} \mid \mathbf{\Phi})$ is also referred to as the *likelihood function*.

MLE assumes the parameters of pdfs are fixed but unknown and aims to find the set of parameters that maximizes the likelihood of generating the observed data. For example, if the pdf $p_n(\mathbf{x} \mid \mathbf{\Phi})$ is assumed to be a Gaussian distribution $N(\boldsymbol{\mu}, \boldsymbol{\Sigma})$, the components of $\mathbf{\Phi}$ will then include exactly the components of mean-vector $\boldsymbol{\mu}$ and covariance matrix $\boldsymbol{\Sigma}$. Since $X_1, X_2, \ldots, X_n$ are independent random variables, the likelihood can be rewritten as follows:

$$p_n(\mathbf{x} \mid \mathbf{\Phi}) = \prod_{k=1}^{n} p(x_k \mid \mathbf{\Phi}) \tag{3.124}$$

The likelihood $p_n(\mathbf{x} \mid \mathbf{\Phi})$ can be viewed as the probability of generating the sample data set $\mathbf{x}$ based on parameter set $\mathbf{\Phi}$. The *maximum likelihood estimator* of $\mathbf{\Phi}$ is denoted as $\mathbf{\Phi}_{MLE}$ that maximizes the likelihood $p_n(\mathbf{x} \mid \mathbf{\Phi})$. That is,

$$\mathbf{\Phi}_{MLE} = \underset{\Phi}{\operatorname{argmax}}\, p_n(\mathbf{x} \mid \mathbf{\Phi}) \tag{3.125}$$

This estimation method is called the *maximum likelihood estimation* method and is often abbreviated as MLE. Since the logarithm function is a monotonically increasing function, the parameter set $\mathbf{\Phi}_{MLE}$ that maximizes the log-likelihood should also maximize the likelihood. If $p_n(\mathbf{x} \mid \mathbf{\Phi})$ is differentiable function of $\mathbf{\Phi}$, $\mathbf{\Phi}_{MLE}$ can be attained by taking the partial derivative with respect to $\mathbf{\Phi}$ and setting it to zero. Specifically, let $\mathbf{\Phi}$ be a k-component parameter vector $\mathbf{\Phi} = (\Phi_1, \Phi_2, \ldots, \Phi_k)^t$ and $\nabla_{\mathbf{\Phi}}$ be the gradient operator:

$$\nabla_{\mathbf{\Phi}} = \begin{bmatrix} \frac{\partial}{\partial \Phi_1} \\ \vdots \\ \frac{\partial}{\partial \Phi_k} \end{bmatrix} \tag{3.126}$$

The log-likelihood becomes:

$$l(\mathbf{\Phi}) = \log p_n(\mathbf{x} \mid \mathbf{\Phi}) = \sum_{k=1}^{n} \log p(x_k \mid \mathbf{\Phi}) \tag{3.127}$$

and its partial derivative is:

$$\nabla_{\mathbf{\Phi}}\ l(\mathbf{\Phi}) = \sum_{k=1}^{n} \nabla_{\mathbf{\Phi}} \log p(x_k \mid \mathbf{\Phi}) \tag{3.128}$$

Thus, the maximum likelihood estimate of $\mathbf{\Phi}$ can be obtained by solving the following set of k equations:

$$\nabla_{\mathbf{\Phi}}\ l(\mathbf{\Phi}) = 0 \tag{3.129}$$

Example 3.1

Let's take a look at the maximum likelihood estimator of a univariate Gaussian pdf, given as the following equation:

$$p(x \mid \mathbf{\Phi}) = \frac{1}{\sqrt{2\pi}\sigma} \exp\left[-\frac{(x-\mu)^2}{2\sigma^2} \right] \tag{3.130}$$

where μ and σ^2 are the mean and the variance respectively. The parameter vector $\mathbf{\Phi}$ denotes (μ, σ^2). The log-likelihood is:

$$\begin{aligned} \log p_n(\mathbf{x} \mid \mathbf{\Phi}) &= \sum_{k=1}^{n} \log p(x_k \mid \mathbf{\Phi}) \\ &= \sum_{k=1}^{n} \log\left(\frac{1}{\sqrt{2\pi}\sigma} \exp\left[-\frac{(x_k-\mu)^2}{2\sigma^2} \right] \right) \\ &= -\frac{n}{2}\log(2\pi\sigma^2) - \frac{1}{2\sigma^2}\sum_{k=1}^{n}(x_k-\mu)^2 \end{aligned} \tag{3.131}$$

and the partial derivative of the above expression is:

$$\begin{aligned}\frac{\partial}{\partial \mu}\log p_n(x \mid \mathbf{\Phi}) &= \sum_{k=1}^{n}\frac{1}{\sigma^2}(x_k-\mu)\\ \frac{\partial}{\partial \sigma^2}\log p_n(x \mid \mathbf{\Phi}) &= -\frac{n}{2\sigma^2}+\sum_{k=1}^{n}\frac{(x_k-\mu)^2}{2\sigma^4}\end{aligned} \tag{3.132}$$

We set the two partial differential derivatives to zero,

$$\begin{aligned}&\sum_{k=1}^{n}\frac{1}{\sigma^2}(x_k-\mu)=0\\ &-\frac{n}{\sigma^2}+\sum_{k=1}^{n}\frac{(x_k-\mu)^2}{\sigma^4}=0\end{aligned} \tag{3.133}$$

The maximum likelihood estimates for μ and σ^2 are obtained by solving the above equations:

$$\begin{aligned}\mu_{MLE} &= \frac{1}{n}\sum_{k=1}^{n} x_k = E(x)\\ \sigma^2_{MLE} &= \frac{1}{n}\sum_{k=1}^{n}(x_k-\mu_{MLE})^2 = E\left[(x-\mu_{MLE})^2\right]\end{aligned} \tag{3.134}$$

Equation (3.134) indicates that the maximum likelihood estimation for mean and variance is just the sample mean and variance.

Example 3.2
For the multivariate Gaussian pdf $p(\mathbf{x})$

$$p(\mathbf{x} \mid \mathbf{\Phi}) = \frac{1}{(2\pi)^{d/2}|\mathbf{\Sigma}|^{1/2}}\exp\left[-\frac{1}{2}(\mathbf{x}-\boldsymbol{\mu})^t\Sigma^{-1}(\mathbf{x}-\boldsymbol{\mu})\right] \tag{3.135}$$

The maximum likelihood estimates of $\boldsymbol{\mu}$ and Σ can be obtained by a similar procedure.

$$\begin{aligned}\hat{\boldsymbol{\mu}}_{MLE} &= \frac{1}{n}\sum_{k=1}^{n}\mathbf{x}_k\\ \hat{\mathbf{\Sigma}}_{MLE} &= \frac{1}{n}\sum_{k=1}^{n}(\mathbf{x}_k-\hat{\boldsymbol{\mu}}_{MLE})(\mathbf{x}_k-\hat{\boldsymbol{\mu}}_{MLE})^t = E\left[(\mathbf{x}_k-\hat{\boldsymbol{\mu}}_{MLE})(\mathbf{x}_k-\hat{\boldsymbol{\mu}}_{MLE})^t\right]\end{aligned} \tag{3.136}$$

Once again, the maximum likelihood estimation for mean vector and covariance matrix is the sample mean vector and sample covariance matrix.

In some situations, maximum likelihood estimation of $\mathbf{\Phi}$ may not exist, or the maximum likelihood estimator may not be uniquely defined, i.e., there may be more than one MLE of $\mathbf{\Phi}$ for a specific set of sample values. Fortunately, according to Fisher's theorem, for most practical problems with a well-behaved family of distributions, the MLE exists and is uniquely defined [4, 25, 26].

In fact, the maximum likelihood estimator can be proven to be sound under certain conditions. As mentioned before, the estimator $\theta(\mathbf{X})$ is a function of the vector of random variables $\mathbf{X}$ that represent the sample data. $\theta(\mathbf{X})$ itself is also a random variable, with a distribution determined by joint distributions of $\mathbf{X}$. Let $\tilde{\mathbf{\Phi}}$ be the parameter vector of true distribution $p(x \mid \mathbf{\Phi})$ from which the samples are drawn. If the following three conditions hold:

1. The sample $\mathbf{x}$ is a drawn from the assumed family of distribution,
2. The family of distributions is well behaved,
3. The sample $\mathbf{x}$ is large enough,

then maximum likelihood estimator, $\mathbf{\Phi}_{MLE}$, has a Gaussian distribution with a mean $\tilde{\mathbf{\Phi}}$ and a variance of the form $1/nB_{\mathbf{x}}^2$ [26], where n is the size of sample and $B_{\mathbf{x}}$ is the *Fisher information*, which is determined solely by $\tilde{\mathbf{\Phi}}$ and $\mathbf{x}$. An estimator is said to be *consistent*, iff the estimate will converge to the true distribution when there is infinite number of training samples.

$$\lim_{n \to \infty} \mathbf{\Phi}_{MLE} = \tilde{\mathbf{\Phi}} \tag{3.137}$$

$\mathbf{\Phi}_{MLE}$ is a consistent estimator based on the analysis above. In addition, it can be shown that no consistent estimator has a lower variance than $\mathbf{\Phi}_{MLE}$. In other words, no estimator provides a closer estimate of the true parameters than the maximum likelihood estimator.

3.2.3. Bayesian Estimation and MAP Estimation

Bayesian estimation has a different philosophy than maximum likelihood estimation. While MLE assumes that the parameter $\mathbf{\Phi}$ [3] is fixed but unknown, Bayesian estimation assumes that the parameter $\mathbf{\Phi}$ itself is a random variable with a prior distribution $p(\mathbf{\Phi})$. Suppose we observe a sequence of random samples $\mathbf{x} = \{x_1, x_2, \ldots, x_n\}$, which are i.i.d. with a pdf $p(x \mid \mathbf{\Phi})$. According to Bayes' rule, we have the posterior distribution of $\mathbf{\Phi}$ as:

$$p(\mathbf{\Phi} \mid \mathbf{x}) = \frac{p(\mathbf{x} \mid \mathbf{\Phi}) p(\mathbf{\Phi})}{p(\mathbf{x})} \propto p(\mathbf{x} \mid \mathbf{\Phi}) p(\mathbf{\Phi}) \tag{3.138}$$

[3] For simplicity, we assume the parameter $\mathbf{\Phi}$ is a scalar instead of a vector here. However, the extension to a parameter vector $\mathbf{\Phi}$ can be derived according to a similar procedure.

In Eq. (3.138), we dropped the denominator $p(\mathbf{x})$ here because it is independent of the parameter Φ. The distribution in Eq. (3.138) is called the posterior distribution of Φ because it is the distribution of Φ after we observed the values of random variables $X_1, X_2, \ldots, X_n$.

3.2.3.1. Prior and Posterior Distributions

For mathematical tractability, conjugate priors are often used in Bayesian estimation. Suppose a random sample is taken of a known distribution with pdf $p(\mathbf{x} \mid \Phi)$. A conjugate prior for the random variable (or vector) is defined as the prior distribution for the parameters of the probability density function of the random variable (or vector), such that the class-conditional pdf $p(\mathbf{x} \mid \Phi)$, the posterior distribution $p(\Phi \mid \mathbf{x})$, and the prior distribution $p(\Phi)$ belong to the same distribution family. For example, it is well known that the conjugate prior for the mean of a Gaussian pdf is also a Gaussian pdf [4]. Now, let's derive such a posterior distribution $p(\Phi \mid \mathbf{x})$ from the widely used Gaussian conjugate prior.

Example 3.3

Suppose $X_1, X_2, \ldots, X_n$ are drawn from a Gaussian distribution for which the mean Φ is a random variable and the variance σ^2 is known. The likelihood function $p(\mathbf{x} \mid \Phi)$ can be written as:

$$p(\mathbf{x} \mid \Phi) = \frac{1}{(2\pi)^{n/2}\sigma^n} \exp\left[-\frac{1}{2}\sum_{i=1}^{n}\left(\frac{x_i - \Phi}{\sigma}\right)^2\right] \propto \exp\left[-\frac{1}{2}\sum_{i=1}^{n}\left(\frac{x_i - \Phi}{\sigma}\right)^2\right] \tag{3.139}$$

To further simply Eq. (3.139), we could use Eq. (3.140)

$$\sum_{i=1}^{n}(x_i - \Phi)^2 = n(\Phi - \bar{x}_n)^2 + \sum_{i=1}^{n}(x_i - \bar{x}_n)^2 \tag{3.140}$$

where $\bar{x}_n = \frac{1}{n}\sum_{i=1}^{n} x_i =$ the sample mean of $\mathbf{x} = \{x_1, x_2, \ldots, x_n\}$.

Let's rewrite $p(\mathbf{x} \mid \Phi)$ in Eq. (3.139) into Eq. (3.141):

$$p(\mathbf{x} \mid \Phi) \propto \exp\left[-\frac{n}{2\sigma^2}(\Phi - \bar{x}_n)^2\right] \exp\left[-\frac{1}{2\sigma^2}\sum_{i=1}^{n}(x_i - \bar{x}_n)^2\right] \tag{3.141}$$

Now suppose the prior distribution of Φ is also a Gaussian distribution with mean μ and variance ν^2, i.e., the prior distribution $p(\Phi)$ is given as follows:

$$p(\Phi) = \frac{1}{(2\pi)^{1/2}\nu} \exp\left[-\frac{1}{2}\left(\frac{\Phi - \mu}{\nu}\right)^2\right] \propto \exp\left[-\frac{1}{2}\left(\frac{\Phi - \mu}{\nu}\right)^2\right] \tag{3.142}$$

By combining Eqs. (3.141) and (3.142) while dropping the second term in Eq. (3.141) we could attain the posterior pdf $p(\Phi \mid \mathbf{x})$ in the following equation:

$$p(\Phi \mid \mathbf{x}) \propto \exp\left\{-\frac{1}{2}\left[\frac{n}{\sigma^2}(\Phi - \bar{x}_n)^2 + \frac{1}{v^2}(\Phi - \mu)^2\right]\right\} \tag{3.143}$$

Now if we define ρ and τ as follows:

$$\rho = \frac{\sigma^2 \mu + n v^2 \bar{x}_n}{\sigma^2 + n v^2} \tag{3.144}$$

$$\tau^2 = \frac{\sigma^2 v^2}{\sigma^2 + n v^2} \tag{3.145}$$

We can rewrite Eq. (3.143) as:

$$p(\Phi \mid \mathbf{x}) \propto \exp\left\{-\frac{1}{2}\left[\frac{1}{\tau^2}(\Phi - \rho)^2 + \frac{n}{\sigma^2 + n v^2}(\bar{x}_n - \mu)^2\right]\right\} \tag{3.146}$$

Since the second term in Eq. (3.146) does not depend on Φ, it can be absorbed in the constant factor. Finally, we have the posterior pdf in the following form:

$$p(\Phi \mid \mathbf{x}) = \frac{1}{\sqrt{2\pi}\tau} \exp\left[\frac{-1}{2\tau^2}(\Phi - \rho)^2\right] \tag{3.147}$$

Equation (3.147) shows that the posterior pdf $p(\Phi \mid \mathbf{x})$ is a Gaussian distribution with mean ρ and variance τ^2 as defined in Eqs. (3.144) and (3.145). The Gaussian prior distribution defined in Eq. (3.142) is a conjugate prior.

3.2.3.2. General Bayesian Estimation

The foremost requirement of a good estimator θ is that it can yield an estimate of Φ ($\theta(\mathbf{X})$) which is close to the real value Φ. In other words, a good estimator is one for which it is highly probable that the error $\theta(\mathbf{X}) - \Phi$ is close to 0. In general, we can define a loss function[4] $R(\Phi, \bar{\Phi})$. It measures the loss or cost associated with the fact that the true value of the parameter is Φ while the estimate is $\bar{\Phi}$. When only the prior distribution

[4] Bayesian estimation and loss functions are based on Bayes' decision theory, described in Chapter 4.

$p(\Phi)$ is available and no sample data has been observed, if we choose one particular estimate $\bar{\Phi}$, the expected loss is:

$$E\left[R(\Phi,\bar{\Phi})\right] = \int R(\Phi,\bar{\Phi})p(\Phi)d\Phi \tag{3.148}$$

The fact that we could derive posterior distribution from the likelihood function and the prior distribution [as shown in the derivation of Eq. (3.147)] is very important here because it allows us to compute the expected posterior loss after sample vector $\mathbf{x}$ is observed. The expected posterior loss associated with estimate $\bar{\Phi}$ is:

$$E\left[R(\Phi,\bar{\Phi}) \mid \mathbf{x}\right] = \int R(\Phi,\bar{\Phi})p(\Phi \mid \mathbf{x})d\Phi \tag{3.149}$$

The Bayesian estimator of Φ is defined as the estimator that attains minimum Bayes risk, that is, minimizes the expected posterior loss function (3.149). Formally, the Bayesian estimator is chosen according to:

$$\theta_{Bayes}(\mathbf{x}) = \underset{\theta}{\operatorname{argmin}}\ E\left[R(\Phi,\theta(\mathbf{x})) \mid \mathbf{x}\right] \tag{3.150}$$

The Bayesian estimator of Φ is the estimator θ_{Bayes} for which Eq. (3.150) is satisfied for every possible value of $\mathbf{x}$ of random vector $\mathbf{X}$. Therefore, the form of the Bayesian estimator θ_{Bayes} should depend only on the loss function and the prior distribution, but not the sample value $\mathbf{x}$.

One of the most common loss functions used in statistical estimation is the mean squared error function [20]. The mean squared error function for Bayesian estimation should have the following form:

$$R(\Phi,\theta(\mathbf{x})) = (\Phi - \theta(\mathbf{x}))^2 \tag{3.151}$$

In order to find the Bayesian estimator, we are seeking θ_{Bayes} to minimize the expected posterior loss function:

$$E\left[R(\Phi,\theta(\mathbf{x})) \mid \mathbf{x}\right] = E\left[(\Phi - \theta(\mathbf{x}))^2 \mid \mathbf{x}\right] = E(\Phi^2 \mid \mathbf{x}) - 2\theta(\mathbf{x})E(\Phi \mid \mathbf{x}) - \theta(\mathbf{x})^2 \tag{3.152}$$

The minimum value of this function can be obtained by taking the partial derivative of Eq. (3.152) with respect to $\theta(\mathbf{x})$. Since the above equation is simply a quadratic function of $\theta(\mathbf{x})$, it can be shown that the minimum loss can be achieved when θ_{Bayes} is chosen based on the following equation:

$$\theta_{Bayes}(\mathbf{x}) = E(\Phi \mid \mathbf{x}) \tag{3.153}$$

Equation (3.153) translates into the fact that the Bayesian estimate of the parameter Φ for mean squared error function is equal to the mean of the posterior distribution of Φ. In the following section, we discuss another popular loss function (MAP estimation) that also generates the same estimate for certain distribution functions.

3.2.3.3. MAP Estimation

One intuitive interpretation of Eq. (3.138) is that a prior pdf $p(\Phi)$ represents the relative likelihood before the values of $X_1, X_2, \ldots, X_n$ have been observed; while the posterior pdf $p(\Phi \mid \mathbf{x})$ represents the relative likelihood after the values of $X_1, X_2, \ldots, X_n$ have been observed. Therefore, choosing an estimate $\overline{\Phi}$ that maximizes the posterior probability is consistent without intuition. This estimator is in fact the *maximum posterior probability* (MAP) estimator and is the most popular Bayesian estimator.

The loss function associated with the MAP estimator is the so-called uniform loss function [20]:

$$R(\Phi, \theta(\mathbf{x})) = \begin{cases} 0, & \text{if } |\theta(\mathbf{x}) - \Phi| \le \Delta \\ 1, & \text{if } |\theta(\mathbf{x}) - \Phi| > \Delta \end{cases} \quad \text{where } \Delta > 0 \tag{3.154}$$

Now let's see how this uniform loss function results in MAP estimation. Based on the loss function defined above, the expected posterior loss function is:

$$\begin{aligned} E(R(\Phi, \theta(\mathbf{x})) \mid \mathbf{x}) &= P(|\theta(\mathbf{x}) - \Phi| > \Delta \mid \mathbf{x}) \\ &= 1 - P(|\theta(\mathbf{x}) - \Phi| \le \Delta \mid \mathbf{x}) = 1 - \int_{\theta(\mathbf{x})-\Delta}^{\theta(\mathbf{x})+\Delta} p(\Phi \mid \mathbf{x}) \end{aligned} \tag{3.155}$$

The quantity in Eq. (3.155) is minimized by maximizing the shaded area under $p(\Phi \mid \mathbf{x})$ over the interval $[\theta(\mathbf{x}) - \Delta, \theta(\mathbf{x}) + \Delta]$ in Figure 3.16. If $p(\Phi \mid \mathbf{x})$ is a smooth curve and Δ is small enough, the shaded area can be computed roughly as:

$$\int_{\theta(\mathbf{x})-\Delta}^{\theta(\mathbf{x})+\Delta} p(\Phi \mid \mathbf{x}) \cong 2\Delta p(\Phi \mid \mathbf{x})\Big|_{\Phi=\theta(\mathbf{x})} \tag{3.156}$$

Thus, the shaded area can be approximately maximized by choosing $\theta(\mathbf{x})$ to be the maximum point of $p(\Phi \mid \mathbf{x})$. This concludes our proof the using the error function in Eq. (3.154) indeed will generate MAP estimator.

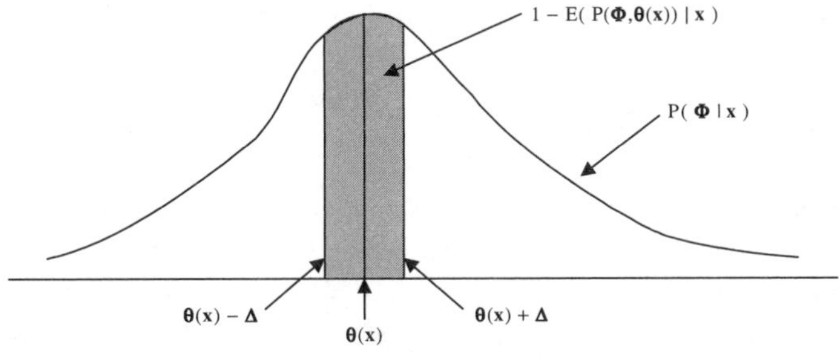

Figure 3.16 Illustration of the minimum expected posterior loss function for MAP estimation [20].

MAP estimation is to find the parameter estimate Φ_{MAP} or estimator $\theta_{MAP}(\mathbf{x})$ that maximizes the posterior probability,

$$\Phi_{MAP} = \theta_{MAP}(\mathbf{x}) = \underset{\Phi}{\text{argmax}}\ p(\Phi \mid \mathbf{x}) = \underset{\Phi}{\text{argmax}}\ p(\mathbf{x} \mid \Phi) p(\Phi) \tag{3.157}$$

Φ_{MAP} can also be specified in the logarithm form as follows:

$$\Phi_{MAP} = \underset{\Phi}{\text{argmax}}\ \left[\log p(\mathbf{x} \mid \Phi) + \log p(\Phi)\right] \tag{3.158}$$

Φ_{MAP} can be attained by solving the following partial differential equation:

$$\frac{\partial \log p(\mathbf{x} \mid \Phi)}{\partial \Phi} + \frac{\partial \log p(\Phi)}{\partial \Phi} = 0 \tag{3.159}$$

Thus the MAP equation for finding Φ_{MAP} can be established.

$$\left.\frac{\partial \log p(\mathbf{x} \mid \Phi)}{\partial \Phi}\right|_{\Phi=\Phi_{MAP}} = \left.\frac{-\partial \log p(\Phi)}{\partial \Phi}\right|_{\Phi=\Phi_{MAP}} \tag{3.160}$$

There are interesting relationships between MAP estimation and MLE estimation. The prior distribution is viewed as the knowledge of the statistics of the parameters of interest before any sample data is observed. For the case of MLE, the parameter is assumed to be fixed but unknown. That is, there is no preference (knowledge) of what the values of parameters should be. The prior distribution $p(\Phi)$ can only be set to constant for the entire parameter space, and this type of prior information is often referred to as *non-informative prior* or *uniform prior*. By substituting $p(\Phi)$ with a uniform distribution in Eq. (3.157), MAP estimation is identical to MLE. In this case, the parameter estimation is solely determined by the observed data. A sufficient amount of training data is often a requirement for MLE. On the other hand, when the size of the training data is limited, the use of the prior density becomes valuable. If some prior knowledge of the distribution of the parameters can be obtained, MAP estimation provides a way of incorporating prior information in the parameter learning process.

Example 3.4

Now, let's formulate MAP estimation for Gaussian densities. As described in Section 3.2.3.1, the conjugate prior distribution for a Gaussian density is also a Gaussian distribution. Similarly, we assumed random variables $X_1, X_2, \ldots, X_n$ drawn from a Gaussian distribution for which the mean Φ is unknown and the variance σ^2 is known, while the conjugate prior distribution of Φ is a Gaussian distribution with mean μ and variance v^2. It is shown in Section 3.2.3.1 that the posterior pdf can be formulated as in Eq. (3.147). The MAP estimation for Φ can be solved by taking the derivative of Eq. (3.147) with respect to Φ:

$$\Phi_{MAP} = \rho = \frac{\sigma^2 \mu + n\nu^2 \overline{x}_n}{\sigma^2 + n\nu^2} \tag{3.161}$$

where n is the total number of training samples and $\overline{x}_n$ the sample mean.

The MAP estimate of the mean Φ is a weighted average of the sample mean $\overline{x}_n$ and the prior mean. When n is zero (when there is no training data at all), the MAP estimate is simply the prior mean μ. On the other hand, when n is large ($n \to \infty$), the MAP estimate will converge to the maximum likelihood estimate. This phenomenon is consistent with our intuition and is often referred to as *asymptotic equivalence* or *asymptotic convergence*. Therefore, in practice, the difference between MAP estimation and MLE is often insignificant when a large amount of training data is available. When the prior variance ν^2 is very large (e.g., $\nu^2 >> \sigma^2 / n$), the MAP estimate will converge to the ML estimate because a very large ν^2 translates into a non-informative prior.

It is important to note that the requirement of learning prior distributions for MAP estimation is critical. In some cases, the prior distribution is very difficult to estimate and MLE is still an attractive estimation method. As mentioned before, the MAP estimation framework is particularly useful for dealing with sparse data, such as parameter adaptation. For example, in speaker adaptation, the speaker-independent (or multiple speakers) database can be used to first estimate the prior distribution [9]. The model parameters are adapted to a target speaker through a MAP framework by using limited speaker-specific training data as discussed in Chapter 9.

3.3. SIGNIFICANCE TESTING

Significance testing is one of the most important theories and methods of statistical *inference*. A problem of statistical inference, or, more simply, a statistics problem, is one in which data that have been generated in accordance with some unknown probability distribution must be analyzed, and some type of inference about the unknown distribution must be made. Hundreds of test procedures have developed in statistics for various kinds of hypotheses testing. We focus only on tests that are used in spoken language systems.

The selection of appropriate models for the data or systems is essential for spoken language systems. When the distribution of certain sample data is unknown, it is usually appropriate to make some assumptions about the distribution of the data with a distribution function whose properties are well known. For example, people often use Gaussian distributions to model the distribution of background noise in spoken language systems. One important issue is how good our assumptions are, and what the appropriate values of the parameters for the distributions are, even when we can use the methods in Section 3.2 to estimate parameters from sample data. Statistical tests are often applied to determine if the distribution with specific parameters is appropriate to model the sample data. In this section, we describe the most popular testing method for the goodness of distribution fitting – the χ^2 goodness-of-fit test.

Another important type of statistical tests is designed to evaluate the excellence of two different methods or algorithms for the same tasks when there is uncertainty regarding the results. To assure that the two systems are evaluated on the same or similar conditions, experimenters often carefully choose similar or even the exactly same data sets for testing. This is why we refer to this type of statistical test as a *paired observations* test. In both speech recognition and speech synthesis, the paired observations test is a very important tool for interpreting the comparison results.

3.3.1. Level of Significance

We now consider statistical problems involving a parameter ϕ whose value is unknown but must lie in a certain parameter space Ω. In statistical tests, we let H_0 denote the hypothesis that $\phi \in \Omega_0$ and let H_1 denote the hypothesis that $\phi \in \Omega_1$. The subsets Ω_0 and Ω_1 are disjoint and $\Omega_0 \cup \Omega_1 = \Omega$, so exactly one of the hypotheses H_0 and H_1 must be true. We must now decide whether to accept H_0 or H_1 by observing a random sample $X_1, \cdots, X_n$ drawn from a distribution involving the unknown parameter ϕ. A problem like this is called hypotheses testing. A procedure for deciding whether to accept H_0 or H_1 is called a *test procedure* or simply a *test*. The hypothesis H_0 is often referred to as the *null hypothesis* and the hypothesis H_1 as the *alternative hypothesis*. Since there are only two possible decisions, accepting H_0 is equivalent to rejecting H_1 and rejecting H_0 is equivalent to accepting H_1. Therefore, in testing hypotheses, we often use the terms *accepting or rejecting the null hypothesis* H_0 as the only decision choices.

Usually we are presented with a random sample $\mathbf{X} = (X_1, \cdots, X_n)$ to help us in making the test decision. Let *S* denote the sample space of *n*-dimensional random vector $\mathbf{X}$. The testing procedure is equivalent to partitioning the sample space *S* into two subsets. One subset specifies the values of $\mathbf{X}$ for which one will accept H_0 and the other subset specifies the values of $\mathbf{X}$ for which one will reject H_0. The second subset is called the *critical region* and is often denoted as *C*.

Since there is uncertainty associated with the test decision, for each value of $\phi \in \Omega$, we are interested in the probability $\rho(\phi)$ that the testing procedure rejects H_0. The function $\rho(\phi)$ is called the *power function* of the test and can be specified as follows:

$$\rho(\phi) = P(\mathbf{X} \in C \mid \phi) \tag{3.162}$$

For $\phi \in \Omega_0$, the decision to reject H_0 is incorrect. Therefore, if $\phi \in \Omega_0$, $\rho(\phi)$ is the probability that the statistician will make an incorrect decision (false rejection). In statistical tests, an upper bound α_0 $(0 < \alpha_0 < 1)$ is specified, and we only consider tests for which $\rho(\phi) \le \alpha_0$ for every value of $\phi \in \Omega_0$. The upper bound α_0 is called the *level of significance*. The smaller α_0 is, the less likely it is that the test procedure will reject H_0. Since α_0 specifies the upper bound for false rejection, once a hypothesis is rejected by the test procedure, we can be $(1-\alpha_0)$ confident the decision is correct. In most applications, α_0 is set to be 0.05 and the test is said to be carried out at the 0.05 level of significance or 0.95 level of confidence.

We define the size α of a given test as the maximum probability, among all the values of ϕ which satisfy the null hypothesis, of making an incorrect decision.

$$\alpha = \max_{\theta \in \Omega_0} \rho(\phi) \tag{3.163}$$

Once we obtain the value of α, the test procedure is straightforward. First, the statistician specifies a certain level of significance α_0 in a given problem of testing hypotheses, then he or she rejects the null hypothesis if the size α is such that $\alpha \le \alpha_0$.

The size α of a given test is also called the *tail area* or the *p-value* corresponding to the observed value of data sample **X** because it corresponds to tail area of the distribution. The hypothesis will be rejected if the level of significance α_0 is such that $\alpha_0 > \alpha$ and should be accepted for any value of $\alpha_0 < \alpha$. Alternatively, we can say the observed value of **X** is *just significant* at the level of significance α without using the level of significance α_0. Therefore, if we had found that the observed value of one data sample **X** was just significant at the level of 0.0001, while the other observed value of data sample **Y** was just significant at the level of 0.001, then we can conclude the sample **X** provides much stronger evidence against H_0. In statistics, an observed value of one data sample **X** is generally said to be statistically significant if the corresponding tail area is smaller than the traditional value 0.05. For cases requiring more significance (confidence), 0.01 can be used.

A statistically significant observed data sample **X** that provides strong evidence against H_0 does not necessary provide strong evidence that the actual value of ϕ is significantly far away from parameter set Ω_0. This situation can arise, particularly when the size of random data sample is large, because a test with larger sample size will in general reject hypotheses with more confidence, unless the hypothesis is indeed the true one.

3.3.2. Normal Test (Z-Test)

Suppose we need to find whether a coin toss is fair or not. Let p be the probability of heads. The hypotheses to be tested are as follows:

$H_0 : p = ½$

$H_1 : p \ne ½$

We assume that a random sample size n is taken, and let random variable M denote the number of times we observe heads as the result. The random variable M has a binomial distribution $B(n, ½)$. Because of the shape of binomial distribution, M can lie on either side of the mean. This is why it is called a typical *two-tailed test*. The tail area or p-value for the observed value k can be computed as:

$$p = \begin{cases} 2P(k \le M \le n) & \text{for } k > n/2 \\ 2P(0 \le M \le k) & \text{for } k < n/2 \\ 1.0 & \text{for } k = n/2 \end{cases} \tag{3.164}$$

The *p*-value in Eq. (3.164) can be computed directly using the binomial distribution. The test procedure will reject H_0 when *p* is less than the significance level α_0.

In many situations, the *p*-value for the distribution of data sample **X** is difficult to obtain due to the complexity of the distribution. Fortunately, if some statistic *Z* of the data sample **X** has some well-known distribution, the test can then be done in the *Z* domain instead. If *n* is large enough ($n > 50$), a *normal test* (or *Z-test*) can be used to approximate a binomial probability. Under H_0, the mean and variance for *M* are $E(M) = n/2$ and $Var(M) = n/4$. The new random variable *Z* is defined as,

$$Z = \frac{|M - n/2| - 1/2}{\sqrt{n/4}} \tag{3.165}$$

which can be approximated as standard Gaussian distribution $N(0,1)$ under H_0. The *p*-value can now be computed as $p = 2P(Z \geq z)$ where *z* is the realized value of *Z* after *M* is observed. Thus, H_0 is rejected if $p < \alpha_0$, where α_0 is the level of significance.

3.3.3. χ^2 Goodness-of-Fit Test

The normal test (Z-test) can be extended to test the hypothesis that a given set of data came from a certain distribution with all parameters specified. First let's look at the case of discrete distribution fitting.

Suppose that a large population consists of items of *k* different types and let p_i be the probability that a random selected item belongs to type *i*. Now, let $q_1, \ldots, q_k$ be a set of specific numbers satisfying the probabilistic constraint ($q_i \geq 0$ for $i = 1, \ldots, k$ and $\sum_{i=1}^{k} q_i = 1$). Finally, suppose that the following hypotheses are to be tested:

$H_0 : p_i = q_i$ for $i = 1, \ldots, k$

$H_1 : p_i \neq q_i$ for at least one value of *i*

Assume that a random sample of size *n* is to be taken from the given population. For $i = 1, \ldots, k$, let N_i denote the number of observations in the random sample which are of type *i*. Here, $N_1, \ldots, N_k$ are nonnegative numbers and $\sum_{i=1}^{k} N_i = n$. Random variables $N_1, \ldots, N_k$ have a multinomial distribution. Since the *p*-value for the multinomial distribution is hard to obtain, instead we use another statistic about $N_1, \ldots, N_k$. When H_0 is true, the expected number of observations of type *i* is nq_i. In other words, the difference between the actual number of observations N_i and the expected number nq_i should be small when H_0 is true. It seems reasonable to base the test on the differences $N_i - nq_i$ and to reject H_0 when the differences are large. It can be proved [14] that the following random variable λ

$$\lambda = \sum_{i=1}^{k} \frac{(N_i - nq_i)^2}{nq_i} \tag{3.166}$$

converges to the χ^2 distribution with $k-1$ degrees of freedom as the sample size $n \to \infty$.

A χ^2 test of goodness-of-fit can be carried out in the following way. Once a level of significance α_0 is specified, we can use the following p-value function to find critical point c:[5]

$$P(\lambda > c) = 1 - F_{\chi^2}(x = c) = \alpha_0 \tag{3.167}$$

where $F_{\chi^2}(x)$ is the distribution function for χ^2 distribution. The test procedure simply rejects H_0 when the realized value λ is such that $\lambda > c$. Empirical results show that the χ^2 distribution will be a good approximation to the actual distribution of λ as long as the value of each expectation nq_i is not too small (≥ 5). The approximation should still be satisfactory if $nq_i \geq 1.5$ for $i = 1,\ldots,k$.

For continuous distributions, a modified χ^2 goodness-of-fit test procedure can be applied. Suppose that we would like to hypothesize a null hypothesis H_0 in which continuous random sample data $X_1,\ldots,X_k$ are drawn from a certain continuous distribution with all parameters specified or estimated. Also, suppose the observed values of random sample $x_1,\ldots,x_k$ are bounded within interval Ω. First, we divide the range of the hypothesized distribution into m subintervals within interval Ω such that the expected number of values, say E_i, in each interval is at least 5. For $i = 1,\ldots,k$, we let N_i denote the number of observations in the ith subintervals. As in Eq. (3.166), one can prove that the following random variable λ

$$\lambda = \sum_{i=1}^{m} \frac{(N_i - E_i)^2}{E_i} \tag{3.168}$$

converges to the χ^2 distribution with $m-k-1$ degrees of freedom as the sample size $n \to \infty$, where k is the number of parameters that must be estimated from the sample data in order to calculate the expected number of values, E_i. Once the χ^2 distribution is established, the same procedure can be used to find the critical c in Eq. (3.167) to make test decisions.

Example 3.5

Suppose we are given a random variable X of sample size 100 points and we want to determine whether we can reject the following hypothesis:

$$H_0 : X \sim N(0,1) \tag{3.169}$$

To perform χ^2 goodness-of-fit test, we first divide the range of X into 10 subintervals. The corresponding probability falling in each subinterval, the expected number of points falling in each subinterval and the actual number of points falling in each subintervals [10] are illustrated in Table 3.1.

---

[5] Since χ^2 pdf is a monotonic function, the test is a one-tail test. Thus, we only need to calculate one tail area.

Table 3.1 The probability falling in each subinterval of an $N(0,1)$, and 100 sample points, the expected number of points falling in each subinterval, and the actual number of points falling in each subinterval [10].

Subinterval I_i	$P(X \in I_i)$	$E_i = 100P(X \in I_i)$	N_i
[-∞, -1.6]	0.0548	5.48	2
[-1.6, -1.2]	0.0603	6.03	9
[-1.2, -0.8]	0.0968	9.68	6
[-0.8, -0.4]	0.1327	13.27	11
[-0.4, 0.0]	0.1554	15.54	19
[0.0, 0.4]	0.1554	15.54	25
[0.4, 0.8]	0.1327	13.27	17
[0.8, 1.2]	0.0968	9.68	2
[1.2, 1.6]	0.0603	6.03	6
[-1.6, ∞]	0.0548	5.48	3

The value for λ can then be calculated as follows:

$$\lambda = \sum_{i=1}^{m} \frac{(N_i - E_i)^2}{E_i} = 18.286$$

Since λ can be approximated as a χ^2 distribution with $m-k-1=10-0-1=9$ degrees of freedom, the critical point c at the 0.05 level of significance is calculated[6] to be 16.919 according to Eq. (3.167). Thus, we should reject the hypothesis H_0 because the calculated λ is greater than the critical point c.

The χ^2 goodness-of-fit test at the 0.05 significance level is in general used to determine when a hypothesized distribution is not an adequate distribution to use. To accept the diswtribution as a good fit, one needs to make sure the hypothesized distribution cannot be rejected at the 0.4 to 0.5 level-of-significance. The alternative is to use the χ^2 goodness-of-fit test for a number of potential distributions and select the one with smallest calculated χ^2.

When all the parameters are specified (instead of estimated), the Kolmogorov-Smirnov test [5] can also be used for the goodness-of-fit test. The Kolmogorov-Smirnov test in general is a more powerful test procedure when the sample size is relatively small.

3.3.4. Matched-Pairs Test

In this section, we discuss experiments in which two different methods (or systems) are to be compared to learn which one is better. To assure the two methods are evaluated under

[6] In general, we use a cumulative distribution function table to find the point with specific desired cumulative probability for complicated distributions, like the χ^2 distribution.

similar conditions, two closely resemble data samples or ideally the same data sample should be used to evaluate both methods. This type of hypotheses test is called *matched-paired* test [5].

3.3.4.1. The Sign Test

For $i = 1,\ldots,n$, let p_i denote the probability that method *A* is better than method *B* when testing on the ith paired data sample. We shall assume that the probability p_i has the same value *p* for each of the *n* pairs. Suppose we wish to test the null hypothesis that method *A* is no better than method *B*. That is, the hypotheses to be tested have the following form:

$$H_0 : p \le \tfrac{1}{2}$$

$$H_1 : p > \tfrac{1}{2}$$

Suppose that, for each pair of data samples, either one method or the other will appear to be better, and the two methods cannot tie. Under these assumptions, the *n* pairs represent *n* Bernoulli trials, for each of which there is probability *p* that method *A* yields better performance. Thus the number of pairs *M* in which method *A* yields better performance will have a binomial distribution $B(n, p)$. For the simple sign test where one needs to decide which method is better, *p* will be set to $1/2$. Hence a reasonable procedure is to reject H_0 if $M > c$, where *c* is a critical point. This procedure is called a signed test. The critical point can be found according to

$$P(M > c) = 1 - F_B(x = c) = \alpha_0 \tag{3.170}$$

where $F_B(x)$ is the distribution for binomial distribution. Thus, for observed values $M > c$, we will reject H_0.

3.3.4.2. Magnitude-Difference Test

The only information that the sign test utilizes from each pair of data samples, is the sign of the difference between two performances. To do a sign test, one does not need to obtain a numeric measurement of the magnitude of the difference between the two performances. However, if the measurement of magnitude of the difference for each pair is available, a test procedure based on the relative magnitudes of the differences can be used [11].

We assume now that the performance of each method can be measured for any data samples. For $i = 1,\ldots,n$, let A_i denote the performance of the method *A* on the ith pair of data samples and B_i denote the performance of the method *B* on the ith pair of data sample. Moreover, we shall let $D_i = A_i - B_i$. Since $D_1,\ldots,D_n$ are generated on *n* different pairs of data samples, they should be independent random variables. We also assume that $D_1,\ldots,D_n$ have the same distribution. Suppose now we are interested in testing the null hypothesis that method *A* and method *B* have on the average the same performance on the *n* pairs of data samples.

Let μ_D be the mean of D_i. The MLE estimate of μ_D is:

$$\mu_D = \sum_{i=1}^{n} \frac{D_i}{n} \tag{3.171}$$

The test hypotheses are:
$H_0 : \mu_D = 0$
$H_1 : \mu_D \neq 0$
The MLE estimate of the variance of D_i is

$$\sigma_D^2 = \frac{1}{n} \sum_{i=1}^{n} (D_i - \mu_D)^2 \tag{3.172}$$

We define a new random variable Z as follows:

$$Z = \frac{\mu_D}{\sigma_D / \sqrt{n}} \tag{3.173}$$

If n is large enough (> 50), Z is proved to have a standard Gaussian distribution $N(0,1)$. The normal test procedure described in Section 3.3.2 can be used to test H_0. This type of matched-paired tests usually depends on having enough pairs of data samples for the assumption that Z can be approximated with a Gaussian distribution. It also requires enough data samples to estimate the mean and variance of the D_i.

3.4. Information Theory

Transmission of information is a general definition of what we call communication. Claude Shannon's classic paper of 1948 gave birth to a new field in information theory that has become the cornerstone for coding and digital communication. In the paper titled "A Mathematical Theory of Communication," he wrote [21]:

> *The fundamental problem of communication is that of reproducing at one point either exactly or approximately a message selected at another point.*

Information theory is a mathematical framework for approaching a large class of problems related to encoding, transmission, and decoding information in a systematic and disciplined way. Since speech is a form of communication, information theory has served as the underlying mathematical foundation for spoken language processing.

3.4.1. Entropy

Three interpretations can be used to describe the quantity of *information*: (1) the amount of uncertainty before seeing an event, (2) the amount of surprise when seeing an event, and (3) the amount of information after seeing an event. Although these three interpretations seem slightly different, they are virtually the same under the framework of information theory.

According to information theory, the information derivable from outcome x_i depends on its probability. If the probability $P(x_i)$ is small, we can derive a large degree of information, because the outcome that it has occurred is very rare. On the other hand, if the probability is large, the information derived will be small, because the outcome is well expected. Thus, the amount of information is defined as follows:

$$I(x_i) = \log \frac{1}{P(x_i)} \tag{3.174}$$

The reason to use a logarithm can be interpreted as follows. The information for two independent events to occur (where the joint probability is the multiplication of both individual probabilities) can be simply carried out by the addition of the individual information of each event. When the logarithm base is 2, the unit of information is called a *bit*. This means that one bit of information is required to specify the outcome. In this probabilistic framework, the amount of information represents uncertainty. Suppose X is a discrete random variable taking value x_i (referred to as a symbol) from a finite or countable infinite sample space $S = \{x_1, x_2, \ldots, x_i, \ldots\}$ (referred to as an alphabet). The symbol x_i is produced from an information source with alphabet S, according to the probability distribution of the random variable X. One of the most important properties of an information source is the entropy $H(S)$ of the random variable X, defined as the average amount of information (expected information):

$$H(X) = E[I(X)] = \sum_S P(x_i) I(x_i) = \sum_S P(x_i) \log \frac{1}{P(x_i)} = E\left[-\log P(X)\right] \tag{3.175}$$

This entropy $H(X)$ is the amount of information required to specify what kind of symbol has occurred on average. It is also the average uncertainty for the symbol. Suppose that the sample space S has an alphabet size $\|S\| = N$. The entropy $H(X)$ attains the maximum value when the pf has a uniform distribution, i.e.:

$$P(x_i) = P(x_j) = \frac{1}{N} \quad \text{for all } i \text{ and } j \tag{3.176}$$

Equation (3.176) can be interpreted to mean that *uncertainty* reaches its maximum level when no outcome is more probable than any other. It can be proved that the entropy $H(X)$ is nonnegative and becomes zero only if the probability function is a deterministic one, i.e.,

$$H(X) \geq 0 \text{ with equality i.f.f. } P(x_i) = 1 \text{ for some } x_i \in S \tag{3.177}$$

There is another very interesting property for the entropy. If we replace the pf of generating symbol x_i in Eq. (3.175) with any other arbitrary pf, the new value is no smaller than the original entropy. That is,

$$H(X) \le E[-\log Q(X)] = -\sum_S P(x_i)\log Q(x_i) \tag{3.178}$$

Equation (3.178) has a very important meaning. It shows that we are more uncertain about the data if we misestimate the distribution governing the data source. The equality for Eq. (3.178) occurs if and only if $P(x_i) = Q(x_i) \;\; 1 \le i \le N$. Equation (3.178), often referred to as *Jensen's inequality*, is the basis for the proof of EM algorithm in Chapter 4. Similarly, Jensen's inequality can be extended to continuous pdf:

$$-\int f_x(x)\log f_x(x)dx \le -\int g_x(x)\log f_x(x)dx \tag{3.179}$$

with equality occurring if and only if $f_x(x) = g_x(x) \;\; \forall x$.

The proof of Eq. (3.178) follows from the fact $\log(x) \le x-1, \; \forall x$, so the following quantity must have an non-positive value.

$$\sum_S P(x_i)\log\frac{Q(x_i)}{P(x_i)} \le \sum_S P(x_i)\left[\frac{Q(x_i)}{P(x_i)} - 1\right] = 0 \tag{3.180}$$

Based on this property, the negation of the quantity in Eq. (3.180) can be used for the measurement of the distance of two probability distributions. Specifically, the *Kullback-Leibler* (KL) *distance* (*relative entropy*, *discrimination*, or *divergence*) is defined as:

$$KL(P \| Q) = E\left[\log\frac{P(X)}{Q(X)}\right] = \sum_S P(x_i)\log\frac{P(x_i)}{Q(x_i)} \tag{3.181}$$

As discussed in Chapter 11, the branching factor of a grammar or language is an important measure of degree of difficulty of a particular task in spoken language systems. This relates to the size of the word list from which a speech recognizer or a natural language processor needs to disambiguate in a given context. According to the entropy definition above, this branching factor estimate (or average choices for an alphabet) is defined as follows:

$$PP(X) = 2^{H(X)} \tag{3.182}$$

$PP(X)$ is called the *perplexity* of source *X*, since it describes how confusing the grammar (or language) is. The value of perplexity is equivalent to the size of an imaginary equivalent list, whose words are equally probable. The bigger the perplexity, the higher branching factor. To find out the perplexity of English, Shannon devised an ingenious way [22] to estimate the entropy and perplexity of English words and letters. His method is similar to a guessing game where a human subject guesses sequentially the words of a text hidden from him, using the relative frequencies of her/his guesses as the estimates of the probability distribution underlying the source of the text. Shannon's perplexity estimate of English comes out to be about 2.39 for English letters and 130 for English words. Chapter 11 has a detailed description on the use of perplexity for language modeling.

3.4.2. Conditional Entropy

Now let us consider transmission of symbols through an information channel. Suppose the input alphabet is $X=(x_1,x_2,\ldots,x_s)$, the output alphabet is $Y=(y_1,y_2,\ldots,y_t)$, and the information channel is defined by the channel matrix $M_{ij}=P(y_j \mid x_i)$, where $P(y_j \mid x_i)$ is the conditional probability of receiving output symbol y_j when input symbol x_i is sent. Figure 3.17 shows an example of an information channel.

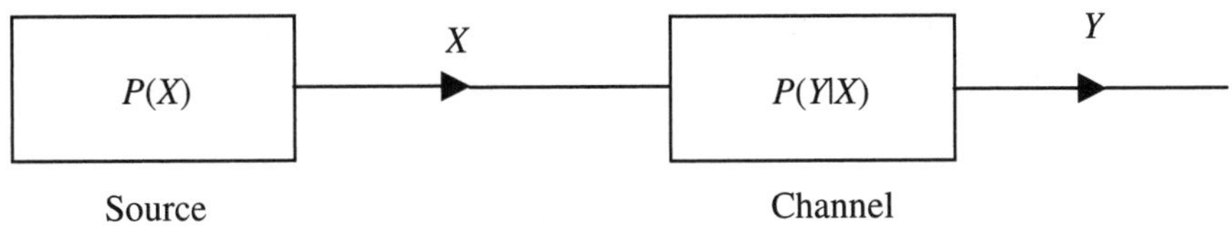

Figure 3.17 Example of information channel. The source is described by source pf $P(X)$ and the channel is characterized by the conditional pf $P(Y|X)$.

Before transmission, the average amount of information, or the uncertainty of the input alphabet X, is the prior entropy $H(\mathrm{X})$.

$$H(X)=\sum_X P(X=x_i)\log\frac{1}{P(X=x_i)} \tag{3.183}$$

where $P(x_i)$ is the prior probability. After transmission, suppose y_j is received; then the average amount of information, or the uncertainty of the input alphabet A, is reduced to the following *posterior* entropy:

$$H(X \mid Y=y_j)=-\sum_X P(X=x_i \mid Y=y_j)\log P(X=x_i \mid Y=y_j) \tag{3.184}$$

where the $P(x_i \mid y_j)$ are the posterior probabilities. Averaging the posterior entropy $H(X \mid y_j)$ over all output symbols y_j leads to the following equation:

$$\begin{aligned} H(X \mid Y) &= \sum_Y P(Y=y_j)H(X \mid Y=y_j) \\ &= -\sum_Y P(Y=y_j)\sum_X P(X=x_i \mid Y=y_j)\log P(X=x_i \mid Y=y_j) \\ &= -\sum_X\sum_Y P(X=x_i,Y=y_j)\log P(X=x_i \mid Y=y_j) \end{aligned} \tag{3.185}$$

This *conditional entropy,* defined in Eq. (3.185), is the average amount of information or the uncertainty of the input alphabet X given the outcome of the output event Y. Based on the definition of conditional entropy, we derive the following equation:

$$\begin{aligned} H(X,Y) &= -\sum_X \sum_Y P(X = x_i, Y = y_i) \log P(X = x_i, Y = y_i) \\ &= -\sum_X \sum_Y P(X = x_i, Y = y_i) \{\log P(X = x_i) + \log P(Y = y_i \mid X = x_i)\} \\ &= H(X) + H(Y \mid X) \end{aligned} \quad (3.186)$$

Equation (3.186) has an intuitive meaning – the uncertainty about two random variables equals the sum of uncertainty about the first variable and the conditional entropy for the second variable given the first variable is known. Equations (3.185) and (3.186) can be generalized to random vectors **X** and **Y** where each contains several random variables.

It can be proved that the chain rule [Eq. (3.16)] applies to entropy.

$$H(X_1, \cdots, X_n) = H(X_n \mid X_1, \cdots, X_{n-1}) + \cdots + H(X_2 \mid X_1) + H(X_1) \quad (3.187)$$

Finally, the following inequality can also be proved:

$$H(X \mid Y, Z) \leq H(X \mid Y) \quad (3.188)$$

with equality i.f.f. X and Z being independent when conditioned on Y. Equation (3.188) basically confirms the intuitive belief that uncertainty decreases when more information is known.

3.4.3. The Source Coding Theorem

Information theory is the foundation for data compressing. In this section we describe *Shannon's source coding theorem*, also known as the *first coding theorem*. In source coding, we are interested in *lossless* compression, which means the compressed information (or symbols) can be recovered (decoded) perfectly. The entropy serves as the upper bound for a source lossless compression.

Consider an information source with alphabet $S = \{0, 1, \ldots, N-1\}$. The goal of data compression is to *encode* the output symbols into a string of binary symbols. An interesting question arises: *What is the minimum number of bits required, on the average, to encode the output symbols of the information source?*

Let's assume we have a source that can emit four symbols {0,1,2,3} with equal probability $P(0) = P(1) = P(2) = P(3) = 1/4$. Its entropy is 2 bits as illustrated in Eq. (3.189):

$$H(S) = \sum_{i=0}^{3} P(i) \log_2 \frac{1}{P(i)} = 2 \quad (3.189)$$

It is obvious that 2 bits per symbol is good enough to encode this source. A possible binary code for this source is {00, 01, 10, 11}. It could happen, though some symbols are more likely than others, for example, $P(0) = 1/2, \ P(1) = 1/4, \ P(2) = 1/8, \ P(3) = 1/8$. In this case the entropy is only 1.75 bits. One obvious idea is to use fewer bits for lower values that are frequently used and more bits for larger values that are rarely used. To represent this source we can use a variable-length code $\{0,10,110,111\}$, where no codeword is a prefix for the rest and thus a string of 0s and 1s can be uniquely broken into those symbols. The encoding scheme with such a property is called *uniquely decipherable* (or instantaneous) coding, because as soon as the decoder observes a sequence of codes, it can decisively determine the sequence of the original symbols. If we let $r(x)$ be the number of bits (length) used to encode symbol x, the average rate R of bits per symbol used for encoding the information source is:

$$R = \sum_x r(x)P(x) \tag{3.190}$$

In our case, R is 1.75 bits as shown in Eq. (3.191):

$$R = 0.5\times1 + 0.25\times2 + 0.125\times3 + 0.125\times3 = 1.75 \tag{3.191}$$

Such variable-length coding strategy is called *Huffman coding*. Huffman coding belongs to *entropy coding* because it matches the entropy of the source. In general, *Shannon's source coding* theorem says that a source cannot be coded with fewer bits than its entropy. We will skip the proof here. Interested readers can refer to [3, 15, 17] for the detailed proof. This theorem is consistent with our intuition because the entropy measure is exactly the information content of the information measured in bits. If the entropy increases, then uncertainty increases, resulting in a large amount of information. Therefore, it takes more bits to encode the symbols. In the case above, we are able to match this rate, but, in general, this is impossible, though we can get arbitrarily close to it. The Huffman code for this source offers a compression rate of 12.5% relative to the code designed for the uniform distribution.

Shannon's source coding theorem establishes not only the lower bound for lossless compression but also the upper bound. Let $\lceil x \rceil$ denote the smallest integer that greater or equal to x. As in the similar procedure above, we can make the code length assigned to source output x equal to

$$l(x) = \lceil -\log P(x) \rceil \tag{3.192}$$

The average length L satisfies the following inequality:

$$L = \sum_x l(x)P(x) < \sum_x [1 - \log P(x)]P(x) = 1 + H(X) \tag{3.193}$$

Equation (3.193) means that the average rate R only exceeds the value of entropy by less than one bit.

L can be made arbitrarily close to the entropy by block coding. Instead of encoding single output symbols of the information source, one can encode each block of length n. Let's assume the source is memoryless, so $X_1, X_2, \ldots, X_n$ are independent. According to Eq. (3.193), the average rate R for this block code satisfies:

$$L < 1 + H(X_1, X_2, \ldots, X_n) = 1 + nH(X) \tag{3.194}$$

This makes the average number of bits per output symbol, L/n, satisfy

$$\lim_{n\to\infty}\frac{1}{n}L \le H(X) \tag{3.195}$$

In general, Huffman coding arranges the symbols in order of decreasing probability, assigns the bit 0 to the symbol of highest probability and the bit 1 to what is left, and proceeds the same way for the second highest probability value (which now has a code 10) and iterate. This results in 2.25 bits for the uniform distribution case, which is higher than the 2 bits we obtain with equal-length codes.

Lempel-Ziv coding is a coding strategy that uses correlation to encode *strings* of symbols that occur frequently. Although it can be proved to converge to the entropy, its convergence rate is much slower [27]. Unlike Huffman coding, Lempel-Ziv coding is independent of the distribution of the source; i.e., it needs not be aware of the distribution of the source before encoding. This type of coding scheme is often referred to as *universal* encoding scheme.

3.4.4. Mutual Information and Channel Coding

Let's review the information channel illustrated in Figure 3.17. An intuitively plausible measure of the average amount of information provided by the random event *Y* about the random event *X* is the average difference between the number of bits it takes to specify the outcome of *X* when the outcome of *Y* is not known and the outcome of Y is known. *Mutual information* is defined as the difference in the entropy of *X* and the conditional entropy of *X* given *Y*:

$$\begin{aligned}
I(X;Y) &= H(X) - H(X\mid Y) \\
&= \sum_X P(x_i)\log\frac{1}{P(x_i)} - \sum_X\sum_Y P(x_i,y_j)\log\frac{1}{P(x_i\mid y_j)} \\
&= \sum_X\sum_Y P(x_i,y_j)\log\frac{P(x_i\mid y_j)}{P(x_i)} = \sum_X\sum_Y P(x_i,y_j)\log\frac{P(x_i,y_j)}{P(x_i)P(y_i)} \\
&= E\left[\log\frac{P(X,Y)}{P(X)P(Y)}\right]
\end{aligned} \tag{3.196}$$

$I(X;Y)$ is referred to as the mutual information between *X* and *Y*. $I(X;Y)$ is symmetrical; i.e., $I(X;Y)=I(Y;X)$. The quantity $P(x,y)/P(x)P(y)$ is often referred to as the *mutual information between symbol x and y*. $I(X;Y)$ is bounded:

$$0 \le I(X;Y) \le \min\left[H(X),H(Y)\right] \tag{3.197}$$

$I(X;Y)$ reaches the minimum value (zero) when the random variables *X* and *Y* are independent.

Mutual information represents the information obtained (or the reduction in uncertainty) through a channel by observing an output symbol. If the information channel is noiseless, the input symbol can be determined definitely by observing an output symbol. In this case, the conditional entropy $H(X|Y)$ equals zero and it is called a *noiseless channel.* We obtain the maximum mutual information $I(X;\ Y) = H(X)$. However, the information channel is generally *noisy* so that the conditional entropy $H(X|Y)$ is not zero. Therefore, maximizing the mutual information is equivalent to obtaining a low-noise information channel, which offers a closer relationship between input and output symbols.

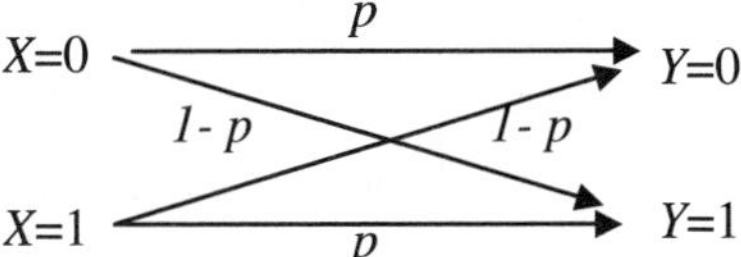

Figure 3.18 A binary channel with two symbols.

Let's assume that we have a binary channel, a channel with a binary input and output as shown in Figure 3.18. Associated with each output are a probability p that the output is correct, and a probability $(1-p)$ that it is not, so that the channel is *symmetric.*

If we observe a symbol $Y = 1$ at the output, we don't know for sure what symbol X was transmitted, though we know $P(X=1 \mid Y=1) = p$ and $P(X=0 \mid Y=1) = (1-p)$, so that we can measure our uncertainty about X by its conditional entropy:

$$H(X \mid Y=1) = -p\log p - (1-p)\log(1-p) \tag{3.198}$$

If we assume that our source X has a uniform distribution, $H(X \mid Y) = H(X \mid Y=1)$ as shown in Eq. (3.198) and $H(X) = 1$. The mutual information between X and Y is given by

$$I(X,Y) = H(X) - H(X \mid Y) = 1 + p\log p + (1-p)\log(1-p) \tag{3.199}$$

It measures the information that Y carries by about X. The channel capacity C is the maximum of the mutual information over all distributions of X. That is,

$$C = \max_{P(x)} I(X;Y) \tag{3.200}$$

The channel capacity C can be attained by varying the distribution of the information source until the mutual information is maximized for the channel. The channel capacity C can be regarded as a channel that can transmit at most C bits of information per unit of time. *Shannon's channel coding* theorem says that for a given channel there exists a code that permits error-free transmission across the channel, provided that $R \le C$, where R is the rate of the communication system, which is defined as the number of bits per unit of time being transmitted by the communication system. Shannon's channel coding theorem states the fact that *arbitrarily reliable communication is possible at any rate below channel capacity*.

Figure 3.19 illustrates a transmission channel with the source encoder and destination decoder. The source encoder will encode the source symbol sequence $\mathbf{x} = x_1, x_2, \ldots, x_n$ into channel input sequence $y_1, y_2, \ldots, y_k$. The destination decoder takes the output sequence $z_1, z_2, \ldots, z_k$ from the channel and converts it into the estimates of the source output $\bar{\mathbf{x}} = \bar{x}_1, \bar{x}_2, \ldots, \bar{x}_n$. The goal of this transmission is to make the probability of correct decoding $P(\bar{\mathbf{x}} = \mathbf{x})$ asymptotically close to 1 while keeping the compression ratio $\Re = n/k$ as large as possible. *Shannon's source-channel coding* theorem (also referred to as *Shannon's second coding* theorem) says that it is possible to find an encoder-decoder pair of rate $\Re$ for a noisy information channel, provided that $\Re \times H(X) \le C$.

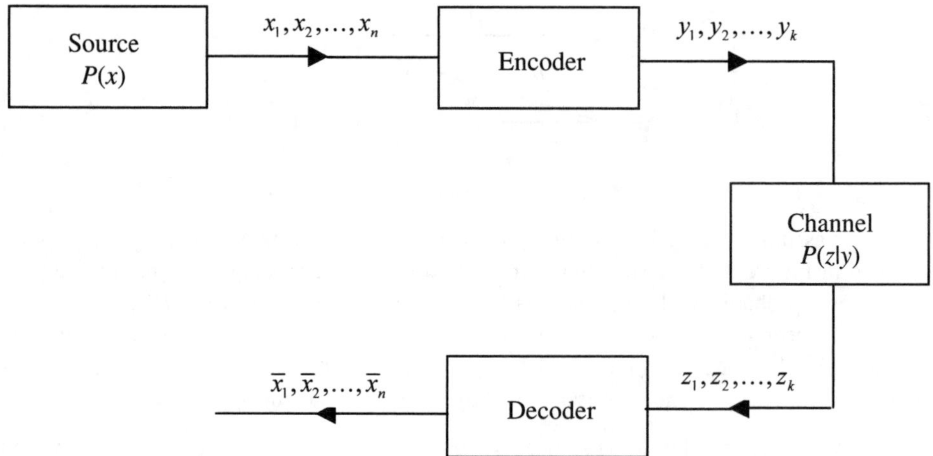

Figure 3.19 Transmission of information through a noisy channel [15].

Because of channel errors, speech coders need to provide error correction codes that will decrease the bit rate allocated to the speech. In practice, there is a tradeoff between the bit rate used for source coding and the bit rate for channel coding. In Chapter 7 we will describe speech coding in great detail.

3.5. Historical Perspective and Further Reading

The idea of uncertainty and probability can be traced all the way back to about 3500 B.C., when games of chance played with bone objects were developed in Egypt. Cubical dice with markings virtually identical to modern dice have been found in Egyptian tombs dating around 2000 B.C. Gambling with dice played an important part in the early development of probability theory. Modern mathematical theory of probability is believed to have been started by the French mathematicians Blaise Pascal (1623-1662) and Pierre Fermat (1601-1665) when they worked on certain gambling problems involving dice. English mathematician Thomas Bayes (1702-1761) was first to use probability inductively and established a mathematical basis for probability inference, leading to what is now known as Bayes' theorem. The theory of probability has developed steadily since then and has been widely applied in diverse fields of study. There are many good textbooks on probability theory. The book by DeGroot [6] is an excellent textbook for both probability and statistics which covers all the necessary elements for engineering majors. The authors also recommend [14], [19], or [24] for interested readers.

Estimation theory is a basic subject in statistics covered in textbooks. The books by DeGroot [6], Wilks [26] and Hoel [13] offer excellent discussions of estimation theory. They all include comprehensive treatments for maximum likelihood estimation and Bayesian estimation. Maximum likelihood estimation was introduced in 1912 by R. A. Fisher (1890-1962) and has been applied to various domains. It is arguably the most popular parameter estimation method due to its intuitive appeal and excellent performance with large training samples. The EM algorithm in Chapter 4 and the estimation of hidden Markov models in Chapter 8 are based on the principle of MLE. The use of prior distribution in Bayesian estimation is very controversial in statistics. Some statisticians adhere to the Bayesian philosophy of statistics by taking the Bayesian estimation view of the parameter $\mathbf{\Phi}$ having a probability distribution. Others, however, believe that in many problems $\mathbf{\Phi}$ is not a random variable but rather a fixed number whose value is unknown. Those statisticians believe that a prior distribution can be assigned to a parameter $\mathbf{\Phi}$ only when there is extensive prior knowledge of the past; thus the non-informative priors are completely ruled out. Both groups of statisticians agree that whenever a meaningful prior distribution can be obtained, the theory of Bayesian estimation is applicable and useful. The books by DeGroot [6] and Poor [20] are excellent for learning the basics of Bayesian and MAP estimations. Bayesian and MAP adaptation are particularly powerful when the training samples are sparse. Therefore, they are often used for adaptation where the knowledge of prior distribution can help to adapt the model to a new but limited training set. The speaker adaptation work done by Brown et al. [2] first applied Bayesian estimation to speech recognition and [9] is another good paper on using MAP for hidden Markov models. References [4], [16] and [14] have extensive studies of different conjugate prior distributions for various standard distributions. Finally, [1] has an extensive reference for Bayesian estimation.

Significance testing is an essential tool for statisticians to interpret all the statistical experiments. Neyman and Pearson provided some of the most important pioneering work in hypotheses testing [18]. There are many different testing methods presented in most statistics books. The χ^2 test, invented in 1900 by Karl Pearson, is arguably the most widely used testing method. Again, the textbook by DeGroot [6] is an excellent source for the basics of testing and various testing methods. The authors recommend [7] as an interesting book that uses many real-world examples to explain statistical theories and methods, particularly the significance testing.

Information theory first appeared in Claude Shannon's historical paper: *A Mathematical Theory of Communication* [21]. In it, Shannon, analyzed communication as the transmission of a message from a source through a channel to a receiver. In order to solve the problem he created a new branch of applied mathematics – *information and coding theory*. IEEE published a collection of Shannon's papers [23] containing all of his published works, as well as many that have never been published. Those published include his classic papers on information theory and switching theory. Among the unpublished works are his once-secret wartime reports, his Ph.D. thesis on population genetics, unpublished Bell Labs memoranda, and a paper on the theory of juggling. The textbook by McEliece [17] is excellent for learning all theoretical aspects of information and coding theory. However, it might be out of print now. Instead, the books by Hamming [12] and Cover [3] are two current great references for information and coding theory. Finally, F. Jelinek's *Statistical Methods for Speech Recognition* [15] approaches the speech recognition problem from an information-theoretic aspect. It is a useful book for people interested in both topics.

REFERENCES

[1] Bernardo, J.M. and A.F.M. Smith, *Bayesian Theory*, 1996, New York, John Wiley.

[2] Brown, P., C.-H. Lee, and J. Spohrer, "Bayesian Adaptation in Speech Recognition," *Proc. of the IEEE Int. Conf. on Acoustics, Speech and Signal Processing*, 1983, Boston, MA, pp. 761-764.

[3] Cover, T.M. and J.A. Thomas, *Elements of Information Theory*, 1991, New York, John Wiley and Sons.

[4] DeGroot, M.H., *Optimal Statistical Decisions*, 1970, New York, NY, McGraw-Hill.

[5] DeGroot, M.H., *Probability and Statistics*, Addison-Wesley Series in Behavioral Science: Quantitive Methods, F. Mosteller, ed., 1975, Reading, MA, Addison-Wesley Publishing Company.

[6] DeGroot, M.H., *Probability and Statistics*, 2nd ed, Addison-Wesley Series in Behavioral Science: Quantitive Methods, F. Mosteller, ed., 1986, Reading, MA, Addison-Wesley Publishing Company.

[7] Freedman, D., *et al.*, *Statistics*, 2nd ed, 1991, New York, W. W. Norton & Company, Inc.

[8] Gales, M.J., *Model Based Techniques for Noise Robust Speech Recognition*, PhD Thesis in Engineering Department 1995, Cambridge University.

[9] Gauvain, J.L. and C.H. Lee, "Maximum a Posteriori Estimation for Multivariate Gaussian Mixture Observations of Markov Chains," *IEEE Trans. on Speech and Audio Processing*, 1994, **2**(2), pp. 291-298.

[10] Gillett, G.E., *Introduction to Operations Research: A Computer-Oriented Algorithmic Approach*, McGraw-Hill Series in Industrial Engineering and Management Science, J. Riggs, ed., 1976, New York, McGraw-Hill.

[11] Gillick, L. and S.J. Cox, "Some Statistical Issues in the Comparison of Speech Recognition Algorithms," *IEEE Int. Conf. on Acoustics, Speech and Signal Processing*, 1989, Glasgow, Scotland, UK, IEEE, pp. 532-535.

[12] Hamming, R.W., *Coding and Information Theory*, 1986, Englewood Cliffs, NJ, Prentice-Hall.

[13] Hoel, P.G., *Introduction to Mathematical Statistics*, 5th ed., 1984, John Wiley & Sons.

[14] Jeffreys, H., *Theory of Probability*, 1961, Oxford University Press.

[15] Jelinek, F., *Statistical Methods for Speech Recognition*, Language, Speech, and Communication, 1998, Cambridge, MA, MIT Press.

[16] Lindley, D.V., "The Use of Prior Probability Distributions in Statistical Inference and Decision," *Fourth Berkeley Symposium on Mathematical Statistics and Probability*, 1961, Berkeley, CA, Univ. of California Press.

[17] McEliece, R., *The Theory of Information and Coding*, Encyclopedia of Mathematics and Its Applications, eds. R. Gian-Carlo. Vol. 3, 1977, Reading, MA, Addison-Wesley Publishing Company.

[18] Neyman, J. and E.S. Pearson, "On the Problem of the Most Efficient Tests of Statistical Hypotheses," *Philosophical Trans. of Royal Society*, 1928, **231**, pp. 289-337.

[19] Papoulis, A., *Probability, Random Variables, and Stochastic Processes*, 3rd ed., 1991, New York, McGraw-Hill.

[20] Poor, H.V., *An Introduction to Signal Detection and Estimation*, Springer Tests in Electrical Engineering, J.B. Thomas, ed., 1988, New York, Springer-Verlag.

[21] Shannon, C., "A Mathematical Theory of Communication," *Bell System Technical Journal*, 1948, **27**, pp. 379-423, 623-526.

[22] Shannon, C.E., "Prediction and Entropy of Printed English," *Bell System Technical Journal*, 1951, pp. 50-62.

[23] Shannon, C.E., *Claude Elwood Shannon: Collected Papers*, 1993, IEEE.

[24] Viniotis, Y., *Probability and Random Processes for Electrical Engineering*, Outline Series in Electronics & Electrical Engineering, Schaum, ed., 1998, New York, WCB McGraw-Hill.

[25] Wald, A., "Note of Consistency of Maximum Likelihood Estimate," *Ann. Mathematical Statistics*, 1949(20), pp. 595-601.

[26] Wilks, S.S., *Mathematical Statistics*, 1962, New York, John Wiley and Sons.

[27] Ziv, J. and A. Lempel, "A Universal Algorithm for Sequential Data Compression," *IEEE Trans. on Information Theory*, 1997, **IT-23**, pp. 337-343.

CHAPTER 4

Pattern Recognition

Spoken language processing relies heavily on pattern recognition, one of the most challenging problems for machines. In a broader sense, the ability to recognize patterns forms the core of our intelligence. If we can incorporate the ability to reliably recognize patterns in our work and life, we can make machines much easier to use. The process of human pattern recognition is not well understood.

Due to the inherent variability of spoken language patterns, we emphasize the use of statistical approaches in this book. The decision for pattern recognition is based on appropriate probabilistic models of the patterns. This chapter presents several mathematical fundamentals for statistical pattern recognition and classification. In particular, Bayes' decision theory and estimation techniques for parameters of classifiers are introduced. Bayes' decision theory, which plays a central role for statistical pattern recognition, introduces the concept of decision-making based on both *posterior* knowledge obtained from specific observation data, and *prior* knowledge of the categories. To build such a classifier or predictor, it is critical to estimate *prior* class probabilities and the class-conditional probabilities for a Bayes' classifier.

Supervised learning has class information for the data. Only the probabilistic structure needs to be learned. Maximum likelihood estimation (MLE) and maximum posterior probability estimation (MAP) that we discussed in Chapter 3 are two most powerful methods. Both MLE and MAP aim to maximize the likelihood function. The MLE criterion does not necessarily minimize the recognition error rate. Various discriminant estimation methods are introduced for that purpose. *Maximum mutual information estimation* (MMIE) is based on criteria to achieve maximum model separation (the model for the correct class is well separated from other competing models) instead of likelihood criteria. The MMIE criterion is one step closer but still is not directly related to minimizing the error rate. Other discriminant estimation methods, such as *minimum error-rate estimation*, use the ultimate goal of pattern recognition – minimizing the classification errors. *Neural networks* are one class of discriminant estimation methods.

The EM algorithm is an iterative algorithm for unsupervised learning in which class information is unavailable or only partially available. The EM algorithm forms the theoretical basis for training hidden Markov models (HMM) as described in Chapter 8. To better understand the relationship between MLE and EM algorithms, we first introduce *vector quantization* (VQ), a widely used source-coding technique in speech analysis. The well-known *k-means* clustering algorithm best illustrates the relationship between MLE and the EM algorithm. We close this chapter by introducing a powerful binary prediction and regression technique, classification and regression trees (CART). CART represents an important technique that combines rule-based expert knowledge and statistical learning.

4.1. BAYES' DECISION THEORY

Bayes' decision theory forms the basis of statistical pattern recognition. The theory is based on the assumption that the decision problem can be specified in probabilistic terms and that all of the relevant probability values are known. Bayes' decision theory can be viewed as a formalization of a common-sense procedure, i.e., the aim to achieve minimum-error-rate classification. This common-sense procedure can be best observed in the following real-world decision examples.

Consider the problem of making predictions for the stock market. We use the Dow Jones Industrial average index to formulate our example, where we have to decide tomorrow's Dow Jones Industrial average index in one of the three categories (events): *Up, Down*, or *Unchanged*. The available information is the probability function $P(\omega)$ of the three categories. The variable ω is a discrete random variable with the value $\omega = \omega_i \; (i = 1,2,3)$. We call the probability $P(\omega_i)$ a *prior* probability, since it reflects prior knowledge of tomorrow's Dow Jones Industrial index. If we have to make a decision based only on the *prior* probability, the most plausible decision may be made by selecting the class ω_i with the highest prior probability $P(\omega_i)$. This decision is unreasonable, in that we always make the same decision even though we know that all three categories of Dow Jones Industrial index changes will possibly appear. If we are given further observable data, such as the federal-funds interest rate or the jobless rate, we can make a more informed decision. Let x be a continuous random variable whose value is the federal-fund interest rate, and $f_{x|\omega}(x \mid \omega)$ be a class-conditional pdf For simplicity, we denote the pdf $f_{x|\omega}(x \mid \omega)$ as $p(x \mid \omega_i)$, where $i = 1$, 2, 3 unless there is ambiguity. The class-conditional probability density function is often referred to as the *likelihood* function as well, since it measures how likely it is that the underlying parametric model of class ω_i will generate the data sample x. Since we know the prior probability $P(\omega_i)$ and class-conditional pdf $p(x \mid \omega_i)$, we can compute the conditional probability $P(\omega_i \mid x)$ using Bayes' rule:

$$P(\omega_i \mid x) = \frac{p(x \mid \omega_i)P(\omega_i)}{p(x)} \tag{4.1}$$

where $p(x) = \sum_{i=1}^{3} p(x \mid \omega_i)P(\omega_i)$.

The probability term in the left-hand side of Eq. (4.1) is called the *posterior* probability as it is the probability of class ω_i after observing the federal-funds interest rate x. An intuitive decision rule would be choosing the class ω_k with the greatest posterior probability. That is,

$$k = \arg\max_i P(\omega_i \mid x) \tag{4.2}$$

In general, the denominator $p(x)$ in Eq. (4.1) is unnecessary because it is a constant term for all classes. Therefore, Eq. (4.2) becomes

$$k = \arg\max_i P(\omega_i \mid x) = \arg\max_i p(x \mid \omega_i)P(\omega_i) \tag{4.3}$$

The rule in Eq. (4.3) is referred to as Bayes' decision rule. It shows how the observed data x changes the decision based on the *prior* probability $P(\omega_i)$ to one based on the posterior probability $P(\omega_i \mid x)$. Decision making based on the *posterior* probability is more reliable, because it employs prior knowledge together with the present observed data. As a matter of fact, when the prior knowledge is non-informative ($P(\omega_1) = P(\omega_2) = P(\omega_3) = 1/3$), the observed data fully control the decision. On the other hand, when observed data are ambiguous, then prior knowledge controls the decision. There are many kinds of decision rules based on posterior probability. Our interest is to find the decision rule that leads to minimum overall risk, or minimum error rate in decision.

4.1.1. Minimum-Error-Rate Decision Rules

Bayes' decision rule is designed to minimize the overall risk involved in making a decision. Bayes' decision based on posterior probability $P(\omega_i \mid x)$ instead of prior probability $P(\omega_i)$ is a natural choice. Given an observation x, if $P(\omega_k \mid x) \ge P(\omega_i \mid x)$ for all $i \ne k$, we can decide that the true class is ω_k. To justify this procedure, we show such a decision results in minimum decision error.

Let $\Omega = \{\omega_1, \ldots, \omega_s\}$ be the finite set of s possible categories to be predicted and $\Delta = \{\delta_1, \ldots, \delta_t\}$ be a finite set of t possible decisions. Let $l(\delta_i \mid \omega_j)$ be the loss function incurred for making decision δ_i when the true class is ω_j. Using the *prior* probability $P(\omega_i)$ and class-conditional pdf $p(x \mid \omega_i)$, the posterior probability $P(\omega_i \mid x)$ is computed by Bayes' rule as shown in Eq. (4.1). Since the posterior probability $P(\omega_j \mid x)$ is the probability that the true class is ω_j after observing the data x, the expected loss associated with making decision δ_i is:

$$R(\delta_i \mid x) = \sum_{j=1}^{s} l(\delta_i \mid \omega_j) P(\omega_j \mid x) \tag{4.4}$$

In decision-theoretic terminology, the above expression is called *conditional risks*. The overall risk R is the expected loss associated with a given decision rule. The decision rule is employed as a decision function $\delta(x)$ that maps the data x to one of the decisions $\Delta = \{\delta_1, \ldots, \delta_t\}$. Since $R(\delta_i \mid x)$ is the conditional risk associated with decision δ_i, the overall risk is given by:

$$R = \int_{-\infty}^{\infty} R(\delta(x) \mid x) p(x) dx \tag{4.5}$$

If the decision function $\delta(x)$ is chosen so that the conditional risk $R(\delta(x) \mid x)$ is minimized for every x, the overall risk is minimized. This leads to the Bayes' decision rule: To minimize the overall risk, we compute the conditional risk shown in Eq. (4.4) for $i = 1, \ldots, t$ and select the decision δ_i for which the conditional risk $R(\delta_i \mid x)$ is minimum. The resulting minimum overall risk is known as *Bayes' risk* that has the best performance possible.

The loss function $l(\delta_i \mid \omega_j)$ in the Bayes' decision rule can be defined as:

$$l(\delta_i \mid \omega_j) = \begin{cases} 0 & i = j \\ 1 & i \neq j \end{cases} \quad i, j = 1, \ldots, s \tag{4.6}$$

This loss function assigns no loss to a correct decision where the true class is ω_i and the decision is δ_i, which implies that the true class must be ω_i. It assigns a unit loss to any error where $i \neq j$; i.e., all errors are equally costly. This type of loss function is known as a *symmetrical* or *zero-one* loss function. The risk corresponding to this loss function equals the classification error rate, as shown in the following equation.

$$\begin{aligned} R(\delta_i \mid x) &= \sum_{j=1}^{s} l(\delta_i \mid \omega_j) P(\omega_j \mid x) = \sum_{j \neq i} P(\omega_j \mid x) \\ &= \sum_{j=1}^{s} P(\omega_j \mid x) - P(\omega_i \mid x) = 1 - P(\omega_i \mid x) \end{aligned} \tag{4.7}$$

Here $P(\omega_i \mid x)$ is the conditional probability that decision δ_i is correct after observing the data x. Therefore, in order to minimize classification error rate, we have to choose the decision of class i that maximizes the posterior probability $P(\omega_i \mid x)$. Furthermore, since $p(x)$ is a constant, the decision is equivalent to picking the class i that maximizes $p(x \mid \omega_i)P(\omega_i)$. The Bayes' decision rule can be formulated as follows:

$$\delta(x) = \arg\max_i P(\omega_i \mid x) = \arg\max_i P(x \mid \omega_i)P(\omega_i) \tag{4.8}$$

This decision rule, which is based on the maximum of the posterior probability $P(\omega_i \mid x)$, is called the *minimum-error-rate decision rule*. It minimizes the classification error rate. Although our description is for random variable x, Bayes' decision rule is applicable to multivariate random vector **x** without loss of generality.

A pattern classifier can be regarded as a device for partitioning the feature space into decision regions. Without loss of generality, we consider a two-class case. Assume that the classifier divides the space $\Re$ into two regions, $\Re_1$ and $\Re_2$. To compute the likelihood of errors, we need to consider two cases. In the first case, x falls in $\Re_1$, but the true class is ω_2. In the other case, x falls in $\Re_2$, but the true class is ω_1. Since these two cases are mutually exclusive, we have

$$\begin{aligned} P(error) &= P(x \in \Re_1, \omega_2) + P(x \in \Re_2, \omega_1) \\ &= P(x \in \Re_1 \mid \omega_2)P(\omega_2) + P(x \in \Re_2 \mid \omega_1)P(\omega_1) \\ &= \int_{\Re_1} P(x \mid \omega_2)P(\omega_2)dx + \int_{\Re_2} P(x \mid \omega_1)P(\omega_1)dx \end{aligned} \tag{4.9}$$

Figure 4.1 illustrates the calculation of the classification error in Eq. (4.9). The two terms in the summation are merely the tail areas of the function $P(x \mid \omega_i)P(\omega_i)$. It is clear

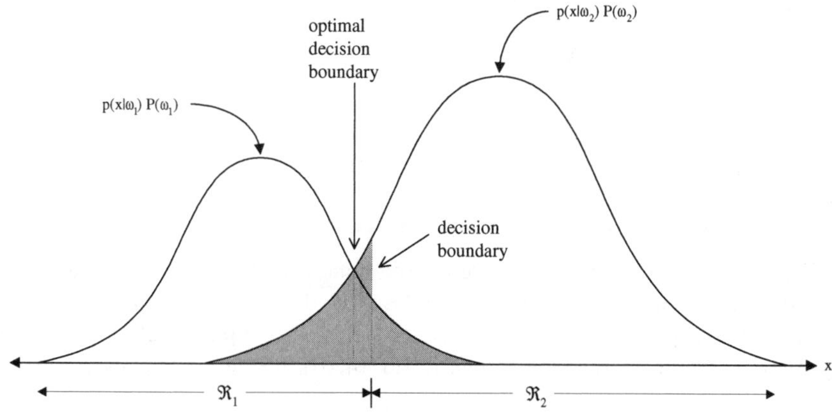

Figure 4.1 Calculation of the likelihood of classification error [22]. The shaded area represents the integral value in Eq. (4.9).

that this decision boundary is not optimal. If we move the decision boundary a little bit to the left, so that the decision is made to choose the class i based on the maximum value of $P(x \mid \omega_i)P(\omega_i)$, the tail integral area $P(error)$ becomes minimum, which is the Bayes' decision rule.

4.1.2. Discriminant Functions

The decision problem above can also be viewed as a pattern classification problem where unknown data $\mathbf{x}$[1] are classified into known categories, such as the classification of sounds into phonemes using spectral data $\mathbf{x}$. A classifier is designed to classify data $\mathbf{x}$ into s categories by using s discriminant functions, $d_i(\mathbf{x})$, computing the similarities between the unknown data $\mathbf{x}$ and each class ω_i and assigning $\mathbf{x}$ to class ω_j if

$$d_j(\mathbf{x}) > d_i(\mathbf{x}) \ \forall i \neq j \tag{4.10}$$

This representation of a classifier is illustrated in Figure 4.2.

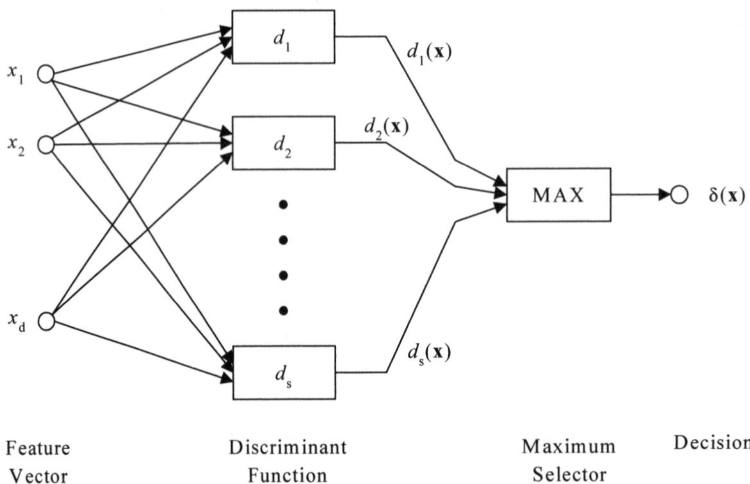

Figure 4.2 Block diagram of a classifier based on discriminant functions [22].

A Bayes' classifier can be represented in the same way. Based on the Bayes' classifier, unknown data $\mathbf{x}$ are classified on the basis of Bayes' decision rule, which minimizes the conditional risk $R(\delta_i \mid \mathbf{x})$. Since the classification decision of a pattern classifier is based on the maximum discriminant function shown in Eq. (4.10), we define our discriminant function as:

$$d_i(\mathbf{x}) = -R(\delta_i \mid \mathbf{x}) \tag{4.11}$$

[1] Assuming $\mathbf{x}$ is a d-dimensional vector.

As such, the maximum discriminant function corresponds to the minimum conditional risk. In the minimum-error-rate classifier, the decision rule is to maximize the posterior probability $P(\omega_i \mid \mathbf{x})$. Thus, the discriminant function can be written as follows:

$$d_i(\mathbf{x}) = P(\omega_i \mid \mathbf{x}) = \frac{p(\mathbf{x} \mid \omega_i)P(\omega_i)}{p(\mathbf{x})} = \frac{p(\mathbf{x} \mid \omega_i)P(\omega_i)}{\sum_{j=1}^{s} p(\mathbf{x} \mid \omega_j)P(\omega_j)} \tag{4.12}$$

There is a very interesting relationship between Bayes' decision rule and the hypotheses testing method described in Chapter 3. For a two-class pattern recognition problem, the Bayes' decision rule in Eq. (4.2) can be written as follows:

$$p(\mathbf{x} \mid \omega_1)P(\omega_1) \underset{\omega_2}{\overset{\omega_1}{\gtrless}} p(\mathbf{x} \mid \omega_2)P(\omega_2) \tag{4.13}$$

Eq. (4.13) can be rewritten as:

$$\ell(x) = \frac{p(\mathbf{x} \mid \omega_1)}{p(\mathbf{x} \mid \omega_2)} \underset{\omega_2}{\overset{\omega_1}{\gtrless}} \frac{P(\omega_2)}{P(\omega_1)} \tag{4.14}$$

The term $\ell(x)$ is called *likelihood ratio* and is the basic quantity in hypothesis testing [73]. The term $P(\omega_2)/P(\omega_1)$ is called the *threshold value* of the likelihood ratio for the decision. Often it is convenient to use the log-likelihood ratio instead of the likelihood ratio for the decision rule. Namely, the following single discriminant function can be used instead of $d_1(x)$ and $d_2(x)$ for:

$$d(\mathbf{x}) = \log \ell(\mathbf{x}) = \log p(\mathbf{x} \mid \omega_1) - \log p(\mathbf{x} \mid \omega_2) \underset{\omega_2}{\overset{\omega_1}{\gtrless}} \log P(\omega_2) - \log P(\omega_1) \tag{4.15}$$

As the classifier assigns data $\mathbf{x}$ to class ω_i, the data space is divided into s regions, $\Re_1^d, \Re_2^d, \ldots, \Re_s^d$, called decision regions. The boundaries between decision regions are called decision boundaries and are represented as follows (if they are contiguous):

$$d_i(\mathbf{x}) = d_j(\mathbf{x}) \quad i \neq j \tag{4.16}$$

For points on the decision boundary, the classification can go either way. For a Bayes' classifier, the conditional risk associated with either decision is the same and how to break the tie does not matter. Figure 4.3 illustrates an example of decision boundaries and regions for a three-class classifier on a scalar data sample x.

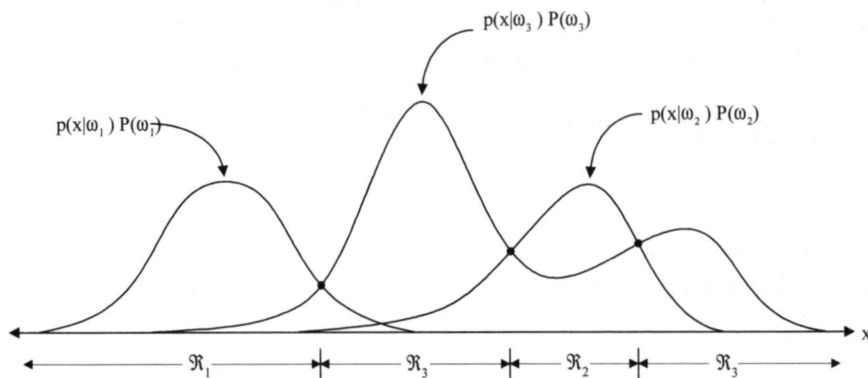

Figure 4.3 An example of decision boundaries and regions. For simplicity, we use scalar variable x instead of a multi-dimensional vector [22].

4.2. How To Construct Classifiers

In the Bayes' classifier, or the minimum-error-rate classifier, the prior probability $P(\omega_i)$ and class-conditional pdf $p(\mathbf{x} \mid \omega_i)$ are known. Unfortunately, in pattern recognition, we rarely have complete knowledge of class-conditional pdfs and/or prior probability. They often must be estimated or learned from the training data. In practice, the estimation of the prior probabilities is relatively easy. Estimation of the class-conditional pdf is more complicated. There is always concern to have sufficient training data relative to the tractability of the huge dimensionality of the sample data **x**. In this chapter we focus on estimation methods for the class-conditional pdf.

The estimation of the class-conditional pdfs can be nonparametric or parametric. In nonparametric estimation, no model structure is assumed and the pdf is directly estimated from the training data. When large amounts of sample data are available, nonparametric learning can accurately reflect the underlying probabilistic structure of the training data. However, available sample data are normally limited in practice, and parametric learning can achieve better estimates if valid model assumptions are made. In parametric learning, some general knowledge about the problem space allows one to parameterize the class-conditional pdf, so the severity of sparse training data can be reduced significantly. Suppose the pdf $p(\mathbf{x} \mid \omega_i)$ is assumed to have a certain probabilistic structure, such as the Gaussian pdf In such cases, only the mean vector $\boldsymbol{\mu}_i$ (or mean μ_i) and covariance matrix Σ_i (or variance σ^2) need to be estimated.

When the observed data x only takes discrete values from a finite set of N values, the class-conditional pdf is often assumed nonparametric, so there will be $N-1$ free parameters in the probability function $p(\mathbf{x}|\omega_i)$.[2] When the observed data $\mathbf{x}$ takes continuous values, parametric approaches are usually necessary. In many systems, the continuous class-conditional pdf (likelihood) $p(\mathbf{x}|\omega_i)$ is assumed to be a Gaussian distribution or a mixture of Gaussian distributions.

In pattern recognition, the set of data samples, which is often collected to estimate the parameters of the recognizer (including the prior and class-conditional pdf), is referred to as the *training set*. In contrast to the training set, the *testing set* is referred to the independent set of data samples, which is used to evaluate the recognition performance of the recognizer.

For parameter estimation or learning, it is also important to distinguish between *supervised learning* and *unsupervised learning*. Let's denote the pair $(\mathbf{x},\omega)$ as a sample, where $\mathbf{x}$ is the observed data and ω is the class from which the data $\mathbf{x}$ comes. From the definition, it is clear that $(\mathbf{x},\omega)$ are jointly distributed random variables. In supervised learning, ω, information about the class of the sample data $\mathbf{x}$ is given. Such sample data are usually called labeled data or complete data, in contrast to incomplete data where the class information ω is missing for unsupervised learning. Techniques for parametric unsupervised learning are discussed in Section 4.4.

In Chapter 3 we introduced two most popular parameter estimation techniques – *maximum likelihood estimation* (MLE) and *maximum a posteriori probability estimation* (MAP). Both MLE and MAP are supervised learning methods since the class information is required. MLE is the most widely used because of its efficiency. The goal of MLE is to find the set of parameters that maximizes the probability of generating the training data actually observed. The class-conditional pdf is typically parameterized. Let $\mathbf{\Phi}_i$ denote the parameter vector for class i. We can represent the class-conditional pdf as a function of $\mathbf{\Phi}_i$ as $p(\mathbf{x}|\omega_i,\mathbf{\Phi}_i)$. As stated earlier, in supervised learning, the class name ω_i is given for each sample data in training set $\{\mathbf{x}_1,\mathbf{x}_2,\ldots,\mathbf{x}_n\}$. We need to make an assumption[3] that samples in class ω_i give no information about the parameter vector $\mathbf{\Phi}_j$ of the other class ω_j. This assumption allows us to deal with each class independently, since the parameter vectors for different categories are functionally independent. The class-conditional pdf can be rewritten as $p(\mathbf{x}|\mathbf{\Phi})$, where $\mathbf{\Phi}=\{\mathbf{\Phi}_1,\mathbf{\Phi}_i,\ldots,\mathbf{\Phi}_n\}$. If a set of random samples $\{\mathbf{X}_1,\mathbf{X}_2,\ldots,\mathbf{X}_n\}$ is drawn independently according to a pdf $p(\mathbf{x}|\mathbf{\Phi})$, where the value of the parameter $\mathbf{\Phi}$ is unknown, the MLE method described in Chapter 3 can be directly applied to estimate $\mathbf{\Phi}$.

Similarly, MAP estimation can be applied to estimate $\mathbf{\Phi}$ if knowledge about a prior distribution is available. In general, MLE is used for estimating parameters from scratch without any prior knowledge, and MAP estimation is used for parameter adaptation where the behavior of a prior distribution is known and only a small amount of adaptation data is available. When the amount of adaptation data increases, MAP estimation converges to MLE.

---

[2] Since all the probabilities need to add up to one.

[3] This assumption is only true for non-discriminative estimation. Samples in class ω_i may affect parameter vector $\mathbf{\Phi}_i$ of the other classes in discriminative estimation methods as described in Section 4.3

4.2.1. Gaussian Classifiers

A *Gaussian classifier* is a Bayes' classifier where class-conditional probability density $p(\mathbf{x} \mid \omega_i)$ for each class ω_i is assumed to have a Gaussian distribution:[4]

$$p(\mathbf{x} \mid \omega_i) = \frac{1}{(2\pi)^{d/2} |\Sigma_i|^{1/2}} \exp\left[-\frac{1}{2} (\mathbf{x} - \boldsymbol{\mu}_i)^t \Sigma_i^{-1} (\mathbf{x} - \boldsymbol{\mu}_i) \right] \tag{4.17}$$

As discussed in Chapter 3, the parameter estimation techniques are well suited for the Gaussian family. The MLE of the Gaussian parameter is just its sample mean and variance (or co-variance matrix). A Gaussian classifier is equivalent to the one using a quadratic discriminant function. As noted in Eq. (4.12), the discriminant function for a Bayes' decision rule is the posterior probability $p(\omega_i \mid \mathbf{x})$ or $p(\mathbf{x} \mid \omega_i)P(\omega_i)$. Assuming $p(\mathbf{x} \mid \omega_i)$ is a multivariate Gaussian density as shown in Eq. (4.17), a discriminant function can be written as follows:

$$\begin{aligned} d_i(\mathbf{x}) &= \log p(\mathbf{x} \mid \omega_i) P(\omega_i) \\ &= -\frac{1}{2} (\mathbf{x} - \boldsymbol{\mu}_i)^t \Sigma_i^{-1} (\mathbf{x} - \boldsymbol{\mu}_i) + \log P(\omega_i) - \frac{1}{2} \log |\Sigma| - \frac{d}{2} \log 2\pi \end{aligned} \tag{4.18}$$

If we have a uniform prior $P(\omega_i)$, it is clear that the above discriminant function $d_i(\mathbf{x})$ is a quadratic function. Once we have the s Gaussian discriminant functions, the decision process simply assigns data $\mathbf{x}$ to class ω_j if

$$j = \arg\max_i d_i(\mathbf{x}) \tag{4.19}$$

When all the Gaussian pdfs have the same covariance matrix ($\Sigma_i = \Sigma$ for $i = 1, 2, \ldots, s$), the quadratic term $\mathbf{x}^t \Sigma^{-1} \mathbf{x}$ is independent of the class and can be treated as a constant. Thus the following new discriminant function $d_i(\mathbf{x})$ can be used [22]:

$$d_i(\mathbf{x}) = \mathbf{a}_i^t \mathbf{x} + \mathbf{c}_i \tag{4.20}$$

where $\mathbf{a}_i = \Sigma^{-1} \boldsymbol{\mu}_i$ and $\mathbf{c}_i = -\frac{1}{2} \boldsymbol{\mu}_i^t \Sigma_i^{-1} \boldsymbol{\mu}_i + \log P(\omega_i)$. $d_i(\mathbf{x})$ in Eq. (4.20) is a linear discriminant function. For linear discriminant functions, the decision boundaries are hyperplanes. For the two-class case (ω_1 and ω_2), and assuming that data sample $\mathbf{x}$ is a real random vector, the decision boundary can be shown to be the following hyperplane:

---

[4] The Gaussian distribution may include a mixture of Gaussian pdfs.

$$\mathbf{A}^t(\mathbf{x}-\mathbf{b})=0 \tag{4.21}$$

where

$$\mathbf{A}=\Sigma^{-1}(\boldsymbol{\mu}_1-\boldsymbol{\mu}_2) \tag{4.22}$$

and

$$\mathbf{b}=\frac{1}{2}(\boldsymbol{\mu}_1-\boldsymbol{\mu}_2)-\frac{(\boldsymbol{\mu}_1-\boldsymbol{\mu}_2)\log\left[P(\omega_1/\omega_2\right]}{(\boldsymbol{\mu}_1-\boldsymbol{\mu}_2)^t\Sigma^{-1}(\boldsymbol{\mu}_1-\boldsymbol{\mu}_2)} \tag{4.23}$$

Figure 4.4 shows a two dimensional decision boundary for a two-class Gaussian classifier with the same covariance matrix. Please note that the decision hyperplane is generally not orthogonal to the line between the means $\boldsymbol{\mu}_1$ and $\boldsymbol{\mu}_2$, although it does intersect that line at the point **b**, which is halfway between $\boldsymbol{\mu}_1$ and $\boldsymbol{\mu}_2$. The analysis above is based on the case of uniform priors ($p(\omega_1)=p(\omega_2)$). For nonuniform priors, the decision hyperplane moves away from the more likely mean.

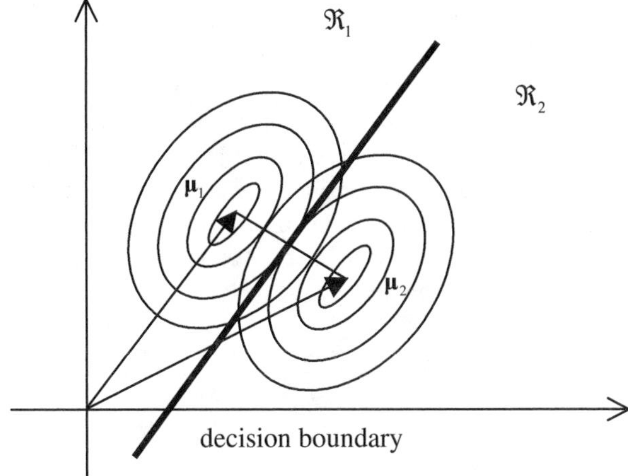

Figure 4.4 Decision boundary for a two-class Gaussian classifier. Gaussian distributions for the two categories have the same covariance matrix Σ. Each ellipse represents the region with the same likelihood probability [22].

Finally, if each dimension of random vector **x** is statistically independent and has the same variance σ^2, i.e., $\Sigma_1=\Sigma_2=\sigma^2\mathbf{I}$, Figure 4.4 becomes Figure 4.5. The ellipse in Figure 4.4 becomes a circle because the variance σ^2 is the same for all dimensions [22].

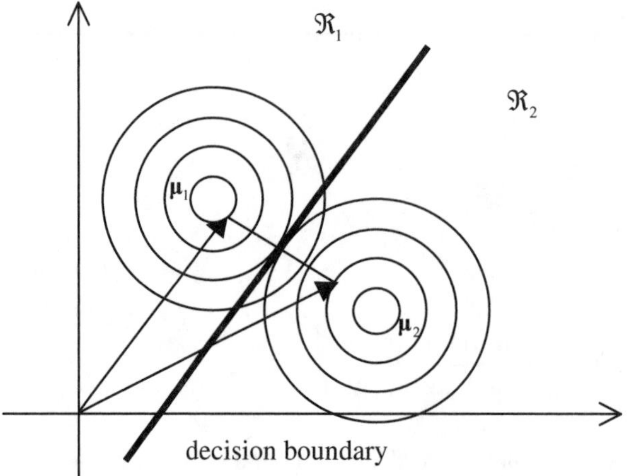

Figure 4.5 Decision boundary for a two-class Gaussian classifier. Gaussian distributions for the two categories have the same covariance matrix $\sigma^2\mathbf{I}$. Each circle represents the region with the same likelihood probability [22].

4.2.2. The Curse of Dimensionality

More features (higher dimensions for sample $\mathbf{x}$) and more parameters for the class-conditional pdf $p(\mathbf{x} \mid \Phi)$ may lead to lower classification error rate. If the features are statistically independent, there are theoretical arguments that support better classification performance with more features. Let us consider a simple two-class Gaussian classifier. Suppose the prior probabilities $p(\omega_i)$ are equal and the class-conditional Gaussian pdfs $p(\mathbf{x} \mid \boldsymbol{\mu}_i, \Sigma)$ share the same covariance matrix Σ. According to Eqs. (4.9) and (4.21), the Bayes' classification error rate is given by:

$$
\begin{aligned}
P(error) &= 2\int_{\Re_2} P(\mathbf{x} \mid \omega_1)P(\omega_1)d\mathbf{x} \\
&= 2\int_{\mathbf{A}^t(\mathbf{x}-\mathbf{b})=0}^{\infty} \frac{1}{(2\pi)^{d/2}|\Sigma|^{1/2}} \exp\left[-\tfrac{1}{2}(\mathbf{x}-\boldsymbol{\mu}_i)^t \Sigma^{-1}(\mathbf{x}-\boldsymbol{\mu}_i)\right] d\mathbf{x} \qquad (4.24) \\
&= \frac{1}{\sqrt{2\pi}}\int_{\frac{r}{2}}^{\infty} e^{-\frac{1}{2}z^2} dz
\end{aligned}
$$

where $r = \sqrt{(\boldsymbol{\mu}_1 - \boldsymbol{\mu}_2)^t \Sigma^{-1} (\boldsymbol{\mu}_1 - \boldsymbol{\mu}_2)}$. When features are independent, the covariance matrix becomes a diagonal one. The following equation shows that each independent feature helps to reduce the error rate:[5]

[5] When the means of a feature for the two classes are exactly the same, adding such a feature does not reduce the Bayes' error. Nonetheless, according to Eq. (4.25), the Bayes' error cannot possibly be increased by incorporating an additional independent feature.

$$r = \sqrt{\sum_{i=1}^{d} \left(\frac{\mu_{1i} - \mu_{2i}}{\sigma_i} \right)^2} \tag{4.25}$$

where μ_{1i} and μ_{2i} are the ith -dimension of mean vectors $\boldsymbol{\mu}_1$ and $\boldsymbol{\mu}_2$ respectively.

Unfortunately, in practice, the inclusion of additional features may lead to worse classification results. This paradox is called *the curse of dimensionality*. The fundamental issue, called *trainability*, refers to how well the parameters of the classifier are trained from the limited training samples. Trainability can be illustrated by a typical curve-fitting (or regression) problem. Figure 4.6 shows a set of eleven data points and several polynomial fitting curves with different orders. Both the first-order (linear) and second-order (quadratic) polynomials shown provide fairly good fittings for these data points. Although the tenth-order polynomial fits the data points perfectly, no one would expect such an under-determined solution to fit the new data well. In general, many more data samples would be necessary to get a good estimate of a tenth-order polynomial than of a second-order polynomial, because reliable interpolation or extrapolation can be attained only with an over-determined solution.

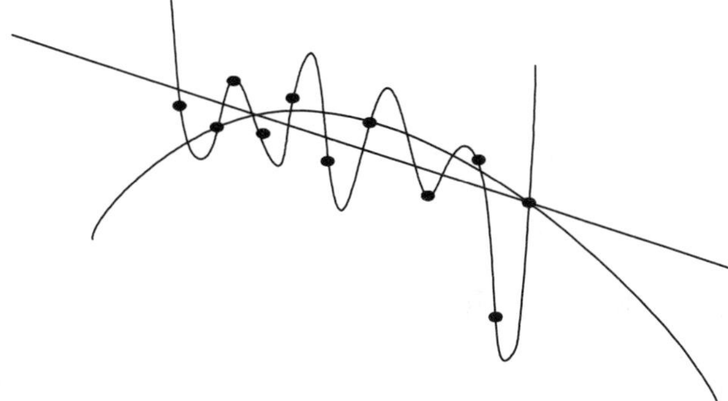

Figure 4.6 Fitting eleven data points with polynomial functions of different orders [22].

Figure 4.7 shows the error rates for two-phonemes (/*ae*/ and /*ih*/) classification where two phonemes are modeled by mixtures of Gaussian distributions. The parameters of mixtures of Gaussian are trained from a varied set of training samples via maximum likelihood estimation. The curve illustrates the classification error rate as a function of the number of training samples and the number of mixtures. For every curve associated with a finite number of samples, there are an optimal number of mixtures. This illustrates the importance of trainability: it is critical to assure there are enough samples for training an increased number of features or parameters. When the size of training data is fixed, increasing the number of features or parameters beyond a certain point is likely to be counterproductive.

When you have an insufficient amount of data to estimate the parameters, some simplification can be made to the structure of your models. In general, the estimation for higher-order statistics, like variances or co-variance matrices, requires more data than that for lower- order statistics, like mean vectors. Thus more attention often is paid to dealing with the estimation of covariance matrices. Some frequently used heuristics for Gaussian distributions include the use of the same covariance matrix for all mixture components [77], diagonal covariance matrix, and shrinkage (also referred to as regularized discriminant analysis), where the covariance matrix is interpolated with the constant covariance matrix [23, 50].

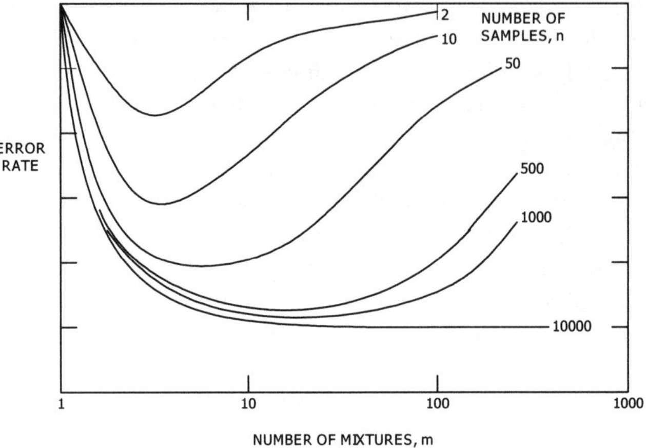

Figure 4.7 Two-phoneme (*/ae/* and */ih/*) classification results as a function of the number of Gaussian mixtures and the number of training samples.

4.2.3. Estimating the Error Rate

Estimating the error rate of a classifier is important. We want to see whether the classifier is good enough to be useful for our task. For example, telephone applications show that some minimum accuracy is required before users would switch from using the touch-tone to the speech recognizer. It is also critical to compare the performance of a classifier (algorithm) against an alternative. In this section we deal with how to estimate the true classification error rate.

One approach is to compute the theoretic error rate from the parametric model as shown in Eq. (4.24). However, there are several problems with this approach. First, such an approach almost always under-estimates, because the parameters estimated from the training samples might not be realistic unless the training samples are representative and sufficient. Second, all the assumptions about models and distributions might be severely wrong. Finally, it is very difficult to compute the exact error rate, as in the simple case illustrated in Eq. (4.24).

Instead, you can estimate the error rate empirically. In general, the recognition error rate on the training set should be viewed only as a lower bound, because the estimate can be made to minimize the error rate on the training data. Therefore, a better estimate of the recognition performance should be obtained on an independent test set. The question now is how representative is the error rate computed from an arbitrary independent test set. The common process of using some of the data samples for design and reserving the rest for test is called the *holdout* or *H* method.

Suppose the true but unknown classification error rate of the classifier is p, and one observes that k out of n independent randomly drawn test samples are misclassified. The random variable K should have a binomial distribution $B(n,p)$. The maximum likelihood estimation for p should be

$$\hat{p} = \frac{k}{n} \tag{4.26}$$

The statistical test for binomial distribution is discussed in Chapter 3. For a 0.05 significance level, we can compute the following equations to get the range (p_1, p_2):

$$2P(k \le m \le n) = 2\sum_{m=k}^{n} \binom{n}{m} (p_1)^m (1-p_1)^{n-m} = 0.05 \quad \text{when } k > np_1 \tag{4.27}$$

$$2P(0 \le m \le k) = 2\sum_{m=0}^{k} \binom{n}{m} (p_2)^m (1-p_2)^{n-m} = 0.05 \quad \text{when } k < np_2 \tag{4.28}$$

Equations (4.27) and (4.28) are cumbersome to solve, so the normal test described in Chapter 3 can be used instead. The null hypothesis H_0 is

$$H_0 : p = \hat{p}$$

We can use the normal test to find the two boundary points p_1 and p_2 at which we would not reject the null hypothesis H_0.

The range (p_1, p_2) is called the 0.95 confidence intervals because one can be 95% confident that the true error rate p falls in the range (p_1, p_2). Figure 4.8 illustrates 95% confidence intervals as a function of $\hat{p}$ and n. The curve certainly agrees with our intuition – the larger the number of test samples n, the more confidence we have in the MLE estimated error rate $\hat{p}$; otherwise, the $\hat{p}$ can be used only with caution.

Based on the description in the previous paragraph, the larger the test set is, the better it represents the recognition performance of possible data. On one hand, we need more training data to build a reliable and consistent estimate. On the other hand, we need a large independent test set to derive a good estimate of the true recognition performance. This creates a contradictory situation for dividing the available data set into training and independent test set. One way to effectively use the available database is *V-fold cross validation*. It first splits the entire database into V equal parts. Each part is used in turn as an independent test set while the remaining (V - 1) parts are used for training. The error rate can then be better estimated by averaging the error rates evaluated on the V different testing sets. Thus, each part can contribute to both training and test sets during V-fold cross validation. This procedure, also called the *leave-one-out* or U method [53], is particularly attractive when the number of available samples are limited.

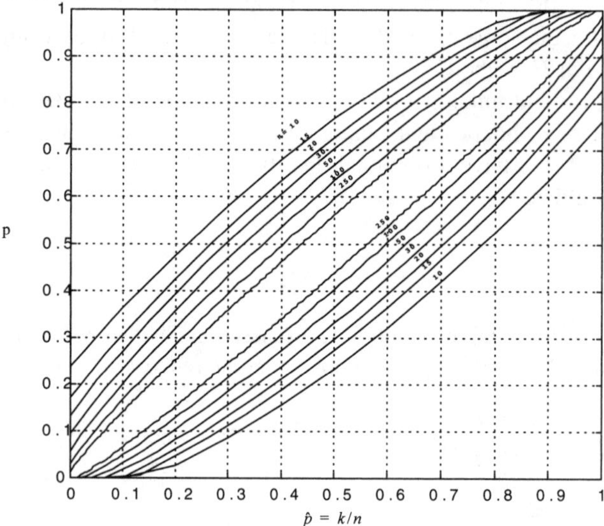

Figure 4.8 95% confidence intervals for classification error rate estimation with normal test.

4.2.4. Comparing Classifiers

Given so many design alternatives, it is critical to compare the performance of different classifiers so that the best classifier can be used for real-world applications. It is common for designers to test two classifiers on some test samples and decide if one is superior to the other. Relative efficacy can be claimed only if the difference in performance is statistically significant. In other words, we establish the null hypothesis H_0 that the two classifiers have the same error rates. Based on the observed error patterns, we decide whether we could reject H_0 at the 0.05 level of significance. The test for different classifiers falls into the category of *matched-pairs* tests described in Chapter 3. Classifiers are compared with the same test samples.

We present an effective matched-pairs test – McNemar's test [66] which is particularly suitable for comparing classification results. Suppose there are two classifiers: Q_1 and Q_2. The estimated classification error rates on the same test set for these two classifiers are p_1 and p_2 respectively. The null hypothesis H_0 is $p_1 = p_2$. The classification performance of the two classifiers can be summarized as in Table 4.1. We define q_{ij} as follows:

$$q_{00} = P(Q_1 \text{ and } Q_2 \text{ classify data sample correctly})$$

$$q_{01} = P(Q_1 \text{ classifies data sample correctly, but } Q_2 \text{ incorrectly})$$

q_{10} = $P(Q_2$ classifies data sample correctly, but Q_1 incorrectly)

q_{11} = $P(Q_1$ and Q_2 classify data sample incorrectly)

Table 4.1 Classification performance table for classifiers Q_1 and Q_2. N_{00} is the number of samples which Q_1 and Q_2 classify correctly, N_{01} is the number of samples which Q_1 classifies correctly, but Q_2 incorrectly, N_{10} is the number of samples which Q_2 classifies correctly, but Q_1 incorrectly, and N_{11} is the number of samples which Q_1 and Q_2 classify incorrectly [30].

		Q_2 Correct	Q_2 Incorrect
Q_1	Correct	N_{00}	N_{01}
	Incorrect	N_{10}	N_{11}

The null hypothesis H_0 is equivalent to H_0^1 : $q_{01} = q_{10}$. If we define $q = q_{10}/(q_{01}+q_{10})$, H_0 is equivalent to H_0^2 : $q = \frac{1}{2}$. H_0^2 represents the hypothesis that, given only one of the classifiers makes an error, it is equally likely to be either one. We can test H_0^2 based on the data samples on which only one of the classifiers made an error. Let $n = N_{01} + N_{10}$. The observed random variable N_{01} should have a binomial distribution $B(n, \frac{1}{2})$. Therefore, the normal test (z-test) described in Chapter 3 can be applied directly to test the null hypothesis H_0^2.

The above procedure is called the *McNemar's test* [66]. If we view the classification results as N (the total number of test samples) independent matched pairs, the sign test as described in Chapter 3 can be directly applied to test the null hypothesis that classifier Q_1 is not better than classifier Q_2, that is, the probability that classifier Q_1 performs better than classifier Q_2, p, is smaller than or equal to ½.

McNemar's test is applicable when the errors made by a classifier are independent among different test samples. Although this condition is true for most static pattern recognition problems, it is not the case for most speech recognition problems. In speech recognition, the errors are highly inter-dependent because of the use of higher-order language models (described in Chapter 11).

The solution is to divide the test data stream into segments in such a way that errors in one segment are statistically independent of errors in any other segment [30]. A natural candidate for such a segment is a sentence or a phrase after which the speaker pauses. Let N_1^i be the number of errors[6] made on the ith segment by classifier Q_1 and N_2^i be the number of errors made on the ith segment by classifier Q_2. Under this formulation, the magnitude-difference test described in Chapter 3 can be applied directly to test the null hypothesis that classifiers Q_1 and Q_2 have on the average the same error rate on the pairs of n independent segments.

[6] The errors for speech recognition include substitutions, insertions and deletions as discussed in Chapter 9.

4.3. Discriminative Training

Both MLE and MAP criteria maximize the probability of the model associated with the corresponding data. Only data labeled as belonging to class ω_i are used to train the parameters. There is no guarantee that the observed data $\mathbf{x}$ from class ω_i actually have a higher likelihood $P(\mathbf{x} | \omega_i)$ than the likelihood $P(\mathbf{x} | \omega_j)$ associated with class j, given $j \neq i$. Models generated by MLE or MAP have a loose discriminant nature. Several estimation methods aim for maximum discrimination among models to achieve best pattern recognition performance.

4.3.1. Maximum Mutual Information Estimation

The pattern recognition problem can be formalized as an information channel, as illustrated in Figure 4.9. The source symbol ω is encoded into data $\mathbf{x}$ and transmitted through an information channel to the observer. The observer utilizes pattern recognition techniques to decode $\mathbf{x}$ into source symbol $\hat{\omega}$. Consistent with the goal of communication channels, the observer hopes the decoded symbol $\hat{\omega}$ is the same as the original source symbol ω. Maximum mutual information estimation tries to improve channel quality between input and output symbols.

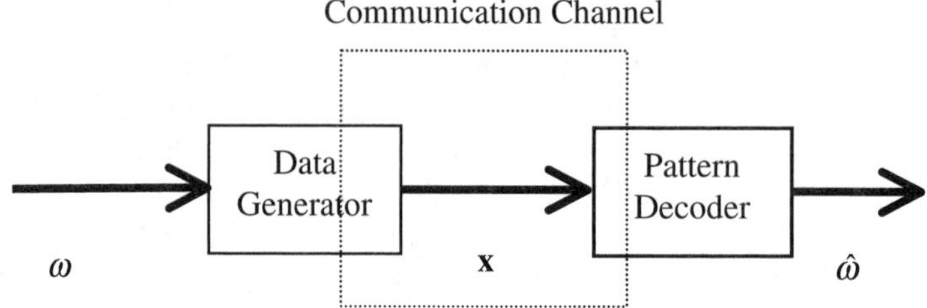

Figure 4.9 An information channel framework for pattern recognition.

As described in Section 4.1.1, the decision rule for the minimum-error-rate classifier selects the class ω_i with maximum posterior probability $P(\omega_i | \mathbf{x})$. It is a good criterion to maximize the posterior probability $P(\omega_i | \mathbf{x})$ for parameter estimation. Recalling Bayes' rule in Section 4.1, the posterior probability $p(\omega_i | \mathbf{x})$ (assuming $\mathbf{x}$ belongs to class ω_i) is:

$$P(\omega_i | \mathbf{x}) = \frac{p(\mathbf{x} | \omega_i) P(\omega_i)}{p(\mathbf{x})} \tag{4.29}$$

and $p(\mathbf{x})$ can be expressed as follows:

$$p(\mathbf{x}) = \sum_k p(\mathbf{x} | \omega_k) p(\omega_k) \tag{4.30}$$

In the classification stage, $p(\mathbf{x})$ can be considered as a constant. However, during training, the value of $p(\mathbf{x})$ depends on the parameters of all models and is different for different **x**. Equation (4.29) is referred to as *conditional likelihood*. A conditional maximum likelihood estimator (CMLE) θ_{CMLE} is defined as follows:

$$\theta_{CMLE}(\mathbf{x}) = \mathbf{\Phi}_{MAP} = \underset{\mathbf{\Phi}}{\operatorname{argmax}}\ p_{\mathbf{\Phi}}(\omega_i \mid \mathbf{x}) \tag{4.31}$$

The summation in Eq. (4.30) extends over all possible classes that include the correct model and all the possible competing models. The parameter vector $\mathbf{\Phi}$ in Eq. (4.31) includes not only the parameter $\mathbf{\Phi}_i$ corresponding to class ω_i, but also those for all other classes.

Note that in Chapter 3, the mutual information between random variable **X** (observed data) and Ω (class label) is defined as:

$$I(\mathbf{X},\Omega) = E\left(\log \frac{p(\mathbf{X},\Omega)}{p(\mathbf{X})P(\Omega)}\right) = E\left(\log \frac{p(\mathbf{X}\mid\Omega)P(\Omega)}{p(\mathbf{X})P(\Omega)}\right) \tag{4.32}$$

Since we don't know the probability distribution for $p(\mathbf{X},\Omega)$, we assume our sample $(\mathbf{x},\omega_i)$ is representative and define the following *instantaneous mutual information*:

$$I(\mathbf{x},\omega_i) = \log \frac{p(\mathbf{x},\omega_i)}{p(\mathbf{x})P(\omega_i)} \tag{4.33}$$

If equal prior $p(\omega_i)$ is assumed for all classes, maximizing the conditional likelihood in Eq. (4.29) is equivalent to maximizing the mutual information defined in Eq.(4.33). CMLE becomes *maximum mutual information estimation* (MMIE). It is important to note that, in contrast to MLE, MMIE is concerned with distributions over all possible classes. Equation (4.30) can be rewritten as two terms, one corresponding to the correct one, and the other corresponding to the competing models:

$$p(\mathbf{x}) = p(\mathbf{x}\mid\omega_i)P(\omega_i) + \sum_{k\neq i} p(\mathbf{x}\mid\omega_k)P(\omega_k) \tag{4.34}$$

Based on the new expression of $p(\mathbf{x})$ shown in Eq. (4.34), the posterior probability $p(\omega_i \mid \mathbf{x})$ in Eq. (4.29) can be rewritten as:

$$P(\omega_i \mid \mathbf{x}) = \frac{p(\mathbf{x}\mid\omega_i)P(\omega_i)}{p(\mathbf{x}\mid\omega_i)P(\omega_i) + \sum_{k\neq i} p(\mathbf{x}\mid\omega_k)P(\omega_k)} \tag{4.35}$$

Now, maximization of the posterior probability $p(\omega_i \mid \mathbf{x})$ with respect to all models leads to a discriminant model.[7] It implies that the contribution of $p(\mathbf{x}\mid\omega_i)P(\omega_i)$ from the true model needs to be enforced, while the contribution of all the competing models, specified by

[7] General minimum-error-rate estimation is described in Section 4.3.2.

$\sum_{k \neq i} p(\mathbf{x} \mid \omega_k) P(\omega_k)$, needs to be minimized. Maximization of Eq. (4.35) can be further rewritten as:

$$P(\omega_i \mid \mathbf{x}) = \frac{1}{1 + \frac{\sum_{k \neq i} p(\mathbf{x} \mid \omega_i) p(\omega_i)}{p(\mathbf{x} \mid \omega_i) p(\omega_i)}} \tag{4.36}$$

Maximization is thus equivalent to maximization of the following term, which is clearly a discriminant criterion between model ω_i and the sum of all other competing models.

$$\frac{p(\mathbf{x} \mid \omega_i) p(\omega_i)}{\sum_{k \neq i} p(\mathbf{x} \mid \omega_k) p(\omega_k)} \tag{4.37}$$

Equation (4.37) also illustrates a fundamental difference between MLE and MMIE. In MLE, only the correct model needs to be updated during training. However, every MMIE model is updated even with one training sample. Furthermore, the greater the prior probability $p(\omega_k)$ for class ω_k, the more effect it has on the maximum mutual information estimator θ_{MMIE}. This makes sense, since the greater the prior probability $p(\omega_k)$, the greater the chance for the recognition system to mis-recognize ω_i as ω_k. MLE is a simplified version of MMIE by restricting the training of model using the data for the model only. This simplification allows the denominator term in Eq. (4.29) to contain the correct model so that it can be dropped as a constant term. Thus, maximization of the posterior probability $p(\omega_i \mid \mathbf{x})$ can be transformed into maximization of the likelihood $p(\mathbf{x} \mid \omega_i)$.

Although likelihood and posterior probability are transformable based on Bayes' rule, MLE and MMIE often generate different results. Discriminative criteria like MMIE attempt to achieve minimum error rate. It might actually produce lower likelihood for the underlying probability density $p(\mathbf{x} \mid \omega_k)$. However, if the assumption of the underlying distributions is correct and there are enough (or infinite) training data, the estimates should converge to the true underlying distributions. Therefore, Bayes' rule should be satisfied and MLE and MMIE should produce the same estimate.

Arthur Nadas [71] showed that if the prior distribution (language model) and the assumed likelihood distribution family are correct, both MLE and MMIE are consistent estimators, but MMIE has a greater variance. However, when some of those premises are not valid, it is desirable to use MMIE to find the estimate that maximizes the mutual information (instead of likelihood) between sample data and its class information. The difference between these two estimation techniques is that MMIE not only aims to increase the likelihood for the correct class, but also tries to decrease the likelihood for the incorrect classes. Thus, MMIE in general possesses more discriminating power among different categories. Although MMIE is theoretically appealing, computationally it is very expensive. Comparing with MLE, every data sample needs to train all the possible models instead of the corresponding model. It also lacks an efficient maximization algorithm. You need to use a gradient descent algorithm.

4.3.1.1. Gradient Descent

To maximize Eq. (4.37) over the entire parameter space $\mathbf{\Phi} = \{\mathbf{\Phi}_1, \mathbf{\Phi}_2, \ldots, \mathbf{\Phi}_S\}$ with S classes, we define the mutual information term in Eq. (4.37) to be a function of $\mathbf{\Phi}$. To fit into the traditional gradient descent framework, we take the inverse of Eq. (4.37) as our optimization function to minimize the following function:[8]

$$F(\mathbf{\Phi}) = \frac{\sum_{k \neq i} p_{\Phi_k}(\mathbf{x} \mid \omega_k) p(\omega_k)}{p_{\Phi_i}(\mathbf{x} \mid \omega_i) p(\omega_i)} \tag{4.38}$$

The gradient descent algorithm starts with some initial estimate $\mathbf{\Phi}^0$ and computes the gradient vector $\nabla F(\mathbf{\Phi})$ (∇ is defined in Chapter 3). It obtains a new estimate $\mathbf{\Phi}^1$ by moving $\mathbf{\Phi}^0$ in the direction of the steepest descent, i.e., along the negative of the gradient. Once it obtains the new estimate, it can perform the same gradient descent procedure iteratively until $F(\mathbf{\Phi})$ converges to the local minimum. In summary, it obtains $\mathbf{\Phi}^{t+1}$ from $\mathbf{\Phi}^t$ by the following formula:

$$\mathbf{\Phi}^{t+1} = \mathbf{\Phi}^t - \varepsilon_t \nabla F(\mathbf{\Phi})|_{\mathbf{\Phi}=\mathbf{\Phi}^t} \tag{4.39}$$

where ε_t is the learning rate (or step size) for the gradient descent.

Why can gradient descent lead $F(\mathbf{\Phi})$ to a local minimum? Based on the definition of gradient vector, $F(\mathbf{\Phi})$ can be approximated by the first order expansion if the correction term $\Delta\mathbf{\Phi}$ is small enough.

$$F(\mathbf{\Phi}^{t+1}) \approx F(\mathbf{\Phi}^t) + \Delta\mathbf{\Phi} * \nabla F(\mathbf{\Phi})|_{\mathbf{\Phi}=\mathbf{\Phi}^t} \tag{4.40}$$

$\Delta\mathbf{\Phi}$ can be expressed as the following term based on Eq. (4.39)

$$\Delta\mathbf{\Phi} = \mathbf{\Phi}^{t+1} - \mathbf{\Phi}^t = -\varepsilon_t \nabla F(\mathbf{\Phi})|_{\mathbf{\Phi}=\mathbf{\Phi}^t} \tag{4.41}$$

Thus, we can obtain the following equation:

$$\begin{aligned} F(\mathbf{\Phi}^{t+1}) - F(\mathbf{\Phi}^t) &= -\varepsilon_t \left\langle \nabla F(\mathbf{\Phi})|_{\mathbf{\Phi}=\mathbf{\Phi}^t}, \nabla F(\mathbf{\Phi})|_{\mathbf{\Phi}=\mathbf{\Phi}^t} \right\rangle \\ &= -\varepsilon_t \left\| \nabla F(\mathbf{\Phi})|_{\mathbf{\Phi}=\mathbf{\Phi}^t} \right\|^2 < 0 \end{aligned} \tag{4.42}$$

where $\langle x, y \rangle$ represents the inner product of two vectors, and $\|x\|$ represents the Euclidean norm of the vector. Equation (4.42) means that the gradient descent finds a new estimate $\mathbf{\Phi}^{t+1}$ that makes the value of the function $F(\mathbf{\Phi})$ decrease.

The gradient descent algorithm needs to go through an iterative hill-climbing procedure to converge to the local minimum (estimate). Gradient descent usually requires many iterations to converge. The algorithm usually stops when the change of the parameter $\Delta\mathbf{\Phi}$ becomes small enough. That is,

[8] You can use the logarithm of the object function to make it easier to compute the derivative in gradient descent.

$$\left|\varepsilon_t \nabla F(\mathbf{\Phi})|_{\mathbf{\Phi}=\mathbf{\Phi}^t}\right| < \lambda \tag{4.43}$$

where λ is a preset threshold.

Based on the derivation above, the learning rate coefficient ε_t must be small enough for gradient descent to converge. However, if ε_t is too small, convergence is needlessly slow. Thus, it is very important to choose an appropriate ε_t. It is proved [47] [48] that gradient converges almost surely if the learning rate coefficient ε_t satisfies the following condition:

$$\sum_{t=0}^{\infty} \varepsilon_t = \infty, \quad \sum_{t=0}^{\infty} \varepsilon_t^{\ 2} < \infty, \quad \text{and } \varepsilon_t > 0 \tag{4.44}$$

One popular choice of ε_t satisfying the above condition is

$$\varepsilon_t = \frac{1}{t+1} \tag{4.45}$$

Another way to find an appropriate ε_t is through the second-order expansion:

$$F(\mathbf{\Phi}^{t+1}) \approx F(\mathbf{\Phi}^t) + \Delta\mathbf{\Phi} \nabla F(\mathbf{\Phi})|_{\mathbf{\Phi}=\mathbf{\Phi}^t} + \frac{1}{2}(\Delta\mathbf{\Phi})^t \mathbf{D} \Delta\mathbf{\Phi} \tag{4.46}$$

where $\mathbf{D}$ is the *Hessian* matrix [23] of the second-order gradient operator where the *i*-th row and *j*-th element $D_{i,j}$ are given by the following partial derivative:

$$D_{i,j} = \frac{\partial^2 F(\mathbf{\Phi})}{\partial \mathbf{\Phi}_i \partial \mathbf{\Phi}_j} \tag{4.47}$$

By substituting $\Delta\mathbf{\Phi}$ from Eq. (4.41) into Eq. (4.46), we can obtain

$$F(\mathbf{\Phi}^{t+1}) \approx F(\mathbf{\Phi}^t) - \varepsilon_t \left\|\nabla F\right\|^2 + \frac{1}{2}\varepsilon_t^2 \nabla F^t \mathbf{D} \nabla F \tag{4.48}$$

From this, it follows that ε_t can be chosen as follows to minimize $F(\mathbf{\Phi})$ [23]:

$$\varepsilon_t = \frac{\left\|\nabla F\right\|^2}{\nabla F^t \mathbf{D} \nabla F} \tag{4.49}$$

Sometimes it is desirable to impose a different learning rate for the correct model vs. competing models. Therefore re-estimation Eq. (4.39) can be generalized to the following form [19, 48]:

$$\mathbf{\Phi}^{t+1} = \mathbf{\Phi}^t - \varepsilon_t U_t \nabla F(\mathbf{\Phi})|_{\mathbf{\Phi}=\mathbf{\Phi}^t} \tag{4.50}$$

where U_t is the learning bias matrix which is a positive definite matrix. One particular choice of U_t is $\mathbf{D}^{-1}$, where $\mathbf{D}$ is the Hessian matrix defined in Eq. (4.47). When the learning rate is set to be 1.0, Eq. (4.50) becomes Newton's algorithm, where the gradient descent is chosen to minimize the second-order expansion. Equation (4.50) becomes:

$$\mathbf{\Phi}^{t+1} = \mathbf{\Phi}^{t} - \mathbf{D}^{-1}\nabla F(\mathbf{\Phi})\,|_{\mathbf{\Phi}=\mathbf{\Phi}^{t}} \tag{4.51}$$

When probabilistic parameters are iteratively re-estimated, probabilistic constraints must be satisfied in each iteration as probability measure, such as:

1. For discrete distributions, all the values of the probability function ought to be nonnegative. Moreover the sum of all discrete probability values needs to be one, i.e., $\sum_i a_i = 1$
2. For continuous distributions (assuming Gaussian density family), the variance needs to be nonnegative. For Gaussian mixtures, the diagonal covariance entries need to be nonnegative and the sum of mixture weights needs to be one, i.e., $\sum_i c_i = 1$

In general, gradient descent is an unconstrained minimization (or maximization) process that needs to be modified to accommodate constrained minimization (or maximization) problems. The tricks to use are parameter transformations that implicitly maintain these constraints during gradient descent. The original parameters are updated through the inverse transform from the transformed parameter space to the original parameter space. The transformation is done in such a way that constraints on the original parameter are always maintained. Some of these transformations are listed as follows [48]:

1. For probabilities which need to be nonnegative and sum to one, like discrete probability function and mixture weight, the following transformation can be performed:

$$a_i = \frac{\exp(\tilde{a}_i)}{\sum_k \exp(\tilde{a}_k)} \tag{4.52}$$

2. For mean μ and variance (or diagonal covariance entries) σ^2, the following transformation can be used.

$$\mu = \tilde{\mu}\sigma \tag{4.53}$$

$$\sigma = \exp(\tilde{\sigma}) \tag{4.54}$$

After the transformations, we can now compute the gradient with respect to the transformed parameters $(\tilde{a}_i, \tilde{\mu}, \tilde{\sigma})$ using the chain rule. Once the new estimate for the transformed parameters is obtained through gradient descent, one can easily transform them back to the original parameter domain.

4.3.2. Minimum-Error-Rate Estimation

Parameter estimation techniques described so far aim to maximize either the likelihood (class-conditional probability) (MLE and MAP) or the posterior probability (MMIE) in Bayes' equation, Eq. (4.1). Although the criteria used in those estimation methods all have their own merit and under some conditions should lead to satisfactory results, the ultimate parameter estimation criterion for pattern recognition should be to minimize the recognition error rate (or the Bayes' risk) directly. *Minimum-error-rate estimation* is also called *minimum-classification-error* (MCE) training, or discriminative training. Similar to MMIE, the algorithm generally tests the classifier using re-estimated models in the training procedure, and subsequently improves the correct models and suppresses mis-recognized or near-miss models.[9] Neural networks are in this class. Although minimum-error-rate estimation cannot be easily applied, it is still attractive that the criterion is identical to the goal of pattern recognition.

We have used the posterior probability $p(\omega_i \mid \mathbf{x})$ in Bayes' rule as the discriminant function. In fact, just about any discriminant function can be used for minimum-error-rate estimation. For example, as described in Section 4.2.1, a Bayes' Gaussian classifier is equivalent to a quadratic discriminant function. The goal now is to find the estimation of parameters for a discriminant function family $\{d_i(\mathbf{x})\}$ to achieve the minimum error rate. One such error measure is defined in Eq. (4.5). The difficulty associated with the discriminative training approach lies in the fact that the error function needs to be consistent with the true error rate measure and also suitable for optimization.[10] Unfortunately, the error function defined in Section 4.1.1 [Eq. (4.5)] is based on a finite set, which is a piecewise constant function of the parameter vector $\mathbf{\Phi}$. It is not suitable for optimization.

To find an alternative smooth error function for MCE, let us assume that the discriminant function family contains s discriminant functions $d_i(\mathbf{x}, \mathbf{\Phi})$, $i = 1, 2, \ldots, s$. $\mathbf{\Phi}$ denotes the entire parameter set for s discriminant functions. We also assume that all the discriminant functions are nonnegative. We define the following error (misclassification) measure:

$$e_i(\mathbf{x}) = -d_i(\mathbf{x}, \mathbf{\Phi}) + \left[\frac{1}{s-1} \sum_{j \neq i} d_j(\mathbf{x}, \mathbf{\Phi})^{\eta} \right]^{1/\eta} \tag{4.55}$$

where η is a positive number. The intuition behind the above measure is the attempt to enumerate the decision rule. For a ω_i class input $\mathbf{x}$, $e_i(\mathbf{x}) > 0$ implies recognition error; while $e_i(\mathbf{x}) \leq 0$ implies correct recognition. The number η can be thought to be a coefficient to select competing classes in Eq. (4.55). When $\eta = 1$, the competing class term is the average of all the competing discriminant function scores. When $\eta \to \infty$, the competing class term becomes $\max_{j \neq i} d_j(\mathbf{x}, \mathbf{\Phi})$ representing the discriminant function score for the top

[9] A near-miss model occurs when the incorrect model has higher likelihood than the correct model.

[10] In general, a function is optimizable if it is a smooth function and has a derivative.

competing class. By varying the value of η, one can take all the competing classes into account based on their individual significance.

To transform $e_i(\mathbf{x})$ into a normalized smooth function, we can use the sigmoid function to embed $e_i(\mathbf{x})$ in a smooth zero-one function. The loss function can be defined as follows:

$$l_i(\mathbf{x};\mathbf{\Phi}) = sigmoid(e_i(\mathbf{x})) \tag{4.56}$$

$$\text{where } sigmoid(x) = \frac{1}{1+e^{-x}} \tag{4.57}$$

When $e_i(\mathbf{x})$ is a big negative number, which indicates correct recognition, the loss function $l_i(\mathbf{x};\mathbf{\Phi})$ has a value close to zero, which implies no loss incurred. On the other hand, when $e_i(\mathbf{x})$ is a positive number, it leads to a value between zero and one that indicates the likelihood of an error. Thus $l_i(\mathbf{x};\mathbf{\Phi})$ essentially represents a soft recognition error count.

For any data sample **x**, the recognizer's loss function can be defined as:

$$l(\mathbf{x},\mathbf{\Phi}) = \sum_{i=1}^{s} l_i(\mathbf{x},\mathbf{\Phi})\delta(\omega = \omega_i) \tag{4.58}$$

where $\delta(\bullet)$ is a Boolean function which will return 1 if the argument is true and 0 if the argument is false. Since **x** is a random vector, the expected loss according to Eq. (4.58) can be defined as:

$$L(\mathbf{\Phi}) = E_{\mathbf{X}}(l(\mathbf{x},\mathbf{\Phi})) = \sum_{i=1}^{s} \int_{\omega=\omega_i} l(\mathbf{x},\mathbf{\Phi})p(\mathbf{x})d\mathbf{x} \tag{4.59}$$

Since $\max_{\mathbf{\Phi}}\left[\int f(\mathbf{x},\mathbf{\Phi})d\mathbf{x}\right] = \int\left[\max_{\mathbf{\Phi}} f(\mathbf{x},\mathbf{\Phi})\right]d\mathbf{x}$, $\mathbf{\Phi}$ can be estimated by gradient descent over $l(\mathbf{x},\mathbf{\Phi})$ instead of expected loss $L(\mathbf{\Phi})$. That is, minimum classification error training of parameter $\mathbf{\Phi}$ can be estimated by first choosing an initial estimate $\mathbf{\Phi}_0$ and the following iterative estimation equation:

$$\mathbf{\Phi}^{t+1} = \mathbf{\Phi}^t - \varepsilon_t \nabla l(\mathbf{x},\mathbf{\Phi})\big|_{\mathbf{\Phi}=\mathbf{\Phi}^t} \tag{4.60}$$

You can follow the gradient descent procedure described in Section 4.3.1.1 to achieve the MCE estimate of $\mathbf{\Phi}$.

Both MMIE and MCE are much more computationally intensive than MLE, owing to the inefficiency of gradient descent algorithms. Therefore, discriminant estimation methods, like MMIE and MCE, are usually used for tasks containing few classes or data samples. A more pragmatic approach is *corrective training* [6], which is based on a very simple error-correcting procedure. First, a labeled training set is used to train the parameters for each corresponding class by standard MLE. For each training sample, a list of confusable classes is created by running the recognizer and kept as its *near-miss* list. Then, the parameters of the correct class are moved in the direction of the data sample, while the parameters of the "near-miss" class are moved in the opposite direction of the data samples. After all training samples have been processed; the parameters of all classes are updated. This procedure is repeated until the parameters for all classes converge. Although there is no theoretical proof that such a process converges, some experimental results show that it outperforms both MLE and MMIE methods [4].

We have described various estimators: minimum mean square estimator, maximum likelihood estimator, maximum posterior estimator, maximum mutual information estimator, and minimum error estimator. Although based on different training criteria, they are all powerful estimators for various pattern recognition problems. Every estimator has its strengths and weaknesses. It is almost impossible always to favor one over the others. Instead, you should study their characteristics and assumptions and select the most suitable ones for the domains you are working on.

In the following section we discuss *neural networks*. Both neural networks and MCE estimations follow a very similar discriminant training framework.

4.3.3. Neural Networks

In the area of pattern recognition, the advent of new learning procedures and the availability of high-speed parallel supercomputers have given rise to a renewed interest in neural networks.[11] Neural networks are particularly interesting for speech recognition, which requires massive constraint satisfaction, i.e., the parallel evaluation of many clues and facts and their interpretation in the light of numerous interrelated constraints. The computational flexibility of the human brain comes from its large number of neurons in a mesh of axons and dendrites. The communication between neurons is via the synapse and afferent fibers. There are many billions of neural connections in the human brain. At a simple level it can be considered that nerve impulses are comparable to the phonemes of speech, or to letters, in that they do not themselves convey meaning but indicate different intensities [95, 101] that are interpreted as meaningful units by *the language of the brain*. Neural networks attempt to achieve real-time response and humanlike performance using many simple processing elements operating in parallel as in biological nervous systems. Models of neural networks use a particular topology for the interactions and interrelations of the connections of the *neural units*. In this section we describe the basics of neural networks, including the multi-layer perceptrons and the back-propagation algorithm for training neural networks.

[11] A neural network is sometimes called an artificial neural network (ANN), a neural net, or a connectionist model.

4.3.3.1. Single-Layer Perceptrons

Figure 4.10 shows a basic *single-layer perceptron*. Assuming there are N inputs, labeled as $x_1, x_2, \ldots, x_N$, we can form a linear function with weights $w_{0j}, w_{1j}, w_{2j}, \ldots, w_{Nj}$ to give the output y_j, defined as

$$y_j = w_{0j} + \sum_{i=1}^{N} w_{ij} x_i = \mathbf{w}_j \mathbf{x}^t = g_j(\mathbf{x}) \tag{4.61}$$

where $\mathbf{w}_j = (w_{0j}, w_{1j}, w_{2j}, \ldots, w_{Nj})$ and $\mathbf{x} = (1, x_1, x_2, \ldots, x_N)$.

For pattern recognition purposes, we associate each class ω_j out of s classes $(\omega_1, \omega_2, \ldots, \omega_s)$ with such a linear discriminant function $g_j(\mathbf{x})$. By collecting all the discriminant functions, we will have the following matrix representation:

$$\mathbf{y} = \mathbf{g}(\mathbf{x}) = \mathbf{W}^t \mathbf{x} \tag{4.62}$$

where $\mathbf{g}(\mathbf{x}) = (g_1(\mathbf{x}), g_2(\mathbf{x}), \ldots, g_s(\mathbf{x}))^t$; $\mathbf{W} = (\mathbf{w}_1^t, \mathbf{w}_2^t, \ldots, \mathbf{w}_s^t)^t$ and $\mathbf{y} = (y_1, y_2, \ldots, y_s)^t$. The pattern recognition decision can then be based on these discriminant functions as in Bayes' decision theory. That is,

$$\mathbf{x} \in \omega_k \quad \text{iff} \quad k = \arg\max_i g_i(\mathbf{x}) \tag{4.63}$$

The *perceptron training algorithm* [68], guaranteed to converge for linearly separable classes, is often used for training the weight matrix $\mathbf{W}$. The algorithm basically divides the sample space $\Re^N$ into regions of corresponding classes. The decision boundary is characterized by hyper-planes of the following form:

$$g_i(\mathbf{x}) - g_j(\mathbf{x}) = 0 \quad \forall i \neq j \tag{4.64}$$

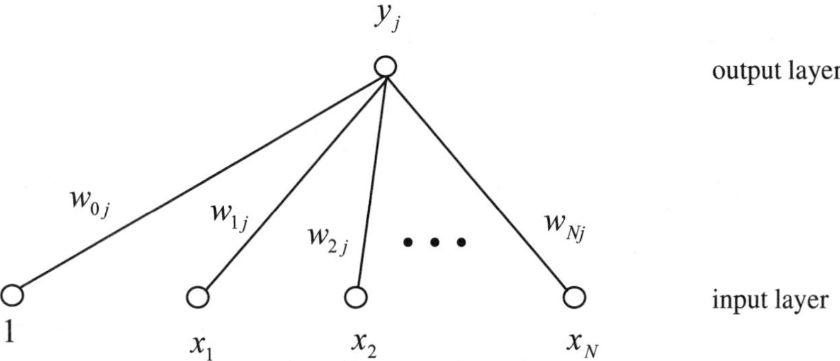

Figure 4.10 A single-layer perceptron.

Unfortunately, for data samples that are not linearly separable, the perceptron algorithm does not converge. However, if we can relax the definition of classification errors in this case, we can still use a powerful algorithm to train the weight matrix $\mathbf{W}$. This approach is the *least square error* (LSE) algorithm described in Chapter 3, which aims at minimizing *sum-of-squared-error* (SSE) criterion, instead of minimizing the classification errors. The sum-of-squared-error is defined as:

$$\text{SSE} = \sum_i \sum_{\mathbf{x}\in\omega_i} \| \mathbf{g}(\mathbf{x}) - \delta_i \|^2 \tag{4.65}$$

where δ_i is an M-dimensional *index vector* with all zero components except that the ith one is 1.0, since the desired output for $\mathbf{g}(\mathbf{x})$ is typically equal to 1.0 if $\mathbf{x} \in \omega_i$ and 0 if $\mathbf{x} \notin \omega_i$.

The use of LSE leads to discriminant functions that have real outputs approximating the values 1 or 0. Suppose there are M input vectors $\mathbf{X} = (\mathbf{x}_1^t, \mathbf{x}_2^t, \ldots, \mathbf{x}_M^t)$ in the training set. Similar to the LSE for linear functions described in Chapter 3 (cf. Section 3.2.1.2), the LSE estimate of weight matrix $\mathbf{W}$ will have the following closed form:

$$\mathbf{W} = ((\mathbf{X}\mathbf{X}^t))^{-1}\mathbf{L}\Sigma \tag{4.66}$$

where $\mathbf{L}$ is a $(N+1)\times s$ matrix where the k-th column is the mean vector $\boldsymbol{\mu}_k = (1, \mu_{k1}, \mu_{k2}, \ldots, \mu_{kN})^t$ of all the vectors classified into class ω_k, and Σ is an $s \times s$ diagonal matrix with diagonal entry $c_{k,k}$ representing the number of vectors classified into ω_k. LSE estimation using linear discriminant functions is equivalent to estimating Bayes' Gaussian densities where all the densities are assumed to share the same covariance matrix [98], as described in Section 4.2.1.

Although the use of LSE algorithm solves the convergence problems, it loses the power of nonlinear logical decision (i.e., minimizing the classification error rate), since it is only approximating the simple logical decision between alternatives. An alternative approach is to use a smooth and differential sigmoid function as the threshold function:

$$\begin{aligned}\mathbf{y} &= sigmoid(\mathbf{g}(\mathbf{x})) = sigmoid((g_1(\mathbf{x}), g_2(\mathbf{x}), \ldots, g_s(\mathbf{x}))^t) \\ &= (sigmoid(g_1(\mathbf{x})), sigmoid(g_2(\mathbf{x})), \ldots, sigmoid(g_s(\mathbf{x})))^t\end{aligned} \tag{4.67}$$

where $sigmoid(x)$ is the sigmoid function defined in Eq. (4.57). With the sigmoid function, the following new sum-of-squared-error term closely tracks the classification error:

$$\text{NSSE} = \sum_i \sum_{\mathbf{x}\in\omega_i} \| sigmoid(\mathbf{g}(\mathbf{x})) - \delta_i \|^2 \tag{4.68}$$

where δ_i is the same index vector defined above. Since there is no analytic way of minimizing a nonlinear function, the use of the sigmoid threshold function requires an iterative gradient descent algorithm, back-propagation, which will be described in the next section.

4.3.3.2. Multi-Layer Perceptron

One of the technical developments sparking the recent resurgence of interest in neural networks has been the popularization of *multi-layer perceptrons* (MLP) [37, 90]. Figure 4.11 shows a multi-layer perceptron. In contrast to a single-layer perceptron, it has two hidden layers. The hidden layers can be viewed as feature extractors. Each layer has the same computation models as the single-layer perceptron; i.e., the value of each node is computed as a linear weighted sum of the input nodes and passed to a sigmoid type of threshold function.

$$\begin{aligned}
\mathbf{h}_1 &= sigmoid(\mathbf{g}_{h1}(\mathbf{x})) = sigmoid(\mathbf{W}_{h1}^t \mathbf{x}) \\
\mathbf{h}_2 &= sigmoid(\mathbf{g}_{h2}(\mathbf{h}_1)) = sigmoid(\mathbf{W}_{h2}^t \mathbf{h}_1) \\
\mathbf{y} &= sigmoid(\mathbf{g}_y(\mathbf{h}_2)) = sigmoid(\mathbf{W}_y^t \mathbf{h}_2)
\end{aligned} \tag{4.69}$$

where $sigmoid(x)$ is the sigmoid function defined in Eq. (4.57).

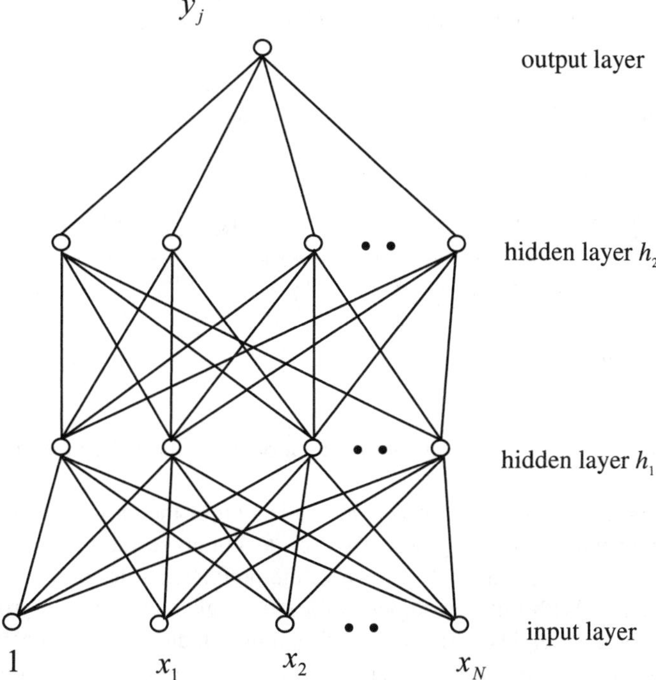

Figure 4.11 A multi-layer perceptron with four total layers. The middle two layers are hidden.

According to Eq. (4.69), we can propagate the computation from input layer to output layer and denote the output layer as a nonlinear function of the input layer.

$$\mathbf{Y} = MLP(\mathbf{x}) \tag{4.70}$$

Let's denote $O(\mathbf{x})$ as the desired output for input vector $\mathbf{x}$. For pattern classification, $O(\mathbf{x})$ will be an s-dimensional vector with the desired output pattern set to one and the remaining patterns set to zero. As we mentioned before, there is no analytic way to minimize the mean square error $E = \sum_{\mathbf{x}} \| MLP(\mathbf{x}) - O(\mathbf{x}) \|^2$. Instead, an iterative gradient descent algorithm called back propagation [89, 90] needs to be used to reduce error. Without loss of generality, we assume there is only one input vector $\mathbf{x} = (1, x_1, x_2, \ldots, x_N)$ with desired output $\mathbf{o} = (o_1, o_2, \ldots, o_s)$. All the layers in the MLP are numbered 0, 1, 2,... upward from the input layer. The back propagation algorithm can then be described as in Algorithm 4.1.

In computing the partial derivative $\frac{\partial E}{\partial w_{ij}^k(t)}$, you need to use the chain rule. w_{ij}^K is the weight connecting the output layer and the last hidden layer; the partial derivative is:

$$\begin{aligned}
\frac{\partial E}{\partial w_{ij}^K} &= \frac{\partial(\sum_{i=1}^{s}(y_i - o_i)^2)}{\partial w_{ij}^K} \\
&= \frac{\partial(\sum_{i=1}^{s}(y_i - o_i)^2)}{\partial y_j} \times \frac{\partial y_j}{\partial(w_{0j}^K + \sum_{i=1}^{N} w_{ij}^K v_i^{K-1})} \times \frac{\partial(w_{0j}^K + \sum_{i=1}^{N} w_{ij}^K v_i^{K-1})}{\partial w_{ij}^K} \\
&= 2(y_i - o_i) y_j (y_i - 1) v_i^{K-1}
\end{aligned} \tag{4.71}$$

For layers $k = K-1, K-2, \cdots$, one can apply chain rules similarly for gradient $\frac{\partial E}{\partial w_{ij}^k(t)}$.

The back propagation algorithm is a generalization of the minimum mean squared error (MMSE) algorithm. It uses a gradient search to minimize the difference between the desired outputs and the actual net outputs, where the optimized criterion is directly related to pattern classification. With initial parameters for the weights, the training procedure is then repeated to update the weights until the cost function is reduced to an acceptable value or remains unchanged. In the algorithm described above, we assume a single training example. In real-world application, these weights are estimated from a large number of training observations in a manner similar to hidden Markov modeling. The weight updates in Step 3 are accumulated over all the training data. The actual gradient is then estimated for the complete set of training data before the beginning of the next iteration. Note that the estimation criterion for neural networks is directly related to classification rather than maximum likelihood.

ALGORITHM 4.1: *THE BACK PROPAGATION ALGORITHM*

Step 1: Initialization: Set $t = 0$ and choose initial weight matrices $\mathbf{W}$ for each layer. Let's denote $w_{ij}^k(t)$ as the weighting coefficients connecting i^{th} input node in layer $k-1$ and j^{th} output node in layer k at time t.

Step 2: Forward Propagation: Compute the values in each node from input layer to output layer in a propagating fashion, for $k = 1$ to K

$$v_j^k = sigmoid(w_{0j}(t) + \sum_{i=1}^{N} w_{ij}^k(t) v_i^{k-1}) \quad \forall j \tag{4.72}$$

where $sigmoid(x) = \dfrac{1}{1+e^{-x}}$ and v_j^k is denoted as the jth node in the kth layer

Step 3: Back Propagation: Update the weights matrix for each layer from output layer to input layer according to:

$$\overline{w}_{ij}^k(t+1) = w_{ij}^k(t) - \alpha \frac{\partial E}{\partial w_{ij}^k(t)} \tag{4.73}$$

where $E = \sum_{i=1}^{s} \| y_i - o_i \|^2$ and $(y_1, y_2, \ldots y_s)$ is the computed output vector in Step 2.

α is referred to as the learning rate and has to be small enough to guarantee convergence. One popular choice is $1/(t+1)$.

Step 4: Iteration: Let $t = t + 1$. Repeat Steps 2 and 3 until some convergence condition is met.

4.4. UNSUPERVISED ESTIMATION METHODS

As described in Section 4.2, in unsupervised learning, information about class ω of the data sample $\mathbf{x}$ is unavailable. Data observed are *incomplete* since the class data ω is missing. One might wonder why we are interested in such an unpromising problem, and whether or not it is possible to learn anything from incomplete data. Interestingly enough, the formal solution to this problem is almost identical to the solution for the supervised learning case – MLE. We discuss vector quantization (VQ), which uses principles similar to the EM algorithm. It is important in its own right in spoken language systems.

4.4.1. Vector Quantization

As described in Chapter 3, source coding refers to techniques that convert the signal source into a sequence of bits that are transmitted over a communication channel and then used to reproduce the original signal at a different location or time. In speech communication, the reproduced sound usually allows some acceptable level of distortion to achieve low bit rate. The goal of source coding is to reduce the number of bits necessary to transmit or store data, subject to a distortion or fidelity criterion, or equivalently, to achieve the minimum possible distortion for a prescribed bit rate. *Vector quantization* (VQ) is one of the most efficient source-coding techniques.

Quantization is the process of approximating continuous amplitude signals by discrete symbols. The quantization of a single signal value or parameter is referred to as scalar quantization. In contrast, joint quantization of multiple signal values or parameters is referred to as vector quantization. Conventional pattern recognition techniques have been used effectively to solve the quantization or data compression problem with successful application to speech coding, image coding, and speech recognition [36, 85]. In both speech recognition and synthesis systems, vector quantization serves an important role in many aspects of the systems, ranging from discrete acoustic prototypes of speech signals for the discrete HMM, to robust signal processing and data compression.

A vector quantizer is described by a codebook, which is a set of fixed *prototype vectors* or reproduction vectors. Each of these prototype vectors is also referred to as a codeword. To perform the quantization process, the input vector is matched against each codeword in the codebook using some *distortion measure*. The input vector is then replaced by the index of the codeword with the smallest distortion. Therefore, a description of the vector quantization process includes:

1. the distortion measure;
2. the generation of each codeword in the codebook.

4.4.1.1. Distortion Measures

Since vectors are replaced by the index of the codeword with smallest distortion, the transmitted data can be recovered only by replacing the code index sequence with the corresponding codeword sequence. This inevitably causes distortion between the original data and the transmitted data. How to minimize the distortion is thus the central goal of vector quantization. This section describes a couple of the most common distortion measures.

Assume that $\mathbf{x} = (x_1, x_2, \ldots, x_d)^t \in R^d$ is a d-dimensional vector whose components $\{x_k, 1 \le k \le d\}$ are real-valued, continuous-amplitude random variables. After vector quantization, the vector $\mathbf{x}$ is mapped (quantized) to another discrete-amplitude d-dimensional vector $\mathbf{z}$.

$$\mathbf{z} = q(\mathbf{x}) \tag{4.74}$$

In Eq. (4.74) $q()$ is the quantization operator. Typically, $\mathbf{z}$ is a vector from a finite set $\mathbf{Z} = \{\mathbf{z}_j, 1 \le j \le M\}$, where $\mathbf{z}_j$ is also a d-dimensional vector. The set $\mathbf{Z}$ is referred to as the codebook, M is the size of the codebook, and $\mathbf{z}_j$ is j^{th} *codeword*. The size M of the codebook is also called the number of partitions (or levels) in the codebook. To design a codebook, the

d-dimensional space of the original random vector $\mathbf{x}$ can be partitioned into M regions or cells $\{C_i, 1 \le i \le M\}$, and each cell C_i is associated with a codeword vector $\mathbf{z}_i$. VQ then maps (quantizes) the input vector $\mathbf{x}$ to codeword $\mathbf{z}_i$ if $\mathbf{x}$ lies in C_i. That is,

$$q(\mathbf{x}) = \mathbf{z}_i \text{ if } \mathbf{x} \in C_i \tag{4.75}$$

An example of partitioning of a two-dimensional space ($d = 2$) for the purpose of vector quantization is shown in Figure 4.12. The shaded region enclosed by the dashed lines is the cell C_i. Any input vector $\mathbf{x}$ that lies in the cell C_i is quantized as $\mathbf{z}_i$. The shapes of the various cells can be different. The positions of the codewords within each cell are shown by dots in Figure 4.12. The codeword $\mathbf{z}_i$ is also referred to as the *centroid* of the cell C_i because it can be viewed as the central point of the cell C_i.

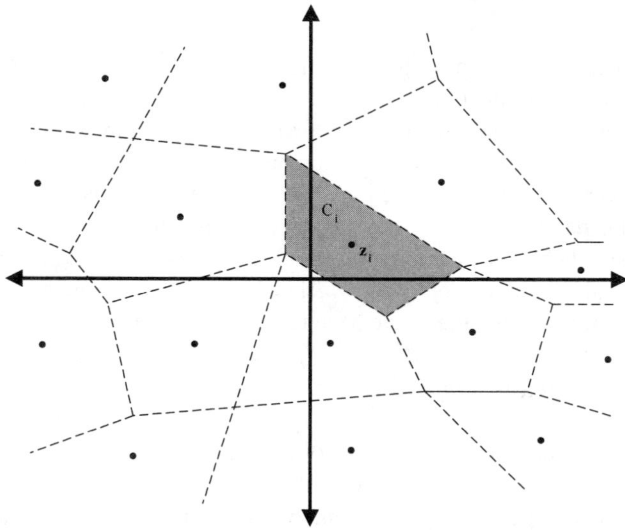

Figure 4.12 Partitioning of a two-dimensional space into 16 cells.

When $\mathbf{x}$ is quantized as $\mathbf{z}$, a quantization error results. A distortion measure $d(\mathbf{x}, \mathbf{z})$ can be defined between $\mathbf{x}$ and $\mathbf{z}$ to measure the quantization quality. Using this distortion measure, Eq. (4.75) can be reformulated as follows:

$$q(\mathbf{x}) = \mathbf{z}_i \text{ if and only if } i = \underset{k}{\operatorname{argmin}}\ d(\mathbf{x}, \mathbf{z}_k) \tag{4.76}$$

The distortion measure between $\mathbf{x}$ and $\mathbf{z}$ is also known as a distance measure in the speech context. The measure must be tractable in order to be computed and analyzed, and also must be subjectively relevant so that differences in distortion values can be used to indicate differences in original and transmitted signals. The most commonly used measure is the Euclidean distortion measure, which assumes that the distortions contributed by quantizing the different parameters are equal. Therefore, the distortion measure $d(\mathbf{x}, \mathbf{z})$ can be defined as follows:

$$d(\mathbf{x},\mathbf{z}) = (\mathbf{x}-\mathbf{z})^t(\mathbf{x}-\mathbf{z}) = \sum_{i=1}^{d}(x_i - z_i)^2 \tag{4.77}$$

The distortion defined in Eq. (4.77) is also known as sum of squared error. In general, unequal weights can be introduced to weight certain contributions to the distortion more than others. One choice for weights that is popular in many practical applications is to use the inverse of the covariance matrix of $\mathbf{z}$.

$$d(\mathbf{x},\mathbf{z}) = (\mathbf{x}-\mathbf{z})^t\Sigma^{-1}(\mathbf{x}-\mathbf{z}) \tag{4.78}$$

This distortion measure, known as the *Mahalanobis* distance, is actually the exponential term in a Gaussian density function.

Another way to weight the contributions to the distortion measure is to use *perceptually*-based distortion measures. Such distortion measures take advantage of subjective judgments of perceptual difference caused by two different signals. A perceptually-based distortion measure has the property that signal changes that make the sounds being perceived different should be associated with large distances. Similarly signal changes that keep the sound perceived the same should be associated with small distances. A number of perceptually based distortion measures have been used in speech coding [3, 75, 76].

4.4.1.2. The *K*-Means Algorithm

To design an M-level codebook, it is necessary to partition d-dimensional space into M cells and associate a quantized vector with each cell. Based on the source-coding principle, the criterion for optimization of the vector quantizer is to minimize overall average distortion over all M-levels of the VQ. The overall average distortion can be defined by

$$\begin{aligned} D = E\left[d(\mathbf{x},\mathbf{z})\right] &= \sum_{i=1}^{M} p(\mathbf{x}\in C_i)E\left[d(\mathbf{x},\mathbf{z}_i) \mid \mathbf{x}\in C_i\right] \\ &= \sum_{i=1}^{M} p(\mathbf{x}\in C_i)\int_{x\in C_i} d(\mathbf{x},\mathbf{z}_i)p(\mathbf{x}\mid \mathbf{x}\in C_i)d\mathbf{x} = \sum_{i=1}^{M} D_i \end{aligned} \tag{4.79}$$

where the integral is taken over all components of vector $\mathbf{x}$; $p(\mathbf{x}\in C_i)$ denotes the prior probability of codeword $\mathbf{z}_i$; $p(\mathbf{x}\mid \mathbf{x}\in C_i)$ denotes the multidimensional probability density function of $\mathbf{x}$ in cell C_i; and D_i is the average distortion in cell C_i. No analytic solution exists to guarantee global minimization of the average distortion measure for a given set of speech data. However, an iterative algorithm, which guarantees a local minimum, exists and works well in practice.

We say a quantizer is optimal if the overall average distortion is minimized over all M-levels of the quantizer. There are two necessary conditions for optimality. The first is that the optimal quantizer is realized by using a nearest-neighbor selection rule as specified by Eq. (4.76). Note that the average distortion for each cell C_i

$$E\left[d(\mathbf{x},\mathbf{z}_i) \mid \mathbf{x} \in C_i\right] \tag{4.80}$$

can be minimized when $\mathbf{z}_i$ is selected such that $d(\mathbf{x},\mathbf{z}_i)$ is minimized for $\mathbf{x}$. This means that the quantizer must choose the codeword that results in the minimum distortion with respect to $\mathbf{x}$. The second condition for optimality is that each codeword $\mathbf{z}_i$ is chosen to minimize the average distortion in cell C_i. That is, $\mathbf{z}_i$ is the vector that minimizes

$$D_i = p(\mathbf{z}_i)E\left[d(\mathbf{x},\mathbf{z}) \mid \mathbf{x} \in C_i\right] \tag{4.81}$$

Since the overall average distortion D is a linear combination of average distortions in C_i, they can be independently computed after classification of $\mathbf{x}$. The vector $\mathbf{z}_i$ is called the centroid of the cell C_i and is written

$$\mathbf{z}_i = \text{cent}(C_i) \tag{4.82}$$

The centroid for a particular region (cell) depends on the definition of the distortion measure. In practice, given a set of training vectors $\{\mathbf{x}_t, 1 \le t \le T\}$, a subset of K_i vectors will be located in cell C_i. In this case, $p(\mathbf{x} \mid \mathbf{z}_i)$ can be assumed to be $1/K_i$, and $p(\mathbf{z}_i)$ becomes K_i/T. The average distortion D_i in cell C_i can then be given by

$$D_i = \frac{1}{T}\sum_{\mathbf{x}\in C_i} d(\mathbf{x},\mathbf{z}_i) \tag{4.83}$$

The second condition for optimality can then be rewritten as follows:

$$\hat{\mathbf{z}}_i = \arg\min_{\mathbf{z}_i} D_i(\mathbf{z}_i) = \arg\min_{\mathbf{z}_i} \frac{1}{T}\sum_{\mathbf{x}\in C_i} d(\mathbf{x},\mathbf{z}_i) \tag{4.84}$$

When the sum of squared error in Eq. (4.77) is used for the distortion measure, the attempt to find such $\hat{\mathbf{z}}_i$ to minimize the sum of squared error is equivalent to least squared error estimation, which was described in Chapter 3. Minimization of D_i in Eq. (4.84) with respect to $\mathbf{z}_i$ is given by setting the derivative of D_i to zero:

$$\begin{aligned}
\nabla_{\mathbf{z}_i} D_i &= \nabla_{\mathbf{z}_i} \frac{1}{T}\sum_{\mathbf{x}\in C_i} (\mathbf{x}-\mathbf{z}_i)^t(\mathbf{x}-\mathbf{z}_i) \\
&= \frac{1}{T}\sum_{\mathbf{x}\in C_i} \nabla_{\mathbf{z}_i}(\mathbf{x}-\mathbf{z}_i)^t(\mathbf{x}-\mathbf{z}_i) \\
&= \frac{-2}{T}\sum_{\mathbf{x}\in C_i} (\mathbf{x}-\mathbf{z}_i) = 0
\end{aligned} \tag{4.85}$$

By solving Eq. (4.85), we obtain the least square error estimate of centroid $\hat{\mathbf{z}}_i$ simply as the sample mean of all the training vectors **x**, quantized to cell C_i:

$$\hat{\mathbf{z}}_i = \frac{1}{K_i} \sum_{\mathbf{x} \in C_i} \mathbf{x} \tag{4.86}$$

If the Mahalanobis distance measure [Eq. (4.78)] is used, minimization of D_i in Eq. (4.84) can be done similarly:

$$\begin{aligned}
\nabla_{\mathbf{z}_i} D_i &= \nabla_{\mathbf{z}_i} \frac{1}{T} \sum_{\mathbf{x} \in C_i} (\mathbf{x} - \mathbf{z}_i)^t \Sigma^{-1} (\mathbf{x} - \mathbf{z}_i) \\
&= \frac{1}{T} \sum_{\mathbf{x} \in C_i} \nabla_{\mathbf{z}_i} (\mathbf{x} - \mathbf{z}_i)^t \Sigma^{-1} (\mathbf{x} - \mathbf{z}_i) \\
&= \frac{-2}{T} \sum_{\mathbf{x} \in C_i} \Sigma^{-1} (\mathbf{x} - \mathbf{z}_i) = 0
\end{aligned} \tag{4.87}$$

and centroid $\hat{\mathbf{z}}_i$ is obtained from

$$\hat{\mathbf{z}}_i = \frac{1}{K_i} \sum_{\mathbf{x} \in C_i} \mathbf{x} \tag{4.88}$$

One can see that $\hat{\mathbf{z}}_i$ is again the sample mean of all the training vectors **x**, quantized to cell C_i. Although Eq. (4.88) is obtained based on the Mahalanobis distance measure, it also works with a large class of Euclidean-like distortion measures [61]. Since the Mahalanobis distance measure is actually the exponential term in a Gaussian density, minimization of the distance criterion can be easily translated into maximization of the logarithm of the Gaussian likelihood. Therefore, similar to the relationship between least square error estimation for the linear discrimination function and the Gaussian classifier described in Section 4.3.3.1, the distance minimization process (least square error estimation) above is in fact a *maximum likelihood estimation*.

According to these two conditions for VQ optimality, one can iteratively apply the nearest-neighbor selection rule and Eq. (4.88) to get the new centroid $\hat{\mathbf{z}}_i$ for each cell in order to minimize the average distortion measure. This procedure is known as the *k-means* algorithm or the *generalized Lloyd* algorithm [29, 34, 56]. In the *k*-means algorithm, the basic idea is to partition the set of training vectors into *M* clusters C_i $(1 \le i \le M)$ in such a way that the two necessary conditions for optimality described above are satisfied. The *k*-means algorithm can be described as in Algorithm 4.2.

ALGORITHM 4.2: *THE K-MEANS ALGORITHM*

Step 1: Initialization: Choose some adequate method to derive initial VQ codewords ($\mathbf{z}_i,\ 1 \le i \le M$) in the codebook.
Step 2: Nearest-neighbor Classification: Classify each training vector $\{\mathbf{x}_k\}$ into one of the cells C_i by choosing the closest codeword $\mathbf{z}_i$ ($\mathbf{x} \in C_i$, i.f.f. $d(\mathbf{x},\mathbf{z}_i) \le d(\mathbf{x},\mathbf{z}_j)$ for all $j \ne i$). This classification is also called minimum-distance classifier.
Step 3: Codebook Updating: Update the codeword of every cell by computing the centroid of the training vectors in each cell according to Eq. (4.84) ($\hat{\mathbf{z}}_i = cent(C_i),\ 1 \le i \le M$).
Step 4: Iteration: Repeat steps 2 and 3 until the ratio of the new overall distortion D at the current iteration relative to the overall distortion at the previous iteration is above a preset threshold.

In the process of minimizing the average distortion measure, the k-means procedure actually breaks the minimization process into two steps. Assuming that the centroid $\mathbf{z}_i$ (or mean) for each cell C_i has been found, then the minimization process is found simply by partitioning all the training vectors into their corresponding cells according to the distortion measure. After all of the new partitions are obtained, the minimization process involves finding the new centroid within each cell to minimize its corresponding within-cell average distortion D_i based on Eq. (4.84). By iterating over these two steps, a new overall distortion D smaller than that of the previous step can be obtained.

Theoretically, the k-means algorithm can converge only to a local optimum [56]. Furthermore, any such solution is, in general, not unique [33]. Initialization is often critical to the quality of the eventual converged codebook. Global optimality may be approximated by repeating the k-means algorithm for several sets of codebook initialization values, and then one can choose the codebook that produces the minimum overall distortion. In the next subsection we will describe methods for finding a decent initial codebook.

4.4.1.3. The LBG Algorithm

Since the initial codebook is critical to the ultimate quality of the final codebook, it has been shown that it is advantageous to design an M-vector codebook in stages. This extended k-means algorithm is known as the LBG algorithm proposed by Linde, Buzo, and Gray [56]. The LBG algorithm first computes a 1-vector codebook, then uses a splitting algorithm on the codewords to obtain the initial 2-vector codebook, and continues the splitting process until the desired M-vector codebook is obtained. The procedure is formally implemented by Algorithm 4.3.

ALGORITHM 4.3: *THE LBG ALGORITHM*

Step 1: Initialization: Set M (number of partitions or cells) =1. Find the centroid of all the training data according to Eq. (4.84).
Step 2: Splitting: Split M into $2M$ partitions by splitting each current codeword by finding two points that are far apart in each partition using a heuristic method, and use these two points as the new centroids for the new $2M$ codebook. Now set $M = 2M$.
Step 3: K-means Stage: Now use the k-means iterative algorithm described in the previous section to reach the best set of centroids for the new codebook.
Step 4: Termination: If M equals the VQ codebook size required, STOP; otherwise go to Step 2.

4.4.2. The EM Algorithm

We introduce the EM algorithm that is important to hidden Markov models and other learning techniques. It discovers model parameters by maximizing the log-likelihood of incomplete data and by iteratively maximizing the expectation of log-likelihood from complete data. The EM algorithm is a generalization of the VQ algorithm described above.

The EM algorithm can also be viewed as a generalization of the MLE method, when the data observed is incomplete. Without loss of generality, we use scalar random variables here to describe the EM algorithm. Suppose we observe training data y. In order to determine the parameter vector $\mathbf{\Phi}$ that maximizes $P(Y=y\,|\,\mathbf{\Phi})$, we would need to know some hidden data x (that is unobserved). For example, x may be a hidden number that refers to component densities of observable data y, or x may be the underlying hidden state sequence in *hidden Markov models* (as discussed in Chapter 8). Without knowing this hidden data x, we could not easily use the maximum likelihood estimation to estimate $\hat{\mathbf{\Phi}}$, which maximizes $P(Y=y\,|\,\mathbf{\Phi})$. Instead, we assume a parameter vector $\mathbf{\Phi}$ and estimate the probability that each x occurred in the generation of y. This way we can pretend that we had in fact observed a complete data pair (x, y), with frequency proportional to the probability $P(X=x,Y=y\,|\,\mathbf{\Phi})$, to compute a new $\bar{\mathbf{\Phi}}$, the maximum likelihood estimate of $\mathbf{\Phi}$. We can then set the parameter vector $\mathbf{\Phi}$ to be this new $\bar{\mathbf{\Phi}}$ and repeat the process to iteratively improve our estimate.

The issue now is whether or not the process (EM algorithm) described above converges. Without loss of generality, we assume that both random variables X (unobserved) and Y (observed) are discrete random variables. According to Bayes' rule,

$$P(X=x,Y=y\,|\,\bar{\mathbf{\Phi}}) = P(X=x\,|\,Y=y,\bar{\mathbf{\Phi}})P(Y=y\,|\,\bar{\mathbf{\Phi}}) \tag{4.89}$$

Our goal is to maximize the log-likelihood of the observable, real data y generated by parameter vector $\bar{\mathbf{\Phi}}$. Based on Eq. (4.89), the log-likelihood can be expressed as follows:

$$\log P(Y=y\,|\,\bar{\mathbf{\Phi}}) = \log P(X=x,Y=y\,|\,\bar{\mathbf{\Phi}}) - \log P(X=x\,|\,Y=y,\bar{\mathbf{\Phi}}) \tag{4.90}$$

Now, we take the conditional expectation of $\log P(Y=y\mid\bar{\Phi})$ over X computed with parameter vector Φ:

$$\begin{aligned} E_\Phi[\log P(Y=y\mid\bar{\Phi})]_{X|Y=y} &= \sum_x \left(P(X=x\mid Y=y,\Phi)\log P(Y=y\mid\bar{\Phi})\right) \\ &= \log P(Y=y\mid\bar{\Phi}) \end{aligned} \tag{4.91}$$

where we denote $E_\Phi[f]_{X|Y=y}$ as the expectation of function f over X computed with parameter vector Φ. Then using Eq. (4.90) and (4.91) , the following expression is obtained:

$$\begin{aligned} \log P(Y=y\mid\bar{\Phi}) &= E_\Phi[\log P(X,Y=y\mid\bar{\Phi})]_{X|Y=y} - E_\Phi[\log P(X\mid Y=y,\bar{\Phi})]_{X|Y=y} \\ &= Q(\Phi,\bar{\Phi}) - H(\Phi,\bar{\Phi}) \end{aligned} \tag{4.92}$$

where

$$\begin{aligned} Q(\Phi,\bar{\Phi}) &= E_\Phi[\log P(X,Y=y\mid\bar{\Phi})]_{X|Y=y} \\ &= \sum_x \left(P(X=x\mid Y=y,\Phi)\log P(X=x,Y=y\mid\bar{\Phi})\right) \end{aligned} \tag{4.93}$$

and

$$\begin{aligned} H(\Phi,\bar{\Phi}) &= E_\Phi[\log P(X\mid Y=y,\bar{\Phi})]_{X|Y=y} \\ &= \sum_x \left(P(X=x\mid Y=y,\bar{\Phi})\log P(X=x\mid Y=y,\bar{\Phi})\right) \end{aligned} \tag{4.94}$$

The convergence of the EM algorithm lies in the fact that if we choose $\bar{\Phi}$ so that

$$Q(\Phi,\bar{\Phi}) \ge Q(\Phi,\Phi) \tag{4.95}$$

then

$$\log P(Y=y\mid\bar{\Phi}) \ge \log P(Y=y\mid\Phi) \tag{4.96}$$

since it follows from Jensen's inequality that $H(\Phi,\bar{\Phi}) \le H(\Phi,\Phi)$ [21]. The function $Q(\Phi,\bar{\Phi})$ is known as the Q-function or auxiliary function. This fact implies that we can maximize the Q-function, which is the expectation of log-likelihood from complete data pair (x, y), to update parameter vector from Φ to $\bar{\Phi}$, so that the incomplete log-likelihood $L(x,\Phi)$ increases monotonically. Eventually, the likelihood will converge to a local maximum if we iterate the process.

The name of the EM algorithm comes from E for expectation and M for maximization. The implementation of the EM algorithm includes the E (expectation) step, which calculates the auxiliary Q-function $Q(\Phi,\bar{\Phi})$ and the M (maximization) step, which maximizes $Q(\Phi,\bar{\Phi})$ over $\bar{\Phi}$ to obtain $\hat{\Phi}$. The general EM algorithm can be described in the following way.

ALGORITHM 4.4: *THE EM ALGORITHM*

Step 1: Initialization: Choose an initial estimate Φ.
Step 2: E-Step: Compute auxiliary *Q*-function $Q(\Phi, \bar{\Phi})$ (which is also the expectation of log-likelihood from complete data) based on Φ.
Step 3: M-Step: Compute $\hat{\Phi} = \arg\max_{\Phi} Q(\Phi, \bar{\Phi})$ to maximize the auxiliary *Q*-function.
Step 4: Iteration: Set $\Phi = \hat{\Phi}$, repeat from Step 2 until convergence.

The M-step of the EM algorithm is actually a maximum likelihood estimation of complete data (assuming we know the unobserved data x based on observed data y and initial parameter vector Φ). The EM algorithm is usually used in applications where no analytic solution exists for maximization of log-likelihood of incomplete data. Instead, the *Q*-function is iteratively maximized to obtain the estimation of parameter vector.

4.4.3. Multivariate Gaussian Mixture Density Estimation

The vector quantization process described in Section 4.4.1 partitions the data space into separate regions based on some distance measure regardless of the probability distributions of the original data. This process may introduce errors in partitions that could potentially destroy the original structure of data. An alternative way for modeling a VQ codebook is to use a family of Gaussian probability density functions, such that each cell will be represented by a (Gaussian) probability density function as shown in Figure 4.13. These probability density functions can then overlap, rather than partition, in order to represent the entire data space. The objective for a mixture Gaussian VQ is to maximize the likelihood of the observed data (represented by the product of the Gaussian mixture scores) instead of minimizing the overall distortion. The centroid of each cell (the mean vectors of each Gaussian pdf) obtained via such a representation may be quite different from that obtained using the traditional *k*-means algorithm, since the distribution properties of the data are taken into account.

There should be an obvious analogy between the EM algorithm and the *k*-means algorithm described in the Section 4.4.1.2. In the *k*-means algorithm, the class information for the observed data samples is hidden and unobserved, so an EM-like algorithm instead of maximum likelihood estimate needs to be used. Therefore, instead of a single process of maximum likelihood estimation, the *k*-means algorithm first uses the old codebook to find the nearest neighbor for each data sample followed by maximum likelihood estimation of the new codebook and iterates the process until the distortion stabilizes. The steps 2 and 3 in the *k*-means algorithm are actually the E and M steps in the EM algorithm respectively.

Mixture density estimation [41] is a typical example of EM estimation. In the mixtures of Gaussian density, the probability density for observable data $\mathbf{y}$ is the weighted sum of each Gaussian component:

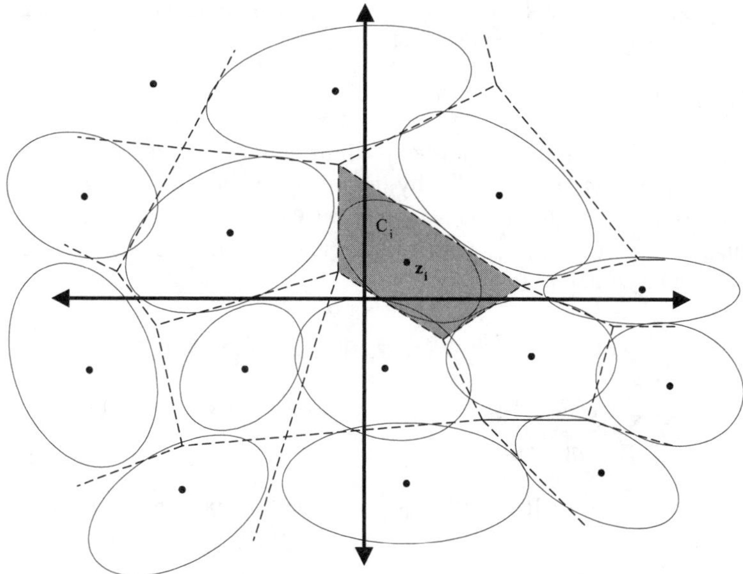

Figure 4.13 Partitioning of a two-dimensional space with 16 Gaussian density functions.

$$p(\mathbf{y} \mid \mathbf{\Phi}) = \sum_{k=1}^{K} c_k p_k(\mathbf{y} \mid \mathbf{\Phi}_k) = \sum_{k=1}^{K} c_k N_k(\mathbf{y} \mid \boldsymbol{\mu}_k, \Sigma_k) \tag{4.97}$$

where $0 \le c_k \le 1$, for $1 \le k \le K$ and $\sum_{k=1}^{K} c_k = 1$.

Unlike the case of a single Gaussian estimation, we also need to estimate the mixture weight c_k. In order to do so, we can assume that observable data $\mathbf{y}$ come from one of the component densities $p_X(\mathbf{y} \mid \mathbf{\Phi}_X)$, where X is a random variable taking value from $\{1,2,\ldots K\}$ to indicate the Gaussian component. It is clear that x is unobserved and used to specify the pdf component $\mathbf{\Phi}_X$. Assuming that the probability density function for complete data $(x,\mathbf{y})$ is given by the joint probability:

$$p(\mathbf{y}, x \mid \mathbf{\Phi}) = P(X = x) p_x(\mathbf{y} \mid \mathbf{\Phi}_x) = P(X = x) N_x(\mathbf{y} \mid \boldsymbol{\mu}_x, \Sigma_x) \tag{4.98}$$

$P(X = x)$ can be regarded as the probability of the unobserved data x used to specify the component density $p_x(\mathbf{y} \mid \mathbf{\Phi}_x)$ from which the observed data $\mathbf{y}$ is drawn. If we assume the number of components is K and $\mathbf{\Phi}$ is the vector of all probability parameters ($P(X)$, $\mathbf{\Phi}_1, \mathbf{\Phi}_2, \ldots, \mathbf{\Phi}_K$), the probability density function of incomplete (observed) data $\mathbf{y}$ can be specified as the following marginal probability:

$$p(\mathbf{y} \mid \mathbf{\Phi}) = \sum_{x} p(\mathbf{y}, x \mid \mathbf{\Phi}) = \sum_{x} P(X = x) p_x(\mathbf{y} \mid \mathbf{\Phi}_x) \tag{4.99}$$

By comparing Eq. (4.97) and (4.99), we can see that the mixture weight is represented as the probability function $P(X = x)$. That is,

$$c_k = P(X = k) \tag{4.100}$$

According to the EM algorithm, the maximization of the logarithm of the likelihood function $\log p(\mathbf{y} \mid \mathbf{\Phi})$ can be performed by iteratively maximizing the conditional expectation of the logarithm of Eq. (4.98), i.e., $\log p(\mathbf{y}, x \mid \mathbf{\Phi})$. Suppose we have observed N independent samples: $\{\mathbf{y}_1, \mathbf{y}_2, \ldots, \mathbf{y}_N\}$ with hidden unobserved data $\{x_1, x_2, \ldots, x_N\}$; the Q-function can then be written as follows:

$$\begin{aligned} Q(\mathbf{\Phi}, \bar{\mathbf{\Phi}}) &= \sum_{i=1}^{N} Q_i(\mathbf{\Phi}, \bar{\mathbf{\Phi}}) = \sum_{i=1}^{N} \sum_{x_i} P(x_i \mid \mathbf{y}_i, \mathbf{\Phi}) \log p(\mathbf{y}_i, x_i \mid \bar{\mathbf{\Phi}}) \\ &= \sum_{i=1}^{N} \sum_{x_i} \frac{p(\mathbf{y}_i, x_i \mid \mathbf{\Phi})}{p(\mathbf{y}_i \mid \mathbf{\Phi})} \log p(\mathbf{y}_i, x_i \mid \bar{\mathbf{\Phi}}) \end{aligned} \tag{4.101}$$

By replacing items in Eq. (4.101) with Eqs. (4.98) and (4.100), the following equation can be obtained:

$$Q(\mathbf{\Phi}, \bar{\mathbf{\Phi}}) = \sum_{k=1}^{K} \gamma_k \log \bar{c}_k + \sum_{k=1}^{K} Q_\lambda(\mathbf{\Phi}, \bar{\mathbf{\Phi}}_k) \tag{4.102}$$

where

$$\gamma_k^i = \frac{c_k p_k(\mathbf{y}_i \mid \mathbf{\Phi}_k)}{P(\mathbf{y}_i \mid \mathbf{\Phi})} \tag{4.103}$$

$$\gamma_k = \sum_{i=1}^{N} \gamma_k^i = \sum_{i=1}^{N} \frac{c_k p_k(\mathbf{y}_i \mid \mathbf{\Phi}_k)}{P(\mathbf{y}_i \mid \mathbf{\Phi})} \tag{4.104}$$

$$Q_\lambda(\Phi, \bar{\mathbf{\Phi}}_k) = \sum_{i=1}^{N} \gamma_k^i \log p_k(\mathbf{y}_i \mid \bar{\mathbf{\Phi}}_k) = \sum_{i=1}^{N} \frac{c_k p_k(\mathbf{y}_i \mid \mathbf{\Phi}_k)}{P(\mathbf{y}_i \mid \mathbf{\Phi})} \log p_k(\mathbf{y}_i \mid \bar{\mathbf{\Phi}}_k) \tag{4.105}$$

Now we can perform a maximum likelihood estimation on the complete data (x, $\mathbf{y}$) during the M-step. By taking the derivative with respect to each parameter and setting it to zero, we obtain the following EM re-estimate of $c_k, \boldsymbol{\mu}_k$, and $\boldsymbol{\Sigma}_k$:

$$\hat{c}_k = \frac{\gamma_k}{\sum_{k=1}^{K} \gamma_k} = \frac{\gamma_k}{N} \tag{4.106}$$

$$\hat{\boldsymbol{\mu}}_k = \frac{\sum_{i=1}^{N} \gamma_k^i \mathbf{y}_i}{\sum_{i=1}^{N} \gamma_k^i} = \frac{\sum_{i=1}^{N} \frac{c_k p_k(\mathbf{y}_i \mid \boldsymbol{\Phi}_k)\mathbf{y}_i}{P(\mathbf{y}_i \mid \boldsymbol{\Phi})}}{\sum_{i=1}^{N} \frac{c_k p_k(\mathbf{y}_i \mid \boldsymbol{\Phi}_k)}{P(\mathbf{y}_i \mid \boldsymbol{\Phi})}} \tag{4.107}$$

$$\hat{\Sigma}_k = \frac{\sum_{i=1}^{N} \gamma_k^i (\mathbf{y}_i - \boldsymbol{\mu}_k)(\mathbf{y}_i - \boldsymbol{\mu}_k)^t}{\sum_{i=1}^{N} \gamma_k^i} = \frac{\sum_{i=1}^{N} \frac{c_k p_k(\mathbf{y}_i \mid \boldsymbol{\Phi}_k)(\mathbf{y}_i - \boldsymbol{\mu}_k)(\mathbf{y}_i - \boldsymbol{\mu}_k)^t}{P(\mathbf{y}_i \mid \boldsymbol{\Phi})}}{\sum_{i=1}^{N} \frac{c_k p_k(\mathbf{y}_i \mid \boldsymbol{\Phi}_k)}{P(\mathbf{y}_i \mid \boldsymbol{\Phi})}} \tag{4.108}$$

The quantity γ_k^i defined in Eq. (4.103) can be interpreted as the posterior probability that the observed data $\mathbf{y}_i$ belong to Gaussian component k ($N_k(\mathbf{y} \mid \boldsymbol{\mu}_k, \Sigma_k)$). This information as to whether the observed data $\mathbf{y}_i$ should belong to Gaussian component k is hidden and can only be observed through the hidden variable x (c_k). The EM algorithm described above is used to uncover how likely the observed data $\mathbf{y}_i$ are expected to be in each Gaussian component. The re-estimation formulas are consistent with our intuition. These MLE formulas calculate the weighted contribution of each data sample according to the mixture posterior probability γ_k^i.

In fact, VQ is an approximate version of EM algorithms. A traditional VQ with the Mahalanobis distance measure is equivalent to a mixture Gaussian VQ with the following conditions:

$$c_k = 1/K \tag{4.109}$$

$$\gamma_k^i = \begin{cases} 1, & \mathbf{y}_i \in C_k \\ 0, & \text{otherwise} \end{cases} \tag{4.110}$$

The difference between VQ and the EM algorithm is that VQ performs a hard assignment of the data sample $\mathbf{y}_i$ to clusters (cells) while the EM algorithm performs a soft assignment of the data sample $\mathbf{y}_i$ to clusters. As discussed in Chapter 8, this difference carries over to the case of the Viterbi algorithm vs. the Baum-Welch algorithm in hidden Markov models.

4.5. CLASSIFICATION AND REGRESSION TREES

Classification and regression trees (CART) [15, 82] have been used in a variety of pattern recognition applications. Binary decision trees, with splitting questions attached to each node, provide an easy representation that interprets and predicts the structure of a set of data. The application of binary decision trees is much like playing the *number-guessing* game, where the examinee tries to deduce the chosen number by asking a series of binary number-comparing questions.

Consider a simple binary decision tree for height classification. Every person's data in the study may consist of several measurements, including race, gender, weight, age, occupation, and so on. The goal of the study is to develop a classification method to assign a person one of the following five height classes: *tall* (T), *medium-tall* (t), *medium* (M), *medium-short*(s) and *short* (S). Figure 4.14 shows an example of such a binary tree structure. With this binary decision tree, one can easily predict the height class for any new person (with all the measured data, but no height information) by traversing the binary trees. Traversing the binary tree is done through answering a series of yes/no questions in the traversed nodes with the measured data. When the answer is *no*, the right branch is traversed next; otherwise, the left branch will be traversed instead. When the path ends at a leaf node, you can use its attached label as the height class for the new person. If you have the average height for each leaf node (computed by averaging the heights from those people who fall in the same leaf node during training), you can actually use the average height in the leaf node to predict the height for the new person.

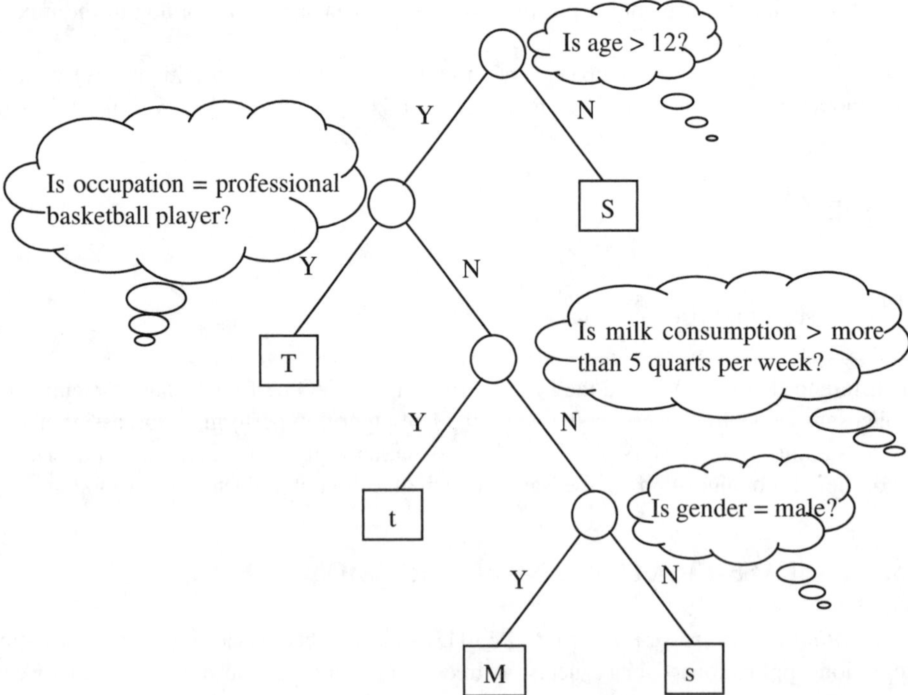

Figure 4.14 A binary tree structure for height classification.

This classification process is similar to a rule-based system where the classification is carried out by a sequence of decision rules. The choice and order of rules applied in a rule-based system is typically designed subjectively by hand through an introspective analysis based on the impressions and intuitions of a limited number of data samples. CART, on the other hand, provides an automatic and data-driven framework to construct the decision process based on objective criteria. Most statistical pattern recognition techniques are designed for data samples having a standard structure with homogeneous variables. CART is designed instead to handle data samples with high dimensionality, mixed data types, and nonstandard data structure. It has the following advantages over other pattern recognition techniques:

- CART can be applied to any data structure through appropriate formulation of the set of potential questions.
- The binary tree structure allows for compact storage, efficient classification, and easily understood interpretation of the predictive structure of the data.
- It often provides, without additional effort, not only classification and recognition, but also an estimate of the misclassification rate for each class.
- It not only handles missing data, but also is very robust to outliers and mislabeled data samples.

To construct a CART from the training samples with their classes (let's denote the set as $\Im$), we first need to find a set of questions regarding the measured variables; e.g., "*Is age > 12*?", "*Is occupation = professional basketball player*?", "*Is gender = male*?" and so on. Once the question set is determined, CART uses a greedy algorithm to generate the decision trees. All training samples $\Im$ are placed in the root of the initial tree. The *best question* is then chosen from the question set to split the root into two nodes. Of course, we need a measurement of how well each question splits the data samples to pick the best question. The algorithm recursively splits the most promising node with the best question until the right-sized tree is obtained. We describe next how to construct the question set, how to measure each split, how to grow the tree, and how to choose the right-sized tree.

4.5.1. Choice of Question Set

Assume that the training data has the following format:

$$\mathbf{x} = (x_1, x_2, \ldots x_d) \tag{4.111}$$

where each variable x_i is a discrete or continuous data type. We can construct a *standard set* of questions Q as follows:

1. Each question is about the value of only a single variable. Questions of this type are called *simple* or *singleton* questions.

2. If x_i is a discrete variable from the set $\{c_1, c_2, \ldots, c_K\}$, Q includes all questions of the following form:

$$\{\text{Is } x_i \in S?\} \tag{4.112}$$

where S is any subset of $\{c_1, c_2, \ldots, c_K\}$

3. If x_i is a continuous variable, Q includes all questions of the following form:

$$\{\text{Is } x_i \leq c?\} \text{ for } c \in (-\infty, \infty) \tag{4.113}$$

The question subset generated from discrete variables (in condition 2 above) is clearly a finite set ($2^{K-1}-1$). On the other hand, the question subset generated from continuous variables (in condition 3 above) seems to be an infinite set based on the definition. Fortunately, since the training data samples are finite, there are only finite number of distinct splits for the training data. For a continuous variable x_i, the data points in $\Im$ contain at most *M* distinct values $v_1, v_2, \ldots, v_M$. There are only at most *M* different splits generated by the set of questions in the form:

$$\{\text{Is } x_i \leq c_n\} \quad n = 1, 2, \ldots, M \tag{4.114}$$

where $c_n = \dfrac{v_{n-1} + v_n}{2}$ and $v_0 = 0$. Therefore, questions related to a continuous variable also form a finite subset. The fact that Q is a finite set allows the enumerating of all possible questions in each node during tree growing.

The construction of a question set is similar to that of rules in a rule-based system. Instead of using the all-possible question set Q, some people use knowledge selectively to pick a subset of Q, which is sensitive to pattern classification. For example, the vowel subset and consonant subset are a natural choice for these sensitive questions for phoneme classification. However, the beauty of CART is the ability to use all possible questions related to the measured variables, because CART has a statistical data-driven framework to determine the decision process (as described in subsequent sections). Instead of setting some constraints on the questions (splits), most CART systems use all the possible questions for Q.

4.5.2. Splitting Criteria

A question in CART framework represents a split (partition) of data samples. All the leaf nodes (L in total) represent L disjoint subsets $A_1, A_2, \ldots, A_L$. Now we have the entire potential question set Q, the task is how to find the best question for a node split. The selection of the best question is equivalent to finding the best split for the data samples of the node.

Since each node t in the tree contains some training samples, we can compute the corresponding class probability density function $P(\omega \mid t)$. The classification process for the node can then be interpreted as a random process based on $P(\omega \mid t)$. Since our goal is classification, the objective of a decision tree is to reduce the uncertainty of the event being decided upon. We want the leaf nodes to be as pure as possible in terms of the class distribution. Let Y be the random variable of classification decision for data sample $\mathbf{X}$. We could define the weighted entropy for any node t as follows:

$$\bar{H}_t(Y) = H_t(Y)P(t) \tag{4.115}$$

$$H_t(Y) = -\sum_i P(\omega_i \mid t)\log P(\omega_i \mid t) \tag{4.116}$$

where $P(\omega_i \mid t)$ is the percentage of data samples for class *i* in node *t*; and $P(t)$ is the prior probability of visiting node *t* (equivalent to the ratio of number of data samples in node *t* and the total number of training data samples). With this weighted entropy definition, the splitting criterion is equivalent to finding the question which gives the greatest entropy reduction, where the entropy reduction for a question *q* to split a node *t* into nodes *l* and *r* can be defined as:

$$\Delta\bar{H}_t(q) = \bar{H}_t(Y) - (\bar{H}_l(Y) + \bar{H}_r(Y)) = \bar{H}_t(Y) - \bar{H}_t(Y \mid q) \tag{4.117}$$

The reduction in entropy is also the mutual information between Y and question q. The task becomes that of evaluating the entropy reduction $\Delta\bar{H}_q$ for each potential question (split), and picking the question with the greatest entropy reduction, that is,

$$q^* = \underset{q}{\operatorname{argmax}}\,(\Delta\bar{H}_t(q)) \tag{4.118}$$

If we define the entropy for a tree, T, as the sum of weighted entropies for all the terminal nodes, we have:

$$\bar{H}(T) = \sum_{t \text{ is terminal}} \bar{H}_t(Y) \tag{4.119}$$

It can be shown that the tree-growing (splitting) process repeatedly reduces the entropy of the tree. The resulting tree thus has a better classification power. For continuous pdf, likelihood gain is often used instead, since there is no straightforward entropy measurement [43]. Suppose one specific split divides the data into two groups, $\mathbf{X}_1$ and $\mathbf{X}_2$, which can then be used to train two Gaussian distributions $N_1(\boldsymbol{\mu}_1, \Sigma_1)$ and $N_2(\boldsymbol{\mu}_2, \Sigma_2)$. The log-likelihoods for generating these two data groups are:

$$L_1(\mathbf{X}_1 \mid N_1) = \log\prod_{\mathbf{x}_1} N(\mathbf{x}_1, \boldsymbol{\mu}_1, \Sigma_1) = -\left(d\log 2\pi + \log\left|\Sigma_1\right| + d\right)a/2 \tag{4.120}$$

$$L_2(\mathbf{X}_2 \mid N_2) = \log\prod_{\mathbf{x}_2} N(\mathbf{x}_2, \text{¼}_2, \Sigma_2) = -\left(d\log 2\pi + \log\left|\Sigma_2\right| + d\right)b/2 \tag{4.121}$$

where d is the dimensionality of the data; and a and b are the sample counts for the data groups $\mathbf{X}_1$ and $\mathbf{X}_2$ respectively. Now if the data $\mathbf{X}_1$ and $\mathbf{X}_2$ are merged into one group and modeled by one Gaussian $N(\boldsymbol{\mu},\Sigma)$, according to MLE, we have

$$\boldsymbol{\mu} = \frac{a}{a+b}\boldsymbol{\mu}_1 + \frac{b}{a+b}\boldsymbol{\mu}_2 \tag{4.122}$$

$$\Sigma = \frac{a}{a+b}\left[\Sigma_1 + (\boldsymbol{\mu}_1 - \boldsymbol{\mu})(\boldsymbol{\mu}_1 - \boldsymbol{\mu})^t\right] + \frac{b}{a+b}\left[\Sigma_2 + (\boldsymbol{\mu}_2 - \boldsymbol{\mu})(\boldsymbol{\mu}_2 - \boldsymbol{\mu})^t\right] \tag{4.123}$$

Thus, the likelihood gain of splitting the data $\mathbf{X}$ into two groups $\mathbf{X}_1$ and $\mathbf{X}_2$ is:

$$\begin{aligned}\Delta \overline{L}_t(q) &= L_1(\mathbf{X}_1 \mid N) + L_2(\mathbf{X}_2 \mid N) - L_{\mathbf{x}}(\mathbf{X} \mid N) \\ &= (a+b)\log|\Sigma| - a\log|\Sigma_1| - b\log|\Sigma_2|\end{aligned} \tag{4.124}$$

For regression purposes, the most popular splitting criterion is the mean squared error measure, which is consistent with the common *least squared* regression methods. For instance, suppose we need to investigate the real height as a regression function of the measured variables in the height study. Instead of finding height classification, we could simply use the average height in each node to predict the height for any data sample. Suppose Y is the actual height for training data $\mathbf{X}$, then overall regression (prediction) error for a node t can be defined as:

$$E(t) = \sum_{\mathbf{X}\in t} | Y - d(\mathbf{X}) |^2 \tag{4.125}$$

where $d(\mathbf{X})$ is the regression (predictive) value of Y.

Now, instead of finding the question with greatest entropy reduction, we want to find the question with largest squared error reduction. That is, we want to pick the question q^* that maximizes:

$$\Delta E_t(q) = E(t) - (E(l) + E(r)) \tag{4.126}$$

where l and r are the leaves of node t. We define the expected square error $V(t)$ for a node t as the overall regression error divided by the total number of data samples in the node.

$$V(t) = E\left(\sum_{\mathbf{X}\in t} | Y - d(\mathbf{X}) |^2\right) = \frac{1}{N(t)}\sum_{\mathbf{X}\in t} | Y - d(\mathbf{X}) |^2 \tag{4.127}$$

Note that $V(t)$ is actually the variance estimate of the height, if $d(\mathbf{X})$ is made to be the average height of data samples in the node. With $V(t)$, we define the weighted squared error $\overline{V}(t)$ for a node t as follows.

$$\overline{V}(t) = V(t)P(t) = \left(\frac{1}{N(t)}\sum_{\mathbf{X}\in t} | Y - d(\mathbf{X}) |^2\right)P(t) \tag{4.128}$$

Finally, the splitting criterion can be rewritten as:

$$\Delta\overline{V}_t(q) = \overline{V}(t) - (\overline{V}(l) + \overline{V}(r)) \tag{4.129}$$

Based on Eqs. (4.117) and (4.129), one can see the analogy between entropy and variance in the splitting criteria for CART. The use of entropy or variance as splitting criteria is under the assumption of uniform misclassification costs and uniform prior distributions. When nonuniform misclassification costs and prior distributions are used, some other splitting might be used for splitting criteria. Noteworthy ones are *Gini index of diversity* and *twoing rule*. Those interested in alternative splitting criteria can refer to [11, 15].

For a wide range of splitting criteria, the properties of the resulting CARTs are empirically insensitive to these choices. Instead, the criterion used to get the right-sized tree is much more important. We discuss this issue in Section 4.5.6.

4.5.3. Growing the Tree

Given the question set Q and splitting criteria $\Delta\overline{H}_t(q)$, the tree-growing algorithm starts from the initial root-only tree. At each node of tree, the algorithm searches through the variables one by one, from x_1 to x_N. For each variable, it uses the splitting criteria to find the best question (split). Then it can pick the best question out of the N best single-variable questions. The procedure can continue splitting each node until either of the following conditions is met for a node:

1. No more splits are possible; that is, all the data samples in the node belong to the same class;
2. The greatest entropy reduction of best question (split) falls below a pre-set threshold β, i.e.:

$$\max_{q \in Q} \Delta\overline{H}_t(q) < \beta \tag{4.130}$$

3. The number of data samples falling in the leaf node t is below some threshold α. This is to assure that there are enough training tokens for each leaf node if one needs to estimate some parameters associated with the node.

When a node cannot be further split, it is declared a terminal node. When all active (nonsplit) nodes are terminal, the tree-growing algorithm stops.

The algorithm is greedy because the question selected for any given node is the one that seems to be the best, without regard to subsequent splits and nodes. Thus, the algorithm constructs a tree that is locally optimal, but not necessarily globally optimal (but hopefully globally *good enough*). This tree-growing algorithm has been successfully applied in many applications [5, 39, 60]. A dynamic programming algorithm for determining global optimality is described in [78]; however, it is suitable only in restricted applications with relatively few variables.

4.5.4. Missing Values and Conflict Resolution

Sometimes, the available data sample $\mathbf{x} = (x_1, x_2, \ldots x_d)$ has some value x_j missing. This missing-value case can be handled by the use of *surrogate questions* (*splits*). The idea is intuitive. We define a similarity measurement between any two questions (splits) q and $\tilde{q}$ of a node t. If the best question of node t is the question q on the variable x_i, we can find the question $\tilde{q}$ that is most similar to q on a variable other than x_i. $\tilde{q}$ is our best surrogate question. Similarly, we find the second-best surrogate question, third-best and so on. The surrogate questions are considered as the backup questions in the case of missing x_i values in the data samples. The surrogate question is used in descending order to continue tree traversing for those data samples. The surrogate question gives CART unique ability to handle the case of missing data. The similarity measurement is basically a measurement reflecting the similarity of the class probability density function [15].

When choosing the best question for splitting a node, several questions on the same variable x_i may achieve the same entropy reduction and generate the same partition. As in rule-based problem solving systems, a *conflict resolution* procedure [99] is needed to decide which question to use. For example, discrete questions q_1 and q_2 have the following format:

$$q_1 \; : \; \{\text{Is } x_i \in S_1 \, ?\} \tag{4.131}$$

$$q_2 \; : \; \{\text{Is } x_i \in S_2 \, ?\} \tag{4.132}$$

Suppose S_1 is a subset of S_2, and one particular node contains only data samples whose x_i value contains only values in S_1, but no other. Now question q_1 or q_2 performs the same splitting pattern and therefore achieves exactly the same amount of entropy reduction. In this case, we call q_1 a sub-question of question q_2, because q_1 is a more specific version.

A *specificity ordering* conflict resolution strategy is used to favor the discrete question with fewer elements because it is more specific to the current node. In other words, if the elements of a question are a subset of the elements of another question with the same entropy reduction, the question with the subset of elements is preferred. Preferring more specific questions will prevent decision trees from over-generalizing. The specificity ordering conflict resolution can be implemented easily by presorting the set of discrete questions by the number of elements they contain in descending order, before applying them to decision trees. A similar specificity ordering conflict resolution can also be implemented for continuous-variable questions.

4.5.5. Complex Questions

One problem with allowing only simple questions is that the data may be over-fragmented, resulting in similar leaves in different locations of the tree. For example, when the best question (rule) to split a node is actually a composite question of the form "*Is* $x_i \in S_1$?" or "*Is* $x_i \in S_2$?", a system allowing only simple questions will generate two separate questions to split the data into three clusters rather than two as shown in Figure 4.15. Also data for which the answer is *yes* are inevitably fragmented across two shaded nodes. This is inefficient and ineffective since these two very similar data clusters may now both contain insufficient training examples, which could potentially handicap future tree growing. Splitting data unnecessarily across different nodes leads to unnecessary computation, redundant clusters, reduced trainability, and less accurate entropy reduction.

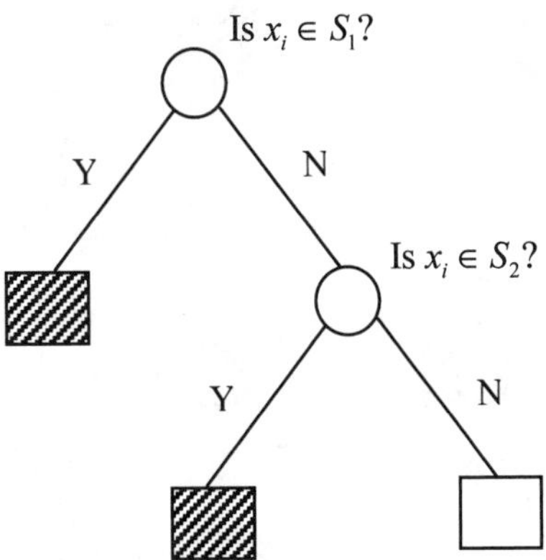

Figure 4.15 An over-split tree for the question "*Is* $x_i \in S_1$?" or "*Is* $x_i \in S_2$?"

We deal with this problem by using a composite-question construction [38, 40]. It involves conjunctive and disjunctive combinations of all questions (and their negations). A composite question is formed by first growing a tree with simple questions only and then clustering the leaves into two sets. Figure 4.16 shows the formation of one composite question. After merging, the structure is still a binary question. To construct the composite question, multiple OR operators are used to describe the composite condition leading to either one of the final clusters, and AND operators are used to describe the relation within a particular route. Finally, a Boolean reduction algorithm is used to simplify the Boolean expression of the composite question.

To speed up the process of constructing composite questions, we constrain the number of leaves or the depth of the binary tree through heuristics. The most frequently used heuristics is the limitation of the depth when searching a composite question. Since composite questions are essentially binary questions, we use the same greedy tree-growing algorithm to find the best composite question for each node and keep growing the tree until the stop criterion is met. The use of composite questions not only enables flexible clustering, but also improves entropy reduction. Growing the sub-tree a little deeper before constructing the composite question may achieve longer-range optimum, which is preferable to the local optimum achieved in the original greedy algorithm that used simple questions only.

The construction of composite questions can also be applied to continuous variables to obtained complex rectangular partitions. However, some other techniques are used to obtain general partitions generated by hyperplanes not perpendicular to the coordinate axes. Questions typically have a linear combination of continuous variables in the following form [15]:

$$\{\text{Is } \sum_i a_i x_i \le c?\} \tag{4.133}$$

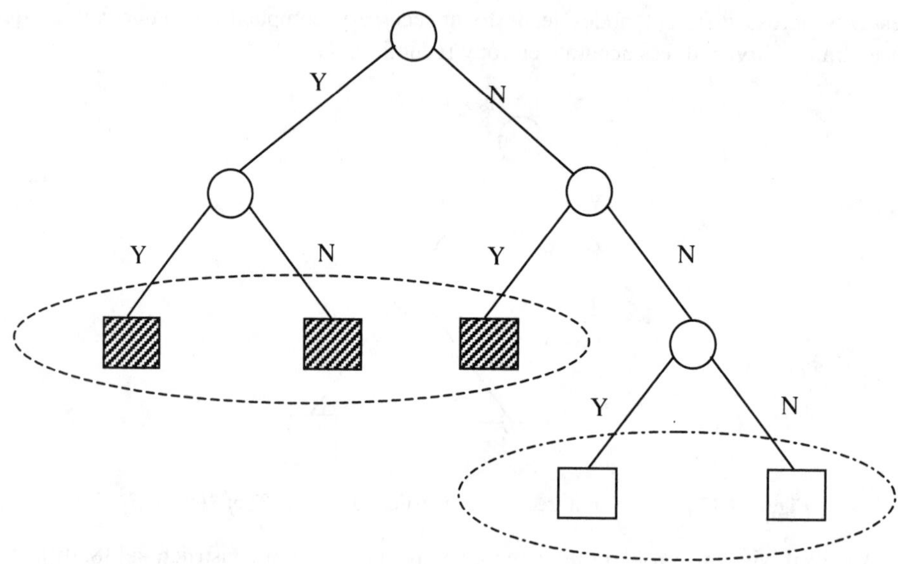

Figure 4.16 The formation of a composite question from simple questions.

4.5.6. The Right-Sized Tree

One of the most critical problems for CART is that the tree may be strictly tailored to the training data and has no generalization capability. When you split a leaf node in the tree to get entropy reduction until each leaf node contains data from one single class, that tree possesses a zero percent classification error on the training set. This is an over-optimistic estimate of the test-set misclassification rate. Independent test sample estimation or cross-validation is often used to prevent decision trees from over-modeling idiosyncrasies of the training data. To get a right-sized tree, you can minimize the misclassification rate for future independent test data.

Before we describe the solution for finding the right sized tree, let's define a couple of useful terms. Naturally we will use the plurality rule $\delta(t)$ to choose the class for a node t:

$$\delta(t) = \underset{i}{\operatorname{argmax}}\, P(\omega_i \mid t) \tag{4.134}$$

Similar to the notation used in Bayes' decision theory, we can define the misclassification rate $R(t)$ for a node t as:

$$R(t) = r(t)P(t) \tag{4.135}$$

where $r(t) = 1 - \max_i P(\omega_i \mid t)$ and $P(t)$ is the frequency (probability) of the data falling in node t. The overall misclassification rate for the whole tree T is defined as:

$$R(T) = \sum_{t \in \tilde{T}} R(t) \tag{4.136}$$

where $\tilde{T}$ represents the set of terminal nodes. If a nonuniform misclassification cost $c(i \mid j)$, the cost of misclassifying class j data as class i data, is used, $r(t)$ is redefined as:

$$r(t) = \min_i \sum_j c(i \mid j) P(j \mid t) \tag{4.137}$$

As we mentioned, $R(T)$ can be made arbitrarily small (eventually reduced to zero) for the training data if we keep growing the tree. The key now is how we choose the tree that can minimize $R^*(T)$, which is denoted as the misclassification rate of independent test data. Almost no tree initially grown can perform well on independent test data. In fact, using more complicated stopping rules to limit the tree growing seldom works, and it is either stopped too soon at some terminal nodes, or continued too far in other parts of the tree. Instead of inventing some clever stopping criteria to stop the tree growing at the right size, we let the tree over-grow (based on rules in Section 4.5.3). We use a pruning strategy to gradually cut back the tree until the minimum $R^*(T)$ is achieved. In the next section we describe an algorithm to prune an over-grown tree, *minimum cost-complexity pruning*.

4.5.6.1. Minimum Cost-Complexity Pruning

To prune a tree, we need to find a subtree (or branch) that makes the least impact in terms of a cost measure, whether it is pruned or not. This candidate to be pruned is called the *weakest subtree*. To define such a weakest subtree, we first need to define the cost measure.

DEFINITION 1: For any sub-tree T of $T_{\max}$ ($T \prec T_{\max}$), let $|\tilde{T}|$ denote the number of terminal nodes in tree T.

DEFINITION 2: Let $\alpha \geq 0$ be a real number called the *complexity parameter*. The cost-complexity measure can be defined as:

$$R_\alpha(T) = R(T) + \alpha |\tilde{T}| \tag{4.138}$$

DEFINITION 3: For each α, define the minimal cost-complexity subtree $T(\alpha) \prec T_{\max}$ that minimizes $R_\alpha(T)$, that is,

$$T(\alpha) = \arg\min_{T \prec T_{\max}} R_\alpha(T) \tag{4.139}$$

Based on Definitions 2 and 3, if α is small, the penalty for having a large tree is small and $T(\alpha)$ will be large. In fact, $T(0) = T_{\max}$ because $T_{\max}$ has a zero misclassification rate, so it will minimize $R_o(T)$. On the other hand, when α increases, $T(\alpha)$ becomes smaller and smaller. For a sufficient large α, $T(\alpha)$ may collapse into a tree with only the root. The increase of α produces a sequence of pruned trees and it is the basis of the pruning process. The pruning algorithm rests on two theorems. The first is given as follows.

THEOREM 1: For every value of α, there exists a unique minimal cost-complexity subtree $T(\alpha)$ as defined in Definition 3.[12]

To progressively prune the tree, we need to find the weakest subtree (node). The idea of a weakest subtree is the following: *if we collapse the weakest subtree into a single terminal node, the cost-complexity measure would increase least.* For any node t in the tree T, let $\{t\}$ denote the subtree containing only the node t, and T_t denote the branch starting at node t. Then we have

$$R_\alpha(T_t) = R(T_t) + \alpha |\tilde{T}_t| \tag{4.140}$$

$$R_\alpha(\{t\}) = R(t) + \alpha \tag{4.141}$$

When α is small, T_t has a smaller cost-complexity than the single-node tree $\{t\}$. However, when α increases to a point where the cost-complexity measures for T_t and $\{t\}$ are the same, it makes sense to collapse T_t into a single terminal node $\{t\}$. Therefore, we decide the critical value for α by solving the following inequality:

$$R_\alpha(T_t) \le R_\alpha(\{t\}) \tag{4.142}$$

We obtain:

$$\alpha \le \frac{R(t) - R(T_t)}{|\tilde{T}_t| - 1} \tag{4.143}$$

Based on Eq. (4.143), we define a measurement $\eta(t)$ for each node t in tree T:

$$\eta(t) = \begin{cases} \dfrac{R(t) - R(T_t)}{|\tilde{T}_t| - 1}, & t \notin \tilde{T} \\ +\infty, & t \in \tilde{T} \end{cases} \tag{4.144}$$

Based on measurement $\eta(t)$, we then define the weakest subtree T_{t_1} as the tree branch starting at the node t_1 such that

[12] You can find the proof to this in [15].

$$t_1 = \arg\min_{t \in T} \eta(t) \tag{4.145}$$

$$\alpha_1 = \eta(t_1) \tag{4.146}$$

As α increases, the node t_1 is the first node such that $R_\alpha(\{t\})$ becomes equal to $R_\alpha(T_t)$. At this point, it would make sense to prune subtree T_{t_1} (collapse T_{t_1} into a single-node subtree $\{t_1\}$), and α_1 is the value of α where the pruning occurs.

Now the tree T after pruning is referred to as T_1, i.e.,

$$T_1 = T - T_{t_1} \tag{4.147}$$

We then use the same process to find the weakest subtree T_{t_2} in T_1 and the new pruning point α_2. After pruning away T_{t_2} from T_1 to form the new pruned tree T_2, we repeat the same process to find the next weakest subtree and pruning point. If we continue the process, we get a sequence of decreasing pruned trees:

$$T \succ T_1 \succ T_2 \succ T_2 \cdots \succ \{r\} \tag{4.148}$$

where $\{r\}$ is the single-node tree containing the root of tree T with corresponding pruning points:

$$\alpha_0 < \alpha_1 < \alpha_2 < \alpha_3 < \cdots \tag{4.149}$$

where $\alpha_0 = 0$.

With the process above, the following theorem (which is basic for the minimum cost-complexity pruning) can be proved.

THEOREM 2 : Let T_0 be the original tree T.

$$\text{For } k \geq 0,\ \alpha_k \leq \alpha < \alpha_{k+1},\ T(\alpha) = T(\alpha_k) = T_k \tag{4.150}$$

4.5.6.2. Independent Test Sample Estimation

The minimum cost-complexity pruning algorithm can progressively prune the over-grown tree to form a decreasing sequence of subtrees $T \succ T_1 \succ T_2 \succ T_2 \cdots \succ \{r\}$, where $T_k = T(\alpha_k)$, $\alpha_0 = 0$ and $T_0 = T$. The task now is simply to choose one of those subtrees as the optimal-sized tree. Our goal is to find the optimal-sized tree that minimizes the misclassification for independent test set $R^*(T)$. When the training set $\Im$ is abundant, we can afford to set aside an independent test set $\Re$ from the training set. Usually $\Re$ is selected as one third of the training set $\Im$. We use the remaining two thirds of the training set $\Im - \Re$ (still abundant) to train the initial tree T and apply the minimum cost-complexity pruning algorithm to attain the decreasing sequence of subtrees $T \succ T_1 \succ T_2 \succ T_2 \cdots \succ \{r\}$. Next, the test set $\Re$ is run through the sequence of subtrees to get corresponding estimates of test-set misclassification $R^*(T), R^*(T_1), R^*(T_2), \cdots, R^*(\{r\})$. The optimal-sized tree T_{k^*} is then picked as the one with minimum test-set misclassification measure, i.e.:

$$k^* = \arg\min_k R^*(T_k) \tag{4.151}$$

The *independent test sample estimation* approach has the drawback that it reduces the effective training sample size. This is why it is used only when there is abundant training data. Under most circumstances where training data is limited, *cross-validation* is often used.

4.5.6.3. Cross-Validation

CART can be pruned via *v-fold cross-validation*. It follows the same principle of cross validation described in Section 4.2.3. First it randomly divides the training set $\Im$ into v disjoint subsets $\Im_1, \Im_2, \cdots, \Im_v$, each containing roughly the same data samples. It then defines the ith training set

$$\Im^i = \Im - \Im_i \quad i = 1, 2, \ldots, v \tag{4.152}$$

so that $\Im^i$ contains the fraction $(v-1)/v$ of the original training set. v is usually chosen to be large, like 10.

In v-fold cross-validation, v auxiliary trees are grown together with the main tree T grown on $\Im$. The ith tree is grown on the ith training set $\Im^i$. By applying minimum cost-complexity pruning, for any given value of the cost-complexity parameter α, we can obtain the corresponding minimum cost-complexity subtrees $T(\alpha)$ and $T^i(\alpha)$, $i = 1, 2, \ldots, v$. According to Theorem 2 in Section 4.5.6.1, those minimum cost-complexity subtrees will form $v+1$ sequences of subtrees:

$$T \succ T_1 \succ T_2 \succ T_2 \cdots \succ \{r\} \text{ and} \tag{4.153}$$

$$T^i \succ T_1^i \succ T_2^i \succ T_3^i \cdots \succ \{r^i\} \quad i = 1, 2, \ldots, v \tag{4.154}$$

ALGORITHM 4.5: *THE CART ALGORITHM*

Step 1: Question Set: Create a standard set of questions Q that consists of all possible questions about the measured variables.
Step 2: Splitting Criterion: Pick a splitting criterion that can evaluate all the possible questions in any node. Usually it is either entropy-like measurement for classification trees or mean square errors for regression trees.
Step 3: Initialization: Create a tree with one (root) node, consisting of all training samples.
Step 4: Split Candidates: Find the best composite question for each terminal node:
- a. Generate a tree with several simple-question splits as described in Section 4.5.3.
- b. Cluster leaf nodes into two classes according to the same splitting criterion.
- c. Based on the clustering done in (b), construct a corresponding composite question.

Step 5: Split: Out of all the split candidates in Step 4, split the one with best criterion.
Step 6: Stop Criterion: If all the leaf nodes containing data samples from the same class or all the potential splits generating improvement fall below a pre-set threshold β, go to Step 7; otherwise go to Step 4.
Step 7: Use *independent test sample estimate* or *cross-validation estimate* to prune the original tree into the optimal size.

The basic assumption of cross-validation is that the procedure is *stable* if v is large. That is, $T(\alpha)$ should have the same classification accuracy as $T^i(\alpha)$. Although we cannot directly estimate the test-set misclassification for the main tree $R^*(T(\alpha))$, we could approximate it via the test-set misclassification measure $R^*(T^i(\alpha))$, since each data sample in $\Im$ occurs in one and only one test set $\Im_i$. The v-fold cross-validation estimate $R^{CV}(T(\alpha))$ can be computed as:

$$R^{CV}(T(\alpha)) = \frac{1}{v}\sum_{i=1}^{v} R^*(T^i(\alpha)) \tag{4.155}$$

Similar to Eq. (4.151), once $R^{CV}(T(\alpha))$ is computed, the optimal v-fold cross-validation tree $T_{k^{CV}}$ can be found through

$$k^{CV} = \arg\min_k R^{CV}(T_k) \tag{4.156}$$

Cross-validation is computationally expensive in comparison with independent test sample estimation, though it makes more effective use of all training data and reveals useful information regarding the stability of the tree structure. Since the auxiliary trees are grown on a smaller training set (a fraction $v-1/v$ of the original training data), they tend to have a higher misclassification rate. Therefore, the cross-validation estimates $R^{CV}(T)$ tend to be an over-estimation of the misclassification rate. The algorithm for generating a CART tree is illustrated in Algorithm 4.5.

4.6. Historical Perspective and Further Reading

Pattern recognition is a multidisciplinary field that comprises a broad body of loosely related knowledge and techniques. Historically, there are two major approaches to pattern recognition – the statistical and the syntactical approaches. Although this chapter is focused on the statistical approach, syntactical pattern recognition techniques, which aim to address the limitations of the statistical approach in handling contextual or structural information, can be complementary to statistical approaches for spoken language processing, such as parsing. Syntactic pattern recognition is based on the analogy that complex patterns can be decomposed recursively into simpler subpatterns, much as a sentence can be decomposed into words and letters. Fu [24] provides an excellent book on syntactic pattern recognition.

The foundation of statistical pattern recognition is Bayesian theory, which can be traced back to the 18th century [9, 54] and its invention by the British mathematician Thomas Bayes (1702-1761). Chow [20] was the first to use Bayesian decision theory for pattern recognition. Statistical pattern recognition has been used successfully in a wide range of applications, from optical/handwritten recognition [13, 96], to speech recognition [7, 86] and to medical/machinery diagnosis [1, 27]. The books by Duda et al. [22] and Fukunaga [25] are two classic treatments of statistical pattern recognition. Duda et al. have a second edition of the classic pattern recognition book [23] that includes many up-to-date topics.

MLE and MAP are two most frequently used estimation methods for pattern recognition because of their simplicity and efficiency. In Chapters 8 and 9, they play an essential role in model parameter estimation. Estimating the recognition performance and comparing different recognition systems are important subjects in pattern recognition. The importance of a large number of test samples was reported in [49]. McNemar's test is dated back to the 1940s [66]. The modification of the test for continuous speech recognition systems presented in this chapter is based on an interesting paper [30] that contains a general discussion on using hypothesis-testing methods for continuous speech recognition.

Gradient descent is fundamental for most discriminant estimation methods, including MMIE, MCE, and neural networks. The history of gradient descent can be traced back to Newton's method for root finding [72, 81]. Both the book by Duda et al. [23] and the paper by Juang et al. [48] provide a good description of gradient descent. MMIE was first proposed in [16, 71] for the speech recognition problem. According to these two works, MMIE is more robust than MLE to incorrect model assumptions. MCE was first formulated by Juang et al. [48] and successfully applied to small-vocabulary speech recognition [47].

The modern era of neural networks was brought to the scientific community by McCulloch and Pitts. In the pioneering paper [64], McCulloch and Pitts laid out the mathematical treatment of the behavior of networks of simple neurons. The most important result they showed is that a network would compute any computable function. John von Neumann was influenced by this paper to use switch-delay elements derived from the McCulloch-Pitts neuron in the construction of the EDVAC (Electronic Discrete Variable Automatic Computer) that was developed based on ENIAC (Electronic Numerical Integrator and Computer) [2, 35]. The ENIAC was the famous first general-purpose electronic computer built at the

Moore School of Electrical Engineering at the University of Pennsylvania from 1943 to 1946 [31]. The two-layer perceptron work [87] by Rosenblatt, was the first to provide rigorous proofs about perceptron convergence. A 1969 book by Minsky and Papert [68] reveals that there are fundamental limits to what single-layer perceptrons can compute. It was not until the 1980s that the discovery of multi-layer perceptrons (with hidden layers and nonlinear threshold functions) and back-propagation [88] reawakened interest in neural networks. The two-volume PDP book [90, 91], *Parallel Distributed Processing: Explorations in the Microstructures of Cognition*, edited by Rummelhart and McClelland, brought the back-propagation learning method to the attention of the widest audience. Since then, various applications of neural networks in diverse domains have been developed, including speech recognition [14, 58], speech production and perception [93, 94], optical and handwriting character recognition [55, 92], visual recognition [26], game playing [97], and natural language processing [63]. There are several good textbooks for neural networks. In particular, the book by Haykin [35] provides a very comprehensive coverage of all foundations of neural networks. Bishop [12] provides a thoughtful treatment of neural networks from the perspective of pattern recognition. Short, concise tutorial papers on neural networks can be found in [44, 57].

Vector quantization originated from speech coding [17, 32, 45, 61]. The *k*-means algorithm was introduced by Lloyd [59]. Over the years, there have been many variations of VQ, including fuzzy VQ [10], learning VQ (LVQ) [51], and supervised VQ [18, 42]. The first published investigation toward the EM-like algorithm for incomplete data learning can be attributed to Pearson [79]. The modern EM algorithm was formalized by Dempster, Laird, and Rubin [21]. McLachlan and Krishnan [65] provide a thorough overview and history of the EM algorithm. The convergence of the EM algorithm is an interesting research topic and Wu [100] has an extensive description of the rate of convergence. The EM algorithm is the basis for all unsupervised learning that includes hidden variables. The famous HMM training algorithm, as described in Chapter 8, is based on the EM algorithm.

CART uses a very intuitive and natural principle of sequential questions and answers, which can be traced back to 1960s [70]. The popularity of CART is attributed to the book by Breiman *et al.* [15]. Quinlan proposed some interesting variants of CART, like ID3 [82] and C4.5 [84]. CART has recently been one of the most popular techniques in machine learning. Mitchell includes a good overview chapter on the latest CART techniques in his machine-learning book [69]. In addition to the strategies of node splitting and pruning mentioned in this chapter, [62] used a very interesting approach for splitting and pruning criteria based on a statistical significance testing of the data's distributions. Moreover, [28] proposed an iterative expansion pruning algorithm which is believed to perform as well as cross-validation pruning and yet is computationally cheaper [52]. CART has been successfully used in a variety of spoken language applications such as letter-to-sound conversion [46, 60], allophone model clustering [8, 38, 39], language models [5], automatic rule generation [83], duration modeling of phonemes [74, 80], and supervised vector quantization [67].

REFERENCES

[1] Albert, A. and E.K. Harris, *Multivariate Interpretation of Clinical Laboratory Data*, 1987, New York, Marcel Dekker.

[2] Aspray, W. and A. Burks, "Papers of John von Neumann on Computing and Computer Theory" in *Charles Babbage Institute Reprint Series for the History of Computing* 1986, Cambridge, MA, MIT Press.

[3] Atal, B.S. and M.R. Schroeder, "Predictive Coding of Speech Signals and Subjective Error Criteria," *IEEE Trans. on Acoustics, Speech and Signal Processing*, 1979, **ASSP-27**(3), pp. 247-254.

[4] Bahl, L.R., *et al.*, "A New Algorithm for the Estimation of Hidden Markov Model Parameters," *Proc. of the IEEE Int. Conf. on Acoustics, Speech and Signal Processing*, 1988, New York, NY, pp. 493-496.

[5] Bahl, L.R., *et al.*, "A Tree-Based Statistical Language Model for Natural Language Speech Recognition," *IEEE Trans. on Acoustics, Speech, and Signal Processing*, 1989, **37**(7), pp. 1001-1008.

[6] Bahl, L.R., *et al.*, "Estimating Hidden Markov Model Parameters so as to Maximize Speech Recognition Accuracy," *IEEE Trans. on Speech and Audio Processing*, 1993, **1**(1), pp. 77-83.

[7] Bahl, L.R., F. Jelinek, and R.L. Mercer, "A Maximum Likelihood Approach to Continuous Speech Recognition," *IEEE Trans. on Pattern Analysis and Machine Intelligence*, 1983, **5**(2), pp. 179-190.

[8] Bahl, L.R., *et al.*, "Decision Trees for Phonological Rules in Continuous Speech" in *Proc. of the IEEE Int. Conf. on Acoustics, Speech and Signal Processing* 1991, Toronto, Canada, pp. 185-188.

[9] Bayes, T., "An Essay Towards Solving a Problem in the Doctrine of Chances," *Philosophical Tansactions of the Royal Society*, 1763, **53**, pp. 370-418.

[10] Bezdek, J., *Pattern Recognition with Fuzzy Objective Function Algorithms*, 1981, New York, NY, Plenum Press.

[11] Bhargava, T.N. and V.R.R. Uppuluri, "Sampling Distribution of Gini's Index of Diversity," *Applied Mathematics and Computation*, 1977, **3**, pp. 1-24.

[12] Bishop, C.M., *Neural Networks for Pattern Recognition*, 1995, Oxford, UK, Oxford University Press.

[13] Blesser, B., *et al.*, "A Theoretical Approach for Character Recognition Based on Phenomenological Attributes," *Int. Journal of Man-Machine Studies*, 1974, **6**(6), pp. 701-714.

[14] Bourlard, H. and N. Morgan, *Connectionist Speech Recognition – A Hybrid Approach*, 1994, Boston, MA, Kluwer Academic Publishers.

[15] Breiman, L., *et al.*, *Classification and Regression Trees*, 1984, Pacific Grove, CA, Wadsworth.

[16] Brown, P.F., *The Acoustic-Modeling Problem in Automatic Speech Recognition*, PhD Thesis in Computer Science Department 1987, Carnegie Mellon University, Pittsburgh, PA.

[17] Buzo, A., *et al.*, "Speech Coding Based upon Vector Quantization," *IEEE Trans. on Acoustics, Speech and Signal Processing*, 1980, **28**(5), pp. 562-574.

[18] Cerf, P.L., W. Ma, and D.V. Compernolle, "Multilayer Perceptrons as Labelers for Hidden Markov Models," *IEEE Trans. on Speech and Audio Processing*, 1994, **2**(1), pp. 185-193.

[19] Chang, P.C. and B.H. Juang, "Discriminative Training of Dynamic Programming Based Speech Recognizers," *IEEE Int. Conf. on Acoustics, Speech and Signal Processing*, 1992, San Fancisco.

[20] Chow, C.K., "An Optimum Character Recognition System Using Decision Functions," *IRE Trans.*, 1957, pp. 247-254.

[21] Dempster, A.P., N.M. Laird, and D.B. Rubin, "Maximum-Likelihood from Incomplete Data via the EM Algorithm," *Journal of Royal Statistical Society ser. B*, 1977, **39**, pp. 1-38.

[22] Duda, R.O. and P.E. Hart, *Pattern Classification and Scene Analysis*, 1973, New York, N.Y., John Wiley and Sons.

[23] Duda, R.O., D.G. Stork, and P.E. Hart, *Pattern Classification and Scene Analysis: Pattern Classification*, 2nd ed, 1999, John Wiley & Sons.

[24] Fu, K.S., *Syntactic Pattern Recognition and Applications*, 1982, Englewood Cliffs, NJ, Prentice Hall.

[25] Fukunaga, K., *Introduction to Statistical Pattern Recognition*, 2nd ed., 1990, Orlando, FL, Academic Press.

[26] Fukushima, K., S. Miykake, and I. Takayuki, "Neocognition: A Neural Network Model for a Mechanism of Visual Pattern Recognition," *IEEE Trans. on Systems, Man and Cybernetics*, 1983, **SMC-13**(5), pp. 826-834.

[27] Gastwirth, J.L., "The Statistical Precision of Medical Screening Procedures: Application to Polygraph and AIDS Antibodies Test Data (with Discussion)," *Statistics Science*, 1987, **2**, pp. 213-238.

[28] Gelfand, S., C. Ravishankar, and E. Delp, "An Iterative Growing and Pruning Algorithm for Classification Tree Design," *IEEE Trans. on Pattern Analysis and Machine Intelligence*, 1991, **13**(6), pp. 163-174.

[29] Gersho, A., "On the Structure of Vector Quantization," *IEEE Trans. on Information Theory*, 1982, **IT-28**, pp. 256-261.

[30] Gillick, L. and S.J. Cox, "Some Statistical Issues in the Comparison of Speech Recognition Algorithms," *IEEE Int. Conf. on Acoustics, Speech and Signal Processing*, 1989, Glasgow, Scotland, UK, IEEE, pp. 532-535.

[31] Goldstine, H., *The Computer from Pascal to von Neumann*, 2nd ed, 1993, Princeton, NJ, Princeton University Press.

[32] Gray, R.M., "Vector Quantization," *IEEE ASSP Magazine*, 1984, **1**(April), pp. 4-29.

[33] Gray, R.M. and E.D. Karnin, "Multiple Local Optima in Vector Quantizers," *IEEE Trans. on Information Theory*, 1982, **IT-28**, pp. 256-261.

[34] Hartigan, J.A., *Clustering Algorithm*, 1975, New York, NY, J. Wiley.

[35] Haykin, S., *Neural Networks: A Comprehensive Foundation*, 2nd ed, 1999, Upper Saddle River, NJ, Prentice-Hall.

[36] Hedelin, P., P. Knagenhjelm, and M. Skoglund, "Vector Quantization for Speech Transmission" in *Speech Coding and Synthesis*, W.B. Kleijn and K.K. Paliwal, eds., 1995, Amsterdam, pp. 311-396, Elsevier.

[37] Hinton, G.E., "Connectionist Learning Procedures," *Artificial Intelligence*, 1989, **40**, pp. 185--234.

[38] Hon, H.W., *Vocabulary-Independent Speech Recognition: The VOCIND System*, Ph.D Thesis in Department of Computer Science 1992, Carnegie Mellon University, Pittsburgh, PA.

[39] Hon, H.W. and K.F. Lee, "On Vocabulary-Independent Speech Modeling," *IEEE Int. Conf. on Acoustics, Speech and Signal Processing*, 1990, Albuquerque, NM, pp. 725-728.

[40] Hon, H.W. and K.F. Lee, "Vocabulary-Independent Subword Modeling and Adaptation," *IEEE Workshop on Speech Recognition*, 1991, Harriman, NY, Arden House.

[41] Huang, X.D., Y. Ariki, and M.A. Jack, *Hidden Markov Models for Speech Recognition*, 1990, Edinburgh, U.K., Edinburgh University Press.

[42] Hunt, M.J., *et al.*, "An Investigation of PLP and IMELDA Acoustic Representations and of Their Potential for Combination" in *Proc. of the IEEE Int. Conf. on Acoustics, Speech and Signal Processing* 1991, Toronto, Canada, pp. 881-884.

[43] Hwang, M.Y. and X.D. Huang, "Dynamically Configurable Acoustic Modelings for Speech Recognition," *IEEE Int. Conf. on Acoustics, Speech and Signal Processing*, 1998, Seattle, WA.

[44] Jain, A., J. Mao, and K.M. Mohiuddin, "Artificial Neural Networks: A Tutorial," *Computer*, 1996, **29**(3), pp. 31-44.

[45] Jayant, N.S. and P. Noll, *Digital Coding of Waveforms*, 1984, Englewood Cliffs, NJ, Prentice-Hall.

[46] Jiang, L., H.W. Hon, and X.D. Huang, "Improvements on a Trainable Letter-to-Sound Converter," *Eurospeech*, 1997, Rhodes, Greece.

[47] Juang, B.H., W. Chou, and C.H. Lee, "Statistical and Discriminative Methods for Speech Recognition" in *Automatic Speech and Speaker Recognition – Advanced Topics*, C.H. Lee, F.K. Soong, and K.K. Paliwal, eds., 1996, Boston, MA, pp. 109-132, Kluwer Academic Publishers.

[48] Juang, B.H. and S. Katagiri, "Discriminative Learning for Minimum Error Classification," *IEEE Trans. on Acoustics, Speech and Signal Processing*, 1992, **SP-40**(12), pp. 3043-3054.

[49] Kanal, L.N. and B. Chandrasekaran, "On Dimensionality and Sample Size in Statistical Pattern Classification," *Proc. of NEC*, 1968, **24**, pp. 2-7.

[50] Kanal, L.N. and N.C. Randall, "Recognition System Design by Statistical Analysis," *ACM Proc. of 19th National Conf.*, 1964, pp. D2.5-1-D2.5-10.

[51] Kohonen, T., *Learning Vector Quantization for Pattern Recognition*, 1986, Helsinki University of Technology, Finland.

[52] Kuhn, R. and R.D. Mori, "The Application of Semantic Classification Trees to Natural Language Understanding," *IEEE Trans. on Pattern Analysis and Machine Intelligence*, 1995(7), pp. 449-460.

[53] Lachenbruch, P.A. and M.R. Mickey, "Estimation of Error Rates in Discriminant Analysis," *Technometrics*, 1968, **10**, pp. 1-11.

[54] Laplace, P.S., *Theorie Analytique des Probabilities*, 1812, Paris, Courcier.

[55] Lee, D.S., S.N. Srihari, and R. Gaborski, "Bayesian and Neural Network Pattern Recognition: A Theoretical Connection and Empirical Results with Handwriting Characters" in *Artificial Neural Network and Statistical Pattern Recognition: Old and New Connections*, I.K. Sethi and A.K. Jain, eds., 1991, Amsterdam, North-Holland.

[56] Linde, Y., A. Buzo, and R.M. Gray, "An Algorithm for Vector Quantizer Design," *IEEE Trans. on Communication*, 1980, **COM-28**(1), pp. 84-95.

[57] Lippmann, R.P., "An Introduction to Computing with Neural Nets," *IEEE ASSP Magazine*, 1987, pp. 4-22.

[58] Lippmann, R.P., "Review of Neural Nets for Speech Recognition," *Neural Computation*, 1989, **1**, pp. 1-38.

[59] Lloyd, S.P., "Least Squares Quantization in PCM," *IEEE Trans. on Information Theory*, 1982, **IT-2**, pp. 129-137.

[60] Lucassen, J.M. and R.L. Mercer, "An Information-Theoretic Approach to the Automatic Determination of Phonemic Baseforms," *Proc. of the IEEE Int. Conf. on Acoustics, Speech and Signal Processing*, 1984, San Diego, California, pp. 42.5.1-42.5.4.

[61] Makhoul, J., S. Roucos, and H. Gish, "Vector Quantization in Speech Coding," *Proc. of the IEEE*, 1985, **73**(11), pp. 1551-1588.

[62] Martin, J.K., "An Exact Probability Metric for Decision Tree Splitting and Stopping," *Articial Intelligence and Statistics*, 1995, **5**, pp. 379-385.

[63] McClelland, J., "The Programmable Blackboard Model" in *Parallel Distributed Processing – Explorations in the Microstructure of Cognition, Volume II: Psychological and Biological Models* 1986, Cambridge, MA, MIT Press.

[64] McCulloch, W.S. and W. Pitts, "A Logical Calculus of Ideas Immanent in Nervous Activity," *Bulletin of Mathematical Biophysics*, 1943.

[65] McLachlan, G. and T. Krishnan, *The EM Algorithm and Extensions*, 1996, New York, NY, Wiley Interscience.

[66] McNemar, E.L., "Note on the Sampling Error of the Difference Between Correlated Proportions or Percentages," *Psychometrika*, 1947, **12**, pp. 153-157.

[67] Meisel, W.S., *et al.*, "The SSI Large-Vocabulary Speaker-Independent Continuous Speech Recognition System," 1991, pp. 337-340.

[68] Minsky, M. and S. Papert, *Perceptrons*, 1969, Cambridge, MA, MIT Press.

[69] Mitchell, T., *Machine Learning*, McGraw-Hill Series in Computer Science, 1997, McGraw-Hill.

[70] Morgan, J.N. and J.A. Sonquist, "Problems in the Analysis of Survey Data and a Proposal," *Journal of American Statistics Association*, 1962, **58**, pp. 415-434.

[71] Nadas, A., "A Decision-Theoretic Formulation of a Training Problem in Speech Recognition and a Comparison of Training by Unconditional Versus Conditional Maximum Likelihood," *IEEE Trans. on Acoustics, Speech and Signal Processing*, 1983, **4**, pp. 814-817.

[72] Newton, I., *Philosophiae Naturalis Principlea Mathematics*, 1687, London, Royal Society Press.

[73] Neyman, J. and E.S. Pearson, "On the Problem of the Most Efficient Tests of Statistical Hypotheses," *Philosophical Trans. of Royal Society*, 1928, **231**, pp. 289-337.

[74] Ostendorf, M. and N. Velleux, "A Hierarchical Stochastic Model for Automatic Prediction of Prosodic Boundary Location," *Computational Linguistics*, 1995, **20**(1), pp. 27-54.

[75] Paliwal, K. and W.B. Kleijn, "Quantization of LPC Parameters" in *Speech Coding and Synthesis*, W.B. Kleijn and K.K. Paliwal, eds., 1995, Amsterdam, pp. 433-466, Elsevier.

[76] Paul, D.B., "An 800 bps Adaptive Vector Quantization in Speech Coding," *IEEE Int. Conf. on Acoustics, Speech and Signal Processing*, 1983, pp. 73-76.

[77] Paul, D.B., "The Lincoln Tied-Mixture HMM Continuous Speech Recognizer" in *Morgan Kaufmann Publishers* 1990, San Mateo, CA, pp. 332-336.

[78] Payne, H.J. and W.S. Meisel, "An Algorithm for Constructing Optimal Binary Decision Trees," *IEEE Trans. on Computers*, 1977, **C-26**(September), pp. 905-916.

[79] Pearson, K., "Contributions to The Mathematical Theorem of Evolution," *Philosophical Trans. of Royal Society*, 1894, **158A**, pp. 71-110.

[80] Pitrelli, J. and V. Zue, "A Hierarchical Model for Phoneme Duration in American English" in *Proc. of Eurospeech,* 1989.

[81] Press, W.H., *et al.*, *Numerical Recipes in C*, 1988, Cambridge, UK, Cambridge University Press.

[82] Quinlan, J.R., "Introduction of Decision Trees" in *Machine Learning: An Artifical Intelligence Approach*, R. Michalski, J. Carbonell, and T. Mitchell, eds., 1986, Boston, M.A., pp. 1-86, Kluwer Academic Publishers.

[83] Quinlan, J.R., "Generating Production Rules from Decision Trees," *Int. Joint Conf. on Artificial Intelligence*, 1987, pp. 304-307.

[84] Quinlan, J.R., *C4.5: Programs for Machine Learning*, 1993, San Francisco, Morgan Kaufmann.

[85] Rabiner, L. and B.H. Juang, "Speech Recognition System Design and Implementation Issues" in *Fundamentals of Speech Recognition*, L. Rabiner and B.H. Juang, eds., 1993, Englewood Cliffs, NJ, Prentice Hall, pp. 242-340.

[86] Reddy, D.R., "Speech Recognition by Machine: A Review," *IEEE Proc.*, 1976, **64**(4), pp. 502-531.

[87] Rosenblatt, F., "The Perceptron – A Probabilistic Model for Information Storage and Organization in the Brain," *Psychological Review*, 1958, **65**, pp. 386-408.

[88] Rummelhart, D.E., G.E. Hinton, and R.J. Williams, "Learning Internal Representations by Error Propagation" in *Parallel Distributed Processing*, D.E. Rumelhart and J.L. McClelland, eds., 1986, Cambridge, MA, MIT Press, pp. 318-362.

[89] Rummelhart, D.E., G.E. Hinton, and R.J. Williams, "Learning Representations by Back-Propagating Errors," *Nature*, 1986, **323**, pp. 533-536.

[90] Rummelhart, D.E. and J.L. McClelland, *Parallel Distributed Processing – Explorations in the Microstructure of Cognition, Volume I: Foundations*, 1986, Cambridge, MA, MIT Press.

[91] Rummelhart, D.E. and J.L. McClelland, *Parallel Distributed Processing – Explorations in the Microstructure of Cognition, Volume II: Psychological and Biological Models*, 1986, Cambridge, MA, MIT Press.

[92] Schalkoff, R.J., *Digital Image Processing and Computer Vision*, 1989, New York, NY, John Wiley & Sons.

[93] Sejnowski, T.J. and C.R. Rosenberg, "Parallel Networks that Learn to Pronounce English Text," *Complex Systems*, 1987, **1**, pp. 145-168.

[94] Sejnowski, T.J., *et al.*, "Combining Visual and Acoustic Speech Signals with a Neural Network Improve Intelligibility" in *Advances in Neural Information Processing Systems* 1990, San Mateo, CA, Morgan Kaufmann, pp. 232-239.

[95] Sparkes, J.J., "Pattern Recognition and a Model of the Brain," *Int. Journal of Man-Machine Studies*, 1969, **1**(3), pp. 263-278.

[96] Tappert, C., C.Y. Suen, and T. Wakahara, "The State of the Art in On-Line Handwriting Recognition," *IEEE Trans. on Pattern Analysis and Machine Intelligence*, 1990, **12**(8), pp. 787-808.

[97] Tesauro, G. and T.J. Sejnowski, "A Neural Network That Learns to Play Backgammon," *Neural Information Processing Systems, American Institute of Physics*, 1988, pp. 794-803.

[98] White, H., "Learning in Artificial Neural Networks: A Statistical Perspective," *Neural Computation*, 1989, **1**(4), pp. 425-464.

[99] Winston, P.H., *Artificial Intelligence*, 1984, Reading, MA, Addison-Wesley.

[100] Wu, C.F.J., "On the Convergence Properties of the EM Algorithm," *The Annals of Statistics*, 1983, **11**(1), pp. 95-103.

[101] Young, J.Z., *Programmes of the Brain*, 1975, Oxford, England, Oxford University Press.

PART II

SPEECH PROCESSING

CHAPTER 5

Digital Signal Processing

One of the most popular ways of characterizing speech is in terms of a *signal* or acoustic waveform. Shown in Figure 5.1 is a representation of the speech signal that ensures that the information content can be easily extracted by human listeners or computers. This is why digital signal processing plays a fundamental role for spoken language processing. We describe here the fundamentals of digital signal processing: digital signals and systems, frequency-domain transforms for both continuous and discrete frequencies, digital filters, the relationship between analog and digital signals, filterbanks, and stochastic processes. In this chapter we set the mathematical foundations of frequency analysis that allow us to develop specific techniques for speech signals in Chapter 6.

The main theme of this chapter is the development of frequency-domain methods computed through the Fourier transform. When we boost the bass knob in our amplifier we are increasing the gain at low frequencies, and when we boost the treble knob we are increasing the gain at high frequencies. Representation of speech signals in the frequency domain is especially useful because the frequency structure of a phoneme is generally unique.

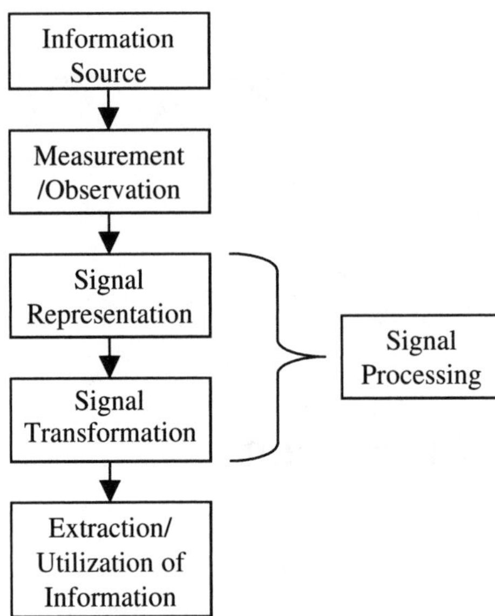

Figure 5.1 Signal processing is both a representation and a transformation that allows a useful information extraction from a source. The representation and transformation are based on a model of the signal, often parametric, that is convenient for subsequent processing.

5.1. DIGITAL SIGNALS AND SYSTEMS

To process speech signals, it is convenient to represent them mathematically as functions of a continuous variable *t*, which represents time. Let us define an *analog signal* $x_a(t)$ as a function varying continuously in time. If we sample the signal *x* with a sampling period *T* (i.e., $t = nT$), we can define a discrete-time signal as $x[n] = x_a(nT)$, also known as *digital signal.*[1] In this book we use parentheses to describe an analog signal and brackets for digital signals. Furthermore we can define the sampling frequency F_s as $F_s = 1/T$, the inverse of the sampling period *T*. For example, for a sampling rate $F_s = 8\,\text{kHz}$, its corresponding sampling period is 125 microseconds. In Section 5.5 it is shown that, under some circumstances, the analog signal $x_a(t)$ can be recovered exactly from the digital signal $x[n]$. Figure 5.2 shows an analog signal and its corresponding digital signal. In subsequent figures, for convenience, we will sometimes plot digital signals as continuous functions.

The term *Digital Signal Processing* (DSP) refers to methods for manipulating the sequence of numbers $x[n]$ in a digital computer. The acronym DSP is also used to refer to a *Digital Signal Processor*, i.e., a microprocessor specialized to perform DSP operations.

[1] Actually the term digital signal is defined as a discrete-time signal whose values are represented by integers within a range, whereas a general discrete-time signal would be represented by real numbers. Since the term digital signal is much more commonly used, we will use that term, except when the distinction between them is necessary.

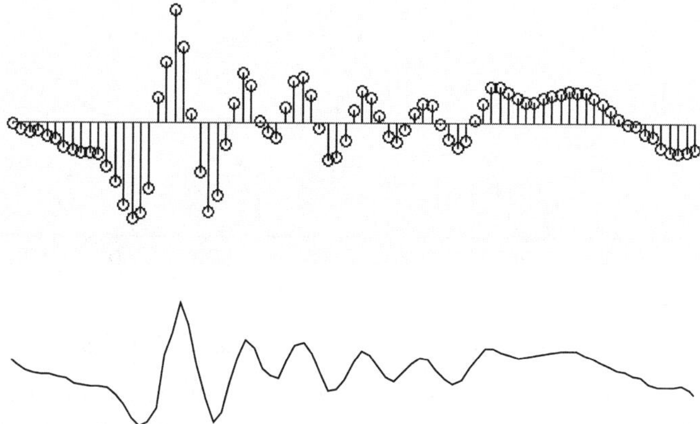

Figure 5.2 Analog signal and its corresponding digital signal.

We start with sinusoidal signals and show they are the fundamental signals for linear systems. We then introduce the concept of convolution and linear time-invariant systems. Other digital signals and nonlinear systems are also introduced.

5.1.1. Sinusoidal Signals

One of the most important signals is the sine wave or *sinusoid*

$$x_0[n] = A_0 \cos(\omega_0 n + \phi_0) \tag{5.1}$$

where A_0 is the sinusoid's amplitude, ω_0 the angular frequency, and ϕ_0 the phase. The angle in the trigonometric functions is expressed in radians, so that the angular frequency ω_0 is related to the normalized linear frequency f_0 by the relation $\omega_0 = 2\pi f_0$, and $0 \le f_0 \le 1$. This signal is *periodic*[2] with period $T_0 = 1/f_0$. In Figure 5.3 we can see an example of a sinusoid with frequency $f_0 = 0.04$, or a period of $T_0 = 25$ samples.

Sinusoids are important because speech signals can be decomposed as sums of sinusoids. When we boost the bass knob in our amplifier we are increasing the gain for sinusoids of low frequencies, and when we boost the treble knob we are increasing the gain for sinusoids of high frequencies.

---

[2] A signal $x[n]$ is periodic with period N if and only if $x[n]=x[n+N]$, which requires $\omega_0 = 2\pi / N$. This means that the digital signal in Eq. (5.1) is not periodic for all values of ω_0, even though its continuous signal counterpart $x(t) = A_0 \cos(\omega_0 t + \phi_0)$ is periodic for all values of ω_0 (see Section 5.5).

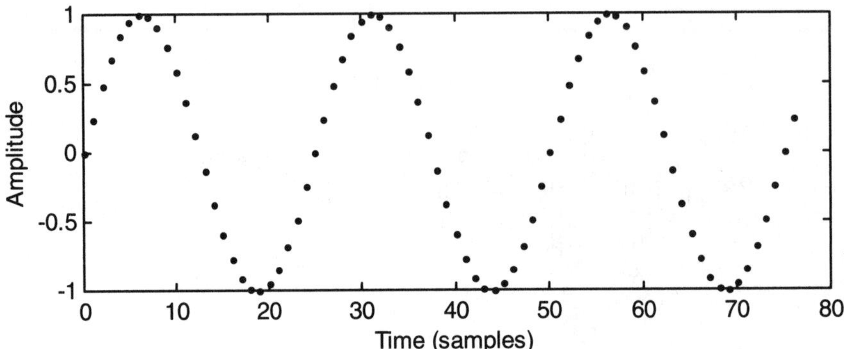

Figure 5.3 A digital sinusoid with a period of 25 samples.

What is the sum of two sinusoids $x_0[n]$ and $x_1[n]$ of the same frequency ω_0 but different amplitudes A_0 and A_1, and phases ϕ_0 and ϕ_1? The answer is another sinusoid of the same frequency but a different amplitude A and phase ϕ. While this can be computed through trigonometric identities, it is somewhat tedious and not very intuitive. For this reason we introduce another representation based on complex numbers, which proves to be very useful when we study digital filters.

A complex number x can be expressed as $z = x+jy$, where $j = \sqrt{-1}$, x is the real part and y is the imaginary part, with both x and y being real numbers. Using Euler's relation, given a real number ϕ, we have

$$e^{j\phi} = \cos\phi + j\sin\phi \tag{5.2}$$

so that the complex number z can also be expressed in polar form as $z = Ae^{j\phi}$, where A is the amplitude and ϕ is the phase. Both representations can be seen in Figure 5.4, where the real part is shown in the abscissa (x-axis) and the imaginary part in the ordinate (y-axis).

Using complex numbers, the sinusoid in Eq. (5.1) can be expressed as the real part of the corresponding complex exponential

$$x_0[n] = A_0 \cos(\omega_0 n + \phi_0) = \mathrm{Re}\{A_0 e^{j(\omega_0 n + \phi_0)}\} \tag{5.3}$$

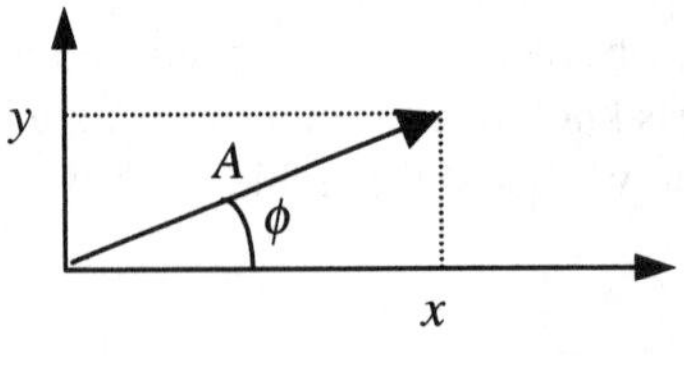

Figure 5.4 Complex number representation in Cartesian form $z = x + jy$ and polar form $z = Ae^{j\phi}$. Thus $x = A\cos\phi$ and $y = A\sin\phi$.

and thus the sum of two complex exponential signals equals

$$A_0 e^{j(\omega_0 n+\phi_0)} + A_1 e^{j(\omega_0 n+\phi_1)} = e^{j\omega_0 n}\left(A_0 e^{j\phi_0} + A_1 e^{j\phi_1}\right) = e^{j\omega_0 n} A e^{j\phi} = A e^{j(\omega_0 n+\phi)} \tag{5.4}$$

Taking the real part in both sides results in

$$A_0 \cos(\omega_0 n + \phi_0) + A_2 \cos(\omega_0 n + \phi_1) = A\cos(\omega_0 n + \phi) \tag{5.5}$$

or in other words, the sum of two sinusoids of the same frequency is another sinusoid of the same frequency.

To compute A and ϕ, dividing Eq. (5.4) by $e^{j\omega_0 n}$ leads to a relationship between the amplitude A and phase ϕ:

$$A_0 e^{j\phi_0} + A_1 e^{j\phi_1} = A e^{j\phi} \tag{5.6}$$

Equating real and imaginary parts in Eq. (5.6) and dividing them we obtain:

$$\tan\phi = \frac{A_0 \sin\phi_0 + A_1 \sin\phi_1}{A_0 \cos\phi_0 + A_1 \cos\phi_1} \tag{5.7}$$

and adding the squared of real and imaginary parts and using trigonometric identities[3]

$$A^2 = A_0^2 + A_1^2 + 2A_0 A_1 \cos(\phi_0 - \phi_1) \tag{5.8}$$

This complex representation of Figure 5.5 lets us analyze and visualize the amplitudes and phases of sinusoids of the same frequency as vectors. The sum of N sinusoids of the same frequency is another sinusoid of the same frequency that can be obtained by adding the real and imaginary parts of all complex vectors. In Section 5.2.1 we show that the output of a linear time-invariant system to a sinusoid is another sinusoid of the same frequency.

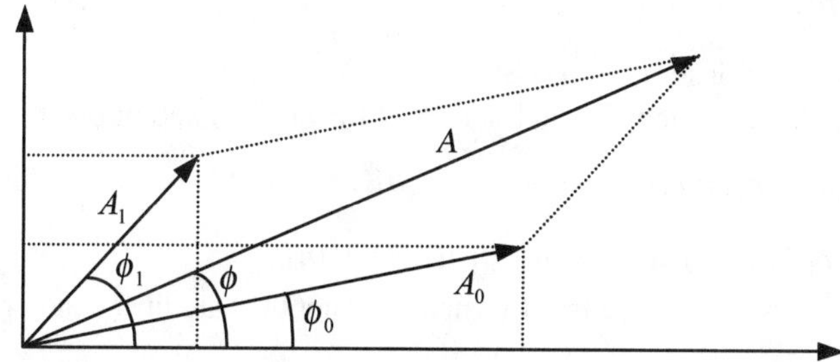

Figure 5.5 Geometric representation of the sum of two sinusoids of the same frequency. It follows the complex number representation in Cartesian form of Figure 5.4.

[3] $\sin^2\phi + \cos^2\phi = 1$ and $\cos(a-b) = \cos a \cos b + \sin a \sin b$.

5.1.2. Other Digital Signals

In the field of digital signal processing there are other signals that repeatedly arise and that are shown in Table 5.1.

Table 5.1 Some useful digital signals: the Kronecker delta, unit step, rectangular signal, real exponential ($a<1$) and real part of a complex exponential ($r<1$).

Kronecker delta, or *unit impulse*	$\delta[n]=\begin{cases}1 & n=0\\ 0 & otherwise\end{cases}$	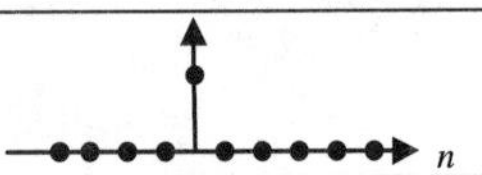 n
Unit step	$u[n]=\begin{cases}1 & n\geq 0\\ 0 & n<0\end{cases}$	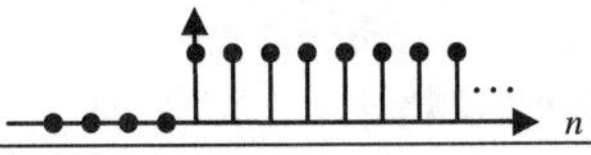 ... n
Rectangular signal	$\text{rect}_N[n]=\begin{cases}1 & 0\leq n<N\\ 0 & otherwise\end{cases}$	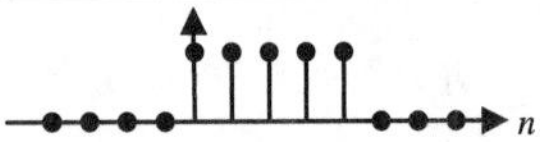 n
Real exponential	$x[n]=a^n u[n]$	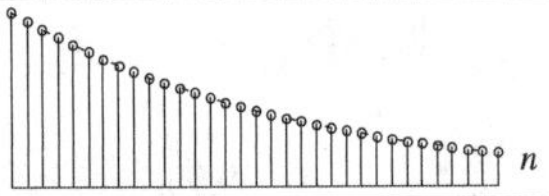 n
Complex exponential	$x[n]=a^n u[n]=r^n e^{jn\omega_0}u[n]$ $=r^n(\cos n\omega_0+j\sin n\omega_0)u[n]$	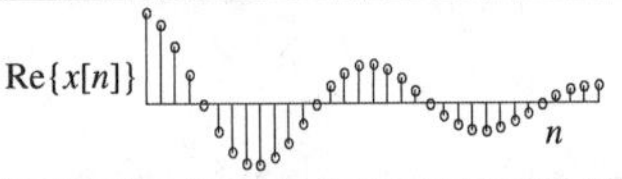 $\text{Re}\{x[n]\}$ n

If $r=1$ and $\omega_0\neq 0$ we have a complex sinusoid as shown in Section 5.1.1. If $\omega_0=0$ we have a real exponential signal, and if $r<1$ and $\omega_0\neq 0$ we have an exponentially decaying oscillatory sequence, also known as a damped sinusoid.

5.1.3. Digital Systems

A digital system is a system that, given an input signal $x[n]$, generates an output signal $y[n]$:

$$y[n]=T\{x[n]\} \tag{5.9}$$

whose input/output relationship can be seen in Figure 5.6.

In general, a digital system T is defined to be linear *iff* (if and only if)

$$T\{a_1x_1[n]+a_2x_2[n]\}=a_1T\{x_1[n]\}+a_2T\{x_2[n]\} \tag{5.10}$$

for any values of a_1, a_2 and any signals $x_1[n]$ and $x_2[n]$.

Here, we study systems according to whether or not they are linear and/or time invariant.

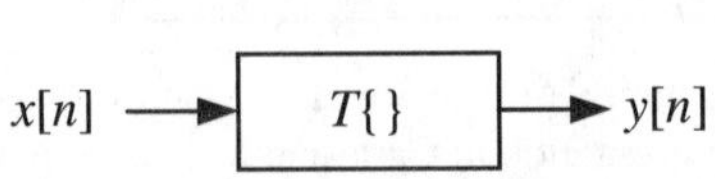

$x[n]$ → $T\{\}$ → $y[n]$

Figure 5.6 Block diagram of a digital system whose input is digital signal $x[n]$, and whose output is digital signal $y[n]$.

5.1.3.1. Linear Time-Invariant Systems

A system is *time-invariant* if given Eq. (5.9), then

$$y[n-n_0] = T\{x[n-n_0]\} \tag{5.11}$$

Linear digital systems of a special type, the so-called *linear time-invariant* (LTI),[4] are described by

$$y[n] = \sum_{k=-\infty}^{\infty} x[k]h[n-k] = x[n]*h[n] \tag{5.12}$$

where $*$ is defined as the *convolution* operator. It is left to the reader to show that the linear system in Eq. (5.12) indeed satisfies Eq. (5.11).

LTI systems are completely characterized by the signal $h[n]$, which is known as the system's *impulse response* because it is the output of the system when the input is an impulse $x[n] = \delta[n]$. Most of the systems described in this book are LTI systems.

Table 5.2 Properties of the convolution operator.

Commutative	$x[n]*h[n] = h[n]*x[n]$
Associative	$x[n]*(h_1[n]*h_2[n]) = (x[n]*h_1[n])*h_2[n] = x[n]*h_1[n]*h_2[n]$
Distributive	$x[n]*(h_1[n]+h_2[n]) = x[n]*h_1[n]+x[n]*h_2[n]$

The convolution operator is commutative, associative and distributive as shown in Table 5.2 and Figure 5.7.

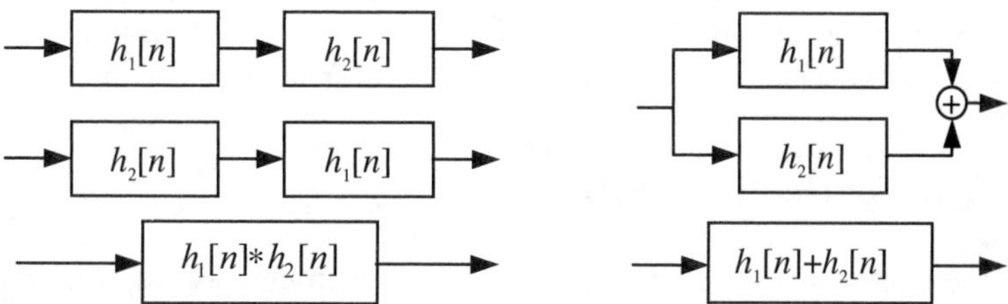

Figure 5.7 The block diagrams on the left, representing the commutative property, are equivalent. The block diagrams on the right, representing the distributive property, are also equivalent.

[4] Actually the term linear time-invariant (LTI) systems is typically reserved for continuous or analog systems, and linear shift-invariant system is used for discrete-time signals, but we will use LTI for discrete-time signals too since it is widely used in this context.

5.1.3.2. Linear Time-Varying Systems

An interesting type of digital systems is that whose output is a linear combination of the input signal at different times:

$$y[n] = \sum_{k=-\infty}^{\infty} x[k]g[n, n-k] \tag{5.13}$$

The digital system in Eq. (5.13) is linear, since it satisfies Eq. (5.10). The Linear Time-Invariant systems of Section 5.1.3.1 are a special case of Eq. (5.13) when $g[n, n-k] = h[n-k]$. The systems in Eq. (5.13) are called *linear time-varying* (LTV) systems, because the weighting coefficients can vary with time.

A useful example of such system is the so-called *amplitude modulator*

$$y[n] = x[n]\cos\omega_0 n \tag{5.14}$$

used in AM transmissions. As we show in Chapter 6, speech signals are the output of LTV systems. Since these systems are difficult to analyze, we often approximate them with linear time-invariant systems.

5.1.3.3. Nonlinear Systems

Many *nonlinear* systems do not satisfy Eq. (5.10). Table 5.3 includes a list of typical nonlinear systems used in speech processing. All these nonlinear systems are memoryless, because the output at time n depends only on the input at time n, except for the *median smoother* of order $(2N + 1)$ whose output depends also on the previous and the following N samples.

5.2. Continuous-Frequency Transforms

A very useful transform for LTI systems is the Fourier transform, because it uses complex exponentials as its basis functions, and its generalization: the z-transform. In this section we cover both transforms, which are continuous functions of frequency, and their properties.

5.2.1. The Fourier Transform

It is instructive to see what the output of a LTI system with impulse response $h[n]$ is when the input is a complex exponential. Substituting $x[n] = e^{j\omega_0 n}$ in Eq. (5.12) and using the commutative property of the convolution we obtain

$$y[n] = \sum_{k=-\infty}^{\infty} h[k]e^{j\omega_0(n-k)} = e^{j\omega_0 n}\sum_{k=-\infty}^{\infty} h[k]e^{-j\omega_0 k} = e^{j\omega_0 n}H(e^{j\omega_0}) \tag{5.15}$$

Table 5.3 Examples of nonlinear systems for speech processing. All of them are memoryless except for the median smoother.

Nonlinear System	**Equation**
Median Smoother of order (2*N*+1)	$y[n] = \text{median}\{x[n-N],\cdots,x[n],\cdots,x[n+N]\}$
Full-Wave Rectifier	$y[n] = \lvert x[n] \rvert$
Half-Wave Rectifier	$y[n] = \begin{cases} x[n] & x[n] \ge 0 \\ 0 & x[n] < 0 \end{cases}$
Frequency Modulator	$y[n] = A\cos\left(\omega_0 + \Delta\omega x[n]\right)n$
Hard-Limiter	$y[n] = \begin{cases} A & x[n] \ge A \\ x[n] & \lvert x[n] \rvert < A \\ -A & x[n] \le -A \end{cases}$
Uniform Quantizer (*L*-bit) with $2N = 2^L$ intervals of width Δ	$y[n] = \begin{cases} (N-1/2)\Delta & x[n] \ge (N-1)\Delta \\ (m+1/2)\Delta & m\Delta \le x[n] < (m+1)\Delta \quad 0 \le m < N-1 \\ (-m+1/2)\Delta & -m\Delta \le x[n] < -(m-1)\Delta \quad 0 < m < N-1 \\ (-N+1/2)\Delta & x[n] < -(N-1)\Delta \end{cases}$

which is another complex exponential of the same frequency and amplitude multiplied by the complex quantity $H(e^{j\omega_0})$ given by

$$H(e^{j\omega}) = \sum_{n=-\infty}^{\infty} h[n]e^{-j\omega n} \tag{5.16}$$

Since the output of a LTI system to a complex exponential is another complex exponential, it is said that complex exponentials are *eigensignals* of LTI systems, with the complex quantity $H(e^{j\omega_0})$ being their *eigenvalue*.

The quantity $H(e^{j\omega})$ is defined as the *discrete-time Fourier transform* of $h[n]$. It is clear from Eq. (5.16) that $H(e^{j\omega})$ is a periodic function of ω with period 2π, and therefore we need to keep only one period to fully describe it, typically $-\pi < \omega < \pi$ (Figure 5.8).

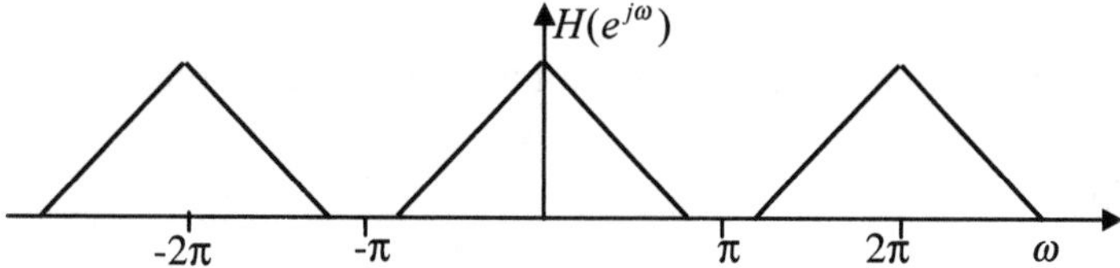

Figure 5.8 $H(e^{j\omega})$ is a periodic function of ω.

$H(e^{j\omega})$ is a complex function of ω which can be expressed in terms of the real and imaginary parts:

$$H(e^{j\omega}) = H_r(e^{j\omega}) + jH_i(e^{j\omega}) \tag{5.17}$$

or in terms of the magnitude and phase as

$$H(e^{j\omega}) = \left|H(e^{j\omega})\right| e^{j\arg[H(e^{j\omega})]} \tag{5.18}$$

Thus if the input to the LTI system is a sinusoid as in Eq. (5.1), the output will be

$$y_0[n] = A_0 \left|H(e^{j\omega_0})\right| \cos\left(\omega_0 n + \phi_0 + \arg\{H(e^{j\omega_0})\}\right) \tag{5.19}$$

according to Eq. (5.15). Therefore if $\left|H(e^{j\omega_0})\right| > 1$, the LTI system will amplify that frequency, and likewise it will attenuate, or *filter* it, if $\left|H(e^{j\omega_0})\right| < 1$. That is one reason why these systems are also called filters. The Fourier transform $H(e^{j\omega})$ of a filter $h[n]$ is called the system's *frequency response* or *transfer function.*

The angular frequency ω is related to the normalized linear frequency f by the simple relation $\omega = 2\pi f$. We show in Section 5.5 that linear frequency f_l and normalized frequency f are related by $f_l = fF_s$, where F_s is the sampling frequency.

The inverse *discrete-time Fourier transform* is defined as

$$h[n] = \frac{1}{2\pi}\int_{-\pi}^{\pi} H(e^{j\omega})e^{j\omega n} d\omega \tag{5.20}$$

The Fourier transform is invertible, and Eq. (5.16) and (5.20) are transform pairs:

$$\begin{aligned} h[n] &= \frac{1}{2\pi}\int_{-\pi}^{\pi} H(e^{j\omega})e^{j\omega n} d\omega = \frac{1}{2\pi}\int_{-\pi}^{\pi}\left(\sum_{m=-\infty}^{\infty} h[m]e^{-j\omega m}\right)e^{j\omega n} d\omega \\ &= \sum_{m=-\infty}^{\infty} h[m]\frac{1}{2\pi}\int_{-\pi}^{\pi} e^{j\omega(n-m)} d\omega = \sum_{m=-\infty}^{\infty} h[m]\delta[n-m] = h[n] \end{aligned} \tag{5.21}$$

since

$$\frac{1}{2\pi}\int_{-\pi}^{\pi} e^{j\omega(n-m)} d\omega = \delta[n-m] \tag{5.22}$$

A sufficient condition for the existence of the Fourier transform is

$$\sum_{n=-\infty}^{\infty} \left|h[n]\right| < \infty \tag{5.23}$$

Although we have computed the Fourier transform of the impulse response of a filter $h[n]$, Eq. (5.16) and (5.20) can be applied to any signal $x[n]$.

5.2.2. Z-Transform

The *z-transform* is a generalization of the Fourier transform. The *z*-transform of a digital signal $h[n]$ is defined as

$$H(z) = \sum_{n=-\infty}^{\infty} h[n] z^{-n} \tag{5.24}$$

where z is a complex variable. Indeed, the Fourier transform of $h[n]$ equals its *z*-transform evaluated at $z = e^{j\omega}$. While the Fourier and *z*-transforms are often used interchangeably, we normally use the Fourier transform to plot the filter's frequency response, and the *z*-transform to analyze more general filter characteristics, given its polynomial functional form. We can also use the *z*-transform for unstable filters, which do not have Fourier transforms.

Since Eq. (5.24) is an infinite sum, it is not guaranteed to exist. A sufficient condition for convergence is:

$$\sum_{n=-\infty}^{\infty} |h[n]| |z|^{-n} < \infty \tag{5.25}$$

which is true only for a *region of convergence* (ROC) in the complex *z*-plane $R_1 < |z| < R_2$ as indicated in Figure 5.9.

For a signal $h[n]$ to have a Fourier transform, its *z*-transform $H(z)$ has to include the unit circle, $|z| = 1$, in its convergence region. Therefore, a sufficient condition for the existence of the Fourier transform is given in Eq. (5.23) by applying Eq. (5.25) to the unit circle.

An LTI system is defined to be *causal* if its impulse response is a causal signal, i.e. $h[n] = 0$ for $n < 0$. Similarly, a LTI system is *anti-causal* if $h[n] = 0$ for $n > 0$. While all physical systems are causal, noncausal systems are still useful since causal systems could be decomposed into causal and anti-causal systems.

A system is defined to be *stable* if for every bounded input it produces a bounded output. A necessary and sufficient condition for an LTI system to be stable is

$$\sum_{n=-\infty}^{\infty} |h[n]| < \infty \tag{5.26}$$

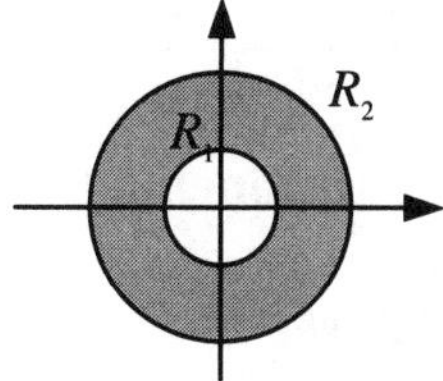

Figure 5.9 Region of convergence of the *z*-transform in the complex plane.

which means, according to Eq. (5.23), that $h[n]$ has a Fourier transform, and therefore that its z-transform includes the unit circle in its region of convergence.

Just as in the case of Fourier transforms, we can use the z-transform for any signal, not just for a filter's impulse response.

The *inverse z-transform* is defined as

$$h[n] = \frac{1}{2\pi j} \oint H(z) z^{n-1} dz \tag{5.27}$$

where the integral is performed along a closed contour that is within the *region of convergence*. Eqs. (5.24) and (5.27) plus knowledge of the region of convergence form a transform pair: *i.e.* one can be exactly determined if the other is known. If the integral is performed along the unit circle (i.e., doing the substitution $z = e^{j\omega}$) we obtain Eq. (5.20), the inverse Fourier transform.

5.2.3. Z-Transforms of Elementary Functions

In this section we compute the z-transforms of the signals defined in Table 5.1. The z-transforms of such signals are summarized in Table 5.4. In particular we compute the z-transforms of left-sided and right-sided complex exponentials, which are essential to compute the inverse z-transform of rational polynomials. As we see in Chapter 6, speech signals are often modeled as having z-transforms that are rational polynomials.

Table 5.4 Z-transforms of some useful signals together with their region of convergence.

Signal	Z-Transform	Region of Convergence
$h_1[n] = \delta[n-N]$	$H_1(z) = z^{-N}$	$z \neq 0$
$h_2[n] = u[n] - u[n-N]$	$H_2(z) = \dfrac{1-z^{-N}}{1-z^{-1}}$	$z \neq 0$
$h_3[n] = a^n u[n]$	$H_3(z) = \dfrac{1}{1-az^{-1}}$	$\lvert a \rvert < \lvert z \rvert$
$h_4[n] = -a^n u[-n-1]$	$H_4(z) = \dfrac{1}{1-az^{-1}}$	$\lvert z \rvert < \lvert a \rvert$

5.2.3.1. Right-Sided Complex Exponentials

A right-sided complex exponential sequence

$$h_3[n] = a^n u[n] \tag{5.28}$$

has a z-transform given by

$$H_3(z) = \sum_{n=0}^{\infty} a^n z^{-n} = \lim_{N\to\infty} \frac{1-(az^{-1})^N}{1-az^{-1}} = \frac{1}{1-az^{-1}} \quad \text{for} \quad |a|<|z| \tag{5.29}$$

by using the sum of the terms of a geometric sequence and making $N \to \infty$. This region of convergence ($|a|<|z|$) is typical of causal signals (those that are zero for $n<0$).

When a z-transform is expressed as the ratio of two polynomials, the roots of the numerator are called *zeros*, and the roots of the denominator are called *poles*. Zeros are the values of z for which the z-transform equals 0, and poles are the values of z for which the z-transform equals infinity.

$H_3(z)$ has a pole at $z=a$, because its value goes to infinity at $z=a$. According to Eq. (5.26), $h_3[n]$ is a stable signal if and only if $|a|<1$, or in other words, if its pole is inside the unit circle. In general, a causal and stable system has all its poles inside the unit circle. As a corollary, a system which has poles outside the unit circle is either noncausal or unstable or both. This is a very important fact, which we exploit throughout the book.

5.2.3.2. Left-Sided Complex Exponentials

A left-sided complex exponential sequence

$$h_4[n] = -a^n u[-n-1] \tag{5.30}$$

has a z-transform given by

$$\begin{aligned} H_4(z) &= -\sum_{n=-\infty}^{-1} a^n z^{-n} = -\sum_{n=1}^{\infty} a^{-n} z^n = 1 - \sum_{n=0}^{\infty} a^{-n} z^n \\ &= 1 - \frac{1}{1-a^{-1}z} = \frac{-a^{-1}z}{1-a^{-1}z} = \frac{1}{1-az^{-1}} \end{aligned} \quad \text{for } |z|<|a| \tag{5.31}$$

This region of convergence ($|z|<|a|$) is typical of noncausal signals (those that are nonzero for $n<0$). Observe that $H_3(z)$ and $H_4(z)$ are functionally identical and only differ in the region of convergence. In general, the region of convergence of a signal that is nonzero for $-\infty<n<\infty$ is $R_1<|z|<R_2$.

5.2.3.3. Inverse Z-Transform of Rational Functions

Integrals in the complex plane such as Eq. (5.27) are not easy to do, but fortunately they are not necessary for the special case of $H(z)$ being a rational polynomial transform. In this case, partial fraction expansion can be used to decompose the signal into a linear combination of signals like $h_1[n]$, $h_3[n]$ and $h_4[n]$ as in Table 5.4.

For example,

$$H_5(z) = \frac{2+8z^{-1}}{2-5z^{-1}-3z^{-2}} \tag{5.32}$$

has as roots of its denominator $z = 3, -1/2$. Therefore it can be decomposed as

$$H_5(z) = \frac{A}{1-3z^{-1}} + \frac{B}{1+(1/2)z^{-1}} = \frac{(2A+2B)+(A-6B)z^{-1}}{2-5z^{-1}-3z^{-2}} \tag{5.33}$$

so that *A* and *B* are the solution of the following set of linear equations:

$$\begin{aligned} 2A+2B &= 2 \\ A-6B &= 8 \end{aligned} \tag{5.34}$$

whose solution is $A=2$ and $B=-1$, and thus Eq. (5.33) is expressed as

$$H_5(z) = 2\left(\frac{1}{1-3z^{-1}}\right) - \left(\frac{1}{1+(1/2)z^{-1}}\right) \tag{5.35}$$

However, we cannot compute the inverse *z*-transform unless we know the region of convergence. If, for example, we are told that the region of convergence includes the unit circle (necessary for the system to be stable), then the inverse transform of

$$H_4(z) = \frac{1}{1-3z^{-1}} \tag{5.36}$$

must have a region of convergence of $|z| < 3$ according to Table 5.4, and thus be a left-sided complex exponential:

$$h_4[n] = -3^n u[-n-1] \tag{5.37}$$

and the transform of

$$H_3(z) = \frac{1}{1+(1/2)z^{-1}} \tag{5.38}$$

must have a region of convergence of $1/2 < |z|$ according to Table 5.4, and thus be a right-sided complex exponential:

$$h_3[n] = (-1/2)^n u[n] \tag{5.39}$$

so that

$$h_5[n] = -2 \cdot 3^n u[-n-1] - (-1/2)^n u[n] \tag{5.40}$$

While we only showed an example here, the method used generalizes to rational transfer functions with more poles and zeros.

5.2.4. Properties of the Z- and Fourier Transforms

In this section we include a number of properties that are used throughout the book and that can be derived from the definition of Fourier and z-transforms (see Table 5.5). Of special interest are the convolution property and Parseval's theorem, which are described below.

5.2.4.1. The Convolution Property

The z-transform of $y[n]$, convolution of $x[n]$ and $h[n]$, can be expressed as a function of their z-transforms:

$$\begin{aligned} Y(z) &= \sum_{n=-\infty}^{\infty} y[n]z^{-n} = \sum_{n=-\infty}^{\infty}\left(\sum_{k=-\infty}^{\infty} x[k]h[n-k]\right)z^{-n} \\ &= \sum_{k=-\infty}^{\infty} x[k]\left(\sum_{n=-\infty}^{\infty} h[n-k]z^{-n}\right) = \sum_{k=-\infty}^{\infty} x[k]\left(\sum_{n=-\infty}^{\infty} h[n]z^{-(n+k)}\right) \\ &= \sum_{k=-\infty}^{\infty} x[k]z^{-k}H(z) = X(z)H(z) \end{aligned} \tag{5.41}$$

which is the fundamental property of LTI systems: "The z-transform of the convolution of two signals is the product of their z-transforms." This is also known as the convolution property. The ROC of $Y(z)$ is now the intersection of the ROCs of $X(z)$ and $H(z)$ and cannot be empty for $Y(z)$ to exist.

Likewise, we can obtain a similar expression for the Fourier transforms:

$$Y(e^{j\omega}) = X(e^{j\omega})H(e^{j\omega}) \tag{5.42}$$

A dual version of the convolution property can be proven for the product of digital signals:

$$x[n]y[n] \leftrightarrow \frac{1}{2\pi} X(e^{j\omega}) * Y(e^{j\omega}) \tag{5.43}$$

whose transform is the continuous convolution of the transforms with a scale factor. The convolution of functions of continuous variables is defined as

$$y(t) = x(t) * h(t) = \int_{-\infty}^{\infty} x(\tau)h(t-\tau)d\tau \tag{5.44}$$

Note how this differs from the discrete convolution of Eq. (5.12).

5.2.4.2. Power Spectrum and Parseval's Theorem

Let's define the *autocorrelation* of signal $x[n]$ as

$$R_{xx}[n] = \sum_{m=-\infty}^{\infty} x[m+n]x^*[m] = \sum_{l=-\infty}^{\infty} x[l]x^*[-(n-l)] = x[n]*x^*[-n] \tag{5.45}$$

where the superscript asterisk (*) means complex conjugate[5] and should not be confused with the convolution operator.

Using the fundamental property of LTI systems in Eq. (5.42) and the symmetry properties in Table 5.5, we can express its Fourier transform $S_{xx}(\omega)$ as

$$S_{xx}(\omega) = X(\omega)X^*(\omega) = |X(\omega)|^2 \tag{5.46}$$

which is the *power spectrum.* The Fourier transform of the autocorrelation is the power spectrum:

$$R_{xx}[n] \leftrightarrow S_{xx}(\omega) \tag{5.47}$$

or alternatively

$$R_{xx}[n] = \frac{1}{2\pi}\int_{-\pi}^{\pi} S_{xx}(\omega)e^{j\omega n}d\omega \tag{5.48}$$

If we set $n = 0$ in Eq. (5.48) and use Eq. (5.45) and (5.46), we obtain

$$\sum_{n=-\infty}^{\infty} |x[n]|^2 = \frac{1}{2\pi}\int_{-\pi}^{\pi} |X(\omega)|^2 \, d\omega \tag{5.49}$$

which is called *Parseval's theorem* and says that we can compute the signal's energy in the time domain or in the frequency domain.

5.3. DISCRETE-FREQUENCY TRANSFORMS

Here we describe transforms, including the DFT, DCT and FFT, that take our discrete-time signal into a discrete frequency representation. Discrete-frequency transforms are the natural transform for periodic signals, though we show in Section 5.7 and Chapter 6 how they are also useful for aperiodic signals such as speech.

---

[5] If $z = x + jy = Ae^{j\phi}$, its complex conjugate is defined as $z^* = x - jy = Ae^{-j\phi}$.

Table 5.5 Properties of the Fourier and *z*-transforms.

Property	Signal	Fourier Transform	*z*-Transform
Linearity	$ax_1[n]+bx_2[n]$	$aX_1(e^{j\omega})+bX_2(e^{j\omega})$	$aX_1(z)+bX_2(z)$
Symmetry	$x[-n]$	$X(e^{-j\omega})$	$X(z^{-1})$
	$x^*[n]$	$X^*(e^{-j\omega})$	$X^*(z^*)$
	$x^*[-n]$	$X^*(e^{j\omega})$	$X^*(1/z^*)$
	$x[n]$ real	$X(e^{j\omega})$ is *Hermitian* $X(e^{-j\omega})=X^*(e^{j\omega})$ $\lvert X(e^{j\omega})\rvert$ is even[6] $\text{Re}\{X(e^{j\omega})\}$ is even $\arg\{X(e^{j\omega})\}$ is odd[7] $\text{Im}\{X(e^{j\omega})\}$ is odd	$X(z^*)=X^*(z)$
	$\text{Even}\{x[n]\}$	$\text{Re}\{X(e^{j\omega})\}$	
	$\text{Odd}\{x[n]\}$	$j\,\text{Im}\{X(e^{j\omega})\}$	
Time-shifting	$x[n-n_0]$	$X(e^{j\omega})e^{-j\omega n_0}$	$X(z)z^{-n_0}$
Modulation	$x[n]e^{j\omega_0 n}$	$X(e^{j(\omega-\omega_0)})$	$X(e^{-j\omega_0}z)$
	$x[n]z_0^n$		$X(z/z_0)$
Convolution	$x[n]*h[n]$	$X(e^{j\omega})H(e^{j\omega})$	$X(z)H(z)$
	$x[n]y[n]$	$\frac{1}{2\pi}X(e^{j\omega})*Y(e^{j\omega})$	
Parseval's Theorem	$R_{xx}[n]=\sum_{m=-\infty}^{\infty}x[m+n]x^*[m]$	$S_{xx}(\omega)=\lvert X(\omega)\rvert^2$	$X(z)X^*(1/z^*)$

A discrete transform of a signal $x[n]$ is another signal defined as

$$X[k]=\text{T}\{x[n]\} \tag{5.50}$$

Linear transforms are special transforms that decompose the input signal $x[n]$ into a linear combination of other signals:

$$x[n]=\sum_{k=-\infty}^{\infty}X[k]\varphi_k[n] \tag{5.51}$$

[6] A function *f*(*x*) is called even if and only if $f(x)=f(-x)$.

[7] A function *f*(*x*) is called odd if and only if $f(x)=-f(-x)$.

where $\varphi_k[n]$ is a set of *orthonormal* functions

$$<\varphi_k[n],\varphi_l[n]>=\delta[k-l] \tag{5.52}$$

with the inner product defined as

$$<\varphi_k[n],\varphi_l[n]>=\sum_{n=-\infty}^{\infty}\varphi_k[n]\varphi_l^*[n] \tag{5.53}$$

With this definition, the coefficients $X[k]$ are the projection of $x[n]$ onto $\varphi_k[n]$:

$$X[k]=<x[n],\varphi_k[n]> \tag{5.54}$$

as illustrated in Figure 5.10.

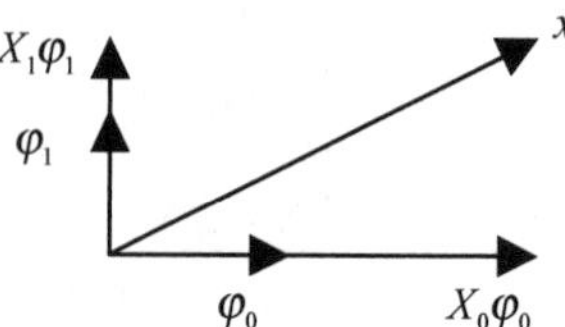

Figure 5.10 Orthonormal expansion of a signal $x[n]$ in a two-dimensional space.

5.3.1. The Discrete Fourier Transform (DFT)

If a $x_N[n]$ signal is periodic with period N then

$$x_N[n]=x_N[n+N] \tag{5.55}$$

and the signal is uniquely represented by N consecutive samples. Unfortunately, since Eq. (5.23) is not met, we cannot guarantee the existence of its Fourier transform. The *Discrete Fourier Transform* (DFT) of a periodic signal $x_N[n]$ is defined as

$$X_N[k]=\sum_{n=0}^{N-1}x_N[n]e^{-j2\pi nk/N} \qquad 0\le k<N \tag{5.56}$$

$$x_N[n]=\frac{1}{N}\sum_{k=0}^{N-1}X_N[k]e^{j2\pi nk/N} \qquad 0\le n<N \tag{5.57}$$

which are transform pairs. Equation (5.57) is also referred as a *Fourier series* expansion.

In Figure 5.11 we see the approximation of a periodic square signal with period $N=100$ as a sum of 19 *harmonic* sinusoids, i.e., we used only the first 19 $X_N[k]$ coefficients in Eq. (5.57).

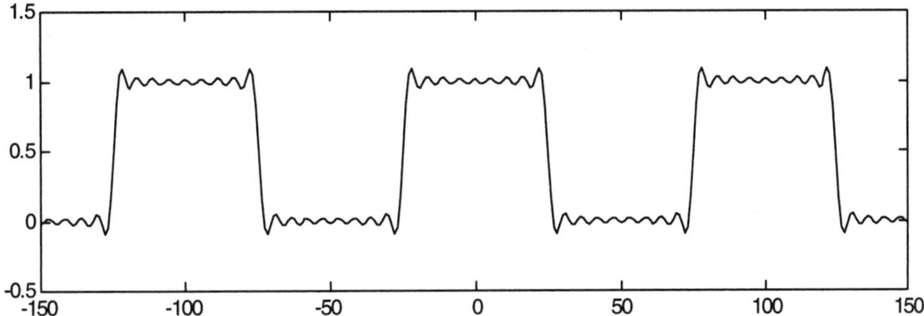

Figure 5.11 Decomposition of a periodic square signal with period 100 samples as a sum of 19 harmonic sinusoids with frequencies $\omega_k = 2\pi k/100$.

$$\tilde{x}_N[n] = \frac{1}{N}\sum_{k=-18}^{18} X_N[k]e^{j2\pi nk/N} = \frac{X_N[0]}{N} + \frac{2}{N}\sum_{k=1}^{18} X_N[k]\cos(2\pi nk/N) \tag{5.58}$$

Had we used 100 harmonic sinusoids, the periodic signal would have been reproduced *exactly*. Nonetheless, retaining a smaller number of sinusoids can provide a decent approximation for a periodic signal.

5.3.2. Fourier Transforms of Periodic Signals

Using the DFT, we now discuss how to compute the Fourier transforms of a complex exponential, an impulse train, and a general periodic signal, since they are signals often used in DSP. We also present a relationship between the continuous-frequency Fourier transform and the discrete Fourier transform.

5.3.2.1. The Complex Exponential

One of the simplest periodic functions is the complex exponential $x[n] = e^{j\omega_0 n}$. Since it has infinite energy, we cannot compute its Fourier transform in its strict sense. Since such signals are so useful, we devise an alternate formulation.

First, let us define the function

$$d_\Delta(\omega) = \begin{cases} 1/\Delta & 0 \le \omega < \Delta \\ 0 & otherwise \end{cases} \tag{5.59}$$

which has the following property

$$\int_{-\infty}^{\infty} d_\Delta(\omega)\,d\omega = 1 \tag{5.60}$$

for all values of $\Delta > 0$.

It is useful to define the continuous delta function $\delta(\omega)$, also known as the *Dirac* delta, as

$$\delta(\omega) = \lim_{\Delta \to 0} d_\Delta(\omega) \tag{5.61}$$

which is a *singular function* and can be seen in Figure 5.12. The Dirac delta is a function of a continuous variable and should not be confused with the Kronecker delta, which is a function of a discrete variable.

Using Eqs. (5.59) and (5.61) we can then see that

$$\int_{-\infty}^{\infty} X(\omega)\delta(\omega)d\omega = \lim_{\Delta \to 0} \int_{-\infty}^{\infty} X(\omega)d_\Delta(\omega)d\omega = X(0) \tag{5.62}$$

and similarly

$$\int_{-\infty}^{\infty} X(\omega)\delta(\omega - \omega_0)d\omega = X(\omega_0) \tag{5.63}$$

so that

$$X(\omega)\delta(\omega - \omega_0) = X(\omega_0)\delta(\omega - \omega_0) \tag{5.64}$$

because the integrals on both sides are identical.

Using Eq. (5.63), we see that the convolution of $X(\omega)$ and $\delta(\omega - \omega_0)$ is

$$X(\omega) * \delta(\omega - \omega_0) = \int_{-\infty}^{\infty} X(u)\delta(\omega - \omega_0 - u)du = X(\omega - \omega_0) \tag{5.65}$$

For the case of a complex exponential, inserting $X(\omega) = e^{j\omega n}$ into Eq. (5.63) results in

$$\int_{-\infty}^{\infty} \delta(\omega - \omega_0)e^{j\omega n}d\omega = e^{j\omega_0 n} \tag{5.66}$$

By comparing Eq. (5.66) with (5.20) we can then obtain

$$e^{j\omega_0 n} \leftrightarrow 2\pi\delta(\omega - \omega_0) \tag{5.67}$$

so that the Fourier transform of a complex exponential is an impulse concentrated at frequency ω_0.

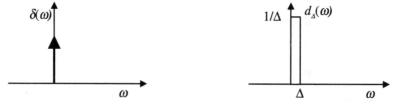

Figure 5.12 Representation of the $\delta(\omega)$ function and its approximation $d_\Delta(\omega)$.

5.3.2.2. The Impulse Train

Since the impulse train

$$p_N[n] = \sum_{k=-\infty}^{\infty} \delta[n-kN] \tag{5.68}$$

is periodic with period N, it can be expanded in Fourier series according to (5.56) as

$$P_N[k] = 1 \tag{5.69}$$

so that using the inverse Fourier series Eq. (5.57), $p_N[n]$ can alternatively be expressed as

$$p_N[n] = \frac{1}{N}\sum_{k=0}^{N-1} e^{j2\pi kn/N} \tag{5.70}$$

which is an alternate expression to Eq. (5.68) as a sum of complex exponentials. Taking the Fourier transform of Eq. (5.70) and using Eq. (5.67) we obtain

$$P_N(e^{j\omega}) = \frac{2\pi}{N}\sum_{k=0}^{N-1} \delta(\omega - 2\pi k/N) \tag{5.71}$$

which is another impulse train in the frequency domain (See Figure 5.13). The impulse train in the time domain is given in terms of the Kronecker delta, and the impulse train in the frequency domain is given in terms of the Dirac delta.

Figure 5.13 An impulse train signal and its Fourier transform, which is also an impulse train.

5.3.2.3. General Periodic Signals

We now compute the Fourier transform of a general periodic signal using the results of Section 5.3.2.2 and show that, in addition to being periodic, the transform is also discrete. Given a periodic signal $x_N[n]$ with period N, we define another signal $x[n]$:

$$x[n] = \begin{cases} x_N[n] & 0 \le n < N \\ 0 & otherwise \end{cases} \tag{5.72}$$

so that

$$x_N[n] = \sum_{k=-\infty}^{\infty} x[n-kN] = x[n] * \sum_{k=-\infty}^{\infty} \delta[n-kN] = x[n] * p_N[n] \tag{5.73}$$

which is the convolution of $x[n]$ with an impulse train $p_N[n]$ as in Eq. (5.68). Since $x[n]$ is of finite length, it has a Fourier transform $X(e^{j\omega})$. Using the convolution property $X_N(e^{j\omega}) = X(e^{j\omega})P_N(e^{j\omega})$, where $P_N(e^{j\omega})$ is the Fourier transform of $p_N[n]$ as given by Eq. (5.71), we obtain another impulse train:

$$X_N(e^{j\omega}) = \frac{2\pi}{N} \sum_{k=-\infty}^{\infty} X(e^{j2\pi k/N})\delta(\omega - 2\pi k/N) \tag{5.74}$$

Therefore the Fourier transform $X_N(e^{j\omega})$ of a periodic signal $x_N[n]$ can be expressed in terms of samples $\omega_k = 2\pi k/N$, spaced $2\pi/N$ apart, of the Fourier transform $X(e^{j\omega})$ of $x[n]$, one period of the signal $x_N[n]$. The relationships between $x[n]$, $x_N[n]$, $X(e^{j\omega})$ and $X_N(e^{j\omega})$ are shown in Figure 5.14.

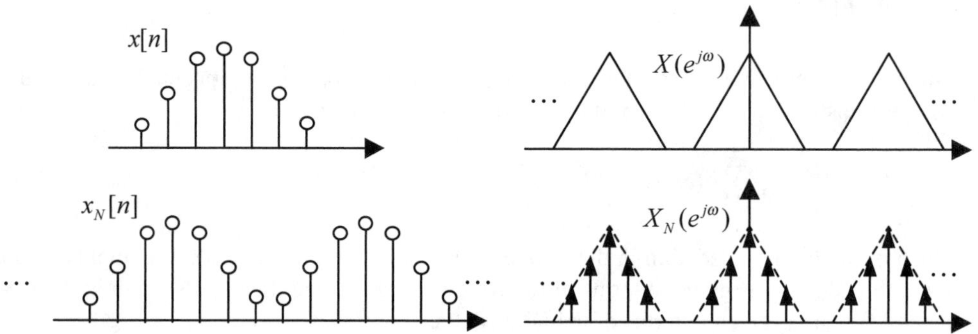

Figure 5.14 Relationships between finite and periodic signals and their Fourier transforms. On one hand, $x[n]$ is a length N discrete signal whose transform $X(e^{j\omega})$ is continuous and periodic with period 2π. On the other hand, $x_N[n]$ is a periodic signal with period N whose transform $X_N(e^{j\omega})$ is discrete and periodic.

5.3.3. The Fast Fourier Transform (FFT)

There is a family of fast algorithms to compute the DFT, which are called Fast Fourier Transforms (FFT). Direct computation of the DFT from Eq. (5.56) requires N^2 operations, assuming that the trigonometric functions have been pre-computed. The FFT algorithm only requires on the order of $N \log_2 N$ operations, so it is widely used for speech processing.

5.3.3.1. Radix-2 FFT

Let's express the discrete Fourier transform of $x[n]$

$$X[k] = \sum_{n=0}^{N-1} x[n]e^{-j2\pi nk/N} = \sum_{n=0}^{N-1} x[n]W_N^{nk} \qquad 0 \le k < N \tag{5.75}$$

where we have defined for convenience

$$W_N = e^{-j2\pi/N} \tag{5.76}$$

Equation (5.75) requires N^2 complex multiplies and adds. Now, let's suppose N is even, and let $f[n] = x[2n]$ represent the even-indexed samples of $x[n]$, and $g[n] = x[2n+1]$ the odd-indexed samples. We can express Eq. (5.75) as

$$X[k] = \sum_{n=0}^{N/2-1} f[n]W_{N/2}^{nk} + W_N^k \sum_{n=0}^{N/2-1} g[n]W_{N/2}^{nk} = F[k] + W_N^k G[k] \tag{5.77}$$

where $F[k]$ and $G[k]$ are the $N/2$ point DFTs of $f[n]$ and $g[n]$, respectively. Since both $F[k]$ and $G[k]$ are defined for $0 \le k < N/2$, we need to also evaluate them for $N/2 \le k < N$, which is straightforward, since

$$F[k + N/2] = F[k] \tag{5.78}$$

$$G[k + N/2] = G[k] \tag{5.79}$$

If $N/2$ is also even, then both $f[n]$ and $g[n]$ can be decomposed into sequences of even and odd indexed samples and therefore its DFT can be computed using the same process. Furthermore, if N is an integer power of 2, this process can be iterated and it can be shown that the number of multiplies and adds is $N \log_2 N$, which is a significant saving from N^2. This is the *decimation-in-time* algorithm and can be seen in Figure 5.15. A dual algorithm called *decimation-in-frequency* can be derived by decomposing the signal into its first $N/2$ and its last $N/2$ samples.

5.3.3.2. Other FFT Algorithms

Although the radix-2 FFT is the best known algorithm, there are other variants that are faster and are more often used in practice. Among those are the radix-4, radix-8, split-radix and prime-factor algorithm.

The same process used in the derivation of the radix-2 decimation-in-time algorithm applies if we decompose the sequences into four sequences: $f_1[n] = x[4n]$, $f_2[n] = x[4n+1]$, $f_3[n] = x[4n+2]$, and $f_4[n] = x[4n+3]$. This is the radix-4 algorithm, which can be applied when N is a power of 4, and is generally faster than an equivalent radix-2 algorithm.

Similarly there are radix-8 and radix-16 algorithms for N being powers of 8 and 16 respectively, which use fewer multiplies and adds. But because of possible additional control logic, it is not obvious that they will be faster, and every algorithm needs to be optimized for a given processor.

There are values of N, such as $N = 128$, for which we cannot use radix-4, radix-8 nor radix-16, so we have to use the less efficient radix-2. A combination of radix-2 and radix-4, called *split-radix* [5], has been shown to have fewer multiplies than both radix-2 and radix-4, and can be applied to N being a power of 2.

Finally, another possible decomposition is $N = p_1 p_2 \cdots p_L$ with p_i being prime numbers. This leads to the *prime-factor algorithm* [2]. While this family of algorithms offers a similar number of operations as the algorithms above, it offers more flexibility in the choice of *N*.

5.3.3.3. FFT Subroutines

Typically, FFT subroutines are computed *in-place* to save memory and have the form `fft (float *xr, float *xi, int n)` where `xr` and `xi` are the real and imaginary parts respectively of the input sequence, before calling the subroutine, and the real and imaginary parts of the output transform, after returning from it. C code that implements a decimation-in-time radix-2 FFT of Figure 5.15 is shown in Figure 5.16.

The first part of the subroutine in Figure 5.16 is doing the so-called *butterflies*, which use the trigonometric factors, also called *twiddle factors*. Normally, those twiddle factors are pre-computed and stored in a table. The second part of the subroutine deals with the fact that the output samples are not linearly ordered (see Figure 5.15); in fact, the indexing has the bits reversed, which is why we need to do *bit reversal*, also called *descrambling*.

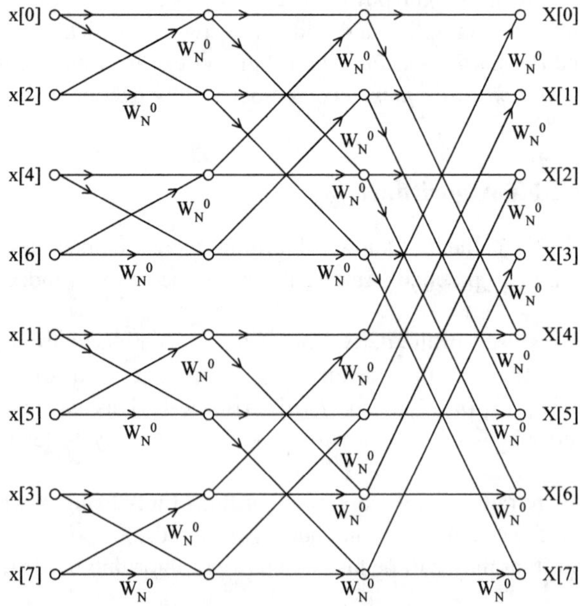

Figure 5.15 Decimation in time radix-2 algorithm for an 8-point FFT.

```
void fft2 (float *x, float *y, int n, int m)
{
     int n1, n2, i, j, k, l;
     float  xt, yt, c, s;
     double e, a;

     /* Loop through all m stages */
     n2 = n;
     for (k = 0; k < m; k++) {
         n1 = n2;
         n2 = n2 / 2;
         e = PI2 / n1;
         for (j = 0; j < n2; j++) {
             /* Compute Twiddle factors */
             a = j * e;
             c = (float) cos (a);
             s = (float) sin (a);

             /* Do the butterflies */
             for (i = j; i < n; i += n1) {
                 l = i + n2;
                 xt = x[i] - x[l];
                 x[i] = x[i] + x[l];
                 yt = y[i] - y[l];
                 y[i] = y[i] + y[l];
                 x[l] = c * xt + s * yt;
                 y[l] = c * yt - s * xt;
             }
         }
     }

     /* Bit reversal: descrambling */
     j = 0;
     for (i = 0; i < n - 1; i++) {
         if (i < j) {
             xt = x[j];
             x[j] = x[i];
             x[i] = xt;
             xt = y[j];
             y[j] = y[i];
             y[i] = xt;
         }
         k = n / 2;
         while (k <= j) {
             j -= k;
             k /= 2;
         }
         j += k;
     }
}
```

Figure 5.16 C source for a decimation-in-time radix-2 FFT. Before calling the subroutine, x and y contain the real and imaginary parts of the input signal respectively. After returning from the subroutine, x and y contain the real and imaginary parts of the Fourier transform of the input signal. n is the length of the FFT and is related to m by $n = 2^m$.

To compute the inverse FFT an additional routine is not necessary; it can be computed with the subroutine above. To see that, we expand the DFT in Eq. (5.56) into its real and imaginary parts:

$$X_R[k]+jX_I[k]=\sum_{n=0}^{N-1}\left(x_R[n]+jx_I[n]\right)e^{-j2\pi nk/N} \tag{5.80}$$

take complex conjugate and multiply by j to obtain

$$X_I[k]+jX_R[k]=\sum_{n=0}^{N-1}\left(x_I[n]+jx_R[n]\right)e^{j2\pi nk/N} \tag{5.81}$$

which has the same functional form as the expanded inverse DFT of Eq. (5.57)

$$x_R[k]+jx_I[k]=\frac{1}{N}\sum_{n=0}^{N-1}\left(X_R[n]+jX_I[n]\right)e^{j2\pi nk/N} \tag{5.82}$$

so that the inverse FFT can be computed by calling `fft (xi, xr, n)` other than the (1/*N*) factor.

Often the input signal $x[n]$ is real, so that we know from the symmetry properties of Table 5.5 that its Fourier transform is Hermitian. This symmetry can be used to compute the length-*N* FFT more efficiently with a length (*N*/2) FFT. One way of doing so is to define $f[n]=x[2n]$ to represent the even-indexed samples of $x[n]$, and $g[n]=x[2n+1]$ the odd-indexed samples. We can then define a length (*N*/2) complex signal $h[n]$ as

$$h[n]=f[n]+jg[n]=x[2n]+jx[2n+1] \tag{5.83}$$

whose DFT is

$$H[k]=F[k]+jG[k]=H_R[k]+jH_I[k] \tag{5.84}$$

Since $f[n]$ and $g[n]$ are real, their transforms are Hermitian and thus

$$H^*[-k]=F^*[-k]-jG^*[-k]=F[k]-jG[k] \tag{5.85}$$

Using Eqs. (5.84) and (5.85), we can obtain $F[k]$ and $G[k]$ as a function of $H_R[k]$ and $H_I[k]$:

$$F[k]=\frac{H[k]+H^*[-k]}{2}=\left(\frac{H_R[k]+H_R[-k]}{2}\right)+j\left(\frac{H_I[k]-H_I[-k]}{2}\right) \tag{5.86}$$

$$G[k]=\frac{H[k]-H^*[-k]}{2j}=\left(\frac{H_I[k]+H_I[-k]}{2}\right)-j\left(\frac{H_R[k]-H_R[-k]}{2}\right) \tag{5.87}$$

As shown in Eq. (5.77), $X[k]$ can be obtained as a function of $F[k]$ and $G[k]$

$$X[k] = F[k] + G[k]W_N^{-k} \tag{5.88}$$

so that the DFT of the real sequence $x[n]$ is obtained through Eqs. (5.83), (5.86), (5.87) and (5.88). The computational complexity is a length ($N/2$) complex FFT plus N real multiplies and $3N$ real adds.

5.3.4. Circular Convolution

The convolution of two periodic signals is not defined according to Eq. (5.12). Given two periodic signals $x_1[n]$ and $x_2[n]$ with period N, we define their *circular convolution* as

$$y[n] = x_1[n] \otimes x_2[n] = \sum_{m=0}^{N-1} x_1[m]x_2[n-m] = \sum_{m=<N>} x_1[m]x_2[n-m] \tag{5.89}$$

where $m =< N >$ in Eq. (5.89) means that the sum lasts only one period. In fact, the sum could be over any N consecutive samples, not just the first N. Moreover, $y[n]$ is also periodic with period N. Furthermore, it is left to the reader to show that

$$Y[k] = X_1[k]X_2[k] \tag{5.90}$$

i.e., the DFT of $y[n]$ is the product of the DFTs of $x_1[n]$ and $x_2[n]$.

An important application of the above result is the computation of a regular convolution using a circular convolution. Let $x_1[n]$ and $x_2[n]$ be two signals such that $x_1[n]=0$ outside $0 \le n < N_1$, and $x_2[n]=0$ outside $0 \le n < N_2$. We know that their regular convolution $y[n] = x_1[n] * x_2[n]$ is zero outside $0 \le N_1 + N_2 - 1$. If we choose an integer N such that $N \ge N_1 + N_2 - 1$, we can define two periodic signals $\tilde{x}_1[n]$ and $\tilde{x}_2[n]$ with period N such that

$$\tilde{x}_1[n] = \begin{cases} x_1[n] & 0 \le n < N_1 \\ 0 & N_1 \le n < N \end{cases} \tag{5.91}$$

$$\tilde{x}_2[n] = \begin{cases} x_2[n] & 0 \le n < N_2 \\ 0 & N_2 \le n < N \end{cases} \tag{5.92}$$

where $x_1[n]$ and $x_2[n]$ have been *zero padded.* It can be shown that the circular convolution $\tilde{y}[n] = \tilde{x}_1[n] \otimes \tilde{x}_2[n]$ is identical to $y[n]$ for $0 \le n < N$, which means that $y[n]$ can be obtained as the inverse DFT of $\tilde{Y}[k] = \tilde{X}_1[k]\tilde{X}_2[k]$. This method of computing the regular convolution of two signals is more efficient than the direct calculation when N is large. While the crossover point will depend on the particular implementations of the FFT and convolution, as well as the processor, in practice this has been found beneficial for $N \ge 1024$.

5.3.5. The Discrete Cosine Transform (DCT)

The Discrete Cosine Transform (DCT) is widely used for speech processing. It has several definitions. The DCT-II $C[k]$ of a real signal $x[n]$ is defined by:

$$C[k] = \sum_{n=0}^{N-1} x[n]\cos\left(\pi k(n+1/2)/N\right) \quad \text{for } 0 \le k < N \tag{5.93}$$

with its inverse given by

$$x[n] = \frac{1}{N}\left\{C[0] + 2\sum_{k=1}^{N-1} C[k]\cos\left(\pi k(n+1/2)/N\right)\right\} \quad \text{for } 0 \le n < N \tag{5.94}$$

The DCT-II can be derived from the DFT by assuming $x[n]$ is a real periodic sequence with period $2N$ and with an even symmetry $x[n] = x[2N-1-n]$. It is left to the reader to show, that $X[k]$ and $C[k]$ are related by

$$X[k] = 2e^{j\pi k/2N}C[k] \quad \text{for } 0 \le k < N \tag{5.95}$$

$$X[2N-k] = 2e^{-j\pi k/2N}C[k] \quad \text{for } 0 \le k < N \tag{5.96}$$

It is left to the reader to prove Eq. (5.94) is indeed the inverse transform using Eqs. (5.57), (5.95), and (5.96). Other versions of the DCT-II have been defined that differ on the normalization constants but are otherwise the same.

There are eight different ways to extend an N-point sequence and make it both periodic and even, such that can be uniquely recovered. The DCT-II is just one of the ways, with three others being shown in Figure 5.17.

The DCT-II is the most often used discrete cosine transform because of its *energy compaction*, which results in its coefficients being more concentrated at low indices than the DFT. This property allows us to approximate the signal with fewer coefficients [10].

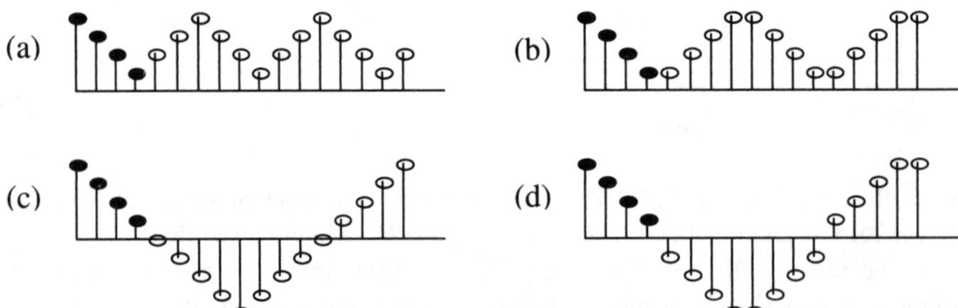

Figure 5.17 Four ways to extend a four-point sequence $x[n]$ to make it both periodic and have even symmetry. The figures in (a), (b), (c) and (d) correspond to the DCT-I, DCT-II, DCT-III and DCT-IV respectively.

From Eq. (5.95) and (5.96) we see that the DCT-II of a real sequence can be computed with a length-2*N* FFT of a real and even sequence, which in turn can be computed with a length (*N*/2) complex FFT and some additional computations. Other fast algorithms have been derived to compute the DCT directly [15], using the principles described in Section 5.3.3.1. Two-dimensional transforms can also be used for image processing.

5.4. DIGITAL FILTERS AND WINDOWS

We describe here the fundamentals of digital filter design and study *finite-impulse response* (FIR) and *infinite-impulse response* (IIR) filters, which are special types of linear time-invariant digital filters. We establish the time-frequency duality and study the ideal low-pass filter (frequency limited) and its dual window functions (time limited).

5.4.1. The Ideal Low-Pass Filter

It is useful to find an impulse response $h[n]$ whose Fourier transform is

$$H(e^{j\omega}) = \begin{cases} 1 & |\omega| < \omega_0 \\ 0 & \omega_0 < |\omega| < \pi \end{cases} \tag{5.97}$$

which is the ideal *low-pass* filter because it lets all frequencies below ω_0 pass through unaffected and completely blocks frequencies above ω_0. Using the definition of Fourier transform, we obtain

$$h[n] = \frac{1}{2\pi}\int_{-\omega_0}^{\omega_0} e^{j\omega n} d\omega = \frac{\left(e^{j\omega_0 n} - e^{-j\omega_0 n}\right)}{2\pi jn} = \frac{\sin \omega_0 n}{\pi n} = \left(\frac{\omega_0}{\pi}\right)\operatorname{sinc}\left(2 f_0 n\right) \tag{5.98}$$

where we have defined the so-called sinc function as

$$\operatorname{sinc}(x) = \frac{\sin \pi x}{\pi x} \tag{5.99}$$

which is a real and even function of x and is plotted in Figure 5.18. Note that the sinc function is *0* when x is a nonzero integer.

Thus, an ideal low-pass filter is noncausal since it has an impulse response with an infinite number of nonzero coefficients.

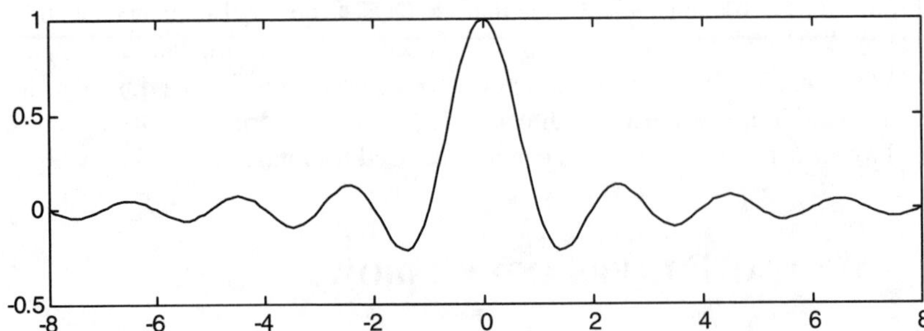

Figure 5.18 A sinc function, which is the impulse response of the ideal low-pass filter with a scale factor.

5.4.2. Window Functions

Window functions are signals that are concentrated in time, often of limited duration. While window functions such as *triangular*, *Kaiser*, *Barlett,* and *prolate spheroidal* occasionally appear in digital speech processing systems, the rectangular, Hanning, and Hamming are the most widely used. Window functions are also concentrated in low frequencies. These window functions are useful in digital filter design and all throughout Chapter 6.

5.4.2.1. The Rectangular Window

The *rectangular* window is defined as

$$h_\pi[n] = u[n] - u[n-N] \tag{5.100}$$

and we refer to it often in this book. Its z-transform is given by

$$H_\pi(z) = \sum_{n=0}^{N-1} z^{-n} \tag{5.101}$$

which results in a polynomial of order $(N-1)$. Multiplying both sides of Eq. (5.101) by z^{-1}, we obtain

$$z^{-1}H_\pi(z) = \sum_{n=1}^{N} z^{-n} = H_\pi(z) - 1 + z^{-N} \tag{5.102}$$

and therefore the sum of the terms of a geometric series can also be expressed as

$$H_\pi(z) = \frac{1-z^{-N}}{1-z^{-1}} \tag{5.103}$$

Although $z=1$ appears to be a pole in Eq. (5.103), it actually isn't because it is canceled by a zero at $z=1$. Since $h_\pi[n]$ has finite length, Eq. (5.25) must be satisfied for $z \neq 0$, so the region of convergence is everywhere but at $z=0$. Moreover, all finite-length sequences have a region of convergence that is the complete z-plane except for possibly $z=0$.

The Fourier transform of the rectangular window is, using Eq. (5.103):

$$\begin{aligned} H_\pi(e^{j\omega}) &= \frac{1-e^{-j\omega N}}{1-e^{-j\omega}} = \frac{\left(e^{j\omega N/2}-e^{-j\omega N/2}\right)e^{-j\omega N/2}}{\left(e^{j\omega/2}-e^{-j\omega/2}\right)e^{-j\omega/2}} \\ &= \frac{\sin \omega N/2}{\sin \omega/2} e^{-j\omega(N-1)/2} = A(\omega)e^{-j\omega(N-1)/2} \end{aligned} \tag{5.104}$$

where $A(\omega)$ is real and even. The function $A(\omega)$, plotted in Figure 5.19 in dB,[8] is 0 for $\omega_k = 2\pi k/N$ with $k \neq \{0, \pm N, \pm 2N, \ldots\}$, and is the discrete-time equivalent of the sinc function.

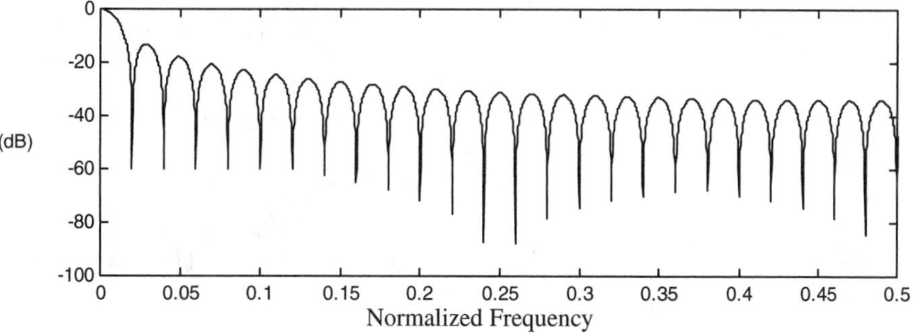

Figure 5.19 Frequency response (magnitude in dB) of the rectangular window with $N = 50$, which is a digital sinc function.

5.4.2.2. The Generalized Hamming Window

The *generalized Hamming window* is defined as

$$h_h[n] = \begin{cases} (1-\alpha) - \alpha\cos(2\pi n/N) & 0 \le n < N \\ 0 & otherwise \end{cases} \tag{5.105}$$

[8] An energy value E is expressed in decibels (dB) as $\overline{E} = 10\log_{10} E$. If the energy value is $2E$, it is therefore 3dB higher. Logarithmic measurements like dB are useful because they correlate well with how the human auditory system perceives volume.

and can be expressed in terms of the rectangular window in Eq. (5.100) as

$$h_h[n] = (1-\alpha)h_\pi[n] - \alpha h_\pi[n]\cos(2\pi n / N) \tag{5.106}$$

whose transform is

$$H_h(e^{j\omega}) = (1-\alpha)H_\pi(e^{j\omega}) - (\alpha/2)H_\pi(e^{j(\omega-2\pi/N)}) - (\alpha/2)H_\pi(e^{j(\omega+2\pi/N)}) \tag{5.107}$$

after using the modulation property in Table 5.5. When $\alpha = 0.5$ the window is known as the *Hanning* window, whereas for $\alpha = 0.46$ it is the *Hamming* window. Hanning and Hamming windows and their magnitude frequency responses are plotted in Figure 5.20.

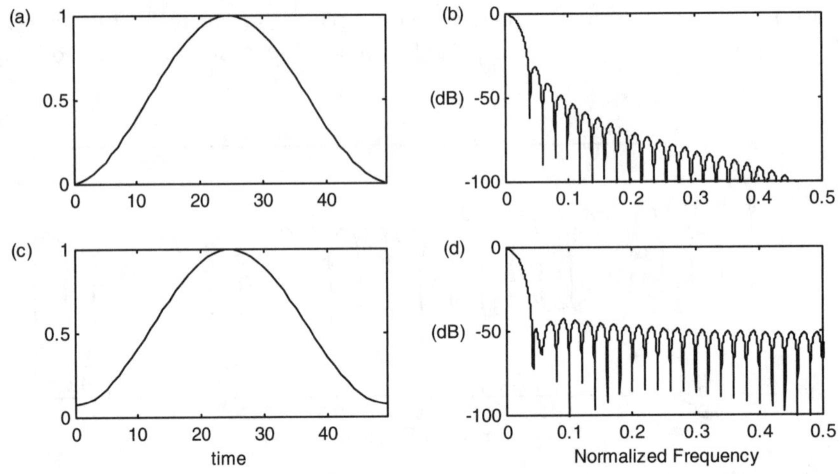

Figure 5.20 (a) Hanning window and (b) the magnitude of its frequency response in dB; (c) Hamming window and (d) the magnitude of its frequency response in dB for $N = 50$.

The main lobe of both Hamming and Hanning is twice as wide as that of the rectangular window, but the attenuation is much greater than that of the rectangular window. The secondary lobe of the Hanning window is 31 dB below the main lobe, whereas for the Hamming window it is 44 dB below. On the other hand, the attenuation of the Hanning window decays with frequency quite rapidly, which is not the case for the Hamming window, whose attenuation stays approximately constant for all frequencies.

5.4.3. FIR Filters

From a practical point of view, it is useful to consider LTI filters whose impulse responses have a limited number of nonzero coefficients:

$$h[n] = \begin{cases} b_n & 0 \le n \le M \\ 0 & otherwise \end{cases} \tag{5.108}$$

These types of LTI filters are called *finite-impulse response* (FIR) filters. The input/output relationship in this case is

$$y[n] = \sum_{r=0}^{M} b_r x[n-r] \tag{5.109}$$

The z-transform of $x[n-r]$ is

$$\sum_{n=-\infty}^{\infty} x[n-r]z^{-n} = \sum_{n=-\infty}^{\infty} x[n]z^{-(n+r)} = z^{-r}X(z) \tag{5.110}$$

Therefore, given that the z-transform is linear, $H(z)$ is

$$H(z) = \frac{Y(z)}{X(z)} = \sum_{r=0}^{M} b_r z^{-r} = Az^{-L}\prod_{r=1}^{M-L}\left(1-c_r z^{-1}\right) \tag{5.111}$$

whose region of convergence is the whole z-plane except for possibly $z=0$. Since $\sum_{r=0}^{M} |b_r|$ is finite, FIR systems are always stable, which makes them very attractive. Several special types of FIR filters will be analyzed below: linear-phase, first-order and low-pass FIR filters.

5.4.3.1. Linear-Phase FIR Filters

Linear-phase filters are important because, other than a delay, the phase of the signal is unchanged. Only the magnitude is affected. Therefore, the temporal properties of the input signal are preserved. In this section we show that linear-phase FIR filters can be built if the filter exhibits symmetry.

Let's explore the particular case of $h[n]$ real, $M = 2L$, an even number, and $h[n] = h[M-n]$ (called a *Type-I* filter). In this case

$$\begin{aligned} H(e^{j\omega}) &= \sum_{n=0}^{M} h[n]e^{-j\omega n} = h[L]e^{-j\omega L} + \sum_{n=0}^{L-1}\left(h[n]e^{-j\omega n} + h[M-n]e^{-j\omega(2L-n)}\right) \\ &= h[L]e^{-j\omega L} + \sum_{n=0}^{L-1} h[n]\left(e^{-j\omega(n-L)} + e^{j\omega(n-L)}\right)e^{-j\omega L} \\ &= \left(h[L] + \sum_{n=1}^{L} 2h[n+L]\cos(\omega n)\right)e^{-j\omega L} = A(\omega)e^{-j\omega L} \end{aligned} \tag{5.112}$$

where $A(\omega)$ is a real and even function of ω, since the cosine is an even function, and $A(\omega)$ is a linear combination of cosines. Furthermore, we see that the phase $\arg\{H(e^{j\omega})\} = L\omega$, which is a linear function of ω, and therefore $h[n]$ is called a *linear-*

phase system. It can be shown that if $h[n] = -h[M-n]$, we also get a linear phase system but $A(\omega)$ this time is a pure imaginary and odd function (Type III filter). It is left to the reader to show that in the case of M being odd the system is still linear phase (Types II and IV filters). Moreover, $h[n]$ doesn't have to be real and:

$$h[n] = \pm h^*[M-n] \tag{5.113}$$

is a sufficient condition for $h[n]$ to be linear phase.

5.4.3.2. First-Order FIR Filters

A special case of FIR filters is the first-order filter:

$$y[n] = x[n] + \alpha x[n-1] \tag{5.114}$$

for real values of α, which, unless $\alpha = 1$, is not linear phase. Its z-transform is

$$H(z) = 1 + \alpha z^{-1} \tag{5.115}$$

It is of interest to analyze the magnitude and phase of its frequency response

$$\begin{aligned} |H(e^{j\omega})|^2 &= |1 + \alpha(\cos\omega - j\sin\omega)|^2 \\ &= (1+\alpha\cos\omega)^2 + (\alpha\sin\omega)^2 = 1 + \alpha^2 + 2\alpha\cos\omega \end{aligned} \tag{5.116}$$

$$\theta(e^{j\omega}) = -\arctan\left(\frac{\alpha\sin\omega}{1+\alpha\cos\omega}\right) \tag{5.117}$$

It is customary to display the magnitude response in decibels (dB):

$$10\log|H(e^{j\omega})|^2 = 10\log\left[(1+\alpha)^2 + 2\alpha\cos\omega\right] \tag{5.118}$$

as shown in Figure 5.21 for various values of α.

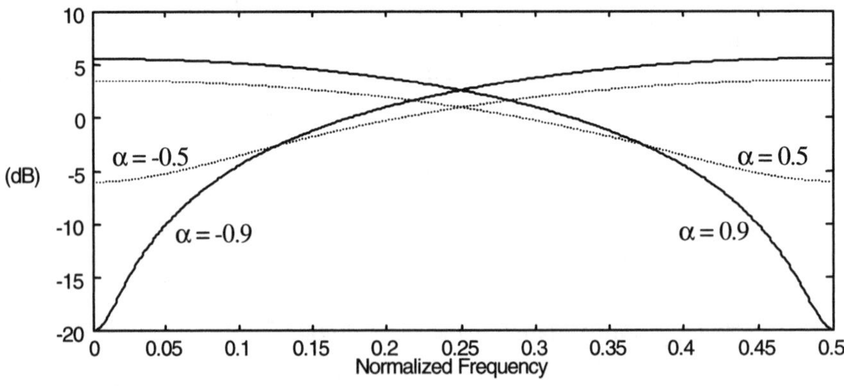

Figure 5.21 Frequency response of the first order FIR filter for various values of α.

We see that for $\alpha > 0$ we have a *low-pass* filter whereas for $\alpha < 0$ it is a *high-pass* filter, also called a *pre-emphasis* filter, since it emphasizes the high frequencies. In general, filters that boost the high frequencies and attenuate the low frequencies are called *high-pass* filters, and filters that emphasize the low frequencies and de-emphasize the high frequencies are called *low-pass* filters. The parameter α controls the slope of the curve.

5.4.3.3. Window Design FIR Lowpass Filters

The ideal lowpass filter lets all frequencies below ω_0 go through and eliminates all energy from frequencies above that range. As we described in Section 5.4.1, the ideal lowpass filter has an infinite impulse response, which poses difficulties for implementation in a practical system, as it requires an infinite number of multiplies and adds.

Since we know that the sinc function decays over time, it is reasonable to assume that a truncated sinc function that keeps a large enough number of samples *N* could be a good approximation to the ideal low-pass filter. Figure 5.22 shows the magnitude of the frequency response of such a truncated sinc function for different values of *N*. While the approximation gets better for larger *N*, the overshoot near ω_0 doesn't go away and in fact stays at about 9% of the discontinuity even for large *N*. This is known as the *Gibbs phenomenon*, since Yale professor Josiah Gibbs first noticed it in 1899.

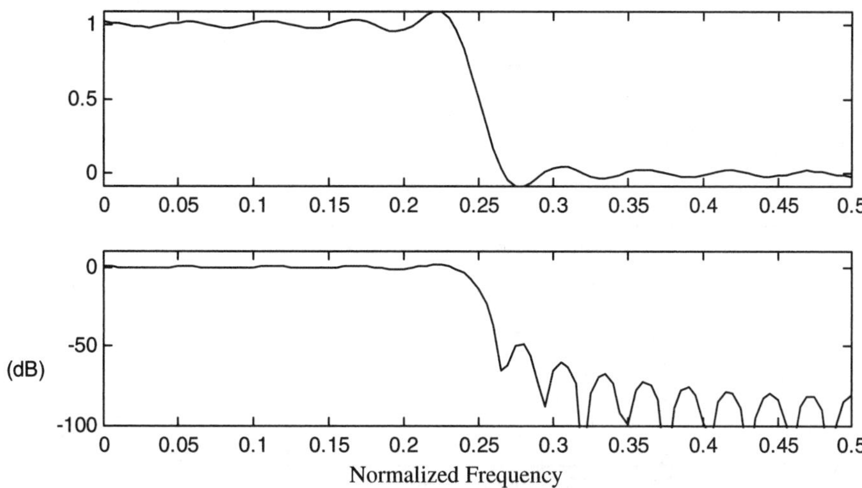

Figure 5.22 Magnitude frequency response of the truncated sinc signal (*N*=200) for $\omega_0 = \pi / 4$. It is an approximation to the ideal low-pass filter, though we see that overshoots are present near the transition. The first graph is linear magnitude and the second is in dB.

In computing the truncated sinc function, we have implicitly multiplied the ideal low-pass filter, the sinc function, by a rectangular window. In the so-called *window design* filter design method, the filter coefficients are obtained by multiplying the ideal sinc function by a tapering window function, such as the Hamming window. The resulting frequency response is the convolution of the ideal lowpass filter function with the transform of the window (shown in Figure 5.23), and it does not exhibit the overshoots in Figure 5.22, at the expense of a slower transition.

5.4.3.4. Parks McClellan Algorithm

While the window design method is simple, it is hard to predict what the final response will be. Other methods have been proposed whose coefficients are obtained to satisfy some constraints. If our constraints are a maximum ripple of δ_p in the *passband* ($0 \le \omega < \omega_p$), and a minimum attenuation of δ_s in the *stopband* ($\omega_s \le \omega < \pi$), the optimal solution is given by the Parks McClellan algorithm [14].

The transformation

$$x = \cos\omega \tag{5.119}$$

maps the interval $0 \le \omega \le \pi$ into $-1 \le x \le 1$. We note that

$$\cos(n\omega) = T_n(\cos\omega) \tag{5.120}$$

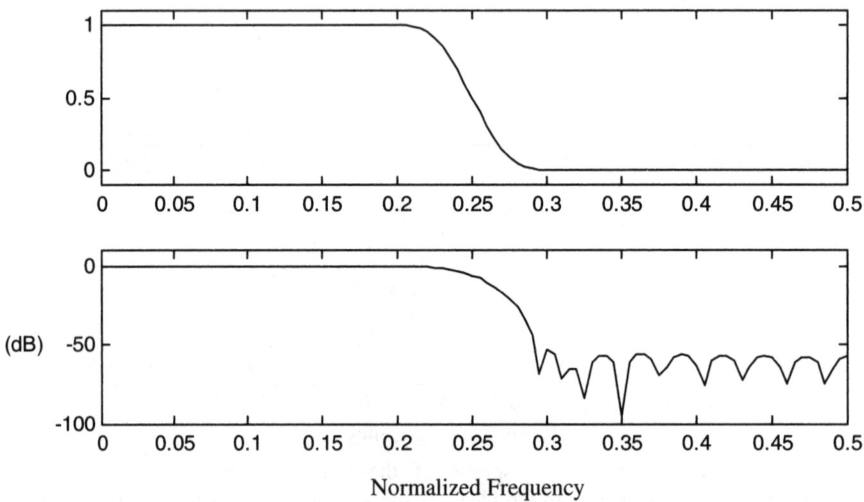

Figure 5.23 Magnitude frequency response of a low-pass filter obtained with the window design method and a Hamming window (N = 200). The first graph is linear magnitude and the second is in dB.

where $T_n(x)$ is the nth-order *Chebychev* polynomial. The first two Chebychev polynomials are given by $T_0(x)=1$ and $T_1(x)=x$. If we add the following trigonometric identities

$$\begin{aligned}\cos(n+1)\omega &= \cos n\omega\cos\omega - \sin n\omega\sin\omega\\ \cos(n-1)\omega &= \cos n\omega\cos\omega + \sin n\omega\sin\omega\end{aligned} \tag{5.121}$$

and use Eqs. (5.119) and (5.120), we obtain the following recursion formula:

$$T_{n+1}(x) = 2xT_n(x) - T_{n-1}(x) \qquad \text{for} \quad n>1 \tag{5.122}$$

Using Eq. (5.120), the magnitude response of a linear phase Type-I filter in Eq. (5.112) can be expressed as an Lth-order polynomial in $\cos\omega$:

$$A(\omega) = \sum_{k=0}^{L} a_k(\cos\omega)^k \tag{5.123}$$

which, using Eq. (5.119) results in a polynomial

$$P(x) = \sum_{k=0}^{L} a_k x^k \tag{5.124}$$

Given that a desired response is $D(x) = D(\cos\omega)$, we define the weighted squared error as

$$E(x) = E(\cos\omega) = W(\cos\omega)[D(\cos\omega) - P(\cos\omega)] = W(x)[D(x) - P(x)] \tag{5.125}$$

where $W(\cos\omega)$ is the weighting in ω. A necessary and sufficient condition for this weighted squared error to be minimized is to have $P(x)$ alternate between minima and maxima. For the case of a low-pass filter,

$$D(\cos\omega) = \begin{cases} 1 & \cos\omega_p \le \cos\omega \le 1 \\ 0 & -1 \le \cos\omega \le \cos\omega_s \end{cases} \tag{5.126}$$

and the weight in the stopband is several times larger than in the passband.

These constraints and the response of a filter designed with such a method are shown in Figure 5.24. We can thus obtain a similar transfer function with fewer coefficients using this method.

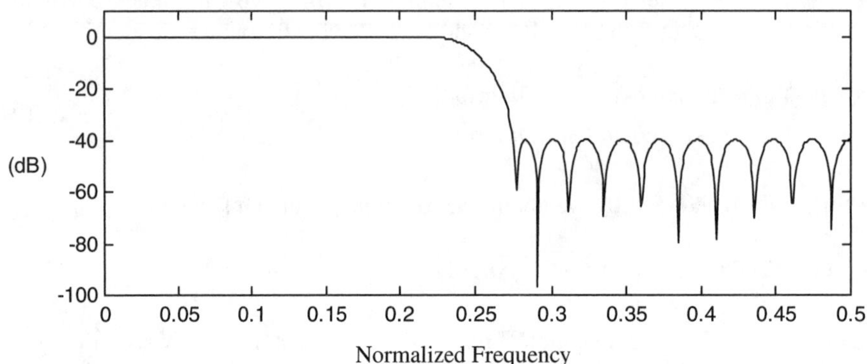

Figure 5.24 Magnitude frequency response of a length-19 lowpass filter designed with the Parks McClellan algorithm.

5.4.4. IIR Filters

Other useful filters are a function of past values of the input and also the output

$$y[n] = \sum_{k=1}^{N} a_k y[n-k] + \sum_{r=0}^{M} b_r x[n-r] \tag{5.127}$$

whose z-transform is given by

$$H(z) = \frac{Y(z)}{X(z)} = \frac{\sum_{r=0}^{M} b_r z^{-r}}{1 - \sum_{k=1}^{N} a_k z^{-k}} \tag{5.128}$$

which in turn can be expressed as a function of the roots of the numerator c_r (called *zeros*), and denominator d_k (called *poles*) as

$$H(z) = \frac{Az^{-L} \prod_{r=1}^{M-L} \left(1 - c_r z^{-1}\right)}{\prod_{k=1}^{N} \left(1 - d_k z^{-1}\right)} \tag{5.129}$$

It is not obvious what the impulse response of such a system is by looking at either Eq. (5.128) or Eq. (5.129). To do that, we can compute the inverse z-transform of Eq. (5.129). If $M < N$ in Eq. (5.129), $H(z)$ can be expanded into partial fractions (see Section 5.2.3.3) as

$$H(z) = \sum_{k=1}^{N} \frac{A_k}{1 - d_k z^{-1}} \tag{5.130}$$

and if $M \geq N$

$$H(z) = \sum_{k=1}^{N} \frac{A_k}{1 - d_k z^{-1}} + \sum_{k=0}^{M-N} B_k z^{-k} \tag{5.131}$$

which we can now compute, since we know that the inverse z-transform of $H_k(z) = A_k / (1 - d_k z^{-1})$ is

$$h_k[n] = \begin{cases} A_k d_k^n u[n] & |d_k| < 1 \\ -A_k d_k^n u[-n-1] & |d_k| > 1 \end{cases} \tag{5.132}$$

so that the convergence region includes the unit circle and therefore $h_k[n]$ is stable. Therefore, a necessary and sufficient condition for $H(z)$ to be stable *and* causal simultaneously is that all its poles be inside the unit circle: i.e., $|d_k| < 1$ for all k, so that its impulse response is given by

$$h[n] = B_n + \sum_{k=1}^{N} A_k d_k^n u[n] \tag{5.133}$$

which has an infinite impulse response, and hence its name.

Since IIR systems may have poles outside the unit circle, they are not guaranteed to be stable and causal like their FIR counterparts. This makes IIR filter design more difficult, since only stable and causal filters can be implemented in practice. Moreover, unlike FIR filters, IIR filters do not have linear phase. Despite these difficulties, IIR filters are popular because they are more efficient than FIR filters in realizing steeper roll-offs with fewer coefficients. In addition, as shown in Chapter 6, they represent many physical systems.

5.4.4.1. First-Order IIR Filters

An important type of IIR filter is the first-order filter of the form

$$y[n] = Ax[n] + \alpha y[n-1] \tag{5.134}$$

for α real. Its transfer function is given by

$$H(z) = \frac{A}{1 - \alpha z^{-1}} \tag{5.135}$$

This system has one pole and no zeros. As we saw in our discussion of z-transforms in Section 5.2.3, a necessary condition for this system to be both stable and causal is that $|\alpha| < 1$. Since for the low-pass filter case $0 < \alpha < 1$, it is convenient to define $\alpha = e^{-b}$ where $b > 0$. In addition, the corresponding impulse response is infinite:

$$h[n] = A\alpha^n u[n] \tag{5.136}$$

whose Fourier transform is

$$H(e^{j\omega}) = \frac{A}{1-\alpha e^{-j\omega}} = \frac{A}{1-e^{-b-j\omega}} \tag{5.137}$$

and magnitude square is given by

$$|H(e^{j\omega})|^2 = \frac{|A|^2}{1+\alpha^2 - 2\alpha\cos\omega} \tag{5.138}$$

which is shown in Figure 5.25 for $\alpha > 0$, which corresponds to a low-pass filter.

The bandwidth of a low-pass filter is defined as the point where its magnitude square is half of its maximum value. Using the first-order Taylor approximation of the exponential function, the following approximation can be used when $b \to 0$:

$$|H(e^{j0})|^2 = \frac{A^2}{|1-e^{-b}|^2} \approx \frac{A^2}{b^2} \tag{5.139}$$

If the bandwidth ω_b is also small, we can similarly approximate

$$|H(e^{j\omega_b})|^2 = \frac{A^2}{|1-e^{-b-j\omega_b}|^2} \approx \frac{A^2}{|b+j\omega_b|^2} = \frac{A^2}{\left(b^2+\omega_b^2\right)} \tag{5.140}$$

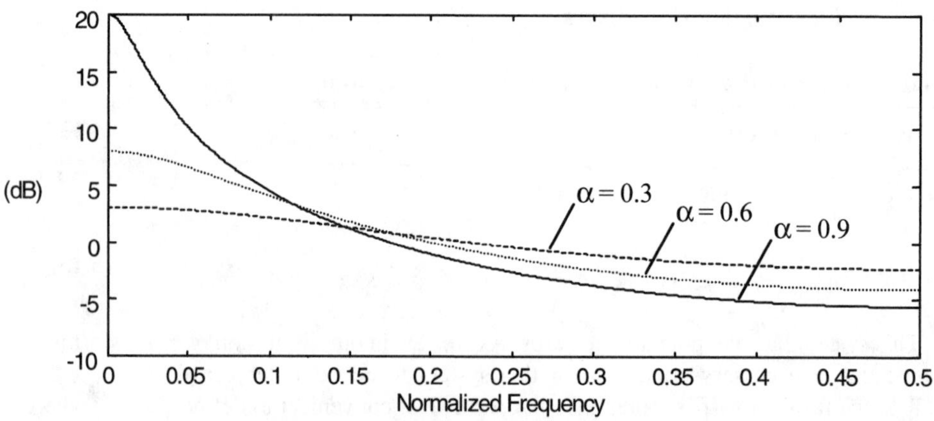

Figure 5.25 Magnitude frequency response of the first-order IIR filter.

so that for $\omega_b = b$ we have $|H(e^{jb})|^2 \approx 0.5|H(e^{j0})|^2$. In other words, the bandwidth of this filter equals b, for small values of b. The relative error in this approximation[9] is smaller than 2% for $b < 0.5$, which corresponds to $0.6 < \alpha < 1$. The relationship with the unnormalized bandwidth B is

$$\alpha = e^{-2\pi B/F_s} \tag{5.141}$$

For $\alpha < 0$ it behaves as a high-pass filter, and a similar discussion can be carried out.

5.4.4.2. Second-Order IIR Filters

An important type of IIR filters is the set of second-order filters of the form

$$y[n] = Ax[n] + a_1 y[n-1] + a_2 y[n-2] \tag{5.142}$$

whose transfer function is given by

$$H(z) = \frac{A}{1 - a_1 z^{-1} - a_2 z^{-2}} \tag{5.143}$$

This system has two poles and no zeros. A special case is when the coefficients A, a_1, and a_2 are real. In this case the two poles are given by

$$z = \frac{a_1 \pm \sqrt{a_1^2 + 4a_2}}{2} \tag{5.144}$$

which for the case of $a_1^2 + 4a_2 > 0$ yields two real roots, and is a degenerate case of two first-order systems. The more interesting case is when $a_1^2 + 4a_2 < 0$. In this case we see that the two roots are complex conjugates of each other, which can be expressed in their magnitude and phase notation as

$$z = e^{-\sigma \pm j\omega_0} \tag{5.145}$$

As we mentioned before, $\sigma > 0$ is a necessary and sufficient condition for the poles to be inside the unit circle and thus for the system to be stable. With those values, the z-transform is given by

$$H(z) = \frac{A}{(1 - e^{-\sigma + j\omega_0} z^{-1})(1 - e^{-\sigma - j\omega_0} z^{-1})} = \frac{A}{1 - 2e^{-\sigma}\cos(\omega_0) z^{-1} + e^{-2\sigma} z^{-2}} \tag{5.146}$$

In Figure 5.26 we show the magnitude of its Fourier transform for a value of σ and ω_0. We see that the response is centered around ω_0 and is more concentrated for smaller values of

[9] The exact value is $\omega_b = \arccos\left[2 - \cosh b\right]$, where $\cosh b = \left(e^b + e^{-b}\right)/2$ is the hyperbolic cosine.

σ. This is a type of *bandpass* filter, since it favors frequencies in a band around ω_0. It is left to the reader as an exercise to show that the bandwidth[10] is approximately 2σ. The smaller the ratio σ/ω_0, the sharper the resonance. The filter coefficients can be expressed as a function of the unnormalized bandwidth B and resonant frequency F and the sampling frequency F_s (all expressed in Hz) as

$$a_1 = 2e^{-\pi B/F_s}\cos\left(2\pi F/F_s\right) \tag{5.147}$$

$$a_2 = -e^{-2\pi B/F_s} \tag{5.148}$$

These types of systems are also known as second-order resonators and will be of great use for speech synthesis (Chapter 16), particularly for formant synthesis.

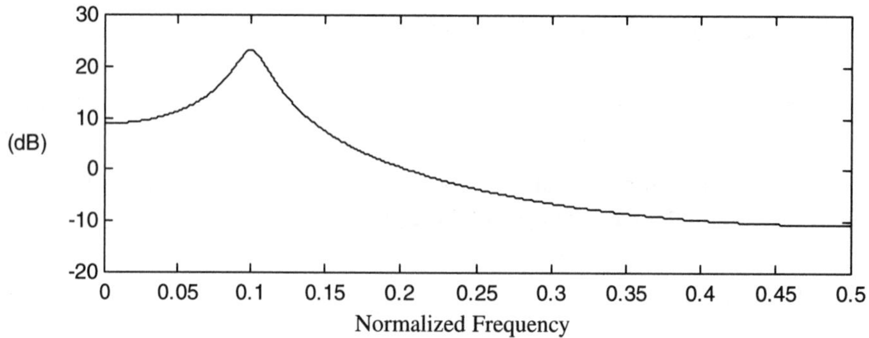

Figure 5.26 Frequency response of the second-order IIR filter for center frequency of $F = 0.1F_s$ and bandwidth $B = 0.01F_s$.

5.5. DIGITAL PROCESSING OF ANALOG SIGNALS

To use digital signal processing methods, it is necessary to convert the speech signal $x(t)$, which is analog, to a digital signal $x[n]$, which is formed by periodically sampling the analog signal $x(t)$ at intervals equally spaced T seconds apart:

$$x[n] = x(nT) \tag{5.149}$$

where T is defined as the sampling period, and its inverse $F_s = 1/T$ as the sampling frequency. In the speech applications considered in this book, F_s can range from 8000 Hz for telephone applications to 44,100 Hz for high-fidelity audio applications. This section explains the sampling theorem, which essentially says that the analog signal $x(t)$ can be

[10] The bandwidth of a bandpass filter is the region between half maximum magnitude squared values.

uniquely recovered given its digital signal $x[n]$ if the analog signal $x(t)$ has no energy for frequencies above the *Nyquist* frequency $F_s/2$.

We not only prove the sampling theorem, but also provide great insight into the analog-digital conversion, which is used in Chapter 7.

5.5.1. Fourier Transform of Analog Signals

The Fourier transform of an analog signal $x(t)$ is defined as

$$X(\Omega)=\int_{-\infty}^{\infty} x(t)e^{-j\Omega t}dt \tag{5.150}$$

with its inverse transform being

$$x(t)=\frac{1}{2\pi}\int_{-\infty}^{\infty} X(\Omega)e^{j\Omega t}d\Omega \tag{5.151}$$

They are transform pairs. You can prove similar relations for the Fourier transform of analog signals as for their digital signals counterpart.

5.5.2. The Sampling Theorem

Let's define $x_p(t)$

$$x_p(t)=x(t)p(t) \tag{5.152}$$

as a sampled version of $x(t)$, where

$$p(t)=\sum_{n=-\infty}^{\infty}\delta(t-nT) \tag{5.153}$$

where $\delta(t)$ is the Dirac delta defined in Section 5.3.2.1. Therefore, $x_p(t)$ can also be expressed as

$$x_p(t)=\sum_{n=-\infty}^{\infty}x(t)\delta(t-nT)=\sum_{n=-\infty}^{\infty}x(nT)\delta(t-nT)=\sum_{n=-\infty}^{\infty}x[n]\delta(t-nT) \tag{5.154}$$

after using Eq. (5.149). In other words, $x_p(t)$ can be uniquely specified given the digital signal $x[n]$.

Using the modulation property of Fourier transforms of analog signals, we obtain

$$X_p(\Omega) = \frac{1}{2\pi} X(\Omega) * P(\Omega) \tag{5.155}$$

Following a derivation similar to that in Section 5.3.2.2, one can show that the transform of the impulse train $p(t)$ is given by

$$P(\Omega) = \frac{2\pi}{T} \sum_{k=-\infty}^{\infty} \delta(\Omega - k\Omega_s) \tag{5.156}$$

where $\Omega_s = 2\pi F_s$ and $F_s = 1/T$, so that

$$X_p(\Omega) = \frac{1}{T} \sum_{k=-\infty}^{\infty} X(\Omega - k\Omega_s) \tag{5.157}$$

From Figure 5.27 it can be seen that if

$$X(\Omega) = 0 \text{ for } |\Omega| > \Omega_s / 2 \tag{5.158}$$

then $X(\Omega)$ can be completely recovered from $X_p(\Omega)$ as follows

$$X(\Omega) = R_{\Omega_s}(\Omega) X_p(\Omega) \tag{5.159}$$

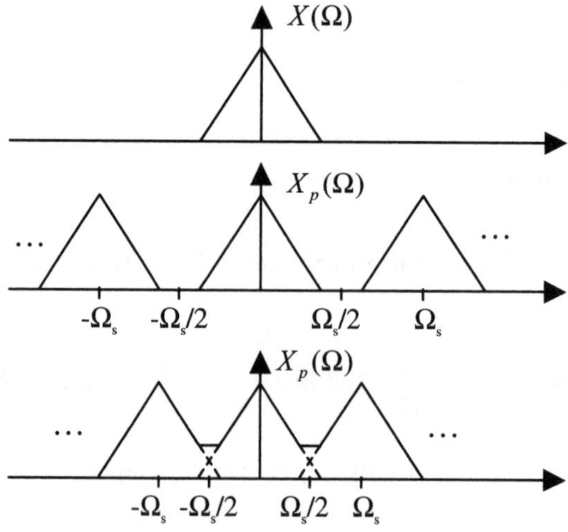

Figure 5.27 $X(\Omega)$, $X_p(\Omega)$ for the case of no aliasing and aliasing.

where

$$R_{\Omega_s}(\Omega) = \begin{cases} T & |\Omega| < \Omega_s / 2 \\ 0 & otherwise \end{cases} \tag{5.160}$$

is an ideal lowpass filter. We can also see that if Eq. (5.158) is not met, then *aliasing* will take place and $X(\Omega)$ can no longer be recovered from $X_p(\Omega)$. Since, in general, we cannot be certain that Eq. (5.158) is true, the analog signal is low-pass filtered with an ideal filter given by Eq. (5.160), which is called anti-aliasing filter, prior to sampling. Limiting the bandwidth of our analog signal is the price we have to pay to be able to manipulate it digitally.

The inverse Fourier transform of Eq. (5.160), computed through Eq. (5.151), is a sinc function

$$r_T(t) = \text{sinc}(t/T) = \frac{\sin(\pi t/T)}{\pi t/T} \tag{5.161}$$

so that using the convolution property in Eq. (5.159) we obtain

$$x(t) = r_T(t) * x_p(t) = r_T(t) * \sum_{k=-\infty}^{\infty} x[k]\delta(t-kT) = \sum_{k=-\infty}^{\infty} x[k] r_T(t-kT) \tag{5.162}$$

The *sampling theorem* states that we can recover the continuous time signal $x(t)$ just from its samples $x[n]$ using Eqs. (5.161) and (5.162). The angular frequency $\Omega_s = 2\pi F_s$ is expressed in terms of the sampling frequency F_S. $T = 1/F_s$ is the sampling period, and $F_s/2$ the *Nyquist* frequency. Equation (5.162) is referred to as *bandlimited interpolation* because $x(t)$ is reconstructed by interpolating $x[n]$ with sinc functions that are bandlimited.

Now let's see the relationship between $X_p(\Omega)$ and $X(e^{j\omega})$, the Fourier transform of the discrete sequence $x[n]$. From Eq. (5.154) we have

$$X_p(\Omega) = \sum_{n=-\infty}^{\infty} x[n] e^{-j\Omega nT} \tag{5.163}$$

so that the continuous transform $X_p(\Omega)$ equals the discrete Fourier transform $X(e^{j\omega})$ at $\omega = \Omega T$.

5.5.3. Analog-to-Digital Conversion

The process of converting an analog signal $x(t)$ into a digital signal $x[n]$ is called *Analog-to-Digital conversion*, or A/D for short, and the device that does it is called an *Analog-to-Digital Converter*. In Section 5.5.2 we saw that an ideal low-pass anti-aliasing filter was required on the analog signal, which of course is not realizable in practice so that an approximation has to be used. In practice, sharp analog filters can be implemented on the same chip using switched capacitor filters, which have attenuations above 60 dB in the stop band so that aliasing tends not to be an important issue for speech signals. The passband is not exactly flat, but this again does not have much significance for speech signals (for other signals, such as those used in modems, this issue needs to be studied more carefully).

Although such sharp analog filters are possible, they can be expensive and difficult to implement. One common solution involves the use of a simple analog low-pass filter with a large attenuation at $MF_s/2$, a multiple of the required cutoff frequency. Then *oversampling* is done at the new rate MF_s, followed by a sharper digital filter with a cut-off frequency of $F_s/2$ and downsampling (see Section 5.6). This is equivalent to having used a sharp analog filter, with the advantage of a lower-cost implementation. This method also allows variable sampling rates with minimal increase in cost and complexity. This topic is discussed in more detail in Chapter 7 in the context of sigma-delta modulators.

In addition, the pulses in Eq. (5.59) cannot be zero length in practice, and therefore the sampling theorem does not hold. However, current hardware allows the pulses to be small enough that the analog signal can be approximately recovered. The signal level is then maintained during T seconds, while the conversion to digital is being carried out.

A real A/D converter cannot provide real numbers for $x[n]$, but rather a set of integers typically represented with 16 bits, which gives a range between –32,768 and 32,767. Such conversion is achieved by comparing the analog signal to a number of different signal levels. This means that *quantization noise* has been added to the digital signal. This is typically not a big problem for speech signals if using 16 bits or more since, as is shown in Chapter 7, other noises will mask the quantization noise anyway. Typically, quantization noise becomes an issue only if 12 or fewer bits are used. A more detailed study of the effects of quantization is presented in Chapter 7.

Finally, A/D subsystems are not exactly linear, which adds another source of distortion. This nonlinearity can be caused by, among things, jitter and drift in the pulses and unevenly spaced comparators. For popular A/D subsystems, such as *sigma-delta* A/D, an offset is typically added to $x[n]$, which in practice is not very important, because speech signals do not contain information at $f = 0$, and thus can be safely ignored.

5.5.4. Digital-to-Analog Conversion

The process of converting the digital signal $x[n]$ back into an analog $x(t)$ is called *digital-to-analog conversion*, or D/A for short. The ideal band-limited interpolation requires ideal sinc functions as shown in Eq. (5.162), which are not realizable. To convert the digital signal to analog, a zero-order hold filter

$$h_0(t) = \begin{cases} 1 & 0 < t < T \\ 0 & \text{otherwise} \end{cases} \tag{5.164}$$

is often used, which produces an analog signal as shown in Figure 5.28. The output of such a filter is given by

$$x_0(t) = h_0(t) * \sum_{n=-\infty}^{\infty} x[n]\delta(t-nT) = \sum_{n=-\infty}^{\infty} x[n]h_0(t-nT) \tag{5.165}$$

The Fourier transform of the zero-hold filter in Eq. (5.164) is, using Eq. (5.150),

$$H_0(\Omega) = \frac{2\sin(\Omega T/2)}{\Omega} e^{-j\Omega T/2} \tag{5.166}$$

and, since we need an ideal lowpass filter to achieve the band-limited interpolation of Eq. (5.162), the signal $x_0(t)$ has to be filtered with a reconstruction filter with transfer function

$$H_r(\Omega) = \begin{cases} \dfrac{\Omega T/2}{\sin(\Omega T/2)} e^{j\Omega T/2} & |\Omega| < \pi/T \\ 0 & |\Omega| > \pi/T \end{cases} \tag{5.167}$$

In practice, the phase compensation is ignored, as it amounts to a delay of *T*/2 seconds. Its magnitude response can be seen in Figure 5.29. In practice, such an analog filter is not realizable and an approximation is made. Since the zero-order hold filter is already low-pass, the reconstruction filter doesn't need to be that sharp.

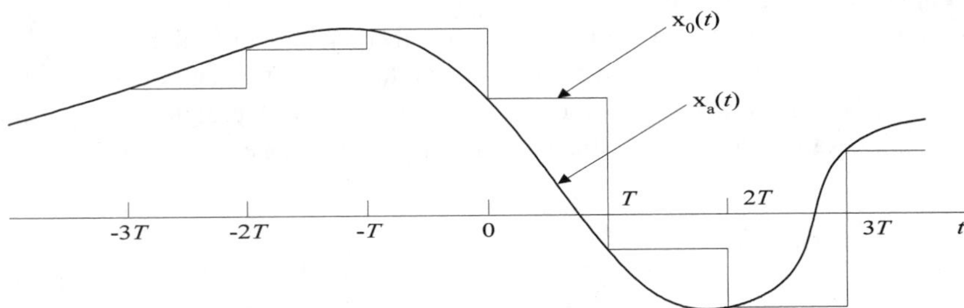

Figure 5.28 Output of a zero-order hold filter.

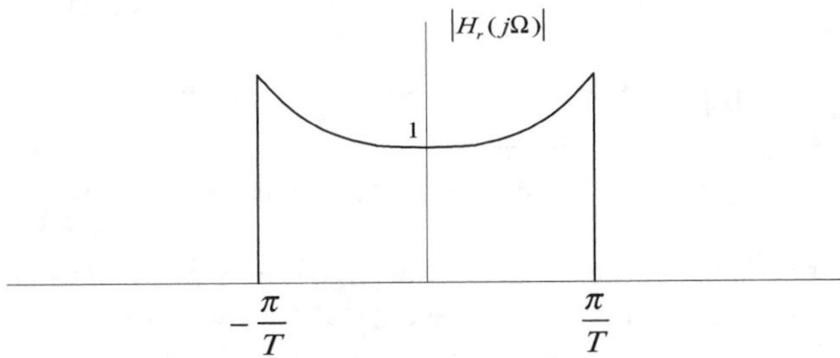

Figure 5.29 Magnitude frequency response of the reconstruction filter used in digital-to-analog converters after a zero-hold filter.

In the above discussion we note that practical A/D and D/A systems introduce distortions, which causes us to wonder whether it is a good idea to go through this process just to manipulate digital signals. It turns out that for most speech processing algorithms described in Chapter 6, the advantages of operating with digital signals outweigh the disadvantage of the distortions described above. Moreover, commercial A/D and D/A systems are such that the errors and distortions can be arbitrarily small. The fact that music in digital format (as in compact discs) has won out over analog format (cassettes) shows that this is indeed the case. Nonetheless, it is important to be aware of the above limitations when designing a system.

5.6. MULTIRATE SIGNAL PROCESSING

The term *Multirate Signal Processing* refers to processing of signals sampled at different rates. A particularly important problem is that of sampling-rate conversion. It is often the case that we have a digital signal $x[n]$ sampled at a sampling rate F_s, and we want to obtain an equivalent signal $y[n]$ but at a different sampling rate F'_s. This often occurs in A/D systems that oversample in order to use smaller quantizers, such as a delta or sigma delta-quantizer (see Chapter 7), and a simpler analog filter, and then have to downsample the signal. Other examples include mixing signals of different sampling rates and downsampling to reduce computation (many signal processing algorithms have a computational complexity proportional to the sampling rate or its square).

A simple solution is to convert the digital signal $x[n]$ into an analog signal $x(t)$ with a D/A system running at F_s and then convert it back to digital with an A/D system running at F'_s. An interesting problem is whether this could be done in the digital domain directly, and the techniques to do so belong to the general class of multi-rate processing.

5.6.1. Decimation

If we want to reduce the sampling rate by a factor of M, i.e., $T' = MT$, we take every M samples. In order to avoid aliasing, we need to lowpass filter the signal to bandlimit it to frequencies $1/T'$. This is shown in Figure 5.30, where the arrow pointing down indicates the decimation.

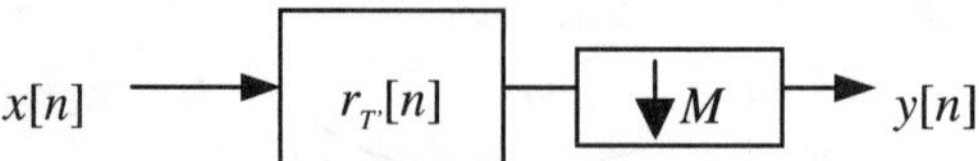

Figure 5.30 Block diagram of the decimation process.

Since the output is not desired at all instants n, but only every M samples, the computation can be reduced by a factor of M over the case where lowpass filtering is done first and *decimation* later. To do this we express the analog signal $x_l(t)$ at the output of the lowpass filter as

$$x_l(t) = \sum_{k=-\infty}^{\infty} x[k] r_{T'}(t - kT) \tag{5.168}$$

and then look at the value $t' = nT'$. The decimated signal $y[n]$ is then given by

$$y[n] = x_l(nT') = \sum_{k=-\infty}^{\infty} x[k] r_{T'}\left(nT' - kT\right) = \sum_{k=-\infty}^{\infty} x[k] \text{sinc}\left(\frac{Mn-k}{M}\right) \tag{5.169}$$

which can be expressed as

$$y[n] = \sum_{k=-\infty}^{\infty} x[k] h\left[Mn - k\right] \tag{5.170}$$

where

$$h[n] = \text{sinc}(n / M) \tag{5.171}$$

In practice, the ideal lowpass filter $h[n]$ is approximated by an FIR filter with a cutoff frequency of $1/(2M)$.

5.6.2. Interpolation

If we want to increase the sampling rate by a factor of N, so that $T' = T / N$, we do not have any aliasing and no further filtering is necessary. In fact we already know one out of every N output samples

$$y[Nn] = x[n] \tag{5.172}$$

and we just need to compute the $(N-1)$ samples in-between. Since we know that $x[n]$ is a bandlimited signal, we can use the sampling theorem in Eq. (5.162) to reconstruct the analog signal as

$$x_l(t) = \sum_{k=-\infty}^{\infty} x[k] r_T(t - kT) \tag{5.173}$$

and thus the interpolated signal $y[n]$ as

$$y[n] = x(nT') = \sum_{k=-\infty}^{\infty} x[k] r_T\left(nT' - kT\right) = \sum_{k=-\infty}^{\infty} x[k] \text{sinc}\left(\frac{n-kN}{N}\right) \tag{5.174}$$

Now let's define

$$x'[k'] = \begin{cases} x[Nk] & k' = Nk \\ 0 & otherwise \end{cases} \tag{5.175}$$

which, inserted into Eq. (5.174), gives

$$y[n] = \sum_{k'=-\infty}^{\infty} x'[k']\text{sinc}\left((n-k')/N\right) \tag{5.176}$$

This can be seen in Figure 5.31, where the block with the arrow pointing up implements Eq. (5.175).

Equation (5.174) can be expressed as

$$y[n] = \sum_{k=-\infty}^{\infty} x[k]h\left[n-kN\right] \tag{5.177}$$

where we have defined

$$h[n] = \text{sinc}(n/N) \tag{5.178}$$

Again, in practice, the ideal low-pass filter $h[n]$ is approximated by an FIR filter with a cutoff frequency of $1/(2N)$.

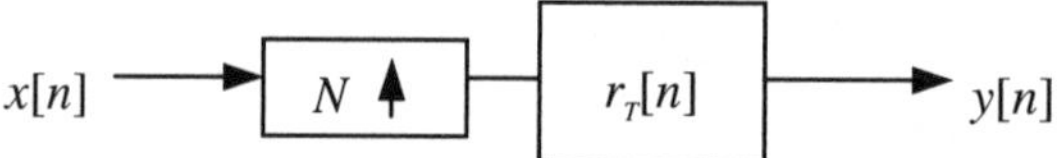

Figure 5.31 Block diagram of the interpolation process.

5.6.3. Resampling

To resample the signal so that $T' = TM/N$, or $F_s' = F_s\left(N/M\right)$, we can first upsample the signal by N and then downsample it by M. However, there is a more efficient way. Proceeding similarly to decimation and interpolation, one can show the output is given by

$$y[n] = \sum_{k=-\infty}^{\infty} x[k]h[nM-kN] \tag{5.179}$$

where

$$h[n] = \text{sinc}\left(\frac{n}{\max(N,M)}\right) \tag{5.180}$$

for the ideal case. In practice, $h[n]$ is an FIR filter with a cutoff frequency of $1/(2\max(N,M))$. We can see that Eq. (5.179) is a superset of Eqs. (5.170) and (5.177).

5.7. FILTERBANKS

A filterbank is a collection of filters that span the whole frequency spectrum. In this section we describe the fundamentals of filterbanks, which are used in speech and audio coding, echo cancellation, and other applications. We first start with a filterbank with two equal bands, then explain multi-resolution filterbanks, and present the FFT as a filterbank. Finally we introduce the concept of lapped transforms and wavelets.

5.7.1. Two-Band Conjugate Quadrature Filters

A two-band filterbank is shown in Figure 5.32 , where the filters $f_0[n]$ and $g_0[n]$ are low-pass filters, and the filters $f_1[n]$ and $g_1[n]$ are high-pass filters, as shown in Figure 5.33. Since the output of $f_0[n]$ has a bandwidth half of that of $x[n]$, we can sample it at half the rate of $x[n]$. We do that by decimation (throwing out every other sample), as shown in Figure 5.32. The output of such a filter plus decimation is $x_0[m]$. Similar results can be shown for $f_1[n]$ and $x_1[n]$.

For reconstruction, we upsample $x_0[m]$, by inserting a 0 between every sample. Then we low-pass filter it with filter $g_0[n]$ to complete the interpolation, as we saw in Section 5.6. A similar process can be done with the high pass filters $f_1[n]$ and $g_1[n]$. Adding the two bands produces $\tilde{x}[n]$, which is identical to $x[n]$ if the filters are ideal.

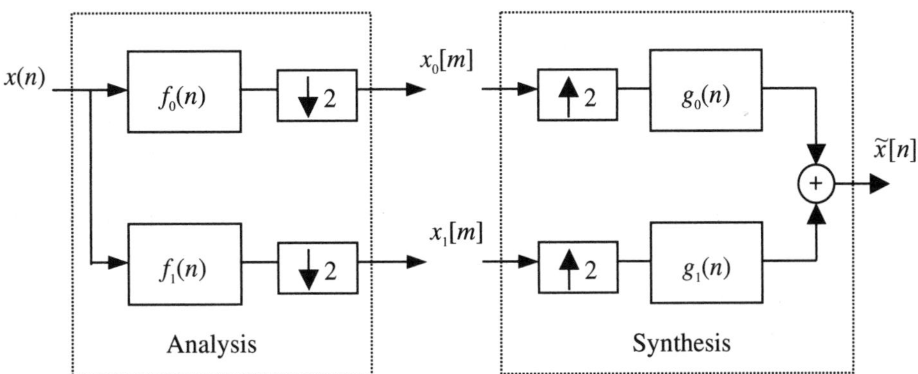

Figure 5.32 Two-band filterbank.

In practice, however, ideal filters such as those in Figure 5.33 are not achievable, so we would like to know if it is possible to build a filterbank that has perfect reconstruction with FIR filters. The answer is affirmative, and in this section we describe conjugate quadrature filters, which are the basis for the solutions.

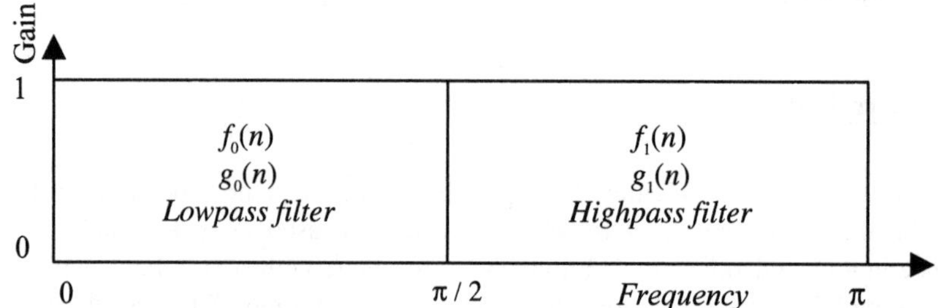

Figure 5.33 Ideal frequency responses of analysis and synthesis filters for the two-band filterbank.

To investigate this, let's analyze the cascade of a downsampler and an upsampler (Figure 5.34). The output $y[n]$ is a signal whose odd samples are zero and whose even samples are the same as those of the input signal $x[n]$.

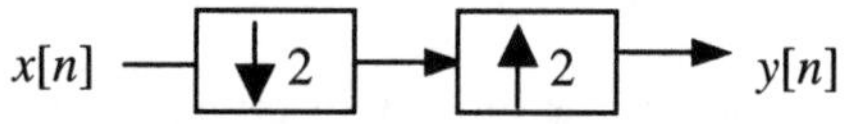

$x[n] \rightarrow \boxed{\downarrow 2} \rightarrow \boxed{\uparrow 2} \rightarrow y[n]$

Figure 5.34 Cascade of a downsampler and an upsampler.

The z-transform of the output is given by

$$\begin{aligned} Y(z) &= \sum_{\substack{n=-\infty \\ n\,even}}^{\infty} x[n]z^{-n} = \frac{1}{2}\sum_{n=-\infty}^{\infty} x[n]z^{-n} + \frac{1}{2}\sum_{n=-\infty}^{\infty} (-1)^n x[n]z^{-n} \\ &= \frac{X(z)+X(-z)}{2} \end{aligned} \tag{5.181}$$

Using Eq. (5.181) and the system in Figure 5.32, we can express the z-transform of the output in Figure 5.32 as

$$\begin{aligned} \tilde{X}(z) &= \left(\frac{F_0(z)G_0(z)+F_1(z)G_1(z)}{2} \right) X(z) \\ &+ \left(\frac{F_0(-z)G_0(z)+F_1(-z)G_1(z)}{2} \right) X(-z) \\ &= \left(\frac{F_0(z)X(z)+F_0(-z)X(-z)}{2} \right) G_0(z) + \left(\frac{F_1(z)X(z)+F_1(-z)X(-z)}{2} \right) G_1(z) \end{aligned} \tag{5.182}$$

which for perfect reconstruction requires the output to be a delayed version of the input, and thus

$$\begin{aligned} F_0(z)G_0(z) + F_1(z)G_1(z) &= 2z^{-(L-1)} \\ F_0(-z)G_0(z) + F_1(-z)G_1(z) &= 0 \end{aligned} \tag{5.183}$$

These conditions are met if we select the so-called *Conjugate Quadrature Filters* (CQF) [17], which are FIR filters that specify $f_1[n]$, $g_0[n]$, and $g_1[n]$ as a function of $f_0[n]$:

$$\begin{aligned} f_1[n] &= (-1)^n f_0[L-1-n] \\ g_0[n] &= f_0[L-1-n] \\ g_1[n] &= f_1[L-1-n] \end{aligned} \tag{5.184}$$

where $f_0[n]$ is an FIR filter of even length L. The z-transforms of Eq. (5.184) are

$$\begin{aligned} F_1(z) &= -z^{-(L-1)} F_0(-z^{-1}) \\ G_0(z) &= z^{-(L-1)} F_0(z^{-1}) \\ G_1(z) &= -F_0(-z) \end{aligned} \tag{5.185}$$

so that the second equation in Eq.(5.183) is met if L is even. In order to analyze the first equation in Eq. (5.183), let's define $P(z)$ as

$$\begin{aligned} P(z) &= F_0(z)F_0(z^{-1}) \\ p[n] &= \sum_m f_0[m] f_0[m+n] \end{aligned} \tag{5.186}$$

then insert Eq. (5.185) into (5.183), use Eq. (5.186), and obtain the following condition:

$$P(z) + P(-z) = 2 \tag{5.187}$$

Taking the inverse z-transform of Eq. (5.187) and using Eq. (5.181), we obtain

$$p[n] = \begin{cases} 1 & n = 0 \\ 0 & n = 2k \end{cases} \tag{5.188}$$

so that all even samples of the autocorrelation of $f_0[n]$ are zero, except for $n = 0$. Since $f_0[n]$ is a half-band low-pass filter, $p[n]$ is also a half-band low-pass filter. The ideal half-band filter $h[n]$

$$h[n] = \frac{\sin(\pi n / 2)}{\pi n} \tag{5.189}$$

satisfies Eq. (5.188), as does any half-band zero-phase filter (a linear phase filter with no delay). Therefore, the steps to build CQF are

1. Design a $(2L - 1)$ tap[11] half-band linear-phase low-pass filter $p[n]$ with any available technique, for an even value of L. For example, one could use the Parks McClellan algorithm, constraining the passband and stopband cutoff frequencies so that $\omega_p = \pi - \omega_s$ and using an error weighting that is the same for the passband and stopband. This results in a half-band linear-phase filter with equal ripple δ in both bands. Another possibility is to multiply the ideal half-band filter in Eq. (5.189) by a window with low-pass characteristics.
2. Add a value δ to $p[0]$ so that we can guarantee that $P(e^{j\omega}) \geq 0$ for all ω and thus is a legitimate power spectral density.
3. Spectrally factor $P(z) = F_0(z)F_0(z^{-1})$ by computing its roots.
4. Compute $f_1[n]$, $g_0[n]$ and $g_1[n]$ from Eq. (5.184).

5.7.2. Multiresolution Filterbanks

While the above filterbank has equal bandwidth for both filters, it may be desirable to have varying bandwidths, since it has been proven to work better in speech recognition systems. In this section we show how to use the two-band conjugate quadrature filters described in the previous section to design a filterbank with more than two bands. In fact, multi-resolution analysis such as that of Figure 5.35, are possible with bands of different band-widths (see Figure 5.36).

One interesting result is that the product of time resolution and frequency resolution is constant (all the tiles in Figure 5.37 have the same area), since filters with smaller band-widths do not need to be sampled as often. Instead of using Fourier basis for decomposition, multi-resolution filterbanks allow more flexibility in the tiling of the time-frequency plane.

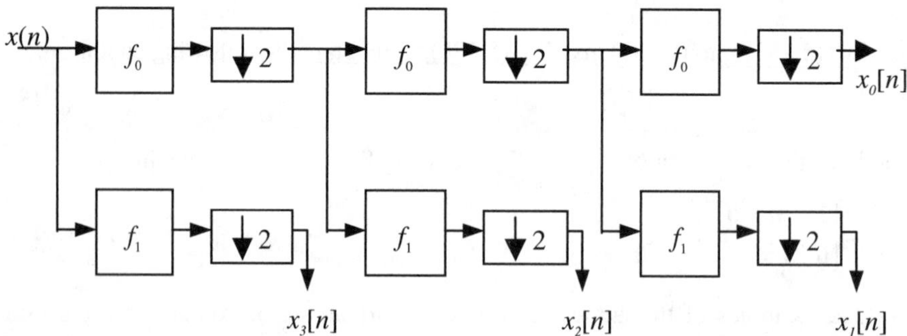

Figure 5.35 Analysis part of a multi-resolution filterbank designed with conjugate quadrature filters. Only $f_0[n]$ needs to be specified.

[11] A filter with N taps is a filter of length N.

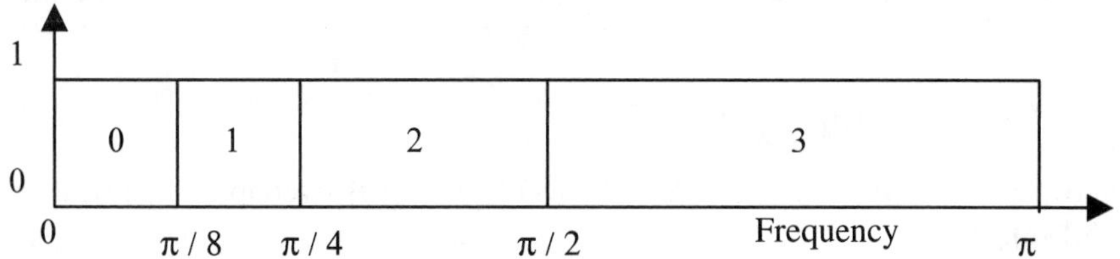

Figure 5.36 Ideal frequency responses of the multi-resolution filterbank of Figure 5.35. Note that $x_0[n]$ and $x_1[n]$ occupy 1/8 of the total bandwidth.

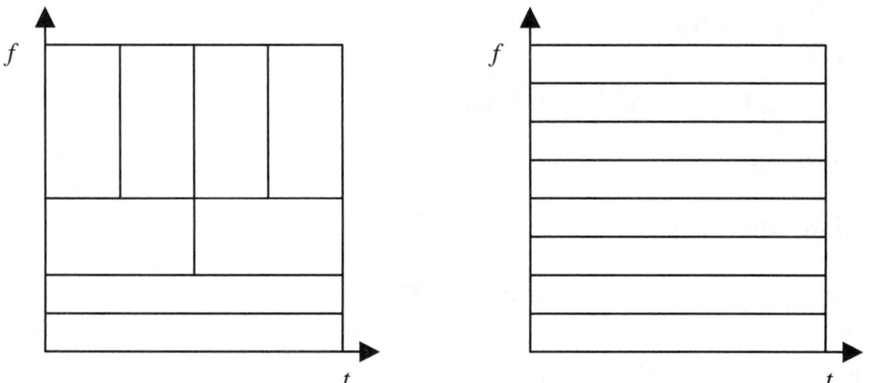

Figure 5.37 Two different time-frequency tilings: the non-uniform filterbank and that obtained through a short-time Fourier transform. Notice that the area of each tile is constant.

5.7.3. The DFT as a Filterbank

It turns out that we can use the Fourier transform to construct a filterbank. To do that, we decompose the input signal $x[n]$ as a sum of *short-time* signals $x_m[n]$

$$x[n] = \sum_{m=-\infty}^{\infty} x_m[n] \tag{5.190}$$

where $x_m[n]$ is obtained as

$$x_m[n] = x[n]w_m[n] \tag{5.191}$$

the product of $x[n]$ by a *window* function $w_m[n]$ of length *N*. From Eqs. (5.190) and (5.191) we see that the window function has to satisfy

$$\sum_{m=-\infty}^{\infty} w_m[n] = 1 \qquad \forall n \tag{5.192}$$

If the short-term signals $x_m[n]$ are spaced M samples apart, we define the window $w_m[n]$ as:

$$w_m[n] = w[n - Mm] \tag{5.193}$$

where $w[n] = 0$ for $n < 0$ and $n > N$. The windows $w_m[n]$ overlap in time while satisfying Eq. (5.192).

Since $x_m[n]$ has N nonzero values, we can evaluate its length-N DFT as

$$\begin{aligned} X_m[k] &= \sum_{l=0}^{N-1} x_m[Mm+l]e^{-j\omega_k l} \\ &= \sum_{l=0}^{N-1} x[Mm+l]w[l]e^{-j\omega_k l} = \sum_{l=0}^{N-1} x[Mm+l]f_k[-l] \end{aligned} \tag{5.194}$$

where $\omega_k = 2\pi k / N$ and the analysis filters $f_k[l]$ are given by

$$f_k[l] = w[-l]e^{j\omega_k l} \tag{5.195}$$

If we define $\tilde{X}_k[n]$ as

$$\tilde{X}_k[n] = x[n] * f_k[n] = \sum_{r=-\infty}^{\infty} x[n-r]f_k[r] = \sum_{l=0}^{N-1} x[n+l]f_k[-l] \tag{5.196}$$

then Eqs. (5.194) and (5.196) are related by

$$X_m[k] = \tilde{X}_k[mM] \tag{5.197}$$

This manipulation is shown in Figure 5.38, so that the DFT output $X_m[k]$ is $\tilde{X}_k[n]$ decimated by M.

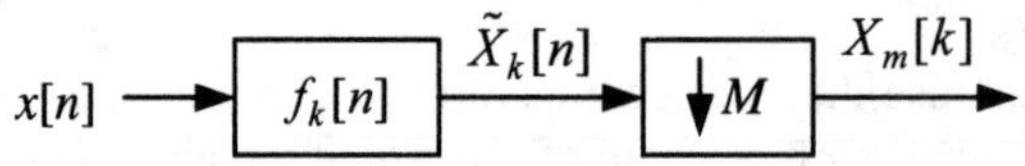

$x[n]$ → $f_k[n]$ → $\tilde{X}_k[n]$ → $\downarrow M$ → $X_m[k]$

Figure 5.38 Fourier analysis used to build a linear filter.

The short-time signal $x_m[n]$ can be recovered through the inverse DFT of $X_m[k]$ as

$$x_m[mM+l] = h[l]\sum_{k=0}^{N-1} X_m[k]e^{j\omega_k l} \tag{5.198}$$

where $h[n]$ has been defined as

$$h[n] = \begin{cases} 1/N & 0 \le n < N \\ 0 & \text{otherwise} \end{cases} \tag{5.199}$$

so that Eq. (5.198) is valid for all values of l, and not just $0 \le l < N$.

Making the change of variables $mM + l = n$ in Eq. (5.198) and inserting it into Eq. (5.190) results in

$$
\begin{aligned}
x[n] &= \sum_{m=-\infty}^{\infty} h[n-mM] \sum_{k=0}^{N-1} X_m[k] e^{j\omega_k (n-mM)} \\
&= \sum_{k=0}^{N-1} \sum_{m=-\infty}^{\infty} X_m[k] g_k[n-mM]
\end{aligned}
\tag{5.200}
$$

where the synthesis filters $g_k[n]$ are defined as

$$
g_k[n] = h[n] e^{j\omega_k n} \tag{5.201}
$$

Now, let's define the upsampled version of $X_m[k]$ as

$$
\hat{X}_k[l] = \begin{cases} X_m[k] & l = mM \\ 0 & \text{otherwise} \end{cases} \tag{5.202}
$$

which, inserted into Eq. (5.200), yields

$$
x[n] = \sum_{k=0}^{N-1} \sum_{l=-\infty}^{\infty} \hat{X}_k[l] g_k[n-l] = \sum_{k=0}^{N-1} \hat{X}_k[n] * g_k[n] \tag{5.203}
$$

Thus, the signal can be reconstructed. The block diagram of the analysis/resynthesis filterbank implemented by the DFT can be seen in Figure 5.39, where $x_k[m] = X_m[k]$ and $\tilde{x}[n] = x[n]$.

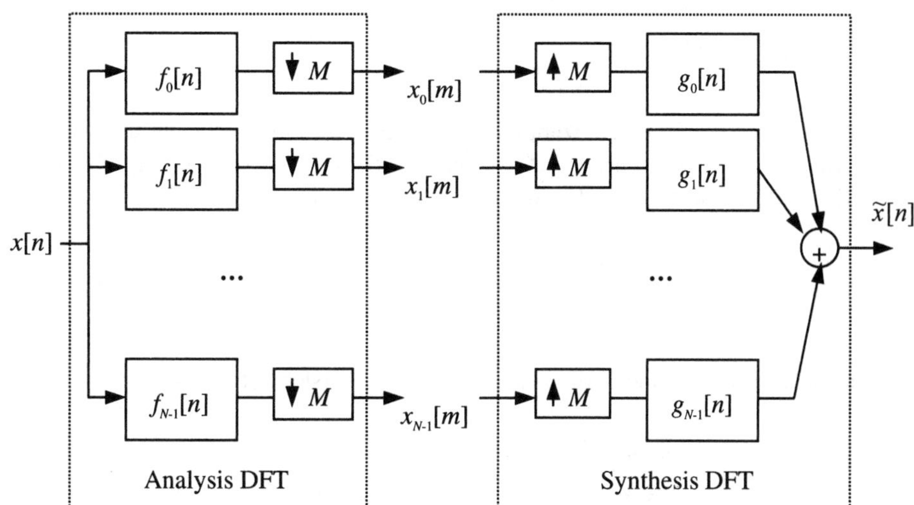

Figure 5.39 A filterbank with N analysis and synthesis filters.

For perfect reconstruction we need $N \geq M$. If $w[n]$ is a rectangular window of length N, the frame rate has to be $M = N$. We can also use overlapping windows with $N = 2M$ (50% overlap), such as Hamming or Hanning windows, and still get perfect reconstruction. The use of such overlapping windows increases the data rate by a factor of 2, but the analysis filters have much less spectral leakage because of the higher attenuation of the Hamming/Hanning window outside the main lobe.

5.7.4. Modulated Lapped Transforms

The filterbank of Figure 5.39 is useful because, as we see in Chapter 7, it is better to quantize the spectral coefficients than the waveform directly. If the DFT coefficients are quantized, there will be some discontinuities at frame boundaries. To solve this problem we can distribute the window $w[n]$ between the analysis and synthesis filters so that

$$w[n] = w_a[n]w_s[n] \tag{5.204}$$

so that the analysis filters are given by

$$f_k[n] = w_a[-n]e^{j\omega_k n} \tag{5.205}$$

and the synthesis filters by

$$g_k[n] = w_s[n]e^{-j\omega_k n} \tag{5.206}$$

This way, if there is a quantization error, the use of a tapering synthesis window will substantially decrease the border effect. A common choice is $w_a[n] = w_s[n]$, which for the case of $w[n]$ being a Hanning window divided by N, results in

$$w_a[n] = w_s[n] = \frac{1}{\sqrt{N}}\sin\left(\frac{\pi n}{N}\right) \qquad \text{for} \quad 0 \leq n < N \tag{5.207}$$

so that the analysis and synthesis filters are the reversed versions of each other:

$$f_k[-n] = g_k[n] = \frac{\sin(\pi n / N)}{\sqrt{N}} e^{j2\pi nk/N}\Pi_N[n] = h_k^N[n] \tag{5.208}$$

whose frequency response can be seen in Figure 5.40.

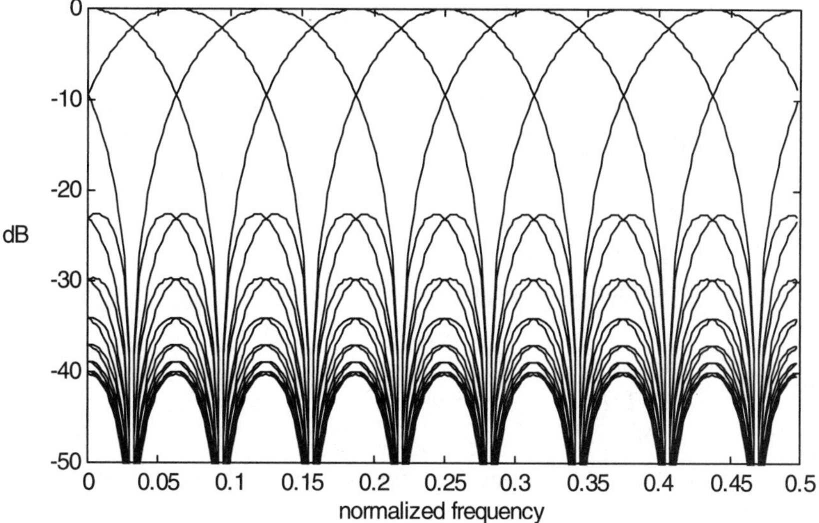

Figure 5.40 Frequency response of the Lapped Orthogonal Transform filterbank.

The functions $h_k^N[n]$ in Eq. (5.208) are sine modulated complex exponentials, which have the property

$$h_k^{N/2}[n] = \sqrt{2} h_k^N[2n] \tag{5.209}$$

which is a property typical of functions called *wavelets*, i.e., they can be obtained from each other by stretching by 2 and scaling them appropriately. Such wavelets can be seen in Figure 5.41.

If instead of modulating a complex exponential we use a cosine sequence, we obtain the *Modulated Lapped Transform* (MLT) [7], also known as the *Modified Discrete Cosine Transform* (MDCT):

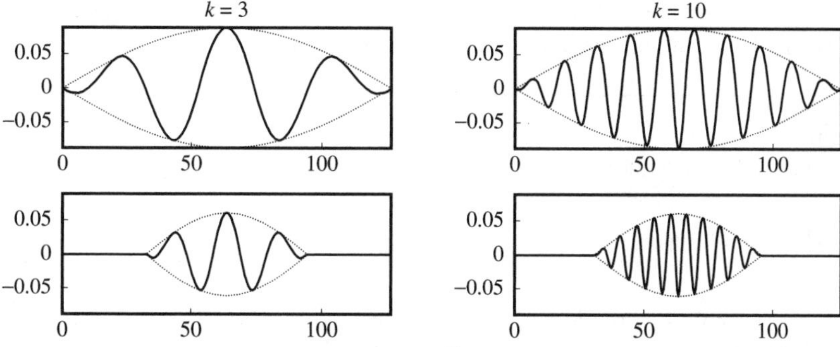

Figure 5.41 Iterations of the wavelet $h_k^N[n]$ for several values of k and N.

$$p_{kn} = f_k[2M-1-n] = g_k[n] = h[n]\sqrt{\frac{2}{M}}\cos\left[\left(k+\frac{1}{2}\right)\left(n+\frac{M+1}{2}\right)\frac{\pi}{M}\right] \tag{5.210}$$

for $k = 0,1,\cdots,M-1$ and $n = 0,1,\cdots,2M-1$. There are M filters with $2M$ taps each, and $h[n]$ is a symmetric window $h[n] = h[2M-1-n]$ that satisfies

$$h^2[n] + h^2[n+M] = 1 \tag{5.211}$$

where the most common choice for $h[n]$ is

$$h[n] = \sin\left[\left(n+\frac{1}{2}\right)\frac{\pi}{2M}\right] \tag{5.212}$$

A fast algorithm can be used to compute these filters based on the DCT, which is called the *Lapped Orthogonal Transform* (LOT).

5.8. STOCHASTIC PROCESSES

While in this chapter we have been dealing with *deterministic signals*, we also need to deal with *noise,* such as the static present in a poorly tuned AM station. To analyze noise signals we need to introduce the concept of stochastic processes, also known as random processes. A *discrete-time* stochastic process $\mathbf{x}[n]$, also denoted by $\mathbf{x}_n$, is a sequence of random variables for each time instant *n*. *Continuous-time* stochastic processes $\mathbf{x}(t)$, random variables for each value of *t*, will not be the focus of this book, though their treatment is similar to that of discrete-time processes. We use bold for random variables and regular text for deterministic signals.

Here, we cover the statistics of stochastic processes, defining stationary and ergodic processes and the output of linear systems to such processes.

Example 5.1

We can define a random process $\mathbf{x}[n]$ as

$$\mathbf{x}[n] = \cos[\omega n + \boldsymbol{\varphi}] \tag{5.213}$$

where $\boldsymbol{\varphi}$ is real random variable with a uniform pdf in the interval $(-\pi,\pi)$. Several realizations of this random process are displayed in Figure 5.42.

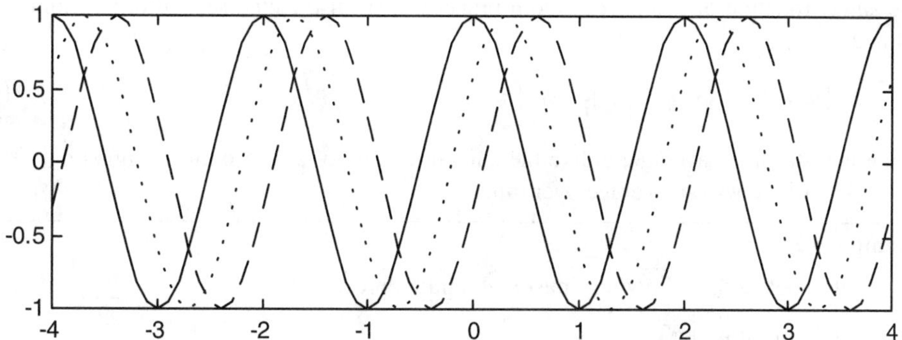

Figure 5.42 Several realizations of a sinusoidal random process with a random phase.

5.8.1. Statistics of Stochastic Processes

In this section we introduce several statistics of stochastic processes such as distribution, density function, mean and autocorrelation. We also define several types of processes depending on these statistics.

For a specific n, $\mathbf{x}[n]$ is a random variable with *distribution*

$$F(x,n) = P\{\mathbf{x}[n] \le x\} \tag{5.214}$$

Its first derivative with respect to x is the first-order density function, or simply the *probability density function* (pdf)

$$f(x,n) = \frac{dF(x,n)}{dx} \tag{5.215}$$

The second-order distribution of the process $\mathbf{x}[n]$ is the joint distribution

$$F(x_1,x_2;n_1,n_2) = P\{\mathbf{x}[n_1] \le x_1, \mathbf{x}[n_2] \le x_2\} \tag{5.216}$$

of the random variables $\mathbf{x}[n_1]$ and $\mathbf{x}[n_2]$. The corresponding density equals

$$f(x_1,x_2;n_1,n_2) = \frac{\partial^2 F(x_1,x_1;n_1,n_2)}{\partial x_1 \partial x_2} \tag{5.217}$$

A complex random process $\mathbf{x}[n] = \mathbf{x}_r[n] + j\mathbf{x}_i[n]$ is specified in terms of the joint statistics of the real processes $\mathbf{x}_r[n]$ and $\mathbf{x}_i[n]$.

The *mean* $\mu[n]$ of $\mathbf{x}[n]$, also called first-order moment, is defined as the expected value of the random variable $\mathbf{x}[n]$ for each value of n:

$$\mu_x[n] = E\{\mathbf{x}[n]\} = \int_{-\infty}^{\infty} \mathbf{x}[n] f(\mathbf{x},n) d\mathbf{x} \tag{5.218}$$

The autocorrelation of complex random process $\mathbf{x}[n]$, also called second-order moment, is defined as

$$R_{xx}[n_1,n_2]=E\left\{\mathbf{x}[n_1]\mathbf{x}^*[n_2]\right\}=R^*_{xx}[n_2,n_1] \tag{5.219}$$

which is a statistical average, unlike the autocorrelation of a deterministic signal defined in Eq. (5.45), which was an average over time.

Example 5.2

Let's look at the following sinusoidal random process

$$\mathbf{x}[n]=\mathbf{r}\cos[\omega n+\boldsymbol{\varphi}] \tag{5.220}$$

where $\mathbf{r}$ and $\boldsymbol{\varphi}$ are independent and $\boldsymbol{\varphi}$ is uniform in the interval $(-\pi,\pi)$. This process is *zero-mean* because

$$\mu_x[n]=E\left\{\mathbf{r}\cos[\omega n+\boldsymbol{\varphi}]\right\}=E\left\{\mathbf{r}\right\}E\left\{\cos[\omega n+\boldsymbol{\varphi}]\right\}=0 \tag{5.221}$$

since $\mathbf{r}$ and $\boldsymbol{\varphi}$ are independent and

$$E\left\{\cos[\omega n+\boldsymbol{\varphi}]\right\}=\int_{-\pi}^{\pi}\cos[\omega n+\boldsymbol{\varphi}]\frac{1}{2\pi}d\boldsymbol{\varphi}=0 \tag{5.222}$$

Its autocorrelation is given by

$$\begin{aligned}R_{xx}[n_1,n_2]&=E\{\mathbf{r}^2\}\int_{-\pi}^{\pi}\cos[\omega n_1+\boldsymbol{\varphi}]\cos[\omega n_2+\boldsymbol{\varphi}]\frac{1}{2\pi}d\boldsymbol{\varphi}\\&=\frac{1}{2}E\{\mathbf{r}^2\}\int_{-\pi}^{\pi}\{\cos[\omega(n_1+n_2)+\boldsymbol{\varphi}]+\cos[\omega(n_2-n_1)]\}\frac{1}{2\pi}d\boldsymbol{\varphi}\\&=\frac{1}{2}E\{\mathbf{r}^2\}\cos[\omega(n_2-n_1)]\end{aligned} \tag{5.223}$$

which only depends on the time difference n_2-n_1.

An important property of a stochastic process is that its autocorrelation $R_{xx}[n_1,n_2]$ is a *positive-definite* function, i.e., for any a_i,a_j

$$\sum_i\sum_j a_i a_j^* R_{xx}[n_i,n_j]\ge 0 \tag{5.224}$$

which is a consequence of the identity

$$0\le E\left\{\left|\sum_i a_i\mathbf{x}[n_i]\right|^2\right\}=\sum_i\sum_j a_i a_j^* E\left\{\mathbf{x}[n_i]\mathbf{x}^*[n_j]\right\} \tag{5.225}$$

Similarly, the autocovariance of a complex random process is defined as

$$C_{xx}[n_1,n_2]=E\left\{\left(\mathbf{x}[n_1]-\mu_x[n_1]\right)\left(\mathbf{x}[n_2]-\mu_x[n_2]\right)^*\right\}=R_{xx}[n_1,n_2]-\mu_x[n_1]\mu_x^*[n_2] \quad (5.226)$$

The correlation coefficient of process $\mathbf{x}[n]$ is defined as

$$r_{xx}[n_1,n_2]=\frac{C_{xx}[n_1,n_2]}{\sqrt{C_{xx}[n_1,n_1]C_{xx}[n_2,n_2]}} \quad (5.227)$$

An important property of the correlation coefficient is that it is bounded by 1:

$$\left|r_{xx}[n_1,n_2]\right|\le 1 \quad (5.228)$$

which is the *Cauchy-Schwarz* inequality. To prove it, we note that for any real number a

$$\begin{aligned}0&\le E\left\{\left|a(\mathbf{x}[n_1]-\mu[n_1])+(\mathbf{x}[n_2]-\mu[n_2])\right|^2\right\}\\&=a^2C_{xx}[n_1,n_1]+2aC_{xx}[n_1,n_2]+C_{xx}[n_2,n_2]\end{aligned} \quad (5.229)$$

Since the quadratic function in Eq. (5.229) is positive for all a, its roots have to be complex, and thus its discriminant has to be negative:

$$C_{xx}^2[n_1,n_2]-C_{xx}[n_1,n_1]C_{xx}[n_2,n_2]\le 0 \quad (5.230)$$

from which Eq. (5.228) is derived.

The cross-correlation of two stochastic processes $\mathbf{x}[n]$ and $\mathbf{y}[n]$ is defined as

$$R_{xy}[n_1,n_2]=E\left\{\mathbf{x}[n_1]\mathbf{y}^*[n_2]\right\}=R_{yx}^*[n_2,n_1] \quad (5.231)$$

where we have explicitly indicated with subindices the random process. Similarly, their cross-covariance is

$$C_{xy}[n_1,n_2]=R_{xy}[n_1,n_2]-\mu_x[n_1]\mu_y^*[n_2] \quad (5.232)$$

Two processes $\mathbf{x}[n]$ and $\mathbf{y}[n]$ are called *orthogonal* iff

$$R_{xy}[n_1,n_2]=0 \qquad \text{for every } n_1 \text{ and } n_2 \quad (5.233)$$

They are called *uncorrelated* iff

$$C_{xy}[n_1,n_2]=0 \qquad \text{for every } n_1 \text{ and } n_2 \quad (5.234)$$

Independent processes. If two processes $\mathbf{x}[n]$ and $\mathbf{y}[n]$ are such that the random variables $\mathbf{x}[n_1], \mathbf{x}[n_2], \cdots, \mathbf{x}[n_m]$, and $\mathbf{y}[n_1'], \mathbf{y}[n_2'], \cdots, \mathbf{y}[n_m']$ are mutually independent, then these processes are called independent. If two processes are independent, then they are also uncorrelated, though the converse is not generally true.

Gaussian processes. A process $\mathbf{x}[n]$ is called Gaussian if the random variables $\mathbf{x}[n_1], \mathbf{x}[n_2], \cdots, \mathbf{x}[n_m]$ are jointly Gaussian for any m and $n_1, n_2, \cdots, n_m$. If two processes are Gaussian and also uncorrelated, then they are also statistically independent.

5.8.2. Stationary Processes

Stationary processes are those whose statistical properties do not change over time. While truly stationary processes do not exist in speech signals, they are a reasonable approximation and have the advantage of allowing us to use the Fourier transforms defined in Section 5.2.1. In this section we define stationarity and analyze some of its properties.

A stochastic process is called *strict-sense stationary* (SSS) if its statistical properties are invariant to a shift of the origin: i.e., both processes $\mathbf{x}[n]$ and $\mathbf{x}[n+l]$ have the same statistics for any l. Likewise, two processes $\mathbf{x}[n]$ and $\mathbf{y}[n]$ are called *jointly strict-sense stationary* if their joint statistics are the same as those of $\mathbf{x}[n+l]$ and $\mathbf{y}[n+l]$ for any l.

From the definition, it follows that the mth-order density of an SSS process must be such that

$$f(x_1, \cdots, x_m; n_1, \cdots, n_m) = f(x_1, \cdots, x_m; n_1 + l, \cdots, n_m + l) \tag{5.235}$$

for any l. Thus the first-order density satisfies $f(x,n) = f(x, n+l)$ for any l, which means that it is independent of n:

$$f(x,n) = f(x) \tag{5.236}$$

or, in other words, the density function is constant with time.

Similarly, $f(x_1, x_2; n_1 + l, n_2 + l)$ is independent of l, which leads to the conclusion

$$f(x_1, x_2; n_1, n_2) = f(x_1, x_2; m) \qquad m = n_1 - n_2 \tag{5.237}$$

or, in other words, the joint density of $\mathbf{x}[n]$ and $\mathbf{x}[n+m]$ is not a function of n, only of m, the time difference between the two samples.

Let's compute the first two moments of a SSS process:

$$E\{x[n]\} = \int x[n] f(x[n]) = \int x f(x) = \mu \tag{5.238}$$

$$E\{x[n+m]x^*[n]\} = \int x[n+m]x^*[n] f(x[n+m], x[n]) = R_{xx}[m] \tag{5.239}$$

or, in other words, its mean is not a function of time and its autocorrelation depends only on m.

A stochastic process $\mathbf{x}[n]$ that obeys Eq. (5.238) and (5.239) is called *wide-sense stationary* (WSS). From this definition, a SSS process is also a WSS process but the converse is not true in general. Gaussian processes are an important exception, and it can be proved that a WSS Gaussian process is also SSS.

For example, the random process of Eq. (5.213) is WSS, because it has zero mean and its autocorrelation function, as given by Eq. (5.223), is only a function of $m = n_1 - n_2$. By setting $m = 0$ in Eq. (5.239) we see that the average power of a WSS stationary process

$$E\{|x[n]|^2\} = R[0] \tag{5.240}$$

is independent of n.

The autocorrelation of a WSS process is a conjugate-symmetric function, also referred to as a *Hermitian* function:

$$R[-m] = E\{x[n-m]x^*[n]\} = E\{x[n]x^*[n+m]\} = R^*[m] \tag{5.241}$$

so that if $x[n]$ is real, $R[m]$ is even.

From Eqs. (5.226), (5.238), and (5.239) we can compute the autocovariance as

$$C[m] = R[m] - |\mu|^2 \tag{5.242}$$

and its correlation coefficient as

$$r[m] = C[m]/C[0] \tag{5.243}$$

Two processes $\mathbf{x}[n]$ and $\mathbf{y}[n]$ are called jointly WSS if both are WSS and their cross-correlation depends only on $m = n_1 - n_2$:

$$R_{xy}[m] = E\{x[n+m]y^*[n]\} \tag{5.244}$$

$$C_{xy}[m] = R_{xy}[m] - \mu_x \mu_y^* \tag{5.245}$$

5.8.2.1. Ergodic Processes

A critical problem in the theory of stochastic processes is the estimation of their various statistics, such as the mean and autocorrelation given that often only one realization of the random process is available. The first approximation would be to replace the expectation in Eq. (5.218) with its *ensemble average*:

$$\mu[n] \cong \frac{1}{M} \sum_{i=0}^{M-1} x_i[n] \tag{5.246}$$

where $x_i[n]$ are different samples of the random process.

As an example, let $\mathbf{x}[n]$ be the frequency-modulated (FM) random process received by a FM radio receiver:

$$\mathbf{x}[n] = a[n] + \mathbf{v}[n] \tag{5.247}$$

which contains some additive noise $\mathbf{v}[n]$. The realization $x_i[n]$ received by receiver i will be different from the realization $x_j[n]$ for receiver j. We know that each signal has a certain level of noise, so one would hope that by averaging them, we could get the mean of the process for a sufficiently large number of radio receivers.

In many cases, however, only one sample of the process is available. According to Eq. (5.246) this would mean that that the sample signal equals the mean, which does not seem very robust. We could also compute the signal's time average, but this may not tell us much about the random process in general. However, for a special type of random processes called ergodic, their ensemble averages equal appropriate time averages.

A process $\mathbf{x}[n]$ with constant mean

$$E\{\mathbf{x}[n]\} = \mu \tag{5.248}$$

is called *mean-ergodic* if, with probability 1, the ensemble average equals the time average when N approaches infinity:

$$\lim_{N \to \infty} \mu_N = \mu \tag{5.249}$$

where μ_N is the time average

$$\mu_N = \frac{1}{N} \sum_{n=-N/2}^{N/2-1} \mathbf{x}[n] \tag{5.250}$$

which, combined with Eq. (5.248), indicates that μ_N is a random variable with mean μ. Taking expectations in Eq. (5.250) and using Eq. (5.248), it is clear that

$$E\{\mu_N\} = \mu \tag{5.251}$$

so that proving Eq. (5.249) is equivalent to proving

$$\lim_{N \to \infty} \sigma_N^2 = 0 \tag{5.252}$$

with σ_N^2 being the variance of μ_N. It can be shown [12] that a process $\mathbf{x}[n]$ is mean ergodic *iff*

$$\lim_{N \to \infty} \frac{1}{N^2} \sum_{n=-N/2}^{N/2-1} \sum_{m=-N/2}^{N/2-1} C_{xx}[n,m] = 0 \tag{5.253}$$

It can also be shown [12] that a sufficient condition for a WSS process to be mean ergodic is to satisfy

$$\lim_{m\to\infty} C_{xx}[m] = 0 \tag{5.254}$$

which means that if the random variables $\mathbf{x}[n]$ and $\mathbf{x}[n+m]$ are uncorrelated for large m, then process $\mathbf{x}[n]$ is mean ergodic. This is true for many regular processes.

A similar condition can be proven for a WSS process to be covariance ergodic. In most cases in this book we assume ergodicity, first because of convenience for mathematical tractability, and second because it is a good approximation to assume that samples that are far apart are uncorrelated. *Ergodicity allows us to compute means and covariances of random processes by their time averages.*

5.8.3. LTI Systems with Stochastic Inputs

If the WSS random process $\mathbf{x}[n]$ is the input to an LTI system with impulse response $h[n]$, the output

$$\mathbf{y}[n] = \sum_{m=-\infty}^{\infty} h[m]\mathbf{x}[n-m] = \sum_{m=-\infty}^{\infty} h[n-m]\mathbf{x}[m] \tag{5.255}$$

is another WSS random process. To prove this we need to show that the mean is not a function of n:

$$\mu_y[n] = E\{\mathbf{y}[n]\} = \sum_{m=-\infty}^{\infty} h[m]E\{\mathbf{x}[n-m]\} = \mu_x \sum_{m=-\infty}^{\infty} h[m] \tag{5.256}$$

The cross-correlation between input and output is given by

$$\begin{aligned} R_{xy}[m] &= E\{\mathbf{x}[n+m]\mathbf{y}*[n]\} = \sum_{l=-\infty}^{\infty} h^*[l]E\{\mathbf{x}[n+m]\mathbf{x}*[n-l]\} \\ &= \sum_{l=-\infty}^{\infty} h^*[l]R_{xx}[m+l] = \sum_{l=-\infty}^{\infty} h^*[-l]R_{xx}[m-l] = h^*[-m]*R_{xx}[m] \end{aligned} \tag{5.257}$$

and the autocorrelation of the output

$$\begin{aligned} R_{yy}[m] &= E\{\mathbf{y}[n+m]\mathbf{y}*[n]\} = \sum_{l=-\infty}^{\infty} h[l]E\{\mathbf{x}[n+m-l]\mathbf{y}*[n]\} \\ &= \sum_{l=-\infty}^{\infty} h[l]R_{xy}[m-l] = h[m]*R_{xy}[m] = h[m]*h^*[-m]*R_{xx}[m] \end{aligned} \tag{5.258}$$

is only a function of m.

5.8.4. Power Spectral Density

The Fourier transform of a WSS random process $\mathbf{x}[n]$ is a stochastic process in the variable ω

$$\mathbf{X}(\omega) = \sum_{n=-\infty}^{\infty} \mathbf{x}[n] e^{-j\omega n} \tag{5.259}$$

whose autocorrelation is given by

$$\begin{aligned} E\left\{\mathbf{X}(\omega+u)\mathbf{X}^*(\omega)\right\} &= E\left\{\sum_{l=-\infty}^{\infty} \mathbf{x}[l] e^{-j(\omega+u)l} \sum_{m=-\infty}^{\infty} \mathbf{x}^*[m] e^{j\omega m}\right\} \\ &= \sum_{n=-\infty}^{\infty} e^{-j(\omega+u)n} \sum_{m=-\infty}^{\infty} E\{\mathbf{x}[m+n]\mathbf{x}^*[m]\} e^{-jum} \end{aligned} \tag{5.260}$$

where we made a change of variables $l = n+m$ and changed the order of expectation and summation. Now, if $\mathbf{x}[n]$ is WSS

$$R_{xx}[n] = E\left\{\mathbf{x}[m+n]\mathbf{x}^*[m]\right\} \tag{5.261}$$

and if we set $u = 0$ in Eq. (5.260) together with Eq. (5.261), then we obtain

$$S_{xx}(\omega) = E\left\{\left|\mathbf{X}(\omega)\right|^2\right\} = \sum_{n=-\infty}^{\infty} R_{xx}[n] e^{-j\omega n} \tag{5.262}$$

$S_{xx}(\omega)$ is called the *power spectral density* of the WSS random process $\mathbf{x}[n]$, and it is the Fourier transform of its autocorrelation function $R_{xx}[n]$, with the inversion formula being

$$R_{xx}[n] = \frac{1}{2\pi} \int_{-\infty}^{\infty} S_{xx}(\omega) e^{j\omega n} d\omega \tag{5.263}$$

Note that Eqs. (5.48) and (5.263) are identical, though in one case we compute the autocorrelation of a signal as a time average, and the other is the autocorrelation of a random process as an ensemble average. For an ergodic process both are the same.

Just as we take Fourier transforms of deterministic signals, we can also compute the power spectral density of a random process as long as it is wide-sense stationary, which is why these wide-sense stationary processes are so useful.

If the random process $\mathbf{x}[n]$ is real then $R_{xx}[n]$ is real and even and, using properties in Table 5.5, $S_{xx}(\omega)$ is also real and even.

Parseval's theorem for random processes also applies here:

$$E\left\{\left|\mathbf{x}[n]\right|^2\right\} = R_{xx}[0] = \frac{1}{2\pi} \int_{-\pi}^{\pi} S_{xx}(\omega) d\omega \tag{5.264}$$

so that we can compute the signal's energy from the area under $S_{xx}(\omega)$. Let's get a physical interpretation of $S_{xx}(\omega)$. In order to do that we can similarly derive the cross-power spectrum $S_{xy}(\omega)$ of two WSS random processes $\mathbf{x}[n]$ and $\mathbf{y}[n]$ as the Fourier transform of their cross-correlation:

$$S_{xy}(\omega) = \sum_{n=-\infty}^{\infty} R_{xy}[n] e^{-j\omega n} \tag{5.265}$$

which allows us, taking Fourier transforms in Eq. (5.257), to obtain the cross-power spectrum between input and output to a linear system as

$$S_{xy}(\omega) = S_{xx}(\omega) H^*(\omega) \tag{5.266}$$

Now, taking the Fourier transform of Eq. (5.258), the power spectrum of the output is thus given by

$$S_{yy}(\omega) = S_{xy}(\omega) H(\omega) = S_{xx}(\omega) |H(\omega)|^2 \tag{5.267}$$

Finally, suppose we filter $\mathbf{x}[n]$ through the ideal bandpass filter

$$H_b(\omega) = \begin{cases} \sqrt{\pi / c} & \omega_0 - c < \omega < \omega_0 + c \\ 0 & otherwise \end{cases} \tag{5.268}$$

The energy of the output process is

$$0 \le E\left\{|\mathbf{y}[n]|^2\right\} = R_{yy}[0] = \frac{1}{2\pi} \int_{-\pi}^{\pi} S_{yy}(\omega) d\omega = \frac{1}{2c} \int_{\omega_0 - c}^{\omega_0 + c} S_{xx}(\omega) d\omega \tag{5.269}$$

so that taking the limit when $c \to 0$ results in

$$0 \le \lim_{c \to 0} \frac{1}{2c} \int_{\omega_0 - c}^{\omega_0 + c} S_{xx}(\omega) d\omega = S_{xx}(\omega_0) \tag{5.270}$$

which is the *Wiener-Khinchin* theorem and says that the power spectrum of a WSS process $\mathbf{x}[n]$, real or complex, is always positive for any ω. Equation (5.269) also explains the name power spectral density, because $S_{xx}(\omega)$ represents the density of power at any given frequency ω.

5.8.5. Noise

A process $\mathbf{x}[n]$ is *white noise* if, and only if, its samples are uncorrelated:

$$C_{xx}[n_1, n_2] = C[n_1] \delta[n_1 - n_2] \tag{5.271}$$

and is zero-mean $\mu_x[n] = 0$.

If in addition $\mathbf{x}[n]$ is WSS, then

$$C_{xx}[n] = R_{xx}[n] = q\delta[n] \tag{5.272}$$

which has a flat power spectral density

$$S_{xx}(\omega) = q \qquad \text{for all } \omega \tag{5.273}$$

The thermal noise phenomenon in metallic resistors can be accurately modeled as white Gaussian noise. White noise doesn't have to be Gaussian (white Poisson impulse noise is one of many other possibilities).

Colored noise is defined as a zero-mean WSS process whose samples are correlated with autocorrelation $R_{xx}[n]$. Colored noise can be generated by passing white noise through a filter $h[n]$ such that $S_{xx}(\omega) = |H(\omega)|^2$. A type of colored noise that is very frequently encountered in speech signals is the so-called *pink noise*, whose power spectral density decays with ω. A more in-depth discussion of noise and its effect on speech signals is included in Chapter 10.

5.9. HISTORICAL PERSPECTIVE AND FURTHER READING

It is impossible to cover the field of Digital Signal Processing in just one chapter. The book by Oppenheim and Schafer [10] is one of the most widely used as a comprehensive treatment. For a more in-depth coverage of digital filter design, you can read the book by Parks and Burrus [13]. A detailed study of the FFT is provided by Burrus and Parks [2]. The theory of signal processing for analog signals can be found in Oppenheim and Willsky [11]. The theory of random signals can be found in Papoulis [12]. Multirate processing is well studied in Crochiere and Rabiner [4]. Razavi [16] covers analog-digital conversion. Software programs, such as MATLAB [1], contain a large number of packaged subroutines. Malvar [7] has extensive coverage of filterbanks and lapped transforms.

The field of Digital Signal Processing has a long history. The greatest advances in the field started in the 17th century. In 1666, English mathematician and physicist Sir *Isaac Newton* (1642-1727) invented differential and integral calculus, which was independently discovered in 1675 by German mathematician *Gottfried Wilhelm Leibniz* (1646-1716). They both developed discrete mathematics and numerical methods to solve such equations when closed-form solutions were not available. In the 18th century, these techniques were further extended. Swiss brothers *Johann* (1667-1748) and *Jakob Bernoulli* (1654-1705) invented the calculus of variations and polar coordinates. French mathematician *Joseph Louis Lagrange* (1736-1813) developed algorithms for numerical integration and interpolation of continuous functions. The famous Swiss mathematician *Leonhard Euler* (1707-1783) developed the theory of complex numbers and number theory so useful in the DSP field, in addition to the first full analytical treatment of algebra, the theory of equations, trigonometry and analytical geometry. In 1748, Euler examined the motion of a vibrating string and discovered that sinusoids are eigenfunctions for linear systems. Swiss scientist *Daniel Bernoulli* (1700-1782), son of Johann Bernoulli, also conjectured in 1753 that all physical motions of a string could be represented by linear combinations of normal modes. However, both Euler and Bernoulli, and later Lagrange, discarded the use of trigonometric series because *it was impossible to represent signals with corners*. The 19th century brought us the theory of harmonic analysis. One of those who contributed most to the field of Digital Signal Processing is *Jean Baptiste Joseph Fourier* (1768-1830), a French mathematician who in 1822 published *The Analytical Theory of Heat*, where he derived a mathematical formulation for the phenomenon of heat conduction. In this treatise, he also developed the concept of Fourier series and harmonic analysis and the Fourier transform. One of Fourier's disciples, the French mathematician *Simeon-Denis Poisson* (1781-1840), studied the convergence of Fourier series together with countryman *Augustin Louis Cauchy* (1789-1857). Nonetheless, it was German *Peter Dirichlet* (1805-1859) who gave the first set of conditions sufficient to guarantee the convergence of a Fourier series. French mathematician *Pierre Simon Laplace* (1749-1827) invented the Laplace transform, a transform for continuous-time signals over the whole complex plane. French mathematician *Marc-Antoine Parseval* (1755-1836) derived the theorem that carries his name. German *Leopold Kronecker* (1823-1891) did work with discrete delta functions. French mathematician *Charles Hermite* (1822-1901) discovered complex conjugate matrices. American *Josiah Willard Gibbs* (1839-1903) studied the phenomenon of Fourier approximations to periodic square waveforms.

Until the early 1950s, all signal processing was analog, including the *long-playing* (LP) record first released in 1948. Pulse Code Modulation (PCM) had been invented by *Paul M. Rainey* in 1926 and independently by *Alan H. Reeves* in 1937, but it wasn't until 1948 when *Oliver, Pierce,* and *Shannon* [9] laid the groundwork for PCM (see Chapter 7 for details). Bell Labs engineers developed a PCM system in 1955, the so-called T-1 carrier system, which was put into service in 1962 as the world's first common-carrier digital communications system and is still used today. The year 1948 also saw the invention of the transistor at Bell Labs and a small prototype computer at Manchester University and marked the birth of modern Digital Signal Processing. In 1958, *Jack Kilby* of Texas Instruments invented the integrated circuit and in 1970, researchers at Lincoln Laboratories developed the first real-time DSP computer, which performed signal processing tasks about 100 times faster than general-purpose computers of the time. In 1978, Texas Instruments introduced *Speak & Spell*™, a toy that included an integrated circuit especially designed for speech synthesis. Intel Corporation introduced in 1971 the 4-bit Intel 4004, the first general-purpose microprocessor chip, and in 1972 they introduced the 8-bit 8008. In 1982 Texas Instruments introduced the TMS32010, the first commercially viable single-chip Digital Signal Processor (DSP), a microprocessor specially designed for fast signal processing operations. At a cost of about $100, the TMS32010 was a 16-bit fixed-point chip with a hardware multiplier built-in that executed 5 million instructions per second (MIPS). Gordon Moore, Intel's founder, came up with the law that carries his name stating that computing power doubles every 18 months, allowing ever faster processors. By the end of the 20th century, DSP chips could perform floating-point operations at a rate over 1000MIPS and had a cost below $5, so that today they are found in many devices from automobiles to cellular phones.

While hardware improvements significantly enabled the development of the field, digital algorithms were also needed. The 1960s saw the discovery of many of the concepts described in this chapter. In 1965, *James W. Cooley* and *John W. Tukey* [3] discovered the FFT, although it was later found [6] that German mathematician *Carl Friedrich Gauss* (1777-1855) had already invented it over a century earlier. The FFT sped up calculations by orders of magnitude, which opened up many possible algorithms for the slow computers of the time. *James F. Kaiser*, *Bernard Gold,* and *Charles Rader* published key papers on digital filtering. *John Stockham* and *Howard Helms* independently discovered fast convolution by doing convolution with FFTs.

An association that has had a large impact on the development of modern Digital Signal Processing is the Institute of Electrical and Electronic Engineers (IEEE), which has over 350,000 members in 150 nations and is the world's largest technical organization. It was founded in 1884 as the American Institute of Electrical Engineers (AIEE). IEEE's other parent organization, the Institute of Radio Engineers (IRE), was founded in 1912, and the two merged in 1963. The IEEE Signal Processing Society is a society within the IEEE devoted to Signal Processing. Originally founded in 1948 as the Institute of Radio Engineers Professional Group on Audio, it was later renamed the IEEE Group on Audio (1964), the IEEE Audio and Electroacoustics group (1965), the IEEE group on Acoustics Speech and Signal Processing (1974), the Acoustic, Speech and Signal Processing Society (1976), and finally IEEE Signal Processing Society (1989). In 1976 the society initiated its practice of holding an annual conference, the International Conference on Acoustic, Speech and Signal Processing (ICASSP), which has been held every year since, and whose proceedings constitute an invaluable reference. Frederik Nebeker [8] provides a history of the society's first 50 years rich in insights from the pioneers.

REFERENCES

[1] Burrus, C.S., *et al.*, *Computer-Based Exercises for Signal Processing Using Matlab*, 1994, Upper Saddle River, NJ, Prentice Hall.

[2] Burrus, C.S. and T.W. Parks, *DFT/FFT and Convolution Algorithms: Theory and Implementation*, 1985, New York, John Wiley.

[3] Cooley, J.W. and J.W. Tukey, "An Algorithm for the Machine Calculation of Complex Fourier Series," *Mathematics of Computation*, 1965, **19**(Apr.), pp. 297-301.

[4] Crochiere, R.E. and L.R. Rabiner, *Multirate Digital Signal Processing*, 1983, Upper Saddle River, NJ, Prentice-Hall.

[5] Duhamel, P. and H. Hollman, "Split Radix FFT Algorithm," *Electronic Letters*, 1984, **20**(January), pp. 14-16.

[6] Heideman, M.T., D.H. Johnson, and C.S. Burrus, "Gauss and the History of the Fast Fourier Transform," *IEEE ASSP Magazine*, 1984, **1**(Oct), pp. 14-21.

[7] Malvar, H., *Signal Processing with Lapped Transforms*, 1992, Artech House.

[8] Nebeker, F., *Fifty Years of Signal Processing: The IEEE Signal Processing Society and Its Technologies*, 1998, IEEE.

[9] Oliver, B.M., J.R. Pierce, and C. Shannon, "The Philosophy of PCM," *Proc. Institute of Radio Engineers*, 1948, **36**, pp. 1324-1331.

[10] Oppenheim, A.V., R.W. Schafer, and J.R. Buck, *Discrete-Time Signal Processing*, 2nd ed., 1999, Prentice-Hall, Upper Saddle River, NJ.

[11] Oppenheim, A.V. and A.S. Willsky, *Signals and Systems*, 1997, Upper Saddle River, NJ, Prentice-Hall.

[12] Papoulis, A., *Probability, Random Variables, and Stochastic Processes*, 3rd ed., 1991, New York, McGraw-Hill.

[13] Parks, T.W. and C.S. Burrus, *Digital Filter Design*, 1987, New York, NY, John Wiley.

[14] Parks, T.W. and J.H. McClellan, "A Program for the Design of Linear Phase Finite Impulse Response Filters," *IEEE Trans. on Audio Electroacoustics*, 1972, **AU-20**(Aug), pp. 195-199.

[15] Rao, K.R. and P. Yip, *Discrete Cosine Transform: Algorithms, Advantages and Applications*, 1990, San Diego, CA, Academic Press.

[16] Razavi, B., *Principles of Data Conversión System Design*, 1995, IEEE Press.

[17] Smith, M.J.T. and T.P. Barnwell, "A Procedure for Designing Exact Reconstruction Filter Banks for Tree Structured Subband Coders," *Int. Conf. on Acoustics, Speech and Signal Processing*, 1984, San Diego, CA, pp. 27.1.1-27.1.4.

CHAPTER 6

Speech Signal Representations

This chapter presents several representations for speech signals useful in speech coding, synthesis, and recognition. The central theme is the decomposition of the speech signal as a source passed through a linear time-varying filter. This filter can be derived from models of speech production based on the theory of acoustics where the source represents the air flow at the vocal cords, and the filter represents the resonances of the vocal tract which change over time. Such a source-filter model is illustrated in Figure 6.1. We describe methods to compute both the source or *excitation* $e[n]$ and the filter $h[n]$ from the speech signal $x[n]$.

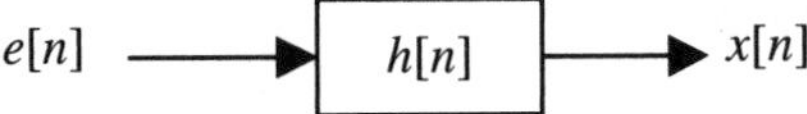

Figure 6.1 Basic source-filter model for speech signals.

To estimate the filter we present methods inspired by speech production models (such as linear predictive coding and cepstral analysis) as well as speech perception models (such as mel-frequency cepstrum). Once the filter has been estimated, the source can be obtained by passing the speech signal through the inverse filter. Separation between source and filter is one of the most difficult challenges in speech processing.

It turns out that phoneme classification (either by human or by machines) is mostly dependent on the characteristics of the filter. Traditionally, speech recognizers estimate the filter characteristics and ignore the source. Many speech synthesis techniques use a source-filter model because it allows flexibility in altering the pitch and the filter. Many speech coders also use this model because it allows a low bit rate.

We first introduce the spectrogram as a representation of the speech signal that highlights several of its properties and describe the short-time Fourier analysis, which is the basic tool to build the spectrograms of Chapter 2. We then introduce several techniques used to separate source and filter: LPC and cepstral analysis, perceptually motivated models, formant tracking, and pitch tracking.

6.1. Short-Time Fourier Analysis

In Chapter 2, we demonstrated how useful *spectrograms* are to analyze phonemes and their transitions. A spectrogram of a time signal is a special two-dimensional representation that displays time in its horizontal axis and frequency in its vertical axis. A gray scale is typically used to indicate the energy at each point (t, f) with white representing low energy and black high energy. In this section we cover short-time Fourier analysis, the basic tool with which to compute them.

The idea behind a spectrogram, such as that in Figure 6.2, is to compute a Fourier transform every 5 milliseconds or so, displaying the energy at each time/frequency point. Since some regions of speech signals shorter than, say, 100 milliseconds often appear to be periodic, we use the techniques discussed in Chapter 5. However, the signal is no longer periodic when longer segments are analyzed, and therefore the exact definition of Fourier transform cannot be used. Moreover, that definition requires knowledge of the signal for infinite time. For both reasons, a new set of techniques called *short-time analysis* are proposed. These techniques decompose the speech signal into a series of short segments, referred to as *analysis frames*, and analyze each one independently.

In Figure 6.2 (a), note the assumption that the signal can be approximated as periodic within X and Y is reasonable. In regions (Z, W) and (H, G), the signal is not periodic and looks like *random noise*. The signal in (Z, W) appears to have different noisy characteristics than those of segment (H, G). The use of an analysis frame implies that the region is short enough for the behavior (periodicity or noise-like appearance) of the signal to be approximately constant. If the region where speech seems periodic is too long, the pitch period is not constant and not all the periods in the region are similar. In essence, the speech region has to be short enough so that the signal is *stationary* in that region: i.e., the signal characteristics (whether periodicity or noise-like appearance) are uniform in that region. A more formal definition of stationarity is given in Chapter 5.

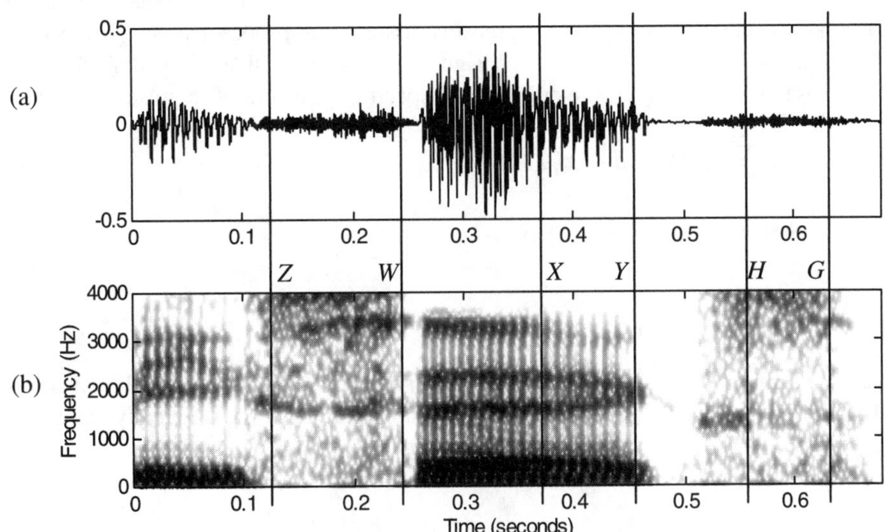

Figure 6.2 (a) Waveform with (b) its corresponding wideband spectrogram. Darker areas mean higher energy for that time and frequency. Note the vertical lines spaced by pitch periods.

Similarly to the filterbanks described in Chapter 5, given a speech signal $x[n]$, we define the short-time signal $x_m[n]$ of frame m as

$$x_m[n] = x[n]w_m[n] \tag{6.1}$$

the product of $x[n]$ by a *window* function $w_m[n]$, which is zero everywhere except in a small region.

While the window function can have different *values* for different frames *m*, a popular choice is to keep it constant for all frames:

$$w_m[n] = w[m-n] \tag{6.2}$$

where $w[n] = 0$ for $|n| > N/2$. In practice, the window length is on the order of 20 to 30 ms.

With the above framework, the short-time Fourier representation for frame *m* is defined as

$$X_m(e^{j\omega}) = \sum_{n=-\infty}^{\infty} x_m[n]e^{-j\omega n} = \sum_{n=-\infty}^{\infty} w[m-n]x[n]e^{-j\omega n} \tag{6.3}$$

with all the properties of Fourier transforms studied in Chapter 5.

In Figure 6.3 we show the short-time spectrum of voiced speech. Note that there are a number of peaks in the spectrum. To interpret this, assume the properties of $x_m[n]$ persist outside the window, and that, therefore, the signal is periodic with period M in the true sense. In this case, we know (see Chapter 5) that its spectrum is a sum of impulses

$$X_m(e^{j\omega}) = 2\pi \sum_{k=-\infty}^{\infty} X_m[k]\delta(\omega - 2\pi k / M) \tag{6.4}$$

Given that the Fourier transform of $w[n]$ is

$$W(e^{j\omega}) = \sum_{n=-\infty}^{\infty} w[n]e^{-j\omega n} \tag{6.5}$$

so that the transform of $w[m-n]$ is $W(e^{-j\omega})e^{-j\omega m}$. Therefore, using the convolution property, the transform of $x[n]w[m-n]$ for fixed m is the convolution in the frequency domain

$$X_m(e^{j\omega}) = \sum_{k=-\infty}^{\infty} X_m[k]W(e^{-j(\omega-2\pi k/N)})e^{-j(\omega-2\pi k/N)m} \tag{6.6}$$

which is a sum of weighted $W(e^{j\omega})$, shifted on every harmonic, the narrow peaks seen in Figure 6.3 (b) with a rectangular window. The short-time spectrum of a periodic signal exhibits peaks (equally spaced $2\pi / M$ apart) representing the harmonics of the signal. We estimate $X_m[k]$ from the short-time spectrum $X_m(e^{j\omega})$, and we see the importance of the length and choice of window.

Equation (6.6) indicates that one cannot recover $X_m[k]$ by simply retrieving $X_m(e^{j\omega})$, although the approximation can be reasonable if there is a small value of λ such that

$$W(e^{j\omega}) \approx 0 \text{ for } |\omega - \omega_k| > \lambda \tag{6.7}$$

which is the case outside the main lobe of the window's frequency response.

Recall from Section 5.4.2.1 that, for a rectangular window of length *N*, $\lambda = 2\pi / N$. Therefore, Eq. (6.7) is satisfied if $N \geq M$, i.e., the rectangular window contains at least one pitch period. The width of the main lobe of the window's frequency response is inversely proportional to the length of the window. The pitch period in Figure 6.3 is $M = 71$ at a sampling rate of 8 kHz. A shorter window is used in Figure 6.3 (c), which results in wider analysis lobes, though still visible.

Also recall from Section 5.4.2.2 that for a Hamming window of length *N*, $\lambda = 4\pi / N$: twice as wide as that of the rectangular window, which entails $N \geq 2M$. Thus, for Eq. (6.7) to be met, a Hamming window must contain at least two pitch periods. The lobes are visible in Figure 6.3 (d) since $N = 240$, but they are not visible in Figure 6.3 (e) since $N = 120$, and $N < 2M$.

In practice, one cannot know what the pitch period is ahead of time, which often means you need to prepare for the lowest pitch period. A low-pitched voice with a

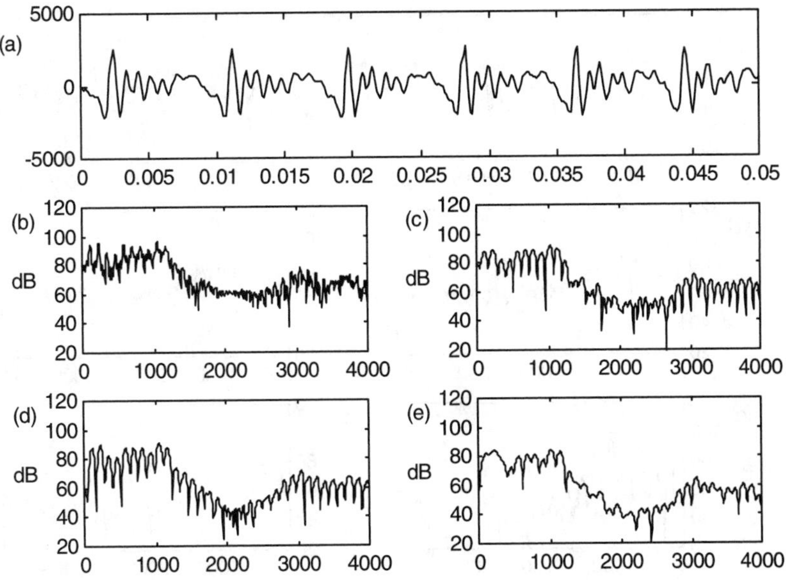

Figure 6.3 Short-time spectrum of male voiced speech (vowel /ah/ with local pitch of 110Hz): (a) time signal, spectra obtained with (b) 30 ms rectangular window and (c) 15 ms rectangular window, (d) 30 ms Hamming window, (e) 15 ms Hamming window. The window lobes are not visible in (e), since the window is shorter than 2 times the pitch period. Note the spectral leakage present in (b).

$F_0 = 50\,\text{Hz}$ requires a rectangular window of at least 20 ms and a Hamming window of at least 40 ms for the condition in Eq. (6.7) to be met. If speech is non-stationary within 40 ms, taking such a long window implies obtaining an average spectrum during that segment instead of several distinct spectra. For this reason, the rectangular window provides better *time resolution* than the Hamming window. Figure 6.4 shows analysis of female speech for which shorter windows are feasible.

But the frequency response of the window is not completely zero outside its main lobe, so one needs to see the effects of this incorrect assumption. From Section 5.4.2.1 note that the second lobe of a rectangular window is only approximately 17 dB below the main lobe. Therefore, for the kth harmonic the value of $X_m(e^{j2\pi k/M})$ contains not $X_m[k]$, but also a weighted sum of $X_m[l]$. This phenomenon is called *spectral leakage* because the amplitude of one harmonic leaks over the rest and masks its value. If the signal's spectrum is white, spectral leakage does not cause a major problem, since the effect of the second lobe on a harmonic is only $10\log_{10}(1+10^{-17/10}) = 0.08\text{dB}$. On the other hand, if the signal's spectrum decays more quickly in frequency than the decay of the window, the spectral leakage results in inaccurate estimates.

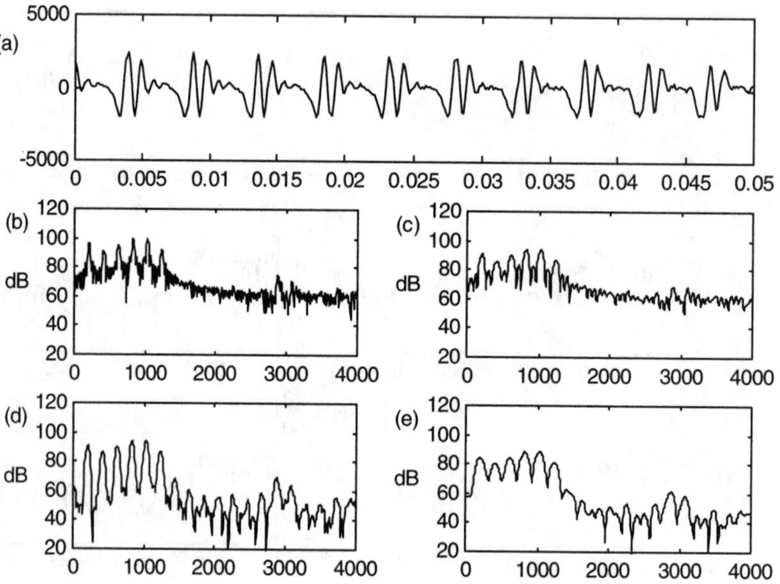

Figure 6.4 Short-time spectrum of female voiced speech (vowel /aa/ with local pitch of 200Hz): (a) time signal, spectra obtained with (b) 30 ms rectangular window and (c) 15 ms rectangular window, (d) 30 ms Hamming window, (e) 15 ms Hamming window. In all cases the window lobes are visible, since the window is longer than 2 times the pitch period. Note the spectral leakage present in (b) and (c).

From Section 5.4.2.2, observe that the second lobe of a Hamming window is approximately 43 dB, which means that the spectral leakage effect is much less pronounced. Other windows, such as Hanning, or triangular windows, also offer less spectral leakage than the rectangular window. This important fact is the reason why, despite their better time resolution, rectangular windows are rarely used for speech analysis. In practice, window lengths are on the order of 20 to 30 ms. This choice is a compromise between the stationarity assumption and the frequency resolution.

In practice, the Fourier transform in Eq. (6.3) is obtained through an FFT. If the window has length *N*, the FFT has to have a length greater than or equal to *N*. Since FFT algorithms often have lengths that are powers of 2 ($L = 2^R$), the windowed signal with length *N* is augmented with $(L - N)$ zeros either before, after, or both. This process is called *zero-padding*. A larger value of *L* provides a finer description of the discrete Fourier transform; but it does not increase the analysis frequency resolution: this is the sole mission of the window length *N*.

In Figure 6.3, observe the broad peaks, resonances or formants, which represent the filter characteristics. For voiced sounds there is typically more energy at low frequencies than at high frequencies, also called *roll-off*. It is impossible to determine exactly the filter characteristics, because we know only samples at the harmonics, and we have no knowledge of the values in between. In fact, the resonances are less obvious in Figure 6.4 because the harmonics sample the spectral envelope less densely. For high-pitched female speakers and children, it is even more difficult to locate the formant resonances from the short-time spectrum.

Figure 6.5 shows the short-time analysis of unvoiced speech, for which no regularity is observed.

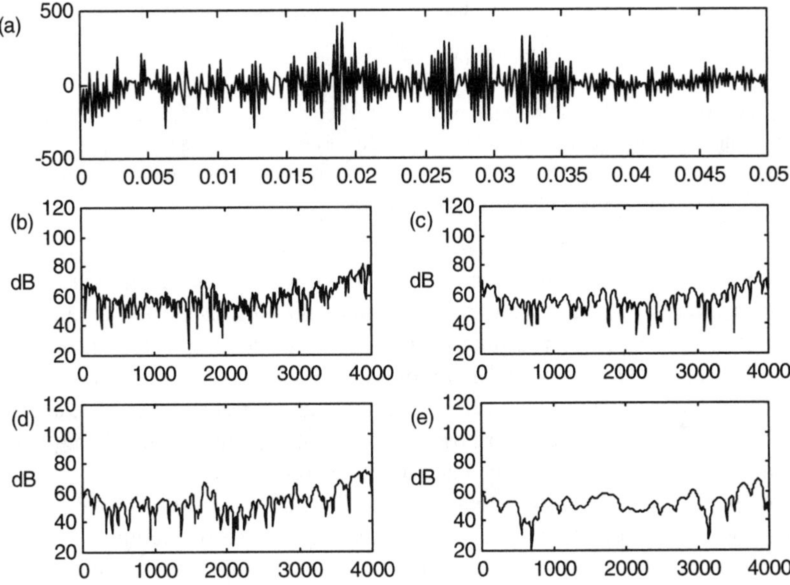

Figure 6.5 Short-time spectrum of unvoiced speech: (a) time signal, (b) 30 ms rectangular window, (c) 15 ms rectangular window, (d) 30 ms Hamming window, (e) 15 ms Hamming window.

6.1.1. Spectrograms

Since the spectrogram displays just the energy and not the phase of the short-term Fourier transform, we compute the energy as

$$\log |X[k]|^2 = \log\left(X_r^2[k] + X_i^2[k]\right) \tag{6.8}$$

with this value converted to a gray scale according to Figure 6.6. Pixels whose values have not been computed are interpolated. The slope controls the contrast of the spectrogram, while the saturation points for white and black control the dynamic range.

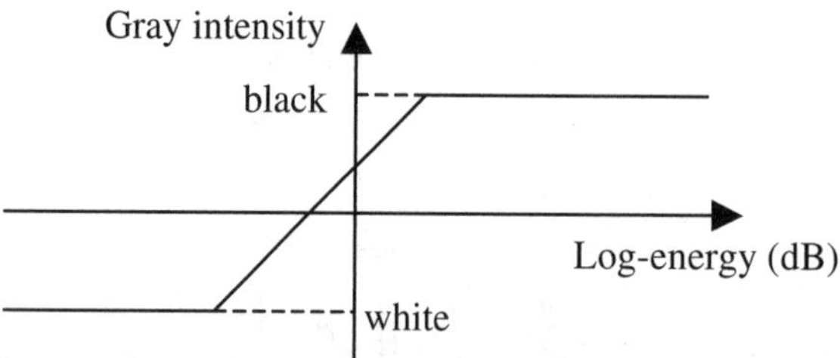

Figure 6.6 Conversion between log-energy values (in the x-axis) and gray scale (in the y-axis). Larger log-energies correspond to a darker gray color. There is a linear region for which more log-energy corresponds to darker gray, but there is saturation at both ends. Typically there is 40 to 60 dB between the pure white and the pure black.

There are two main types of spectrograms: *narrow-band* and *wide-band*. Wide-band spectrograms use relatively short windows (< 10 ms) and thus have good time resolution at the expense of lower frequency resolution, since the corresponding filters have wide bandwidths (> 200 Hz) and the harmonics cannot be seen. Note the vertical stripes in Figure 6.2, due to the fact that some windows are centered at the high part of a pitch pulse, and others in between have lower energy. Spectrograms can aid in determining formant frequencies and fundamental frequency, as well as voiced and unvoiced regions.

Narrow-band spectrograms use relatively long windows (> 20 ms), which lead to filters with narrow bandwidth (< 100 Hz). On the other hand, time resolution is lower than for wide-band spectrograms (see Figure 6.7). Note that the harmonics can be clearly seen, because some of the filters capture the energy of the signal's harmonics, and filters in between have little energy.

Some implementation details also need to be taken into account. Since speech signals are real, the Fourier transform is Hermitian, and its power spectrum is also even. Thus, it is only necessary to display values for $0 \le k \le N/2$ for N even. In addition, while the traditional spectrogram uses a gray scale, a color scale can also be used, or even a 3-D representation. In addition, to make the spectrograms easier to read, sometimes the signal is first pre-emphasized (typically with a first-order difference FIR filter) to boost the high frequencies to counter the roll-off of natural speech.

By inspecting both narrow-band and wide-band spectrograms, we can learn the filter's magnitude response and whether the source is voiced or not. Nonetheless it is very difficult to separate source and filter due to nonstationarity of the speech signal, spectral leakage, and the fact that only the filter's magnitude response can be known at the signal's harmonics.

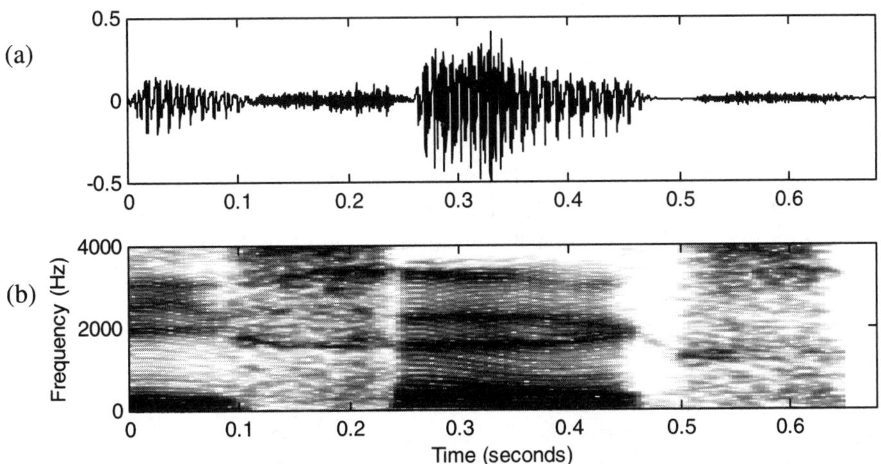

Figure 6.7 Waveform (a) with its corresponding narrowband spectrogram (b). Darker areas mean higher energy for that time and frequency. The harmonics can be seen as horizontal lines spaced by fundamental frequency. The corresponding wideband spectrogram can be seen in Figure 6.2.

6.1.2. Pitch-Synchronous Analysis

In the previous discussion, we assumed that the window length is fixed, and we saw the tradeoffs between a window that contained several pitch periods (narrow-band spectrograms) and a window that contained less than a pitch period (wide-band spectrograms). One possibility is to use a rectangular window whose length is exactly one pitch period; this is called *pitch-synchronous* analysis. To reduce spectral leakage a tapering window, such as Hamming or Hanning, can be used, with the window covering exactly two pitch periods. This latter option provides a very good compromise between time and frequency resolution. In this representation, no stripes can be seen in either time or frequency. The difficulty in computing pitch synchronous analysis is that, of course, we need to know the local pitch period, which, as we see in Section 6.7, is not an easy task.

6.2. Acoustical Model of Speech Production

Speech is a sound wave created by vibration that is propagated in the air. Acoustic theory analyzes the laws of physics that govern the propagation of sound in the vocal tract. Such a theory should consider three-dimensional wave propagation, the variation of the vocal tract shape with time, losses due to heat conduction and viscous friction at the vocal tract walls, softness of the tract walls, radiation of sound at the lips, nasal coupling, and excitation of sound. While a detailed model that considers all of the above is not yet available, some models provide a good approximation in practice, as well as a good understanding of the physics involved.

6.2.1. Glottal Excitation

As discussed in Chapter 2, the vocal cords constrict the path from the lungs to the vocal tract. This is illustrated in Figure 6.8. As lung pressure is increased, air flows out of the lungs and through the opening between the vocal cords (*glottis*). At one point the vocal cords are together, thereby blocking the airflow, which builds up *pressure* behind them. Eventually the pressure reaches a level sufficient to force the vocal cords to open and thus allow air to flow through the glottis. Then, the pressure in the glottis falls and, if the tension in the vocal cords is properly adjusted, the reduced pressure allows the cords to come together, and the cycle is repeated. This condition of sustained oscillation occurs for voiced sounds. The *closed-phase* of the oscillation takes place when the glottis is closed and the *volume velocity* is zero. The *open-phase* is characterized by a non-zero volume velocity, in which the lungs and the vocal tract are coupled.

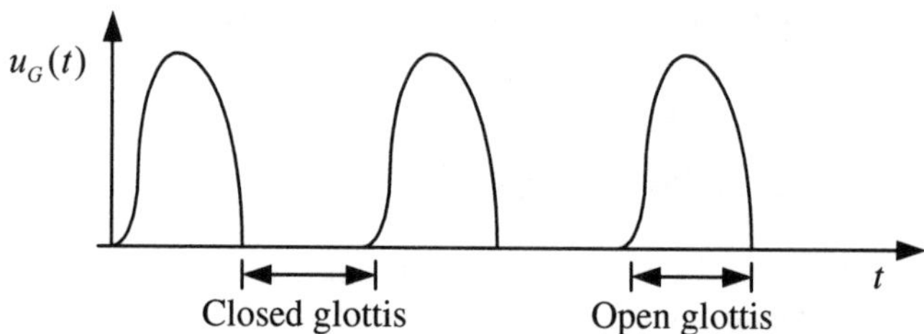

Figure 6.8 Glottal excitation: volume velocity is zero during the closed-phase, during which the vocal cords are closed.

Rosenberg's glottal model [39] defines the shape of the glottal volume velocity with the *open quotient*, or duty cycle, as the ratio of pulse duration to pitch period, and the *speed quotient* as the ratio of the rising to falling pulse durations.

6.2.2. Lossless Tube Concatenation

A widely used model for speech production is based on the assumption that the vocal tract can be represented as a concatenation of lossless tubes, as shown in Figure 6.9. The constant cross-sectional areas $\{A_k\}$ of the tubes approximate the area function $A(x)$ of the vocal tract. If a large number of tubes of short length are used, we reasonably expect the frequency response of the concatenated tubes to be close to those of a tube with continuously varying area function.

For frequencies corresponding to wavelengths that are long compared to the dimensions of the vocal tract, it is reasonable to assume plane wave propagation along the axis of the tubes. If in addition we assume that there are no losses due to viscosity or thermal conduction, and that the area A does not change over time, the sound waves in the tube satisfy the following pair of differential equations:

$$-\frac{\partial p(x,t)}{\partial x} = \frac{\rho}{A}\frac{\partial u(x,t)}{\partial t}$$
$$-\frac{\partial u(x,t)}{\partial x} = \frac{A}{\rho c^2}\frac{\partial p(x,t)}{\partial t} \tag{6.9}$$

where $p(x,t)$ is the sound pressure in the tube at position x and time t, $u(x,t)$ is the volume velocity flow in the tube at position x and time t, ρ is the density of air in the tube, c is the velocity of sound, and A is the cross-sectional area of the tube.

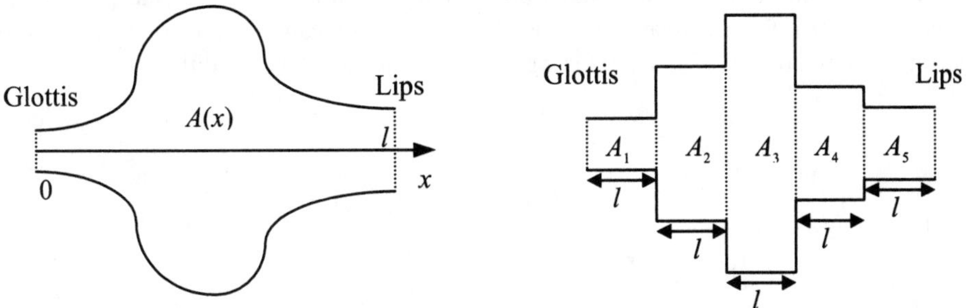

Figure 6.9 Approximation of a tube with continuously varying area $A(x)$ as a concatenation of 5 lossless acoustic tubes.

Since Eqs. (6.9) are linear, the pressure and volume velocity in the kth tube are related by

$$u_k(x,t) = u_k^+(t - x/c) - u_k^-(t + x/c)$$
$$p_k(x,t) = \frac{\rho c}{A_k}\left[u_k^+(t - x/c) + u_k^-(t + x/c)\right] \tag{6.10}$$

where $u_k^+(t - x/c)$ and $u_k^-(t - x/c)$ are the traveling waves in the positive and negative directions respectively and x is the distance measured from the left-hand end of tube k^{th}: $0 \le x \le l$. The reader can prove that this is indeed the solution by substituting Eq. (6.10) into (6.9).

When there is a junction between two tubes, as in Figure 6.10, part of the wave is reflected at the junction, as measured by r_k, the reflection coefficient

$$r_k = \frac{A_{k+1} - A_k}{A_{k+1} + A_k} \tag{6.11}$$

so that the larger the difference between the areas the more energy is reflected. The proof [9] is beyond the scope of this book. Since A_k and A_{k+1} are positive, it is easy to show that r_k satisfies the condition

$$-1 \le r_k \le 1 \tag{6.12}$$

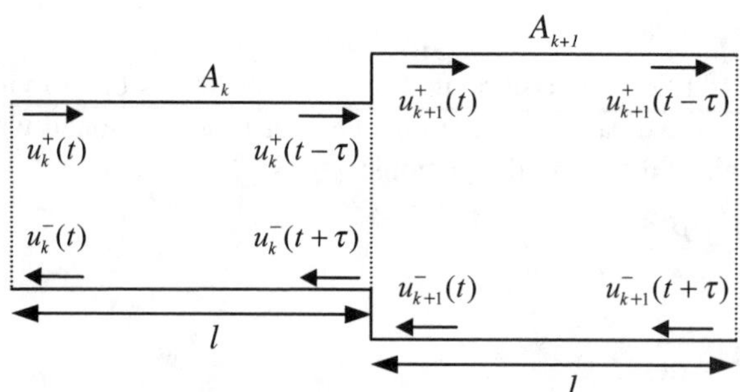

Figure 6.10 Junction between two lossless tubes.

A relationship between the z-transforms of the volume velocity at the glottis $u_G[n]$ and the lips $u_L[n]$ for a concatenation of N lossless tubes can be derived [9] using a discrete-time version of Eq. (6.10) and taking into account boundary conditions for every junction:

$$V(z) = \frac{U_L(z)}{U_G(z)} = \frac{0.5z^{-N/2}(1+r_G)\prod_{k=1}^{N}(1+r_k)}{\begin{bmatrix} 1 & -r_G \end{bmatrix}\left(\prod_{k=1}^{N}\begin{bmatrix} 1 & -r_k \\ -r_k z^{-1} & z^{-1} \end{bmatrix}\right)\begin{bmatrix} 1 \\ 0 \end{bmatrix}} \tag{6.13}$$

where r_G is the reflection coefficient at the glottis and $r_N = r_L$ is the reflection coefficient at the lips. Equation (6.11) is still valid for the glottis and lips, where $A_0 = \rho c / Z_G$ is the equivalent area at the glottis and $A_{N+1} = \rho c / Z_L$ the equivalent area at the lips. Z_G and Z_L are the equivalent impedances at the glottis and lips, respectively. Such impedances relate the volume velocity and pressure, for the lips the expression is

$$U_L(z) = P_L(z) / Z_L \tag{6.14}$$

In general, the concatenation of N lossless tubes results in an N-pole system as shown in Eq. (6.13). For a concatenation of N tubes, there are at most $N/2$ complex conjugate poles, or resonances or formants. These resonances occur when a given frequency gets *trapped* in the vocal tract because it is reflected back at the lips and then again back at the glottis.

Since each tube has length l and there are N of them, the total length is $L = lN$. The propagation delay in each tube $\tau = l/c$, and the sampling period is $T = 2\tau$, the round trip in a tube. We can find a relationship between the number of tubes N and the sampling frequency $F_s = 1/T$:

$$N = \frac{2LF_s}{c} \tag{6.15}$$

For example, for F_s = 8000 kHz, c = 34000 cm/s, and L = 17 cm, the average length of a male adult vocal tract, we obtain N = 8, or alternatively 4 formants. Experimentally, the vocal tract transfer function has been observed to have approximately 1 formant per kilohertz. Shorter vocal tract lengths (females or children) have fewer resonances per kilohertz and vice versa.

The pressure at the lips has been found to approximate the derivative of volume velocity, particularly at low frequencies. Thus, $Z_L(z)$ can be approximated by

$$Z_L(z) \approx R_0(1-z^{-1}) \tag{6.16}$$

which is 0 for low frequencies and reaches R_0 asymptotically. This dependency upon frequency results in a reflection coefficient that is also a function of frequency. For low frequencies, $r_L = 1$, and no loss occurs. At higher frequencies, loss by radiation translates into widening of formant bandwidths.

Similarly, the glottal impedance is also a function of frequency in practice. At high frequencies, Z_G is large and $r_G \approx 1$ so that all the energy is transmitted. For low frequencies, $r_G < 1$, whose main effect is an increase of bandwidth for the lower formants.

Moreover, energy is lost as a result of vibration of the tube walls, which is more pronounced at low frequencies. Energy is also lost, to a lesser extent, as a result of viscous friction between the air and the walls of the tube, particularly at frequencies above 3 kHz. The yielding walls tend to raise the resonance frequencies while the viscous and thermal losses tend to lower them. The net effect in the transfer function is a broadening of the resonances' bandwidths.

Despite thermal losses, yielding walls in the vocal tract, and the fact that both r_L and r_G are functions of frequency, the all-pole model of Eq. (6.13) for $V(z)$ has been found to be a good approximation in practice [13]. In Figure 6.11 we show the measured area function of a vowel and its corresponding frequency response obtained using the approximation as a concatenation of 10 lossless tubes with a constant r_L. The measured formants and corresponding bandwidths match quite well with this model despite all the approximations made. Thus, this concatenation of lossless tubes model represents reasonably well the acoustics inside the vocal tract. Inspired by the above results, we describe in Section 6.3 "Linear Predictive Coding," an all-pole model for speech.

In the production of the nasal consonants, the velum is lowered to trap the nasal tract to the pharynx, whereas a complete closure is formed in the oral tract (/m/ at the lips, /n/ just back of the teeth and /ng/ just forward of the velum itself. This configuration is shown in Figure 6.12, which shows two branches, one of them completely closed. For nasals, the radiation occurs primarily at the nostrils. The set of resonances is determined by the shape and length of the three tubes. At certain frequencies, the wave reflected in the closure cancels the wave at the pharynx, preventing energy from appearing at nostrils. The result is that for nasal sounds, the vocal tract transfer function $V(z)$ has anti-resonances (zeros) in addition to resonances. It has also been observed that nasal resonances have broader bandwidths than non-nasal voiced sounds, due to the greater viscous friction and thermal loss because of the large surface area of the nasal cavity.

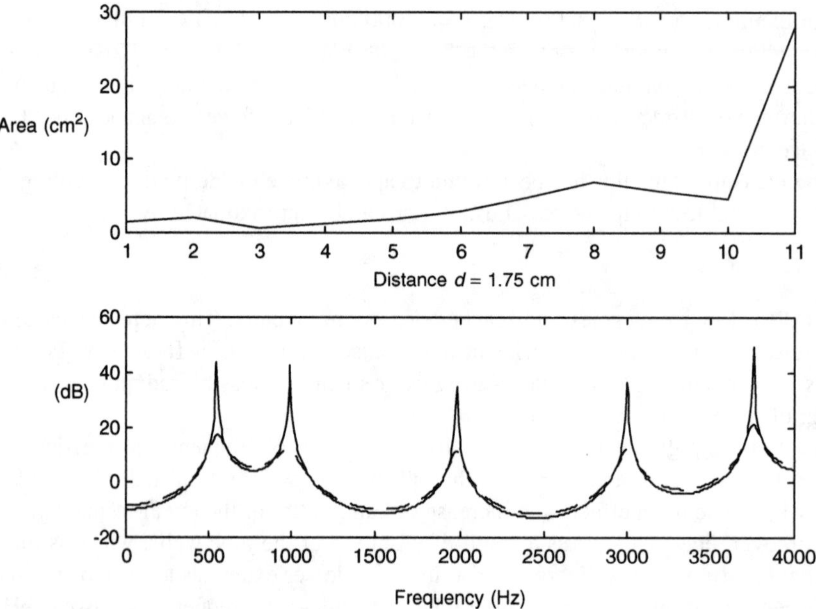

Figure 6.11 Area function and frequency response for vowel /a/ and its approximation as a concatenation of 10 lossless tubes. A reflection coefficient at the load of $k = 0.72$ (dotted line) is displayed. For comparison, the case of $k = 1.0$ (solid line) is also shown.

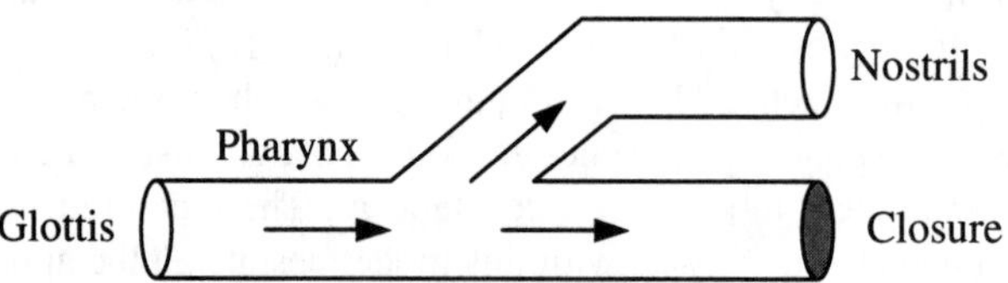

Figure 6.12 Coupling of the nasal cavity with the oral cavity.

6.2.3. Source-Filter Models of Speech Production

As shown in Chapter 10, speech signals are captured by microphones that respond to changes in air pressure. Thus, it is of interest to compute the pressure at the lips $P_L(z)$, which can be obtained as

$$P_L(z) = U_L(z)Z_L(z) = U_G(z)V(z)Z_L(z) \tag{6.17}$$

For voiced sounds we can model $u_G[n]$ as an impulse train convolved with $g[n]$, the glottal pulse (see Figure 6.13). Since $g[n]$ is of finite length, its z-transform is an all-zero system.

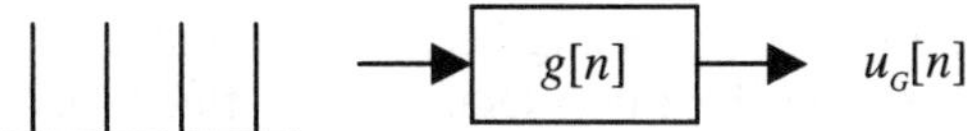

Figure 6.13 Model of the glottal excitation for voiced sounds.

The complete model for both voiced and unvoiced sounds is shown in Figure 6.14. We have modeled $u_G[n]$ in unvoiced sounds as random noise.

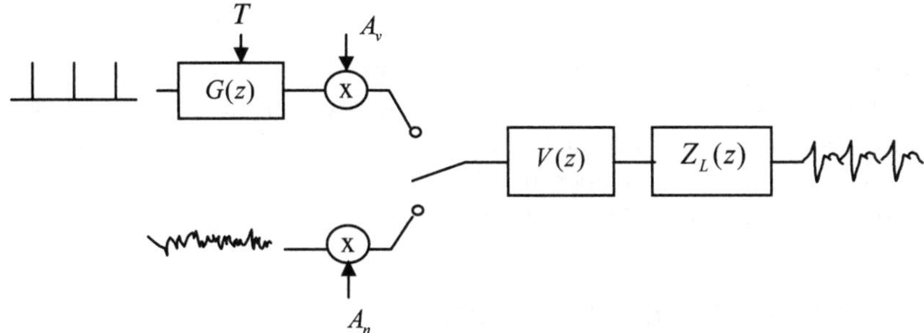

Figure 6.14 General discrete-time model of speech production. The excitation can be either an impulse train with period T and amplitude A_v driving a filter $G(z)$ or random noise with amplitude A_n.

We can simplify the model in Figure 6.14 by grouping $G(z)$, $V(z)$, and $Z_L(z)$ into $H(z)$ for voiced sounds, and $V(z)$ and $Z_L(z)$ into $H(z)$ for unvoiced sounds. The simplified model is shown in Figure 6.15, where we make explicit the fact that the filter changes over time.

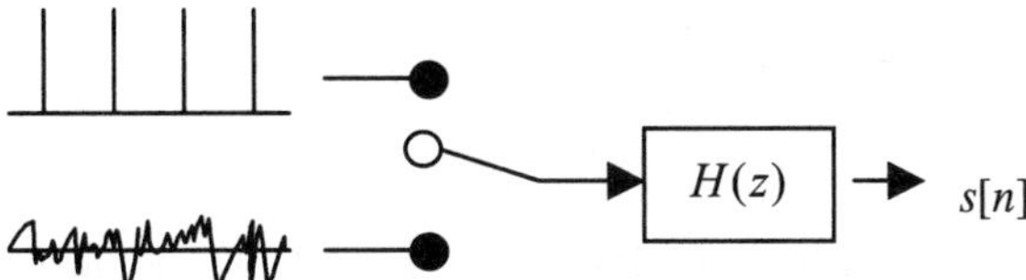

Figure 6.15 Source-filter model for voiced and unvoiced speech.

This model is a decent approximation, but fails on voiced fricatives, since those sounds contain both a periodic component and an aspirated component. In this case, a *mixed excitation* model can be applied, using for voiced sounds a sum of both an impulse train and colored noise (Figure 6.16).

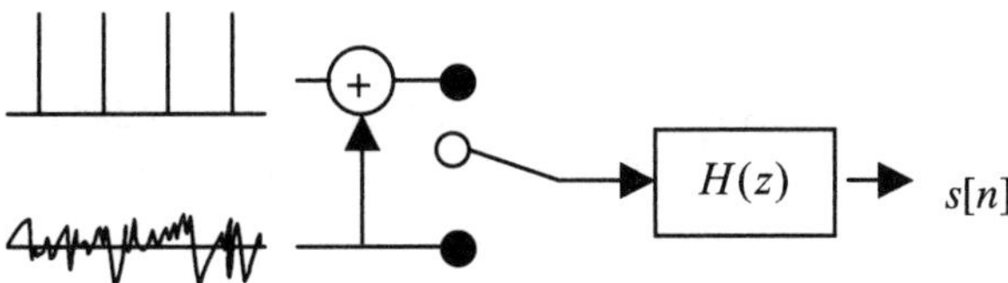

Figure 6.16 A mixed excitation source-filter model of speech.

The model in Figure 6.15 is appealing because the source is white (has a flat spectrum) and all the *coloring* is in the filter. Other source-filter decompositions attempt to model the source as the signal at the glottis, in which the source is definitely not white. Since $G(z)$, $Z_L(z)$ contain zeros, and $V(z)$ can also contain zeros for nasals, $H(z)$ is no longer all-pole. However, recall in Chapter 5, we state that the z-transform of $x[n] = a^n u[n]$ is

$$X(z) = \sum_{n=0}^{\infty} a^n z^{-n} = \frac{1}{1 - az^{-1}} \quad \text{for} \quad |a| < |z| \tag{6.18}$$

so that by inverting Eq. (6.18) we see that a zero can be expressed with infinite poles. This is the reason why all-pole models are still reasonable approximations as long as a large enough number of poles is used. Fant [12] showed that on the average the speech spectrum contains one pole per kHz. Setting the number of poles p to $F_s + 2$, where F_s is the sampling frequency expressed in kHz, has been found to work well in practice.

6.3. Linear Predictive Coding

A very powerful method for speech analysis is based on *linear predictive coding* (LPC) [4, 7, 19, 24, 27], also known as LPC analysis or *auto-regressive* (AR) modeling. This method is widely used because it is fast and simple, yet an effective way of estimating the main parameters of speech signals.

As shown in Section 6.2, an all-pole filter with a sufficient number of poles is a good approximation for speech signals. Thus, we could model the filter $H(z)$ in Figure 6.15 as

$$H(z) = \frac{X(z)}{E(z)} = \frac{1}{1 - \sum_{k=1}^{p} a_k z^{-k}} = \frac{1}{A(z)} \tag{6.19}$$

where p is the order of the LPC analysis. The *inverse filter* $A(z)$ is defined as

$$A(z) = 1 - \sum_{k=1}^{p} a_k z^{-k} \tag{6.20}$$

Taking inverse z-transforms in Eq. (6.19) results in

$$x[n] = \sum_{k=1}^{p} a_k x[n-k] + e[n] \tag{6.21}$$

Linear predictive coding gets its name from the fact that it predicts the current sample as a linear combination of its past p samples:

$$\tilde{x}[n] = \sum_{k=1}^{p} a_k x[n-k] \tag{6.22}$$

The prediction error when using this approximation is

$$e[n]=x[n]-\tilde{x}[n]=x[n]-\sum_{k=1}^{p}a_k x[n-k] \tag{6.23}$$

6.3.1. The Orthogonality Principle

To estimate the predictor coefficients from a set of speech samples, we use the short-term analysis technique. Let's define $x_m[n]$ as a segment of speech selected in the vicinity of sample m:

$$x_m[n]=x[m+n] \tag{6.24}$$

We define the short-term prediction error for that segment as

$$E_m=\sum_n e_m^2[n]=\sum_n \left(x_m[n]-\tilde{x}_m[n]\right)^2=\sum_n\left(x_m[n]-\sum_{j=1}^{p}a_j x_m[n-j]\right)^2 \tag{6.25}$$

In the absence of knowledge about the probability distribution of a_i, a reasonable estimation criterion is minimum mean squared error, introduced in Chapter 4. Thus, given a signal $x_m[n]$, we estimate its corresponding LPC coefficients as those that minimize the total prediction error E_m. Taking the derivative of Eq. (6.25) with respect to a_i and equating to 0, we obtain:

$$<\mathbf{e}_m,\mathbf{x}_m^i>=\sum_n e_m[n]x_m[n-i]=0 \qquad 1\le i\le p \tag{6.26}$$

where we have defined $\mathbf{e}_m$ and $\mathbf{x}_m^i$ as vectors of samples, and their inner product has to be 0. This condition, known as *orthogonality principle,* says that the predictor coefficients that minimize the prediction error are such that the error must be orthogonal to the past vectors, and is seen in Figure 6.17.

Equation (6.26) can be expressed as a set of p linear equations

$$\sum_n x_m[n-i]x_m[n]=\sum_{j=1}^{p}a_j\sum_n x_m[n-i]x_m[n-j] \qquad i=1,2,\ldots,p \tag{6.27}$$

For convenience, we can define the correlation coefficients as

$$\phi_m[i,j]=\sum_n x_m[n-i]x_m[n-j] \tag{6.28}$$

so that Eqs. (6.27) and (6.28) can be combined to obtain the so-called *Yule-Walker* equations:

$$\sum_{j=1}^{p} a_j \phi_m[i,j] = \phi_m[i,0] \qquad i = 1,2,\ldots,p \tag{6.29}$$

Solution of the set of p linear equations results in the p LPC coefficients that minimize the prediction error. With a_i satisfying Eq. (6.29), the total prediction error in Eq. (6.25) takes on the following value:

$$E_m = \sum_n x_m^2[n] - \sum_{j=1}^{p} a_j \sum_n x_m[n] x_m[n-j] = \phi[0,0] - \sum_{j=1}^{p} a_j \phi[0,j] \tag{6.30}$$

It is convenient to define a normalized prediction error $u[n]$ with unity energy

$$\sum_n u_m^2[n] = 1 \tag{6.31}$$

and a gain G, such that

$$e_m[n] = G u_m[n] \tag{6.32}$$

The gain G can be computed from the short-term prediction error

$$E_m = \sum_n e_m^2[n] = G^2 \sum_n u_m^2[n] = G^2 \tag{6.33}$$

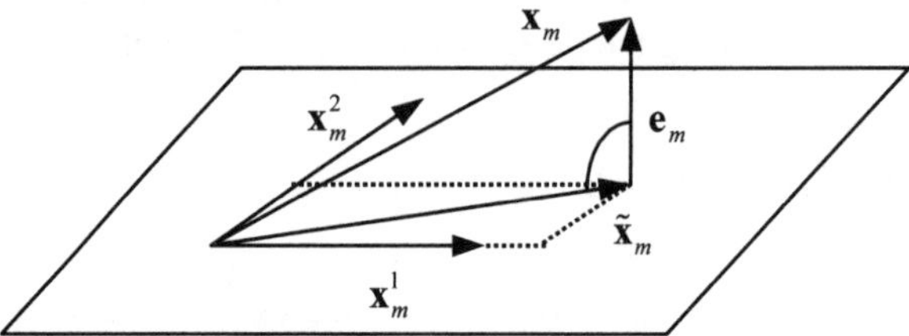

Figure 6.17 The orthogonality principle. The prediction error is orthogonal to the past samples.

6.3.2. Solution of the LPC Equations

The solution of the Yule-Walker equations in Eq. (6.29) can be achieved with any standard matrix inversion package. Because of the special form of the matrix here, some efficient solutions are possible, as described below. Also, each solution offers a different insight so we present three different algorithms: the covariance method, the autocorrelation method, and the lattice method.

6.3.2.1. Covariance Method

The covariance method [4] is derived by defining directly the interval over which the summation in Eq. (6.28) takes place:

$$E_m = \sum_{n=0}^{N-1} e_m^2[n] \tag{6.34}$$

so that $\phi_m[i,j]$ in Eq. (6.28) becomes

$$\phi_m[i,j] = \sum_{n=0}^{N-1} x_m[n-i]x_m[n-j] = \sum_{n=-i}^{N-1-j} x_m[n]x_m[n+i-j] = \phi_m[j,i] \tag{6.35}$$

and Eq. (6.29) becomes

$$\begin{pmatrix} \phi_m[1,1] & \phi_m[1,2] & \phi_m[1,3] & \cdots & \phi_m[1,p] \\ \phi_m[2,1] & \phi_m[2,2] & \phi_m[2,3] & \cdots & \phi_m[2,p] \\ \phi_m[3,1] & \phi_m[3,2] & \phi_m[3,3] & \cdots & \phi_m[3,p] \\ \cdots & \cdots & \cdots & \cdots & \cdots \\ \phi_m[p,1] & \phi_m[p,2] & \phi_m[p,3] & \cdots & \phi_m[p,p] \end{pmatrix} \begin{pmatrix} a_1 \\ a_2 \\ a_3 \\ \cdots \\ a_p \end{pmatrix} = \begin{pmatrix} \phi_m[1,0] \\ \phi_m[2,0] \\ \phi_m[3,0] \\ \cdots \\ \phi_m[p,0] \end{pmatrix} \tag{6.36}$$

which can be expressed as the following matrix equation

$$\mathbf{\Phi a} = \psi \tag{6.37}$$

where the matrix $\mathbf{\Phi}$ in Eq. (6.37) is symmetric and *positive definite*, for which efficient methods are available, such as the Cholesky decomposition. For this method, also called the squared root method, the matrix $\mathbf{\Phi}$ is expressed as

$$\mathbf{\Phi} = \mathbf{VDV}^t \tag{6.38}$$

where $\mathbf{V}$ is a lower triangular matrix (whose main diagonal elements are 1's), and $\mathbf{D}$ is a diagonal matrix. So each element of $\mathbf{\Phi}$ can be expressed as

$$\phi[i,j] = \sum_{k=1}^{j} V_{ik} d_k V_{jk} \qquad 1 \le j < i \tag{6.39}$$

or alternatively

$$V_{ij} d_j = \phi[i,j] - \sum_{k=1}^{j-1} V_{ik} d_k V_{jk} \qquad 1 \le j < i \tag{6.40}$$

and for the diagonal elements

$$\phi[i,i] = \sum_{k=1}^{i} V_{ik} d_k V_{ik} \tag{6.41}$$

or alternatively

$$d_i = \phi[i,i] - \sum_{k=1}^{i-1} V_{ik}^2 d_k, \qquad i \ge 2 \tag{6.42}$$

with

$$d_1 = \phi[1,1] \tag{6.43}$$

The Cholesky decomposition starts with Eq. (6.43) then alternates between Eqs. (6.40) and (6.42). Once the matrices $\mathbf{V}$ and $\mathbf{D}$ have been determined, the LPC coefficients are solved in a two-step process. The combination of Eqs. (6.37) and (6.38) can be expressed as

$$\mathbf{VY} = \psi \tag{6.44}$$

with

$$\mathbf{Y} = \mathbf{DV}^t\mathbf{a} \tag{6.45}$$

or alternatively

$$\mathbf{V}^t\mathbf{a} = \mathbf{D}^{-1}\mathbf{Y} \tag{6.46}$$

Therefore, given matrix $\mathbf{V}$ and Eq. (6.44), $\mathbf{Y}$ can be solved recursively as

$$Y_i = \psi_i - \sum_{j=1}^{i-1} V_{ij} Y_j, \qquad 2 \le i \le p \tag{6.47}$$

with the initial condition

$$Y_1 = \psi_1 \tag{6.48}$$

Having determined $\mathbf{Y}$, Eq. (6.46) can be solved recursively in a similar way

$$a_i = Y_i / d_i - \sum_{j=i+1}^{p} V_{ji} a_j, \qquad 1 \le i < p \tag{6.49}$$

with the initial condition

$$a_p = Y_p / d_p \tag{6.50}$$

where the index i in Eq. (6.49) proceeds backwards.

The term covariance analysis is somewhat of a misnomer, since we know from Chapter 5 that the covariance of a signal is the correlation of that signal with its mean removed. It was so called because the matrix in Eq. (6.36) has the properties of a covariance matrix, though this algorithm is more like a cross-correlation.

6.3.2.2. Autocorrelation Method

The summation in Eq. (6.28) had no specific range. In the autocorrelation method [24, 27], we assume that $x_m[n]$ is 0 outside the interval $0 \le n < N$:

$$x_m[n] = x[m+n]w[n] \tag{6.51}$$

with $w[n]$ being a window (such as a Hamming window) which is 0 outside the interval $0 \le n < N$. With this assumption, the corresponding prediction error $e_m[n]$ is non-zero over the interval $0 \le n < N + p$, and, therefore, the total prediction error takes on the value

$$E_m = \sum_{n=0}^{N+p-1} e_m^2[n] \tag{6.52}$$

With this range, Eq. (6.28) can be expressed as

$$\phi_m[i,j] = \sum_{n=0}^{N+p-1} x_m[n-i]x_m[n-j] = \sum_{n=0}^{N-1-(i-j)} x_m[n]x_m[n+i-j] \tag{6.53}$$

or alternatively

$$\phi_m[i,j] = R_m[i-j] \tag{6.54}$$

with $R_m[k]$ being the autocorrelation sequence of $x_m[n]$:

$$R_m[k] = \sum_{n=0}^{N-1-k} x_m[n]x_m[n+k] \tag{6.55}$$

Combining Eqs. (6.54) and (6.29), we obtain

$$\sum_{j=1}^{p} a_j R_m[|i-j|] = R_m[i] \tag{6.56}$$

which corresponds to the following matrix equation

$$\begin{pmatrix} R_m[0] & R_m[1] & R_m[2] & \cdots & R_m[p-1] \\ R_m[1] & R_m[0] & R_m[1] & \cdots & R_m[p-2] \\ R_m[2] & R_m[1] & R_m[0] & \cdots & R_m[p-3] \\ \cdots & \cdots & \cdots & \cdots & \cdots \\ R_m[p-1] & R_m[p-2] & R_m[p-3] & \cdots & R_m[0] \end{pmatrix} \begin{pmatrix} a_1 \\ a_2 \\ a_3 \\ \cdots \\ a_p \end{pmatrix} = \begin{pmatrix} R_m[1] \\ R_m[2] \\ R_m[3] \\ \cdots \\ R_m[p] \end{pmatrix} \tag{6.57}$$

The matrix in Eq. (6.57) is symmetric and all the elements in its diagonals are identical. Such matrices are called *Toeplitz*. Durbin's recursion exploits this fact resulting in a very efficient algorithm (for convenience, we omit the subscript m of the autocorrelation function), whose proof is beyond the scope of this book:

1. Initialization

$$E^0 = R[0] \tag{6.58}$$

2. Iteration. For $i = 1, \cdots, p$ do the following recursion:

$$k_i = \left(R[i] - \sum_{j=1}^{i-1} a_j^{i-1} R[i-j] \right) / E^{i-1} \tag{6.59}$$

$$a_i^i = k_i \tag{6.60}$$

$$a_j^i = a_j^{i-1} - k_i a_{i-j}^{i-1}, \qquad 1 \le j < i \tag{6.61}$$

$$E^i = (1 - k_i^2) E^{i-1} \tag{6.62}$$

3. Final solution:

$$a_j = a_j^p \qquad 1 \le j \le p \tag{6.63}$$

where the coefficients k_i, called *reflection coefficients*, are bounded between –1 and 1 (see Section 6.3.2.1.3). In the process of computing the predictor coefficients of order p, the recursion finds the solution of the predictor coefficients for all orders less than p.

Replacing $R[j]$ by the normalized autocorrelation coefficients $r[j]$, defined as

$$r[j] = R[j] / R[0] \tag{6.64}$$

results in identical LPC coefficients, and the recursion is more robust to problems with arithmetic precision. Likewise, the normalized prediction error at iteration i is defined by dividing Eq. (6.30) by $R[0]$, which, using Eq.(6.54), results in

$$V^i = \frac{E^i}{R[0]} = 1 - \sum_{j=1}^{i} a_j r[j] \tag{6.65}$$

The normalized prediction error is, using Eqs. (6.62) and (6.65),

$$V^p = \prod_{i=1}^{p} (1 - k_i^2) \tag{6.66}$$

6.3.2.3. Lattice Formulation

In this section we derive the lattice formulation [7, 19], an equivalent algorithm to the Levinson Durbin recursion, which has some precision benefits. It is advantageous to define the *forward prediction error* obtained at stage i of the Levinson Durbin procedure as

$$e^i[n] = x[n] - \sum_{k=1}^{i} a_k^i x[n-k] \tag{6.67}$$

whose z-transform is given by

$$E^i(z) = A^i(z)X(z) \tag{6.68}$$

with $A^i(z)$ being defined by

$$A^i(z) = 1 - \sum_{k=1}^{i} a_k^i z^{-k} \tag{6.69}$$

which, combined with Eq. (6.61), results in the following recursion:

$$A^i(z) = A^{i-1}(z) - k_i z^{-i} A^{i-1}(z^{-1}) \tag{6.70}$$

Similarly, we can define the so-called *backward prediction error* as

$$b^i[n] = x[n-i] - \sum_{k=1}^{i} a_k^i x[n+k-i] \tag{6.71}$$

whose z-transform is

$$B^i(z) = z^{-i} A^i(z^{-1}) X(z) \tag{6.72}$$

Now combining Eqs. (6.68), (6.70), and (6.72), we obtain

$$E^i(z) = A^{i-1}(z)X(z) - k_i z^{-i} A^{i-1}(z^{-1})X(z) = E^{i-1}(z) - k_i z^{-1} B^{i-1}(z) \tag{6.73}$$

whose inverse z-transform is given by

$$e^i[n] = e^{i-1}[n] - k_i b^{i-1}[n-1] \tag{6.74}$$

Also, substituting Eq. (6.70) into (6.72) and using Eq. (6.68), we obtain

$$B^i(z) = z^{-1} B^{i-1}(z) - k_i E^{i-1}(z) \tag{6.75}$$

whose inverse z-transform is given by

$$b^i[n] = b^{i-1}[n-1] - k_i e^{i-1}[n] \tag{6.76}$$

Equations (6.74) and (6.76) define the forward and backward prediction error sequences for an ith-order predictor in terms of the corresponding forward and backward prediction errors of an $(i-1)^{th}$-order predictor. We initialize the recursive algorithm by noting that the 0th-order predictor is equivalent to using no predictor at all; thus

$$e^0[n] = b^0[n] = x[n] \tag{6.77}$$

and the final prediction error is $e[n] = e^p[n]$.

A block diagram of the lattice method is given in Figure 6.18, which resembles a lattice, hence its name.

While the computation of the k_i coefficients can be done through the Levinson Durbin recursion of Eqs. (6.59) through (6.62), it can be shown that an equivalent calculation can be found as a function of the forward and backward prediction errors. To do so we minimize the sum of the forward prediction errors

$$E^i = \sum_{n=0}^{N-1} \left(e^i[n]\right)^2 \tag{6.78}$$

by substituting Eq. (6.74) in (6.78), taking the derivative with respect to k_i, and equating to 0:

$$k_i = \frac{\sum_{n=0}^{N-1} e^{i-1}[n] b^{i-1}[n-1]}{\sum_{n=0}^{N-1} \left(b^{i-1}[n-1]\right)^2} \tag{6.79}$$

Using Eqs. (6.67) and (6.71), it can be shown that

$$\sum_{n=0}^{N-1} \left(e^{i-1}[n]\right)^2 = \sum_{n=0}^{N-1} \left(b^{i-1}[n-1]\right)^2 \tag{6.80}$$

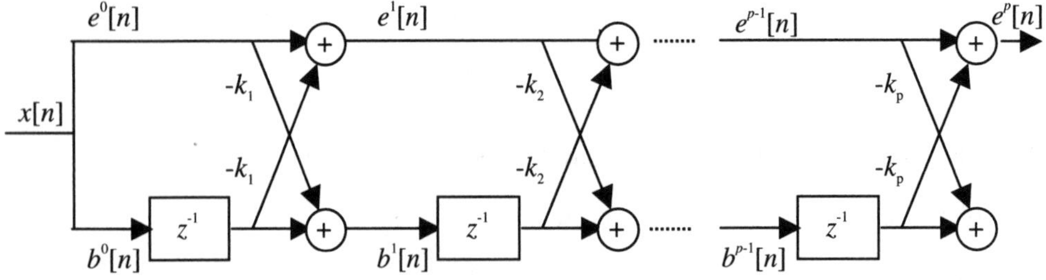

Figure 6.18 Block diagram of the lattice filter.

since minimization of both yields identical Yule-Walker equations. Thus Eq. (6.79) can be alternatively expressed as

$$k_i = \frac{\sum_{n=0}^{N-1} e^{i-1}[n] b^{i-1}[n-1]}{\sqrt{\sum_{n=0}^{N-1} \left(e^{i-1}[n]\right)^2 \sum_{n=0}^{N-1} \left(b^{i-1}[n-1]\right)^2}} = \frac{< \mathbf{e}^{i-1}, \mathbf{b}^{i-1} >}{\left|\mathbf{e}^{i-1}\right| \left|\mathbf{b}^{i-1}\right|} \tag{6.81}$$

where we have defined the vectors $\mathbf{e}^i = \left(e^i[0] \cdots e^i[N-1]\right)$ and $\mathbf{b}^i = \left(b^i[0] \cdots b^i[N-1]\right)$. The inner product of two vectors **x** and **y** is defined as

$$< \mathbf{x}, \mathbf{y} > = \sum_{n=0}^{N-1} x[n] y[n] \tag{6.82}$$

and its norm as

$$\left|\mathbf{x}\right|^2 = < \mathbf{x}, \mathbf{x} > = \sum_{n=0}^{N-1} x^2[n] \tag{6.83}$$

Equation (6.81) has the form of a normalized cross-correlation function, and, therefore, the reflection coefficients are also called *partial correlation coefficients* (PARCOR). As with any normalized cross-correlation function, the k_i coefficients are bounded by

$$-1 \le k_i \le 1 \tag{6.84}$$

This is a necessary and sufficient condition for all the roots of the polynomial $A(z)$ to be inside the unit circle, therefore guaranteeing a stable filter. This condition can be checked to avoid numerical imprecision by stopping the recursion if the condition is not met. The inverse lattice filter can be seen in Figure 6.19, which resembles the lossless tube model. This is why the k_i are also called *reflection coefficients*.

Lattice filters are often used in fixed-point implementation, because lack of precision doesn't result in unstable filters. Any error that may take place – for example due to quantization – is generally not be sufficient to cause k_i to fall outside the range in Eq. (6.84). If, owing to round-off error, the reflection coefficient falls outside the range, the lattice filter can be ended at the previous step.

More importantly, linearly varying coefficients can be implemented in this fashion. While, typically, the reflection coefficients are constant during the analysis frame, we can implement a linear interpolation of the reflection coefficients to obtain the error signal. If the coefficients of both frames are in the range in Eq. (6.84), the linearly interpolated reflection coefficients also have that property, and thus the filter is stable. This is a property that the predictor coefficients don't have.

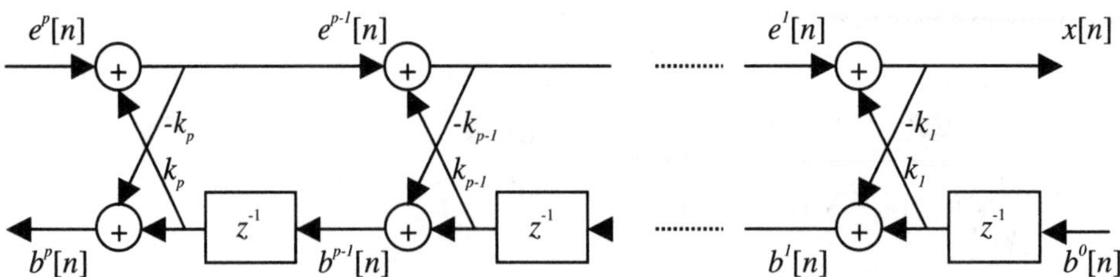

Figure 6.19 Inverse lattice filter used to generate the speech signal, given its residual.

6.3.3. Spectral Analysis via LPC

Let's now analyze the frequency-domain behavior of the LPC analysis by evaluating

$$H(e^{j\omega}) = \frac{G}{1 - \sum_{k=1}^{p} a_k e^{-j\omega k}} = \frac{G}{A(e^{j\omega})} \tag{6.85}$$

which is an *all-pole* or IIR filter. If we plot $H(e^{j\omega})$, we expect to see peaks at the roots of the denominator. Figure 6.20 shows the 14-order LPC spectrum of the vowel of Figure 6.3 (d).

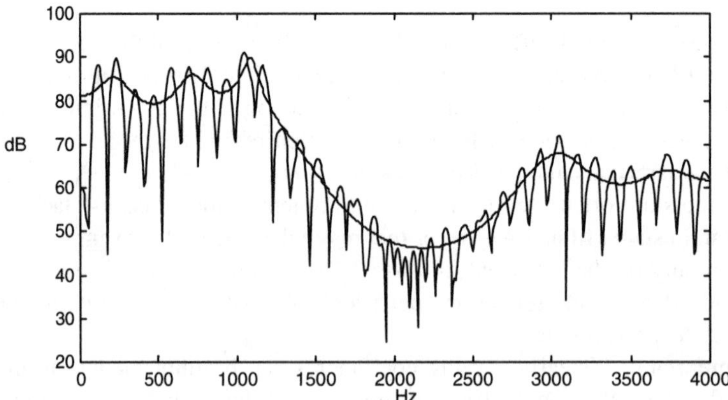

Figure 6.20 LPC spectrum of the */ah/* phoneme in the word *lives* of Figure 6.3. Used here are a 30-ms Hamming window and the autocorrelation method with $p = 14$. The short-time spectrum is also shown.

For the autocorrelation method, the squared error of Eq. (6.52) can be expressed, using Eq. (6.85) and Parseval's theorem, as

$$E_m = \frac{G^2}{2\pi}\int_{-\pi}^{\pi} \frac{|X_m(e^{j\omega})|^2}{|H(e^{j\omega})|^2} d\omega \tag{6.86}$$

Since the integrand in Eq. (6.86) is positive, minimizing E_m is equivalent to minimizing the ratio of the energy spectrum of the speech segment $|X_m(e^{j\omega})|^2$ to the magnitude squared of the frequency response of the linear system $|H(e^{j\omega})|^2$. The LPC spectrum matches more closely the peaks than the valleys (see Figure 6.20), because the regions where $|X_m(e^{j\omega})| > |H(e^{j\omega})|$ contribute more to the error than those where $|H(e^{j\omega})| > |X_m(e^{j\omega})|$.

Even nasals, which have zeros in addition to poles, can be represented with an infinite number of poles. In practice, if *p* is large enough we can approximate the signal spectrum with arbitrarily small error. Figure 6.21 shows different fits for different values of *p*. The higher *p*, the more details of the spectrum are preserved.

The prediction order is not known for arbitrary speech, so we need to set it to balance spectral detail with estimation errors.

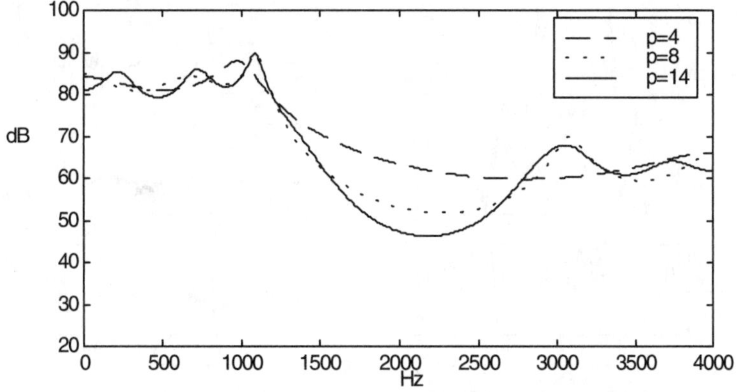

Figure 6.21 LPC spectra of Figure 6.20 for various values of the predictor order *p*.

6.3.4. The Prediction Error

So far, we have concentrated on the filter component of the source-filter model. Using Eq. (6.23), we can compute the prediction error signal, also called the *excitation*, or *residual* signal. For unvoiced speech we expect the residual to be approximately white noise. In practice, this approximation is quite good, and replacement of the residual by white noise followed by the LPC filter typically results in no audible difference. For voiced speech we expect the residual to approximate an impulse train. In practice, this is not the case, because the all-pole assumption is not altogether valid; thus, the residual, although it contains spikes, is far from an impulse train. Replacing the residual by an impulse train, followed by the LPC filter, results in speech that sounds somewhat robotic, partly because real speech is not perfectly periodic (it has a random component as well), and because the zeroes are not modeled with the LPC filter. Residual signals computed from inverse LPC filters for several vowels are shown in Figure 6.22.

How do we choose p? This is an important design question. Larger values of p lead to lower prediction errors (see Figure 6.23). Unvoiced speech has higher error than voiced speech, because the LPC model is more accurate for voiced speech. In general, the normalized error rapidly decreases, and then converges to a value of around 12–14 for 8 kHz speech. If we use a large value of p, we are fitting the individual harmonics; thus the LPC filter is modeling the source, and the separation between source and filter is not going to be so good. The more coefficients we have to estimate, the larger the variance of their estimates, since the number of available samples is the same. A rule of thumb is to use 1 complex pole per kHz plus 2–4 poles to model the radiation and glottal effects.

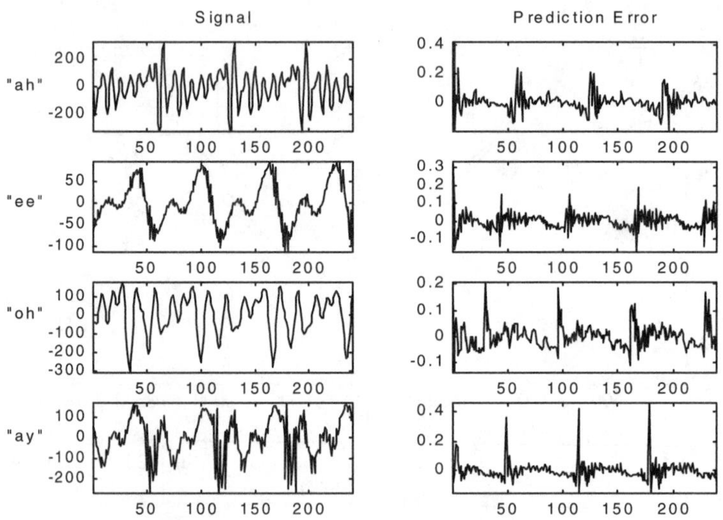

Figure 6.22 LPC prediction error signals for several vowels.

For unvoiced speech, both the autocorrelation and the covariance methods provide similar results. For voiced speech, however, the covariance method can provide better estimates if the analysis window is shorter than the local pitch period and the window only includes samples from the closed phase (when the vocal tract is closed at the glottis and speech signal is due mainly to free resonances). This is called *pitch synchronous* analysis and results in lower prediction error, because the true excitation is close to zero during the whole analysis window. During the open phase, the trachea, the vocal folds, and the vocal tract are acoustically coupled, and this coupling will change the free resonances. Additionally, the prediction error is higher for both the autocorrelation and the covariance methods if samples from the open phase are included in the analysis window, because the prediction during those instants is poor.

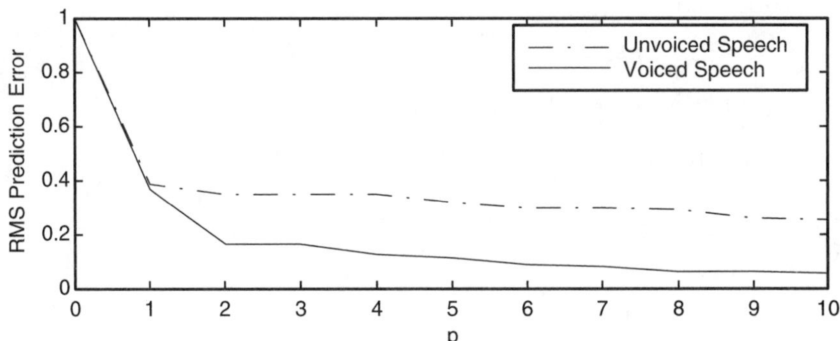

Figure 6.23 Variation of the normalized prediction error with the number of prediction coefficients p for the voiced segment of Figure 6.3 and the unvoiced speech of Figure 6.5. The autocorrelation method was used with a 30 ms Hamming window, and a sampling rate of 8 kHz.

6.3.5. Equivalent Representations

There are a number of alternate useful representations of the predictor coefficients. The most important are the line spectrum frequencies, reflection coefficients, log-area ratios, and the roots of the predictor polynomial.

6.3.5.1. Line Spectral Frequencies

Line Spectral Frequencies (LSF) [18] provide an equivalent representation of the predictor coefficients that is very popular in speech coding. It is derived from computing the roots of the polynomials $P(z)$ and $Q(z)$ defined as

$$P(z) = A(z) + z^{-(p+1)} A(z^{-1}) \tag{6.87}$$

$$Q(z) = A(z) - z^{-(p+1)} A(z^{-1}) \tag{6.88}$$

To gain insight on these roots, look at a second-order predictor filter with a pair of complex roots:

$$A(z) = 1 - a_1 z^{-1} - a_2 z^{-2} = 1 - 2\rho_0 \cos(2\pi f_0) z^{-1} + \rho_0^2 z^{-2} \tag{6.89}$$

where $0 < \rho_0 < 1$ and $0 < f_0 < 0.5$. Inserting Eq. (6.89) into (6.87) and (6.88) results in

$$\begin{aligned} P(z) &= 1-(a_1+a_2)z^{-1}-(a_1+a_2)z^{-2}+z^{-3} \\ Q(z) &= 1-(a_1-a_2)z^{-1}+(a_1-a_2)z^{-2}-z^{-3} \end{aligned} \tag{6.90}$$

From Eq. (6.90) we see that $z=-1$ is a root of *P*(*z*) and $z=1$ a root of *Q*(*z*), which can be divided out and results in

$$\begin{aligned} P(z) &= (1+z^{-1})(1-2\beta_1 z^{-1}+z^{-2}) \\ Q(z) &= (1-z^{-1})(1-2\beta_2 z^{-1}+z^{-2}) \end{aligned} \tag{6.91}$$

where β_1 and β_2 are given by

$$\begin{aligned} \beta_1 &= \frac{a_1+a_2+1}{2} = \rho_0\cos(2\pi f_0)+\frac{1-\rho_0^2}{2} \\ \beta_2 &= \frac{a_1-a_2-1}{2} = \rho_0\cos(2\pi f_0)-\frac{1-\rho_0^2}{2} \end{aligned} \tag{6.92}$$

It can be shown that $|\beta_1|<1$ and $|\beta_2|<1$ for all possible values of f_0 and ρ_0. With this property, the roots of *P*(*z*) and *Q*(*z*) in Eq. (6.91) are complex and given by $\beta_1 \pm j\sqrt{1-\beta_1^2}$ and $\beta_2 \pm j\sqrt{1-\beta_2^2}$, respectively. Because they lie in the unit circle, they can be uniquely represented by their angles

$$\begin{aligned} \cos(2\pi f_1) &= \rho_0\cos(2\pi f_0)+\frac{1-\rho_0^2}{2} \\ \cos(2\pi f_2) &= \rho_0\cos(2\pi f_0)-\frac{1-\rho_0^2}{2} \end{aligned} \tag{6.93}$$

where f_1 and f_2 are the *line spectral frequencies* of *A*(*z*). Since $|\rho_0|<1$, $\cos(2\pi f_2) < \cos(2\pi f_0)$, and thus $f_2 > f_0$. It's also the case that $\cos(2\pi f_1) > \cos(2\pi f_0)$ and thus $f_1 < f_0$. Furthermore, as $\rho_0 \to 1$, we see from Eq. (6.93) that $f_1 \to f_0$ and $f_2 \to f_0$. We conclude that, given a pole at f_0, the two line spectral frequencies bracket it, i.e., $f_1 < f_0 < f_2$, and that they are closer together as the pole of the second-order resonator gets closer to the unit circle.

We have proven that for a second-order predictor, the roots of *P*(*z*) and *Q*(*z*) lie in the unit circle, that ± 1 are roots, and that, once sorted, the roots of *P*(*z*) and *Q*(*z*) alternate. Although we do not prove it here, it can be shown that these conclusions hold for other predictor orders, and, therefore, the *p* predictor coefficients can be transformed into *p* line spectral frequencies. We also know that $z=1$ is always a root of *Q*(*z*), whereas $z=-1$ is a root of *P*(*z*) for even *p* and a root of *Q*(*z*) for odd *p*.

To compute the LSF for $p>2$, we replace $z=\cos(\omega)$ and compute the roots of $P(\omega)$ and $Q(\omega)$ by any available root finding method. A popular technique, given that there are p roots which are real in ω and bounded between 0 and 0.5, is to bracket them by observing changes in sign of both functions in a dense grid. To compute the predictor coefficients from the LSF coefficients we can factor $P(z)$ and $Q(z)$ as a product of second-order filters as in Eq. (6.91), and then $A(z) = \left(P(z) + Q(z)\right)/2$.

In practice, LSF are useful because of *sensitivity* (a quantization of one coefficient generally results in a spectral change only around that frequency) and *efficiency* (LSF result in low spectral distortion). This doesn't occur with other representations. As long as the LSF coefficients are ordered, the resulting LPC filter is stable, though the proof is beyond the scope of this book. LSF coefficients are used extensively in Chapter 7.

6.3.5.2. Reflection Coefficients

For the autocorrelation method, the predictor coefficients may be obtained from the reflection coefficients by the following recursion:

$$\begin{aligned} a_i^i &= k_i & i &= 1, \cdots, p \\ a_j^i &= a_j^{i-1} - k_i a_{i-j}^{i-1} & 1 &\le j < i \end{aligned} \tag{6.94}$$

where $a_i = a_i^p$. Similarly, the reflection coefficients may be obtained from the prediction coefficients using a backward recursion of the form

$$\begin{aligned} k_i &= a_i^i & i &= p, \cdots, 1 \\ a_j^{i-1} &= \frac{a_j^i + a_i^i a_{i-j}^i}{1 - k_i^2} & 1 &\le j < i \end{aligned} \tag{6.95}$$

where we initialize $a_i^p = a_i$.

Reflection coefficients are useful when implementing LPC filters whose values are interpolated over time, because, unlike the predictor coefficients, they are guaranteed to be stable at all times as long as the anchors satisfy Eq. (6.84).

6.3.5.3. Log-Area Ratios

The *log-area ratio* coefficients are defined as

$$g_i = \ln\left(\frac{1 - k_i}{1 + k_i}\right) \tag{6.96}$$

with the inverse being given by

$$k_i = \frac{1 - e^{g_i}}{1 + e^{g_i}} \tag{6.97}$$

The log-area ratio coefficients are equal to the natural logarithm of the ratio of the areas of adjacent sections of a lossless tube equivalent of the vocal tract having the same transfer function. Since for stable predictor filters $-1 < k_i < 1$, we have from Eq. (6.96) that $-\infty < g_i < \infty$. For speech signals, it is not uncommon to have some reflection coefficients close to 1, and quantization of those values can cause a large change in the predictor's transfer function. On the other hand, the log-area ratio coefficients have relatively flat spectral sensitivity (i.e., a small change in their values causes a small change in the transfer function) and thus are useful in coding.

6.3.5.4. Roots of the Polynomial

An alternative to the predictor coefficients results from computing the complex roots of the predictor polynomial:

$$A(z) = 1 - \sum_{k=1}^{p} a_k z^{-k} = \prod_{k=1}^{p} (1 - z_k z^{-1}) \tag{6.98}$$

These roots can be represented as

$$z_k = e^{(-\pi b_k + j2\pi f_k)/F_s} \tag{6.99}$$

where b_k, f_k, and F_s represent the bandwidth, center frequency, and sampling frequency, respectively. Since a_k are real, all complex roots occur in conjugate pairs so that if (b_k, f_k) is a root, so is $(b_k, -f_k)$. The bandwidths b_k are always positive, because the roots are inside the unit circle ($|z_k| < 1$) for a stable predictor. Real roots $z_k = e^{-\pi b_k / F_s}$ can also occur. While algorithms exist to compute the complex roots of a polynomial, in practice there are sometimes numerical difficulties in doing so.

If the roots are available, it is straightforward to compute the predictor coefficients by using Eq. (6.98). Since the roots of the predictor polynomial represent resonance frequencies and bandwidths, they are used in the formant synthesizers of Chapter 16.

6.4. Cepstral Processing

A *homomorphic* transformation $\hat{x}[n] = D(x[n])$ is a transformation that converts a convolution

$$x[n] = e[n] * h[n] \tag{6.100}$$

into a sum

$$\hat{x}[n] = \hat{e}[n] + \hat{h}[n] \tag{6.101}$$

In this section we introduce the *cepstrum* as one homomorphic transformation [32] that allows us to separate the source from the filter. We show that we can find a value N such that the cepstrum of the filter $\hat{h}[n] \approx 0$ for $n \geq N$, and that the cepstrum of the excitation $\hat{e}[n] \approx 0$ for $n < N$. With this assumption, we can approximately recover both $e[n]$ and $h[n]$ from $\hat{x}[n]$ by homomorphic filtering. In Figure 6.24, we show how to recover $h[n]$ with a homomorphic filter:

$$l[n] = \begin{cases} 1 & |n| < N \\ 0 & |n| \geq N \end{cases} \tag{6.102}$$

where D is the cepstrum operator.

The excitation signal can be similarly recovered with a homomorphic filter given by

$$l[n] = \begin{cases} 1 & |n| \geq N \\ 0 & |n| < N \end{cases} \tag{6.103}$$

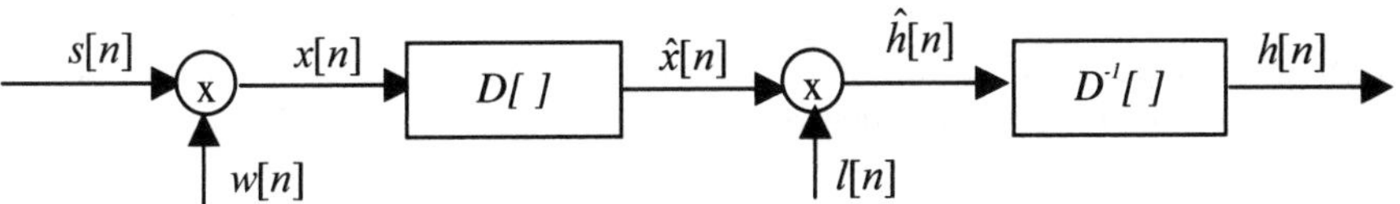

Figure 6.24 Homomorphic filtering to recover the filter's response from a periodic signal. We have used the homomorphic filter of Eq. (6.102).

6.4.1. The Real and Complex Cepstrum

The *real cepstrum* of a digital signal $x[n]$ is defined as

$$c[n] = \frac{1}{2\pi} \int_{-\pi}^{\pi} \ln |X(e^{j\omega})| \, e^{j\omega n} d\omega \tag{6.104}$$

and the *complex cepstrum* of $x[n]$ is defined as

$$\hat{x}[n] = \frac{1}{2\pi} \int_{-\pi}^{\pi} \ln X(e^{j\omega}) \, e^{j\omega n} d\omega \tag{6.105}$$

where the complex logarithm is used:

$$\hat{X}(e^{j\omega}) = \ln X(e^{j\omega}) = \ln |X(e^{j\omega})| + j\theta(\omega) \tag{6.106}$$

and the phase $\theta(\omega)$ is given by

$$\theta(\omega) = \arg\left[X(e^{j\omega})\right] \tag{6.107}$$

You can see from Eqs. (6.104) and (6.105) that both the real and the complex cepstrum satisfy Eq. (6.101) and thus they are homomorphic transformations.

If the signal $x[n]$ is real, both the real cepstrum $c[n]$ and the complex cepstrum $\hat{x}[n]$ are also real signals. Therefore the term complex cepstrum doesn't mean that it is a complex signal but rather that the complex logarithm is taken.

It can easily be shown that $c[n]$ is the even part of $\hat{x}[n]$:

$$c[n] = \frac{\hat{x}[n] + \hat{x}[-n]}{2} \tag{6.108}$$

From here on, when we refer to cepstrum without qualifiers, we are referring to the real cepstrum, since it is the most widely used in speech technology.

The cepstrum was invented by Bogert et al. [6], and its term was coined by reversing the first syllable of the word spectrum, given that it is obtained by taking the inverse Fourier transform of the log-spectrum. Similarly, they defined the term *quefrency* to represent the independent variable *n* in *c*[*n*]. The quefrency has dimension of time.

6.4.2. Cepstrum of Pole-Zero Filters

A very general type of filters are those with rational transfer functions

$$H(z) = \frac{Az^r \prod_{k=1}^{M_i}(1-a_k z^{-1}) \prod_{k=1}^{M_o}(1-u_k z)}{\prod_{k=1}^{N_i}(1-b_k z^{-1}) \prod_{k=1}^{N_o}(1-v_k z)} \tag{6.109}$$

with the magnitudes of a_k, b_k, u_k, and v_k all less than 1. Therefore, $(1-a_k z^{-1})$ and $(1-b_k z^{-1})$ represent the zeros and poles inside the unit circle, whereas $(1-u_k z)$ and $(1-v_k z)$ represent the zeros and poles outside the unit circle, and z^r is a shift from the time origin. Thus, the complex logarithm is

$$\begin{aligned}\hat{H}(z) &= \ln[A] + \ln[z^r] + \sum_{k=1}^{M_i} \ln(1-a_k z^{-1}) \\ &- \sum_{k=1}^{N_i} \ln(1-b_k z^{-1}) + \sum_{k=1}^{M_o} \ln(1-u_k z) - \sum_{k=1}^{N_o} \ln(1-v_k z)\end{aligned} \tag{6.110}$$

where the term $\ln[z^r]$ contributes to the imaginary part of the complex cepstrum only with a term $j\omega r$. Since it just carries information about the time origin, it's typically ignored. We use the Taylor series expansion

$$\ln(1-x) = -\sum_{n=1}^{\infty} \frac{x^n}{n} \tag{6.111}$$

in Eq. (6.110) and take inverse z-transforms to obtain

$$\hat{h}[n] = \begin{cases} \log[A] & n = 0 \\ \sum_{k=1}^{N_i} \frac{b_k^n}{n} - \sum_{k=1}^{M_i} \frac{a_k^n}{n} & n > 0 \\ \sum_{k=1}^{M_o} \frac{u_k^n}{n} - \sum_{k=1}^{N_o} \frac{v_k^n}{n} & n < 0 \end{cases} \tag{6.112}$$

If the filter's impulse response doesn't have zeros or poles outside the unit circle, the so-called *minimum phase* signals, then $\hat{h}[n] = 0$ for $n < 0$. *Maximum phase* signals are those with $\hat{h}[n] = 0$ for $n > 0$. If a signal is minimum phase, its complex cepstrum can be uniquely determined from its real cepstrum:

$$\hat{h}[n] = \begin{cases} 0 & n < 0 \\ c[n] & n = 0 \\ 2c[n] & n > 0 \end{cases} \tag{6.113}$$

It is easy to see from Eq. (6.112) that both the real and complex cepstrum are decaying sequences, which is the reason why, typically, a finite number of coefficients are sufficient to approximate it, and, therefore, people refer to the truncated cepstrum signal as a *cepstrum vector*.

6.4.2.1. LPC-Cepstrum

The case when the rational transfer function in Eq. (6.109) has been obtained with an LPC analysis is particularly interesting, since LPC analysis is such a widely used method. While Eq. (6.112) applies here, too, it is useful to find a recursion which doesn't require us to compute the roots of the predictor polynomial. Given the LPC filter

$$H(z) = \frac{G}{1 - \sum_{k=1}^{p} a_k z^{-k}} \tag{6.114}$$

we take the logarithm

$$\hat{H}(z) = \ln G - \ln\left(1 - \sum_{l=1}^{p} a_l z^{-l}\right) = \sum_{k=-\infty}^{\infty} \hat{h}[k] z^{-k} \tag{6.115}$$

and the derivative of both sides with respect to z

$$\frac{-\sum_{n=1}^{p} n a_n z^{-n-1}}{1 - \sum_{l=1}^{p} a_l z^{-l}} = -\sum_{k=-\infty}^{\infty} k\hat{h}[k] z^{-k-1} \tag{6.116}$$

Multiplying both sides by $-z\left(1 - \sum_{l=1}^{p} a_l z^{-l}\right)$, we obtain

$$\sum_{n=1}^{p} n a_n z^{-n} = \sum_{n=-\infty}^{\infty} n\hat{h}[n] z^{-n} - \sum_{l=1}^{p} \sum_{k=-\infty}^{\infty} k\hat{h}[k] a_l z^{-k-l} \tag{6.117}$$

which, after replacing $l = n - k$, and equating terms in z^{-n}, results in

$$\begin{aligned} n a_n &= n\hat{h}[n] - \sum_{k=1}^{n-1} k\hat{h}[k] a_{n-k} \qquad 0 < n \le p \\ 0 &= n\hat{h}[n] - \sum_{k=n-p}^{n-1} k\hat{h}[k] a_{n-k} \qquad n > p \end{aligned} \tag{6.118}$$

so that the complex cepstrum can be obtained from the LPC coefficients by the following recursion:

$$\hat{h}[n] = \begin{cases} 0 & n < 0 \\ \ln G & n = 0 \\ a_n + \sum_{k=1}^{n-1} \left(\frac{k}{n}\right) \hat{h}[k] a_{n-k} & 0 < n \le p \\ \sum_{k=n-p}^{n-1} \left(\frac{k}{n}\right) \hat{h}[k] a_{n-k} & n > p \end{cases} \tag{6.119}$$

where the value for $n = 0$ can be obtained from Eqs. (6.115) and (6.111). We note that, while there are a finite number of LPC coefficients, the number of cepstrum coefficients is infinite. Speech recognition researchers have shown empirically that a finite number is sufficient: 12–20 depending on the sampling rate and whether or not frequency warping is done. In Chapter 8 we discuss the use of the cepstrum in speech recognition.

This recursion should not be used in the reverse mode to compute the LPC coefficients from *any* set of cepstrum coefficients, because the recursion in Eq. (6.119) assumes an all-pole model with all poles inside the unit circle, and that might not be the case for an arbitrary cepstrum sequence, so that the recursion might yield a set of unstable LPC coefficients. In some experiments it has been shown that quantized LPC-cepstrum can yield unstable LPC coefficients over 5% of the time.

6.4.3. Cepstrum of Periodic Signals

It is important to see what the cepstrum of periodic signals looks like. To do so, let's consider the following signal:

$$x[n] = \sum_{k=0}^{M-1} \alpha_k \delta[n-kN] \tag{6.120}$$

which can be viewed as an impulse train of period N multiplied by an analysis window, so that only M impulses remain. Its z-transform is

$$X(z) = \sum_{k=0}^{M-1} \alpha_k z^{-kN} \tag{6.121}$$

which is a polynomial in z^{-N} rather than z^{-1}. Therefore, $X(z)$ can be expressed as a product of factors of the form $(1-a_k z^{-Nk})$ and $(1-u_k z^{Nk})$. Following the derivation in Section 6.4.2, it is clear that its complex cepstrum is nonzero only at integer multiples of N:

$$\hat{x}[n] = \sum_{k=-\infty}^{\infty} \beta_k \delta[n-kN] \tag{6.122}$$

A particularly interesting case is when $\alpha_k = \alpha^k$ with $0 < \alpha < 1$, so that Eq. (6.121) can be expressed as

$$X(z) = 1 + \alpha z^{-N} + \cdots + (\alpha z^{-N})^{M-1} = \frac{1-(\alpha z^{-N})^M}{1-\alpha z^{-N}} \tag{6.123}$$

so that taking the logarithm of Eq. (6.123) and expanding it in Taylor series using Eq. (6.111) results in

$$\hat{X}(z) = \ln X(z) = \sum_{r=1}^{\infty} \frac{\alpha^r}{r} z^{-rN} - \sum_{l=1}^{\infty} \frac{\alpha^{lM}}{l} z^{-lMN} = \sum_{n=1}^{\infty} \hat{x}[n] z^{-n} \tag{6.124}$$

which lets us compute the complex cepstrum as

$$\hat{x}[n] = \sum_{r=1}^{\infty} \frac{\alpha^r}{r} \delta[n - rN] - \sum_{l=1}^{\infty} \frac{\alpha^{lM}}{l} \delta[n - lMN] \tag{6.125}$$

An infinite impulse train can be obtained by making $\alpha \to 1$ and $M \to \infty$ in Eq. (6.125).

$$\hat{x}[n] = \sum_{r=1}^{\infty} \frac{\delta[n - rN]}{r} \tag{6.126}$$

We see from Eq. (6.126) that the cepstrum of an impulse train goes to 0 as n increases. This justifies our assumption of homomorphic filtering.

6.4.4. Cepstrum of Speech Signals

We can compute the cepstrum of a speech segment by windowing the signal with a window of length N. In practice, the cepstrum is not computed through Eq. (6.112), since root-finding algorithms are slow and offer numerical imprecision for the large values of N used. Instead, we can compute the cepstrum directly through its definition of Eq. (6.105), using the DFT as follows:

$$X_a[k] = \sum_{n=0}^{N-1} x[n] e^{-j2\pi nk/N}, \quad 0 \le k < N \tag{6.127}$$

$$\hat{X}_a[k] = \ln X_a[k], \quad 0 \le k < N \tag{6.128}$$

$$\hat{x}_a[n] = \frac{1}{N} \sum_{n=0}^{N-1} \hat{X}_a[k] e^{-j2\pi nk/N}, \quad 0 \le n < N \tag{6.129}$$

The subscript a means that the new complex cepstrum $\hat{x}_a[n]$ is an aliased version of $\hat{x}[n]$ given by

$$\hat{x}_a[n] = \sum_{r=-\infty}^{\infty} \hat{x}[n + rN] \tag{6.130}$$

which can be derived by using the sampling theorem of Chapter 5, by reversing the concepts of time and frequency.

This aliasing introduces errors in the estimation that can be reduced by choosing a large value for N.

Computation of the complex cepstrum requires computing the complex logarithm and, in turn, the phase. However, given the principal value of the phase $\theta_p[k]$, there are infinite possible values for $\theta[k]$:

$$\theta[k] = \theta_p[k] + 2\pi n_k \tag{6.131}$$

From Chapter 5 we know that if $x[n]$ is real, $\arg\left[X(e^{j\omega})\right]$ is an odd function and also continuous. Thus we can do *phase unwrapping* by choosing n_k to guarantee that $\theta[k]$ is a smooth function, i.e., by forcing the difference between adjacent values to be small:

$$\left|\theta[k] - \theta[k-1]\right| < \pi \tag{6.132}$$

A linear phase term r as in Eq. (6.110), would contribute to the phase difference in Eq. (6.132) with $2\pi r / N$, which may result in errors in the phase unwrapping if $\theta[k]$ is changing sufficiently rapidly. In addition, there could be large changes in the phase difference if $X_a[k]$ is noisy. To guarantee that we can track small phase differences, a value of N several times larger than the window size is required: i.e., the input signal has to be zero-padded prior to the FFT computation. Finally, the delay r in Eq. (6.109), can be obtained by forcing the phase to be an odd function, so that:

$$\theta[N/2] = \pi r \tag{6.133}$$

For unvoiced speech, the unwrapped phase is random, and therefore only the real cepstrum has meaning. In practical situations, even voiced speech has some frequencies at which noise dominates (typically very low and high frequencies), which results in phase $\theta[k]$ that changes drastically from frame to frame. Because of this, the complex cepstrum in Eq. (6.105) is rarely used for real speech signals. Instead, the real cepstrum is used much more often:

$$C_a[k] = \ln\left|X_a[k]\right|, \quad 0 \le k < N \tag{6.134}$$

$$c_a[n] = \frac{1}{N}\sum_{n=0}^{N-1} C_a[k] e^{-j2\pi nk/N}, \quad 0 \le n < N \tag{6.135}$$

Similarly, it can be shown that for the new real cepstrum $c_a[n]$ is an aliased version of $c[n]$ given by

$$c_a[n] = \sum_{r=-\infty}^{\infty} c[n + rN] \tag{6.136}$$

which again has aliasing that can be reduced by choosing a large value for N.

6.4.5. Source-Filter Separation via the Cepstrum

We have seen that, if the filter is a rational transfer function, and the source is an impulse train, the homomorphic filtering of Figure 6.24 can approximately separate them. Because of problems in estimating the phase in speech signals (see Section 6.4.4), we generally compute the real cepstrum using Eqs. (6.127), (6.134), and (6.135), and then compute the complex cepstrum under the assumption of a minimum phase signal according to Eq. (6.113). The result of separating source and filter using this cepstral deconvolution is shown in Figure 6.25 for voiced speech and Figure 6.26 for unvoiced speech.

The real cepstrum of white noise $x[n]$ with an expected magnitude spectrum $|X(e^{j\omega})|=1$ is 0. If colored noise is present, the cepstrum of the observed colored noise $\hat{y}[n]$ is identical to the cepstrum of the coloring filter $\hat{h}[n]$, except for a gain factor. The above is correct if we take an infinite number of noise samples, but in practice, this cannot be done and a limited number have to be used, so that this is only an approximation, though it is often used in speech processing algorithms.

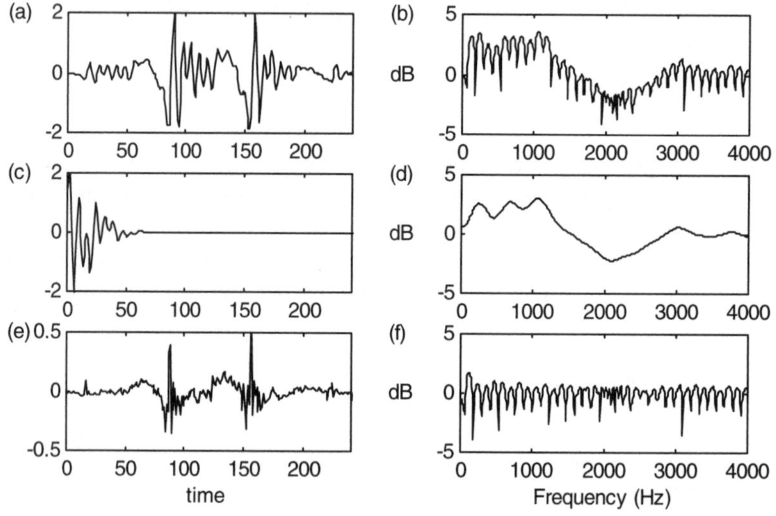

Figure 6.25 Separation of source and filter using homomorphic filtering for voiced speech with the scheme of Figure 6.24 with $N = 20$ in the homomorphic filter of Eq. (6.102) with the real cepstrum: (a) windowed signal, (b) log-spectrum, (c) filter's impulse response, (d) smoothed log-spectrum, (e) windowed excitation signal, (f) log-spectrum of high-part of cepstrum. Note that the windowed excitation is not a windowed impulse train because of the minimum phase assumption.

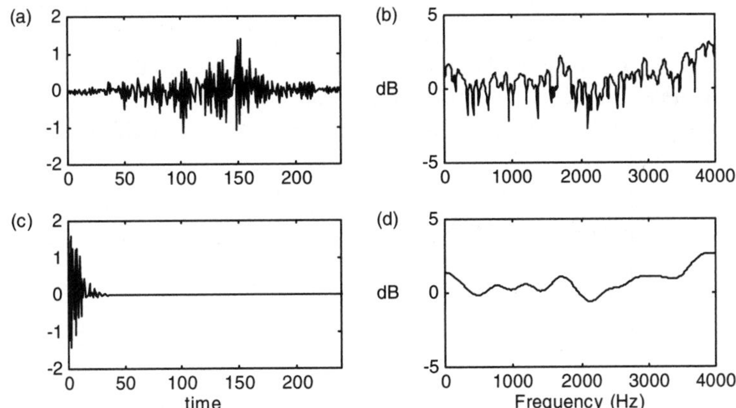

Figure 6.26 Separation of source and filter using homomorphic filtering for unvoiced speech with the scheme of Figure 6.24 with $N = 20$ in the homomorphic filter of Eq. (6.102) with the real cepstrum: (a) windowed signal, (b) log-spectrum, (c) filter's impulse response, (d) smoothed log-spectrum.

6.5. Perceptually Motivated Representations

In this section we describe some aspects of human perception, and methods motivated by the behavior of the human auditory system: bilinearly transformed cepstrum, Mel-Frequency Cepstrum Coefficients (MFCC), and Perceptual Linear Prediction (PLP). These methods have been successfully used in speech recognition.

6.5.1. The Bilinear Transform

The transformation

$$s = \frac{z^{-1} - \alpha}{1 - \alpha z^{-1}} \tag{6.137}$$

for $0 < \alpha < 1$ belongs to the class of *bilinear* transforms. It is a mapping in the complex plane that maps the unit circle onto itself. The frequency transformation is obtained by making the substitution $z = e^{j\omega}$ and $s = e^{j\Omega}$:

$$\Omega = \omega + 2\arctan\left[\frac{\alpha \sin(\omega)}{1 - \alpha \cos(\omega)}\right] \tag{6.138}$$

This transformation is very similar to the Bark and mel scale for an appropriate choice of the parameter α (see Chapter 2). Oppenheim [31] showed that the advantage of this transformation is that it can be used to transform a time sequence in the linear frequency into another time sequence in the warped frequency, as shown in Figure 6.27. This bilinear transform has been successfully applied to cepstral and autocorrelation coefficients.

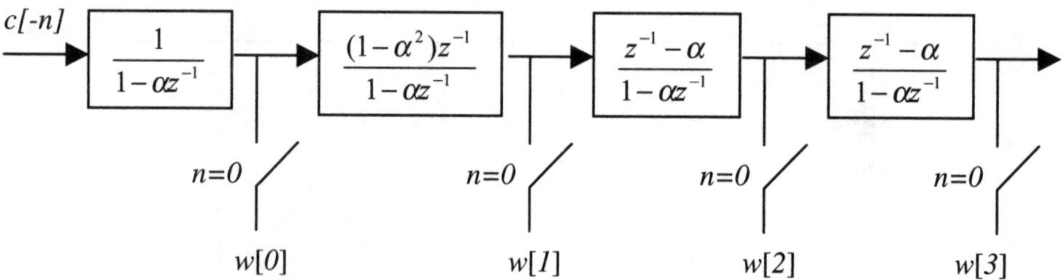

Figure 6.27 Implementation of the frequency-warped cepstral coefficients as a function of the linear-frequency cepstrum coefficients. Both sets of coefficients are causal. The input is the time-reversed cepstrum sequence, and the output can be obtained by sampling the outputs of the filters at time $n = 0$. The filters used for $w[m]$ $m > 2$ are the same. Note that, for a finite-length cepstrum, an infinite-length warped cepstrum results.

For a finite number of cepstral coefficients the bilinear transform in Figure 6.27 results in an infinite number of warped cepstral coefficients. Since truncation is usually done in practice, the bilinear transform is equivalent to a matrix multiplication, where the matrix is a function of the warping parameter α. Shikano [43] showed these warped cepstral coefficients were beneficial for speech recognition.

6.5.2. Mel-Frequency Cepstrum

The *Mel-Frequency Cepstrum Coefficients* (MFCC) is a representation defined as the real cepstrum of a windowed short-time signal derived from the FFT of that signal. The difference from the real cepstrum is that a nonlinear frequency scale is used, which approximates the behavior of the auditory system. Davis and Mermelstein [8] showed the MFCC representation to be beneficial for speech recognition.

Given the DFT of the input signal

$$X_a[k] = \sum_{n=0}^{N-1} x[n] e^{-j2\pi nk/N}, \quad 0 \le k < N \tag{6.139}$$

we define a filterbank with M filters ($m = 1, 2, \cdots, M$), where filter m is triangular filter given by:

$$H_m[k] = \begin{cases} 0 & k < f[m-1] \\ \dfrac{2(k - f[m-1])}{(f[m+1] - f[m-1])(f[m] - f[m-1])} & f[m-1] \le k \le f[m] \\ \dfrac{2(f[m+1] - k)}{(f[m+1] - f[m-1])(f[m+1] - f[m])} & f[m] \le k \le f[m+1] \\ 0 & k > f[m+1] \end{cases} \tag{6.140}$$

Such filters compute the average spectrum around each center frequency with increasing bandwidths, and they are displayed in Figure 6.28.

Alternatively, the filters can be chosen as

$$H'_m[k] = \begin{cases} 0 & k < f[m-1] \\ \dfrac{(k - f[m-1])}{(f[m] - f[m-1])} & f[m-1] \le k \le f[m] \\ \dfrac{(f[m+1] - k)}{(f[m+1] - f[m])} & f[m] \le k \le f[m+1] \\ 0 & k > f[m+1] \end{cases} \tag{6.141}$$

which satisfies $\sum_{m=1}^{M} H'_m[k] = 1$. The mel-cepstrum computed with $H_m[k]$ or $H'_m[k]$ will differ by a constant vector for all inputs, so the choice becomes unimportant when used in a speech recognition system that has been trained with the same filters.

Let's define f_l and f_h to be the lowest and highest frequencies of the filterbank in Hz, F_s the sampling frequency in Hz, M the number of filters, and N the size of the FFT. The boundary points $f[m]$ are uniformly spaced in the mel-scale:

$$f[m] = \left(\frac{N}{F_s}\right) B^{-1}\left(B(f_l) + m\frac{B(f_h) - B(f_l)}{M+1}\right) \tag{6.142}$$

where the mel-scale B is given by Eq. (2.6), and B^{-1} is its inverse

$$B^{-1}(b) = 700\left(\exp(b/1125) - 1\right) \tag{6.143}$$

We then compute the log-energy at the output of each filter as

$$S[m] = \ln\left[\sum_{k=0}^{N-1} |X_a[k]|^2 H_m[k]\right], \qquad 0 < m \le M \tag{6.144}$$

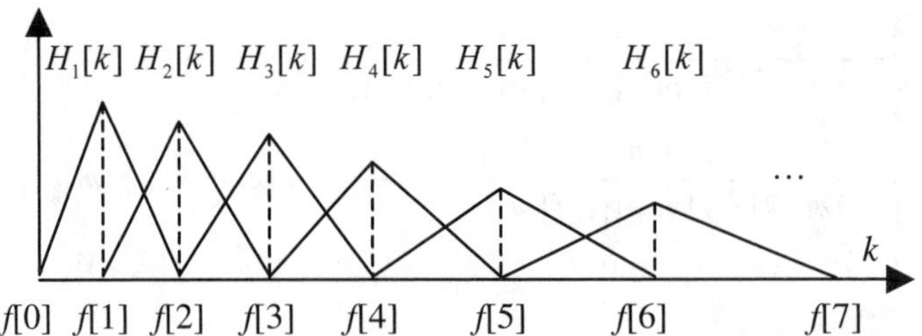

Figure 6.28 Triangular filters used in the computation of the mel-cepstrum using Eq. (6.140).

The mel-frequency cepstrum is then the discrete cosine transform of the *M* filter outputs:

$$c[n] = \sum_{m=0}^{M-1} S[m] \cos\left(\pi n(m-1/2)/M\right) \qquad 0 \le n < M \tag{6.145}$$

where *M* varies for different implementations from 24 to 40. For speech recognition, typically only the first 13 cepstrum coefficients are used. It is important to note that the MFCC representation is no longer a homomorphic transformation. It would be if the order of summation and logarithms in Eq. (6.144) were reversed:

$$S[m] = \sum_{k=0}^{N-1} \ln\left(|X_a[k]|^2 H_m[k]\right) \qquad 0 < m \le M \tag{6.146}$$

In practice, however, the MFCC representation is approximately homomorphic for filters that have a smooth transfer function. The advantage of the MFCC representation using (6.144) instead of (6.146) is that the filter energies are more robust to noise and spectral estimation errors. This algorithm has been used extensively as a feature vector for speech recognition systems.

While the definition of cepstrum in Section 6.4.1 uses an inverse DFT, since $S[m]$ is even, a DCT-II can be used instead (see Chapter 5).

6.5.3. Perceptual Linear Prediction (PLP)

Perceptual Linear Prediction (PLP) [16] uses the standard Durbin recursion of Section 6.3.2.1.2 to compute LPC coefficients, and typically the LPC coefficients are transformed to LPC-cepstrum using the recursion in Section 6.4.2.1. But unlike standard linear prediction, the autocorrelation coefficients are not computed in the time domain through Eq. (6.55).

The autocorrelation $R_x[n]$ is the inverse Fourier transform of the power spectrum $|X(\omega)|^2$ of the signal. We cannot compute the continuous-frequency Fourier transform easily, but we can take an FFT to compute $X[k]$, so that the autocorrelation can be obtained as the inverse Fourier transform of $|X[k]|^2$. Since the discrete Fourier transform is not performing linear convolution but circular convolution, we need to make sure that the FFT size is larger than twice the window length (see Section 5.3.4) for this to hold. This alternate way of computing autocorrelation coefficients, entailing two FFTs and N multiplies and adds, should yield identical results. Since normally only a small number p of autocorrelation coefficients are needed, this is generally not a cost-effective way to do it, unless the first FFT has to be computed for other reasons.

Perceptual linear prediction uses the above method, but replaces $|X[k]|^2$ by a perceptually motivated power spectrum. The most important aspect is the non-linear frequency scaling, which can be achieved through a set of filterbanks similar to those described in Section 6.5.2, so that this critical-band power spectrum can be sampled in approximately 1-Bark intervals. Another difference is that, instead of taking the logarithm on the filterbank energy outputs, a different non-linearity compression is used, often the cubic root. It is reported [16] that the use of this different non-linearity is beneficial for speech recognizers in noisy conditions.

6.6. Formant Frequencies

Formant frequencies are the resonances in the vocal tract and, as we saw in Chapter 2, they convey the differences between different sounds. Expert spectrogram readers are able to recognize speech by looking at a spectrogram, particularly at the formants. It has been argued that they are very useful features for speech recognition, but they haven't been widely used because of the difficulty in estimating them.

One way of obtaining formant candidates at a frame level is to compute the roots of a pth-order LPC polynomial [3, 26]. There are standard algorithms to compute the complex roots of a polynomial with real coefficients [36], though convergence is not guaranteed. Each complex root z_i can be represented as

$$z_i = \exp(-\pi b_i + j2\pi f_i) \tag{6.147}$$

where f_i and b_i are the formant frequency and bandwidth, respectively, of the ith root. Real roots are discarded and complex roots are sorted by increasing f, discarding negative values. The remaining pairs (f_i, b_i) are the formant candidates. Traditional formant trackers discard roots whose bandwidths are higher than a threshold [46], say 200 Hz.

Closed-phase analysis of voiced speech [5] uses only the regions for which the glottis is closed and thus there is no excitation. When the glottis is open, there is a coupling of the vocal tract with the lungs and the resonance bandwidths are somewhat larger. Determination of the closed-phase regions directly from the speech signal is difficult, so often an *electroglottograph* (EGG) signal is used [23]. EGG signals, obtained by placing electrodes at the speaker's throat, are very accurate in determining the times when the glottis is closed. Using samples in the closed-phase covariance analysis can yield accurate results [46]. For female speech, the closed-phase is short, and sometimes non-existent, so such analysis can be a challenge. EGG signals are useful also for pitch tracking and are described in more detail in Chapter 16.

Another common method consists of finding the peaks on a smoothed spectrum, such as that obtained through an LPC analysis [26, 40]. The advantage of this method is that you can always compute the peaks and it is more computationally efficient than extracting the complex roots of a polynomial. On the other hand, this procedure generally doesn't estimate the formant's bandwidth. The first three formants are typically estimated this way for formant synthesis (see Chapter 16), since they are the ones that allow sound classification, whereas the higher formants are more speaker dependent.

Sometimes, the signal goes through some *conditioning*, which includes sampling rate conversion to remove frequencies outside the range we are interested in. For example, if we are interested only in the first three formants, we can safely downsample the input signal to 8 kHz, since we know all three formants should be below 4 kHz. This downsampling reduces computation and the chances of the algorithm to find formant values outside the expected range (otherwise peaks or roots could be chosen above 4 kHz which we know do not correspond to any of the first three formants). Pre-emphasis filtering is also often used to whiten the signal.

Because of the thresholds imposed above, it is possible that the formants are not continuous. For example, when the vocal tract's spectral envelope is changing rapidly, bandwidths obtained through the above methods are overestimates of the true bandwidths, and they may exceed the threshold and thus be rejected. It is also possible for the peak-picking algorithm to classify a harmonic as a formant during some regions where it is much stronger than the other harmonics. Due to the thresholds used, a given frame could have no formants, only one formant (either first, second, or third), two, three, or more. Formant alignment from one frame to another has often been done using heuristics to prevent such discontinuities.

6.6.1. Statistical Formant Tracking

It is desirable to have an approach that does not use any thresholds on formant candidates and uses a probabilistic model to do the tracking instead of heuristics [1]. The formant candidates can be obtained from roots of the LPC polynomial, peaks in the smoothed spectrum, or even from a dense sample of possible points. If the first n formants are desired, and we have $(p/2)$ formant candidates, a maximum of r n-tuples are considered, where r is given by

$$r = \binom{p/2}{n} \tag{6.148}$$

A Viterbi search (see Chapter 8) is then carried out to find the most likely path of formant n-tuples given a model with some a priori knowledge of formants. The prior distribution for formant targets is used to determine which formant candidate to use of all possible choices for the given phoneme (i.e., we know that F1 for an AE should be around 800 Hz).

Formant continuity is imposed through the prior distribution of the formant slopes. This algorithm produces n formants for every frame, including silence.

Since we are interested in obtaining the first three formants (n = 3) and F3 is known to be lower than 4 kHz, it is advantageous to downsample the signal to 8 kHz in order to avoid obtaining formant candidates above 4 kHz and to let us use a lower-order analysis which offers fewer numerical problems when computing the roots. With p = 14, it results in a maximum of r = 35 triplets for the case of no real roots.

Let **X** be a sequence of T feature vectors $\mathbf{x}_t$ of dimension n:

$$\mathbf{X} = (\mathbf{x}_1, \mathbf{x}_2, \ldots, \mathbf{x}_T)' \tag{6.149}$$

where the prime denotes transpose.

We estimate the formants with the knowledge of what sound occurs at that particular time, for example by using a speech recognizer that segments the waveform into different phonemes (see Chapter 9) or states q_t within a phoneme. In this case we assume that the output distribution of each state i is modeled by one Gaussian density function with a mean μ_i and covariance matrix Σ_i. We can define up to N states, with λ being the set of all means and covariance matrices for all:

$$\lambda = (\mu_1, \Sigma_1, \mu_2, \Sigma_2, \cdots, \mu_N, \Sigma_N) \tag{6.150}$$

Therefore, the log-likelihood for **X** is given by

$$\ln p(\mathbf{X} \mid \hat{\mathbf{q}}, \lambda) = -\frac{TM}{2} \ln(2\pi) - \frac{1}{2} \sum_{t=1}^{T} \ln |\Sigma_{q_t}| - \frac{1}{2} \sum_{t=1}^{T} (\mathbf{x}_t - \mu_{q_t})' \Sigma_{q_t}^{-1} (\mathbf{x}_t - \mu_{q_t}) \tag{6.151}$$

Maximizing **X** in Eq. (6.151) leads to the trivial solution $\hat{\mathbf{X}} = (\mu_{q_1}, \mu_{q_2}, \ldots, \mu_{q_T})'$, a piecewise function whose value is that of the best n-tuple candidate. This function has discontinuities at state boundaries and thus is not likely to represent well the physical phenomena of speech.

This problem arises because the slopes at state boundaries do not match the slopes of natural speech. To avoid these discontinuities, we would like to match not only the target formants at each state, but also the formant slopes at each state. To do that, we augment the feature vector $\mathbf{x}_t$ at frame t with the delta vector $\mathbf{x}_t - \mathbf{x}_{t-1}$. Thus, we increase the parameter space of λ with the corresponding means δ_i and covariance matrices Γ_i of these delta parameters, and assume statistical independence among them. The corresponding new log-likelihood has the form

$$\begin{aligned} \ln p(\mathbf{X} \mid \hat{\mathbf{q}}, \lambda) &= K - \frac{1}{2} \sum_{t=1}^{T} \ln |\Sigma_{q_t}| - \frac{1}{2} \sum_{t=2}^{T} \ln |\Gamma_{q_t}| \\ &- \frac{1}{2} \sum_{t=1}^{T} (\mathbf{x}_t - \mu_{q_t})' \Sigma_{q_t}^{-1} (\mathbf{x}_t - \mu_{q_t}) - \frac{1}{2} \sum_{t=2}^{T} (\mathbf{x}_t - \mathbf{x}_{t-1} - \delta_{q_t})' \Gamma_{q_t}^{-1} (\mathbf{x}_t - \mathbf{x}_{t-1} - \delta_{q_t}) \end{aligned} \tag{6.152}$$

Maximization of Eq. (6.152) with respect to $\mathbf{x}_t$ requires solving several sets of linear equations. If Γ_i and Σ_i are diagonal covariance matrices, it results in a set of linear equations for each of the M dimensions

$$\mathbf{BX} = \mathbf{c} \tag{6.153}$$

where **B** is a tridiagonal matrix (all values are zero except for those in the main diagonal and its two adjacent diagonals), which leads to a very efficient solution [36]. For example, the values of **B** and **c** for $T = 3$ are given by

$$\mathbf{B} = \begin{pmatrix} \frac{1}{\sigma_{q_1}^2} + \frac{1}{\gamma_{q_2}^2} & -\frac{1}{\gamma_{q_2}^2} & 0 \\ -\frac{1}{\gamma_{q_2}^2} & \frac{1}{\sigma_{q_2}^2} + \frac{1}{\gamma_{q_2}^2} + \frac{1}{\gamma_{q_3}^2} & -\frac{1}{\gamma_{q_3}^2} \\ 0 & -\frac{1}{\gamma_{q_3}^2} & \frac{1}{\sigma_{q_3}^2} + \frac{1}{\gamma_{q_3}^2} \end{pmatrix} \tag{6.154}$$

$$\mathbf{c} = \begin{pmatrix} \frac{\mu_{q_1}}{\sigma_{q_1}^2} - \frac{\delta_{q_2}}{\gamma_{q_2}^2} & \frac{\mu_{q_2}}{\sigma_{q_2}^2} + \frac{\delta_{q_2}}{\gamma_{q_2}^2} - \frac{\delta_{q_3}}{\gamma_{q_3}^2} & \frac{\mu_{q_3}}{\sigma_{q_3}^2} + \frac{\delta_{q_3}}{\gamma_{q_3}^2} \end{pmatrix}' \tag{6.155}$$

where just one dimension is represented, and the process is repeated for all dimensions with a computational complexity of $O(TM)$.

The maximum likelihood sequence $\hat{\mathbf{x}}_t$ is close to the targets μ_i while keeping the slopes close to δ_i for a given state i, thus estimating a continuous function. Because of the delta coefficients, the solution depends on all the parameters of all states and not just the current state. This procedure can be performed for the formants as well as the bandwidths.

The parameters μ_i, Σ_i, δ_i, and Γ_i can be re-estimated using the EM algorithm described in Chapter 8. In [1] it is reported that two or three iterations are sufficient for speaker-dependent data.

The formant track obtained through this method can be rough, and it may be desired to smooth it. Smoothing without knowledge about the speech signal would result in either blurring the sharp transitions that occur in natural speech, or maintaining ragged formant tracks where the underlying physical phenomena vary slowly with time. Ideally we would like a larger adjustment to the raw formant when the error in the estimate is large relative to the variance of the corresponding state within a phoneme. This can be done by modeling the formant measurement error as a Gaussian distribution. Figure 6.29 shows an utterance from a male speaker with the smoothed formant tracks, and Figure 6.30 compares the raw and smoothed formants. When no real formant is visible from the spectrogram, the algorithm tends to assign a large bandwidth (not shown in the figure).

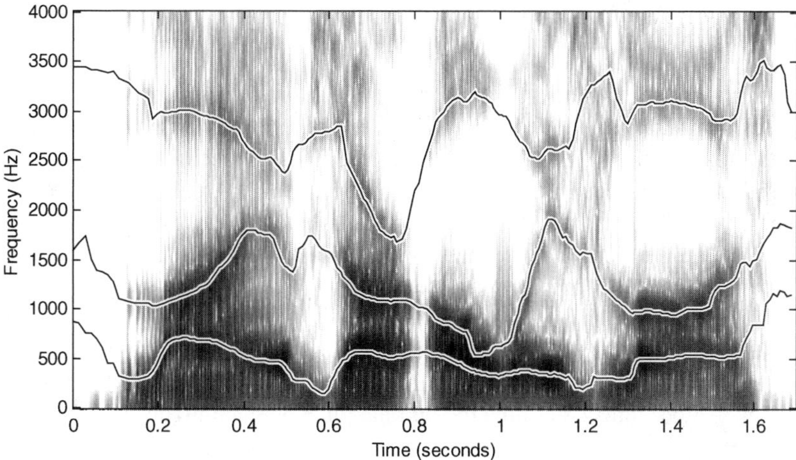

Figure 6.29 Spectrogram and three smoothed formants.

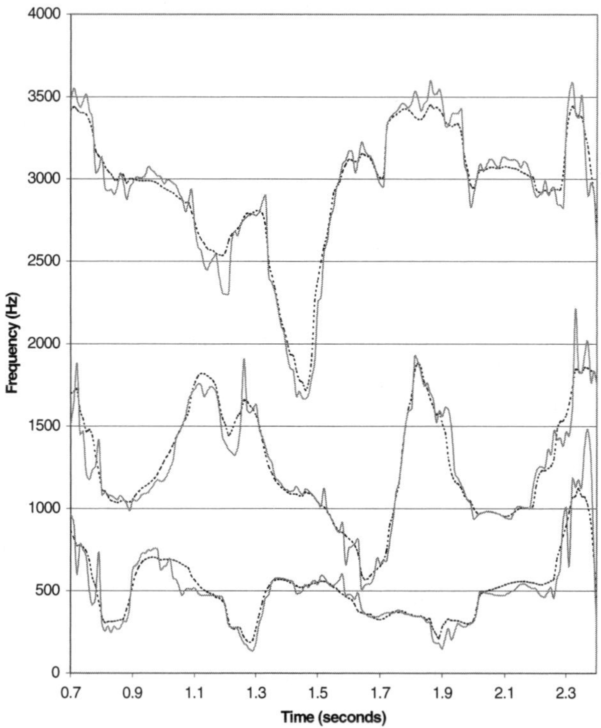

Figure 6.30 Raw formants (ragged gray line) and smoothed formants (dashed line).

6.7. THE ROLE OF PITCH

Pitch determination is very important for many speech processing algorithms. The concatenative speech synthesis methods of Chapter 16 require pitch tracking on the desired speech segments if prosody modification is to be done. Chinese speech recognition systems use pitch tracking for tone recognition, which is important in disambiguating the myriad of homophones. Pitch is also crucial for prosodic variation in text-to-speech systems (see Chapter 15) and spoken language systems (see Chapter 17). While in the previous sections we have dealt with features representing the filter, pitch represents the source of the model illustrated in Figure 6.1.

Pitch determination algorithms also use short-term analysis techniques, which means that for every frame $\mathbf{x}_m$ we get a score $f(T \mid \mathbf{x}_m)$ that is a function of the candidate pitch periods T. These algorithms determine the optimal pitch by maximizing

$$T_m = \arg\max_{T} f(T \mid \mathbf{x}_m) \tag{6.156}$$

We describe several such functions computed through the autocorrelation method and the normalized cross-correlation method, as well as the signal conditioning that is often performed. Other approaches based on cepstrum [28] have also been used successfully. A good summary of techniques used for pitch tracking is provided by [17, 45].

Pitch determination using Eq. (6.156) is error prone, and a smoothing stage is often done. This smoothing, described in Section 6.7.4, takes into consideration that the pitch does not change quickly over time.

6.7.1. Autocorrelation Method

A commonly used method to estimate pitch is based on detecting the highest value of the autocorrelation function in the region of interest. This region must exclude $m = 0$, as that is the absolute maximum of the autocorrelation function [37]. As discussed in Chapter 5, the statistical autocorrelation of a sinusoidal random process

$$\mathbf{x}[n] = \cos(\omega_0 n + \varphi) \tag{6.157}$$

is given by

$$R[m] = E\{\mathbf{x}^*[n]\mathbf{x}[n+m]\} = \frac{1}{2}\cos(\omega_0 m) \tag{6.158}$$

which has maxima for $m = lT_0$, the pitch period and its harmonics, so that we can find the pitch period by computing the highest value of the autocorrelation. Similarly, it can be shown that any WSS periodic process $\mathbf{x}[n]$ with period T_0 also has an autocorrelation $R[m]$ which exhibits its maxima at $m = lT_0$.

In practice, we need to obtain an estimate $\hat{R}[m]$ from knowledge of only N samples. If we use a window $w[n]$ of length N on $\mathbf{x}[n]$ and assume it to be real, the empirical autocorrelation function is given by

$$\hat{R}[m] = \frac{1}{N} \sum_{n=0}^{N-1-|m|} w[n]\mathbf{x}[n]w[n+|m|]\mathbf{x}[n+|m|] \tag{6.159}$$

whose expected value can be shown to be

$$E\{\hat{R}[m]\} = R[m](w[m] * w[-m]) \tag{6.160}$$

where

$$w[m] * w[-m] = \sum_{n=0}^{N-|m|-1} w[n]w[n+|m|] \tag{6.161}$$

which, for the case of a rectangular window of length N, is given by

$$w[m] * w[-m] = \begin{cases} 1 - \dfrac{|m|}{N} & |m| < N \\ 0 & |m| \geq N \end{cases} \tag{6.162}$$

which means that $\hat{R}[m]$ is a biased estimator of $R[m]$. So, if we compute the peaks based on Eq. (6.159), the estimate of the pitch will also be biased. Although the variance of the estimate is difficult to compute, it is easy to see that as m approaches N, fewer and fewer samples of $x[n]$ are involved in the calculation, and thus the variance of the estimate is expected to increase. If we multiply Eq. (6.159) by $N/(N-m)$, the estimate will be unbiased but the variance will be larger.

Using the empirical autocorrelation in Eq. (6.159) for the random process in Eq. (6.157) results in an expected value of

$$E\{\hat{R}[m]\} = \left(1 - \frac{|m|}{N}\right)\frac{\cos(\omega_0 m)}{2}, \quad |m| < N \tag{6.163}$$

whose maximum coincides with the pitch period for $m > m_0$.

Since pitch periods can be as low as 40 Hz (for a very low-pitched male voice) or as high as 600 Hz (for a very high-pitched female or child's voice), the search for the maximum is conducted within a region. This F0 detection algorithm is illustrated in Figure 6.31 where the lag with highest autocorrelation is plotted for every frame. In order to see periodicity present in the autocorrelation, we need to use a window that contains at least two pitch periods, which, if we want to detect a 40 Hz pitch, implies 50 ms (see Figure 6.32). For window lengths so long, the assumption of stationarity starts to fail, because a pitch period at the beginning of the window can be significantly different than at the end of the window.

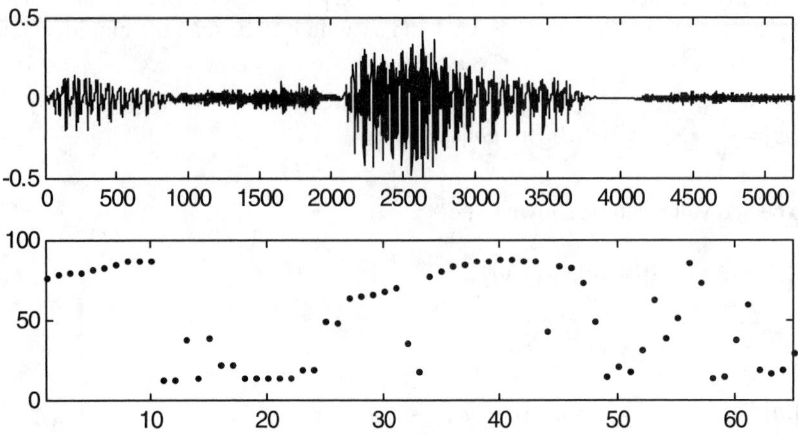

Figure 6.31 Waveform and unsmoothed pitch track with the autocorrelation method. A frame shift of 10 ms, a Hamming window of 30 ms, and a sampling rate of 8 kHz were used. Notice that two frames in the voiced region have an incorrect pitch. The pitch values in the unvoiced regions are essentially random.

One possible solution to this problem is to estimate the autocorrelation function with different window lengths for different lags m.

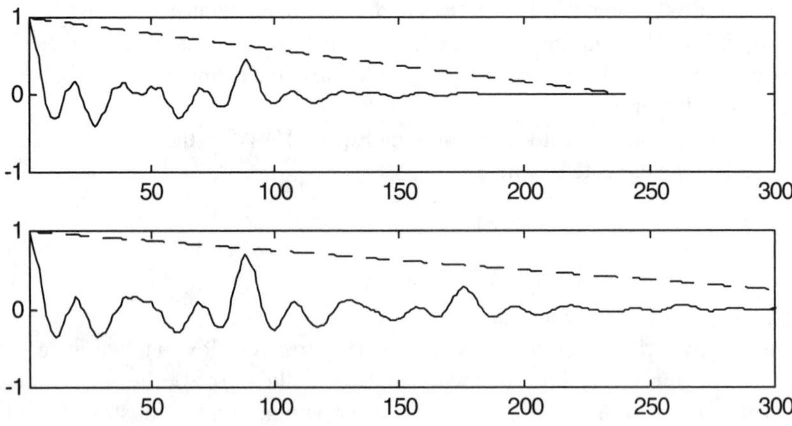

Figure 6.32 Autocorrelation function for frame 40 in Figure 6.31. The maximum occurs at 89 samples. A sampling frequency of 8 kHz and window shift of 10 ms are used. The top figure is using a window length of 30 ms, whereas the bottom one is using 50 ms. Notice the quasi-periodicity in the autocorrelation function.

The candidate pitch periods in Eq. (6.156) can be simply $T_m = m$; i.e., the pitch period is any integer number of samples. For low values of T_m, the frequency resolution is lower than for high values. To maintain a relatively constant frequency resolution, we do not have to search all the pitch periods for large T_m. Alternatively, if the sampling frequency is not high, we may need to use fractional pitch periods (often done in the speech coding algorithms of Chapter 7).

The autocorrelation function can be efficiently computed by taking a signal, windowing it, and taking an FFT and then the square of the magnitude.

6.7.2. Normalized Cross-Correlation Method

A method that is free from these border problems and has been gaining in popularity is based on the *normalized cross-correlation* [2]

$$\alpha_t(T) = \cos(\theta) = \frac{<\mathbf{x}_t, \mathbf{x}_{t-T}>}{|\mathbf{x}_t||\mathbf{x}_{t-T}|} \tag{6.164}$$

where $\mathbf{x}_t = \{x[t-N/2], x[t-N/2+1], \cdots, x[t+N/2-1]\}$ is a vector of N samples centered at time t, and $<\mathbf{x}_t, \mathbf{x}_{t-T}>$ is the inner product between the two vectors defined as

$$<\mathbf{x}_n, \mathbf{y}_l> \sum_{m=-N/2}^{N/2-1} x[n+m]y[l+m] \tag{6.165}$$

so that, using Eq. (6.165), the normalized cross-correlation can be expressed as

$$\alpha_t(T) = \frac{\sum_{n=-N/2}^{N/2-1} x[t+n]x[t+n-T]}{\sqrt{\sum_{n=-N/2}^{N/2-1} x^2[t+n] \sum_{m=-N/2}^{N/2-1} x^2[t+m+T]}} \tag{6.166}$$

where we see that the numerator in Eq. (6.166) is very similar to the autocorrelation in Section 6.7.1, but where N terms are used in the summation for all values of T.

The maximum of the normalized cross-correlation method is shown in Figure 6.33 (b). Unlike the autocorrelation method, the estimate of the normalized cross-correlation is not biased by the term $(1-m/N)$. For perfectly periodic signals, this results in identical values of the normalized cross-correlation function for kT. This can result in pitch halving, where $2T$ can be chosen as the pitch period, which happens in Figure 6.33 (b) at the beginning of the utterance. Using a decaying bias $(1-m/M)$ with $M \gg N$, can be useful in reducing pitch halving, as we see in Figure 6.33 (c).

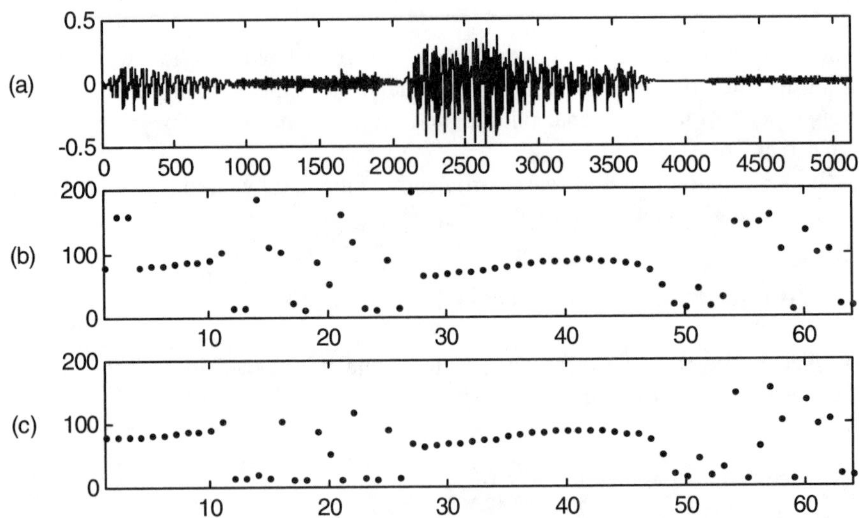

Figure 6.33 (a) Waveform and (b, c) unsmoothed pitch tracks with the normalized cross-correlation method. A frame shift of 10 ms, window length of 10 ms, and sampling rate of 8 kHz were used. (b) is the standard normalized cross-correlation method, whereas (c) has a decaying term. If we compare it to the autocorrelation method of Figure 6.31, the middle voiced region is correctly identified in both (b) and (c), but two frames at the beginning of (b) that have pitch halving are eliminated with the decaying term. Again, the pitch values in the unvoiced regions are essentially random.

Because the number of samples involved in the calculation is constant, this estimate is unbiased and has lower variance than that of the autocorrelation. Unlike the autocorrelation method, the window length could be lower than the pitch period, so that the assumption of stationarity is more accurate and it has more time resolution. While pitch trackers based on the normalized cross-correlation typically perform better than those based on the autocorrelation, they also require more computation, since all the autocorrelation lags can be efficiently computed through 2 FFTs and N multiplies and adds (see Section 5.3.4).

Let's gain some insight about the normalized cross-correlation. If $x[n]$ is periodic with period T, then we can predict it from a vector T samples in the past as:

$$\mathbf{x}_t = \rho \mathbf{x}_{t-T} + \mathbf{e}_t \tag{6.167}$$

where ρ is the prediction gain. The normalized cross-correlation measures the angle between the two vectors, as can be seen in Figure 6.34, and since it is a cosine, it has the property that $-1 \le \alpha_n(P) \le 1$.

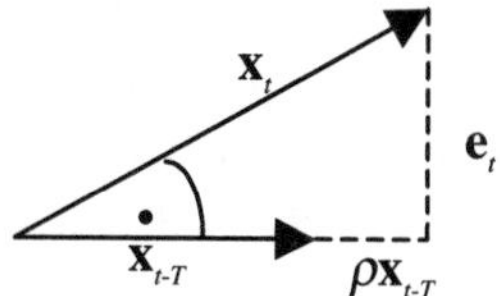

Figure 6.34 The prediction of $\mathbf{x}_t$ with $\mathbf{x}_{t-T}$ results in an error $\mathbf{e}_t$.

If we choose the value of the prediction gain ρ so as to minimize the prediction error

$$|\mathbf{e}_t|^2 = |\mathbf{x}_t|^2 - |\mathbf{x}_t|^2 \cos^2(\theta) = |\mathbf{x}_t|^2 - |\mathbf{x}_t|^2 \alpha_t^2(T) \tag{6.168}$$

and assume $\mathbf{e}_t$ is a zero-mean Gaussian random vector with a standard deviation $\sigma | \mathbf{x}_t |$, then

$$\ln f(\mathbf{x}_t \mid T) = K + \frac{\alpha_t^2(T)}{2\sigma^2} \tag{6.169}$$

so that the maximum likelihood estimate corresponds to finding the value T with highest normalized cross-correlation. Using Eq. (6.166), it is possible that $\alpha_t(T) < 0$. In this case, there is negative correlation between $\mathbf{x}_t$ and $\mathbf{x}_{t-T}$, and it is unlikely that T is a good choice for pitch. Thus, we need to force $\rho > 0$, so that Eq. (6.169) is converted into

$$\ln f(\mathbf{x}_t \mid T) = K + \frac{\left(\max(0, \alpha_t(T))\right)^2}{2\sigma^2} \tag{6.170}$$

The normalized cross-correlation of Eq. (6.164) predicts the current frame with a frame that occurs T samples before. Voiced speech may exhibit low correlation with a previous frame at a spectral discontinuity, such as those appearing at stops. To account for this, an enhancement can be done to consider not only the *backward* normalized cross-correlation, but also the *forward* normalized cross-correlation, by looking at a frame that occurs T samples ahead of the current frame, and taking the highest of both.

$$\ln f(\mathbf{x}_t \mid T) = K + \frac{\left(\max(0, \alpha_t(T), \alpha_t(-T))\right)^2}{2\sigma^2} \tag{6.171}$$

6.7.3. Signal Conditioning

Noise in the signal tends to make pitch estimation less accurate. To reduce this effect, signal conditioning or pre-processing has been proposed prior to pitch estimation [44]. Typically this involves bandpass filtering to remove frequencies above 1 or 2 kHz, and below 100 Hz or so. High frequencies do not have much voicing information and have significant noise energy, whereas low frequencies can have 50/60 Hz interference from power lines or non-linearities from some A/D subsystems that can also mislead a pitch estimation algorithm.

In addition to the noise in the very low frequencies and aspiration at high bands, the stationarity assumption is not as valid at high frequencies. Even a slowly changing pitch, say, nominal 100 Hz increasing 5 Hz in 10 ms, results in a fast-changing harmonic: the 30th harmonic at 3000 Hz changes 150 Hz in 10 ms. The corresponding short-time spectrum no longer shows peaks at those frequencies.

Because of this, it is advantageous to filter out such frequencies prior to the computation of the autocorrelation or normalized cross-correlation. If an FFT is used to compute the autocorrelation, this filter is easily done by setting to 0 the undesired frequency bins.

6.7.4. Pitch Tracking

Pitch tracking using the above methods typically fails in several cases:

- *Sub-harmonic errors*. If a signal is periodic with period T, it is also periodic with period $2T$, $3T$, etc. Thus, we expect the scores also to be high for the multiples of T, which can mislead the algorithm. Because the signal is never perfectly stationary, those multiples, or sub-harmonics, tend to have slightly lower scores than the fundamental. If the pitch is identified as $2T$, pitch halving is said to occur.
- *Harmonic errors*. If harmonic M dominates the signal's total energy, the score at pitch period T/M will be large. This can happen if the harmonic falls in a formant frequency that boosts its amplitude considerably compared to that of the other harmonics. If the pitch is identified as $T/2$, pitch doubling is said to occur.
- *Noisy conditions*. When the SNR is low, pitch estimates are quite unreliable for most methods.
- *Vocal fry*. While pitch is generally continuous, for some speakers it can suddenly change and even halve, particularly at the end of an unstressed voiced region. The pitch here is really not well defined and imposing smoothness constraints can hurt the system.
- *F0 jumps* up or down by an octave occasionally.
- *Breathy-voiced speech* is difficult to distinguish from periodic background noise.
- *Narrow-band filtering* of unvoiced excitations by certain vocal tract configurations can lead to signals that appear periodic.

For these reasons, pitch trackers do not determine the pitch value at frame m based exclusively on the signal at that frame. For a frame where there are several pitch candidates with similar scores, the fact that pitch does not change abruptly with time is beneficial in disambiguation, because the following frame possibly has a clearer pitch candidate, which can help.

To integrate the normalized cross-correlation into a probabilistic framework, you can combine tracking with the use of a priori information [10]. Let's define $\mathbf{X} = \{\mathbf{x}_0, \mathbf{x}_1, \ldots, \mathbf{x}_{M-1}\}$ as a sequence of input vectors for M consecutive frames centered at equally spaced time instants, say 10 ms. Furthermore, if we assume that the $\mathbf{x}_i$ are independent of each other, the joint distribution takes on the form:

$$f(\mathbf{X} \mid \mathbf{T}) = \prod_{i=0}^{M-1} f(\mathbf{x}_i \mid T_i) \tag{6.172}$$

where $\mathbf{T} = \{T_0, T_1, \ldots, T_{M-1}\}$ is the pitch track for the input. The *maximum a posteriori* (MAP) estimate of the pitch track is:

$$\mathbf{T}_{MAP} = \max_{\mathbf{T}} f(\mathbf{T} \mid \mathbf{X}) = \max_{\mathbf{T}} \frac{f(\mathbf{T}) f(\mathbf{X} \mid \mathbf{T})}{f(\mathbf{X})} = \max_{\mathbf{T}} f(\mathbf{T}) f(\mathbf{X} \mid \mathbf{T}) \tag{6.173}$$

according to Bayes' rule, with the term $f(\mathbf{X} \mid \mathbf{T})$ being given by Eq. (6.172) and $f(\mathbf{x}_i \mid T_i)$ by Eq. (6.169), for example.

The function $f(\mathbf{T})$ constitutes the a priori statistics for the pitch and can help disambiguate the pitch, by avoiding pitch doubling or halving given knowledge of the speaker's average pitch, and by avoiding rapid transitions given a model of how pitch changes over time. One possible approximation is given by assuming that the a priori probability of the pitch period at frame i depends only on the pitch period for the previous frame:

$$f(\mathbf{T}) = f(T_0, T_1, \ldots, T_{M-1}) = f(T_{M-1} \mid T_{M-2}) f(T_{M-2} \mid T_{M-3}) \cdots f(T_1 \mid T_0) f(T_0) \tag{6.174}$$

One possible choice for $f(T_t \mid T_{t-1})$ is to decompose it into a component that depends on T_t and another that depends on the difference $(T_t - T_{t-1})$. If we approximate both as Gaussian densities, we obtain

$$\ln f(T_t \mid T_{t-1}) = K' - \frac{(T_t - \mu)^2}{2\beta^2} - \frac{(T_t - T_{t-1} - \delta)^2}{2\gamma^2} \tag{6.175}$$

so that when Eqs. (6.170) and (6.175) are combined, the log-probability of transitioning to T_i at time t from pitch T_j at time $t-1$ is given by

$$S_t(T_i, T_j) = \frac{(\max(0, \alpha_t(T_i)))^2}{2\sigma^2} - \frac{(T_i - \mu)^2}{2\beta^2} - \frac{(T_i - T_j - \delta)^2}{2\gamma^2} \tag{6.176}$$

so that the log-likelihood in Eq. (6.173) can be expressed as

$$\ln f(\mathbf{T}) f(\mathbf{X} \mid \mathbf{T}) = (\max(0, \alpha_0(T_0)))^2 + \max_{i_t} \sum_{t=1}^{M-1} S_t(T_{i_t}, T_{i_{t-1}}) \tag{6.177}$$

which can be maximized through dynamic programming. For a region where pitch is not supposed to change, $\delta = 0$, the term $(T_i - T_j)^2$ in Eq. (6.176) acts as a penalty that keeps the pitch track from jumping around. A mixture of Gaussians can be used instead to model different rates of pitch change, as in the case of Mandarin Chinese with four tones characterized by different slopes. The term $(T_i - \mu)^2$ attempts to get the pitch close to its expected value to avoid pitch doubling or halving, with the average μ being different for male and female speakers. Pruning can be done during the search without loss of accuracy (see Chapter 12).

Pitch trackers also have to determine whether a region of speech is voiced or unvoiced. A good approach is to build a statistical classifier with techniques described in Chapter 8 based on energy and the normalized cross-correlation described above. Such classifiers, i.e., an HMM, penalize jumps between voiced and unvoiced frames to avoid voiced regions having isolated unvoiced frames inside and vice versa. A threshold can be used on the a posteriori probability to distinguish voiced from unvoiced frames.

6.8. Historical Perspective and Further Reading

In 1978, Lawrence R. Rabiner and Ronald W. Schafer [38] wrote a book summarizing the work to date on digital processing of speech, which remains a good source for the reader interested in further reading in the field. The book by Deller, Hansen, and Proakis [9] includes more recent work and is also an excellent reference. O'Shaughnessy [33] also has a thorough description of the subject. Malvar [25] covers filterbanks and lapped transforms extensively.

The extensive wartime interest in sound spectrography led Koenig and his colleagues at Bell Laboratories [22] in 1946 to the invaluable development of a tool that has been used for speech analysis since then: the spectrogram. Potter et al. [35] showed the usefulness of the analog spectrogram in analyzing speech. The spectrogram facilitated research in the field and led Peterson and Barney [34] to publish in 1952 a detailed study of formant values of different vowels. The development of computers and the FFT led Oppenheim, in 1970 [30], to develop digital spectrograms, which imitated the analog counterparts.

The MIT Acoustics Lab started work in speech in 1948 with Leo R. Beranek, who in 1954 published the seminal book *Acoustics*, where he studied sound propagation in tubes. In

1950, Kenneth N. Stevens joined the lab and started work on speech perception. Gunnar Fant visited the lab at that time and as a result started a strong speech production effort at KTH in Sweden.

The 1960s marked the birth of digital speech processing. Two books, Gunnar Fant's *Acoustical Theory of Speech Production* [13] in 1960 and James Flanagan's *Speech Analysis: Synthesis and Perception* [14] in 1965, had a great impact and sparked interest in the field. The advent of the digital computer prompted Kelly and Gertsman to create in 1961 the first digital speech synthesizer [21]. Short-time Fourier analysis, cepstrum, LPC analysis, and pitch and formant tracking were the fruit of that decade.

Short-time frequency analysis was first proposed for analog signals by Fano [11] in 1950 and later by Schroeder and Atal [42].

The mathematical foundation behind linear predictive coding dates to the auto-regressive models of George Udny Yule (1927) and Gilbert Walker (1931), which led to the well-known Yule-Walker equations. These equations resulted in a Toeplitz matrix, named after Otto Toeplitz (1881-1940) who studied it extensively. N. Levinson suggested in 1947 an efficient algorithm to invert such a matrix, which J. Durbin refined in 1960 and is now known as the Levinson-Durbin recursion. The well-known LPC analysis consisted of the application of the above results to speech signals, as developed by Bishnu Atal [4], J. Burg [7], Fumitada Itakura and S. Saito [19] in 1968, and Markel [27] and John Makhoul [24] in 1973.

The cepstrum was first proposed in 1964 by Bogert, Healy, and John Tukey [6] and further studied by Alan V. Oppenheim [29] in 1965. The popular mel-frequency cepstrum was proposed by Davis and Mermelstein [8] in 1980, combining the advantages of cepstrum with knowledge of the non-linear perception of frequency by the human auditory system that had been studied by E. Zwicker [47] in 1961.

Formant tracking was first investigated by Ken Stevens and James Flanagan in the late 1950s, with the foundations for most modern techniques being developed by Schafer and Rabiner [40], Itakura [20], and Markel [26]. Pitch tracking through digital processing was first studied by B. Gold [15] in 1962 and then improved by A. M. Noll [28], M. Schroeder [41], and M. Sondhi [44] in the late 1960s.

REFERENCES

[1] Acero, A., "Formant Analysis and Synthesis Using Hidden Markov Models," *Eurospeech*, 1999, Budapest pp. 1047-1050.

[2] Atal, B.S., *Automatic Speaker Recognition Based on Pitch Contours*, PhD Thesis, 1968, Polytechnic Institute of Brooklyn.

[3] Atal, B.S. and L. Hanauer, "Speech Analysis and Synthesis by Linear Prediction of the Speech Wave," *Journal of the Acoustical Society of America*, 1971, **50**, pp. 637-655.

[4] Atal, B.S. and M.R. Schroeder, "Predictive Coding of Speech Signals," *Report of the 6th Int. Congress on Acoustics*, 1968, Tokyo, Japan.

[5] Berouti, M.G., D.G. Childers, and A. Paige, "Glottal Area versus Glottal Volume Velocity," *Int. Conf. on Acoustics, Speech and Signal Processing*, 1977, Hartford, Conn, pp. 33-36.

[6] Bogert, B., M. Healy, and J. Tukey, "The Quefrency Alanysis of Time Series for Echoes," *Proc. Symp. on Time Series Analysis*, 1963, New York, J. Wiley, pp. 209-243.

[7] Burg, J., "Maximum Entropy Spectral Analysis," *Proc. of the 37th Meeting of the Society of Exploration Geophysicists*, 1967.

[8] Davis, S. and P. Mermelstein, "Comparison of Parametric Representations for Monosyllable Word Recognition in Continuously Spoken Sentences," *IEEE Trans. on Acoustics, Speech and Signal Processing*, 1980, **28**(4), pp. 357-366.

[9] Deller, J.R., J.H.L. Hansen, and J.G. Proakis, *Discrete-Time Processing of Speech Signals*, 2000, IEEE Press.

[10] Droppo, J. and A. Acero, "Maximum a Posteriori Pitch Tracking," *Int. Conf. on Spoken Language Processing*, 1998, Sydney, Australia, pp. 943-946.

[11] Fano, R.M., "Short-time Autocorrelation Functions and Power Spectra," *Journal of the Acoustical Society of America*, 1950, **22**(Sep), pp. 546-550.

[12] Fant, G., "On the Predictability of Formant Levels and Spectrum Envelopes from Formant Frequencies" in *For Roman Jakobson*, M. Halle, ed., 1956, The Hague, NL, Mouton & Co., pp. 109-120.

[13] Fant, G., *Acoustic Theory of Speech Production*, 1970, The Hague, NL, Mouton.

[14] Flanagan, J., *Speech Analysis Synthesis and Perception*, 1972, New York, Springer-Verlag.

[15] Gold, B., "Computer Program for Pitch Extraction," *Journal of the Acoustical Society of America*, 1962, **34**(7), pp. 916-921.

[16] Hermansky, H., "Perceptual Linear Predictive (PLP) Analysis of Speech," *Journal of the Acoustical Society of America*, 1990, **87**(4), pp. 1738-1752.

[17] Hess, W., *Pitch Determination of Speech Signals*, 1983, New York, Springer-Verlag.

[18] Itakura, F., "Line Spectrum Representation of Linear Predictive Coefficients," *Journal of the Acoustical Society of America*, 1975, **57**(4), pp. 535.

[19] Itakura, F. and S. Saito, "Analysis Synthesis Telephony Based on the Maximum Likelihood Method," *Proc. 6th Int. Congress on Acoustics*, 1968, Tokyo, Japan.

[20] Itakura, F. and S. Saito, "A Statistical Method for Estimation of Speech Spectral Density and Formant Frequencies," *Elec. and Comm. in Japan*, 1970, **53-A**(1), pp. 36-43.

[21] Kelly, J.L. and L.J. Gerstman, "An Artificial Talker Driven From Phonetic Input," *Journal of Acoustical Society of America*, 1961, **33**, pp. 835.

[22] Koenig, R., H.K. Dunn, and L.Y. Lacy, "The Sound Spectrograph," *Journal of the Acoustical Society of America*, 1946, **18**, pp. 19-49.

[23] Krishnamurthy, A.K. and D.G. Childers, "Two Channel Speech Analysis," *IEEE Trans. on Acoustics, Speech and Signal Processing*, 1986, **34**, pp. 730-743.

[24] Makhoul, J., "Spectral Analysis of Speech by Linear Prediction," *IEEE Trans. on Acoustics, Speech and Signal Processing*, 1973, **21**(3), pp. 140-148.

[25] Malvar, H., *Signal Processing with Lapped Transforms*, 1992, Artech House.

[26] Markel, J.D., "Digital Inverse Filtering—A New Tool for Formant Trajectory Estimation," *IEEE Trans. on Audio and Electroacoustics*, 1972, **AU-20**(June), pp. 129-137.

[27] Markel, J.D. and A.H. Gray, "On Autocorrelation Equations as Applied to Speech Analysis," *IEEE Trans. on Audio and Electroacoustics*, 1973, **AU-21**(April), pp. 69-79.

[28] Noll, A.M., "Cepstrum Pitch Determination," *Journal of the Acoustical Society of America*, 1967, **41**, pp. 293-309.

[29] Oppenheim, A.V., *Superposition in a Class of Nonlinear Systems*, 1965, Research Lab. of Electronics, MIT, Cambridge, Massachusetts.

[30] Oppenheim, A.V., "Speech Spectrograms Using the Fast Fourier Transform," *IEEE Spectrum*, 1970, **7**(Aug), pp. 57-62.

[31] Oppenheim, A.V. and D.H. Johnson, "Discrete Representation of Signals," *The Proc. of the IEEE*, 1972, **60**(June), pp. 681-691.

[32] Oppenheim, A.V., R.W. Schafer, and T.G. Stockham, "Nonlinear Filtering of Multiplied and Convolved Signals," *Proc. of the IEEE*, 1968, **56**, pp. 1264-1291.

[33] O'Shaughnessy, D., *Speech Communication—Human and Machine*, 1987, Addison-Wesley.

[34] Peterson, G.E. and H.L. Barney, "Control Methods Used in a Study of the Vowels," *Journal of the Acoustical Society of America*, 1952, **24**(2), pp. 175-184.

[35] Potter, R.K., G.A. Kopp, and H.C. Green, *Visible Speech*, 1947, New York, D. Van Nostrand Co. Republished by Dover Publications, Inc., 1966.

[36] Press, W.H., *et al.*, *Numerical Recipes in C*, 1988, New York, NY, Cambridge University Press.

[37] Rabiner, L.R., "On the Use of Autocorrelation Analysis for Pitch Detection," *IEEE Trans. on Acoustics, Speech and Signal Processing*, 1977, **25**, pp. 24-33.

[38] Rabiner, L.R. and R.W. Schafer, *Digital Processing of Speech Signals*, 1978, Englewood Cliffs, NJ, Prentice-Hall.

[39] Rosenberg, A.E., "Effect of Glottal Pulse Shape on the Quality of Natural Vowels," *Journal of the Acoustical Society of America*, 1971, **49**, pp. 583-590.

[40] Schafer, R.W. and L.R. Rabiner, "System for Automatic Formant Analysis of Voiced Speech," *Journal of the Acoustical Society of America*, 1970, **47**, pp. 634-678.

[41] Schroeder, M., "Period Histogram and Product Spectrum: New Methods for Fundamental Frequency Measurement," *Journal of the Acoustical Society of America*, 1968, **43**(4), pp. 829-834.

[42] Schroeder, M.R. and B.S. Atal, "Generalized Short-Time Power Spectra and Autocorrelation," *Journal of the Acoustical Society of America*, 1962, **34**(Nov), pp. 1679-1683.

[43] Shikano, K., K.-F. Lee, and R. Reddy, "Speaker Adaptation through Vector Quantization," *IEEE Int. Conf. on Acoustics, Speech and Signal Processing*, 1986, Tokyo, Japan, pp. 2643-2646.

[44] Sondhi, M.M., "New Methods for Pitch Extraction," *IEEE Trans. on Audio and Electroacoustics*, 1968, **16**(June), pp. 262-268.

[45] Talkin, D., "A Robust Algorithm for Pitch Tracking" in *Speech Coding and Synthesis*, W.B. Kleijn and K.K. Paliwal, eds., 1995, Amsterdam, Elsevier, pp. 485-518, .

[46] Yegnanarayana, B. and R.N.J. Veldhuis, "Extraction of Vocal-Tract System Characteristics from Speech Signals," *IEEE Trans. on Speech and Audio Processing*, 1998, **6**(July), pp. 313-327.

[47] Zwicker, E., "Subdivision of the Audible Frequency Range into Critical Bands," *Journal of the Acoustical Society of America*, 1961, **33**(Feb), p. 248.

CHAPTER 7

Speech Coding

Transmission of speech using data networks requires the speech signal to be digitally encoded. Voice over IP has become very popular because of the Internet, where bandwidth limitations make it necessary to compress the speech signal. Digital storage of audio signals, which can result in higher quality and smaller size than the analog counterpart, is commonplace in compact discs, digital video discs, and MP3 files. Many spoken language systems also use coded speech for efficient communication. For these reasons we devote a chapter to speech and audio coding techniques.

Rather than exhaustively cover all the existing speech and audio coding algorithms, we uncover their underlying technology and enumerate some of the most popular standards. The coding technology discussed in this chapter has a strong link to both speech recognition and speech synthesis. For example, the speech synthesis algorithms described in Chapter 16 use many techniques described here.

7.1. SPEECH CODERS ATTRIBUTES

How do we compare different speech or audio coders? We can refer to a number of factors, such as signal bandwidth, bit rate, quality of reconstructed speech, noise robustness, computational complexity, delay, channel-error sensitivity, and standards.

Speech signals can be bandlimited to 10 kHz without significantly affecting the hearer's perception. The telephone network limits the bandwidth of speech signals to between 300 and 3400 Hz, which gives *telephone speech* a lower quality. Telephone speech is typically sampled at 8 kHz. The term *wideband speech* is used for a bandwidth of 50–7000 Hz and a sampling rate of 16 kHz. Finally, *audio coding* is used in dealing with high-fidelity audio signals, in which case the signal is sampled at 44.1 kHz.

Reduction in bit rate is the primary purpose of speech coding. The previous bit stream can be compressed to a lower rate by removing redundancy in the signal, resulting in savings in storage and transmission bandwidth. If only redundancy is removed, the original signal can be recovered exactly (*lossless* compression). In *lossy* compression, the signal cannot be recovered exactly, though hopefully it will sound similar to the original.

Depending on system and design constraints, fixed-rate or variable-rate speech coders can be used. Variable-rate coders are used for non-real time applications, such as voice storage (silence can be coded with fewer bits than fricatives, which in turn use fewer bits than vowels), or for packet voice transmissions, such as CDMA cellular for better channel utilization. Transmission of coded speech through a noisy channel may require devoting more bits to channel coding and fewer to source coding. For most real-time communication systems, a maximum bit rate is specified.

The quality of the reconstructed speech signal is a fundamental attribute of a speech coder. Bit rate and quality are intimately related: the lower the bit rate, the lower the quality. While the bit rate is inherently a number, it is difficult to quantify the quality. The most widely used measure of quality is the *Mean Opinion Score* (MOS) [25], which is the result of averaging opinion scores for a set of between 20 and 60 untrained subjects. Each listener characterizes each set of utterances with a score on a scale from 1 (unacceptable quality) to 5 (excellent quality), as shown in Table 7.1. An MOS of 4.0 or higher defines *good* or *toll* quality, where the reconstructed speech signal is generally indistinguishable from the original signal. An MOS between 3.5 and 4.0 defines *communication* quality, which is sufficient for telephone communications. We show in Section 7.2.1 that if each sample is quantized with 16 bits, the resulting signal has *toll* quality (essentially indistinguishable from the unquantized signal). See Chapter 16 for more details on perceptual quality measurements.

Table 7.1 Mean Opinion Score (MOS) is a numeric value computed as an average for a number of subjects, where each number maps to the above subjective quality.

Excellent	Good	Fair	Poor	Bad
5	4	3	2	1

Another measure of quality is the *signal-to-noise ratio* (SNR), defined as the ratio between the signal's energy and the noise's energy in terms of dB:

$$SNR = \frac{\sigma_x^2}{\sigma_e^2} = \frac{E\{x^2[n]\}}{E\{e^2[n]\}} \tag{7.1}$$

The MOS rating of a codec on noise-free speech is often higher than its MOS rating for noisy speech. This is generally caused by specific assumptions in the speech coder that tend to be violated when a significant amount of noise is present in the signal. This phenomenon is more accentuated for lower-bit-rate coders that need to make more assumptions.

The computational complexity and memory requirements of a speech coder determine the cost and power consumption of the hardware on which it is implemented. In most cases, real-time operation is required at least for the decoder. Speech coders can be implemented in inexpensive *Digital Signal Processors* (DSP) that form part of many consumer devices, such as answering machines and DVD players, for which storage tends to be relatively more expensive than processing power. DSPs are also used in cellular phones because bit rates are limited.

All speech coders have some delay, which, if excessive, can affect the dynamics of a two-way communication. For instance, delays over 150 ms can be unacceptable for highly interactive conversations. Coder delay is the sum of different types of delay. The first is the *algorithmic delay* arising because speech coders usually operate on a block of samples, called a *frame,* which needs to be accumulated before processing can begin. Often the speech coder requires some additional *look-ahead* beyond the frame to be encoded. The *computational delay* is the time that the speech coder takes to process the frame. For real-time operation, the computational delay has to be smaller than the algorithmic delay. A block of bits is generally assembled by the encoder prior to transmission, possibly to add error-correction properties to the bit stream, which cause *multiplexing delay.* Finally, there is the *transmission delay,* due to the time it takes for the frame to traverse the channel. The decoder will incur a *decoder delay* to reconstruct the signal. In practice, the total delay of many speech coders is at least three frames.

If the coded speech needs to be transmitted over a channel, we need to consider possible channel errors, and our speech decoder should be insensitive to at least some of them. There are two types of errors: random errors and burst errors, and they could be handled differently. One possibility to increase the robustness against such errors is to use channel coding techniques, such as those proposed in Chapter 3. Joint source and channel coding allows us to find the right combination of bits to devote to speech coding with the right amount devoted to channel coding, adjusting this ratio adaptively depending on the channel. Since channel coding will only reduce the number of errors, and not eliminate them, graceful degradation of speech quality under channel errors is typically a design factor for speech coders. When the channel is the Internet, complete frames may be missing because they have not arrived in time. Therefore, we need techniques that degrade gracefully with missing frames.

7.2. Scalar Waveform Coders

In this section we describe several waveform coding techniques, such as linear PCM, μ-law, and A-law PCM, APCM, DPCM, DM, and ADPCM, that quantize each sample using scalar quantization. These techniques attempt to approximate the waveform, and, if a large enough bit rate is available, will get arbitrarily close to it.

7.2.1. Linear Pulse Code Modulation (PCM)

Analog-to-digital converters perform both sampling and quantization simultaneously. To better understand how this process affects the signal it's better to study them separately. We analyzed the effects of sampling in Chapter 5, so now we analyze the effects of quantization, which encodes each sample with a fixed number of bits. With B bits, it is possible to represent 2^B separate quantization levels. The output of the quantizer $\hat{x}[n]$ is given by

$$\hat{x}[n] = Q\{x[n]\} \tag{7.2}$$

Linear *Pulse Code Modulation* (PCM) is based on the assumption that the input discrete signal $x[n]$ is bounded

$$|x[n]| \le X_{\max} \tag{7.3}$$

and that we use *uniform quantization* with quantization step size Δ which is constant for all levels x_i

$$x_i - x_{i-1} = \Delta \tag{7.4}$$

The input/output characteristics are shown by Figure 7.1 for the case of a 3-bit uniform quantizer. The so-called *mid-riser* quantizer has the same number of positive and negative levels, whereas the *mid-tread* quantizer has one more negative than positive levels. The code $c[n]$ is expressed in two's complement representation, which for Figure 7.1 varies between –4 and +3. For the mid-riser quantizer the output $\hat{x}[n]$ can be obtained from the code $c[n]$ through

$$\hat{x}[n] = sign(c[n])\frac{\Delta}{2} + c[n]\Delta \tag{7.5}$$

and for the mid-tread quantizer

$$\hat{x}[n] = c[n]\Delta \tag{7.6}$$

which is often used in computer systems that use two's complement representation.

There are two independent parameters for a uniform quantizer: the number of levels $N = 2^B$, and the step size Δ. Assuming Eq. (7.3), we have the relationship

$$2X_{\max} = \Delta 2^B \tag{7.7}$$

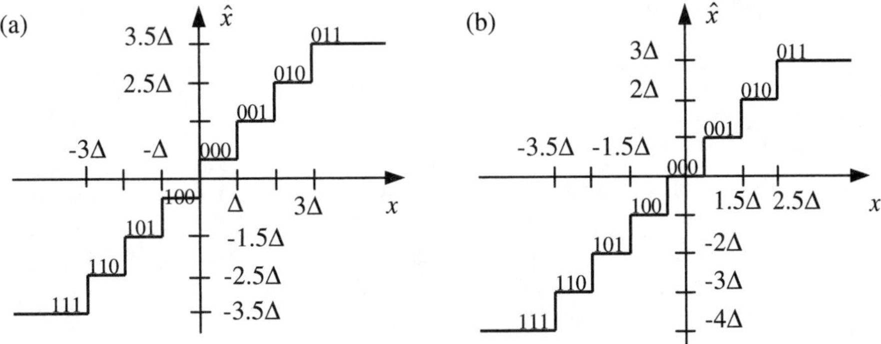

Figure 7.1 Three-bit uniform quantization characteristics: (a) mid-riser, (b) mid-tread.

In quantization, it is useful to express the relationship between the unquantized sample $x[n]$ and the quantized sample $\hat{x}[n]$ as

$$\hat{x}[n] = x[n] + e[n] \tag{7.8}$$

with $e[n]$ being the quantization noise. If we choose Δ and B to satisfy Eq. (7.7), then

$$-\frac{\Delta}{2} \le e[n] \le \frac{\Delta}{2} \tag{7.9}$$

While there is obviously a deterministic relationship between $e[n]$ and $x[n]$, it is convenient to assume a probabilistic model for the quantization noise:

1. $e[n]$ is white: $E\{e[n]e[n+m]\} = \sigma_e^2 \delta[m]$
2. $e[n]$ and $x[n]$ are uncorrelated: $E\{x[n]e[n+m]\} = 0$
3. $e[n]$ is uniformly distributed in the interval $(-\Delta/2, \Delta/2)$

These assumptions are unrealistic for some signals, except in the case of speech signals, which rapidly fluctuate between different quantization levels. The assumptions are reasonable if the step size Δ is small enough, or alternatively the number of levels is large enough (say, more than 2^6).

The variance of such uniform distribution (see Chapter 3) is

$$\sigma_e^2 = \frac{\Delta^2}{12} = \frac{X_{\max}^2}{3 \times 2^{2B}} \tag{7.10}$$

after using Eq. (7.7). The SNR is given by

$$SNR(dB) = 10\log_{10}\left(\frac{\sigma_x^2}{\sigma_e^2}\right) = (20\log_{10} 2)B + 10\log_{10} 3 - 20\log_{10}\left(\frac{X_{\max}}{\sigma_x}\right) \tag{7.11}$$

which implies that each bit contributes to 6 dB of *SNR*, since $20\log_{10} 2 \cong 6$.

Speech samples can be approximately described as following a *Laplacian* distribution [40]

$$p(x) = \frac{1}{\sqrt{2}\sigma_x} e^{-\frac{\sqrt{2}|x|}{\sigma_x}} \tag{7.12}$$

and the probability of x falling outside the range $(-4\sigma_x, 4\sigma_x)$ is 0.35%. Thus, using $X_{\max} = 4\sigma_x$, $B = 7$ bits in Eq. (7.11) results in an *SNR* of 35 dB, which would be acceptable in a communications system. Unfortunately, signal energy can vary over 40 dB, due to variability from speaker to speaker as well as variability in transmission channels. Thus, in practice, it is generally accepted that 11 bits are needed to achieve an SNR of 35 dB while keeping the clipping to a minimum.

Digital audio stored in computers (Windows WAV, Apple AIF, Sun AU, and SND formats among others) use 16-bit linear PCM as their main format. The *Compact Disc-Digital Audio* (CD-DA or simply CD) also uses 16-bit linear PCM. Invented in the late 1960s by James T. Russell, it was launched commercially in 1982 and has become one of the most successful examples of consumer electronics technology: there were about 700 million audio CD players in 1997. A CD can store up to 74 minutes of music, so the total amount of digital data that must be stored on a CD is 44,100 samples/(channel*second) * 2 bytes/sample * 2 channels * 60 seconds/minute * 74 minutes = 783,216,000 bytes. This 747 MB are stored in a disk only 12 centimeters in diameter and 1.2 mm thick. CD-ROMs can record only 650 MB of computer data because they use the remaining bits for error correction.

7.2.2. µ-law and A-law PCM

Human perception is affected by SNR, because adding noise to a signal is not as noticeable if the signal energy is large enough. Ideally, we want SNR to be constant for all quantization levels, which requires the step size to be proportional to the signal value. This can be done by using a logarithmic *compander*[1]

$$y[n] = \ln|x[n]| \tag{7.13}$$

followed by a uniform quantizer on $y[n]$ so that

$$\hat{y}[n] = y[n] + \varepsilon[n] \tag{7.14}$$

and, thus,

$$\hat{x}[n] = \exp\{\hat{y}[n]\}\text{sign}\{x[n]\} = x[n]\exp\{\varepsilon[n]\} \tag{7.15}$$

[1] A compander is a nonlinear function that compands one part of the x-axis.

after using Eqs. (7.13) and (7.14). If $\varepsilon[n]$ is small, then Eq. (7.15) can be expressed as

$$\hat{x}[n] \cong x[n](1+\varepsilon[n]) = x[n] + x[n]\varepsilon[n] \tag{7.16}$$

and, thus, the $SNR = 1/\sigma_\varepsilon^2$ is constant for all levels. This type of quantization is not practical, because an infinite number of quantization steps would be required. An approximation is the so-called μ-law [51]:

$$y[n] = X_{\max} \frac{\log\left[1+\mu\dfrac{|x[n]|}{X_{\max}}\right]}{\log[1+\mu]} \operatorname{sign}\{x[n]\} \tag{7.17}$$

which is approximately logarithmic for large values of $x[n]$ and approximately linear for small values of $x[n]$. A related compander called A-law is also used

$$y[n] = X_{\max} \frac{1+\log\left[\dfrac{A|x[n]|}{X_{\max}}\right]}{1+\log A} \operatorname{sign}\{x[n]\} \tag{7.18}$$

which has greater resolution than μ-law for small sample values, but a range equivalent to 12 bits. In practice, they both offer similar quality. The μ-law curve can be seen in Figure 7.2.

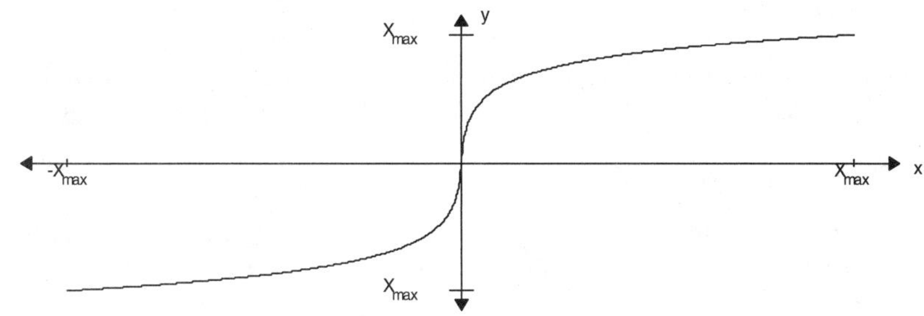

Figure 7.2 Nonlinearity used in the μ-law compression.

In 1972 the ITU-T[2] recommendation G.711 standardized telephone speech coding at 64 kbps for digital transmission of speech through telephone networks. It uses 8 bits per sample and an 8-kHz sampling rate with either μ-law or A-law. In North America and Japan, μ-law with μ = 255 is used, whereas, in the rest of the world, A-law with A = 87.56 is used. Both compression characteristics are very similar and result in an approximate SNR of 35 dB. Without the logarithmic compressor, a uniform quantizer requires approximately 12 bits

[2] The International Telecommunication Union (ITU) is a part of the United Nations Economic, Scientific and Cultural Organization (UNESCO). ITU-T is the organization within ITU responsible for setting global telecommunication standards. Within ITU-T, Study Group 15 (SG15) is responsible for formulating speech coding standards. Prior to 1993, telecommunication standards were set by the *Comité Consultatif International Téléphonique et Télégraphique* (CCITT), which was reorganized into the ITU-T that year.

per sample to achieve the same level of quality. All the speech coders for telephone speech described in this chapter use G.711 as a baseline reference, whose quality is considered *toll*, and an MOS of about 4.3. G.711 is used by most digital central office switches, so that when you make a telephone call using your plain old telephone service (POTS), your call is encoded with G.711.

7.2.3. Adaptive PCM

When quantizing speech signals we confront a dilemma. On the one hand, we want the quantization step size to be large enough to accommodate the maximum peak-to-peak range of the signal and avoid clipping. On the other hand, we need to make the step size small to minimize the quantization noise. One possible solution is to adapt the step size to the level of the input signal.

The basic idea behind *Adaptive PCM* (APCM) is to let the step size $\Delta[n]$ be proportional to the standard deviation of the signal $\sigma[n]$:

$$\Delta[n] = \Delta_0 \sigma[n] \tag{7.19}$$

An equivalent method is to use a fixed quantizer but have a time-varying gain $G[n]$, which is inversely proportional to the signal's standard deviation

$$G[n] = G_0 / \sigma[n] \tag{7.20}$$

Estimation of the signal's variance, or short-time energy, is typically done by low-pass filtering $x^2[n]$. With a first-order IIR filter, the variance $\sigma^2[n]$ is computed as

$$\sigma^2[n] = \alpha\sigma^2[n-1] + (1-\alpha)x^2[n-1] \tag{7.21}$$

with α controlling the time constant of the filter $T = -1/(F_s \ln\alpha)$, F_s the sampling rate, and $0 < \alpha < 1$. In practice, α is chosen so that the time constant ranges between 1 ms ($\alpha = 0.88$ at 8 kHz) and 10 ms ($\alpha = 0.987$ at 8 kHz).

Alternatively, $\sigma^2[n]$ can be estimated from the past M samples:

$$\sigma^2[n] = \frac{1}{M}\sum_{m=n-M}^{n-1} x^2[m] \tag{7.22}$$

In practice, it is advantageous to set limits on the range of values of $\Delta[n]$ and $G[n]$:

$$\Delta_{\min} \le \Delta[n] \le \Delta_{\max} \tag{7.23}$$

$$G_{\min} \le G[n] \le G_{\max} \tag{7.24}$$

with the ratios $\Delta_{\max}/\Delta_{\min}$ and $G_{\max}/G_{\min}$ determining the dynamic range of the system. If our objective is to obtain a relatively constant SNR over a range of 40 dB, these ratios can be 100.

Feedforward adaptation schemes require us to transmit, in addition to the quantized signal, either the step size $\Delta[n]$ or the gain $G[n]$. Because these values evolve slowly with time, they can be sampled and quantized at a low rate. The overall rate will be the sum of the bit rate required to transmit the quantized signal plus the bit rate required to transmit either the gain or the step size.

Another class of adaptive quantizers use *feedback adaptation* to avoid having to send information about the step size or gain. In this case, the step size and gain are estimated from the quantizer output, so that they can be recreated at the decoder without any extra information. The corresponding short-time energy can then be estimated through a first-order IIR filter as in Eq. (7.21) or a rectangular window as in Eq. (7.22), but replacing $x^2[n]$ by $\hat{x}^2[n]$.

Another option is to adapt the step size

$$\Delta[n] = P\Delta[n-1] \tag{7.25}$$

where $P > 1$ if the previous codeword corresponds to the largest positive or negative quantizer level, and $P < 1$ if the previous codeword corresponds to the smallest positive or negative quantizer level. A similar process can be done for the gain.

APCM exhibits an improvement between 4–8 dB over μ-law PCM for the same bit rate.

7.2.4. Differential Quantization

Speech coding is about finding redundancy in the signal and removing it. We know that there is considerable correlation between adjacent samples, because on the average the signal doesn't change rapidly from sample to sample. A simple way of capturing this is to quantize the difference $d[n]$ between the current sample $x[n]$ and its predicted value $\tilde{x}[n]$

$$d[n] = x[n] - \tilde{x}[n] \tag{7.26}$$

with its quantized value represented as

$$\hat{d}[n] = Q\{d[n]\} = d[n] + e[n] \tag{7.27}$$

where $e[n]$ is the quantization error. Then, the quantized signal is the sum of the predicted signal $\tilde{x}[n]$ and the quantized difference $\hat{d}[n]$

$$\hat{x}[n] = \tilde{x}[n] + \hat{d}[n] = x[n] + e[n] \tag{7.28}$$

If the prediction is good, Eq. (7.28) tells us that the quantization error will be small. Statistically, we need the variance of $e[n]$ to be lower than that of $x[n]$ for differential coding to provide any gain. Systems of this type are generically called *Differential Pulse Code Modulation* (DPCM) [11] and can be seen in Figure 7.3.

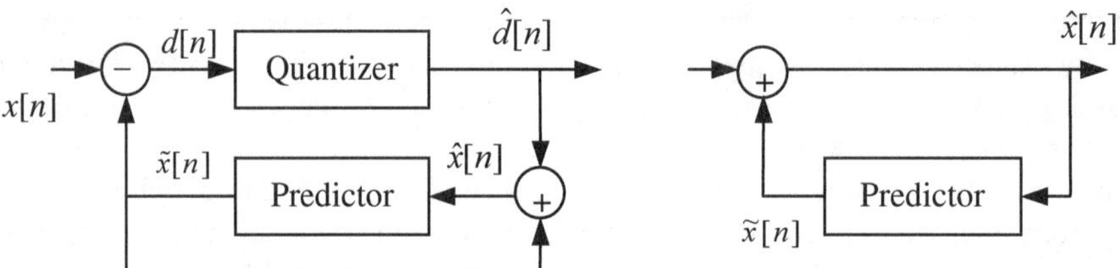

Figure 7.3 Block diagram of a DPCM encoder and decoder with feedback prediction.

Delta Modulation (DM) [47] is a 1-bit DPCM, which predicts the current sample to be the same as the past sample:

$$\tilde{x}[n] = x[n-1] \tag{7.29}$$

so that we transmit whether the current sample is above or below the previous sample.

$$d[n] = \begin{cases} \Delta & x[n] > x[n-1] \\ -\Delta & x[n] \le x[n-1] \end{cases} \tag{7.30}$$

with Δ being the step size. If Δ is too small, the reconstructed signal will not increase as fast as the original signal, a condition known as *slope overload distortion*. When the slope is small, the step size Δ also determines the peak error; this is known as *granular noise*. Both quantization errors can be seen in Figure 7.4. The choice of Δ that minimizes the mean squared error will be a tradeoff between slope overload and granular noise.

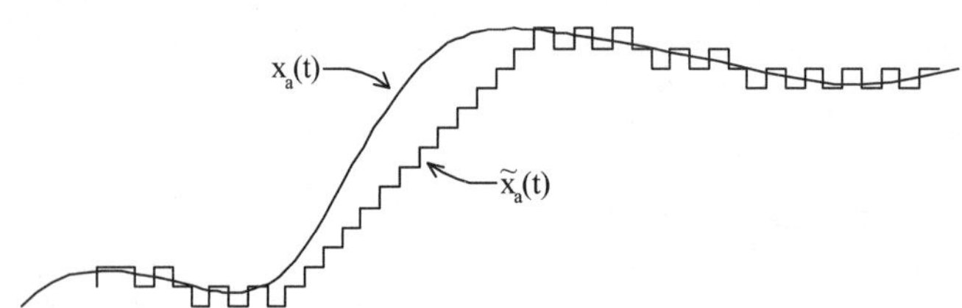

Figure 7.4 An example of slope overload distortion and granular noise in a DM encoder.

If the signal is oversampled by a factor *N*, and the step size is reduced by the same amount (i.e., Δ/N), the slope overload will be the same, but the granular noise will decrease by a factor *N*. While the coder is indeed very simple, sampling rates of over 200 kbps are needed for SNRs comparable to PCM, so DM is rarely used as a speech coder.

However, delta modulation is useful in the design of analog-digital converters, in a variant called sigma-delta modulation [44] shown in Figure 7.5. First the signal is lowpass filtered with a simple analog filter, and then it is oversampled. Whenever the predicted signal $\tilde{x}[n]$ is below the original signal $x[n]$, the difference $d[n]$ is positive. This difference $d[n]$

is averaged over time with a digital integrator whose output is $e[n]$. If this situation persists, the accumulated error $e[n]$ will exceed a positive value A, which causes a 1 to be encoded into the stream $q[n]$. A digital-analog converter is used in the loop which increments by one the value of the predicted signal $\tilde{x}[n]$. The system acts in the opposite way if the predicted signal $\tilde{x}[n]$ is above the original signal $x[n]$ for an extended period of time. Since the signal is oversampled, it changes very slowly from one sample to the next, and this quantization can be accurate. The advantages of this technique as an analog-digital converter are that inexpensive analog filters can be used and only a simple 1-bit A/D is needed. The signal can next be low-passed filtered with a more accurate digital filter and then downsampled.

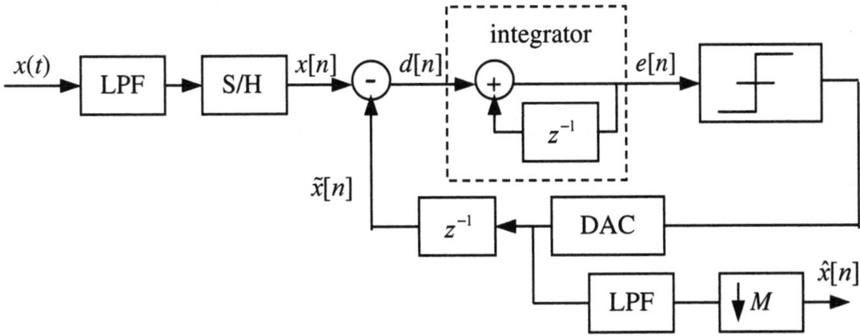

Figure 7.5 A sigma-delta modulator used in an oversampling analog-digital converter.

Adaptive Delta Modulation (ADM) combines ideas from adaptive quantization and delta modulation with the so-called *Continuously Variable Slope Delta Modulation* (CVSDM) [22] having a step size that increases

$$\Delta[n] = \begin{cases} \alpha\Delta[n-1] + k_1 & \text{if } e[n], e[n-1] \text{ and } e[n-2] \text{ have same sign} \\ \alpha\Delta[n-1] + k_2 & \text{otherwise} \end{cases} \tag{7.31}$$

with $0 < \alpha < 1$ and $0 < k_2 << k_1$. The step size increases if the last three errors have the same sign and decreases otherwise.

Improved DPCM is achieved through linear prediction in which $\tilde{x}[n]$ is a linear combination of past quantized values $\hat{x}[n]$

$$\tilde{x}[n] = \sum_{k=1}^{p} a_k \hat{x}[n-k] \tag{7.32}$$

DPCM systems with fixed prediction coefficients can provide from 4 to 11 dB improvement over direct linear PCM, for prediction orders up to $p = 4$, at the expense of increased computational complexity. Larger improvements can be obtained by adapting the prediction coefficients. The coefficients can be transmitted in a feedforward fashion or not transmitted if the feedback scheme is selected.

ADPCM [6] combines differential quantization with adaptive step-size quantization. ITU-T Recommendation G.726 uses ADPCM at bit rates of 40, 32, 24, and 16 kbps, with 5, 4, 3, and 2 bits per sample, respectively. It employs an adaptive feedback quantizer and an adaptive feedback pole-zero predictor. Speech at bit rates of 40 and 32 kbps offer toll quality, while the other rates don't. G.727 is called embedded ADPCM because the 2-bit quantizer is embedded into the 3-bit quantizer, which is embedded into the 4-bit quantizer, and into the 5-bit quantizer. This makes it possible for the same codec to use a lower bit rate, with a graceful degradation in quality, if channel capacity is temporarily limited. Earlier standards G.721 [7, 13] (created in 1984) and G.723 have been subsumed by G.726 and G.727. G.727 has a MOS of 4.1 for 32 kbps and is used in submarine cables. The Windows WAV format also supports a variant of ADPCM. These standards are shown in Table 7.2.

Table 7.2 Common scalar waveform standards used.

Standard	Bit Rate (kbits/sec)	MOS	Algorithm	Sampling Rate (kHz)
Stereo CD Audio	1411	5.0	16-bit linear PCM	44.1
WAV, AIFF, SND	Variable	-	16/8-bit linear PCM	8, 11.025, 16, 22.05, 44.1, 48
G.711	64	4.3	µ-law/A-law PCM	8
G.727	40, 32, 24, 16	4.2 (32k)	ADPCM	8
G.722	64, 56, 48		Subband ADPCM	16

Wideband speech (50–7000 Hz) increases intelligibility of fricatives and overall perceived quality. In addition, it provides more subject presence and adds a feeling of transparent communication. ITU-T Recommendation G.722 encodes wideband speech with bit rates of 48, 56, and 64 kbps. Speech is divided into two subbands with QMF filters (see Chapter 5). The upper band is encoded using a 16-kbps ADPCM similar to the G.727 standard. The lower band is encoded using a 48-kbps ADPCM with the 4- and 5-bit quantizers embedded in the 6-bit quantizer. The quality of this system scores almost 1 MOS higher than that of telephone speech.

7.3. SCALAR FREQUENCY DOMAIN CODERS

Frequency domain is advantageous because:

1. The samples of a speech signal have a great deal of correlation among them, whereas frequency domain components are approximately uncorrelated and

2. The perceptual effects of masking described in Chapter 2 can be more easily implemented in the frequency domain. These effects are more pronounced for high-bandwidth signals, so frequency-domain coding has been mostly used for CD-quality signals and not for 8-kHz speech signals.

7.3.1. Benefits of Masking

As discussed in Chapter 2, masking is a phenomenon by which human listeners cannot perceive a sound if it is below a certain level. The consequence is that we don't need to encode such sound. We now illustrate how this masked threshold is computed for MPEG[3]-1 layer 1. Given an input signal $s[n]$ quantized with b bits, we obtain the normalized signal $x[n]$ as

$$x[n] = \frac{s[n]}{N2^{b-1}} \tag{7.33}$$

where $N = 512$ is the length of the DFT. Then, using a Hanning window,

$$w[n] = 0.5 - 0.5\cos(2\pi n / N) \tag{7.34}$$

we obtain the log-power spectrum as

$$P[k] = P_0 + 10\log_{10}\left|\sum_{n=0}^{N-1} w[n]x[n]e^{-j2\pi nk/N}\right|^2 \tag{7.35}$$

where P_0 is the playback SPL, which, in the absence of any volume information, is defined as 90 dB.

Tonal components are identified in Eq. (7.35) as local maxima, which exceed neighboring components within a certain Bark distance by at least 7 dB. Specifically, bin k is tonal if and only if

$$P[k] > P[k \pm 1] \tag{7.36}$$

and

$$P[k] > P[k \pm l] + 7dB \tag{7.37}$$

where $1 < l \le \Delta_k$, and Δ_k is given by

$$\Delta_k = \begin{cases} 2 & 2 < k < 63 \quad (170\,\text{Hz} - 5.5\,\text{kHz}) \\ 3 & 63 \le k < 127 \quad (5.5\,\text{kHz}, 11\,\text{kHz}) \\ 6 & 127 \le k \le 256 \quad (11\,\text{kHz}, 22\,\text{kHz}) \end{cases} \tag{7.38}$$

[3] MPEG (Moving Picture Experts Group) is the nickname given to a family of International Standards for coding audiovisual information.

so that the power of that tonal masker is computed as the sum of the power in that bin and its left and right adjacent bins:

$$P_{TM}[k] = 10\log_{10}\left(\sum_{j=-1}^{j} 10^{0.1P[k+j]}\right) \tag{7.39}$$

The noise maskers are computed as the sum of power spectrum of the remaining frequency bins $\bar{k}$ in a critical band not within a neighborhood Δ_k of the tonal maskers:

$$P_{NM}[\bar{k}] = 10\log_{10}\left(\sum_{j} 10^{0.1P[j]}\right) \tag{7.40}$$

where j spans a critical band.

To compute the overall masked threshold we need to sum all masking thresholds contributed by each frequency bin i, which is approximately equal to the maximum (see Chapter 2):

$$T[k] = \max\left(T_h[k], \max_i\left(T_i[k]\right)\right) \tag{7.41}$$

In Chapter 2 we saw that whereas temporal postmasking can last from 50 to 300 ms, temporal premasking tends to last about 5 ms. This is also important because when a frequency transform is quantized, the blocking effects of transform's coders can introduce noise above the temporal premasking level that can be audible, since 1024 points corresponds to 23 ms at a 44-kHz sampling rate. To remove this pre-echo distortion, audible in the presence of castanets and other abrupt transient signals, subband filtering has been proposed, whose time constants are well below the 5-ms premasking time constant.

7.3.2. Transform Coders

We now use the *Adaptive Spectral Entropy Coding* (ASPEC) of High Quality Music Signals algorithm, which is the basis for the MPEG1 Layer 1 audio coding standard [24], to illustrate how transform coders work. The DFT coefficients are grouped into 128 subbands, and 128 scalar quantizers are used to transmit all the DFT coefficients. It has been empirically found that a difference of less than 1 dB between the original amplitude and the quantized value cannot be perceived. Each subband j has a quantizer having k_j levels and step size of T_j as

$$k_j = 1 + 2 \times \text{rnd}\left(P_j / T_j\right) \tag{7.42}$$

where T_j is the quantized JND threshold, P_j is the quantized magnitude of the largest real or imaginary component of the jth subband, and rnd( ) is the nearest integer rounding function. Entropy coding (see Chapter 3) is used to encode the coefficients of that subband. Both

T_j and P_j are quantized on a dB scale using 8-bit uniform quantizers with a 170-dB dynamic range, thus with a step size of 0.66 dB. Then they are transmitted as side information.

There are two main methods of obtaining a frequency-domain representation:

1. Through subband filtering via a filterbank (see Chapter 5). When a filterbank is used, the bandwidth of each band is chosen to increase with frequency following a perceptual scale, such as the Bark scale. As shown in Chapter 5, such filterbanks yield perfect reconstruction in the absence of quantization.
2. Through frequency-domain transforms. Instead of using a DFT, higher efficiency can be obtained by the use of an MDCT (see Chapter 5).

The exact implementation of the MPEG1 Layer 1 standard is much more complicated and beyond the scope of this book, though it follows the main ideas described here; the same is true for the popular MPEG1 Layer III, also known as MP3. Implementation details can be found in [42].

7.3.3. Consumer Audio

Dolby Digital, MPEG, DTS, and the Perceptual Audio Coder (PAC) [28] are all audio coders based on frequency-domain coding. Except for MPEG-1, which supports only stereo signals, the rest support multichannel.

Dolby Digital is multichannel digital audio, using lossy AC-3 [54] coding technology from original PCM with a sample rate of 48 kHz at up to 24 bits. The bit rate varies from 64 to 448 kbps, with 384 being the normal rate for 5.1 channels and 192 the normal rate for stereo (with or without surround encoding). Most Dolby Digital decoders support up to 640 kbps. Dolby Digital is the format used for audio tracks on almost all Digital Video/Versatile Discs (DVD). A DVD-5 with only one surround stereo audio stream (at 192 kbps) can hold over 55 hours of audio. A DVD-18 can hold over 200 hours.

MPEG was established in 1988 as part of the joint ISO (International Standardization Organization) / IEC (International Electrotechnical Commission) Technical Committee on Information Technology. MPEG-1 was approved in 1992 and MPEG-2 in 1994. Layers I to III define several specifications that provide better quality at the expense of added complexity. MPEG-1 audio is limited to 384 kbps. MPEG1 Layer III audio [23], also known as MP3, is very popular on the Internet, and many compact players exist.

MPEG-2 audio, one of the audio formats used in DVD, is multichannel digital audio, using lossy compression from 16-bit linear PCM at 48 kHz. Tests have shown that for nearly all types of speech and music, at a data rate of 192 kbps and over, on a stereo channel, scarcely any difference between original and coded versions was observable (ranking of coded item > 4.5), with the original signal needing 1.4 Mbps on a CD (reduction by a factor of 7). One advantage of the MPEG audio technique is that future findings regarding psychoacoustic effects can be incorporated later, so it can be expected that today's quality level using 192 kbps will be achievable at lower data rates in the future. A variable bit rate of 32 to 912 kbps is supported for DVDs.

DTS (Digital Theater Systems) Digital Surround is another multi-channel (5.1) digital audio format, using lossy compression derived from 20-bit linear PCM at 48 kHz. The compressed data rate varies from 64 to 1536 kbps, with typical rates of 768 and 1536 kbps.

7.3.4. Digital Audio Broadcasting (DAB)

Digital Audio Broadcasting (DAB) is a means of providing current AM and FM listeners with a new service that offers: sound quality comparable to that of compact discs, increased service availability (especially for reception in moving vehicles), flexible coverage scenarios, and high spectrum efficiency.

Different approaches have been considered for providing listeners with such a service. Currently, the most advanced system is one commonly referred to as Eureka 147 DAB, which has been under development in Europe under the Eureka Project EU147 since 1988. Other approaches include various American in-band systems (IBOC, IBAC, IBRC, FMDigital, and FMeX) still in development, as well as various other systems promising satellite delivery, such as WorldSpace and CD Radio, still in development as well. One satellite-delivery system called MediaStar (formerly Archimedes) proposes to use the Eureka 147 DAB signal structure, such that a single receiver could access both terrestrial and satellite broadcasts.

DAB has been under development since 1981 at the Institut für Rundfunktechnik (IRT) and since 1987 as part of a European research project (Eureka 147). The Eureka 147 DAB specification was standardized by the European Telecommunications Standards Institute (ETSI) in February 1995 as document ETS 300 401, with a draft second edition issued in June 1996. In December 1994, the International Telecommunication Union—Radiocommunication (ITU-R) recommended that this technology, referred to as Digital System A, be used for implementing DAB services.

The Eureka 147 DAB signal consists of multiple carriers within a 1.536-MHz channel bandwidth. Four possible modes of operation define the channel coding configuration, specifying the total number of carriers, the carrier spacing, and also the guard interval duration. Each channel provides a raw data rate of 2304 kbps; after error protection, a useful data rate of anywhere between approximately 600 kbps up to 1800 kbps is available to the service provider, depending on the user-specified multiplex configuration. This useful data rate can be divided into an infinite number of possible configurations of audio and data programs. All audio programs are individually compressed using MUSICAM (MPEG-1 Layer II).

For each useful bit, 1 1/3 ... 4 bits are transmitted. This extensive redundancy makes it possible to reconstruct the transmitted bit sequence in the receiver, even if part of it is disrupted during transmission (FEC—forward error correction). In the receiver, error concealment can be carried out at the audio reproduction stage, so that residual transmission errors which could not be corrected do not always cause disruptive noise.

7.4. CODE EXCITED LINEAR PREDICTION (CELP)

The use of linear predictors removes redundancy in the signal, so that coding of the residual signal can be done with simpler quantizers. We first introduce the LPC vocoder and then introduce coding of the residual signal with a very popular technique called CELP.

7.4.1. LPC Vocoder

A typical model for speech production is shown in Figure 7.6, which has a source, or excitation, driving a linear time-varying filter. For voiced speech, the excitation is an impulse train spaced P samples apart. For unvoiced speech, the source is white random noise. The filter $h_m[n]$ for frame m changes at regular intervals, say every 10 ms. If this filter is represented with linear predictive coding, it is called an *LPC vocoder* [3].

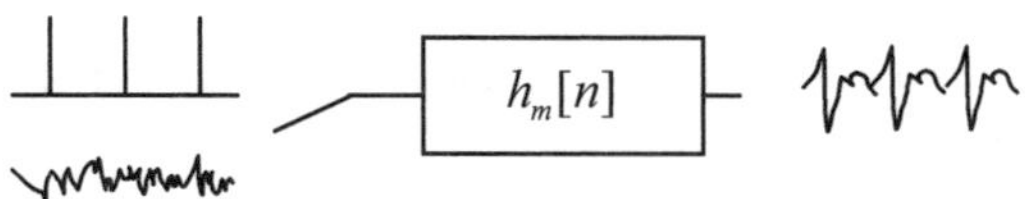

Figure 7.6 Block diagram of an LPC vocoder.

In addition to transmitting the gain and LPC coefficients, the encoder has to determine whether the frame is voiced or unvoiced, as well as the pitch period P for voiced frames.

The LPC vocoder produces reasonable quality for unvoiced frames, but often results in somewhat mechanical sound for voiced sounds, and a buzzy quality for voiced fricatives. More importantly, the LPC vocoder is quite sensitive to voicing and pitch errors, so that an accurate pitch tracker is needed for reasonable quality. The LPC vocoder also performs poorly in the presence of background noise. Nonetheless, it can be highly intelligible. The Federal Standard 1015 [55], proposed for secure communications, is based on a 2.4-kbps LPC vocoder.

It's also possible to use linear predictive coding techniques together with Huffman coding [45] to achieve lossless compression of up to 50%.

7.4.2. Analysis by Synthesis

Code Excited Linear Prediction (CELP) [5] is an umbrella term for a family of techniques that quantize the LPC residual using VQ, thus the term *code excited*, using analysis by synthesis. In addition CELP uses the fact that the residual of voiced speech has periodicity and can be used to predict the residual of the current frame. In CELP coding the LPC coefficients are quantized and transmitted (feedforward prediction), as well as the codeword index. The prediction using LPC coefficients is called *short-term prediction*. The prediction of the residual based on pitch is called *long-term prediction*. To compute the quantized coefficients we use an *analysis-by-synthesis* technique, which consists of choosing the combination of parameters whose reconstructed signal is closest to the analysis signal. In practice, not all coefficients of a CELP coder are estimated in an analysis-by-synthesis manner.

We first estimate the pth-order LPC coefficients from the samples $x[n]$ for frame t using the autocorrelation method, for example. We then quantize the LPC coefficients to $(a_1, a_2, \cdots, a_p)$ with the techniques described in Section 7.4.5. The residual signal $e[n]$ is obtained by inverse filtering $x[n]$ with the quantized LPC filter

$$e[n] = x[n] - \sum_{i=1}^{p} a_i x[n-i] \tag{7.43}$$

Given the transfer function of the LPC filter

$$H(z) = \frac{1}{A(z)} = \frac{1}{1 - \sum_{i=1}^{p} a_i z^{-i}} = \sum_{i=0}^{\infty} h_i z^{-i} \tag{7.44}$$

we can obtain the first M coefficients of the impulse response $h[n]$ of the LPC filter by driving it with an impulse as

$$h[n] = \begin{cases} 1 & n = 0 \\ \sum_{i=1}^{n} a_i h[n-i] & 0 < n < p \\ \sum_{i=1}^{p} a_i h[n-i] & p \le n < M \end{cases} \tag{7.45}$$

so that if we quantize a frame of M samples of the residual $\mathbf{e} = (e[0], e[1], \cdots e[M-1])^T$ to $\mathbf{e}_i = (e_i[0], e_i[1], \cdots e_i[M-1])^T$, we can compute the reconstructed signal $\hat{x}_i[n]$ as

$$\hat{x}_i[n] = \sum_{m=0}^{n} h[m] e_i[n-m] + \sum_{m=n+1}^{\infty} h[m] e[n-m] \tag{7.46}$$

where the second term in the sum depends on the residual for previous frames, which we already have. Let's define signal $r_0[n]$ as the second term of Eq. (7.46):

$$r_0[n] = \sum_{m=n+1}^{\infty} h[m] e[n-m] \tag{7.47}$$

which is the output of the LPC filter when there is no excitation for frame t. The important thing to note is that $r_0[n]$ does not depend on $e_i[n]$.

It is convenient to express Eqs. (7.46) and (7.47) in matrix form as

$$\hat{\mathbf{x}}_i = \mathbf{H}\mathbf{e}_i + \mathbf{r}_0 \tag{7.48}$$

where matrix $\mathbf{H}$ corresponds to the LPC filtering operation with its memory set to 0:

$$\mathbf{H} = \begin{bmatrix} h_0 & 0 & \cdots & 0 & 0 \\ h_1 & h_0 & \cdots & 0 & 0 \\ \cdots & \cdots & \cdots & \cdots & \cdots \\ h_{M-1} & h_{M-2} & \cdots & h_0 & 0 \\ h_M & h_{M-1} & \cdots & h_1 & h_0 \end{bmatrix} \tag{7.49}$$

Given the large dynamic range of the residual signal, we use gain-shape quantization, where we quantize the gain and the gain-normalized residual separately:

$$\mathbf{e}_i = \lambda \mathbf{c}_i \tag{7.50}$$

where λ is the gain and $\mathbf{c}_i$ is the codebook entry i. This codebook is known as the *fixed codebook* because its vectors do not change from frame to frame. Usually the size of the codebook is selected as 2^N so that full use is made of all N bits. Codebook sizes typically vary from 128 to 1024. Combining Eq. (7.48) with Eq. (7.50), we obtain

$$\hat{\mathbf{x}}_i = \lambda \mathbf{H}\mathbf{c}_i + \mathbf{r}_0 \tag{7.51}$$

The error between the original signal $\mathbf{x}$ and the reconstructed signal $\hat{\mathbf{x}}_i$ is

$$\boldsymbol{\varepsilon}_i = \mathbf{x} - \hat{\mathbf{x}}_i \tag{7.52}$$

The optimal gain λ and codeword index i are the ones that minimize the squared error between the original signal and the reconstructed[4] signal:

$$E(i,\lambda) = \left|\mathbf{x} - \hat{\mathbf{x}}_i\right|^2 = \left|\mathbf{x} - \lambda\mathbf{H}\mathbf{c}_i - \mathbf{r}_0\right|^2 = \left|\mathbf{x} - \mathbf{r}_0\right|^2 + \lambda^2 \mathbf{c}_i^T \mathbf{H}^T \mathbf{H}\mathbf{c}_i - 2\lambda \mathbf{c}_i^T \mathbf{H}^T (\mathbf{x} - \mathbf{r}_0) \tag{7.53}$$

where the term $\left|\mathbf{x} - \mathbf{r}_0\right|^2$ does not depend on λ or i and can be neglected in the minimization. For a given $\mathbf{c}_i$, the gain λ_i that minimizes Eq. (7.53) is given by

$$\lambda_i = \frac{\mathbf{c}_i^T \mathbf{H}^T (\mathbf{x} - \mathbf{r}_0)}{\mathbf{c}_i^T \mathbf{H}^T \mathbf{H}\mathbf{c}_i} \tag{7.54}$$

Inserting Eq. (7.54) into (7.53) lets us compute the index j as the one that minimizes

$$j = \arg\min_i \left\{ -\frac{\left(\mathbf{c}_i^T \mathbf{H}^T (\mathbf{x} - \mathbf{r}_0)\right)^2}{\mathbf{c}_i^T \mathbf{H}^T \mathbf{H}\mathbf{c}_i} \right\} \tag{7.55}$$

So we first obtain the codeword index j according to Eq. (7.55) and then the gain λ_j according to Eq. (7.54), which is scalarly quantized to $\hat{\lambda}_j$. Both codeword index j and $\hat{\lambda}_j$ are transmitted. In the algorithm described here, we first chose the quantized LPC coeffi-

[4] A beginner's mistake is to find the codebook index that minimizes the squared error of the residual. This does not minimize the difference between the original signal and the reconstructed signal.

cients $(a_1, a_2, \cdots, a_p)$ independently of the gains and codeword index, and then we chose the codeword index independently of the quantized gain $\hat{\lambda}_j$. This procedure is called *open-loop* estimation, because some parameters are obtained independently of the others. This is shown in Figure 7.7. *Closed-loop* estimation [49] means that all possible combinations of quantized parameters are explored. Closed-loop is more computationally expensive but yields lower squared error.

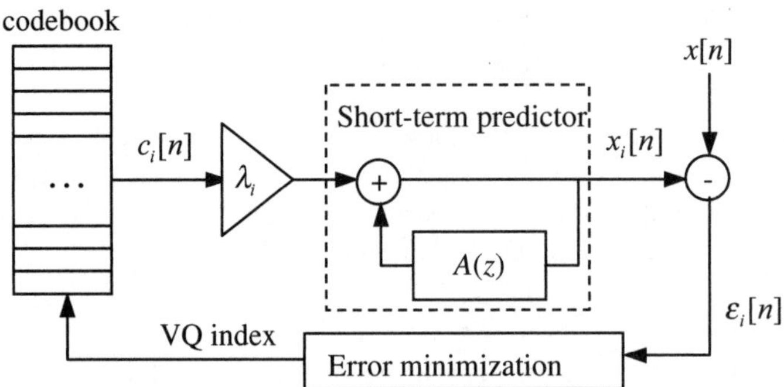

Figure 7.7 Analysis-by-synthesis principle used in a basic CELP.

7.4.3. Pitch Prediction: Adaptive Codebook

The fact that speech is highly periodic during voiced segments can also be used to reduce redundancy in the signal. This can be done by predicting the residual signal $e[n]$ at the current vector with samples from the past residual signal shifted a pitch period t:

$$e[n] = \lambda_t^a e[n-t] + \lambda_i^f c_i^f[n] = \lambda_t^a c_t^a[n] + \lambda_i^f c_i^f[n] \tag{7.56}$$

Using the matrix framework we described before, Eq. (7.56) can be expressed as

$$\mathbf{e}_{ti} = \lambda_t^a \mathbf{c}_t^a + \lambda_i^f \mathbf{c}_i^f \tag{7.57}$$

where we have made use of an *adaptive codebook* [31], where $\mathbf{c}_t^a$ is the adaptive codebook entry j with corresponding gain λ^a, and $\mathbf{c}_i^f$ is the fixed or stochastic codebook entry i with corresponding gain λ^f. The adaptive codebook entries are segments of the recently synthesized excitation signal

$$\mathbf{c}_t^a = (e[-t], e[1-t], \cdots, e[M-1-t])^T \tag{7.58}$$

where t is the delay which specifies the start of the adaptive codebook entry t. The range of t is often between 20 and 147, since this can be encoded with 7 bits. This corresponds to a range in pitch frequency between 54 and 400 Hz for a sampling rate of 8 kHz.

The contribution of the adaptive codebook is much larger than that of the stochastic codebook for voiced sounds. So we generally search for the adaptive codebook first, using Eq. (7.58) and a modified version of Eqs. (7.55) and (7.54), replacing i by t. Closed-loop search of both t and gain here often yields a much larger error reduction.

7.4.4. Perceptual Weighting and Postfiltering

The objective of speech coding is to reduce the bit rate while maintaining a perceived level of quality; thus, minimization of the error is not necessarily the best criterion. A perceptual weighting filter tries to shape the noise so that it gets masked by the speech signal (see Chapter 2). This generally means that most of the quantization noise energy is located in spectral regions where the speech signal has most of its energy. A common technique [4] consists in approximating this perceptual weighting with a linear filter

$$W(z) = \frac{A(z/\beta)}{A(z/\gamma)} \tag{7.59}$$

where $A(z)$ is the predictor polynomial

$$A(z) = 1 - \sum_{i=1}^{p} a_i z^{-i} \tag{7.60}$$

Choosing γ and β so that and $0 < \gamma < \beta \le 1$, implies that the roots of $A(z/\beta)$ and $A(z/\gamma)$ will move closer to the origin of the unit circle than the roots of $A(z)$, thus resulting in a frequency response with wider resonances. This perceptual filter therefore deemphasizes the contribution of the quantization error near the formants. A common choice of parameters is $\beta = 1.0$ and $\gamma = 0.8$, since it simplifies the implementation. This filter can easily be included in the matrix **H**, and a CELP coder incorporating the perceptual weighting is shown in Figure 7.8.

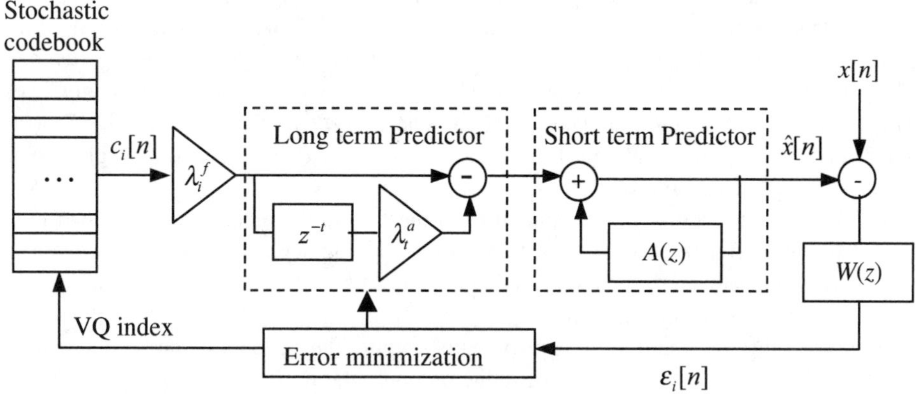

Figure 7.8 Diagram of a CELP coder. Both long-term and short-term predictors are used, together with a perceptual weighting.

Despite the perceptual weighting filter, the reconstructed signal still contains audible noise. This filter reduces the noise in those frequency regions that are perceptually irrelevant without degrading the speech signal. The postfilter generally consists of a short-term postfilter to emphasize the formant structure and a long-term postfilter to enhance the periodicity of the signal [10]. One possible implementation follows Eq. (7.59) with values of $\beta = 0.5$ and $\gamma = 0.75$.

7.4.5. Parameter Quantization

To achieve a low bit rate, all the coefficients need to be quantized. Because of its coding efficiency, vector quantization is the compression technique of choice to quantize the predictor coefficients. The LPC coefficients cannot be quantized directly, because small errors produced in the quantization process may result in large changes in the spectrum and possibly unstable filters. Thus, equivalent representations that guarantee stability are used, such as reflection coefficients, log-area ratios, and the line spectral frequencies (LSF) described in Chapter 6. LSF are used most often, because it has been found empirically that they behave well when they are quantized and interpolated [2]. For 8 kHz, 10 predictor coefficients are often used, which makes using a single codebook impractical because of the large dimension of the vector. Split-VQ [43] is a common choice, where the vectors are divided into several subvectors, and each is vector quantized. Matrix quantization can also be used to exploit the correlation of these subvectors across consecutive time frames. *Transparent quality*, defined as average spectral distortion below 1 dB with no frames above 4 dB, can be achieved with fewer than 25 bits per frame.

A frame typically contains around 20 to 30 milliseconds, which at 8 kHz represents 160–240 samples. Because of the large vector dimension, it is impractical to quantize a whole frame with a single codebook. To reduce the dimensionality, the frame is divided into four or more nonoverlapping sub-frames. The LSF coefficients for each subframe are linearly interpolated between the two neighboring frames.

The typical range of the pitch prediction for an 8-kHz sampling rate goes from 54 to 400 Hz, from 20 to 147 samples, and from 2.5 ms to 18.375 ms, which can be encoded with 7 bits. An additional bit is often used to encode fractional delays for the lower pitch periods. These fractional delays can be implemented through upsampling as described in Chapter 5. The subframe gain of the adaptive codebook can be effectively encoded with 3 or 4 bits. Alternatively, the gains of all sub-frames within a frame can be encoded through VQ, resulting in more efficient compression.

The fixed codebook can be trained from data using the techniques described in Chapter 4. This will offer the lowest distortion for the training set but doesn't guarantee low distortion for mismatched test signals. Also, it requires additional storage, and full search increases computation substantially.

Since subframes should be approximately white, the codebook can be populated from samples of a white process. A way of reducing computation is to let those noise samples be only +1, 0, or –1, because only additions are required. Codebooks of a specific type, known as *algebraic codebooks* [1], offer even more computational savings because they contain many 0s. Locations for the 4 pulses per subframe under the G.729 standard are shown in Table 7.3.

Full search can efficiently be done with this codebook structure. Algebraic codebooks can provide almost as low distortion as trained codebooks can, with low computational complexity.

Table 7.3 Algebraic codebooks for the G.729 standard. Each of the four codebooks has one pulse in one possible location indicated by 3 bits for the first three codebooks and 4 bits for the last codebook. The sign is indicated by an additional bit. A total of 17 bits are needed to encode a 40-sample subframe.

Amplitude	Positions
±1	0, 5, 10, 15, 20, 25, 30, 35
±1	1, 6, 11, 16, 21, 26, 31, 36
±1	2, 7, 12, 17, 22, 27, 32, 37
±1	3, 8, 13, 18, 23, 28, 33, 38 4, 9, 14, 19, 24, 29, 34, 39

7.4.6. CELP Standards

There are many standards for speech coding based on CELP, offering various points in the bit-rate/quality plane, mostly depending on when they were created and how refined the technology was at that time.

Voice over Internet Protocol (Voice over IP) consists of transmission of voice through data networks such as the Internet. H.323 is an umbrella standard which references many other ITU-T recommendations. H.323 provides the system and component descriptions, call model descriptions, and call signaling procedures. For audio coding, G.711 is mandatory, while G.722, G.728, G.723.1, and G.729 are optional. G.728 is a low-delay CELP coder that offers toll quality at 16 kbps [9], using a feedback 50th-order predictor, but no pitch prediction. G.729 [46] offers toll quality at 8 kbps, with a delay of 10 ms. G.723.1, developed by DSP Group, including Audiocodes Ltd., France Telecom, and the University of Sherbrooke, has slightly lower quality at 5.3 and 6.3 kbps, but with a delay of 30 ms. These standards are shown in Table 7.4.

Table 7.4 Several CELP standards used in the H.323 specification used for teleconferencing and voice streaming through the Internet.

Standard	Bit Rate (kbps)	MOS	Algorithm	H.323	Comments
G.728	16	4.0	No pitch prediction	Optional	Low-delay
G.729	8	3.9	ACELP	Optional	
G.723.1	5.3, 6.3	3.9	ACELP for 5.3k	Optional	

In 1982, the Conference of European Posts and Telegraphs (CEPT) formed a study group called the Groupe Spécial Mobile (GSM) to study and develop a pan-European public land mobile system. In 1989, GSM responsibility was transferred to the European Telecommunication Standards Institute (ETSI), and the phase I GSM specifications were published in 1990. Commercial service was started in mid 1991, and by 1993 there were 36 GSM networks in 22 countries, with 25 additional countries considering or having already selected GSM. This is not only a European standard; South Africa, Australia, and many Middle and Far East countries have chosen GSM. The acronym GSM now stands for Global System for Mobile telecommunications. The GSM group studied several voice coding algorithms on the basis of subjective speech quality and complexity (which is related to cost, processing delay, and power consumption once implemented) before arriving at the choice of a Regular Pulse Excited–Linear Predictive Coder (RPE-LPC) with a Long Term Predictor loop [56]. Neither the original full-rate at 13 kbps [56] nor the half-rate at 5.6 kbps [19] achieves toll quality, though the enhanced full-rate (EFR) standard based on ACELP [26] has toll quality at the same rates.

The *Telecommunication Industry Association* (TIA) and the *Electronic Industries Alliance* (EIA) are organizations accredited by the *American National Standards Institute* (ANSI) to develop voluntary industry standards for a wide variety of telecommunication products. TR-45 is the working group within TIA devoted to mobile and personal communication systems. Time Division Multiple Access (TDMA) is a digital wireless technology that divides a narrow radio channel into framed time slots (typically 3 or 8) and allocates a slot to each user. The TDMA Interim Standard 54, or TIA/EIA/IS54, was released in early 1991 by both TIA and EIA. It is available in North America at both the 800-MHz and 1900-MHz bands. IS54 [18] at 7.95 kbps is used in North America's TDMA (Time Division Multiple Access) digital telephony and has quality similar to the original full-rate GSM. TDMA IS-136 is an update released in 1994.

Table 7.5 CELP standards used in cellular telephony.

Standard	Bit Rate (kbps)	MOS	Algorithm	Cellular	Comments
Full-rate GSM	13	3.6	VSELP RTE-LTP	GSM	
EFR GSM	12.2	4.5	ACELP	GSM	
IS-641	7.4	4.1	ACELP	PCS1900	
IS-54	7.95	3.9	VSELP	TDMA	
IS-96a	max 8.5	3.9	QCELP	CDMA	Variable-rate

Code Division Multiple Access (CDMA) is a form of *spread spectrum*, a family of digital communication techniques that have been used in military applications for many years. The core principle is the use of noiselike carrier waves, and, as the name implies, bandwidths much wider than that required for simple point-to-point communication at the same data rate. Originally there were two motivations: either to resist enemy efforts to jam the communications (anti-jam, or AJ) or to hide the fact that communication was even taking place, sometimes called low probability of intercept (LPI). The service started in 1996 in the United States, and by the end of 1999 there were 50 million subscribers worldwide. IS-96 QCELP [14], used in North America's CDMA, offers variable-rate coding at 8.5, 4, 2, and 0.8 kbps. The lower bit rate is transmitted when the coder detects background noise. TIA/EIA/IS-127-2 is a standard for an enhanced variable-rate codec, whereas TIA/EIA/IS-733-1 is a standard for high-rate. Standards for CDMA, TDMA, and GSM are shown in Table 7.5.

Third generation (3G) is the generic term used for the next generation of mobile communications systems. 3G systems will provide enhanced services to those—such as voice, text, and data—predominantly available today. The Universal Mobile Telecommunications System (UMTS) is a part of ITU's International Mobile Telecommunications (IMT)-2000 vision of a global family of third-generation mobile communications systems. It has been assigned to the frequency bands 1885–2025 and 2110–2200 MHz. The first networks are planned to launch in Japan in 2001, with European countries following in early 2002. A major part of 3G is General Packet Radio Service (GPRS), under which carriers charge by the packet rather than by the minute. The speech coding standard for CDMA2000, the umbrella name for the third-generation standard in the United States, gained approval for its first phase in 2000. An adaptive multi-rate wideband speech codec has also been proposed for the GSM's 3G [16], which has five modes of operation from 24 kbps down to 9.1 kbps.

While most of the work described above uses a sampling rate of 8 kHz, there has been growing interest in using CELP techniques for high bandwidth and particularly in a scalable way so that a basic layer contains the lower frequency and the higher layer either is a full-band codec [33] or uses a parametric model [37].

7.5. Low-Bit Rate Speech Coders

In this section we describe a number of low-bit-rate speech coding techniques including the mixed-excitation LPC vocoder, harmonic coding, and waveform interpolation. These coding techniques are also used extensively in speech synthesis.

Waveform-approximating coders are designed to minimize the difference between the original signal and the coded signal. Therefore, they produce a reconstructed signal whose SNR goes to infinity as the bit rate increases, and they also behave well when the input signal is noisy or music. In this category we have the scalar waveform coders of Section 7.2, the frequency-domain coders of Section 7.3, and the CELP coders of Section 7.4.

Low-bit-rate coders, on the other hand, do not attempt to minimize the difference between the original signal and the quantized signal. Since these coders are designed to operate at low bit rates, their SNR does not generally approach infinity even if a large bit rate is used. The objective is to compress the original signal with another one that is perceptually equivalent. Because of the reliance on an inaccurate model, these low-bit-rate coders often distort the speech signal even if the parameters are not quantized. In this case, the distortion can consist of more than quantization noise. Furthermore, these coders are more sensitive to the presence of noise in the signal, and they do not perform as well on music.

In Figure 7.9 we compare the MOS of waveform approximating coders and low-bit-rate coders as a function of the bit rate. CELP uses a model of speech to obtain as much prediction as possible, yet allows for the model not to be exact, and thus is a waveform-approximating coder. CELP is a robust coder that works reasonably well when the assumption of only a clean speech signal breaks either because of additive noise or because there is music in the background. Researchers are working on the challenging problem of creating more scalable coders that offer best performance at all bit rates.

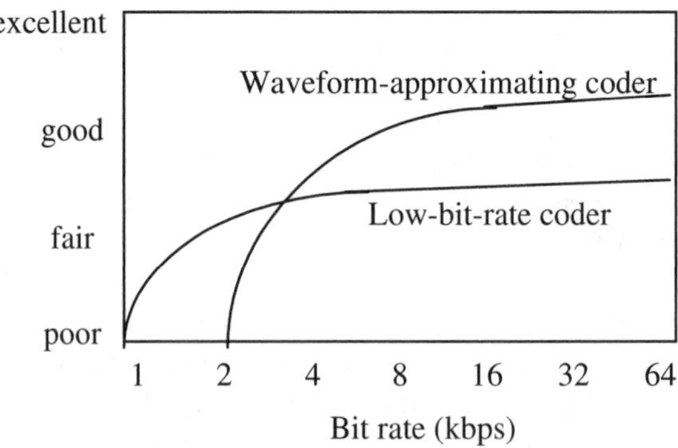

Figure 7.9 Typical subjective performance of waveform-approximating and low-bit-rate coders as a function of the bit rate. Note that waveform-approximating coders are a better choice for bit rates higher than about 3 kbps, whereas parametric coders are a better choice for lower bit rates. The exact cutoff point depends on the specific algorithms compared.

7.5.1. Mixed-Excitation LPC Vocoder

The main weakness of the LPC vocoder is the binary decision between voiced and unvoiced speech, which results in errors especially for noisy speech and voiced fricatives. By having a separate voicing decision for each of a number of frequency bands, the performance can be enhanced significantly [38]. The new proposed U.S. Federal Standard at 2.4 kbps is a Mixed Excitation Linear Prediction (MELP) LPC vocoder [39], which has a MOS of about 3.3. This exceeds the quality of the older 4800-bps Federal Standard 1016 [8] based on CELP. The bit rate of the proposed standard can be reduced while maintaining the same quality by jointly quantizing several frames together [57]. A hybrid codec that uses MELP in strongly voiced regions and CELP in weakly voiced and unvoiced regions [53] has shown to yield lower bit rates. MELP can also be combined with the waveform interpolation technique of Section 7.5.3 [50].

7.5.2. Harmonic Coding

Sinusoidal coding decomposes the speech signal [35] or the LP residual signal [48] into a sum of sinusoids. The case where these sinusoids are harmonically related is of special interest for speech synthesis (see Chapter 16), so we will concentrate on it in this section, even though a similar treatment can be followed for the case where the sinusoids are not harmonically related. In fact, a combination of harmonically related and nonharmonically related sinusoids can also be used [17]. We show in Section 7.5.2.2 that we don't need to transmit the phase of the sinusoids, only the magnitude.

As shown in Chapter 5, a periodic signal $\tilde{s}[n]$ with period T_0 can be expressed as a sum of T_0 harmonic sinusoids

$$\tilde{s}[n] = \sum_{l=0}^{T_0-1} A_l \cos(nl\omega_0 + \phi_l) \tag{7.61}$$

whose frequencies are multiples of the fundamental frequency $\omega_0 = 2\pi / T_0$, and where A_l and ϕ_l are the sinusoid amplitudes and phases, respectively. If the pitch period T_0 has fractional samples, the sum in Eq. (7.61) includes only the integer part of T_0 in the summation. Since a real signal $s[n]$ will not be perfectly periodic in general, we have a modeling error

$$e[n] = s[n] - \tilde{s}[n] \tag{7.62}$$

We can use short-term analysis to estimate these parameters from the input signal $s[n]$ at frame k, in the neighborhood of $t = kN$, where N is the frame shift:

$$s_k[n] = s[n]w_k[n] = s[n]w[kN - n] \tag{7.63}$$

if we make the assumption that the sinusoid parameters for frame k (ω_0^k, A_l^k and ϕ_l^k) are constant within the frame.

At resynthesis time, there will be discontinuities at unit boundaries, due to the block processing, unless we specifically smooth the parameters over time. One way of doing this is with overlap-add method between frames $(k - 1)$ and k:

$$\hat{s}[n] = w[n]\tilde{s}_{k-1}[n] + w[n-N]\tilde{s}_k[n-N], \quad 0 \le n < N \tag{7.64}$$

where the window $w[n]$ must be such that

$$w[n] + w[n-N] = 1, \quad 0 \le n < N \tag{7.65}$$

to achieve perfect reconstruction. This is the case for the common Hamming and Hanning windows.

This harmonic model [35] is similar to the classic filterbank, though rather than the whole spectrum we transmit only the fundamental frequency ω_0 and the amplitudes A_l and phases ϕ_l of the harmonics. This reduced representation doesn't result in loss of quality for a frame shift N that corresponds to 12 ms or less. For unvoiced speech, using a default pitch of 100 Hz results in acceptable quality.

7.5.2.1. Parameter Estimation

For simplicity in the calculations, let's define $\tilde{s}[n]$ as a sum of complex exponentials

$$\tilde{s}[n] = \sum_{l=0}^{T_0-1} A_l \exp\{j(nl\omega_0 + \phi_l)\} \quad (7.66)$$

and perform short-time Fourier transform with a window $w[n]$

$$\tilde{S}_W(\omega) = \sum_{l=0}^{T_0-1} A_l e^{j\phi_l} W(\omega - l\omega_0) \quad (7.67)$$

where $W(\omega)$ is the Fourier transform of the window function. The goal is to estimate the sinusoid parameters as those that minimize the squared error:

$$E = |S(\omega) - \tilde{S}_W(\omega)|^2 \quad (7.68)$$

If the main lobes of the analysis window do not overlap, we can estimate the phases ϕ_l as

$$\phi_l = \arg S(l\omega_0) \quad (7.69)$$

and the amplitudes A_l as

$$A_l = \frac{|S(l\omega_0)|}{W(0)} \quad (7.70)$$

For example, the Fourier transform of a $(2N + 1)$ point rectangular window centered around the origin is given by

$$W(\omega) = \frac{\sin((2N+1)\omega/2)}{\sin(\omega/2)} \quad (7.71)$$

whose main lobes will not overlap in Eq. (7.67) if $2T_0 < 2N+1$: i.e., the window contains at least two pitch periods. The implicit assumption in the estimates of Eqs. (7.69) and (7.70) is that there is no spectral leakage, but a rectangular window does have significant spectral leakage, so a different window is often used in practice. For windows such as Hanning or Hamming, which reduce the leakage significantly, it has been found experimentally that these estimates are acceptable if the window contains at least two and a half pitch periods.

Typically, the window is centered around 0 (nonzero in the interval $-N \le n \le N$) to avoid numerical errors in estimating the phases.

Another implicit assumption in Eqs. (7.69) and (7.70) is that we know the fundamental frequency ω_0 ahead of time. Since, in practice, this is not the case, we can estimate it as the one which minimizes Eq. (7.68). This pitch-estimation method can generate pitch doubling or tripling when a harmonic falls within a formant that accounts for the majority of the signal's energy.

Voiced/unvoiced decisions can be computed from the ratio between the energy of the signal and that of the reconstruction error

$$SNR = \frac{\sum_{n=-N}^{N} |s[n]|^2}{\sum_{n=-N}^{N} |s[n] - \tilde{s}[n]|^2} \tag{7.72}$$

where it has been empirically found that frames with SNR higher than 13 dB are generally voiced and lower than 4 dB unvoiced. In between, the signal is considered to contain a mixed excitation. Since speech is not perfectly stationary within the analysis frame, even noise-free periodic signals will yield finite SNR.

For unvoiced speech, a good assumption is to default to a pitch of 100 Hz. The use of fewer sinusoids leads to perceptual artifacts.

Improved quality can be achieved by using an analysis-by-synthesis framework [17, 34] since the closed-loop estimation is more robust to pitch-estimation and voicing decision errors.

7.5.2.2. Phase Modeling

An impulse train $e[n]$, a periodic excitation, can be expressed as a sum of complex exponentials

$$e[n] = T_0 \sum_{k=-\infty}^{\infty} \delta[n - n_0 - kT_0] = \sum_{l=0}^{T_0-1} e^{j(n-n_0)\omega_0 l} \tag{7.73}$$

which, if passed through a filter $H(\omega) = A(\omega)\exp\Phi(\omega)$, will generate

$$s[n] = \sum_{l=0}^{T_0-1} A(l\omega_0)\exp\{j[(n-n_0)\omega_0 l + \Phi(l\omega_0)]\} \tag{7.74}$$

Comparing Eq. (7.66) with (7.74), the phases of our sinusoidal model are given by

$$\phi_l = -n_0\omega_0 l + \Phi(l\omega_0) \tag{7.75}$$

Since the sinusoidal model has too many parameters to lead to low-rate coding, a common technique is to not encode the phases. In Chapter 6 we show that if a system is considered minimum phase, the phases can be uniquely recovered from knowledge of the magnitude spectrum.

The magnitude spectrum is known at the pitch harmonics, and the remaining values can be filled in by interpolation: e.g., linear or cubic splines [36]. This interpolated magnitude spectrum can be approximated through the real cepstrum:

$$|\tilde{A}(\omega)| = c_0 + 2\sum_{k=1}^{K} c_k \cos(k\omega) \tag{7.76}$$

and the phase, assuming a minimum phase system, is given by

$$\tilde{\Phi}(\omega) = -2\sum_{k=1}^{K} c_k \sin(k\omega) \tag{7.77}$$

The phase $\phi_0(t)$ of the first harmonic between frames $(k-1)$ and k can be obtained from the instantaneous frequency $\omega_0(t)$

$$\phi_0(t) = \phi_0((k-1)N) + \int_{(k-1)N}^{t} \omega_0(t)dt \tag{7.78}$$

if we assume the frequency $\omega_0(t)$ in that region to vary linearly between frames $(k-1)$ and k:

$$\omega_0(t) = \omega_0^{k-1} + \frac{\omega_0^k - \omega_0^{k-1}}{N} t \tag{7.79}$$

and insert Eq. (7.79) into (7.78), evaluating at $t = kN$, to obtain

$$\phi_0^k = \phi_0(kN) = \phi_0((k-1)N) + (\omega_0^{k-1} + \omega_0^k)(N/2) \tag{7.80}$$

the phase of the sinusoid at ω_0 as a function of the fundamental frequencies at frames $(k-1)$, k and the phase at frame $(k-1)$:

$$\phi_l^k = \Phi^k(l\omega_0) + l\phi_0^k \tag{7.81}$$

The phases computed by Eqs. (7.80) and (7.81) are a good approximation in practice for perfectly voiced sounds. For unvoiced sounds, random phases are needed, or else the reconstructed speech sounds buzzy. Voiced fricatives and many voiced sounds have an aspiration component, so that a mixed excitation is needed to represent them. In these cases, the source is split into different frequency bands and each band is classified as either voiced or unvoiced. Sinusoids in voiced bands use the phases described above, whereas sinusoids in unvoiced bands have random phases.

7.5.2.3. Parameter Quantization

To quantize the sinusoid amplitudes, we can use an LPC fitting and then quantize the line spectral frequencies. Also we can do a cepstral fit and quantize the cepstral coefficients. To be more effective, a mel scale should be used.

While these approaches help in reducing the number of parameters and in quantizing those parameters, they are not the most effective way of quantizing the sinusoid amplitudes. A technique called *Variable-Dimension Vector Quantization* (VDVQ) [12] has been devised to address this. Each codebook vector $\mathbf{c}_i$ has a fixed dimension N determined by the length of the FFT used. The vector of sinusoid amplitudes $\mathbf{A}$ has a dimension l that depends on the number of harmonics and thus the pitch of the current frame. To compute the distance between $\mathbf{A}$ and $\mathbf{c}_i$, the codebook vectors are resampled to a size l and the distance is computed between two vectors of dimension l. Euclidean distance of the log-amplitudes is often used.

In this method, only the distance at the harmonics is evaluated instead of the distance at the points in the envelope that are not actually present in the signal. Also, this technique does not suffer from inaccuracies of the model used, such as the inability of linear predictive coding to model nasals.

7.5.3. Waveform Interpolation

The main idea behind waveform interpolation (WI) [29] is that the pitch pulse changes slowly over time for voiced speech. During voiced segments, the speech signal is nearly periodic. WI coders can operate as low as 2.4 kbps.

Starting at an arbitrary time instant, it is easy to identify a first pitch cycle $x_1[n]$, a second $x_2[n]$, a third $x_3[n]$, and so on. We then express our signal $x[n]$ as a function of these pitch cycle waveforms $x_m[n]$

$$x[n] = \sum_{m=-\infty}^{\infty} x_m[n - t_m] \tag{7.82}$$

where $P_m = t_m - t_{m-1}$ is the pitch period at time t_m in samples, and the pitch cycle is a windowed version of the input

$$x_m[n] = w_m[n]x[n] \tag{7.83}$$

for example, with a rectangular window. To transmit the signal in a lossless fashion we need to transmit all pitch waveforms $x_m[n]$.

If the signal is perfectly periodic, we need to transmit only one pitch waveform $x_m[n]$ and the pitch period P. In practice, voiced signals are not perfectly periodic, so that we need to transmit more than just one pitch waveform. On the other hand, voiced speech is nearly periodic, and consecutive pitch waveforms are very similar. Thus, we probably do not need to transmit all, and we could send every other pitch waveform, for example.

It is convenient to define a two-dimensional surface $u[n,l]$ (shown in Figure 7.10) such that the pitch waveform $x_m[n]$ can be obtained as

$$x_m[n] = u[n, t_m] \tag{7.84}$$

so that $u[n,l]$ is defined for $l = t_m$, with the remaining points having been computed through interpolation. A frequency representation of the pitch cycle can also be used instead of the time pitch cycle.

This surface can then be sampled at regular time intervals $l = sT$. It has been shown empirically that transmitting the pitch waveform $x_s[n]$ about 40 times per second (a 25-ms interval is equivalent to T = 200 samples for an $F_s = 8000$ Hz sampling rate) is sufficient for voiced speech. The so-called *slowly evolving waveform* (SEW) $\tilde{u}[n,l]$ can be generated by low-pass filtering $u[n,l]$ along the l-axis:

$$x_s[n] = \tilde{u}[n, sT] = \frac{\sum_m h[sT - t_m]u[n, t_m]}{\sum_m h[sT - t_m]} \tag{7.85}$$

where $h[n]$ is a low-pass filter and $x_s[n]$ is a sampled version of $\tilde{u}[n,l]$.

The decoder has to reconstruct each pitch waveform $x_m[n]$ from the SEW $x_s[n]$ by interpolation between adjacent pitch waveforms, and thus the name *waveform interpolation (WI) coding*:

$$\tilde{x}_m[n] = \tilde{u}[n,t_m] = \frac{\sum_s h[t_m - sT]x_s[n]}{\sum_s h[t_m - sT]} \tag{7.86}$$

If the sampling period is larger than the local pitch period ($T > P_m$), perfect reconstruction will not be possible, and there will be some error in the approximation

$$x_m[n] = \tilde{x}_m[n] + \hat{x}_m[n] \tag{7.87}$$

or alternatively in the two-dimensional representation

$$u[n,l] = \tilde{u}[n,l] + \hat{u}[n,l] \tag{7.88}$$

where $\hat{x}_m[n]$ and $\hat{u}[n,l]$ represent the *rapidly evolving waveforms* (REW).

Since this technique can also be applied to unvoiced speech, where the concept of pitch waveform doesn't make sense, the more general term *characteristic waveform* is used instead. For unvoiced speech, an arbitrary *period* of around 100 Hz can be used.

For voiced speech, we expect the rapidly varying waveform $\hat{u}[n,l]$ in Eq. (7.88) to have much less energy than the slowly evolving waveform $\tilde{u}[n,l]$. For unvoiced speech the converse is true: $\hat{u}[n,l]$ has more energy than $\tilde{u}[n,l]$. For voiced fricatives, both components may be comparable and thus we want to transmit both.

In Eqs. (7.85) and (7.86) we need to average characteristic waveforms that have, in general, different lengths. To handle this, all characteristic waveforms are typically normalized in length prior to the averaging operation. This length normalization is done by padding with zeros $x_m[n]$ to a certain length M, or truncating $x_m[n]$ if $P_m > M$. Another possible normalization is done via linear resampling. This decomposition is shown in Figure 7.10.

Another representation uses the Fourier transform of $x_m[n]$. This case is related to the harmonic model of Section 7.5.2. In the harmonic model, a relatively long window is needed to average the several pitch waveforms within the window, whereas this waveform interpolation method has higher time resolution. In constructing the characteristic waveforms we have implicitly used a rectangular window of length one pitch period, but other windows can be used, such as a Hanning window that covers two pitch periods. This frequency-domain representation offers advantages in coding both the SEW and the REW, because properties of the human auditory system can help reduce the bit rate. This decomposition is often done on the LPC residual signal.

In particular, the REW $\hat{u}[n,l]$ has the characteristics for noise, and as such only a rough description of its power spectral density is needed. At the decoder, random noise is generated with the transmitted power spectrum. The spectrum of $\hat{u}[n,l]$ can be vector quantized to as few as eight shapes with little or no degradation.

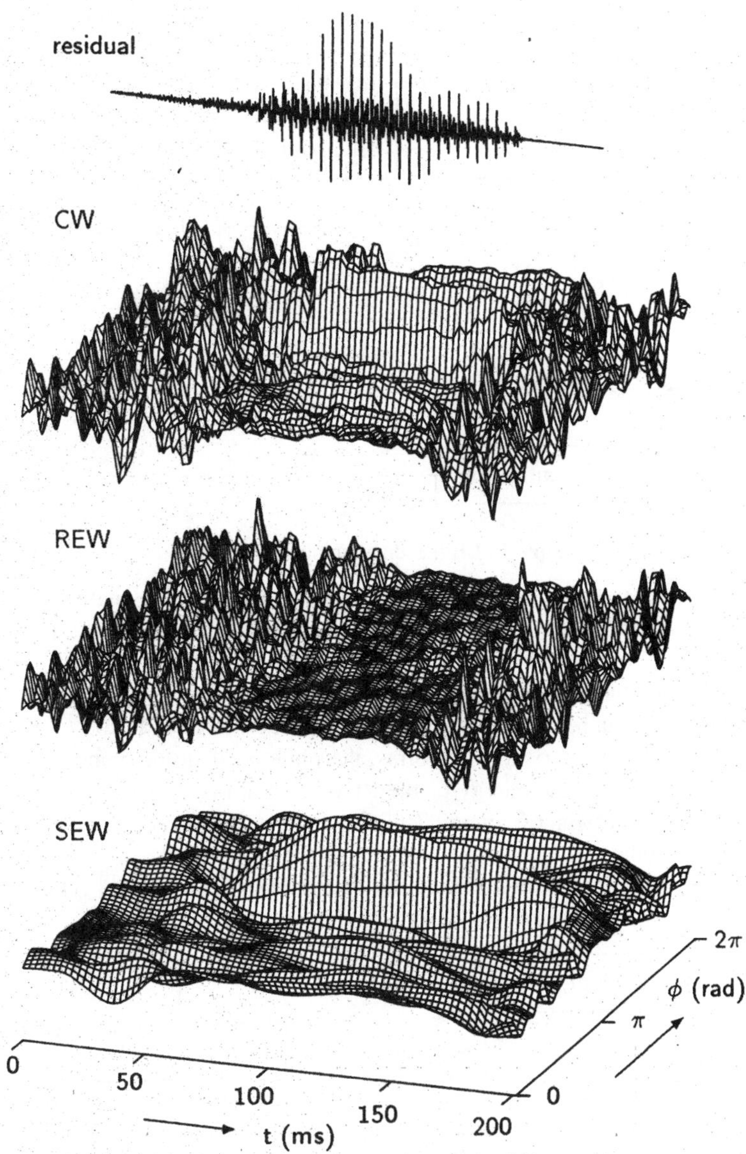

Figure 7.10 LP residual signal and its associated characteristic waveform (CW) $u(t,\phi)$. In the ϕ axis we have a normalized pitch pulse at every given time t. Decomposition of the surface into a slowly evolving waveform (SEW) and a rapidly evolving waveform (REW). (After Kleijn and Haagen [30], reprinted by permission of IEEE).

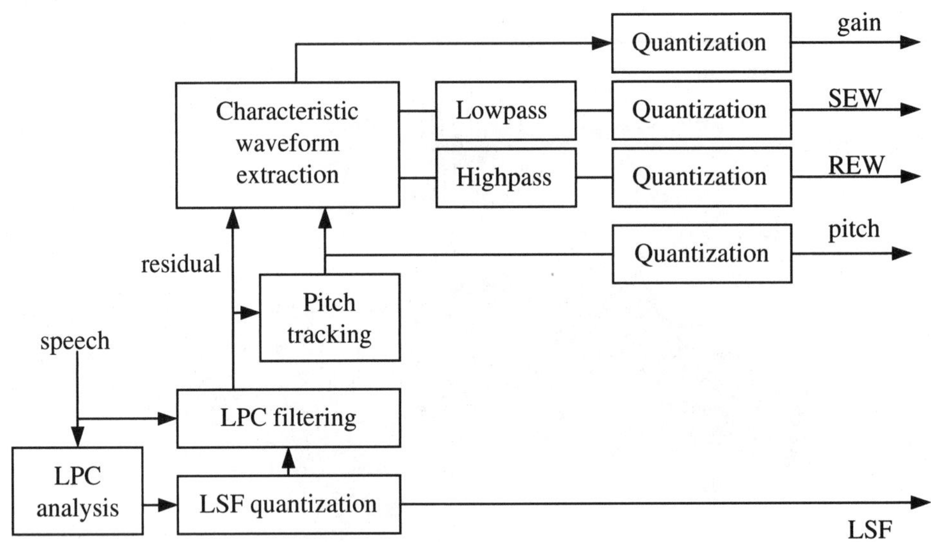

Figure 7.11 Block diagram of the WI encoder.

The SEW $\tilde{u}[n,l]$ is more important perceptually, and for high quality the whole shape needs to be transmitted. Higher accuracy is desired at lower frequencies so that a perceptual frequency scale (mel or Bark) is often used. Since the magnitude of $\tilde{u}[n,l]$ is perceptually more important than the phase, for low bit rates the phase of the SEW is not transmitted. The magnitude spectrum can be quantized with the VDVQ described in Section 7.5.2.3.

To obtain the characteristic waveforms, the pitch needs to be computed. We can find the pitch period such that the energy of the REW is minimized. To do this we use the approaches described in Chapter 6. Figure 7.11 shows a block diagram of the encoder and Figure 7.12 of the decoder.

Parameter estimation using an analysis-by-synthesis framework [21] can yield better results than the open-loop estimation described above.

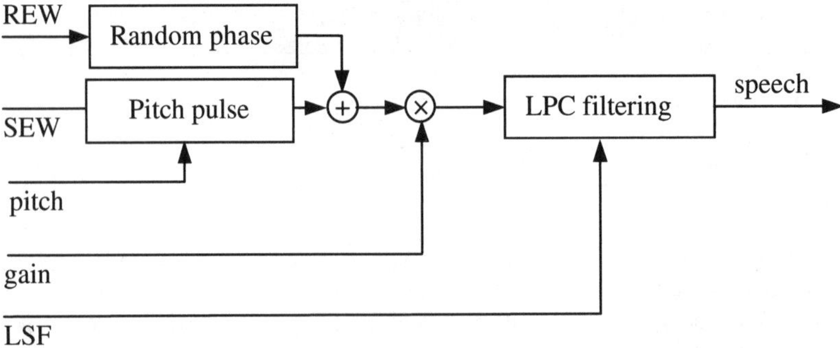

Figure 7.12 Block diagram of the WI decoder.

7.6. HISTORICAL PERSPECTIVE AND FURTHER READING

This chapter is only an introduction to speech and audio coding technologies. The reader is referred to [27, 32, 41, 52] for coverage in greater depth. A good source of the history of speech coding can be found in [20].

In 1939, Homer Dudley of AT&T Bell Labs first proposed the channel vocoder [15], the first analysis-by-synthesis system. This vocoder analyzed slowly varying parameters for both the excitation and the spectral envelope. Dudley thought of the advantages of bandwidth compression and information encryption long before the advent of digital communications.

PCM was first conceived in 1937 by Alex Reeves at the Paris Laboratories of AT&T, and it started to be deployed in the United States Public Switched Telephone Network in 1962. The digital compact disc, invented in the late 1960s by James T. Russell and introduced commercially in 1984, also uses PCM as coding standard. The use of μ-law encoding was proposed by Smith [51] in 1957, but it wasn't standardized for telephone networks (G.711) until 1972. In 1952, Schouten et al. [47] proposed delta modulation and Cutler [11] invented differential PCM. ADPCM was developed by Barnwell [6] in 1974.

Speech coding underwent a fundamental change with the development of linear predictive coding in the early 1970s. Atal [3] proposed the LPC vocoder in 1971, and then CELP [5] in 1984. The majority of coding standards for speech signals today use a variation on CELP.

Sinusoidal coding [35] and waveform interpolation [29] were developed in 1986 and 1991, respectively, for low-bit-rate telephone speech. Transform coders such as MP3 [23], MPEG II, and Perceptual Audio Coder (PAC) [28] have been used primarily in audio coding for high-fidelity applications.

Recently, researchers have been improving the technology for cellular communications by trading off source coding and channel coding. For poor channels more bits are allocated to channel coding and fewer to source coding to reduce dropped calls. Scalable coders that have different layers with increased level of precision, or bandwidth, are also of great interest.

REFERENCES

[1] Adoul, J.P., *et al.*, "Fast CELP Coding Based on Algebraic Codes," *Int. Conf. on Acoustics, Speech and Signal Processing*, 1987, Dallas, TX, pp. 1957-1960.

[2] Atal, B.S., R.V. Cox, and P. Kroon, "Spectral Quantization and Interpolation for CELP Coders," *Int. Conf. on Acoustics, Speech and Signal Processing*, 1989, Glasgow, pp. 69-72.

[3] Atal, B.S. and L. Hanauer, "Speech Analysis and Synthesis by Linear Prediction of the Speech Wave," *Journal of the Acoustical Society of America*, 1971, **50**, pp. 637-655.

[4] Atal, B.S. and M.R. Schroeder, "Predictive Coding of Speech Signals and Subjective Error Criteria," *IEEE Trans. on Acoustics, Speech and Signal Processing*, 1979, **ASSP-27**(3), pp. 247-254.

[5] Atal, B.S. and M.R. Schroeder, "Stochastic Coding of Speech at Very Low Bit Rates," *Proc. Int. Conf. on Comm.*, 1984, Amsterdam, pp. 1610-1613.

[6] Barnwell, T.P., *et al.*, *Adaptive Differential PCM Speech Transmission*, 1974, Rome Air Development Center.

[7] Benvenuto, N., G. Bertocci, and W.R. Daumer, "The 32-kbps ADPCM Coding Standard," *AT&T Technical Journal*, 1986, **65**, pp. 12-22.

[8] Campbell, J.P., T.E. Tremain, and V.C. Welch, "The DoD 4.8 kbps Standard (Proposed Federal Standard 1016)" in *Advances in Speech Coding*, B. Atal, V. Cuperman, and A. Gersho, eds. 1991, Kluwer Academic Publishers, pp. 121-133.

[9] Chen, J.H., *et al.*, "A Low-Delay CELP Coder for the CCITT 16 kbps Speech Coding Standard," *IEEE Journal on Selected Areas Communcations*, 1992, **10**(5), pp. 830-849.

[10] Chen, J.H. and A. Gersho, "Adaptive Postfiltering for Quality Enhancement of Coded Speech," *IEEE Trans. on Speech and Audio Processing*, 1995, **3**(1), pp. 59-71.

[11] Cutler, C.C., *Differential Quantization for Communication Signals*, 1952, US Patent 2,605,361.

[12] Das, A. and A. Gersho, "Variable Dimension Vector Quantization," *IEEE Signal Processing Letters*, 1996, **3**(7), pp. 200-202.

[13] Daumer, W.R., *et al.*, "Overview of the 32kbps ADPCM Algorithm," *Proc. IEEE Global Telecomm*, 1984, pp. 774-777.

[14] DeJaco, P.J.A., W. Gardner, and C. Lee, "QCELP: The North American CDMA Digital Cellular Variable Speech Coding Standard," *Proc. Workshop on Speech Coding for Telecommunications*, 1993, Sainte Adele, Quebec, pp. 5-6.

[15] Dudley, H., "The Vocoder," *Bell Labs Record*, 1939, **17**, pp. 122-126.

[16] Erdmann, C., *et al.*, "An Adaptive Rate Wideband Speech Codec with Adaptive Gain Re-Quantization," *IEEE Workshop on Speech Coding*, 2000, Delavan, Wisconsin.

[17] Etemoglu, C.O., V. Cuperman, and A. Gersho, "Speech Coding with an Analysis-by-Synthesis Sinusoidal Model," *Int. Conf. on Acoustics, Speech and Signal Processing*, 2000, Istanbul, Turkey, pp. 1371-1374.

[18] Gerson, I.A. and M.A. Jasiuk, "Vector Sum Excited Linear Prediction (VSELP)" in *Advances in Speech Coding*, B.S. Atal, V. Cuperman, and A. Gersho, eds. 1991, Boston, MA, Kluwer Academic Publishers, pp. 69-79.

[19] Gerson, I.A. and M.A. Jasiuk., "Techniques for Improving the Performance of CELP-type Speech Coders," *IEEE Journal Selected Areas Communications*, 1991, **10**(5), pp. 858-865.

[20] Gold, B. and N. Morgan, *Speech and Audio Signal Processing: Processing and Perception of Speech and Music*, 2000, New York, John Wiley.

[21] Gottesman, O. and A. Gersho, "High Quality Enhanced Waveform Interpolative Coding at 2.8 kbps," *Int. Conf. on Acoustics, Speech and Signal Processing*, 2000, Istanbul, Turkey, pp. 1363-1366.

[22] Greefkes, J.A., "A Digitally Companded Delta Modulation Modem for Speech Transmission," *Proc. Int. Conf. on Communications*, 1970, pp. 7.33-7.48.

[23] ISO, *Coding of Moving Pictures and Associated Audio—Audio Part*, 1993, Int. Standards Organization.

[24] ISO/IEC, *Information Technology—Coding of Moving Pictures and Associated Audio for Digital Storage Media at up to about 1.5 Mbps, Part 3: Audio (MPEG-1)*, 1992, Int. Standards Organization.

[25] ITU-T, *Methods for Subjective Determination of Transmission Quality*, 1996, Int. Telecommunication Unit.

[26] Jarvinen, K., *et al.*, "GSM Enhanced Full Rate Speech Codec," *Int. Conf. on Acoustics, Speech and Signal Processing*, 1997, Munich, Germany, pp. 771-774.

[27] Jayant, N.S. and P. Noll, *Digital Coding of Waveforms*, 1984, Upper Saddle River, NJ, Prentice Hall.

[28] Johnston, J.D., *et al.*, "ATT Perceptual Audio Coding (PAC)" in *Audio Engineering Society (AES) Collected Papers on Digital Audio Bit Rate Reduction*, N. Gilchrist and C. Grewin, eds. 1996, pp. 73-82.

[29] Kleijn, W.B., "Continuous Representations in Linear Predictive Coding," *Int. Conf. on Acoustics, Speech and Signal Processing*, 1991, Toronto, Canada, pp. 201-204.

[30] Kleijn, W.B. and J. Haagen, "Transformation and Decomposition of the Speech Signal for Coding," *IEEE Signal Processing Letters*, 1994, **1**, pp. 136-138.

[31] Kleijn, W.B., D.J. Krasinski, and R.H. Ketchum, "An Efficient Stochastically Excited Linear Predictive Coding Algorithm for High Quality Low Bit Rate Transmission of Speech," *Speech Communication*, 1988, **7**, pp. 305-316.

[32] Kleijn, W.B. and K.K. Paliwal, *Speech Coding and Synthesis*, 1995, Amsterdam, Netherlands, Elsevier.

[33] Koishida, K., V. Cuperman, and A. Gersho, "A 16-KBIT/S Bandwidth Scalable Audio Coder Based on the G.729 Standard," *Int. Conf. on Acoustics, Speech and Signal Processing*, 2000, Istanbul, Turkey, pp. 1149-1152.

[34] Li, C. and V. Cuperman, "Analysis-by-Synthesis Multimode Harmonic Speech Coding at 4 kbps," *Int. Conf. on Acoustics, Speech and Signal Processing*, 2000, Istanbul, Turkey, pp. 1367-1370.

[35] McAulay, R.J. and T.F. Quateri, "Speech Analysis/Synthesis Based on a Sinusoidal Representation," *IEEE Trans. on Acoustics, Speech and Signal Processing*, 1986, **34**, pp. 744-754.

[36] McAulay, R.J. and T.F. Quateri, "Sinusoidal Coding" in *Speech Coding and Synthesis*, W.B. Kleijn and K.K. Paliwal, eds. 1995, Elsevier, pp. 121-174.

[37] McCree, A., "A 14 kbps Wideband Speech Coder with a Parametric Highband Model," *Int. Conf. on Acoustics, Speech and Signal Processing*, 2000, Istanbul, Turkey, pp. 1153-1156.

[38] McCree, A.V. and T.P. Barnwell, "Improving the Performance of a Mixed-Excitation LPC Vocoder in Acoustic Noise," *Int. Conf. on Acoustics, Speech and Signal Processing*, 1992, San Francisco, pp. II-137-138.

[39] McCree, A.V., *et al.*, "A 2.4 kbit/s MELP Coder Candidate for the New U.S. Federal Standard," *Int. Conf. on Acoustics, Speech and Signal Processing*, 1996, Atlanta, GA, pp. 200-203.

[40] Paez, M.D. and T.H. Glisson, "Minimum Squared-Error Quantization in Speech," *IEEE Trans. on Comm*, 1972, **20**, pp. 225-230.

[41] Painter, T. and A. Spanias, "A Review of Algorithms for Perceptual Coding of Digital Audio Signals," *Proc. Int. Conf. on DSP*, 1997, pp. 179-205.

[42] Painter, T. and A. Spanias, "Perceptual Coding of Digital Audio," *Proc. of IEEE*, 2000(April), pp. 451-513.

[43] Paliwal, K.K. and B. Atal, "Efficient Vector Quantization of LPC Parameters at 24 Bits/Frame," *IEEE Trans. on Speech and Audio Processing*, 1993, **1**(1), pp. 3-14.

[44] Prevez, M.A., H.V. Sorensen, and J.V.D. Spiegel, "An Overview of Sigma-Delta Converters," *IEEE Signal Processing Magazine*, 1996, **13**(1), pp. 61-84.

[45] Robinson, T., *Simple Lossless and Near-Lossless Waveform Compression*, 1994, Cambridge University Engineering Department.

[46] Salami, R., C. Laflamme, and B. Bessette, "Description of ITU-T Recommendation G.729 Annex A: Reduced Complexity 8 kbps CS-ACELP Codec," *Int. Conf. on Acoustics, Speech and Signal Processing*, 1997, Munich, Germany, pp. 775-778.

[47] Schouten, J.S., F.E. DeJager, and J.A. Greefkes, *Delta Modulation, a New Modulation System for Telecommunications*, 1952, Phillips, pp. 237-245.

[48] Shlomot, E., V. Cuperman, and A. Gersho, "Combined Harmonic and Waveform Coding of Speech at Low Bit Rates," *Int. Conf. on Acoustics, Speech and Signal Processing*, 1998, Seattle, WA, pp. 585-588.

[49] Singhal, S. and B.S. Atal, "Improving Performance of Multi-Pulse LPC Coders at Low Bit Rates," *Int. Conf. on Acoustics, Speech and Signal Processing*, 1984, San Diego, pp. 1.3.1-1.3.4.

[50] Skoglund, J., R. Cox, and J. Collura, "A Combined WI and MELP Coder at 5.2KBPS," *Int. Conf. on Acoustics, Speech and Signal Processing*, 2000, Istanbul, Turkey, pp. 1387-1390.

[51] Smith, B., "Instantaneous Companding of Quantized Signals," *Bell Systems Technical Journal*, 1957, **36**(3), pp. 653-709.

[52] Spanias, A.S., "Speech Coding: A Tutorial Review," *Proc. of the IEEE*, 1994, **82**(10), pp. 1441-1582.

[53] Stachurski, J. and A. McCree, "A 4 kbps Hybrid MELP/CELP Coder with Alignment Phase Encoding and Zero Phase Equalization," *Int. Conf. on Acoustics, Speech and Signal Processing*, 2000, Istanbul, Turkey, pp. 1379-1382.

[54] Todd, C., "AC-3: Flexible Perceptual Coding for Audio Transmission and Storage," *Audio Engineering Society 96th Convention*, 1994.

[55] Tremain, T.E., *The Government Standard Linear Predictive Coding Algorithm*, in *Speech Technology Magazine*, 1982, pp. 40-49.

[56] Vary, P., *et al.*, "A Regular-Pulse Excited Linear Predictive Code," *Speech Communication*, 1988, **7**(2), pp. 209-215.

[57] Wang, T., *et al.*, "A 1200 BPS Speech Coder Based on MELP," *Int. Conf. on Acoustics, Speech and Signal Processing*, 2000, Istanbul, Turkey, pp. 1375-1378.

PART III

SPEECH RECOGNITION

CHAPTER 8

Hidden Markov Models

The hidden Markov model (HMM) is a very powerful statistical method of characterizing the observed data samples of a discrete-time series. Not only can it provide an efficient way to build parsimonious parametric models, but can also incorporate the dynamic programming principle in its core for a unified pattern segmentation and pattern classification of time-varying data sequences. The data samples in the time series can be discretely or continuously distributed; they can be scalars or vectors. The underlying assumption of the HMM is that the data samples can be well characterized as a parametric random process, and the parameters of the stochastic process can be estimated in a precise and well-defined framework. The basic HMM theory was published in a series of classic papers by Baum and his colleagues [4]. The HMM has become one of the most powerful statistical methods for modeling speech signals. Its principles have been successfully used in automatic speech recognition, formant and pitch tracking, speech enhancement, speech synthesis, statistical language modeling, part-of-speech tagging, spoken language understanding, and machine translation [3, 4, 8, 10, 12, 18, 23, 37].

8.1. THE MARKOV CHAIN

A *Markov chain* models a class of random processes that incorporates a minimum amount of memory without being completely memoryless. In this subsection we focus on the discrete-time Markov chain only.

Let $\mathbf{X} = X_1, X_2, \ldots X_n$ be a sequence of random variables from a finite discrete alphabet $O = \{o_1, o_2, \ldots, o_M\}$. Based on the Bayes' rule, we have

$$P(X_1, X_2, \ldots, X_n) = P(X_1) \prod_{i=2}^{n} P(X_i \mid X_1^{i-1}) \tag{8.1}$$

where $X_1^{i-1} = X_1, X_2, \ldots, X_{i-1}$. The random variables X are said to form a first-order Markov chain, provided that

$$P(X_i | X_1^{i-1}) = P(X_i | X_{i-1}) \tag{8.2}$$

As a consequence, for the first-order Markov chain, Eq. (8.1) becomes

$$P(X_1, X_2, \ldots, X_n) = P(X_1) \prod_{i=2}^{n} P(X_i \mid X_{i-1}) \tag{8.3}$$

Equation (8.2) is also known as the *Markov assumption*. This assumption uses very little memory to model dynamic data sequences: the probability of the random variable at a given time depends only on the value at the preceding time. The Markov chain can be used to model time-invariant (stationary) events if we discard the time index *i*,

$$P(X_i = s | X_{i-1} = s') = P(s | s') \tag{8.4}$$

If we associate X_i to a state, the Markov chain can be represented by a finite state process with transition between states specified by the probability function $P(s|s')$. Using this finite state representation, the Markov assumption [Eq. (8.2)] is translated to the following: the probability that the Markov chain will be in a particular state at a given time depends only on the state of the Markov chain at the previous time.

Consider a Markov chain with *N* distinct states labeled by $\{1, \ldots, N\}$, with the state at time *t* in the Markov chain denoted as s_t; the parameters of a Markov chain can be described as follows:

$$a_{ij} = P(s_t = j \mid s_{t-1} = i) \quad 1 \le i, j \le N \tag{8.5}$$

$$\pi_i = P(s_1 = i) \quad 1 \le i \le N \tag{8.6}$$

where a_{ij} is the transition probability from state *i* to state *j*; and π_i is the initial probability that the Markov chain will start in state *i*. Both transition and initial probabilities are bound to the constraints:

$$\sum_{j=1}^{N} a_{ij} = 1; \quad 1 \le i \le N$$
$$\sum_{j=1}^{N} \pi_j = 1 \tag{8.7}$$

The Markov chain described above is also called the observable Markov model because the output of the process is the set of states at each time instance t, where each state corresponds to an observable event X_i. In other words, there is one-to-one correspondence between the observable event sequence $\mathbf{X}$ and the Markov chain state sequence $\mathbf{S} = s_1, s_2, \ldots s_n$. Consider a simple three-state Markov chain for the Dow Jones Industrial average as shown in Figure 8.1. At the end of each day, the Dow Jones Industrial average may correspond to one of the following states:

state 1 – *up* (in comparison to the index of previous day)
state 2 – *down* (in comparison to the index of previous day)
state 3 – *unchanged* (in comparison to the index of previous day)

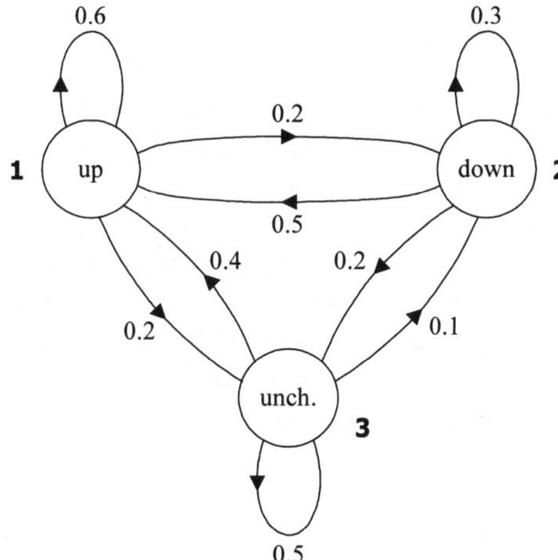

Figure 8.1 A Markov chain for the Dow Jones Industrial average. Three states represent *up*, *down*, and *unchanged*, respectively.

The parameter for this Dow Jones Markov chain may include a state-transition probability matrix

$$A = \{a_{ij}\} = \begin{bmatrix} 0.6 & 0.2 & 0.2 \\ 0.5 & 0.3 & 0.2 \\ 0.4 & 0.1 & 0.5 \end{bmatrix}$$

and an initial state probability matrix

$$\boldsymbol{\pi} = (\pi_i)^t = \begin{pmatrix} 0.5 \\ 0.2 \\ 0.3 \end{pmatrix}$$

Suppose you would like to find out the probability for five consecutive *up* days. Since the observed sequence of *up-up-up-up-up* corresponds to the state sequence of (1, 1, 1, 1, 1), the probability will be

$$P(5 \text{ consecutive } up \text{ days}) = P(1,1,1,1,1)$$
$$= \pi_1 a_{11} a_{11} a_{11} a_{11} = 0.5 \times (0.6)^4 = 0.0648$$

8.2. Definition of the Hidden Markov Model

In the Markov chain, each state corresponds to a deterministically observable event, i.e., the output of such sources in any given state is not random. A natural extension to the Markov chain introduces a non-deterministic process that generates output observation symbols in any given state. Thus, the observation is a probabilistic function of the state. This new model is known as a *hidden Markov model,* which can be viewed as a double-embedded stochastic process with an underlying stochastic process (the state sequence) not directly observable. This underlying process can only be probabilistically associated with another observable stochastic process producing the sequence of features we can observe.

A hidden Markov model is basically a Markov chain where the output observation is a random variable X generated according to a output probabilistic function associated with each state. Figure 8.2 shows a revised hidden Markov model for the Dow Jones Industrial average. You see that each state now can generate all three output observations: *up*, *down*, and *unchanged*, according to its output pdf. This means that there is no longer a one-to-one correspondence between the observation sequence and the state sequence, so you cannot unanimously determine the state sequence for a given observation sequence, i.e., the state sequence is not observable and therefore hidden. This is why the world *hidden* is placed in front of *Markov models*. Although the state of an HMM is hidden, it often contains salient information about the data we are modeling. For example, the first state in Figure 8.2 indicates a *bull* market, and the second state indicates a *bear* market as specified by the output probability in each state. Formally speaking, a hidden Markov model is defined by:

- $\mathrm{O} = \{o_1, o_2, \ldots, o_M\}$—An output observation alphabet.[1] The observation symbols correspond to the physical output of the system being modeled. In the case of the Dow Jones Industrial average HMM, the output observation alphabet is the collection of three categories—$O = \{up, down, unchanged\}$.

[1] Although we use the discrete output observation here, we can extend it to the continuous case with a continuous pdf. You can also use vector quantization to map a continuous vector variable into a discrete alphabet set.

- $\Omega = \{1, 2, \ldots, N\}$—A set of states representing the state space. Here s_t is denoted as the state at time t. In the case of the Dow Jones Industrial average HMM, the state may indicate a bull market, a bear market, and a stable market.
- $\mathbf{A} = \{a_{ij}\}$—A transition probability matrix, where a_{ij} is the probability of taking a transition from state *i* to state *j*, i.e.,

$$a_{ij} = P(s_t = j | s_{t-1} = i) \tag{8.8}$$

- $\mathbf{B} = \{b_i(k)\}$—An output probability matrix,[2] where $b_i(k)$ is the probability of emitting symbol o_k when state *i* is entered. Let $\mathbf{X} = X_1, X_2, \ldots, X_t, \ldots$ be the observed output of the HMM. The state sequence $S = s_1, s_2, \ldots, s_t, \ldots$ is not observed (hidden), and $b_i(k)$ can be rewritten as follows:

$$b_i(k) = P(X_t = o_k \mid s_t = i) \tag{8.9}$$

- $\boldsymbol{\pi} = \{\pi_i\}$—A initial state distribution where

$$\pi_i = P(s_0 = i) \quad 1 \le i \le N \tag{8.10}$$

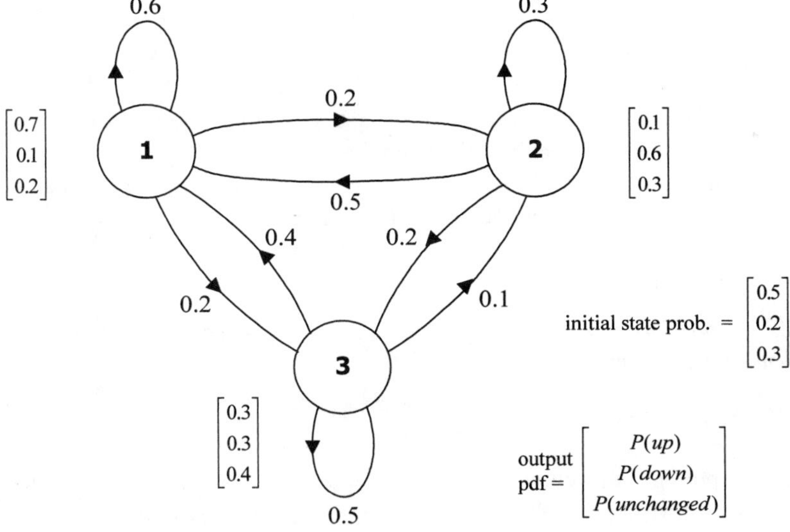

Figure 8.2 A hidden Markov model for the Dow Jones Industrial average. The three states no longer have deterministic meanings as in the Markov chain illustrated in Figure 8.1.

[2] The output distribution can also be transition-dependent. Although these two formulations look different, the state-dependent one can be reformulated as a transition-dependent one with the constraint of all the transitions entering the same state sharing the same output distribution.

Since a_{ij}, $b_i(k)$, and π_i are all probabilities, they must satisfy the following properties:

$$a_{ij} \geq 0, \quad b_i(k) \geq 0, \quad \pi_i \geq 0 \quad \forall \text{ all } i, j, k \tag{8.11}$$

$$\sum_{j=1}^{N} a_{ij} = 1 \tag{8.12}$$

$$\sum_{k=1}^{M} b_i(k) = 1 \tag{8.13}$$

$$\sum_{i=1}^{N} \pi_i = 1 \tag{8.14}$$

To sum up, a complete specification of an HMM includes two constant-size parameters, N and M, representing the total number of states and the size of observation alphabets, observation alphabet O, and three sets (matrices) of probability measures $\mathbf{A}$, $\mathbf{B}$, $\boldsymbol{\pi}$. For convenience, we use the following notation

$$\Phi = (\mathbf{A}, \mathbf{B}, \boldsymbol{\pi}) \tag{8.15}$$

to indicate the whole parameter set of an HMM and sometimes use the parameter set Φ to represent the HMM interchangeably without ambiguity.

In the first-order hidden Markov model discussed above, there are two assumptions. The first is the *Markov assumption* for the Markov chain.

$$P(s_t | s_1^{t-1}) = P(s_t | s_{t-1}) \tag{8.16}$$

where s_1^{t-1} represents the state sequence $s_1, s_2, \ldots, s_{t-1}$. The second is the *output-independence assumption*:

$$P(X_t | X_1^{t-1}, s_1^t) = P(X_t | s_t) \tag{8.17}$$

where X_1^{t-1} represents the output sequence $X_1, X_2, \ldots, X_{t-1}$. The output-independence assumption states that the probability that a particular symbol is emitted at time t depends only on the state s_t and is conditionally independent of the past observations.

You might argue that these assumptions limit the memory of the first-order hidden Markov models and may lead to model deficiency. However, in practice, they make evaluation, decoding, and learning feasible and efficient without significantly affecting the modeling capability, since those assumptions greatly reduce the number of parameters that need to be estimated.

Given the definition of HMMs above, three basic problems of interest must be addressed before they can be applied to real-world applications.

1. **The Evaluation Problem**—Given a model Φ and a sequence of observations $\mathbf{X} = (X_1, X_2, \ldots, X_T)$, what is the probability $P(\mathbf{X}|\Phi)$, i.e., the probability that the model generates the observations?
2. **The Decoding Problem**—Given a model Φ and a sequence of observations $\mathbf{X} = (X_1, X_2, \ldots, X_T)$, what is the most likely state sequence $\mathbf{S} = (s_0, s_1, s_2, \ldots, s_T)$ in the model that produces the observations?
3. **The Learning Problem**—Given a model Φ and a set of observations, how can we adjust the model parameter $\hat{\Phi}$ to maximize the joint probability (likelihood) $\prod_{\mathbf{X}} P(\mathbf{X} \mid \Phi)$?

If we could solve the *evaluation* problem, we would have a way of evaluating how well a given HMM matches a given observation sequence. Therefore, we could use HMM to do pattern recognition, since the likelihood $P(\mathbf{X}|\Phi)$ can be used to compute posterior probability $P(\Phi|\mathbf{X})$, and the HMM with highest posterior probability can be determined as the desired pattern for the observation sequence. If we could solve the decoding problem, we could find the best matching state sequence given an observation sequence, or, in other words, we could uncover the hidden state sequence. As discussed in Chapters 12 and 13, these are the basics for the decoding in continuous speech recognition. Last but not least, if we could solve the learning problem, we would have the means to automatically estimate the model parameter Φ from an ensemble of training data. These three problems are tightly linked under the same probabilistic framework. The efficient implementation of these algorithms shares the principle of dynamic programming that we briefly discuss next.

8.2.1. Dynamic Programming and DTW

The dynamic programming concept, also known as *dynamic time warping* (DTW) in speech recognition [40], has been widely used to derive the overall distortion between two speech templates. In these template-based systems, each speech template consists of a sequence of speech vectors. The overall distortion measure is computed from the accumulated distance between two feature vectors that are aligned between two speech templates with minimal overall distortion. The DTW method can warp two speech templates ($\mathbf{x}_1\mathbf{x}_2\ldots\mathbf{x}_N$) and ($\mathbf{y}_1\mathbf{y}_2\ldots\mathbf{y}_M$) in the time dimension to alleviate nonlinear distortion as illustrated in Figure 8.3.

This is roughly equivalent to the problem of finding the minimum distance in the trellis between these two templates. Associated with every pair (*i, j*) is a distance $d(i, j)$ between two speech vectors $\mathbf{x}_i$ and $\mathbf{y}_j$. To find the optimal path between starting point (1, 1) and end point (*N, M*) from left to right, we need to compute the optimal accumulated distance $D(N, M)$. We can enumerate all possible accumulated distance from (1,1) to (*N, M*) and identify the one that has the minimum distance. Since there are *M* possible moves for each step from left to right in Figure 8.3, all the possible paths from (1, 1) to (*N, M*) will be exponential. Dynamic programming principles can drastically reduce the amount of computation by avoiding the enumeration of sequences that cannot possibly be optimal. Since the same optimal path after each step must be based on the previous step, the minimum distance $D(i, j)$ must satisfy the following equation:

$$D(i,j) = \min_k \left[D(i-1,k) + d(k,j) \right] \tag{8.18}$$

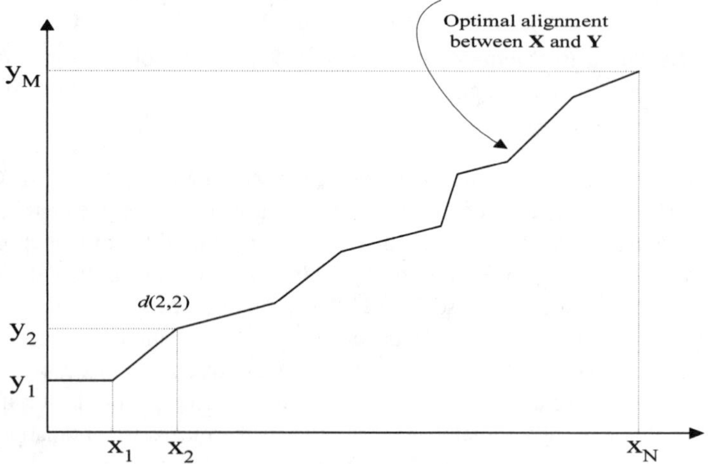

Figure 8.3 Direct comparison between two speech templates $\mathbf{X}=\mathbf{x}_1\mathbf{x}_2...\mathbf{x}_N$ and $\mathbf{Y}=\mathbf{y}_1\mathbf{y}_2...\mathbf{y}_M$.

Equation (8.18) indicates you only need to consider and keep only the best move for each pair although there are M possible moves. The recursion allows the optimal path search to be conducted incrementally from left to right. In essence, dynamic programming delegates the solution recursively to its own sub-problem. The computation proceeds from the small sub-problem ($D(i-1,k)$) to the larger sub-problem ($D(i,j)$). We can identify the optimal match $\mathbf{y}_j$ with respect to $\mathbf{x}_i$ and save the index in a back pointer table $B(i, j)$ as we move forward. The optimal path can be traced back after the optimal path is identified. The algorithm is described in Algorithm 8.1.

The advantage of the dynamic programming lies in the fact that once a sub-problem is solved, the partial result can be stored and never needs to be recalculated. This is a very important principle that you see again and again in building practical spoken language systems.

Speech recognition based on DTW is simple to implement and fairly effective for small-vocabulary speech recognition. Dynamic programming can temporally align patterns to account for differences in speaking rates across talkers as well as across repetitions of the word by the same talker. However, it does not have a principled way to derive an averaged template for each pattern from a large amount of training samples. A multiplicity of reference training tokens is typically required to characterize the variation among different utterances. As such, the HMM is a much better alternative for spoken language processing.

ALGORITHM 8.1: *THE DYNAMIC PROGRAMMING ALGORITHM*

Step 1: Initialization

$D(1,1) = d(1,1), B(1,1) = 1$, for $j = 2,\ldots,M$ compute $D(1,j) = \infty$

Step 2: Iteration

for $i = 2,\ldots,N$ {

for $j = 1,\ldots,M$ compute {

$$D(i,j) = \min_{1\le p\le M}\left[D(i-1,p)+d(p,j)\right]$$

$$B(i,j) = \arg\min_{1\le p\le M}\left[D(i-1,p)+d(p,j)\right] \;\}\}$$

Step 3: Backtracking and Termination

The optimal (minimum) distance is $D(N,M)$ and the optimal path is $(s_1, s_2, \ldots, s_N)$ where $s_N = M$ and $s_i = B(i+1, s_{i+1})$, $i = N-1, N-2, \ldots, 1$

8.2.2. How to Evaluate an HMM—The Forward Algorithm

To calculate the probability (likelihood) $P(\mathbf{X}|\Phi)$ of the observation sequence $\mathbf{X} = (X_1, X_2, \ldots, X_T)$, given the HMM Φ, the most intuitive way is to sum up the probabilities of all possible state sequences:

$$P(\mathbf{X} \mid \Phi) = \sum_{\text{all } \mathbf{S}} P(\mathbf{S} \mid \Phi) P(\mathbf{X} \mid \mathbf{S}, \Phi) \tag{8.19}$$

In other words, to compute $P(\mathbf{X}|\Phi)$, we first enumerate all possible state sequences $\mathbf{S}$ of length T, that generate observation sequence $\mathbf{X}$, and then sum all the probabilities. The probability of each path $\mathbf{S}$ is the product of the state sequence probability (first factor) and the joint output probability (the second factor) along the path.

For one particular state sequence $\mathbf{S} = (s_1, s_2, \ldots, s_T)$, where s_1 is the initial state, the state-sequence probability in Eq. (8.19) can be rewritten by applying Markov assumption:

$$P(\mathbf{S} \mid \Phi) = P(s_1 \mid \Phi) \prod_{t=2}^{T} P(s_t | s_{t-1}, \Phi) = \pi_{s_1} a_{s_1 s_2} \ldots a_{s_{T-1} s_T} = a_{s_0 s_1} a_{s_1 s_2} \ldots a_{s_{T-1} s_T} \tag{8.20}$$

where $a_{s_0 s_1}$ denotes π_{s_1} for simplicity. For the same state sequence $\mathbf{S}$, the joint output probability along the path can be rewritten by applying the output-independent assumption:

$$\begin{aligned} P(\mathbf{X} \mid \mathbf{S}, \Phi) &= P(X_1^T \mid S_1^T, \Phi) = \prod_{t=1}^{T} P(X_t | s_t, \Phi) \\ &= b_{s_1}(X_1) b_{s_2}(X_2) \ldots b_{s_T}(X_T) \end{aligned} \tag{8.21}$$

Substituting Eqs. (8.20) and (8.21) into (8.19), we get:

$$\begin{aligned} P(\mathbf{X} \mid \Phi) &= \sum_{\text{all } \mathbf{S}} P(\mathbf{S} \mid \Phi) P(\mathbf{X} \mid \mathbf{S}, \Phi) \\ &= \sum_{\text{all } \mathbf{S}} a_{s_0 s_1} b_{s_1}(X_1) a_{s_1 s_2} b_{s_2}(X_2) \ldots a_{s_{T-1} s_T} b_{s_T}(X_T) \end{aligned} \tag{8.22}$$

Equation (8.22) can be interpreted as follows. First we enumerate all possible state sequence with length *T*. For any given state sequence, we start from initial state s_1 with probability π_{s_1} or $a_{s_0 s_1}$. We take a transition from s_{t-1} to s_t with probability $a_{s_{t-1} s_t}$ and generate the observation X_t with probability $b_{s_t}(X_t)$ until we reach the last transition.

However, direct evaluation of Eq. (8.22) according to the interpretation above requires enumeration of $O(N^T)$ possible state sequences, which results in exponential computational complexity. Fortunately, a more efficient algorithm can be used to calculate Eq. (8.22). The trick is to store intermediate results and use them for subsequent state-sequence calculations to save computation. This algorithm is known as the *forward algorithm.*

Based on the HMM assumptions, the calculation of $P(s_t \mid s_{t-1}, \Phi) P(X_t \mid s_t, \Phi)$ involves only s_{t-1}, s_t, and X_t, so, it is possible to compute the likelihood with $P(\mathbf{X}|\Phi)$ with recursion on *t*. Let's define forward probability:

$$\alpha_t(i) = P(X_1^t, s_t = i | \Phi) \tag{8.23}$$

$\alpha_t(i)$ is the probability that the HMM is in state *i* at time t having generated partial observation X_1^t (namely $X_1 X_2 \ldots X_t$). $\alpha_t(i)$ can be calculated inductively as illustrated in Algorithm 8.2. This inductive procedure shown in Eq. (8.24) can be illustrated in a *trellis*. Figure 8.4 illustrates the computation of forward probabilities α via a trellis framework for the Dow Jones Industrial average HMM shown in Figure 8.2. Given two consecutive *up* days for the Dow Jones Industrial average, we can find the forward probability α based on the model of Figure 8.2. An arrow in Figure 8.4 indicates a transition from its origin state to its destination state. The number inside each cell indicates the forward probability α. We start the α cells from $t = 0$, where the α cells contains exactly the initial probabilities. The other cells are computed in a *time-synchronous* fashion from left to right, where each cell for time *t* is completely computed before proceeding to time *t*+1. When the states in the last column have been computed, the sum of all probabilities in the final column is the probability of generating the observation sequence. For most speech problems, we need to have the HMM end in some particular exit state (a.k.a final state, S_F), and we thus have $P(\mathbf{X} \mid \Phi) = \alpha_T(s_F)$.

It is easy to show that the complexity for the forward algorithm is $O(N^2 T)$ rather than exponential. This is because we can make full use of partially computed probabilities for the improved efficiency.

ALGORITHM 8.2: *THE FORWARD ALGORITHM*

Step 1: Initialization

$\alpha_1(i) = \pi_i b_i(\mathrm{X}_1) \qquad 1 \le i \le N$

Step 2: Induction

$$\alpha_t(j) = \left[\sum_{i=1}^{N} \alpha_{t-1}(i) a_{ij} \right] b_j(X_t) \qquad 2 \le t \le T; \; 1 \le j \le N \tag{8.24}$$

Step 3: Termination

$P(\mathbf{X}|\Phi) = \sum_{i=1}^{N} \alpha_T(i)$ If it is required to end in the final state, $P(\mathbf{X} | \Phi) = \alpha_T(s_F)$

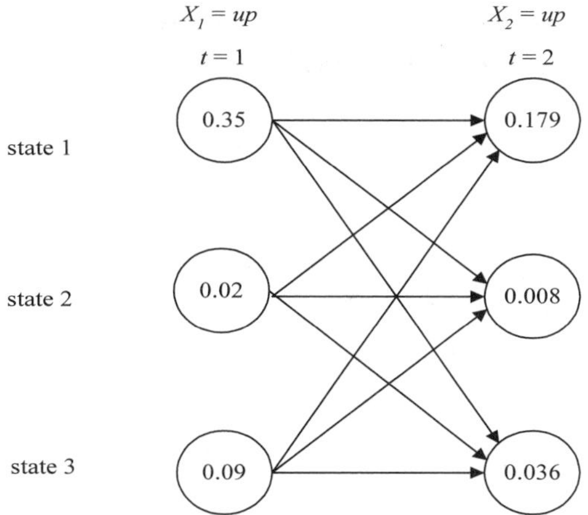

Figure 8.4 The forward trellis computation for the HMM of the Dow Jones Industrial average.

8.2.3. How to Decode an HMM—The Viterbi Algorithm

The forward algorithm, in the previous section, computes the probability that an HMM generates an observation sequence by summing up the probabilities of all possible paths, so it does not provide the best path (or state sequence). In many applications, it is desirable to find such a path. As a matter of fact, finding the best path (state sequence) is the cornerstone for searching in continuous speech recognition. Since the state sequence is hidden (unobserved) in the HMM framework, the most widely used criterion is to find the state sequence that has the highest probability of being taken while generating the observation sequence. In other words, we are looking for the state sequence $\mathbf{S} = (s_1, s_2, \ldots, s_T)$ that maximizes $P(\mathbf{S}, \mathbf{X} | \Phi)$. This problem is very similar to the optimal-path problem in dynamic programming. As a consequence, a formal technique based on dynamic programming, known as *Viterbi* algorithm [43], can be used to find the best state sequence for an HMM. In practice, the same method can be used to evaluate HMMs that offers an approximate solution close to the case obtained using the forward algorithm described in Section 8.2.2.

The Viterbi algorithm can be regarded as the dynamic programming algorithm applied to the HMM or as a modified forward algorithm. Instead of summing up probabilities from different paths coming to the same destination state, the Viterbi algorithm picks and remembers the best path. To define the best-path probability:

$$V_t(i) = P(X_1^t, S_1^{t-1}, s_t = i \mid \Phi)$$

$V_t(i)$ is the probability of the most likely state sequence at time t, which has generated the observation X_1^t (until time t) and ends in state i. A similar induction procedure for the Viterbi algorithm can be described in Algorithm 8.3.

ALGORITHM 8.3: *THE VITERBI ALGORITHM*

Step 1: Initialization

$V_1(i) = \pi_i b_i(X_1) \qquad 1 \le i \le N$

$B_1(i) = 0$

Step 2: Induction

$$V_t(j) = \underset{1\le i\le N}{Max}\left[V_{t-1}(i)a_{ij}\right]b_j(X_t) \quad 2 \le t \le T; \quad 1 \le j \le N \tag{8.25}$$

$$B_t(j) = \underset{1\le i\le N}{Arg\max}\left[V_{t-1}(i)a_{ij}\right] \quad 2 \le t \le T; \quad 1 \le j \le N \tag{8.26}$$

Step 3: Termination

The best score $= \underset{1\le i\le N}{Max}\left[V_t(i)\right]$

$s_T^* = \underset{1\le i\le N}{Arg\max}\left[B_T(i)\right]$

Step 4: Backtracking

$s_t^* = B_{t+1}(s_{t+1}^*) \quad t = T-1, T-2, \ldots, 1$

$\mathbf{S}^* = (s_1^*, s_2^*, \ldots, s_T^*)$ is the best sequence

This Viterbi algorithm can also be illustrated in a trellis framework similar to the one for the forward algorithm shown in Figure 8.4. Instead of summing up all the paths, Figure 8.5 illustrates the computation of V_t by picking the best path in each cell. The number inside each cell indicates the best score V_t and the best path leading to each cell is indicated by solid lines while the rest of the paths are indicated by dashed line. Again, the computation is done in a *time-synchronous* fashion from left to right. The complexity for the Viterbi algorithm is also $O(N^2T)$.

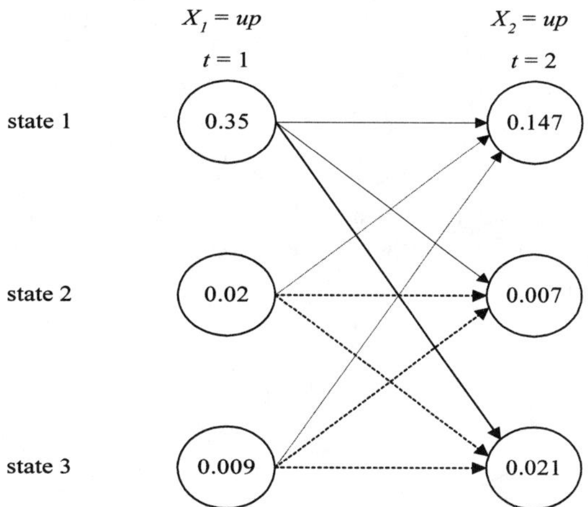

Figure 8.5 The Viterbi trellis computation for the HMM of the Dow Jones Industrial average.

8.2.4. How to Estimate HMM Parameters—Baum-Welch Algorithm

It is very important to estimate the model parameters $\mathbf{\Phi} = (\mathbf{A}, \mathbf{B}, \boldsymbol{\pi})$ to accurately describe the observation sequences. This is by far the most difficult of the three problems, because there is no known analytical method that maximizes the joint probability of the training data in a closed form. Instead, the problem can be solved by the iterative *Baum-Welch* algorithm, also known as the *forward-backward* algorithm.

The HMM learning problem is a typical case of unsupervised learning discussed in Chapter 4, where the data is incomplete because of the hidden state sequence. The EM algorithm is perfectly suitable for this problem. As a matter of fact, Baum and colleagues used the same principle as that of the EM algorithm. Before we describe the formal Baum-Welch algorithm, we first define a few useful terms. In a manner similar to the forward probability, we define backward probability as:

$$\beta_t(i) = P(X_{t+1}^T | s_t = i, \mathbf{\Phi}) \tag{8.27}$$

where $\beta_t(i)$ is the probability of generating partial observation X_{t+1}^T (from t+1 to the end) given that the HMM is in state i at time t, $\beta_t(i)$ can then be calculated inductively;

Initialization:

$$\beta_T(i) = 1/N \qquad 1 \le i \le N$$

Induction:

$$\beta_t(i) = \left[\sum_{j=1}^{N} a_{ij} b_j(X_{t+1}) \beta_{t+1}(j) \right] \quad t=T-1\ldots1; \quad 1 \le i \le N \tag{8.28}$$

The relationship of adjacent α and β (α_{t-1} & α_t and β_t & β_{t+1}) can be best illustrated as shown in Figure 8.6. α is computed recursively from left to right, and β recursively from right to left.

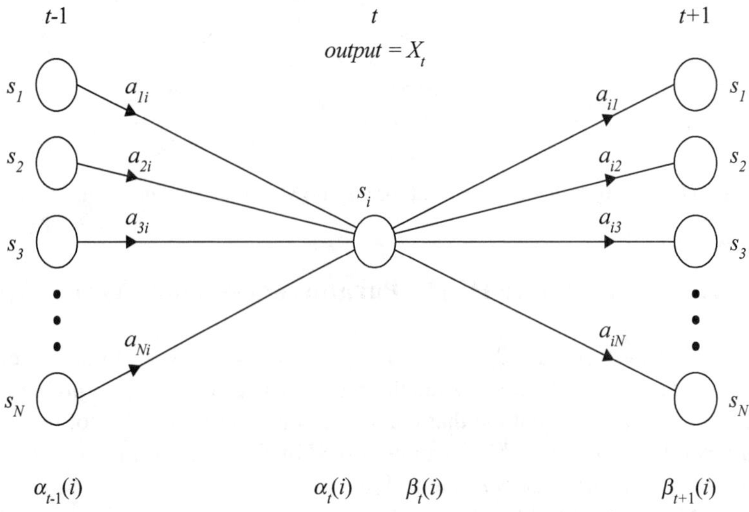

Figure 8.6 The relationship of α_{t-1} and α_t and β_t and β_{t+1} in the forward-backward algorithm.

Next, we define $\gamma_t(i,j)$, which is the probability of taking the transition from state i to state j at time t, given the model and observation sequence, i.e.,

$$\begin{aligned} \gamma_t(i,j) &= P(s_{t-1}=i, s_t=j \mid X_1^T, \Phi) \\ &= \frac{P(s_{t-1}=i, s_t=j, X_1^T \mid \Phi)}{P(X_1^T \mid \Phi)} \\ &= \frac{\alpha_{t-1}(i) a_{ij} b_j(X_t) \beta_t(j)}{\sum_{k=1}^{N} \alpha_T(k)} \end{aligned} \tag{8.29}$$

The equation above can be best illustrated as shown in Figure 8.7.

We can iteratively refine the HMM parameter vector $\Phi = (\mathbf{A}, \mathbf{B}, \boldsymbol{\pi})$ by maximizing the likelihood $P(\mathbf{X}|\Phi)$ for each iteration. We use $\hat{\Phi}$ to denote the new parameter vector derived from the parameter vector Φ in the previous iteration. According to the EM algorithm of Chapter 4, the maximization process is equivalent to maximizing the following Q-function:

$$Q(\Phi,\hat{\Phi}) = \sum_{\text{all } \mathbf{S}} \frac{P(\mathbf{X},\mathbf{S}\mid\Phi)}{P(\mathbf{X}\mid\Phi)} \log P(\mathbf{X},\mathbf{S}\mid\hat{\Phi}) \tag{8.30}$$

where $P(\mathbf{X},\mathbf{S}|\Phi)$ and $\log P(\mathbf{X},\mathbf{S}|\hat{\Phi})$ can be expressed as:

$$P(\mathbf{X},\mathbf{S}\mid\Phi) = \prod_{t=1}^{T} a_{s_{t-1}s_t} b_{s_t}(X_t) \tag{8.31}$$

$$\log P(\mathbf{X},\mathbf{S}\mid\Phi) = \sum_{t=1}^{T} \log a_{s_{t-1}s_t} + \sum_{t=1}^{T} \log b_{s_t}(X_t) \tag{8.32}$$

Equation (8.30) can thus be rewritten as

$$Q(\Phi,\hat{\Phi}) = Q_{\mathbf{a}_i}(\Phi,\hat{\mathbf{a}}_i) + Q_{\mathbf{b}_j}(\Phi,\hat{\mathbf{b}}_j) \tag{8.33}$$

where

$$Q_{\mathbf{a}_i}(\Phi,\hat{\mathbf{a}}_i) = \sum_i \sum_j \sum_t \frac{P(\mathbf{X}, s_{t-1}=i, s_t=j\mid\Phi)}{P(\mathbf{X}\mid\Phi)} \log \hat{a}_{ij} \tag{8.34}$$

$$Q_{\mathbf{b}_j}(\Phi,\hat{\mathbf{b}}_j) = \sum_j \sum_k \sum_{t\in X_t=o_k} \frac{P(\mathbf{X}, s_t=j\mid\Phi)}{P(\mathbf{X}\mid\Phi)} \log \hat{b}_j(k) \tag{8.35}$$

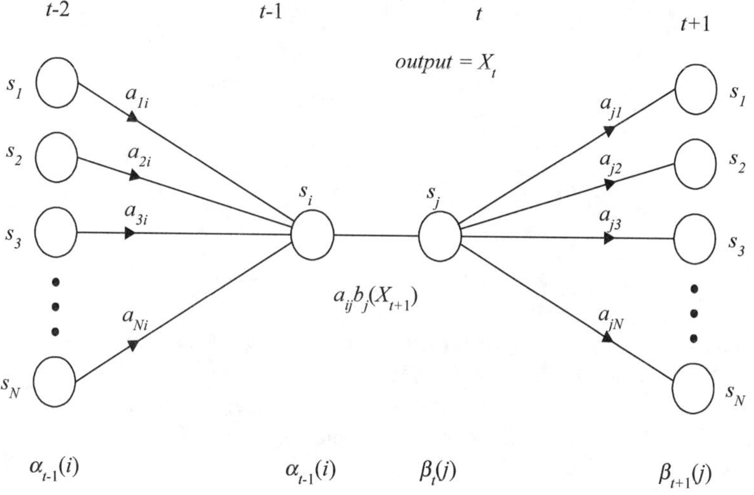

Figure 8.7 Illustration of the operations required for the computation of $\gamma_t(i,j)$, which is the probability of taking the transition from state i to state j at time t.

Since we separate the Q-function into two independent terms, the maximization procedure on $Q(\Phi,\hat{\Phi})$ can be done by maximizing the individual terms separately, subject to probability constraints.

$$\sum_{j=1}^{N} a_{ij} = 1 \quad \forall \text{ all } i \tag{8.36}$$

$$\sum_{k=1}^{M} b_j(k) = 1 \quad \forall \text{ all } j \tag{8.37}$$

Moreover, all these terms in Eqs. (8.34) and (8.35) have the following form:

$$F(x) = \sum_i y_i \log x_i \tag{8.38}$$

where $\sum_i x_i = 1$

By using the Lagrange multipliers, the function above can be proved to achieve maximum value at

$$x_i = \frac{y_i}{\sum_i y_i} \tag{8.39}$$

Using this formation, we obtain the model estimate as[3]:

$$\hat{a}_{ij} = \frac{\dfrac{1}{P(\mathbf{X}|\Phi)}\sum_{t=1}^{T} P(\mathbf{X}, s_{t-1}=i, s_t=j|\Phi)}{\dfrac{1}{P(\mathbf{X}|\Phi)}\sum_{t=1}^{T} P(\mathbf{X}, s_{t-1}=i|\Phi)} = \frac{\sum_{t=1}^{T}\gamma_t(i,j)}{\sum_{t=1}^{T}\sum_{k=1}^{N}\gamma_t(i,k)} \tag{8.40}$$

$$\hat{b}_j(k) = \frac{\dfrac{1}{P(\mathbf{X}|\Phi)}\sum_{t=1}^{T} P(\mathbf{X}, s_t=j|\Phi)\delta(X_t, o_k)}{\dfrac{1}{P(\mathbf{X}|\Phi)}\sum_{t=1}^{T} P(\mathbf{X}, s_t=j|\Phi)} = \frac{\sum_{t\in X_t=o_k}\sum_i \gamma_t(i,j)}{\sum_{t=1}^{T}\sum_i \gamma_t(i,j)} \tag{8.41}$$

By carefully examining the HMM re-estimation Eqs. (8.40) and (8.41), you can see that Eq. (8.40) is essentially the ratio between the expected number of transitions from state i to state j and the expected number of transitions from state i. For the output probability re-estimation Eq. (8.41), the numerator is the expected number of times the observation data

[3] Notice that the initial probability $\hat{\pi}_i$ can be derived as a special case of the transition probability. $\hat{\pi}_i$ is often fixed (i.e., $\hat{\pi}_i = 1$ for the initial state) for most speech applications.

emitted from state j is the observation symbol o_k, and the denominator is the expected number of times the observation data is emitted from state j.

According to the EM algorithm, the forward-backward (Baum-Welch) algorithm guarantees a monotonic likelihood improvement on each iteration, and eventually the likelihood converges to a local maximum. The forward-backward algorithm can be described in a way similar to the general EM algorithm as shown in Algorithm 8.4.

ALGORITHM 8.4: *THE FORWARD-BACKWARD ALGORITHM*

Step 1: Initialization: Choose an initial estimate Φ.

Step 2: E-step: Compute auxiliary function $Q(\Phi, \hat{\Phi})$ based on Φ.

Step 3: M-step: Compute $\hat{\Phi}$ according to the re-estimation Eqs. (8.40) and (8.41) to maximize the auxiliary Q-function.

Step 4: Iteration: Set $\Phi = \hat{\Phi}$, repeat from step 2 until convergence.

Although the forward-backward algorithm described above is based on one training observation sequence, it can be easily generalized to multiple training observation sequences under the independence assumption between these sequences. To train an HMM from M data sequences is equivalent to finding the HMM parameter vector Φ that maximizes the joint likelihood:

$$\prod_{i=1}^{M} P(\mathbf{X}_i \mid \Phi) \tag{8.42}$$

The training procedure performs the forward-backward algorithm on each independent observation sequence to calculate the expectations (or sometimes referred to as counts) in Eqs. (8.40) and (8.41). These counts in the denominator and numerator, respectively, can be added across M data sequences respectively. Finally, all the model parameters (probabilities) are normalized to make them sum up to one. This constitutes one iteration of Baum-Welch re-estimation; iteration continues until convergence. This procedure is practical and useful because it allows you to train a good HMM in a typical speech recognition scenario where a large amount of training data is available.

For example, if we let $\gamma_t^m(i,j)$ denote the $\gamma_t(i,j)$ from the mth data sequence and T^m denote the corresponding length, Eq. (8.40) can be extended as:

$$\hat{a}_{ij} = \frac{\sum_{m=1}^{M}\sum_{t=1}^{T^m} \gamma_t^m(i,j)}{\sum_{m=1}^{M}\sum_{t=1}^{T^m}\sum_{k=1}^{N} \gamma_t^m(i,k)} \tag{8.43}$$

8.3. CONTINUOUS AND SEMICONTINUOUS HMMS

If the observation does not come from a finite set, but from a continuous space, the discrete output distribution discussed in the previous sections needs to be modified. The difference between the discrete and the continuous HMM lies in a different form of output probability functions. For speech recognition, use of continuous HMMs implies that the quantization procedure to map observation vectors from the continuous space to the discrete space for the discrete HMM is no longer necessary. Thus, the inherent quantization error can be eliminated.

8.3.1. Continuous Mixture Density HMMs

In choosing continuous output probability density functions $b_j(\mathbf{x})$, the first candidate is multivariate Gaussian mixture density functions. This is because they can approximate any continuous probability density function, as discussed in Chapter 3. With M Gaussian mixture density functions, we have:

$$b_j(\mathbf{x}) = \sum_{k=1}^{M} c_{jk} N(\mathbf{x}, \boldsymbol{\mu}_{jk}, \boldsymbol{\Sigma}_{jk}) = \sum_{k=1}^{M} c_{jk} b_{jk}(\mathbf{x}) \tag{8.44}$$

where $N(\mathbf{x}, \boldsymbol{\mu}_{jk}, \boldsymbol{\Sigma}_{jk})$ or $b_{jk}(\mathbf{x})$ denotes a single Gaussian density function with mean vector $\boldsymbol{\mu}_{jk}$ and covariance matrix $\boldsymbol{\Sigma}_{jk}$ for state j, M denotes the number of mixture-components, and c_{jk} is the weight for the kth mixture component satisfying

$$\sum_{k=1}^{M} c_{jk} = 1 \tag{8.45}$$

To take the same *divide and conquer* approach as Eq. (8.33), we need to express $b_j(\mathbf{x})$ with respect to each single mixture component as:

$$p(\mathbf{X}, \mathbf{S} \mid \Phi) = \prod_{t=1}^{T} a_{s_{t-1}s_t} b_{s_t}(\mathbf{x}_t) = \sum_{k_1=1}^{M} \sum_{k_2=1}^{M} \cdots \sum_{k_T=1}^{M} \{ \prod_{t=1}^{T} a_{s_{t-1}s_t} b_{s_t k_t}(\mathbf{x}_{s_t}) c_{s_t k_t} \} \tag{8.46}$$

Equation (8.46) can be considered as the summation of densities with all the possible state sequences $\boldsymbol{S}$ and all the possible mixture components $\boldsymbol{K}$, defined in Ω^T as the Tth Cartesian product of $\Omega = \{1, 2, ..., M\}$, as follows:

$$p(\mathbf{X}, \mathbf{S}, \mathbf{K} \mid \Phi) = \prod_{t=1}^{T} a_{s_{t-1}s_t} b_{s_t k_t}(\mathbf{x}_t) c_{s_t k_t} \tag{8.47}$$

Therefore, the joint probability density is

$$p(\mathbf{X}|\Phi) = \sum_{\mathbf{S}} \sum_{\mathbf{K}\in\Omega^T} p(\mathbf{X},\mathbf{S},\mathbf{K}|\Phi) \tag{8.48}$$

An auxiliary function $Q(\Phi,\hat{\Phi})$ of two model points, Φ and $\hat{\Phi}$, given an observation **X**, can be written as:

$$Q(\Phi,\hat{\Phi}) = \sum_{\mathbf{S}} \sum_{\mathbf{K}\in\Omega^T} \frac{p(\mathbf{X},\mathbf{S},\mathbf{K}|\Phi)}{p(\mathbf{X}|\Phi)} \log p(\mathbf{X},\mathbf{S},\mathbf{K}|\hat{\Phi}) \tag{8.49}$$

From (8.47), the following decomposition can be shown:

$$\begin{aligned} &\log p(\mathbf{X},\mathbf{S},\mathbf{K}|\hat{\Phi}) \\ &= \sum_{t=1}^{T} \log \hat{a}_{s_{t-1}s_t} + \sum_{t=1}^{T} \log \hat{b}_{s_t k_t}(\mathbf{x}_t) + \sum_{t=1}^{T} \log \hat{c}_{s_t k_t} \end{aligned} \tag{8.50}$$

Maximization of the likelihood by way of re-estimation can be accomplished on individual parameter sets owing to the separability shown in (8.50). The separation of (8.50) is the key to the increased versatility of a re-estimation algorithm in accommodating mixture observation densities. The auxiliary function can be rewritten in a separated form in a similar manner as Eq. (8.33):

$$\begin{aligned} Q(\Phi,\hat{\Phi}) &= \sum_{\mathbf{S}} \sum_{\mathbf{K}} \frac{p(\mathbf{X},\mathbf{S},\mathbf{K}|\Phi)}{p(\mathbf{X}|\Phi)} \log p(\mathbf{X},\mathbf{S},\mathbf{K}|\hat{\Phi}) \\ &= \sum_{i=1}^{N} Q_{\mathbf{a}_i}(\Phi,\hat{\mathbf{a}}_i) + \sum_{j=1}^{N} \sum_{k=1}^{M} Q_{\mathbf{b}_{jk}}(\Phi,\hat{\mathbf{b}}_{jk}) + \sum_{j=1}^{N} \sum_{k=1}^{M} Q_{\mathbf{c}_{jk}}(\Phi,\hat{\mathbf{c}}_{jk}) \end{aligned} \tag{8.51}$$

The only difference we have is:

$$Q_{\mathbf{b}_{jk}}(\Phi,\hat{\mathbf{b}}_{jk}) = \sum_{t=1}^{T} p(s_t = j, k_t = k|\mathbf{X},\Phi) \log \hat{b}_{jk}(\mathbf{x}_t), \tag{8.52}$$

and

$$Q_{\mathbf{c}_{jk}}(\Phi,\hat{\mathbf{c}}_{jk}) = \sum_{t=1}^{T} p(s_t = j, k_t = k|\mathbf{X},\Phi) \log \hat{c}_{jk} \tag{8.53}$$

The optimization procedure is similar to what is discussed in the discrete HMM. The only major difference is $Q_{\mathbf{b}_{jk}}(\Phi,\hat{\mathbf{b}}_{jk})$. Maximization of $Q_{\mathbf{b}_{jk}}(\Phi,\hat{\mathbf{b}}_{jk})$ with respect to $\hat{\mathbf{b}}_{jk}$ is obtained through differentiation with respect to $\{\boldsymbol{\pi}_{jk}, \boldsymbol{\Sigma}_{jk}\}$ that satisfies:

$$\nabla_{\hat{\mathbf{b}}_{jk}} Q_{\mathbf{b}_{ijk}}(\Phi,\hat{\mathbf{b}}_{jk}) = 0 \tag{8.54}$$

The solutions are:

$$\hat{\boldsymbol{\mu}}_{jk} = \frac{\sum_{t=1}^{T} \frac{p(\mathbf{X}, s_t = j, k_t = k \mid \Phi)}{p(\mathbf{X} \mid \Phi)} \mathbf{x}_t}{\sum_{t=1}^{T} \frac{p(\mathbf{X}, s_t = j, k_t = k \mid \Phi)}{p(\mathbf{X} \mid \Phi)}} = \frac{\sum_{t=1}^{T} \zeta_t(j,k)\mathbf{x}_t}{\sum_{t=1}^{T} \zeta_t(j,k)} \tag{8.55}$$

$$\begin{aligned} \hat{\boldsymbol{\Sigma}}_{jk} &= \frac{\sum_{t=1}^{T} \frac{p(\mathbf{X}, s_t = j, k_t = k \mid \Phi)}{p(\mathbf{X} \mid \Phi)} (\mathbf{x}_t - \hat{\boldsymbol{\mu}}_{jk})(\mathbf{x}_t - \hat{\boldsymbol{\mu}}_{jk})^t}{\sum_{t=1}^{T} \frac{p(\mathbf{X}, s_t = j, k_t = k \mid \Phi)}{p(\mathbf{X} \mid \Phi)}} \\ &= \frac{\sum_{t=1}^{T} \zeta_t(j,k)(\mathbf{x}_t - \hat{\boldsymbol{\mu}}_{jk})(\mathbf{x}_t - \hat{\boldsymbol{\mu}}_{jk})^t}{\sum_{t=1}^{T} \zeta_t(j,k)} \end{aligned} \tag{8.56}$$

where $\zeta_t(j,k)$ is computed as:

$$\zeta_t(j,k) = \frac{p(\mathbf{X}, s_t = j, k_t = k \mid \Phi)}{p(\mathbf{X} \mid \Phi)} = \frac{\sum_{i=1}^{N} \alpha_{t-1}(i) a_{ij} c_{jk} b_{jk}(\mathbf{x}_t) \beta_t(j)}{\sum_{i=1}^{N} \alpha_T(i)} \tag{8.57}$$

In a similar manner to the discrete HMM, we can derive the reestimation equation for c_{jk} as follows:

$$\hat{c}_{jk} = \frac{\sum_{t=1}^{T} \zeta_t(j,k)}{\sum_{t=1}^{T} \sum_{k=1}^{M} \zeta_t(j,k)} \tag{8.58}$$

8.3.2. Semicontinuous HMMs

Traditionally, the discrete and the continuous mixture density HMMs have been treated separately. In fact, the gap between them can be bridged under some minor assumptions with the so-called semicontinuous or tied-mixture HMM. It assumes the mixture density functions are tied together across all the models to form a set of shared kernels. In the discrete HMM, a VQ codebook is typically used to map the continuous input feature vector $\mathbf{x}$ to o_k, so we can use the discrete output probability distribution $b_j(k)$. The codebook can be essentially regarded as one of such shared kernels. Accordingly, Eq. (8.44) can be modified as:

$$b_j(\mathbf{x}) = \sum_{k=1}^{M} b_j(k) f(\mathbf{x} \mid o_k) = \sum_{k=1}^{M} b_j(k) N(\mathbf{x}, \boldsymbol{\mu}_k, \boldsymbol{\Sigma}_k) \tag{8.59}$$

where o_k is the kth codeword, $b_j(k)$ is the same output probability distribution in the discrete HMM or the mixture weights for the continuous mixture density function, and $N(\mathbf{x}, \boldsymbol{\mu}_k, \boldsymbol{\Sigma}_k)$ are assumed to be independent of the Markov model and they are shared across all the Markov models with a very large number of mixtures M.

From the discrete HMM point of view, the needed VQ codebook consists of M continuous probability density functions, and each *codeword* has a mean vector and a covariance matrix. Typical quantization produces a codeword index that has minimum distortion to the given continuous observation **x**. In the semicontinuous HMM, the quantization operation produces values of continuous probability density functions $f(\mathbf{x} \mid o_k)$ for all the codewords o_k. The structure of the semicontinuous model can be roughly the same as that of the discrete one. However, the output probabilities are no longer used directly as in the discrete HMM. In contrast, the VQ codebook density functions, $N(\mathbf{x}, \boldsymbol{\mu}_k, \boldsymbol{\Sigma}_k)$, are combined with the discrete output probability as in Eq. (8.59). This is a combination of *discrete* model-dependent weighting coefficients with the *continuous* codebook probability density functions. Such a representation can be used to re-estimate the original VQ codebook together with the HMM.

The semicontinuous model also resembles the M-mixture continuous HMM with all the continuous output probability density functions shared among all Markov states. Compared with the continuous mixture HMM, the semicontinuous HMM can maintain the modeling ability of large-mixture probability density functions. In addition, the number of free parameters and the computational complexity can be reduced, because all the probability density functions are tied together, thus providing a good compromise between detailed acoustic modeling and trainability.

In practice, because M is large, Eq. (8.59) can be simplified by using the L most significant values $f(\mathbf{x} \mid o_k)$ for each **x** without affecting the performance. Experience has shown that values of L in the range of 1-3% of M are adequate. This can be conveniently obtained during the VQ operations by sorting the VQ output and keeping the L most significant values. Let $\eta(\mathbf{x})$ denote the set of L VQ codewords that has the largest $f(\mathbf{x} \mid o_k)$ for the given **x**. Then we have:

$$b_j(\mathbf{x}) \cong \sum_{k \in \eta(\mathbf{x})} f(\mathbf{x} \mid o_k) b_j(k) \tag{8.60}$$

Since the number of mixture components in $\eta(\mathbf{x})$ is of lower order than M, Eq. (8.60) can significantly reduce the amount of computation. In fact, $\eta(\mathbf{x})$ is the key to bridge the gap between the continuous and discrete HMM. If $\eta(\mathbf{x})$ contains only the most significant

$f(\mathbf{x} \mid o_k)$ (i.e., only the closest codeword to $\mathbf{x}$), the semicontinuous HMM degenerates to the discrete HMM. On the other hand, a large VQ codebook can be used such that each Markov state could contain a number of its own codewords (a mixture of probability density functions). The discrete output probability $b_{ij}(k)$ thus becomes the mixture weights for each state. This would go to the other extreme, a standard continuous mixture density model. We can also define $\eta(\mathbf{x})$ in such a way that we can have partial tying of $f(\mathbf{x} \mid o_k)$ for different phonetic classes. For example, we can have a phone-dependent codebook.

When we have a tied VQ codebook, re-estimation of these mean vectors and covariance matrices of different models will involve interdependencies. If any observation $\mathbf{x}_t$ (no matter what model it is designated for) has a large value of posterior probability $\zeta_t(j,k)$, it will have a large contribution on re-estimation of parameters of codeword o_k. We can compute the posterior probability for each codeword from $\zeta_t(j,k)$ as defined in Eq. (8.57).

$$\zeta_t(k) = p(\mathbf{x}_t = o_k \mid \mathbf{X}, \Phi) = \sum_j \zeta_t(j,k) \tag{8.61}$$

The re-estimation formulas for the tied mixture can be written as:

$$\hat{\boldsymbol{\mu}}_k = \frac{\sum_{t=1}^{T} \zeta_t(k)\mathbf{x}_t}{\sum_{t=1}^{T} \zeta_t(k)} \tag{8.62}$$

$$\hat{\Sigma}_k = \frac{\sum_{t=1}^{T} \zeta_t(k)(\mathbf{x}_t - \hat{\boldsymbol{\mu}}_k)(\mathbf{x}_t - \hat{\boldsymbol{\mu}}_k)^t}{\sum_{t=1}^{T} \zeta_t(k)} \tag{8.63}$$

8.4. Practical Issues in Using HMMs

While the HMM provides a solid framework for speech modeling, there are a number of issues you need to understand to make effective use of spoken language processing. In this section we point out some of the key issues related to practical applications. For expedience, we mostly use the discrete HMM as our example here.

8.4.1. Initial Estimates

In theory, the re-estimation algorithm of the HMM should reach a local maximum for the likelihood function. A key question is how to choose the right initial estimates of the HMM parameters so that the local maximum becomes the global maximum.

In the discrete HMM, if a probability is initialized to be zero, it will remain zero forever. Thus, it is important to have a reasonable set of initial estimates. Empirical study has shown that, for discrete HMMs, you can use a uniform distribution as the initial estimate. It works reasonably well for most speech applications, though good initial estimates are always helpful to compute the output probabilities.

If continuous mixture density HMMs are used, good initial estimates for the Gaussian density functions are essential. There are a number of ways to obtain such initial estimates:

- You can use the k-means clustering procedure, as used in vector quantization clustering. The Markov state segmentation can be derived from the discrete HMM, since it is not very sensitive to the initial parameters. Based on the segmented data, you can use the k-means algorithm to derive needed Gaussian mean and covariance parameters. The mixture coefficients can be based on the uniform distribution.
- You can estimate the semicontinuous HMM from the discrete HMM. You simply need to estimate an additional covariance matrix for each VQ codeword and run an additional four or five iterations to refine the semi-continuous HMM based on the discrete HMM, which typically requires four or five iterations from the uniform distribution. When the semi-continuous HMM is trained, you take the top *M codewords,* and in each Markov state use them as the initial Gaussian density functions for the continuous density mixture model.
- You can start training a single mixture Gaussian model. You can compute the parameters from previously segmented data. You can then iteratively split the Gaussian density function in a way similar to VQ codebook generation. You typically need two or three iterations to refine the continuous density after each splitting.

8.4.2. Model Topology

Speech is a time-evolving nonstationary signal. Each HMM state has the ability to capture some quasi-stationary segment in the non-stationary speech signal. A left-to-right topology, as illustrated in Figure 8.8, is a natural candidate to model the speech signal. It has a self-transition to each state that can be used to model contiguous speech features belonging to the same state. When the quasi-stationary speech segment evolves, the left-to-right transition enables a natural progression of such evolution. In such a topology, each state has a state-dependent output probability distribution that can be used to interpret the observable speech signal. This topology is, in fact, one of the most popular HMM structures used in state-of-the-art speech recognition systems. The output probability distribution can be either discrete distributions or a mixture of continuous density functions.

For the left-to-right HMM, the most important parameter in determining the topology is the number of states. The choice of model topology depends on available training data and what the model is used for. If each HMM is used to represent a phone, you need to have at least three to five output distributions. If such a model is used to represent a word, more states are generally required, depending on the pronunciation and duration of the word. For example, the word *tetrahydrocannabino* should have a large number of states in comparison to the word *a*. You may use at least 24 states for the former and three states for the latter. If you have the number of states depending on the duration of the signal, you may want to use 15 to 25 states for each second of speech signal. One exception is that, for silence, you may want to have a simpler topology. This is because silence is stationary, and one or two states will be sufficient.

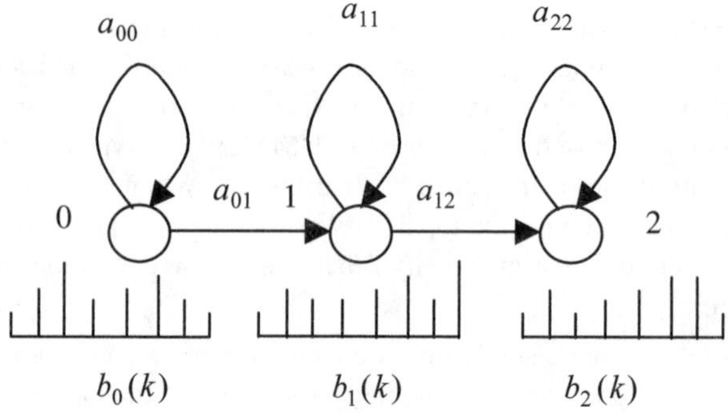

Figure 8.8 A typical hidden Markov model used to model phonemes. There are three states (0-2) and each state has an associated output probability distribution.

In practice, it is convenient to define a *null transition*. This is convenient if we want to simply traverse the HMM without consuming any observation symbol. To incorporate the null arc, you need to slightly modify the basic forward-backward or Viterbi probability equations, provided that no loops of empty transitions exist. If we denote the empty transition between state i and j as a_{ij}^{ε}, they need to satisfy the following constraints:

$$\sum_j a_{ij} + a_{ij}^{\varepsilon} = 1, \forall i \tag{8.64}$$

The forward probability can be augmented as follows:

$$\alpha_t(j) = \left[\sum_{i=1}^{N} \alpha_{t-1}(i) a_{ij} b_i(\mathbf{x}_t) + \sum_{i=1}^{N} \alpha_t(i) a_{ij}^{\varepsilon} \right] \quad 1 \le t \le T; \quad 1 \le j \le N \tag{8.65}$$

Equation (8.65) appears to have a recursion, but it actually uses the value of the same time column of $\alpha_t(i)$, provided that i is already computed, which is easily achievable if we have left-to-right empty transitions without loops of empty transitions.

8.4.3. Training Criteria

The argument for maximum likelihood estimation (MLE) is based on an assumption that the true distribution of speech is a member of the family of distributions used. This amounts to the assertion that the observed speech is genuinely produced by the HMM being used, and the only unknown parameters are the values. However, this can be challenged. Typical HMMs make many inaccurate assumptions about the speech production process, such as the output-independence assumption, the Markov assumption, and the continuous probability density assumption. Such inaccurate assumptions substantially weaken the rationale for maximum likelihood criteria. For instance, although maximum likelihood estimation is consistent (convergence to the true value), it is meaningless to have such a property if the wrong model is used. The true parameters in such cases will be the true parameters of the wrong models. Therefore, an estimation criterion that can work well in spite of these inaccurate assumptions should offer improved recognition accuracy compared with the maximum likelihood criterion. These alternative criteria include the MCE and MMIE, as discussed in Chapter 4. Finally, if we have prior knowledge about the model distribution, we can employ the Bayes' method such as MAP that can effectively combine both the prior and posterior distributions in a consistent way, which is particularly suitable for adaptation or dealing with insufficient training data.

Among all these criteria, MLE remains one of the most widely used, because of its simplicity and superior performance when appropriate assumptions are made about the system design. MCE and MMIE work well for small- to medium-vocabulary speech recognition [2, 26, 36]. You can train a number of other iterations based on the ML estimates. Neither MCE nor MMIE has been found extremely effective for large-vocabulary speech recognition. However, it is possible to combine the MMIE or MCE model with the MLE model for improved performance. This is because the error patterns generated from these different models are not the same. We can decode the test utterance with these different models and vote for the most consistent results [15, 25]. We discuss MAP methods in Chapter 9, since it is mostly helpful for speaker adaptive speech recognition.

8.4.4. Deleted Interpolation

For improved robustness, it is often necessary to combine well trained general models (such as speaker-independent) with those that are less well trained but more detailed (such as speaker-dependent). For example, you can typically improve speech recognition accuracy with speaker-dependent training. Nevertheless, you may not have sufficient data for a particular speaker so it is desirable to use a speaker-independent model that is more general but less accurate to smooth the speaker-dependent model. One effective way to achieve robustness is to interpolate both models with a technique called deleted interpolation, in which the interpolation weights are estimated using cross-validation data. The objective function is to maximize the probability of the model generating the held-out data.

Now, let us assume that we want to interpolate two sets of models, $P_A(\mathbf{x})$ and $P_B(\mathbf{x})$, which can be either discrete probability distributions or continuous density functions, to form an interpolated model $P_{DI}(\mathbf{x})$. The interpolation procedure can be expressed as follows:

$$P_{DI}(\mathbf{x}) = \lambda P_A(\mathbf{x}) + (1-\lambda)P_B(\mathbf{x}) \tag{8.66}$$

where the interpolation weight λ is what we need to derive from the training data.

Consider that we want to interpolate a speaker-independent model $P_A(\mathbf{x})$ with a speaker-dependent model $P_B(\mathbf{x})$. If we use speaker-independent data to estimate the interpolation weight, we may not capture needed speaker-specific information that should be reflected in the interpolation weights. What is worse is that the interpolation weight for the speaker-independent model should be equal to 1.0 if we use the same speaker-independent data from which the model was derived to estimate the interpolation weight. This is because of the MLE criterion. If we use speaker-dependent data instead, we have the weight for the speaker-dependent model equal 1.0 without achieving the desired smoothing effect. Thus the interpolation weights need to be trained using different data or *deleted* data with the so called cross-validation method.

We can have the training data normally divided into M parts, and train a set of $P_A(\mathbf{x})$ and $P_B(\mathbf{x})$ models using the standard EM algorithm from each combination of M-1 parts, with the deleted part serving as the unseen data to estimate the interpolation weights λ. These M sets of interpolation weights are then averaged to obtain the final weights.

ALGORITHM 8.5: *DELETED INTERPOLATION PROCEDURE*

Step 1: Initialize λ with a guessed estimate.

Step 2: Update λ by the following formula:

$$\hat{\lambda} = \frac{1}{M}\sum_{j=1}^{M}\sum_{t=1}^{n_j}\frac{\lambda P_{A-j}(\mathbf{x}_t^j)}{\lambda P_{A-j}(\mathbf{x}_t^j) + (1-\lambda)P_{B-j}(\mathbf{x}_t^j)} \tag{8.67}$$

where $P_{A-j}(\mathbf{x}_t^j)$ and $P_{B-j}(\mathbf{x}_t^j)$ is $P_A(\mathbf{x})$ and $P_B(\mathbf{x})$ estimated by the entire training corpus except part *j*, the deleted part, respectively; n_j is the total number of data points in part *j* that have been aligned to estimate the model; and $\mathbf{x}_t^j$ indicates the *t*-th data point in the *j*-th set of the aligned data.

Step 3: If the new value $\hat{\lambda}$ is sufficiently close to the previous value λ, stop. Otherwise, go to Step 2.

In fact, the interpolation weights in Eq. (8.66) are similar to the Gaussian mixture weights, although $P_A(\mathbf{x})$ and $P_B(\mathbf{x})$ may not be Gaussian density functions. When we have M sets of data, we can use the same EM algorithm to estimate the interpolation weights as illustrated in Algorithm 8.5.

The deleted interpolation procedure described above can be applied after each training iteration. Then, for the following iteration of training, the learned interpolation weights can be used as illustrated in Eq. (8.66) to compute the forward-backward paths or the Viterbi maximum path. We can also have more than two distributions interpolated together. Deleted interpolation has been widely used in both acoustic and language modeling where smoothing is needed.

8.4.5. Parameter Smoothing

One simple reality for probabilistic modeling is that as many observations as possible are required to reliably estimate model parameters. However, in reality, only a finite amount of training data is available. If the training data are limited, this will result in some parameters being inadequately trained, and classification based on poorly trained models will result in higher recognition error rate. There are many possible solutions to address the problem of insufficient training data:

- You can increase the size of the training data. *There is no data like more data.*
- You can reduce the number of free parameters to be re-estimated. This has its limitations, because a number of significant parameters are always needed to model physical events.
- You can interpolate one set of parameter estimates with another set of parameter estimates, for which an adequate amount of training data exists. Deleted interpolation, discussed in Section 8.4.4, can be used effectively. In the discrete HMM, one simple approach is to set a floor to both the transition probability and the output probability in order to eliminate possible zero estimates. The same principle applies to the SCHMM as well as the mixing coefficients of the continuous density HMM. Parameter flooring can be regarded as a special case of interpolation with the uniform distribution.
- You can tie parameters together to reduce the number of free parameters. The SCHMM is a typical example of such parameter-tying techniques.

For the continuous mixture HMM, you need to pay extra attention to smoothing the covariance matrices. There are a number of techniques you can use:

- You can interpolate the covariance matrix with those that are better trained or a priori via the MAP method.

- You can tie the Gaussian covariance matrices across different mixture components or across different Markov states. A very general shared Gaussian density model is discussed in [20].
- You can use the diagonal covariance matrices if the correlation among feature coefficients is weak, which is the case if you use uncorrelated features such as the MFCC.
- You can combine these methods together.

In practice, we can reduce the speech recognition error rate by 5-20% with various smoothing techniques, depending on the available amount of training data.

8.4.6. Probability Representations

When we compute the forward and backward probabilities in the forward-backward algorithm, they will approach zero in exponential fashion if the observation sequence length, T, becomes large enough. For sufficiently large T, the dynamic range of these probabilities will exceed the precision range of essentially any machine. Thus, in practice, it will result in underflow on the computer if probabilities are represented directly. We can resolve this implementation problem by scaling these probabilities with some scaling coefficient so that they remain within the dynamic range of the computer. All of these scaling coefficients can be removed at the end of the computation without affecting the overall precision.

For example, let $\alpha_t(i)$ be multiplied by a scaling coefficient, S_t:

$$S_t = 1 / \sum_i \alpha_t(i) \tag{8.68}$$

so that $\sum_i S_t \alpha_t(i) = 1$ for $1 \le t \le T$. $\beta_t(i)$ can also be multiplied by S_t for $1 \le t \le T$. The recursion involved in computing the forward and backward variables can be scaled at each stage of time t by S_t. Notice that $\alpha_t(i)$ and $\beta_t(i)$ are computed recursively in exponential fashion; therefore, at time t, the total scale factor applied to the forward variable $\alpha_t(i)$ is

$$Scale_\alpha(t) = \prod_{k=1}^{t} S_k \tag{8.69}$$

and the total scale factor applied to the backward variable $\beta_t(i)$ is

$$Scale_\beta(t) = \prod_{k=t}^{T} S_k \tag{8.70}$$

This is because the individual scaling factors are multiplied together in the forward and backward recursion. Let $\alpha'_t(i)$, $\beta'_t(i)$, and $\gamma'_t(i,j)$ denote their corresponding scaled variables, respectively. Note that

$$\sum_i \alpha'_T(i) = Scale_\alpha(T)\sum_i \alpha_T(i) = Scale_\alpha(T)P(\mathbf{X}\mid\Phi) \tag{8.71}$$

The scaled intermediate probability, $\gamma'_t(i,j)$, can then be written as:

$$\gamma'_t(i,j) = \frac{Scale_\alpha(t-1)\alpha_{t-1}(i)a_{ij}b_j(X_t)\beta_t(j)Scale_\beta(t)}{Scale_\alpha(T)\sum_{i=1}^{N}\alpha_T(i)} = \gamma_t(i,j) \tag{8.72}$$

Thus, the intermediate probabilities can be used in the same way as the unscaled probabilities, because the scaling factor is cancelled out in Eq. (8.72). Therefore, re-estimation formulas can be kept exactly except that $P(\mathbf{X}\mid\Phi)$ should be computed according to

$$P(\mathbf{X}\mid\Phi) = \sum_i \alpha'_T(i) / Scale_\alpha(T) \tag{8.73}$$

In practice, the scaling operation need not be performed at every observation time. It can be used at any scaling interval for which the underflow is likely to occur. In the unscaled interval, $Scale_\alpha$ can be kept as unity.

An alternative way to avoid underflow is to use a logarithmic representation for all the probabilities. This not only ensures that scaling is unnecessary, as underflow cannot happen, but also offers the benefit that integers can be used to represent the logarithmic values, thereby changing floating point operations to fixed point ones, which is particularly suitable for Viterbi-style computation, as Eq. (8.25) requires no probability addition.

In the forward-backward algorithm we need to have probability addition. We can keep a table on $\log_b P_2 - \log_b P_1$. If we represent probability P by $\log_b P$, more precision can be obtained by setting b closer to unity. Let us assume that we want to add P_1 and P_2 and that $P_1 \ge P_2$. We have:

$$\log_b(P_1 + P_2) = \log_b P_1 + \log_b(1 + b^{\log_b P_2 - \log_b P_1}) \tag{8.74}$$

If P_2 is many orders of magnitude smaller than P_1, adding the two numbers will just result in P_1. We could store all possible values of $\log_b(1+b^x)$ in a table. Using logarithms introduces errors for addition operation. In practice, double-precision float representation can be used to minimize the impact of the precision problems.

8.5. HMM Limitations

There are a number of limitations in the conventional HMMs. For example, HMMs assume the duration follows an exponential distribution, the transition probability depends only on the origin and destination, and all observation frames are dependent only on the state that generated them, not on neighboring observation frames. Researchers have proposed a number of techniques to address these limitations, albeit these solutions have not significantly improved speech recognition accuracy for practical applications.

8.5.1. Duration Modeling

One major weakness of conventional HMMs is that they do not provide an adequate representation of the temporal structure of speech. This is because the probability of state occupancy decreases exponentially with time as shown in Eq. (8.75). The probability of t consecutive observations in state i is the probability of taking the self-loop at state i for t times, which can be written as

$$d_i(t) = a_{ii}^t(1 - a_{ii}) \tag{8.75}$$

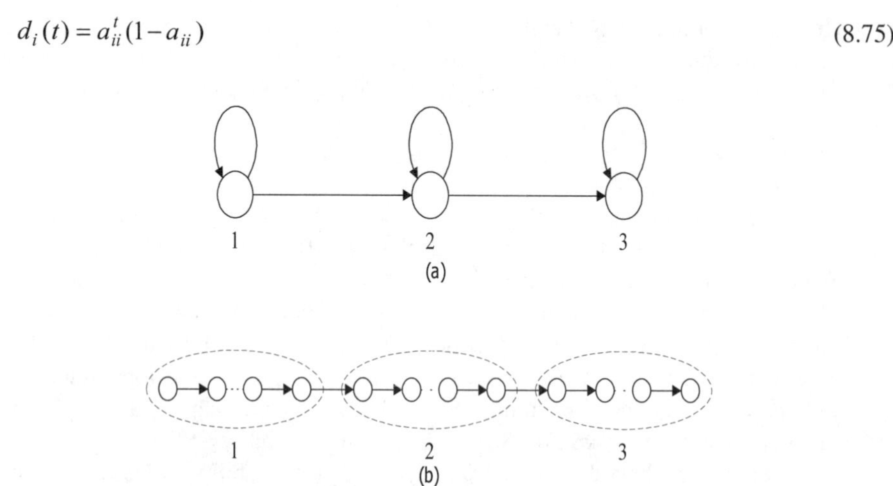

Figure 8.9 A standard HMM (a) and its corresponding explicit duration HMM (b) where the self transitions are replaced with the explicit duration probability distribution for each state.

An improvement to the standard HMM results from the use of HMMs with an explicit time duration distribution for each state [30, 39]. To explain the principle of time duration modeling, a conventional HMM with exponential state duration density and a time duration HMM with specified state duration densities (which can be either a discrete distribution or a continuous density) are illustrated in Figure 8.9. In (a), the state duration probability has an exponential form as in Eq. (8.75). In (b), the self-transition probabilities are replaced with an explicit duration probability distribution. At time t, the process enters state i for duration τ with probability density $d_i(\tau)$, during which the observations $X_{t+1}, X_{t+2} \ldots X_{t+\tau}$ are generated. It then transfers to state j with transition probability a_{ij} only after the appropriate τ observations have occurred in state i. Thus, by setting the time duration probability density to be the exponential density of Eq. (8.75) the time duration HMM can be made equivalent to the standard HMM. The parameters $d_i(\tau)$ can be estimated from observations along with the other parameters of the HMM. For expedience, the duration density is usually truncated at a maximum duration value T_d. To re-estimate the parameters of the HMM with time duration modeling, the forward recursion must be modified as follows:

$$\alpha_t(j) = \sum_{\tau} \sum_{i, i \neq j} \alpha_{t-\tau}(i) a_{ij} d_j(\tau) \prod_{l=1}^{\tau} b_j(X_{t-\tau+l}) \tag{8.76}$$

where the transition from state i to state j depends not only upon the transition probability a_{ij} but also upon all the possible time periods τ that may occur in state j. Intuitively, Eq. (8.76) illustrates that when state j is reached from previous states i, the observations may stay in state j for a period of τ with duration density $d_j(\tau)$, and each observation emits its own output probability. All possible durations must be considered, which leads to summation with respect to τ. The independence assumption of observations results in the Π term of the output probabilities. Similarly, the backward recursion can be written as:

$$\beta_t(i) = \sum_{\tau} \sum_{j, j \neq i} a_{ij} d_j(\tau) \prod_{l=1}^{\tau} b_j(X_{t+l}) \beta_{t+\tau}(j) \tag{8.77}$$

The modified Baum-Welch algorithm can then be used based on Eq. (8.76) and (8.77). The proof of the re-estimation algorithm can be based on the modified Q-function except that $P(\mathbf{X}, \mathbf{S}|\Phi)$ should be replaced with $P(\mathbf{X}, \mathbf{S}, \mathbf{T}|\Phi)$, which denotes the joint probability of observation, $\mathbf{X}$, state sequence, $\mathbf{S} = \{s_1, s_2 \ldots, s_k \ldots s_{N_s}\}$ in terms of state s_k with time duration τ_k, and the corresponding duration sequence, $\mathbf{T} = \{\tau_1, \tau_2, \ldots \tau_k \ldots \tau_{N_s}\}$.

$$Q(\Phi, \hat{\Phi}) = \frac{1}{P(\mathbf{X}|\Phi)} \sum_{\mathbf{T}} \sum_{\mathbf{S}} P(\mathbf{X}, \mathbf{S}, \mathbf{T}|\Phi) \log P(\mathbf{X}, \mathbf{S}, \mathbf{T}|\hat{\Phi}) \tag{8.78}$$

In a manner similar to the standard HMM, $\gamma_{t,\tau}(i, j)$ can be defined as the transition probability from state i at time t to state j with time duration τ in state j. $\gamma_{t,\tau}(i, j)$ can be written as:

$$\gamma_{t,\tau}(i, j) = \alpha_t(i) a_{ij} d_j(\tau) \prod_{l=1}^{\tau} b_j(X_{t+l}) \beta_{t+\tau}(j) / \sum_{k=1}^{N} \alpha_T(k) \tag{8.79}$$

Similarly, the probability of being in state j at time t with duration τ can be computed as:

$$\gamma_{t,\tau}(j) = \sum_{i} \gamma_{t,\tau}(i, j) \tag{8.80}$$

The re-estimation algorithm can be derived from Eq. (8.80), the Viterbi decoding algorithm can be used for the time duration model, and the optimal path can be determined according to:

$$V_t(j) = \underset{i}{Max}\, \underset{\tau}{Max}[V_{t-\tau}(i) a_{ij} d_j(\tau) \prod_{l=1}^{\tau} b_j(X_{t-\tau+l})] \tag{8.81}$$

There are drawbacks to the use of the time duration modeling discussed here. One is the great increase in computational complexity by a factor of $O(D^2)$, where D is the time duration distribution length. Another problem is the large number of additional parameters D that must be estimated. One proposed remedy is to use a continuous density function instead of the discrete distribution $d_j(\tau)$.

In practice, duration models offer only modest improvement for speaker-independent continuous speech recognition. Many systems even eliminate the transition probability completely because the output probabilities are so dominant. Nevertheless, duration information is very effective for pruning unlikely candidates during the large-vocabulary speech recognition decoding process.

8.5.2. First-Order Assumption

As you can see from the previous section, the duration of each stationary segment captured by a single state is inadequately modeled. Another way to alleviate the duration problem is to eliminate the first-order transition assumption and to make the underlying state sequence a second-order Markov chain [32]. As a result, the transition probability between two states at time t depends on the states in which the process was at time t-1 and t-2. For a given state sequence $\mathbf{S} = \{s_1, s_2, \ldots s_T\}$, the probability of the state should be computed as:

$$P(\mathbf{S}) = \prod_t a_{s_{t-2}s_{t-1}s_t} \tag{8.82}$$

where $a_{s_{t-2}s_{t-1}s_t} = P(s_t | s_{t-2}s_{t-1})$ is the transition probability at time t, given the two-order state history. The re-estimation procedure can be readily extended based on Eq. (8.82). For example, the new forward probability can be re-defined as:

$$\alpha_t(j,k) = P(X_1^t, s_{t-1} = j, s_t = k | \lambda) = \sum_i \alpha_{t-1}(i,j) a_{ijk} b_k(X_t) \tag{8.83}$$

where $a_{ijk} = P(s_t = k | s_{t-2} = i, s_{t-1} = j)$. Similarly, we can define the backward probability as:

$$\beta_t(i,j) = P(X_{t+1}^T | s_{t-1} = i, s_t = j, \lambda) = \sum_k a_{ijk} b_k(X_{t+1}) \beta_{t+1}(j,k) \tag{8.84}$$

With Eq. (8.83) and (8.84), the MLE estimates can be derived easily based on the modified $\gamma_t(i,j,k)$:

$$\begin{aligned} \gamma_t(i,j,k) &= P(s_{t-1} = i, s_t = j, s_{t+1} = k, \mathbf{X} | \Phi) \\ &= \alpha_t(i,j) a_{ijk} b_k(X_{t+1}) \beta_{t+1}(j,k) / P(\mathbf{X} | \Phi) \end{aligned} \tag{8.85}$$

In practice, the second-order model is computationally very expensive as we have to consider the increased state space, which can often be realized with an equivalent first-order hidden Markov model on the two-fold product state space. It has not offered significantly improved accuracy to justify its increase in computational complexity for most applications.

8.5.3. Conditional Independence Assumption

The third major weakness in HMMs is that all observation frames are dependent only on the state that generated them, not on neighboring observation frames. The conditional independence assumption makes it hard to effectively handle nonstationary frames that are strongly correlated. There are a number of ways to alleviate the conditional independence assumption [34]. For example, we can assume the output probability distribution depends not only on the state but also on the previous frame. Thus, the probability of a given state sequence can be rewritten as:

$$P(\mathbf{X}|\mathbf{S},\Phi) = \prod_{t=1}^{T} P(X_t|X_{t-1}, s_t, \Phi) \tag{8.86}$$

As the parameter space becomes huge, we often need to quantize X_{t-1} into a smaller set of codewords so that we can keep the number of free parameters under control. Thus, Eq. (8.86) can be simplified as:

$$P(\mathbf{X} \mid \mathbf{S},\Phi) = \prod_{t=1}^{T} P(X_t \mid \Re(X_{t-1}), s_t, \Phi) \tag{8.87}$$

where $\Re(\,)$ denotes the quantized vector that has a small codebook size, L. Although this can dramatically reduce the space of the free conditional output probability distributions, the total number of free parameters will still increase by L times.

The re-estimation for conditional dependent HMMs can be derived with the modified Q-function, as discussed in the previous sections. In practice, it has not demonstrated convincing accuracy improvement for large-vocabulary speech recognition.

8.6. HISTORICAL PERSPECTIVE AND FURTHER READING

The Markov chain was named after Russian scientist A. Markov for his pioneering work in analyzing the letter sequence in the text of a literary work in 1913 [33]. In the 1960s, Baum and others further developed efficient methods for training the model parameters [4, 5]. When the observation is real valued, the use of continuous or semi-continuous HMMs can improve the overall performance. Baum et al. also developed the method to use continuous density functions that are strictly log concave [5], which was relaxed by Liporace [31] and expanded by Juang to include mixture density functions [27].

The Viterbi algorithm shares the same concept that was independently discovered by researchers in many separate fields [28], including Vintsyuk [42], Needleman and Wunsch [35], Sankoff [41], Sakoe and Chiba [40], and Wagner and Fischer [44].

Jim Baker did his Ph.D. thesis under Raj Reddy at Carnegie Mellon using HMMs for speech recognition [3]. At the same time Fred Jelinek and his colleagues at IBM Research pioneered widespread applications [23]. Since the 1980s, partly because of the DARPA-funded speech projects, HMMs have become a mainstream technique for modeling speech, as exemplified by advanced systems developed at BBN, Bell Labs, Carnegie Mellon, IBM, Microsoft, SRI, and others [9, 17, 29, 46]. The Ph.D. theses from Kai-Fu Lee [29], Hsiao-Wuen Hon [16], and Mei-Yuh Hwang [22] at Carnegie Mellon addressed many important practical issues in using HMMs for speech recognition. There are also a number of good books on the practical use of HMMs [18, 24, 38, 45].

The choice of different output probabilities depends on a number of factors such as the availability of training data, the feature characteristics, the computational complexity, and the number of free parameters [19] [34]. The semi-continuous model, also known as the tied-mixture model, was independently proposed by Huang and Jack [21] and Bellegarda and Nahamoo [6]. Other improvements include explicit duration modeling [1, 11, 13, 14, 30, 39], high-order and conditional models [7, 32, 34], which have yet to be shown effective for practical speech recognition.

Both Carnegie Mellon University's open speech software[4] and Cambridge University's HTK[5] are a good starting point for those interested in using the existing tools for running experiments.

HMMs have become the most prominent techniques for speech recognition today. Most of the state-of-the-art speech recognition systems on the market are based on HMMs described in this chapter.

REFERENCES

[1] Anastasakos, A., R. Schwartz, and H. Sun, "Duration Modeling in Large Vocabulary Speech Recognition" in *Proc. of the IEEE Int. Conf. on Acoustics, Speech and Signal Processing*, 1995, Detroit, MI, pp. 628-631.

[2] Bahl, L.R., *et al.*, "Speech Recognition with Continuous-Parameter Hidden Markov Models," *Computer Speech and Language*, 1987, **2**, pp. 219-234.

[3] Baker, J.K., "The DRAGON System—An Overview," *Trans. on Acoustics, Speech and Signal Processing*, 1975, **23**(1), pp. 24-29.

[4] Baum, L.E. and J.A. Eagon, "An Inequality with Applications to Statistical Estimation for Probabilistic Functions of Markov Processes and to a Model for Ecology," *Bulletin of American Mathematical Society*, 1967, **73**, pp. 360-363.

[5] Baum, L.E., *et al.*, "A Maximization Technique Occurring in the Statistical Analysis of Probabilistic Functions of Markov Chains," *Annals of Mathematical Statistics*, 1970, **41**, pp. 164-171.

[6] Bellegarda, J.R. and D. Nahamoo, "Tied Mixture Continuous Parameter Models for Large Vocabulary Isolated Speech Recognition," *Int. Conf. on Acoustics, Speech and Signal Processing*, 1989, pp. 13-16.

[7] Brown, P., *The Acoustic-Modeling Problem in Automatic Speech Recognition*, Ph.D. Thesis in *Computer Science*, 1987, Carnegie Mellon University, Pittsburgh.

---

[4] http://www.speech.cs.cmu.edu/sphinx/

[5] http://htk.eng.cam.ac.uk/

[8] Brown, P.F., *et al.*, "The Mathematics of Statistical Machine Translation: Parameter Estimation," *Computational Linguistics*, 1995, **19**(2), pp. 263-312.

[9] Chou, W., C.H. Lee, and B.H. Juang, "Minimum Error Rate Training of Inter-Word Context Dependent Acoustic Model Units in Speech Recognition" in *Proc. of the Int. Conf. on Spoken Language Processing*, 1994, Yokohama, Japan, pp. 439-442.

[10] Church, K., "A Stochastic Parts Program and Noun Phrase Parser for Unrestricted Text," *Proc. of the Second Conf. on Applied Natural Language Processing*, 1988, Austin, Texas, pp. 136-143.

[11] Deng, L., M. Lennig, and P. Mermelstein, "Use of Vowel Duration Information in a Large Vocabulary Word Recognizer," *Journal of the Acoustical Society of America*, 1989, **86**(2 August), pp. 540-548.

[12] DeRose, S.J., "Grammatical Category Disambiguation by Statistical Optimization," *Computational Linguistics*, 1988(1), pp. 31-39.

[13] Dumouchel, P. and D. Shaughnessy, "Segmental Duration and HMM Modeling," *Proc. of the European Conf. on Speech Communication and Technology*, 1995, Madrid, Spain, pp. 803-806.

[14] Ferguson, J.D., "Variable Duration Models for Speech" in *Proc. of the Symposium on the Application of Hidden Markov Models to Text and Speech*, J.D. Ferguson, ed., 1980, Princeton, New Jersey, pp. 143-179.

[15] Fiscus, J., "A Post-Processing System to Yield Reduced Word Error Rates: Recognizer Output Voting Error Reduction (ROVER)," *IEEE Workshop on Automatic Speech Recognition and Understanding*, 1997, Santa Barbara, CA, pp. 347-352.

[16] Hon, H.W., *Vocabulary-Independent Speech Recognition: The VOCIND System*, Ph.D Thesis in *Department of Computer Science* 1992, Carnegie Mellon University, Pittsburgh.

[17] Huang, X.D., *et al.*, "The SPHINX-II Speech Recognition System: An Overview," *Computer Speech and Language*, 1993, pp. 137-148.

[18] Huang, X.D., Y. Ariki, and M.A. Jack, *Hidden Markov Models for Speech Recognition*, 1990, Edinburgh, U.K., Edinburgh University Press.

[19] Huang, X.D., *et al.*, "A Comparative Study of Discrete, Semicontinuous, and Continuous Hidden Markov Models," *Computer Speech and Language*, 1993, **7**(4), pp. 359-368.

[20] Huang, X.D., *et al.*, "Deleted Interpolation and Density Sharing for Continuous Hidden Markov Models," *IEEE Int. Conf. on Acoustics, Speech and Signal Processing*, 1996.

[21] Huang, X.D. and M.A. Jack, "Semi-Continuous Hidden Markov Models with Maximum Likelihood Vector Quantization" in *IEEE Workshop on Speech Recognition*, 1988.

[22] Hwang, M., *Subphonetic Modeling in HMM-based Speech Recognition Systems*, Ph.D. Thesis in *Computer Science*, 1994, Carnegie Mellon University, Pittsburgh.

[23] Jelinek, F., "Continuous Speech Recognition by Statistical Methods," *Proc. of the IEEE*, 1976, **64**(4), pp. 532-556.

[24] Jelinek, F., *Statistical Methods for Speech Recognition*, 1998, Cambridge, MA, MIT Press.

[25] Jiang, L. and X. Huang, "Unified Decoding and Feature Representation for Improved Speech Recognition," *Proc. of the 6th European Conf. on Speech Communication and Technology*, 1999, Budapest, Hungary, pp. 1331-1334.

[26] Juang, B.H., W. Chou, and C.H. Lee, "Statistical and Discriminative Methods for Speech Recognition" in *Automatic Speech and Speaker Recognition—Advanced Topics*, C.H. Lee, F.K. Soong, and K.K. Paliwal, eds., 1996, Boston, pp. 109-132, Kluwer Academic Publishers.

[27] Juang, B.H., S.E. Levinson, and M.M. Sondhi, "Maximum Likelihood Estimation for Multivariate Mixture Observations of Markov Chains," *IEEE Trans. on Information Theory*, 1986, **IT-32**(2), pp. 307-309.

[28] Kruskal, J., "An Overview of Sequence Comparison" in *Time Warps, String Edits, and Macromolecules: The Theory and Practice of Sequence Comparison*, D. Sankoff and J. Kruskal, eds. 1983, Reading, MA., Addison-Wesley, pp. 1-44.

[29] Lee, K.F., *Large-Vocabulary Speaker-Independent Continuous Speech Recognition: The SPHINX System*, Ph.D. Thesis in Computer Science Dept., 1988, Carnegie Mellon University, Pittsburgh.

[30] Levinson, S.E., "Continuously Variable Duration Hidden Markov Models for Automatic Speech Recognition," *Computer Speech and Language*, 1986, pp. 29-45.

[31] Liporace, L.R., "Maximum Likelihood Estimation for Multivariate Observations of Markov Sources," *IEEE Trans. on Information Theory*, 1982, **28**, pp. 729-734.

[32] Mari, J., J. Haton, and A. Kriouile, "Automatic Word Recognition Based on Second-Order Hidden Markov Models," *IEEE Trans. on Speech and Audio Processing*, 1977, **5**(1), pp. 22-25.

[33] Markov, A.A., "An Example of Statistical Investigation in the Text of 'Eugene Onyegin', Illustrating Coupling of Tests in Chains," *Proc. of the Academy of Sciences of St. Petersburg*, 1913, Russia, pp. 153-162.

[34] Ming, J. and F. Smith, "Modelling of the Interframe Dependence in an HMM Using Conditional Gaussian Mixtures," *Computer Speech and Language*, 1996, **10**(4), pp. 229-247.

[35] Needleman, S. and C. Wunsch, "A General Method Applicable to the Search for Similarities in the Amino-acid Sequence of Two Proteins," *Journal of Molecular Biology*, 1970, **48**, pp. 443-453.

[36] Normandin, Y., "Maximum Mutual Information Estimation of Hidden Markov Models" in *Automatic Speech and Speaker Recognition*, C.H. Lee, F.K. Soong, and K.K. Paliwal, eds., 1996, Norwell, MA, Kluwer Academic Publishers.

[37] Rabiner, L.R., "A Tutorial on Hidden Markov Models and Selected Applications in Speech Recognition," *Proc. of IEEE*, 1989, **77**(2), pp. 257-286.

[38] Rabiner, L.R. and B.H. Juang, *Fundamentals of Speech Recognition*, Prentice Hall Signal Processing Series, ed. A.V. Oppenheim, 1993, Englewood Cliffs, NJ, Prentice-Hall.

[39] Russell, M.J. and R.K. Moore, "Explicit Modeling of State Occupancy in Hidden Markov Models for Automatic Speech Recognition," *Int. Conf. on Acoustics, Speech and Signal Processing*, 1985, pp. 5-8.

[40] Sakoe, H. and S. Chiba, "Dynamic Programming Algorithm Optimization for Spoken Word Recognition," *IEEE Trans. on Acoustics, Speech and Signal Processing*, 1978, **26**(1), pp. 43-49.

[41] Sankoff, D., "Matching Sequences under Deletion-Insertion Constraints," *Proc. of the National Academy of Sciences*, 1972, **69**, pp. 4-6.

[42] Vintsyuk, T.K., "Speech Discrimination by Dynamic Programming," *Cybernetics*, 1968, **4**(1), pp. 52-57.

[43] Viterbi, A.J., "Error Bounds for Convolutional Codes and an Asymptotically Optimum Decoding Algorithm," *IEEE Trans. on Information Theory*, 1967, **13**(2), pp. 260-269.

[44] Wagner, R. and M. Fischer, "The String-to-String Correction Problem," *Journal of the ACM*, 1974, **21**, pp. 168-173.

[45] Waibel, A.H. and K.F. Lee, *Readings in Speech Recognition*, 1990, San Mateo, CA, Morgan Kaufman Publishers.

[46] Young, S.J. and P.C. Woodland, "The Use of State Tying in Continuous Speech Recognition," *Proc. of Eurospeech*, 1993, Berlin, pp. 2203-2206.

CHAPTER 9

Acoustic Modeling

After years of research and development, accuracy of automatic speech recognition remains one of the most important research challenges. A number of well-known factors determines accuracy; those most noticeable are variations in context, in speaker, and in environment. Acoustic modeling plays a critical role in improving accuracy and is arguably the central part of any speech recognition system.

For the given acoustic observation $\mathbf{X} = X_1X_2...X_n$, the goal of speech recognition is to find out the corresponding word sequence $\hat{\mathbf{W}} = w_1w_2...w_m$ that has the maximum posterior probability $P(\mathbf{W}|\mathbf{X})$ as expressed by Eq. (9.1).

$$\hat{\mathbf{W}} = \arg\max_{\mathbf{w}} P(\mathbf{W}|\mathbf{X}) = \arg\max_{\mathbf{w}} \frac{P(\mathbf{W})P(\mathbf{X}|\mathbf{W})}{P(\mathbf{X})} \tag{9.1}$$

Since the maximization of Eq. (9.1) is carried out with the observation **X** fixed, the above maximization is equivalent to maximization of the following equation:

$$\hat{\mathbf{W}} = \arg\max_{\mathbf{w}} P(\mathbf{W})P(\mathbf{X} \mid \mathbf{W}) \tag{9.2}$$

The practical challenge is how to build accurate acoustic models, $P(\mathbf{X} \mid \mathbf{W})$, and language models, $P(\mathbf{W})$, that can truly reflect the spoken language to be recognized. For large-vocabulary speech recognition, since there are a large number of words, we need to decompose a word into a subword sequence. Thus $P(\mathbf{X} \mid \mathbf{W})$ is closely related to phonetic modeling. $P(\mathbf{X} \mid \mathbf{W})$ should take into account speaker variations, pronunciation variations, environmental variations, and context-dependent phonetic coarticulation variations. Last, but not least, any static acoustic or language model will not meet the needs of real applications. So it is vital to dynamically adapt both $P(\mathbf{W})$ and $P(\mathbf{X} \mid \mathbf{W})$ to maximize $P(\mathbf{W} \mid \mathbf{X})$ while using spoken language systems. The decoding process of finding the best word sequence **W** to match the input speech signal **X** in speech recognition systems is more than a simple pattern recognition problem, since in continuous speech recognition you have an infinite number of word patterns to search, as discussed in detail in Chapters 12 and 13.

In this chapter we focus on discussing solutions that work well in practice. To highlight solutions that are effective, we use the Whisper speech recognition system [49] developed at Microsoft Research as a concrete example to illustrate how to build a working system and how various techniques can help to reduce speech recognition errors.[1] We hope that by studying what worked well in the past we can illuminate the possibilities for further improvement of the state of the art.

The hidden Markov model we discussed in Chapter 8 is the underpinning for acoustic phonetic modeling. It provides a powerful way to integrate segmentation, time warping, pattern matching, and context knowledge in a unified manner. The underlying technologies are undoubtedly evolving, and the research community is aggressively searching for more powerful solutions. Most of the techniques discussed in this chapter can be readily derived from the fundamentals discussed in earlier chapters.

9.1. Variability in the Speech Signal

The research community has produced technologies that, with some constraints, can accurately recognize spoken input. Admittedly, today's state-of-the-art systems still cannot match humans' performance. Although we can build a very accurate speech recognizer for a particular speaker, in a particular language and speaking style, in a particular environment, and limited to a particular task, it remains a research challenge to build a recognizer that can essentially understand anyone's speech, in any language, on any topic, in any free-flowing style, and in almost any speaking environment.

[1] Most of the experimental results used here are based on a development test set for the 60,000-word speaker-independent continuous dictation task. The training set consists of 35,000 utterances from about 300 speakers. The test set consists of 410 utterances from 10 speakers that were not used in the training data. The language model is derived from 2 billion words of English text corpora.

Accuracy and robustness are the ultimate measures for the success of speech recognition algorithms. There are many reasons why existing algorithms or systems did not deliver what people want. In the sections that follow we summarize the major factors involved.

9.1.1. Context Variability

Spoken language interaction between people requires knowledge of word meanings, communication context, and common sense. Words with widely different meanings and usage patterns may have the same phonetic realization. Consider the challenge represented by the following utterance:

> *Mr. Wright should write to Ms. Wright right away about his Ford or four door Honda.*

For a given word with the same pronunciation, the meaning could be dramatically different, as indicated by *Wright*, *write*, and *right*. What makes it even more difficult is that *Ford or* and *Four Door* are not only phonetically identical, but also semantically relevant. The interpretation is made within a given word boundary. Even with smart linguistic and semantic information, it is still impossible to decipher the correct word sequence, unless the speaker pauses between words or uses intonation to set apart these semantically confusable phrases.

In addition to the context variability at word and sentence level, you can find dramatic context variability at the phonetic level. As illustrated in Figure 9.1, the acoustic realization of phoneme /ee/ for word *peat* and *wheel* depends on its left and right context. The dependency becomes more important in fast speech or spontaneous speech conversations, since many phonemes are not fully realized.

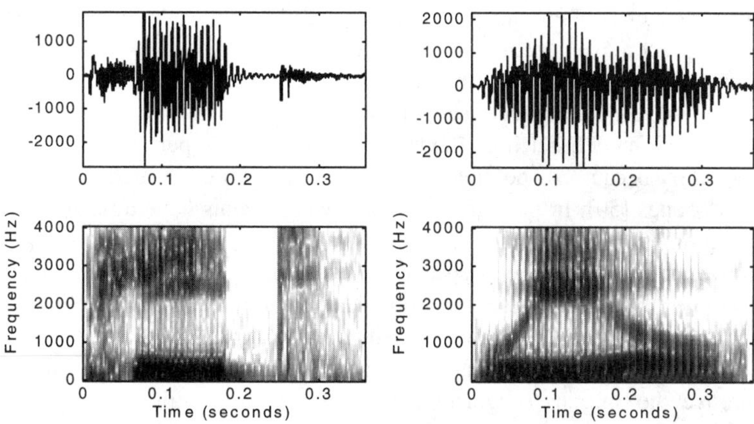

Figure 9.1 Waveforms and spectrograms for words *peat* (left) and *wheel* (right). The phoneme /ee/ is illustrated with two different left and right contexts. This illustrates that different contexts may have different effects on a phone.

9.1.2. Style Variability

To deal with acoustic realization variability, a number of constraints can be imposed on the use of the speech recognizer. For example, we can have an *isolated* speech recognition system, in which users have to pause between each word. Because the pause provides a clear boundary for the word, we can easily eliminate errors such as *Ford or* and *four door.* In addition, isolated speech provides a correct silence context to each word so that it is easier to model and decode the speech, leading to a significant reduction in computational complexity and error rate. In practice, the word-recognition error rate of an isolated speech recognizer can typically be reduced by more than a factor of three (from 7% to 2%) as compared with to a comparable continuous speech recognition system [5]. The disadvantage is that such an isolated speech recognizer is unnatural to most people. The throughput is also significantly lower than that for continuous speech.

In continuous speech recognition, the error rate for casual, spontaneous speech, as occurs in our daily conversation, is much higher than for carefully articulated read-aloud speech. The rate of speech also affects the word recognition rate. It is typical that the higher the *speaking rate* (words/minute), the higher the error rate. If a person whispers, or shouts, to reflect his or her emotional changes, the variation increases even more significantly.

9.1.3. Speaker Variability

Every individual speaker is different. The speech he or she produces reflects the physical vocal tract size, length and width of the neck, a range of physical characteristics, age, sex, dialect, health, education, and personal style. As such, one person's speech patterns can be entirely different from those of another person. Even if we exclude these interspeaker differences, the same speaker is often unable to precisely produce the same utterance. Thus, the shape of the vocal tract movement and rate of delivery may vary from utterance to utterance, even with dedicated effort to minimize the variability.

For *speaker-independent* speech recognition, we typically use more than 500 speakers to build a combined model. Such an approach exhibits large performance fluctuations among new speakers because of possible mismatches in the training data between exiting speakers and new ones [50]. In particular, speakers with accents have a tangible error-rate increase of 2 to 3 times.

To improve the performance of a speaker-independent speech recognizer, a number of constraints can be imposed on its use. For example, we can have a user enrollment that requires the user to speak for about 30 minutes. With the *speaker-dependent* data and training, we may be able to capture various speaker-dependent acoustic characteristics that can significantly improve the speech recognizer's performance. In practice, speaker-dependent speech recognition offers not only improved accuracy but also improved speed, since decoding can be more efficient with an accurate acoustic and phonetic model. A typical speaker-dependent speech recognition system can reduce the word recognition error by more than 30% as compared with a comparable speaker-independent speech recognition system.

The disadvantage of speaker-dependent speech recognition is that it takes time to collect speaker-dependent data, which may be impractical for some applications such as an automatic telephone operator. Many applications have to support walk-in speakers, so speaker-independent speech recognition remains an important feature. When the amount of speaker-dependent data is limited, it is important to make use of both speaker-dependent and speaker-independent data using *speaker-adaptive* training techniques, as discussed in Section 9.6. Even for speaker-independent speech recognition, you can still use speaker-adaptive training based on recognition results to quickly adapt to each individual speaker during usage.

9.1.4. Environment Variability

The world we live in is full of sounds of varying loudness from different sources. When we interact with computers, we may have people speaking in the background. Someone may slam the door, or the air conditioning may start humming without notice. If speech recognition is embedded in mobile devices, such as PDAs (personal digital assistants) or cellular phones, the spectrum of noises varies significantly because the owner moves around. These external parameters, such as the characteristics of the environmental noise and the type and placement of the microphone, can greatly affect speech recognition system performance. In addition to the background noises, we have to deal with noises made by speakers, such as lip smacks and noncommunication words. Noise may also be present from the input device itself, such as microphone and A/D interference noises.

In a similar manner to speaker-independent training, we can build a system by using a large amount of data collected from a number of environments; this is referred to as *multistyle training* [70]. We can use adaptive techniques to normalize the mismatch across different environment conditions in a manner similar to speaker-adaptive training, as discussed in Chapter 10. Despite the progress being made in the field, environment variability remains as one of the most severe challenges facing today's state-of-the-art speech systems.

9.2. How to Measure Speech Recognition Errors

It is critical to evaluate the performance of speech recognition systems. The word recognition error rate is widely used as one of the most important measures. When you compare different acoustic modeling algorithms, it is important to compare their relative error reduction. Empirically, you need to have a test data set that contains more than 500 sentences (with 6 to 10 words for each sentence) from 5 to 10 different speakers to reliably estimate the recognition error rate. Typically, you need to have more than 10% relative error reduction to consider adopting a new algorithm.

As a sanity check, you may want to use a small sample from the training data to measure the performance of the *training set*, which is often much better than what you can get from testing new data. Training-set performance is useful in the development stage to identify potential implementation bugs. Eventually, you need to use a *development set* that typically consists of data never used in training. Since you may tune a number of parameters with your development set, it is important to evaluate performance of a *test set* after you decide the optimal parameter setting. The test set should be completely new with respect to both training and parameter tuning.

There are typically three types of word recognition errors in speech recognition:

- *Substitution*: an incorrect word was substituted for the correct word
- *Deletion*: a correct word was omitted in the recognized sentence
- *Insertion*: an extra word was added in the recognized sentence[2]

For instance, a speech recognition system may produce an incorrect result as follows, where substitutions are ***bold***, insertions are <u>underlined</u>, and deletions are denoted as **. There are four errors in this example.

Correct: *Did mob mission area of the Copeland ever go to m4 in nineteen eighty one*
Recognized: *Did mob mission area ** the* ***copy*** *<u>land</u> ever go to m4 in nineteen* ***east*** *one*

To determine the minimum error rate, you can't simply compare two word sequences one by one. For example, suppose you have utterance *The effect is clear* recognized as <u>*Effect is not clear*</u>. If you compare word to word, the error rate is 75% (*The* vs. <u>*Effect*</u>, *effect* vs. <u>*is*</u>, *is* vs. <u>*not*</u>). In fact, the error rate is only 50% with one deletion (*The*) and one insertion (<u>*not*</u>). In general, you need to align a recognized word string against the correct word string and compute the number of substitutions (*Subs*), deletions (*Dels*), and insertions (*Ins*). The *Word Error Rate* is defined as:

$$\text{Word Error Rate} = 100\% \times \frac{Subs + Dels + Ins}{\text{No. of words in the correct sentenc}} \tag{9.3}$$

This alignment is also known as the *maximum substring matching* problem, which can be easily handled by the dynamic programming algorithm discussed in Chapter 8.

Let the correct word string be $w_1 w_2 \cdots w_n$, where w_i denotes the ith word in the correct word string, and the recognized word string be $\hat{w}_1 \hat{w}_2 \cdots \hat{w}_m$, where $\hat{w}_i$ denotes the ith word in the recognized word string. We denote $R[i, j]$ as the minimum error of aligning substring $w_1 w_2 \cdots w_n$ against substring $\hat{w}_1 \hat{w}_2 \cdots \hat{w}_m$. The optimal alignment and the associated word error rate $R[n, m]$ for correct word string $w_1 w_2 \cdots w_n$ and the recognized word string $\hat{w}_1 \hat{w}_2 \cdots \hat{w}_m$ are obtained via the dynamic programming algorithm illustrated in Algorithm 9.1. The accumulated cost function $R[i, j]$ progresses from $R[1, 1]$ to $R[n, m]$ corresponding to the minimum distance from (1, 1) to (*n*, *m*). We store the back pointer information $B[i, j]$ as we move along. When we reach the final grid (*n*, *m*), we back trace along the optimal path to find out if there are substitutions, deletions, or insertions on the matched path, as stored in $B[i, j]$.

[2] Even for isolated speech recognition, you may still have the insertion error, since the word boundary needs to be detected in most applications. It is possible that one isolated utterance is recognized as two words.

ALGORITHM 9.1: *ALGORITHM TO MEASURE THE WORD ERROR RATE*

Step 1: *Initialization* $R[0,0]=0 \;\; R[i,j]=\infty \text{ if } (i<0) \text{ or } (j<0) \;\; B[0,0]=0$

Step 2: *Iteration*

for $i=1,\ldots,n$ {

for $j=1,\ldots,m$ {

$$R[i,j]=\min\begin{bmatrix} R[i-1,j]+1 \text{ (deletion)} \\ R[i-1,j-1] \text{ (match)} \\ R[i-1,j-1]+1 \text{ (substitution)} \\ R[i,j-1]+1 \text{ (insertion)} \end{bmatrix}$$

$$B[i,j]=\begin{cases} 1 & \text{if deletion} \\ 2 & \text{if insertion} \\ 3 & \text{if match} \\ 4 & \text{if substitution} \end{cases} \quad \}\,\}$$

Step 3: *Backtracking and termination*

word error rate $=100\% \times \dfrac{R(n,m)}{n}$

optimal backward path = $(s_1, s_2, \ldots, 0)$

where $s_1 = B[n,m]$, $s_t = \begin{bmatrix} B[i-1,j] \text{ if } s_{t-1}=1 \\ B[i,j-1] \text{ if } s_{t-1}=2 \\ B[i-1,j-1] \text{ if } s_{t-1}=3 \text{ or } 4 \end{bmatrix}$ for $t=2,\ldots$ until $s_t=0$

For applications involved with rejection, such as word confidence measures as discussed in Section 9.7, you need to measure both false rejection rate and false acceptance rate. In speaker or command verification, the false acceptance of a valid user/command is also referred to as Type I error, as opposed to the false rejection of a valid user/command (Type II) [17]. A higher false rejection rate generally leads to a lower false acceptance rate. A plot of the false rejection rate versus the false acceptance rate, widely used in communication theory, is called the *receiver operating characteristic* (ROC) curve.

9.3. SIGNAL PROCESSING—EXTRACTING FEATURES

The role of a signal processing module, as illustrated in Figure 1.2, is to reduce the data rate, to remove noises, and to extract salient features that are useful for subsequent acoustic matching. Using as building blocks the topics we discussed in earlier chapters, we briefly illustrate here what is important in modeling speech to deal with variations we must address. More advanced environment normalization techniques are discussed in Chapter 10.

9.3.1. Signal Acquisition

Today's computers can handle most of the necessary speech signal acquisition tasks in software. For example, most PC sound cards have direct memory access, and the speech can be digitized to memory without burdening the CPU with input/output interrupts. The operating system can correctly handle most of the necessary AD/DA functions in real time.

To perform speech recognition, a number of components—such as digitizing speech, feature extraction and transformation, acoustic matching, and language model-based search—can be pipelined time-synchronously from left to right. Most operating systems can supply mechanisms for organizing pipelined programs in a multitasking environment. Buffers must be appropriately allocated so that you can ensure time-synchronous processing of each component. Large buffers are generally required on slow machines because of potential delays in processing an individual component. The right buffer size can be easily determined by experimentally tuning the system with different machine load situations to find a balance between resource use and relative delay.

For speech signal acquisition, the needed buffer typically ranges from 4 to 64 kB with 16-kHz sampling rate and 16-bit A/D precision. In practice, 16-kHz sampling rate is sufficient for the speech bandwidth (8 kHz). Reduced bandwidth, such as telephone channel, generally increases speech recognition error rate. Table 9.1 shows some empirical relative word recognition error increase using a number of different sampling rates. If we take the 8-kHz sampling as our baseline, we can reduce the word recognition error with a comparable recognizer by about 10% if we increase the sampling rate to 11 kHz. If we further increase the sampling rate to 16 kHz, the word recognition error rate can be further reduced by an additional 10%. Further increasing the sampling rate to 22 kHz does not have any additional impact on the word recognition errors, because most of the salient speech features are within an 8-kHz bandwidth.

Table 9.1 Relative error rate reduction with different sampling rates. The reduction is relative to that of the preceding row.

Sampling Rate	Relative Error-Rate Reduction
8 kHz	Baseline
11 kHz	+10%
16 kHz	+10%
22 kHz	+0%

9.3.2. End-Point Detection

To activate speech signal capture, you can use a number of modes including either *push to talk* or *continuously listening*. The push-to-talk mode uses a special push event to activate or deactivate speech capture, which is immune to the potential background noise and can eliminate unnecessary use of processing resources to detect speech events. This mode sometimes also requires you to *push and hold while talking*. You push to indicate speech's beginning and then release to indicate the end of speech capture. The disadvantage is the necessity to activate the application each time the person speaks.

The continuously listening model listens all the time and automatically detects whether there is a speech signal or not. It needs a so-called *speech end-point detector*, which is typically based on an extremely efficient two-class pattern classifier. Such a classifier is used to filter out obvious silence, but the ultimate decision on the utterance boundary is left to the speech recognizer. In comparison to the push-to-talk mode, the continuously listening mode requires more processing resources, also with potential classification errors.

The endpoint detector is often based on an energy threshold that is a function of time. The logarithm of the energy threshold can be dynamically generated based on energy profiles across a certain period of time. Constraints on word duration can also be imposed to better classify a sequence of frames so that extremely short spikes can be eliminated.

It is not critical for the automatic end-point detector to offer exact end-point accuracy. The key feature required of it is a low rejection rate (i.e., the automatic end-point detector should not interpret speech segments as silence/noise segments). Any false rejection leads to an error in the speech recognizer. On the other hand, a possible false acceptance (i.e., the automatic end-point detector interprets noise segments as speech segments) may be rescued by the speech recognizer later if the recognizer has appropriate noise models, such as specific models for clicks, lip smacks, and background noise.

Explicit end-point detectors work reasonably well with recordings exhibiting a signal-to-noise ratio of 30 dB or greater, but they fail considerably on noisier speech. As discussed, speech recognizers can be used to determine the end points by aligning the vocabulary words preceded and followed by a silence/noise model. This scheme is generally much more reliable than any threshold-based explicit end-point detection, because recognition can jointly detect both the end points and words or other explicit noise classes, but requires more computational resources. A compromise is to use a simple adaptive two-class (speech vs. silence/noise) classifier to locate speech activities (with enough buffers at both ends) and notify the speech recognizer for subsequent processing. For the two-class classifier, we can use both the log-energy and delta log-energy as the feature. Two Gaussian density functions, $\{\Phi_1, \Phi_2\} = \Phi$, can be used to model the background stationary noise and speech, respectively. The parameters of the Gaussian density can be estimated using the labeled speech and noise data or estimated in an unsupervised manner.

When enough frames are classified as speech segments by the efficient two-class classifier, the speech recognizer is notified to start recognizing the signal. As shown in Figure 9.2, we should include enough frames before the beginning frame, t_b, for the speech recognizer to minimize the possible detection error. In the same manner, when enough noise/silence frames are detected at t_e, we should keep providing the speech recognizer with enough frames for processing before declaring that the end of the utterance has been reached.

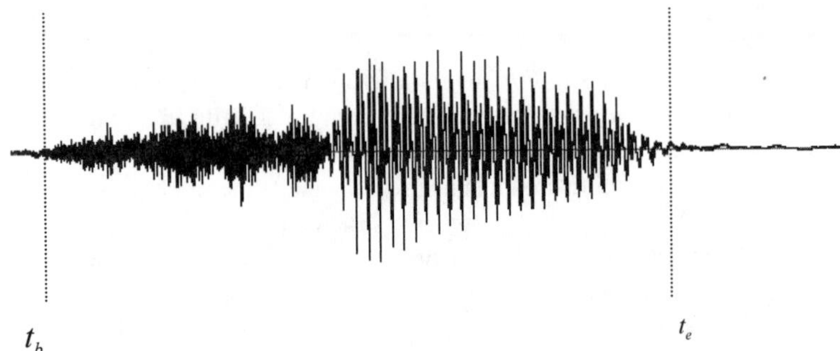

Figure 9.2 End-point detection boundary t_b and t_e may need extra buffering for subsequent speech recognition.

Since there are only two classes, these parameters can be dynamically adapted using the EM algorithm during runtime. As discussed in Chapter 4, the EM algorithm can iteratively estimate the Gaussian parameters without having a precise segmentation between speech and noise segments. This is very important, because we need to keep the parameters dynamic for robust end-point detection in constantly changing environments.

To track the varying background noises, we use an exponential window to give weight to the most recent signal:

$$w_k = \exp(-\alpha k) \tag{9.4}$$

where α is a constant that controls the adaptation rate, and k is the index of the time. In fact, you could use different rates for noise and speech when you use the EM algorithm to estimate the two-class Gaussian parameters. It is advantageous to use a smaller time constant for noise than for speech. With such a weighting window, the means of the Gaussian density, as discussed in Chapter 4, can be rewritten as:

$$\hat{\boldsymbol{\mu}}_k = \frac{\sum_{i=-\infty}^{t} w_i \frac{c_k P(\mathbf{x}_i \mid \Phi_k)\mathbf{x}_i}{\sum_{k=1}^{2} P(\mathbf{x}_i \mid \Phi_k)}}{\sum_{i=-\infty}^{t} w_i \frac{c_k P(\mathbf{x}_i \mid \Phi_k)}{\sum_{k=1}^{2} P(\mathbf{x}_i \mid \Phi_k)}}, k \in \{0,1\} \tag{9.5}$$

9.3.3. MFCC and Its Dynamic Features

The extraction of reliable features is one of the most important issues in speech recognition. There are a large number of features we can use. However, as discussed in Chapter 4, the *curse-of-dimensionality* problem reminds us that the amount of training data is always limited. Therefore, incorporation of additional features may not lead to any measurable error reduction. This does not necessarily mean that the additional features are poor ones, but rather that we may have insufficient data to reliably model those features.

The first feature we use is the speech waveform itself. In general, time-domain features are much less accurate than frequency-domain features such as the mel-frequency cepstral coefficients (MFCC) discussed in Chapter 6 [23]. This is because many features such as formants, useful in discriminating vowels, are better characterized in the frequency domain with a low-dimension feature vector.

As discussed in Chapter 2, temporal changes in the spectra play an important role in human perception. One way to capture this information is to use *delta coefficients* that measure the change in coefficients over time. Temporal information is particularly complementary to HMMs, since HMMs assume each frame is independent of the past, and these dynamic features broaden the scope of a frame. It is also easy to incorporate new features by augmenting the static feature.

When 16-kHz sampling rate is used, a typical state-of-the-art speech system can be build based on the following features.

- 13th-order MFCC $\mathbf{c}_k$
- 13th-order 40-msec 1st-order delta MFCC computed from $\Delta\mathbf{c}_k = \mathbf{c}_{k+2} - \mathbf{c}_{k-2}$
- 13th-order 2nd-order delta MFCC computed from $\Delta\Delta\mathbf{c}_k = \Delta\mathbf{c}_{k+1} - \Delta\mathbf{c}_{k-1}$

A short-time analysis Hamming window of 25 ms is typically used to compute the MFCC $\mathbf{c}_k$. The window shift is typically 10 ms. Please note that $c_k[0]$ is included in the feature vector, which has a role similar to that of the log power. The feature vector used for speech recognition is typically a combination of these features

$$\mathbf{x}_k = \begin{pmatrix} \mathbf{c}_k \\ \Delta\mathbf{c}_k \\ \Delta\Delta\mathbf{c}_k \end{pmatrix} \tag{9.6}$$

The relative error reduction with a typical speech recognition system is illustrated in Table 9.2. As you can see from the table, the 13th-order MFCC outperforms 13th-order LPC cepstrum coefficients, which indicates that perceptually motivated mel-scale representation indeed helps recognition. In a similar manner, perceptually based LPC features such as PLP can achieve similar improvement. The MFCC order has also been studied experimentally for speech recognition. The higher-order MFCC does not further reduce the error rate in comparison with the 13th-order MFCC, which indicates that the first 13 coefficients already contain most salient information needed for speech recognition. In addition to mel-scale representation, another perceptually motivated feature such as the first- and second-order delta features can significantly reduce the word recognition error, while the higher-order delta features provide no further information.

Feature extraction in these experiments is typically optimized together with the classifier, since there are a number of modeling assumptions, such as the diagonal covariance in the Gaussian density function, that are closely related to what features to use. It is possible that these relative error reductions would vary if a different speech recognizer were used.

Table 9.2 Relative error reduction with different features. The reduction is relative to that of the preceding row.

Feature Set	Relative Error Reduction
13th-order LPC cepstrum coefficients	Baseline
13th-order MFCC	+10%
16th-order MFCC	+0%
+1st- and 2nd-order dynamic features	+20%
+3rd-order dynamic features	+0%

9.3.4. Feature Transformation

Before you use feature vectors such as MFCC for recognition, you can preprocess or transform them into a new space that alleviates environment noise, channel distortion, and speaker variations. You can also transform the features that are most effective for preserving class separability so that you can further reduce the recognition error rate. Since we devote Chapter 10 completely to environment and channel normalization, we briefly discuss here how we can transform the feature vectors to improve class separability.

To further reduce the dimension of the feature vector, you can use a number of dimension reduction techniques to map the feature vector into more effective representations. If the mapping is linear, the mapping function is well defined and you can find the coefficients of the linear function so as to optimize your objective functions. For example, when you combine the first- and second-order dynamic features with the static MFCC vector, you can use *principal-component analysis* (PCA) (also known as *Karhunen-Loeve* transform) [32] to map the combined feature vector into a smaller dimensional vector. The optimum basis vectors of the principal-component analysis are the eigenvectors of the covariance matrix of a given distribution. In practice, you can compute the eigenvectors of the autocorrelation matrix as the basis vectors. The effectiveness of the transformed vector, in terms of representing the original feature vector, is determined by the corresponding eigenvalue of each value in the vector. You can discard the feature with the smallest eigenvalue, since the mean-square error between the transformed vector and the original vector is determined by the eigenvalue of each feature in the vector. In addition, the transformed feature vector is uncorrelated. This is particularly suitable for the Gaussian probability density function with a diagonal covariance matrix.

The recognition error is the best criterion for deciding what feature sets to use. However, it is hard to obtain such an estimate to evaluate feature sets systematically. A simpler approach is to use within-class and between-class scatter matrices to formulate criteria of class separability, which is also called *Linear Discriminant Analysis* (LDA) transformation. We can compute the within-class scatter matrix as:

$$S_w = \sum_{\mathbf{x}\in\omega_i} P(\omega_i)E\{(\mathbf{x}-\boldsymbol{\mu}_i)(\mathbf{x}-\boldsymbol{\mu}_i)^t \mid \omega_i\} = \sum_{\mathbf{x}\in\omega_i} P(\omega_i)\Sigma_i \tag{9.7}$$

where the sum is for all the data $\mathbf{x}$ within the class ω_i. This is the scatter of samples around their respective class mean. On the other hand, the between-class scatter matrix is the scatter of the expected vectors around the mixture mean:

$$S_B = \sum_{\boldsymbol{\mu}_i\in\omega_i} P(\omega_i)(\boldsymbol{\mu}_i - \mathbf{m}_0)(\boldsymbol{\mu}_i - \mathbf{m}_0)^t \tag{9.8}$$

where $\mathbf{m}_0$ represents the expected mean vector of the mixture distribution:

$$\mathbf{m}_0 = E\{\mathbf{x}\} = \sum_{\omega_i} P(\omega_i)\mathbf{m}_i \tag{9.9}$$

To formulate criteria to transform feature vector $\mathbf{x}$, we need to derive the linear transformation matrix $\mathbf{A}$. One of the measures can be the trace of $S_w^{-1}S_B$:

$$J = tr(S_w^{-1}S_B) \tag{9.10}$$

The trace is the sum of the eigenvalues of $S_w^{-1}S_B$ and hence the sum of the variances in the principal directions. The number is larger when the between-class scatter is large or the within-class scatter is small. You can derive the transformation matrix based on the eigenvectors of $S_w^{-1}S_B$. In a manner similar to PCA, you can reduce the dimension of the original input feature vector by discarding the smallest eigenvalues [16, 54].

Researchers have used the LDA method to measure the effectiveness of several feature vectors for speaker normalization [41]. Other feature processing techniques designed for speaker normalization include neural-network-based speaker mapping [51], frequency warping for vocal tract normalization (VTN) via mel-frequency scaling [67, 100], and bilinear transformation [2].

To reduce interspeaker variability by a speaker-specific frequency warping, you can simply shift the center frequencies of the mel-spaced filter bank. Let $k\Delta f_{mel}$, k = 1, ..., K, denote the center frequencies in mel-scale. Then the center frequencies in hertz for a warping factor of α are computed by Eq. (9.11) before the cosine transformation of the MFCC feature vector.

$$f_{Hz}^{\alpha}(k\Delta f_{mel}) = 700(10^{k\Delta f_{mel}/2595} - 1)/\alpha \tag{9.11}$$

The warping factor is estimated for each speaker by computing the likelihood of the training data for feature sets obtained with different warping factors using the HMM. The relative error reduction based on the feature transformation method has been limited, typically under 10%.

9.4. PHONETIC MODELING—SELECTING APPROPRIATE UNITS

As discussed in Chapter 2, the phonetic system is related to a particular language. We focus our discussion on language-independent technologies but use English in our examples to illustrate how we can use the language-independent technologies to model the salient phonetic information in the language. For general-purpose large-vocabulary speech recognition, it is difficult to build whole-word models because:

- Every new task contains novel words without any available training data, such as proper nouns and newly invented jargons.
- There are simply too many words, and these different words may have different acoustic realizations, as illustrated in Chapter 2. It is unlikely that we have sufficient repetitions of these words to build context-dependent word models.

How to select the most basic units to represent salient acoustic and phonetic information for the language is an important issue in designing a workable system. At a high level, there are a number of issues we must consider in choosing appropriate modeling units.

- The unit should be *accurate,* to represent the acoustic realization that appears in different contexts.
- The unit should be *trainable*. We should have enough data to estimate the parameters of the unit. Although words are accurate and representative, they are the least trainable choice in building a working system, since it is nearly impossible to get several hundred repetitions for all the words, unless we are using a speech recognizer that is domain specific, such as a recognizer designed for digits only.
- The unit should be *generalizable,* so that any new word can be derived from a predefined unit inventory for task-independent speech recognition. If we have a fixed set of word models, there is no obvious way for us to derive the new word model.

A practical challenge is how to balance these selection criteria for speech recognition. In this section we compare a number of units and point out their strengths and weaknesses in practical applications.

9.4.1. Comparison of Different Units

What is a unit of language? In English, words are typically considered as a principal carrier of meaning and are seen as the smallest unit that is capable of independent use. As the most natural unit of speech, whole-word models have been widely used for many speech recognition systems. A distinctive advantage of using word models is that we can capture the phonetic coarticulation inherent within these words. When the vocabulary is small, we can create word models that are context dependent.

For example, if the vocabulary is English digits, we can have different word models for the word *one* to represent the word in different contexts. Thus each word model is dependent on its left and right context. If someone says *three one two*, the recognizer uses the word model *one* that specifically depends on the left context *three* and right context *two*. Since the vocabulary is small (10), we need to have only 10*10*10=1000 word models, which is achievable when you collect enough training data. With context-dependent, or even context-independent, word models, a wide range of phonological variations can be automatically accommodated. When these word models are adequately trained, they usually yield the best recognition performance in comparison to other modeling units. Therefore, for small vocabulary recognition, whole-word models are widely used, since they are both *accurate* and *trainable*, and there is no need to be *generalizable*.

While words are suitable units for small-vocabulary speech recognition, they are not a practical choice for large-vocabulary continuous speech recognition. First, each word has to be treated individually, and data cannot be shared across word models; this implies a prohibitively large amount of training data and storage. Second, for some task configurations, the recognition vocabulary may consist of words that never appeared in the training data. As a result, some form of word-model composition technique is required to generate word models. Third, it is very expensive to model interword coarticulation effects or adapt a word-based system for a new speaker, a new channel, or new context usage.

To summarize, word models are *accurate* if enough data are available. Thus, they are *trainable* only for small tasks. They are typically not *generalizable*.

Alternatively, there are only about 50 phones in English, and they can be sufficiently trained with just a few hundred sentences. Unlike word models, phonetic models provide no training problem. Moreover, they are also vocabulary independent by nature and can be trained on one task and tested on another. Thus, phones are more *trainable* and *generalizable*. However, the phonetic model is inadequate because it assumes that a phoneme in any context is identical. Although we may try to say each word as a concatenated sequence of independent phonemes, these phonemes are not produced independently, because our articulators cannot move instantaneously from one position to another. Thus, the realization of a phoneme is strongly affected by its immediately neighboring phonemes. For example, if context-independent phonetic models are used, the same model for *t* must capture various events, such as flapping, unreleased stops, and realizations in */t s/* and */t r/*. Then, if */t s/* is the only context in which *t* occurs in the training, while */t r/* is the only context in the testing, the model used is highly inappropriate. While word models are not generalizable, phonetic models overgeneralize and, thus, lead to less accurate models.

A compromise between the word and phonetic model is to use larger units such as *syllables*. These units encompass phone clusters that contain the most variable contextual effects. However, while the central portions of these units have no contextual dependencies, the beginning and ending portions are still susceptible to some contextual effects. There are only about 1200 tone-dependent syllables in Chinese and approximately 50 syllables in Japanese, which makes syllable a suitable unit for these languages. Unfortunately, the large number of syllables (over 30,000) in English presents a challenge in terms of trainability.

9.4.2. Context Dependency

If we make units context dependent, we can significantly improve the recognition accuracy, provided there are enough training data to estimate these context-dependent parameters. Context-dependent phonemes have been widely used for large-vocabulary speech recognition, thanks to its significantly improved accuracy and trainability. A context usually refers to the immediate left and/or right neighboring phones.

A *triphone* model is a phonetic model that takes into consideration both the left and the right neighboring phones. If two phones have the same identity but different left or right contexts, they are considered different triphones. We call different realizations of a phoneme *allophones*. Triphones are an example of allophones.

The left and right contexts used in triphones, while important, are only two of many important contributing factors that affect the realization of a phone. Triphone models are powerful because they capture the most important coarticulatory effects. They are generally much more consistent than context-independent phone models. However, as context-dependent models generally have increased parameters, trainability becomes a challenging issue. We need to balance trainability and accuracy with a number of parameter-sharing techniques.

Modeling interword context-dependent phones is complicated. For example, in the word *speech*, pronounced */s p iy ch/*, both left and right contexts for */p/* and */iy/* are known, while the left context for */s/* and the right context for */ch/* are dependent on the preceding and following words in actual sentences. The juncture effect on word boundaries is one of the most serious coarticulation phenomena in continuous speech, especially with short function words like *the* or *a*. Even with the same left and right context identities, there may be significantly different realizations for a phone at different word positions (the *beginning, middle,* or *end* of a word). For example, the phone */t/* in *that rock* is almost extinct, while the phone */t/* in the middle of *theatrical* sounds like */ch/*. This implies that different word positions have an effect on the realization of the same triphone.

In addition to the context, stress also plays an important role in the realization of a particular phone. Stressed vowels tend to have longer duration, higher pitch, and more intensity, while unstressed vowels appear to move toward a neutral, central *schwa*-like phoneme. Agreement about the phonetic identity of a syllable has been reported to be greater in stressed syllables for both humans and automatic phone recognizers. In English, *word-level* stress is referred to as *free stress,* because the stressed syllable can take on any position within a word, in contrast to *bound stress* found in languages such as French and Polish, where the position of the stressed syllable is fixed within a word. Therefore, stress in English can be used as a constraint for lexical access. In fact, stress can be used as a unique feature to distinguish a set of word pairs, such as **_im_**_port_ vs. _im_**_port_**, and **_ex_**_port_ vs. _ex_**_port_**. For example, the phone set used for Whisper, such as */er/-/axr/* and */ah/-/ix/-/ax/*, describes these stressed and unstressed vowels. One example illustrating how stress can significantly affect the realization of phone is demonstrated in Figure 9.3, where phone */t/* in word *Italy* vs. *Italian* is pronounced differently in American English due the location of the stress, albeit the triphone context is identical for both words.

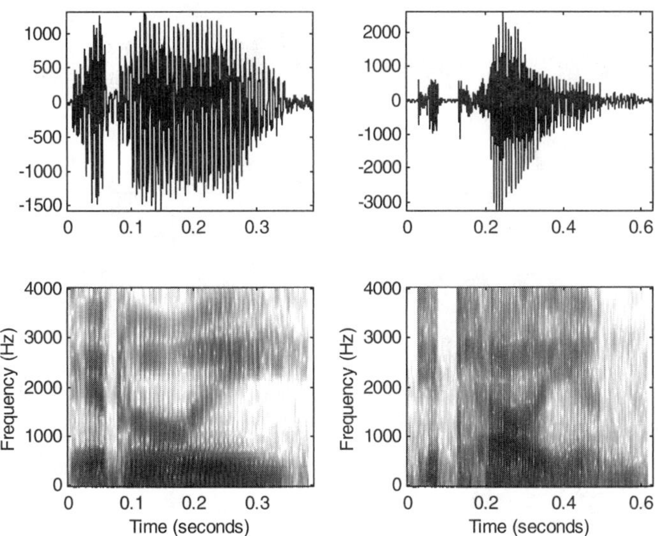

Figure 9.3 The importance of stress is illustrated in *Italy* vs. *Italian* for phone */t/*. The realizations are quite different, even though they share the same left and right context.

Sentence-level stress, on the other hand, represents the overall stress pattern of continuous speech. While sentence-level stress does not change the meaning of any particular lexicon item, it usually increases the relative prominence of portions of the utterance for the purpose of contrast or emphasis. *Contrastive stress* is normally used to coordinate constructions such as *there are* ***import*** *records and there are* ***domestic*** *ones*, as well as for the purpose of correction, as in *I said* ***import****, not* ***export.*** *Emphatic* stress is commonly used to distinguish a sentence from its negation, e.g., *I* ***did*** *have dinner*. Sentence-level stress is very hard to model without incorporating high-level semantic and pragmatic knowledge. In most state-of- the-art speech recognition systems, only word-level stress is used for creating allophones.

9.4.3. Clustered Acoustic-Phonetic Units

Triphone modeling assumes that every triphone context is different. Actually, many phones have similar effects on the neighboring phones. The position of our articulators has an important effect on how we pronounce neighboring vowels. For example, /b/ and /p/ are both labial stops and have similar effects on the following vowel, while /r/ and /w/ are both liquids and have similar effects on the following vowel. Contrary to what we illustrate in Figure 9.1, Figure 9.4 illustrates this phenomenon. It is desirable to find instances of similar contexts and merge them. This would lead to a much more manageable number of models that can be better trained.

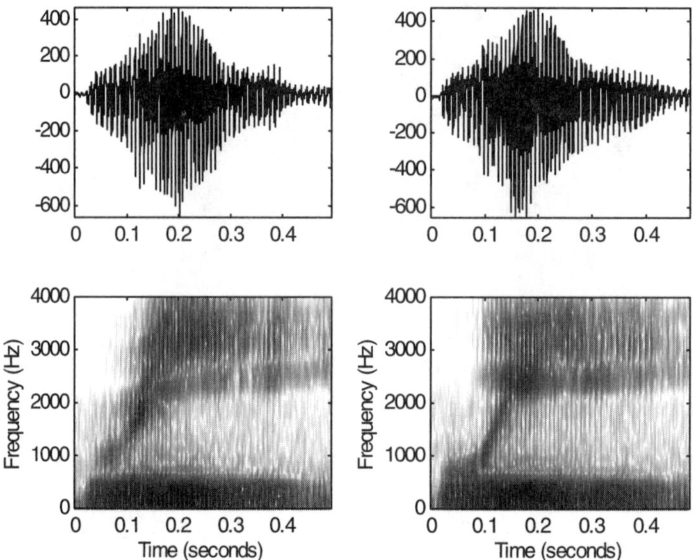

Figure 9.4 The spectrograms for the phoneme /iy/ with two different *left-contexts* are illustrated. Note that /r/ and /w/ have similar effects on /iy/. This illustrates that different left-contexts may have similar effects on a phone.

The trainability and accuracy balance between phonetic and word models can be generalized further to model subphonetic events. In fact, both phonetic and subphonetic units have the same benefits, as they share parameters at the unit level. This is the key benefit in comparison to the word units. Papers by [11, 45, 57, 66, 111] provide examples of the application of this concept to cluster hidden Markov models. For subphonetic modeling, we can treat the state in phonetic HMMs as the basic subphonetic unit. Hwang and Huang further generalized clustering to the state-dependent output distributions across different phonetic models [57]. Each cluster thus represents a set of similar Markov states and is called a *senone* [56]. A subword model is thus composed of a sequence of senones after the clustering is finished. The optimal number of senones for a system is mainly determined by the available training corpus and can be tuned on a development set.

Each allophone model is an HMM made of states, transitions, and probability distributions. To improve the reliability of the statistical parameters of these models, some distributions can be tied. For example, distributions for the central portion of an allophone may be tied together to reflect the fact that they represent the stable (context-independent) physical realization of the central part of the phoneme, uttered with a stationary configuration of the vocal tract. Clustering at the granularity of the state rather than the entire model can keep the dissimilar states of two models apart while the other corresponding states are merged, thus leading to better parameter sharing.

Figure 9.5 illustrates how state-based clustering can lead to improved representations. These two HMMs come from the same phone class with a different right context, leading to very different output distributions in the last state. As the left contexts are identical, the first and second output distributions are almost identical. If we measure the overall model similarity based on the accumulative overall output distribution similarities of all states, these two models may be clustered, leading to a very inaccurate distribution for the last state. Instead, we cluster the first two output distributions while leaving the last one intact.

There are two key issues in creating trainable context-dependent phonetic or subphonetic units:

- We need to enable better parameter sharing and smoothing. As Figure 9.4 illustrates, many phones have similar effects on neighboring phones. If the acoustic realization is indeed identical, we tie them together to improve trainability and efficiency.
- Since the number of triphones in English is very large (over 100,000), there are many new or unseen triphones that are in the test set but not in the training set. It is important to map these unseen triphones into appropriately trained triphones.

As discussed in Chapter 4, a decision tree is a binary tree to classify target objects by asking binary questions in a hierarchical manner. Modeling unseen triphones is particularly important for *vocabulary independence*, since it is difficult to collect a training corpus which covers enough occurrences of every possible subword unit. We need to find models that are accurate, trainable, and especially generalizable. The senonic decision tree classifies Markov states of triphones represented in the training corpus by asking linguistic questions composed of conjunctions, disjunctions, and/or negations of a set of predetermined simple categorical linguistic questions. Examples of these simple categorical questions are: *Is the left-context phone a fricative? Is the right-context phone a front vowel?* The typical question set used in Whisper to generate the senone tree is shown in Table 9.3. So, for each node in the tree, we check whether its left or right phone belongs to one of the categories. As discussed in Chapter 4, we measure the corresponding entropy reduction or likelihood increase for each question and select the question that has the largest entropy decrease to split the node.

Thus, the tree can be automatically constructed by searching, for each node, the question that renders the maximum entropy decrease. Alternatively, complex questions can be formed for each node for improved splitting.

When we grow the tree, it needs to be pruned using cross-validation as discussed in Chapter 4. When the algorithm terminates, the leaf nodes of the tree represent the senones to be used. Figure 9.6 shows an example tree we built to classify the second state of all /k/ triphones seen in a training corpus. After the tree is built, it can be applied to the second state of *any* /k/ triphone, thanks to the generalizability of the binary tree and the general linguistic questions. Figure 9.6 indicates that the second state of the /k/ triphone in *welcome* is mapped to the second senone, no matter whether this triphone occurs in the training corpus or not.

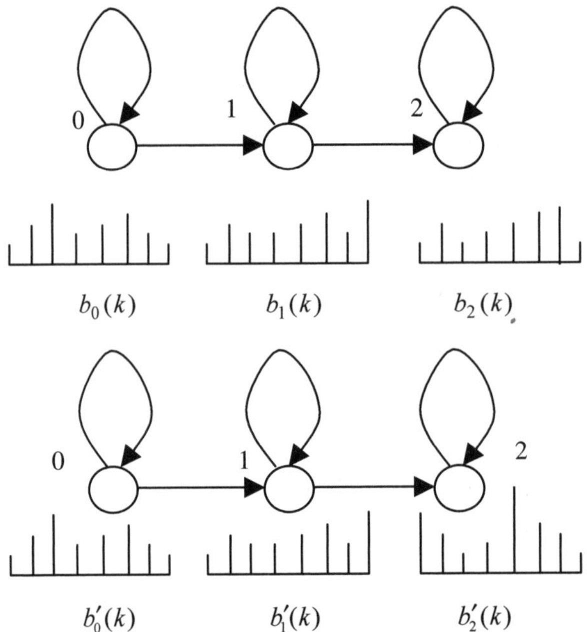

Figure 9.5 State-based vs. model-based clustering. These two models are very similar, as both the first and the second output distributions are almost identical. The key difference is the output distribution of the third state. If we measure the overall model similarity, which is often based on the accumulative output distribution similarities of all states, these two models may be clustered, leading to a very inaccurate distribution for the last state. If we cluster output distributions at state level, we can cluster the first two output distributions while leaving the last ones intact, leading to more accurate representations.

Table 9.3 Some example questions used in building senone trees.

Questions	Phones in Each Question Category
Aspseg	*hh*
Sil	*sil*
Alvstp	*d t*
Dental	*dh th*
Labstp	*b p*
Liquid	*l r*
Lw	*l w*
S/Sh	*s sh*
Sylbic	*er axr*
Velstp	*g k*
Affric	*ch jh*
Lqgl-B	*l r w*
Nasal	*m n ng*
Retro	*r er axr*
Schwa	*ax ix axr*
Velar	*ng g k*
Fric2	*th s sh f*
Fric3	*dh z zh v*
Lqgl	*l r w y*
S/Z/Sh/Zh	*s z sh zh*
Wglide	*uw aw ow w*
Labial	*w m b p v*
Palatl	*y ch jh sh zh*
Yglide	*iy ay ey oy y*
High	*ih ix iy uh uw y*
Lax	*eh ih ix uh ah ax*
Low	*ae aa ao aw ay oy*
Orstp2	*p t k*
Orstp3	*b d g*
Alvelr	*n d t s z*
Diph	*uw aw ay ey iy ow oy*
Fric1	*dh th s sh z zh v f*
Round	*uh ao uw ow oy w axr er*
Frnt-R	*ae eh ih ix iy ey ah ax y aw*
Tense	*iy ey ae uw ow aa ao ay oy aw*
Back-L	*uh ao uw ow aa er axr l r w aw*
Frnt-L	*ae eh ih ix iy ey ah ax y oy ay*
Back-R	*uh ao uw ow aa er axr oy l r w ay*
Orstp1	*b d g p t k ch jh*
Vowel	*ae eh ih ix iy uh ah ax aa ao uw aw ay ey ow oy er axr*
Son	*ae eh ih ix iy ey ah ax oy ay uh ao uw ow aa er axr aw l r w y*
Voiced	*ae eh ih ix iy uh ah ax aa ao uw aw ay ey ow oy l r w y er axr m n ng jh b d dh g v z zh*

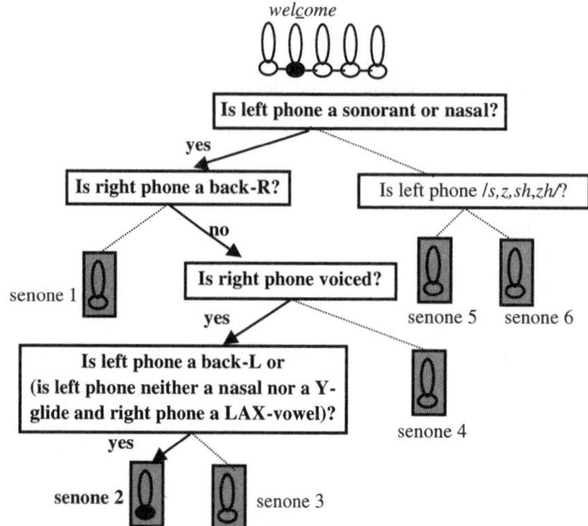

Figure 9.6 A decision tree for classifying the second state of K-triphone HMMs [48].

In practice, senone models significantly reduce the word recognition error rate in comparison with model-based clustered triphone models, as illustrated in Table 9.4. It is the senonic model's significant reduction of the overall system parameters that enables the continuous mixture HMMs to perform well for large-vocabulary speech recognition [56].

Table 9.4 Relative error reductions for different modeling units. The reduction is relative to that of the preceding row.

Units	Relative Error Reductions
Context-independent phone	Baseline
Context-dependent phone	+25%
Clustered triphone	+15%
Senone	+24%

9.4.4. Lexical Baseforms

When appropriate subword units are used, we must have the correct pronunciation for each word so that concatenation of subword units can accurately represent the word to be recognized. The dictionary represents the standard pronunciation used as a starting point for building a workable speech recognition system. We also need to provide alternative pronunciations for words such as *tomato* that may have very different pronunciations. For example, the COMLEX dictionary from LDC has about 90,000 baseforms that cover most words used in many years of *The Wall Street Journal*. The CMU Pronunciation Dictionary, which was optimized for continuous speech recognition, has about 100,000 baseforms.

In continuous speech recognition, we must also use phonologic rules to modify inter-word pronunciations or to have reduced sounds. Assimilation is a typical coarticulation phenomenon—a change in a segment to make it more like a neighboring segment. Typical examples include phrases such as *did you /d ih jh y ah/*, *set you /s eh ch er/*, *last year / l ae s ch iy r/*, *because you've /b iy k ah zh uw v/*, etc. Deletion is also common in continuous speech. For example, /t/ and /d/ are often deleted before a consonant. Thus, in conversational speech, you may find examples like *find him /f ay n ix m/*, *around this /ix r aw n ih s/*, and *Let me in /l eh m eh n/*.

Dictionaries often don't include proper names. For example, the 20,000 names included in the COMLEX dictionary are a small fraction of 1–2 million names in the USA. To deal with these new words, we often have to derive their pronunciation automatically. These new words have to be added on the fly, either by the user or through an interface from speech-aware applications. Unlike Spanish or Italian, rule-based letter-to-sound (LTS) conversion for English is often impractical, since so many words in English don't follow phonological rules. A trainable LTS converter is attractive, since its performance can be improved by constantly learning from examples so that it can generalize rules for the specific task. Trainable LTS converters can be based on neural networks, HMMs, or the CART described in Chapter 4. In practice, CART-based LTS has a very accurate performance [10, 61, 71, 89].

When CART is used, the basic YES-NO question for LTS conversion looks like: *Is the second right letter 'p'?* or: *Is the first left output phone /ay/?* The question for letters and phones can be on either the left or the right side. The range of question positions should be long enough to cover the most important phonological variations. Empirically, a 10-letter window (5 for left letter context and 5 for right letter context) and 3-phone window context is generally sufficient. A primitive set of questions can include all the singleton questions about each letter or phone identity. If we allow the node to have a complex question—that is, a combination of primitive questions—the depth of the tree can be greatly reduced and performance improved. For example, a complex question: *Is the second left letter 't' and the first left letter 'i' and the first right letter 'n'?* can capture *o* in the common suffix *tion* and convert it to the correct phone. Complex questions can also alleviate possible data-fragmentation problems caused by the greedy nature of the CART algorithm.

Categorical questions can be formed in both the letter and phone domains with our common linguistic knowledge. For example, the most often used set includes the letter or phone clusters for vowels, consonants, nasals, liquids, fricatives, and so on. In growing the decision tree, the context distance also plays a major role in the overall quality. It is very important to weight the entropy reduction according to the distance (either letter or phoneme) to avoid overgeneralization, which forces the tree to look more carefully at the *nearby* context than at the *far-away* context. Each leaf of the tree has a probability distribution for letter-to-phoneme mapping.

There are a number of ways to improve the effectiveness of the decision tree. First, pruning controls the tree's depth. For example, certain criteria have to be met for a node to be split. Typically splitting requires a minimum number of counts and a minimum entropy reduction. Second, the distribution at the leaves can be smoothed. For example, a leaf distribution can be interpolated with the distributions of its ancestor nodes using deleted-interpolation. Finally, we can partition the training data and build multiple trees with different prediction capabilities. These trees accommodate different phonological rules with different language origins.

When the decision tree is used to derive the phonetic pronunciation, the phonetic conversion error is about 8% for the *Wall Street Journal* newspaper text corpora [61]. These errors can be broadly classified into two categories. The first includes errors of proper nouns and foreign words. For example, *Pacino* can be mistakenly converted to */p ax s iy n ow /* instead of */p ax ch iy n ow/.* The second category includes generalization errors. For example, *shier* may be converted to */sh ih r/* instead of the correct pronunciation */sh ay r/* if the word *cashier /k ae sh ih r/* appears in the training data. The top three phone confusion pairs are */ix/ax/, /dx/t/,* and */ae/ax/.* The most confusing pair is */ix/ax/.* This is not surprising, because */ix/ax/* is among the most inconsistent transcriptions in most of the published dictionaries. There is no consensus for */ix/ax/* transcription among phoneticians.

Although automatic LTS conversion has a reasonable accuracy, it is hardly practical if you don't use an exception dictionary. This is especially true for proper nouns. In practice, you can often ask the person who knows how to pronounce the word to either speak or write down the correct phonetic pronunciation, updating the exception dictionary if the correct one disagrees with what the LTS generates. When acoustic examples are available, you can use the decision tree to generate multiple results and use these results as a language model to perform phone recognition on the acoustic examples. The best overall acoustic and LTS probability can be used as the most likely candidate in the exception dictionary. Since there may be many ways to pronounce a word, you can keep multiple pronunciations in the dictionary with a probability for each possible one. If the pronunciation probability is inaccurate, an increase in multiple pronunciations essentially increases the size and confusion of the vocabulary, leading to increased speech recognition error rate.

Even if you have accurate phonetic baseforms, pronunciations in spontaneous speech differ significantly from the standard baseform. Analysis of manual phonetic transcription of conversational speech reveals a large number (> 20%) of cases of genuine ambiguity: instances where human labelers disagree on the identity of the surface form [95]. For example, the word *because* has more than 15 different pronunciation variations, such as */b iy k ah z/, /b ix k ah z/, /k ah z/, /k ax z/, /b ix k ax z/, /b ax k ah z/, /b ih k ah z/, /k s/, /k ix z/, /k ih z/,/b iy k ah s/, /b iy k ah/, /b iy k ah zh/, /ax z/*, etc., in the context of conversational speech [39]. To characterize the acoustic evidence in the context of this ambiguity, you can partly resolve the ambiguity by deriving a suitable phonetic baseform from speech data [29, 95, 97]. This is because the widespread variation can be due either to a lexical fact (such as that the word *because* can be *'cause* in informal speech) or to the dialect differences. African American vernacular English has many vowels different from general American English.

To incorporate widespread pronunciations, we can use a probabilistic finite state machine to model each word's pronunciation variations, as shown in Figure 9.7. The probability with each arc indicates how likely that path is to be taken, with all the arcs that leave a node summing to 1. As with HMMs, these weights can be estimated from a real corpus for improved speech recognition [20, 85, 102, 103, 110]. In practice, the relative error reduction of using probabilistic finite state machines is very modest (5–10%).

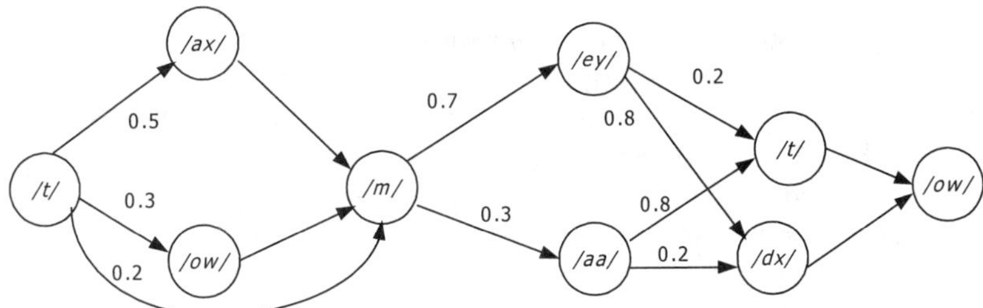

Figure 9.7 A possible pronunciation network for word *tomato*. The vowel /ey/ is more likely to flap, thereby having a higher transition probability into /dx/.

9.5. ACOUSTIC MODELING—SCORING ACOUSTIC FEATURES

After feature extraction, we have a sequence of feature vectors, **X**, such as the MFCC vector, as our input data. We need to estimate the probability of these acoustic features, given the word or phonetic model, **W**, so that we can recognize the input data for the correct word. This probability is referred to as acoustic probability, $P(\mathbf{X} \mid \mathbf{W})$. In this section we focus our discussion on the HMM. As discussed in Chapter 8, it is the most successful method for acoustic modeling. Other emerging techniques are discussed in Section 9.8.

9.5.1. Choice of HMM Output Distributions

As discussed in Chapter 8, you can use discrete, continuous, or semicontinuous HMMs. When the amount of training data is sufficient, parameter tying becomes unnecessary. A continuous model with a large number of mixtures offers the best recognition accuracy, although its computational complexity also increases linearly with the number of mixtures. On the other hand, the discrete model is computationally efficient, but has the worst performance among the three models. The semicontinuous model provides a viable alternative between system robustness and trainability.

When either the discrete or the semicontinuous HMM is employed, it is helpful to use multiple codebooks for a number of features for significantly improved performance. Each codebook then represents a set of different speech parameters. One way to combine these multiple output observations is to assume that they are independent, computing the output probability as the product of the output probabilities of each codebook. For example, the semicontinuous HMM output probability of multiple codebooks can be computed as the product of each codebook:

$$b_i(\mathbf{x}) = \prod_m \sum_{k=1}^{L^m} f^m(\mathbf{x}^m \mid o_k^m) b_i^m(o_k^m) \tag{9.12}$$

where superscript m denotes the codebook-m related parameters. Each codebook consists of L^m-mixture continuous density functions.

Following our discussion in Chapter 8, the re-estimation algorithm for the multiple-codebook-based HMM could be extended. Since multiplication of the output probability density of each codebook leads to several independent terms in the Q-function, for codebook m, $\zeta_t(j,k^m)$ can be modified as follows:

$$\zeta_t(j,k^m) = \frac{\sum_i \alpha_{t-1}(i) a_{ij} b_j^m(k^m) f^m(\mathbf{x}_t \mid v_k^m) \prod_{m \neq n} \sum_k b_j^n(k^n) f^n(\mathbf{x}_t \mid v_k^n) \beta_t(j)}{\sum_k \alpha_T^m(k)} \tag{9.13}$$

Other intermediate probabilities can also be computed in a manner similar to what we discussed in Chapter 8.

Multiple codebooks can dramatically increase the representation power of the VQ codebook and can substantially improve speech recognition accuracy. You can typically build a codebook for $\mathbf{c}_k$, $\Delta\mathbf{c}_k$, and $\Delta\Delta\mathbf{c}_k$, respectively. As energy has a very different dynamic range, you can further improve the performance by building a separate codebook for $c_k[0]$, $\Delta c_k[0]$, and $\Delta\Delta c_k[0]$. In comparison to building a single codebook for $\mathbf{x}_k$ as illustrated in Eq. (9.6), the multiple-codebook system can reduce the error rate by more than 10%.

In practice, the most important parameter for the output probability distribution is the number of mixtures or the size of the codebooks. When there are sufficient training data, relative error reductions with respect to the discrete HMM are those shown in Figure 9.8.

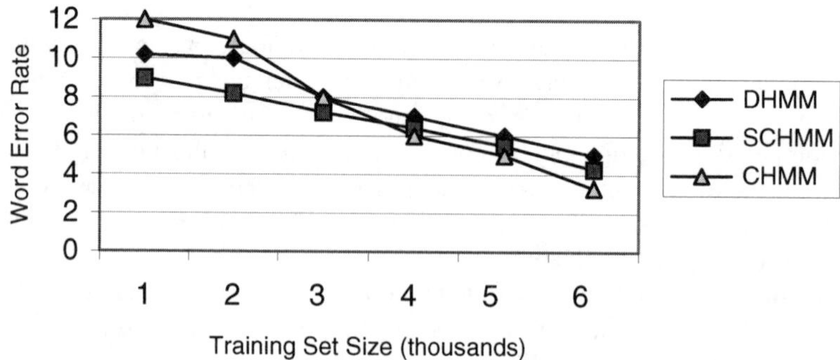

Figure 9.8 Continuous speaker-independent word recognition error rates of the discrete HMM (DHMM), SCHMM, and the continuous HMM (CHMM) with respect to the training set sizes (thousands of training sentences). Both the DHMM and SCHMM have multiple codebooks. The CHMM has 20 mixture diagonal Gaussian density functions.

As you can see from Figure 9.8, the SCHMM offers improved accuracy in comparison with the discrete HMM or the continuous HMM when the amount of training data is limited. When we increase the training data size, the continuous mixture density HMM starts to outperform both the discrete and the semicontinuous HMM, since the need to share model parameters becomes less critical.

Performance is also a function of the number of mixtures. With a small number of mixtures, the continuous HMM lacks the modeling power and it actually performs worse than the discrete HMM across the board. Only after we dramatically increase the number of mixtures does the continuous HMM start to offer improved recognition accuracy. The SCHMM can typically reduce the discrete HMM error rate by 10–15% across the board. The continuous HMM with 20 diagonal Gaussian density functions performed worse than either the discrete or the SCHMM when the size of training data was small. It outperformed either the discrete HMM or the SCHMM when sufficient amounts of training data became available. When the amount of training data is sufficiently large, it can reduce the error rate of the semicontinuous HMM by 15–20%.

9.5.2. Isolated vs. Continuous Speech Training

If we build a word HMM for each word in the vocabulary for isolated speech recognition, the training or recognition can be implemented directly, using the basic algorithms introduced in Chapter 8. To estimate model parameters, examples of each word in the vocabulary are collected. The model parameters are estimated from all these examples using the forward-backward algorithm and the reestimation formula. It is not necessary to have precise end-point detection, because the silence model automatically determines the boundary if we concatenate silence models with the word model in both ends.

If subword units,[3] such as phonetic models, are used, we need to share them across different words for large-vocabulary speech recognition. These subword units are concatenated to form a word model, possibly adding silence models at the beginning and end, as illustrated in Figure 9.9.

To concatenate subword units to form a word model, you can have a null transition from the final state of the previous subword HMM to the initial state of the next subword HMM, as indicated by the dotted line in Figure 9.9. As described in Chapter 8, you can estimate the parameters of the concatenated HMM accordingly. Please notice that the added null transition arc should satisfy the probability constraint with the transition probability of each phonetic HMM. The self-loop transition probability of the last state in each individual HMM has the topology illustrated in Figure 9.9. If we estimate these parameters with the concatenated model, the null arc transition probability, a_{ij}^{ε}, should satisfy the constraint $\sum_j (a_{ij} + a_{ij}^{\varepsilon}) = 1$ such that the self-loop transition probability of the last state is no longer equal to 1. For interword concatenation or concatenation involving multiple pronunciations, you can use multiple null arcs to concatenate individual models together.

[3] We have a detailed discussion on word models vs. subword models in Section 9.4.1.

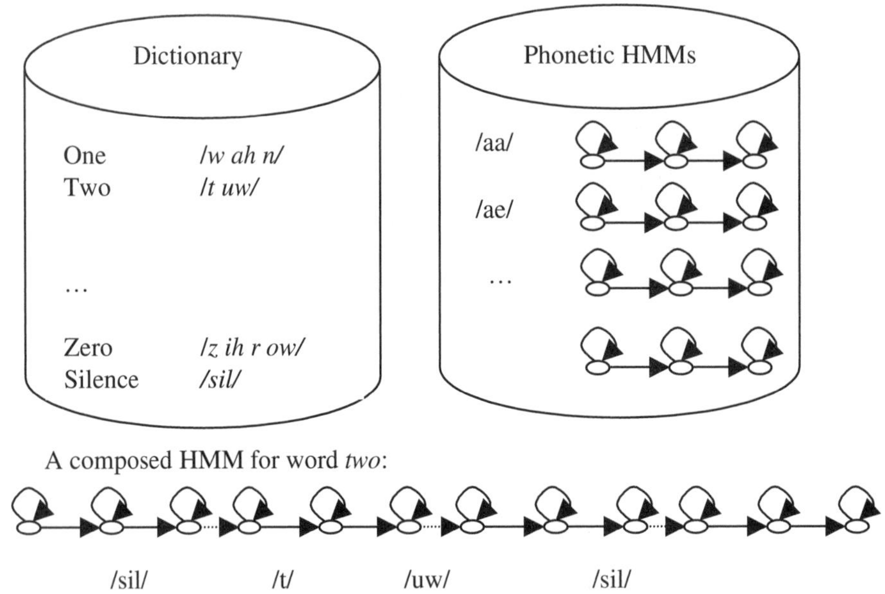

Figure 9.9 The construction of an isolated word model by concatenating multiple phonetic models based on the pronunciation dictionary.

In the example given in Figure 9.9, we have ten English digits in the vocabulary. We build an HMM for each English phone. The dictionary provides information on each word's pronunciation. We have a special word, *Silence*, that maps to a */sil/* HMM that has the same topology as the standard phonetic HMM. For each word in the vocabulary we first derive the phonetic sequence for each word from the dictionary. We link these phonetic models together to form a word HMM for each word in the vocabulary. The link between two phonetic models is shown in the figure as the dotted arrow.

For example, for word *two*, we create a word model based on the beginning silence */sil/*, phone */t/*, phone */uw/*, and ending silence */sil/*. The concatenated word model is then treated in the same manner as a standard large composite HMM. We use the standard forward-backward algorithm to estimate the parameters of the composite HMM from multiple sample utterances of the word *two*. After several iterations, we automatically get the HMM parameters for */sil/*, */t/*, and */uw/*. Since a phone can be shared across different words, the phonetic parameters may be estimated from acoustic data in different words.

The ability to automatically align each individual HMM to the corresponding unsegmented speech observation sequence is one of the most powerful features in the forward-backward algorithm. When the HMM concatenation method is used for continuous speech, you need to compose multiple words to form a sentence HMM based on the transcription of the utterance. In the same manner, the forward-backward algorithm absorbs a range of possible word boundary information of models automatically. There is no need to have a precise segmentation of the continuous speech.

In general, to estimate the parameters of the HMM, each word is instantiated with its concatenated word model (which may be a concatenation of subword models). The words in the sentence are concatenated with optional silence models between them. If there is a need to modify interword pronunciations due to interword pronunciation change, such as *want you*, you can add a different optional phonetic sequence for ***t-y*** in the concatenated sentence HMM.

In the digit recognition example, if we have a continuous training utterance *one three*, we compose a sentence HMM, as shown in Figure 9.10, where we have an optional silence HMM between the words *one* and *three*, linked with a null transition from the last state of the word model *one* to the first state of the word model *three*. There is also a direct null arc connection between the models *one* and *three* because a silence may not exist in the training example. These optional connections ensure that all the possible acoustic realizations of the natural continuous speech are considered, so that the forward-backward algorithm can automatically discover the correct path and accurately estimate the corresponding HMM from the given speech observation.

In general, the concatenated sentence HMM can be trained using the forward-backward algorithm with the corresponding observation sequence. Since the entire sentence HMM is trained on the entire observation sequence for the corresponding sentence, most possible word boundaries are inherently considered. Parameters of each model are based on those state-to-speech alignments. It does not matter where the word boundaries are. Such a training method allows complete freedom to align the sentence model against the observation, and no explicit effort is needed to find word boundaries.

In speech decoding, a word may begin and end anywhere within a given speech signal. As word boundaries cannot be detected accurately, all possible beginning and end points have to be accounted for. This converts a linear search (as for isolated word recognition) to a tree search, and a polynomial recognition algorithm to an exponential one. How to design an efficient decoder is discussed in Chapters 12 and 13.

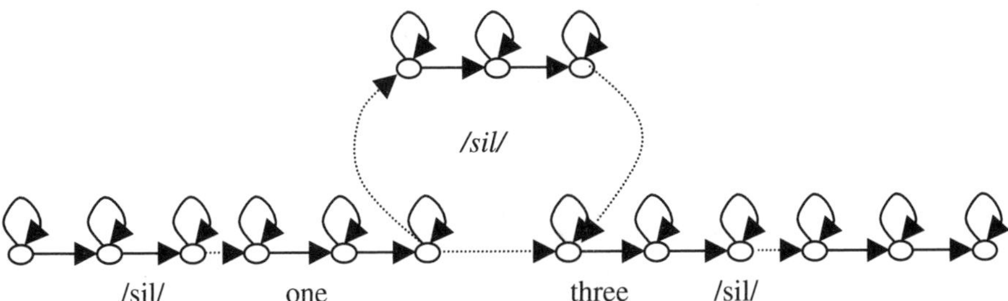

Figure 9.10 A composite sentence HMM. Each word can be a word HMM or a composite phonetic word HMM, as illustrated in Figure 9.9.

9.6. Adaptive Techniques—Minimizing Mismatches

As Figure 1.2 illustrated, it is important to adapt both acoustic models and language models for new situations. A decent model can accommodate a wide range of variabilities. However, the mismatch between the model and operating conditions always exists. One of the most important factors in making a speech system usable is to minimize the possible mismatch dynamically with a small amount of *calibration data*. Adaptive techniques can be used to modify system parameters to better match variations in microphone, transmission channel, environment noise, speaker, style, and application contexts. As a concrete example, speaker-dependent systems can provide a significant word error-rate reduction in comparison to speaker-independent systems if a large amount of speaker-dependent training data exists [50]. Speaker-adaptive techniques can bridge the gap between these two configurations with a small fraction of the speaker-specific training data needed to build a full speaker-dependent system. These techniques can also be used incrementally as more speech is available from a particular speaker. When speaker-adaptive models are built, you can have not only improved accuracy but also improved speed and potentially reduced model parameter sizes because of accurate representations, which is particularly appealing for practical speech recognition.

There are a number of ways to use adaptive techniques to minimize mismatches. You can have a nonintrusive adaptation process that works in the background all the time. This is typically unsupervised, using only the outcome of the recognizer (with a high confidence score, as discussed in Section 9.7) to guide the model adaptation. This approach can continuously modify the model parameters so that any nonstationary mismatches can be eliminated. As discussed in Chapter 13, systems that are required to transcribe speech in a non-real-time fashion may use multiple recognition passes. You can use unsupervised adaptation on the test data to improve the models after each pass to improve performance for a subsequent recognition pass.

Since the use of recognition results may be imperfect, there is a possibility of divergence if the recognition error rate is high. If the error rate is low, the adaptation results may still not be as good as supervised adaptation in which the correct transcription is provided for the user to read, a process referred to as the *enrollment process*. In this process you can check a wide range of parameters as follows:

- Check the background noise by asking the user not to speak.
- Adjust the microphone gain by asking the user to speak normally.
- Adapt the acoustic parameters by asking the user to read several sentences.
- Change the decoder parameters for the best speed with no loss of accuracy.
- Compose dynamically new enrollment sentences based on the user-specific error patterns.

The challenge for model adaptation is that we can use only a small amount of observable data to modify model parameters. This constraint requires different modeling strategies from the ones we discussed in building the baseline system, as the amount of training data is generally sufficient for offline training. In this section we focus on a number of adaptive techniques that can be applied to compensate either speaker or environment variations. Most of these techniques are model-based, since the acoustic model parameters rather than the acoustic feature vectors are adapted. We use speaker-adaptation examples to illustrate how these techniques can be used to improve system performance. We can generalize to environment adaptation by using environment-specific adaptation data and a noise-compensation model, which we discuss in Chapter 10. In a similar manner, we can modify the language model as discussed in Chapter 11.

9.6.1. Maximum a Posteriori (MAP)

Maximum a posteriori (MAP) estimation, as discussed in Chapter 4, can effectively deal with data-sparse problems, as we can take advantage of prior information about existing models. We can adjust the parameters of pretrained models in such a way that limited new training data would modify the model parameters guided by the prior knowledge to compensate for the adverse effect of a mismatch [35]. The prior density prevents large deviations of the parameters unless the new training data provide strong evidence.

More specifically, we assume that an HMM is characterized by a parameter vector $\mathbf{\Phi}$ that is a random vector, and that prior knowledge about the random vector is available and characterized by a prior probability density function $p(\mathbf{\Phi})$, whose parameters are to be determined experimentally.

With the observation data $\mathbf{X}$, the MAP estimate is expressed as follows:

$$\hat{\Phi} = \underset{\Phi}{\arg\max}[p(\Phi \mid \mathbf{X})] = \underset{\Phi}{\arg\max}[p(\mathbf{X} \mid \Phi)p(\Phi)] \tag{9.14}$$

If we have no prior information, $p(\mathbf{\Phi})$ is the uniform distribution, and the MAP estimate becomes identical to the ML estimate. We can use the EM algorithm as the ML to estimate the parameters of HMMs. The corresponding Q-function can be defined as:

$$Q_{MAP}(\Phi,\hat{\Phi}) = \log p(\hat{\Phi}) + Q(\Phi,\hat{\Phi}) \tag{9.15}$$

The EM algorithm for the ML criterion can be applied here directly. The actual expression depends on the assumptions made about the prior density. For the widely used continuous Gaussian mixture HMM, there is no joint conjugate prior density. We can assume different components of the HMM to be mutually independent, so that the optimization can be split into different subproblems involving only a single component of the parameter set. For example, the prior density function for the mixture Gaussian can be as follows:

$$p_{b_i}(\mathbf{c}_i, \boldsymbol{\mu}_i, \boldsymbol{\Sigma}_i) = p_{c_i}(\mathbf{c}_i)\prod_k p_{b_{ik}}(\boldsymbol{\mu}_{ik}, \boldsymbol{\Sigma}_{ik}) \tag{9.16}$$

where $p_{c_i}(\mathbf{c}_i)$ is a Dirichlet prior density for the mixing coefficient vector of all mixture components in the Markov state i, and $p_{b_{ik}}(\boldsymbol{\mu}_{ik}, \boldsymbol{\Sigma}_{ik})$ denotes the prior density for parameters of the kth Gaussian component in the state i. The Dirichlet prior density $p_{c_i}(\mathbf{c}_i)$ is characterized by a vector υ_i of positive hyperparameters such that:

$$p_{c_i}(\mathbf{c}_i) \propto \prod_k c_{ik}^{\upsilon_{ik}-1} \tag{9.17}$$

For full covariance D-dimensional Gaussian densities, the prior density can be a normal-Wishart density parameterized by two values $\eta > D-1, \tau > 0$, the vector $\boldsymbol{\mu}_{nw}$, and the symmetric positive definite matrix $\mathbf{S}$ as follows:

$$\begin{aligned} & p_{b_{ik}}(\boldsymbol{\mu}_{ik}, \boldsymbol{\Sigma}_{ik}) \propto \\ & \sqrt{\det(\boldsymbol{\Sigma}_{ik})^{D-\eta}} \exp(-\frac{\tau}{2}(\boldsymbol{\mu}_{ik} - \boldsymbol{\mu}_{nw})\boldsymbol{\Sigma}_{ik}^{-1}(\boldsymbol{\mu}_{ik} - \boldsymbol{\mu}_{nw})^t - \frac{1}{2} tr(\mathbf{S}\boldsymbol{\Sigma}_{ik}^{-1})) \end{aligned} \tag{9.18}$$

We can apply the same procedure as the MLE Baum-Welch reestimation algorithm. For example, with the Q-function defined in Eq. (9.15), we can apply the Lagrange method to derive the mixture coefficients as follows:

$$\begin{cases} \dfrac{\partial}{\partial \hat{c}_{ik}}(\log p_{c_i}(\hat{\mathbf{c}}_i) + \sum_k \sum_t \xi_t(i,k) \log \hat{c}_{ik}) + \lambda = 0, \forall k \\ \sum_k \hat{c}_{ik} = 1 \end{cases} \tag{9.19}$$

Based on Eqs. (9.17) and (9.19), the solution is:

$$\hat{c}_{ik} = \frac{\upsilon_{ik} - 1 + \sum_t \xi_t(i,k)}{\sum_l (\upsilon_{il} - 1 + \sum_t \xi_t(i,l))} \tag{9.20}$$

A comparison between Eq. (9.20) and the ML estimate Eq. (8.58) shows that the MAP estimate is a weighted average between the mode of the prior density and the ML estimate with proportions given by $\upsilon_{ijk} - 1$ and $\sum \xi_t(i,k)$, respectively.

We can optimize Eq. (9.15) with respect to mean and covariance parameters in a similar fashion. For example, the solution of these estimates is:

$$\hat{\boldsymbol{\mu}}_{ik} = \frac{\tau_{ik}\boldsymbol{\mu}_{nw_{ik}} + \sum_{t=1}^{T} \zeta_t(i,k)\mathbf{x}_t}{\tau_{ik} + \sum_{t=1}^{T} \zeta_t(i,k)} \tag{9.21}$$

$$\hat{\boldsymbol{\Sigma}}_{ik} = \frac{\mathbf{S}_{ik} + \tau_{ik}(\hat{\boldsymbol{\mu}}_{ik} - \boldsymbol{\mu}_{nw_{ik}})(\hat{\boldsymbol{\mu}}_{ik} - \boldsymbol{\mu}_{nw_{ik}})^t + \sum_{t=1}^{T} \zeta_t(i,k)(\mathbf{x} - \hat{\boldsymbol{\mu}}_{ik})(\mathbf{x} - \hat{\boldsymbol{\mu}}_{ik})^t}{\eta_{ik} - D + \sum_{t=1}^{T} \zeta_t(i,k)} \tag{9.22}$$

where τ_{ik} is the parameter in the normal-gamma density for the corresponding state i.

Thus, the reestimation formula for the Gaussian mean is a weighted sum of the prior mean with the ML mean estimate $\sum_{t=1}^{T}\zeta_t(i,k)\mathbf{x}_t / \sum_{t=1}^{T}\zeta_t(i,k)$. τ_{ik} is a balancing factor between prior mean and the ML mean estimate. When τ_{ik} is large, the value of the prior knowledge is small and the value of the mean $\boldsymbol{\mu}_{nw_{ik}}$ is assumed to have high certainty, leading to the dominance of the final estimate. When the amount of adaptation data increases, the MAP estimate approaches the ML estimate, as the adaptation data overwrite any important prior that may influence the final estimate. Similarly, the covariance estimation formula has the same interpretation of the balance between the prior and new data.

One major limitation of the MAP-based approach is that it requires an accurate initial guess for the prior $p(\Phi)$, which is often difficult to obtain. We can use the already trained initial models that embody some characteristics of the original training conditions. A typical way to generate an initial Gaussian prior is to cluster the initial training data based on speaker or environment similarity measures. We can derive a set of models based on the partition, which can be seen as a set of observations drawn from a distribution having $p(\Phi)$. We can, thus, estimate the prior based on the sample moments to derive the corresponding prior parameters.

Another major limitation is that the MAP-based approach is a local approach to updating the model parameters. Namely, only model parameters that are observed in the adaptation data can be modified from the prior value. When the system has a large number of free parameters, the adaptation can be very slow. Thus in practice we need to find correlations between the model parameters, so that the unobserved or poorly adapted parameters can be altered [3, 22]. Another possibility is to impose structural information so the model parameters can be shared for improved adaptation speed [96].

The MAP training can be iterative, too, which requires an initial estimate of model parameters. A careful initialization for the Gaussian densities is also very important. Unlike the discrete distributions, there is no such a thing as a uniform density for a total lack of information about the value of the parameters. We need to use the same initialization procedure as discussed in Chapter 8.

For speaker-adaptive speech recognition, it has been experimentally found that τ_{ik} can be a fixed constant value for all the Gaussian components across all the dimensions. Thus the MAP HMM can be regarded as an interpolated model between the speaker-independent and speaker-dependent HMM. Both are derived from the standard ML forward-backward algorithm. Experimental performance of MAP training is discussed in Section 9.6.3.

9.6.2. Maximum Likelihood Linear Regression (MLLR)

When the continuous HMM is used for acoustic modeling, the most important parameter set to adapt is the output Gaussian density parameters, i.e., the mean vector and the covariance matrix. We can use a set of linear regression transformation functions to map both means and covariances in order to maximize the likelihood of the adaptation data [68]. The maximum likelihood linear regression (MLLR) mapping is consistent with the underlying criterion for building the HMM while keeping the number of free parameters under control. Since the transformation parameters can be estimated from a relatively small amount of adaptation data, it is very effective for rapid adaptation. MLLR has been widely used to obtain adapted models for either a new speaker or a new environment condition.

More specifically, in the mixture Gaussian density functions, the kth mean vector $\boldsymbol{\mu}_{ik}$ for each state i can be transformed using following equation:

$$\tilde{\boldsymbol{\mu}}_{ik} = \mathbf{A}_c \boldsymbol{\mu}_{ik} + \mathbf{b}_c \tag{9.23}$$

where $\mathbf{A}_c$ is a regression matrix and $\mathbf{b}_c$ is an additive bias vector associated with some broad class c, which can be either a broad phone class or a set of tied Markov states. The goal of Eq. (9.23) is to map the mean vector into a new space such that the mismatch can be eliminated. Because the amount of adaptation data is small, we need to make sure the number of broad classes c is small so we have only a small number of free parameters to estimate. Equation (9.23) can be simplified into:

$$\tilde{\boldsymbol{\mu}}_{ik} = \mathbf{W}_c \boldsymbol{\mu}_{ik} \tag{9.24}$$

where $\boldsymbol{\mu}_{ik}$ is extended as $[1, \boldsymbol{\mu}_{ik}^t]^t$ and $\mathbf{W}_c$ is the extended transform, $[\mathbf{b}_c, \mathbf{A}_c]$.

This mapping approach is based on the assumption that $\mathbf{W}_c$ can be tied for a wide range of broad phonetic classes so that the overall number of free parameters is significantly less than the number of mean vectors. Therefore, the same transformation can be used for several distributions if they represent similar acoustic characteristics.

To estimate these transformation parameters in the MLE framework, we can use the same Q-function we discussed in Chapter 8. We need to optimize only

$$\sum_i \sum_{k=1}^{M} Q_{b_i}(\Phi, \hat{\mathbf{b}}_{ik}) \tag{9.25}$$

with respect to $\mathbf{W}_c$. Maximization of $Q_{b_i}(\Phi, \hat{\mathbf{b}}_{ik})$ with respect to $\mathbf{W}_c$ can be achieved by computing the partial derivatives. For the Gaussian mixture density function, the partial derivative with respect to $\mathbf{W}_c$ is:

$$\frac{\partial \hat{b}_{ik}(\mathbf{x})}{\partial \mathbf{W}_c} = N(\mathbf{x}, \tilde{\boldsymbol{\mu}}_{ik}, \hat{\boldsymbol{\Sigma}}_{ik}) \boldsymbol{\Sigma}_{ik}^{-1} (\mathbf{x} - \mathbf{W}_c \boldsymbol{\mu}_{ik}) \boldsymbol{\mu}_{ik}^t \tag{9.26}$$

Let us denote the set of Gaussian components forming the broad transformation classes as C; we use $b_{ik} \in C$ to denote that the kth Gaussian density in state i belongs to the class. We can expand the Q-function with the partial derivatives and set it to zero, leading to the following equation:

$$\sum_{t=1}^{T} \sum_{b_{ik} \in C} \zeta_t(i,k) \boldsymbol{\Sigma}_{ik}^{-1} \mathbf{x}_t \boldsymbol{\mu}_{ik}^t = \sum_{t=1}^{T} \sum_{b_{ik} \in C} \zeta_t(i,k) \boldsymbol{\Sigma}_{ik}^{-1} \mathbf{W}_c \boldsymbol{\mu}_{ik} \boldsymbol{\mu}_{ik}^t \tag{9.27}$$

We can rewrite Eq. (9.27) as:

$$\mathbf{Z} = \sum_{b_{ik} \in C} \mathbf{V}_{ik} \mathbf{W}_c \mathbf{D}_{ik} \tag{9.28}$$

where

$$\mathbf{Z} = \sum_{t=1}^{T} \sum_{b_{ik} \in C} \zeta_t(i,k) \boldsymbol{\Sigma}_{ik}^{-1} \mathbf{x}_t \boldsymbol{\mu}_{ik}^t, \tag{9.29}$$

$$\mathbf{V}_{ik} = \sum_{t=1}^{T} \zeta_t(i,k) \boldsymbol{\Sigma}_{ik}^{-1}, \tag{9.30}$$

and

$$\mathbf{D}_{ik} = \boldsymbol{\mu}_{ik} \boldsymbol{\mu}_{ik}^t. \tag{9.31}$$

Estimating $\mathbf{W}_c$ for Eq. (9.28) is computationally expensive, as it requires solving simultaneous equations. Nevertheless, if we assume that the covariance matrix is diagonal, we can have a closed-form solution that is computationally efficient. Thus, we can define

$$\mathbf{G}_q = \sum_{b_{ijk} \in C} v_{qq} \mathbf{D}_{ik} \tag{9.32}$$

where v_{qq} denotes the qth diagonal element of matrix $\mathbf{V}_{ik}$. The transformation matrix can be computed row by row. So for the qth row of the transformation matrix $\mathbf{W}_q$, we can derive it from the qth row of $\mathbf{Z}_q$ [defined in Eq. (9.29)] as follows:

$$\mathbf{W}_q = \mathbf{Z}_q \mathbf{G}_q^{-1} \tag{9.33}$$

Since $\mathbf{G}_q$ may be a singular matrix, we need to make sure we have enough training data for the broad class. Thus, if the amount of training data is limited, we must tie a number of transformation classes together.

We can run several iterations to maximize the likelihood for the given adaptation data. At each iteration, transformation matrices can be initialized to identity transformations. We can iteratively repeat the process to update the means until convergence is achieved. We can also incrementally adapt the mean vectors after each observation sequence or set of observation sequences while the required statistics are accumulated over time. Under the assumption that the alignment of each observation sequence against the model is reasonably accurate, we can accumulate these estimated counts over time and use them incrementally. In order to deal with the tradeoff between specificity and robust estimation, we can dynamically generate regression classes according to the senone tree. Thus, we can incrementally increase the number of regression classes when more and more data become available.

MLLR adaptation can be generalized to include the variances with the ML framework, although the additional gain after mean transformation is often less significant (less than relative 2% error reduction). When the user donates about 15 sentences for enrollment training, Table 9.5 illustrates how the MLLR adaptation technique can be used to further reduce the word recognition error rate for a typical dictation application. Here, there is only one context-independent phonetic class for all the context-dependent Gaussian densities. As we can see, most of the error reduction came from adapting the mean vectors.

We can further extend MLLR to *speaker-adaptive training* (SAT) [6, 74]. In conventional speaker-independent training, we simply use data from different speakers to build a speaker-independent model. An inherent difficulty in this approach is that spectral variations of different speakers give the speaker-independent acoustic model higher variance than the corresponding speaker-dependent model. We can include MLLR transformation in the process of training to derive the MLLR parameters for each individual speaker. Thus the training data are transformed to maximize the likelihood for the overall speaker-independent model. This process can be run iteratively to reduce mismatches of different speakers. By explicitly accounting for the interspeaker variations during training and decoding, SAT reduces the error rate by an additional 5–10%.

Table 9.5 Relative error reductions with MLLR methods. The reduction is relative to that of the preceding row.

Models	Relative Error Reduction
CHMM	Baseline
MLLR on mean only	+12%
MLLR on mean and variance	+2%
MLLR SAT	+8%

9.6.3. MLLR and MAP Comparison

The MLLR method can be combined with MAP. This guarantees that with the increased amount of training data, we can have not only a set of compact MLLR transformation functions for rapid adaptation, but also directly modified model parameters that converge to the ML estimates. We can use MAP to adapt the model parameters and then add MLLR to transform these adapted models. It is also possible to incorporate the MAP principle directly into MLLR [18, 19].

As an example, the result of a 60,000-word dictation application using various adaptation methods is shown in Figure 9.11.[4] The speaker-dependent model used 1000 utterances. Also included as a reference is the speaker-independent result, which is used as the starting point for adaptive training. When the speaker-independent model is adapted with about 200 utterances, the speaker-adaptive model has already outperformed both speaker-independent and speaker-dependent systems. The results clearly demonstrate that we have insufficient training data for speaker-dependent speech recognition, as MAP-based outperform ML-based models. This also illustrates that we can make effective use of speaker-independent data for speaker-dependent speech recognition. Also, notice that the MLLR method has a faster adaptation rate than the MAP method. The MLLR method has context-independent phonetic classes. So, when the amount of adaptation data is limited, the MLLR method offers better overall performance.

However, the MAP becomes more accurate when the amount of adaptation data increases to 600 per speaker. This is because we can modify all the model parameters with the MAP training, and the MLLR transformation can never have the same degrees of freedom as the MAP method. When the MLLR is combined with MAP, we can have not only rapid adaptation but also superior performance over either the MLLR or MAP method across a wide range of adaptation data points. There are a number of different ways to combine both MLLR and MAP for improved performance [4, 98].

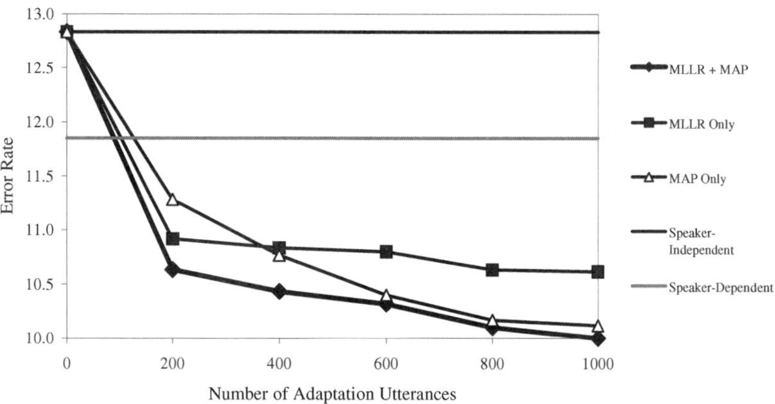

Figure 9.11 Comparison of Whisper with MLLR, MAP, and combined MLLR and MAP. The error rate is shown for a different amount of adaptation data. The speaker-dependent and speaker-independent models are also included. The speaker-dependent model was trained with 1000 sentences.

[4] In practice, if a large well-trained, speaker-independent model is used, the baseline performance may be very good, and hence, the relative error reduction from speaker adaptation may be smaller than for smaller and simpler models.

9.6.4. Clustered Models

Both MAP and MLLR techniques are based on using an appropriate initial model for adaptive modeling. How accurate we make the initial model directly affects the overall performance. An effective way to minimize the mismatch is, thus, to cluster similar speakers and environments in the training data, building a set of models for each cluster that has minimal mismatch for different conditions. When we have enough training data, and enough coverage for a wide range of conditions, this approach ensures significantly improved robustness.

For example, we often need a set of clustered models for different telephone channels, including different cellular phone standards. We also need to build gender-dependent models or speaker-clustered models for improved performance. In fact, when we construct speaker-clustered models, we can apply MLLR transformations or neural networks to minimize speaker variations such that different speakers can be mapped to the same golden speaker that is the representative of the cluster.

Speaker clusters can be created based on the information of each speaker-dependent HMM. The clustering procedure is similar to the decision-tree procedure discussed in Section 9.4.3. Using clustered models increases the amount of computational complexity. It also fragments the training data. Clustering is often needed to combine other smoothing techniques, such as deleted interpolation or MLLR transformation, in order to create clustered models from the pooled model. We can also represent a speaker as a weighted sum of individual speaker cluster models with the cluster adaptive training [33] or eigenvoice techniques [64].

When we select an appropriate model, we can compute the likelihood of the test speech against all the models and select the model that has the highest likelihood. Alternatively, we can compute likelihoods as part of the decoding process and prune away less promising models dynamically without significantly increasing the computational load. When multiple models are plausible, we can compute the weighted sum of the clustered models with pretrained mixing coefficients for different clusters, much as we train the deleted interpolation weights.

Traditionally speaker clustering is performed across different speakers without considering phonetic similarities across different speakers. In fact, clustered speaker groups may have very different degrees of variations for different phonetic classes. You can further generalize speaker clustering to the subword or subphonetic level [62]. With multiple instances derived from clustering for each subword or subphonetic unit, you can model speaker variation explicitly across different subword or subphonetic models.

In practice, gender-dependent models can reduce the word recognition error by 10%. More refined speaker-clustered models can further reduce the error rate, but not as much as the gain from gender-dependent models, unless we have a large number of clustered speakers. If the new user happens to be similar to one of these speaker clusters, we can approach speaker-dependent speech recognition without enrollment. For environment-dependent models, clustering is more critical. The challenge is to anticipate the kind of environment or channel distortions the system will have to deal with. Since this is often unpredictable, we need to use adaptive techniques such as MAP and MLLR to minimize the mismatch. We discuss this in more detail in Chapter 10.

9.7. CONFIDENCE MEASURES: MEASURING THE RELIABILITY

One of the most critical components in a practical speech recognition system is a reliable *confidence measure*. With an accurate confidence measure for each recognized word, the conversational back end can repair potential speech recognition errors, can reject out-of-vocabulary words, and can identify key words (perform word spotting) that are relevant to the back end. In a speaker-dependent or speaker-adaptive system, the confidence measure can help user enrollment (to eliminate mispronounced words). It is also critical for unsupervised speaker adaptation, allowing selective use of recognition results so that transcriptions with lower confidence can be discarded for adaptation.

In theory, an accurate estimate of $P(\mathbf{W} \mid \mathbf{X})$, the posterior probability, is itself a good confidence measure for word $\mathbf{W}$ given the acoustic input $\mathbf{X}$. Most practical speech recognition systems simply ignore $P(\mathbf{X})$, as it is a constant in evaluating $P(\mathbf{W})P(\mathbf{X} \mid \mathbf{W}) / P(\mathbf{X})$ across different words. $P(\mathbf{W} \mid \mathbf{X})$ can be expressed:

$$P(\mathbf{W} \mid \mathbf{X}) = \frac{P(\mathbf{W})P(\mathbf{X} \mid \mathbf{W})}{P(\mathbf{X})} = \frac{P(\mathbf{W})P(\mathbf{X} \mid \mathbf{W})}{\sum_{\mathbf{W}} P(\mathbf{W})P(\mathbf{X} \mid \mathbf{W})} \tag{9.34}$$

Equation (9.34) essentially provides a solid framework for measuring confidence levels. It is the ratio between the score for the word hypothesis $P(\mathbf{W})P(\mathbf{X} \mid \mathbf{W})$ and the acoustic probability $\sum_{\mathbf{W}} P(\mathbf{W})P(\mathbf{X} \mid \mathbf{W})$. In the sections that follow we discuss a number of ways to model and use such a ratio in practical systems.

9.7.1. Filler Models

You can compute $P(\mathbf{X})$ in Eq. (9.34) with a general-purpose recognizer. It should be able to recognize anything such that it can *fill the holes* of the grammar in the normal speech recognizer. The filler model has various forms [7, 63]. One of the most widely used is the so-called all-phone network, in which all the possible phonetic and nonspeech HMMs are connected to each other, and with which any word sequence can be recognized.

In addition to evaluating $P(\mathbf{W})P(\mathbf{X} \mid \mathbf{W})$ as needed in normal speech recognition, a separate decoding process is used to evaluate $\sum_{\mathbf{W}} P(\mathbf{W})P(\mathbf{X} \mid \mathbf{W})$. Here $\mathbf{W}$ is either a phonetic or a word model. You can also apply phonetic n-gram probabilities that are derived from a lexicon targeted for possible new words. The best path from the all-phone network is compared with the best path from the normal decoder. The ratio between the two, as expressed in Eq. (9.34), is used to measure the confidence for either word or phone. In the decoding process (see Chapters 12 and 13), you can accumulate the phonetic ratio derived from Eq. (9.34) on a specific word. If the accumulative $P(\mathbf{W} \mid \mathbf{X})$ for the word is less than a predetermined threshold, the word is rejected as either a new word or a nonspeech event.

Both context-independent and context-dependent phonetic models can be used for the fully connected network. When context-dependent phonetic models are used, you need to make sure that only correct contextual phonetic connections are made. Although context-dependent models offer significant improvement for speech recognition, the filler phonetic network seems to be insensitive to context-dependency in empirical experiments.

There are *word-spotting* applications that need to spot just a small number of key words. You can use the filler models described here for word spotting. You can also build antiword models trained with all the data that are not associated with the key words of interest. Empirical experiments indicate that large-vocabulary speech recognition is the most suitable choice for word spotting. You can use a general-purpose *n*-gram (see Chapter 11) to generate recognition results and identify needed key words from the word lattice. This is because a large-vocabulary system provides a better estimate of $\sum_{\mathbf{W}} P(\mathbf{W})P(\mathbf{X} \mid \mathbf{W})$ with a more accurate language model probability. In practice, we don't need to use all the hypotheses to compute $\sum_{\mathbf{W}} P(\mathbf{W})P(\mathbf{X} \mid \mathbf{W})$. Instead, *n*-best lists and scores [40] can be used to provide an effective estimate of $\sum_{\mathbf{W}} P(\mathbf{W})P(\mathbf{X} \mid \mathbf{W})$.

9.7.2. Transformation Models

To determine the confidence level for each word, subword confidence information is often helpful. Different phones have different impacts on our perception of words. The weight for each subword confidence score can be optimized from the real training data. If a word *w* has *N* phones, we can compute the confidence score of the word as follows:

$$CS(w) = \sum_{i=1}^{N} \wp_i(x_i) / N \tag{9.35}$$

where *CS(w)* is the confidence score for word *w*, x_i is the confidence score for subword unit *i* in word *w*, and $\wp_i$ is the mapping function that may be tied across a number of subword units. The transformation function can be defined as:

$$\wp_i(x) = ax + b \tag{9.36}$$

We can use discriminative training to optimize the parameters *a* and *b*, respectively. A cost function can be defined as a sigmoid function of *CS*(*w*). As shown in Figure 9.12, the optimal transformation parameters vary substantially across different phones. The weight for consonants is also typically larger than that of vowels.

The transformation function can be context dependent. Figure 9.13 illustrates the ROC curve of the context-dependent transformation model in comparison with the corresponding phonetic filler model. The filler model essentially has a uniform weight across all the phones in a given word. The estimated transformation model has 15–40% false-acceptance error reduction at various fixed false-rejection rates. The false-acceptance rate of the transformation model is consistently lower than that of the filler model [63].

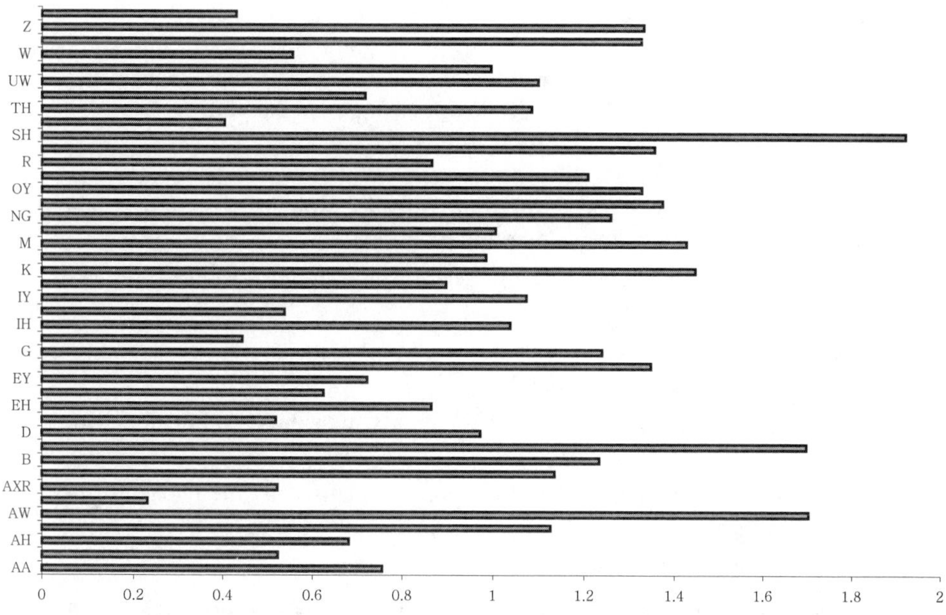

Figure 9.12 Transformation parameter a for each context-independent phone class. The weight of consonants is typically larger than that of vowels [63].

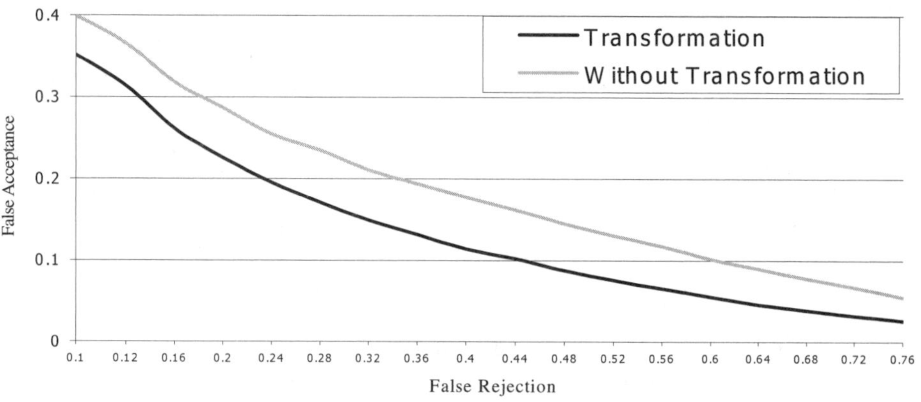

Figure 9.13 The ROC curve of phonetic filler models with and without optimal feature transformation [63].

9.7.3. Combination Models

In practical systems, there are a number of features you can use to improve the performance of confidence measures. For example, the following features are helpful:

- Word stability ratio from different language model weights (*WrdStabRatio*). This is obtained by applying different language weights to see how stably each word shows up in the recognition *n*-best list.
- Logarithm of the average number of active words around the ending frame of the word (*WrdCntEnd*).
- Acoustic score per frame within the word normalized by the corresponding active senone scores (*AscoreSen*).
- Logarithm of the average number of active words within the word (*WrdCntW*).
- Acoustic score per frame within the word normalized by the phonetic filler model (*AscoreFiller*).
- Language model score (*LMScore*).
- Language model back-off (trigram, bigram, or unigram hit) for the word (*LMBackOff*).
- Logarithm of the average number of active states within the word (*StateCnt*).
- Number of phones in the word (*Nphones*).
- Logarithm of the average number of active words around the beginning frame of the word (*WrdCntBeg*).
- Whether the word has multiple pronunciations (*Mpron*).
- Word duration (*WordDur*).

To clarify each feature's relative importance, Table 9.6 shows its linear correlation coefficient against the correct/incorrect tag for each word in the training set. Word stability ratio (*WrdStabRatio*) has the largest correlation value.

Several kinds of classifiers can be used to compute the confidence scores. Previous research has shown that the difference between classifiers, such as linear classifiers, generalized linear models, decision trees, and neural networks, is insignificant. The simplest linear classifier based on discriminative training performs well in practice. As some features are highly correlated, you can iteratively remove features to combat the curse of dimensionality. The combination model can have up to 40–80% false-acceptance error reduction at fixed false-rejection rate in comparison to the single-feature approach.

Table 9.6 Correlation coefficients of several features against correct/incorrect tag.

Feature	Correlation
WrdStabRatio	0.590
WrdCntW	–0.223
LMBackOff	0.171
AscoreSen	0.250
LMScore	0.175
Nphones	0.091
WordDur	0.012
WrdCntEnd	–0.321
AscoreFiller	0.219
StateCnt	–0.155
Mpron	0.057
WrdCntBeg	–0.067

9.8. OTHER TECHNIQUES

In addition to HMMs, a number of interesting alternative techniques are being actively investigated by researchers. We briefly review two promising methods here.

9.8.1. Neural Networks

You have seen both single-layer and multilayer neural nets in Chapter 4 for dealing with static patterns. In dealing with nonstationary signals, you need to address how to map an input sequence properly to an output sequence when two sequences are not synchronous, which should include proper alignment, segmentation, and classification. The basic neural networks are not well equipped to address these problems in a unified way.

Recurrent neural networks have an internal state that is a function of the current input and the previous internal state. A number of them use time-step delayed recurrent loops on the hidden or output units of a feedforward network, as discussed in earlier chapters. For sequences of finite numbers of delays, we can transform these networks into equivalent feedforward networks by unfolding them over the time period. They can be trained with the standard back propagation procedure, with the following modifications:

- The desired outputs are functions of time, and error functions have to be computed for every copy of the output layer. This requires the selection of an appropriate time-dependent target function, which is often difficult to define.
- All copies of the unfolded weights are constrained to be identical during the training. We can compute the correction terms separately for each weight and use the average to update the final estimate.

In most of these networks, you can have a partially recurrent network that has feedback of the hidden and output units to the input layer. For example, the feedforward network can be used in a set of local feedforward connections with one time-step delay. These networks are usually implemented by extending the input field with additional feedback units containing both the hidden and output values generated by the preceding input. You can encode the past nonstationary information that is often required to generate the correct output, given the current input, as illustrated in Figure 9.14.

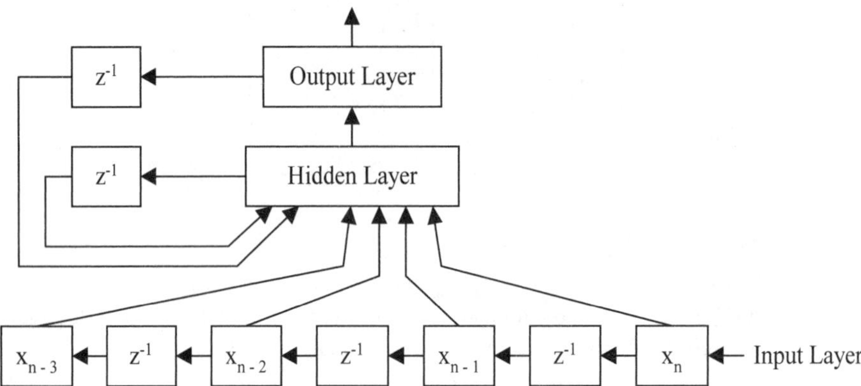

Figure 9.14 A recurrent network with contextual inputs, hidden vector feedback, and output vector feedback.

One of the popular neural networks is the *Time Delay Neural Network* (TDNN) [105]. Like static networks, the TDNN can be trained to recognize a sequence of predefined length (defined by the width of the input window). The activation in the hidden layer is computed from the current and multiple time-delayed values of the preceding layer, and the output units are activated only when a complete speech segment has been processed. A typical TDNN is illustrated in Figure 9.15. The TDNN has been successfully used to classify presegmented phonemes.

All neural networks have been shown to yield good performance for small-vocabulary speech recognition. Sometimes they are better than HMMs for short, isolated speech units. By recurrence and the use of temporal memory, they can perform some kind of integration over time. It remains a challenge for neural networks to demonstrate that they can be as effective as HMMs for dealing with nonstationary signals, as is often required for large-vocabulary speech recognition.

To deal with continuous speech, the most effective solution is to integrate neural nets with HMMs [91, 113]. The neural network can be used as the output probabilities to replace the Gaussian mixture densities. Comparable results can be obtained with the integrated approach. These HMM output probabilities could be estimated by applying the Bayes' rule to the output of neural networks that have been trained to classify HMM state categories. The neural networks can consist either of a single large trained network or of a group of separately trained small networks [21, 31, 75, 90].

A number of techniques have been developed to improve the performance of training these networks. Training can be embedded in an EM-style process. For example, dynamic programming can be used to segment the training data. The segmented data are then used to retrain the network. It is also possible to have Baum-Welch style training [14, 42].

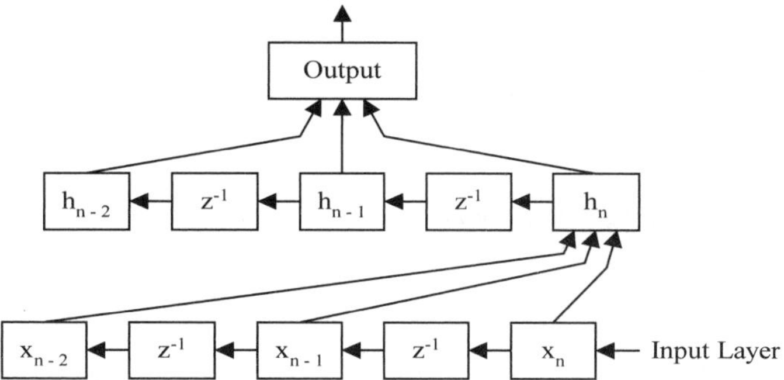

Figure 9.15 A time-delay neural network (TDNN), where the box h_t denotes the hidden vector at time *t*, the box x_t denotes the input vector at time *t*, and the box z^{-1} denotes a delay of one sample.

9.8.2. Segment Models

As discussed in Chapter 8, the HMM *output-independence assumption* results in a piecewise stationary process within an HMM state. Although the nonstationary speech may be modeled sufficiently with a large number of states, the states in which salient speech features are present are far from stationary [25, 99]. While the use of time-derivative features (e.g., delta and/or delta-delta features) alleviates these limitations, the use of such longer-time-span features may invalidate the conditional independence assumption.

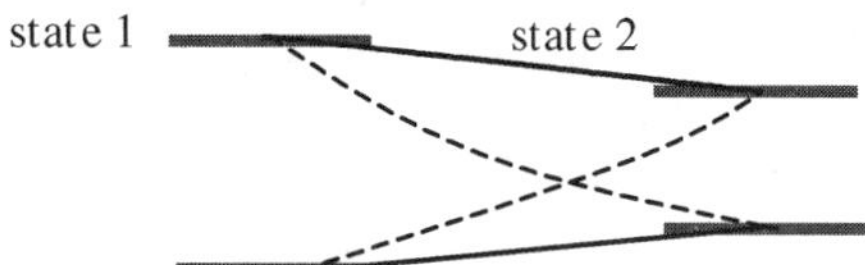

Figure 9.16 Diagram illustrating that HMM's output observation can hop between two unexpected quasi-stationary states [46].

The use of Gaussian mixtures for continuous or semicontinuous HMMs, as described in Chapter 8, could introduce another potential problem, where arbitrary transitions among the Gaussian mixture components between adjacent HMM states are allowed [59]. Figure 9.16 illustrates two HMM states with two mixture components. The solid lines denote the valid trajectories actually observed in the training data. However, in modeling these two trajectories, the Gaussian mixtures inadvertently allow two *phantom* trajectories, shown in dashed lines in Figure 9.16, because no constraint is imposed on the mixture transitions across the state. It is possible that such phantom trajectories degrade recognition performance, because the models can overrepresent speech signals that should be modeled by other acoustic units. Segment models can alleviate such HMM modeling deficiencies [77, 79].

In the standard HMM, the output probability distribution is modeled by a quasi-stationary process, i.e.,

$$P(\mathbf{x}_1^L \mid s) = \prod_{i=1}^{L} b_s(\mathbf{x}_i) \tag{9.37}$$

For the segment model (SM), the output observation distribution is modeled by two stochastic processes:

$$P(\mathbf{x}_1^L \mid s) = P(\mathbf{x}_1^L \mid s, L)P(L \mid s) \tag{9.38}$$

The first term of Eq. (9.38) is no longer decomposable in the absence of the output-independence assumption. The second term is similar to the duration model described in Chapter 8. In contrast to the HMM whose quasi-stationary process for each state s generates one frame $\mathbf{x}_i$, a state in a segment model can generate a variable-length observation sequence $\{\mathbf{x}_1, \mathbf{x}_2, \cdots \mathbf{x}_L\}$ with random length L.

Since the likelihood evaluation of segment models cannot be decomposed, the computation of the evaluation is not shareable between different segments (even for the case where two segments differ only by one frame). This results in a significant increase in computation for both training and decoding [77]. In general, the search state space is increased by a factor of $L_{\max}$, the maximum segment duration. If segment models are used for phone segments, $L_{\max}$ could be as large as 60. On top of this large increase in search state space, the evaluation of segment models is usually an order of magnitude more expensive than for HMM, since the evaluation involves several frames. Thus, the segment model is often implemented in a multipass search framework, as described in Chapter 13.

Segment models have produced encouraging performance for small-vocabulary or isolated recognition tasks [25, 44, 79]. However, their effectiveness on large-vocabulary continuous speech recognition remains an open issue because of necessary compromises to reduce the complexity of implementation.

9.8.2.1. Parametric Trajectory Models

Parametric trajectory models [25, 37] were first proposed to model a speech segment with curve-fitting parameters. They approximate the D-dimensional acoustic observation vector $\mathbf{X} = (\mathbf{x}_1, \mathbf{x}_2, \cdots, \mathbf{x}_T)$ by a polynomial function. Specifically, the observation vector $\mathbf{x}_t$ can be represented as

$$\mathbf{x}_t = \mathbf{C} \times F_t + \mathbf{e}_t(\Sigma) = \begin{pmatrix} c_1^0 & c_1^1 & \cdots & c_1^N \\ c_2^0 & c_2^1 & \cdots & c_2^N \\ \vdots & \vdots & \vdots & \vdots \\ c_D^0 & c_D^1 & c_D^2 & c_D^N \end{pmatrix} \begin{pmatrix} f_0(t) \\ f_1(t) \\ \vdots \\ f_N(t) \end{pmatrix} + \mathbf{e}_t(\Sigma) \tag{9.39}$$

where the matrix $\mathbf{C}$ is the trajectory parameter matrix, F_t is the family of Nth-order polynomial functions, and $\mathbf{e}_t(\Sigma)$ is the residual fitting error. Equation (9.39) can be regarded as modeling the time-varying mean in the output distribution for an HMM state. To simplify computation, the distribution of the residual error is often assumed to be an independent and identically distributed random process with a normal distribution $N(0,\Sigma)$. To accommodate diverse durations for the same segment, the relative linear time sampling of the fixed trajectory is assumed [37].

Each segment M is characterized by a trajectory parameter matrix $\mathbf{C}_m$ and covariance matrix Σ_m. The likelihood for each frame can be specified [46] as

$$P(\mathbf{x}_t \mid \mathbf{C}_m, \Sigma_m) = \frac{\exp\left\{-tr\left[(\mathbf{x}_t - \mathbf{C}_m F_t)\Sigma_m^{-1}(\mathbf{x}_t - \mathbf{C}_m F_t)^t\right]/2\right\}}{(2\pi)^{D/2} \mid \Sigma_m \mid^{1/2}} \tag{9.40}$$

If we let $\mathbf{F}_T = (F_0, F_1, \cdots, F_{T-1})^t$, then the likelihood for the whole acoustic observation vector can be expressed as

$$P(\mathbf{X} \mid \mathbf{C}_m, \Sigma_m) = \frac{\exp\left\{-tr\left[(\mathbf{X} - \mathbf{C}_m \mathbf{F}_T)\Sigma_m^{-1}(\mathbf{X} - \mathbf{C}_m \mathbf{F}_T)^t\right]/2\right\}}{(2\pi)^{DT/2} \mid \Sigma_m \mid^{T/2}} \tag{9.41}$$

Multiple mixtures can also be applied to SM. Suppose segment M is modeled by K trajectory mixtures. The likelihood for the acoustic observation vector $\mathbf{X}$ becomes

$$\sum_{k=1}^{K} w_k p_k(\mathbf{X} \mid \mathbf{C}_k, \Sigma_k) \tag{9.42}$$

Hon et al. [47] showed that only a handful of target trajectories are needed for speaker-independent recognition, in contrast to the many mixtures required for continuous Gaussian HMMs. This should support the phantom-trajectory argument involved in Figure 9.16.

The estimation of segment parameters can be accomplished by the EM algorithm described in Chapter 4. Assume a sample of L observation segments $\mathbf{X}_1, \mathbf{X}_2, \cdots, \mathbf{X}_L$, with corresponding duration $T_1, T_2, \cdots, T_L$, are generated by the segment model M. The MLE formulae using the EM algorithm are:

$$\gamma_k^i = \frac{w_k p_k(\mathbf{X}_i \mid \mathbf{C}_k, \Sigma_k)}{P(\mathbf{X}_i \mid \Phi_m)} \tag{9.43}$$

$$\hat{w}_k = \frac{1}{L}\sum_{i=1}^{L}\frac{w_k p_k(\mathbf{X}_i \mid \mathbf{C}_k, \Sigma_k)}{P(\mathbf{X}_i \mid \Phi_{\mathrm{m}})} \tag{9.44}$$

$$\hat{\mathbf{C}}_k = \left[\sum_{i=1}^{L}\gamma_k^i \mathbf{X}_i \mathbf{F}_{T_i}^t\right] / \left[\sum_{i=1}^{L}\gamma_k^i \mathbf{F}_{T_i}\mathbf{F}_{T_i}^t\right] \tag{9.45}$$

$$\hat{\Sigma}_k = \sum_{i=1}^{L}\gamma_k^i(\mathbf{X}_i - \mathbf{C}_k\mathbf{F}_{T_i})(\mathbf{X}_i - \mathbf{C}_k\mathbf{F}_{T_i})^t \Big/ \sum_{i=1}^{L}\gamma_k^i T_i \tag{9.46}$$

Parametric trajectory models have been successfully applied to phone classification [25, 46] and word spotting [37], and offer a modestly improved performance over HMMs.

9.8.2.2. Unified Frame- and Segment-Based Models

The strengths of the HMM and the segment-model approaches are complementary. HMMs are very effective in modeling the subtle details of speech signals by using one state for each quasi-stationary region. However, the transitions between quasi-stationary regions are largely neglected by HMMs because of their short durations. In contrast, segment models are powerful in modeling the transitions and longer-range speech dynamics, but might need to give up the detailed modeling to assure trainability and tractability. It is possible to have a unified framework to combine both methods [47].

In the unified complementary framework, the acoustic model $p(\mathbf{X}_1^T \mid \mathbf{W})$ can be considered as two joint hidden processes, as in the following equation:

$$\begin{aligned} p(\mathbf{X} \mid \mathbf{W}) &= \sum_{\mathbf{q}^h}\sum_{\mathbf{q}^s} p(\mathbf{X}, \mathbf{q}^h, \mathbf{q}^s \mid \mathbf{W}) \\ &= \sum_{\mathbf{q}^h}\sum_{\mathbf{q}^s} p(\mathbf{X} \mid \mathbf{q}^h, \mathbf{q}^s) p(\mathbf{q}^s \mid \mathbf{q}^h) p(\mathbf{q}^h \mid \mathbf{W}) \end{aligned} \tag{9.47}$$

where $\mathbf{q}^h$ represents the hidden process of the HMM and $\mathbf{q}^s$, the segment model. The conditional probability of the acoustic signal $p(\mathbf{X} \mid \mathbf{q}^s, \mathbf{q}^h)$ can be further decomposed into two separate terms:

$$p(\mathbf{X} \mid \mathbf{q}^s, \mathbf{q}^h) = p(\mathbf{X} \mid \mathbf{q}^h) p(\mathbf{X} \mid \mathbf{q}^s)^a \tag{9.48}$$

where a is a constant that is called *segment-model weight*. The first term is the contribution from normal frame-based HMM evaluation. We further assume for the second term that recognition of segment units can be performed by detecting and decoding a sequence of salient events in the acoustic stream that are statistically independent. In other words,

$$p(\mathbf{X} \mid \mathbf{q}^s) = \prod_i p(\mathbf{X}_i \mid q_i^s) \tag{9.49}$$

where $\mathbf{X}_i$ denotes the ith segment.

We assume that the phone sequence and the phone boundaries hypothesized by HMMs and segment models agree with each other. Based on the independent-segment assumption, this leads to a segment duration model as

$$p(\mathbf{q}^s \mid \mathbf{q}^h) = \prod_i p(t_i, t_{i+1} - 1 \mid \mathbf{X}_i) \tag{9.50}$$

By treating the combination as a hidden-data problem, we can apply the decoding and iterative EM reestimation techniques here. This unified framework enables both frame- and segment-based models to *jointly* contribute to optimal segmentations, which leads to more efficient pruning during the search. The inclusion of the segment models does not require massive revisions in the decoder, because the segment model scores can be handled in the same manner as the language model scores; whereas the segment evaluation is performed at each segment boundary.

Since subphonetic units are often used in HMMs to model the detailed quasi-stationary speech region, the segment units should be used to model long-range transition. As studies have shown that phone transitions play an essential role in humans' perception, the phone-pair segment unit that spans over two adjacent phones can be used [47]. Let e_i and t_i denote the phone and the starting time of the ith segment, respectively. For a phone-pair (e_{i-1}, e_i) segment between t_i and t_{i+1}, the segment likelihood can be computed as follows:

$$p(\mathbf{X}_i \mid q_i^s) = p(\mathbf{x}_{t_{i-1}}^{t_{i+1}} \mid e_{i-1}, e_i) \tag{9.51}$$

Rather than applying segment evaluation for every two phones, an *overlapped evaluation* scheme can be used, as shown in Figure 9.17 (a), where a phone-pair segment model evaluation is applied at each phone boundary. The overlapped evaluation implies that each phone is evaluated twice in the overall score. Most importantly, the overlapped evaluation places constraints on overlapped regions to assure consistent trajectory transitions. This is an important feature, as trajectory mixtures prohibit *phantom* trajectories within a segment unit, but there is still no mechanism to prevent arbitrary trajectory transitions between adjacent segment units.

Some phone-pairs might not have sufficient training data. Units containing silence might also have obscure trajectories due to the arbitrary duration of silence. As a result, a

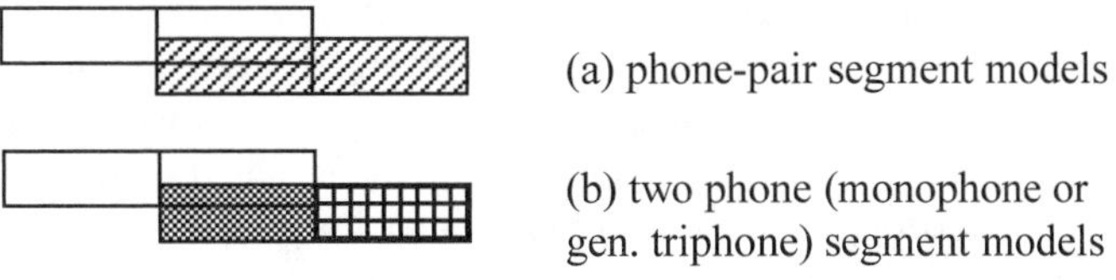

Figure 9.17 Overlapped evaluation using (a) a phone-pair segment model, or (b) back-off to two phone units when the phone-pair (e_{i-1}, e_i) segment model does not exist [47].

phone-pair unit (e_{i-1}, e_i) can be backed off with two phone units as shown in Figure 9.17 (b). The phone units can be context independent or context dependent [46]. Thus, the back-off segment-model evaluation becomes:

$$p(\mathbf{X}_i \mid q_i^s) = \beta * p(\mathbf{x}_{t_{i-1}}^{t_i} \mid e_{i-1}) p(\mathbf{x}_{t_i}^{t_{i+1}} \mid e_i) \tag{9.52}$$

where β is the back-off weight, generally smaller than 1.0. The use of back-off weight has the effect of giving more preference to phone-pair segment models than to two-phone-based back-off segment models.

The phone-pair segment model outperformed the phone-pair HMM by more than 20% in a phone-pair classification experiment [46]. The unified framework achieved about 8% word-error-rate reduction on the WSJ dictation task in comparison to the HMM-based Whisper [47].

9.9. Case Study: Whisper

Microsoft's Whisper engine offers general-purpose speaker-independent continuous speech recognition [49]. Whisper can be used for command and control, dictation, and conversational applications. Whisper offers many features such as continuous speech recognition, speaker-independence with adaptation, and dynamic vocabulary. Whisper has a unified architecture that can be scaled to meet different application and platform requirements.

The Whisper system uses MFCC representations (see Chapter 6) and both first- and second-order delta MFCC coefficients. Two-mean cepstral normalization discussed in Chapter 10 is used to eliminate channel distortion for improved robustness.

The HMM topology is a three-state left-to-right model for each phone. Senone models discussed in Section 9.4.3 are derived from both inter- and intraword context-dependent phones. The generic shared density function architecture can support either semicontinuous or continuous density hidden Markov models.

The SCHMM has a multiple-feature front end. Independent codebooks are built for the MFCC, first-order delta MFCC, second-order delta MFCC, and power and first and second power, respectively. Deleted interpolation is used to interpolate output distributions of context-dependent and context-independent senones. All codebook means and covariance matrices are reestimated together with the output distributions except the power covariance matrices, which are fixed.

The overall senone models can reduce the error rate significantly in comparison to the triphone or clustered triphone model. The shared Markov state also makes it possible to use continuous-density HMMs efficiently for large-vocabulary speech recognition. When a sufficient amount of training data becomes available, the best performance is obtained with the continuous-density mixture HMM. Each senone has 20 mixtures, albeit such an error reduction came at the cost of significantly increased computational complexity.

We can further generalize sharing to the level of each individual Gaussian probability density function. Each Gaussian function is treated as the basic unit to be shared across any Markov state. At this extreme, there is no need to use senones or shared states any more, and the shared probability density functions become the acoustic kernels that can be used to form any mixture function for any Markov state with appropriate mixture weights. Parameter sharing is, thus, advanced from a phone unit to a Markov state unit (senones) to a density component unit.

Regarding lexicon modeling, most words have one pronunciation in the lexicon. For words that are not in the dictionary, the LTS conversion is based on the decision tree that is trained from the existing lexicon. For the purpose of efficiency, the dictionary is used to store the most frequently used words. The LTS is only used for new words that need to be added on the fly.

For speaker adaptation, the diagonal variances and means are adapted using the MAP method. Whisper also uses MLLR to modify the mean vectors only. The MLLR classes are phone dependent. The transition probabilities are context independent and they are not modified during the adaptation stage.

The language model used in Whisper can be either the trigram or the context-free grammar. The difference is largely related to the decoder algorithm, as discussed in Chapter 13. The trigram lexicon has the 60,000 most-frequent words extracted from a large text corpus. Word selection is based on both the frequency and the word's part-of-speech information. For example, verbs and the inflected forms have a higher weight than proper nouns in the selection process.

Whisper's overall word recognition error rate for speaker-independent continuous speech recognition is about 7% for the standard DARPA business-news dictation test set. For isolated dictation with similar materials, the error rate is less than 3%. If speaker-dependent data are available, it can further reduce the error rate by 15–30%, with less than 30 minutes' speech from each person. The performance can be obtained real-time on today's PC systems.

9.10. Historical Perspective and Further Reading

The first machine to recognize speech was a commercial toy named Radio Rex manufactured in the 1920s. Fueled by increased computing resources, acoustic-phonetic modeling has progressed significantly since then. Relative word error rates have been reduced by 10% every year, as illustrated in Figure 9.18, thanks to the use of HMMs, the availability of large speech and text corpora, the establishment of standards for performance evaluation, and advances in computer technology. Before the HMM was established as the standard, there were many competing techniques, which can be traced back to the 1950s. Gold and Morgan's book provides an excellent historical perspective [38].

The HMM is powerful in that, with the availability of training data, the parameters of the model can be estimated and adapted automatically to give optimal performance. There are many HMM-based state-of-the-art speech recognition systems [1, 12, 27, 34, 49, 55, 72, 73, 93, 108, 109, 112]. Alternatively, we can first identify speech segments, then classify the segments and use the segment scores to recognize words. This approach has produced competitive recognition performance that is similar to HMM-based systems in several small- to medium-vocabulary tasks [115].

Speech recognition systems attempt to model the sources of variability in several ways. At the level of signal representation, in addition to MFCC, researchers have developed representations that emphasize perceptually important speaker-independent features of the signal, and deemphasize speaker-dependent characteristics [43]. Other methods based on linear discriminant analysis to improve class separability [28, 54] and speaker normalization transformation to minimize speaker variations [51, 67, 86, 106, 107, 114] have achieved limited success. Linear discriminant analysis can be traced back to *Fisher's linear discriminant* [30], which projects a dimensional vector onto a single line that is oriented to maximize class separability. Its extension to speech recognition can be found in [65].

At the level of acoustic-phonetic modeling, we need to provide an accurate distance measure of the input feature vectors against the phonetic or word models from the signal-processing front end. Before the HMM was used, the most successful acoustic-phonetic model was based on the speech template where the feature vectors are stored as the model and dynamic-programming-based time warping was used to measure the distance between the input feature vectors and the word or phonetic templates [88, 94]. The biggest problem for template-based systems is that they are not as trainable as HMMs, since it is difficult to generate a template that is as representative as all the speech samples we have for the particular units of interest.

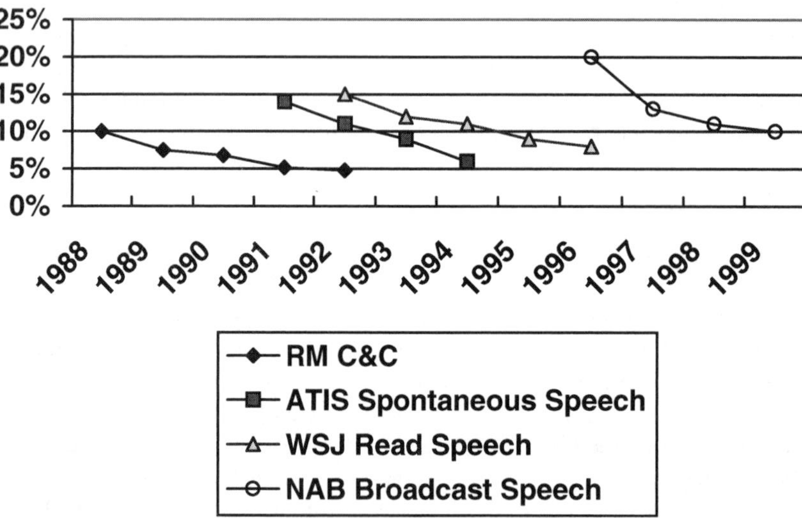

Figure 9.18 History of DARPA speech recognition word-error-rate benchmark evaluation results from 1988 to 1999. There are four major tasks: the Resource Management command and control task (RM C&C, 1000 words), the Air Travel Information System spontaneous speech understanding task (ATIS, 2000 words), the *Wall Street Journal* dictation task (WSJ, 20,000 words), and the Broadcast News Transcription Task (NAB, 60,000 words) [80-84].

Another approach that attracted many researchers is the knowledge-based one that originated from the Artificial Intelligence research community. This approach requires extensive knowledge engineering, which often led to many inconsistent rules. Due to the complexity of speech recognition, rule-based approaches generally cannot match the performance of data-driven approaches such as HMMs, which can automatically extract salient rules from a large amount of training data [105].

Senones are now widely used in many state-of-the-art systems. Word models or allophone models can also be built by concatenation of basic structures made by states, transitions, and distributions such as *fenones* [8, 9] or *senones* [58].

Segment models, as proposed by Roucos and Ostendorf [79, 92], assume that each variable-length segment is mapped to a fixed number of representative frames. The resulting model is very similar to the HMM with a large number of states. Ostendorf published a comprehensive survey paper [77] on segment models. The parametric trajectory segment model was introduced by Deng et al. [25] and Gish et al. [37] independently. Gish's work is very similar to our description in Section 9.8.2.1, which is based on Hon et al. [46, 47], where the evaluation and estimation are more efficient because no individual polynomial fitting is required for likelihood computation. In addition to the phone-pair units described in this chapter, segment models have also been applied to phonetic units [25], subphonetic units [25], diphones [36], and syllables [78]. The dynamic model [24, 26] is probably the most aggressive attempt to impose a global transition constraint on the speech model. It uses the phonetic target theories on unobserved vocal-tract parameters, which are fed to an MLP to produce the observed acoustic data.

Today, it is not uncommon to have tens of thousands of sentences available for system training and testing. These corpora permit researchers to quantify the acoustic cues important for phonetic contrasts and to determine parameters of the recognizers in a statistically meaningful way. While many of these corpora were originally collected under the sponsorship of the U.S. Defense Advanced Research Projects Agency (ARPA) to spur human language technology development among its contractors [82], they have nevertheless gained international acceptance. Recognition of the need for shared resources led to the creation of the Linguistic Data Consortium (LDC)[5] in the United States in 1992 to promote and support the widespread development and sharing of resources for human language technology. The LDC supports various corpus development activities and distributes corpora obtained from a variety of sources. Currently, LDC distributes about twenty different speech corpora including those cited above, comprising many hundreds of hours of speech. The availability of a large body of data in the public domain, coupled with the specification of evaluation standards, has resulted in uniform documentation of test results, thus contributing to greater reliability in monitoring progress.

To further improve the performance of acoustic-phonetic models, we need a robust system so that performance degrades gracefully (rather than catastrophically) as conditions diverge from those under which it was trained. The best approach is likely to have systems continuously adapted to changing conditions (new speakers, microphone, task, etc.). Such adaptation can occur at many levels in systems, subword models, word pronunciations, language models, and so on. We also need to make the system portable, so that we can rapidly design, develop, and deploy systems for new applications. At present, systems tend to suffer

[5] http://www.cis.upenn.edu/ldc

significant degradation when moved to a new task. In order to retain peak performance, they must be trained on examples specific to the new task, which is time consuming and expensive. In the new task, system users may not know exactly which words are in the system vocabulary. This leads to a certain percentage of out-of-vocabulary words in natural conditions. Currently, systems lack a very robust method of detecting such out-of-vocabulary words. These words often are inaccurately mapped into the words in the system, causing unacceptable errors.

An introduction to all aspects of acoustic modeling can be found in *Spoken Dialogues with Computers* [76] and *Fundamentals of Speech Recognition* [87]. A good treatment of HMM-based speech recognition is given in [52, 60, 105]. Bourlard and Morgan's book [15] is a good introduction to speech recognition based on neural networks. There are a range of applications such as predictive networks that estimate each frame's acoustic vector, given the history [69, 104] and nonlinear transformation of observation vectors [13, 53, 101].

You can find tools to build acoustic models from Carnegie Mellon University's speech open source Web site.[6] This site contains the release of CMU's Sphinx acoustic modeling toolkit and documentation. A version of Microsoft's Whisper system can be found in the Microsoft Speech SDK.[7]

REFERENCES

[1] Abrash, V., *et al.*, "Acoustic Adaptation Using Nonlinear Transformations of HMM Parameters" in *Proc. of the IEEE Int. Conf. on Acoustics, Speech and Signal Processing* 1996, Atlanta, pp. 729-732.

[2] Acero, A., *Acoustical and Environmental Robustness in Automatic Speech Recognition*, 1993, Boston, Kluwer Academic Publishers.

[3] Ahadi-Sarkani, S.M., *Bayesian and Predictive Techniques for Speaker Adaptation*, Ph. D. Thesis, 1996, Cambridge University, .

[4] Ahn, S., S. Kang, and H. Ko, "Effective Speaker Adaptations for Speaker Verification," *IEEE Int. Conf. on Acoustics, Speech and Signal Processing*, 2000, Istanbul, Turkey, pp. 1081-1084.

[5] Alleva, F., *et al.*, "Can Continuous Speech Recognizers Handle Isolated Speech?," *Speech Communication*, 1998, **26**, pp. 183-189.

[6] Anastasakos, T., *et al.*, "A Compact Model for Speaker Adaptive Training," *Int. Conf. on Spoken Language Processing*, 1996, Philadelphia, pp. 1137-1140.

[7] Asadi, A., R. Schwartz, and J. Makhoul, "Automatic Modeling for Adding New Words to a Large-Vocabulary Continuous Speech Recognition System," in *Proc. of the IEEE Int. Conf. on Acoustics, Speech and Signal Processing*, 1991, Toronto, pp. 305-308.

[8] Bahl, L.R., *et al.*, "Multonic Markov Word Models for Large Vocabulary Continuous Speech Recognition," *IEEE Trans. on Speech and Audio Processing*, 1993, **1**(3), pp. 334-344.

[6] http://www.speech.cs.cmu.edu/sphinx/

[7] http://www.microsoft.com/speech

[9] Bahl, L.R., *et al.*, "A Method for the Construction of Acoustic Markov Models for Words," *IEEE Trans. on Speech and Audio Processing*, 1993, **1**(4), pp. 443-452.

[10] Bahl, L.R., *et al.*, "Automatic Phonetic Baseform Determination," *Proc. IEEE Int. Conf. on Acoustics, Speech and Signal Processing*, 1991, Toronto, pp. 173-176.

[11] Bahl, L.R., *et al.*, "Decision Trees for Phonological Rules in Continuous Speech," *Proc. IEEE Int. Conf. on Acoustics, Speech and Signal Processing*, 1991, Toronto, Canada, pp. 185-188.

[12] Bellegarda, J.R., *et al.*, "Experiments Using Data Augmentation for Speaker Adaptation," *Proc. of the IEEE Int. Conf. on Acoustics, Speech and Signal Processing*, 1995, Detroit, pp. 692-695.

[13] Bengio, Y., *et al.*, "Global Optimization of a Neural Network-Hidden Markov Model Hybrid," *IEEE Trans. on Neural Networks*, 1992, **3**(2), pp. 252-259.

[14] Bourlard, H., "A New Training Algorithm for Statistical Sequence Recognition with Applications to Transition-Based Speech Recognition," *IEEE Signal Processing Letters*, 1996, **3**, pp. 203-205.

[15] Bourlard, H. and N. Morgan, *Connectionist Speech Recognition - A Hybrid Approach*, 1994, Boston, MA, Kluwer Academic Publishers.

[16] Brown, P.F., *The Acoustic-Modeling Problem in Automatic Speech Recognition*, PhD Thesis in Computer Science Department, 1987, Carnegie Mellon University, Pittsburgh, PA.

[17] Campbell, J., "Speaker Recognition: A Tutorial," *Proc. of the IEEE*, 1997, **85**(9), pp. 1437-1462.

[18] Chesta, C., O. Siohan, and C.H. Lee, "Maximum a Posteriori Linear Regression for Hidden Markov Model Adaptation," *Eurospeech*, 1999, Budapest, pp. 211-214.

[19] Chou, W., "Maximum a Posteriori Linear Regression with Elliptically Symmetric Matrix Priors," *Eurospeech*, 1999, Budapest, pp. 1-4.

[20] Cohen, M., *Phonological Structures for Speech Recognition*, Ph.D. Thesis 1989, University of California, Berkeley.

[21] Cook, G. and A. Robinson, "Transcribing Broadcast News with the 1997 Abbot System," *Int. Conf. on Acoustics, Speech and Signal Processing*, 1998, Seattle, WA, pp. 917-920.

[22] Cox, S., "Predictive Speaker Adaptation in Speech Recognition," *Computer Speech and Language*, 1995, **9**, pp. 1-17.

[23] Davis, S. and P. Mermelstein, "Comparison of Parametric Representations for Monosyllable Word Recognition in Continuously Spoken Sentences," *IEEE Trans. on Acoustics, Speech and Signal Processing*, 1980, **28**(4), pp. 357-366.

[24] Deng, L., "A Dynamic, Feature-based Approach to the Interface Between Phonology and Phonetics for Speech Modeling and Recognition," *Speech Communication*, 1998, **24**(4), pp. 299-323.

[25] Deng, L., *et al.*, "Speech Recognition Using Hidden Markov Models with Polynomial Regression Functions as Nonstationary States," *IEEE Trans. on Speech and Audio Processing*, 1994, **2**(4), pp. 507-520.

[26] Digalakis, V., *Segment-based Stochastic Models of Spectral Dynamics for Continuous Speech Recognition*, Ph.D. Thesis in *Electrical Computer System Engineering*, 1992, Boston University.

[27] Digalakis, V. and H. Murveit, "Genones: Optimizing the Degree of Mixture Tying in a Large Vocabulary Hidden Markov Model Based Speech Recognizer" in *Proc. of the IEEE Int. Conf. on Acoustics, Speech and Signal Processing*, 1994, Adelaide, Australia, pp. 537-540.

[28] Doddington, G.R., "Phonetically Sensitive Discriminants for Improved Speech Recognition," *Int. Conf. on Acoustics, Speech and Signal Processing*, 1989, Glasgow, Scotland, pp. 556-559.

[29] Eichner, M. and M. Wolff, "Data-Driven Generation of Pronunciation Dictionaries in the German Verbmobil Project—Discussion of Experimental Results," *IEEE Int. Conf. on Acoustics, Speech and Signal Processing*, 2000, Istanbul, Turkey, pp. 1687-1690.

[30] Fisher, R., "The Use of Multiple Measurements in Taxonomic Problems," *Annals of Eugenics*, 1936, **7**(1), pp. 179-188.

[31] Fritsch, J. and M. Finke, "ACID/HNN: Clustering Hierarchies of Neural Networks for Context-Dependent Connectionist Acoustic Modeling," *Int. Conf. on Acoustics, Speech and Signal Processing*, 1998, Seattle, WA, pp. 505-508.

[32] Fukunaga, K., *Introduction to Statistical Pattern Recognition*, 2nd ed, 1990, Orlando, FL, Academic Press.

[33] Gales, M., "Cluster Adaptive Training for Speech Recognition," *Int. Conf. on Spoken Language Processing*, 1998, Sydney, Australia, pp. 1783-1786.

[34] Gauvain, J.L., L. Lamel, and M. Adda-Decker, "Developments in Continuous Speech Dictation using the ARPA WSJ Task," *Proc. of the IEEE Int. Conf. on Acoustics, Speech and Signal Processing*, 1995, Detroit, MI, pp. 65-68.

[35] Gauvain, J.L. and C.H. Lee, "Bayesian Learning of Gaussian Mixture Densities for Hidden Markov Models," *Proc. of the DARPA Speech and Natural Language Workshop*, 1991, Palo Alto, CA, pp. 272-277.

[36] Ghitza, O. and M.M. Sondhi, "Hidden Markov Models with Templates as Non-Stationary States: An Application to Speech Recognition," *Computer Speech and Language*, 1993, **7**(2), pp. 101-120.

[37] Gish, H. and K. Ng, "A Segmental Speech Model with Applications to Word Spotting," *Proc. of the IEEE Int. Conf. on Acoustics, Speech and Signal Processing*, 1993, Minneapolis, MN, pp. 447-450.

[38] Gold, B. and N. Morgan, *Speech and Audio Signal Processing: Processing and Perception of Speech and Music*, 2000, New York, John Wiley.

[39] Greenberg, S., D. Ellis, and J. Hollenback, "Insights into Spoken Language Gleaned from Phonetic Transcription of the Switchboard Corpus," *Int. Conf. on Spoken Language Processing*, 1996, Philadelphia, PA, pp. addendum 24-27.

[40] Gunawardana, A., H.W. Hon, and L. Jiang, "Word-Based Acoustic Confidence Measures for Large-Vocabulary Speech Recognition," *Int. Conf. on Spoken Language Processing*, 1998, Sydney, Australia, pp. 791-794.

[41] Haeb-Umbach, R., "Investigations on Inter-Speaker Variability in the Feature Space," *IEEE Int. Conf. on Acoustics, Speech and Signal Processing*, 1999, Phoenix, AZ.

[42] Hennebert, J., *et al.*, "Estimation of Global Posteriors and Forward-Backward Training of Hybrid HMM/ANN Systems," *Proc. of the Eurospeech Conf.*, 1997, Rhodes, Greece, pp. 1951-1954.

[43] Hermansky, H., "Perceptual Linear Predictive (PLP) Analysis of Speech," *Journal of the Acoustical Society of America*, 1990, **87**(4), pp. 1738-1752.

[44] Holmes, W. and M. Russell, "Probabilistic-Trajectory Segmental HMMs," *Computer Speech and Language*, 1999, **13**, pp. 3-37.

[45] Hon, H.W. and K.F. Lee, "CMU Robust Vocabulary-Independent Speech Recognition System," *Proc. of the IEEE Int. Conf. on Acoustics, Speech and Signal Processing*, 1991, Toronto, pp. 889-892.

[46] Hon, H.W. and K. Wang, "Combining Frame and Segment Based Models for Large Vocabulary Continuous Speech Recognition," *IEEE Workshop on Automatic Speech Recognition and Understanding*, 1999, Keystone, CO.

[47] Hon, H.-W. and K. Wang, "Unified Frame and Segment Based Models for Automatic Speech Recognition," *Int. Conf. on Acoustic, Signal and Speech Processing*, 2000, Istanbul, Turkey, IEEE, pp. 1017-1020.

[48] Huang, X., *et al.*, "From Sphinx II to Whisper: Making Speech Recognition Usable" in *Automatic Speech and Speaker Recognition*, C.H. Lee, F.K. Soong, and K.K. Paliwal, eds. 1996, Norwell, MA, pp. 481-508, Kluwer Academic Publishers.

[49] Huang, X., *et al.*, "From Sphinx-II to Whisper - Make Speech Recognition Usable" in *Automatic Speech and Speaker Recognition*, C.H. Lee, F.K. Soong, and K.K. Paliwal, eds. 1996, Norwell, MA, Kluwer Academic Publishers.

[50] Huang, X. and K.-F. Lee, "On Speaker-Independent, Speaker-Dependent, and Speaker-Adaptive Speech Recognition," *IEEE Trans. on Speech and Audio Processing*, 1993, **1**(2), pp. 150-157.

[51] Huang, X.D., "Speaker Normalization for Speech Recognition," *Proc. of the IEEE Int. Conf. on Acoustics, Speech and Signal Processing*, 1992, San Francisco pp. 465-468.

[52] Huang, X.D., Y. Ariki, and M.A. Jack, *Hidden Markov Models for Speech Recognition*, 1990, Edinburgh, U.K., Edinburgh University Press.

[53] Huang, X.D., K. Lee, and A. Waibel, "Connectionist Speaker Normalization and its Applications to Speech Recognition," *IEEE Workshop on Neural Networks for Signal Processing*, 1991, New York, pp. 357-366.

[54] Hunt, M.J., *et al.*, "An Investigation of PLP and IMELDA Acoustic Representations and of Their Potential for Combination," *Proc. of the IEEE Int. Conf. on Acoustics, Speech and Signal Processing*, 1991, Toronto, pp. 881-884.

[55] Huo, Q., C. Chan, and C.-H. Lee, "On-Line Adaptation of the SCHMM Parameters Based on the Segmental Quasi-Bayes Learning for Speech Recognition," *IEEE Trans. on Speech and Audio Processing*, 1996, **4**(2), pp. 141-144.

[56] Hwang, M.Y., X. Huang, and F. Alleva, "Predicting Unseen Triphones with Senones," *Proc. of the IEEE Int. Conf. on Acoustics, Speech and Signal Processing*, 1993, Minneapolis, pp. 311-314.

[57] Hwang, M.Y. and X.D. Huang, "Acoustic Classification of Phonetic Hidden Markov Models" in *Proc. of Eurospeech* 1991.

[58] Hwang, M.Y. and X.D. Huang, "Shared-Distribution Hidden Markov Models for Speech Recognition," *IEEE Trans. on Speech and Audio Processing*, 1993, **1**(4), pp. 414-420.

[59] Iyer, R., *et al.*, "Hidden Markov Models for Trajectory Modeling," *Int. Conf. on Spoken Language Processing*, 1998, Sydney, Australia.

[60] Jelinek, F., *Statistical Methods for Speech Recognition*, 1998, Cambridge, MA, MIT Press.

[61] Jiang, L., H.W. Hon, and X. Huang, "Improvements on a Trainable Letter-to-Sound Converter," *Proc. of Eurospeech*, 1997, Rhodes, Greece, pp. 605-608.

[62] Jiang, L. and X. Huang, "Subword-Dependent Speaker Clustering for Improved Speech Recognition," *Int Conf. on Spoken Language Processing*, 2000, Beijing, China.

[63] Jiang, L. and X.D. Huang, "Vocabulary-Independent Word Confidence Measure Using Subword Features," *Int. Conf. on Spoken Language Processing*, 1998, Syndey, Australia.

[64] Kuhn, R., *et al.*, "Eigenvoices for Speaker Adaptation," *Int. Conf. on Spoken Language Processing*, 1998, Sydney, Australia, pp. 1771-1774.

[65] Kumar, N. and A. Andreou, "Heteroscedastic Discriminant Analysis and Reduced Rank HMMs for Improved Speech Recognition," *Speech Communication*, 1998, **26**, pp. 283-297.

[66] Lee, K.F., *Large-Vocabulary Speaker-Independent Continuous Speech Recognition: The SPHINX System*, Ph.D. Thesis in *Computer Science Dept.* 1988, Carnegie Mellon University, Pittsburgh.

[67] Lee, L. and R. Rose, "Speaker Normalization Using Efficient Frequency Warping Procedures," *IEEE Int. Conf. on Acoustics, Speech and Signal Processing*, 1996, Atlanta, GA, pp. 353-356.

[68] Leggetter, C.J. and P.C. Woodland, "Maximum Likelihood Linear Regression for Speaker Adaptation of Continuous Density Hidden Markov Models," *Computer Speech and Language*, 1995, **9**, pp. 171-185.

[69] Levin, E., "Word Recognition Using Hidden Control Neural Architecture," *Proc. of the IEEE Int. Conf. on Acoustics, Speech and Signal Processing*, 1990, Albuquerque, NM, pp. 433-436.

[70] Lippmann, R.P., E.A. Martin, and D.P. Paul, "Multi-Style Training for Robust Isolated-Word Speech Recognition," *Int. Conf. on Acoustics, Speech and Signal Processing*, 1987, Dallas, TX, pp. 709-712.

[71] Lucassen, J.M. and R.L. Mercer, "An Information-Theoretic Approach to the Automatic Determination of Phonemic Baseforms," *Proc. of the IEEE Int. Conf. on Acoustics, Speech and Signal Processing*, 1984, San Diego, pp. 42.5.1-42.5.4.

[72] Lyu, R.Y., *et al.*, "Golden Mandarin (III) - A User-Adaptive Prosodic Segment-Based Mandarin Dictation Machine for Chinese Language with Very Large Vocabulary" in *Proc. of the IEEE Int. Conf. on Acoustics, Speech and Signal Processing*, 1995, Detroit, MI, pp. 57-60.

[73] Matsui, T., T. Matsuoka, and S. Furui, "Smoothed N-best Based Speaker Adaptation for Speech Recognition" in *Proc. of the IEEE Int. Conf. on Acoustics, Speech and Signal Processing*, 1997, Munich, Germany, pp. 1015-1018.

[74] McDonough, J., *et al.*, "Speaker-Adapted Training on the Switchboard Corpus" in *Proc. of the IEEE Int. Conf. on Acoustics, Speech and Signal Processing* 1997, Munich, Germany, pp. 1059-1062.

[75] Morgan, N. and H. Bourlard, *Continuous Speech Recognition: An Introduction to Hybrid HMM/Connectionist Approach*, in *IEEE Signal Processing Magazine*, 1995, pp. 25-42.

[76] Mori, R.D., *Spoken Dialogues with Computers*, 1998, London, Academic Press.

[77] Ostendorf, M., V.V. Digalakis, and O.A. Kimball, "From HMM's to Segment Models: a Unified View of Stochastic Modeling for Speech Recognition," *IEEE Trans. on Speech and Audio Processing*, 1996, **4**(5), pp. 360-378.

[78] Ostendorf, M. and K. Ross, "A Multi-Level Model for Recognition of Intonation Labels" in *Computing Prosody*, Y. Sagisaka, W.N. Campell, and N. Higuchi, eds., 1997, New York, pp. 291-308, Springer Verlag.

[79] Ostendorf, M. and S. Roukos, "A Stochastic Segment Model for Phoneme-Based Continuous Speech Recognition," *IEEE Trans. on Acoustics, Speech and Signal Processing*, 1989, **37**(1), pp. 1857-1869.

[80] Pallett, D., J.G. Fiscus, and J.S. Garofolo, "DARPA Resource Management Benchmark Test Results June 1990," in *Proc. of the DARPA Speech and Natural Language Workshop* 1990, Hidden Valley, PA, pp. 298-305, Pallett.

[81] Pallett, D., J.G. Fiscus, and J.S. Garofolo, "DARPA Resource Management Benchmark Test Results," *Proc. of the DARPA Speech and Natural Language Workshop*, 1991, Morgan Kaufmann Publishers, pp. 49-58.

[82] Pallett, D.S., *et al.*, "The 1994 Benchmark Tests for the ARPA Spoken Language Program" in *Proc. of the ARPA Spoken Language Technology Workshop*, 1995, Austin, TX, pp. 5-38.

[83] Pallett, D.S., *et al.*, "1997 Broadcast News Benchmark Test Results: English and Non-English," *Proc. of the Broadcast News Transcription and Understanding Workshop*, 1998, Landsdowne, Virginia, Morgan Kaufmann Publishers.

[84] Pallett, D.S., J.G. Fiscus, and M.A. Przybocki, "1996 Preliminary Broadcast News Benchmark Tests," *Proc. of the DARPA Speech Recognition Workshop*, 1997, Chantilly, VA, Morgan Kaufmann Publishers.

[85] Pereira, F., M. Riley, and R. Sproat, "Weighted Rational Transductions and Their Application to Human Language Processing," *Proc. of the ARPA Human Language Technology Workshop*, 1994, Plainsboro, NJ, pp. 249-254.

[86] Pye, D. and P.C. Woodland, "Experiments in Speaker Normalization and Adaptation for Large Vocabulary Speech Recognition," in *Proc. of the IEEE Int. Conf. on Acoustics, Speech and Signal Processing* 1997, Munich, Germany, pp. 1047-1050.

[87] Rabiner, L.R. and B.H. Juang, *Fundamentals of Speech Recognition*, May, 1993, Prentice-Hall.

[88] Rabiner, L.R. and S.E. Levinson, "Isolated and Connected Word Recognition - Theory and Selected Applications," *IEEE Trans. on Communication*, 1981, **COM-29**(5), pp. 621-659.

[89] Riley, M. and A. Ljolje, eds. *Automatic Generation of Detailed Pronunciation Lexicons*, in Automatic Speech and Speaker Recognition, ed. C. Lee, F. Soong, and K. Paliwal, 1996, Kluwer Academic Publishers.

[90] Robinson, A., "An Application of Recurrent Nets to Phone Probability Estimation," *IEEE Trans. on Neural Networks*, 1994, **5**, pp. 298-305.

[91] Robinson, A.J., *et al.*, "A Neural Network Based, Speaker Independent, Large Vocabulary," *Proc. of the European Conf. on Speech Communication and Technology*, 1999, Berlin pp. 1941-1944.

[92] Roucos, S., *et al.*, "Stochastic Segment Modeling Using the Estimate-Maximize Algorithm," *Int. Conf. on Acoustic, Speech and Signal Processing*, 1988, New York, pp. 127-130.

[93] Sagisaka, Y. and L.S. Lee, "Speech Recognition of Asian Languages" in *Proc. IEEE Automatic Speech Recognition Workshop*, 1995, Snowbird, UT, pp. 55-57.

[94] Sakoe, H. and S. Chiba, "Dynamic Programming Algorithm Optimization for Spoken Word Recognition," *IEEE Trans. on Acoustics, Speech and Signal Processing*, 1978, **26**(1), pp. 43-49.

[95] Saraclar, M. and S. Khudanpur, "Pronunciation Ambiguity vs. Pronunciation Variability in Speech Recognition," *IEEE Int. Conf. on Acoustics, Speech and Signal Processing*, 2000, Istanbul, Turkey, pp. 1679-1682.

[96] Shinoda, K. and C. Lee, "Structural MAP Speaker Adaptation Using Hierarchical Priors," *IEEE Workshop on Automatic Speech Recognition and Understanding*, 1997, Santa Barbara, CA, pp. 381-388.

[97] Singh, R., B. Raj, and R. Stern, "Automatic Generation of Phone Sets and Lexical Transcriptions," *IEEE Int. Conf. on Acoustics, Speech and Signal Processing*, 2000, Istanbul, Turkey, pp. 1691-1694.

[98] Siohan, O., C. Chesta, and C. Lee, "Joint Maximum a Posteriori Estimation of Transformation and Hidden Markov Model Parameters," *IEEE Int. Conf. on Acoustics, Speech and Signal Processing*, 2000, Istanbul, Turkey, pp. 965-968.

[99] Siu, M., *et al.*, "Parametric Trajectory Mixtures for LVCSR," *Int. Conf. on Spoken Language Processing*, 1998, Sydney, Australia.

[100] Sixtus, A., *et al.*, "Recent Improvements of the RWTH Large Vocabulary Speech Recognition System on Spontaneous Speech," *IEEE Int. Conf. on Acoustics, Speech and Signal Processing*, 2000, Istanbul, Turkey, pp. 1671-1674.

[101] Sorenson, H., "A Cepstral Noise Reduction Multi-Layer Network," *Int. Conf. on Acoustics, Speech and Signal Processing*, 1991, Toronto, pp. 933-936.

[102] Sproat, R. and M. Riley, "Compilation of Weighted Finite-State Transducers from Decision Trees," *ACL-96*, 1996, Santa Cruz, pp. 215-222.

[103] Tajchman, G., E. Fosler, and D. Jurafsky, "Building Multiple Pronunciation Models for Novel Words Using Exploratory Computational Phonology," *Eurospeech*, 1995, pp. 2247-2250.

[104] Tebelskis, J. and A. Waibel, "Large Vocabulary Recognition Using Linked Predictive Neural Networks," *Int. Conf. on Acoustics, Speech and Signal Processing*, 1990, Albuquerque, NM, pp. 437-440.

[105] Waibel, A.H. and K.F. Lee, *Readings in Speech Recognition*, 1990, San Mateo, CA, Morgan Kaufman Publishers.

[106] Watrous, R., "Speaker Normalization and Adaptation Using Second-Order Connectionist Networks," *IEEE Trans. on Neural Networks*, 1994, **4**(1), pp. 21-30.

[107] Welling, L., S. Kanthak, and H. Ney, "Improved Methods for Vocal Tract Normalization," *IEEE Int. Conf. on Acoustics, Speech and Signal Processing*, 1999, Phoenix, AZ.

[108] Wilpon, J.G., C.H. Lee, and L.R. Rabiner, "Connected Digit Recognition Based on Improved Acoustic Resolution," *Computer Speech and Language*, 1993, **7**(1), pp. 15-26.

[109] Woodland, P.C., *et al.*, "The 1994 HTK Large Vocabulary Speech Recognition System," *Proc. of the IEEE Int. Conf. on Acoustics, Speech and Signal Processing*, 1995, Detroit, pp. 73-76.

[110] Wooters, C. and A. Stolcke, "Multiple-Pronunciation Lexical Modeling in a Speaker Independent Speech Understanding System," *Proc. of the Int. Conf. on Spoken Language Processing*, 1994, Yokohama, Japan, pp. 1363-1366.

[111] Young, S.J. and P.C. Woodland, "The Use of State Tying in Continuous Speech Recognition," *Proc. of Eurospeech*, 1993, Berlin pp. 2203-2206.

[112] Zavaliagkos, G., R. Schwartz, and J. Makhoul, "Batch, Incremental and Instantaneous Adaptation Techniques for Speech Recognition," *Proc. of the IEEE Int. Conf. on Acoustics, Speech and Signal Processing*, 1995, Detroit, pp. 676-679.

[113] Zavaliagkos, G., *et al.*, "A Hybrid Segmental Neural Net/Hidden Markov Model System for Continuous Speech Recognition," *IEEE Trans. on Speech and Audio Processing*, 1994, **2**, pp. 151-160.

[114] Zhan, P. and M. Westphal, "Speaker Normalization Based on Frequency Warping" in *Proc. of the IEEE Int. Conf. on Acoustics, Speech and Signal Processing* 1997, Munich, Germany, pp. 1039-1042.

[115] Zue, V., *et al.*, "The MIT SUMMIT System: A Progress Report," *Proc. of DARPA Speech and Natural Language Workshop*, 1989, pp. 179-189.

CHAPTER 10

Environmental Robustness

A speech recognition system trained in the lab with clean speech may degrade significantly in the real world if the clean speech used in training doesn't match real-world speech. If its accuracy doesn't degrade very much under mismatched conditions, the system is called *robust*. There are several reasons why real-world speech may differ from clean speech; in this chapter we focus on the influence of the *acoustical environment*, defined as the transformations that affect the speech signal from the time it leaves the mouth until it is in digital format.

Chapter 9 discussed a number of variability factors that are critical to speech recognition. Because the acoustical environment is so important to practical systems, we devote this chapter to ways of increasing the environmental robustness, including microphone, echo cancellation, and a number of methods that enhance the speech signal, its spectrum, and the corresponding acoustic model in a speech recognition system.

10.1. The Acoustical Environment

The acoustical environment is defined as the set of transformations that affect the speech signal from the time it leaves the speaker's mouth until it is in digital form. Two main sources of distortion are described here: additive noise and channel distortion. Additive noise, such as a fan running in the background, door slams, or other speakers' speech, is common in our daily life. Channel distortion can be caused by reverberation, the frequency response of a microphone, the presence of an electrical filter in the A/D circuitry, the response of the local loop of a telephone line, a speech codec, etc. Reverberation, caused by reflections of the acoustical wave in walls and other objects, can also dramatically alter the speech signal.

10.1.1. Additive Noise

Additive noise can be stationary or nonstationary. Stationary noise, such as that made by a computer fan or air conditioning, has a power spectral density that does not change over time. Nonstationary noise, caused by door slams, radio, TV, and other speakers' voices, has statistical properties that change over time. A signal captured with a close-talking microphone has little noise and reverberation, even though there may be lip smacks and breathing noise. A microphone that is not close to the speaker's mouth may pick up a lot of noise and/or reverberation.

As described in Chapter 5, a signal $x[n]$ is defined as white noise if its power spectrum is flat, $S_{xx}(f) = q$, a condition equivalent to different samples being uncorrelated, $R_{xx}[n] = q\delta[n]$. Thus, a white noise signal has to have zero mean. This definition tells us about the second-order moments of the random process, but not about its distribution. Such noise can be generated synthetically by drawing samples from a distribution $p(x)$; thus we could have uniform white noise if $p(x)$ is uniform, or Gaussian white noise if $p(x)$ is Gaussian. While typically subroutines are available that generate uniform white noise, we are often interested in white Gaussian noise, as it resembles better the noise that tends to occur in practice. See Algorithm 10.1 for a method to generate white Gaussian noise. Variable x is normally continuous, but it can also be discrete.

White noise is useful as a conceptual entity, but it seldom occurs in practice. Most of the noise captured by a microphone is *colored*, since its spectrum is not flat. *Pink* noise is a particular type of colored noise that has a low-pass nature, as it has more energy at the low frequencies and rolls off at higher frequencies. The noise generated by a computer fan, an air conditioner, or an automobile engine can be approximated by pink noise. We can synthesize pink noise by filtering white noise with a filter whose magnitude squared equals the desired power spectrum.

A great deal of additive noise is nonstationary, since its statistical properties change over time. In practice, even the noises from a computer, an air conditioning system, or an automobile are not perfectly stationary. Some nonstationary noises, such as keyboard clicks, are caused by physical objects. The speaker can also cause nonstationary noises such as lip smacks and breath noise. The *cocktail party effect* is the phenomenon under which a human listener can focus onto one conversation out of many in a cocktail party. The noise of the conversations that are not focused upon is called *babble* noise. When the nonstationary noise is correlated with a known signal, the adaptive echo-canceling (AEC) techniques of Section 10.3 can be used.

ALGORITHM 10.1: *WHITE NOISE GENERATION*

To generate white noise in a computer, we can first generate a random variable ρ with a Rayleigh distribution:

$$p_\rho(\rho) = \rho e^{-\rho^2/2} \tag{10.1}$$

from another random variable *r* with a uniform distribution between (0, 1), $p_r(r) = 1$, by simply equating the probability mass $p_\rho(\rho)|d\rho| = p_r(r)|dr|$ so that $\left|\frac{dr}{d\rho}\right| = \rho e^{-\rho^2/2}$; with integration, it results in $r = e^{-\rho^2/2}$ and the inverse is given by

$$\rho = \sqrt{-2\ln r} \tag{10.2}$$

If *r* is uniform between (0, 1), and ρ is computed through Eq. (10.2), it follows a Rayleigh distribution as in Eq. (10.1). We can then generate Rayleigh white noise by drawing independent samples from such a distribution.

If we want to generate white Gaussian noise, the method used above does not work, because the integral of the Gaussian distribution does not exist in closed form. However, if ρ follows a Rayleigh distribution as in Eq. (10.1), obtained using Eq. (10.2) where *r* is uniform between (0, 1), and θ is uniformly distributed between (0, 2π), then the white Gaussian noise can be generated as the following two variables *x* and *y*:

$$\begin{aligned} x &= \rho\cos(\theta) \\ y &= \rho\sin(\theta) \end{aligned} \tag{10.3}$$

They are independent Gaussian random variables with zero mean and unity variance, since the Jacobian of the transformation is given by

$$J = \begin{vmatrix} \dfrac{\partial p_x}{\partial \rho} & \dfrac{\partial p_x}{\partial \theta} \\ \dfrac{\partial p_y}{\partial \rho} & \dfrac{\partial p_y}{\partial \theta} \end{vmatrix} = \begin{vmatrix} \cos\theta & -\rho\sin\theta \\ \sin\theta & \rho\cos\theta \end{vmatrix} = \rho \tag{10.4}$$

and the joint density $p(x, y)$ is given by

$$\begin{aligned} p(x,y) &= \frac{p(\rho,\theta)}{J} = \frac{p(\rho)p(\theta)}{\rho} = \frac{1}{2\pi}e^{-\rho^2/2} \\ &= \frac{1}{2\pi}e^{-(x^2+y^2)/2} = N(x,0,1)N(y,0,1) \end{aligned} \tag{10.5}$$

The presence of additive noise can sometimes change the way the speaker speaks. The *Lombard effect* [40] is a phenomenon by which a speaker increases his vocal effort in the presence of background noise. When a large amount of noise is present, the speaker tends to shout, which entails not only a higher amplitude, but also often higher pitch, slightly different formants, and a different coloring of the spectrum. It is very difficult to characterize these transformations analytically, but recently some progress has been made [36].

10.1.2. Reverberation

If both the microphone and the speaker are in an *anechoic*[1] chamber or in free space, a microphone picks up only the direct acoustic path. In practice, in addition to the direct acoustic path, there are reflections of walls and other objects in the room. We are well aware of this effect when we are in a large room, which can prevent us from understanding if the reverberation time is too long. Speech recognition systems are much less robust than humans and they start to degrade with shorter reverberation times, such as those present in a normal office environment.

As described in Chapter 2, the signal level at the microphone is inversely proportional to the distance r from the speaker for the direct path. For the kth reflected sound wave, the sound has to travel a larger distance r_k, so that its level is proportionally lower. This reflection also takes time $T_k = r_k / c$ to arrive, where c is the speed of sound in air.[2] Moreover, some energy absorption a takes place each time the sound wave hits a surface. The impulse response of such filter looks like

$$h[n] = \sum_{k=0}^{\infty} \frac{\rho_k}{r_k} \delta[n - T_k] = \frac{1}{c} \sum_{k=0}^{\infty} \frac{\rho_k}{T_k} \delta[n - T_k] \tag{10.6}$$

where ρ_k is the combined attenuation of the kth reflected sound wave due to absorption. Anechoic rooms have $\rho_k \approx 0$. In general ρ_k is a (generally decreasing) function of frequency, so that instead of impulses $\delta[n]$ in Eq. (10.6), other (low-pass) impulse responses are used.

Often we have available a large amount of speech data recorded with a close-talking microphone, and we would like to use the speech recognition system with a far field microphone. To do that we can filter the clean-speech training database with a filter $h[n]$, so that the filtered speech resembles speech collected with the far field microphone, and then retrain the system. This requires estimating the impulse response $h[n]$ of a room. Alternatively, we can filter the signal from the far field microphone with an inverse filter to make it resemble the signal from the close-talking microphone.

---

[1] An anechoic chamber is a room that has walls made of special fiberglass or other sound-absorbing materials so that it absorbs all echoes. It is equivalent to being in free space, where there are neither walls nor reflecting surfaces.

[2] In air at standard atmospheric pressure and humidity the speed of sound is $c = 331.4 + 0.6T \ (m/s)$. It varies with different media and different levels of humidity and pressure.

One way to estimate the impulse response is to play a white noise signal $x[n]$ through a loudspeaker or artificial mouth; the signal $y[n]$ captured at the microphone is given by

$$y[n] = x[n] * h[n] + v[n] \tag{10.7}$$

where $v[n]$ is the additive noise present at the microphone. This noise is due to sources such as air conditioning and computer fans and is an obstacle to measuring $h[n]$. The impulse response can be estimated by minimizing the error over N samples

$$E = \frac{1}{N}\sum_{n=0}^{N-1}\left(y[n] - \sum_{m=0}^{M-1} h[m]x[n-m] \right)^2 \tag{10.8}$$

which, taking the derivative with respect to $h[m]$ and equating to 0, results in our estimate $\hat{h}[l]$:

$$\begin{aligned}
\left.\frac{\partial E}{\partial h[l]}\right|_{h[l]=\hat{h}[l]} &= \frac{1}{N}\sum_{n=0}^{N-1}\left(y[n] - \sum_{m=0}^{M-1} \hat{h}[m]x[n-m] \right)x[n-l] \\
&= \frac{1}{N}\sum_{n=0}^{N-1} y[n]x[n-l] - \sum_{m=0}^{M-1}\hat{h}[m]\left(\frac{1}{N}\sum_{n=0}^{N-1} x[n-m]x[n-l] \right) \\
&= \frac{1}{N}\sum_{n=0}^{N-1} y[n]x[n-l] - \hat{h}[l] - \sum_{m=0}^{M-1}\hat{h}[m]\left(\frac{1}{N}\sum_{n=0}^{N-1} x[n-m]x[n-l] - \delta[m-l] \right) = 0
\end{aligned} \tag{10.9}$$

Since we know our white process is ergodic, it follows that we can replace time averages by ensemble averages as $N \to \infty$:

$$\lim_{N\to\infty} \frac{1}{N}\sum_{n=0}^{N-1} x[n-m]x[n-l] = E\{x[n-m]x[n-l]\} = \delta[m-l] \tag{10.10}$$

so that we can obtain a reasonable estimate of the impulse response as

$$\hat{h}[l] = \frac{1}{N}\sum_{n=0}^{N-1} y[n]x[n-l] \tag{10.11}$$

Inserting Eq. (10.7) into Eq. (10.11) ,we obtain

$$\hat{h}[l] = h[l] + e[l] \tag{10.12}$$

where the estimation error $e[n]$ is given by

$$e[l] = \frac{1}{N}\sum_{n=0}^{N-1} v[n]x[n-l] + \sum_{m=0}^{M-1} h[m]\left(\frac{1}{N}\sum_{n=0}^{N-1} x[n-m]x[n-l] - \delta[m-l] \right) \tag{10.13}$$

If $v[n]$ and $x[n]$ are independent processes, then $E\{e[l]\} = 0$, since $x[n]$ is zero-mean, so that the estimate of Eq. (10.11) is unbiased. The covariance matrix decreases to 0 as

$N \to \infty$, with the dominant term being the noise $v[n]$. The choice of N for a low-variance estimate depends on the filter length M and the noise level present in the room.

The filter $h[n]$ could also be estimated by playing sine waves of different frequencies or a chirp[3] [52]. Since playing a white noise signal or sine waves may not be practical, another method is based on collecting stereo recordings with a close-talking microphone and a far field microphone. The filter $h[n]$ of length M is estimated so that when applied to the close-talking signal $x[n]$ it minimizes the squared error with the far field signal $y[n]$, which results in the following set of M linear equations:

$$\sum_{m=0}^{M-1} h[m] R_{xx}[m-n] = R_{xy}[n] \tag{10.14}$$

which is a generalization of Eq. (10.11) when $x[n]$ is not a white noise signal.

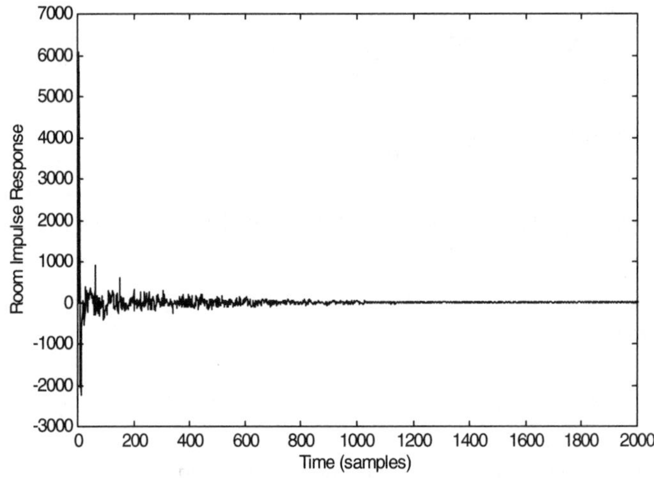

Figure 10.1 Typical impulse response of an average office. Sampling rate was 16 kHz. It was estimated by driving a 4-minute segment of white noise through an artificial mouth and using Eq. (10.11). The filter length is about 125 ms.

It is not uncommon to have reverberation times of over 100 milliseconds in office rooms. In Figure 10.1 we show the typical impulse response of an average office.

10.1.3. A Model of the Environment

A widely used model of the degradation encountered by the speech signal when it gets corrupted by both additive noise and channel distortion is shown in Figure 10.2. We can derive

[3] A chirp function continuously varies its frequency. For example, a linear chirp varies its frequency linearly with time: $\sin(n(\omega_0 + \omega_1 n))$.

the relationships between the clean signal and the corrupted signal both in power-spectrum and cepstrum domains based on such a model [2].

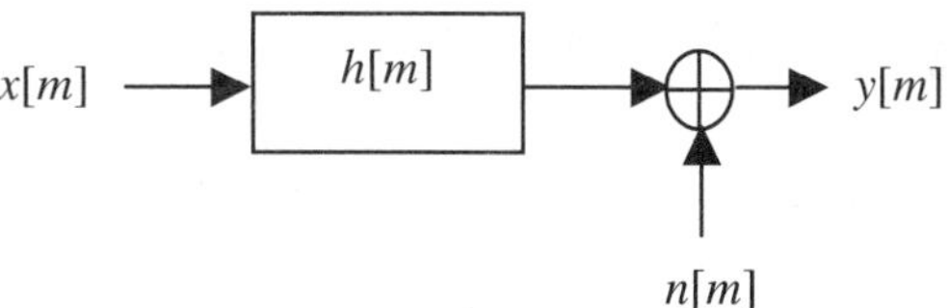

Figure 10.2 A model of the environment.

In the time domain, additive noise and linear filtering results in

$$y[m] = x[m] * h[m] + n[m] \tag{10.15}$$

It is convenient to express this in the frequency domain using the short-time analysis methods of Chapter 6. To do that, we window the signal, take a 2*K*-point DFT in Eq. (10.15) and then the magnitude squared:

$$\begin{aligned} |Y(f_k)|^2 &= |X(f_k)|^2 |H(f_k)|^2 + |N(f_k)|^2 + 2\,\mathrm{Re}\{X(f_k)H(f_k)N^*(f_k)\} \\ &= |X(f_k)|^2 |H(f_k)|^2 + |N(f_k)|^2 + 2|X(f_k)||H(f_k)||N(f_k)|\cos(\theta_k) \end{aligned} \tag{10.16}$$

where $k = 0, 1, \cdots, K$, we have used upper case for frequency domain linear spectra, and θ_k is the angle between the filtered signal and the noise for bin *k*.

The expected value of the *cross-term* in Eq. (10.16) is zero, since $x[m]$ and $n[m]$ are statistically independent. In practice, this term is not zero for a given frame, though it is small if we average over a range of frequencies, as we often do when computing the popular mel-cepstrum (see Chapter 6). When using a filterbank, we can obtain a relationship for the energies at each of the *M* filters:

$$|Y(f_i)|^2 \approx |X(f_i)|^2 |H(f_i)|^2 + |N(f_i)|^2 \tag{10.17}$$

where it has been shown experimentally that this assumption works well in practice.

Equation (10.17) is also implicitly assuming that the length of $h[n]$, the filter's impulse response, is much shorter than the window length 2*N*. That means that for filters with long reverberation times, Eq. (10.17) is inaccurate. For example, for $|N(f)|^2 = 0$, a window shift of *T*, and a filter's impulse response $h[n] = \delta[n-T]$, we have $Y_t[f_m] = X_{t-1}[f_m]$, i.e., the output spectrum at frame *t* does not depend on the input spectrum at that frame. This is a more serious assumption, which is why speech recognition systems tend to fail under long reverberation times.

By taking logarithms in Eq. (10.17), and after some algebraic manipulation, we obtain

$$\begin{aligned} \ln|Y(f_i)|^2 &\approx \ln|X(f_i)|^2 + \ln|H(f_i)|^2 \\ &+ \ln\left(1+\exp\left(\ln|N(f_i)|^2 - \ln|X(f_i)|^2 - \ln|H(f_i)|^2\right)\right) \end{aligned} \tag{10.18}$$

Since most speech recognition systems use cepstrum features, it is useful to see the effect of the additive noise and channel distortion directly on the cepstrum. To do that, let's define the following length-(M + 1) cepstrum vectors:

$$\begin{aligned} \mathbf{x} &= \mathbf{C}\left(\ln|X(f_0)|^2 \quad \ln|X(f_1)|^2 \quad \cdots \quad \ln|X(f_M)|^2\right) \\ \mathbf{h} &= \mathbf{C}\left(\ln|H(f_0)|^2 \quad \ln|H(f_1)|^2 \quad \cdots \quad \ln|H(f_M)|^2\right) \\ \mathbf{n} &= \mathbf{C}\left(\ln|N(f_0)|^2 \quad \ln|N(f_1)|^2 \quad \cdots \quad \ln|N(f_M)|^2\right) \\ \mathbf{y} &= \mathbf{C}\left(\ln|Y(f_0)|^2 \quad \ln|Y(f_1)|^2 \quad \cdots \quad \ln|Y(f_M)|^2\right) \end{aligned} \tag{10.19}$$

where **C** is the DCT matrix and we have used lower-case bold to represent cepstrum vectors.

Combining Eqs. (10.18) and (10.19) results in

$$\mathbf{y} = \mathbf{x} + \mathbf{h} + \mathbf{g}(\mathbf{n} - \mathbf{x} - \mathbf{h}) \tag{10.20}$$

where the nonlinear function **g**(**z**) is given by

$$\mathbf{g}(\mathbf{z}) = \mathbf{C}\ln\left(1 + e^{\mathbf{C}^{-1}\mathbf{z}}\right) \tag{10.21}$$

Equations (10.20) and (10.21) say that we can compute the cepstrum of the corrupted speech if we know the cepstrum of the clean speech, the cepstrum of the noise, and the cepstrum of the filter. In practice, the DCT matrix **C** is not square, so that the dimension of the cepstrum vector is much smaller than the number of filters. This means that we are losing resolution when going back to the frequency domain, and thus Eqs. (10.20) and (10.21) represent only an approximation, though it has been shown to work reasonably well.

As discussed in Chapter 9, the distribution of the cepstrum of **x** can be modeled as a mixture of Gaussian densities. Even if we assume that **x** follows a Gaussian distribution, **y** in Eq. (10.20) is no longer Gaussian because of the nonlinearity in Eq. (10.21).

It is difficult to visualize the effect on the distribution, given the nonlinearity involved. To provide some insight, let's consider the frequency-domain version of Eq. (10.18) when no filtering is done, i.e., $H(f)=1$:

$$y = x + \ln\left(1 + \exp(n - x)\right) \tag{10.22}$$

where x, n, and y represent the log-spectral energies of the clean signal, noise, and noisy signal, respectively, for a given frequency. Using simulated data, not real speech, we can analyze the effect of this transformation. Let's assume that both x and n are Gaussian random variables. We can use Monte Carlo simulation to draw a large number of points from those two Gaussian distributions and obtain the corresponding noisy values y using Eq. (10.22). Figure 10.3 shows the resulting distribution for several values of σ_x. We fixed $\mu_n = 0\,\text{dB}$, since it is only a relative level, and set $\sigma_n = 2\,\text{dB}$, a typical value. We also set $\mu_x = 25\,\text{dB}$ and see that the resulting distribution can be bimodal when σ_x is very large. Fortunately, for modern speech recognition systems that have many Gaussian components, σ_x is never that large and the resulting distribution is unimodal.

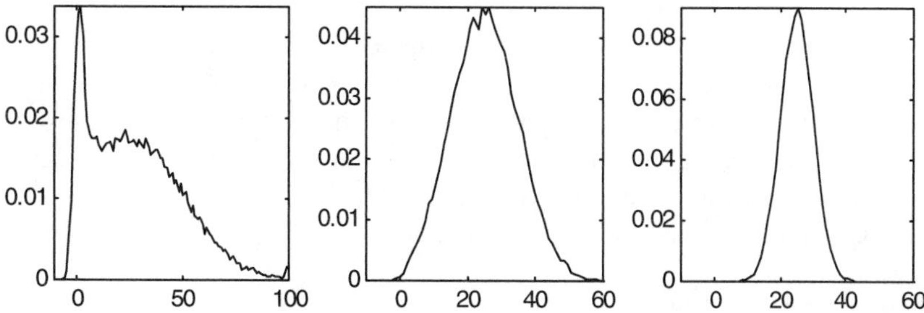

Figure 10.3 Distributions of the corrupted log-spectra y of Eq. (10.22) using simulated data. The distribution of the noise log-spectrum n is Gaussian with mean 0 dB and standard deviation of 2 dB. The distribution of the clean log-spectrum x is Gaussian with mean 25 dB and standard deviations of 25, 10, and 5 dB, respectively (the x-axis is expressed in dB). The first distribution is bimodal, whereas the other two are approximately Gaussian. Curves are plotted using Monte Carlo simulation.

Figure 10.4 shows the distribution of y for two values of μ_x, given the same values for the noise distribution, $\mu_n = 0\,\text{dB}$ and $\sigma_n = 2\,\text{dB}$, and a more realistic value for $\sigma_x = 5\,\text{dB}$. We see that the distribution is always unimodal, though not necessarily symmetric, particularly for low SNR ($\mu_x - \mu_n$).

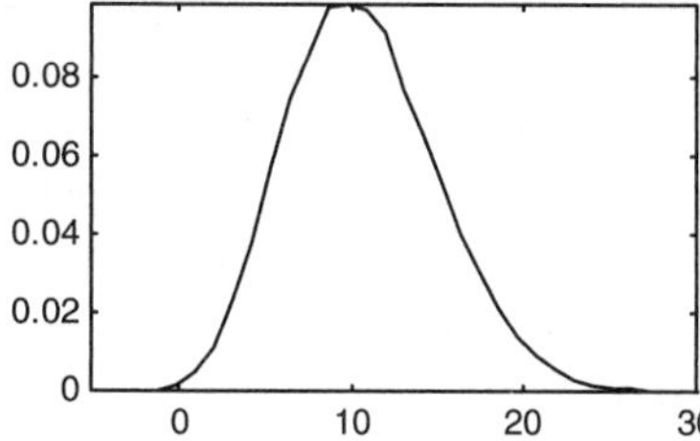

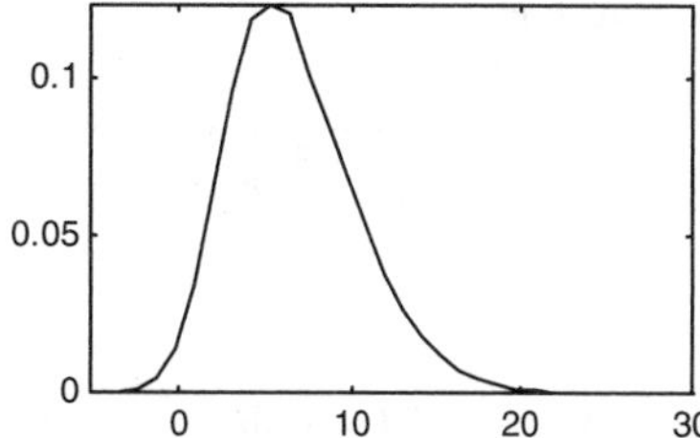

Figure 10.4 Distributions of the corrupted log-spectra y of Eq. (10.22) using simulated data. The distribution of the noise log-spectrum n is Gaussian with mean 0 dB and standard deviation of 2 dB. The distribution of the clean log-spectrum is Gaussian with standard deviation of 5 dB and means of 10 and 5 dB, respectively. The first distribution is approximately Gaussian while the second is nonsymmetric. Curves are plotted using Monte Carlo simulation.

The distributions used in an HMM are mixtures of Gaussians so that, even if each Gaussian component is transformed into a non-Gaussian distribution, the composite distribution can be modeled adequately by another mixture of Gaussians. In fact, if you retrain the model using the standard Gaussian assumption on corrupted speech, you can get good results, so this approximation is not bad.

10.2. Acoustical Transducers

Acoustical transducers are devices that convert the acoustic energy of sound into electrical energy (microphones) and vice versa (loudspeakers). In the case of a microphone this transduction is generally realized with a diaphragm, whose movement in response to sound pressure varies the parameters of an electrical system (a variable-resistance conductor, a condenser, etc.), producing a variable voltage that constitutes the microphone output. We focus on microphones because they play an important role in designing speech recognition systems.

There are near field or close-talking microphones, and far field microphones. Close-talking microphones, either head-mounted or telephone handsets, pick up much less background noise, though they are more sensitive to throat clearing, lip smacks, and breath noise. Placement of such a microphone is often very critical, since, if it is right in front of the mouth, it can produce pops in the signal with plosives such as */p/*. Far field microphones can be lapel mounted or desktop mounted and pick up more background noise than near field microphones. Having a small but variable distance to the microphone could be worse than a larger but more consistent distance, because the corresponding HMM may have lower variability.

When used in speech recognition systems, the most important measurement is the signal-to-noise ratio (SNR), since the lower the SNR the higher the error rate. In addition, different microphones have different transfer functions, and even the same microphone offers different transfer functions depending on the distance between mouth and microphone. Varying noise and channel conditions are a challenge that speech recognition systems have to address, and in this chapter we present some techniques to combat them.

The most popular type of microphone is the condenser microphone. We shall study in detail its directionality patterns, frequency response, and electrical characteristics.

10.2.1. The Condenser Microphone

A *condenser microphone* has a capacitor consisting of a pair of metal plates separated by an insulating material called a dielectric (see Figure 10.5). Its capacitance C is given by

$$C = \varepsilon_0 \pi b^2 / h \tag{10.23}$$

where ε_0 is a constant, b is the width of the plate, and h is the separation between the plates. If we polarize the capacitor with a voltage V_{cc}, it acquires a charge Q given by

$$Q = CV_{cc} \tag{10.24}$$

One of the plates is free to move in response to changes in sound pressure, which results in a change in the plate separation Δh, thereby changing the capacitance and producing a change in voltage $\Delta V = \Delta h V_{cc} / h$. Thus, the sensitivity[4] of the microphone depends on the polarizing voltage V_{cc}, which is why this voltage can often be 100 V or more.

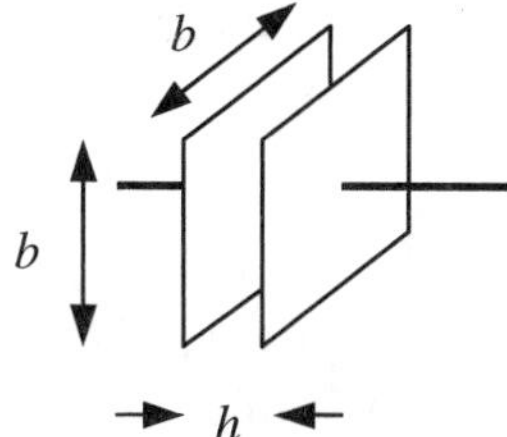

Figure 10.5 A diagram of a condenser microphone.

Electret microphones are a type of condenser microphones that do not require a special polarizing voltage V_{cc}, because a charge is impressed on either the diaphragm or the back plate during manufacturing and it remains for the life of the microphone. Electret microphones are light and, because of their small size, they offer good responses at high frequencies.

From the electrical point of view, a microphone is equivalent to a voltage source $v(t)$ with an impedance Z_M, as shown in Figure 10.6. The microphone is connected to a preamplifier which has an equivalent impedance R_L.

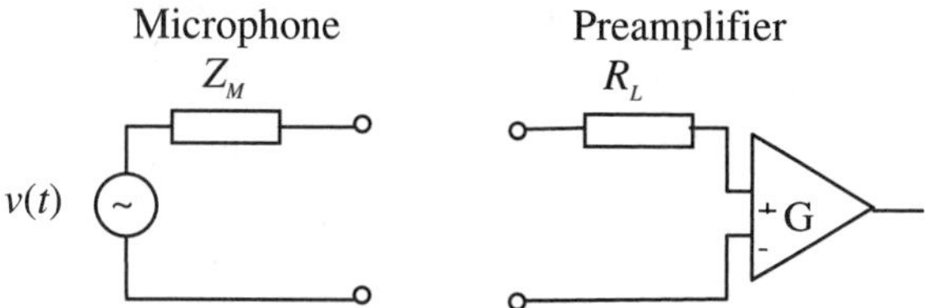

Figure 10.6 Electrical equivalent of a microphone.

[4] The sensitivity of a microphone measures the *open-circuit voltage* of the electric signal the microphone delivers for a sound wave for a given sound pressure level, often 94 dB SPL, when there is no load or a high impedance. This voltage is measured in dBV, where the 0-dB reference is 1 V rms.

From Figure 10.6 we can see that the voltage on R_L is

$$v_R(t) = v(t)\frac{R_L}{(R_M + R_L)} \tag{10.25}$$

Maximization of $v_R(t)$ in Eq. (10.25) results in $R_L = \infty$, or in practice $R_L >> R_M$, which is called *bridging*. Thus, for highest sensitivity the impedance of the amplifier has to be at least 10 times higher than that of the microphone. If the microphone is connected to an amplifier with lower impedance, there is a *load loss* of signal level. Most low-impedance microphones are labeled as 150 ohms, though the actual values may vary between 100 and 300. Medium impedance is 600 ohms and high impedance is 600–10,000 ohms. In practice, the microphone impedance is a function of frequency. Signal power is measured in dBm, where the 0-dB reference corresponds to 1 mW dissipated in a 600-ohm resistor. Thus, 0 dBm is equivalent to 0.775 V.

Since the output impedance of a condenser microphone is very high (~ 1 Mohm), a JFET transistor must be coupled to lower the equivalent impedance. Such a transistor needs to be powered with DC voltage through a different wire, as in Figure 10.7. A standard sound card has a jack with the audio on the tip, ground on the sleeve, DC bias V_{DD} on the ring, and a medium impedance. When using *phantom power*, the V_{CC} bias is provided directly in the audio signal, which must be *balanced* to ground.

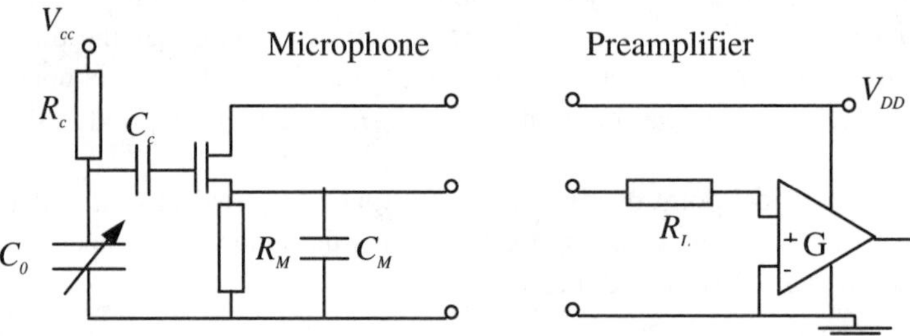

Figure 10.7 Equivalent circuit for a condenser microphone with DC bias on a separate wire.

It is important to understand how noise affects the signal of a microphone. If thermal noise arises in the resistor R_L, it will have a power

$$P_N = 4kTB \tag{10.26}$$

where $k = 1.38 \times 10^{-23}$ J/K is the Bolzmann's constant, T is the temperature in °K, and B is the bandwidth in Hz. The thermal noise in Eq. (10.26) at room temperature ($T = 297$°K) and for a bandwidth of 4 kHz is equivalent to –132 dBm. In practice, the noise is significantly higher than this because of preamplifier noise, radio-frequency noise and electromagnetic interference (poor grounding connections). It is, thus, important to keep the signal path between the microphone and the preamp as short as possible to avoid extra noise. It is desirable to have a microphone with low impedance to decrease the effect of noise due to radio-frequency interference, and to decrease the signal loss if long cables are used. Most microphones specify their SNR and range where they are linear (dynamic range). For condenser microphones, a power supply is necessary (DC bias required). Microphones with balanced output (the signal appears across two inner wires not connected to ground, with the shield of the cable connected to ground) are more resistant to radio frequency interference.

10.2.2. Directionality Patterns

A microphone's directionality pattern measures its sensitivity to a particular direction. Microphones may also be classified by their directional properties as *omnidirectional* (or *non-directional*) and *directional*, the latter subdivided into bidirectional and unidirectional, based upon their response characteristics.

10.2.2.1. Omnidirectional Microphones

By definition, the response of an omnidirectional microphone is independent of the direction from which the encroaching sound wave is coming. Figure 10.8 shows the polar response of an omnidirectional mike. A microphone's *polar response*, or *pickup pattern*, graphs its output voltage for an input sound source with constant level at various angles around the mic. Typically, a polar response assumes a preferred direction, called the major axis or front of the microphone, which corresponds to the direction at which the microphone is most sensitive. The front of the mike is labeled as zero degrees on the polar plot, but since an omnidirectional mic has no particular direction at which it is the most sensitive, the omnidirectional mike has no true front and hence the zero-degree axis is arbitrary. Sounds coming from any direction around the microphone are picked up equally. Omnidirectional microphones provide no noise cancellation.

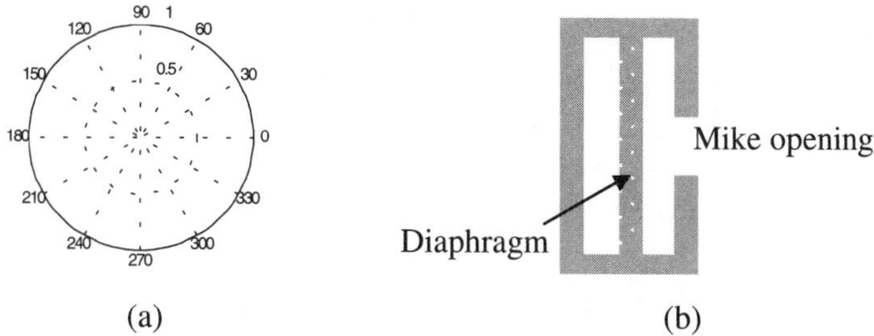

Figure 10.8 (a) Polar response of an ideal omnidirectional microphone and (b) its cross section.

Figure 10.8 shows the mechanics of the ideal[5] omnidirectional condenser microphone. A sound wave creates a pressure all around the microphone. The pressure enters the opening of the mike and the diaphragm moves. An electrical circuit converts the diaphragm move-

[5] Ideal omnidirectional microphones do not exist.

ment into an electrical voltage, or response. Sound waves impinging on the mike create a pressure at the opening regardless of the direction from which they are coming; therefore we have a nondirectional, or omnidirectional, microphone. As we have seen in Chapter 2, if the source signal is $Be^{j\omega t}$, the signal at a distance r is given by $(A/r)e^{j\omega t}$ independently of the angle.

This is the most inexpensive of the condenser microphones, and it has the advantage of a flat frequency response that doesn't change with the angle or distance to the microphone. On the other hand, because of its uniform polar pattern, it picks up not only the desired signal but also noise from any direction. For example, if a pair of speakers is monitoring the microphone output, the sound from the speakers can reenter the microphone and create an undesirable sound called *feedback.*

10.2.2.2. Bidirectional Microphones

The bidirectional microphone is a *noise-canceling* microphone; it responds less to sounds incident from the sides. The bidirectional mike utilizes the properties of a gradient microphone to achieve its noise-canceling polar response. You can see how this is accomplished by looking at the diagram of a simplified gradient bidirectional condenser microphone, as shown in Figure 10.9. A sound impinging upon the front of the microphone creates a pressure at the front opening. A short time later, this same sound pressure enters the back of the microphone. The sound pressure never arrives at the front and back at the same time. This creates a displacement of the diaphragm and, just as with the omnidirectional mike, a corresponding electrical signal. For sounds impinging from the side, however, the pressure from an incident sound wave at the front opening is identical to the pressure at the back. Since both openings lead to one side of the diaphragm, there is no displacement of the diaphragm, and the sound is not reproduced.

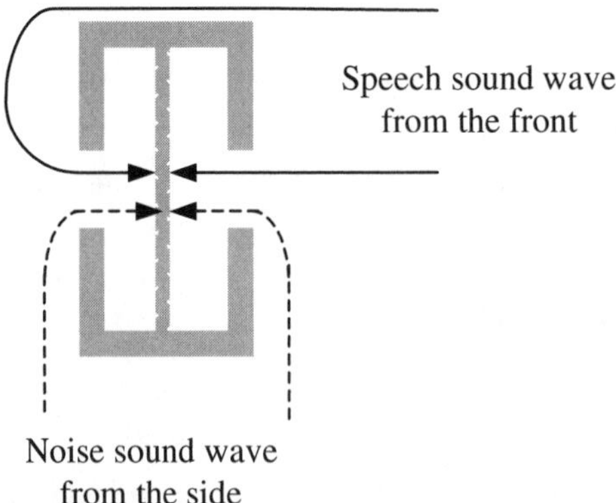

Figure 10.9 Cross section of an ideal bidirectional microphone.

To compute the polar response of this gradient microphone let's make the approximation of Figure 10.10, where the microphone signal is the difference between the signal at the front and rear of the diaphragm, the separation between plates is $2d$, and r is the distance between the source and the center of the microphone.

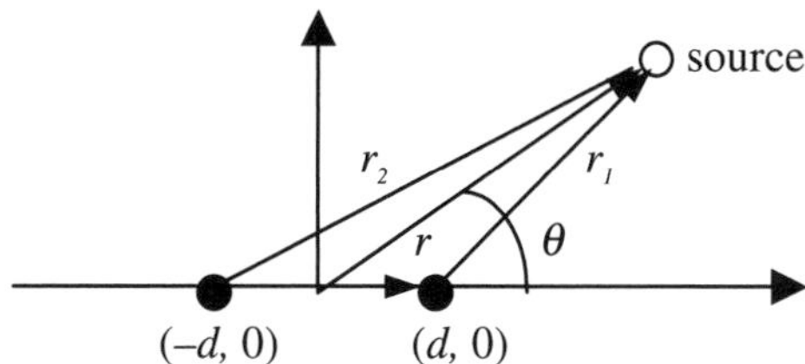

Figure 10.10 Approximation to the noise-canceling microphone of Figure 10.9.

You can see that r_1, the distance between the source and the front of the diaphragm, is the norm of the vector specifying the source location minus the vector specifying the location of the front of the diaphragm

$$r_1 = \left| re^{j\theta} - d \right| \tag{10.27}$$

Similarly, you obtain the distance between the source and the rear of the diaphragm

$$r_2 = \left| re^{j\theta} + d \right| \tag{10.28}$$

The source arrives at the front of the diaphragm with a delay $\delta_1 = r_1 / c$, where c is the speed of sound in air. Similarly, the delay to the rear of the diaphragm is $\delta_2 = r_2 / c$. If the source is a complex exponential $e^{j\omega t}$, the difference signal between the front and rear is given by

$$x(t) = \frac{A}{r_1} e^{j2\pi f(t-\delta_1)} - \frac{A}{r_2} e^{j2\pi f(t-\delta_2)} = \frac{A}{r} e^{j2\pi ft} G(f,\theta) \tag{10.29}$$

where A is a constant and, using Eqs. (10.27), (10.28) and (10.29), the gain $G(f,\theta)$ is given by

$$G(f,\theta) = \frac{e^{-j2\pi\left|e^{j\theta}-\lambda\right|\tau f}}{\left|e^{j\theta} - \lambda\right|} - \frac{e^{-j2\pi\left|e^{j\theta}+\lambda\right|\tau f}}{\left|e^{j\theta} + \lambda\right|} \tag{10.30}$$

where we have defined $\lambda = d / r$ and $\tau = r / c$.

The magnitude of Eq. (10.30) is used to plot the polar response of Figure 10.11. As can be seen by the plot, the pattern resembles a figure eight. The bidirectional mike has an interchangeable front and back, since the response has a maximum in two opposite directions. In practice, this bidirectional microphone is an ideal case, and the polar response has to be measured empirically.

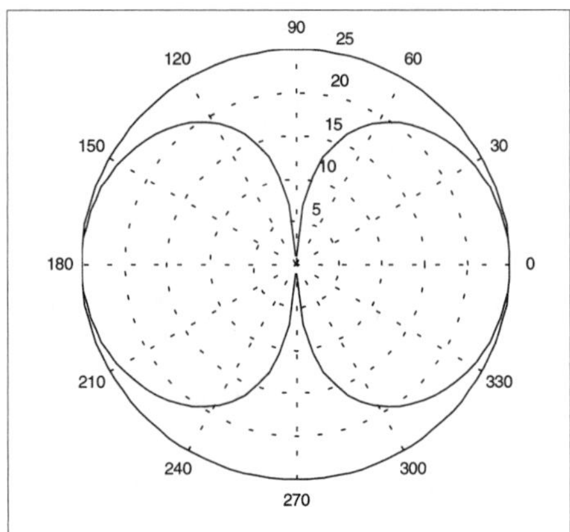

Figure 10.11 Polar response of a bidirectional microphone obtained through Eq. (10.30) with $d = 1$ cm, $r = 50$ cm, $c = 33{,}000$ cm/s, and $f = 1000$ Hz.

According to the idealized model, the frequency response of omnidirectional microphones is constant with frequency, and this approximately holds in practice for real omnidirectional microphones. On the other hand, the polar pattern of directional microphones is not constant with frequency. Clearly it is a function of frequency, as can be seen in Eq. (10.30). In fact, the frequency response of a bidirectional microphone at 0° is shown in Figure 10.12 for both near field and far field conditions.

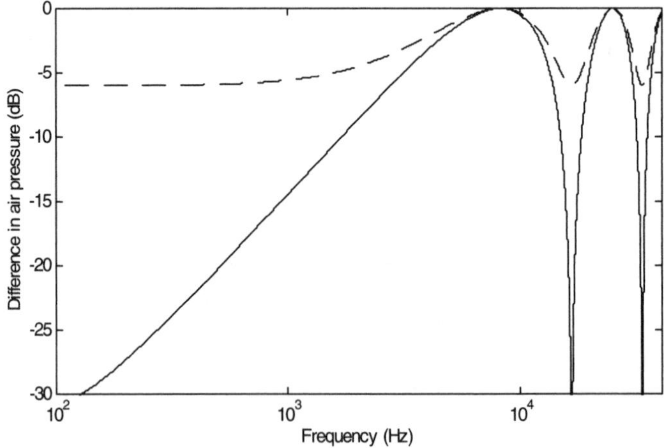

Figure 10.12 Frequency response of a bidirectional microphone with $d = 1\,\text{cm}$ at 0° obtained through Eq. (10.30). The larger the distance between plates, the lower the frequency of the maxima. The highest values are obtained for 8250 Hz and 24,750 Hz and the null for 16,500 Hz. The solid line corresponds to far field conditions ($\lambda = 0.02$) and the dotted line to near field conditions ($\lambda = 0.5$).

It can be shown, after taking the derivative of $G(f,0)$ in Eq. (10.30) and equating to zero, that the maxima are given by

$$f_n = \frac{c}{4d}(2n-1) \tag{10.31}$$

with $n = 1, 2, \cdots$. We can observe from Eq. (10.31) that the larger the width of the diaphragm, the lower the first maximum.

The increase in frequency response, or sensitivity, in the near field, compared to the far field, is a measure of noise cancellation. Consequently the microphone is said to be noise canceling. The microphone is also referred to as a differential or gradient microphone, since it measures the gradient (difference) in sound pressure between two points in space. The boost in low-frequency response in the near field is also referred to as the *proximity effect*, often used by singers to boost their bass levels by getting the microphone closer to their mouths.

By evaluating Eq. (10.30) it can be seen that low-frequency sounds in a bidirectional microphone are not reproduced as well as higher frequencies, leading to a *thin* sounding mike.

Let's interpret Figure 10.12. The net sound pressure between these two points, separated by a distance $D = 2d$, is influenced by two factors: phase shift and inverse square law.

The influence of the sound-wave phase shift is less at low frequencies than at high, because the distance D between the front and rear port entries becomes a small fraction of the low-frequency wavelength. Therefore, there is little phase shift between the ports at low frequencies, as the opposite sides of the diaphragm receive nearly equal amplitude and phase. The result is slight diaphragm motion and a weak microphone output signal. At higher frequencies, the distance D between sound ports becomes a larger fraction of the wavelength. Therefore, more phase shift exists across the diaphragm. This causes a higher microphone output.

The pressure difference caused by phase shift rises with frequency at a rate of 20 dB per decade. As the frequency rises to where the microphone port spacing D equals half a wavelength, the net pressure is at its maximum. In this situation, the diaphragm movement is also at its maximum, since the front and rear see equal amplitude but in opposite polarities of the wave front. This results in a peak in the microphone frequency response, as illustrated in Figure 10.12. As the frequency continues to rise to where the microphone port spacing D equals one complete wavelength, the net pressure is at its minimum. Here, the diaphragm does not move at all, since the front and rear sides see equal amplitude at the same polarity of the wave front. This results in a dip in the microphone frequency response, as shown in Figure 10.12.

A second factor creating a net pressure difference across the diaphragm is the impact of the inverse square law. If the sound-pressure difference between the front and rear ports of a noise-canceling microphone were measured near the sound source and again further from the source, the near field measurement would be greater than the far field. In other words, the microphone's net pressure difference and, therefore, output signal, is greater in the near sound field than in the far field. The inverse-square-law effect is independent of frequency. The net pressure that causes the diaphragm to move is a combination of both the phase shift and inverse-square-law effect. These two factors influence the frequency response of the microphone differently, depending on the distance to the sound source. For distant sound, the influence of the net pressure difference from the inverse-square-law effect is weaker than the phase-shift effect; thus, the rising 20-dB-per-decade frequency response dominates the total frequency response. As the microphone is moved closer to the sound source, the influence of the net pressure difference from the inverse square law is greater than that of the phase shift; thus the total microphone frequency response is largely flat.

The difference in near field to far field frequency response is a characteristic of all noise-canceling microphones and applies equally to both acoustic and electronic types.

10.2.2.3. Unidirectional Microphones

Unidirectional microphones are designed to pick-up the speaker's voice by directing the audio reception toward the speaker, focusing on the desired input and rejecting sounds emanating from other directions that can negatively impact clear communications, such as computer noise from fans or other sounds.

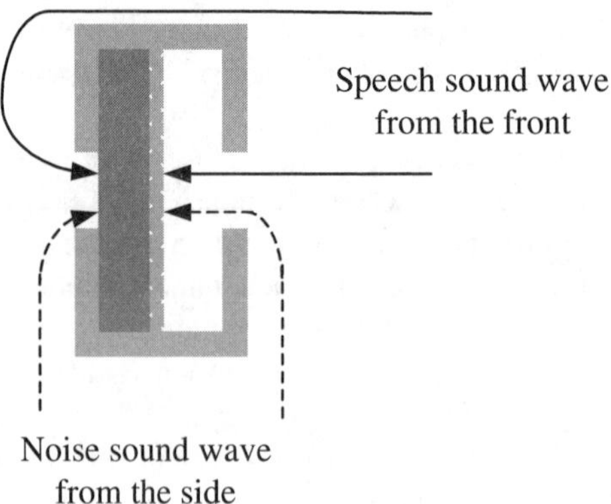

Figure 10.13 Cross section of a unidirectional microphone.

Figure 10.13 shows the cross-section of a unidirectional microphone, which also relies upon the principles of a gradient microphone. Notice that the unidirectional mic looks similar to the bidirectional, except that there is a resistive material (often cloth or foam) between the diaphragm and the opening of one end. The material's resistive properties *slow down* the pressure on its path from the back opening to the diaphragm. If the additional delay through the back plate is given by τ_0, the gain can be given by

$$G(f,\theta)=\frac{e^{-j2\pi\left|e^{j\theta}-\lambda\right|\tau f}}{\left|e^{j\theta}-\lambda\right|}-\frac{e^{-j2\pi\left(\tau_0+\left|e^{j\theta}+\lambda\right|\tau\right)f}}{\left|e^{j\theta}+\lambda\right|} \tag{10.32}$$

which was obtained by modifying Eq. (10.30). Unidirectional microphones have the greatest response to sound waves impinging from one direction, typically referred to as the front, or major axis of the microphone. One typical response of a unidirectional microphone is the *cardioid* pattern shown in the polar plot of Figure 10.14, plotted from Eq. (10.32). The frequency response at 0° is similar to that of Figure 10.12. Because the cardioid pattern of polar response is so popular among them, unidirectional mikes are often referred to as cardioid mikes.

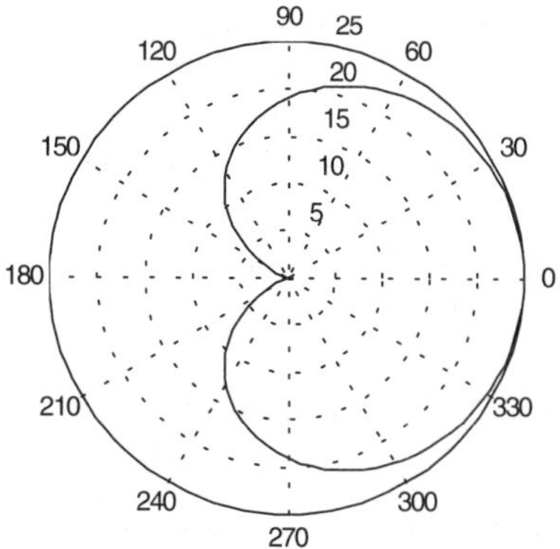

Figure 10.14 Polar response of a unidirectional microphone. The polar response was obtained through Eq. (10.32) with d = 1 cm, r = 50 cm, c = 33,000 cm/s, f = 1 kHz, and $\tau_0 = 0.06$ ms.

Equation (10.32) was derived under a simplified schematic based on Figure 10.10, which is an idealized model so that, in practice, the polar response of a real microphone has to be measured empirically. The frequency response and polar pattern of a commercial microphone are shown in Figure 10.15.

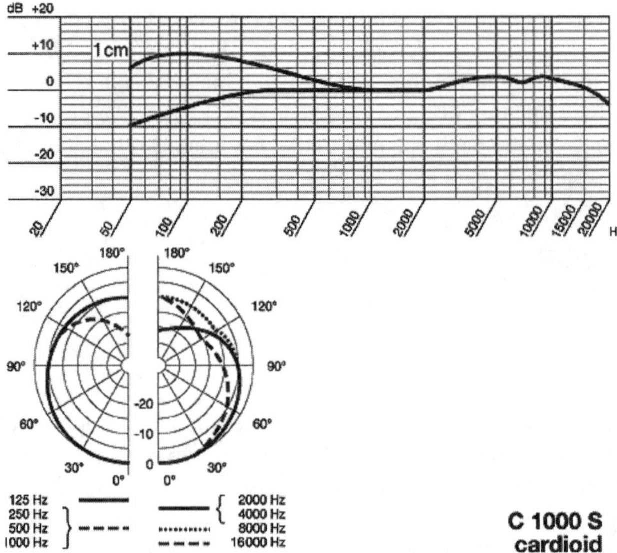

Figure 10.15 Characteristics of an AKG C1000S cardioid microphone: (top) frequency response for near and far field conditions (note the proximity effect) and (bottom) polar pattern for different frequencies.

Although this noise cancellation decreases the overall response to sound pressure (*sensitivity*) of the microphone, the directional and frequency-response improvements far outweigh the lessened sensitivity. It is particularly well suited for use as a desktop mic or as part of an embedded microphone in a laptop or desktop computer. Unidirectional microphones achieve superior noise-rejection performance over omnidirectionals. Such performance is necessary for clean audio input and for audio signal processing algorithms such as acoustic echo cancellation, which form the core of speakerphone applications.

10.2.3. Other Transduction Categories

In a *passive microphone*, sound energy is directly converted to electrical energy, whereas an *active microphone* requires an external energy source that is modulated by the sound wave. Active transducers thus require phantom power, but can have higher sensitivity.

We can also classify microphones according to the physical property to which the sound wave responds. A *pressure microphone* has an electrical response that corresponds to the pressure in a sound wave, while a *pressure gradient microphone* has a response corresponding to the difference in pressure across some distance in a sound wave. A pressure microphone is a fine reproducer of sound, but a gradient microphone typically has a response greatest in the direction of a desired signal or talker and rejects undesired background sounds. This is particularly beneficial in applications that rely upon the reproduction of only a desired signal, where any undesired signal entering the reproduction severely degrades performance. Such is the case in voice recognition or speakerphone applications.

In terms of the mechanism by which they create an electrical signal corresponding to the sound wave they detect, microphones are classified as *electromagnetic*, *electrostatic*, and *piezoelectric*. *Dynamic* microphones are the most popular type of electromagnetic microphone and *condenser* microphones the most popular type of electrostatic microphone.

Electromagnetic microphones induce voltage based on a varying magnetic field. *Ribbon microphones* are a type of electromagnetic microphones that employ a thin metal ribbon suspended between the poles of a magnet. *Dynamic* microphones are electromagnetic microphones that employ a moving coil suspended by a light diaphragm (see Figure 10.16), acting like a speaker but in reverse. The diaphragm moves with changes in sound pressure, which in turns moves the coil, which causes current to flow as lines of flux from the magnet are cut. Dynamic microphones need no batteries or power supply, but they deliver low signal levels that need to be preamplified.

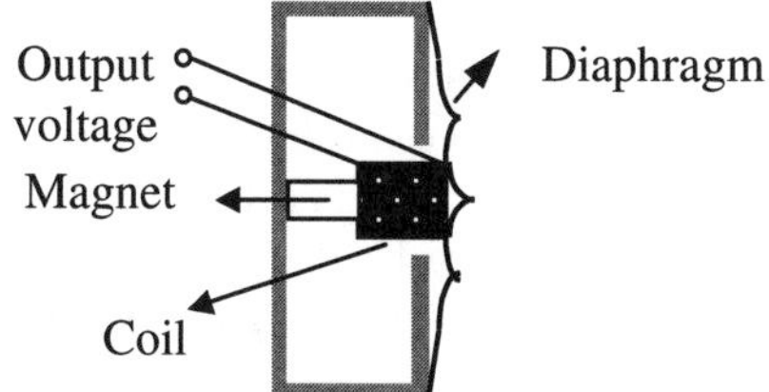

Figure 10.16 Dynamic microphone schematics.

Piezoresistive and piezoelectric microphones are based on the variation of electric resistance of their sensor induced by changes in sound pressure. *Carbon button microphones* consist of a small cylinder packed with tiny granules of carbon that, when compacted by sound pressure, reduce the electric resistance. Such microphones, often used in telephone handsets, offer a worse frequency response than condenser microphones, and lower dynamic range.

10.3. ADAPTIVE ECHO CANCELLATION (AEC)

If a spoken language system allows the user to talk while speech is being output through the loudspeakers, the microphone picks up not only the user's voice, but also the speech from the loudspeaker. This problem may be avoided with a *half-duplex* system that does not listen when a signal is being played through the loudspeaker, though such systems offer an unnatural user experience. On the other hand, a *full-duplex* system that allows *barge-in* by the user to interrupt the system offers a better user experience. For barge-in to work, the signal played through the loudspeaker needs to be canceled. This is achieved with echo cancellation (see Figure 10.17), as discussed in this section.

In hands-free conferencing the local user's voice is output by the remote loudspeaker, whose signal is captured by the remote microphone and after some delay is output by the local loudspeaker. People are tolerant to these echoes if either they are greatly attenuated or the delay is short. Perceptual studies have shown that the longer the delay, the greater the attenuation needed for user acceptance.

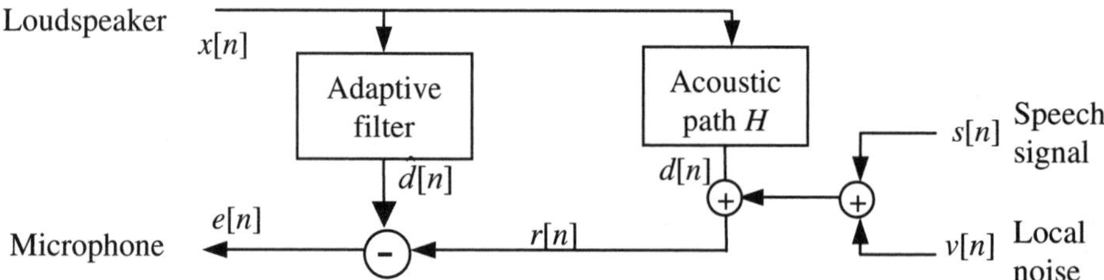

Figure 10.17 Block diagram of an echo-canceling application. $x[n]$ represents the signal from the loudspeaker, $s[n]$ the speech signal, $v[n]$ the local background noise, and $e[n]$ the signal that goes to the microphone.

The use of *echo cancellation* is mandatory in telephone communications and hands-free conferencing when it is desired to have full-duplex voice communication. This is particularly important when the call is routed through a satellite that can have delays larger than 200 ms. A block diagram is shown in Figure 10.18.

In Figure 10.17, the return signal $r[n]$, assuming no local noise, is the sum

$$r[n] = d[n] + s[n] \tag{10.33}$$

where $s[n]$ is the speech signal and $d[n]$ is the attenuated and possibly distorted version of the loudspeaker's signal $x[n]$. The purpose of the echo canceler is to remove the echo $d[n]$ from the return signal $r[n]$, which is done by means of an adaptive FIR filter whose coefficients are computed to minimize the energy of the canceled signal $e[n]$. The filter coefficients are reestimated *adaptively* to track slowly changing line conditions.

This problem is essentially that of adaptive filtering *only* when $s[n] = 0$, or in other words when the user is silent. For this reason, you have to implement a *double-talk detection* module that detects when the speaker is silent. This is typically feasible because the echo $d[n]$ is usually small, and if the return signal $r[n]$ has high energy it means that the user is

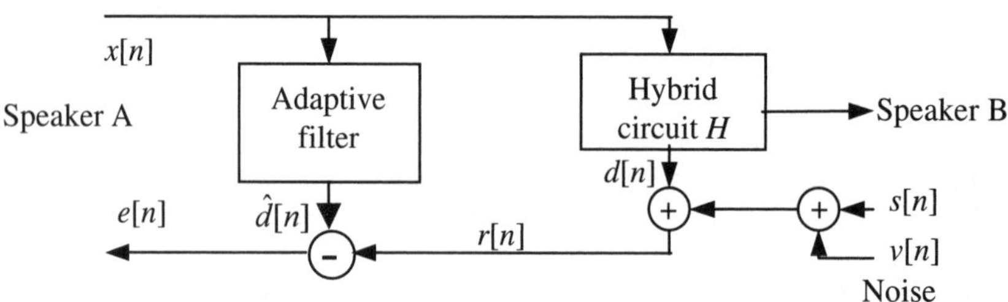

Figure 10.18 Block diagram of echo canceling for a telephone communication. $x[n]$ represents the remote call signal, $s[n]$ the local outgoing signal. The hybrid circuit H does a 2-4 wire conversion and is nonideal because of impedance mismatches.

not silent. Errors in double-talk detection result in divergence of the filter, so it is generally preferable to be conservative in the decision and when in doubt not adapt the filter coefficients. Initialization could be done by sending a known signal with white spectrum. The quality of the filtering is measured by the so-called *echo-return loss enhancement* (ERLE):

$$ERLE(dB) = 10\log_{10}\frac{E\{d^2[n]\}}{E\{(d[n]-\hat{d}[n])^2\}} \tag{10.34}$$

The filter coefficients are chosen to maximize the ERLE. Since the telephone-line characteristics, or the acoustic path (due to speaker movement), can change over time, the filter is often adaptive. Another reason for adaptive filters is that reliable ERLE maximization requires a large number of samples, and such a delay is not tolerable.

In the following sections, we describe the fundamentals of adaptive filtering. While there are some nonlinear adaptive filters, the vast majority are linear FIR filters, with the LMS algorithm being the most important. We introduce the LMS algorithm, study its convergence properties, and present two extensions: the normalized LMS algorithm and transform-domain LMS algorithms.

10.3.1. The LMS Algorithm

Let's assume that a desired signal $d[n]$ is generated from an input signal $x[n]$ as follows

$$d[n] = \sum_{k=0}^{L-1} g_k x[n-k] + u[n] = \mathbf{G}^T\mathbf{X}[n] + u[n] \tag{10.35}$$

with $\mathbf{G} = \{g_0, g_1, \cdots g_{L-1}\}$, the input signal vector $\mathbf{X}[n] = \{x[n], x[n-1], \cdots x[n-L+1]\}$, and $u[n]$ being noise that is independent of $x[n]$.

We want to estimate $d[n]$ in terms of the sum of previous samples of $x[n]$. To do that we define the estimate signal $y[n]$ as

$$y[n] = \sum_{k=0}^{L-1} w_k[n]x[n-k] = \mathbf{W}^T[n]\mathbf{X}[n] \tag{10.36}$$

where $\mathbf{W}[n] = \{w_0[n], w_1[n], \cdots w_{L-1}[n]\}$ is the time-dependent coefficient vector. The instantaneous error between the desired and the estimated signal is given by

$$e[n] = d[n] - \mathbf{W}^T[n]\mathbf{X}[n] \tag{10.37}$$

The *least mean square* (LMS) algorithm updates the value of the coefficient vector in the steepest descent direction

$$\mathbf{W}[n+1] = \mathbf{W}[n] + \varepsilon e[n]\mathbf{X}[n] \tag{10.38}$$

where ε is the step size. This algorithm is very popular because of its simplicity and effectiveness [58].

10.3.2. Convergence Properties of the LMS Algorithm

The choice of ε is important: if it is too small, the adaptation rate will be slow and it might not even track the nonstationary trends of $x[n]$, whereas if ε is too large, the error might actually increase. We analyze the conditions under which the LMS algorithm converges.

Let's define the error in the coefficient vector $\mathbf{V}[n]$ as

$$\mathbf{V}[n] = \mathbf{G} - \mathbf{W}[n] \tag{10.39}$$

and combine Eqs. (10.35), (10.37), (10.38), and (10.39) to obtain

$$\mathbf{V}[n+1] = \mathbf{V}[n] - \varepsilon \mathbf{X}[n]\mathbf{X}^T[n]\mathbf{V}[n] - \varepsilon u[n]\mathbf{X}[n] \tag{10.40}$$

Taking expectations in Eq. (10.40) results in

$$E\{\mathbf{V}[n+1]\} = E\{\mathbf{V}[n]\} - \varepsilon E\{\mathbf{X}[n]\mathbf{X}^T[n]\mathbf{V}[n]\} \tag{10.41}$$

where we have assumed that $u[n]$ and $x[n]$ are independent and that either is a zero-mean process. Finally, we express the autocorrelation of $\mathbf{X}[n]$ as

$$\mathbf{R}_{xx} = E\{\mathbf{X}[n]\mathbf{X}^T[n]\} = \mathbf{Q}\Lambda\mathbf{Q}^T \tag{10.42}$$

where $\mathbf{Q}$ is a matrix of its eigenvectors and Λ is a diagonal matrix of its eigenvalues $\{\lambda_0, \lambda_1, \cdots, \lambda_{L-1}\}$, which are all real valued because of the symmetry of $\mathbf{R}_{xx}$.

Although we know that $\mathbf{X}[n]$ and $\mathbf{V}[n]$ are not statistically independent, we assume in this section that they are, so that we can obtain some insight on the convergence properties. With this assumption, Eq. (10.41) can be expressed as

$$E\{\mathbf{V}[n+1]\} = E\{\mathbf{V}[n]\}(1 - \varepsilon \mathbf{R}_{xx}) \tag{10.44}$$

which, applied recursively, leads to

$$E\{\mathbf{V}[n+1]\} = E\{\mathbf{V}[0]\}(1 - \varepsilon \mathbf{R}_{xx})^n \tag{10.45}$$

Using Eqs. (10.39) and (10.42) in (10.45), we can express the $(i+1)$th element of $E\{\mathbf{W}[n]\}$ as

$$E\{w_i[n]\} = g_i + \sum_{j=0}^{L-1} q_{ij}(1-\varepsilon\lambda_j)^n E\{\tilde{v}_i[0]\} \tag{10.46}$$

where q_{ij} is the $(i+1, j+1)$th element of the eigenvector matrix $\mathbf{Q}$, and $\tilde{v}_i[n]$ is the rotated coefficient error vector defined as

$$\tilde{\mathbf{V}}[n] = \mathbf{Q}^T\mathbf{V}[n] \tag{10.47}$$

From Eq. (10.46) we see that the mean value of the LMS filter coefficients converges exponentially to the true value if

$$0 < \varepsilon < 1/\lambda_j \tag{10.48}$$

so that the adaptation constant ε must be determined from the largest eigenvalue of $\mathbf{X}[n]$ for the mean LMS algorithm to converge.

10.3.3. Normalized LMS Algorithm

In practice, mean convergence doesn't tell us the nature of the fluctuations that the coefficients experience. Analysis of the variance of $\mathbf{V}[n]$ together with some more approximations result in mean-squared convergence if

$$0 < \varepsilon < \frac{K}{L\sigma_x^2} \tag{10.49}$$

with $\sigma_x^2 = E\{x^2[n]\}$ being the input signal power and K a constant that depends weakly on the nature of the input signal statistics but not on its power.

Because of the inaccuracies of the independence assumptions above, a rule of thumb used in practice to determine the adaptation constant ε is

$$0 < \varepsilon < \frac{0.1}{L\sigma_x^2} \tag{10.50}$$

The choice of largest value for ε in Eq. (10.49) makes the LMS algorithm track nonstationary variations in x fastest, and achieve faster convergence. On the other hand, the misadjustment of the filter coefficients increases as both the filter length L and adaptation constant ε increase. For this reason, often the adaptation constant can be made a function of

n ($\varepsilon[n]$), with larger values at first and smaller values once convergence has been determined.

The *normalized LMS algorithm* (NLMS) uses the result of Eq. (10.49) and, therefore, defines a normalized step size

$$\varepsilon[n] = \frac{\varepsilon}{\delta + L\hat{\sigma}_x^2[n]} \tag{10.51}$$

where the constant δ avoids a division by 0 and $\hat{\sigma}_x^2[n]$ is an estimate of the input signal power, which is typically done with an exponential window

$$\hat{\sigma}_x^2[n] = (1-\beta)\hat{\sigma}_x^2[n-1] + \beta x^2[n] \tag{10.52}$$

or a sliding rectangular window

$$\hat{\sigma}_x^2[n] = \frac{1}{N}\sum_{i=0}^{N-1} x^2[n-i] = \hat{\sigma}_x^2[n-1] + \frac{1}{N}\left(x^2[n] - x^2[n-N]\right) \tag{10.53}$$

where both β and N control the effective memory of the estimators in Eqs. (10.52) and (10.53), respectively. Finally, we need to pick ε so that $0 < \varepsilon < 2$ to assure convergence. Choice of the NLMS algorithm simplifies the selection of ε, and the NLMS often converges faster than the LMS algorithm in practical situations.

10.3.4. Transform-Domain LMS Algorithm

As discussed in Section 10.3.2, convergence of the LMS algorithm is determined by the largest eigenvalue of the input. Since complex exponentials are approximate eigenvectors for LTI systems, the LMS algorithm's convergence is dominated by the frequency band with largest energy, and convergence in other frequency bands is generally much slower. This is the rationale for the *subband LMS algorithm,* which performs independent LMS algorithms for different frequency bands, as proposed by Boll [14].

The *block LMS* (BLMS) algorithm keeps the coefficients unchanged for a block k of L samples

$$\mathbf{W}[k+1] = \mathbf{W}[k] + \varepsilon \sum_{m=0}^{L-1} e[kL+m]\mathbf{X}[kL+m] \tag{10.54}$$

which is represented by a linear convolution and therefore can be implemented efficiently using length-$2N$ FFTs according to overlap-save method of Figure 10.19. Notice that implementing a linear convolution with a circular convolution operator such as the FFT requires the use of the dashed box.

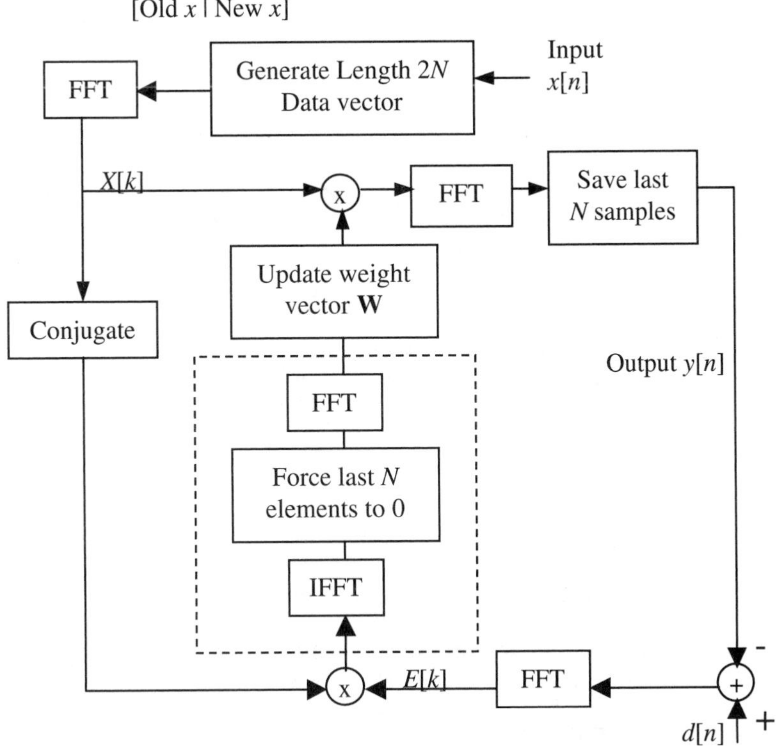

Figure 10.19 Block diagram of the constrained frequency-domain block LMS algorithm. The unconstrained version of this algorithm eliminates the computation inside the dashed box.

An unconstrained frequency-domain LMS algorithm can be implemented by removing the constraint in Figure 10.19, therefore implementing a circular instead of a linear convolution. While this is not exact, the algorithm requires only three FFTs instead of five. In some practical applications, there is no difference in convergence between the constrained and unconstrained cases.

10.3.5. The RLS Algorithm

The search for the optimum filter can be accelerated when the gradient vector is properly deviated toward the minimum. This approach uses the Newton-Raphson method to iteratively compute the root of $f(x)$ (see Figure 10.20) so that the value at iteration $i + 1$ is given by

$$x_{i+1} = x_i - \frac{f(x_i)}{f'(x_i)} \tag{10.55}$$

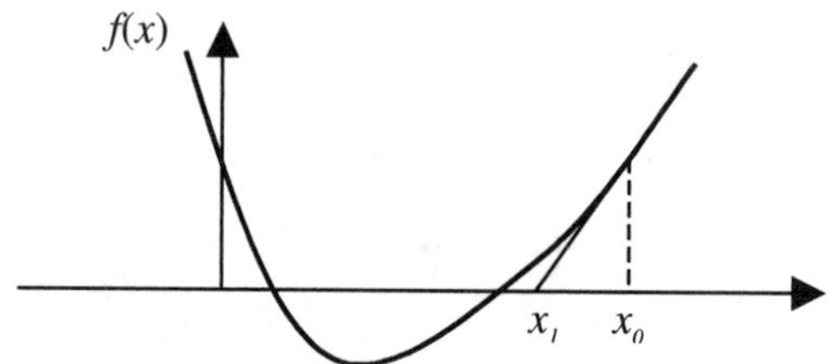

Figure 10.20 Newton-Raphson method to compute the roots of a function.

To minimize function $f(x)$ we thus compute the roots of $f'(x)$ through the above method:

$$x_{i+1} = x_i - \frac{f'(x_i)}{f''(x_i)} \tag{10.56}$$

In the case of a vector, Eq. (10.56) is transformed into

$$\mathbf{w}_{i+1} = \mathbf{w}_i - \varepsilon[n]\left(\nabla^2 e(\mathbf{w}_i)\right)^{-1} \nabla e(\mathbf{w}_i) \tag{10.57}$$

where we add a step size $\varepsilon[n]$, and where $\nabla^2 e(\mathbf{w}_i)$ is the *Hessian* of the least-squares function which, for Eq. (10.37), equals the autocorrelation of $\mathbf{x}$:

$$\nabla^2 e(\mathbf{w}_i) = \mathbf{R}[n] = E\{\mathbf{x}[n]\mathbf{x}^T[n]\} \tag{10.58}$$

The *recursive least squares* (RLS) algorithm specifies a method of estimating Eq. (10.58) using an exponential window:

$$\mathbf{R}[n] = \lambda\mathbf{R}[n-1] + \mathbf{x}[n]\mathbf{x}^T[n] \tag{10.59}$$

While the RLS algorithm converges faster than the LMS algorithm, it also is more computationally expensive, as it requires a matrix inversion for every sample. Several algorithms have been derived to speed it up [54].

10.4. MULTIMICROPHONE SPEECH ENHANCEMENT

The use of more than one microphone is motivated by the human auditory system, in which the use of both ears has been shown to enhance detection of the direction of arrival, as well as increase SNR when one ear is covered. The methods the human auditory system uses to accomplish this task are still not completely known, and the techniques described in this section do not mimic that behavior.

Microphone arrays use multiple microphones and knowledge of the microphone locations to predict delays and thus create a beam that focuses on the direction of the desired speaker and rejects signals coming from other angles. Reverberation, as discussed in Section 10.1.2, can be combated with these techniques. Blind source separation techniques are another family of statistical techniques that typically do not use spatial constraints, but rather statistical independence between different sources.

While in this section we describe only linear processing, i.e., the output speech is a linearly filtered version of the microphone signals, we could also combine these techniques with the nonlinear methods of Section 10.5.

10.4.1. Microphone Arrays

The goals of microphone arrays are twofold: finding the position of a sound source in a room, and improving the SNR of the received signal. *Steering* is helpful in videoconferencing, where a camera has to follow the current speaker. Since the speaker is typically far away from the microphone, the received signal likely contains a fair amount of additive noise. Microphone arrays can also be used to increase the SNR.

Let $x[n]$ be the signal at the source S. Microphone i picks up a signal

$$y_i[n] = x[n] * g_i[n] + v_i[n] \tag{10.60}$$

that is a filtered version of the source plus additive noise $v_i[n]$. If we have N such microphones, we can attempt to recover $s[n]$ because all the signals $y_i[n]$ should be correlated.

A typical assumption made is that all the filters $g_i[n]$ are delayed versions of the same filter $g[n]$

$$g_i[n] = g[n - D_i] \tag{10.61}$$

with the delay $D_i = d_i / c$, d_i being the distance between the source S and microphone i, and c the speed of sound in air. We cannot recover signal $x[n]$ without knowledge of $g[n]$ or the signal itself, so the goal is to obtain the filtered signal $y[n]$

$$y[n] = x[n] * g[n] \tag{10.62}$$

so that, combining Eqs. (10.60), (10.61), and (10.62),

$$y_i[n] = y[n - D_i] + v_i[n] \tag{10.63}$$

Assuming $v_i[n]$ are independent and Gaussianly distributed, the optimal estimate of $x[n]$ is given by

$$\tilde{y}[n] = \frac{1}{N} \sum_{i=0}^{N-1} y_i[n + D_i] = y[n] + v[n] \tag{10.64}$$

which is the so-called *delay-and-sum beamformer* [24, 29], where the residual noise $v[n]$

$$v[n] = \frac{1}{N}\sum_{i=0}^{N-1} v_i[n+D_i] \tag{10.65}$$

has a variance that decreases as the number of microphones N increases, since the noises $v_i[n+D_i]$ are uncorrelated.

Equation (10.64) requires estimation of the delays D_i. To attenuate the additive noise $v[n]$, it is not necessary to identify the absolute delays, but rather the delays relative to one reference microphone (for example, the center microphone). It can be shown that the maximum likelihood solution consists in maximizing the energy of $\tilde{y}[n]$ in Eq. (10.64), which is the sum of cross-correlations:

$$D_i = \arg\max_{D_i} \left(\sum_{i=0}^{N-1} \sum_{j=0}^{N-1} R_{ij}[D_i - D_j] \right) \qquad 0 \le i < N \tag{10.66}$$

This approach assumes that we know nothing about the geometry of the microphone placement. In fact, given a point source and assuming no reflections, we can compute the delay based on the distance between the source and the microphone. The use of geometry allows us to reduce the number of parameters to estimate from (N - 1) to a maximum of 3, in case we desire to estimate the exact location. This location is often described in spherical coordinates (φ, θ, ρ) with φ being the direction of arrival, θ the elevation angle, and ρ the distance to the reference microphone, as shown in Figure 10.21.

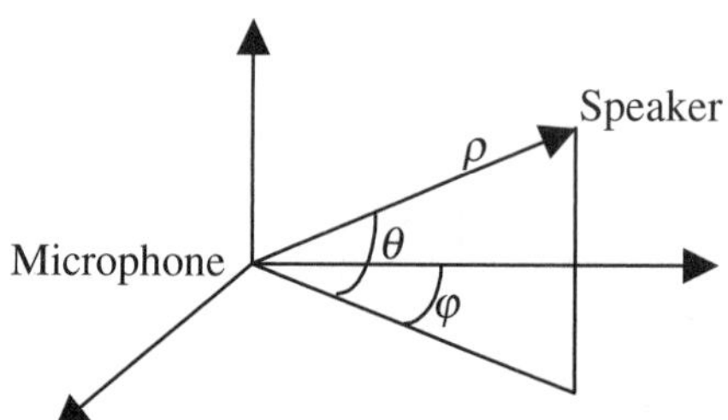

Figure 10.21 Spherical coordinates (φ, θ, ρ) with φ being the direction of arrival, θ the elevation angle, and ρ the distance to the reference microphone.

While 2-D and 3-D microphone configurations can be used, which would allow us to determine not just the steering angle φ, but also distance to the origin ρ and azimuth θ, linear microphone arrays are the most widely used configurations because they are the simplest. In a linear array all the microphones are placed on a line (see Figure 10.22). In this case, we cannot determine the elevation angle θ. To determine both φ and ρ we need at least two microphones in the array.

If the microphones are relatively close to each other compared to the distance to the source, the angle of arrival φ is approximately the same for all signals. With this assumption, the normalized delay $\bar{D}_i$ with respect to the reference microphone is given by

$$\bar{D}_i = -a_i \sin(\varphi)/c \tag{10.67}$$

where a_i is the *y*-axis coordinate in Figure 10.22 for microphone *i*, where the reference microphone has $a_0 = 0$ and also $\bar{D}_0 = 0$.

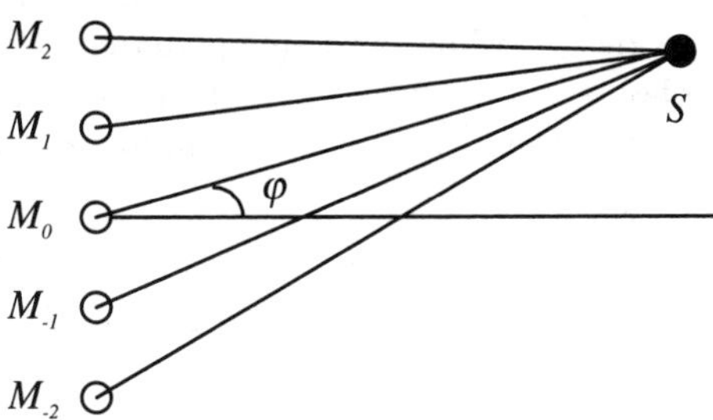

Figure 10.22 Linear microphone array (five microphones). The source signal arrives at each microphone with a different delay, which allows us to find the correct angle of arrival.

With approximation, we define $\bar{D}_i(\varphi)$, the relative delay of the signal at microphone *i* to the reference microphone, as a function of the direction of arrival angle φ and independent of ρ. The optimal direction of arrival φ is then that which maximizes the energy of the estimated signal $\tilde{x}[n]$ over a set of samples

$$\begin{aligned}\varphi &= \arg\max_{\varphi} \sum_n \left(\frac{1}{N} \sum_{i=0}^{N-1} y_i[n + \bar{D}_i(\varphi)] \right)^2 \\ &= \arg\max_{\varphi} \sum_n \left(\frac{1}{N} \sum_{i=0}^{N-1} y_i[n - \frac{a_i}{c} \sin(\varphi)] \right)^2\end{aligned} \tag{10.68}$$

The term *beamforming* entails that this array favors a specific direction of arrival φ and that sources arriving from other directions are not in phase and therefore are attenuated. Since the source can move over time, maximization of Eq. (10.68) can be done in an adaptive fashion.

As the beam is steered away from the broadside, the system exhibits a reduction in spatial discrimination because the beam pattern broadens. Furthermore, beamwidth varies with frequency, so an array has an approximate bandwidth given by the upper f_u and lower f_l frequencies

$$\begin{aligned}f_u &= \frac{c}{d \max_{\phi,\phi'} |\cos\varphi - \cos\varphi'|} \\ f_l &= \frac{f_u}{N}\end{aligned} \tag{10.69}$$

with d being the sensor spacing, φ' the steering angle measured with respect to the axis of the array, and φ the direction of the source. For a desired range of $\pm 30^\circ$ and five sensors spaced 5 cm apart, the range is approximately 880 to 4400 Hz. We see in Figure 10.23 that at very low frequencies the response is essentially omnidirectional, since the microphone spacing is small compared to the large wavelength. At high frequencies more lobes start appearing, and the array steers toward not only the preferred direction but others as well. For speech signals, the upshot is that we either need a lot of microphones to provide a directional polar pattern at low frequencies, or we need them to be spread far enough apart, or both.

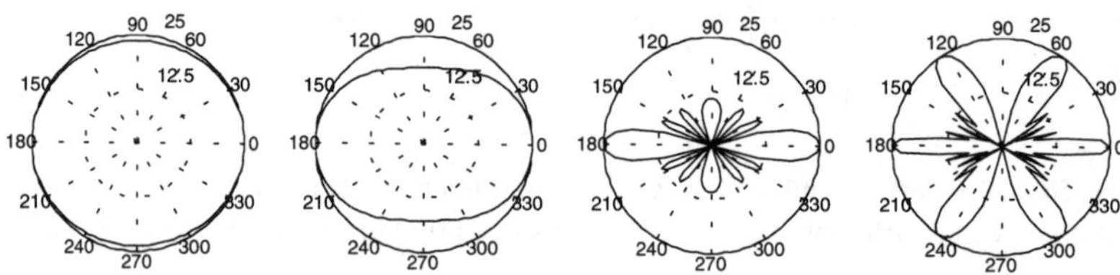

Figure 10.23 Polar pattern of a microphone array with steering angle of $\varphi' = 0$, five microphones spaced 5 cm apart for 400, 880, 4400, and 8000 Hz from left to right, respectively, for a source located at 5 m.

The polar pattern in Figure 10.23 was computed as follows:

$$P(f, r, \varphi) = \sum_{i=1}^{N} \frac{e^{-j2\pi f\left[a_i \sin\varphi' + \left|re^{j\varphi} - ja_i\right|\right]/c}}{\left|re^{j\varphi} - ja_i\right|} \tag{10.70}$$

though the sensors could be spaced nonuniformly, as in Figure 10.24, allowing for better behavior across the frequency spectrum.

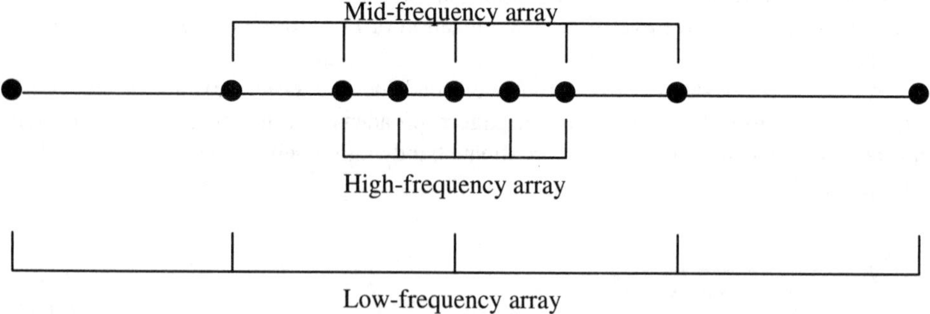

Figure 10.24 Nonuniform linear microphone array containing three subarrays for the high, mid, and low frequencies.

Once a microphone array has been steered towars a direction φ', it attenuates noise source coming from other directions. The beamwidth depends not only on the frequency of the signal, but also on the steering direction. If the beam is steered toward a direction φ', then the direction of the source for which the beam response fall to half its power has been found empirically to be

$$\varphi_{3dB}(f) = \cos^{-1}\left\{\cos\varphi' \pm \frac{K}{Ndf}\right\} \tag{10.71}$$

with K being a constant. Equation (10.71) shows that the smaller the array, the wider the beam, and that lower frequencies yield wider beams also. Figure 10.25 shows that the bandwidth of the array when steering toward a 30° direction is lower than when steering at 0°.

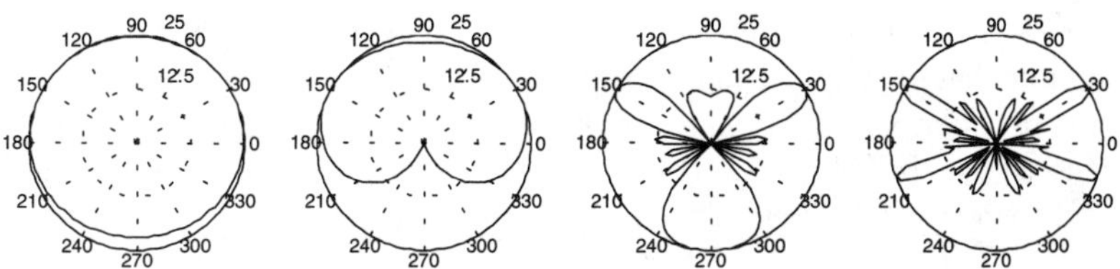

Figure 10.25 Polar pattern of a microphone array with steering angle of $\varphi' = 30^\circ$, five microphones spaced 5 cm apart for 400, 880, 3000, and 4400 Hz from left to right, respectively, for a source located at 5 m.

Microphone arrays have been shown to improve recognition accuracy when the microphones and the speaker are far apart [51]. Several companies are commercializing microphone arrays for teleconferencing or speech recognition applications.

Only in anechoic chambers does the assumption in Eq. (10.61) hold, since in practice many reflections take place, which are also different for different microphones. In addition, the assumption of a common direction of arrival for all microphones may not hold either. For this case of reverberant environments, single beamformers typically fail. While computing the direction of arrival is much more difficult in this case, the SNR can still be improved.

Let's define the desired signal $d[n]$ as that obtained in the reference microphone. We can estimate the vector $\mathbf{H}[n] = \{h_{11}, \cdots, h_{1L}, h_{21}, \cdots, h_{2L}, \cdots, h_{(N-1)1}, \cdots, h_{(N-1)L}\}$ for the (N-1) L-tap filters that minimizes the error array [25]

$$e[n] = d[n] - \mathbf{H}[n]\mathbf{Y}[n] \tag{10.72}$$

where the $(N-1)$ microphone signals are represented in the vector

$$\mathbf{Y}[n] = \{y_1[n], \cdots, y_1[n-L-1], y_2[n], \cdots, y_2[n-L-1], \cdots, y_{N-1}[n], \cdots, y_{N-1}[n-L-1]\}$$

The filter coefficients $\mathbf{G}[n]$ can be estimated through the adaptive filtering techniques described in Section 10.3. The clean signal is then estimated as

$$\hat{x}[n] = \frac{1}{2}\left(d[n] + \mathbf{H}[n]\mathbf{Y}[n]\right) \tag{10.73}$$

This last method does not assume anything about the geometry of the microphone array.

10.4.2. Blind Source Separation

The problem of separating the desired speech from interfering sources, the *cocktail party effect* [15], has been one of the holy grails in signal processing. *Blind source separation* (BSS) is a set of techniques that assume no information about the mixing process or the sources, apart from their mutual statistical independence, hence is termed blind. *Independent component analysis* (ICA), developed in the last few years [19, 38], is a set of techniques to solve the BSS problem that estimate a set of linear filters to separate the mixed signals under the assumption that the original sources are statistically independent.

Let's first consider *instantaneous mixing*. Let's assume that R microphone signals $y_i[n]$, denoted by $\mathbf{y}[n] = \left(y_1[n], y_2[n], \cdots, y_R[n]\right)$, are obtained by a linear combination of R unobserved source signals $x_i[n]$, denoted by $\mathbf{x}[n] = \left(x_1[n], x_2[n], \cdots, x_R[n]\right)$:

$$\mathbf{y}[n] = \mathbf{G}\mathbf{x}[n] \tag{10.74}$$

for all n, with $\mathbf{G}$ being the R x R mixing matrix. This mixing is termed instantaneous, since the sensor signals at time n depend on the sources at the same, but no earlier, time point. Had the mixing matrix been given, its inverse could have been applied to the sensor signals to recover the sources by $\mathbf{x}[n] = \mathbf{G}^{-1}\mathbf{y}[n]$. In the absence of any information about the mixing, the blind separation problem consists of estimating a separating matrix $\mathbf{H} = \mathbf{G}^{-1}$ from the observed microphone signals alone. The source signals can then be recovered by

$$\mathbf{x}[n] = \mathbf{H}\mathbf{y}[n] \tag{10.75}$$

We'll use here the probabilistic formulation of ICA, though alternate frameworks for ICA have been derived also [18]. Let $p_{\mathbf{x}}(\mathbf{x}[n])$ be the probability density function (pdf) of the source signals, so that the pdf of microphone signals $\mathbf{y}[n]$ is given by

$$p_{\mathbf{y}}(\mathbf{y}[n]) = |\mathbf{H}|\, p_{\mathbf{x}}(\mathbf{H}\mathbf{y}[n]) \tag{10.76}$$

and if we furthermore assume the sources $\mathbf{x}[n]$ are independent from themselves in time, $\mathbf{x}[n+i]$ $i \neq 0$, then the joint probability is given by

$$p_{\mathbf{y}}(\mathbf{y}[0], \mathbf{y}[1], \cdots, \mathbf{y}[N-1]) = \prod_{n=0}^{N-1} p_{\mathbf{y}}(\mathbf{y}[n]) = |\mathbf{H}|^N \prod_{n=0}^{N-1} p_{\mathbf{x}}(\mathbf{H}\mathbf{y}[n]) \tag{10.77}$$

whose normalized log-likelihood is given by

$$\Psi = \frac{1}{N} \ln p_{\mathbf{y}}(\mathbf{y}[0], \mathbf{y}[1], \cdots, \mathbf{y}[N-1]) = \ln |\mathbf{H}| + \frac{1}{N} \sum_{n=0}^{N-1} \ln p_{\mathbf{x}}(\mathbf{H}\mathbf{y}[n]) \tag{10.78}$$

It can be shown that

$$\frac{\partial \ln |\mathbf{H}|}{\partial \mathbf{H}} = \left(\mathbf{H}^T\right)^{-1} \tag{10.79}$$

so that that the gradient of Ψ [38] in Eq. (10.78) is given by

$$\frac{\partial \Psi}{\partial \mathbf{H}} = \left(\mathbf{H}^T\right)^{-1} + \frac{1}{N} \sum_{n=0}^{N-1} \phi(\mathbf{H}\mathbf{y}[n]) \left(\mathbf{y}[n]\right)^T \tag{10.80}$$

where $\phi(\mathbf{x})$ is expressed as

$$\phi(\mathbf{x}) = \frac{\partial \ln p_{\mathbf{x}}(\mathbf{x})}{\partial \mathbf{x}} \tag{10.81}$$

If we further assume the distribution is a zero mean Gaussian distribution with standard deviation σ, then Eq. (10.81) results in

$$\phi(\mathbf{x}) = -\frac{\mathbf{x}}{\sigma^2} \tag{10.82}$$

which inserted into Eq. (10.80) yields

$$\frac{\partial \Psi}{\partial \mathbf{H}} = \left(\mathbf{H}^T\right)^{-1} - \frac{\mathbf{H}}{\sigma^2} \left(\frac{1}{N} \sum_{n=0}^{N-1} \mathbf{y}[n] \left(\mathbf{y}[n]\right)^T \right) = \left(\mathbf{H}^T\right)^{-1} - \frac{\mathbf{H}}{\sigma^2} \mathbf{R} \tag{10.83}$$

with $\mathbf{R}$ being the matrix of cross-correlations, i.e.,

$$R_{ij} = \frac{1}{N} \sum_{n=0}^{N-1} y_i[n] y_j[n] \tag{10.84}$$

Setting Eq. (10.83) to $\mathbf{0}$ results in maximization of Eq. (10.78) under the Gaussian assumption:

$$\mathbf{H}^T \mathbf{H} = \sigma^2 \mathbf{R}^{-1} \tag{10.85}$$

which can be solved using the Cholesky decomposition described in Chapter 6.

Since σ is generally not known, it can be shown from Eq. (10.85) that the sources can be recovered only to within a scaling factor [17]. Scaling is in general not a big problem, since speech recognition systems perform automatic gain control (AGC). Moreover, the sources can be recovered to within a permutation. To see this, let's define a two-dimensional matrix $\mathbf{A}$

$$\mathbf{A}=\begin{pmatrix} 0 & 1 \\ 1 & 0 \end{pmatrix} \tag{10.86}$$

which is orthogonal:

$$\mathbf{A}^T\mathbf{A}=\mathbf{I} \tag{10.87}$$

If **H** is a solution of Eq. (10.85), then **AH** is also a solution. Thus, a permutation of the sources yields the same correlation matrix in Eq. (10.84). Although we have shown it only under the Gaussian assumption, separation up to a scaling factor and source permutation is a general result in blind source separation [17].

Unfortunately, the Gaussian assumption does not guarantee separation. To see this, we can define a two-dimensional rotation matrix **A**

$$\mathbf{A}=\begin{pmatrix} \cos\theta & -\sin\theta \\ \sin\theta & \cos\theta \end{pmatrix} \tag{10.88}$$

which is also orthogonal, so that if **H** is a solution of Eq. (10.85), then **AH** is also a solution.

The Gaussian assumption entails considering only second-order statistics, and to ensure separation we could consider higher-order statistics. Since speech signals do not follow a Gaussian distribution, we could use a Laplacian distribution, as we saw in Chapter 7:

$$p_x(x)=\frac{\beta}{2}e^{-\beta|x|} \tag{10.89}$$

which, using Eq. (10.81), results in

$$\phi(x)=\begin{cases} -\beta & x>0 \\ \beta & x<0 \end{cases} \tag{10.90}$$

and thus a nonlinear function of **H** for Eq. (10.80). Since a closed-form solution is not possible, a common solution in this case is gradient descent, where the gradient is given by

$$\frac{\partial \Psi}{\partial \mathbf{H}_n}=\left(\mathbf{H}_n^T\right)^{-1}+\phi(\mathbf{H}_n\mathbf{y}[n])(\mathbf{y}[n])^T \tag{10.91}$$

and the update formula by

$$\mathbf{H}_{n+1}=\mathbf{H}_n-\varepsilon\frac{\partial \Psi}{\partial \mathbf{H}_n}=\mathbf{H}_n-\varepsilon\left[\left(\mathbf{H}_n^T\right)^{-1}+\phi(\mathbf{H}_n\mathbf{y}[n])(\mathbf{y}[n])^T\right] \tag{10.92}$$

which is the so-called *infomax* rule [10].

Often the nonlinearity in Eq. (10.90) is replaced by a sigmoid[6] function:

$$\phi(x) = -\beta \tanh(\beta x) \tag{10.93}$$

which implies a density function

$$p_x(x) = \frac{\beta}{2\pi \cosh(\beta x)} \tag{10.94}$$

The sigmoid converges to the Laplacian as $\beta \to \infty$. Nonlinear functions in Eqs. (10.90) and (10.93) can be expanded in Taylor series so that all the moments of the observed signals are used and not just the second order, as in the case of the Gaussian assumption. These nonlinearities have been shown to be more effective in separating the sources. The use of more accurate density functions for $p_{\mathbf{x}}(\mathbf{x})$, such as a mixture of Gaussians [9], also results in nonlinear $\phi(\mathbf{x})$ functions that have shown better separation.

A problem of Eq. (10.92) is that it requires a matrix inversion at every iteration. The so-called *natural gradient* [7] was suggested to avoid this, also providing faster convergence. To do this we can multiply the gradient of Eq. (10.91) by a positive definite matrix, the inverse of the Fisher's information matrix $\mathbf{H}_n^T \mathbf{H}_n$, for example, to whiten the signal:

$$\mathbf{H}_{n+1} = \mathbf{H}_n - \varepsilon \frac{\partial \Psi}{\partial \mathbf{H}} \mathbf{H}_n^T \mathbf{H}_n \tag{10.95}$$

which, combined with Eq. (10.91), results in

$$\mathbf{H}_{n+1} = \mathbf{H}_n - \varepsilon \left[\mathbf{I} + \phi(\hat{\mathbf{x}}[n])(\hat{\mathbf{x}}[n])^T \right] \mathbf{H}_n \tag{10.96}$$

where the estimated sources are given by

$$\hat{\mathbf{x}}[n] = \mathbf{H}_n \mathbf{y}[n] \tag{10.97}$$

Notice the similarity of this approach to the RLS algorithm of Section 10.3.5. Similarly to most Newton-Raphson methods, the convergence of this approach is quadratic instead of linear as long as we are close enough to the maximum.

Another way of overcoming the lack of separation under the independent Gaussian assumption is to make use of temporal information, which we know is important for speech signals. If the model of Eq. (10.74) is extended to contain additive noise

$$\mathbf{y}[n] = \mathbf{G}\mathbf{x}[n] + \mathbf{v}[n] \tag{10.98}$$

[6] The sigmoid function can be expressed in terms of the hyperbolic tangent $\tanh(x) = \sinh(x)/\cosh(x)$, where $\sinh(x) = (e^x - e^{-x})/2$ and $\cosh(x) = (e^x + e^{-x})/2$.

we can compute the autocorrelation of $\mathbf{y}[n]$ as

$$\mathbf{R}_{\mathbf{y}}[n] = \mathbf{G}\mathbf{R}_{\mathbf{x}}[n]\mathbf{G}^T + \mathbf{R}_{\mathbf{v}}[n] \tag{10.99}$$

or, after some manipulation,

$$\mathbf{R}_{\mathbf{x}}[n] = \mathbf{H}(\mathbf{R}_{\mathbf{y}}[n] - \mathbf{R}_{\mathbf{v}}[n])\mathbf{H}^T \tag{10.100}$$

which we know must be diagonal because the sources $\mathbf{x}$ are independent, and thus $\mathbf{H}$ can be estimated to minimize the squared error of the off-diagonal terms of $\mathbf{R}_{\mathbf{x}}[n]$ for several values of n [11]. Equation (10.100) is a generalization of Eq. (10.85) when considering temporal correlation and additive noise.

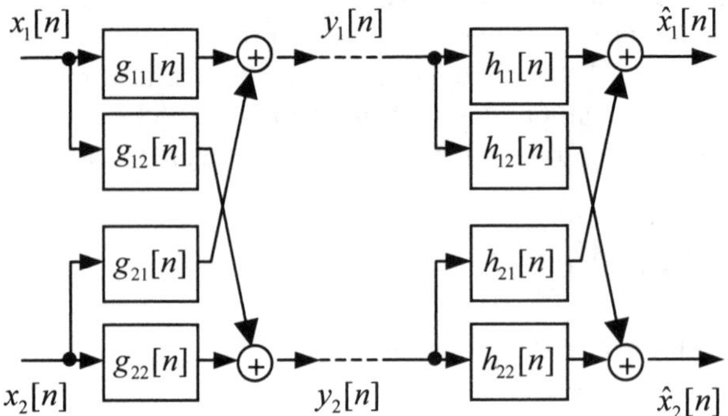

Figure 10.26 Convolutional model for the case of two microphones.

The case of instantaneous mixing is not realistic, as we need to consider the transfer functions between the sources and the microphones created by the room acoustics. It can be shown that the reconstruction filters $h_{ij}[n]$ in Figure 10.26 will completely recover the original signals $x_i[n]$ *if and only if* their z-transforms are the inverse of the z-transforms of the mixing filters $g_{ij}[n]$:

$$\begin{aligned}\begin{pmatrix} H_{11}(z) & H_{12}(z) \\ H_{21}(z) & H_{22}(z) \end{pmatrix} &= \begin{pmatrix} G_{11}(z) & G_{12}(z) \\ G_{21}(z) & G_{22}(z) \end{pmatrix}^{-1} \\ &= \frac{1}{G_{11}(z)G_{22}(z) - G_{12}(z)G_{21}(z)} \begin{pmatrix} G_{11}(z) & G_{12}(z) \\ G_{21}(z) & G_{22}(z) \end{pmatrix}\end{aligned} \tag{10.101}$$

If the matrix in Eq. (10.101) is not invertible, separability is impossible. This can happen if both microphones pick up the same signal, which could happen if either the two microphones are too close to each other or the two sources are too close to each other. It's reasonable to assume the mixing filters $g_{ij}[n]$ to be FIR filters, whose length will generally depend on the reverberation time, which in turn depends on the room size, microphone position, wall absorbance, and so on. In general this means that the reconstruction filters $h_{ij}[n]$ have an infinite impulse response. In addition, the filters $g_{ij}[n]$ may have zeroes outside the unit circle, so that perfect reconstruction filters would need to have poles outside the unit circle. For this reason it is not possible, in general, to recover the original signals exactly.

In practice, it's convenient to assume such filters to be FIR of length q, which means that the original signals $x_1[n]$ and $x_2[n]$, will not be recovered exactly. Thus the problem consists in estimating the reconstruction filters $h_{ij}[n]$ directly from the microphone signals $y_1[n]$ and $y_2[n]$, so that the estimated signals $\hat{x}_i[n]$ are as close as possible to the original signals. Often we are satisfied if the resulting signals are separated, even if they contain some amount of reverberation.

An approach commonly used to combat this problem consists of taking a filterbank and assuming instantaneous mixing within each filter [38]. This approach can separate real sources much more effectively, but it suffers from the problem of permutations, which in this case is more severe because frequencies from different sources can be mixed together. To avoid this, we may need a probabilistic model of the sources that takes into account correlations across frequencies [3]. Another problem occurs when the number of sources is larger than the number of microphones.

10.5. ENVIRONMENT COMPENSATION PREPROCESSING

The goal of this section is to present a number of techniques used to *clean up* the signal of additive noise and/or channel distortions prior to the speech recognition system. Although the techniques presented here are developed for the case of one microphone, they can be generalized to the case where several microphones are available using the approaches described in Section 10.4. These techniques can also be used to *enhance* the signal captured with a speakerphone or a desktop microphone in teleconferencing applications.

Since the use of human auditory system is so robust to changes in acoustical environment, many researchers have attempted to develop signal processing schemes that mimic the functional organization of the peripheral auditory system [27, 49]. The PLP cepstrum described in Chapter 6 has also been shown to be very effective in combating noise and channel distortions [60].

Another alternative is to consider the feature vector as an integral part of the recognizer, and thus researchers have investigated its design so as to maximize recognition accuracy, as discussed in Chapter 9. Such approaches include LDA [34] and neural networks [45]. These discriminatively trained features can also be optimized to operate better under noisy conditions, thus possibly beating the standard mel-cepstrum, especially when several independent features are combined [50]. The mel-cepstrum is the most popular feature vector for speech recognition. In this context we present a number of techniques that have been proposed over the years to compensate for the effects of additive noise and channel distortions on the cepstrum.

10.5.1. Spectral Subtraction

The basic assumption in this section is that the desired clean signal $x[m]$ has been corrupted by additive noise $n[m]$:

$$y[m] = x[m] + n[m] \tag{10.102}$$

and that both $x[m]$ and $n[m]$ are statistically independent, so that the power spectrum of the output $y[m]$ can be approximated as the sum of the power spectra:

$$|Y(f)|^2 \approx |X(f)|^2 + |N(f)|^2 \tag{10.103}$$

with equality if we take expected values, as the expected value of the cross term vanishes (see Section 10.1.3).

Although we don't know $|N(f)|^2$, we can obtain an estimate using the average periodogram over M frames that are known to be just noise (i.e., when no signal is present) as long as the noise is stationary

$$|\hat{N}(f)|^2 = \frac{1}{M}\sum_{i=0}^{M-1}|Y_i(f)|^2 \tag{10.104}$$

Spectral subtraction supplies an intuitive estimate for $|X(f)|$ using Eqs. (10.103) and (10.104) as

$$|\hat{X}(f)|^2 = |Y(f)|^2 - |\hat{N}(f)|^2 = |Y(f)|^2\left(1 - \frac{1}{SNR(f)}\right) \tag{10.105}$$

where we have defined the frequency-dependent signal-to-noise ratio $SNR(f)$ as

$$SNR(f) = \frac{|Y(f)|^2}{|\hat{N}(f)|^2} \tag{10.106}$$

Equation (10.105) describes the magnitude of the Fourier transform but not the phase. This is not a problem if we are interested in computing the mel-cepstrum as discussed in Chapter 6. We can just modify the magnitude and keep the original phase of $Y(f)$ using a filter $H_{ss}(f)$:

$$\hat{X}(f) = Y(f)H_{ss}(f) \tag{10.107}$$

where, according to Eq. (10.105), $H_{ss}(f)$ is given by

$$H_{ss}(f) = \sqrt{1 - \frac{1}{SNR(f)}} \tag{10.108}$$

Since $\left|\hat{X}(f)\right|^2$ is a power spectral density, it has to be positive, and therefore

$$SNR(f) \geq 1 \tag{10.109}$$

but we have no guarantee that $SNR(f)$, as computed by Eq. (10.106), satisfies Eq. (10.109). In fact, it is easy to see that noise frames do not comply. To enforce this constraint, Boll [13] suggested modifying Eq. (10.108) as follows:

$$H_{ss}(f) = \sqrt{\max\left(1 - \frac{1}{SNR(f)}, a\right)} \tag{10.110}$$

with $a \geq 0$, so that the quantity within the square root is always positive, and where $f_{ss}(x)$ is given by

$$f_{ss}(x) = \sqrt{\max\left(1 - \frac{1}{x}, a\right)} \tag{10.111}$$

It is useful to express $SNR(f)$ in dB so that

$$\bar{x} = 10\log_{10} SNR \tag{10.112}$$

and the gain of the filter in Eq. (10.111) also in dB:

$$g_{ss}(\bar{x}) = 20\log_{10} f_{ss}(\bar{x}) \tag{10.113}$$

Using Eqs. (10.111) and (10.112), we can express Eq. (10.113) by

$$g_{ss}(\bar{x}) = \max\left(10\log_{10}\left(1 - 10^{-\bar{x}/10}\right), -A\right) \tag{10.114}$$

after expressing the attenuation a in dB:

$$a = 10^{-A/10} \tag{10.115}$$

Equation (10.114) is plotted in Figure 10.27 for A = 10 dB.

The spectral subtraction rule in Eq. (10.111) is quite intuitive. To implement it we can do a short-time analysis, as shown in Chapter 6, by using overlapping windowed segments, zero-padding, computing the FFT, modifying the magnitude spectrum, taking the inverse FFT, and adding the resulting windows.

This implementation results in output speech that has significantly less noise, though it exhibits what is called *musical noise* [12]. This is caused by frequency bands f for which $|Y(f)|^2 \approx |\hat{N}(f)|^2$. As shown in Figure 10.27, a frequency f_0 for which $|Y(f_0)|^2 < |\hat{N}(f_0)|^2$ is attenuated by A dB, whereas a neighboring frequency f_1, where $|Y(f_1)|^2 > |\hat{N}(f_1)|^2$, has a much smaller attenuation. These rapid changes with frequency introduce tones at varying frequencies that appear and disappear rapidly.

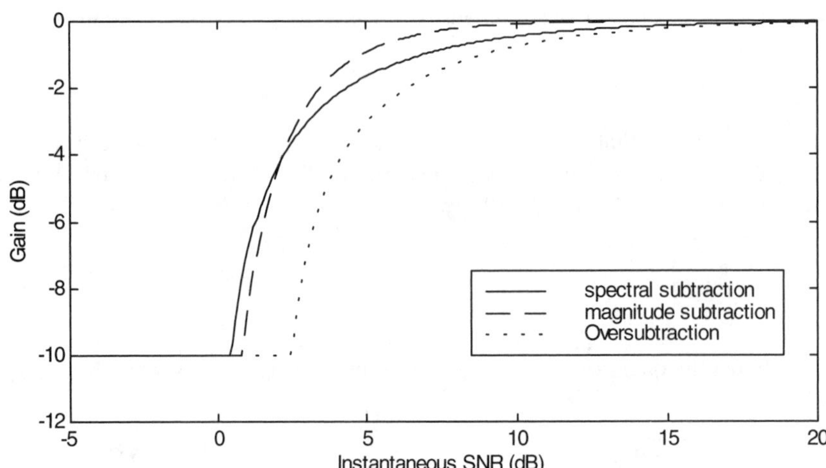

Figure 10.27 Magnitude of the spectral subtraction filter gain as a function of the input instantaneous SNR for A = 10 dB, for the spectral subtraction of Eq. (10.114), magnitude subtraction of Eq. (10.118), and oversubtraction of Eq. (10.119) with β = 2 dB.

The main reason for the presence of musical noise is that the estimates of $SNR(f)$ through Eqs. (10.104) and (10.106) are poor. This is partly because $SNR(f)$ is computed independently for each frequency, whereas we know that $SNR(f_0)$ and $SNR(f_1)$ are correlated if f_0 and f_1 are close to each other. Thus, one possibility is to smooth the filter in Eq. (10.114) over frequency. This approach suppresses a smaller amount of noise, but it does not distort the signal as much, and thus may be preferred by listeners. Similarly, smoothing over time

$$SNR(f,t) = \gamma SNR(f,t-1) + (1-\gamma)\frac{|Y(f)|^2}{|\hat{N}(f)|^2} \tag{10.116}$$

can also be done to reduce the distortion, at the expense of a smaller noise attenuation. Smoothing over both time and frequency can be done to obtain more accurate SNR measurements and thus less distortion. As shown in Figure 10.28, use of spectral subtraction can reduce the error rate.

Additionally, the attenuation A can be made a function of frequency. This is useful when we want to suppress more noise at one frequency than another, which is a tradeoff between noise reduction and nonlinear distortion of speech.

Other enhancements to the basic algorithm have been proposed to reduce the musical noise. Sometimes Eq. (10.111) is generalized to

$$f_{ms}(x) = \left(\max\left(1 - \frac{1}{x^{\alpha/2}}, a \right) \right)^{1/\alpha} \tag{10.117}$$

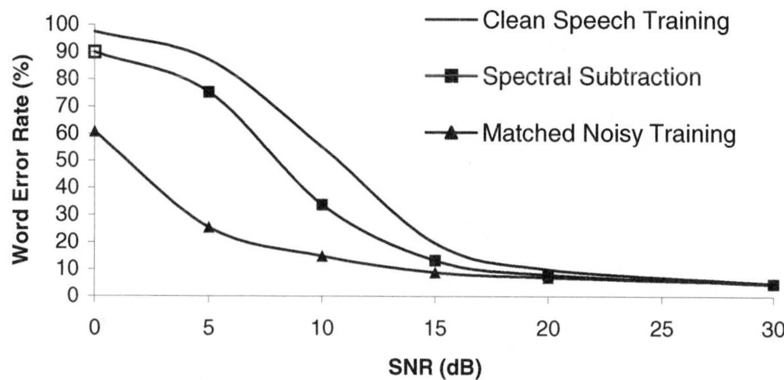

Figure 10.28 Word error rate as a function of SNR (dB) using Whisper on the *Wall Street Journal* 5000-word dictation task. White noise was added at different SNRs. The solid line represents the baseline system trained with clean speech, the line with squares the use of spectral subtraction with the previous clean HMMs. They are compared to a system trained on the same speech with the same SNR as the speech tested on.

where $\alpha = 2$ corresponds to the *power spectral subtraction* rule in Eq. (10.111), and $\alpha = 1$ corresponds to the *magnitude subtraction* rule (plotted in Figure 10.27 for A = 10 dB):

$$g_{ms}(\bar{x}) = \max\left(20\log_{10}\left(1-10^{-\bar{x}/5}\right), -A\right) \tag{10.118}$$

Another variation, called oversubtraction, consists of multiplying the estimate of the noise power spectral density $\left|\hat{N}(f)\right|^2$ in Eq. (10.104) by a constant $10^{\beta/10}$, where $\beta > 0$, which causes the power spectral subtraction rule in Eq. (10.114) to be transformed to another function

$$g_{os}(\bar{x}) = \max\left(10\log_{10}\left(1-10^{-(\bar{x}-\beta)/10}\right), -A\right) \tag{10.119}$$

This causes $|Y(f)|^2 < |\hat{N}(f)|^2$ to occur more often than $|Y(f)|^2 > |\hat{N}(f)|^2$ for frames for which $|Y(f)|^2 \approx |\hat{N}(f)|^2$, and thus reduces the musical noise.

10.5.2. Frequency-Domain MMSE from Stereo Data

You have seen that several possible functions, such as Eqs. (10.114), (10.118), or (10.119), can be used to attenuate the noise, and it is not clear that any one of them is better than the others, since each has been obtained through different assumptions. This opens the possibility of estimating the curve $g(\bar{x})$ using a different criterion, and, thus, different approximations than those used in Section 10.5.1.

One interesting possibility occurs when we have pairs of stereo utterances that have been recorded simultaneously in noise-free conditions in one channel and noisy conditions in the other channel. In this case, we can estimate $f(x)$ using a minimum mean squared criterion (Porter and Boll [47], Ephraim and Malah [23]), so that

$$\hat{f}(x) = \underset{f(x)}{\arg\min} \left\{ \sum_{i=0}^{N-1} \sum_{j=0}^{M-1} \left(X_i(f_j) - f\left(SNR(f_j)\right) Y_i(f_j) \right)^2 \right\} \tag{10.120}$$

or $g(x)$ as

$$\hat{g}(x) = \underset{g(x)}{\arg\min} \left\{ \sum_{i=0}^{N-1} \sum_{j=0}^{M-1} \left(10\log_{10} \left| X_i(f_j) \right|^2 - g\left(SNR(f_j)\right) - 10\log_{10} \left| Y_i(f_j) \right|^2 \right)^2 \right\} \tag{10.121}$$

which can be solved by discretizing $f(x)$ and $g(x)$ into several bins and summing over all M frequencies and N frames. This approach results in a curve that is smoother and thus offers less musical noise and lower distortion. Stereo utterances of noise-free and noisy speech are needed to estimate $f(x)$ and $g(x)$ through Eqs. (10.120) and (10.121) for any given acoustical environment and can be collected with two microphones, or the noisy speech can be obtained by adding to the clean speech artificial noise from the testing environment.

Another generalization of this approach is to use a different function $f(x)$ or $g(x)$ for every frequency [2] as shown in Figure 10.29. This also allows for a lower squared error at the expense of having to store more data tables. In the experiments of Figure 10.29, we note that more subtraction is needed at lower frequencies than at higher frequencies in this case.

If such stereo data is available to estimate these curves, it makes the enhanced speech sound better [23] than does spectral subtraction. When used in speech recognition systems, it also leads to higher accuracies [2].

10.5.3. Wiener Filtering

Let's reformulate Eq. (10.102) from the statistical point of view. The process $\mathbf{y}[n]$ is the sum of random process $\mathbf{x}[n]$ and the additive noise $\mathbf{v}[n]$ process:

$$\mathbf{y}[n] = \mathbf{x}[n] + \mathbf{v}[n] \tag{10.122}$$

We wish to find a linear estimate $\hat{\mathbf{x}}[n]$ in terms of the process $\mathbf{y}[n]$:

$$\hat{\mathbf{x}}[n] = \sum_{m=-\infty}^{\infty} h[m]\mathbf{y}[n-m] \tag{10.123}$$

which is the result of a linear time-invariant filtering operation. The MMSE estimate of $h[n]$ in Eq. (10.123) minimizes the squared error

$$E\left\{ \left[\mathbf{x}[n] - \sum_{m=-\infty}^{\infty} h[m]\mathbf{y}[n-m] \right]^2 \right\} \tag{10.124}$$

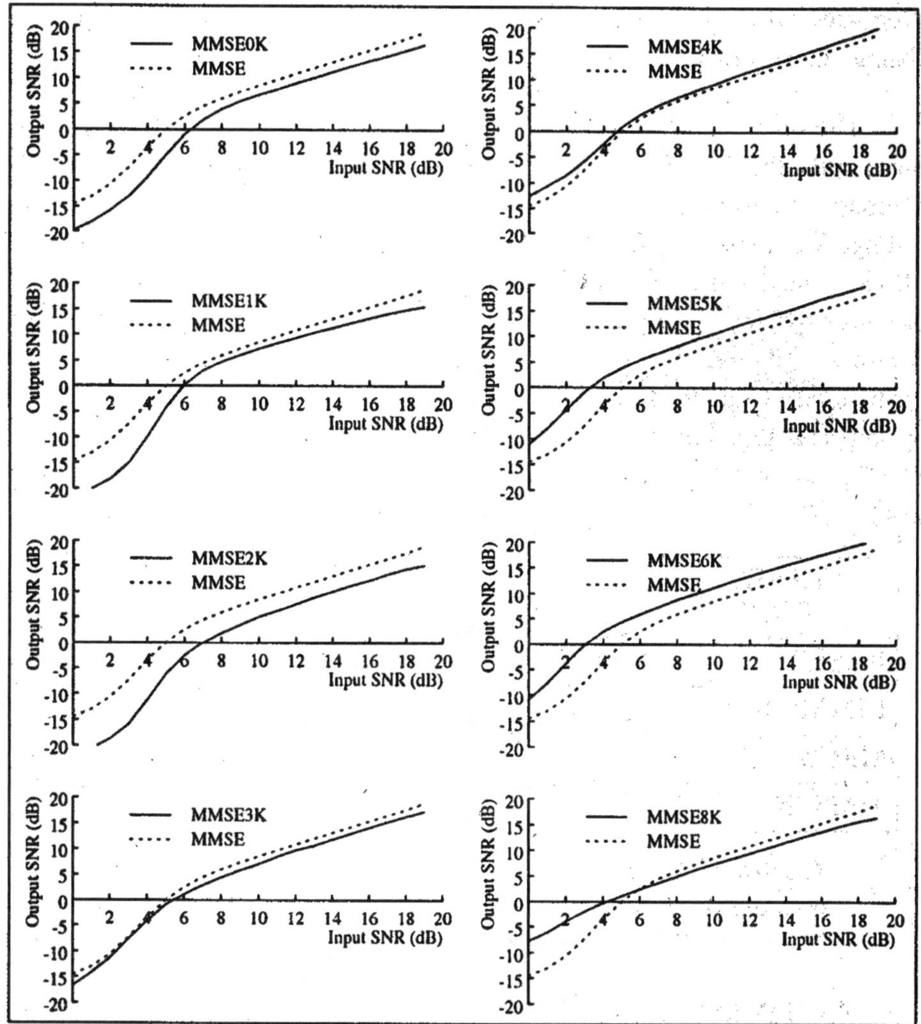

Figure 10.29 Empirical curves for input-to-output instantaneous SNR. Eight different curves for 0, 1, 2, 3, 4, 5, 6, 7 and 8 kHz are obtained following Eq. (10.121) [2] using speech recorded simultaneously from a close-talking microphone and a desktop microphone.

which results in the famous *Wiener-Hopf* equation

$$R_{xy}[l] = \sum_{m=-\infty}^{\infty} h[m] R_{yy}[l-m] \tag{10.125}$$

so that, taking Fourier transforms, the resulting filter can be expressed in the frequency domain as

$$H(f) = \frac{S_{xy}(f)}{S_{yy}(f)} \tag{10.126}$$

If the signal $\mathbf{x}[n]$ and the noise $\mathbf{v}[n]$ are orthogonal, which is often the case, then

$$S_{xy}(f) = S_{xx}(f) \text{ and } S_{yy}(f) = S_{xx}(f) + S_{vv}(f) \tag{10.127}$$

so that Eq. (10.126) is given by

$$H(f) = \frac{S_{xx}(f)}{S_{xx}(f) + S_{vv}(f)} \tag{10.128}$$

Equation (10.128) is called the *noncausal Wiener filter*. This can be realized only if we know the power spectra of both the noise and the signal. Of course, if $S_{xx}(f)$ and $S_{vv}(f)$ do not overlap, then $H(f) = 1$ in the band of the signal and $H(f) = 0$ in the band of the noise.

In practice, $S_{xx}(f)$ is unknown. If it were known, we could compute its mel-cepstrum, which would coincide exactly with the mel-cepstrum before noise addition. To solve this chicken-and-egg problem, we need some kind of model. Ephraim [22] proposed the use of an HMM where, if we know what state the current frame falls under, we can use its mean spectrum as $S_{xx}(f)$. In practice we do not know what state each frame falls into either, so he proposed to weigh the filters for each state by the a posterior probability that the frame falls into each state. This algorithm, when used in speech enhancement, results in gains of 7 dB or more.

A causal version of the Wiener filter can also be derived. A dynamical state model algorithm called the Kalman filter (see [42] for details) is also an extension of the Wiener filter.

10.5.4. Cepstral Mean Normalization (CMN)

Different microphones have different transfer functions, and even the same microphone has a varying transfer function depending on the distance to the microphone and the room acoustics. In this section we describe a powerful and simple technique that is designed to handle convolutional distortions and, thus, increases the robustness of speech recognition systems to unknown linear filtering operations.

Given a signal $x[n]$, we compute its cepstrum through short-time analysis, resulting in a set of T cepstral vectors $\mathbf{X} = \{\mathbf{x}_0, \mathbf{x}_1, \cdots, \mathbf{x}_{T-1}\}$. Its sample mean $\overline{\mathbf{x}}$ is given by

$$\overline{\mathbf{x}} = \frac{1}{T}\sum_{t=0}^{T-1}\mathbf{x}_t \tag{10.129}$$

Cepstral mean normalization (CMN) (Atal [8]) consists of subtracting $\bar{\mathbf{x}}$ from each vector $\mathbf{x}_t$ to obtain the normalized cepstrum vector $\hat{\mathbf{x}}_t$:

$$\hat{\mathbf{x}}_t = \mathbf{x}_t - \bar{\mathbf{x}} \tag{10.130}$$

Let's now consider a signal $y[n]$, which is the output of passing $x[n]$ through a filter $h[n]$. We can compute another sequence of cepstrum vectors $\mathbf{Y} = \{\mathbf{y}_0, \mathbf{y}_1, \cdots, \mathbf{y}_{T-1}\}$. Now let's further define a vector $\mathbf{h}$ as

$$\mathbf{h} = \mathbf{C}\left(\ln\left|H(\omega_0)\right|^2 \quad \cdots \quad \ln\left|H(\omega_M)\right|^2 \right) \tag{10.131}$$

where $\mathbf{C}$ is the DCT matrix. We can see that

$$\mathbf{y}_t = \mathbf{x}_t + \mathbf{h} \tag{10.132}$$

and thus the sample mean $\bar{\mathbf{y}}_t$ equals

$$\bar{\mathbf{y}} = \frac{1}{T}\sum_{t=0}^{T-1} \mathbf{y}_t = \frac{1}{T}\sum_{t=0}^{T-1} (\mathbf{x}_t + \mathbf{h}) = \bar{\mathbf{x}} + \mathbf{h} \tag{10.133}$$

and its normalized cepstrum is given by

$$\hat{\mathbf{y}}_t = \mathbf{y}_t - \bar{\mathbf{y}}_t = \hat{\mathbf{x}}_t \tag{10.134}$$

which indicates that cepstral mean normalization is immune to linear filtering operations. This procedure is performed on every utterance for both training and testing. Intuitively, the mean vector $\bar{\mathbf{x}}$ conveys the spectral characteristics of the current microphone and room acoustics. In the limit, when $T \to \infty$ for each utterance, we should expect means from utterances from the same recording environment to be the same. Use of CMN to the cepstrum vectors does not modify the delta or delta-delta cepstrum.

Let's analyze the effect of CMN on a short utterance. Assume that our utterance contains a single phoneme, say /s/. The mean $\bar{\mathbf{x}}$ will be very similar to the frames in this phoneme, since /s/ is quite stationary. Thus, after normalization, $\hat{\mathbf{x}}_t \approx 0$. A similar result will happen for other fricatives, which means that it would be impossible to distinguish these ultrashort utterances, and the error rate will be very high. If the utterance contains more than one phoneme but is still short, this problem is not insurmountable, but the confusion among phonemes is still higher than if no CMN had been applied. Empirically, it has been found that this procedure does not degrade the recognition rate on utterances from the same acoustical environment, as long as they are longer than 2–4 seconds. Yet the method provides significant robustness against linear filtering operations. In fact, for telephone recordings, where each call has a different frequency response, the use of CMN has been shown to provide as much as 30% relative decrease in error rate. When a system is trained on one microphone and tested on another, CMN can provide significant robustness.

Interestingly enough, it has been found in practice that the error rate for utterances within the same environment is actually somewhat lower, too. This is surprising, given that there is no mismatch in channel conditions. One explanation is that, even for the same microphone and room acoustics, the distance between the mouth and the microphone varies for different speakers, which causes slightly different transfer functions, as we studied in Section 10.2. In addition, the cepstral mean characterizes not only the channel transfer function, but also the average frequency response of different speakers. By removing the long-term speaker average, CMN can act as sort of speaker normalization.

One drawback of CMN is it does not discriminate silence and voice in computing the utterance mean. An extension to CMN consists in computing different means for noise and speech [5]:

$$\begin{aligned} \mathbf{h}^{(j+1)} &= \frac{1}{N_s} \sum_{t \in q_s} \mathbf{x}_t - \mathbf{m}_s \\ \mathbf{n}^{(j+1)} &= \frac{1}{N_n} \sum_{t \in q_n} \mathbf{x}_t - \mathbf{m}_n \end{aligned} \tag{10.135}$$

i.e., the difference between the average vector for speech frames in the utterance and the average vector $\mathbf{m}_s$ for speech frames in the training data, and similarly for the noise frames $\mathbf{m}_n$. Speech/noise discrimination could be done by classifying frames into speech frames and noise frames, computing the average cepstra for each, and subtracting them from the average in the training data. This procedure works well as long as the speech/noise classification is accurate. It's best done by the recognizer, since other speech detection algorithms can fail in high background noise (see Section 10.6.2). To avoid errors in transitions between speech and noise, delta and delta-delta can be computed prior to this speech/noise mean normalization so that they are unaffected. As shown in Figure 10.30, this algorithm has been shown to improve robustness not only to varying channels but also to noise.

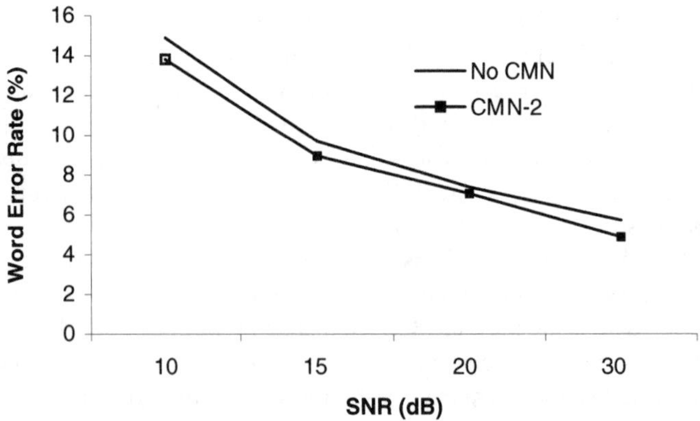

Figure 10.30 Word error rate as a function of SNR (dB) for both no CMN and CMN-2 [5]. White noise was added at different SNRs and the system was trained with speech with the same SNR. The Whisper system is used on the 5000-word *Wall Street Journal* task using a bigram language model.

10.5.5. Real-Time Cepstral Normalization

CMN requires the complete utterance to compute the cepstral mean; thus, it cannot be used in a real-time system, and an approximation needs to be used. In this section we discuss a modified version of CMN that can address this problem, as well as a set of techniques called RASTA that attempt to do the same thing.

We can interpret CMN as the operation of subtracting a low-pass filter $d[n]$, where all the T coefficients are identical and equal $1/T$, which is a high-pass filter with a cutoff frequency ω_c that is arbitrarily close to 0. This interpretation indicates that we can implement other types of high-pass filters. One that has been found to work well in practice is the exponential filter, so the cepstral mean $\overline{\mathbf{x}}_t$ is a function of time

$$\overline{\mathbf{x}}_t = \alpha \mathbf{x}_t + (1-\alpha)\overline{\mathbf{x}}_{t-1} \tag{10.136}$$

where α is chosen so that the filter has a time constant[7] of at least 5 seconds of speech.

Other types of filters have been proposed in the literature. In fact, a popular approach consists of an IIR bandpass filter with the transfer function:

$$H(z) = 0.1z^4 * \frac{2 + z^{-1} - z^{-3} - 2z^{-4}}{1 - 0.98z^{-1}} \tag{10.137}$$

which is used in the so-called relative spectral processing or RASTA [32]. As in CMN, the high-pass portion of the filter is expected to alleviate the effect of convolutional noise introduced in the channel. The low-pass filtering helps to smooth some of the fast frame-to-frame spectral changes present. Empirically, it has been shown that the RASTA filter behaves similarly to the real-time implementation of CMN, albeit with a slightly higher error rate. Both the RASTA filter and real-time implementations of CMN require the filter to be properly initialized. Otherwise, the first utterance may use an incorrect cepstral mean. The original derivation of RASTA includes a few stages prior to the bandpass filter, and this filter is performed on the spectral energies, not the cepstrum.

10.5.6. The Use of Gaussian Mixture Models

Algorithms such as spectral subtraction of Section 10.5.1 or the frequency-domain MMSE of Section 10.5.2 implicitly assume that different frequencies are uncorrelated from each other. Because of that, the spectrum of the enhanced signal may exhibit abrupt changes across frequency and not look like spectra of real speech signals. Using the model of the

[7] The time constant τ of a low-pass filter is defined as the value for which the output is cut in half. For an exponential filter of parameter α and sampling rate F_s, $\alpha = \ln 2/(TF_s)$.

environment of Section 10.1.3, we can express the clean-speech cepstral vector $\mathbf{x}$ as a function of the observed noisy cepstral vector $\mathbf{y}$ as

$$\mathbf{x} = \mathbf{y} - \mathbf{h} - \mathbf{C}\ln\left(1 - e^{\mathbf{C}^{-1}(\mathbf{n}-\mathbf{y})}\right) \tag{10.138}$$

where the noise cepstral vector $\mathbf{n}$ is a random vector. The MMSE estimate of $\mathbf{x}$ is given by

$$\hat{\mathbf{x}}_{MMSE} = E\{\mathbf{x} \mid \mathbf{y}\} = \mathbf{y} - \mathbf{h} - \mathbf{C}E\left\{\ln\left(1 - e^{\mathbf{C}^{-1}(\mathbf{n}-\mathbf{y})}\right) \mid \mathbf{y}\right\} \tag{10.139}$$

where the expectation uses the distribution of $\mathbf{n}$. Solution to Eq. (10.139) results in a nonlinear function which can be learned, for example, with a neural network [53].

A popular model to attack this problem consists in modeling the probability distribution of the noisy speech $\mathbf{y}$ as a mixture of K Gaussians:

$$p(\mathbf{y}) = \sum_{k=0}^{K-1} p(\mathbf{y} \mid k)P[k] = \sum_{k=0}^{K-1} \mathrm{N}(\mathbf{y}, \boldsymbol{\mu}_k, \boldsymbol{\Sigma}_k)P[k] \tag{10.140}$$

where $P[k]$ is the prior probability of each Gaussian component k. If $\mathbf{x}$ and $\mathbf{y}$ are jointly Gaussian within class k, then $p(\mathbf{x} \mid \mathbf{y}, k)$ is also Gaussian [42] with mean:

$$E\{\mathbf{x} \mid \mathbf{y}, k\} = \boldsymbol{\mu}_{\mathbf{x}}^k + \boldsymbol{\Sigma}_{\mathbf{xy}}^k \left(\boldsymbol{\Sigma}_{\mathbf{y}}^k\right)^{-1} (\mathbf{y} - \boldsymbol{\mu}_{\mathbf{y}}^k) = \mathbf{C}_k\mathbf{y} + \mathbf{r}_k \tag{10.141}$$

so that the joint distribution of $\mathbf{x}$ and $\mathbf{y}$ is given by

$$\begin{aligned} p(\mathbf{x}, \mathbf{y}) &= \sum_{k=0}^{K-1} p(\mathbf{x}, \mathbf{y} \mid k)P[k] = \sum_{k=0}^{K-1} p(\mathbf{x} \mid \mathbf{y}, k)p(\mathbf{y} \mid k)P[k] \\ &= \sum_{k=0}^{K-1} \mathrm{N}(\mathbf{x}, \mathbf{C}_k\mathbf{y} + \mathbf{r}_k, \boldsymbol{\Gamma}_k)\mathrm{N}(\mathbf{y}, \boldsymbol{\mu}_k, \boldsymbol{\Sigma}_k)P[k] \end{aligned} \tag{10.142}$$

where $\mathbf{r}_k$ is called the *correction vector*, $\mathbf{C}_k$ is the *rotation matrix*, and the matrix $\boldsymbol{\Gamma}_k$ tells us how uncertain we are about the compensation.

A maximum likelihood estimate of $\mathbf{x}$ maximizes the joint probability in Eq. (10.142). Assuming the Gaussians do not overlap very much (as in the FCDCN algorithm [2]):

$$\hat{\mathbf{x}}_{ML} \approx \arg\max_k p(\mathbf{x}, \mathbf{y}, k) = \arg\max_k \mathrm{N}(\mathbf{y}, \boldsymbol{\mu}_k, \boldsymbol{\Sigma}_k)\mathrm{N}(\mathbf{x}, \mathbf{C}_k\mathbf{y} + \mathbf{r}_k, \boldsymbol{\Gamma}_k)P[k] \tag{10.143}$$

whose solution is

$$\hat{\mathbf{x}}_{ML} = \mathbf{C}_{\hat{k}}\mathbf{y} + \mathbf{r}_{\hat{k}} \tag{10.144}$$

where

$$\hat{k} = \arg\max_{k} N(\mathbf{y}, \boldsymbol{\mu}_k, \boldsymbol{\Sigma}_k) P[k] \tag{10.145}$$

It is often more robust to compute the MMSE estimate of **x** (as in the CDCN [2] and RATZ [43] algorithms):

$$\hat{\mathbf{x}}_{MMSE} = E\{\mathbf{x} \mid \mathbf{y}\} = \sum_{k=0}^{K-1} p(k \mid \mathbf{y}) E\{\mathbf{x} \mid \mathbf{y}, k\} = \sum_{k=0}^{K-1} p(k \mid \mathbf{y})\left(\mathbf{C}_k\mathbf{y} + \mathbf{r}_k\right) \tag{10.146}$$

as a weighted sum for all mixture components, where the posterior probability $p(k \mid \mathbf{y})$ is given by

$$p(k \mid \mathbf{y}) = \frac{p(\mathbf{y} \mid k) P[k]}{\sum_{k=0}^{K-1} p(\mathbf{y} \mid k) P[k]} \tag{10.147}$$

where the rotation matrix $\mathbf{C}_k$ in Eq. (10.144) can be replaced by **I** with a modest degradation in performance in return for faster computation [21].

A number of different algorithms [2, 43] have been proposed that vary in how the parameters $\boldsymbol{\mu}_k$, $\boldsymbol{\Sigma}_k$, $\mathbf{r}_k$, and $\boldsymbol{\Gamma}_k$ are estimated. If stereo recordings are available from both the clean signal and the noisy signal, then we can estimate $\boldsymbol{\mu}_k$, $\boldsymbol{\Sigma}_k$ by fitting a mixture Gaussian model to **y** as described in Chapter 3. Then $\mathbf{C}_k$, $\mathbf{r}_k$ and $\boldsymbol{\Gamma}_k$ can be estimated directly by linear regression of **x** and **y**. The FCDCN algorithm [2, 6] is a variant of this approach when it is assumed that $\boldsymbol{\Sigma}_k = \sigma^2\mathbf{I}$, $\boldsymbol{\Gamma}_k = \gamma^2\mathbf{I}$, and $\mathbf{C}_k = \mathbf{I}$, so that $\boldsymbol{\mu}_k$ and $\mathbf{r}_k$ are estimated through a VQ procedure and $\mathbf{r}_k$ is the average difference $(\mathbf{y} - \mathbf{x})$ for vectors **y** that belong to mixture component *k*. An enhancement is to use the instantaneous SNR of a frame, defined as the difference between the log-energy of that frame and the average log-energy of the background noise. It is advantageous to use different correction vectors for different instantaneous SNR levels. The log-energy can be replaced by the zeroth-order cepstral coefficient with little change in recognition accuracy.

Often, stereo recordings are not available and we need other means of estimating parameters $\boldsymbol{\mu}_k$, $\boldsymbol{\Sigma}_k$, $\mathbf{r}_k$, and $\boldsymbol{\Gamma}_k$. CDCN [6] is one such algorithm that has a model of the environment as described in Section 10.1.3, which defines a nonlinear relationship between **x**, **y** and the environmental parameters **n** and **h** for the noise and channel. This method also uses an MMSE approach where the correction vector is a weighted average of the correction vectors for all classes. An extension of CDCN using a vector Taylor series approximation [44] for that nonlinear function has been shown to offer improved results. Other methods that do not require stereo recordings or a model of the environment are presented in [43].

10.6. Environmental Model Adaptation

We describe a number of techniques that achieve compensation by adapting the HMM to the noisy conditions. The most straightforward method is to retrain the whole HMM with the speech from the new acoustical environment. Another option is to apply standard adaptive techniques discussed in Chapter 9 to the case of environment adaptation. We consider a model of the environment that allows constrained adaptation methods for more efficient adaptation in comparison to the general techniques.

10.6.1. Retraining on Corrupted Speech

If there is a mismatch between acoustical environments, it is sensible to retrain the HMM. This is done in practice for telephone speech where only telephone speech, and no clean high-bandwidth speech, is used in the training phase.

Unfortunately, training a large-vocabulary speech recognizer requires a very large amount of data, which is often not available for a specific noisy condition. For example, it is difficult to collect a large amount of training data in a car driving at 50 mph, whereas it is much easier to record it at idle speed. Having a small amount of matched-conditions training data could be worse than a large amount of mismatched-conditions training data. Often we want to adapt our model given a relatively small sample of speech from the new acoustical environment.

One option is to take a noise waveform from the new environment, add it to all the utterances in our database, and retrain the system. If the noise characteristics are known ahead of time, this method allows us to adapt the model to the new environment with a relatively small amount of data from the new environment, yet use a large amount of training data. Figure 10.31 shows the benefit of this approach over a system trained on clean speech for the case of additive white noise. If the target acoustical environment also has a different channel, we can also filter all the utterances in the training data prior to retraining. This method allows us to adapt the model to the new environment with a relatively small amount of data from the new environment.

If the noise sample is available offline, this simple technique can provide good results at no cost during recognition. Otherwise the noise addition and model retraining would need

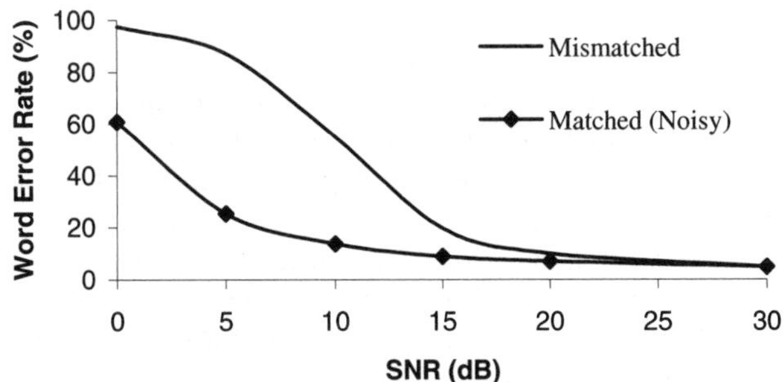

Figure 10.31 Word error rate as a function of the testing data SNR (dB) for Whisper trained on clean data and a system trained on noisy data at the same SNR as the testing set as in Figure 10.30. White noise at different SNRs is added.

to occur at runtime. This is feasible for speaker-dependent small-vocabulary systems where the training data can be kept in memory and where the retraining time can be small, but it is generally not feasible for large-vocabulary speaker-independent systems because of memory and computational limitations.

One possibility is to create a number of artificial acoustical environments by corrupting our clean database with noise samples of varying levels and types, as well as varying channels. Then all those waveforms from multiple acoustical environments can be used in training. This is called *multistyle training* [39], since our training data comes from different conditions. Because of the diversity of the training data, the resulting recognizer is more robust to varying noise conditions. In Figure 10.32 we see that, though generally the error-rate curve is above the matched-condition curve, particularly for clean speech, multistyle training does not require knowledge of the specific noise level and thus is a viable alternative to the theoretical lower bound of matched conditions.

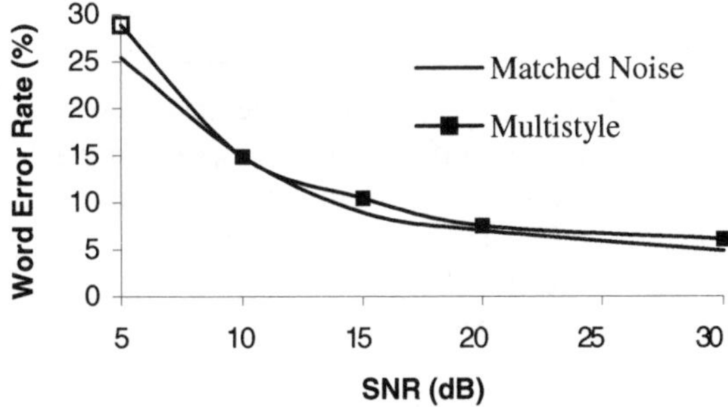

Figure 10.32 Word error rates of multistyle training compared to matched-noise training as a function of the SNR in dB for additive white noise. Whisper is trained as in Figure 10.30. The error rate of multistyle training is between 12% (for low SNR) and 25% (for high SNR) higher in relative terms than that of matched-condition training. Nonetheless, multistyle training does better than a system trained on clean data for all conditions other than clean speech.

10.6.2. Model Adaptation

We can also use the standard adaptation methods used for speaker adaptation, such as MAP or MLLR described in Chapter 9. Since MAP is an unstructured method, it can offer results similar to those of matched conditions, but it requires a significant amount of adaptation data. MLLR can achieve reasonable performance with about a minute of speech for minor mismatches [41]. For severe mismatches, MLLR also requires a large number of transformations, which, in turn, require a larger amount of adaptation data as discussed in Chapter 9.

Let's analyze the case of a single MLLR transform, where the affine transformation is simply a bias. In this case the MLLR transform consists only of a vector $\mathbf{h}$ that, as in the case of CMN described in Section 10.5.4, can be estimated from a single utterance. Instead of estimating $\mathbf{h}$ as the average cepstral mean, this method estimates $\mathbf{h}$ as the maximum likelihood estimate, given a set of sample vectors $\mathbf{X} = \{\mathbf{x}_0, \mathbf{x}_1, \cdots, \mathbf{x}_{T-1}\}$ and an HMM model λ [48], and it is a version of the EM algorithm where all the vector means are tied together (see Algorithm 10.2). This procedure for estimating the cepstral bias has a very slight reduction in error rates over CMN, although the improvement is larger for short utterances [48].

ALGORITHM 10.2: *MLE SIGNAL BIAS REMOVAL*

Step 1: Initialize $\mathbf{h}^{(0)} = \mathbf{0}$ at iteration $j = 0$

Step 2: Obtain model $\lambda^{(j)}$ by updating the means from $\mathbf{m}_k$ to $\mathbf{m}_k + \mathbf{h}^{(j)}$, for all Gaussians *k*.

Step 3: Run recognition with model $\lambda^{(j)}$ on the current utterance and determine a state segmentation $\theta[t]$ for each frame *t*.

Step 4: Estimate $\mathbf{h}^{(j+1)}$ as the vector that maximizes the likelihood, which, using covariance matrices $\mathbf{\Sigma}_k$, is given by:

$$\mathbf{h}^{(j+1)} = \left(\sum_{t=0}^{T-1} \mathbf{\Sigma}_{\theta[t]}^{-1} \right)^{-1} \sum_{t=0}^{T-1} \mathbf{\Sigma}_{\theta[t]}^{-1} \left(\mathbf{x}_t - \mathbf{m}_{\theta[t]} \right) \tag{10.148}$$

Step 5: If converged, stop; otherwise, increment *j* and go to Step 2. In practice two iterations are often sufficient.

If both additive noise and linear filtering are applied, the cepstrum for the noise and that for most speech frames are affected differently. The *speech/noise mean normalization* [5] algorithm can be extended similarly, as shown in Algorithm 10.3. The idea is to estimate a vector $\overline{\mathbf{n}}$ and $\overline{\mathbf{h}}$, such that all the Gaussians associated to the noise model are shifted by $\overline{\mathbf{n}}$, and all remaining Gaussians are shifted by $\overline{\mathbf{h}}$.

We can make Eq. (10.150) more efficient by tying all the covariance matrices. This transforms Eq. (10.150) into

$$\begin{aligned} \mathbf{h}^{(j+1)} &= \frac{1}{N_s} \sum_{t \in q_s} \mathbf{x}_t - \mathbf{m}_s \\ \mathbf{n}^{(j+1)} &= \frac{1}{N_n} \sum_{t \in q_n} \mathbf{x}_t - \mathbf{m}_n \end{aligned} \tag{10.149}$$

i.e., the difference between the average vector for speech frames in the utterance and the average vector $\mathbf{m}_s$ for speech frames in the training data, and similarly for the noise frames $\mathbf{m}_n$. This is essentially the same equation as in the speech-noise cepstral mean normalization described in Section 10.5.4. The difference is that the speech/noise discrimination is done by the recognizer instead of by a separate classifier. This method is more accurate in high-background-noise conditions where traditional speech/noise classifiers can fail. As a compromise, a codebook with considerably fewer Gaussians than a recognizer can be used to estimate $\overline{\mathbf{n}}$ and $\overline{\mathbf{h}}$.

ALGORITHM 10.3: *SPEECH/NOISE MEAN NORMALIZATION*

Step 1: Initialize $\mathbf{h}^{(0)} = \mathbf{0}$, $\mathbf{n}^{(0)} = \mathbf{0}$ at iteration $j = 0$

Step 2: Obtain model $\lambda^{(j)}$ by updating the means of speech Gaussians from $\mathbf{m}_k$ to $\mathbf{m}_k + \mathbf{h}^{(j)}$, and of noise Gaussians from $\mathbf{m}_l$ to $\mathbf{m}_l + \mathbf{n}^{(j)}$.

Step 3: Run recognition with model $\lambda^{(j)}$ on the current utterance and determine a state segmentation $\theta[t]$ for each frame *t*.

Step 4: Estimate $\mathbf{h}^{(j+1)}$ and $\mathbf{n}^{(j+1)}$ as the vectors that maximize the likelihood for speech frames ($t \in q_s$) and noise frames ($t \in q_n$), respectively:

$$\mathbf{h}^{(j+1)} = \left(\sum_{t \in q_s} \mathbf{\Sigma}_{\theta[t]}^{-1} \right)^{-1} \sum_{t \in q_s} \mathbf{\Sigma}_{\theta[t]}^{-1} \left(\mathbf{x}_t - \mathbf{m}_{\theta[t]} \right)$$
$$\mathbf{n}^{(j+1)} = \left(\sum_{t \in q_n} \mathbf{\Sigma}_{\theta[t]}^{-1} \right)^{-1} \sum_{t \in q_n} \mathbf{\Sigma}_{\theta[t]}^{-1} \left(\mathbf{x}_t - \mathbf{m}_{\theta[t]} \right) \quad (10.150)$$

Step 5: If converged, stop; otherwise, increment *j* and go to Step 2.

10.6.3. Parallel Model Combination

By using the clean-speech models and a noise model, we can approximate the distributions obtained by training a HMM with corrupted speech. The memory requirements for the algorithm are then significantly reduced, as the training data is not needed online. *Parallel model combination* (PMC) is a method to obtain the distribution of noisy speech given the distribution of clean speech and noise as mixture of Gaussians. As discussed in Section 10.1.3, if the clean-speech cepstrum follows a Gaussian distribution and the noise cepstrum follows another Gaussian distribution, the noisy speech has a distribution that is no longer Gaussian. The PMC method nevertheless makes the assumption that the resulting distribution is Gaussian whose mean and covariance matrix are the mean and covariance matrix of the resulting non-Gaussian distribution. If it is assumed that the distribution of clean speech is a mixture of *N* Gaussians, and the distribution of the noise is a mixture of *M* Gaussians, the distribution of the noisy speech contains *NM* Gaussians. The feature vector is often composed of the cepstrum, delta cepstrum, and delta-delta cepstrum. The model combination can be seen in Figure 10.33.

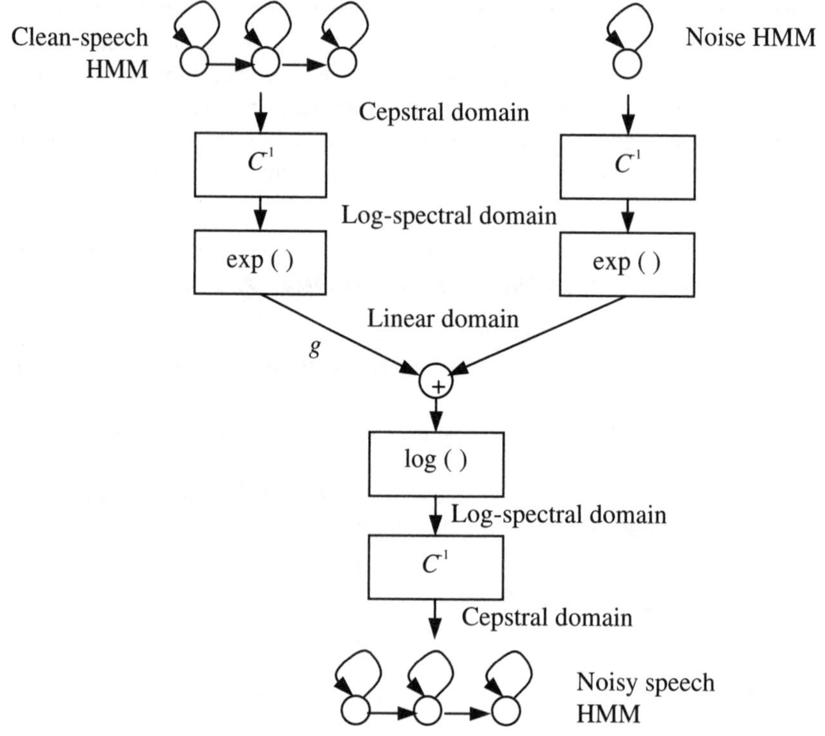

Figure 10.33 Parallel model combination for the case of one-state noise HMM.

If the mean and covariance matrix of the cepstral noise vector **n** are given by $\boldsymbol{\mu}_{\mathbf{n}}^{c}$ and $\boldsymbol{\Sigma}_{\mathbf{n}}^{c}$, respectively, we first compute the mean and covariance matrix in the log-spectral domain:

$$\begin{aligned} \boldsymbol{\mu}_{\mathbf{n}}^{l} &= \mathbf{C}^{-1}\boldsymbol{\mu}_{\mathbf{n}}^{c} \\ \boldsymbol{\Sigma}_{\mathbf{n}}^{l} &= \mathbf{C}^{-1}\boldsymbol{\Sigma}_{\mathbf{n}}^{c}(\mathbf{C}^{-1})^{T} \end{aligned} \tag{10.151}$$

In the linear domain $\mathbf{N} = e^{\mathbf{n}}$, the distribution is lognormal, whose mean vector $\boldsymbol{\mu}_{\mathbf{N}}$ and covariance matrix $\boldsymbol{\Sigma}_{\mathbf{N}}$ can be shown (see Chapter 3) to be given by

$$\begin{aligned} \boldsymbol{\mu}_{\mathbf{N}}[i] &= \exp\left\{\boldsymbol{\mu}_{\mathbf{n}}^{l}[i] + \boldsymbol{\Sigma}_{\mathbf{n}}^{l}[i,i]/2\right\} \\ \boldsymbol{\Sigma}_{\mathbf{N}}[i,j] &= \boldsymbol{\mu}_{\mathbf{N}}[i]\boldsymbol{\mu}_{\mathbf{N}}[j]\left(\exp\left\{\boldsymbol{\Sigma}_{\mathbf{n}}^{l}[i,j]\right\} - 1\right) \end{aligned} \tag{10.152}$$

with expressions similar to Eqs. (10.151) and (10.152) for the mean and covariance matrix of **X**.

Using the model of the environment with no filter is equivalent to obtaining a random linear spectral vector **Y** given by (see Figure 10.33)

$$\mathbf{Y} = \mathbf{X} + \mathbf{N} \tag{10.153}$$

and, since **X** and **N** are independent, we can obtain the mean and covariance matrix of **Y** as

$$\begin{aligned} \boldsymbol{\mu}_{\mathbf{Y}} &= \boldsymbol{\mu}_{\mathbf{X}} + \boldsymbol{\mu}_{\mathbf{N}} \\ \boldsymbol{\Sigma}_{\mathbf{Y}} &= \boldsymbol{\Sigma}_{\mathbf{X}} + \boldsymbol{\Sigma}_{\mathbf{N}} \end{aligned} \tag{10.154}$$

Although the sum of two lognormal distributions is not lognormal, the popular *log-normal approximation* [26] consists in assuming that **Y** is lognormal. In this case we can apply the inverse formulae of Eq. (10.152) to obtain the mean and covariance matrix in the log-spectral domain:

$$\begin{aligned} \boldsymbol{\Sigma}_{\mathbf{y}}^{l}[i,j] &= \ln\left\{\frac{\boldsymbol{\Sigma}_{\mathbf{Y}}[i,j]}{\boldsymbol{\mu}_{\mathbf{Y}}[i]\boldsymbol{\mu}_{\mathbf{Y}}[j]}+1\right\} \\ \boldsymbol{\mu}_{\mathbf{y}}^{l}[i] &= \ln \boldsymbol{\mu}_{\mathbf{Y}}[i] - \frac{1}{2}\ln\left\{\frac{\boldsymbol{\Sigma}_{\mathbf{Y}}[i,j]}{\boldsymbol{\mu}_{\mathbf{Y}}[i]\boldsymbol{\mu}_{\mathbf{Y}}[j]}+1\right\} \end{aligned} \tag{10.155}$$

and finally return to the cepstrum domain applying the inverse of Eq. (10.151):

$$\begin{aligned} \boldsymbol{\mu}_{\mathbf{y}}^{c} &= \mathbf{C}\boldsymbol{\mu}_{\mathbf{y}}^{l} \\ \boldsymbol{\Sigma}_{\mathbf{y}}^{c} &= \mathbf{C}\boldsymbol{\Sigma}_{\mathbf{y}}^{l}\mathbf{C}^{T} \end{aligned} \tag{10.156}$$

The lognormal approximation cannot be used directly for the delta and delta-delta cepstrum. Another variant that can be used in this case and is more accurate than the lognormal approximation is the *data-driven parallel model combination* (DPMC) [26], which uses Monte Carlo simulation to draw random cepstrum vectors from both the clean-speech HMM and noise distribution to create cepstrum of the noisy speech by applying Eqs. (10.20) and (10.21) to each sample point. These composite cepstrum vectors are not kept in memory, only their means and covariance matrices are, therefore reducing the required memory though still requiring a significant amount of computation. The number of vectors drawn from the distribution was at least 100 in [26]. A way of reducing the number of random vectors needed to obtain good Monte Carlo simulations is proposed in [56]. A version of PMC using numerical integration, which is very computationally expensive, yielded the best results.

Figure 10.34 and Figure 10.35 compare the values estimated through the lognormal approximation to the true value, where for simplicity we deal with scalars. Thus x, n, and y represent the log-spectral energies of the clean signal, noise, and noisy signal, respectively, for a given frequency. Assuming x and n to be Gaussian with means μ_x and μ_n and variances σ_x and σ_n respectively, we see that the lognormal approximation is accurate when the standard deviations σ_x and σ_n are small.

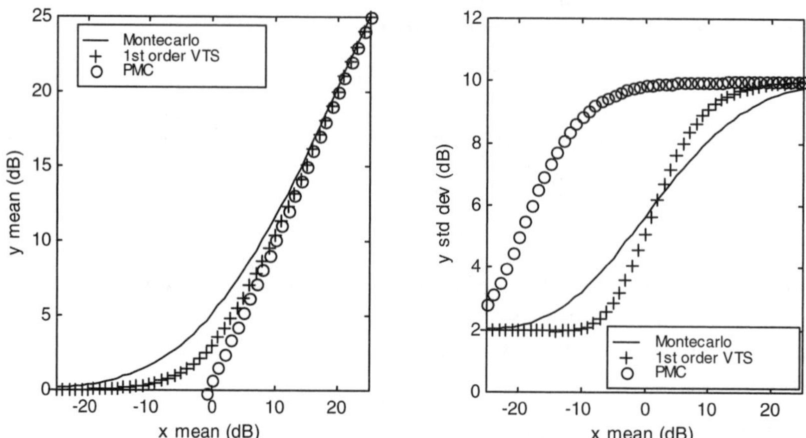

Figure 10.34 Means and standard deviation of noisy log-spectrum y in dB according to Eq. (10.165). The distribution of the noise log-spectrum n is Gaussian with mean 0 dB and standard deviation 2 dB. The distribution of the clean log-spectrum x is Gaussian, having a standard deviation of 10 dB and a mean varying from –25 to 25 dB. Both the mean and the standard deviation of y are more accurate in first-order VTS than in PMC.

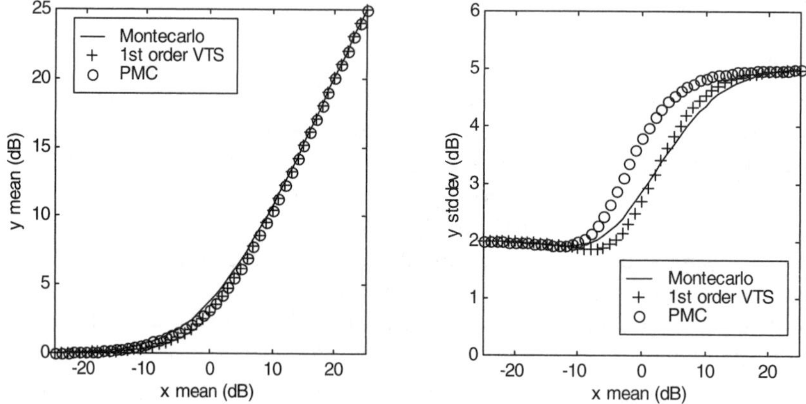

Figure 10.35 Means and standard deviation of noisy log-spectrum y in dB according to Eq. (10.165). The distribution of the noise log-spectrum n is Gaussian with mean 0 dB and standard deviation of 2 dB. The distribution of the clean log-spectrum x is Gaussian with a standard deviation of 5 dB and a mean varying from –25 dB to 25 dB. The mean of y is well estimated in both PMC and first-order VTS. The standard deviation of y is more accurate in first-order VTS than in PMC.

10.6.4. Vector Taylor Series

The model of the acoustical environment described in Section 10.1.3 describes the relationship between the cepstral vectors **x**, **n**, and **y** of the clean speech, noise, and noisy speech, respectively:

$$\mathbf{y} = \mathbf{x} + \mathbf{h} + \mathbf{g}(\mathbf{n} - \mathbf{x} - \mathbf{h}) \tag{10.157}$$

where **h** is the cepstrum of the filter, and the nonlinear function **g(z)** is given by

$$\mathbf{g}(\mathbf{z}) = \mathbf{C}\ln\left(1 + e^{\mathbf{C}^{-1}\mathbf{z}}\right) \tag{10.158}$$

Moreno [44] suggests the use of Taylor series to approximate the nonlinearity in Eq. (10.158), though he applies it in the spectral instead of the cepstral domain. We follow that approach to compute the mean and covariance matrix of **y** [4].

Assume that **x**, **h**, and **n** are Gaussian random vectors with means $\boldsymbol{\mu}_x$, $\boldsymbol{\mu}_h$, and $\boldsymbol{\mu}_n$ and covariance matrices $\boldsymbol{\Sigma}_x$, $\boldsymbol{\Sigma}_h$, and $\boldsymbol{\Sigma}_n$, respectively, and furthermore that **x**, **h**, and **n** are independent. After algebraic manipulation it can be shown that the Jacobian of Eq. (10.157) with respect to **x**, **h**, and **n** evaluated at $\boldsymbol{\mu} = \boldsymbol{\mu}_\mathbf{n} - \boldsymbol{\mu}_\mathbf{x} - \boldsymbol{\mu}_\mathbf{h}$ can be expressed as

$$\begin{aligned} \left.\frac{\partial \mathbf{y}}{\partial \mathbf{x}}\right|_{(\boldsymbol{\mu}_n,\boldsymbol{\mu}_x,\boldsymbol{\mu}_h)} &= \left.\frac{\partial \mathbf{y}}{\partial \mathbf{h}}\right|_{(\boldsymbol{\mu}_n,\boldsymbol{\mu}_x,\boldsymbol{\mu}_h)} = \mathbf{A} \\ \left.\frac{\partial \mathbf{y}}{\partial \mathbf{n}}\right|_{(\boldsymbol{\mu}_n,\boldsymbol{\mu}_x,\boldsymbol{\mu}_h)} &= \mathbf{I} - \mathbf{A} \end{aligned} \tag{10.159}$$

where the matrix **A** is given by

$$\mathbf{A} = \mathbf{C}\mathbf{F}\mathbf{C}^{-1} \tag{10.160}$$

and **F** is a diagonal matrix whose elements are given by vector $\mathbf{f}(\boldsymbol{\mu})$, which in turn is given by

$$\mathbf{f}(\boldsymbol{\mu}) = \frac{1}{1 + e^{\mathbf{C}^{-1}\boldsymbol{\mu}}} \tag{10.161}$$

Using Eq. (10.159) we can then approximate Eq. (10.157) by a first-order Taylor series expansion around $(\boldsymbol{\mu}_\mathbf{n}, \boldsymbol{\mu}_\mathbf{x}, \boldsymbol{\mu}_\mathbf{h})$ as

$$\begin{aligned} \mathbf{y} &\approx \boldsymbol{\mu}_x + \boldsymbol{\mu}_h + \mathbf{g}(\boldsymbol{\mu}_n - \boldsymbol{\mu}_x - \boldsymbol{\mu}_h) \\ &+ \mathbf{A}(\mathbf{x} - \boldsymbol{\mu}_x) + \mathbf{A}(\mathbf{h} - \boldsymbol{\mu}_h) + (\mathbf{I} - \mathbf{A})(\mathbf{n} - \boldsymbol{\mu}_n) \end{aligned} \tag{10.162}$$

The mean of **y**, $\boldsymbol{\mu}_y$, can be obtained from Eq. (10.162) as

$$\boldsymbol{\mu}_y \approx \boldsymbol{\mu}_x + \boldsymbol{\mu}_h + \mathbf{g}(\boldsymbol{\mu}_n - \boldsymbol{\mu}_x - \boldsymbol{\mu}_h) \tag{10.163}$$

and its covariance matrix $\boldsymbol{\Sigma}_\mathbf{y}$ by

$$\boldsymbol{\Sigma}_\mathbf{y} \approx \mathbf{A}\boldsymbol{\Sigma}_\mathbf{x}\mathbf{A}^T + \mathbf{A}\boldsymbol{\Sigma}_\mathbf{h}\mathbf{A}^T + (\mathbf{I}-\mathbf{A})\boldsymbol{\Sigma}_\mathbf{n}(\mathbf{I}-\mathbf{A})^T \tag{10.164}$$

so that even if $\boldsymbol{\Sigma}_x$, $\boldsymbol{\Sigma}_h$, and $\boldsymbol{\Sigma}_n$ are diagonal, $\boldsymbol{\Sigma}_\mathbf{y}$ is no longer diagonal. Nonetheless, we can assume it to be diagonal, because this way we can transform a clean HMM to a corrupted HMM that has the same functional form and use a decoder that has been optimized for diagonal covariance matrices.

It is difficult to visualize how good the approximation is, given the nonlinearity involved. To provide some insight, let's consider the frequency-domain version of Eqs. (10.157) and (10.158) when no filtering is done:

$$y = x + \ln\left(1 + \exp\left(n - x\right)\right) \tag{10.165}$$

where x, n, and y represent the log-spectral energies of the clean signal, noise, and noisy signal, respectively, for a given frequency. In Figure 10.34 we show the mean and standard deviation of the noisy log-spectral energy y in dB as a function of the mean of the clean log-spectral energy x with a standard deviation of 10 dB. The log-spectral energy of the noise n is Gaussian with mean 0 dB and standard deviation 2 dB. We compare the correct values obtained through Monte Carlo simulation (or DPMC) with the values obtained through the lognormal approximation of Section 10.6.3 and the first-order VTS approximation described here. We see that the VTS approximation is more accurate than the lognormal approximation for the mean and especially for the standard deviation of y, assuming the model of the environment described by Eq. (10.165).

Figure 10.35 is similar to Figure 10.34 except that the standard deviation of the clean log-energy x is only 5 dB, a more realistic number in speech recognition systems. In this case, both the lognormal approximation and the first-order VTS approximation are good estimates of the mean of y, though the standard deviation estimated through the lognormal approximation in PMC is not as good as that obtained through first-order VTS, again assuming the model of the environment described by Eq. (10.165). The overestimate of the variance in the lognormal approximation might, however, be useful if the model of the environment is not accurate.

To compute the means and covariance matrices of the delta and delta-delta parameters, let's take the derivative of the approximation of $\mathbf{y}$ in Eq. (10.162) with respect to time:

$$\frac{\partial \mathbf{y}}{\partial t} \approx \mathbf{A}\frac{\partial \mathbf{x}}{\partial t} \tag{10.166}$$

so that the delta-cepstrum computed through $\Delta\mathbf{x}_t = \mathbf{x}_{t+2} - \mathbf{x}_{t-2}$, is related to the derivative [28] by

$$\Delta \mathbf{x} \approx 4 \frac{\partial \mathbf{x}_t}{\partial t} \tag{10.167}$$

so that

$$\boldsymbol{\mu}_{\Delta \mathbf{y}} \approx \mathbf{A} \boldsymbol{\mu}_{\Delta \mathbf{x}} \tag{10.168}$$

and similarly

$$\boldsymbol{\Sigma}_{\Delta \mathbf{y}} \approx \mathbf{A} \boldsymbol{\Sigma}_{\Delta \mathbf{x}} \mathbf{A}^T + (\mathbf{I} - \mathbf{A}) \boldsymbol{\Sigma}_{\Delta \mathbf{n}} (\mathbf{I} - \mathbf{A})^T \tag{10.169}$$

where we assumed that $\mathbf{h}$ is constant within an utterance, so that $\Delta \mathbf{h} = 0$.

Similarly, for the delta-delta cepstrum, the mean is given by

$$\boldsymbol{\mu}_{\Delta^2 \mathbf{y}} \approx \mathbf{A} \boldsymbol{\mu}_{\Delta^2 \mathbf{x}} \tag{10.170}$$

and the covariance matrix by

$$\boldsymbol{\Sigma}_{\Delta^2 \mathbf{y}} \approx \mathbf{A} \boldsymbol{\Sigma}_{\Delta^2 \mathbf{x}} \mathbf{A}^T + (\mathbf{I} - \mathbf{A}) \boldsymbol{\Sigma}_{\Delta^2 \mathbf{n}} (\mathbf{I} - \mathbf{A})^T \tag{10.171}$$

where we again assumed that $\mathbf{h}$ is constant within an utterance, so that $\Delta^2 \mathbf{h} = 0$.

Equations (10.163), (10.168), and (10.170) resemble the MLLR adaptation formulae of Chapter 9 for the means, though in this case the matrix is different for each Gaussian and is heavily constrained.

We are interested in estimating the environmental parameters $\boldsymbol{\mu}_{\mathbf{n}}$, $\boldsymbol{\mu}_{\mathbf{h}}$, and $\boldsymbol{\Sigma}_{\mathbf{n}}$, given a set of T observation frames $\mathbf{y}_t$. This estimation can be done iteratively using the EM algorithm on Eq. (10.162). If the noise process is stationary, $\boldsymbol{\Sigma}_{\Delta \mathbf{n}}$ could be approximated, assuming independence between $\mathbf{n}_{t+2}$ and $\mathbf{n}_{t-2}$, by $\boldsymbol{\Sigma}_{\Delta \mathbf{n}} = 2\boldsymbol{\Sigma}_{\mathbf{n}}$. Similarly, $\boldsymbol{\Sigma}_{\Delta^2 \mathbf{n}}$ could be approximated, assuming independence between $\Delta \mathbf{n}_{t+1}$ and $\Delta \mathbf{n}_{t-1}$, by $\boldsymbol{\Sigma}_{\Delta^2 \mathbf{n}} = 4\boldsymbol{\Sigma}_{\mathbf{n}}$. If the noise process is not stationary, it is best to estimate $\boldsymbol{\Sigma}_{\Delta \mathbf{n}}$ and $\boldsymbol{\Sigma}_{\Delta^2 \mathbf{n}}$ from input data directly.

If the distribution of $\mathbf{x}$ is a mixture of N Gaussians, each Gaussian is transformed according to the equations above. If the distribution of $\mathbf{n}$ is also a mixture of M Gaussians, the composite distribution has NM Gaussians. While this increases the number of Gaussians, the decoder is still functionally the same as for clean speech. Because normally you do not want to alter the number of Gaussians of the system when you do noise adaptation, it is often assumed that $\mathbf{n}$ is a single Gaussian.

10.6.5. Retraining on Compensated Features

We have discussed adapting the HMM to the new acoustical environment using the standard front-end features, in most cases the mel-cepstrum. Section 10.5 dealt with cleaning the noisy feature without retraining the HMMs. It's logical to consider a combination of both, where the features are cleaned to remove noise and channel effects and then the HMMs are retrained to take into account that this processing stage is not perfect. This idea is illustrated in Figure 10.36, where we compare the word error rate of the standard matched-noise-condition training with the matched-noise-condition training after it has been compensated by a variant of the mixture Gaussian algorithms described in Section 10.5.6 [21]. An improvement is obtained by retraining on compensated features, which beats the unprocessed matched-condition training.

The low error rates of both curves in Figure 10.36 are hard to obtain in practice, because they assume we know exactly what the noise level and type are ahead of time, which in general is hard to do. On the other hand, this could be combined with the multistyle training discussed in Section 10.6.1 or with a set of clustered models discussed in Chapter 9.

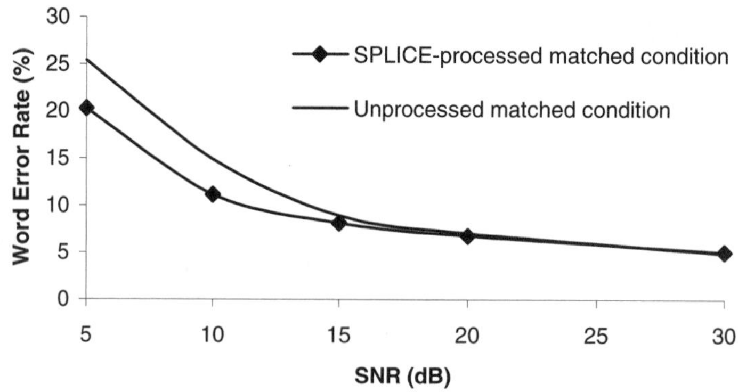

Figure 10.36 Word error rates of matched-noise training without feature preprocessing and with the SPLICE algorithm [21] as a function of the SNR in dB for additive white noise. Whisper is trained as in Figure 10.30. Error rate with the mixture Gaussian model is up to 30% lower than that of standard noisy matched conditions for low SNRs while it is about the same for high SNRs.

10.7. Modeling Nonstationary Noise

The previous sections deal mostly with stationary noise. In practice, there are many nonstationary noises that often match a random word in the system's lexicon better than the silence model. In this case, the benefit of using speech recognition vanishes quickly.

The most typical types of noise present in desktop applications are mouth noise (lip smacks, throat clearings, coughs, nasal clearings, heavy breathing, uhms and uhs, etc), computer noise (keyboard typing, microphone adjustment, computer fan, disk head seeking, etc.), and office noise (phone rings, paper rustles, shutting door, interfering speakers, etc.). We can use a simple method that has been successful in speech recognition [57], as shown in Algorithm 10.4. This method consists of adding noise words modeled with HMMs to absorb these nonstationary noises.

In practice, updating the transcription turns out to be important, because human labelers often miss short noises that the system can uncover. Since the noise training data are often limited in terms of coverage, some noises can be easily matched to short word models, such as: *if*, *two*. Due to the unique characteristics of noise rejection, we often need to further augment confidence measures such as those described in Chapter 9. In practice, we need an additional classifier to provide more detailed discrimination between speech and noise. We can use a two-level classifier for this purpose. The ratio between the *all-speech model* score (fully connected context-independent phone models) and the *all-noise model score* (fully connected silence and noise phone models) can be used.

Another approach [55] consists of having an HMM for noise with several states to deal with nonstationary noises. The decoder needs to conduct a three-dimensional Viterbi search which evaluates at each frame every possible speech state as well as every possible noise state to achieve the *speech/noise decomposition* (see Figure 10.37). The computational complexity of such an approach is very large, though it can handle nonstationary noises quite well in theory.

ALGORITHM 10.4: *EXPLICIT NOISE MODELING*

Step 1: Augmenting the vocabulary with *noise words* (such as ++SMACK++), each composed of a single *noise phoneme* (such as +SMACK+), which are thus modeled with a single HMM. These noise words have to be labeled in the transcriptions so that they can be trained.
Step 2: Training noise models, as well as the other models, using the standard HMM training procedure.
Step 3: Updating the transcription. To do that, convert the transcription into a network, where the noise words can be optionally inserted between each word in the original transcription. A forced alignment segmentation is then conducted with the current HMM optional noise words inserted. The segmentation with the highest likelihood is selected, thus yielding an optimal transcription.
Step 4: If converged, stop; otherwise go to Step 2.

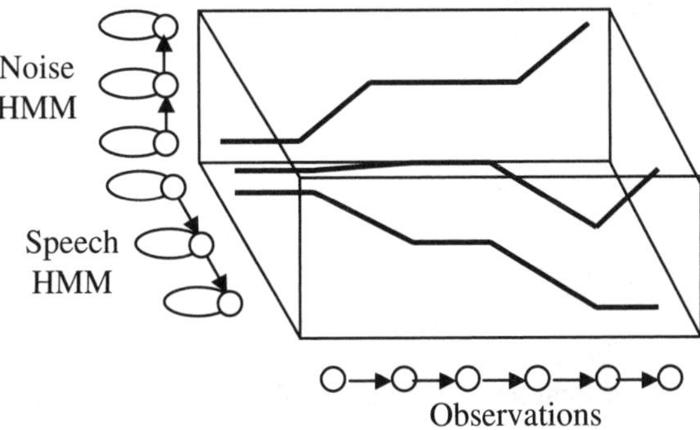

Figure 10.37 Speech noise decomposition and a three-dimensional Viterbi decoder.

10.8. HISTORICAL PERSPECTIVE AND FURTHER READING

This chapter contains a number of diverse topics that are often described in different fields; no single reference covers it all. For further reading on adaptive filtering, you can check the books by Widrow and Stearns [59] and Haykin [30]. Theodoridis and Bellanger provide [54] a good summary of adaptive filtering, and Breining et al. [16] a good summary of echo-canceling techniques. Lee [38] has a good summary of independent component analysis for blind source separation. Deller et al. [20] provide a number of techniques for speech enhancement. Juang [35] and Junqua [37] survey techniques used in improving the robustness of speech recognition systems to noise. Acero [2] compares a number of feature transformation techniques in the cepstral domain and introduces the model of the environment used in this chapter.

Adaptive filtering theory emerged early in the 1900s. The Wiener and LMS filters were derived by Wiener and Widrow in 1919 and 1960, respectively. Norbert Wiener joined the MIT faculty in 1919 and made profound contributions to generalized harmonic analysis, the famous Wiener-Hopf equation, and the resulting Wiener filter. The LMS algorithm was developed by Widrow and his colleagues at Stanford University in the early 1960s.

From a practical point of view, the use of gradient microphones (Olsen [46]) has proven to be one of the more important contributions to increased robustness. Directional microphones are commonplace today in most speech recognition systems.

Boll [13] first suggested the use of spectral subtraction. This has been the cornerstone for noise suppression, and many systems nowadays still use a variant of Boll's original algorithm.

The Cepstral mean normalization algorithm was proposed by Atal [8] in 1974, although it wasn't until the early 1990s that it became commonplace in most speech recognition systems evaluated in the DARPA speech programs [33]. Hermansky proposed PLP [31] in 1990. The work of Rich Stern's robustness group at CMU (especially the Ph.D. thesis work of Acero [1] and Moreno [43]) and the Ph.D. thesis of Gales [26] also represented advances in the understanding of the effect of noise in the cepstrum.

Bell and Sejnowski [10] gave the field of independent component analysis a boost in 1995 with their infomax rule. The field of source separation is a promising alternative to improve the robustness of speech recognition systems when more than one microphone is available.

REFERENCES

[1] Acero, A., Acoustical and Environmental Robustness in Automatic Speech Recognition, PhD Thesis in Electrical and Computer Engineering 1990, Carnegie Mellon University, Pittsburgh.

[2] Acero, A., Acoustical and Environmental Robustness in Automatic Speech Recognition, 1993, Boston, Kluwer Academic Publishers.

[3] Acero, A., S. Altschuler, and L. Wu, "Speech/Noise Separation Using Two Microphones and a VQ Model of Speech Signals," Int. Conf. on Spoken Language Processing, 2000, Beijing, China.

[4] Acero, A., et al., "HMM Adaptation Using Vector Taylor Series for Noisy Speech Recognition," Int. Conf. on Spoken Language Processing, 2000, Beijing, China.

[5] Acero, A. and X.D. Huang, "Augmented Cepstral Normalization for Robust Speech Recognition," Proc. of the IEEE Workshop on Automatic Speech Recognition, 1995, Snowbird, UT.

[6] Acero, A. and R. Stern, "Environmental Robustness in Automatic Speech Recognition," Int. Conf. on Acoustics, Speech and Signal Processing, 1990, Albuquerque, NM, pp. 849-852.

[7] Amari, S., A. Cichocki, and H.H. Yang, eds. A New Learning Algorithm for Blind Separation, Advances in Neural Information Processing Systems, 1996, Cambridge, MA, MIT Press.

[8] Atal, B.S., "Effectiveness of Linear Prediction Characteristics of the Speech Wave for Automatic Speaker Identification and Verification," Journal of the Acoustical Society of America, 1974, **55**(6), pp. 1304-1312.

[9] Attias, H., "Independent Factor Analysis," Neural Computation, 1998, **11**, pp. 803-851.

[10] Bell, A.J. and T.J. Sejnowski, "An Information Maximisation Approach to Blind Separation and Blind Deconvolution," Neural Computation, 1995, **7**(6), pp. 1129-1159.

[11] Belouchrani, A., et al., "A Blind Source Separation Technique Using Second Order Statistics," IEEE Trans. on Signal Processing, 1997, **45**(2), pp. 434-444.

[12] Berouti, M., R. Schwartz, and J. Makhoul, "Enhancement of Speech Corrupted by Acoustic Noise," Proc. of the IEEE Int. Conf. on Acoustics, Speech and Signal Processing, 1979, pp. 208-211.

[13] Boll, S.F., "Suppression of Acoustic Noise in Speech Using Spectral Subtraction," IEEE Trans. on Acoustics, Speech and Signal Processing, 1979, **27**(Apr.), pp. 113-120.

[14] Boll, S.F. and D.C. Pulsipher, "Suppression of Acoustic Noise in Speech Using Two Microphone Adaptive Noise Cancellation," IEEE Trans. on Acoustics Speech and Signal Processing, 1980, **28**(December), pp. 751-753.

[15] Bregman, A.S., Auditory Scene Analysis, 1990, Cambridge MA, MIT Press.

[16] Breining, C., Acoustic Echo Control, in IEEE Signal Processing Magazine, 1999. pp. 42-69.

[17] Cardoso, J., "Blind Signal Separation: Statistical Principles," Proc. of the IEEE, 1998, **9**(10), pp. 2009-2025.

[18] Cardoso, J.F., "Infomax and Maximum Likelihood for Blind Source Separation," IEEE Signal Processing Letters, 1997, **4**, pp. 112-114.

[19] Comon, P., "Independent Component Analysis: A New Concept," Signal Processing, 1994, **36**, pp. 287-314.

[20] Deller, J.R., J.H.L. Hansen, and J.G. Proakis, Discrete-Time Processing of Speech Signals, 2000, IEEE Press.

[21] Deng, L., et al., "Large-Vocabulary Speech Recognition Under Adverse Acoustic Environments," Int. Conf. on Spoken Language Processing, 2000, Beijing, China.

[22] Ephraim, Y., "Statistical Model-Based Speech Enhancement System," Proc. of the IEEE, 1992, **80**(1), pp. 1526-1555.

[23] Ephraim, Y. and D. Malah, "Speech Enhancement Using Minimum Mean Square Error Short Time Spectral Amplitude Estimator," IEEE Trans. on Acoustics, Speech and Signal Processing, 1984, **32**(6), pp. 1109-1121.

[24] Flanagan, J.L., et al., "Computer-Steered Microphone Arrays for Sound Transduction in Large Rooms," Journal of the Acoustical Society of America, 1985, **78**(5), pp. 1508-1518.

[25] Frost, O.L., "An Algorithm for Linearly Constrained Adaptive Array Processing," Proc. of the IEEE, 1972, **60**(8), pp. 926-935.

[26] Gales, M.J., Model Based Techniques for Noise Robust Speech Recognition, PhD Thesis in Engineering Department, 1995, Cambridge University.

[27] Ghitza, O., "Robustness against Noise: The Role of Timing-Synchrony Measurement," Proc. of the IEEE Int. Conf. on Acoustics, Speech and Signal Processing, 1987, pp. 2372-2375.

[28] Gopinath, R.A., et al., "Robust Speech Recognition in Noise—Performance of the IBM Continuous Speech Recognizer on the ARPA Noise Spoke Task," Proc. ARPA Workshop on Spoken Language Systems Technology, 1995, pp. 127-133.

[29] Griffiths, L.J. and C.W. Jim, "An Alternative Approach to Linearly Constrained Adaptive Beamforming," IEEE Trans. on Antennas and Propagation, 1982, **30**(1), pp. 27-34.

[30] Haykin, S., Adaptive Filter Theory, 2nd ed, 1996, Upper Saddle River, NJ, Prentice-Hall.

[31] Hermansky, H., "Perceptual Linear Predictive (PLP) Analysis of Speech," Journal of the Acoustical Society of America, 1990, **87**(4), pp. 1738-1752.

[32] Hermansky, H. and N. Morgan, "RASTA Processing of Speech," IEEE Trans. on Speech and Audio Processing, 1994, **2**(4), pp. 578-589.

[33] Huang, X.D., et al., "The SPHINX-II Speech Recognition System: An Overview," Computer Speech and Language, 1993, pp. 137-148.

[34] Hunt, M. and C. Lefebre, "A Comparison of Several Acoustic Representations for Speech Recognition with Degraded and Undegraded Speech," Int. Conf. on Acoustic, Speech and Signal Processing, 1989, pp. 262-265.

[35] Juang, B.H., "Speech Recognition in Adverse Environments," Computer Speech and Language, 1991, **5**, pp. 275-294.

[36] Junqua, J.C., "The Lombard Reflex and Its Role in Human Listeners and Automatic Speech Recognition," Journal of the Acoustical Society of America, 1993, **93**(1), pp. 510-524.

[37] Junqua, J.C. and J.P. Haton, Robustness in Automatic Speech Recognition, 1996, Kluwer Academic Publishers.

[38] Lee, T.W., Independent Component Analysis: Theory and Applications, 1998, Kluwer Academic Publishers.

[39] Lippmann, R.P., E.A. Martin, and D.P. Paul, "Multi-Style Training for Robust Isolated-Word Speech Recognition," Int. Conf. on Acoustics, Speech and Signal Processing, 1987, Dallas, TX, pp. 709-712.

[40] Lombard, E., "Le Signe de l'élévation de la Voix," Ann. Maladies Oreille, Larynx, Nez, Pharynx, 1911, **37**, pp. 101-119.

[41] Matassoni, M., M. Omologo, and D. Giuliani, "Hands-Free Speech Recognition Using a Filtered Clean Corpus and Incremental HMM Adaptation," Proc. Int. Conf. on Acoustics, Speech and Signal Processing, 2000, Istanbul, Turkey, pp. 1407-1410.

[42] Mendel, J.M., Lessons in Estimation Theory for Signal Processing, Communications, and Control, 1995, Upper Saddle River, NJ, Prentice Hall.

[43] Moreno, P., Speech Recognition in Noisy Environments, PhD Thesis in Electrical and Computer Engineering 1996, Carnegie Mellon University, Pittsburgh.

[44] Moreno, P.J., B. Raj, and R.M. Stern, "A Vector Taylor Series Approach for Environment Independent Speech Recognition," Int. Conf. on Acoustics, Speech and Signal Processing, 1996, Atlanta, pp. 733-736.

[45] Morgan, N. and H. Bourlard, Continuous Speech Recognition: An Introduction to Hybrid HMM/Connectionist Approach, in IEEE Signal Processing Magazine, 1995, pp. 25-42.

[46] Olsen, H.F., "Gradient Microphones," Journal of the Acoustical Society of America, 1946, **17**,(3), pp. 192-198.

[47] Porter, J.E. and S.F. Boll, "Optimal Estimators for Spectral Restoration of Noisy Speech," Proc. of the IEEE Int. Conf. on Acoustics, Speech and Signal Processing, 1984, San Diego, CA, pp. 18.A.2.1-4.

[48] Rahim, M.G. and B.H. Juang, "Signal Bias Removal by Maximum Likelihood Estimation for Robust Telephone Speech Recognition," IEEE Trans. on Speech and Audio Processing, 1996, **4**(1), pp. 19-30.

[49] Seneff, S., "A Joint Synchrony/Mean-Rate Model of Auditory Speech Processing," Journal of Phonetics, 1988, **16**(1), pp. 55-76.

[50] Sharma, S., et al., "Feature Extraction Using Non-Linear Transformation for Robust Speech Recognition on the Aurora Database," Int. Conf. on Acoustics, Speech and Signal Processing, 2000, Istanbul, Turkey, pp. 1117-1120.

[51] Sullivan, T.M. and R.M. Stern, "Multi-Microphone Correlation-Based Processing for Robust Speech Recognition," Int. Conf. on Acoustics, Speech and Signal Processing, 1993, Minneapolis, pp. 2091-2094.

[52] Suzuki, Y., et al., "An Optimum Computer-Generated Pulse Signal Suitable for the Measurement of Very Long Impulse Responses," Journal of the Acoustical Society of America, 1995, **97**(2), pp. 1119-1123.

[53] Tamura, S. and A. Waibel, "Noise Reduction Using Connectionist Models," Int. Conf. on Acoustics, Speech and Signal Processing, 1988, New York, pp. 553-556.

[54] Theodoridis, S. and M.G. Bellanger, Adaptive Filters and Acoustic Echo Control, in IEEE Signal Processing Magazine, 1999, pp. 12-41.

[55] Varga, A.P. and R.K. Moore, "Hidden Markov Model Decomposition of Speech and Noise," Proc. of the IEEE Int. Conf. on Acoustics, Speech and Signal Processing, 1990 pp. 845-848.

[56] Wan, E.A., R.V.D. Merwe, and A.T. Nelson, "Dual Estimation and the Unscented Transformation" in Advances in Neural Information Processing Systems, S.A. Solla, T.K. Leen, and K.R. Muller, eds. 2000, Cambridge, MA, MIT Press, pp. 666-672.

[57] Ward, W., "Modeling Non-Verbal Sounds for Speech Recognition," Proc. Speech and Natural Language Workshop, 1989, Cape Cod, MA, Morgan Kauffman, pp. 311-318.

[58] Widrow, B. and M.E. Hoff, "Adaptive Switching Algorithms," IRE Wescon Convention Record, 1960, pp. 96-104.

[59] Widrow, B. and S.D. Stearns, Adaptive Signal Processing, 1985, Upper Saddle River, NJ, Prentice Hall.

[60] Woodland, P.C., "Improving Environmental Robustness in Large Vocabulary Speech Recognition," Int. Conf. on Acoustics, Speech and Signal Processing, 1996, Atlanta, Georgia, pp. 65-68.

CHAPTER 11

Language Modeling

Acoustic pattern matching, as discussed in Chapter 9, and knowledge about language are equally important in recognizing and understanding natural speech. Lexical knowledge (i.e., vocabulary definition and word pronunciation) is required, as are the syntax and semantics of the language (the rules that determine what sequences of words are grammatically well-formed and meaningful). In addition, knowledge of the pragmatics of language (the structure of extended discourse, and what people are likely to say in particular contexts) can be important to achieving the goal of spoken language understanding systems. In practical speech recognition, it may be impossible to separate the use of these different levels of knowledge, since they are often tightly integrated.

In this chapter we review the basic concept of Chomsky's formal language theory and the probabilistic language model. For the formal language model, two things are fundamental: the grammar and the parsing algorithm. The *grammar* is a formal specification of the permissible structures for the language. The *parsing* technique is the method of analyzing the sentence to see if its structure is compliant with the grammar. With the advent of bodies of text (*corpora*) that have had their structures hand-annotated, it is now possible to generalize the formal grammar to include accurate probabilities. Furthermore, the probabilistic relationship among a sequence of words can be directly derived and modeled from the corpora with the so-called stochastic language models, such as *n*-gram, avoiding the need to create broad coverage formal grammars. Stochastic language models play a critical role in building a working spoken language system, and we discuss a number of important issues associated with them.

11.1. Formal Language Theory

In constructing a syntactic grammar for a language, it is important to consider the generality, the selectivity, and the understandability of the grammar. The *generality* and *selectivity* basically determine the range of sentences the grammar accepts and rejects. The *understandability* is important, since it is up to the authors of the system to create and maintain the grammar. For SLU systems described in Chapter 17, we need to have a grammar that covers and generalizes to most of the typical sentences for an application. The system also needs to distinguish the kind of sentences for different actions in a given application. Without understandability, it is almost impossible to improve a practical SLU system since it typically involves a large number of developers to maintain and refine the grammar.

The most common way of representing the grammatical structure of a sentence, "*Mary loves that person,*" is by using a tree, as illustrated in Figure 11.1. The node labeled *S* is the parent node of the nodes labeled *NP* and *VP* for noun phrase and verb phrase, respectively. The *VP* node is the parent node of node *V*—for verb. Each leaf is associated with the word

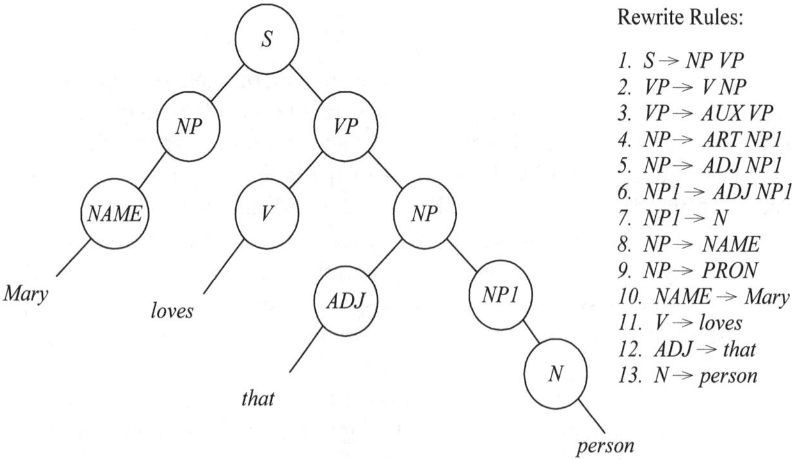

Figure 11.1 A tree representation of a sentence and its corresponding grammar.

in the sentence to be analyzed. To construct such a tree for a sentence, we must know the structure of the language so that a set of rewrite rules can be used to describe what tree structures are allowable. These rules, as illustrated in Figure 11.1, determine that a certain symbol may be expanded in the tree by a sequence of symbols. The grammatical structure helps in determining the meaning of the sentence. It tells us that *that* in the sentence modifies *person*. "Mary loves *that* person."

11.1.1. Chomsky Hierarchy

In Chomsky's *formal language theory* [1, 14, 15], a grammar is defined as $G = (V, T, P, S)$, where V and T are finite sets of *non-terminals* and *terminals*, respectively. V contains all the *non-terminal* symbols. We often use upper-case symbols to denote them. In the example discussed here, *S, NP, NP1, VP, NAME, ADJ, N*, and *V* are non-terminal symbols. The *terminal* set T contains *Mary, loves, that*, and *person*, which are often denoted with lower-case symbols. P is a finite set of *production (rewrite) rules*, as illustrated in the rewrite rules in Figure 11.1. S is a special non-terminal, called the *start symbol*.

The language to be analyzed is essentially a string of terminal symbols, such as "*Mary loves that person.*" It is produced by applying production rules sequentially to the start symbol. The production rule is of the form $\alpha \rightarrow \beta$, where α and β are arbitrary strings of grammar symbols V and T, and the α must not be empty. In formal language theory, four major languages and their associated grammars are hierarchically structured. They are referred to as the Chomsky hierarchy [1] as defined in Table 11.1. There are four kinds of automata that can accept the languages produced by these four types of grammars. Among these automata, the finite-state automaton is not only the mathematical device used to implement the regular grammar but also one of the most significant tools in computational linguistics. Variations of automata such as finite-state transducers, hidden Markov models, and *n*-gram models are important examples in spoken language processing.

These grammatical formulations can be compared according to their generative capacity, i.e., the range that the formalism can cover. While there is evidence that natural languages are at least weakly context sensitive, the context-sensitive requirements are rare in practice. The context-free grammar (CFG) is a very important structure for dealing with both machine language and natural language. CFGs are not only powerful enough to describe most of the structure in spoken language,[1] but also restrictive enough to have efficient parsers to analyze natural sentences. Since CFGs offer a good compromise between parsing efficiency and power in representing the structure of the language, they have been widely applied to natural language processing. Alternatively, regular grammars, as represented with a finite-state machine, can be applied to more restricted applications. Since finite-state grammars are a subset of the more general context-free grammar, we focus our discussion on context-free grammars only, although the parsing algorithm for finite-state grammars can be more efficient.

[1] The effort to prove natural languages are not context-free is summarized in Pullum and Gazdar [54].

Table 11.1 Chomsky hierarchy and the corresponding machine that accepts the language.

Types	Constraints	Automata
Phrase structure grammar	$\alpha \to \beta$. This is the most general grammar.	Turing machine
Context-sensitive grammar	A subset of the phrase structure grammar. $\lvert\alpha\rvert \le \lvert\beta\rvert$, where $\lvert.\rvert$ indicates the length of the string.	Linear bounded automata
Context-free grammar (CFG)	A subset of the context sensitive grammar. The production rule is $A \to \beta$, where *A* is a non-terminal. This production rule is shown to be equivalent to Chomsky normal form: $A \to w$ and $A \to BC$, where *w* is a terminal and *B, C* are non-terminals.	Push down automata
Regular grammar	A subset of the CFG. The production rule is expressed as: $A \to w$ and $A \to wB$.	Finite-state automata

As discussed in Section 11.1.2, a parsing algorithm offers a procedure that searches through various ways of combining grammatical rules to find a combination that generates a tree to illustrate the structure of the input sentence, which is similar to the search problem in speech recognition. The result of the parsing algorithm is a parse tree,[2] which can be regarded as a record of the CFG rules that account for the structure of the sentence. In other words, if we parse the sentence, working either top-down from *S* or bottom-up from each word, we automatically derive something that is similar to the tree representation, as illustrated in Figure 11.1.

A push-down automaton is also called a *recursive transition network* (RTN), which is an alternative formalism to describe context-free grammars. A transition network consists of nodes and labeled arcs. One of the nodes is specified as the initial state *S*. Starting at the initial state, we traverse an arc if the current word in the sentence is in the category on the arc. If the arc is followed, the current word is updated to the next word. A phrase can be parsed if there is a path from the starting node to a *pop* arc that indicates a complete parse for all the words in the phrase. Simple transition networks without recursion are often called *finite-state machines* (FSM). Finite-state machines are equivalent in expressive power to regular grammars and, thus, are not powerful enough to describe all languages that can be described by CFGs. Chapter 12 has a more detailed discussion on RTNs and FSMs used in speech recognition.

[2] The result can be more than one parse tree since natural language sentences are often ambiguous. In practice, a parsing algorithm should not only consider all the possible parse trees but also provide a ranking among them, as discussed in Chapter 17.

11.1.2. Chart Parsing for Context-Free Grammars

Since Chomsky introduced the notion of context-free grammars in the 1950s, a vast literature has arisen on the parsing algorithms. Most parsing algorithms were developed in computer science to analyze programming languages that are not ambiguous in the way that spoken language is [1, 32]. We discuss only the most relevant materials that are fundamental to building spoken language systems, namely the chart parser for the context-free grammar. This algorithm has been widely used in state-of-the-art spoken language understanding systems.

11.1.2.1. Top Down or Bottom Up?

Parsing is a special case of the search problem generally encountered in speech recognition. A parsing algorithm offers a procedure that searches through various ways of combining grammatical rules to find a combination that generates a tree to describe the structure of the input sentence, as illustrated in Figure 11.1. The search procedure can start from the root of the tree with the *S* symbol, attempting to rewrite it into a sequence of terminal symbols that matches the words in the input sentence, which is based on *goal-directed search*. Alternatively, the search procedure can start from the words in the input sentence and identify a word sequence that matches some non-terminal symbol. The bottom-up procedure can be repeated with partially parsed symbols until the root of the tree or the start symbol *S* is identified. This *data-directed search* has been widely used in practical SLU systems.

A top-down approach starts with the *S* symbol, then searches through different ways to rewrite the symbols until the input sentence is generated, or until all possibilities have been examined. A grammar is said to accept a sentence if there is a sequence of rules that allow us to rewrite the start symbol into the sentence. For the grammar in Figure 11.1, a sequence of rewrite rules can be illustrated as follows:

```
S
→ NP VP (rewriting S using S→NP)
→NAME VP (rewriting NP using NP→NAME)
→Mary VP (rewriting NAME using NAME→Mary)
...
→Mary loves that person (rewriting N using N→person)
```

Alternatively, we can take a bottom-up approach to start with the words in the input sentence and use the rewrite rules backward to reduce the sequence of symbols until it becomes S. The left-hand side of each rule is used to rewrite the symbol on the right-hand side as follows:

```
→NAME loves that person (rewriting Mary using NAME→Mary)
→NAME V that person (rewriting loves using V→loves)
...
→NP VP
→S (rewriting NP using S→NP VP)
```

A parsing algorithm must systematically explore every possible state that represents the intermediate node in the parsing tree. If a mistake occurs early on in choosing the rule that rewrites *S*, the intermediate parser results can be quite wasteful if the number of rules becomes large.

The main difference between top-down and bottom-up parsers is the way the grammar rules are used. For example, consider the rule *NP→ADJ NP1*. In a top-down approach, the rule is used to identify an *NP* by looking for the sequence *ADJ NP1*. Top-down parsing can be very predictive. A phrase or a word may be ambiguous in isolation. The top-down approach may prevent some ungrammatical combinations from consideration. It never wastes time exploring trees that cannot result in an *S*. On the other hand, it may predict many different constituents that do not have a match to the input sentence and rebuild large constituents again and again. For example, when the grammar is *left-recursive* (i.e., it contains a non-terminal category that has a derivation that includes itself anywhere along its leftmost branch), the top-down approach can lead a top-down, depth-first left-to-right parser to recursively expand the same non-terminal over again in exactly the same way. This causes an infinite expansion of trees. In contrast, a bottom-up parser takes a sequence *ADJ NP1* and identifies it as an *NP* according to the rule. The basic operation in bottom-up parsing is to take a sequence of symbols and match it to the right-hand side of the rules. It checks the input only once, and only builds each constituent exactly once. However, it may build up trees that have no hope of leading to *S* since it never suggests trees that are not at least locally grounded in the actual input. Since bottom-up parsing is similar to top-down parsing in terms of overall performance and is particularly suitable for robust spoken language processing as described in Chapter 17, we use the bottom-up method as our example to understand the key concept in the next section.

11.1.2.2. Bottom-Up Chart Parsing

As a standard search procedure, the state of the search consists of a symbol list, starting with the words in the sentence. Successor states can be generated by exploring all possible ways to replace a sequence of symbols that matches the right-hand side of a grammar rule with its left-hand side symbol. A simple-minded solution enumerates all the possible matches, leading to prohibitively expensive computational complexity. To avoid this problem, it is necessary to store partially parsed results of the matching, thereby eliminating duplicate work. This is the same technique that has been widely used in dynamic programming, as described in Chapter 8. Since chart parsing does not need to be from left to right, it is more efficient than the graph search algorithm discussed in Chapter 12, which can be used to parse the input sentence from left to right.

A data structure, called a *chart,* is used to allow the parser to store the partial results of the matching. The chart data structure maintains not only the records of all the constituents derived from the sentence so far in the parse tree, but also the records of rules that have matched partially but are still incomplete. These are called *active arcs*. Here, matches are always considered from the point of view of some *active constituents*, which represent the subparts that the input sentence can be divided into according to the rewrite rules. Active constituents are stored in a data structure called an *agenda*. To find grammar rules that match a string involving the active constituent, we need to identify rules that start with the active constituent or rules that have already been started by earlier active constituents and require the current constituent to complete the rule or to extend the rule. The basic operation of a chart-based parser involves combining these partially matched records (active arcs) with a completed constituent to form either a new completed constituent or a new partially matched (but incomplete) constituent that is an extension of the original partially matched constituent. Just like the graph search algorithm, we can use either a depth-first or breadth-first search strategy, depending on how the agenda is implemented. If we use probabilities or other heuristics, we take the best-first strategy discussed in Chapter 12 to select constituents from the agenda. The chart-parser process is defined more precisely in Algorithm 11.1. It is possible to combine both top-down and bottom-up. The major difference is how the constituents are used.

ALGORITHM 11.1: *A BOTTOM-UP CHART PARSER*

Step1: Initialization: Define a list called chart to store active arcs, and a list called an agenda to store active constituents until they are added to the chart.
Step 2: Repeat: Repeat Step 2 to 7 until there is no input left.
Step 3: Push and pop the agenda: If the agenda is empty, look up the interpretations of the next word in the input and push them to the agenda. Pop a constituent *C* from the agenda. If *C* corresponds to position from w_i to w_j of the input sentence, we denote it *C[i,j]*.
Step 4: Add *C* to the chart: Insert *C[i,j]* into the chart.
Step 5: Add key-marked active arcs to the chart: For each rule in the grammar of the form $X \rightarrow C\ Y$, add to the chart an active arc (partially matched constituent) of the form $X[i,j] \rightarrow {}^\circ C Y$, where $^\circ$ denotes the critical position called the key that indicates that everything before $^\circ$ has been seen, but things after $^\circ$ are yet to be matched (incomplete constituent).
Step 6: Move $^\circ$ forward: For any active arc of the form $X[1,j] \rightarrow Y \ldots {}^\circ C \ldots Z$ (everything before w_i) in the chart, add a new active arc of the form $X[1,j] \rightarrow Y \ldots C^\circ \ldots Z$ to the chart.
Step 7: Add new constituents to the agenda: For any active arc of the form $X[1,I] \rightarrow Y \ldots {}^\circ C$, add a new constituent of type *X[1,j]* to the agenda.
Step 8: Exit: If *S[1,n]* is in the chart, where *n* is the length of the input sentence, we can exit successfully unless we want to find all possible interpretations of the sentence. The chart may contain many *S* structures covering the entire set of positions.

Let us look at an example to see how the chart parser parses the sentence *Mary loves that person* using the grammar specified in Figure 11.1. We first create the chart and agenda data structure as illustrated in Figure 11.2 (a), in which the leaves of the tree-like chart data structure corresponds to the position of each input word. The parent of each block in the chart covers from the position of the left child's corresponding starting word position to the right child's corresponding ending word position. Thus, the root block in the chart covers the whole sentence from the first word *Mary* to the last word *person*. The chart parser scans through the input words to match against possible rewrite rules in the grammar. For the first word, the rule *Name→Mary* can be matched, so it is added to the agenda according to Step 3 in Algorithm 11.1. In Step 4, *Name→Mary* is added to the chart from the agenda. After the word Mary is processed, we have *Name→Mary, NP→Name*, and *S→NP°VP* in the chart, as illustrated in Figure 11.2 (b). *NP°VP* in the chart indicates that ° has reached the point at which everything before ° has been matched (in this case *Mary* matched *NP*) but everything after ° is yet to be parsed. The completed parsed chart is illustrated in Figure 11.2 (c).

A parser may assign one or more parsed structures to the sentence in the language it defines. If any sentence is assigned more than one such structure, the grammar is said to be ambiguous. Spoken language is, of course, ambiguous by nature.[3] For example, we can have a sentence like *Mary sold the student bags*. It is unclear whether *student* should be the modifier for *bags* or whether it means that *Mary* sold the *bags* to *the student*.

Chart parsers can be fairly efficient simply because the same constituent is never constructed more than once. In the worst case, the chart parser builds every possible constituent between every possible pair of positions, leading to the worst-case computational complexity of $O(n^3)$, where n is the length of the input sentence. This is still far more efficient than a straightforward brute-force search.

In many practical tasks, we need only a partial parse or shallow parse of the input sentence. You can use cascades of finite-state automata instead of CFGs. Relying on simple finite-state automata rather than full parsing makes such systems more efficient, although finite-state systems cannot model certain kinds of recursive rules, so that efficiency is traded for a certain lack of coverage.

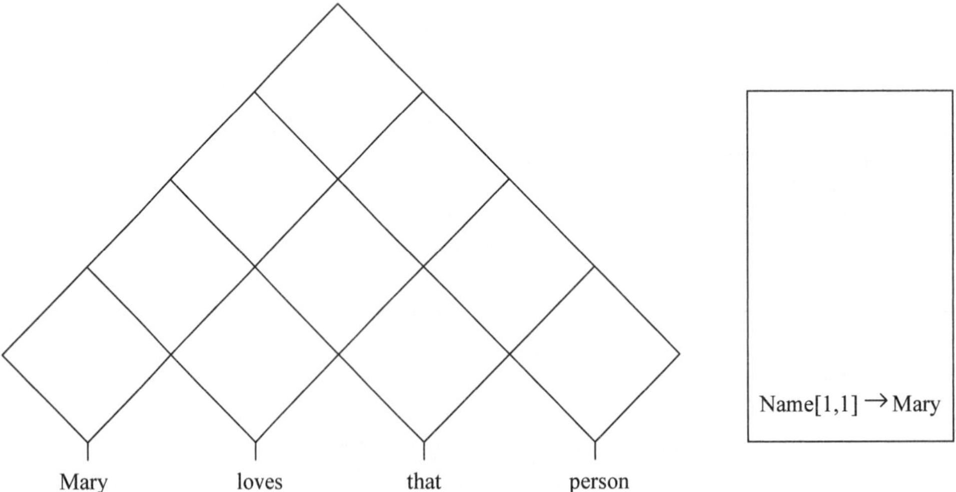

(a) The chart is illustrated on the left, and the agenda is on the right. The agenda now has one rule in it according to Step 3, since the agenda is empty.

[3] The same parse tree can also mean multiple things, so a parse tree itself does not define meaning. " *Mary loves that person*" could be sarcastic and mean something different.

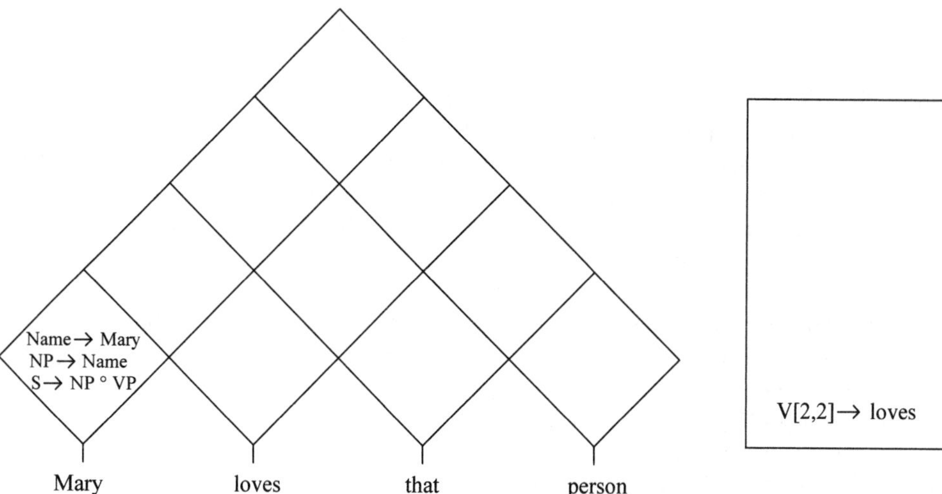

(b) After Mary, the chart now has rules *Name→Mary, NP→Name*, and *S→NP°VP.*

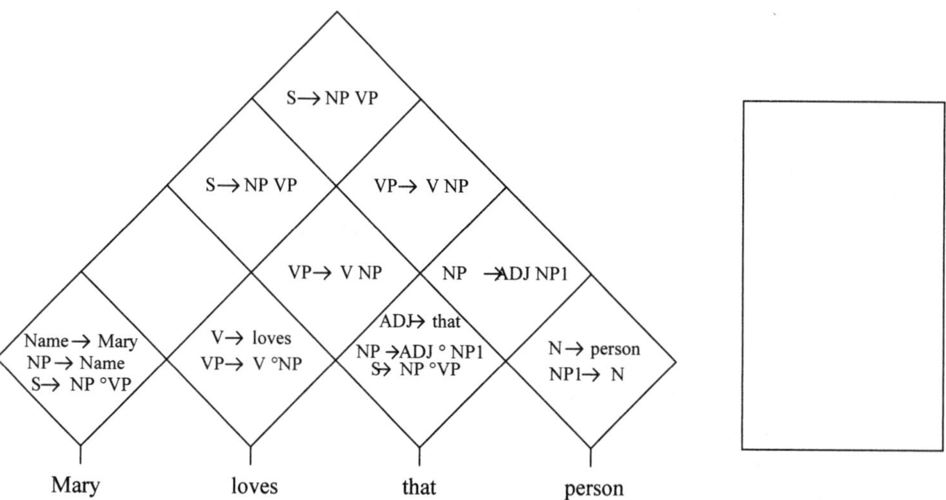

(c) The chart after the whole sentence is parsed. S→ NP VP covers the whole sentence, indicating that the sentence is parsed successfully by the grammar.

Figure 11.2 An example of a chart parser with the grammar illustrated in Figure 11.1. Parts (a) and (b) show the initial chart and agenda to parse the first word; part (c) shows the chart after the sentence is completely parsed.

11.2. STOCHASTIC LANGUAGE MODELS

Stochastic language models (SLM) take a probabilistic viewpoint of language modeling. We need to accurately estimate the probability *P*(**W**) for a given word sequence $\mathbf{W} = w_1 w_2 ... w_n$. In the formal language theory discussed in Section 11.1, *P*(**W**) can be regarded as 1 or 0 if the word sequence is accepted or rejected, respectively, by the grammar. This may be inappropriate for spoken language systems, since the grammar itself is unlikely to have a complete coverage, not to mention that spoken language is often ungrammatical in real conversational applications.

The key goal of SLM is to provide adequate probabilistic information so that likely word sequences should have a higher probability. This not only makes speech recognition more accurate but also helps to dramatically constrain the search space for speech recognition (see Chapters 12 and 13). Notice that SLM can have a wide coverage on all the possible word sequences, since probabilities are used to differentiate different word sequences. The most widely used SLM is the so call *n*-gram model discussed in this chapter. In fact, the CFG can be augmented as the bridge between the *n*-gram and the formal grammar if we can incorporate probabilities into the production rules, as discussed in the next section.

11.2.1. Probabilistic Context-Free Grammars

The CFG can be augmented with probability for each production rule. The advantages of probabilistic CFGs (PCFGs) lie in their ability to more accurately capture the embedded usage structure of spoken language to minimize syntactic ambiguity. The use of probability becomes increasingly important to discriminate many competing choices when the number of rules is large.

In the PCFG, we have to address the parallel problems we discussed for HMMs in Chapter 8. The *recognition problem* is concerned with the computation of the probability of the start symbol *S* generating the word sequence $\mathbf{W} = w_{1,} w_{2.} \ldots w_T$, given the grammar *G*:

$$P(S \Rightarrow \mathbf{W} | G) \tag{11.1}$$

where $\Rightarrow$ denotes a derivation sequence consisting of one or more steps. This is equivalent to the chart parser augmented with probabilities, as discussed in Section 11.1.2.2.

The *training problem* is concerned with determining a set of rules *G* based on the training corpus and estimating the probability of each rule. If the set of rules is fixed, the simplest approach to deriving these probabilities is to count the number of times each rule is used in a corpus containing parsed sentences. We denote the probability of a rule $A \rightarrow \alpha$ by $P(A \rightarrow \alpha | G)$. For instance, if there are *m* rules for left-hand side non-terminal node A: $A \rightarrow \alpha_1, A \rightarrow \alpha_2, \ldots A \rightarrow \alpha_m$, we can estimate the probability of these rules as follows:

$$P(A \rightarrow \alpha_j \mid G) = C(A \rightarrow \alpha_j) / \sum_{i=1}^{m} C(A \rightarrow \alpha_i) \tag{11.2}$$

where *C(.)* denotes the number of times each rule is used.

When you have hand-annotated corpora, you can use the maximum likelihood estimation as illustrated by Eq. (11.2) to derive the probabilities. When you don't have hand-annotated corpora, you can extend the EM algorithm (see Chapter 4) to derive these probabilities. The algorithm is also known as the *inside-outside* algorithm. As we discussed in Chapter 8, you can develop algorithms similar to the Viterbi algorithm to find the most likely parse tree that could have generated the sequence of words $P(\mathbf{W})$ after these probabilities are estimated.

We can make certain independence assumptions about rule usage. Namely, we assume that the probability of a constituent being derived by a rule is independent of how the constituent is used as a subconstituent. For instance, we assume that the probabilities of *NP* rules are the same no matter whether the *NP* is used for the subject or the object of a verb, although the assumptions are not valid in many cases. More specifically, let the word sequence $\mathbf{W}=w_1, w_2 \ldots w_T$ be generated by a PCFG *G*, with rules in Chomsky normal form as discussed in Section 11.1.1:

$$A_i \rightarrow A_m A_n \text{ and } A_i \rightarrow w_l \tag{11.3}$$

where A_m and A_n are two possible non-terminals that expand A_i at different locations. The probability for these rules must satisfy the following constraint:

$$\sum_{m,n} P(A_i \rightarrow A_m A_n \mid G) + \sum_l P(A_i \rightarrow w_l \mid G) = 1, \quad \text{for all } i \tag{11.4}$$

Equation (11.4) simply means that all non-terminals can generate either pairs of non-terminal symbols or a single terminal symbol, and all these production rules should satisfy the probability constraint. Analogous to the HMM forward and backward probabilities discussed in Chapter 8, we can define the inside and outside probabilities to facilitate the estimation of these probabilities from the training data.

A non-terminal symbol A_i can generate a sequence of words $w_j w_{j+1} \ldots w_k$; we define the probability of $Inside(j, A_i, k) = P(A_i \Rightarrow w_j w_{j+1} \ldots w_k \mid G)$ as the *inside constituent probability*, since it assigns a probability to the word sequence inside the constituent. The inside probability can be computed recursively. When only one word is emitted, the transition rule of the form $A_i \rightarrow w_m$ applies. When there is more than one word, rules of the form $A_i \rightarrow A_m A_n$ must apply. The inside probability of $inside(j, A_i, k)$ can be expressed recursively as follows:

$$\begin{aligned} inside(j, A_i, k) &= P(A_i \Rightarrow w_j w_{j+1} \ldots w_k) \\ &= \sum_{n,m} \sum_{l=j}^{k-1} P(A_i \rightarrow A_m A_n) P(A_m \Rightarrow w_j \ldots w_l) P(A_n \Rightarrow w_{l+1} \ldots w_k) \\ &= \sum_{n,m} \sum_{l=j}^{k-1} P(A_i \rightarrow A_m A_n) inside(j, A_m, l) inside(l+1, A_n, k) \end{aligned} \tag{11.5}$$

The inside probability is the sum of the probabilities of all derivations for the section over the span of j to k. One possible derivation of the form can be drawn as a parse tree shown in Figure 11.3.

Another useful probability is the *outside* probability for a non-terminal node A_i covering w_s to w_t, in which they can be derived from the start symbol S, as illustrated in Figure 11.4, together with the rest of the words in the sentence:

$$outside(s, A_i, t) = P(S \Rightarrow w_1 \ldots w_{s-1}\ A_i\ w_{t+1} \ldots w_T) \tag{11.6}$$

After the inside probabilities are computed bottom-up, we can compute the outside probabilities top-down. For each non-terminal symbol A_i, there are one of two possible configurations $A_m \rightarrow A_n A_i$ or $A_m \rightarrow A_i A_n$ as illustrated in Figure 11.5. Thus, we need to consider all the possible derivations of these two forms as follows:

$$\begin{aligned}
outside(s, A_i, t) &= P(S \Rightarrow w_1 \ldots w_{s-1}\ A_i\ w_{t+1} \ldots w_T) \\
&= \sum_{m,n} \left\{ \begin{array}{l} \sum_{l=1}^{s-1} P(A_m \rightarrow A_n A_i) P(A_n \Rightarrow w_l \ldots w_{s-1}) P(S \Rightarrow w_1 \ldots w_{l-1}\ A_m\ w_{t+1} \ldots w_T) + \\ + \sum_{l=t+1}^{T} P(A_m \rightarrow A_i A_n) P(A_n \Rightarrow w_{t+1} \ldots w_l) P(S \Rightarrow w_1 \ldots w_{s-1}\ A_m\ w_{l+1} \ldots w_T) \end{array} \right\} \\
&= \sum_{m,n} \left\{ \begin{array}{l} \sum_{l=1}^{s-1} P(A_m \rightarrow A_n A_i) inside(l, A_n, s-1) outside(l, A_m, t) + \\ + \sum_{l=t+1}^{T} P(A_m \rightarrow A_i A_n) inside(t+1, A_n, l) outside(s, A_m, l) \end{array} \right\}
\end{aligned} \tag{11.7}$$

The inside and outside probabilities are used to compute the sentence probability as follows:

$$P(S \Rightarrow w_1 \ldots w_T) = \sum_i inside(s, A_i, t) outside(s, A_i, t) \qquad \text{for any } s \le t \tag{11.8}$$

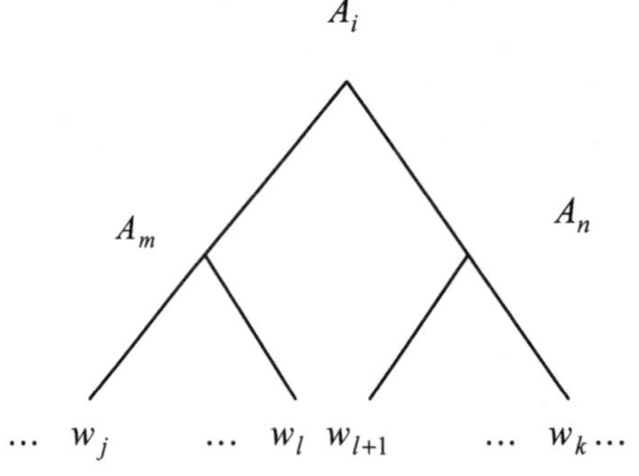

Figure 11.3 Inside probability is computed recursively as sum of all the derivations.

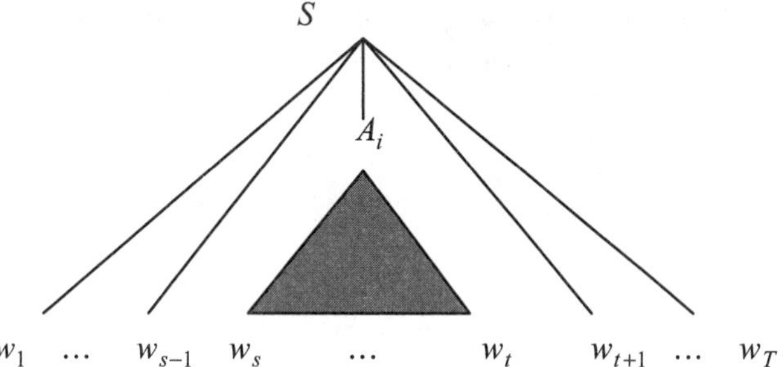

Figure 11.4 Definition of the outside probability.

Since *outside*$(1, A_i, T)$ is equal to 1 for the starting symbol only, the probability for the whole sentence can be conveniently computed using the inside probability alone as

$$P(S \Rightarrow \mathbf{W}|G) = inside(1, S, T) \tag{11.9}$$

We are interested in the probability that a particular rule, $A_i \rightarrow A_m A_n$ is used to cover a span $w_s \ldots w_t$, given the sentence and the grammar:

$$\begin{aligned}\xi(i,m,n,s,t) &= P(A_i \Rightarrow w_s ... w_t, A_i \rightarrow A_m A_n \mid S \Rightarrow \mathbf{W}, G) \\ &= \frac{1}{P(S \Rightarrow \mathbf{W} \mid G)} \sum_{k=s}^{t-1} P(A_i \rightarrow A_m A_n \mid G) inside(s, A_m, k) inside(k+1, A_n, t) outside(s, A_i, t)\end{aligned} \tag{11.10}$$

These conditional probabilities form the basis of the inside-outside algorithm, which is similar to the forward-backward algorithm discussed in Chapter 8. We can start with some initial probability estimates. For each sentence of training data, we determine the inside and outside probabilities in order to compute, for each production rule, how likely it is that the production rule is used as part of the derivation of that sentence. This gives us the number of counts for each production rule in each sentence. Summing these counts across sentences gives us an estimate of the total number of times each production rule is used to produce the

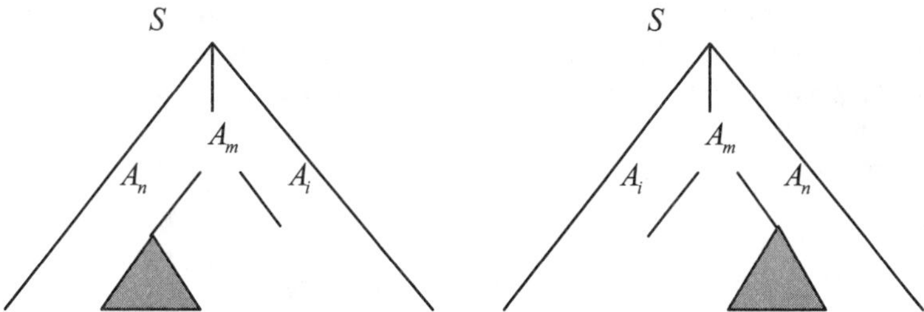

Figure 11.5 Two possible configurations for a non-terminal node A_i.

sentences in the training corpus. Dividing by the total counts of productions used for each non-terminal gives us a new estimate of the probability of the production in the MLE framework. For example, we have:

$$P(A_i \to A_m A_n | G) = \frac{\sum_{s=1}^{T-1} \sum_{t=s+1}^{T} \xi(i,m,n,s,t)}{\sum_{m,n} \sum_{s=1}^{T-1} \sum_{t=s+1}^{T} \xi(i,m,n,s,t)} \tag{11.11}$$

In a similar manner, we can estimate $P(A_i \to w_m | G)$. It is also possible to let the inside-outside algorithm formulate all the possible grammar production rules so that we can select rules with sufficient probability values. If there is no constraint, we may have too many *greedy symbols* that serve as possible non-terminals. In addition, the algorithm is guaranteed only to find a local maximum. It is often necessary to use prior knowledge about the task and the grammar to impose strong constraints to avoid these two problems. The chart parser discussed in Section 11.1.2 can be modified to accommodate PCFGs [29, 45].

One problem with the PCFG is that it assumes that the expansion of any one non-terminal is independent of the expansion of other non-terminals. Thus each PCFG rule probability is multiplied together without considering the location of the node in the parse tree. This is against our intuition since there is a strong tendency toward the context-dependent expansion. Another problem is its lack of sensitivity to words, although lexical information plays an important role in selecting the correct parsing of an ambiguous prepositional phrase attachment. In the PCFG, lexical information can only be represented via the probability of pre-terminal nodes, such as verb or noun, to be expanded lexically. You can add lexical dependencies to PCFGs and make PCFG probabilities more sensitive to surrounding syntactic structure [6, 11, 19, 31, 45].

11.2.2. *N*-gram Language Models

As covered earlier, a language model can be formulated as a probability distribution $P(\mathbf{W})$ over word strings $\mathbf{W}$ that reflects how frequently a string $\mathbf{W}$ occurs as a sentence. For example, for a language model describing spoken language, we might have *P(hi)* = 0.01, since perhaps one out of every hundred sentences a person speaks is *hi*. On the other hand, we would have *P(lid gallops Changsha pop)* = 0, since it is extremely unlikely anyone would utter such a strange string.

$P(\mathbf{W})$ can be decomposed as

$$\begin{aligned} P(\mathbf{W}) &= P(w_1, w_2, \ldots, w_n) \\ &= P(w_1)P(w_2|w_1)P(w_3|w_1,w_2)\cdots P(w_n|w_1,w_2,\ldots,w_{n-1}) \\ &= \prod_{i=1}^{n} P(w_i|w_1,w_2,\ldots,w_{i-1}) \end{aligned} \tag{11.12}$$

where $P(w_i | w_1, w_2, \ldots, w_{i-1})$ is the probability that w_i will follow, given that the word sequence $w_1, w_2, \ldots, w_{i-1}$ was presented previously. In Eq. (11.12), the choice of w_i thus depends on the entire past history of the input. For a vocabulary of size v there are v^{i-1} different histories and so, to specify $P(w_i | w_1, w_2, \ldots, w_{i-1})$ completely, v^i values would have to be estimated. In reality, the probabilities $P(w_i | w_1, w_2, \ldots, w_{i-1})$ are impossible to estimate for even moderate values of i, since most histories $w_1, w_2, \ldots, w_{i-1}$ are unique or have occurred only a few times. A practical solution to the above problems is to assume that $P(w_i | w_1, w_2, \ldots, w_{i-1})$ depends only on some equivalence classes. The equivalence class can be simply based on the several previous words $w_{i-N+1}, w_{i-N+2}, \ldots, w_{i-1}$. This leads to an *n-gram* language model. If the word depends on the previous two words, we have a *trigram*: $P(w_i | w_{i-2}, w_{i-1})$. Similarly, we can have *unigram:* $P(w_i)$, or *bigram:* $P(w_i | w_{i-1})$ language models. The trigram is particularly powerful, as most words have a strong dependence on the previous two words, and it can be estimated reasonably well with an attainable corpus.

In bigram models, we make the approximation that the probability of a word depends only on the identity of the immediately preceding word. To make $P(w_i | w_{i-1})$ meaningful for $i = 1$, we pad the *beginning of the sentence* with a distinguished token <s>; that is, we pretend w_0 = <s>. In addition, to make the sum of the probabilities of all strings equal 1, it is necessary to place a distinguished token </s> at the *end of the sentence*. For example, to calculate *P(Mary loves that person)* we would take

P(Mary loves that person) =
P(Mary|<s>*)P(loves*|*Mary)P(that*|*loves)P(person*|*that)P(*</s>|*person)*

To estimate $P(w_i | w_{i-1})$, the frequency with which the word w_i occurs given that the last word is w_{i-1}, we simply count how often the sequence (w_{i-1}, w_i) occurs in some text and normalize the count by the number of times w_{i-1} occurs.

In general, for a trigram model, the probability of a word depends on the two preceding words. The trigram can be estimated by observing the frequencies or counts of the word pair $C(w_{i-2}, w_{i-1})$ and triplet $C(w_{i-2}, w_{i-1}, w_i)$ as follows:

$$P(w_i | w_{i-2}, w_{i-1}) = \frac{C(w_{i-2}, w_{i-1}, w_i)}{C(w_{i-2}, w_{i-1})} \tag{11.13}$$

The text available for building a model is called a training corpus. For n-gram models, the amount of training data used is typically many millions of words. The estimate of Eq. (11.13) is based on the maximum likelihood principle, because this assignment of probabilities yields the trigram model that assigns the highest probability to the training data of all possible trigram models.

We sometimes refer to the value n of an n-gram model as its order. This terminology comes from the area of Markov models, of which n-gram models are an instance. In particular, an n-gram model can be interpreted as a Markov model of order n-1.

Consider a small example. Let our training data *S* be comprised of the three sentences *John read her book. I read a different book. John read a book by Mulan*. and let us calculate *P(John read a book)* for the maximum likelihood bigram model. We have

$$P(John \mid <s>) = \frac{C(<s>, John)}{C(<s>)} = \frac{2}{3}$$

$$P(read \mid John) = \frac{C(John, read)}{C(John)} = \frac{2}{2}$$

$$P(a \mid read) = \frac{C(read, a)}{C(read)} = \frac{2}{3}$$

$$P(book \mid a) = \frac{C(a, book)}{C(a)} = \frac{1}{2}$$

$$P(</s> \mid book) = \frac{C(book, </s>)}{C(book)} = \frac{2}{3}$$

These trigram probabilities help us estimate the probability for the sentence as:

$$\begin{aligned} & P(John\ read\ a\ book) \\ & = P(John \mid <s>)P(read \mid John)P(a \mid read)P(book \mid a)P(</s> \mid book) \\ & \approx 0.148 \end{aligned} \qquad (11.14)$$

If these three sentences are all the data we have available to use in training our language model, the model is unlikely to generalize well to new sentences. For example, the sentence "*Mulan read her book*" should have a reasonable probability, but the trigram will give it a zero probability simply because we do not have a reliable estimate for *P(read|Mulan)*.

Unlike linguistics, grammaticality is not a strong constraint in the *n*-gram language model. Even though the string is ungrammatical, we may still assign it a high probability if *n* is small.

11.3. Complexity Measure of Language Models

Language can be thought of as an information source whose outputs are words w_i belonging to the vocabulary of the language. The most common metric for evaluating a language model is the word recognition error rate, which requires the participation of a speech recognition system. Alternatively, we can measure the probability that the language model assigns to test word strings without involving speech recognition systems. This is the derivative measure of cross-entropy known as test-set *perplexity*.

The measure of cross-entropy is discussed in Chapter 3. Given a language model that assigns probability $P(\mathbf{W})$ to a word sequence $\mathbf{W}$, we can derive a compression algorithm that encodes the text $\mathbf{W}$ using $-\log_2 P(\mathbf{W})$ bits. The cross-entropy $H(\mathbf{W})$ of a model

$P(w_i|w_{i-n+1}\dots w_{i-1})$ on data $\mathbf{W}$, with a sufficiently long word sequence, can be simply approximated as

$$H(\mathbf{W}) = -\frac{1}{N_{\mathbf{W}}}\log_2 P(\mathbf{W}) \tag{11.15}$$

where $N_{\mathbf{W}}$ is the length of the text $\mathbf{W}$ measured in words.

The perplexity $PP(\mathbf{W})$ of a language model $P(\mathbf{W})$ is defined as the reciprocal of the (geometric) average probability assigned by the model to each word in the test set $\mathbf{W}$. This is a measure, related to cross-entropy, known as test-set perplexity:

$$PP(\mathbf{W}) = 2^{H(\mathbf{W})} \tag{11.16}$$

The perplexity can be roughly interpreted as the geometric mean of the branching factor of the text when presented to the language model. The perplexity defined in Eq. (11.16) has two key parameters: a language model and a word sequence. The test-set[4] perplexity evaluates the generalization capability of the language model. The training-set perplexity measures how the language model fits the training data, like the likelihood. It is generally true that lower perplexity correlates with better recognition performance. This is because the perplexity is essentially a statistically weighted word branching measure on the test set. The higher the perplexity, the more branches the speech recognizer needs to consider statistically.

While the perplexity [Eqs. (11.16) and (11.15)] is easy to calculate for the *n*-gram [Eq. (11.12)], it is slightly more complicated to compute for a probabilistic CFG. We can first parse the word sequence and use Eq. (11.9) to compute $P(\mathbf{W})$ for the test-set perplexity. The perplexity can also be applied to nonstochastic models such as CFGs. We can assume they have a uniform distribution in computing $P(\mathbf{W})$.

A language with higher perplexity means that the number of words branching from a previous word is larger on average. In this sense, perplexity is an indication of the complexity of the language if we have an accurate estimate of $P(\mathbf{W})$. For a given language, the difference between the perplexity of a language model and the true perplexity of the language is an indication of the quality of the model. The perplexity of a particular language model can change dramatically in terms of the vocabulary size, the number of states of grammar rules, and the estimated probabilities. A language model with perplexity *X* has roughly the same difficulty as another language model in which every word can be followed by *X* different words with equal probabilities. Therefore, in the task of continuous digit recognition, the perplexity is 10. Clearly, lower perplexity will generally have less confusion in recognition. Typical perplexities yielded by *n*-gram models on English text range from about 50 to almost 1000 (corresponding to cross-entropies from about 6 to 10 bits/word), depending on the type of text. In the task of 5,000-word continuous speech recognition for the *Wall Street Journal*, the test-set perplexities of the trigram grammar and the bigram grammar are re-

---

[4] We often distinguish between the word sequence from the unseen test data and that from the training data to derive the language model.

ported to be about 128 and 176 respectively.[5] In the tasks of 2000-word conversational Air Travel Information System (ATIS), the test-set perplexity of the word trigram model is typically less than 20.

Since perplexity does not take into account acoustic confusability, we eventually have to measure speech recognition accuracy. For example, if the vocabulary of a speech recognizer contains the E-set of English alphabet: *B, C, D, E, G, P*,and *T*, we can define a CFG that has a low perplexity value of 7. Such a low perplexity does not guarantee we will have good recognition performance, because of the intrinsic acoustic confusability of the E-set.

11.4. *N*-GRAM SMOOTHING

One of the key problems in *n*-gram modeling is the inherent data sparseness of real training data. If the training corpus is not large enough, many actually possible word successions may not be well observed, leading to many extremely small probabilities. For example, with several-million-word collections of English text, more than 50% of trigrams occur only once, and more than 80% of trigrams occur less than five times. Smoothing is critical to make estimated probabilities robust for unseen data. If we consider the sentence *Mulan read a book* in the example we discussed in Section 11.2.2, we have:

$$P(read|\,Mulan) = \frac{C(Mulan, read)}{\sum_w C(Mulan, w)} = \frac{0}{1}$$

giving us *P*(*Mulan read a book*) = 0.

Obviously, this is an underestimate for the probability of "*Mulan read a book*" since there is *some* probability that the sentence occurs in some test set. To show why it is important to give this probability a nonzero value, we turn to the primary application for language models, speech recognition. In speech recognition, if *P*(**W**) is zero, the string **W** will never be considered as a possible transcription, regardless of how unambiguous the acoustic signal is. Thus, whenever a string ***W*** such that *P*(**W**) = 0 occurs during a speech recognition task, an error will be made. Assigning all strings a nonzero probability helps prevent errors in speech recognition. This is the core issue of smoothing. Smoothing techniques adjust the maximum likelihood estimate of probabilities to produce more robust probabilities for unseen data, although the likelihood for the training data may be hurt slightly.

The name smoothing comes from the fact that these techniques tend to make distributions flatter, by adjusting low probabilities such as zero probabilities upward, and high probabilities downward. Not only do smoothing methods generally prevent zero probabilities,

---

[5] Some experimental results show that the test-set perplexities for different languages are comparable. For example, French, English, Italian and German have a bigram test-set perplexity in the range of 95 to 133 for newspaper corpora. Italian has a much higher perplexity reduction (a factor of 2) from bigram to trigram because of the high number of function words. The trigram perplexity of Italian is among the lowest in these languages [34].

but they also attempt to improve the accuracy of the model as a whole. Whenever a probability is estimated from few counts, smoothing has the potential to significantly improve the estimation so that it has better generalization capability.

To give an example, one simple smoothing technique is to pretend each bigram occurs once more than it actually does, yielding

$$P(w_i|w_{i-1}) = \frac{1+C(w_{i-1},w_i)}{\sum_{w_i}(1+C(w_{i-1},w_i))} = \frac{1+C(w_{i-1},w_i)}{V+\sum_{w_i}C(w_{i-1},w_i)} \tag{11.17}$$

where V is the size of the vocabulary. In practice, vocabularies are typically fixed to be tens of thousands of words or less. All words not in the vocabulary are mapped to a single word, usually called the *unknown word*. Let us reconsider the previous example using this new distribution, and let us take our vocabulary V to be the set of all words occurring in the training data S, so that we have $V = 11$ (with both <s> and </s>).

For the sentence *John read a book*, we now have

$$\begin{aligned} &P(John\ read\ a\ book) \\ &= P(John|</s>)P(read\,|\,John)P(a\,|\,read)P(book\,|\,a)P(</s>|\,book) \\ &\approx 0.00035 \end{aligned} \tag{11.18}$$

In other words, we estimate that the sentence *John read a book* occurs about once every three thousand sentences. This is more reasonable than the maximum likelihood estimate of 0.148 of Eq. (11.14). For the sentence *Mulan read a book*, we have

$$\begin{aligned} &P(Mulan\ read\ a\ book) \\ &= P(Mulan|</s>)P(read\,|\,Mulan)P(a\,|\,read)P(book\,|\,a)P(</s>|\,book) \\ &\approx 0.000084 \end{aligned} \tag{11.19}$$

Again, this is more reasonable than the zero probability assigned by the maximum likelihood model. In general, most existing smoothing algorithms can be described with the following equation:

$$\begin{aligned} &P_{smooth}(w_i\,|\,w_{i-n+1}...w_{i-1}) \\ &= \begin{cases} \alpha(w_i\,|\,w_{i-n+1}...w_{i-1}) & \text{if } C(w_{i-n+1}...w_i)>0 \\ \gamma(w_{i-n+1}...w_{i-1})P_{smooth}(w_i\,|\,w_{i-n+2}...w_{i-1}) & \text{if } C(w_{i-n+1}...w_i)=0 \end{cases} \end{aligned} \tag{11.20}$$

That is, if an n-gram has a nonzero count we use the distribution $\alpha(w_i|w_{i-n+1}\cdots w_{i-1})$. Otherwise, we backoff to the lower-order n-gram distribution $P_{smooth}(w_i|w_{i-n+2}\cdots w_{i-1})$, where the scaling factor $\gamma(w_{i-n+1}\cdots w_{i-1})$ is chosen to make the conditional distribution sum to one. We refer to algorithms that fall directly in this framework as *backoff models*.

Several other smoothing algorithms are expressed as the linear interpolation of higher- and lower-order n-gram models as:

$$\begin{aligned} & P_{smooth}(w_i \mid w_{i-n+1} \ldots w_{i-1}) \\ & = \lambda P_{ML}(w_i \mid w_{i-n+1} \ldots w_{i-1}) + (1-\lambda) P_{smooth}(w_i \mid w_{i-n+2} \ldots w_{i-1}) \end{aligned} \tag{11.21}$$

where λ is the interpolation weight that depends on $w_{i-n+1} \ldots w_{i-1}$. We refer to models of this form as interpolated models.

The key difference between backoff and interpolated models is that for the probability of n-grams with nonzero counts, interpolated models use information from lower-order distributions while backoff models do not. In both backoff and interpolated models, lower-order distributions are used in determining the probability of n-grams with zero counts. Now, we discuss several backoff and interpolated smoothing methods. Performance comparison of these techniques in real speech recognition applications is discussed in Section 11.4.4.

11.4.1. Deleted Interpolation Smoothing

Consider the case of constructing a bigram model on training data where we have that *C(enliven you)* = 0 *and C(enliven thou)* = 0. Then, according to both additive smoothing of Eq. (11.17), we have *P(you|enliven)* = *P(thou|enliven)*. However, intuitively we should have *P(you|enliven)* > *P(thou|enliven)*, because the word *you* is much more common than the word *thou* in modern English. To capture this behavior, we can interpolate the bigram model with a unigram model. A unigram model conditions the probability of a word on no other words, and just reflects the frequency of that word in text. We can linearly interpolate a bigram model and a unigram model as follows:

$$P_I(w_i|w_{i-1}) = \lambda P(w_i|w_{i-1}) + (1-\lambda) P(w_i) \tag{11.22}$$

where $0 \le \lambda \le 1$. Because *P(you|enliven)* = *P(thou|enliven)*=0 while presumably *P(you)* > *P(thou)*, we will have that $P_I(you \mid enliven) > P_I(thou \mid enliven)$ as desired.

In general, it is useful to interpolate higher-order n-gram models with lower-order n-gram models, because when there is insufficient data to estimate a probability in the higher-order model, the lower-order model can often provide useful information. An elegant way of performing this interpolation is given as follows

$$\begin{aligned} & P_I(w_i|w_{i-n+1} \cdots w_{i-1}) \\ & = \lambda_{w_{i-n+1} \ldots w_{i-1}} P(w_i|w_{i-n+1} \cdots w_{i-1}) + (1-\lambda_{w_{i-n+1} \ldots w_{i-1}}) P_I(w_i|w_{i-n+2} \cdots w_{i-1}) \end{aligned} \tag{11.23}$$

That is, the nth-order smoothed model is defined *recursively* as a linear interpolation between the nth-order maximum likelihood model and the (n-1)th-order smoothed model. To end the recursion, we can take the smoothed first-order model to be the maximum likelihood distribution (unigram), or we can take the smoothed zeroth-order model to be the uniform distribution. Given a fixed $P(w_i|w_{i-n+1}\ldots w_{i-1})$, it is possible to search efficiently for the interpolation parameters using the deleted interpolation method discussed in Chapter 9.

Notice that the optimal $\lambda_{w_{i-n+1}\ldots w_{i-1}}$ is different for different histories $w_{i-n+1}\ldots w_{i-1}$. For example, for a context we have seen thousands of times, a high λ will be suitable, since the higher-order distribution is very reliable; for a history that has occurred only once, a lower λ is appropriate. Training each parameter $\lambda_{w_{i-n+1}\ldots w_{i-1}}$ independently can be harmful; we need an enormous amount of data to train so many independent parameters accurately. One possibility is to divide the $\lambda_{w_{i-n+1}\ldots w_{i-1}}$ into a moderate number of partitions or buckets, constraining all $\lambda_{w_{i-n+1}\ldots w_{i-1}}$ in the same bucket to have the same value, thereby reducing the number of independent parameters to be estimated. Ideally, we should tie together those $\lambda_{w_{i-n+1}\ldots w_{i-1}}$ that we have a prior reason to believe should have similar values.

11.4.2. Backoff Smoothing

Backoff smoothing is attractive because it is easy to implement for practical speech recognition systems. The Katz backoff model is the canonical example we discuss in this section. It is based on the Good-Turing smoothing principle.

11.4.2.1. Good-Turing Estimates and Katz Smoothing

The Good-Turing estimate is a smoothing technique to deal with infrequent *n*-grams. It is not used by itself for *n*-gram smoothing, because it does not include the combination of higher-order models with lower-order models necessary for good performance. However, it is used as a tool in several smoothing techniques. The basic idea is to partition *n*-grams into groups depending on their frequency (i.e. how many time the *n*-grams appear in the training data) such that the parameter can be smoothed based on *n*-gram frequency.

The Good-Turing estimate states that for any *n*-gram that occurs *r* times, we should pretend that it occurs r^* times as follows:

$$r^* = (r+1)\frac{n_{r+1}}{n_r} \tag{11.24}$$

where n_r is the number of *n*-grams that occur exactly *r* times in the training data. To convert this count to a probability, we just normalize: for an *n*-gram *a* with *r* counts, we take

$$P(a) = \frac{r^*}{N} \tag{11.25}$$

where $N = \sum_{r=0}^{\infty} n_r r^*$. Notice that $N = \sum_{r=0}^{\infty} n_r r^* = \sum_{r=0}^{\infty}(r+1)n_{r+1} = \sum_{r=0}^{\infty} n_r r$, i.e., N is equal to the original number of counts in the distribution [28].

Katz smoothing extends the intuitions of the Good-Turing estimate by adding the combination of higher-order models with lower-order models [38]. Take the bigram as our example, Katz smoothing suggested using the Good-Turing estimate for nonzero counts as follows:

$$C^*(w_{i-1}w_i) = \begin{cases} d_r r & \text{if } r>0 \\ \alpha(w_{i-1})P(w_i) & \text{if } r=0 \end{cases} \tag{11.26}$$

where d_r is approximately equal to r^*/r. That is, all bigrams with a nonzero count r are *discounted* according to a discount ratio d_r, which implies that the counts subtracted from the nonzero counts are distributed among the zero-count bigrams according to the next lower-order distribution, e.g., the unigram model. The value $\alpha(w_{i-1})$ is chosen to equalize the total number of counts in the distribution, i.e., $\sum_{w_i} C^*(w_{i-1}w_i) = \sum_{w_i} C(w_{i-1}w_i)$. The appropriate value for $\alpha(w_{i-1})$ is computed so that the smoothed bigram satisfies the probability constraint:

$$\alpha(w_{i-1}) = \frac{1-\sum_{w_i:C(w_{i-1}w_i)>0} P^*(w_i|w_{i-1})}{\sum_{w_i:C(w_{i-1}w_i)=0} P(w_i)} = \frac{1-\sum_{w_i:C(w_{i-1}w_i)>0} P^*(w_i|w_{i-1})}{1-\sum_{w_i:C(w_{i-1}w_i)>0} P(w_i)} \tag{11.27}$$

To calculate $P^*(w_i|w_{i-1})$ from the corrected count, we just normalize:

$$P^*(w_i \mid w_{i-1}) = \frac{C^*(w_{i-1}w_i)}{\sum_{w_k} C^*(w_{i-1}w_k)} \tag{11.28}$$

In Katz implementation, the d_r are calculated as follows: large counts are taken to be reliable, so they are not discounted. In particular, Katz takes $d_r = 1$ for all $r > k$ for some k, say k in the range of 5 to 8. The discount ratios for the lower counts $r \le k$ are derived from the Good-Turing estimate applied to the global bigram distribution; that is, n_r in Eq. (11.24) denotes the total number of bigrams that occur exactly r times in the training data. These d_r are chosen such that

- the resulting discounts are proportional to the discounts predicted by the Good-Turing estimate, and
- the total number of counts discounted in the global bigram distribution is equal to the total number of counts that should be assigned to bigrams with zero counts according to the Good-Turing estimate.

The first constraint corresponds to the following equation:

$$d_r = \mu \frac{r^*}{r} \tag{11.29}$$

for $r \in \{1, \ldots k\}$ with some constant μ. The Good-Turing estimate predicts that the total mass assigned to bigrams with zero counts is $n_0 \frac{n_1}{n_0} = n_1$, and the second constraint corresponds to the equation

$$\sum_{r=1}^{k} n_r (1 - d_r) r = n_1 \tag{11.30}$$

Based on Eq. (11.30), the unique solution is given by:

$$d_r = \frac{\frac{r^*}{r} - \frac{(k+1)n_{k+1}}{n_1}}{1 - \frac{(k+1)n_{k+1}}{n_1}} \tag{11.31}$$

Katz smoothing for higher-order n-gram models is defined analogously. The Katz n-gram backoff model is defined in terms of the Katz (n-1)-gram model. To end the recursion, the Katz unigram model is taken to be the maximum likelihood unigram model. It is usually necessary to smooth n_r when using the Good-Turing estimate, e.g., for those n_r that are very low. However, in Katz smoothing this is not essential because the Good-Turing estimate is used only for small counts $r<=k$, and n_r is generally fairly high for these values of r. The procedure of Katz smoothing can be summarized as in Algorithm 11.2.

In fact, the Katz backoff model can be expressed in terms of the interpolated model defined in Eq. (11.23), in which the interpolation weight is obtained via Eq. (11.26) and (11.27).

ALGORITHM 11.2: *KATZ SMOOTHING*

$$P_{Katz}(w_i \mid w_{i-1}) = \begin{cases} C(w_{i-1}w_i)/C(w_{i-1}) & \text{if } r > k \\ d_r C(w_{i-1}w_i)/C(w_{i-1}) & \text{if } k \geq r > 0 \\ \alpha(w_{i-1})P(w_i) & \text{if } r = 0 \end{cases}$$

where $d_r = \frac{\frac{r^*}{r} - \frac{(k+1)n_{k+1}}{n_1}}{1 - \frac{(k+1)n_{k+1}}{n_1}}$ and $\alpha(w_{i-1}) = \frac{1 - \sum_{w_i:r>0} P_{Katz}(w_i \mid w_{i-1})}{1 - \sum_{w_i:r>0} P(w_i)}$

11.4.2.2. Alternative Backoff Models

In a similar manner to the Katz backoff model, there are other ways to discount the probability mass. For instance, *absolute discounting* involves subtracting a fixed discount $D <= 1$ from each nonzero count. If we express the absolute discounting in term of interpolated models, we have the following:

$$
\begin{aligned}
&P_{abs}(w_i | w_{i-n+1} \cdots w_{i-1}) \\
&= \frac{\max\{C(w_{i-n+1} \cdots w_i) - D, 0\}}{\sum_{w_i} C(w_{i-n+1} \cdots w_i)} + (1 - \lambda_{w_{i-n+1} \ldots w_{i-1}}) P_{abs}(w_i | w_{i-n+2} \cdots w_{i-1})
\end{aligned}
\tag{11.32}
$$

To make this distribution sum to 1, we normalize it to determine $\lambda_{w_{i-n+1} \cdots w_{i-1}}$. Absolute discounting is explained with the Good-Turing estimate. Empirically the average Good-Turing discount $r - r^*$ associated with *n*-grams of larger counts (*r* over 3) is largely constant over *r*.

Consider building a bigram model on data where there exists a word that is very common, say *Francisco*, that occurs only after a single word, say *San*. Since *C(Francisco)* is high, the unigram probability *P(Francisco)* will be high, and an algorithm such as absolute discounting or Katz smoothing assigns a relatively high probability to occurrence of the word *Francisco* after novel bigram histories. However, intuitively this probability should not be high, since in the training data the word *Francisco* follows only a single history. That is, perhaps *Francisco* should receive a low unigram probability, because the only time the word occurs is when the last word is *San*, in which case the bigram probability models its probability well.

Extending this line of reasoning, perhaps the unigram probability used should not be proportional to the number of occurrences of a word, but instead to the number of different words that it follows. To give an intuitive argument, imagine traversing the training data sequentially and building a bigram model on the preceding data to predict the current word. Then, whenever the current bigram does not occur in the preceding data, the unigram probability becomes a large factor in the current bigram probability. If we assign a count to the corresponding unigram whenever such an event occurs, then the number of counts assigned to each unigram is simply the number of different words that it follows. In Kneser-Ney smoothing [40], the lower-order *n*-gram is not proportional to the number of occurrences of a word, but instead to the number of *different* words that it follows. We summarize the Kneser-Ney backoff model in Algorithm 11.3.

Kneser-Ney smoothing is an extension of other backoff models. Most of the previous models used the lower-order *n*-grams trained with ML estimation. Kneser-Ney smoothing instead considers a lower-order distribution as a significant factor in the combined model such that they are optimized together with other parameters. To derive the formula, more generally, we express it in terms of the interpolated model specified in Eq. (11.23) as:

$$P_{KN}(w_i \mid w_{i-n+1}...w_{i-1}) \\ = \frac{\max\{C(w_{i-n+1}...w_i) - D, 0\}}{\sum_{w_i} C(w_{i-n+1}...w_i)} + (1 - \lambda_{w_{i-n+1}...w_{i-1}}) P_{KN}(w_i \mid w_{i-n+2}...w_{i-1}) \tag{11.33}$$

To make this distribution sum to 1, we have:

$$1 - \lambda_{w_{i-n+1}...w_{i-1}} = \frac{D}{\sum_{w_i} C(w_{i-n+1}...w_i)} \mathbb{C}(w_{i-n+1}...w_{i-1}\bullet) \tag{11.34}$$

where $\mathbb{C}(w_{i-n+1}...w_{i-1}\bullet)$ is the number of unique words that follow the history $w_{i-n+1}...w_{i-1}$. This equation enables us to interpolate the lower-order distribution with all words, not just with words that have zero counts in the higher-order distribution.

ALGORITHM 11.3: *KNESER-NEY BIGRAM SMOOTHING*

$$P_{KN}(w_i \mid w_{i-1}) = \begin{cases} \dfrac{\max\{C(w_{i-1}w_i) - D, 0\}}{C(w_{i-1})} & \text{if } C(w_{i-1}w_i) > 0 \\ \alpha(w_{i-1}) P_{KN}(w_i) & \text{otherwise} \end{cases}$$

where $P_{KN}(w_i) = \mathbb{C}(\bullet w_i) / \sum_{w_i} \mathbb{C}(\bullet w_i)$, $\mathbb{C}(\bullet w_i)$ is the number of unique words preceding w_i.

$\alpha(w_{i-1})$ is chosen to make the distribution sum to 1 so that we have:

$$\alpha(w_{i-1}) = \frac{1 - \sum_{w_i : C(w_{i-1}w_i) > 0} \dfrac{\max\{C(w_{i-1}w_i) - D, 0\}}{C(w_{i-1})}}{1 - \sum_{w_i : C(w_{i-1}w_i) > 0} P_{KN}(w_i)}$$

Now, take the bigram case as an example. We need to find a unigram distribution $P_{KN}(w_i)$ such that the marginal of the bigram smoothed distributions should match the marginal of the training data:

$$\frac{C(w_i)}{\sum_{w_i} C(w_i)} = \sum_{w_{i-1}} P_{KN}(w_{i-1}w_i) = \sum_{w_{i-1}} P_{KN}(w_i \mid w_{i-1}) P(w_{i-1}) \tag{11.35}$$

For $P(w_{i-1})$, we simply take the distribution found in the training data

$$P(w_{i-1}) = \frac{C(w_{i-1})}{\sum_{w_{i-1}} C(w_{i-1})} \tag{11.36}$$

We substitute Eq. (11.33) in Eq. (11.35). For the bigram case, we have:

$$
\begin{aligned}
& C(w_i) \\
&= \sum_{w_{i-1}} C(w_{i-1})\left[\frac{\max\{C(w_{i-1}w_i)-D,0\}}{\sum_{w_i} C(w_{i-1}w_i)} + \frac{D}{\sum_{w_i} C(w_{i-1}w_i)}\mathbb{C}(w_{i-1}\bullet)P_{KN}(w_i)\right] \\
&= \sum_{w_{i-1}:C(w_{i-1}w_i)>0} C(w_{i-1})\frac{C(w_{i-1}w_i)-D}{C(w_{i-1})} + \sum_{w_{i-1}} C(w_{i-1})\frac{D}{C(w_{i-1})}\mathbb{C}(w_{i-1}\bullet)P_{KN}(w_i) \\
&= C(w_i) - \mathbb{C}(\bullet w_{i-1})D + DP_{KN}(w_i) + DP_{KN}(w_i)\sum_{w_{i-1}} \mathbb{C}(w_{i-1}\bullet)
\end{aligned}
\tag{11.37}
$$

Solving the equation, we get

$$
P_{KN}(w_i) = \frac{\mathbb{C}(\bullet w_i)}{\sum_{w_i} \mathbb{C}(\bullet w_i)} \tag{11.38}
$$

which can be generalized to higher-order models:

$$
P_{KN}(w_i \mid w_{i-n+2}...w_{i-1}) = \frac{\mathbb{C}(\bullet w_{i-n+2}...w_i)}{\sum_{w_i} \mathbb{C}(\bullet w_{i-n+2}...w_i)} \tag{11.39}
$$

where $\mathbb{C}(\bullet w_{i-n+2}...w_i)$ is the number of different words that precede $w_{i-n+2}...w_i$.

In practice, instead of using a single discount D for all nonzero counts as in Kneser-Ney smoothing, we can have a number of different parameters (D_i) that depend on the range of counts:

$$
\begin{aligned}
& P_{KN}(w_i \mid w_{i-n+1}...w_{i-1}) \\
&= \frac{C(w_{i-n+1}...w_i) - D(C(w_{i-n+1}...w_i))}{\sum_{w_i} C(w_{i-n+1}...w_i)} + \\
& +\gamma(w_{i-n+1}...w_{i-1})P_{KN}(w_i \mid w_{i-n+2}...w_{i-1})
\end{aligned}
\tag{11.40}
$$

This modification is motivated by evidence that the ideal average discount for n-grams with one or two counts is substantially different from the ideal average discount for n-grams with higher counts.

11.4.3. Class *N*-grams

As discussed in Chapter 2, we can define classes for words that exhibit similar semantic or grammatical behavior. This is another effective way to handle the data sparsity problem.

Class-based language models have been shown to be effective for rapid adaptation, training on small data sets, and reduced memory requirements for real-time speech applications.

For any given assignment of a word w_i to class c_i, there may be many-to-many mappings, e.g., a word w_i may belong to more than one class, and a class c_i may contain more than one word. For the sake of simplicity, assume that a word w_i can be uniquely mapped to only one class c_i. The n-gram model can be computed based on the previous n-1 classes:

$$P(w_i|c_{i-n+1}\cdots c_{i-1}) = P(w_i \mid c_i)P(c_i|c_{i-n+1}\cdots c_{i-1}) \quad (11.41)$$

where $P(w_i|c_i)$ denotes the probability of word w_i given class c_i in the current position, and $P(c_i|c_{i-n+1}\cdots c_{i-1})$ denotes the probability of class c_i given the class history. With such a model, we can learn the class mapping $w \rightarrow c$ from either a training text or task knowledge we have about the application. In general, we can express the class trigram as:

$$P(\mathbf{W}) = \sum_{c_1 \ldots c_n} \prod_i P(w_i \mid c_i)P(c_i \mid c_{i-2,}c_{i-1}) \quad (11.42)$$

If the classes are nonoverlapping, i.e. a word may belong to only one class, then Eq. (11.42) can be simplified as:

$$P(\mathbf{W}) = \prod_i P(w_i \mid c_i)P(c_i \mid c_{i-2,}c_{i-1}) \quad (11.43)$$

If we have the mapping function defined, we can easily compute the class n-gram. We can estimate the empirical frequency of each word $C(w_i)$, and of each class $C(c_i)$. We can also compute the empirical frequency that a word from one class will be followed immediately by a word from another $C(c_{i-1}c_i)$. As a typical example, the bigram probability of a word given the prior word (class) can be estimated as

$$P(w_i \mid w_{i-1}) = P(w_i \mid c_{i-1}) = P(w_i \mid c_i)P(c_i \mid c_{i-1}) = \frac{C(w_i)}{C(c_i)}\frac{C(c_{i-1}c_i)}{C(c_{i-1})} \quad (11.44)$$

For general-purpose large vocabulary dictation applications, class-based n-grams have not significantly improved recognition accuracy. They are mainly used as a backoff model to complement the lower-order n-grams for better smoothing. Nevertheless, for limited domain speech recognition, the class-based n-gram is very helpful as the class can efficiently encode semantic information for improved key word spotting and speech understanding accuracy.

11.4.3.1. Rule-Based Classes

There are a number of ways to cluster words together based on the syntactic-semantic information that exists for the language and the task. For example, part-of-speech can be generally used to produce a small number of classes although this may lead to significantly increased perplexity. Alternatively, if we have domain knowledge, it is often advantageous to cluster together words that have a similar semantic functional role. For example, if we need to build a conversational system for air travel information systems, we can group the name of different airlines such as *United Airlines, KLM,* and *Air China*, into a broad *airline class.* We can do the same thing for the names of different airports such as *JFK*, *Narita*, and *Heathrow*, the names of different cities like *Beijing, Pittsburgh*, and *Moscow,* and so on. Such an approach is particularly powerful, since the amount of training data is always limited. With generalized broad classes of semantically interpretable meaning, it is easy to add a new airline such as *Redmond Air* into the classes if there is indeed a start-up airline named *Redmond Air* that the system has to incorporate. The system is now able to assign a reasonable probability to a sentence like "*Show me all flights of Redmond Air from Seattle to Boston*" in a similar manner as "*Show me all flights of United Airlines from Seattle to Boston.*" We only need to estimate the probability of *Redmond Air,* given the airline class c_i. We can use the existing class n-gram model that contains the broad structure of the air travel information system as it is.

Without such a broad interpretable class, it would be extremely difficult to deal with new names the system needs to handle, although these new names can always be mapped to the special class of the unknown word or proper noun classes. For these new words, we can alternatively map them into a word that has a similar syntactic and semantic role. Thus, the new word inherits all the possible word trigram relationships that may be very similar to those of the existing word observed with the training data.

11.4.3.2. Data-driven Classes

For a general-purpose dictation application, it is impractical to derive functional classes in the same manner as a domain-specific conversational system that focuses on a narrow task. Instead, data-driven clustering algorithms have been used to generalize the concept of word similarities, which is in fact a search procedure to find a class label for each word with a predefined objective function. The set of words with the same class label is called a cluster. We can use the maximum likelihood criterion as the objective function for a given training corpus and a given number of classes, which is equivalent to minimizing the perplexity for the training corpus. Once again, the EM algorithm can be used here. Each word can be initialized to a random cluster (class label). At each iteration, every word is moved to the class that produces the model with minimum perplexity [9, 48]. The perplexity modifications can be calculated independently, so that each word is evaluated as if all other word classes were held fixed. The algorithm converges when no single word can be moved to another class in a way that reduces the perplexity of the clustered n-gram model.

One special kind of class n-gram models is based on the decision tree as discussed in Chapter 4. We can use it to create equivalent classes for words in the history, so that can we have a compact long-distance n-gram language model [2]. The sequential decomposition, as expressed in Eq. (11.12), is approximated as:

$$P(\mathbf{W}) = \prod_{i=1}^{n} P(w_i | E(w_1, w_2, \ldots, w_{i-1})) = \prod_{i=1}^{n} P(w_i | E(\mathbf{h})) \tag{11.45}$$

where $E(\mathbf{h})$ denotes a many-to-one mapping function that groups word histories $\mathbf{h}$ into some equivalence classes. It is important to have a scheme that can provide adequate information about the history so it can serve as a basis for prediction. In addition, it must yield a set of classes that can be reliably estimated. The decision tree method uses entropy as a criterion in developing the equivalence classes that can effectively incorporate long-distance information. By asking a number of questions associated with each node, the decision tree can classify the history into a small number of equivalence classes. Each leaf of the tree, thus, has the probability $P(w_i | E(w_1 \ldots w_{i-1}))$ that is derived according to the number of times the word w_i is found in the leaf. The selection of questions in building the tree can be infinite. We can consider not only the syntactic structure, but also semantic meaning to derive permissible questions from which the entropy criterion would choose. A full-fledged question set that is based on detailed analysis of the history is beyond the limit of our current computing resources. As such, we often use the membership question to check each word in the history.

11.4.4. Performance of *N*-gram Smoothing

The performance of various smoothing algorithms depends on factors such as the training-set sizes. There is a strong correlation between the test-set perplexity and word error rate. Smoothing algorithms leading to lower perplexity generally result in a lower word error rate. Among all the methods discussed here, the Kneser-Ney method slightly outperforms other algorithms over a wide range of training-set sizes and corpora, and for both bigram and trigram models. Albeit the difference is not large, the good performance of the Kneser-Ney smoothing is due to the modified backoff distributions. The Katz algorithms and deleted interpolation smoothing generally yield the next best performance. All these three smoothing algorithms perform significantly better than the n-gram model without any smoothing. The deleted interpolation algorithm performs slightly better than the Katz method in sparse data situations, and the reverse is true when data are plentiful. Katz's algorithm is particularly good at smoothing larger counts; these counts are more prevalent in larger data sets.

Class n-grams offer different kind of smoothing. While clustered n-gram models often offer no significant test-set perplexity reduction in comparison to the word n-gram model, it is beneficial to smooth the word n-gram model via either backoff or interpolation methods.

For example, the decision-tree based long-distance class language model does not offer significantly improved speech recognition accuracy until it is interpolated with the word trigram. They are effective as a domain-specific language model if the class can accommodate domain-specific information.

Smoothing is a fundamental technique for statistical modeling, important not only for language modeling but for many other applications as well. Whenever data sparsity is an issue, smoothing can help performance, and data sparsity is almost always an issue in statistical modeling. In the extreme case, where there is so much training data that all parameters can be accurately trained without smoothing, you can almost always expand the model, such as by moving to a higher-order n-gram model, to achieve improved performance. With more parameters, data sparsity becomes an issue again, but a proper smoothing model is usually more accurate than the original model. Thus, no matter how much data you have, smoothing can almost always help performance, and for a relatively small effort.

11.5. Adaptive Language Models

Dynamic adjustment of the language model parameter, such as n-gram probabilities, vocabulary size, and the choice of words in the vocabulary, is important, since the topic of conversation is highly nonstationary [4, 33, 37, 41, 46]. For example, in a typical dictation application, a particular set of words in the vocabulary may suddenly burst forth and then become dormant later, based on the current conversation. Because the topic of the conversation may change from time to time, the language model should be dramatically different based on the topic of the conversation. We discuss several adaptive techniques that can improve the quality of the language model based on the real usage of the application.

11.5.1. Cache Language Models

To adjust word frequencies observed in the current conversation, we can use a dynamic *cache* language model [41]. The basic idea is to accumulate word n-grams dictated so far in the current document and use these to create a local dynamic n-gram model such as bigram $P_{cache}(w_i|w_{i-1})$. Because of limited data and nonstationary nature, we should use a lower-order language model that is no higher than a trigram model $P_{cache}(w_i|w_{i-2}w_{i-1})$, which can be interpolated with the dynamic bigram and unigram. Empirically, we need to normally give a high weight to the unigram cache model, because it is better trained with the limited data in the cache.

With the cache trigram, we interpolate it with the static n-gram model $P_s(w_i|w_{i-n+1}\ldots w_{i-1})$. The interpolation weight can be made to vary with the size of the cache.

$$\begin{aligned}&P_{cache}(w_i \mid w_{i-n+1}\ldots w_{i-1})\\&=\lambda_c P_s(w_i \mid w_{i-n+1}\ldots w_{i-1})+(1-\lambda_c)P_{cache}(w_i \mid w_{i-2}w_{i-1})\end{aligned} \tag{11.46}$$

The cache model is desirable in practice because of its impressive empirical performance improvement. In a dictation application, we often encounter new words that are not in the static vocabulary. The same words also tend to be repeated in the same article. The cache model can address this problem effectively by adjusting the parameters continually as recognition and correction proceed for incrementally improved performance. A noticeable benefit is that we can better predict words belonging to fixed phrases such as *Windows NT* and *Bill Gates*.

11.5.2. Topic-Adaptive Models

The topic can change over time. Such topic or style information plays a critical role in improving the quality of the static language model. For example, the prediction of whether the word following the phrase *the operating* is *system* or *table* can be improved substantially by knowing whether the topic of discussion is related to computing or medicine.

Domain or topic-clustered language models split the language model training data according to topic. The training data may be divided using the known category information or using automatic clustering. In addition, a given segment of the data may be assigned to multiple topics. A topic-dependent language model is then built from each cluster of the training data. Topic language models are combined using linear interpolation or other methods such as maximum entropy techniques discussed in Section 11.5.3.

We can avoid any pre-defined clustering or segmentation of the training data. The reason is that the best clustering may become apparent only when the current topic of discussion is revealed. For example, when the topic is hand-injury to baseball player, the pre-segmented clusters of topic *baseball & hand-injuries* may have to be combined. This leads to a union of the two clusters, whereas the ideal dataset is obtained by the intersection of these clusters. In general, various combinations of topics lead to a combinatorial explosion in the number of compound topics, and it appears to be a difficult task to anticipate all the needed combinations beforehand.

We base our determination of the most suitable language model data to build a model upon the particular history of a given document. For example, we can use it as a query against the entire training database of documents using *information retrieval* techniques [57]. The documents in the database can be ranked by relevance to the query. The most relevant documents are then selected as the adaptation set for the topic-dependent language model. The process can be repeated as the document is updated.

There are two major steps we need to consider here. The first involves using the available document history to retrieve similar documents from the database. The second consists of using the similar document set retrieved in the first step to adapt the general or topic-independent language model. Available document history depends upon the design and the requirements of the recognition system. If the recognition system is designed for live-mode application, where the recognition results must be presented to the user with a small delay, the available document history will be the history of the document user created so far. On the other hand, in a recognition system designed for batch operation, the amount of time taken by the system to recognize speech is of little consequence to the user. In the batch mode, therefore, a multi-pass recognition system can be used, and the document history will be the recognizer transcript produced in the current pass.

The well-known information retrieval measure called *TFIDF* can be used to locate similar documents in the training database [57]. The term frequency (TF) tf_{ij} is defined as the frequency of the *j*th term in the document D_i, the unigram count of the term j in the document D_i. The inverse document frequency (IDF) idf_j is defined as the frequency of the *j*th term over the entire database of documents, which can be computed as:

$$idf_j = \frac{\text{Total number of documents}}{\text{Number of documents containing term } j} \tag{11.47}$$

The combined TF-IDF measure is defined as:

$$TFIDF_{ij} = tf_{ij} \log(idf_i) \tag{11.48}$$

The combination of TF and IDF can help to retrieve similar documents. It highlights words of particular interest to the query (via TF), while de-emphasizing common words that appear across different documents (via IDF). Each document including the query itself, can be represented by the TFIDF vector. Each element of the vector is the TFIDF value that corresponds to a word (or a term) in the vocabulary. Similarity between the two documents is then defined to be the cosine of the angle between the corresponding vectors. Therefore, we have:

$$Similarity(D_{i,}D_j) = \frac{\sum_k tfidf_{ik} * tfidf_{jk}}{\sqrt{\sum_k (tfidf_{ik})^2 * \sum_k (tfidf_{jk})^2}} \tag{11.49}$$

All the documents in the training database are ranked by the decreasing similarity between the document and the history of the current document dictated so far, or by a topic of particular interest to the user. The most similar documents are selected as the adaptation set for the topic-adaptive language model [46].

11.5.3. Maximum Entropy Models

The language model we have discussed so far combines different *n*-gram models via linear interpolation. A different way to combine sources is the maximum entropy approach. It constructs a single model that attempts to capture all the information provided by the various knowledge sources. Each such knowledge source is reformulated as a set of constraints that the desired distribution should satisfy. These constraints can be, for example, marginal distributions of the combined model. Their intersection, if not empty, should contain a set of probability functions that are consistent with these separate knowledge sources. Once the desired knowledge sources have been incorporated, we make no other assumption about other constraints, which leads to choosing the flattest of the remaining possibilities, the one with the highest entropy. The maximum entropy principle can be stated as follows:

- Reformulate different information sources as constraints to be satisfied by the target estimate.
- Among all probability distributions that satisfy these constraints, choose the one that has the highest entropy.

Given a general event space $\{\mathbf{X}\}$, let $P(\mathbf{X})$ denote the combined probability function. Each constraint is associated with a characteristic function of a subset of the sample space, $f_i(\mathbf{X})$. The constraint can be written as:

$$\sum_{\mathbf{X}} P(\mathbf{X}) f_i(\mathbf{X}) = E_i \tag{11.50}$$

where E_i is the corresponding desired expectation for $f_i(\mathbf{X})$, typically representing the required marginal probability of $P(\mathbf{X})$. For example, to derive a word trigram model, we can reformulate Eq. (11.50) so that constraints are introduced for unigram, bigram, and trigram probabilities. These constraints are usually set only where marginal probabilities can be estimated from a corpus. For example, the unigram constraint can be expressed as

$$f_{w_1}(w) = \begin{cases} 1 & \text{if } w = w_1 \\ 0 & \text{otherwise} \end{cases} \tag{11.51}$$

The desired value E_{w_1} can be the empirical expectation in the training data, $\sum_{w \in training\ data} f_{w_1}(w)/N$, and the associated constraint is

$$\sum_{\mathbf{h}} P(\mathbf{h}) \sum_{w} P(w \mid \mathbf{h}) f_{w_1}(w) = E_{w_1} \tag{11.52}$$

where $\mathbf{h}$ is the word history preceding word w.

We can choose $P(\mathbf{X})$ to diverge minimally from some other known probability function $Q(\mathbf{X})$, that is, to minimize the divergence function:

$$\sum_{\mathbf{X}} P(\mathbf{X}) \log \frac{P(\mathbf{X})}{Q(\mathbf{X})} \tag{11.53}$$

When $Q(\mathbf{X})$ is chosen as the uniform distribution, the divergence is equal to the negative of entropy with a constant. Thus minimizing the divergence function leads to maximizing the entropy. Under a minor consistent assumption, a unique solution is guaranteed to exist in the form [20]:

$$P(\mathbf{X}) \propto \prod_i \mu_i^{f_i(\mathbf{X})} \tag{11.54}$$

where μ_i is an unknown constant to be found. To search the exponential family defined by Eq. (11.54) for the μ_i that make $P(\mathbf{X})$ satisfy all the constraints, an iterative algorithm called generalized iterative scaling exists [20]. It guarantees to converge to the solution with some arbitrary initial μ_i. Each iteration creates a new estimate $P(\mathbf{X})$, which is improved in the sense that it matches the constraints better than its previous iteration [20]. One of the most effective applications of the maximum entropy model is to integrate the cache constraint into the language model directly, instead of interpolating the cache n-gram with the static n-gram. The new constraint is that the marginal distribution of the adapted model is the same as the lower-order n-gram in the cache [56]. In practice, the maximum entropy method has not offered any significant improvement in comparison to the linear interpolation.

11.6. PRACTICAL ISSUES

In a speech recognition system, every string of words $\mathbf{W} = w_1 w_2 \ldots w_n$ taken from the prescribed vocabulary can be assigned a probability, which is interpreted as the a priori probability to guide the recognition process and is a contributing factor in the determination of the final transcription from a set of partial hypothesis. Without language modeling, the entire vocabulary must be considered at every decision point. It is impossible to eliminate many candidates from consideration, or alternatively to assign higher probabilities to some candidates than others to considerably reduce recognition costs and errors.

11.6.1. Vocabulary Selection

For most speech recognition systems, an inflected form is considered as a different word. This is because these inflected forms typically have different pronunciations, syntactic roles, and usage patterns. So the words *work*, *works*, *worked*, and *working* are counted as four different words in the vocabulary.

We prefer to have a smaller vocabulary size, since this eliminates potential confusable candidates in speech recognition, leading to improved recognition accuracy. However, the limited vocabulary size imposes a severe constraint on the users and makes the system less flexible. In practice, the percentage of the Out-Of-Vocabulary (OOV) word rate directly affects the perceived quality of the system. Thus, we need to balance two kinds of errors, the OOV rate and the word recognition error rate. We can have a larger vocabulary to minimize the OOV rate if the system resources permit. We can minimize the expected OOV rate of the test data with a given vocabulary size. A corpus of text is used in conjunction with dictionaries to determine appropriate vocabularies.

The availability of various types and amounts of training data, from various time periods, affects the quality of the derived vocabulary. Given a collection of training data, we can create an ordered word list with the lowest possible OOV curve, such that, for any desired vocabulary size *V*, a minimum-OOV-rate vocabulary can be derived by taking the most frequent *V* words in that list. Viewed this way, the problem becomes one of estimating unigram probabilities of the test distribution, and then ordering the words by these estimates.

As illustrated in Figure 11.6, the perplexity generally increases with the vocabulary size, albeit it really does not make much sense to compare the perplexity of different vocabulary sizes. There are generally more competing words for a given context when the vocabulary size becomes big, which leads to increased recognition error rate. In practice, this is offset by the OOV rate, which decreases with the vocabulary size as illustrated in Figure 11.7. If we keep the vocabulary size fixed, we need more than 200,000 words in the vocabulary to have 99.5% English words coverage. For more inflectional languages such as German, larger vocabulary sizes are required to achieve coverage similar to that of English.[6]

In practice, it is far more important to use data from a specific topic or domain, if we know in what domain the speech recognizer is used. In general, it is also important to consider coverage of a specific time period. We should use training data from that period, or as close to it as possible. For example, if we know we will talk only about air travel, we benefit from using the air-travel related vocabulary and language model. This point is well illustrated by the fact that the perplexity of the domain-dependent bigram can be reduced by more than a factor of five over the general-purpose English trigram.

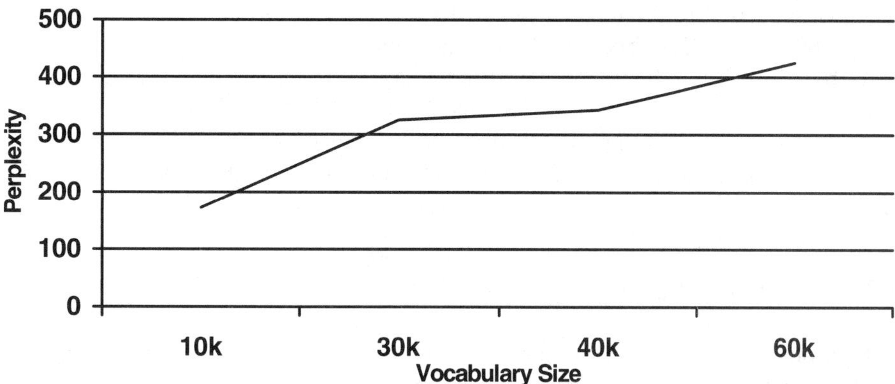

Figure 11.6 The perplexity of bigram with different vocabulary sizes. The training set consists of 500 million words derived from various sources, including newspapers and email. The test set comes from the whole Microsoft Encarta, an encyclopedia that has a wide coverage of different topics.

---

[6] The OOV rate of German is about twice as high as that of English with a 20k-word vocabulary [34].

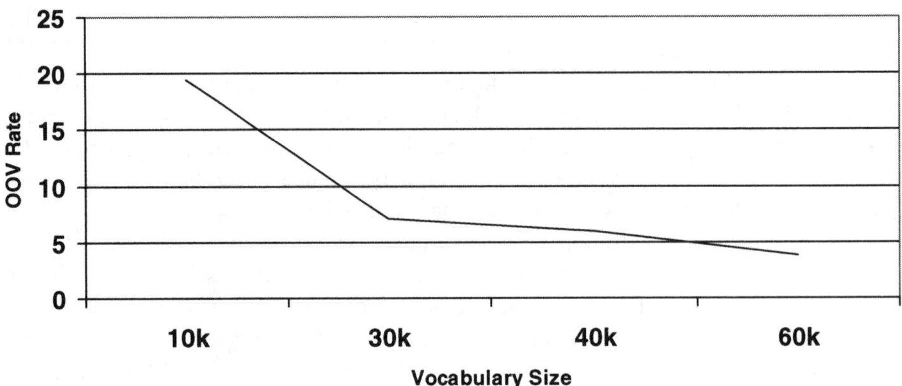

Figure 11.7 The OOV rate with different vocabulary size. The training set consists of 500 million words derived from various sources including newspaper and email. The test set came from the whole Microsoft Encarta encyclopedia.

For a user of a speech recognition system, a more personalized vocabulary can be much more effective than a general fixed vocabulary. The coverage can be dramatically improved as customized new words are added to a starting static vocabulary of 20,000. Typically, the coverage of such a system can be improved from 93% to more than 98% after 1000-4000 customized words are added to the vocabulary [18].

In North American general business English, the least frequent words among the most frequent 60,000 have a frequency of about 1:7,000,000. In optimizing a 60,000-word vocabulary we need to distinguish words with frequency of 1:7,000,000 from those that are slightly less frequent. To differentiate somewhat reliably between a 1:7,000,000 word and, say, a 1:8,000,000 word, we need to observe them enough times for the difference in their counts to be statistically reliable. For constructing a decent vocabulary, it is important that most such words are ranked correctly. We may need 100,000,000 words to estimate these parameters. This agrees with the empirical results, in which as more training data is used, the OOV curve improves rapidly up to 50,000,000 words and then more slowly beyond that point.

11.6.2. *N*-gram Pruning

When high order *n*-gram models are used, the model sizes typically become too large for practical applications. It is necessary to prune parameters from *n*-gram models such that the relative entropy between the original and the pruned model is minimized. You can choose *n*-grams so as to maximize performance (i.e., minimize perplexity) while minimizing the model size [39, 59, 64].

The criterion to prune *n*-grams can be based on some well-understood information-theoretic measure of language model quality. For example, the pruning method by Stolcke [64] removes some *n*-gram estimates while minimizing the performance loss. After pruning, the retained explicit n-gram probabilities are unchanged, but backoff weights are recomputed. Stolcke pruning uses the criterion that minimizes the distance between the distribution embodied by the original model and that of the pruned model based on the *Kullback-Leibler distance* defined in Eq. (3.181). Since it is infeasible to maximize over all possible subsets of n-grams, Stolcke prunning assumes that the n-grams affect the relative entropy roughly independently, and compute the distance due to each individual n-gram. The n-grams are thus ranked by their effect on the model entropy, and those that increase relative entropy the least are pruned accordingly. The main approximation is that we do not consider possible interactions between selected n-grams, and prune based solely on relative entropy due to removing a single n-gram. This avoids searching the exponential space of n-gram subsets.

To compute the relative entropy, $KL(p \parallel p')$, between the original and pruned n-gram models p and p', there is no need to sum over the vocabulary. By plugging in the terms for the backoff estimates, the sum can be factored as shown in Eq. (11.55) for a more efficient computation.

$$\begin{aligned} KL(p \parallel p') = -P(h)\{&P(w \mid h)[\log P(w \mid h') + \log\alpha'(h) - \log P(w \mid h)] \\ &+[\log\alpha'(h) - \log\alpha(h)](1 - \sum_{w_i \in \neg Backoff(w_i h)} P(w_i \mid h))\} \end{aligned} \tag{11.55}$$

where the sum in $\sum_{w_i \in \neg Backoff(w_i h)} P(w_i \mid h)$ is over all non-backoff estimates. To compute the revised backoff weights $\alpha'(h)$, you can simply drop the term for the pruned n-gram from the summation (backoff weight computation is illustrated in Algorithm 11.1).

In practice, pruning is highly effective. Stolcke reported that the trigram model can be compressed by more than 25% without degrading recognition performance. Comparing the pruned 4-gram model to the unpruned trigram model, it is better to use pruned 4-grams than to use a much larger number of trigrams.

11.6.3. CFG vs. *N*-gram Models

This chapter has discussed two major language models. While CFGs remain one of the most important formalisms for interpreting natural language, word n-gram models are surprisingly powerful for domain-independent applications. These two formalisms can be unified for both speech recognition and spoken language understanding. To improve portability of the domain-independent n-gram, it is possible to incorporate domain-specific CFGs into the domain-independent n-gram that can improve generalizability of the CFG and specificity of the n-gram.

The CFG is not only powerful enough to describe most of the structure in spoken language, but also restrictive enough to have efficient parsers. $P(\mathbf{W})$ is regarded as 1 or 0 depending upon whether the word sequence is accepted or rejected by the grammar. The problem is that the grammar is *almost always incomplete*. A CFG-based system is good only when you know what sentences to speak, which diminishes the system's value and usability of the system. The advantage of CFG's structured analysis is, thus, nullified by the poor coverage in most real applications. On the other hand, the *n*-gram model is trained with a large amount of data, the *n*-word dependency can often accommodate both syntactic and semantic structure seamlessly. The prerequisite of this approach is that we have enough training data. The problem for *n*-gram models is that we need a lot of data and the model may not be specific enough.

It is possible to take advantage of both rule-based CFGs and data-driven *n*-grams. Let's consider the following training sentences:

```
Meeting at three with Zhou Li.
Meeting at four PM with Derek.
```

If we use a word trigram, we estimate *P(Zhou|three with)* and *P(Derek|PM with)*, etc. There is no way we can capture needed long-span semantic information in the training data. A unified model has a set of CFGs that can capture the semantic structure of the domain. For the example listed here, we have a CFG for {name} and {time}, respectively. We can use the CFG to parse the training data to spot all the potential semantic structures in the training data. The training sentences now look like:

```
Meeting {at three:TIME} with {Zhou Li:NAME}
Meeting {at four PM:TIME} with {Derek: NAME}
```

With analyzed training data, we can estimate our *n*-gram probabilities as usual. We have probabilities, such as P({name}|{time} with), instead of *P(Zhou|three with)*, which is more meaningful and accurate. Inside each CFG we also derive P("*Zhou Li*"|{name}) and P("*four PM*"|{time}) from the existing *n*-gram (*n*-gram probability inheritance) so that they are normalized. If we add a new name to the existing {name} CFG, we use the existing *n*-gram probabilities to renormalize our CFGs for the new name. The new approach can be regarded as a standard *n*-gram in which the vocabulary consists of words and structured classes, as discussed in Section 11.4.3. The structured class can be very simple, such as {date}, {time}, and {name}, or can be very complicated, such as a CFG that contains deep structured information. The probability of a word or class depends on the previous words or CFG classes.

It is possible to inherit probability from a word *n*-gram LM. Let's take word trigram as our example here. An input utterance $\mathbf{W} = w_1 w_2 ... w_n$ can be segmented into a sequence $\mathbf{T} = t_1 t_2 ... t_m$, where each t_i is either a word in **W** or a CFG non-terminal that covers a sequence of words $\bar{u}_{t_i}$ in **W**. The likelihood of **W** under the segmentation **T** is, therefore,

$$P(\mathbf{W}, \mathbf{T}) = \prod_i P(t_i \mid t_{i-1}, t_{i-2}) \prod_i P(\bar{u}_{t_i} \mid t_i) \tag{11.56}$$

$P(\overline{u}_{t_i} \mid t_i)$, the likelihood of generating a word sequence $\overline{u}_{t_i} = [u_{t_i 1} u_{t_i 2} \ldots u_{t_i k}]$ from the CFG non-terminal t_i, can be inherited from the domain-independent word trigram. We can essentially use the CFG constraint to condition the domain-independent trigram into a domain-specific trigram. Such a unified language model can dramatically improve cross-domain performance using domain-specific CFGs [66].

In summary, the CFG is widely used to specify the permissible word sequences in natural language processing when training corpora are unavailable. It is suitable for dealing with structured command and control applications in which the vocabulary is small and the semantics of the task is well defined. The CFG either accepts the input sentence or rejects it. There is a serious coverage problem associated with CFGs. In other words, the accuracy for the CFG can be extremely high when the test data are covered by the grammar. Unfortunately, unless the task is narrow and well-defined, most users speak sentences that may not be accepted by the CFG, leading to word recognition errors.

Statistical language models such as trigrams assign an estimated probability to any word that can follow a given word history without parsing the structure of the history. Such an approach contains some limited syntactic and semantic information, but these probabilities are typically trained from a large corpus. Speech recognition errors are much more likely to occur within trigrams and (especially) bigrams that have not been observed in the training data. In these cases, the language model typically relies on lower-order statistics. Thus, increased *n*-gram coverage translates directly into improved recognition accuracy, but usually at the cost of increased memory requirements.

It is interesting to compute the true entropy of the language so that we understand what a solid lower bound is for the language model. For English, Shannon [60] used human subjects to guess letters by looking at how many guesses it takes people to derive the correct one based on the history. We can thus estimate the probability of the letters and hence the entropy of the sequence. Shannon computed the per-letter entropy of English with an entropy of 1.3 bits for 26 letters plus space. This may be an underestimate, since it is based on a single text. Since the average length of English written words (including space) is about 5.5 letters, the Shannon estimate of 1.3 bits per letter corresponds to a per-word perplexity of 142 for general English.

Table 11.2 summarizes the performance of several different *n*-gram models on a 60,000-word continuous speech dictation application. The experiments used about 260 million words from a newspaper such as the *Wall Street Journal*. The speech recognizer is based on Whisper described in Chapter 9. As you can see from the table, when the amount of training data is sufficient, both Katz and Kneser-Ney smoothing offer comparable recognition performance, although Kneser-Ney smoothing offers a modest improvement when the amount of training data is limited.

In comparison to Shannon's estimate of general English word perplexity, the trigram language for the *Wall Street Journal* is lower (91.4 vs. 142). This is because the text is mostly business oriented with a fairly homogeneous style and word usage pattern. For example, if we use the trigram language for data from a new domain that is related to personal information management, the test-set word perplexity can increase to 378 [66].

Table 11.2 *N*-gram perplexity and its corresponding speaker-independent speech recognition word error rate.

Models	Perplexity	Word Error Rate
Unigram Katz	1196.45	14.85%
Unigram Kneser-Ney	1199.59	14.86%
Bigram Katz	176.31	11.38%
Bigram Kneser-Ney	176.11	11.34%
Trigram Katz	95.19	9.69%
Trigram Kneser-Ney	91.47	9.60%

11.7. HISTORICAL PERSPECTIVE AND FURTHER READING

There is a large and active area of research in both speech and linguistics. These two distinctive communities worked on the problem with very different paths, leading to the stochastic language models and the formal language theory. The linguistics community has developed tools for tasks like parsing sentences, assigning semantic relations to the parts of a sentence, and so on. Most of these parser algorithms have the same characteristics, that is, they tabulate each sub-derivation and reuse it in building any derivation that shares that sub-derivation with appropriate grammars [22, 65, 67]. They have polynomial complexity with respect to sentence length because of *dynamic programming* principles to search for optimal derivations with respect to appropriate evaluation functions on derivations. There are three well-known dynamic programming parsers with a worst-case behavior of $O(n^3)$, where n is the number of words in the sentence: the Cocke-Younger-Kasami (CYK) algorithm (a bottom-up parser, proposed by J. Cocke, D. Younger, and T. Kasami) [32, 67], the Graham-Harrison-Ruzzo algorithm (bottom-up) [30], and the Earley algorithm (top-down) [21].

On the other hand, the speech community has developed tools to predict the next word on the basis of what has been said, in order to improve speech recognition accuracy [35]. Neither approach has been completely successful. The formal grammar and the related parsing algorithms are too brittle for comfort and require a lot of human retooling to port from one domain to another. The lack of structure and deep understanding has taken its toll on statistical technology's ability to choose the right words to guide speech recognition.

In addition to those discussed in this chapter, many alternative formal techniques are available. Augmented context-free grammars are used for natural language to capture grammatical natural languages such as agreement and subcategorization. Examples include generalized phrase structure grammars and head-driven phrase structure grammars [26, 53]. You can further generalize the augmented context-free grammar to the extent that the requirement of *context free* becomes unnecessary. The entire grammar, known as the *unification grammar,* can be specified as a set of constraints between feature structures [62]. Most of these grammars have only limited success when applied to spoken language systems. In fact, no practical domain-independent parser of unrestricted text has been developed for spoken language systems, partly because disambiguation requires the specification of detailed semantic information. Analysis of the Susanne Corpus with a crude parser suggests that over 80% of sentences are structurally ambiguous. More recently, large *treebanks* of parsed texts have given impetus to statistical approaches to parsing. Probabilities can be estimated from treebanks or plain text [6, 8, 24, 61] to efficiently rank analyses produced by modified chart parsing algorithms. These systems have yielded results of around 75% accuracy in assigning analyses to (unseen) test sentences from the same source as the unambiguous training material. Attempts have also been made to use statistical induction to *learn* the correct grammar for a given corpus of data [7, 43, 51, 58]. Nevertheless, these techniques are limited to simple grammars with category sets of a dozen or so non-terminals, or to training on manually parsed data. Furthermore, even when parameters of the grammar and control mechanism can be learned automatically from training corpora, the required corpora do not exist or are too small for proper training. In practice, we can devise grammars that specify directly how relationships relevant to the task may be expressed. For instance, one may use a phrase-structure grammar in which nonterminals stand for task concepts and relationships and rules specify possible expressions of those concepts and relationships. Such *semantic grammars* have been widely used for spoken language applications as discussed in Chapter 17.

It is worthwhile to point out that many natural language parsing algorithms are *NP-complete*, a term for a class of problems that are suspected to be particularly difficult to process. For example, maintaining lexical and agreement features over a potentially infinite-length sentence causes the unification-based formalisms to be NP-complete [3].

Since the predictive power of a general-purpose grammar is insufficient for reasonable performance, *n*-gram language models continue to be widely used. A complete proof of Good-Turing smoothing was presented by Church *et al.* [17]. Chen and Goodman [13] provide a detailed study on different *n*-gram smoothing algorithms. Jelinek's Eurospeech tutorial paper [35] provides an interesting historical perspective on the community's efforts to improve trigrams. Mosia and Giachin's paper [48] has detailed experimental results on class-based language models. Class-based model may be based on parts of speech or morphology [10, 16, 23, 47, 63]. More detailed discussion of the maximum entropy language model can be found in [5, 36, 42, 44, 52, 55, 56].

One interesting research area is to combine both n-grams and the structure that is present in language. A concerted research effort to explore structure-based language model may be the key for significant progress to occur in language modeling. This can be done as annotated data becomes available. Nasr *et al.* [50] have considered a new unified language model composed of several local models and a general model linking the local models together. The local model used in their system is based on the stochastic FSA, which is estimated from the training corpora. Other efforts to incorporate structured information are described in [12, 25, 27, 49, 66].

You can find tools to build *n*-gram language models at the CMU open source Web site[7] and SRI's language modeling toolkit Web site.[8] Both contain language modeling toolkits and documentation.

---

[7] http://www.speech.cs.cmu.edu/sphinx/

[8] http://www.speech.sri.com/projects/srilm/download.html

REFERENCES

[1] Aho, A.V. and J.D. Ullman, *The Theory of Parsing, Translation and Compiling*, 1972, Englewood Cliffs, NJ, Prentice-Hall.

[2] Bahl, L.R., *et al.*, "A Tree-Based Statistical Language Model for Natural Language Speech Recognition," *IEEE Trans. on Acoustics, Speech, and Signal Processing*, 1989, **37**(7), pp. 1001-1008.

[3] Barton, G., R. Berwick, and E. Ristad, *Computational Complexity and Natural Language*, 1987, Cambridge, MA, MIT Press.

[4] Bellegarda, J., "A Latent Semantic Analysis Framework for Large-Span Language Modeling," *Eurospeech*, 1997, Rhodes, Greece, pp. 1451-1454.

[5] Berger, A., S. DellaPietra, and V. DellaPietra, "A Maximum Entropy Approach to Natural Language Processing," *Computational Linguistics*, 1996, **22**(1), pp. 39-71.

[6] Black, E., *et al.*, "Towards History-based Grammars: Using Richer Models for Probabilistic Parsing," *Proc. of the Annual Meeting of the Association for Computational Linguistics*, 1993, Columbus, Ohio, USA, pp. 31-37.

[7] Briscoe, E.J., ed. *Prospects for Practical Parsing: Robust Statistical Techniques*, in Corpus-based Research into Language: A Feschrift for Jan Aarts, ed. P.d. Haan and N. Oostdijk, 1994, Amsterdam. 67-95, Rodopi.

[8] Briscoe, E.J. and J. Carroll, "Generalized Probabilistic LR Parsing of Natural Language (Corpora) with Unification-based Grammars," *Computational Linguistics*, 1993, **19**, pp. 25-59.

[9] Brown, P.F., *et al.*, "Class-Based *N*-gram Models of Natural Language," *Computational Linguistics*, 1992(4), pp. 467-479.

[10] Cerf-Danon, H. and M. El-Bèze, "Three Different Probabilistic Language Models: Comparison and Combination," *Proc. of the IEEE Int. Conf. on Acoustics, Speech and Signal Processing*, 1991, Toronto, Canada, pp. 297-300.

[11] Charniak, E., "Statistical Parsing with a Context-Free Grammar and Word Statistics," *AAAI-97*, 1997, Menlo Park, pp. 598-603.

[12] Chelba, C., A. Corazza, and F. Jelinek, "A Context Free Headword Language Model" in *Proc. of IEEE Automatic Speech Recognition Workshop"* 1995, Snowbird, Utah, pp. 89-90.

[13] Chen, S. and J. Goodman, "An Empirical Study of Smoothing Techniques for Language Modeling," *Proc. of Annual Meeting of the ACL*, 1996, Santa Cruz, CA.

[14] Chomsky, N., *Syntactic Structures*, 1957, The Hague: Mouton.

[15] Chomsky, N., *Aspects of the Theory of Syntax*, 1965, Cambridge, MIT Press.

[16] Church, K., "A Stochastic Parts Program and Noun Phrase Parser for Unrestricted Text," *Proc. of 2nd Conf. on Applied Natural Language Processing*, 1988, Austin, Texas, pp. 136-143.

[17] Church, K.W. and W.A. Gale, "A Comparison of the Enhanced Good-Turing and Deleted Estimation Methods for Estimating Probabilities of English Bigrams," *Computer Speech and Language*, 1991, pp. 19-54.

[18] Cole, R., *et al.*, *Survey of the State of the Art in Human Language Technology*, eds. http://cslu.cse.ogi.edu/HLTsurvey/HLTsurvey.html, 1996, Cambridge University Press.

[19] Collins, M., "A New Statistical Parser Based on Bigram Lexical Dependencies," *ACL-96*, 1996, pp. 184-191.

[20] Darroch, J.N. and D. Ratcliff, "Generalized Iterative Scaling for Log-Linear Models," *The Annals of Mathematical Statistics*, 1972, **43**(5), pp. 1470-1480.

[21] Earley, J., *An Efficient Context-Free Parsing Algorithm*, PhD Thesis, 1968, Carnegie Mellon University, Pittsburgh.

[22] Earley, J., "An Efficient Context-Free Parsing Algorithm," *Communications of the ACM*, 1970, **6**(8), pp. 451-455.

[23] El-Bèze, M. and A.-M. Derouault, "A Morphological Model for Large Vocabulary Speech Recognition," *Proc. of the IEEE Int. Conf. on Acoustics, Speech and Signal Processing*, 1990, Albuquerque, NM, pp. 577-580.

[24] Fujisaki, T., *et al.*, "A probabilistic parsing method for sentence disambiguation," *Proc. of the Int. Workshop on Parsing Technologies*, 1989, Pittsburgh.

[25] Galescu, L., E.K. Ringger, and A.F. Allen, "Rapid Language Model Development for New Task Domains," *Proc. of the ELRA First Int. Conf. on Language Resources and Evaluation (LREC)*, 1998, Granada, Spain.

[26] Gazdar, G., *et al.*, *Generalized Phrase Structure Grammars*, 1985, Cambridge, MA, Harvard University Press.

[27] Gillett, J. and W. Ward, "A Language Model Combining Trigrams and Stochastic Context-Free Grammars," *Int. Conf. on Spoken Language Processing*, 1998, Sydney, Australia.

[28] Good, I.J., "The Population Frequencies of Species and the Estimation of Population Parameters," *Biometrika*, 1953, pp. 237-264.

[29] Goodman, J., *Parsing Inside-Out*, PhD Thesis in Computer Science, 1998, Harvard University, Cambridge.

[30] Graham, S.L., M.A. Harrison, and W. L.Ruzzo, "An Improved Context-Free Recognizer," *ACM Trans. on Programming Languages and Systems*, 1980, **2**(3), pp. 415-462.

[31] Hindle, D. and M. Rooth, "Structural Ambiguity and Lexical Relations," *DARPA Speech and Natural Language Workshop*, 1990, Hidden Valley, PA, Morgan Kaufmann.

[32] Hopcroft, J.E. and J.D. Ullman, *Introduction to Automata Theory, Languages, and Computation*, 1979, Reading, MA, Addision Wesley.

[33] Iyer, R., M. Ostendorf, and J.R. Rohlicek, "Language Modeling with Sentence-Level Mixtures," *Proc. of the ARPA Human Language Technology Workshop*, 1994, Plainsboro, NJ, pp. 82-86.

[34] Jardino, M., "Multilingual Stochastic *N*-gram Class Language Models," *Proc. of the IEEE Int. Conf. on Acoustics, Speech and Signal Processing*, 1996, Atlanta, GA, pp. 161-163.

[35] Jelinek, F., "Up From Trigrams! The Struggle for Improved Language Models" in *Proc. of the European Conf. on Speech Communication and Technology*, 1991, Genoa, Italy, pp. 1037-1040.

[36] Jelinek, F., *Statistical Methods for Speech Recognition*, 1998, Cambridge, MA, MIT Press.

[37] Jelinek, F., *et al.*, "A dynamic language model for speech recognition" in *Proc. of the DARPA Speech and Natural Language Workshop*, 1991, Asilomar, CA.

[38] Katz, S.M., "Estimation of Probabilities from Sparse Data for the Language Model Component of a Speech Recognizer," *IEEE Trans. Acoustics, Speech and Signal Processing*, 1987(3), pp. 400-401.

[39] Kneser, R., "Statistical Language Modeling using a Variable Context" in *Proc. of the Int. Conf. on Spoken Language Processing*, 1996, Philadelphia, PA, p. 494.

[40] Kneser, R. and H. Ney, "Improved Backing-off for N-gram Language Modeling" in *Proc. of the IEEE Int. Conf. on Acoustics, Speech and Signal Processing* 1995, Detroit, MI, pp. 181-184.

[41] Kuhn, R. and R.D. Mori, "A Cache-Based Natural Language Model for Speech Recognition," *IEEE Trans. on Pattern Analysis and Machine Intelligence*, 1990(6), pp. 570-582.

[42] Lafferty, J.D. and B. Suhm, "Cluster Expansions and Iterative Scaling for Maximum Entropy Language Models" in *Maximum Entropy and Bayesian Methods*, K. Hanson and R. Silver, eds., 1995, Kluwer Academic Publishers.

[43] Lari, K. and S.J. Young, "Applications of Stochastic Context-free Grammars Using the Inside-Outside Algorithm," *Computer Speech and Language*, 1991, **5**(3), pp. 237-257.

[44] Lau, R., R. Rosenfeld, and S. Roukos, "Trigger-Based Language Models: A Maximum Entropy Approach," *Int. Conf. on Acoustics, Speech and Signal Processing*, 1993, Minneapolis, MN, pp. 108-113.

[45] Magerman, D.M. and M.P. Marcus, "Pearl: A Probabilistic Chart Parser," *Proc. of the Fourth DARPA Speech and Natural Language Workshop*, 1991, Pacific Grove, California.

[46] Mahajan, M., D. Beeferman, and X.D. Huang, "Improved Topic-Dependent Language Modeling Using Information Retrieval Techniques," *IEEE Int. Conf. on Acoustics, Speech and Signal Processing*, 1999, Phoenix, AZ, pp. 541-544.

[47] Maltese, G. and F. Mancini, "An Automatic Technique to Include Grammatical and Morphological Information in a Trigram-based Statistical Language Model," *Proc. of the IEEE Int. Conf. on Acoustics, Speech and Signal Processing*, 1992, San Francisco, CA, pp. 157-160.

[48] Moisa, L. and E. Giachin, "Automatic Clustering of Words for Probabilistic Language Models" in *Proc. of the European Conf. on Speech Communication and Technology* 1995, Madrid, Spain, pp. 1249-1252.

[49] Moore, R., *et al.*, "Combining Linguistic and Statistical Knowledge Sources in Natural-Language Processing for ATIS," *Proc. of the ARPA Spoken Language Sys-*

tems Technology Workshop, 1995, Austin, Texas, Morgan Kaufmann, Los Altos, CA.

[50] Nasr, A., *et al.*, "A Language Model Combining *N*-grams and Stochastic Finitie State Automata," *Proc. of the Eurospeech*, 1999, Budapest, Hungary, pp. 2175-2178.

[51] Pereira, F.C.N. and Y. Schabes, "Inside-Outside Reestimation from Partially Bracketed Corpora," *Proc. of the 30th Annual Meeting of the Association for Computational Linguistics*, 1992, pp. 128-135.

[52] Pietra, S.A.D., *et al.*, "Adaptive Language Model Estimation using Minimum Discrimination Estimation," *Proc. of the IEEE Int. Conf. on Acoustics, Speech and Signal Processing*, 1992, San Francisco, CA, pp. 633-636.

[53] Pollard, C. and I.A. Sag, *Head-Driven Phrase Structure Grammar*, 1994, Chicago, University of Chicago Press.

[54] Pullum, G. and G. Gazdar, "Natural Languages and Context-Free Languages," *Linguistics and Philosophy*, 1982, **4**, pp. 471-504.

[55] Ratnaparkhi, A., S. Roukos, and R.T. Ward, "A Maximum Entropy Model for Parsing," *Proc. of the Int. Conf. on Spoken Language Processing*, 1994, Yokohama, Japan, pp. 803-806.

[56] Rosenfeld, R., *Adaptive Statistical Language Modeling: A Maximum Entropy Approach*, Ph.D. Thesis in School of Computer Science, 1994, Carnegie Mellon University, Pittsburgh, PA.

[57] Salton, G. and M.J. McGill, *Introduction to Modern Information Retrieval*, 1983, New York, McGraw-Hill.

[58] Schabes, Y., M. Roth, and R. Osborne, "Parsing the *Wall Street Journal* with the Inside-Outside Algorithm," *Proc. of the Sixth Conf. of the European Chapter of the Association for Computational Linguistics*, 1993, pp. 341-347.

[59] Seymore, K. and R. Rosenfeld, "Scalable Backoff Language Models," *Proc. of the Int. Conf. on Spoken Language Processing*, 1996, Philadelphia, PA, pp. 232.

[60] Shannon, C.E., "Prediction and Entropy of Printed English," *Bell System Technical Journal*, 1951, pp. 50-62.

[61] Sharman, R., F. Jelinek, and R.L. Mercer, "Generating a Grammar for Statistical Training," *Proc. of the Third DARPA Speech and Natural Language Workshop*, 1990, Hidden Valley, Pennsylvania, pp. 267-274.

[62] Shieber, S.M., *An Introduction to Unification-Based Approaches to Grammars*, 1986, Cambridge, UK, CSLI Publication, Leland Stanford Junior University.

[63] Steinbiss, V., *et al.*, "A 10,000-word Continuous Speech Recognition System," *Proc. of the IEEE Int. Conf. on Acoustics, Speech and Signal Processing*, 1990, Albuquerque, NM, pp. 57-60.

[64] Stolcke, A., "Entropy-based Pruning of Backoff Language Models," *DARPA Broadcast News Transcription and Understanding Workshop*, 1998, Lansdowne, VA.

[65] Tomita, M., "An Efficient Augmented-Context-Free Parsing Algorithm," *Computational Linguistics*, 1987, **13**(1-2), pp. 31-46.

[66] Wang, Y., M. Mahajan, and X. Huang, "A Unified Context-Free Grammar and *N*-Gram Model for Spoken Language Processing," *Int. Conf. on Acoustics, Speech and Signal Processing*, 2000, Istanbul, Turkey, pp. 1639-1642.

[67] Younger, D.H., "Recognition and Parsing of Context-Free Languages in Time n^3," *Information and Control*, 1967, **10**, pp. 189-208.

CHAPTER 12

Basic Search Algorithms

Continuous speech recognition (CSR) is both a pattern recognition and search problem. As described in previous chapters, the acoustic and language models are built upon a statistical pattern recognition framework. In speech recognition, making a search decision is also referred to as decoding. In fact, decoding got its name from information theory (see Chapter 3) where the idea is to *decode* a signal that has presumably been encoded by the source process and has been transmitted through the communication channel, as depicted in Chapter 1, Figure 1.1. In this chapter, we first review the general decoder architecture that is based on such a source-channel model.

The decoding process of a speech recognizer is to find a sequence of words whose corresponding acoustic and language models best match the input signal. Therefore, the process of such a decoding process with trained acoustic and language models is often referred to as just a *search* process. Graph search algorithms have been explored extensively in the fields of artificial intelligence, operation research, and game theory. In this chapter first we present several basic search algorithms, which serve as the basic foundation for CSR.

The complexity of a search algorithm is highly correlated with the search space, which is determined by the constraints imposed by the language models. We discuss the impact of different language models, including finite-state grammars, context-free grammars, and *n*-grams.

Speech recognition search is usually done with the Viterbi or A* stack decoders. The reasons for choosing the Viterbi decoder involve arguments that point to speech as a left-to-right process and to the efficiencies afforded by a time-synchronous process. The reasons for choosing a stack decoder involve its ability to more effectively exploit the A* criteria, which holds out the hope of performing an optimal search as well as the ability to handle huge search spaces. Both algorithms have been successfully applied to various speech recognition systems. The relative merits of both search algorithms were quite controversial in the 1980s. Lately, with the help of efficient pruning techniques, Viterbi beam search has been the preferred method for almost all speech recognition tasks. Stack decoding, on the other hand, remains an important strategy to uncover the *n*-best and lattice structures.

12.1. BASIC SEARCH ALGORITHMS

Search is a subject of interest in artificial intelligence and has been well studied for expert systems, game playing, and information retrieval. We discuss several general graph search methods that are fundamental to spoken language systems. Although the basic concept of graph search algorithms is independent of any specific task, the efficiency often depends on how we exploit domain-specific knowledge.

The idea of search implies moving around, examining things, and making decisions about whether the sought object has yet been found. In general, search problems can be represented using the *state-space search* paradigm. It is defined by a triplet (S, O, G), where S is a set of initial states, O a set of operators (or rules) applied on a state to generate a transition with its corresponding cost to another state, and G a set of goal states. A solution in the state-space search paradigm consists in finding a path from an initial state to a goal state. The state-space representation is commonly identified with a directed graph in which each node corresponds to a state and each arc to an application of an operator (or a rule), which transitions from one state to another. Thus, the state-space search is equivalent to searching through the graph with some objective function.

Before we present any graph search algorithms, we need to remind the readers of the importance of the dynamic programming algorithm described in Chapter 8. Dynamic programming should be applied whenever possible and as early as possible because (1) unlike any heuristics, it will not sacrifice optimality; (2) it can transform an exponential search into a polynomial search.

12.1.1. General Graph Searching Procedures

Although dynamic programming is a powerful polynomial search algorithm, many interesting problems cannot be handled by it. A classical example is the traveling salesman's problem. We need to find a shortest-distance tour, starting at one of many cities, visiting each city exactly once, and returning to the starting city. This is one of the most famous problems in the *NP*-hard class [1, 32]. Another classical example is the *N*-queens problem (typically 8-queens), where the goal is to place *N* queens on an $N \times N$ chessboard in such a way that no queen can capture any other queen, i.e., there is no more than one queen in any given row, column, or diagonal. Many of these puzzles have the same characteristics. As we know, the best algorithms currently known for solving the *NP*-hard problem are exponential in the problem size. Most graph search algorithms try to solve those problems using heuristics to avoid or moderate such a combinatorial explosion.

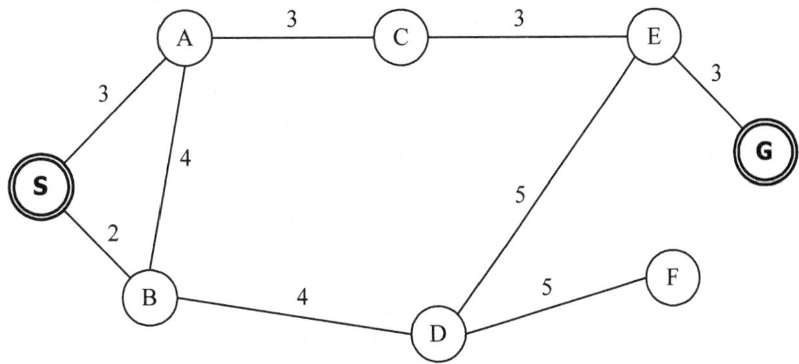

Figure 12.1 A highway distance map for cities S, A, B, C, D, E, F, and G. The salesman needs to find a path to travel from city S to city G [42].

Let's start our discussion of graph search procedure with a simple city-traveling problem [42]. Figure 12.1 shows a highway distance map for all the cities. A salesman named John needs to travel from the starting city S to the end city G. One obvious way to find a path is to derive a graph that allows orderly exploration of all possible paths. Figure 12.2 shows the graph that traces out all possible paths in the city-distance map shown in Figure 12.1. Although the city-city connection is bi-directional, we should note that the search graph in this case must not contain cyclic paths, because they would not lead to any progress in this scenario.

If we define the search space as the potential number of nodes (states) in the graph search procedure, the search space for finding the optimal state sequence in the Viterbi algorithm (described in Chapter 8) is $N \times T$, where *N* is the number of states for the HMM and *T* is the length of the observation. Similary, the search space for John's traveling problem will be 27.

Another important measure for a search graph is the *branching factor,* defined as the average number of successors for each node. Since the number of nodes of a search graph (or tree) grows exponentially with base equal to this branching factor, we certainly need to watch out for search graphs (or trees) with a large branching factor. Sometimes they can be too big to handle (even infinite, as in game playing). We often trade the optimal solution for improved performance and feasibility. That is, the goal for such search problems is to find one satisfactory solution instead of the optimal one. In fact, most AI (artifical intelligence) search problems belong to this category.

The search tree in Figure 12.2 may be implemented either explicitly or implicitly. In an explicit implementation, the nodes and arcs with their corresponding distances (or costs) are explicitly specified by a table. However, an explicit implementation is clearly impractical for large search graphs and impossible for those with infinite nodes. In practice, most parts of the graph may never be explored before a solution is found. Therefore, a sensible strategy is to dynamically generate the search graph. The part that becomes explicit is often referred to as an *active* search space. Throughout the discussion here, it is important to keep in mind this distinction between the implicit search graph that is specified by the start node S and the explicit partial search graphs that are actually constructed by the search algorithm.

To expand the tree, the term *successor operator* (or *move generator*, as it is often called in game search) is defined as an operator that is applied to a node to generate all of the successors of that node and to compute the distance associated with each arc. The successor operator obviously depends on the topology (or rules) of the problem space. Expanding the starting node *S*, and successors of *S*, ad infinitum, gradually makes the implicitly

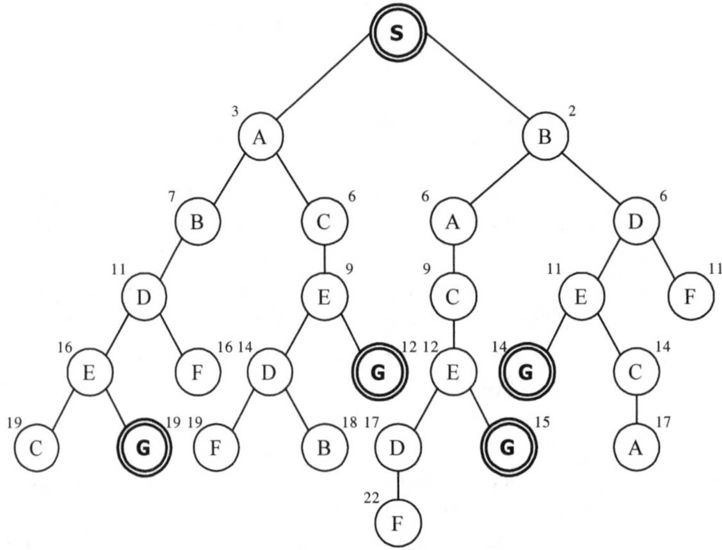

Figure 12.2 The search tree (graph) for the salesman problem illustrated in Figure 12.1. The number next to each node is the accumulated distance from start city to end city [42].

defined graph explicit. This recursive procedure is straightforward, and the search graph (tree) can be constructed without the extra bookkeeping. However, this process would only generate a search tree where the same node might be generated as a part of several possible paths.

For example, node E is being generated in four different paths. If we are interested in finding an optimal path to travel from S to G, it is more efficient to merge those different paths that lead to the same node E. We can pick the shortest path up to C, since everything following E is the same for the rest of the paths. This is consistent with the dynamic programming principle—when looking for the best path from S to G, all partial paths from S to any node E, other than the best path from S to E, should be discarded. The dynamic programming merge also eliminates cyclic paths implicitly, since a cyclic path cannot be the shortest path. Performing this extra bookkeeping (merging different paths leading into the same node) generates a search graph rather than a search tree.

Although a graph search has the potential advantage over a tree search of being more efficient, it does require extra bookkeeping. Whether this effort is justified depends on the individual problem one has to address.

Most search strategies search in a forward direction, i.e., build the search graph (or tree) by starting with the initial configuration (the starting state S) from the root. In the general AI literature, this is referred to as *forward reasoning* [43], because it performs rule-base reasoning by matching the left side of rules first. However, for some specific problem domains, it might be more efficient to use *backward reasoning* [43], where the search graph is built from the bottom up (the goal state G). Possible scenarios include:

- *There are more initial states than goal states.* Obviously it is easy to start with a small set of states and search for paths leading to one of the bigger sets of states. For example, suppose the initial state S is the hometown for John in the city-traveling problem in Figure 12.1 and the goal state G is an unfamiliar city for him. In the absence of a map, there are certainly more locations (neighboring cities) that John can identify as being close[1] to his home city S than those he can identify as being close to an unfamiliar location. In a sense, all of those locations being identified as close to John's home city S are equivalent to the initial state S. This means John might want to consider reasoning backward from the unfamiliar goal city G for the trip planning.
- *The branching factor for backward reasoning is smaller than that for forward reasoning.* In this case it makes sense to search in the direction with lower branching factor.

It is in principle possible to search from both ends simultaneously, until two partial paths meet somewhere in the middle. This strategy is called *bi-directional search* [43]. Bidirectional search seems particularly appealing if the number of nodes at each step grows

[1] *Being close* means that, once John reaches one of those neighboring cities, he can easily remember the best path to return home. It is similar to the killer book for chess play. Once the player reaches a particular board configuration, he can follow the killer book for moves that can guarantee a victory.

exponentially with the depth that needs to be explored. However, sometimes bi-directional search can be devastating. The two searches may cross each other, as illustrated in Figure 12.3.

The process of explicitly generating part of an implicitly defined graph forms the essence of our general graph search procedure. The procedure is summarized in Algorithm 12.1. It maintains two lists: *OPEN*, which stores the nodes waiting for expansion, and *CLOSE*, which stores the already expanded nodes. Steps 6a and 6b are basically the bookkeeping process to merge different paths going into the same node by picking the one that has the minimum distance. Step 6a handles the case where v is in the *OPEN* list and thus is not expanded. The merging process is straightforward, with a single comparison and change of traceback pointer if necessary. However, when v is in the *CLOSE* list and thus is already expanded in Step 6b, the merging requires additional forward propagation of the new score if the current path is found to be better than the best subpath already in the *CLOSE* list. This forward propagation could be very expensive. Fortunately, most of the search strategy can avoid such a procedure if we know that the already expanded node must belong in the best path leading to it. We discuss this in Section 12.5.

As described earlier, it may not be worthwhile to perform bookkeeping for a graph search, so Steps 6a and 6b are optional. If both steps are omitted, the graph search algorithm described above becomes a tree search algorithm. To illustrate different search strategies, tree search is used as the basic graph search algorithm in the sections that follows. However, you should note that all the search methods described here could be easily extended to graph search with the extra bookkeeping (merging) process as illustrated in Steps 6a and 6b of Algorithm 12.1.

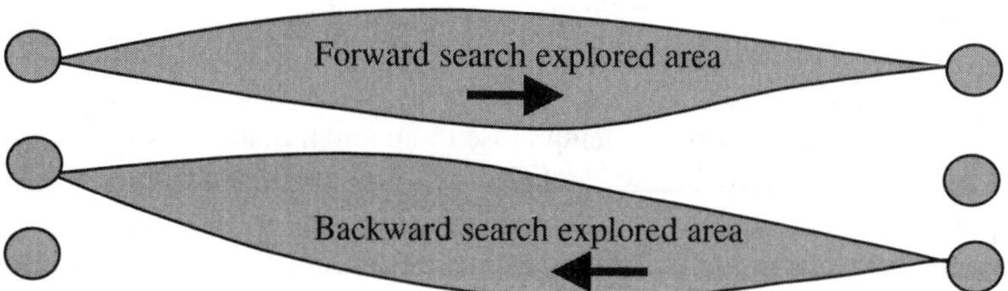

Figure 12.3 A bad case for bi-directional search, where the forward search and the backward search crossed each other [42].

ALGORITHM 12.1: *THE GRAPH-SEARCH ALGORITHM*

Step 1: Initialization: Put S in the *OPEN* list and create an initially empty *CLOSE* list .
Step 2: If the *OPEN* list is empty, exit and declare failure.
Step 3: Pop up the first node N in the *OPEN* list, remove it from the *OPEN* list and put it into the *CLOSE* list.
Step 4: If node N is a goal node, exit successfully with the solution obtained by tracing back the path along the pointers from N to S.
Step 5: Expand node N by applying the successor operator to generate the successor set $SS(N)$ of node N. Be sure to eliminate the ancestors of N from $SS(N)$.
Step 6: $\forall v \in SS(N)$ do

6a. (optional) If $v \in OPEN$ and the accumulated distance of the new path is smaller than that for the one in the *OPEN* list, do
(i) change the traceback (parent) pointer of v to N and adjust the accumulated distance for v.
(ii) go to Step 7.

6b. (optional) If $v \in CLOSE$ and the accumulated distance of the new path is smaller than the partial path ending at v in the *CLOSE* list, do
(i) change the traceback (parent) pointer of v to N and adjust the accumulated distance for all paths that contain v.
(ii) go to Step 7.

6c. Create a pointer pointing to N and push it into the *OPEN* list.

Step 7: Reorder the *OPEN* list according to search strategy or some heuristic measurement.
Step 8: Go to Step 2.

12.1.2. Blind Graph Search Algorithms

If the aim of the search problem is to find an acceptable path instead of the best path, blind search is often used. *Blind search* treats every node in the *OPEN* list the same and blindly decides the order to be expanded without using any domain knowledge. Since blind search treats every node equally, it is often referred to as *uniform search* or *exhaustive search*, because it exhaustively tries out all possible paths. In AI, people are typically not interested in blind search. However, it does provide a lot of insight into many sophisticated heuristic search algorithms. You should note that blind search does not expand nodes randomly. Instead, it follows some systematic way to explore the search graph. Two popular types of blind search are depth-first search and breadth-first search.

12.1.2.1. Depth-First Search

When we are in a maze, the most natural way to find a way out is to mark the branch we take whenever we reach a branching point. The marks allow us to go back to a choice point with an unexplored alternative, withdraw the most recently made choice and undo all consequences of the withdrawn choice whenever a dead-end is reached. Once the alternative choice is selected and marked, we go forward based on the same procedure. This intuitive search strategy is called *backtracking*. The famous *N*-queens puzzle [32] can be handily solved by the backtracking strategy.

Depth-first search picks an arbitrary alternative at every node visited. The search sticks with this partial path and works forward from the partial path. Other alternatives at the same level are ignored completely (for the time being) in the hope of finding a solution based on the current choice. This strategy is equivalent to ordering the nodes in the *OPEN* list by their depth in the search graph (tree). The deepest nodes are expanded first and nodes of equal depth are ordered arbitrarily.

Although depth-first search hopes the current choice leads to a solution, sometimes the current choice could lead to a dead-end (a node which is neither a goal node nor can be expanded further). In fact, it is desirable to have many short dead-ends. Otherwise, the algorithm may search for a very long time before it reaches a dead-end, or it might not ever reach a solution if the search space is infinite. When the search reaches a dead-end, it goes back to the last decision point and proceeds with another alternative.

Figure 12.4 shows all the nodes being expanded under the depth-first search algorithm for the city-traveling problem illustrated in Figure 12.1. The only differences between the graph search and the depth-first search algorithms are:

1. The graph search algorithm generates all successors at a time (although all except one are ignored first), while depth-first search generates only one successor at a time.
2. The graph search, when successfully finding a path, saves only one path from the starting node to the goal node, while depth-first search in general saves the entire record of the search graph.

Depth-first search could be dangerous because it might search an impossible path that is actually an infinite dead-end. To prevent exploring of paths that are too long, a depth bound can be placed to constrain the nodes to be expanded, and any node reaching that depth limit is treated as a terminal node (as if it had no successor).

The general graph search algorithm can be modified into a depth-first search algorithm as illustrated in Algorithm 12.2.

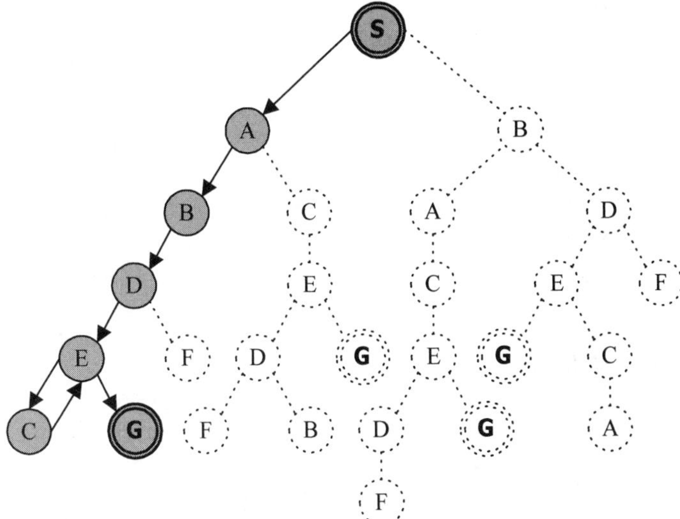

Figure 12.4 The node-expanding procedure of the depth-first search for the path search problem in Figure 12.1. When it fails to find the goal city in node *C*, it backtracks to the parent and continues the search until it finds the goal city. The gray nodes are those that are explored. The dotted nodes are not visited during the search [42].

ALGORITHM 12.2: *THE DEPTH-FIRST SEARCH ALGORITHM*

Step 1: Initialization: Put S in the *OPEN* list and create an initially empty the CLOSE list.
Step 2: If the *OPEN* list is empty, exit and declare failure.
Step 3: Pop up the first node N in the *OPEN* list, remove it from the *OPEN* list and put it into the *CLOSE* list.
Step 4: If node N is a goal node, exit successfully with the solution obtained by tracing back the path along the pointers from N to S.
 4a. If the depth of node N is equal to the depth bound, go to Step 2.
Step 5: Expand node N by applying the successor operator to generate the successor set SS(N) of node N. Be sure to eliminate the ancestors of N from SS(N).
Step 6: $\forall v \in SS(N)$ do
 6c. Create a pointer pointing to N and push it into the *OPEN* list.
Step 7: Reorder the the *OPEN* list in descending order of the depth of the nodes.
Step 8: Go to Step 2.

12.1.2.2. Breadth-First Search

One natural alternative to the depth-first search strategy is breadth-first search. *Breadth-first search* examines all the nodes on one level before considering any of the nodes on the next level (depth). As shown in Figure 12.5, node *B* would be examined just after node *A*. The search moves on level-by-level, finally discovering G on the fourth level.

Breadth-first search is guaranteed to find a solution if one exists, assuming that a finite number of successors (branches) always follow any node. The proof is straightforward. If there is a solution, its path length must be finite. Let's assume the length of the solution is *M*. Breadth-first search explores all paths of the same length increasingly. Since the number of paths of fixed length *N* is always finite, it eventually explores all paths of length *M*. By that time it should find the solution.

It is also easy to show that a breadth-first search can work on a search tree (graph) with infinite depth on which an unconstrained depth-first search will fail. Although a breadth-first might not find a shortest-distance path for the city-travel problem, it is guaranteed to find the one with fewest cities visited (minimum-length path). In some cases, it is a very desirable solution. On the other hand, a breadth-first search may be highly inefficient when all solutions leading to the goal node are at approximately the same depth. The breadth-first search algorithm is summarized in Algorithm 12.3.

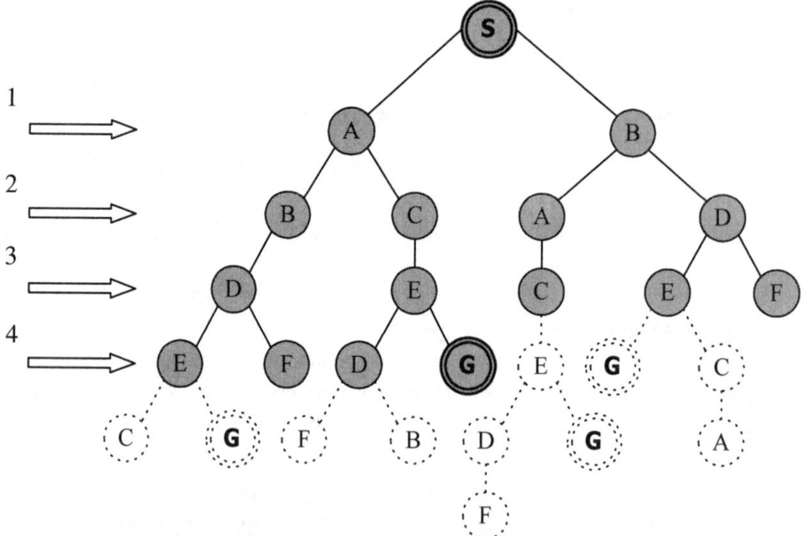

Figure 12.5 The node-expanding procedure of a breadth-first search for the path search problem in Figure 12.1. It searches through each level until the goal is identified. The gray nodes are those that are explored. The dotted nodes are not visited during the search [42].

ALGORITHM 12.3: *THE BREADTH-FIRST SEARCH ALGORITHM*

Step 1: Initialization: Put *S* in the *OPEN* list and create an initially empty the *CLOSE* list.
Step 2: If the *OPEN* list is empty, exit and declare failure.
Step 3: Pop up the first node *N* in the *OPEN* list, remove it from the *OPEN* list and put it into the *CLOSE* list.
Step 4: If node *N* is a goal node, exit successfully with the solution obtained by tracing back the path along the pointers from *N* to *S*.
Step 5: Expand node *N* by applying the successor operator to generate the successor set *SS*(*N*) of node *N*. Be sure to eliminate the ancestors of *N*, from *SS*(*N*).
Step 6: $\forall v \in SS(N)$ do

6c. Create a pointer pointing to *N* and push it into the *OPEN* list.

Step 7: Reorder the *OPEN* list in increasing order of the depth of the nodes.
Step 8. Go to Step 2.

12.1.3. Heuristic Graph Search

Blind search methods, like depth-first search and breadth-first search, have no sense (or guidance) of where the goal node lies ahead. Consequently, they often spend a lot of time searching in hopeless directions. If there is guidance, the search can move in the direction that is more likely to lead to the goal. For example, you may want to find a driving route to the World Trade Center in New York. Without a map at hand, you can still use a straight-line distance estimated by eye as a hint to see if you are closer to the goal (World Trade Center). This *hill-climbing* style of guidance can help you to find the destination much more efficiently.

Blind search finds only one arbitrary solution instead of the optimal solution. To find the optimal solution with depth-first or breadth-first search, you must not stop searching when the first solution is discovered. Instead, the search needs to continue until it reaches all the solutions, so you can compare them to pick the best. This strategy for finding the optimal solution is called *British Museum search* or *brute-force search*. Obviously, it is unfeasible when the search space is large. Again, to conduct selective search and yet still be able to find the optimal solution, some guidance on the search graph is necessary.

The guidance obviously comes from domain-specific knowledge. Such knowledge is usually referred to as *heuristic* information, and search methods taking advantage of it are called *heuristic search* methods. There is usually a wide variety of different heuristics for the problem domain. Some heuristics can reduce search effort without sacrificing optimality, while other can greatly reduce search effort but provide only sub-optimal solutions. In most practical problems, the choice of different heuristics is usually a tradeoff between the quality of the solution and the cost of finding the solution.

Heuristic information works like an evaluation function $h(N)$ that maps each node N to a real number, and which serves to indicate the relative goodness (or cost) of continuing the search path from that node. Since in our city-travel problem, straight-line distance is a natural way of measuring the goodness of a path, we can use the heuristic function $h(N)$ for the distance evaluation as:

$$h(N)\text{=Heuristic estimate of the remaining distance from node } N \text{ to goal } G \quad (12.1)$$

Since $g(N)$, the distance of the partial path to the current node N, is generally known, we have:

$$g(N)\text{=The distance of the partial path already traveled from root } S \text{ to node } N \quad (12.2)$$

We can define a new heuristic function, $f(N)$, which estimates the total distance for the path (not yet finished) going through node N.

$$f(N) = g(N) + h(N) \quad (12.3)$$

A heuristic search method basically uses the heuristic function $f(N)$ to re-order the *OPEN* list in the Step 7 of Algorithm 12.1. The node with the best heuristic value is explored first (expanded first). Some heuristic search strategies also prune some unpromising partial paths forever to save search space. This is why heuristic search is often referred to as heuristic pruning.

The choice of the heuristic function is critical to the search results. If we use one that overestimates the distance of some nodes, the search results may be suboptimal. Therefore, heuristic functions that do not overestimate the distance are often used in search methods aiming to find the optimal solution.

To close this section, we describe two of the most popular heuristic search methods: best-first (or A^* Search) [32, 43] and beam search [43]. They are widely used in many components of spoken language systems.

12.1.3.1. Best-First (A^* Search)

Once we have a reasonable heuristic function to evaluate the goodness of each node in the *OPEN* list, we can explore the best node (the node with smallest $f(N)$ value) first, since it offers the best hope of leading to the best path. This natural search strategy is called *best-first search*. To implement best-first search based on the Algorithm 12.1, we need to first evaluate $f(N)$ for each successor before putting the successors in the *OPEN* list in Step 6. We also need to sort the elements in the *OPEN* list based on $f(N)$ in Step 7, so that the best node is in the front-most position waiting to be expanded in Step 3. The modified procedure for performing best-first search is illustrated in Algorithm 12.4. To avoid duplicating nodes in the *OPEN* list, we include Steps 6a and 6b to take advantage of the dynamic programming principle. They perform the needed bookkeeping process to merge different paths leading into the same node.

ALGORITHM 12.4: *THE BEST-FIRST SEARCH ALGORITHM*

Step 1: Initialization: Put *S* in the *OPEN* list and create an initially empty the *CLOSE* list.
Step 2: If the *OPEN* list is empty, exit and declare failure.
Step 3. Pop up the first node *N* in the *OPEN* list, remove it from the *OPEN* list and put it into the *CLOSE* list.
Step 4: If node *N* is a goal node, exit successfully with the solution obtained by tracing back the path along the pointers from *N* to *S*.
Step 5: Expand node *N* by applying the successor operator to generate the successor set *SS*(*N*) of node *N*. Be sure to eliminate the ancestors of *N*, from *SS*(*N*).
Step 6: $\forall v \in SS(N)$ do

6a. (optional) If $v \in OPEN$ and the accumulated distance of the new path is smaller than that for the one in the the *OPEN* list, do

(i) Change the traceback (parent) pointer of v to *N* and adjust the accumulated distance for v.

(ii) Evaluate heuristic function $f(v)$ for v and go to Step 7.

6b. (optional) If $v \in CLOSE$ and the accumulated distance of the new path is small than the partial path ending at v in the the *CLOSE* list,

(i) Change the traceback (parent) pointer of v to *N* and adjust the accumulated distance and heuristic function f for all the paths containing v.

(ii) go to Step 7.

6c. Create a pointer pointing to *N* and push it into the *OPEN* list.

Step 7: Reorder the the *OPEN* list in the increasing order of the heuristic function $f(N)$.
Step 8: Go to Step 2.

A search algorithm is said to be *admissible* if it can guarantee to find an optimal solution, if one exists. Now we show that if the heuristic function $h(N)$ of estimating the remaining distance from N to goal node G is an underestimate[2] of the true distance from N to goal node G, the best-first search illustrated in Algorithm 12.4 is admissible. In fact, when $h(N)$ satisfies the above criterion, the best-first algorithm is called A^* (pronounced as /*eh*/-star) Search.

The proof can be carried out informally as follows. When the frontmost node in the *OPEN* list is the goal node G in Step 4, it immediately implies that

$$\forall v \in OPEN \quad f(v) \geq f(G) = g(G) + h(G) = g(G) \tag{12.4}$$

[2] For admissibility, we actually require only that the heuristic function not overestimate the distance from *N* to *G*. Since it is very rare to have an exact estimate, we use underestimate throughout this chapter without loss of generality. Sometimes we refer to an underestimate function as a lower-bound estimate of the true value.

Equation (12.4) says that the distance estimate of any incomplete path is no shorter than the first found complete path. Since the distance estimate for any incomplete path is underestimated, the first found complete path in Step 4 must be the optimal path. A similar argument can also be used to prove that the Step 6b is actually not necessary for admissible heuristic functions; that is, there cannot be another path with a shorter distance from the starting node to a node that has been expanded. This is a very important feature since Step 6b is, in general, very expensive and it requires significant updates of many already expanded paths.

The A^* search method is actually a family of search algorithms. When $h(N) = 0$ for all N, the search degenerates into an uninformed search[3] [40]. In fact, this type of uninformed search is the famous *branch-and-bound search* algorithm that is often used in many *operations research* problems. Branch-and-bound search always expands the shortest path leading into an open node until there is a path reaching the goal that is of a length no longer than all incomplete paths terminating at open nodes. When $g(N)$ is defined as the depth of the node N, the use of heuristic function $f(N)$ makes the search method identical to breadth-first search. In Section 12.1.2.2, we mention that breadth-first search is guaranteed to find a minimum length path. This can certainly be derived from the admissibility of the A^* search method.

When the heuristic function is close to the true remaining distance, the search can usually find the optimal solution without too much effort. In fact, when the true remaining distances for all nodes are known, the search can be done in a totally greedy fashion without any search at all, i.e., the only path explored is the solution. Any non-zero heuristic function is then called an informed heuristic function, and the search using such a function is called informed search. A heuristic function h_1 is said to be more informed than a heuristic function h_2 if the estimate h_1 is everywhere larger than h_2 and yet still admissible (underestimate). Finding an informed admissible heuristic function (guaranteed to underestimate for all nodes) is, in general, a difficult task. The heuristic often requires extensive analysis of the domain-specific knowledge and knowledge representation.

Let's look at a simple example—the 8-puzzle problem. The 8-puzzle consists of eight numbered, movable tiles set in a 3×3 frame. One cell of this frame is always empty, so it is possible to move an adjacent numbered tile into the empty cell. A solution for the 8-puzzle is to find a sequence of moves to change the initial configuration into a given goal configuration as shown in Figure 12.6. One choice for an informed admissible heuristic function h_1 is the number of misplaced tiles associated with the current configuration. Since each misplaced tile needs to move at least once to be in the right position, this heuristic function is clearly a lower bound of the true movements remaining. Based on this heuristic function, the value for the initial configuration will be 7 in Figure 12.7. If we examine this problem further, a more informed heuristic function h_2 can be defined as the sum of all row and column distances of all misplaced tiles and their goal positions. For example, the row and column distance between the tile 8 in the initial configuration and the goal position is 2 + 1= 3,

[3] In some literature an uninformed search is referred to as uniform-cost search.

Figure 12.6 Initial and goal configurations for the 8-puzzle problem.

which indicates that one must move tile 8 at least 3 times in order for it to be in the right position. Based on the heuristic function h_2, the value for the initial configuration will be 16 in Figure 12.6. h_2 is again admissible.

In our city-travel problem, one natural choice for the underestimating heuristic function of the remaining distance between node N and goal G is the straight-line distance since the true distance must be no shorter than the straight-line distance.

Figure 12.7 shows an augmented city-distance map with straight-line distance to goal node attached to each node. Accordingly, the heuristic search tree can be easily constructed for improved efficiency. Figure 12.8 shows the search progress of applying the A^* search algorithm for the city-traveling problem by using the straight-line distance heuristic function to estimate the remaining distances.

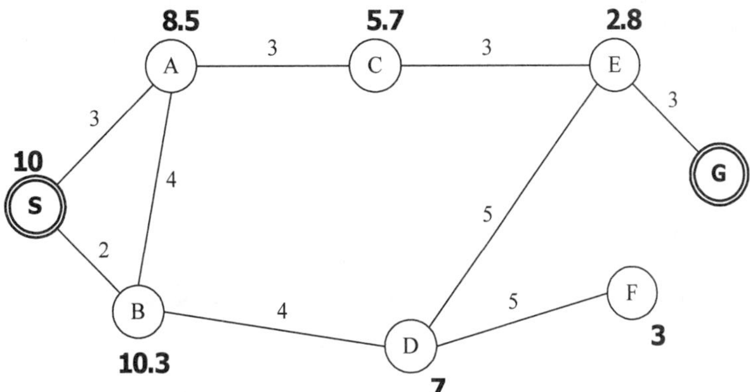

Figure 12.7 The city-travel problem augmented with heuristic information. The numbers beside each node indicate the straight-line distance to the goal node G [42].

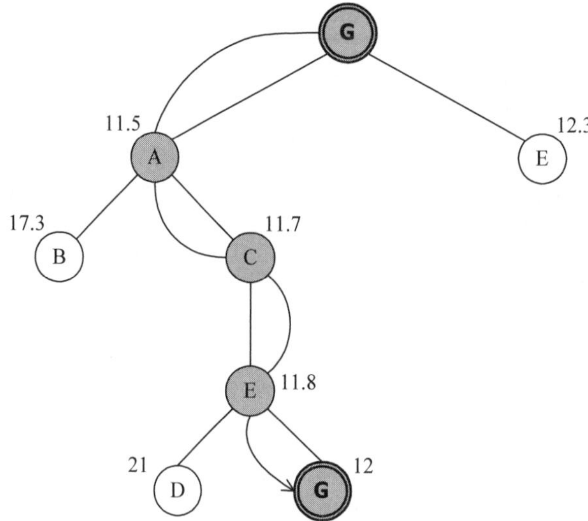

Figure 12.8 The search progress of applying A* search for the city-travel problem. The search determines that path S-A-C-E-G is the optimal one. The number beside the node is *f* values on which the sorting of the *OPEN* list is based [42].

12.1.3.2. Beam Search

Sometimes, it is impossible to find any effective heuristic estimate, as required in A* search, particularly when there is very little (or no) information about the remaining paths. For example, in real-time speech recognition, there is little information about what the speaker will utter for the remaining speech. Therefore, an efficient uninformed search strategy is very important to tackle this type of problem.

Breadth-first style search is an important strategy for heuristic search. A breadth-first search virtually explores all the paths with the same depth before exploring deeper paths. In practice, paths of the same depth are often easier to compare. It requires fewer heuristics to rank the goodness of each path. Even with uninformed heuristic function ($h(N) = 0$), the direct comparison of g (distance so far) of the paths with the same length should be a reasonable choice.

Beam search is a widely used search technique for speech recognition systems [26, 31, 37]. It is a breadth-first style search and progresses along with the depth. Unlike traditional breadth-first search, however, beam search only expands nodes that are likely to succeed at each level. Only these nodes are kept in the beam, and the rest are ignored (pruned) for improved efficiency.

In general, a beam search only keeps up to w best paths at each stage (level), and the rest of the paths are discarded. The number w is often referred to as beam width. The number of nodes explored remains manageable in beam search even if the whole search space is gigantic. If a beam width w is used in a beam search with an average branching factor b, only $w \times b$ nodes need to be explored at any depth, instead of the exponential number needed for breadth-first search. Suppose that a beam width of 2 is used for the city-travel problem. Figure 12.9 illustrates how beam search progresses to find the path. We can also see that the beam search saved a large number of unneeded nodes, as shown by the dotted nodes.

The beam search algorithm can be easily modified from the breadth-first search algorithm and is illustrated in Algorithm 12.5. For simplicity, we do not include the merging step here. In Algorithm 12.5, Step 4 obviously requires sorting, which is time-consuming if the number $w \times b$ is huge. In practice, the beam is usually implemented as a flexible list where nodes are expanded if their heuristic functions $f(N)$ are within some threshold (a.k.a., beam threshold) of the best node (the smallest value) at the same level. Thus, we only need to identify the best node and then prune away nodes that are outside of the threshold. Although this makes the beam size change dynamically, it significantly reduces the effort for sorting of the *Beam-Candidate* list. In fact, by adjusting the beam threshold, the beam size can be controlled indirectly and yet kept manageable.

Unlike A^* search, beam search is an approximate heuristic search method that is not admissible. However, it has a number of unique merits. Because of its simplicity in both its search strategy and its requirement of domain-specific heuristic information, it has become one of the most popular methods for complicated speech recognition problems. It is particularly attractive when integration of different knowledge sources is required in a time-synchronous fashion. It has the advantages of providing a consistent way of exploring nodes level by level and of offering minimally needed communication between different paths. It is also very suitable for parallel implementation because of its breadth-first search nature.

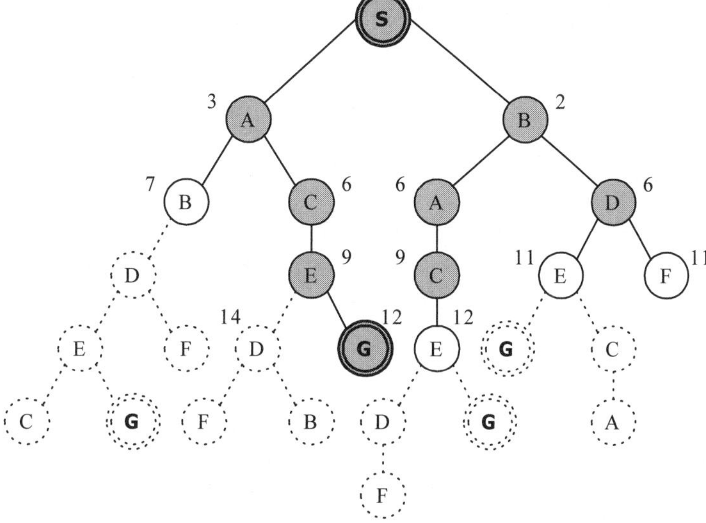

Figure 12.9 Beam search for the city-travel problem. The nodes with gray color are the ones kept in the beam. The transparent nodes were explored but pruned because of higher cost. The dotted nodes indicate all the savings because of pruning [42].

ALGORITHM 12.5: *THE BEAM SEARCH ALGORITHM*

Step 1: Initialization: Put *S* in the *OPEN* list and create an initially empty *CLOSE* list.
Step 2: If the *OPEN* list is empty, exit and declare failure.
Step 3: $\forall N \in OPEN$ do

3a. Pop up node *N* in the *OPEN* list, remove it from the *OPEN* list and put it into the *CLOSE* list.

3b. If node *N* is a goal node, exit successfully with the solution obtained by tracing back the path along the pointers from *N* to *S*.

3c. Expand node *N* by applying a successor operator to generate the successor set *SS*(*N*) of node *N*. Be sure to eliminate the successors, which are ancestors of *N*, from *SS*(*N*).

3d. $\forall v \in SS(N)$ Create a pointer pointing to *N* and push it into *Beam-Candidate* list.

Step 4: Sort the *Beam-Candidate* list according to the heuristic function $f(N)$ so that the best *w* nodes can be pushed into the the *OPEN* list. Prune the rest of nodes in the *Beam-Candidate* list.
Step 5: Go to Step 2.

12.2. SEARCH ALGORITHMS FOR SPEECH RECOGNITION

As described in Chapter 9, the decoder is basically a search process to uncover the word sequence $\hat{\mathbf{W}} = w_1 w_2 ... w_m$ that has the maximum posterior probability $P(\mathbf{W}|\mathbf{X})$ for the given acoustic observation $\mathbf{X} = X_1 X_2 ... X_n$. That is,

$$\hat{\mathbf{W}} = \arg\max_{\mathbf{w}} P(\mathbf{W} \mid \mathbf{X}) = \arg\max_{\mathbf{w}} \frac{P(\mathbf{W})P(\mathbf{X} \mid \mathbf{W})}{P(\mathbf{X})} = \arg\max_{\mathbf{w}} P(\mathbf{W})P(\mathbf{X} \mid \mathbf{W}) \qquad (12.5)$$

One obvious way is to search all possible word sequences and select the one with the best posterior probability score.

The unit of acoustic model $P(\mathbf{X}|\mathbf{W})$ is not necessary a word model. For large-vocabulary speech recognition systems, subword models, which include phonemes, demisyllables, and syllables are often used. When subword models are used, the word model $P(\mathbf{X}|\mathbf{W})$ is then obtained by concatenating the subword models according to the pronunciation transcription of the words in a lexicon or dictionary.

When word models are available, speech recognition becomes a search problem. The goal for speech recognition is thus to find a sequence of word models that best describes the input waveform against the word models. As neither the number of words nor the boundary of each word or phoneme in the input waveform is known, appropriate search strategies to deal with these variable-length nonstationary patterns are extremely important.

When HMMs are used for speech recognition systems, the states in the HMM can be expanded to form the state-search space in the search. In this chapter, we use HMMs as our speech models. Although the HMM framework is used to describe the search algorithms, all techniques mentioned in this and the following chapter can be used for systems based on other modeling techniques, including template matching and neural networks. In fact, many search techniques had been invented before HMMs were applied to speech recognition. Moreover, the HMMs state transition network is actually general enough to represent the general search framework for all modeling approaches.

12.2.1. Decoder Basics

The lessons learned from dynamic programming or the Viterbi algorithm introduced in Chapter 8 tell us that the exponential blind search can be avoided if we can store some intermediate optimal paths (results). Those intermediate paths are used for other paths without being recomputed each time. Moreover, the beam search described in the previous section shows us that efficient search is possible if appropriate pruning is employed to discard highly unlikely paths. In fact, all the search techniques use two strategies: sharing and pruning. *Sharing* means that intermediate results can be kept, so that they can be used by other paths without redundant re-computation. *Pruning* means that unpromising paths can be discarded reliably without wasting time in exploring them further.

Search strategies based on dynamic programming or the Viterbi algorithm with the help of clever pruning, have been applied successfully to a wide range of speech recognition tasks [31], ranging from small-vocabulary tasks, like digit recognition, to unconstraint large-vocabulary (more than 60,000 words) speech recognition. All the efficient search algorithms we discuss in this chapter and the next are considered as variants of dynamic programming or the Viterbi search algorithm.

In Section 12.1, cost (distance) is used as the measure of goodness for graph search algorithms. With Bayes' formulation, searching the minimum-cost path (word sequence) is equivalent to finding the path with maximum probability. For the sake of consistency, we use the inverse of Bayes' posterior probability as our objective function. Furthermore, logarithms are used on the inverse posterior probability to avoid multiplications. That is, the following new criterion is used to find the optimal word sequence $\hat{\mathbf{W}}$:

$$C(\mathbf{W} \mid \mathbf{X}) = \log\left[\frac{1}{P(\mathbf{W})P(\mathbf{X} \mid \mathbf{W})}\right] = -\log\left[P(\mathbf{W})P(\mathbf{X} \mid \mathbf{W})\right] \tag{12.6}$$

$$\hat{\mathbf{W}} = \underset{\mathbf{W}}{\arg\min}\; C(\mathbf{W} \mid \mathbf{X}) \tag{12.7}$$

For simplicity, we also define the following cost measures to mirror the likelihood for acoustic models and language models:

$$C(\mathbf{X} \mid \mathbf{W}) = -\log\left[P(\mathbf{X} \mid \mathbf{W})\right] \tag{12.8}$$

$$C(\mathbf{W}) = -\log\left[P(\mathbf{W})\right] \tag{12.9}$$

12.2.2. Combining Acoustic and Language Models

Although Bayes' equation [Eq. (12.5)] suggests that the acoustic model probability (conditional probability) and language model probability (prior probability) can be combined through simple multiplication, in practice some weighting is desirable. For example, when HMMs are used for acoustic models, the acoustic probability is usually underestimated, owing to the fallacy of the Markov and independence assumptions. Combining the language model probability with an underestimated acoustic model probability according to Eq. (12.5) would give the language model too little weight. Moreover, the two quantities have vastly different dynamic ranges particularly when continuous HMMs are used. One way to balance the two probability quantities is to add a *language model weight* LW to raise the language model probability $P(\mathbf{W})$ to that power $P(\mathbf{W})^{LW}$ [4, 25]. The language model weight LW is typically determined empirically to optimize the recognition performance on a development set. Since the acoustic model probabilities are underestimated, the language model weight LW is typically >1.

Language model probability has another function as a penalty for inserting a new word (or existing words). In particular, when a uniform language model (every word has an equal probability for any condition) is used, the language model probability here can be viewed as purely the penalty of inserting a new word. If this penalty is large, the decoder will prefer fewer longer words in general, and if this penalty is small, the decoder will prefer a greater number of shorter words instead. Since varying the language model weight to match the underestimated acoustic model probability will have some side effect of adjusting the penalty of inserting a new word, we sometimes use another independent *insertion penalty* to adjust the issue of longer or short words. Thus the language model contribution becomes:

$$P(\mathbf{W})^{LW} IP^{N(\mathbf{W})} \tag{12.10}$$

where IP is the insertion penalty (generally $0 < IP \le 1.0$) and $N(\mathbf{W})$ is the number of words in sentence $\mathbf{W}$. According to Eq. (12.10), insertion penalty is generally a constant that is added to the negative-logarithm domain when extending the search to another new word. In Chapter 9, we described how to compute errors in a speech recognition system and introduced three types of error: substitutions, deletions and insertions. Insertion penalty is so named because it usually affects only insertions. Similar to language model weight, the insertion penalty is determined empirically to optimize the recognition performance on a development set.

12.2.3. Isolated Word Recognition

With isolated word recognition, word boundaries are known. If word HMMs are available, the acoustic model probability $P(\mathbf{X}|\mathbf{W})$ can be computed using the forward algorithm introduced in Chapter 8. The search becomes a simple pattern recognition problem, and the word

$\hat{\mathbf{W}}$ with highest forward probability is then chosen as the recognized word. When subword models are used, word HMMs can be easily constructed by concatenating corresponding phoneme HMMs or other types of subword HMMs according to the procedure described in Chapter 9.

12.2.4. Continuous Speech Recognition

Search in continuous speech recognition is rather complicated, even for a small vocabulary, since the search algorithm has to consider the possibility of each word starting at any arbitrary time frame. Some of the earliest speech recognition systems took a two-stage approach towards continuous speech recognition, first hypothesizing the possible word boundaries and then using pattern matching techniques for recognizing the segmented patterns. However, due to significant cross-word co-articulation, there is no reliable segmentation algorithm for detecting word boundaries other than doing recognition itself.

Let's illustrate how you can extend the isolated-word search technique to continuous speech recognition by a simple example, as shown in Figure 12.10. This system contains only two words, w_1 and w_2. We assume the language model used here is an uniform unigram ($P(w_1) = P(w_2) = 1/2$).

It is important to represent the language structures in the same HMM framework. In Figure 12.10, we add one starting state *S* and one collector state *C*. The starting state has a null transition to the initial state of each word HMM with corresponding language model probability (1/2 in this case). The final state of each word HMM has a null transition to the collector state. The collector state then has a null transition back to the starting state in order to allow recursion. Similar to the case of embedding the phoneme (subword) HMMs into the word HMM for isolated speech recognition, we can embed the word HMMs for w_1 and w_2 into a new HMM corresponding to structure in Figure 12.10. Thus, the continuous speech search problem can be solved by the standard HMM formulations.

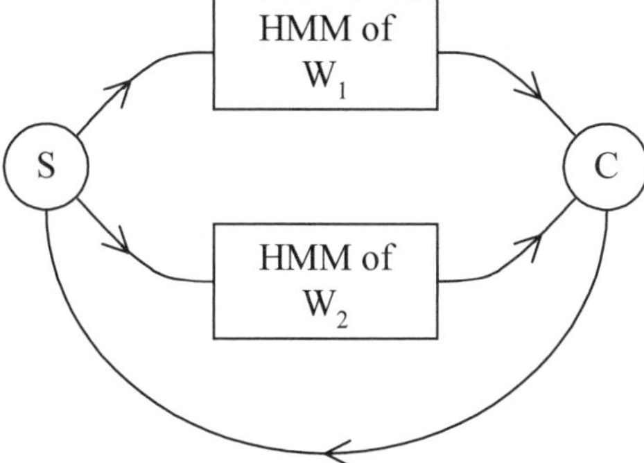

Figure 12.10 A simple example of continuous speech recognition task with two words w_1 and w_2. A uniform unigram language model is assumed for these words. State S is the starting state while state C is a collector state to save fully expanded links between every word pair.

The composite HMMs shown in Figure 12.10 can be viewed as a stochastic finite state network with transition probabilities and output distributions. The search algorithm is essentially producing a match between the acoustic observation **X** and a path[4] in the stochastic finite state network. Unlike isolated word recognition, continuous speech recognition needs to find the optimal word sequence $\hat{\mathbf{W}}$. The Viterbi algorithm is clearly a natural choice for this task since the optimal state sequence $\hat{\mathbf{S}}$ corresponds to the optimal word sequence $\hat{\mathbf{W}}$. Figure 12.11 shows the HMM Viterbi trellis computation for the two-word continuous speech recognition example in Figure 12.10. There is a cell for each state in the stochastic finite state network and each time frame t in the trellis. Each cell $C_{s,t}$ in the trellis can be connected to a cell corresponding to time t or $t+1$ and to states in the stochastic finite state network that can be reached from s. To make a word transition, there is a null transition to connect the final state of each word HMM to the initial state of the next word HMM that can be followed. The trellis computation is done *time-synchronously* from left to right, i.e., each cell for time t is completely computed before proceeding to time $t+1$.

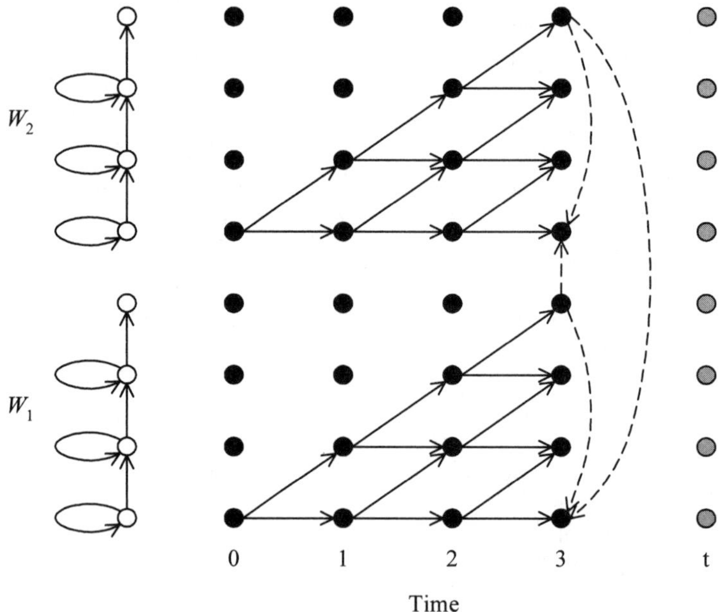

Figure 12.11 HMM trellis for continuous speech recognition example in Figure 12.10. When the final state of the word HMM is reached, a null arc (indicated by a dashed line) is linked from it to the initial state of the following word.

[4] A path here means a sequence of states and transitions.

12.3. LANGUAGE MODEL STATES

The state-space is a good indicator of search complexity. Since the HMM representation for each word in the lexicon is fixed, the state-space is determined by the language models. According to Chapter 11, every language model (grammar) is associated with a state machine (automata). Such a state machine is expanded to form the state-space for the recognizer. The states in such a state machine are referred to as *language models states*. For simplicity, we will use the concepts of state-space and language model states interchangeably. The expansion of language model states to HMM states will be done implicitly. The language model states for isolated word recognition are trivial. They are just the union of the HMM states of each word. In this section we look at the language model states for various grammars for continuous speech recognition.

12.3.1. Search Space with FSM and CFG

As described in Chapter 8, the complexity for the Viterbi algorithm is $O(N^2T)$, where N is the total number of states in the composite HMM and T is the length of input observation. A full time-synchronous Viterbi search is quite efficient for moderate tasks (vocabulary $\leq$ 500). We have already demonstrated in Figure 12.11 how to search for a two-word continuous speech recognition task with a uniform language model. The uniform language model, which allows all words in the vocabulary to follow every word with the same probability, is suitable for connected-digit task. In fact, most small vocabulary tasks in speech recognition applications usually use a finite state grammar (FSG).

Figure 12.12 shows a simple example of an FSM. Similar to the process described in Sections 12.2.3 and 12.2.4, each of the word arcs in an FSG can be expanded as a network of phoneme (subword) HMMs. The word HMMs are connected with null transitions with the grammar state. A large finite state HMM network that encodes all the legal sentences can be constructed based on the expansion procedure. The decoding process is achieved by performing a time-synchronous Viterbi search on this composite finite state HMM.

In practice, FSGs are sufficient for simple tasks. However, when an FSG is made to satisfy the constraints of sharing of different sub-grammars for compactness and support for dynamic modifications, the resulting non-deterministic FSG is very similar to context-free grammar (CFG) in terms of implementation. The CFG grammar consists of a set of productions or rules, which expand nonterminals into a sequence of terminals and nonterminals. Nonterminals in the grammar tend to refer to high-level task-specific concepts such as dates, names, and commands. The terminals are words in the vocabulary. A grammar also has a non-terminal designated as its start state.

Although efficient parsing algorithms, like chart parsing (described in Chapter 11), are available for CFG, they are not suitable for speech recognition, which requires left-to-right processing. A context-free grammar can be formulated with a recursive transition network (RTN). RTNs are more powerful and complicated than the finite state machines described in

Chapter 11 because they allow arc labels to refer to other networks as well as words. We use Figure 12.13 to illustrate how to embed HMMs into a recursive transition network.

Figure 12.13 is an RTN representation of the following CFG:

```
S→ NP VP
NP→ sam | sam davis
VP → VERB tom
VERB → likes | hates
```

There are three types of arcs in an RTN, as shown in Figure 12.13: CAT(x), PUSH (x), and POP(x). The CAT(x) arc indicates that x is a terminal node (which is equivalent to a word arc). Therefore, all the CAT(x) arcs can be expanded by the HMM network for x. The word HMM can again be a composite HMM built from phoneme (or subword) HMMs. Similar to the finite state grammar case in Figure 12.12, each grammar state acts as a state with incoming and outgoing null transitions to connect word HMMs in the CFG.

During decoding, the search pursues several paths through the CFG at the same time. Associated with each of the paths is a grammar state that describes completely how the path can be extended further. When the decoder hypothesizes the end of the current word of a path, it asks the CFG module to extend the path further by one word. There may be several alternative successor words for the given path. The decoder considers all the successor word possibilities. This may cause the path to be extended to generate several more paths to be considered, each with its own grammar state.

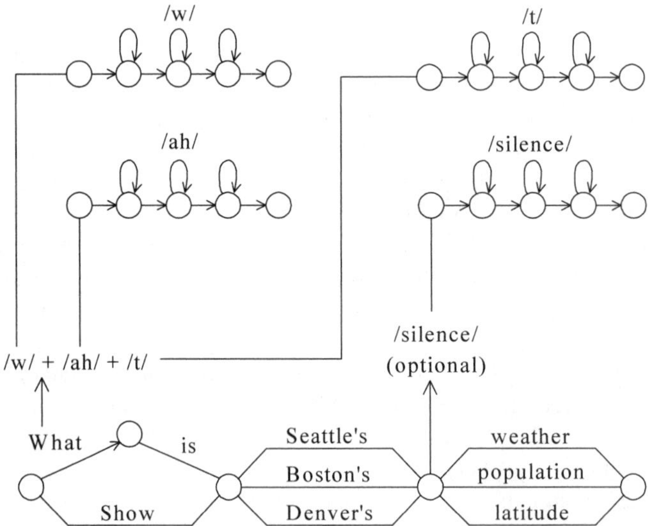

Figure 12.12 An illustration of how to compile a speech recognition task with finite state grammar into a composite HMM.

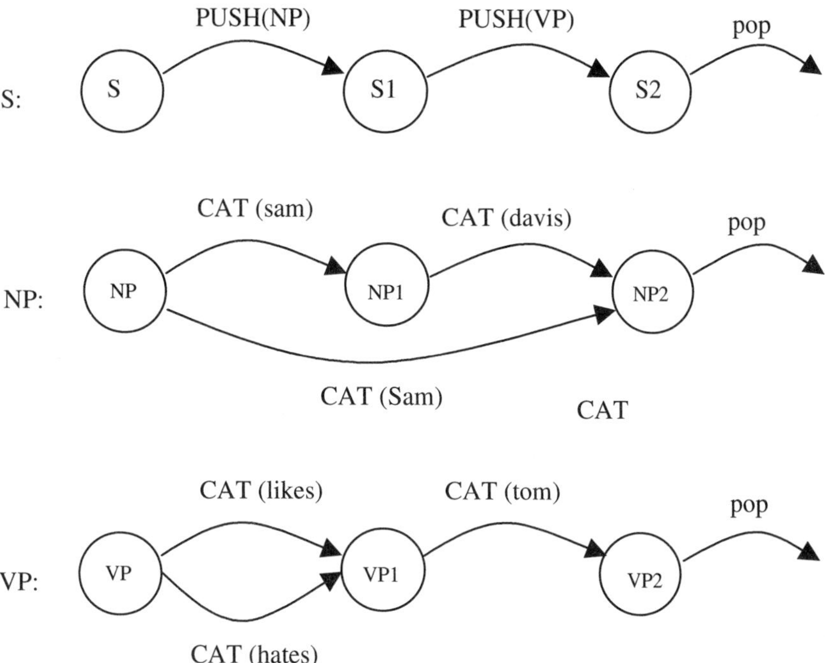

Figure 12.13 A simple RTN example with three types of arcs: CAT(x), PUSH (x), POP.

Readers should note that the same word might be under consideration by the decoder in the context of different paths and grammar states at the same time. For example, there are two word arcs CAT (Sam) in Figure 12.13. Their HMM states should be considered as distinct states in the trellis because they are in completely different grammar states. Two different states in the trellis also means that different paths going into these two states cannot be merged. Since these two partial paths will lead to different successive paths, the search decision needs to be postponed until the end of search. Therefore, when embedding HMMs into word arcs in the grammar network, the HMM state will be assigned a new state identity, although the HMM parameters (transition probabilities and output distributions) can still be shared across different grammar arcs.

Each path consists of a stack of production rules. Each element of the stack also contains the position within the production rule of the symbol that is currently being explored. The search graph (trellis) started from the initial state of CFG (state S). When the path needs to be extended, we look at the next arc (symbol in CFG) in the production. When the search enters a CAT(x) arc (terminal), the path gets extended with the terminal, and the HMM trellis computation is performed on the CAT(x) arc to match the model x against the acoustic data. When the final state of the HMM for x is reached, the search moves on via the null transition to the destination of the CAT(x) arc. When the search enters a PUSH(x) arc, it indicates a nonterminal symbol x is encountered. In effect, the search is about to enter a subnetwork of x; the destination of the PUSH(x) arc is stored in a last-in first-out (LIFO) stack. When the search reaches a POP arc that signals the end of the current network, the control should jump back to the calling network. In other words, the search returns to the state extracted from the top of the LIFO stack. Finally, when we reach the end of the production rule at the very bottom of the stack, we have reached an accepting state in which we have seen a complete grammatical sentence. For our decoding purpose, that is the state we want to pick as the best score at the end of time frame T to get the search result.

The problem of connected word recognition by finite state or context-free grammars is that the number of states increases enormously when it is applied to more complex grammars. Moreover it remains a challenge to generate such FSGs or CFGs from a large corpus, either manually or automatically. As mentioned in Chapter 11, it is questionable whether FSG or CFG is adequate to describe natural languages or unconstrained spontaneous languages. Instead, n-gram language models are often used for natural languages or unconstrained spontaneous languages. In the next section we investigate how to integrate various n-grams into continuous speech recognition.

12.3.2. Search Space with the Unigram

The simplest n-gram is the unigram that is memory-less and depends only on the current word.

$$P(\mathbf{W}) = \prod_{i=1}^{n} P(w_i) \tag{12.11}$$

Figure 12.14 shows such a unigram grammar network. The final state of each word HMM is connected to the collector state by a null transition, with probability 1.0. The collector state is then connected to the starting state by another null transition, with transition probability equal to 1.0. For word expansion, the starting state is connected to the initial state of each word HMM by a null transition, with transition probability equal to the corresponding unigram probability. Using the collector state and starting state for word expansion allows efficient expansion because it first merges all the word-ending paths[5] (only the best one survives) before expansion. It can cut the total cross-word expansion from N^2 to N.

[5] In graph search, a partial path still under consideration is also referred to as a theory, although we will use paths instead of theories in this book.

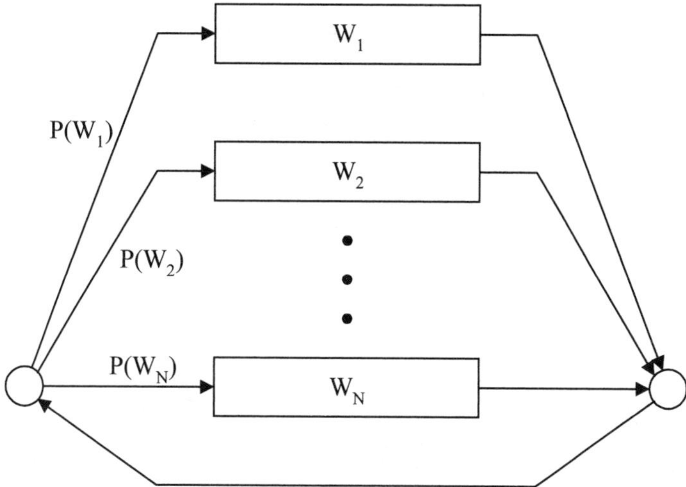

Figure 12.14 A unigram grammar network where the unigram probability is attached as the transition probability from starting state *S* to the first state of each word HMM.

12.3.3. Search Space with Bigrams

When the bigram is used, the probability of a word depends only on the immediately preceding word. Thus, the language model score is:

$$P(\mathbf{W}) = P(w_1 \mid \text{<s>}) \prod_{i=2}^{n} P(w_i \mid w_{i-1}) \tag{12.12}$$

where <s> represents the symbol of starting of a sentence.

Figure 12.15 shows a grammar network using a bigram language model. Because of the bigram constraint, the merge-and-expand framework for unigram search no longer applies here. Instead, the bigram search needs to perform expand-and-merge. Thus, bigram expansion is more expensive than unigram expansion. For a vocabulary size N, the bigram would need N^2 word-to-word transitions in comparison to N for the unigram. Each word transition has a transition probability equal to the corresponding bigram probability. Fortunately, the total number of states for bigram search is still proportional to the vocabulary size N.

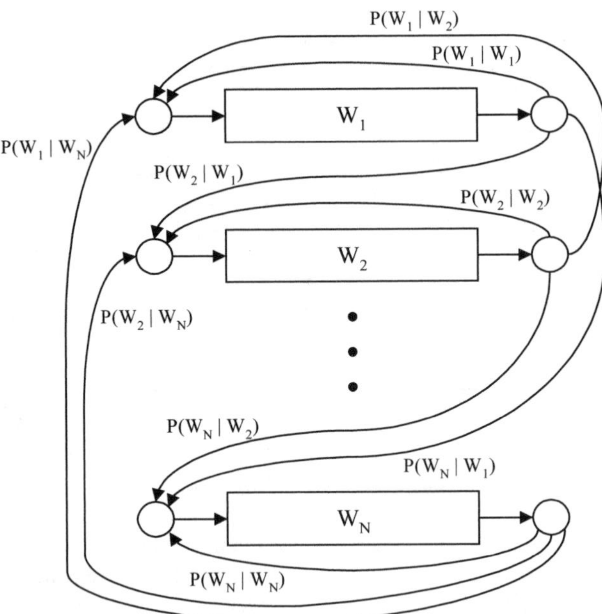

Figure 12.15 A bigram grammar network where the bigram probability $P(w_j \mid w_i)$ is attached as the transition probability from word w_i to w_j [19].

Because the search space for bigram is kept manageable, bigram search can be implemented very efficiently. Bigram search is a good compromise between efficient search and effective language models. Therefore, bigram search is arguably the most widely used search technique for unconstrained large-vocabulary continuous speech recognition. Particularly for the multiple-pass search techniques described in Chapter 13, a bigram search is often used in the first pass search.

12.3.3.1. Backoff Paths

When the vocabulary size N is large, the total bigram expansion N^2 can become computationally prohibitive. As described in Chapter 11, only a limited number of bigrams are observable in any practical corpora for a large vocabulary size. Suppose the probabilities for unseen bigrams are obtained through Katz's backoff mechanism. That is, for unseen bigram $P(w_j \mid w_i)$,

$$P(w_j \mid w_i) = \alpha(w_i)P(w_j) \tag{12.13}$$

where $\alpha(w_i)$ is the backoff weight for word w_i.

Using the backoff mechanism for unseen bigrams, the bigram expansion can be significantly reduced [12]. Figure 12.16 shows the new word expansion scheme. Instead of full bigram expansion, only observed bigrams are connected by direct word transitions with correspondent bigram probabilities. For backoff bigrams, the last state of word w_i is first connected to a central backoff node with transition probability equal to backoff weight $\alpha(w_i)$. The backoff node is then connected to the beginning of each word w_j with transition probability equal to its corresponding unigram probability $P(w_j)$. Readers should note that there are now two paths from w_i to w_j for an observed bigram $P(w_j \mid w_i)$. One is the direct link representing the observable bigram $P(w_j \mid w_i)$, and the other is the two-link backoff path representing $\alpha(w_i)P(w_j)$. For a word pair whose bigram exists, the two-link backoff path is likely to be ignored since the backoff unigram probability is almost always smaller than the observed bigram $P(w_j \mid w_i)$. Suppose there are only N_b different observable bigrams, this scheme requires $N_b + 2N$ instead of N^2 word transitions. Since under normal circumstance $N_b << N^2$, this backoff scheme significantly reduces the cost of word expansion.

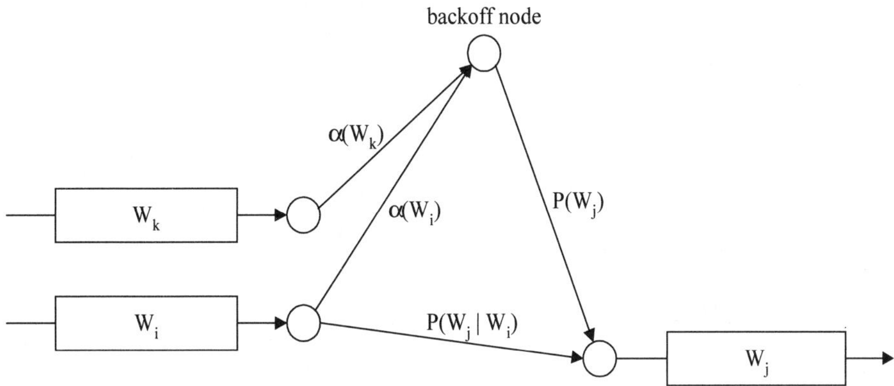

Figure 12.16 Reducing bigram expansion in a search by using the backoff node. In addition to normal bigram expansion arcs for all observed bigrams, the last state of word w_i is first connected to a central backoff node with transition probability equal to backoff weight $\alpha(w_i)$. The backoff node is then connected to the beginning of each word w_j with its corresponding unigram probability $P(w_j)$ [12].

12.3.4. Search Space with Trigrams

For a trigram language model, the language model score is:

$$P(\mathbf{W}) = P(w_1 \mid \text{<s>})P(w_2 \mid \text{<s>}, w_1)\prod_{i=3}^{n} P(w_i \mid w_{i-2}, w_{i-1}) \tag{12.14}$$

The search space is considerably more complex, as shown in Figure 12.17. Since the equivalence grammar class is the previous two words w_i and w_j, the total number of grammar states is N^2. From each of these grammar states, there is a transition to the next word [19].

Obviously, it is very expensive to implement large-vocabulary trigram search given the complexity of the search space. It becomes necessary to dynamically generate the trigram search graph (trellis) via a graph search algorithm. The other alternative is to perform a multiple-pass search strategy, in which the first-pass search uses less detailed language models, like bigrams, to generate an *n*-best list or word lattice, and then a second-pass detailed search can use trigrams on a much smaller search space. Multiple-pass search strategy is discussed in Chapter 13.

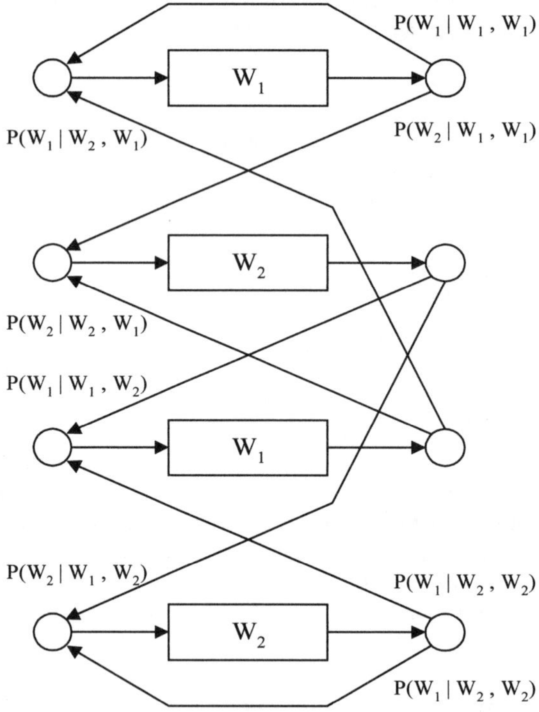

Figure 12.17 A trigram grammar network where the trigram probability $P(w_k | w_i, w_j)$ is attached to transition from grammar state word w_i, w_j to the next word w_k. Illustrated here is a two-word vocabulary, so there are four grammar states in the trigram network [19].

12.3.5. How to Handle Silences Between Words

In continuous speech recognition, there are unavoidable pauses (silences) between words or sentences. The pause is often referred to as silence or a non-speech event in continuous speech recognition. Acoustically, the pause is modeled by a silence model[6] that models background acoustic phenomena. The silence model is usually modeled with a topology flexible enough to accommodate a wide range of lengths, since the duration of a pause is arbitrary.

It can be argued that silences are actually linguistically distinguishable events, which contribute to prosodic and meaning representation. For example, people are likely to pause more often in phrasal boundaries. However, these patterns are so far not well understood for unconstrained natural speech (particularly for spontaneous speech). Therefore, the design of almost all automatic speech recognition systems today allows silences occurring just about anywhere between two lexical tokens or between sentences. It is relatively safe to assume that people pause a little bit between sentences to catch breath, so the silences between sentences are assumed mandatory while silences between words are optional. In most systems, silence is often modeled as a special lexicon entry with special language model probability. This special language model probability is also referred to as silence insertion penalty that is set to adjust the likelihood of inserting such an optional silence between words.

It is relatively straightforward to handle the optional silence between words. We need only to replace all the grammar states connecting words with a small network like the one shown in Figure 12.18. This arrangement is similar to that of the optional silence in training continuous speech, described in Chapter 9. The small network contains two parallel paths. One is the original null transition acting as the direct transition from one word to another, while the other path will need to go through a silence model with the silence insertion penalty attached in the transition probability before going to the next word.

One thing to clarify in the implementation of Figure 12.18 is that this silence expansion needs to be done for every grammar state connecting words. In the unigram grammar network of Figure 12.14, since there is only one collector node to connect words, the silence expansion is required only for this collector node. On the other hand, in the bigram grammar network of Figure 12.15, there is a collector node for every word before expanding to the next word. In this case, the silence expansion is required for every collector node. For a vocabulary size $|V|$, this means there are $|V|$ numbers of silence networks in the grammar search network. This requirement lies in the fact that in bigram search we cannot merge paths before expanding into the next word. Optional silence can then be regarded as part of the search effort for the previous word, so the word expansion needs to be done after finishing the optional silence. Therefore, we treat each word as having two possible pronunciations, one with the silence at the end and one without. This viewpoint integrates silence in the word pronunciation network like the example shown in Figure 12.19.

---

[6] Some researchers extend the context-dependent modeling to silence models. In that case, there are several silence models based on surrounding contexts.

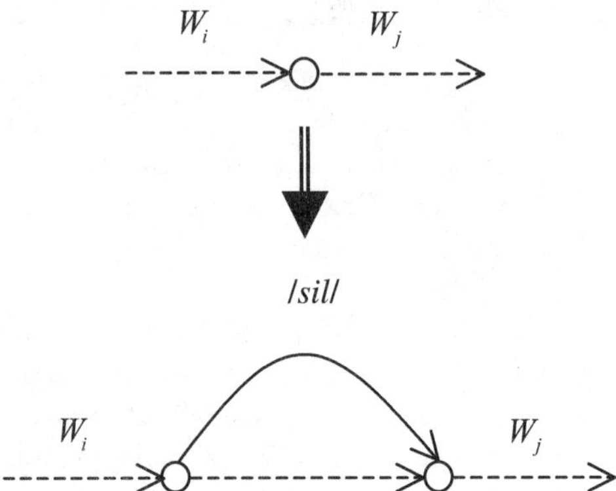

Figure 12.18 Incorporating optional silence (a non-speech event) in the grammar search network where the grammar state connecting different words is laced by two parallel paths. One is the original null transition directly from one word to the other, while the other first goes through the silence word to accommodate the optional silence.

For efficiency reasons, a single silence is sometimes used for large-vocabulary continuous speech recognition using higher order n-gram language model. Theoretically, this could be a source of pruning errors.[7] However, the error could turn out to be so small as to be negligible because there are, in general, very few pauses between word for continuous speech. On the other hand, the overhead of using multiple silences should be very minimal because it is less likely to visit those silence models at the end of words due to pruning.

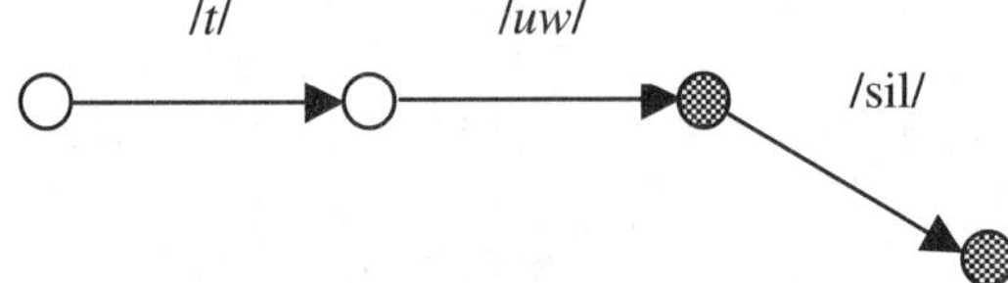

Figure 12.19 An example of treating silence as of the pronunciation network of word TWO. The shaded nodes represent possible word-ending nodes: one without silence and the other one with silence.

12.4. Time-Synchronous Viterbi Beam Search

When HMMs are used for acoustic models, the acoustic model score (likelihood) used in search is by definition the forward probability. That is, all possible state sequences must be considered. Thus,

[7] Speech recognition errors due to sub-optimal search or heuristic pruning are referred to as *pruning errors*, which will be described in detail in Chapter 13.

$$P(\mathbf{X} \mid \mathbf{W}) = \sum_{all \text{ possible } s_0^T} P(\mathbf{X}, s_0^T \mid \mathbf{W}) \tag{12.15}$$

where the summation is to be taken over all possible state sequences **S** with the word sequence **W** under consideration. However, under the trellis framework (as in Figure 12.11), more bookkeeping must be performed since we cannot add scores with different word sequence history. Since the goal of decoding is to uncover the best word sequence, we could approximate the summation with the maximum to find the best state sequence instead. The Bayes' decision rule, Eq. (12.5), becomes

$$\hat{\mathbf{W}} = \arg\max_{\mathbf{w}} P(\mathbf{W})P(\mathbf{X} \mid \mathbf{W}) \cong \arg\max_{\mathbf{w}} \left\{ P(\mathbf{W}) \max_{s_0^T} P(\mathbf{X}, s_0^T \mid \mathbf{W}) \right\} \tag{12.16}$$

Equation (12.16) is often referred to as the *Viterbi approximation*. It can be literally translated to "the *most likely word sequence* is approximated by the *most likely state sequence*." Viterbi search is then sub-optimal. Although the search results by using forward probability and Viterbi probability could, in principle, be different, in practice this is rarely the case. We use this approximation for the rest of this chapter.

The Viterbi search has already been discussed as a solution to one of the three fundamental HMM problems in Chapter 8. It can be executed very efficiently via the same trellis framework. To briefly reiterate, the Viterbi search is a time-synchronous search algorithm that completely processes time t before going on to time $t+1$. For time t, each state is updated by the best score (instead of the sum of all incoming paths) from all states in at time $t-1$. This is why it is often called *time-synchronous Viterbi search*. When one update occurs, it also records the backtracking pointer to remember the most probable incoming state. At the end of search, the most probable state sequence can be recovered by tracing back these backtracking pointers. The Viterbi algorithm provides an optimal solution for handling nonlinear time warping between hidden Markov models and acoustic observation, word boundary detection and word identification in continuous speech recognition. This unified Viterbi search algorithm serves as the basis for all search algorithms as described in the rest of the chapter.

It is necessary to clarify the backtracking pointer for time-synchronous Viterbi search for continuous word recognition. We are generally not interested in the optimal state sequence for speech recognition.[8] Instead, we are only interested in the optimal word sequence indicated by Eq. (12.16). Therefore, we use the backtrack pointer just to remember the word history for the current path, so the optimal word sequence can be recovered at the end of search. To be more specific, when we reach the final state of a word, we create a history node containing the word identity and current time index and append this history node to the existing backtrack pointer. This backtrack pointer is then passed onto the successor node if it

[8] While we are not interested in optimal state sequences for ASR, they are very useful in deriving phonetic segmentation, which could provide important information for developing ASR systems.

is the optimal path leading to the successor node for both intra-word and inter-word transition. The side benefit of keeping this backtrack pointer is that we no longer need to keep the entire trellis during the search. Instead, we only need space to keep two successive time slices (columns) in the trellis computation (the previous time slice and the current time slice) because all the backtracking information is now kept in the backtrack pointer. This simplification is a significant benefit in the implementation of a time-synchronous Viterbi search.

Time-synchronous Viterbi search can be considered as a *breadth-first search* with dynamic programming. Instead of performing a tree search algorithm, the dynamic programming principle helps create a search graph where multiple paths leading to the same search state are merged by keeping the best path (with minimum cost). The Viterbi trellis is a representation of the search graph. Therefore, all the efficient techniques for graph search algorithms can be applied to time-synchronous Viterbi search. Although so far we have described the trellis in an explicit fashion—the whole search space needs to be explored before the optimal path can be found—it is not necessary to do so. When the search space contains an enormous number of states, it becomes impractical to pre-compile the composite HMM entirely and store it in the memory. It is preferable to dynamically build and allocate portions of the search space sufficient to search the promising paths. By using the graph search algorithm described in Section 12.1.1, only part of the entire Viterbi trellis is generated explicitly. By constructing the search space dynamically, the computation cost of the search is proportional only to the number of active hypotheses, independent of the overall size of the potential search space. Therefore, dynamically generated trellises are key to heuristic Viterbi search for efficient large-vocabulary continuous speech recognition, as described in Chapter 13.

12.4.1. The Use of Beam

Based on Chapter 8, the search space for Viterbi search is $O(NT)$ and the complexity is $O(N^2T)$, where N is the total number of HMM states and T is the length of the utterance. For large-vocabulary tasks these numbers are astronomically large even with the help of dynamic programming. In order to avoid examining the overwhelming number of possible cells in the HMM trellis, a heuristic search is clearly needed. Different heuristics generate or explore portions of the trellis in different ways.

A simple way to prune the search space for breadth-first search is the beam search described in Section 12.1.3.2. Instead of retaining all candidates (cells) at every time frame, a threshold T is used to keep only a subset of promising candidates. The state at time t with the lowest cost $D_{\min}$ is first identified. Then each state at time t with cost $> D_{\min} + T$ is discarded from further consideration before moving on to the next time frame t+1. The use of the beam alleviates the need to process all the cells. In practice, it can lead to substantial savings in computation with little or no loss of accuracy.

Although beam search is a simple idea, the combination of time-synchronous Viterbi and beam search algorithms produces the most powerful search strategy for large-vocabulary speech recognition. Comparing paths with equal length under a time-synchronous search framework makes beam search possible. That is, for two different word sequences $\mathbf{W}_1$ and $\mathbf{W}_2$, the posterior probabilities $P(\mathbf{W}_1 \mid \mathbf{x}_0^t)$ and $P(\mathbf{W}_2 \mid \mathbf{x}_0^t)$ are always compared based on the same partial acoustic observation $\mathbf{x}_0^t$. This makes the comparison straightforward because the denominator $P(\mathbf{x}_0^t)$ in Eq. (12.5) is the same for both terms and can be ignored. Since the score comparison for each time frame is fair, the only assumption of beam search is that an optimal path should have a good enough partial-path score for each time frame to survive under beam pruning.

The time-synchronous framework is one of the aspects of Viterbi beam search that is critical to its success. Unlike the time-synchronous framework, time-asynchronous search algorithms such as stack decoding require the normalization of likelihood scores over feature streams of different time lengths. This, as we will see in Section 12.5, is the Achilles' heel of that approach.

The straightforward time-synchronous Viterbi beam search is ineffective in dealing with the gigantic search space of high perplexity tasks. However, with a better understanding of the linguistic search space and the advent of techniques for obtaining *n*-best lists from time-synchronous Viterbi search, described in Chapter 13, time-synchronous Viterbi beam search has turned out to be surprisingly successful in handling tasks of all sizes and all different types of grammars, including FSG, CFG, and *n*-gram [2, 14, 18, 28, 34, 38, 44]. Therefore, it has become the predominant search strategy for continuous speech recognition.

12.4.2. Viterbi Beam Search

To explain the time-synchronous Viterbi beam search in a formal way [31], we first define some quantities:

$D(t;s_t;w) \equiv$ total cost of the best path up to time t that ends in state s_t of grammar word state w.

$h(t;s_t;w) \equiv$ backtrack pointer for the best path up to time t that ends in state s_t of grammar word state w.

Readers should be aware that w in the two quantities above represents a grammar word state in the search space. It is different from just the word identity since the same word could occur in many different language model states, as in the trigram search space shown in Figure 12.17.

There are two types of dynamic programming (DP) transition rules [30], namely intra-word and inter-word transition. The intra-word transition is just like the Viterbi rule for HMMs and can be expressed as follows:

$$D(t;s_t;w) = \min_{s_{t-1}} \left\{ d(\mathbf{x}_t, s_t \mid s_{t-1}; w) + D(t-1; s_{t-1}; w) \right\} \tag{12.17}$$

$$h(t;s_t;w) = h(t-1, b_{\min}(t;s_t;w); w) \tag{12.18}$$

where $d(\mathbf{x}_t, s_t \mid s_{t-1}; w)$ is the cost associated with taking the transition from state s_{t-1} to state s_t while generating output observation $\mathbf{x}_t$, and $b_{\min}(t; s_t; w)$ is the optimal predecessor state of cell $D(t; s_t; w)$. To be specific, they can be expressed as follows:

$$d(\mathbf{x}_t, s_t \mid s_{t-1}; w) = -\log P(s_t \mid s_{t-1}; w) - \log P(\mathbf{x}_t \mid s_t; w) \tag{12.19}$$

$$b_{\min}(t; s_t; w) = \arg\min_{s_{t-1}} \left\{ d(\mathbf{x}_t, s_t \mid s_{t-1}; w) + D(t-1; s_{t-1}; w) \right\} \tag{12.20}$$

The inter-word transition is basically a null transition without consuming any observation. However, it needs to deal with creating a new history node for the backtracking pointer. Let's define $F(w)$ as the final state of word HMM w and $I(w)$ as the initial state of word HMM w. Moreover, state η is denoted as the pseudo initial state. The inter-word transition can then be expressed as follows:

$$D(t; \eta; w) = \min_{v} \left\{ \log P(w \mid v) + D(t; F(v); v) \right\} \tag{12.21}$$

$$h(t; \eta; w) = \langle v_{\min}, t \rangle :: h(t, F(v_{\min}); v_{\min}) \tag{12.22}$$

where $v_{\min} = \arg\min_{v} \left\{ \log P(w \mid v) + D(t; F(v); v) \right\}$ and :: is a link appending operator.

The time-synchronous Viterbi beam search algorithm assumes that all the intra-word transitions are evaluated before inter-word null transitions take place. The same time index is used intentionally for inter-word transition since the null language model state transition does not consume an observation vector. Since the initial state $I(w)$ for word HMM w could have a self-transition, the cell $D(t; I(w); w)$ might already have an active path. Therefore, we need to perform the following check to advance the inter-word transitions.

$$\begin{aligned} &\text{if } D(t; \eta; w) < D(t; I(w); w) \\ &\quad D(t; I(w); w) = D(t; \eta; w) \text{ and } h(t; I(w); w) = h(t; \eta; w) \end{aligned} \tag{12.23}$$

The time-synchronous Viterbi beam search can be summarized as in Algorithm 12.6. For large-vocabulary speech recognition, the experimental results show that only a small percentage of the entire search space (the beam) needs to be kept for each time interval t without increasing error rates. Empirically, the beam size has typically been found to be between 5% and 10% of the entire search space. In Chapter 13 we describe strategies of using different level of beams for more effectively pruning.

12.5. STACK DECODING (A* SEARCH)

If some reliable heuristics are available to guide the decoding, the search can be done in a depth-first fashion around the best path early on, instead of wasting efforts on unpromising paths via the time-synchronous beam search. Stack decoding represents the best attempt to

ALGORITHM 12.6: *TIME-SYNCHRONOUS VITERBI BEAM SEARCH*

Step 1: Initialization: For all the grammar word states w which can start a sentence,

$$D(0;I(w);w)=0$$

$$h(0;I(w);w)=null$$

Step 2: Induction: For time $t=1$ to T do

For all active states do

Intra-word transitions according to Eq. (12.17) and (12.18)

$$D(t;s_t;w)=\min_{s_{t-1}}\{d(\mathbf{x}_t,s_t \mid s_{t-1};w)+D(t-1;s_{t-1};w)\}$$

$$h(t;s_t;w)=h(t-1,b_{\min}(t;s_t;w);w)$$

For all active word-final states do

Inter-word transitions according to Eq. (12.21), (12.22) and (12.23)

$$D(t;\eta;w)=\min_{v}\{\log P(w \mid v)+D(t;F(v);v)\}$$

$$h(t;\eta;w)=\langle v_{\min},t\rangle :: h(t,F(v_{\min});v_{\min})$$

if $D(t;\eta;w)<D(t;I(w);w)$

$$D(t;I(w);w)=D(t;\eta;w) \text{ and } h(t;I(w);w)=h(t;\eta;w)$$

Pruning: Find the cost for the best path and decide the beam threshold

Prune unpromising hypotheses

Step 3: Termination: Pick the best path among all the possible final states of grammar at time T. Obtain the optimal word sequence according to the backtracking pointer $h(t;\eta;w)$

use A* search instead of time-synchronous beam search for continuous speech recognition. Unfortunately, as we will discover in this section, such a heuristic function $h(\bullet)$ (defined in Section 12.1.3) is very difficult to attain in continuous speech recognition, so search algorithms based on A* search are in general less efficient than time-synchronous beam search.

Stack decoding is a variant of the heuristic A* search based on the forward algorithm, where the evaluation function is based on the forward probability. It is a tree search algorithm, which takes a slightly different viewpoint than the time-synchronous Viterbi search. Time-synchronous beam search is basically a breadth-first search, so it is crucial to control the number of all possible language model states as described in Section 12.3. In a typical large-vocabulary Viterbi search with n-gram language models, this number is determined by the equivalent classes of language model histories. On the other hand, stack decoding as a tree search algorithm treats the search as a task for finding a path in a tree whose branches correspond to words in the vocabulary V, non-terminal nodes correspond to incomplete sentences, and terminal nodes correspond to complete sentences. The search tree has a constant branching factor of $|V|$, if we allow every word to be followed by every word. Figure 12.20 illustrates such a search tree for a vocabulary with three words [19].

An important advantage of stack decoding is its consistency with the forward-backward training algorithm. Viterbi search is a graph search, and paths cannot be easily summed because they may have different word histories. In general, the Viterbi search finds the optimal state sequence instead of optimal word sequence. Therefore, Viterbi approximation is necessary to make the Viterbi search feasible, as described in Section 12.4. Stack decoding is a tree search, so each node has a unique history, and the forward algorithm can be used within word model evaluation. Moreover, all possible beginning and ending times (shaded areas in Figure 12.21) are considered [24]. With stack decoding, it is possible to use an objective function that searches for the optimal word string, rather than the optimal state sequence. Furthermore, it is in principle natural for stack decoding to accommodate long-range language models if the heuristics can guide the search to avoid exploring the overwhelmingly large unpromising grammar states.

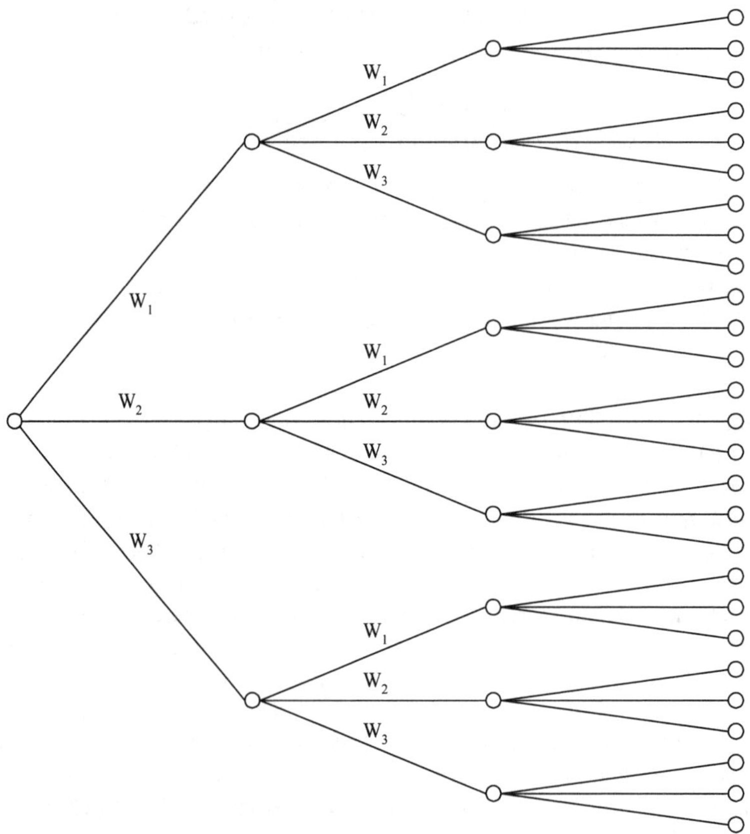

Figure 12.20 A stack decoding search tree for a vocabulary size of three [19].

By formulating stack decoding in a tree search framework, the graph search algorithms described in Section 12.1 can be directly applied to stack decoding. Obviously, blind-search methods, like depth-first and breadth-first search, that do not take advantage of the goodness measurement of how close we are getting to the goal, are usually computationally infeasible in practical speech recognition systems. A* search is clearly attractive for speech recognition, given the hope of a sufficient heuristic function to guide the tree search in a favorable direction without exploring too many unpromising branches and nodes. In contrast to the Viterbi search, it is not time-synchronous and extends paths of different lengths.

The search begins by adding all possible one-word hypotheses to the *OPEN* list. Then the best path is removed from the *OPEN* list, and all paths from it are extended, evaluated, and placed back in the *OPEN* list. This search continues until a complete path that is guaranteed to be better than all paths in the *OPEN* list has been found.

Unlike Viterbi search, where the acoustic probabilities being compared are always based on the same partial input, it is necessary to compare the goodness of partial paths of different lengths to direct the A* tree search. Moreover, since stack decoding is done asynchronously, we need an effective mechanism to determine when to end a phone/word evaluation and move on to the next phone/word. Therefore, the heart and soul of the stack decoding are clearly in

1. Finding an effective and efficient heuristic function for estimating the future remaining input feature stream and
2. Determining when to extend the search to the next word/phone.

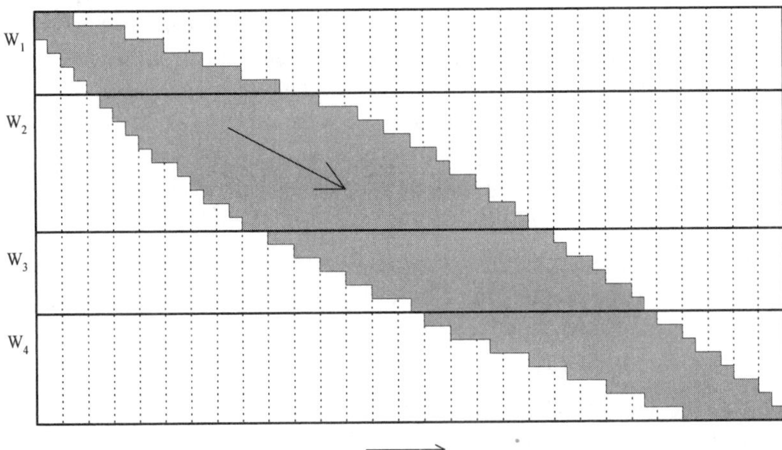

Figure 12.21 The forward trellis space for stack decoding. Each grid point corresponds to a trellis cell in the forward computation. The shaded area represents the values contributing to the computation of the forward score for the optimal word sequence $w_1, w_2, w_3, \ldots$ [24].

In the following section we describe these two critical components. Readers will note that the solutions to these two issues are virtually the same—using a normalization scheme to compare paths of different lengths.

12.5.1. Admissible Heuristics for Remaining Path

The key issue in heuristic search is the selection of an evaluation function. As described in Section 12.1.3, the heuristic function of the path H_N going through node *N* includes the cost up to the node and the estimate of the cost to the target node from node *N*. Suppose path H_N is going through node *N* at time *t*; then the evaluation for path H_N can be expressed as follows:

$$f(H_N^t) = g(H_N^t) + h(H_N^{t,T}) \tag{12.24}$$

where $g(H_N^t)$ is the evaluation function for the partial path of H_N up to time *t*, and $h(H_N^{t,T})$ is the heuristic function of the remaining path from $t+1$ to T for path H_N. The challenge for stack decoders is to devise an admissible function for $h(\bullet)$.

According to Section 12.1.3.1, an admissible heuristic function is one that always underestimates the true cost of the remaining path from $t+1$ to T for path H_N. A trivially admissible function is the zero function. In this case, it results in a very large OPEN list. In addition, since the longer paths tend to have higher cost because of the gradually accumulated cost, the search is likely to be conducted in a breadth-first fashion, which functions very much like a plain Viterbi search. The evaluation function $g(\bullet)$ can be obtained easily by using the HMM forward score as the true cost up to current time *t*. However, how can we find an admissible heuristic function $h(\bullet)$? We present the basic concept here [19, 35].

The goal of $h(\bullet)$ is to find the expected cost for the remaining path. If we can obtain the expected cost per frame ψ for the remaining path, the total expected cost, $(T-t)*\psi$, is simply the product of ψ and the length of the remaining path. One way to find such expected cost per frame is to gather statistics empirically from training data.

1. After the final training iteration, perform Viterbi forced alignment[9] with each training utterance to get an optimal time alignment for each word.
2. Randomly select an interval to cover the number of words ranging from two to ten. Denote this interval as $[i \ldots j]$.
3. Compute the average acoustic cost per frame within this selected interval according to the following formula and save the value in a set Λ:

[9] Viterbi forced alignment means that the Viterbi is performed on the HMM model constructed from the known word transcription. The term "forced" is used because the Viterbi alignment is forced to be performed on the correct model. Viterbi forced alignment is a very useful tool in spoken language processing as it can provide the optimal state-time alignment with the utterances. This detailed alignment can then be used for different purposes, including discriminant training, concatenated speech synthesis, etc.

$$\frac{-1}{j-i}\log P(\mathbf{x}_i^j \mid \mathbf{w}_{i\ldots j}) \tag{12.25}$$

where $\mathbf{w}_{i\ldots j}$ is the word string corresponding to interval $[i \ldots j]$.

4. Repeat Steps 2 and 3 for the entire training set.
5. Define $\psi_{\min}$ and ψ_{avg} as the minimum and average value found in set Λ.

Clearly, $\psi_{\min}$ should be a good under-estimate of the expected cost per frame for the future unknown path. Therefore, the heuristic function $h(H_N^{t,T})$ can be derived as:

$$h(H_N^{t,T}) = (T-t)\psi_{\min} \tag{12.26}$$

Although $\psi_{\min}$ is obtained empirically, stack decoding based on Eq. (12.26) will generally find the optimal solution. However, the search using $\psi_{\min}$ usually runs very slowly, since Eq. (12.26) always under-estimates the true cost for any portion of speech. In practice, a heuristic function like ψ_{avg} that may over-estimate has to be used to prune more hypotheses. This speeds up the search at the expense of possible search errors, because ψ_{avg} should represent the average cost per frame for any portion of speech. In fact, there is an argument that one might be able to use a heuristic function even more than ψ_{avg}. The argument is that ψ_{avg} is derived from the correct path (training data) and the average cost per frame for all paths during search should be more than ψ_{avg} because the paths undoubtedly include correct and incorrect ones.

12.5.2. When to Extend New Words

Since stack decoding is executed asynchronously, it becomes necessary to detect when a phone/word ends, so that the search can extend to the next phone/word. If we have a cost measure that indicates how well an input feature vector of any length matches the evaluated model sequence, this cost measure should drop slowly for the correct phone/word and rise sharply for an incorrect phone/word. In order to do so, it implies we must be able to compare hypotheses of different lengths.

The first thing that comes to mind for this cost measure is simply the forward cost $-\log P(\mathbf{x}_1^t, s_t \mid w_1^k)$, which represents the likelihood of producing acoustic observation $\mathbf{x}_1^t$ based on word sequence w_1^k and ending at state s_t. However, it is definitely not suitable because it is deemed to be smaller for a shorter acoustic input vector. This causes the search to almost always prefer short phones/words, resulting in many insertion errors. Therefore, we must derive some normalized score that satisfies the desired property described above. The normalized cost $\hat{C}(\mathbf{x}_1^t, s_t \mid w_1^k)$ can be represented as follows [6, 24]:

$$\hat{C}(\mathbf{x}_1^t, s_t \mid w_1^k) = -\log\left[\frac{P(\mathbf{x}_1^t, s_t \mid w_1^k)}{\gamma^t}\right] = -\log\left[P(\mathbf{x}_1^t, s_t \mid w_1^k)\right] + t\log\gamma \tag{12.27}$$

where γ ($0 < \gamma < 1$) is a constant normalization factor.

Suppose the search is now evaluating a particular word w_k ; we can define $\hat{C}_{\min}(t)$ as the minimum cost for $\hat{C}(\mathbf{x}_1^t, s_t \mid w_1^k)$ for all the states of w_k, and $\alpha_{\max}(t)$ as the maximum forward probability for $P(\mathbf{x}_1^t, s_t \mid w_1^k)$ for all the states of w_k. That is,

$$\hat{C}_{\min}(t) = \min_{s_t \in w_k} \left[\hat{C}(\mathbf{x}_1^t, s_t \mid w_1^k) \right] \tag{12.28}$$

$$\alpha_{\max}(t) = \max_{s_t \in w_k} \left[P(\mathbf{x}_1^t \mid w_1^k, s_t) \right] \tag{12.29}$$

We want $\hat{C}_{\min}(t)$ to be near 0 just as long as the phone/word we are evaluating is the correct one and we have not gone beyond its end. On the other hand, if the phone/word we are evaluating is the incorrect one or we have already passed its end, we want the $\hat{C}_{\min}(t)$ to be rising sharply. Similar to the procedure of finding the admissible heuristic function, we can set the normalized factor γ empirically during training so that $\hat{C}_{\min}(T) = 0$ when we know the correct word sequence $\mathbf{W}$ that produces acoustic observation sequence $\mathbf{x}_1^T$. Based on Eq. (12.27), γ should be set to:

$$\gamma = \sqrt[T]{\alpha_{\max}(T)} \tag{12.30}$$

Figure 12.22 shows a plot of $\hat{C}_{\min}(t)$ as a function of time for correct match. In addition, the cost for the final state $FS(w_k)$ of word w_k, $\hat{C}(\mathbf{x}_1^t, s_t = FS(w_k) \mid w_1^k)$, which is the score for w_k-ending path, is also plotted. There should be a valley centered around 0 for $\hat{C}(\mathbf{x}_1^t, s_t = FS(w_k) \mid w_1^k)$, which indicates the region of possible ending time for the correct phone/word. Sometimes a stretch of acoustic observations may match better than the average cost, pushing the curve below 0. Similarly, a stretch of acoustic observations may match worse than the average cost, pushing the curve above 0.

There is an interesting connection between the normalized factor γ and the heuristic estimate of the expected cost per frame, ψ, defined in Eq. (12.25). Since the cost is simply the logarithm on the inverse posterior probability, we get the following equation:

$$\psi = \frac{-1}{T} \log P(\mathbf{x}_1^T \mid \hat{\mathbf{W}}) = -\log\left[\alpha_{\max}(T)^{1/T}\right] = -\log \gamma \tag{12.31}$$

Equation (12.31) reveals that these two quantities are basically the same estimate. In fact, if we subtract the heuristic function $f(H_N^t)$ defined in Eq. (12.24) by the constant $\log\left(\gamma^T\right)$, we get exactly the same quantity as the one defined in Eq. (12.27). Decisions on which path to extend first based on the heuristic function and when to extend the search to the next word/phone are basically centered on comparing partial theories with different lengths. Therefore, the normalized cost $\hat{C}(\mathbf{x}_1^t, s_t \mid w_1^k)$ can be used for both purposes.

Based on the connection we have established, the heuristic function, $f(H_N^t)$, which estimates the goodness of a path is simply replaced by the normalized evaluation function $\hat{C}(\mathbf{x}_1^t, s_t \mid w_1^k)$. If we plot the un-normalized cost $C(\mathbf{x}_1^t, s_t \mid w_1^k)$ for the optimal path and other

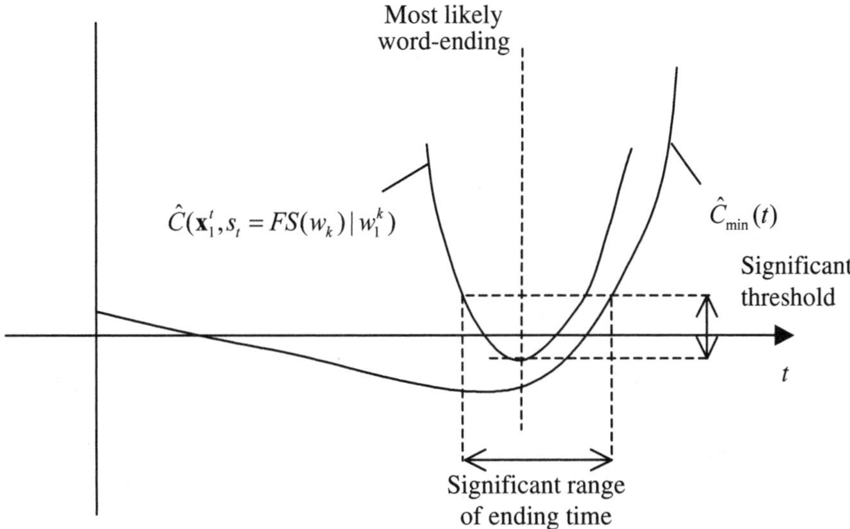

Figure 12.22 $\hat{C}_{\min}(t)$ and $\hat{C}(\mathbf{x}_1^t, s_t = FS(w_k) \mid w_1^k)$ as functions of time t. The valley region represents possible ending times for the correct phone/word.

competing paths as the function time t, the cost values increase as paths get longer (illustrated in Figure 12.23) because every frame adds some non-negative cost to the overall cost. It is clear that using un-normalized cost function $C(\mathbf{x}_1^t, s_t \mid w_1^k)$ generally results in a breadth-first search. What we want is an evaluation that decreases slightly along the optimal path, and hopefully increases along other competing paths. Clearly, the normalized cost function $\hat{C}(\mathbf{x}_1^t, s_t \mid w_1^k)$ fulfills this role, as shown in Figure 12.24.

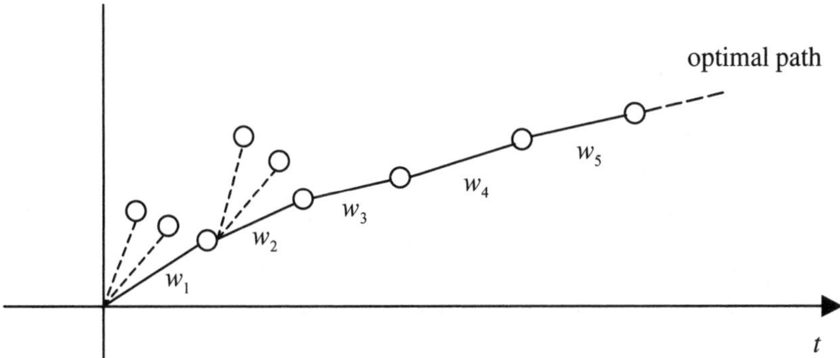

Figure 12.23 Unnormalized cost $C(\mathbf{x}_1^t, s_t \mid w_1^k)$ for optimal path and other competing paths as a function of time.

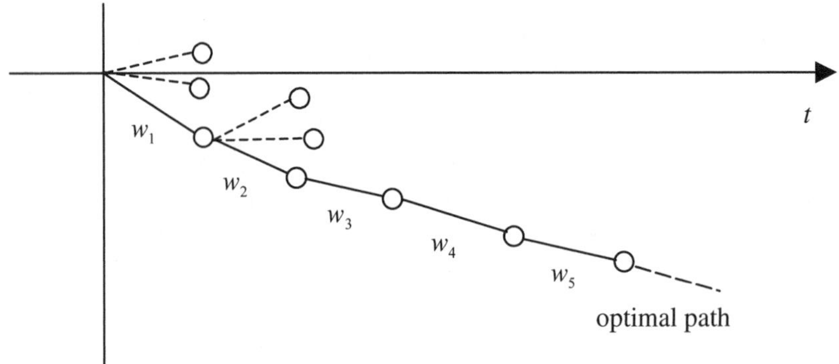

Figure 12.24 Normalized cost $\hat{C}(\mathbf{x}_1^t, s_t \mid w_1^k)$ for the optimal path and other competing paths as a function of time.

Equation (12.30) is a context-less estimation of the normalized factor, which is also referred to as zero-order estimate. To improve the accuracy of the estimate, you can use context-dependent higher-order estimates like [24]:

$\gamma_i = \gamma(\mathbf{x}_i)$ first-order estimate

$\gamma_i = \gamma(\mathbf{x}_i, \mathbf{x}_{i-1})$ second-order estimate

$\gamma_i = \gamma(\mathbf{x}_i, \mathbf{x}_{i-1}, \ldots, \mathbf{x}_{i-N+1})$ n-order estimate

Since the normalized factor γ is estimated from the training data that is also used to train the parameters of the HMMs, the normalized factor γ_i tends to be an overestimate. As a result, $\alpha_{\max}(t)$ might rise slowly for test data even when the correct phone/word model is evaluated. This problem is alleviated by introducing some other scaling factor $\delta < 1$ so that $\alpha_{\max}(t)$ falls slowly for test data for when evaluating the correct phone/word model. The best solution for this problem is to use an independent data set other than the training data to derive the normalized factor γ_i.

12.5.3. Fast Match

Even with an efficient heuristic function and mechanism to determine the ending time for a phone/word, stack decoding could still be too slow for large-vocabulary speech recognition tasks. As described in Section 12.5.1, an effective underestimated heuristic function for the remaining portion of speech is very difficult to derive. On the other hand, a heuristic estimate for the immediate short segment that usually corresponds to the next phone or word may be feasible to attain. In this section, we describe the fast-match mechanism that reduces phone/word candidates for detailed match (expansion).

In asynchronous stack decoding, the most expensive step is to extend the best subpath. For a large-vocabulary search, it implies the calculation of $P(\mathbf{x}_t^{t+k} \mid w)$ over the entire vocabulary size $|V|$. It is desirable to have a fast computation to quickly reduce the possible words starting at a given time t to reduce the search space. This process is often referred to as *fast match* [15, 35]. In fact, fast match is crucial to stack decoding, of which it becomes an integral part. Fast match is a method for the rapid computation of a list of candidates that constrain successive search phases. The expensive *detailed match* can then be performed after fast match. In this sense, fast match can be regarded as an additional pruning threshold to meet before new word/phone can be started.

Fast match, by definition, needs to use only a small amount of computation. However, it should also be accurate enough not to prune away any word/phone candidates that participate in the best path eventually. Fast match is, in general, characterized by the approximations that are made in the acoustic/language models in order to reduce computation. There is an obvious trade-off between these two objectives. Fortunately, many systems [15] have demonstrated that one needs to sacrifice very little accuracy in order to speed up the computation considerably.

Similar to *admissibility* in A^* search, there is also an *admissibility* property in fast match. A fast match method is called admissible if it never prunes away the word/phone candidates that participate in the optimal path. In other words, a fast match is admissible if the recognition errors that appear in a system using the fast match followed by a detailed match are those that would appear if the detailed match were carried out for all words/phones in the vocabulary. Since fast match can be applied to either word or phone level, as we describe in the next section, we explain the admissibility for the case of word-level fast match for simplicity. The same principle can be easily extended to phone-level fast match.

Let V be the vocabulary and $C(\mathbf{X} \mid w)$ be the cost of a detailed match between input $\mathbf{X}$ and word w. Now $F(\mathbf{X} \mid w)$ is an estimator of $C(\mathbf{X} \mid w)$ that is accurate enough and fast to compute. A word list selected by fast match estimator can be attained by first computing $F(\mathbf{X} \mid w)$ for each word w of the vocabulary. Suppose w_b is the word for which the fast match has a minimum cost value:

$$w_b = \arg\min_{w \in V} F(\mathbf{X} \mid w) \tag{12.32}$$

After computing $C(\mathbf{X} \mid w_b)$, the detailed match cost for w_b, we form the fast match word list, Λ, from the word w in the vocabulary such that $F(\mathbf{X} \mid w)$ is no greater than $C(\mathbf{X} \mid w_b)$. In other words,

$$\Lambda = \{ w \in V \mid F(\mathbf{X} \mid w) \le C(\mathbf{X} \mid w_b) \} \tag{12.33}$$

Similar to the admissibility condition for A^* search [3, 33], the fast match estimator $F(\bullet)$ conducted in the way described above is admissible if and only if $F(\mathbf{X} \mid w)$ is always an under-estimator (lower bound) of detailed match $C(\mathbf{X} \mid w)$. That is,

$$F(\mathbf{X} \mid w) \le C(\mathbf{X} \mid w) \qquad \forall \mathbf{X}, w \tag{12.34}$$

The proof is straightforward. If the word w_c has a lower detailed match cost $C(\mathbf{X} \mid w_c)$, you can prove that it must be included in the fast match list Λ because

$$C(\mathbf{X} \mid w_c) \le C(\mathbf{X} \mid w_b) \text{ and } F(\mathbf{X} \mid w_c) \le C(\mathbf{X} \mid w_c) \Rightarrow F(\mathbf{X} \mid w_c) \le C(\mathbf{X} \mid w_b)$$

Therefore, based on the definition of Λ, $w_c \in \Lambda$.

Now the task is to find an admissible fast match estimator. Bahl et al. [6] proposed one fast match approximation for discrete HMMs. As we will see later, this fast match approximation is indeed equivalent to a simplification of the HMM structure. Given the HMM for word w and an input sequence x_1^T of codebook symbols describing the input signal, the probability that the HMM w produces the VQ sequence x_1^T is given by (according to Chapter 8):

$$P(x_1^T \mid w) = \sum_{s_1, s_2, \ldots s_T} \left[P_w(s_1, s_2, \ldots s_T) \prod_{i=1}^{T} P_w(x_i \mid s_i) \right] \tag{12.35}$$

Since we often use Viterbi approximation instead of the forward probability, the equation above can be approximated by:

$$P(x_1^T \mid w) \cong \max_{s_1, s_2, \ldots s_T} \left[P_w(s_1, s_2, \ldots s_T) \prod_{i=1}^{T} P_w(x_i \mid s_i) \right] \tag{12.36}$$

The detailed match cost $C(\mathbf{X} \mid w)$ can now be represented as:

$$C(\mathbf{X} \mid w) = \min_{s_1, s_2, \ldots s_T} \left\{ -\log \left[P_w(s_1, s_2, \ldots s_T) \prod_{i=1}^{T} P_w(x_i \mid s_i) \right] \right\} \tag{12.37}$$

Since the codebook size is finite, it is possible to compute, for each model w, the highest output probability for every VQ label c among all states s_k in HMM w. Let's define $m_w(c)$ to be the following:

$$m_w(c) = \max_{s_k \in w} P_w(c \mid s_k) = \max_{s_k \in w} b_k(c) \tag{12.38}$$

We can further define the $q_{\max}(w)$ as the maximum state sequence with respect to T, i.e., the maximum probability of any complete path in HMM w.

$$q_{\max}(w) = \max_T \left[P_w(s_1, s_2, \ldots s_T) \right] \tag{12.39}$$

Now let's define the fast match estimator $F(\mathbf{A} \mid w)$ as the following:

$$F(\mathbf{X} \mid w) = -\log \left[q_{\max}(w) \prod_{i=1}^{T} m_w(x_i) \right] \tag{12.40}$$

It is easy to show the fast match estimator $F(\mathbf{X} \mid w) \le C(\mathbf{X} \mid w)$ is admissible based on Eq. (12.38) to Eq. (12.40).

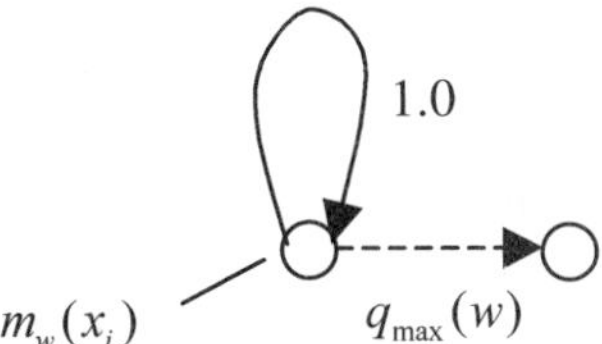

Figure 12.25 The equivalent one-state HMM corresponding to fast match computation defined in Eq. (12.40) [15].

The fast match estimator defined in Eq. (12.40) requires $T+1$ additions for a vector sequence of length T. The operation can be viewed as equivalent to the forward computation with a one-state HMM of the form shown in Figure 12.25. This correspondence can be interpreted as a simplification of the original multiple-state HMM into such a one-state HMM. It thus explains why fast match can be computed much faster than detailed match. Readers should note that this HMM is not actually a true HMM by strict definition, because the output probability distribution $m_w(c)$ and the transition probability distribution do not add up to one.

The fast match computation defined in Eq. (12.40) discards the sequence information with the model unit since the computation is independent of the order of input vectors. Therefore, one needs to decide the acoustic unit for fast match. In general, the longer the unit, the faster the computation is, and, therefore, the smaller the under-estimated cost $F(\mathbf{X} \mid w)$ is. It thus becomes a trade-off between accuracy and speed.

Now let's analyze the real speedup by using fast match to reduce the vocabulary V to the list Λ, followed by the detailed match. Let $|V|$ and $|\Lambda|$ be the sizes for the vocabulary V and the fast match short list Λ. Suppose t_f and t_d are the times required to compute one fast match score and one detailed match score for one word, respectively. Then, the total time required for the fast match followed by the detailed match is $t_f |V| + t_d |\Lambda|$, whereas the time required in doing the detailed match alone for the entire vocabulary is $t_d |V|$. The speed-up ratio is then given as follows:

$$\frac{1}{\left(\frac{t_f}{t_d} + \frac{|\Lambda|}{|V|}\right)} \tag{12.41}$$

We need t_f to be much smaller than t_d and $|\Lambda|$ to be much smaller than $|V|$ to have a significant speed-up using fast match. Using our admissible fast match estimator in Eq. (12.40), the time complexity of the computation for $F(\mathbf{X} \mid w)$ is T instead of N^2T for $C(\mathbf{X} \mid w)$, where N is the number of states in the detailed acoustic model. Therefore, the t_f / t_d saving is about N^2.

In general, in order to make $|\Lambda|$ much smaller than $|V|$, one needs a very accurate fast match estimator that could result in $t_f \approx t_d$. This is why we often relax the constraint of admissibility, although it is a nice principle to adhere to. In practice, most real-time speech recognition systems don't necessarily obey the admissibility principle with the fast match. For example, Bahl et al. [10], Laface et al., [22] and Roe et al., [36] used several techniques to construct off-line groups of acoustically similar words. Armed with this grouping, they can use an aggressive fast match to select only a very short list of words, and words acoustically similar to the words in this list are added to form the short word list Λ for further detailed match processing. By doing so, they are able to report a very efficient fast match method that misses the correct word only 2% of the time. When non-admissible fast match is used, one needs to minimize the additional search error introduced by fast match empirically.

Bahl et al. [6] use a one-state HMM as their fast match units and a tree-structure lexicon similar to the lexical tree structures introduced in Chapter 13 to construct the short word list Λ for next-word expansion in stack decoding. Since the fast match tree search is also done in an asynchronous way, the ending time of each phone is detected using normalized scores similar to those described in Section 12.5.2. It is based on the same idea that this normalized score rises slowly for the correct phone, while it drops rapidly once the end of phone is encountered (so the model is starting to go toward the incorrect phones). During the asynchronous lexical tree search, the unpromising hypotheses are also pruned away by a pruning threshold that is constantly changing once a complete hypothesis (a leaf node) is obtained. On a 20,000-word dictation task, such a fast match scheme was about 100 times faster than detailed match and achieved real-time performance on a commercial workstation with only 0.34% increase in the word error rate being introduced by the fast match process.

12.5.4. Stack Pruning

Even with efficient heuristic functions, the mechanism to determine the ending time for phone/word, and fast match, stack decoding might still be too slow for large-vocabulary speech recognition tasks. A beam within the stack, which saves only a small number of promising hypotheses in the *OPEN* list, is often used to reduce search effort. This *stack pruning* is very similar to beam search. A predetermined threshold ε is used to eliminate hypotheses whose cost value is much worse than the best path so far.

Both fast match and stack pruning could introduce search errors where the eventual optimal path is thrown away prematurely. However, the impact could be reduced to a minimum by empirically adjusting the thresholds in both methods.

The implementation of stack decoding is, in general, more complicated, particularly when some inevitable pruning strategies are incorporated to make the search more efficient. The difficulty of devising both an effectively admissible heuristic function for $h(\bullet)$ and an effective estimation of normalization factors for boundary determination has limited the advantage that stack decoders have over Viterbi decoders. Unlike stack decoding, time-synchronous Viterbi beam search can use an easy comparison of same-length path without heuristic determination of word boundaries. As described in the earlier sections, these simple and unified features of Viterbi beam search allow researchers to incorporate various sound techniques to improve the efficiency of search. Therefore, time-synchronous Viterbi Beam search enjoys a much broader popularity in the speech community. However, the principle of stack decoding is essential particularly for n-best and lattice search. As we describe in Chapter 13, stack decoding plays a very crucial part in multiple-pass search strategies for n-best and lattice search because the early pass is able to establish a near-perfect estimate of the remaining path.

12.5.5. Multistack Search

Even with the help of normalized factor γ or heuristic function $h(\bullet)$, it is still more effective to compare hypotheses of the same length than those of different lengths, because hypotheses with the same length are compared based on the true forward matching score. Inspired by the time-synchronous principle in Viterbi beam search, researchers [8, 35] propose a variant stack decoding based on multiple stacks.

Multistack search is equivalent to a best-first search algorithm running on multiple stacks time-synchronously. Basically, the search maintains a separate stack for each time frame t, so it never needs to compare hypotheses of different lengths. The search runs time-synchronously from left to right just like time-synchronous Viterbi search. For each time frame t, multistack search extracts the best path out of the t-stack, computes one-word extensions, and places all the new paths into the corresponding stacks. When the search finishes, the top path in the last stack is our optimal path. Algorithm 12.7 illustrates the multistack search algorithm.

This time-synchronous multistack search is designed based on the fact that by the time the tth stack is extended, it already contains the best paths that could ever be placed into it. This phenomenon is virtually a variant of the dynamic programming principle introduced in Chapter 8. To make multistack more efficient, some heuristic pruning can be applied to reduce the computation. For example, when the top path of each stack is extended for one more word, we could only consider extensions between minimum and maximum duration. On the other hand, when some heuristic pruning is integrated into the multistack search, one might need to use a small beam in Step 2 of Algorithm 12.7 to extend more than just the best path to guarantee the admissibility.

ALGORITHM 12.7: *MULTISTACK SEARCH*

Step 1: Initialization: for each word v in vocabulary V

for $t = 1, 2, \ldots, T$

Compute $C(\mathbf{x}_1^t \mid v)$ and insert it to t^{th} stack

Step 2: Iteration: for $t = 1, 2, \ldots, T-1$

Sort the tth stack and pop the top path $C(\mathbf{x}_1^t \mid w_1^k)$ out of the stack

for each word v in vocabulary V

for $\tau = t+1, t+2, \ldots, T$

Extend the path $C(\mathbf{x}_1^t \mid w_1^k)$ by word v to get $C(\mathbf{x}_1^\tau \mid w_1^{k+1})$

where $w_1^{k+1} = w_1^k \,||\, v$ and || means string concatenation

Place $C(\mathbf{x}_1^\tau \mid w_1^{k+1})$ in τ^{th} stack

Step 3: Termination: Sort the Tth stack and the top path is the optimal word sequence

12.6. Historical Perspective and Further Reading

Search has been one of the most important topics in artificial intelligence since the origins of the field. It plays the central role in general problem solving [29] and computer games. [43], Nilsson's *Principles of Artificial Intelligence* [32] and Barr and Feigenbaum's *The Handbook of Artificial Intelligence* [11] contain a comprehensive introduction to state-space search algorithms. A* search was first proposed by Hart et al. [17]. A* was thought to be derived from Dijkstra's algorithm [13] and Moore's algorithm [27]. A* search is similar to the *branch-and-bound* algorithm [23, 39], widely used in *operations research*. The proof of admissibility of A* search can be found in [32].

The application of *beam search* in speech recognition was first introduced by the HARPY system [26]. It wasn't widely popular until BBN used it for their BYBLOS system [37]. There are some excellent papers with detailed description of the use of time-synchronous Viterbi beam search for continuous speech recognition [24, 31]. Over the years, many efficient implementations and improvements have been introduced for time-synchronous Viterbi beam search, so real-time large-vocabulary continuous speech recognition can be realized on a general-purpose personal computer.

On the other hand, stack decoding was first developed by IBM [9]. It is successfully used in IBM's large-vocabulary continuous speech recognition systems [3, 16]. Lacking a time-synchronous framework, comparing theories of different lengths and extending theories are more complex as described in this chapter. Because of the complexity of stack decoding, far fewer publications and systems are based on it than on Viterbi beam search [16, 19, 20, 35]. With the introduction of multistack search [8], stack decoding in essence has actually come very close to time-synchronous Viterbi beam search.

Stack decoding is typically integrated with fast match methods to improve its efficiency. Fast match was first implemented for isolated word recognition to obtain a list of potential word candidates [5, 7]. The paper by Gopalakrishnan et al. [15] contains a comprehensive description of fast match techniques to reduce the word expansion for stack decoding. Besides the fast match techniques described in this chapter, there are a number of alternative approaches [5, 21, 41]. Waast's fast match [41], for example, is based on a binary classification tree built automatically from data that comprise both phonetic transcription and acoustic sequence.

References

[1] Aho, A., J. Hopcroft, and J. Ullman, *The Design and Analysis of Computer Algorithms*, 1974, Addison-Wesley Publishing Company.

[2] Alleva, F., X. Huang, and M. Hwang, "An Improved Search Algorithm for Continuous Speech Recognition," *Int. Conf. on Acoustics, Speech and Signal Processing*, 1993, Minneapolis, MN, pp. 307-310.

[3] Bahl, L.R. and et. al., "Large Vocabulary Natural Language Continuous Speech Recognition," *Proc. of the IEEE Int. Conf. on Acoustics, Speech and Signal Processing*, 1989, Glasgow, Scotland, pp. 465-467.

[4] Bahl, L.R., *et al.*, "Language-Model/Acoustic Channel Balance Mechanism," *IBM Technical Disclosure Bulletin*, 1980, **23**(7B), pp. 3464-3465.

[5] Bahl, L.R., *et al.*, "Obtaining Candidate Words by Polling in a Large Vocabulary Speech Recognition System," *Proc. of the IEEE Int. Conf. on Acoustics, Speech and Signal Processing*, 1988, pp. 489-492.

[6] Bahl, L.R., *et al.*, "A Fast Approximate Acoustic Match for Large Vocabulary Speech Recognition," *IEEE Trans. on Speech and Audio Processing*, 1993(1), pp. 59-67.

[7] Bahl, L.R., *et al.*, "Matrix Fast Match: a Fast Method for Identifying a Short List of Candidate Words for Decoding," *Proc. of the IEEE Int. Conf. on Acoustics, Speech and Signal Processing*, 1989, Glasgow, Scotland, pp. 345-347.

[8] Bahl, L.R., P.S. Gopalakrishnan, and R.L. Mercer, "Search Issues in Large Vocabulary Speech Recognition," *Proc. of the 1993 IEEE Workshop on Automatic Speech Recognition*, 1993, Snowbird, UT.

[9] Bahl, L.R., F. Jelinek, and R. Mercer, "A Maximum Likelihood Approach to Continuous Speech Recognition," *IEEE Trans. on Pattern Analysis and Machine Intelligence*, 1983(2), pp. 179-190.

[10] Bahl, L.R., *et al.*, "Constructing Candidate Word Lists Using Acoustically Similar Word Groups," *IEEE Trans. on Signal Processing*, 1992(1), pp. 2814-2816.

[11] Barr, A. and E. Feigenbaum, *The Handbook of Artificial Intelligence: Volume I*, 1981, Addison-Wesley.

[12] Cettolo, M., R. Gretter, and R.D. Mori, "Knowledge Integration" in *Spoken Dialogues with Computers*, R.D. Mori, Editor 1998, London, Academic Press, pp. 231-256.

[13] Dijkstra, E.W., "A Note on Two Problems in Connection with Graphs," *Numerische Mathematik*, 1959, **1**, pp. 269-271.

[14] Gauvain, J.L., *et al.*, "The LIMSI Speech Dictation System: Evaluation on the ARPA Wall Street Journal Corpus," *Proc. of the IEEE Int. Conf. on Acoustics, Speech and Signal Processing*, 1994, Adelaide, Australia, pp. 129-132.

[15] Gopalakrishnan, P.S. 2and L.R. Bahl, "Fast Match Techniques," in *Automatic Speech and Speaker Recognition*, C.H. Lee, F.K. Soong, and K.K. Paliwal, eds., 1996, Norwell, MA, Kluwer Academic Publishers, pp. 413-428.

[16] Gopalakrishnan, P.S., L.R. Bahl, and R.L. Mercer, "A Tree Search Strategy for Large-Vocabulary Continuous Speech Recognition," *Proc. of the IEEE Int. Conf. on Acoustics, Speech and Signal Processing*, 1995, Detroit, MI, pp. 572-575.

[17] Hart, P.E., N.J. Nilsson, and B. Raphael, "A Formal Basis for the Heuristic Determination of Minimum Cost Paths," *IEEE Trans. on Systems Science and Cybernetics*, 1968, **4**(2), pp. 100-107.

[18] Huang, X., *et al.*, "Microsoft Windows Highly Intelligent Speech Recognizer: Whisper," *IEEE Int. Conf. on Acoustics, Speech and Signal Processing*, 1995, pp. 93-96.

[19] Jelinek, F., *Statistical Methods for Speech Recognition*, 1998, Cambridge, MA, MIT Press.

[20] Kenny, P., *et al.*, "A*—Admissible Heuristics for Rapid Lexical Access," *IEEE Trans. on Speech and Audio Processing*, 1993, **1**, pp. 49-58.

[21] Kenny, P., *et al.*, "A New Fast Match for Very Large Vocabulary Continuous Speech Recognition," *IEEE Int. Conf. on Acoustics, Speech and Signal Processing*, 1993, Minneapolis, MN, pp. 656-659.

[22] Laface, P., L. Fissore, and F. Ravera, "Automatic Generation of Words toward Flexible Vocabulary Isolated Word Recognition," *Proc. of the Int. Conf. on Spoken Language Processing*, 1994, Yokohama, Japan, pp. 2215-2218.

[23] Lawler, E.W. and D.E. Wood, "Branch-and-Bound Methods: A Survey," *Operations Research*, 1966(14), pp. 699-719.

[24] Lee, K.F. and F.A. Alleva, "Continuous Speech Recognition" in *Recent Progress in Speech Signal Processing*, S. Furui and M. Sondhi, eds., 1990, Marcel Dekker, Inc.

[25] Lee, K.F., H.W. Hon, and R. Reddy, "An Overview of the SPHINX Speech Recognition System," *IEEE Trans. on Acoustics, Speech and Signal Processing*, 1990, **38**(1), pp. 35-45.

[26] Lowerre, B.T., *The HARPY Speech Recognition System*, PhD Thesis in Computer Science Department, 1976, Carnegie Mellon University.

[27] Moore, E.F., "The Shortest Path Through a Maze," *Int. Symp. on the Theory of Switching*, 1959, Cambridge, MA, Harvard University Press, pp. 285-292.

[28] Murveit, H., *et al.*, "Large Vocabulary Dictation Using SRI's DECIPHER Speech Recognition System: Progressive Search Techniques," *Proc. of the IEEE Int. Conf. on Acoustics, Speech and Signal Processing*, 1993, Minneapolis, MN, pp. 319-322.

[29] Newell, A. and H.A. Simon, *Human Problem Solving*, 1972, Englewood Cliffs, NJ, Prentice Hall.

[30] Ney, H. and X. Aubert, "Dynamic Programming Search: From Digit Strings to Large Vocabulary Word Graphs," in *Automatic Speech and Speaker Recognition*, C.H. Lee, F. Soong and K.K. Paliwal, eds., 1996, Boston, Kluwer Academic Publishers, pp. 385-412.

[31] Ney, H. and S. Ortmanns, *Dynamic Programming Search for Continuous Speech Recognition*, in *IEEE Signal Processing Magazine*, 1999, pp. 64-83.

[32] Nilsson, N.J., *Principles of Artificial Intelligence*, 1982, Berlin, Germany, Springer Verlag.

[33] Nilsson, N.J., *Artificial Intelligence: A New Synthesis*, 1998, Academic Press/Morgan Kaufmann.

[34] Normandin, Y., R. Cardin, and R.D. Mori, "High-Performance Connected Digit Recognition Using Maximum Mutual Information Estimation," *IEEE Trans. on Speech and Audio Processing*, 1994, **2**(2), pp. 299-311.

[35] Paul, D.B., "An Efficient A* Stack Decoder Algorithm for Continuous Speech Recognition with a Stochastic Language Model," *Proc. of the IEEE Int. Conf. on Acoustics, Speech and Signal Processing*, 1992, San Francisco, California, pp. 25-28.

[36] Roe, D.B. and M.D. Riley, "Prediction of Word Confusabilities for Speech Recognition," *Proc. of the Int. Conf. on Spoken Language Processing*, 1994, Yokohama, Japan, pp. 227-230.

[37] Schwartz, R., *et al.*, "Context-Dependent Modeling for Acoustic-Phonetic Recognition of Speech Signals," *Proc. of the IEEE Int. Conf. on Acoustics, Speech and Signal Processing*, 1985, Tampa, FLA, pp. 1205-1208.

[38] Steinbiss, V., *et al.*, "The Philips Research System for Large-Vocabulary Continuous-Speech Recognition," *Proc. of the European Conf. on Speech Communication and Technology*, 1993, Berlin, Germany, pp. 2125-2128.

[39] Taha, H.A., *Operations Research: An Introduction*, 6th ed, 1996, Prentice Hall.

[40] Tanimoto, S.L., *The Elements of Artificial Intelligence : An Introduction Using Lisp*, 1987, Computer Science Press, Inc.

[41] Waast, C. and L.R. Bahl, "Fast Match Based on Decision Tree," *Proc. of the European Conf. on Speech Communication and Technology*, 1995, Madrid, Spain, pp. 909-912.

[42] Winston, P.H., *Artificial Intelligence*, 1984, Reading, MA, Addison-Wesley.

[43] Winston, P.H., *Artificial Intelligence*, 3rd ed, 1992, Reading, MA, Addison-Wesley.

[44] Woodland, P.C., *et al.*, "Large Vocabulary Continuous Speech Recognition Using HTK," *Proc. of the IEEE Int. Conf. on Acoustics, Speech and Signal Processing*, 1994, Adelaide, Australia, pp. 125-128.

CHAPTER 13

Large-Vocabulary Search Algorithms

Chapter 12 discussed the basic search techniques for speech recognition. However, the search complexity for large-vocabulary speech recognition with high-order language models is still difficult to handle. In this chapter we describe efficient search techniques in the context of time-synchronous Viterbi beam search, which becomes the choice for most speech recognition systems because it is very efficient. We use Microsoft Whisper as our case study to illustrate the effectiveness of various search techniques. Most of the techniques discussed here can also be applied to stack decoding.

With the help of beam search, it is unnecessary to explore the entire search space or the entire trellis. Instead, only the promising search state-space needs to be explored. Please keep in mind the distinction between the implicit search graph specified by the grammar network and the explicit partial search graph that is actually constructed by the Viterbi beam search algorithm.

In this chapter we first introduce the most critical search organization for large-vocabulary speech recognition—tree lexicons. Tree lexicons significantly reduce potential search space, although they introduce many practical problems. In particular, we need to address problems such as reentrant lexical trees, factored language model probabilities, subtree optimization, and subtree polymorphism.

Various other efficient techniques also are introduced. Most of these techniques aim for clever pruning with the hope of sparing the correct paths. For more effective pruning, different layers of beams are usually used. While fast match techniques described in Chapter 12 are typically required for stack decoding, similar concepts and techniques can be applied to Viterbi beam search. In practice, the look-ahead strategy is equally effective for Viterbi beam search.

Although it is always desirable to use all the knowledge sources (KSs) in the search algorithm, some are difficult to integrate into the left-to-right time-synchronous search framework. One alternative strategy is to first produce an ordered list of sentence hypotheses (a.k.a. *n-best list*), or a lattice of word hypotheses (a.k.a. *word lattice*) using relatively inexpensive KSs. More expensive KSs can be used to rescore the n-best list or the word lattice to obtain the refined result. Such a multipass strategy has been explored in many large-vocabulary speech recognition systems. Various algorithms to generate sufficient n-best lists or the word lattices are described in the section on multipass search strategies.

Most of the techniques described in this chapter rely on nonadmissible heuristics. Thus, it is critical to derive a framework to evaluate different search strategies and pruning parameters.

13.1. Efficient Manipulation of a Tree Lexicon

The lexicon entry is the most critical component for large-vocabulary speech recognition, since the search space grows linearly along with increased linear vocabulary. Thus an efficient framework for handling large vocabulary undoubtedly becomes the most critical issue for efficient search performance.

13.1.1. Lexical Tree

The search space for n-gram discussed in Chapter 12 is organized based on a straightforward linear lexicon, i.e., each word is represented as a linear sequence of phonemes, independent of other words. For example, the phonetic similarity between the words *task* and *tasks* is not leveraged. In a large-vocabulary system, many words may share the same beginning phonemes. A tree structure is a natural representation for a large-vocabulary lexicon, as many phonemes can be shared to eliminate redundant acoustic evaluations. The lexical tree-based search is thus essential for building a real-time[1] large-vocabulary speech recognizer.

---

[1] The term *real-time* means the decoding process takes no longer than the duration of the speech. Since the decoding process can take place as soon as the speech starts, such a real-time decoder can provide real instantaneous responses after speakers finish talking.

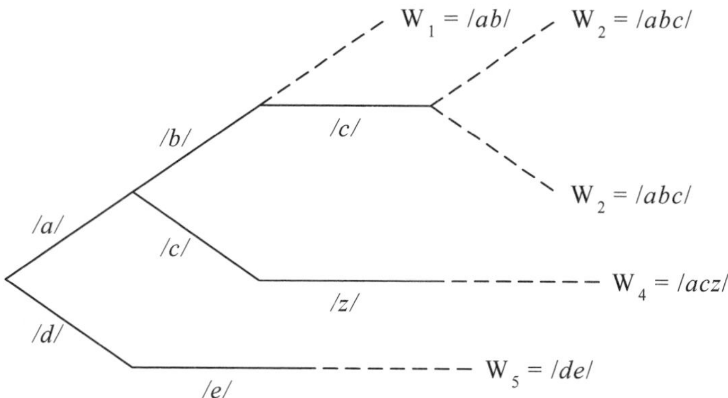

Figure 13.1 An example of a lexical tree, where each branch corresponds to a shared phoneme and the leaf corresponds to a word.

Figure 13.1 shows an example of such a lexical tree, where common beginning phonemes are shared. Each leaf corresponds to a word in the vocabulary. Please note that an extra null arc is used to form the leaf node for each word. This null arc has the following two functions:

1. When the pronunciation transcription of a word is a prefix of other ones, the null arc can function as one branch to end the word.
2. When there are homophones in the lexicon, the null arcs can function as linguistic branches to represent different words such as *two* and *to*.

The advantage of using such a lexical tree representation is obvious: it can effectively reduce the state search space of the trellis. Ney et al. [32] reported that a lexical tree representation of a 12,306-word lexicon with only 43,000 phoneme arcs had a saving of a factor of 2.5 over the linear lexicon with 100,800 phoneme arcs. Lexical trees are also referred to as *prefix trees*, since they are efficient representations of lexicons with sharing among lexical entries that have a common prefix. Table 13.1 shows the distribution of phoneme arcs for this 12,306-word lexical tree. As one can see, even in the fifth level the number of phoneme arcs is only about one-third of the total number of words in the lexicon.

Table 13.1 Distribution of the tree phoneme arcs and active tree phoneme arc for a 12,306-word lexicon using a lexical tree representation [32].

Level	**1**	**2**	**3**	**4**	**5**	**6**	**≥7**
Phoneme arcs	28	331	1511	3116	4380	4950	29.200
Average active arcs	23	233	485	470	329	178	206

The saving by using a lexical tree is substantial, because it not only results in considerable memory saving for representing state-search space but also saves tremendous time by searching far fewer potential paths. Ney et al. [32] report that a tree organization of the lexicon reduces the total search effort by a factor of 7 over the linear lexicon organization. This is because the lion's share of hypotheses during a typical large-vocabulary search is on the first and second phonemes of a word. Haeb-Umbach et al. [23] report that for a 12,306-word dictation task, 79% and 16% of the state hypotheses are in the first and second phonemes, when analyzing the distribution of the state hypotheses over the state position within a word. Obviously, the effect is caused by the ambiguities at the word boundaries. The lexical tree representation reduces that effort by evaluating common phonetic prefixes only once. Table 13.1 also shows the average number of active phoneme arcs in the layers of the lexical tree [32]. Based on this table, you can expect that the overall search cost is far less than the size of the vocabulary. This is the key reason why lexical tree search is widely used for large-vocabulary continuous speech recognition systems.

The lexical tree search requires a sophisticated implementation because of a fundamental deficiency—*a branch in a lexical tree representation does not correspond to a single word with the exception of branches ending in a leaf.* This deficiency translates to the fact that a unique word identity is not determined until a leaf of the tree is reached. This means that any decision about the word identity needs to be delayed until the leaf node is reached, which results in the following complexities.

- Unlike a linear lexicon, where the language model score can be applied when starting the acoustic search of a new word, the lexical tree representation has to delay the application of the language model probability until the leaf is reached. This may result in an increased search effort, because the pruning needs to be done on a less reliable measure, unless a factored language model is used, as discussed in Section 13.1.3.
- Because of the delay of language model contribution by one word, we need to keep a separate copy of an entire lexical tree for each unique language model history.

13.1.2. Multiple Copies of Pronunciation Trees

A simple lexical tree is sufficient if no language model or a unigram is used. This is because the decision at time t depends on the current word only. However, for higher-order n-gram models, the linguistic state cannot be determined locally. A tree copy is required for each language model state. For bigrams, a tree copy is required for each predecessor word. This may seem to be astonishing, because the potential search space is increased by the vocabulary size. Fortunately, experimental results show only a small number of tree copies are required, because efficient pruning can eliminate most of the unneeded ones. Ney et al. [32] report that the search effort using bigrams is increased by only a factor of 2 over the unigram case. In general, when more detailed (better) acoustic and/or language models are used, the effect of a potentially increased search space is often compensated by a more focused beam search from the use of more accurate models. In other words, although the static search space might increase significantly by using more accurate models, the dynamic search space can be under control (sometimes even smaller), thanks to improved evaluation functions.

To deal with tree copies [19, 23, 37], you can create redundant subtrees. When copies of lexical trees are used to disambiguate active linguistic contexts, many of the active state hypotheses correspond to the same redundant unigram state, due to the postponed application of language models. To apply the language model sooner, and to eliminate redundant unigram state computations, a successor tree, T_i, can be created for each linguistic context i. T_i encodes the nonzero n-grams of the linguistic context i as an isomorphic subgraph of the unigram tree, T_0. Figure 13.2 shows the organization of such successor trees and unigram tree for bigram search. For each word w a successor tree, T_w is created with the set of successor words that have nonzero bigram probabilities. Suppose u is a successor of w; the bigram probability $P(u \mid w)$ is attached to the transition connecting the leaf corresponding to u in the successor tree T_w, with the root of the successor tree T_u. The unigram tree is a full-size lexical tree and is shared by all words as the back-off lexical tree. Each leaf of the unigram tree corresponds to one of $|V|$ words in the vocabulary and is linked to the root of its bigram successor tree (T_u) by an arc with the corresponding unigram probability $P(u)$. The backoff weight, $\alpha(u)$, of predecessor u is attached to the arc which links the root of successor tree T_u to the root of the unigram tree.

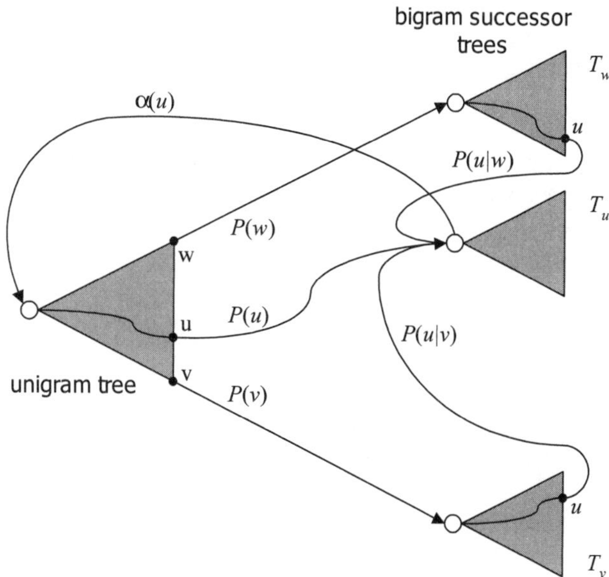

Figure 13.2 Successor trees and unigram trees for bigram search [13].

A careful search organization is required to avoid computational overhead and to guarantee a linear time complexity for exploring state hypotheses. In the following sections we describe techniques to achieve efficient lexical tree recognizers. These techniques include factorization of language model probabilities, tree optimization, and exploiting subtree dominance.

13.1.3. Factored Language Probabilities

As mentioned in Section 13.1.2, search is more efficient if a detailed knowledge source can be applied at an early stage. The idea of *factoring* the language model probabilities across the tree is one such example [4, 19]. When more than one word shares a phoneme arc, the upper bound of their probability can be associated to that arc.[2] The factorization can be applied to both the full lexical tree (unigram) and successor trees (bigram or other higher-order language models).

An unfactored tree only has language model probabilities attached to the leaf nodes, and all the internal nodes have probability 1.0. The procedure for factoring the probabilities across the tree computes the maximum of each node n in the tree according to Eq. (13.1). The tree can then be factored according to Eq. (13.2) so when you traverse the tree you can multiply $F^*(n)$ along the path to get the needed language probability.

$$P^*(n) = \max_{x \in child(n)} P(x) \tag{13.1}$$

$$F^*(n) = \frac{P^*(n)}{P^*(parent(n))} \tag{13.2}$$

An illustration of the factored probabilities is shown in Table 13.2. Using this lexicon, we create the tree depicted in Figure 13.3(a). In this figure the unlabeled internal nodes have a probability of 1.0. We distribute the probabilities according to Eq. (13.1) in Figure 13.3(b), which is factored according to Eq. (13.2), resulting in Figure 13.3(c).

Table 13.2 Sample probabilities $P(w_j)$ and their pseudoword pronunciations [4].

w_j	**Pronunciation**	**P(w_j)**
w_0	/a b c/	0.1
w_1	/a b c/	0.4
w_2	/a c z/	0.3
w_3	/d e/	0.2

[2] The choice of upper bound is because it is an admissible estimate of the path no matter which word will be chosen later.

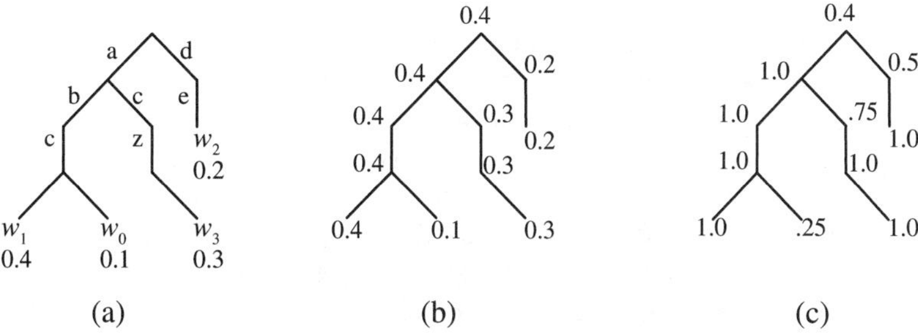

Figure 13.3 (a) Unfactored lexical tree; (b) distributed probabilities with computed $P^*(n)$; (c) factored tree $F^*(n)$ [4].

Using the upper bounds in the factoring algorithm is not an approximation, since the correct language model probabilities are calculated by the product of values traversed along each path from the root to the leaves. However, you should note that the probabilities of all the branches of a node do not sum to one. This can solved by replacing the upper-bound (max) function in Eq. (13.1) with the sum.

$$P^*(n) = \sum_{x \in child(n)} P(x) \tag{13.3}$$

To guarantee that all the branches sum to one, Eq. (13.2) should also be replaced by the following equation:

$$F^*(n) = \frac{P^*(n)}{\sum_{x \in child(parent(n))} P^*(x)} \tag{13.4}$$

A new illustration of the distribution of LM probabilities by using sum instead of upper bound is shown in Figure 13.4. Experimental results have shown that the factoring method with either sum or upper bound has comparable search performance.

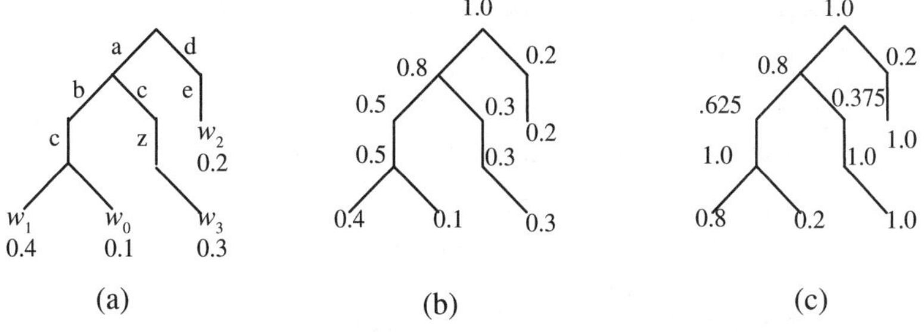

Figure 13.4 Using sum instead of upper bound when factoring tree, the corresponding (a) unfactored lexical tree; (b) distributed probabilities with computed $P^*(n)$; (c) factored tree with computed $F^*(n)$ [4].

One interesting observation is that the language model score can be regarded as a heuristic function to estimate the linguistic expectation of the current word to be searched. In a linear representation of the pronunciation lexicon, application of the linguistic expectation was straightforward, since each state is associated with a unique word. Therefore, given the context defined by the hypothesis under consideration, the expectation for the first phone of word w_i is just $P(w_i \mid w_1^{i-1})$. After the first phone, the expectation for the rest of the phones becomes 1.0, since there is only one possible phone sequence when searching the word w_i. However, for the tree lexicon, it is necessary to compute $E(p_j \mid p_1^{j-1}, w_i^{i-1})$, the expectation of phone p_j given the phonetic prefix p_1^{j-1} and the linguistic context w_1^{i-1}. Let $\phi(j, w_k)$ denote the phonetic prefix of length *j* for w_k. Based on Eqs. (13.1) and (13.2), we can compute the expectation as:

$$E(p_j \mid p_1^{j-1}, w_1^{i-1}) = \frac{P(w_c \mid w_1^{i-1})}{P(w_p \mid w_1^{i-1})} \tag{13.5}$$

where $c = \arg\max_k (w_k \mid w_1^{i-1}, \phi(j, w_k) = p_1^j)$ and $p = \arg\max_k (w_k \mid w_1^{i-1}, \phi(j-1, w_k) = p_1^{j-1})$. Based on Eq. (13.5), an arbitrary *n*-gram model or even a stochastic context-free grammar can be factored accordingly.

13.1.3.1. Efficient Memory Organization of Factored Lexical Trees

A major drawback to the use of successor trees is the large memory overhead required to store the additional information that encodes the structure of the tree and the factored linguistic probabilities. For example, the 5.02 million bigrams in the 1994 NABN (North American Business News) model require 18.2 million nodes. Given a compact binary tree representation that uses 4 bytes of memory per node, 72.8 million bytes are required to store the predecessor-dependent lexical trees. Furthermore, this tree representation is not as amenable to data compression techniques as the linear bigram representation.

The factored probability of successor trees can be encoded as efficiently as the *n*-gram model based on Algorithm 13.1, i.e., one *n*-gram record results in one constant-sized record. Step 3 is illustrated in Figure 13.5(b), where the heavy line ends at the most recently visited node that is not a direct ancestor. The encoding result is shown in Table 13.3.

ALGORITHM 13.1: *ENCODING THE LEXICAL SUCCESSOR TREES (LST)*

For each linguistic context:
Step 1: Distribute the probabilities according to Eq. (13.1).
Step 2: Factor the probabilities according to Eq. (13.2).
Step 3: Perform a depth-first traversal of the LST and encode each leaf record,
(a) the depth of the most recently visited node that is not a direct ancestor,
(b) the probability of the direct ancestor at the depth in (a),
(c) the word identity.

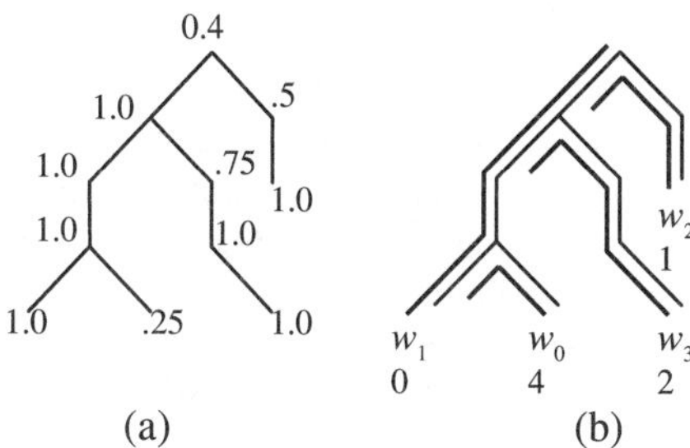

Figure 13.5 (a) Factored tree; (b) tree with common prefix-length annotation.

Clearly the new data structure meets the requirements set forth, and, in fact, it only requires additional log(n) bits per record (n is the depth of the tree). These bits encode the common prefix length for each word. Naturally this requires some modification to the decoding procedure. In particular, the decoder must scan a portion of the n-gram successor list in order to determine which tree nodes should be activated. Depending on the structure of the tree (which is determined by the acoustic model, the lexicon, and language model), the tree structure can be interpreted at runtime or cached for rapid access if memory is available.

Table 13.3 Encoded successor lexical tree; each record corresponds to one augmented factored n-gram.

w_j	**Depth**	$\mathbf{F}^*(w_j)$
w_1	0	0.4
w_0	4	0.25
w_3	2	0.75
w_2	1	0.5

13.1.4. Optimization of Lexical Trees

We now investigate ways to handle the huge search network formed by the multiple copies of lexical trees in different linguistic contexts. The factorization of lexical trees actually makes it easier to search. First, after the factorization of the language model, the intertree transitions shown in Figure 13.2 no longer have the language model scores attached because they are already applied completely before leaving the leaves. Moreover, as illustrated in Figure 13.3, many transitions toward the end of a single-word path now have an associated transition probability that is equal to 1. This observation implies that there could be many duplicated subtrees in the network. Those duplicated subtrees can then be merged to save both space and computation by eliminating redundant (unnecessary) state evaluation. Unlike pruning, this saving is based on the dynamic programming principle, without introducing any potential error.

13.1.4.1. Optimization of Finite State Network

One way to compress the lexical tree network is to use a similar algorithm for optimizing the number of states in a deterministic finite state automaton. The optimization algorithm is based on the *indistinguishable* property of states in a finite state automaton. Suppose that s_1 and s_2 are the initial states for automata T_1 and T_2, then s_1 and s_2 are said to be *indistinguishable* if the languages accepted by automata T_1 and T_2 are exactly the same. If we consider our lexical tree network as a finite state automaton, the symbol emitted from the transition arc includes not only the phoneme identity, but also the factorized language model probability.

The general set-partitioning algorithm [1] can be used for the reduction of finite state automata. The algorithm starts with an initial partition of the automaton states and iteratively refines the partition so that two states s_1 and s_2 are put in the same block B_i if and only if $f(s_1)$ and $f(s_2)$ are both in the same block B_j. For our purpose, $f(s_1)$ and $f(s_2)$ can be defined as the destination state given a phone symbol (in the factored trees, the pair <*phone*, *LM-probability*> can be used). Each time a block is partitioned, the smaller subblock is used for further partitioning. The algorithm stops when all the states that transit to some state in a particular block with arcs labeled with the same symbol are in the same block. When the algorithm halts, each block of the resulting partition is composed of *indistinguishable* states, and those states within each block can then be merged. The algorithm is guaranteed to find the automaton with the minimum number of states. The algorithm has a time complexity of $O(MN \log N)$, where M is the maximum number of branching (fan-out) factors in the lexical tree and N is the number of states in the original tree network.

Although the above algorithm can give optimal finite state networks in terms of number of states, such an optimized network may be difficult to maintain, because the original lexical tree structure could be destroyed and it may be troublesome to add any new word into the tree network [1].

13.1.4.2. Subtree Isomorphism

The finite state optimization algorithm described above does not take advantage of the tree structure of the finite state network, though it generates a network with a minimum number of states. Since our finite state network is a network of trees, the indistinguishability property is actually the same as the definition of subtree isomorphism. Two subtrees are said to be *isomorphic* to each other if they can be made equivalent by permuting the successors. It should be straightforward to prove that two states are indistinguishable, if and only if their subtrees are isomorphic.

There are efficient algorithms [1] to detect whether two subtrees are isomorphic. For all possible pairs of states u and v, if the subtrees starting at u and v, $ST(u)$ and $ST(v)$, are isomorphic, v is merged into u and $ST(v)$ can be eliminated. Note that only internal nodes need to be considered for subtree isomorphism check. The time complexity for this algorithm is $O(N^2)$ [1].

13.1.4.3. Sharing Tails

A *linear tail* in a lexical tree is defined as a subpath ending in a leaf and going through states with a unique successor. It is often referred as a *single-word subpath*. It can be proved that such a linear tail has unit probability attached to its arcs according to Eqs. (13.1) and (13.2). This is because LM probability factorization *pushes forward* the LM probability attached to the last arc of the linear tail, leaving arcs with unit probability. Since all the tails corresponding to the same word w in different successor trees are linked to the root of successor tree T_w,[3] the subtree starting from the first state of each linear tail is isomorphic to the subtree starting from one of the states forming the longest linear tail of w. A simple algorithm to take advantage of this share-tail topology can be employed to reduce the lexical tree network.

Figure 13.6 and Figure 13.7 show a lexical tree network before and after shared-tail optimization. For each word, only the longest linear tail is kept. All other tails can be removed by linking them to an appropriate state in the longest tail, as shown in Figure 13.7.

Shared-tail optimization is not global optimization, because it considers only some special topology optimization. However, there are some advantages associated with shared-tail optimization. First, in practice, duplicated linear tails account for most of the redundancy in lexical tree networks [12]. Moreover, shared-tail optimization has a nice property of maintaining the basic lexical tree structure for the optimized tree network.

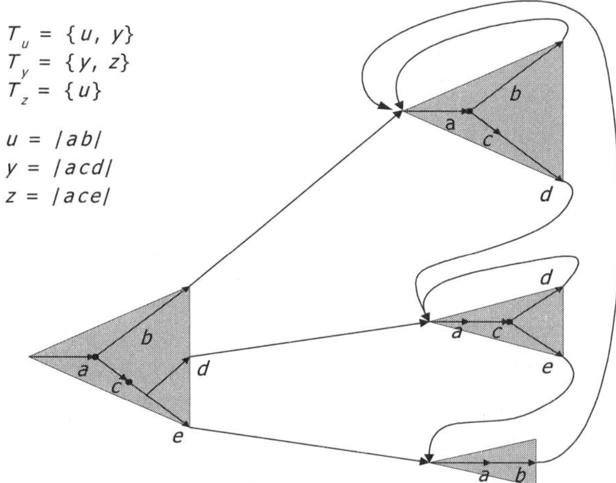

Figure 13.6 An example of a lexical tree network without shared-tail optimization [12]. The vocabulary includes three words, u, y, and z. T_u, T_y, and T_z are the successor trees for u, y, and z, respectively [13].

[3] We assume bigram is used in the discussion of "sharing tails."

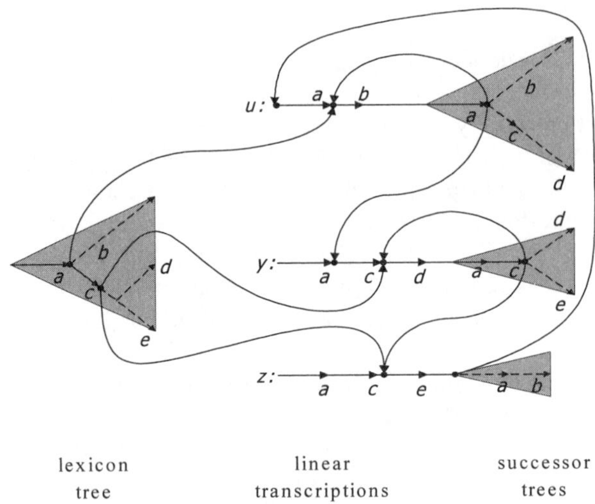

Figure 13.7 The lexical tree network in Figure 13.6 after shared-tail optimization [12].

13.1.5. Exploiting Subtree Polymorphism

The techniques of optimizing the network of successor lexical trees can only eliminate identical subtrees in the network. However, there are still many subtrees that have the same nodes and topology but with different language model scores attached to the arcs. The acoustic evaluation for those subtrees is unnecessarily duplicated. In this section we exploit *subtree dominance* for additional saving.

A subtree instance is *dominated* when the best outcome in that subtree is not better than the worst outcome in another instance of that subtree. The evaluation becomes redundant for the dominated subtree instance. Subtree isomorphism and shared-tail are cases of subtree dominance, but they require prearrangement of the lexical tree network as described in the previous section.

If we need to implement lexical tree search dynamically, the network optimization algorithms are not suitable. Although subtree dominance can be computed using minimax search [35] during runtime, this requires that information regarding subtree isomorphism be available for all corresponding pairs of states for each successor tree T_w. Unfortunately, it is not practical in terms of either computation or space.

In place of computing strict subtree dominance, a *polymorphic* linguistic context assignment to reduce redundancy is employed by estimating subtree dominance based on local information and ignoring the subgraph isomorphism problem. Polymorphic context assignment involves keeping a single copy of the lexical tree and allowing each state to assume the linguistic context of the most promising history. The advantage of this approach is that it employs maximum sharing of data structures and information, so each node in the tree is evaluated, at most, once. However, the use of local knowledge to determine the dominant context could introduce significant errors because of premature pruning. Whisper [4] reports a 65.7% increase in error rate when only the dominant context is kept, based on local knowledge.

To recover the errors created by using local linguistic information to estimate subtree dominance, you need to delay the decision regarding which linguistic context is most promising. This can be done by keeping a heap of contexts at each node in the tree. The heap maintains all contexts (linguistic paths) whose probabilities are within a constant threshold ε, of that of the best global path. The effect of the ε-heap is that more contexts are retained for high-probability states in the lexical tree. The pseudocode fragment in Algorithm 13.2 [3] illustrates a transition from state s_n in context c to state s_m. The terminology used in Algorithm 13.2 is listed as follows:

- $(-\log P(s_m \mid s_n, c))$ is the cost associated with applying acoustic model matching and language model probability of state s_m transited from s_n in context c.
- $InHeap(s_m, c)$ is true if context c is in the heap corresponding to state s_m.
- $Cost(s_m, c)$ is the cost for context c in state s_m.
- $StateInfo(s_m, c)$ is the auxiliary state information associated with context c in state s_m.
- $Add(s_m, c)$ adds context c to the state s_m heap.
- $Delete(s_m, c)$ deletes context c from state s_m heap.
- $WorstContext(s_m)$ retrieves the worst context from the heap of state s_m.

ALGORITHM 13.2: *HANDLING MULTIPLE LINGUISTIC CONTEXTS IN A LEXICAL TREE*

1. $d = Cost(s_n, c) + (-\log P(s_m \mid s_n, c))$

2. **if** $InHeap(s_m, c)$ **then**

    if $d < Cost(s_m, c)$ **then**

        $Cost(s_m, c) = d$

        $StateInfo(s_m, c) = StateInfo(s_n, c)$

else if $d < BestCost(s_m) + \varepsilon$ **then**

    $Add(s_m, c)$; $StateInfo(s_m, c) = StateInfo(s_n, c)$

    $Cost(s_m, c) = d$

    else

    $w = WorstContext(s_m)$

    if $d < Cost(s_m, w)$ **then**

        $Delete(s_m, w)$

        $Add(s_m, c)$; $StateInfo(s_m, c) = StateInfo(s_n, c)$

        $Cost(s_m, c) = d$

When higher-order n-gram is used for lexical tree search, the potential heap size for lexical tree nodes (some also refer to *prefix nodes*) could be unmanageable. With decent acoustic models and efficient pruning, as illustrated in Algorithm 13.2, the average heap size for active nodes in the lexical tree is actually very modest. For example, Whisper's average heap size for active nodes in the 20,000-word WSJ lexical tree decoder is only about 1.6 [3].

13.1.6. Context-Dependent Units and Inter-Word Triphones

So far, we have implicitly assumed that context-independent models are used in the lexical tree search. When context-dependent phonetic or subphonetic models, as discussed in Chapter 9, are used for better acoustic models, the construction and use of a lexical tree become more complicated.

Since senones represent both subphonetic and context-dependent acoustic models, this presents additional difficulty for use in lexical trees. Let's assume that a three-state context-dependent HMM is formed from three senones, one for each state. Each senone is context-dependent and can be shared by different allophones. If we use allophones as the units for lexical tree, the sharing may be poor and fan-out unmanageable. Fortunately, each HMM is uniquely identified by the sequence of senones used to form the HMM. In this way, different context-dependent allophones that share the same *senone sequence* can be treated as the same. This is especially important for lexical tree search, since it reduces the order of the fan-out in the tree.

Interword triphones that require significant fan-ins for the first phone of a word and fan-outs for the last phones usually present an implementation challenge for large-vocabulary speech recognition. A common approach is to delay full interword modeling until a subsequent rescoring phase.[4] Given a sufficiently rich lattice or word graph, this is a reasonable approach, because the static state space in the successive search has been reduced significantly. However, as pointed out in Section 13.1.2, the size of the dynamic state space can remain under control when detailed models are used to allow effective pruning. In addition, a multipass search requires an augmented set of acoustic models to effectively model the biphone contexts used at word boundaries for the first pass. Therefore, it might be desirable to use genuine interword acoustic models in the single-pass search.

Instead of expanding all the fan-ins and fan-outs for inter-word context-dependent phone units in the lexical tree, three *metaunits* are created.

1. The first metaunit, which has a known right context corresponding to the second phone in the word, but uses open left context for the first phone of a word (sometimes referred to as the *word-initial unit*). In this way, the fan-in is represented as a subgraph shared by all words with the same initial left-context-dependent phone.

[4] Multipass search strategy is described in Section 13.3.5.

2. Another metaunit, which has a known left context corresponding to the second-to-last phone of the word, but uses open right context for the last phone of a word (sometimes referred to as the *word-final unit*). Again, the fan-out is represented as a subgraph shared by all words with the same final right-context-dependent phone.
3. The third metaunit, which has both open left and right contexts, and is used for single-phone word unit.

By using these metaunits we can keep the states for the lexical trees under control, because the fan-in and fan-out are now represented as a single node.

During recognition, different left or right contexts within the same metaunit are handled using Algorithm 13.2, where the different acoustic contexts are treated similarly as different linguistic contexts. The open left-context metaunit (fan-ins) can be dealt with in a straightforward way using Aglorithm 13.2, because the left context is always known (the last phone of the previous word) when it is initiated. On the other hand, the open right-context metaunit (fan-out) needs to explore all possible right contexts because the next word is not known yet. To reduce unnecessary computation, fast match algorithms (described in Section 13.2.3) can be used to provide both expected acoustic and language scores for different context-dependent units to result in early pruning of unpromising contexts.

13.2. Other Efficient Search Techniques

Tree structured lexicon represents an efficient framework of manipulation of search space. In this section we present some additional implementation techniques, which can be used to further improve the efficiency of search algorithms. Most of these techniques can be applied to both Viterbi beam search and stack decoding. They are essential ingredients for a practical large-vocabulary continuous speech recognizer.

13.2.1. Using Entire HMM as a State in Search

The state in state-search space based on HMM-trellis computation is, by definition, a Markov state. Phonetic HMM models are the basic unit in most speech recognizers. Even though subphonetic HMMs, like senones, might be used for such a system, the search is often based on phonetic HMMs.

Treating the entire phonetic HMM as a state in state-search has many advantages. The first obvious advantage is that the number of states the search program needs to deal with is smaller. Note that using the entire phonetic HMM does not in effect reduce the number of states in the search. The entire search space is unchanged. All the states within a phonetic HMM are now bundled together. This means that all of them are either kept in the beam, if the phonetic HMM is regarded as promising, or all of them are pruned away. For any given time, the minimum cost among all the states within the phonetic HMM is used as the cost for the phonetic HMM. For pruning purposes, this cost is used to determine the promising degree of this phonetic HMM, i.e., the fate of all the states within this phonetic HMM. Although this does not actually reduce the beam beyond normal pruning, it has the effect of processing fewer candidates in the beam. In programming, this means less checking and bookkeeping, so some computation savings can be expected.

You might wonder if this organization might be ineffective for beam search, since it forces you to keep or prune all the states within a phonetic HMM. In theory, it is possible that only one or two states in the phonetic HMM need to be kept, while other states can be pruned due to high cost score. However, this is, in reality, very rare, since a phone is a small unit and all the states within a phonetic HMM should be relatively promising when the search is near the acoustic region corresponding to the phone.

During the trellis computation, all the phonetic HMM states need to advance one time step when processing one input vector. By performing HMM computation for all states together, the new organization can reduce memory accesses and improve cache locality, since the output and transition probabilities are held in common by all states. Combining this organization strategy with lexical tree search further enhances the efficiency. In lexical tree search, each hypothesis in the beam is associated with a particular node in the lexical tree. These hypotheses are linked together in the heap structure described in Algorithm 13.2 for the purposes of efficient evaluation and heuristic pruning. Since the node corresponds to a phonetic HMM, the HMM evaluation is guaranteed to execute once for each hypothesis sharing this node.

In summary, treating the entire phonetic HMM as a state in state-search space allows you to explore the effective data structure for better sharing and improved memory locality.

13.2.2. Different Layers of Beams

Because of the complexity of search, it often requires pruning of various levels of search to make search feasible. Most systems thus employ different pruning thresholds to control what states participate. The most frequently used thresholds are listed below:

- τ_s controls what states (either phone states or senone states) to retain. This is the most fundamental beam threshold.
- τ_p controls whether the next phone is extended. Although this might not be necessary for both stack decoding and linear Viterbi beam search, it is crucial for lexical tree search, because pruning unpromising phonetic prefixes in the lexical trees could improve search efficiency significantly.
- τ_h controls whether hypotheses are extended for the next word. Since the branching factor for word boundaries is very large, we need this threshold to limit search to only the promising ones.
- τ_c controls where a linguistic context is created in a lexical tree search using higher-order language models. This is also known as ε-heap in Algorithm 13.2.

Pruning can introduce search errors if a state is pruned that would have been on the globally best path. The principle applied here is that the more constraints you have available, the more aggressively you decide whether this path will participate in the globally best path. In this case, at the state level, you have the least constraints. At the phonetic level there are more, and there are the most at the word level. In general, the number of word hypotheses tends to drop significantly at word boundaries. Different thresholds for different levels allow the search designer to fine-tune those thresholds for their tasks to achieve best search performance without significant increase in error rates.

13.2.3. Fast Match

As described in Chapter 12, fast match is a crucial part of stack decoding, which mainly reduces the number of possible word expansions for each path. Similarly, fast match can be applied to the most expensive part—extending the phone HMM fan-outs within or between lexical trees. Fast match is a method for rapidly deriving a list of candidates that constrain successive search phases in which a computationally expensive *detailed match* is performed. In this sense, fast match can be regarded as an additional pruning threshold to meet before a new word/phone can be started.

Fast match is typically characterized by the approximations that are made in the acoustic/language models to reduce computation. The factorization of language model scores among tree branches in lexical trees described in Section 13.1.3 can be viewed as fast match using a language model. The factorized method is also an admissible estimate of the language model scores for the future word. In this section we focus on acoustic model fast match.

13.2.3.1. Look-Ahead Strategy

Fast match, when applied in time-synchronous search, is also called *look-ahead* strategy. since it basically searches ahead of the time-synchronous search by a few frames to determine which words or phones are likely to extend. Typically the look-ahead frames are fixed, and the fast match is also done in time-synchronous fashion with another specialized beam for efficient pruning. You can also use simplified models, like the one-state HMMs or context-independent models [4, 32]. Some systems [21, 22] have tried to simplify the level of details in the input feature vectors by aggregating information from several frames into one. A straightforward way for compressing the feature stream is to skip every other frame of speech for fast match. This allows a longer-range look-ahead, while keeping computation under control. The approach of simplifying the input feature stream instead of simplifying the acoustic models can reuse the fast match results for detailed match.

Whisper [4] uses phoneme look-ahead fast match in lexical tree search, in which pruning is applied based on the estimation of the score of possible phone fan-outs that may follow a given phone. A context-independent phone-net is searched synchronously with the search process but offset N frames into the future. In practice, significant savings can be obtained in search efforts without increase in error rates.

The performance of word and phoneme look-ahead clearly depends on the length of the look-ahead frames. In general, the larger the look-ahead window, the longer is the computation and the shorter the word/phone Λ list. Empirically, the window is a few tens of milliseconds for phone look-ahead and a few hundreds of milliseconds for word look-ahead.

13.2.3.2. The Rich-Get-Richer Strategy

For systems employing continuous-density HMMs, tens of mixtures of Gaussians are often used for the output probability distribution for each state. The computation of the mixtures is one of the bottlenecks when many context-dependent models are used. For example, Whisper uses about 120,000 Gaussians. In addition to using various beam pruning thresholds in the search, there could be significant savings if we have a strategy to limit the number of Gaussians to be computed.

The *Rich-Get-Richer* (RGR) strategy enables us to focus on most promising paths and treat them with detailed acoustic evaluations and relaxed path-pruning thresholds. On the contrary, the less promising paths are extended with less expensive acoustic evaluations and less forgiving path-pruning thresholds. In this way, locally optimal candidates continue to receive the maximum attention while less optimal candidates are retained but evaluated using less precise (computationally expensive) acoustic and/or linguistic models. The RGR strategy gives us finer control in the creation of new paths that has potential to grow exponentially.

RGR is used to control the level of acoustic details in the search. The goal is to reduce the number of context-dependent senone probability (Gaussian) computations required. The context-dependent senones associated with a phone instance p would be evaluated according to the following condition:

$$\begin{aligned}
&Min\left[ci(p)\right] * \alpha + LookAhead\left[ci(p)\right] < \text{ threshold} \\
&\text{where } Min\left[ci(p)\right] = \min_s \left\{\cos t(s) \mid s \in ci_phone(p)\right\} \\
&\text{and } \quad LookAhead\left[ci(p)\right] = \text{look-ahead estimate of } ci(p)
\end{aligned} \tag{13.6}$$

These conditions state that the context-dependent senones associated with p should be evaluated if there exists a state s corresponding to p, whose cost in linear combination with a look-ahead cost score corresponding to p falls within a threshold. In the event that p does not fall within the threshold, the senone scores corresponding to p are estimated using the context-independent senones corresponding to p. This means the context-dependent senones are evaluated only if the corresponding context-independent senones and the look-ahead start showing promise. RGR strategy should save significant senone computation for clearly unpromising paths. Whisper [26] reports that 80% of senone computation can be avoided without introducing significant errors for a 20,000-word WSJ dictation task.

13.3. *N*-BEST AND MULTIPASS SEARCH STRATEGIES

Ideally, a search algorithm should consider all possible hypotheses based on a unified probabilistic framework that integrates all *knowledge sources* (KSs).[5] These KSs, such as acoustic models, language models, and lexical pronunciation models, can be integrated in an HMM state search framework. It is desirable to use the most detailed models, such as context-dependent models, interword context-dependent models, and high-order *n*-grams, in the search as early as possible. When the explored search space becomes unmanageable, due to the increasing size of vocabulary or highly sophisticated KSs, search might be infeasible to implement.

As we develop more powerful techniques, the complexity of models tends to increase dramatically. For example, language understanding models in Chapter 17 require long-distance relationships. In addition, many of these techniques are not operating in a left-to-right manner. A possible alternative is to perform a multipass search and apply several KSs at different stages, in the proper order to constrain the search progressively. In the initial pass, the most discriminant and computationally affordable KSs are used to reduce the number of hypotheses. In subsequent passes, progressively reduced sets of hypotheses are examined, and more powerful and expensive KSs are then used until the optimal solution is found.

The early passes of multipass search can be considered fast match that eliminates those unlikely hypotheses. Multipass search is, in general, not admissible because the optimal word sequence could be wrongly pruned prematurely, due to the fact that not all KSs are used in the earlier passes. However, for complicated tasks, the benefits of computation complexity reduction usually outweigh the nonadmissibility. In practice, multipass search strategy using progressive KSs could generate better results than a search algorithm forced to use less powerful models due to computation and memory constraints.

The most straightforward multipass search strategy is the so-called *n*-best search paradigm. The idea is to use affordable KSs to first produce a list of *n* most probable word sequences in a reasonable time. Then these *n* hypotheses are rescored using more detailed models to obtain the most likely word sequence. The idea of the *n*-best list can be further extended to create a more compact hypotheses representation—namely word lattice or graph. A word lattice is a more efficient way to represent alternative hypotheses. *N*-best or lattice search is used for many large-vocabulary continuous speech recognition systems [20, 30, 44].

In this section we describe the representation of the *n*-best list and word lattice. Several algorithms to generate such an *n*-best-list or word lattice are discussed.

---

[5] In the field of artificial intelligence, the process of performing search through an integrated network of various knowledge sources is called *constraint satisfaction*.

13.3.1. *N*-best Lists and Word Lattices

Table 13.4 shows an example *n*-best (10-best) list generated for a North American Business (NAB) sentence. *N*-best search framework is effective only for *n* of the order of tens or hundreds. If the short *n*-best list that is generated by using less optimal models does not include the correct word sequence, the successive rescoring phases have no chance to generate the correct answer. Moreover, in a typical *n*-best list like the one shown in Table 13.4, many of the different word sequences are just one-word variations of each other. This is not surprising, since similar word sequences should achieve similar scores. In general, the number of *n*-best hypotheses might grow exponentially with the length of the utterance. Word lattices and word graphs are thus introduced to replace *n*-best list with a more compact representation of alternative hypotheses.

Word lattices are composed by word hypotheses. Each word hypothesis is associated with a score and an explicit time interval. Figure 13.8 shows an example of a word lattice corresponding to the *n*-best list example in Table 13.4. It is clear that a word lattice is more efficient representation. For example, suppose the spoken utterance contains 10 words and there are 2 different word hypotheses for each word position. The *n*-best list would need to have $2^{10} = 1024$ different sentences to include all the possible permutations, whereas the word lattice requires only 20 different word hypotheses.

Word graphs, on the other hand, resemble finite state automata, in which arcs are labeled with words. Temporal constraints between words are implicitly embedded in the topology. Figure 13.9 shows a word graph corresponding to the *n*-best list example in Table 13.4. Word graphs in general have an explicit specification of word connections that don't allow overlaps or gaps along the time axis. Nonetheless, word lattices and graphs are similar, and we often use these terms interchangeably.[6] Since an *n*-best list can be treated as a simple word lattice, word lattices are a more general representation of alternative hypotheses. *N*-best lists or word lattices are generally evaluated on the following two parameters:

Table 13.4 An example 10-best list for a North American Business sentence.

1.	I will tell you would I think in my office
2.	I will tell you what I think in my office
3.	I will tell you when I think in my office
4.	I would sell you would I think in my office
5.	I would sell you what I think in my office
6.	I would sell you when I think in my office
7.	I will tell you would I think in my office
8.	I will tell you why I think in my office
9.	I will tell you what I think on my office
10.	I Wilson you I think on my office

[6] We will use the term *word lattice* in the rest of this chapter..

- *Density:* In the n-best case, it is measured by how many alternative word sequences are kept in the n-best list. In the word lattice case, it is measured by the number of word hypotheses or word arcs per uttered word. Obviously, we want the density to be as small as possible for successive rescoring modules, provided the correct word sequence is included in the n-best list or word lattice.
- *The lower bound word error rate:* It is the lowest word error rate for any word sequence in the n-best list or the word lattice.

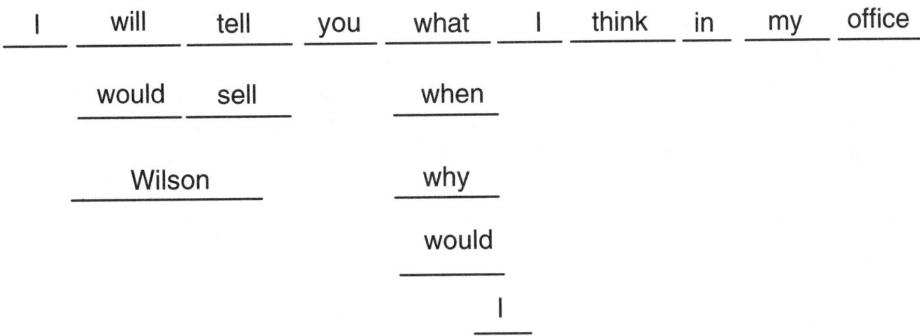

Figure 13.8 A word lattice example. Each word has an explicit time interval associated with it.

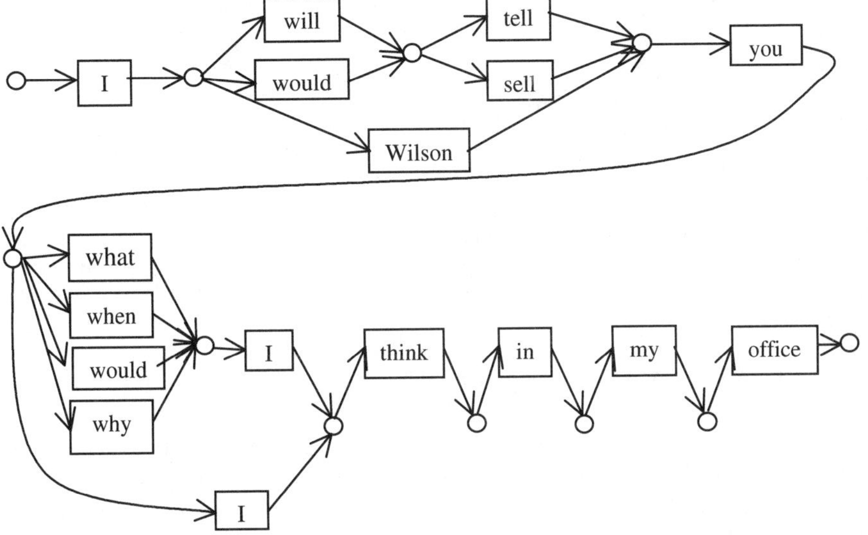

Figure 13.9 A word graph example for the n-best list in Table 13.4. Temporal constraints are implicit in the topology.

Rescoring with highly similar n-best alternatives duplicates computation on common parts. The compact representation of word lattices allows both data structure and computation sharing of the common parts among similar alternative hypotheses, so it is generally computationally less expensive to rescore the word lattice.

Figure 13.10 illustrates the general n-best/lattice search framework. Those KSs providing most constraints, at a lesser cost, are used first to generate the n-best list or word lattice. The n-best list or word lattice is then passed to the rescoring module, which uses the remaining KSs to select the optimal path. You should note that the n-best and word-lattice generators sometimes involve several phases of search mechanisms to generate the n-best list or word lattice. Therefore, the whole search framework in Figure 13.10 could involve several (> 2) phases of search mechanism.

Does the compact n-best or word-lattice representation impose constraints on the complexity of the acoustic and language models applied during successive rescoring modules? The word lattice can be expanded for higher-order language models and detailed context-dependent models, like inter-word triphone models. For example, to use higher-order language models for word lattice entails copying each word in the appropriate context of preceding words (in the trigram case, the two immediately preceding words). To use inter-word triphone models entails replacing the triphones for the beginning and ending phone of each word with appropriate interword triphones. The expanded lattice can then be used with detailed acoustic and language models. For example, Murveit et al. [30] report this can achieve trigram search without exploring the enormous trigram search space.

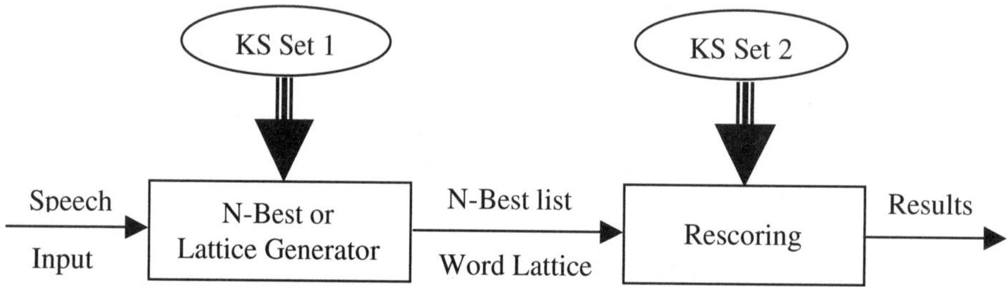

Figure 13.10 N-best/lattice search framework. The most discriminant and inexpensive knowledge sources (KSs 1) are used first to generate the n-best/lattice. The remaining knowledge sources (KSs 2, usually expensive to apply) are used in the rescoring phase to pick up the optimal solution [40].

13.3.2. The Exact *N*-best Algorithm

Stack decoding is the choice of generating n-best candidates because of its best-first principle. We can keep it generating results until it finds n complete paths; these n complete sentences form the n-best list. However, this algorithm usually cannot generate the n best candidates efficiently. The efficient n-best algorithm for time-synchronous Viterbi search was first introduced by Schwartz and Chow [39]. It is a simple extension of time-synchronous Viterbi search. The fundamental idea is to maintain separate records for paths with distinct histories. The history is defined as the whole word sequence up to the current time t and word w. This exact n-best algorithm is also called *sentence-dependent* n-best algorithm. When two or more paths come to the same state at the same time, paths having the same history are merged and their probabilities are summed together; otherwise, only the n-best paths are retained for each state. As commonly used in speech recognition, a typical HMM state has 2 or 3 predecessor states within the word HMM. Thus, for each time frame and each state, the n-best search algorithm needs to compare and merge 2 or 3 sets of n paths into n new paths. At the end of the search, the n paths in the final state of the trellis are simply re-ordered to obtain the n-best word sequences.

This straightforward n-best algorithm can be proved to be admissible[7] in normal circumstances [40]. The complexity of the algorithm is proportional to $O(n)$, where n is the number of paths kept at each state. This is often too slow for practical systems.

13.3.3. Word-Dependent *N*-best and Word-Lattice Algorithm

Since many of the different entries in the n-best list are just one-word variations of each other, as shown in Table 13.4, one efficient algorithm can be derived from the normal 1-best Viterbi algorithm to generate the n-best hypotheses. The algorithm runs just like the normal time-synchronous Viterbi algorithm for all within-word transitions. However for each time frame t, and each word-ending state, the algorithm stores all the different words that can end at current time t and their corresponding scores in a *traceback* list. At the same time, the score of the best hypothesis at each grammar state is passed forward, as in the normal time-synchronous Viterbi search. This obviously requires almost no extra computation above the normal time-synchronous Viterbi search. At the end of search, you can simply search through the stored traceback list to get all the permutations of word sequences with their corresponding scores. If you use a simple threshold, the traceback can be implemented very efficiently to only uncover the word sequences with accumulated cost scores below the threshold. This algorithm is often referred as *traceback*-based n-best algorithm [29, 42] because of the use of the traceback list in the algorithm.

However, there is a serious problem associated with this algorithm. It could easily miss some low-cost hypotheses. Figure 13.11 illustrates an example in which word w_k can be preceded by two different words w_i and w_j in different time frames. Assuming path w_i-w_k has a lower cost than path $w_j - w_k$ when both paths meet during the trellis search of w_k, the path $w_j - w_k$ will be pruned away. During traceback for finding the n-best word sequences, there is only one best starting time for word w_k, determined by the best boundary between the best preceding word w_i and it. Even though path $w_j - w_k$ might have a very low cost (let's say only marginally higher than that of $w_i - w_k$), it could be completely overlooked, since the path has a different starting time for word w_k.

[7] Although one can show, in the worst case, when paths with different histories have near identical scores for each state, the search actually needs to keep all paths (> N) in order to guarantee absolute admissibility. Under this worst case, the admissible algorithm is clearly exponential in the number of words for the utterance, since all permutations of word sequences for the whole sentence need to be kept.

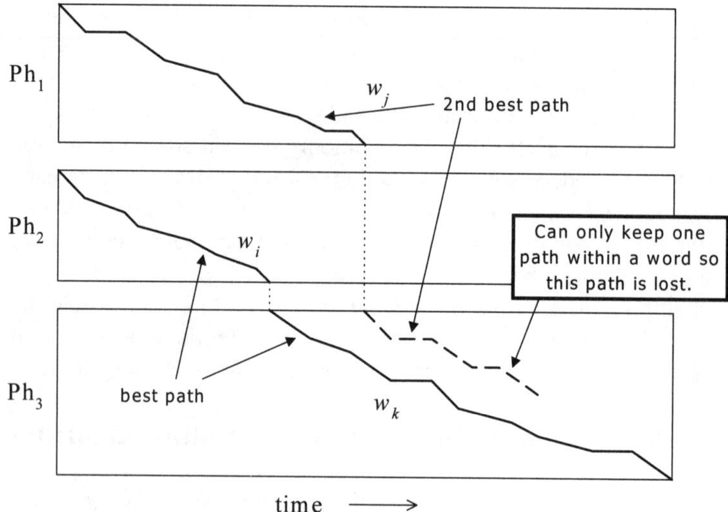

Figure 13.11 Deficiency in traceback-based n-best algorithm. The best subpath, w_i - w_k , will prune away subpath w_j - w_k while searching the word w_k ; the second-best subpath cannot be recovered [40].

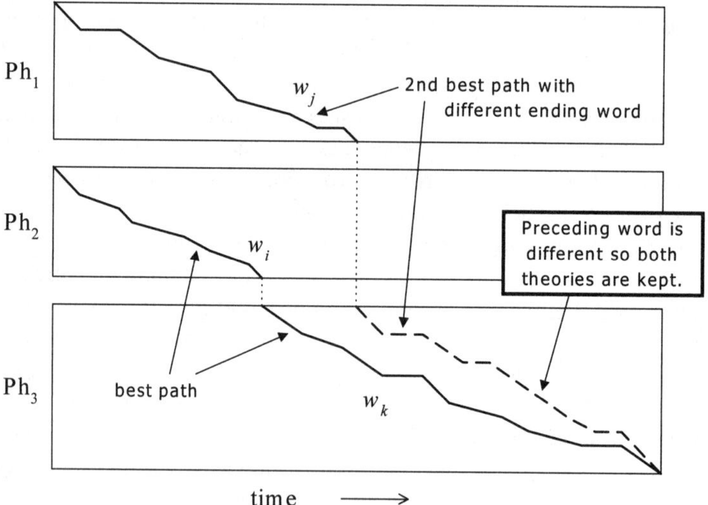

Figure 13.12 Word-dependent n-best algorithm. Both subpaths w_i - w_k and w_j - w_k are kept under the word-dependent assumption [40].

The *word-dependent* n-best algorithm [38] can alleviate the deficiency of the traceback-based n-best algorithm, in which only one starting time is kept for each word, so the starting time is independent of the preceding words. On the other hand, in the sentence-dependent n-best algorithm, the starting time for a word depends on all the preceding words, since different histories are kept separately. A good compromise is the so-called word-dependent assumption: *The starting time of a word depends only on the immediate preceding word.* That is, given a word pair and its ending time, the boundary between these two words is independent of further predecessor words.

In the word-dependent assumption, the history to be considered for a different path is no longer the entire word sequence; instead, it is only the immediately preceding word. This allows you to keep k ($<< n$) different records for each state and each time frame in Viterbi search. Differing slightly from the exact n-best algorithm, a traceback must be performed to find the n-best list at the end of search. The algorithm is illustrated in Figure 13.12. A word-dependent n-best algorithm has a time complexity proportional to k. However, it is no longer admissible because of the word-dependent approximation. In general, this approximation is quite reasonable if the preceding word is long. The loss it entails is insignificant [6].

13.3.3.1. One-Pass *N*-best and Word-Lattice Algorithm

As presented in Section 13.1, one-pass Viterbi beam search can be implemented very efficiently using a tree lexicon. Section 13.1.2 states that multiple copies of lexical trees are necessary for incorporating language models other than the unigram. When bigram is used in lexical tree search, the successor lexical tree is predecessor-dependent. This predecessor-dependent property immediately translates into the word-dependent property,[8] as defined in Section 13.3.3, because the starting time of a word clearly depends on the immediately preceding word. This means that different word-dependent partial paths are automatically saved under the framework of predecessor-dependent successor trees. Therefore, one-pass predecessor-dependent lexical tree search can be modified slightly to output n-best lists or word graphs.

Ney et al. [31] used a word graph builder with a one-pass predecessor-dependent lexical tree search. The idea is to exploit the word-dependent property inherited from the predecessor-dependent lexical tree search. During predecessor-dependent lexical tree search, two additional quantities are saved whenever a word ending state is processed.

$\tau(t;w_i,w_j)$—Representing the optimal word boundary between word w_i and w_j, given word w_j ending at time t.

$h(w_j;\tau(t;w_i,w_j),t) = -\log P(\mathbf{x}_\tau^t \mid w_j)$—Representing the cumulative cost that word w_j produces acoustic vector $\mathbf{x}_\tau, \mathbf{x}_{\tau+1}, \cdots \mathbf{x}_t$.

---

[8] When higher order n-gram models are used, the boundary dependence will be even more significant. For example, when trigrams are used, the boundary for a word juncture depends on the previous two words. Since we generally want a fast method of generating word lattices/graphs, bigram is often used instead of higher order n-gram to generate word lattices/graphs.

At the end of the utterance, the word lattice or n-best list is constructed by tracing back all the permutations of word pairs recorded during the search. The algorithm is summarized in Algorithm 13.3.

ALGORITHM 13.3: *ONE-PASS PREDECESSOR-DEPENDENT LEXICAL TREE SEARCH FOR N-BEST OR WORD-LATTICE CONSTRUCTION*

Step 1: For $t = 1..T$,
1-best predecessor-dependent lexical tree search;
$\forall (w_i, w_j)$ ending at t
record word-dependent crossing time $\tau(t; w_i, w_j)$;
record cumulative word score $h(w_j; \tau(t; w_i, w_j), t)$;
Step 2: Output 1-best result;
Step 3: Construct n-best or word-lattice by tracing back the word-pair records (τ and h).

13.3.4. The Forward-Backward Search Algorithm

As described Chapter 12, the ability to predict how well the search fares in the future for the remaining portion of the speech helps to reduce the search effort significantly. The one-pass search strategy, in general, has very little chance of predicting the cost for the portion that it has not seen. This difficulty can be alleviated by multipass search strategies. In successive phases the search should be able to provide good estimates for the remaining paths, since the entire utterance has been examined by the earlier passes. In this section we investigate a special type of multipass search strategy—forward-backward search.

The idea is to first perform a forward search, during which partial forward scores α for each state can be stored. Then perform a second pass search backward—that is, the second pass starts by taking the final frame of speech and searches its way back until it reaches the start of the speech. During the backward search, the partial forward scores α can be used as an accurate estimate of the heuristic function or the fast match score for the remaining path. Even though different KSs might be used in forward and backward phases, this estimate is usually close to perfect, so the search effort for the backward phase can be significantly reduced.

The forward search must be very fast and is generally a time-synchronous Viterbi search. As in the multipass search strategy, simplified acoustic and language models are often used in forward search. For backward search, either time-synchronous search or time-asynchronous A* search can be employed to find the n-best word sequences or word lattice.

13.3.4.1. Forward-Backward Search

Stack decoding, as described in Chapter 12, is based on the admissible A* search, so the first complete hypothesis found with a cost below that of all the hypotheses in the stack is guaranteed to be the best word sequence. It is straightforward to extend stack decoding to produce the n-best hypotheses by continuing to extend the partial hypotheses according to the same A* criterion until n different hypotheses are found. These n different hypotheses are destined to be the n-best hypotheses under a proof similar to that presented in Chapter 12. Therefore, stack decoding is a natural choice for producing the n-best hypotheses.

However, as described in Chapter 12, the difficulty of finding a good heuristic function that can accurately under-estimate the remaining path has limited the use of stack decoding. Fortunately, this difficulty can be alleviated by *tree-trellis forward-backward search* algorithms [41]. First, the search performs a time-synchronous forward search. At each time frame t, it records the score of the final state of each word ending. The set of words whose final states are active (surviving in the beam) at time t is denoted as Δ_t. The score of the final state of each word w in Δ_t is denoted as $\alpha_t(w)$, which represents the sum of the cost of matching the utterance up to time t given the most likely word sequence ending with word w and the cost of the language model score for that word sequence. At the end of the forward search, the best cost is obtained and denoted as α^T.

After the forward pass is completed, the second search is run in reverse (backward), i.e., considering the last frame T as the beginning one and the first frame as the final one. Both the acoustic models and language models need to be reversed. The backward search is based on A* search. At each time frame t, the best path is removed from the stack and a list of possible one-word extensions for that path is generated. Suppose this best path at time t is ph_{w_j}, where w_j is the first word of this partial path (the last expanded during backward A* search). The exit score of path ph_{w_j} at time t, which now corresponds to the score of the initial state of the word HMM w_j, is denoted as $\beta_t(ph_{w_j})$.

Let us now assume we are concerned about the one-word extension of word w_i for path ph_{w_j}. Remember that there are two fundamental issues for the implementation of A* search algorithm—(1) finding an effective and efficient heuristic function for estimating the future remaining input feature stream and (2) finding the best crossing time between w_i and w_j.

The stored forward score α can be used for solving both issues effectively and efficiently. For each time t, the sum $\alpha_t(w_i)+\beta_t(ph_{w_j})$ represents the cost score of the best complete path including word w_i and partial path ph_{w_j}. $\alpha_t(w_i)$ clearly represents a very good heuristic estimate of the remaining path from the start of the utterance until the end of the word w_i, because it is indeed the best score computed in the forward path for the same quantity. Moreover, the optimal crossing time t^* between w_i and w_j can be easily computed by the following equation:

$$t^* = \arg\min_t \left[\alpha_t(w_i) + \beta_t(ph_{w_j})\right] \tag{13.7}$$

Finally, the new path ph', including the one-word (w_i) extension, is inserted into the stack, ordered by the cost score $\alpha_{t'}(w_i)+\beta_{t'}(ph_{w_j})$. The heuristic function (forward scores α) allows the backward A* search to concentrate search on extending only a few truly promising paths.

As a matter of fact, if the same acoustic and language models are used in both the forward and backward search, this heuristic estimate (forward scores α) is indeed a perfect estimate of the best score the extended path will achieve. The first complete hypothesis generated by backward A* search coincides with the best one found in the time-synchronous forward search and is truly the best hypothesis. Subsequent complete hypotheses correspond sequentially to the n-best list, as they are generated in increasing order of cost. Under this condition, the size of the stack in the backward A* search need only be N. Since the estimate of future is exact, the $(N+1)^{\text{th}}$ path in the stack has no chance to become part of the n-best list. Therefore, the backward search is executed very efficiently to obtain the n-best hypotheses without exploring many unpromising branches. Of course, tree-trellis forward-backward search can also be used like most other multipass search strategies—inexpensive KSs are used in the forward search to get an estimate of α, and more expensive KSs are used in the backward A* search to generate the n-best list.

The same idea of using forward score α can be applied to time-synchronous Viterbi search in the backward search instead of backward A* search [7, 34]. For large-vocabulary tasks, the backward search can run 2 to 3 orders of magnitude faster than a normal Viterbi beam search. To obtain the n-best list from time-synchronous forward-backward search, the backward search can also be implemented in a similar way as a time-synchronous word-dependent n-best search.

13.3.4.2. Word-Lattice Generation

The forward-backward n-best search algorithm can be easily modified to generate word lattices instead of n-best lists. A forward time-synchronous Viterbi search is performed first to compute $\alpha_t(\omega)$, the score of each word ω ending at time t. At the end of the search, this best score α^T is also recorded to establish the global pruning threshold. Then, a backward time-synchronous Viterbi search is performed to compute $\beta_t(\omega)$, the score of each word ω beginning at time t. To decide whether to include word juncture $\omega_i-\omega_j$ in the word lattice/graph at time t, we can check whether the forward-backward score is below a global pruning threshold. Specifically, supposed bigram probability $P(\omega_j \mid \omega_i)$ is used, if

$$\alpha_t(\omega_i)+\beta_t(\omega_j)+\left[-\log P(\omega_j \mid \omega_i)\right]<\alpha^T+\theta \tag{13.8}$$

where θ is the pruning threshold, we will include $\omega_i-\omega_j$ in the word lattice/graph at time t. Once word juncture $\omega_i-\omega_j$ is kept, the search continues looking for the next word-pair, where the first word ω_i will be the second word of the next word-pair.

The above formulation is based on the assumption of using the same acoustic and language models in both forward and backward search. If different KSs are used in forward and backward search, the normalized α and β scores should be used instead.

13.3.5. One-Pass vs. Multipass Search

There are several real-time one-pass search engines [4, 5]. Is it necessary to build a multipass search engine based on n-best or word-lattice rescoring? We address this issue by discussing the disadvantages and advantages of multipass search strategies.

One criticism of multipass search strategies is that they are not suitable for real-time applications. No matter how fast the first pass is, the successive (backward) passes cannot start until users finish speaking. Thus, the search results need to be delayed for at least the time required to execute the successive (backward) passes. This is why the successive passes must be extremely fast in order to shorten the delay. Fortunately, it is possible to keep the delays minimum (under one second) with clever implementation of multipass search algorithms, as demonstrated by Nguyen et al. [18].

Another criticism for multipass search strategies is that each pass has the potential to introduce inadmissible pruning, because decisions made in earlier passes are based on simplified models (KSs). Search is a constraint-satisfaction problem. When a pruning decision in each search pass is made on a subset of constraints (KSs), pruning error is inevitable and is unrecoverable by successive passes. However, inadmissible pruning, like beam pruning and fast match, is often necessary to implement one-pass search in order to cope with the large active search space caused jointly by complex KSs and large-vocabulary tasks. Thus, the problem of inadmissibility is actually shared by both real-time one-pass search and multipass search for different reasons. Fortunately, in both cases, search errors can be reduced to a minimum by clever implementation and by empirically designing all the pruning thresholds carefully, as demonstrated in various one-pass and multipass systems [4, 5, 18].

Despite these concerns regarding multipass search strategies, they remain important components in developing spoken language systems. We list here several important aspects:

1. It might be necessary to use multipass search strategies to incorporate very expensive KSs. Higher-order n-gram, long-distance context-dependent models, and natural language parsing are examples that make the search space unmanageable for one-pass search. Multipass search strategies might be compelling even for some small-vocabulary tasks. For example, there are only a couple of million legal credit card numbers for the authentication task of 16-digit credit card numbers. However, it is very expensive to incorporate all the legal numbers explicitly in the recognition grammar. To first reduce search space down to an n-best list or word lattice/graph might be a desirable approach.
2. Multipass search strategies could be very compelling for spoken language understanding systems. It is problematic to incorporate most natural language

understanding technologies in one-pass search. On the other hand, *n*-best lists or word lattices provide a trivial interface between speech recognition and natural language understanding modules. Such an interface also provides a convenient mechanism for integrating different KSs in a modular way. This is important because the KSs could come from different modalities (like video or pen) that make one-pass integration almost infeasible. This high degree of modality allows different component subsystems to be optimized and implemented independently.

3. *N*-best lists or word lattices are very powerful offline tools for developing new algorithms for spoken language systems. It is often a significant task to fully integrate new modeling techniques, such as segment models, into a one-pass search. The complexity could sometimes slow down the progress of the development of such techniques, since recognition experiments are difficult to conduct. Rescoring of *n*-best list and lattice provides a quick and convenient alternative for running recognition experiments. Moreover, the computation and storage complexity can be kept relatively constant for offline *n*-best or word lattice/graph search strategies even when experimenting with highly expensive new modeling techniques. New modeling techniques can be experimented with using *n*-best/word-graph framework first, being integrated into the system only after significant improvement is demonstrated.
4. Besides being an alternative search strategy, *n*-best generation is also essential for discriminant training. Discriminant training techniques, like MMIE, and MCE described in Chapter 4, often need to compute statistics of all possible rival hypotheses. For isolated word recognition using word models, it is easy to enumerate all the word models as the rival hypotheses. However, for continuous speech recognition, one needs to use an all-phone or all-word model to generate all possible phone sequences or all possible word sequences during training. Obviously, that is too expensive. Instead, one can use *n*-best search to find all the near-miss sentence hypotheses that we want to discriminate against [15, 36].

13.4. Search-Algorithm Evaluation

Throughout this chapter we are careful in following dynamic programming principles, using admissible criteria as much as possible. However, many heuristics are still unavoidable to implement large-vocabulary continuous speech recognition in practice. Those nonadmissible heuristics include:

- Viterbi score instead of forward score described in Chapter 12.
- Beam pruning or stack pruning described in Section 13.2.2 and Chapter 12.

- Subtree dominance pruning described in Section 13.1.5.
- Fast match pruning described in Section 13.2.3.
- Rich-get-richer pruning described in Section 13.2.3.2.
- Multipass search strategies described in Section 13.3.5.

Nonadmissible heuristics generate suboptimal searches where the found path is not necessarily the path with the minimum cost. The question is, how different is this suboptimal from the true optimal path? Unfortunately, there is no way to know the optimal path unless an exhaustive search is conducted. The practical question is whether the suboptimal search hurts the search result. In a test condition where the true result is specified, you can easily compare the search result with the true result to find whether any error occurs. Errors could be due to inaccurate models (including acoustic and language models), suboptimal search, or end-point detection. The error caused by a suboptimal search algorithm is referred to as *search error* or *pruning error*.

How can we find out whether the search commits a pruning error? One of the procedures most often used is straightforward. Let $\hat{\mathbf{W}}$ be the recognized word sequence from the recognizer and $\tilde{\mathbf{W}}$ be the true word sequence. We need to compare the cost for these two word sequences:

$$-\log P(\hat{\mathbf{W}} \mid \mathbf{X}) \propto -\log\left[P(\hat{\mathbf{W}})P(\mathbf{X} \mid \hat{\mathbf{W}})\right] \tag{13.9}$$

$$-\log P(\tilde{\mathbf{W}} \mid \mathbf{X}) \propto -\log\left[P(\tilde{\mathbf{W}})P(\mathbf{X} \mid \tilde{\mathbf{W}})\right] \tag{13.10}$$

The quantity in Eq. (13.9) is supposed to be minimum among all possible word sequences if the search is admissible. Thus, if the quantity in Eq. (13.10) is greater than that in Eq. (13.9), the error is not attributed to search pruning. On the other hand, if the quantity in Eq. (13.10) is smaller than that in Eq. (13.9), there is a search error. The rationale behind the procedure described here is obvious. In the case of search errors, the suboptimal search (or nonadmissible pruning) has obviously pruned the correct path, because the cost of the correct path is smaller than the one found by the recognizer. Although we can conclude that search errors are found in this case, it does not guarantee that the search result is correct if the search can be made optimal. The reason is simply that there might be one pruned path with an incorrect word sequence and lower cost under the same suboptimal search. Therefore, the search errors represent only the upper bound that one can improve on if an optimal search is carried out. Nonetheless, finding search errors by comparing quantities in Eqs. (13.9) and (13.10) is a good measure in different search algorithms.

During the development of a speech recognizer, it is a good idea to always include the correct path in the search space. By including such a path, and some bookkeeping, one can use the correct path to help in determining all the pruning thresholds. If the correct path is pruned away during search by some threshold, some adjustment can be made to relax such a threshold to retain the correct path. For example, one can adjust the pruning threshold for fast match if a word in $\tilde{\mathbf{W}}$ fails to appear on the list supplied by the fast match.

13.5. CASE STUDY—MICROSOFT WHISPER

We use the decoder of Microsoft's Whisper [26, 27] discussed in Chapter 9 as a case study for reviewing the search techniques we have presented in this chapter. Whisper can handle both context-free grammars for small-vocabulary tasks and n-gram language models for large-vocabulary tasks. We describe these two different cases.

13.5.1. The CFG Search Architecture

Although context-free grammars (CFGs) have the disadvantage of being too restrictive and unforgiving, particularly with novice users, they are still one of the most popular configurations for building limited-domain applications because of the following advantages:

- Compact representation results in a small memory footprint.
- Efficient operation during decoding in terms of both space and time.
- Ease of grammar creation and modification for new tasks.

As mentioned in Chapter 12, the CFG grammar consists of a set of productions or rules that expand nonterminals into a sequence of terminals and nonterminals. Nonterminals in the grammar tend to refer to high-level task-specific concepts such as dates, font names, and commands. The terminals are words in the vocabulary. A grammar also has a nonterminal designated as its start state. Whisper also allows some regular expression operators on the right-hand side of the production for notational convenience. These operators are: or '|'; repeat zero or more times '*'; repeat one or more times '+'; and optional ([ ]). The following is a simple CFG example for *binary number:*

```
%start BINARY_NUMBER
BINARY_NUMBER: (zero | one)*
```

Without losing generality, Whisper disallows the left recursion for ease of implementation [2]. The grammar is compiled into a binary linked list format. The binary format currently has a direct one-to-one correspondence with the text grammar components, but is more compact. The compiled format is used by the search engine during decoding. The binary representation consists of variable-sized nodes linked together. The grammar format achieves sharing of subgrammars through the use of shared nonterminal definition rules.

The CFG search is conducted according to RTN framework (see Chapter 12). During decoding, the search engine pursues several paths through the CFG at the same time. Associated with each of the paths is a grammar state that describes completely how the path can be extended further. When the decoder hypothesizes the end of the current word of a path, it asks the grammar module to extend the path further by one word. There may be several alternative successor words for the given path. The decoder considers all the successor word possibilities. This may cause the path to be extended to generate several more paths to be considered, each with its own grammar state. Also note that the same word might be under consideration by the decoder in the context of different paths and grammar states at the same time.

The decoder uses beam search to prune unpromising paths with three different beam thresholds. The state pruning threshold τ_s and new phone pruning threshold τ_p work as described in Section 13.2.2. When extending a path, if the score of the extended path does not exceed the threshold τ_h, the path to be extended is put into a pool. At each frame, for each word in the vocabulary, a winning path that extends to that word is picked from the pool, based on the score. All the remaining paths in the pool are pruned. This level of pruning gives us finer control in the creation of new paths that have potential to grow exponentially.

When two paths representing different word sequences thus far reach the end of the current word with the same grammar state at the same time, only the better path of the two is allowed to continue on. This optimization is safe, except that it does not take into account the effect of different interword left acoustic contexts on the scores of the new word that is started.

Besides beam pruning, the RGR strategy, described in Section 13.2.3.2, is used to avoid unnecessary senone computation. The basic idea is to use the linear combination of context-independent senone score and context-independent look-ahead score to determine whether the context-dependent senone evaluation is worthwhile to pursue.

All of these pruning techniques enable Whisper to perform typical 100- to 200-word CFG tasks in real time running on a 486 PC with 2 MB RAM. Readers might think it is not critical to make CFG search efficient on such a low-end platform.[9] However, it is indeed important to keep the CFG engine fast and lean. The speech recognition engine is eventually only part of an integrated application. The application will benefit if the resources (both CPU and memory) used by the speech decoder are kept as small as possible, so there are more resources left for the application module to use. Moreover, in recognition server applications, several channels of speech recognition can be performed on a single server platform if each speech recognition engine takes only a small portion of the total resources.

13.5.2. The *N*-gram Search Architecture

The CFG decoder is best suited for limited domain command and control applications. For dictation or natural conversational systems, a stochastic grammar such as *n*-grams provides a more natural choice. Using bigrams or trigrams leads to a large number of states to be considered by the search process, requiring an alternative search architecture.

---

[9] Thanks to the progress predicted by Moore's law, the current mainstream PC configuration is an order of magnitude more powerful than the configuration we list here (486 PC with 2 MB RAM) in both speed and memory.

Whisper's n-gram search architecture is based on lexical tree search as described in Section 13.1. To keep the runtime memory[10] as small as possible, Whisper does not need to allocate the entire lexical tree network statically. Instead, it dynamically builds only the portion that needs to be active. To cope with the problem of delayed application of language model scores, Whisper uses the factorization algorithm described in Section 13.1.3 to distribute the language model probabilities through the tree branches. To reduce the memory overhead of the factored language model probabilities, an efficient data structure is used for representing the lexical tree as described in Section 13.1.3.1. This data structure allows Whisper to encode factored language model probabilities in no more than the space required for the original n-gram probabilities. Thus, there is absolutely no storage overhead for using factored lexical trees.

The basic acoustic subword model in Whisper is a context-dependent senone. It also incorporates inter-word triphone models in the lexical tree search as described in Section 13.1.6. Table 13.5 shows the distribution of phoneme arcs for 20,000-word WSJ lexical tree using senones as acoustic models. Context-dependent units certainly prohibit more prefix sharing when compared with Table 13.1. However, the overall arcs in the lexical tree still represent quite a saving when compared with a linear lexicon with about 140,000 phoneme arcs. Most importantly, similar to the case in Table 13.1, most sharing is realized in the beginning prefixes where most computation resides. Moreover, with the help of context-dependent and interword senone models, the search is able to use more reliable knowledge to perform efficient pruning. Therefore, lexical tree with context-dependent models can still enjoy all the benefits associated with lexical tree search.

The search organization is evaluated on the 1992 development test set for the *Wall Street Journal* corpus with a back-off trigram language model. The trigram language model has on the order of 10^7 linguistic equivalent classes, but the number of classes generated is far fewer due to the constraints provided by the acoustic model. Figure 13.13(a) illustrates that the relative effort devoted to the trigram, bigram, and unigram is constant regardless of total search effort, across a set of test utterances. This is because the ratio of states in the language model is constant. The language model is using ~2×10^6 trigrams, ~2×10^6 bigrams, and 6×10^4 unigrams. Figure 13.13(b) illustrates different relative order when word hypotheses are considered. The most common context for word hypotheses is the unigram context, followed by the bigram and trigram contexts. The reason for the reversal from the state-level transitions is the partially overlapping evaluations required by each bigram context. The trigram context is more common than the bigram context for utterances that generate few hypotheses overall. This is likely because the language model models those utterances well.

[10] Here the runtime memory means the virtual memory for the decoder that is the entire image of the decoder.

Table 13.5 Configuration of the first seven levels of the 20,000-word WSJ (*Wall Street Journal*) tree; the large initial fan-out is due to the use of context-dependent acoustic models [4].

Tree Level	Number of Nodes	Fan-Out
1	655	655.0
2	3174	4.85
3	9388	2.96
4	13,703	1.46
5	14,918	1.09
6	13,907	0.93
7	11,389	0.82

To improve efficiency in dealing with tree copies due to the use of higher-order n-gram, one needs to reduce redundant computations in subtrees that are not explicitly part of the given linguistic context. One solution is to use successor trees to include only nonzero successors, as described in Section 13.1.2. Since Whisper builds the search space dynamically, it is not effective for Whisper to use the optimization techniques of the successor-tree network, such as FSN optimization, subtree isomorphism, and sharing tail optimization. Instead, Whisper uses polymorphic linguistic context assignment to reduce redundancy, as described in Section 13.1.5. This involves keeping a single copy of the lexical tree, so that each node in the tree is evaluated at most once. To avoid early inadmissible pruning of different linguistic contexts, an ε-heap of storing paths of different linguistic contexts is created for each node in the tree. The operation of such ε-heaps is in accordance with Algorithm 13.2. The depth of each heap varies dynamically according to a changing threshold that allows more contexts to be retained for promising nodes.

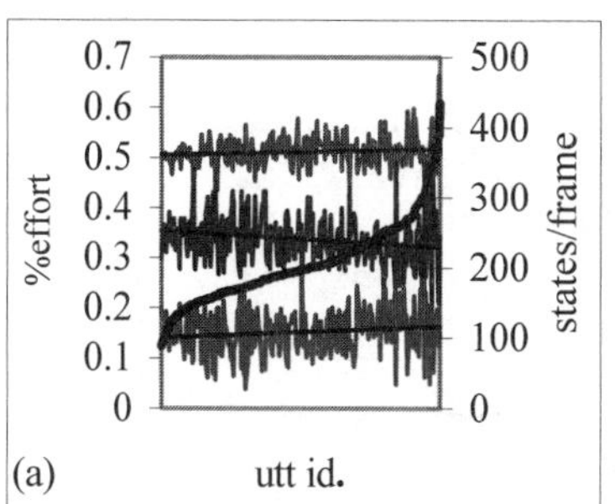

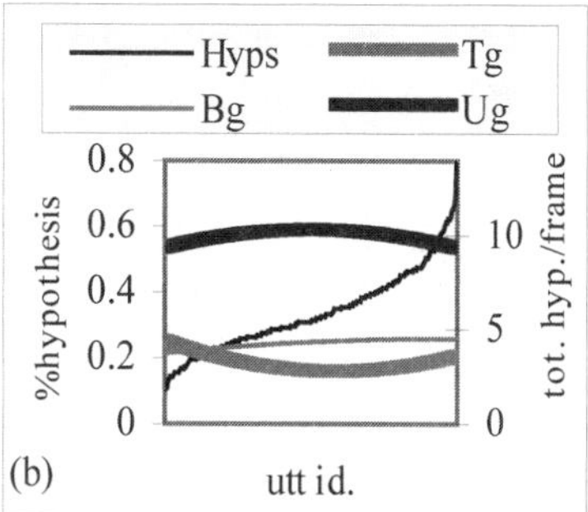

Figure 13.13 (a) Search effort for different linguistic contexts measured by number of active states in each of the three different linguistic contexts. The top series is for the bigram, then the unigram and trigram. The remaining series is the effort per utterance and is plotted on the secondary y-axis. (b) The distribution of word hypotheses with respect to their context. The top line is the unigram context, then the bigram and trigram. The remaining series is the average number of hypotheses per frame for each utterance and is plotted on the secondary y-axis [3].

Table 13.6 illustrates how the depth of the ε-heap, the active states per frame of speech, word error rate, and search time change when the value of threshold ε increases for the 20,000-word WSJ dictation task. As we can see from the table, the average heap size for active nodes is only about 1.6 for the most accurate configuration. Figure 13.14(a) illustrates the distribution of stack depths for a large data set, showing that the stack depth is small even for tree initial nodes. Figure 13.14(b) illustrates the profile of the average stack depth for a sample utterance, showing that the average stack depth remains small across an utterance.

Whisper also employs look-ahead techniques to further reduce the search effort. The acoustic look-ahead technique described in Section 13.2.3.1 attempts to estimate the probability that a phonetic HMM will participate in the final result [3]. Whisper implements acoustic look-ahead by running a CI phone-net synchronously with the search process but offset N frames in the future. One side effect of the acoustic look-ahead is to provide information for the RGR strategy, as described in Section 13.2.3.2, so the search can avoid unnecessary Gaussian computation. Figure 13.15 demonstrates the effectiveness of varying the frame look-ahead from 0 to N frames in terms of states evaluated.

When the look-ahead is increased from 0 to 3 frames, the search effort, in terms of real time, is reduced by ~40% with no loss in accuracy; however, most of that is due to reducing the number of states evaluated per frame. There is no effect on the number of Gaussians evaluated per frame (the system using continuous density) until we begin to negatively impact error rate, indicating that the acoustic space represented by the pruned states is redundant and adequately covered by the retained states prior to the introduction of search errors.

With the techniques discussed here, Whisper is able to achieved real-time performance for the continuous WSJ dictation task (60,000-word) on Pentium-class PCs. The recognition accuracy is identical to that of a standard Viterbi beam decoder with a linear lexicon.

Table 13.6 Effect of heap threshold on contexts/node, states/frame-of-speech (fos), word error rate, and search time [4].

ε	Context / node	states / fos	%error	search time
0	1.000	8805	16.4	1.0x
1.0	1.001	8808	15.5	1.0x
2.0	1.008	8898	14.4	1.0x
3.0	1.018	9252	12.4	1.07x
4.0	1.056	10224	10.5	1.16x
5.0	1.147	11832	10.3	1.36x
6.0	1.315	13749	10.0	1.60x
7.0	1.528	15342	9.9	1.81x
8.0	1.647	15984	9.9	1.86x

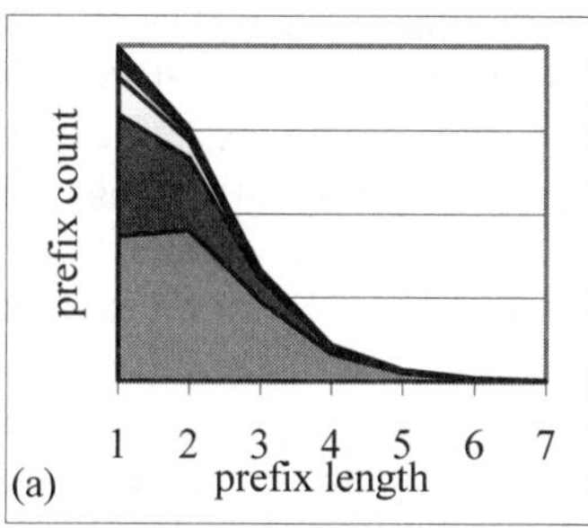

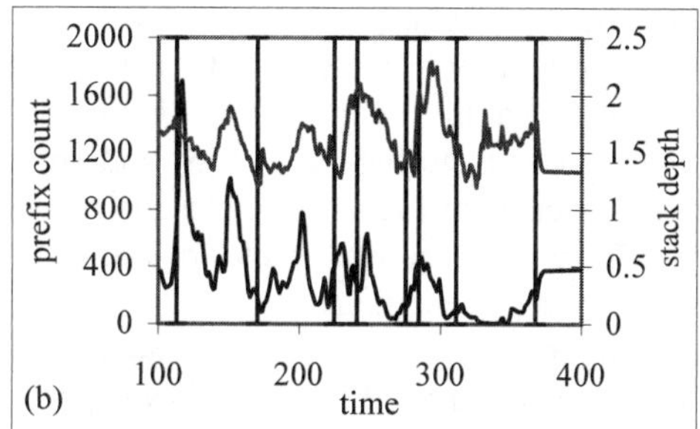

Figure 13.14 (a) A cumulative graph of the prefix count for each stack depth, starting with depth 1, showing the distribution according to prefix length. (b) The prefix count and the average stack depth with respect to one utterance. The vertical bars show the word boundaries [3].

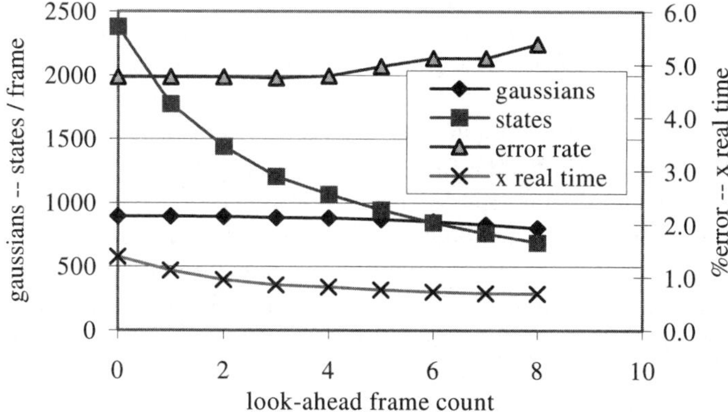

Figure 13.15 Search effort, percent error rate, and real-time factor as a function of the acoustic look-ahead. Note that search effort is the number of Gaussians evaluated per frame and the number of states evaluated per frame [3].

13.6. Historical Perspective and Further Reading

Large-vocabulary continuous speech recognition is a computationally intensive task. Real-time systems started to emerge in the late 1980s. Before that, most systems achieved real-time performance with the help of special hardware [11, 16, 25, 28]. Thanks to Moore's law and various efficient search techniques, real-time systems became a reality on a single-chip general-purpose personal computer in the 1990s [4, 34, 43].

Common wisdom in 1980s saw stack decoding as more efficient for large-vocabulary continuous speech recognition with higher-order *n*-grams. Time-synchronous Viterbi beam search, as described in Sections 13.1 and 13.2, emerged as the most efficient search framework. It has become the most widely used search technique today. The lexical tree representation was first used by IBM as part of its allophonic fast match system [10]. Ney proposed the use of the lexical tree as the primary representation for the search space [32]. The ideas of language model factoring [4, 19] [5] and subtree polymorphism [4] enabled real-time single-pass search with higher-order language models (bigrams and trigrams). Alleva [3] and Ney [33] are two excellent articles regarding the detailed Viterbi beam search algorithm with lexical tree representation.

As mentioned in Chapter 12, fast match was first invented to speed up stack decoding [8, 9]. Ney and Ortmanns [33] and Alleva [3] extended the fast match idea to phone look-ahead in time-synchronous search by using context-independent model evaluation. In Haeb-Umbach et al. [22], a word look-ahead is implemented for a 12.3k-word speaker-dependent continuous speech recognition task. The look-ahead is performed on a lexical tree, with beam search executed every other frame. The results show a factor of 3–5 times of reduction for search space compared to the standard Viterbi beam search, while only 1–2% extra errors are introduced by word look-ahead.

The idea of multipass search strategy has long existed for knowledge-based speech recognition systems [17], where first a phone recognizer is performed, then a lexicon hypothesizer is used to locate all the possible words to form a word lattice, and finally a language model is used to search for the most possible word sequence. However, HMM's popularity predominantly shifted the focus to the unified search approach to achieve global optimization. Computation concerns led many researchers to revisit the multipass search strategy. The first *n*-best algorithm, described in Section 13.3.2, was published by researchers at BBN [39]. Since then, *n*-best and word-lattice based multipass search strategies have become important search frameworks for rapid system deployment, research tools, and spoken language understanding systems. Schwartz et al.'s paper [40] is a good tutorial on the *n*-best or word-lattice generation algorithms. Most of the *n*-best search algorithms can be made to generate word lattices/graphs with minor modifications. Other excellent discussions of multipass search can be found in [14, 24, 30].

REFERENCES

[1] Aho, A., J. Hopcroft, and J. Ullman, *The Design and Analysis of Computer Algorithms*, 1974, Addison-Wesley Publishing Company.

[2] Aho, A.V., R. Sethi, and J.D. Ullman, *Compilers: Principles, Techniques, and Tools*, 1985, Addison-Wesley.

[3] Alleva, F., "Search Organization in the Whisper Continuous Speech Recognition System," *IEEE Workshop on Automatic Speech Recognition*, 1997.

[4] Alleva, F., X. Huang, and M.Y. Hwang, "Improvements on the Pronunciation Prefix Tree Search Organization," *Proc. of the IEEE Int. Conf. on Acoustics, Speech and Signal Processing*, 1996, Atlanta, Georgia, pp. 133-136.

[5] Aubert, X., *et al.*, "Large Vocabulary Continuous Speech Recognition of *Wall Street Journal* Corpus," *Proc. of the IEEE Int. Conf. on Acoustics, Speech and Signal Processing*, 1994, Adelaide, Australia, pp. 129-132.

[6] Aubert, X. and H. Ney, "Large Vocabulary Continuous Speech Recognition Using Word Graphs," *Proc. of the IEEE Int. Conf. on Acoustics, Speech and Signal Processing*, 1995, Detroit, MI, pp. 49-52.

[7] Austin, S., R. Schwartz, and P. Placeway, "The Forward-Backward Search Algorithm for Real-Time Speech Recognition," *Proc. of the IEEE Int. Conf. on Acoustics, Speech and Signal Processing*, 1991, Toronto, Canada, pp. 697-700.

[8] Bahl, L.R., *et al.*, "Obtaining Candidate Words by Polling in a Large Vocabulary Speech Recognition System," *Proc. of the IEEE Int. Conf. on Acoustics, Speech and Signal Processing*, 1988, pp. 489-492.

[9] Bahl, L.R., *et al.*, "Matrix Fast Match: a Fast Method for Identifying a Short List of Candidate Words for Decoding," *Proc. of the IEEE Int. Conf. on Acoustics, Speech and Signal Processing*, 1989, Glasgow, Scotland, pp. 345-347.

[10] Bahl, L.R., P.S. Gopalakrishnan, and R.L. Mercer, "Search Issues in Large Vocabulary Speech Recognition," *Proc. of the 1993 IEEE Workshop on Automatic Speech Recognition*, 1993, Snowbird, UT.

[11] Bisiani, R., T. Anantharaman, and L. Butcher, "BEAM: An Accelerator for Speech Recognition," *Int. Conf. on Acoustics, Speech and Signal Processing*, 1989, pp. 782-784.

[12] Brugnara, F. and M. Cettolo, "Improvements in Tree-Based Language Model Representation," *Proc. of the European Conf. on Speech Communication and Technology*, 1995, Madrid, Spain, pp. 1797-1800.

[13] Cettolo, M., R. Gretter, and R.D. Mori, "Knowledge Integration" in *Spoken Dialogues with Computers*, R.D. Mori, ed., Academic Press, 1998, London, pp. 231-256.

[14] Cettolo, M., R. Gretter, and R.D. Mori, "Search and Generation of Word Hypotheses" in *Spoken Dialogues with Computers*, R.D. Mori, ed., 1998, London, Academic Press, pp. 257-310.

[15] Chou, W., C.H. Lee, and B.H. Juang, "Minimum Error Rate Training Based on N-best String Models," *IEEE Int. Conf. on Acoustics, Speech and Signal Processing*, 1993, Minneapolis, MN, pp. 652-655.

[16] Chow, Y.L., *et al.*, "BYBLOS: The BBN Continuous Speech Recognition System," *Proc. of the IEEE Int. Conf. on Acoustics, Speech and Signal Processing*, 1987, pp. 89-92.

[17] Cole, R.A., *et al.*, "Feature-Based Speaker Independent Recognition of English Letters," *Int. Conf. on Acoustics, Speech and Signal Processing*, 1983, pp. 731-734.

[18] Davenport, J.C., R. Schwartz, and L. Nguyen, "Towards A Robust Real-Time Decoder," *IEEE Int. Conf. on Acoustics, Speech and Signal Processing*, 1999, Phoenix, Arizona, pp. 645-648.

[19] Federico, M., *et al.*, "Language Modeling for Efficient Beam-Search," *Computer Speech and Language*, 1995, pp. 353-379.

[20] Gauvain, J.L., L. Lamel, and M. Adda-Decker, "Developments in Continuous Speech Dictation using the ARPA WSJ Task," *Proc. of the IEEE Int. Conf. on Acoustics, Speech and Signal Processing*, 1995, Detroit, MI, pp. 65-68.

[21] Gillick, L.S. and R. Roth, "A Rapid Match Algorithm for Continuous Speech Recognition," *Proc. of the Speech and Natural Language Workshop*, 1990, Hidden Valley, PA, pp. 170-172.

[22] Haeb-Umbach, R. and H. Ney, "A Look-Ahead Search Technique for Large Vocabulary Continuous Speech Recognition," *Proc. of the European Conf. on Speech Communication and Technology*, 1991, Genova, Italy, pp. 495-498.

[23] Haeb-Umbach, R. and H. Ney, "Improvements in Time-Synchronous Beam-Search for 10000-Word Continuous Speech Recognition," *IEEE Trans. on Speech and Audio Processing*, 1994, **2**(4), pp. 353-365.

[24] Hetherington, I.L., *et al.*, "A* Word Network Search for Continuous Speech Recognition," *Proc. of the European Conf. on Speech Communication and Technology*, 1993, Berlin, Germany, pp. 1533-1536.

[25] Hon, H.W., *A Survey of Hardware Architectures Designed for Speech Recognition*, 1991, Carnegie Mellon University, Pittsburgh, PA.

[26] Huang, X., *et al.*, "From Sphinx II to Whisper: Making Speech Recognition Usable," in *Automatic Speech and Speaker Recognition*, C.H. Lee, F.K. Soong, and K.K. Paliwal, eds. 1996, Norwell, MA, Kluwer Academic Publishers, pp. 481-508.

[27] Huang, X., *et al.*, "Microsoft Windows Highly Intelligent Speech Recognizer: Whisper," *IEEE Int. Conf. on Acoustics, Speech and Signal Processing*, 1995, pp. 93-96.

[28] Jelinek, F., "The Development of an Experimental Discrete Dictation Recognizer," *Proc. of the IEEE*, 1985, **73**(1), pp. 1616-1624.

[29] Marino, J. and E. Monte, "Generation of Multiple Hypothesis in Connected Phonetic-Unit Recognition by a Modified One-Stage Dynamic Programming Algorithm," *Proc. of EuroSpeech*, 1989, Paris, pp. 408-411.

[30] Murveit, H., *et al.*, "Large Vocabulary Dictation Using SRI's DECIPHER Speech Recognition System: Progressive Search Techniques," *Proc. of the IEEE Int. Conf. on Acoustics, Speech and Signal Processing*, 1993, Minneapolis, MN, pp. 319-322.

[31] Ney, H. and X. Aubert, "A Word Graph Algorithm for Large Vocabulary," *Proc. of the Int. Conf. on Spoken Language Processing*, 1994, Yokohama, Japan, pp. 1355-1358.

[32] Ney, H., *et al.*, "Improvements in Beam Search for 10000-Word Continuous Speech Recognition," *Proc. of the IEEE Int. Conf. on Acoustics, Speech and Signal Processing*, 1992, San Francisco, California, pp. 9-12.

[33] Ney, H. and S. Ortmanns, *Dynamic Programming Search for Continuous Speech Recognition*, in *IEEE Signal Processing Magazine*, 1999, pp. 64-83.

[34] Nguyen, L., *et al.*, "Search Algorithms for Software-Only Real-Time Recognition with Very Large Vocabularies," *Proc. of ARPA Human Language Technology Workshop*, 1993, Plainsboro, NJ, pp. 91-95.

[35] Nilsson, N.J., *Problem-Solving Methods in Artificial Intelligence*, 1971, New York, McGraw-Hill.

[36] Normandin, Y., "Maximum Mutual Information Estimation of Hidden Markov Models" in *Automatic Speech and Speaker Recognition*, C.H. Lee, F.K. Soong, and K.K. Paliwal, eds. 1996, Norwell, MA, Kluwer Academic Publishers.

[37] Odell, J.J., *et al.*, "A One Pass Decoder Design for Large Vocabulary Recognition," *Proc. of the ARPA Human Language Technology Workshop*, 1994, Plainsboro, NJ, pp. 380-385.

[38] Schwartz, R. and S. Austin, "A Comparison of Several Approximate Algorithms for Finding Multiple (N-BEST) Sentence Hypotheses," *Proc. of the IEEE Int. Conf. on Acoustics, Speech and Signal Processing*, 1991, Toronto, Canada, pp. 701-704.

[39] Schwartz, R. and Y.L. Chow, "The N-Best Algorithm: an Efficient and Exact Procedure for Finding the N Most Likely Sentence Hypotheses," *Proc. of the IEEE Int. Conf. on Acoustics, Speech and Signal Processing*, 1990, Albuquerque, New Mexico, pp. 81-84.

[40] Schwartz, R., L. Nguyen, and J. Makhoul, "Multiple-Pass Search Strategies" in *Automatic Speech and Speaker Recognition*, C.H. Lee, F.K. Soong, and K.K. Paliwal, eds., 1996, Norwell, MA, Klewer Academic Publishers, pp. 57-81.

[41] Soong, F.K. and E.F. Huang, "A Tree-Trellis Based Fast Search for Finding the N Best Sentence Hypotheses in Continuous Speech Recognition," *Proc. of the IEEE Int. Conf. on Acoustics, Speech and Signal Processing*, 1991, Toronto, Canada, pp. 705-708.

[42] Steinbiss, V., "Sentence-Hypotheses Generation in a Continuous Speech Recognition," *Proc. of EuroSpeech*, 1989, Paris, pp. 51-54.

[43] Steinbiss, V., *et al.*, "The Philips Research System for Large-Vocabulary Continuous-Speech Recognition," *Proc. of the European Conf. on Speech Communication and Technology*, 1993, Berlin, Germany, pp. 2125-2128.

[44] Woodland, P.C., *et al.*, "Large Vocabulary Continuous Speech Recognition Using HTK," *Proc. of the IEEE Int. Conf. on Acoustics, Speech and Signal Processing*, 1994, Adelaide, Australia, pp. 125-128.

PART IV

TEXT-TO-SPEECH SYSTEMS

CHAPTER 14

Text and Phonetic Analysis

Text-to-speech can be viewed as a speech coding system that yields an extremely high compression ratio coupled with a high degree of flexibility in choosing style, voice, rate, pitch range, and other playback effects. In this view of TTS, the text file that is input to a speech synthesizer is a form of coded speech. Thus, TTS subsumes coding technologies discussed in Chapter 7 with the following goals:

- *Compression ratios superior to digitized wave files*—Compression yields benefits in many areas, including fast Internet transmission of spoken messages.
- *Flexibility in output characteristics*—Flexibility includes easy change of gender, average pitch, pitch range, etc., enabling application developers to give their systems' spoken output a unique individual personality. Flexibility also implies easy change of message content; it is generally easier to retype text than it is to record and deploy a digitized speech file.
- *Ability for perfect indexing between text and speech forms*—Preservation of the correspondence between textual representation and the speech wave form allows synchronization with other media and output modes, such as word-by-word reverse video highlighting in a literacy tutor reading aloud.

- *Alternative access of text content*—TTS is the most effective alternative access of text for the blind, hands-free/eyes-free and displayless scenarios.

At first sight, the process of converting text into speech looks straightforward. However, when we analyze how complicated speakers read a text aloud, this simplistic view quickly falls apart. First, we need to convert words in written forms into speakable forms. This process is clearly nontrivial. Second, to sound natural, the system needs to convey the intonation of the sentences properly. This second process is clearly an extremely challenging one. One good analogy is to think how difficult it is to drop a foreign accent when speaking a second language—a process still not quite understood by human beings.

The ultimate goal of simulating the speech of an understanding, effective human speaker from plain text is as distant today as the corresponding Holy Grail goals of the fields of speech recognition and machine translation. This is because such humanlike rendition depends on common-sense reasoning about the world and the text's relation to it, deep knowledge of the language itself in all its richness and variability, and even knowledge of the actual or expected audience—its goals, assumptions, presuppositions, and so on. In typical audio books or recordings for the visually challenged today, the human reader has enough familiarity with and understanding of the text to make appropriate choices for rendition of emotion, emphasis, and pacing, as well as handling both dialog and exposition. While computational power is steadily increasing, there remains a substantial knowledge gap that must be closed before fully human-sounding simulated voices and renditions can be created.

While no TTS system to date has approached optimal quality in the Turing test,[1] a large number of experimental and commercial systems have yielded fascinating insights. Even the relatively limited-quality TTS systems of today have found practical applications.

The basic TTS system architecture is illustrated in Chapter 1. In the present chapter we discuss text analysis and phonetic analysis whose objective is to convert words into speakable phonetic representation. The techniques discussed here are relevant to what we discussed for language modeling in Chapter 11 (like text normalization before computing *n*-gram) and for pronunciation modeling in Chapter 9. The next two modules—prosodic analysis and speech synthesis—are treated in the next two chapters.

14.1. Modules and Data Flow

The text analysis component, guided by presenter controls, is typically responsible for determining document structure, conversion of nonorthographic symbols, and parsing of language structure and meaning. The phonetic analysis component converts orthographic words to phones (unambiguous speech sound symbols). Some TTS systems assume dependency between text analysis, phonetic analysis, prosodic analysis, and speech synthesis, particularly systems based on very large databases containing long stretches of original, unmodified

[1] A test proposed by British mathematician Allan Turing of the ability of a computer to flawlessly imitate human performance on a given speech or language task [29].

digitized speech with their original pitch contours. We discuss our high-level linguistic description of those modules, based on modularity, transparency, and reusability of components, although some aspects of text and phonetic analysis may be unnecessary for some particular systems.

We assume that the entire text (word, sentence, paragraph, document) to be spoken is contained in a single, wholly visible buffer. Some systems may be faced with special requirements for continuous flow-through or visibility of only small (word, phrase, sentence) *chunks* at a time, or extremely complex timing and synchronization requirements. The basic functional processes within the text and phonetic analysis are shown schematically in Figure 14.1.

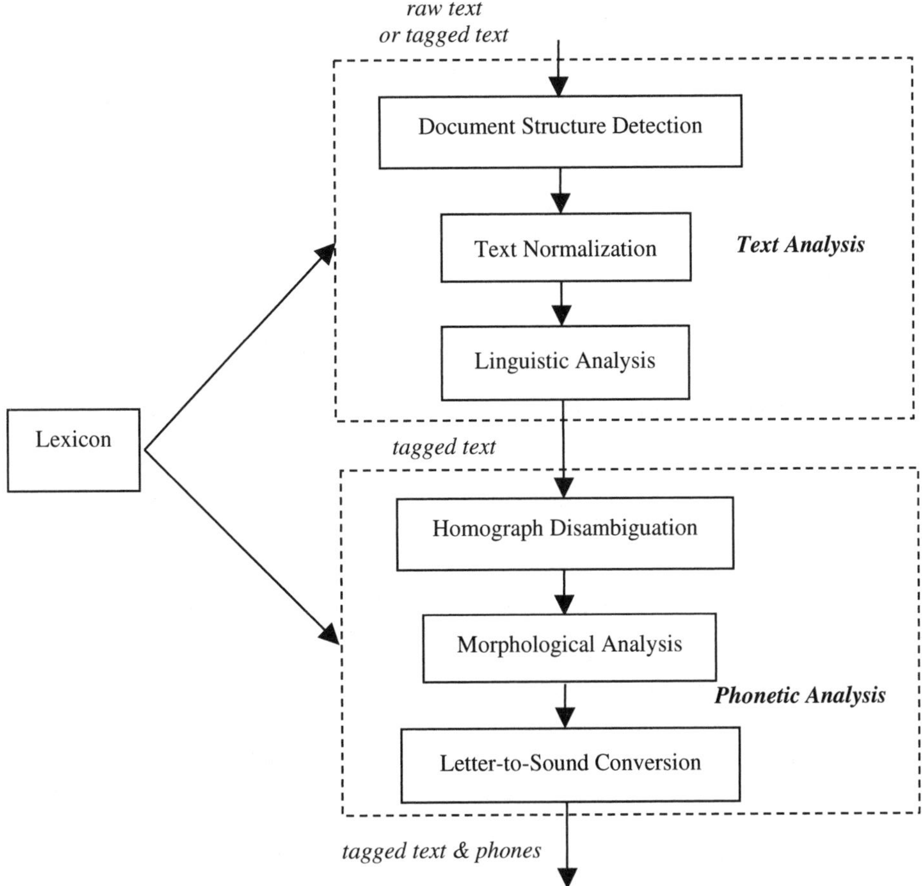

Figure 14.1 Modularized functional blocks for text and phonetic analysis components.

The architecture in Figure 14.1 brings the standard benefits of modularity and transparency. Modularity in this case means that the analysis at each level can be supplied by the most expert knowledge source, or a variety of different sources, as long as the markup conventions for expressing the analysis are uniform. Transparency means that the results of each stage could be reused by other processes for other purposes.

14.1.1. Modules

The *text analysis module* (TAM) is responsible for indicating all knowledge about the text or message that is not specifically phonetic or prosodic in nature. Very simple systems do little more than convert nonorthographic items, such as numbers, into words. More ambitious systems attempt to analyze whitespaces and punctuations to determine document structure, and perform sophisticated syntax and semantic analysis on sentences to determine attributes that help the phonetic analysis to generate correct phonetic representation and prosodic generation to construct superior pitch contours. As shown in Figure 14.1, text analysis for TTS involves three related processes:

- *Document structure detection*—Document structure is important to provide a context for all later processes. In addition, some elements of document structure, such as sentence breaking and paragraph segmentation, may have direct implications for prosody.
- *Text normalization*—Text normalization is the conversion from the variety of symbols, numbers, and other nonorthographic entities of text into a common orthographic transcription suitable for subsequent phonetic conversion.
- *Linguistic analysis*—Linguistic analysis recovers the syntactic constituency and semantic features of words, phrases, clauses, and sentences, which is important for both pronunciation and prosodic choices in the successive processes.

The task of the phonetic analysis is to convert lexical orthographic symbols to phonemic representation along with possible diacritic information, such as stress placement. Phonetic analysis is thus often referred to as grapheme-to-phoneme conversion. The purpose is obvious, since phonemes are the basic units of sound, as described in Chapter 2. Even though future TTS systems might be based on word sounding units with increasing storage technologies, homograph disambiguation and phonetic analysis for new words (either true new words being invented over time or morphologically transformed words) are still necessary for systems to correctly utter every word.

Grapheme-to-phoneme conversion is trivial for languages where there is a simple relationship between orthography and phonology. Such a simple relationship can be well captured by a handful of rules. Languages such as Spanish and Finnish belong to this category and are referred to as *phonetic languages*. English, on the other hand, is remote from phonetic language because English words often have many distinct origins. It is generally believed that the following three services are necessary to produce accurate pronunciations.

- *Homograph disambiguation*—It is important to disambiguate words with different senses to determine proper phonetic pronunciations, such as object (*/ah b jh eh k t/*) as a verb or as a noun (*/aa b jh eh k t/*).
- *Morphological analysis*—Analyzing the component morphemes provides important cues to attain the pronunciations for inflectional and derivational words.
- *Letter-to-sound conversion*—The last stage of the phonetic analysis generally includes general letter-to-sound rules (or modules) and a dictionary lookup to produce accurate pronunciations for any arbitrary word.

All the processes in text and phonetic analysis phases above need not to be deterministic, although most TTS systems today have deterministic processes. What we mean by *not deterministic* is that each of the above processes can generate multiple hypotheses with the hope that the later process can disambiguate those hypotheses by using more knowledge. For example, often it might not be trivial to decide whether the punctuation "." is a sentence ending mark or abbreviation mark during document structure detection. The document structure detection process can pass both hypotheses to the later processes, and the decision can then be delayed until there is enough information to make an informed decision in later modules, such as the text normalization or linguistic analysis phases. When generating multiple hypotheses, the process can also assign probabilistic information if it comprehends the underlying probabilistic structure. This flexible pipeline architecture avoids the mistakes made by early processes based on insufficient knowledge.

Much of the work done by the text/phonetic analysis phase of a TTS system mirrors the processing attempted by *natural language process* (NLP) systems for other purposes, such as automatic proofreading, machine translation, database document indexing, and so on. Increasingly sophisticated NL analysis is needed to make certain TTS processing decisions in the examples illustrated in Table 14.1. Ultimately all decisions are context driven and probabilistic in nature, since, for example, dogs might be cooked and eaten in some cultures.

Table 14.1 Examples of several ambiguous text normalization cases.

Examples	Alternatives	Techniques
Dr. Smith	*doctor* or *drive*?	abbreviation analysis, case analysis
Will you go?	yes-no or wh-question?	syntactic analysis
I ate a hot dog.	accent on *dog*?	semantic, verb/direct object likelihood
I saw a hot dog.	accent on *dog*?	discourse, pragmatic analysis

Most TTS systems today employ specialized natural language processing modules for front-end analysis. In the future, it is likely that less emphasis will be placed on construction of TTS-specific text/phonetic analysis components such as those described in [27], while more resources will likely go to general-purpose NLP systems with cross-functional potential [23]. In other words, all the modules above only perform simple processing and pass all possible hypotheses to the later modules. At the end of the text/phonetic phase, a unified NLP module then performs extensive syntactic/semantic analysis for the best decisions. The necessity for such an architectural approach is already visible in markets where language issues have forced early attention to common lexical and tokenization resources, such as Japan. Japanese system services and applications can usually expect to rely on common cross-functional linguistic resources, and many benefits are reaped, including elimination of bulk, reduction of redundancy and development time, and enforcement of systemwide consistent behavior. For example, under Japanese architectures, TTS, recognition, sorting, word processing, database, and other systems are expected to share a common language and dictionary service.

14.1.2. Data Flows

It is arguable that text input alone does not give the system enough information to express and render the intention of the text producer. Thus, more and more TTS systems focus on providing an infrastructure of standard set of markups (tags), so that the text producer can better express their semantic intention with these markups in addition to plain text. These kinds of markups have different levels of granularity, ranging from simple speed settings specified in *words per minute* up to elaborate schemes for semantic representation of concepts that may bypass the ordinary text analysis module altogether.[2] The markup can be done by internal proprietary conventions or by some standard markup, such as XML (Extensible Markup Language [35]). Some of these markup capabilities will be discussed in Sections 14.3 and 14.4.

For example, an application may know a lot about the structure and content of the text to be spoken, and it can apply this knowledge to the text, using common markup conventions, to greatly improve spoken output quality. On the other hand, some applications may have certain broad requirements such as rate, pitch, callback types, etc. For engines providing such supports, the text and/or phonetic analysis phase can be skipped, in whole or in part. Whether the application or the system has provided the text analysis markup, the structural conventions should be identical and must be sufficient to guide the phonetic analysis. The phonetic analysis module should be presented only with markup tags indicating structure or functions of textual chunks, and words in standard orthography. The similar phonetic markups could also be presented to the phonetic analysis module, the module could be skipped.

[2] This latter type of system is sometimes called *concept-to-speech* or *message-to-speech*, which is described in Chapter 17. It generally generates better speech rendering when domain-specific knowledge is provided to the system.

Internal architectures, data structures, and interfaces may vary widely from system to system. However, most modern TTS systems initially construct a simple description of an utterance or paragraph based on observable attributes, typically text words and punctuation, perhaps augmented by control annotations. This minimal initial *skeleton* is then augmented with many layers of structure hypothesized by the TTS system's internal analysis modules. Beginning with a surface stream of words, punctuation, and other symbols, typical layers of detected structure that may be added include:

- Phonemes
- Syllables
- Morphemes
- Words derived from nonwords (such as dates like "9/10/99")
- Syntactic constituents
- Relative importance of words and phrases
- Prosodic phrasing
- Accentuation
- Duration controls
- Pitch controls

We can now consider how the information needed to support synthesis of a sentence is developed in processing an example sentence such as: "A skilled electrician reported."

In Figure 14.2, the information that must be inferred from text is diagrammed. The flow proceeds as follows:

- **W(ords) ➔ Σ, C(ontrols)**: the syllabic structure (Σ) and the basic phonemic form of a word are derived from lexical lookup and/or the application of rules. The Σ tier shows the syllable divisions (written in text form for convenience). The C tier, at this stage, shows the basic phonemic symbols for each word's syllables.
- **W(ords) ➔ S(yntax/semantics)**: The word stream from text is used to infer a syntactic and possibly semantic structure (S tier) for an input sentence. Syntactic and semantic structure above the word would include syntactic constituents such as Noun Phrase (NP), Verb Phrase (VP), etc. and any semantic features that can be recovered from the current sentence or analysis of other contexts that may be available (such as an entire paragraph or document). The lower-level phrases such as NP and VP may be grouped into broader constituents such as Sentence (S), depending on the parsing architecture.
- **S(yntax/semantics) ➔ P(rosody)**: The P(rosodic) tier is also called the symbolic prosodic module. If a word is semantically *important* in a sentence, that importance can be reflected in speech with a little extra phonetic prominence, called an accent. Some synthesizers begin building a prosodic structure by

placing metrical foot boundaries to the left of every accented syllable. The resulting metrical foot structure is shown as F1, F2, etc. in Figure 14.2 (some *feet* lack an accented head and are 'degenerate'). Over the metrical foot structure, higher-order prosodic constituents, with their own characteristic relative pitch ranges, boundary pitch movements, etc. can be constructed, shown in the figure as intonational phrases IP1, IP2. The details of prosodic analysis, including the meaning of those symbols, are described in Chapter 15.

The final phonetic form of the words to be spoken will reflect not only the original phonetics, but decisions made in the S and P tiers as well. For example, the P(rosody) tier adds detailed pitch and duration controls to the C(ontrol) specification that is passed to the voice synthesis component. Obviously, there can be a huge variety of particular architectures and components involved in the conversion process. Most systems, however, have some analog to each of the components presented above.

S	*S* [f1, f2, ..., fn]								
	NP [f1, f2, ..., fn]						*VP* [f1, f2, ..., fn]		
W	W1	W2	W3				W4		
Σ	A	**skilled**	e	lec	**tri**	cian	re	**por**	ted
C	ax	s k ih l d	ih	l eh k	t r ih	sh ax n	r iy	p ao r	t ax d
P	F1	F2			F3		F4	F5	
	IP1 [f1, f2, ..., fn]						*IP2* [f1, f2, ... , fn]		
	U [f1, f2, ..., fn]								

Figure 14.2 Annotation tiers indicating incremental analysis based on an input (text) sentence "A skilled electrician reported." Flow of incremental annotation is indicated by arrows on the left side.

14.1.3. Localization Issues

A major issue in the text and phonetic analysis components of a TTS system is internationalization and localization. While most of the language processing technologies in this book are exemplified by English case studies, an internationalized TTS architecture enabling minimal expense in localization is highly desirable. From a technological point of view, the text conventions and writing systems of language communities may differ substantially in arbitrary ways, necessitating serious effort in both specifying an internationalized architecture for text and phonetic analysis, and localizing that architecture for any particular language.

For example, in Japanese and Chinese, the unit of *word* is not clearly identified by spaces in text. In French, interword dependencies in pronunciation realization exist (liaison). Conventions for writing numerical forms of dates, times, money, etc. may differ across languages. In French, number groups separated by spaces may need to be integrated as single amounts, which rarely occurs in English. Some of these issues may be more serious for certain types of TTS architectures than others. In general, it is best to specify a rule architecture for text processing and phonetic analysis based on some fundamental formalism that allows for language-particular data tables, and which is powerful enough to handle a wide range of relations and alternatives.

14.2. Lexicon

The most important resource for text and phonetic analysis is the TTS system lexicon (also referred to as a dictionary). As illustrated in Figure 14.1, the TTS system lexicon is shared with almost all components. The lexical service should provide the following kinds of content in order to support a TTS system:

- Inflected forms of lexicon entries
- Phonetic pronunciations (support multiple pronunciations), stress and syllabic structure features for each lexicon entry
- Morphological analysis capability
- Abbreviation and acronym expansion and pronunciation
- Attributes indicating word status, including proper-name tagging, and other special properties
- List of speakable names of all common single characters. Under modern operating systems, the characters should include all Unicode characters.
- Word part-of-speech (POS) and other syntactic/semantic attributes
- Other special features, e.g., how likely a word is to be accented, etc.

It should be clear that the requirements for a TTS system lexical service overlap heavily with those for more general-purpose NLP.

Traditionally, TTS systems have been rule oriented, in particular for grapheme-to-phoneme conversion. Often, tens of so called *letter-to-sound* (LTS) rules (described in detail in Section 14.8) are used first for grapheme-to-phoneme conversion, and the role of the lexicon has been minimized as an *exception list*, whose pronunciations cannot be predicted on the basis of such LTS rules. However, this view of the lexicon's role has increasingly been adjusted as the requirement of a sophisticated NLP analysis for high-quality TTS systems has become apparent. There are a number of ways to optimize a dictionary system. For a good overview of lexical organization issues, please see [4].

To expose different contents about a lexicon entry listed above for different TTS module, it calls for a consistent mechanism. It can be done either through a database query or a function call in which the caller sends a key (usually the orthographic representation of a word) and the desired attribute. For example, a TTS module can use the following function call to look up a particular attribute (like phonetic pronunciations or POS) by passing the attribute *att* and the result will be stored in the pointer *val* upon successful lookup. Moreover, when the lookup is successful (the word is found in the dictionary) the function returns true, otherwise it will return false instead.

```
BOOLEAN DictLookup (string word, ATTTYPE att, (VOID *) val)
```

We should also point out that this functional view of dictionary could further expand the physical dictionary as a service. The morphological analysis and letter-to-sound modules (described in Sections 14.7 and 14.8) can all be incorporated into the same lexical service. That is, underneath dictionary lookup, operation and analysis is encapsulated from users to form a uniform service.

Another consideration in the system's runtime dictionary is compression. While many standard compression algorithms exist, and should be judiciously applied, the organization and extent of the vocabulary itself can also be optimized for small space and quick search. The kinds of American English vocabulary relevant to a TTS system include:

- Grammatical function words (closed class)—about several hundred
- Very common vocabulary—about 5,000 or more
- College-level core vocabulary base forms—about 60,000 or more
- College-level core vocabulary inflected form—about 120,000 or more
- Scientific and technical vocabulary, by field—e.g., legal, medical, engineering, etc.
- Personal names—e.g., family, given, male, female, national origin, etc.
- Place names—e.g., countries, cities, rivers, mountains, planets, stars, etc.
- Slang
- Archaisms

The typical sizes of reasonably complete lists of the above types of vocabulary run from a few hundred function or closed-class words (such as prepositions and pronouns) to 120,000 or so inflected forms of college-level vocabulary items, up to several million surnames and place names. Careful analysis of the likely needs of typical target applications can potentially reduce the size of the runtime dictionary. In general, most TTS systems maintain a system dictionary with a size between 5000 and 200,000 entries. With advanced technologies in database and hashing, search is typically a nonissue for dictionary lookup. In addition, since new forms are constantly produced by various creative processes, such as acronyms, borrowing, slang acceptance, compounding, and morphological manipulation, some means of analyzing words that have not been stored must be provided. This is the topic of Sections 14.7 and 14.8.

14.3. Document Structure Detection

For the purpose of discussion, we assume that all input to the TAM is an XML document, though perhaps largely unmarked, and the output is also a (more extensively marked) XML document. That is to say, all the knowledge recovered during the TAM phase is to be expressed as XML markup. This confirms the independence of the TAM from phonetic and prosodic considerations, allowing a variety of resources, some perhaps not crafted with TTS in mind, to be brought to bear by the TAM on the text. It also implies that that output of the TAM is potentially usable by other, non-TTS processes, such as normalization of language-model training data for building statistical language models (see Chapter 11). This fully modular and transparent view of TTS allows the greatest flexibility in document analysis, provides for direct *authoring* of structure and other customization, while allowing a split between expensive, multipurpose natural language analysis and the core TTS functionality. Although other text format or markup language, such as Adobe Acrobat or Microsoft Word, can be used for the same purpose, the choice of XML is obvious because it is the widely open standard, particularly for the Internet.

XML is a set of conventions for indicating the semantics and scope of various entities that combine to constitute a document. It is conceptually somewhat similar to Hypertext Markup Language (HTML), which is the exchange code for the World Wide Web. In these markup systems, properties are identified by tags with explicit scope, such as "`<b>make this phrase bold</b>`" to indicate a heavy, dark print display. XML in particular attempts to enforce a principled separation between document structure and content, on one hand, and the detailed formatting or presentation requirements of various *uses* of documents, on the other. Since we cannot provide a tutorial on XML here, we freely introduce example tags that indicate document and linguistic structure. The interpretations of these are intuitive to most readers, though, of course, the analytic knowledge underlying decisions to insert tags may be very sophisticated. It will be some time before commercial TTS engines come to a common understanding on the wide variety of text attributes that should be marked, and accept a common set of conventions. Nevertheless, it is reasonable to adopt the idea that TAM should be independent and reusable, thus allowing XML documents (which are expected to proliferate) to function for speech just as for other modalities, as indicated schematically in Figure 14.3.

TTS is regarded in Figure 14.3 as a factored process, with the text analysis perhaps carried out by human editors or by natural language analysis systems. The role of the TTS engine per se may eventually be reduced to the interpretation of structural tags and provision of phonetic information. While commercial engines of the present day are not structured with these assumptions in mind, modularity and transparency are likely to become increasingly important. The increasing acceptance of the basic ideas underlying an XML documentcentric approach to text and phonetic analysis for TTS can be seen in the recent proliferation of XML-like speech markup proposals [24, 33]. While not presenting any of these in detail, in the discussion below we adopt informal conventions that reflect and extend their basic assumptions. The structural markup exploited by the TTS systems of the future may be imposed by XML authoring systems at document creation time, or may be inserted by independent analytical procedures. In any case the distinction between purely automatic structure creation/detection and human annotation and authoring will increasingly blur—just as in natural language translation and information retrieval domains, the distinction between machine-produced results and human-produced results has begun to blur.

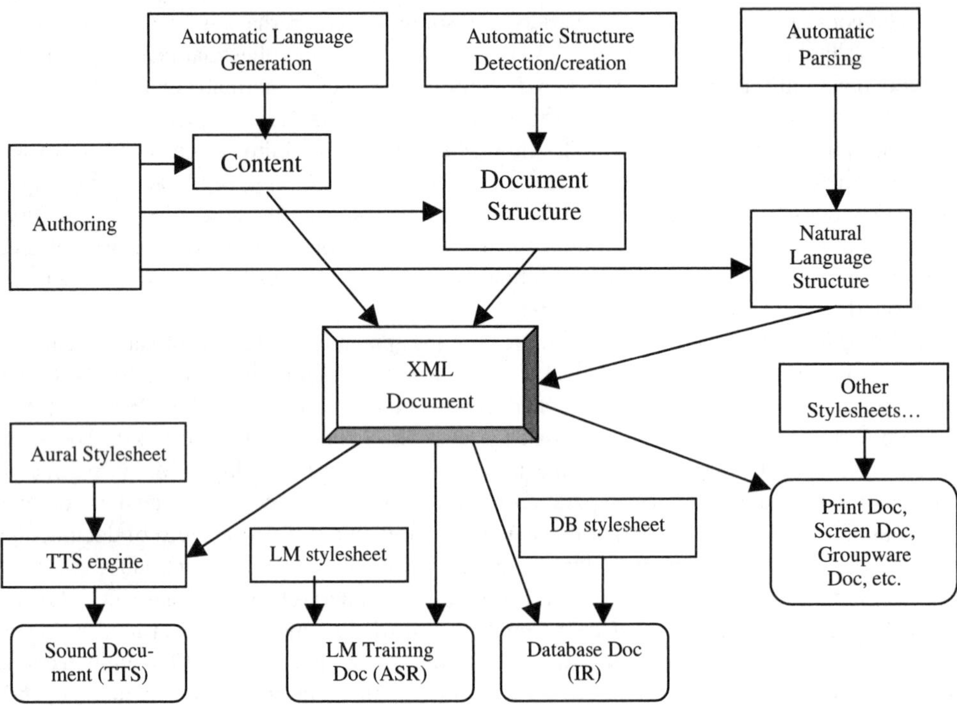

Figure 14.3 A documentcentric view of TTS.

14.3.1. Chapter and Section Headers

Section headers are a standard convention in XML document markup, and TTS systems can use the structural indications to control prosody and to regulate prosodic style, just as a professional reader might treat chapter headings differently. Increasingly, a document created on computer or intended for any kind of electronic circulation incorporates structural markup, and the TTS and audio human-computer-interface systems of the future learn to exploit this (in longer documents, the document structure markup assists in audio navigation, speedup, and skipping). For example, the XML annotation of a book at a high level might follow conventions as shown in Figure 14.4. Viewing a document in this way might lead a TTS system to insert pauses and emphasis correctly, in accordance with the structure marked. Furthermore, an audio interface system would work jointly with a TTS system to allow easy navigation and orientation within such a structure. If future documents are marked up in this fashion, the concept of audio books, for example, would change to rely less on unstructured prerecorded speech and more on smart, XML-aware, high-quality audio navigation and TTS systems, with the output customization flexibility they provide.

For documents without explicit markup information for section and chapter headers, it is in general a nontrivial task to detect them automatically. Therefore, most TTS systems today do not make such an attempt.

```
<Book>
    <Title>The Pity of War</Title>
      <Subtitle>Explaining World War I</Subtitle>
    <Author>Niall Ferguson</Author>
    <TableOfContents>...</TableOfContents>
    <Introduction>
            <Para>...</Para>
            ...
    </Introduction>
    <Chapter>
    <ChapterTitle>The Myths of Militarism</ChapterTitle>
            <Section>
                    <SectionTitle>Prophets</SectionTitle>
                    <Para> ... </Para>
                    ...
            </Section>
    </Chapter>
    ...
</Book>
```

Figure 14.4 An example of the XML annotation of a book.

14.3.2. Lists

Lists or bulleted items may be rendered with distinct intonational contours to indicate aurally their special status. This kind of structure might be indicated in XML as shown in Figure 14.5. Again, TTS engine designers need to get used to the idea of accepting such markup for interpretation, or incorporating technologies that can detect and insert such markup as needed by the downstream phonetic processing modules. Similar to chapter and section headers, most TTS systems today do not make an attempt to detect list structures automatically.

```
<UL>
<LI>compression</LI>
<LI>flexibility</LI>
<LI>text-waveform correspondence</LI>
</UL>
<Caption>The advantages of TTS</Caption>
```

Figure 14.5 An example of a list marked by XML.

14.3.3. Paragraphs

The paragraph has been shown to have direct and distinctive implications for pitch assignment in TTS [26]. The pitch range of good readers or speakers in the first few clauses at the start of a new paragraph is typically substantially higher than that for mid-paragraph sentences, and it narrows further in the final few clauses, before resetting for the next paragraph. Thus, to mimic a high-quality reading style in future TTS systems, the paragraph structure has to be detected from XML tagging or inferred from inspection of raw formatting. Obviously, relying on independently motivated XML tagging is, as always, the superior option, especially since this is a very common structural annotation in XML documents.

In contrast to other document structure information, paragraphs are probably among the easiest to detect automatically. The character <CR> (carriage return) or <NL> (new line) is usually a reliable clue for paragraphs.

14.3.4. Sentences

While sentence breaks are not normally indicated in XML markup today, there is no reason to exclude them, and knowledge of the sentence unit can be crucial for high-quality TTS. In fact, some XML-like conventions for text markup of documents to be rendered by synthesizers (e.g., SABLE) provide for a DIV (division) tag that could take paragraph, sentence, clause, etc. as attribute [24]. If we define sentence broadly as a primal linguistic unit that makes up paragraphs, attributes could be added to a `Sent` tag to express whatever linguistic knowledge exists about the type of the sentence as a whole:

```
<Sent type="yes-no question">
Is life so dear, or peace so sweet, as to be purchased at the price of chains and slavery?
</Sent>
```

Again, as emphasized throughout this section, such annotation could be either applied during creation of the XML documents (of the future) or inserted by independent processes. Such structure-detection processes may be motivated by a variety of needs and may exist outside the TTS system per se.

If no independent markup of sentence structure is available from an external, independently motivated document analysis or natural language system, a TTS system typically relies on simple internal heuristics to guess at sentence divisions. In email and other relatively informal written communications, sentence boundaries may be very hard to detect. In contrast to English, sentence breaking could be trivial for some other written languages. In Chinese, there is a designated symbol (a small circle 。) for marking the end of a sentence, so the sentence breaking could be done in a totally straightforward way. However, for most Asian languages, such as Chinese, Japanese, and Thai, there is in general no space within a sentence. Thus, tokenization is an important issue for Asian languages.

In more formal English writing, sentence boundaries are often signaled by terminal punctuation from the set: {. ! ?} followed by whitespaces and an upper-case initial word. Sometimes additional punctuation may trail the '?' and '!' characters, such as close quotation marks and/or close parenthesis. The character '.' is particularly troubling, because it is, in programming terms, heavily *overloaded*. Apart from its uses in numerical expressions and Internet addresses, its other main use is as a marker of abbreviation, itself a difficult problem for text normalization (see Section 14.4). Consider this pathological jumble of potentially ambiguous cases:

> Mr. Smith came by. He knows that it costs $1.99, but I don't know when he'll be back (he didn't ask, "when should I return?")... His Web site is www.mrsmithhhhhh.com. The car is 72.5 in. long (we don't know which parking space he'll put his car in.) but he said "...and the truth shall set you free," an interesting quote.

Some of these can be resolved in the linguistic analysis module. However for some cases, only probabilistic guesses can be made, and even a human reader may have difficulty. The ambiguous sentence breaking can also be resolved in an abbreviation-processing module (described in Section 14.4.1). Any period punctuation that is not taken to signal an abbreviation and is not part of a number can be taken as end-of-sentence. Of course, as we have seen above, abbreviations are also confusable with words that can naturally end sentences, e.g., "*in*." For the measure abbreviations, an examination of the left context (checking for numeric) may be sufficient. In any case, the complexity of sentence breaking illustrates the value of passing multiple hypotheses and letting later, more knowledgeable modules (such as an abbreviation or linguistic analysis module) make decisions. Algorithm 14.1 shows a simple sentence-breaking algorithm that should be able to handle most cases.

For advanced sentence breakers, a weighted combination of the following kinds of considerations may be used in constructing algorithms for determining sentence boundaries (ordered from easiest/most common to most sophisticated):

- Abbreviation processing—Abbreviation processing is one of the most important tasks in text normalization and will be described in detail in Section 14.4.
- Rules or CART built (Chapter 4) upon features based on: document structure, whitespace, case conventions, etc.
- Statistical frequencies on sentence-initial word likelihood

- Statistical frequencies of typical lengths of sentences for various genres
- Streaming syntactic/semantic (linguistic) analysis—Syntactic/semantic analysis is also essential for providing critical information for phonetic and prosodic analysis. Linguistic analysis will be described in Section 14.5.

As you can see, a deliberate sentence breaking requires a fair amount of linguistic processing, like abbreviation processing and syntactic/semantic analysis. Since this type of analysis is typically included in the later modules (text normalization or linguistic analysis), it might be a sensible decision to delay the decision for sentence breaking until later modules, either text normalization or linguistic analysis. In effect, this arrangement can be treated as the document structure module passing along multiple hypotheses of sentence boundaries, and it allows later modules with deeper linguistic knowledge (text normalization or linguistic analysis) to make more intelligent decisions.

Finally, if a long buffer of unpunctuated words is presented, TTS systems may impose arbitrary limits on the length of a sentence for later processing. For example, the writings of the French author Marcel Proust contain some sentences that are several hundred words long (average sentence length for ordinary prose is about 15 to 25 words).

ALGORITHM 14.1: *A SIMPLE SENTENCE-BREAKING ALGORITHM*

1. **If** found punctuation ./!/? advance one character and **goto 2.**
   else advance one character and **goto 1**.
2. **If** not found whitespace advance one character and **goto 1**.
3. **If** the character is period (.) **goto 4.**
   else goto 5.
4. Perform abbreviation analysis.
   If not an abbreviation **goto 5.**
   else advance one character and **goto 1.**
5. Declare a sentence boundary and sentence type ./!/?
   Advance one character and **goto 1.**

14.3.5. Email

TTS could be ideal for reading email over the phone or in an eyes-busy situation such as when driving a motor vehicle. Here again we can speculate that XML-tagged email structure, minimally something like the example in Figure 14.6, will be essential for high-quality prosody, and for controlling the audio interface, allowing skips and speedups of areas the user has defined as less critical, and allowing the system to announce the function of each block. For example, the `sig` (signature) portion of email certainly has a different semantic function than the main message text and should be clearly identified as such, or skipped, at the listener's discretion. Modern email systems are providing increasingly sophisticated support for structure annotation such as that exemplified in Figure 14.6. Obviously, the email document structure can be detected only with appropriate tags (like XML). It is very difficult for a TTS system to detect it automatically.

```
<message>
    <header>
            <date>11 June 1998</date>
            <from>Leslie</from>
            <to>Jo</to>
            <subject>Surf's Up!</subject>
    </header>
    <body> … </body>
    <sig>Freedom's just another word for nothing left to lose</sig>
</message>
```

Figure 14.6 An example of email marked by XML.

14.3.6. Web Pages

All the comments about TTS reliance on XML markup of document structure can be applied to the case of HTML-marked Web page content as well. In addition to sections, headers, lists, paragraphs, etc., the TTS systems should be aware of XML/HTML conventions such as links (<a href="…">link name</a>) and perhaps apply some distinctive voice quality or prosodic pitch contour to highlight these. The size and color of the section of text also provides useful hints for emphasis. Moreover, the TTS system should also integrate the rendering of audio and video contents on the Web page to create a genuine multimedia experience for the users. More could be said about the rendition of Web content, whether from underlying XML documents or HTML-marked documents prepared specifically for Web presentation. In addition, the World Wide Web Consortium has begun work on standards for aural stylesheets that can work in conjunction with standard HTML to provide special direction in aural rendition [33].

14.3.7. Dialog Turns and Speech Acts

Not all text to be rendered by a TTS system is standard written prose. The more expressive TTS systems could be tasked with rendering natural conversation and dialog in a spontaneous style. As with written documents, the TTS system has to be guided by XML markup of its input. Various systems for marking *dialog turns* (change of speaker) and *speech acts* (the mood and functional intent of an utterance)[3] are used for this purpose, and these annotations will trigger particular phonetic and prosodic rules in TTS systems. The speech act coding

[3] Dialog modeling and the concepts of *dialog turns* and *speech acts* are described in detail in Chapter 17.

schemes can help, for example, in identifying the speaker's intent with respect to an utterance, as opposed to the utterance's structural attributes. The prosodic contour and voice quality selected by the TTS system might be highly dependent on this functional knowledge.

For example, a syntactically well-formed question might be used as information solicitation, with the typical utterance-final pitch upturn as shown in the following:

<REQUEST_INFO>Can you hand me the wrench?</REQUEST_INFO>

But if the same utterance is used as a command, the prosody may change drastically.

<DIRECTIVE>Can you hand me the wrench.</DIRECTIVE>

Research on speech act markup-tag inventories (see Chapter 17) and automatic methods for speech act annotation of dialog is ongoing, and this research has the property considered desirable here, in that it is independently motivated (useful for enhancing speech recognition and language understanding systems). Thus, an advanced TTS system should be expected to exploit dialog and speech act markups extensively.

14.4. TEXT NORMALIZATION

Text often include abbreviations (e.g., FDA for *Food and Drug Administration*) and acronyms (SWAT for *Special Weapons And Tactics*). Novels and short stories may include *spoken* dialog interspersed with exposition; technical manuals may include mathematical formulae, graphs, figures, charts and tables, with associated captions and numbers; email may require interpretation of special conventional symbols such as *emoticons* [e.g., :-) means smileys], as well as Web and Internet address formats, and special abbreviations (e.g., IMHO means *in my humble opinion*). Again, any text source may include part numbers, stock quotes, dates, times, money and currency, and mathematical expressions, as well as standard ordinal and cardinal formats. Without context analysis or prior knowledge, even a human reader would sometimes be hard pressed to give a perfect rendition of every sequence of nonalphabetic characters or of every abbreviation. *Text normalization* (TN) is the process of generating normalized orthography (or, for some systems, direct generation of phones) from text containing words, numbers, punctuation, and other symbols. For example, a simple example is given as follows:

The 7% Solution ➔ THE SEVEN PER CENT SOLUTION

Text normalization is an essential requirement not only for TTS, but also for the preparation of training text corpora for acoustic-model and language-model construction.[4] In addition, speech dictation systems face an analogous problem of *inverse* text normalization for document creation from recognized words, and such systems may depend on knowledge sources similar to those described in this section. The example of an inverse text normalization for the example above is given as follows:

[4] For details of acoustic and language modeling, please refer to Chapters 9 and 11.

THE SEVEN PER CENT SOLUTION ➔ The 7% Solution

Modular text normalization components, which may produce output for multiple downstream consumers, mark up the exemplary text along the following lines:

The <tn snor="SEVEN PER CENT">7%</tn> Solution

The `snor` tag stands for *Standard Normalized Orthographic Representation.*[5] For TTS, input text may include multisentence paragraphs, numbers, dates, times, punctuation, symbols of all kinds, as well as interpretive annotations in a TTS markup language, such as tags for word emphasis or pitch range. Text analysis for TTS is the work of converting such text into a stream of normalized orthography, with all relevant input tagging preserved and new markup added to guide the subsequent modules. Such interpretive annotations added by text analysis are critical for phonetic and prosodic generation phases to produce desired output. The output of the text normalizer may be deterministic, or may preserve a full set of interpretations and processing history with or without probabilistic information to be passed along to later stages. We once again assume that XML markup is an appropriate format for expressing knowledge that can be created by a variety of external processes and exploited by a number of technologies in addition to TTS.

Since today's TTS systems typically cannot expect that their input be independently marked up for text normalization, they incorporate internal technology to perform this function. Future systems may piggyback on full natural language processing solutions developed for independent purposes. Presently, many incorporate minimal, TTS-specific hand-written rules [1], while others are loose agglomerations of modular, task-specific statistical evaluators [3].

For some purposes, an architecture that allows for a set or lattice of possible alternative expansions may be preferable to deterministic text normalization, like the *n-best* lists or *word graph* offered by the speech recognizers described in Chapter 13. Alternatives known to the system can be listed and ranked by probabilities that may be learnable from data. Later stages of processing (linguistic analysis or speech synthesis) can either add knowledge to the lattice structure or recover the best alternative, if needed. Consider the fragment "*at 8 am I . . .* " in some informal writing such as email. Given the flexibility of writing conventions for pronunciation, *am* could be realized as either *A. M.* (the numeric context seems to cue at times) or the auxiliary verb *am*. Both alternatives could be noted in a descriptive lattice of covering interpretations, with confidence measures if known (Table 14.2).

Table 14.2 Two alternative interpretations for sentence fragment "*At 8 am I ...*".

At 8 am I ...	At <time> eight am </time> I ...
At 8 am I ...	At <number> eight </number> am I ...

[5] SNOR, or Standard Normalized Orthographic Representation, is a uniform way of writing words and sentences that corresponds to spoken rendition. SNOR-format sentence texts are required as reference material for many Defense Advanced Research Project Agency and National Institutes of Standards and Technology-sponsored standard speech technology evaluation procedures.

If the potential ambiguity in the interpretation of *am* in the above pair of examples is simply noted, and the alternatives retained rather than suppressed, the choice can be made by a later stage of syntactic/semantic processing. Note another feature of this example—the rough irregular abbreviation form for antemeridian, which by prescriptive convention hopes that high-quality TTS processing can rely entirely on *standard* stylistic conventions. That observation also applies to the *obligatory* use of "?" for all questions.

Specific architectures for the text normalization component of TTS may be highly variable, depending on the system architect's answers to the following questions:

- Are cross-functional language processing resources mandated, or available?
- If so, are phonetic forms, with stress or accent, and normalized orthography, available?
- Is a full syntactic and semantic analysis of input text mandated, or available?
- Can the presenting application add interpretive knowledge to structure the input (text)?
- Are there interface or pipelining requirements that preclude lattice alternatives at every stage?

Because of this variability in requirements and resources, we do not attempt to formally specify a single, all-purpose architectural solution here. Rather, we concentrate on describing the text normalization challenges any system has to face. We note where solutions to these challenges are more readily realized under particular architectural assumptions.

All text normalization consists of two phases: identification of type, and expansion to SNOR or other unambiguous representation. Much of the identification phase, dealing with phenomena of sentence boundary determination, abbreviation expansion, number spell-out, etc., can be modeled as regular expression (see Chapter 11). This raises an interesting architectural issue. Imagine a system based entirely on regular *finite state transducers* (FST, see Chapter 11), as in [27], which enforces an appealing uniformity of processing mechanism and internal structure description. The FST permits a lattice-style representation that does not require premature resolution of any structural choice. An entire text analysis system can be based on such a representation. However, as long as a system confines its attention to issues that commonly come under the heading of text normalization, such as number formats, abbreviations, and sentence breaking, a simpler regular-expression-based uniform mechanism for rule specification and structure representation may be adequate.

Alternatively, TTS systems could make use of advanced tools such as, for example, the lex and yacc tools [17], which provide frameworks for writing customized lexical analyzers and context-free grammar parsers, respectively. In the discussion of typical text normalization requirements below, examples will be provided and then a fragment of Perl pattern-matching code will be shown that allows matching of the examples given. Perl notation [36] is used as a convenient short-hand representing any equivalent regular expressionparsing system and can be regarded as a subset of the functionality provided by any regular expression, FST, or context-free grammar tool set that a TTS software architect may choose to employ. Only a small subset of the simple, fairly standard Perl conventions

to employ. Only a small subset of the simple, fairly standard Perl conventions for regular expression matching are used, and comments are provided in our discussion of text normalization.

A text normalization system typically adds identification information to assist subsequent stages in their tasks. For example, if the TN subsystem has determined with some confidence that a given digit string is a phone number, it can associate XML-like tags with its output, identifying the corresponding normalized orthographic chunk as a candidate for special phone-number intonation. In addition, the identification tags can guide the lexical disambiguation of terms for other processes, like phonetic analysis in TTS systems and training data preparation for speech recognition.

Table 14.3 shows some examples of input fragments with a relaxed form of output normalized orthography. It illustrates a possible ambiguity in TN output. In the (contrived) example, the ambiguity is between a place name and a hypothetical individual named perhaps *Steve* or *Samuel* Asia. Two questions arise in such cases. The first is format of specification. The data between submodules in a TTS system can be passed (or be placed in a centrally viewable *blackboard* location) as tagged text or in a binary format. This is an implementation detail. Most important is that all possibilities known to the TN system be specified in the output, and that confidence measures from the TN, if any, be represented. For example, in many contexts, *South Asia* is the more likely spell-out of *S. Asia*, and this should be indicated implicitly by ordering output strings, or explicitly with probability numbers. The decision could then be delayed until one has enough information in the later module (like linguistic analysis) to make the decision in an informed manner.

Table 14.3 Examples of the normalized output using XML-like tags for text normalization.

Dr. King	<title> DOCTOR </title> KING
7%	<number>SEVEN<ratio>PERCENT</ratio> </number>
S. Asia	<toponym> SOUTH ASIA </toponym> OR <psn_name><initial>S</initial>ASIA</psn_name>

14.4.1. Abbreviations and Acronyms

As noted above, a period is an important but not completely reliable clue to the presence of an abbreviation. Periods may be omitted or misplaced in text for a variety of reasons. For similar reasons of stylistic variability and a writer's (lack of) care and skill, capitalization, another potentially important clue, can be variable as well. For example, all the representations of the abbreviation for *post script* listed below have been observed in actual mail and email. A system must therefore combine knowledge from a variety of contextual sources, such as document structure and origin, when resolving abbreviations:

PS. Don't forget your hat.
Ps. Don't forget your hat.
P.S. Don't forget your hat.
P.s. Don't forget your hat.

And *P.S.*, when examined out of context, could be personal name initials as well. Of course, a given TTS system's user may be satisfied with the simple spoken output */p iy ae s/* in cases such as the above, obviating the need for full interpretation. But at a minimum, when *fallback to letter pronunciation* is chosen, the TTS system must attempt to ensure that some obvious spell-out is not being overlooked. For example, a system should not render the title in *Dr. Jones* as letter names */d iy aa r/.*

Actually, any abbreviation is potentially ambiguous, and there are several distinct types of ambiguity. For example, there are abbreviations, typically quantity and measure terms, which can be realized in English as either plural or singular depending on their numeric coefficient, such as *mm* for *millimeter(s).* This type of ambiguity can get especially tricky in the context of conventionally frozen items. For example, *9mm ammunition* is typically spoken as *nine millimeter ammunition* rather than *nine millimeters ammunition.*

Next, there are forms that can, with appropriate syntactic context, be interpreted either as abbreviations or as simple English words, such as in (inches), particularly at the end of sentences.

Finally, many, perhaps most, abbreviations have entirely different abbreviation spell-outs depending on semantic context, such as *DC* for *direct current* or *District of Columbia.* This variability makes it unlikely that any system ever performs perfectly. However, with sufficient training data, some statistical guidelines for interpretation of common abbreviations in context can be derived. Table 14.4 shows a few more examples of this most difficult type of ambiguity.

An advanced TTS system should attempt to convert reliably at least the following abbreviations:

- Title—Dr., MD, Mr., Mrs., Ms., St. (Saint), … etc.
- Measure—ft., in., mm, cm (centimeter), kg (kilogram), … etc.
- Place names—CO, LA, CA, DC, USA, St. (street), Dr. (drive), … etc.

Table 14.4 Some ambiguous abbreviations.

CO	Colorado	commanding officer
	conscientious objector	carbon monoxide
IRA	Individual Retirement Account	Irish Republican Army
MD	Maryland	doctor of medicine
	muscular dystrophy	

Abbreviation disambiguation usually can be resolved by POS (part-of-speech) analysis. For example, whether *Dr.* is *Doctor* or *Drive* can be resolved by examining the POS features of the previous and following words. If the abbreviation is followed by a capitalized personal name, it can be expanded as *Doctor*, whereas if the abbreviation is preceded by a capitalized place name, a number, or an alphanumeric (like 120th), it will be expanded as *Drive*. Although the example above is resolved via a series of heuristic rules, the disambiguation (POS analysis) can also be done by a statistical approach. In [6], the POS tags are determined based on the most likely POS sequence using POS trigram and lexical-POS unigram. Since an abbreviation can often be distinguished by its POS feature, the most likely POS sequence of the sentence discovered by the trigram search then provides the best guess of the POS (thus the usage) for abbreviations. We describe POS tagging in more detail in Section 14.5.

Other than POS information, the lexical entries for abbreviations should include all features and alternatives necessary to generate a lattice of possible analyses. For example, a typical abbreviation's entry might include information as to whether it could be a word (like *in*), whether period(s) are optional or required, whether plural variants must be generated and if so under what circumstances, whether numerical specification is expected or required, etc.

Acronyms are words created from the first letters or parts of other words. For example, SCUBA is an acronym for *self-contained underwater breathing apparatus*. Generally, to qualify as a true acronym, a letter sequence should reflect normal language phonotactics, such as a reasonable alternation of consonants and vowels. From a TTS system's point of view, the distinctions between acronyms, abbreviations, and plain new or unknown words can be unclear. Many acronyms can be entered into the TTS system lexicon just as ordinary words would be. However, unknown acronyms (not listed in the lexicon) may occasionally be encountered. Although an acronym's case property can be a significant clue to identification, it is often unclear how to speak a given sequence of upper-case letters. Most TTS systems, failing to locate the sequence in the acronym dictionary, spell it out letter-by-letter. Other systems attempt to determine whether the sequence is inherently *speakable*. For example, DEC might be inherently speakable, while FCC is not formed according to normal word phonotactics. When something speakable is found, it is processed via the normal letter-to-sound rules, while something *unspeakable* would be spelled out letter-by-letter. Yet other systems might simply feed the sequence directly to the letter-to-sound rules (see Section 14.8), just as they would any other unknown word. As with all such problems, a larger lexicon usually provides superior results.

The general algorithm for abbreviations and acronyms expansion in text normalization is summarized in Algorithm 14.2. The algorithm assumes that tokenization and POS tagging have been done for the whole sentence. Abbreviation expansion is determined by the POS tags of the potential abbreviation candidates. Acronym expansion is done exclusively by table lookup, and letter-by-letter spell-out is used when acronyms cannot be found in the acronym table.

ALGORITHM 14.2: *ABBREVIATIONS AND ACRONYMS EXPANSION*

1. **If** word token *w* is not in abbreviation table and *w* contains only capital letters **goto 3.**
2. **Abbreviation Expansion**
   If the POS tag of *w* and the correspondent abbreviation match
   Abbreviation expansion by inserting SNOR and interpretive annotation tags
   Advance one word and **goto 1.**
3. **Acronym Expansion**
   If *w* is in the predefined acronym table
   Acronym expansion by inserting SNOR and interpretive annotation tags
   according to acronym expansion table
   else spell out *w* letter-by-letter
4. Advance one word and **goto 1.**

14.4.2. Number Formats

Numbers occur in a wide variety of formats and have a wide variety of contextually dependent reading styles. For example, the digits 370 in the context of the product name *IBM 370 mainframe computer* typically are read as *three seventy*, while in other contexts 370 would be read as *three hundred seventy* or *three hundred and seventy*. In a phone number, such as 370-1111, the string would normally be read as *three seven oh*, while in still other contexts it might be rendered as *three seven zero*. A text analysis system can incorporate rules, perhaps augmented by probabilities, for these situations, but might never achieve perfection in all cases. Phone numbers are a practical place to start, and their treatment illustrates some of the general issues relevant to the other number formats which are covered below.

14.4.2.1. Phone Numbers

Phone numbers may include prefixes and area codes and may have dashes and parentheses as separators. Examples are shown in Table 14.5.

The first two examples have *prefix* codes, while the next four have area codes with minor formatting differences. The final two examples are possible international-format phone numbers. A basic Perl regular expression pattern to subsume the commonality in all the local domestic numbers can be defined as follows:

```
$us_basic = '([0-9]{3}\-[0-9]{4})';
```

This defines a pattern subpart to match 3 digits, followed by a separator dash, followed by another 4 digits. Then the pattern to match the prefix type would be:

```
/([0-9]{1})[\/ -]($us_basic)/
```

Table 14.5 Some different written representations of phone numbers.

9-999-4118
9 345-5555
(617) 932-9209
(617) 932-9209
716-123-4568
409/845-2274
+49 (228) 550-381
+49-228-550-381

In the first example above, this leaves the system pattern variable $1 (corresponding to the first set of capture parentheses in the pattern) set to 9, and $2 (the second set of capture parentheses) set to 999-4118. Then a separate set of tables, indexed by the rule name and the pattern variable contents, could provide orthographic spell-outs for the digits. Clearly a balance has to be struck between the number of pattern variables provided in the expression and the overall complexity of the expression, vis-à-vis the complexity and sophistication of the indexing scheme of the spell-out tables. For example, the $us_basic could be defined to incorporate parentheses capture on the first three digits and the remaining four separately, which might lead to a simpler spell-out table in some cases.

The pattern to match the area code types could be:

```
/(\([0-9]{3}\))[\/ -]($us_basic)/
```

These patterns could be endlessly refined, expanded, and layered to match strings of almost arbitrary complexity. A balance has to be struck between number and complexity of distinct patterns. In any case, no matter how sophisticated the matching mechanism, arbitrary or at best probabilistic decisions have to be made in constructing a TTS system. For example, in matching an area code type, the rule architect must decide how much and what kind of whitespace separation the matching system tolerates between the area code and the rest of the number before a phone-number match is considered unlikely. Or, as another example, does the rule architect allow new lines or other formatting characters to appear between the area code and the basic phone number? These kinds of decisions must be explicitly considered, or made by default, and should be specified to a reasonable degree in user documentation. There are a great many other phone number formats and issues that are beyond the scope of this treatment.

Once a certain type of pattern requires a conversion to normalized orthography, the question of how to perform the conversion arises. The conversion characters can be aligned with the identification, so that conversion occurs implicitly during the pattern matching process. Another way is to separate the conversion from the identification phase. This may or may not lead to gains in efficiency and elimination of redundancy, depending on the overall architecture of the system and whether and how components are expected to be reused. A version of this second approach is sketched here.

Suppose that the pattern match variable $1 has been set to 617 by one of the identification-phase pattern matches described above. Another list can provide pointers to conversion tables, indexed by the rule name or number and the variable name. So for the rule that can match area codes, the relevant entry would be:

```
Identification rule    Variable     Spellout table
Area-Phone             $1           LITERAL_DIGIT
```

The LITERAL_DIGIT spell-out rule set, when presented with the 617 character sequence (the value of $1), simply generates the normalized orthography *six one seven*, by table lookup. In this simple and straightforward approach, spell-out tables such as LITERAL_DIGIT can be reused for portions of a wide variety of identification rules. Other simple numeric spell-out tables would cover different styles of numeric reading, such as *pairwise* style (e.g., *six seventeen*), full decimal with tens, hundreds, thousands units (*six hundred seventeen*), and so on. Some spellout tables may require processing code to supplement the basic table lookup. Additional examples of spell-out tables are not provided for the various other types of text normalization entities exemplified below, but would function similarly.

14.4.2.2. Dates

Dates may be specified in a wide variety of formats, sometimes with a mixture of orthographic and numeric forms. Note that dates in TTS suffer from a mild form of the century-date-change uncertainty (the infamous Y2K bug), so a form such as 5/7/37 may in the future be ambiguous, in its full form, between 1937 and 2037. The safest course is to say as little as possible, i.e., "*five seven thirty seven*", or even "*May seventh, thirty seven*", rather than attempt "*May seventh, nineteen thirty seven*". Table 14.6 shows a variety of date formats and associated normalized orthography.

Table 14.6 Various date formats.

12/19/94 (US)	December nineteenth ninety four
19/12/94 (European)	December nineteenth ninety four
04/27/1992	April twenty seventh nineteen ninety two
May 27, 1995	May twenty seventh nineteen ninety five
July 4, 94	July fourth ninety four
1,994	one thousand nine hundred and ninety four
1994	nineteen ninety four

One issue that comes up with certain number formats, including dates, is range checking. A form like 13/19/94 is basically uninterpretable as a date. This kind of checking, if included in the initial pattern matching, may be slow and may increase formal requirements for power of the pattern matching system. Therefore, range checking can be done at spell-out time (see below) during normalized orthography generation, as long as a backtracking or redo option is present. If range checking is desired as part of the basic identification phase of text normalization, some regular expression matching systems allow for extensions. For example, the following pattern variable matches only numbers less than or equal to 12, the valid month specifications. It can be included as part of a larger, more complex date matching *pattern:*

```
$month = '/(0[123456789]/1[012]/'
```

14.4.2.3. Times

Times may include hours, minute, seconds, and duration specifications as shown in Table 14.7. Time formats exemplify yet another area where linguistic concerns have to intersect with architecture. If simple, flat normalized orthography is generated during a text normalization phase, a later stage may still find a form like *am* ambiguous in pronunciation. If a lattice of alternative interpretations is provided, it should be supplemented with interpretive information on the linguistic status of the alternative text analyses. Alternatively, a single best guess can be made, but even in this case, some kind of interpretive information indicating the status of the choice as, e.g., a time expression, should be provided for later stages of syntactic, semantic, and prosodic interpretation. This reiterates the importance of TTS text analysis systems to generate interpretive annotations tags for subsequent modules' use whenever possible, as discussed in Section 14.4. In some cases, unique text formatting of the choice, corresponding to the system's lexical contents, may be sufficient. That is, in some systems, generation of *A.M.*, for example, may uniquely correspond to the lexicon's entry for that portion of a time expression, which specifies the desired pronunciation and grammatical treatment.

Table 14.7 Several examples for time expressions.

11:15	eleven fifteen
8:30 pm	eight thirty pm
5:20 am	five twenty am
12:15:20	twelve hours fifteen minutes and twenty seconds
07:55:46	seven hours fifty-five minutes and forty-six seconds

14.4.2.4. Money and Currency

As illustrated in Table 14.8, money and currency processing should correctly handle at least the currency indications $, £, DM, ¥, and €, standing for dollars, British pounds, deutsche marks, Japanese yen, and euros, respectively. In general, $ and £ have to precede the numeral; DM, ¥, and € have to follow the numeral. Other currencies are often written in full words and have to follow the numeral, though abbreviations for these are sometimes found, such as *100 francs* and *20 lira.*

Table 14.8 Several money and currency expressions.

$40	forty dollars
£200	two hundred pounds
5 ¥	five yen
25 DM	twenty five deutsche marks
300 €	three hundred euros

14.4.2.5. Account Numbers

Account numbers may refer to bank accounts or social security numbers. Commercial product part numbers often have these kinds of formats as well. In some cases these cannot be readily distinguished from mathematical expressions or even phone numbers. Some examples are shown below:

```
123456-987-125456
000-1254887-87
049-85-5489
```

The other popular number format is that of credit card number, such as

```
4446-2289-2465-7065
3745-122267-22465
```

To process formats like these, it may eventually be desirable for TTS systems to provide customization capabilities analogous to the pronunciation customization features for words found in current TTS systems. Regular expression formalisms of the type exemplified above for phone number, would, if exposed to applications and developers through suitable editors, be adequate for most such needs.

14.4.2.6. Ordinal Numbers

Ordinal numbers are those referring to rank or placement in a series. Examples include:

1st, 2nd, 3rd, 4th, 10th, 11th, 12th, 20th, 100th, 1000th, etc.
1st, 2nd, 3rd, 4th, 10th, 11th, 12th, 20th, 21st, 32nd, 100th, 1000th, etc.

The system's ordinal processing may also be used to generate the denominators of fractions, except for halves, as shown in Table 14.9. Notice that the ordinal must be plural for numerators other than 1.

Table 14.9 Some examples of fractions.

1/2	one half
1/3	one third
1/4	one quarter or one fourth
1/10	one tenth
3/10	three tenths

14.4.2.7. Cardinal Numbers

Cardinal numbers are, loosely speaking, those forms used in simple counting or the statement of amounts. If a given sequence of digits fails to fit any of the more complex formats above, it may be a simple cardinal number. These may be explicitly negative or positive or assumed positive. They may include decimal or fractional specifications. They may be read in several different styles, depending on context and/or aesthetic preferences. Table 14.10 gives some examples of cardinal numbers and alternatives for normalized orthography.

The number-expansion algorithm is summarized in Algorithm 14.3. In this algorithm the text normalization module maintains an extensive pattern table. Each pattern in the table contains its associated pattern in regular expression or Perl format along with a pointer to a rule in the conversion table, which guides the expansion process.

Table 14.10 Some cardinal number types.

123	one two three	one hundred (and) twenty three
1,230	one thousand two hundred (and) thirty	
2426	two four two six	twenty four twenty six
	two thousand four hundred (and) twenty six	

A regular expression to match well-formed cardinals with commas grouping chunks of three digits of the type from 1,000,000 to 999,999,999 might appear as:

```
if ($item =~ /^([0-9]{1,3}),([0-9]{3}),([0-9]{3})/
 {      $NewFrame->{"millions"} = $1;
   $NewFrame->{"thousands"} = $2;
   $NewFrame->{"hundreds"} = $3;
    print "Grouped cardinal found: $item\n";
   return $NewFrame;   }
```

ALGORITHM 14.3: *NUMBER EXPANSION*

1. **Pattern Matching**
   If a match is found **goto 2.**
   else goto 3.
2. **Number Expansion**
   Insert SNOR and interpretive annotation tags according to the associated rule
   Advance the pointer to the right of the match pattern and **goto 1.**
3. **Finish**

14.4.3. Domain-Specific Tags

In keeping with the theme of this section—that is, the increasing importance of independently generated precise markup of text entities—we present a little-used but interesting example.

14.4.3.1. Mathematical Expressions

Mathematical expressions are regarded by some systems as the domain of special-purpose processors. It is a serious question how far to go in mathematical expression parsing, since providing some capability in this area may raise users' expectations to an unrealistic level. The World Wide Web Consortium has developed MathML (mathematical markup language) [34], which provides a standard way of describing math expressions. MathML is an XML extension for describing mathematical expression structure and content to enable mathematics to be served, received, and processed on the Web, similar to the function HTML has performed for text. As XML becomes increasingly pervasive, MathML could possibly be used to guide interpretation of mathematical expressions. For the notation $(x + 2)^2$ a possible MathML representation such as that below might serve as an initial guide for a spoken rendition.

```
<EXPR>
 <EXPR>
 x
 <PLUS/>
 2
 </EXPR>
 <POWER/>
 2
</EXPR>
```

This might be generated by an application or by a specialized preprocessor within the TTS system itself. Prosodic rules or data tables appropriate for math expressions could then be triggered.

14.4.3.2. Chemical Formulae

As XML becomes increasingly common and exploitable by TTS text normalization, other areas follow. For example, Chemical Markup Language (CML [22]) now provides a standard way to describe molecular structure or chemical formulae. CML is an example of how standard conventions for text markup are expected increasingly to replace ad hoc, TTS-internal heuristics.

In CML, the chemical formula C_2OCOH_4 would appear as:

```
<FORMULA>
   <XVAR BUILTIN="STOICH">
   C C O C O H H H H
   </XVAR>
</FORMULA>
```

It seems reasonable to expect that TTS engines of the future will be increasingly devoted to interpreting such precise conventions in high-quality speech renditions rather than endlessly replicating NL heuristics that fail as often as they succeed in guessing the identity of raw text strings.

14.4.4. Miscellaneous Formats

A random list illustrating the range of other types of phenomena for which an English-oriented TTS text analysis module must generate normalized orthography might include:

- Approximately/tilde: The symbol ~ is spoken as *approximately* before (Arabic) numeral or currency amount, otherwise it is the character named *tilde*.

- Folding of accented Roman characters to *nearest* plain version: If the TTS system has no knowledge of dealing with foreign languages, like French or German, a table of folding characters can be provided so that for a term such as *Über-mensch*, rather than spell out the word *Über*, or ignore it, the system can convert it to its *nearest* English-orthography equivalent: *Uber*. The ultimate way to process such foreign words should integrate a language identification module with a multi-lingual TTS system, so that language-specific knowledge can be utilized to produce appropriate text normalization of all text.
- Rather than simply ignore high ASCII characters in English (characters from 128 to 255), the text analysis lexicon can incorporate a table that gives *character names* to all the printable high ASCII characters. These names are either the full Unicode character names, or an abbreviated form of the Unicode names. This would allow speaking the names of characters like © (*copyright sign*), ™ (*trademark*), @ (*at*), ® (*registered mark*), and so on.
- Asterisk: in email, the symbol '*' may be used for emphasis and for setting off an item for special attention. The text analysis module can introduce a little pause to indicate possible emphasis when this situation is detected. For the example of "*Larry has *never* been here*," this may be suppressed for asterisks spanning two or more words. In some texts, a word or phrase appearing completely in UPPER CASE may also be a signal for special emphasis.
- Emoticons: There are several possible emoticons (emotion icons).

  1. :-) or :) SMILEY FACE (humor, laughter, friendliness, sarcasm)
  2. :-(or :(FROWNING FACE (sadness, anger, or disapproval)
  3. ;-) or ;) WINKING SMILEY FACE (naughty)
  4. :-D OPEN-MOUTHED SMILEY FACE (laughing out loud)

  Smileys, of which there are dozens of types, may be tacked onto word start or word end or even occur interword without spaces, as in the following examples.

  :)hi!
  Hi:)
  Hi:)Hi!

14.5. LINGUISTIC ANALYSIS

Linguistic analysis (sometimes also referred to as syntactic and semantic parsing) of natural language (NL) constitutes a major independent research field. Often commercial TTS systems incorporate some minimal parsing heuristics developed strictly for TTS. Alternatively, the TTS systems can also take advantage of independently motivated natural language processing (NLP) systems, which can produce structural and semantic information about sentences. Such linguistically analyzed documents can be used for many purposes other than TTS—information retrieval, machine translation system training, etc.

Provision of some parsing capability is useful to TTS systems in several areas. Parsers may be used in disambiguating the text normalization alternatives described above. Additionally, syntactic/semantic analysis can help to resolve grammatical features of individual words that may vary in pronunciation according to sense or abstract inflection, such as *read*. Finally, parsing can lay a foundation for derivation of a prosodic structure useful in determining segmental duration and pitch contour.

The fundamental types of information desired for TTS from a parsing analysis are summarized below:

- Word part of speech (POS) or word type, e.g., proper name or verb.
- Word sense, e.g., river *bank* vs. money *bank*.
- Phrasal cohesion of words, such as idioms, syntactic phrases, clauses, sentences.
- Modification relations among words.
- Anaphora (co-reference) and synonymy among words and phrases.
- Syntactic type identification, such as questions, quotes, commands, etc.
- Semantic focus identification (emphasis).
- Semantic type and speech act identification, such as requesting, informing, narrating, etc.
- Genre and style analysis.

Here we confine ourselves to discussion of the kind of information that a good parser could, in principle, provide to enable the TTS-specific functionality.

Linguistic analysis supports the phonetic analysis and prosodic generation phases. The modules of phonetic analysis are covered in Sections 14.6, 14.7, and 14.8. A linguistic parser can contribute in several ways to the process of generating (symbolic) phonetic forms from orthographic words found in text. One function of a parser is to provide accurate part-of-speech (POS) labels. This can aid in resolving the pronunciation of several hundred American English homographs, such as *object* and *absent*. Homographs are discussed in greater detail in Section 14.6. Parsers can also aid in identifying names and other special classes of vocabulary for which specialized pronunciation rule sets may exist [32].

Prosody generation deals mainly with assignment of segmental duration and pitch contour that have close relationship with prosodic phrasing (pause placement) and accentuation. Parsing can contribute useful information, such as the syntactic type of an utterance. (e.g., yes/no question contours typically differ from *wh*-question contours, though both are marked simply by '?' in text), as well as semantic relations of synonymy, anaphora, and focus that may affect accentuation and prosodic phrasing. Information from discourse analysis and text genre characterization may affect pitch range and voice quality settings. Further examination of the contribution of parsing specifically to prosodic phrasing, accentuation, and other prosodic interpretation is provided in Chapter 15.

As mentioned earlier, TTS can employ either a general-purpose NL analysis engine or a pipeline of a number of very narrowly targeted, special-purpose NL modules together for the requirement of TTS linguistic analysis. Although we focus on linguistic information for supporting phonetic analysis and prosody generation here, a lot of the information and services are beneficial to document structure detection and text normalization described in previous sections.

The minimum requirement for such a linguistic analysis module is to include a lexicon of the *closed-class* function words, of which only several hundred exist in English (at most), and perhaps homographs. In addition, a minimal set of modular functions or services would include:

- *Sentence breaking*—Sentence breaking has been discussed in Section 14.3.4 above.
- *POS tagging*—POS tagging can be regarded as a two-stage process. The first is POS guessing, which is the process of determining, through a combination of a (possibly small) dictionary and some morphological heuristics or a specialized morphological parser, the POS categories that might be appropriate for a given input term in isolation. The second is POS choosing—that is, the resolution of the POS in context, via local short-window syntactic rules, perhaps combined with probabilistic distribution for the POS guesses of a given word. Sometimes the guessing and choosing functions are combined in a single statistical framework. In [6], lexical probabilities are unigram frequencies of assignments of categories to words estimated from corpora. In the original formulation of the model, the lexical probabilities [$P(c_i \mid w_i)$, where c_i is the hypothesized POS for word w_i], were estimated from the hand-tagged Brown corpus [8]. For Example, the word *see* appeared 771 times as a verb and once as an interjection. Thus the probability that *see* is a verb is estimated to be 771/772 or 0.99. Trigrams are used for contextual probability [$P(c_i \mid c_{i-1}c_{i-2}\cdots c_1) = P(c_i \mid c_{i-1}c_{i-2})$]. Lexical probabilities and trigrams over category sequences are used to score all possible assignments of categories to words for a given input word sequence. The entire set of possible assignments of categories to words in sequence is calculated, and the best-scoring sequence is used. Likewise, simple methods have been used to detect noun phrases (NPs), which can be useful in assigning pronunciation, stress, and prosody. The method described in [6] relies on a table of probabilities for inserting an NP begin bracket '[' between any two POS categories, and similarly for an NP end bracket ']'. This was also trained on the POS-labeled Brown corpus, with further augmentation for the NP labels. For example, the probability of inserting an NP begin bracket after an article was found to be

much lower than that of begin-bracket insertion between a verb and a noun, thus automatically replicating human intuition.

- *Homograph disambiguation*—Homograph disambiguation in general refers to the case of words with the same orthographic representation (written form) but having different semantic meanings and sometimes even different pronunciations. Sometimes it is also referred as sense disambiguation. Examples include "The boy used the *bat* to hit a home run" vs. "We saw a large *bat* in the zoo" (the pronunciation is the same for two *bat*) and "You *record* your voice" vs. "I'd like to buy that *record*" (the pronunciations are different for the two *record*). The linguistic analysis module should at least try to resolve the ambiguity for the case of different pronunciations because it is absolutely required for correct phonetic rendering. Typically, the ambiguity can be resolved based on POS and lexical features. Homograph disambiguation is described in detail in Section 14.6.
- *Noun phrase (NP) and clause detection*—Basic NP and clause information could be critical for a prosodic generation module to generate segmental durations. It also provides useful cues to introduce necessary pauses for intelligibility and naturalness. Phrase and clause structure are well covered in any parsing techniques.
- *Sentence type identification*—Sentence types (declarative, yes-no question, etc.) are critical for macro-level prosody for the sentence. Typical techniques for identifying sentence types have been covered in Section 14.3.4.

If a more sophisticated parser is available, a richer analysis can be derived. A so-called *shallow parse* is one that shows syntactic bracketing and phrase type, based on the POS of words contained in the phrases. A training corpus of shallow-parsed sentences has been created for the Linguistic Data Consortium [16]. The following example illustrates a shallow parse for sentence : "For six years, Marshall Hahn Jr. has made corporate acquisitions in the George Bush mode: kind and gentle."

```
For/IN[six/CD years/NNS],/,[T./NNP Marshall/NNP
Hahn/NNP Jr./NNP]has/VBZ made/VBN[corporate/JJ acquisi-
tions/NNS]in/IN[the/DT George/NNP Bush/NNP mode/NN]
:/:[kind/JJ]and/CC[gentle/JJ]./.
```

The POS labels used in this example are described in Chapter 2 (Table 2.14). A TTS system uses the POS labels in the parse to decide alternative pronunciations and to assign differing degrees of prosodic prominence. Additionally, the bracketing might assist in deciding where to place pauses for great intelligibility. A fuller parse would incorporate more higher-order structure, including sentence type identification, and more semantic analysis, including co-reference.

14.6. HOMOGRAPH DISAMBIGUATION

For written languages, *sense* ambiguities occur when words have different syntactic/semantic meanings. Those words with different senses are called *polysemous* words. For example, *bat* could mean either a kind of animal or the equipment to hit a baseball. Since the pronunciations for the two different senses of *bat* are identical, we are in general only concerned[6] about the other type of polysemous words that are homographs (spelled alike but vary in pronunciation), such as *bass* for a kind of fish (*/b ae s/*) or an instrument (*/b ey s/*).

Homograph variation can often be resolved on POS (grammatical) category. Examples include *object, minute, bow, bass, absent*, etc. Unfortunately, correct determination of POS (whether by a parsing system or statistical methods) is not always sufficient to resolve pronunciation alternatives. For example, simply knowing that the form *bow* is a noun does not allow us to distinguish the pronunciation appropriate for the instrument of archery from that for the front part of a boat. Even more subtle is the pronunciation of *read* in "If you *read* the book, he'll be angry." Without contextual clues, even human readers cannot resolve the pronunciation of *read* from the given sentence alone. Even though the past tense is more likely in some sense, deep semantic and/or discourse analysis would be required to resolve the tense ambiguity.

Several hundred English homographs extracted from the 1974 *Oxford Advanced Learners Dictionary* are listed in [10]. Here are some examples:

- Stress homographs: noun with front-stress vowel, verb with end-stress vowel
  "an *absent* boy" vs. "Do you choose to *absent* yourself?"
- Voicing: noun/verb or adjective/verb distinction made by voice final consonant
  "They will *abuse* him." vs. "They won't take *abuse*."
- –ate words: noun/adjective sense uses schwa, verb sense uses a full vowel
  "He will *graduate*." vs. "He is a *graduate*."
- Double stress: front-stressed before noun, end-stressed when final in phrase
  "an *overnight* bag" vs. "Are you staying *overnight*?"
- -ed adjectives with matching verb past tenses
  "He is a *learned* man." vs. "He *learned* to play piano."
- Ambiguous abbreviations: already described in Section 14.4.1
  in, am, SAT (Saturday vs. Standard Aptitude Test)
- Borrowed words from other languages—They could sometimes be distinguishable based on capitalization.
  "El Camino *Real* road in California" vs. "*real* world"
  "*polish* shoes" vs. "*Polish* accent"

---

[6] Sometimes, a polysemous word with the same pronunciation could have impact for prosodic generation because different semantic properties could have different accentuation effects. Therefore, a high-quality TTS system can definitely be benefited from word-sense disambiguation beyond homograph disambiguation.

- Miscellaneous
  "The *sewer* overflowed." vs. "a *sewer* is not a tailor."
  "He *moped* since his parents refused to buy a *moped*."
  "*Agape* is a Greek word." vs. "His mouth was *agape*."

As discussed earlier, abbreviation/acronym expansion and linguistic analysis described in Sections 14.4.1 and 14.5 are two main sources of information for TTS systems to resolve homograph ambiguities.

We close this section by introducing two special sources of pronunciation ambiguity that are not fully addressed by current TTS systems. The first one is a variation of dialects (or even personal dialect—*idiolect*). For example, some might say *tom[ey]to*, while some others might say *tom[aa]to*. Another example is that Boston natives tend to reduce the */r/* sound in sentences like "Pa*r*k your ca*r* in Ha*r*va*r*d ya*r*d." Similarly, some people use the spelling pronunciation *in-ter-es-ting* as opposed to *intristing*. Finally, speech rate and formality level can influence pronunciation. For example, the */g/* sound in *recognize* may be omitted in faster speech. It might be a sensible decision to output all possible pronunciations as a multiple pronunciation list and hope the synthesis back end picks the one with better acoustic/prosodic voice rendition. While true homographs may be resolved by linguistic and discourse analysis, achieving a consistent presentation of dialectal and stylistic variation is an even more difficult research challenge.

The other special source of ambiguity in TTS is somewhat different from what we have considered so far, but may be a concern in some markets. Most borrowed or foreign single words and place names are realized naturally with pronunciation normalized to the main presentation language. Going beyond that, language detection refers to the ability of a TTS system to recognize the intended language of a multiword stretch of text. For example, consider the fragment "Well, as for the next department head, that is simply *une chose entendue*." The French phrase "*une chose entendue*" (something clearly understood) might be realized in a proper French accent and phone pronunciation by a bilingual English/French reader. For a TTS system to mimic the best performance, the system must have:

- language identification capability
- dictionaries and rules for both languages
- voice rendition capability for both languages

14.7. Morphological Analysis

General issues in morphology are covered in Chapter 2. Here, we consider issues of relating a surface orthographic form to its pronunciation by analyzing its component morphemes, which are minimal, meaningful elements of words, such as prefixes, suffixes, and stem words themselves. This decomposition process is referred as *morphological analysis* [28]. When a dictionary does not list a given orthographic form explicitly, it is sometimes possible to analyze the new word in terms of shorter forms already present. These shorter forms may combine as prefixes, one or more stems or roots, and suffixes to generate new forms. If a word can be so analyzed, the listed pronunciations of the pieces can be combined, perhaps with some adjustment (phonological rules), to yield a phonetic form for the word as a whole.

The prefixes and suffixes are generally considered bound, in the sense that they cannot stand alone but must combine with a stem. A stem, however, can stand alone. A word such as *establishment* may be decomposed into a "stem" *establish* and a suffix *-ment*. In practice, it is not always clear where this kind of analysis should stop. That is, should a system attempt to further decompose the stem *establish* into *establ* and *-ish*? These kinds of questions ultimately belong to etymology, the study of word origins, and there is no final answer. However, for practical purposes, having three classes of entries corresponding to prefixes, stems, and suffixes, where the uses of the affixes are intuitively obvious to educated native speakers, is usually sufficient. In practical language engineering, a difference that makes no difference is no difference, and unless there is a substantial gain in compression or analytical power, it is best to be conservative and list only obvious and highly productive affixes.

The English language presents numerous genuine puzzles in morphological analysis. For example, there is the issue of abstraction: is the word *geese* one morpheme, or two (base *goose* + abstract pluralizing morpheme)? For practical TTS systems, relying on large dictionaries, it is generally best to deal with concrete, observable forms where possible. In such a lexically oriented system, the word geese probably should appear in the lexicon as such, with attached grammatical features including plurality. Likewise, it is simpler to include *children* in the lexical listing rather than create a special pluralizing suffix *-ren* whose use is restricted to the single base *child*.

The morphological analyzer must attempt to cover an input word in terms of the affixes and stems listed in the morphological lexicon. The covering(s) proposed must be legal sequences of forms, so that often a word grammar is supplied to express the allowable patterns of combinations. A word grammar might, for example, restrict suffixation to the final or rightmost stem of a compound, thus allowing plurality on the final element of *businessmen* but not in the initial stem (*businessesman*). In support of the word grammar, all stems and affixes in the lexicon would be listed with morphological combinatory class specifications, usually subtyped in accordance with the base POS categories of the lexicon entries. That is, verbs would typically accept a different set of affixes than nouns or adjectives. In addition, spelling changes that sometimes accompany affixation must be recognized and undone during analysis. For example, the word *stopping* has undergone final consonant doubling as part of accommodating the suffix *ing*.

A morphological analysis system might be as simple as a set of *suffix-stripping* rules for English. If a word cannot be found in the lexicon, a suffix-stripping rule can be applied to first strip out the possible suffix, including *–s, -'s, -ing, -ed, -est, -ment*, etc. If the stripped form can be found in the lexicon, a morphological decomposition is attained. Similarly, *prefix-stripping* rules can be applied to find prefix-stem decomposition for prefixes like *in-, un-, non-, pre-, sub-, etc.*, although in general prefix stripping is less reliable.

Suffix and prefix stripping gives an analysis for many common inflected and some derived words such as *helped, cats, establishment, unsafe, predetermine, subword*, etc. It helps in saving system storage. However, it does not account for compounding, issues of legality of sequence (word grammar), or spelling changes. It can also make mistakes (from a synchronic point of view: *basement* is not *base* + *-ment*), some of which will have consequences in TTS rendition. A more sophisticated version could be constructed by adding elements such as POS type on each suffix/prefix for a rudimentary legality check on combinations. However, a truly robust morphological capability would require more powerful formal machinery and a more thorough analysis. Therefore, adding irregular morphological formation into a system dictionary is always a desirable solution.

Finally, sometimes in commercial product names the compounding structure is signaled by word-medial case differences, e.g., AltaVista™, which can aid phonetic conversion algorithms. These can be treated as two separate words and will often sound more natural if rendered with two separate main stresses. This type of decomposition can be expanded to find compound words that are formed by two separate nouns. Standard morphological analysis algorithms employing suffix/prefix stripping and compound word decomposition are summarized in Algorithm 14.4. Note that the algorithm can be easily modified to handle words constructed by a combination of prefix, suffix, and compound.

ALGORITHM 14.4: *MORPHOLOGICAL ANALYSIS*

1. **Dictionary Lookup**
   Look up word w in lexicon
   If found
     Output attributes of the found lexical entry and exit
2. **Suffix Stripping**
   If word ends in *-s, -'s, -ing, -ed, -est, -ment,* etc.
     Strip the suffix from word w to form u
   If stripped form u found in lexicon
     Output attributes of the stem and suffix and exit
3. **Prefix Stripping**
   If word begins with *in-, un-, non-, pre-, sub-,* etc.
     Strip the prefix from word w to form u
   If stripped form u found in lexicon
     Output attributes of the prefix and stem and exit
4. **Compound word decomposition**
   If detect word-medial case differences within word w
     Break word w into a multiple words u_1, u_2, u_3, ...according to case changes
     For words u_1, u_2, u_3, **goto 1**.
   Else if word w can be decomposed into two nouns u_1, u_2 in lexicon
     Output attributes of the u_1, u_2 and exit
5. Pass word w to letter-to-sound module

14.8. LETTER-TO-SOUND CONVERSION

The best resource for generating (symbolic) phonetic forms from words is an extensive word list. The accuracy and efficiency of such a solution is limited only by the time, effort, and knowledge brought to bear on the dictionary construction process. As described in Section 14.2, a general lexicon service is a critical resource for the TTS system. Thus, the first and the most reliable way for grapheme-to-phoneme conversion is via dictionary lookup.

Where direct dictionary lookup fails, rules may be used to generate phonetic forms. Under earlier naïve assumptions about the regularity and coverage of simple descriptions of English orthography, rules have traditionally been viewed as the primary source of phonetic conversion knowledge, since no dictionary covers every input form and the TTS system must always be able to speak any word. A general *letter-to-sound* (LTS) conversion is thus required to provide phonetic pronunciation for any sequence of letters.

Inspired by the phonetic languages, letter-to-sound conversion is usually carried out by a set of rules. These rules can be thought of as dictionaries of fragments with some special conventions about lookup and context. Typically, rules for phonetic conversion have mimicked phonological rewriting in phonological theory [5], including conventions of ordering, such as *most specific first*. In phonological rules, a target is given and the rewrite is indicated, with context following. For example, a set of rules that changes orthographic *k* to a velar plosive */k/* except when the *k* is word-initial ('[') followed by *n* might appear as:

```
k -> /sil/ % [ _ n
k -> /k/
```

The rule above reads that *k* is rewritten as (phonetic) silence when in word initial position and followed by *n*, otherwise *k* is rewritten as (phonetic) */k/*. The underscore in the first line is a placeholder for the *k* itself in specifying the context. This little set properly treats *k* in *knight*, *darkness*, and *kitten*. These are formally powerful, context-sensitive rules. Generally a TTS system require hundreds or even thousands of such rules to cover words not appearing in the system dictionary or *exception list*. Typically rules are specified in terms of single-letter targets, such as the example for *k* above. However, some systems may have rules for longer fragments, such as the special vowel and consonant combinations in words like *neighbor* and *weigh*. In practice, a binary format for compression, a corresponding fragment matching capability, and a rule index must be defined for efficient system deployment.

Rules of this type are tedious to develop manually. As with any *expert system*, it is difficult to anticipate all possible relevant cases and sometimes hard to check for rule interference and redundancy. In any case, the rules must be verified over a test list of words with known transcriptions. Generally, if prediction of main stress location is not attempted, such rules might account for up to 70% of the words in a test corpus of general English. If prediction of main stress is attempted, the percentage of correct phonetic pronunciations is much lower, perhaps below 50%. The correct prediction of stress depends in part on morphology, which is not typically explicitly attempted in this type of simple rule system (though fragments corresponding to affixes are frequently used, such as *tion* -> */ah ax n/*). Certainly, such rules can be made to approach dictionary accuracy, as longer and more explicit morphological fragments are included. One extreme case is to create one specific rule (containing exact contexts for the whole word) for each word in the dictionary. Obviously this is not desirable, since it is equivalent to putting the word along with its phonetic pronunciation in the dictionary.

In view of how costly it is to develop LTS rules, particularly for a new language, attempts have been made recently to automate the acquisition of LTS conversion rules. These self-organizing methods believe that, given a set of words with correct phonetic transcriptions (the offline dictionary), an automated learning system could capture significant generalizations. Among them, classification and regression trees (CART) have been demonstrated to give satisfactory performances for letter-to-sound conversion. For basic and theoretic description of CART, please refer to Chapter 4.

In the system described in [14], CART methods and phoneme trigrams were used to construct an accurate conversion procedure. All of the experiments were carried on two databases. The first is the NETALK [25], which has hand-labeled alignment between letter and phoneme transcriptions. The second is the CMU dictionary, which does not have any alignment information. The NETALK database consists of 19,940 entries, of which 14,955 were randomly selected as a training set and the remaining 4951 were reserved for testing. Those 4951 words correspond to 4985 entries in the database because of multiple pronunciations. The hand-labeled alignments were used directly to train the CART for LTS conversion. The CMU dictionary has more than 100,000 words, of which the top 60,000 words were selected based on unigram frequencies trained from North American Business News. Among them, 52,415 were used for training and 9719 reserved for testing. Due to multiple pronunciations, those 9719 words have 10,520 entries in the dictionary. Due to lack of alignment information, dynamic programming was used to align each letter to the corresponding phoneme before training the LTS CART.

The basic CART component includes a set of *yes-no* questions and a procedure to select the best question at each node to grow the tree from the root. The basic *yes-no* question for LTS conversion looks like "*Is the second right letter 'p'?*" or "*Is the first left output phoneme /ay/?*" The questions for letters could be on either the left or the right side. For phones, only questions on the left side were used, for simplicity. The range of question positions must be long enough to cover the long-distance phonological variations. It was found that the 11-letter window (5 for left letter context and 5 for right letter context) and 3-phoneme window for left phoneme context are generally sufficient. A primitive set of questions would be the set of all the singleton questions about each letter or phoneme identity. When growing the tree, the question that had the best entropy reduction was chosen at each node. We observed that if we allow the node to have a complex question that is a combination of primitive questions, the depth of the tree will be greatly reduced and the performance improved. For example, the complex question "*Is the second left letter 't' and the first left letter 'i' and the first right letter 'n'?*" can capture '*o*' in common suffix "*tion*" and convert it to the right phoneme. Complex questions can also alleviate the data fragment problem caused by greedy nature of the CART algorithm. This way of finding such complex questions is similar to those used in Chapter 4. The baseline system built using the above techniques has error rates as listed in Table 14.11.

Table 14.11 LTS baseline results using CART [13].

Database	Phoneme	Word
CMU Lexicon	9.7%	35.0%
NETTALK	9.5%	42.3%

The CART LTS system [14] further improved the accuracy of the system via the following extensions and refinements:

- *Phoneme trigram rescoring*: A statistical model of phoneme co-occurrence, or phonotactics, was constructed over the training set. A phonemic trigram was generated from the training samples with back-off smoothing, and this was used to rescore the n-best list generated by LTS.
- *Multiple tree combination:* The training data was partitioned into two parts and two trees were trained. When the performance of these two trees was tested, it was found that they had a great overlap but also behaved differently, as each had a different focus region. Combining them together greatly improved the coverage. To get a better overall accuracy, the tree trained by all the samples was used together with two other trees, each trained by half of the samples. The leaf distributions of three trees were interpolated together with equal weights and then phonemic trigram was used to rescore the n-best output lists.

By incrementally experimenting with addition of these extensions and refinements, the results improved, as shown in Table 14.12.

These experiments did not include prediction of stress location. Stress prediction is a difficult problem, as we pointed out earlier. It requires information beyond the letter string. In principle, one can incorporate more lexical information, including POS and morphologic information, into the CART LTS framework, so it can be more powerful to learn the phonetic correspondence between the letter string and lexical properties.

Table 14.12 LTS using multiple trees and phonemic trigram rescoring [13].

Database	Phoneme	Word
CMU Lexicon	8.2%	26.9%
NETTALK	8.1%	34.2%

14.9. EVALUATION

Ever since the early days of TTS research [21, 31], evaluation has been considered an integral part of the development of TTS systems. End users and application developers are mostly interested in the end-to-end evaluation of TTS systems. This *monolithic* type of whole-system evaluation is often referred to as *black-box* evaluation. On the other hand, *modular* (component) testing is more appropriate for TTS researchers when working with isolated components of the TTS system, for diagnosis or regression testing. We often refer to this type of evaluation as *glass-box* evaluation. We discuss the modular evaluations in each modular TTS chapter, while leaving the evaluation of the whole system to Chapter 16.

For text and phonetic analysis, automated, analytic, and objective evaluation is usually feasible, because the input and output of such module is relatively well defined. The evaluation focuses mainly on symbolic and linguistic level in contrast to the acoustic level, with which prosodic generation and speech synthesis modules need to deal. Such tests usually involve establishing a test corpus of correctly tagged examples of the tested materials, which can be automatically checked against the output of a text analysis module. It is not particularly productive to discuss such testing in the abstract, since the test features must closely track each system's design and implementation. Nevertheless, a few typical areas for testing can be noted. In general, tests are simultaneously testing the linguistic model and content as well as the software implementation of a system, so whenever a discrepancy arises, both possible sources of error must be considered.

For automatic detection of document structures, the evaluation typically focuses on sentence breaking and sentence type detection. Since the definitions of these two types of document structures are straightforward, a standard evaluation database can be easily established.

In the basic level, the evaluation for the text normalization component should include large regression test databases of text micro-entities: addresses, Internet and email entities, numbers in many formats (ordinal, cardinal, mathematical, phone, currency, etc.), titles, and abbreviations in a variety of contexts. These would be paired with the correct reference forms in something like the SNOR used in ASR output evaluation. In its simplest form, this would consist of a database of automatically checkable paired entries like 7% vs. *seven percent*, and $1.20 vs. *one dollar and twenty cents*. If you want to evaluate the semantic capability of text normalization, the regression database might include markups for semantic tags, so that we have 7% vs. "<number>SEVEN<ratio>PERCENT</ratio></number>", and $1.20 vs. "<money>ONE DOLLAR AND TWENTY CENTS</money>". The regression database could include domain-specific entries. This implies some dependence on the system's API—its markup capabilities or tag set. In the examples given in Table 14.13, the first one is a desirable output for domain-independent input, while the second one is suitable for normalization of the same expression in mathematical formula domain.

Table 14.13 Two examples to test domain independent/dependent text normalization.

3-4	*three to four*
	three four
<math_exp> 3-4 </math_exp>	*three minus four*

Some systems may not have a discrete level of orthographic or SNOR representation that easily lends itself to the type of evaluation described in this section. Such systems may have to evaluate their text normalization component in terms of LTS conversion.

An automated test framework for the LTS conversion analysis minimally includes a set of test words and their phonetic transcriptions for automated lookup and comparison tests. The problem is the infinite nature of language: there are always *new words* that the system does not convert correctly, and many of these will initially lack a transcription of record even to allow systematic checking. Therefore, a comprehensive test program for test of phonetic conversion accuracy needs to be paired with a data development effort. The data effort has two goals: to secure a continuous source of potential new words, such as a 24-hour newswire feed, and to maintain and construct an offline test dictionary, where reference transcriptions for new words are constantly created and maintained by human experts. This requirement illustrates the codependence of automated and manual aspects of evaluation. Different types and sources of vocabulary need to be considered separately, and they may have differing testing requirements, depending, again, on the nature of the particular system to be evaluated. For example, some systems have elaborate subsystems targeted specifically for name conversion. Such systems may depend on other kinds of preprocessing technologies, such as name identification modules, that might be tested independently.

The correct phonetic representation of a word usually depends on its sentence and even discourse contexts, as described in Section 14.6. Therefore, the adequacy of LTS conversion should not, in principle, be evaluated on the basis of isolated word pronunciations. However, a list of isolated word pronunciations is often used in LTS conversion because of its simplicity. Discourse contexts are, in general, difficult to represent unless specific applications and markup tags are available to the evaluation database. A reasonable compromise is to use a list of independent sentences with their corresponding phonetic representation for the evaluation of grapheme-to-phoneme conversion.

Error analysis should be treated as equally important as the evaluation itself. For example, if a confusability matrix shows that a given system frequently confuses central and schwa-like unstressed vowels, this may be viewed as less serious than other kinds of errors. Other subareas of LTS conversion that could be singled out for special diagnosis and testing include morphological analysis and stress placement. Of course, testing with phonemic transcriptions is the ultimate *unit test* in the sense that it contains nothing to insure that the correctly transcribed words, when spoken by the system's artificial voice and prosody, are intelligible or pleasant to hear. Phone transcription accuracy is, thus, a necessary but not a sufficient condition of quality.

14.10. CASE STUDY: FESTIVAL

The University of Edinburgh's *Festival* [3] has been designed to take advantage of modular subcomponents for various standard functions. Festival provides a complete text and phonetic analysis with modules organized in sequence roughly equivalent to Figure 14.1. Festival outputs speech of quality comparable to many commercial synthesizers. While default routines are provided for each stage of processing, the system is architecturally designed to accept alternative routines in modular fashion, as long as the data transfer protocols are followed. This variant of the traditional TTS architecture is particularly attractive for commercial purposes (development, maintenance, testing, scalability) as well as research. Festival can be called in various ways with a variety of switches and filters, set from a variety of sanctioned programming and scripting languages. These control options are beyond the scope of this overview.

14.10.1. Lexicon

Festival employs phonemes as the basic sounding units, which are used not only as the atoms of word transcriptions in the lexicons, but also as the organizing principle for unit selection (see Chapter 16) in the synthesizer itself. Festival can support a number of distinct phone sets and it supports mapping from one to another. A phone defined in a set can have various associated phonological features, such as vowel, high, low, etc.

The Festival lexicon, which may contain several components, provides pronunciations for words. The *addenda* is an optional list of words that are unique to a particular user, document, or application. The addenda is searched linearly. The main system lexicon is expected to be large enough to require compression and is assumed to reside on a disk or other external storage. It is accessed via binary search. The lexical entry also contains POS information, which can be modified according to the preference of the system configurer. A typical lexical entry consists of the word key, a POS tag, and phonetic pronunciation (with stress and possible syllabification indicated in parentheses):

```
("walkers" N ((( w ao ) 1) (( k er z ) 0)) )
```

If the syllables structure is not shown with parentheses, a syllabification rule component can be invoked. Separate entry lines are used for words with multiple pronunciations and/or POS, which can be resolved by later processing.

14.10.2. Text Analysis

Festival has been partially integrated with research on the use of automatic identification of document and discourse structures. The discourse tagging is done by a separate component, called SOLE [11]. The tags produced by SOLE indicate features that may have relevance for pitch contour and phrasing in later stages of synthesis (see Chapter 15). These must be recognized and partially interpreted at the text analysis phrase. The SOLE tags tell Festival when the text is comparing or contrasting two objects, when it's referring to old or new information, when it's using a parenthetical or starting a new paragraph, etc., and Festival will decide, based on this information, that it needs to pause, to emphasize or deemphasize, to modify its pitch range, etc.

Additionally, as discussed in Section 14.3, when document creators have knowledge about the structure or content of documents, they can express the knowledge through an XML-based synthesis markup language. A document to be spoken is first analyzed for all such tags, which can indicate alternative pronunciations, semantic or quasi-semantic attributes (different uses of numbers by context for example), as well as document structures, such as explicit sentence or paragraph divisions. The kinds of information potentially supplied by the SABLE tags[7] are exemplified in Figure 14.7.

```
<SABLE>
    <SPEAKER NAME="male1">
    The boy saw the girl in the park <BREAK/> with the telescope.
    The boy saw the girl <BREAK/> in the park with the telescope.

    Good morning <BREAK /> My name is Stuart, which is spelled
    <RATE SPEED="-40%">
    <SAYAS MODE="literal">stuart</SAYAS> </RATE>
    though some people pronounce it
    <PRON SUB="stoo art">stuart</PRON>. My telephone number
    is <SAYAS MODE="literal">2787</SAYAS>.

    I used to work in <PRON SUB="Buckloo">Buccleuch</PRON> Place,
    but no one can pronounce that.
    </SPEAKER>
</SABLE>
```

Figure 14.7 A document fragment augmented with SABLE tags can be processed by the Festival system [3].

For untagged input, or for input inadequately tagged for text division (`<BREAK/>`), sentence breaking is performed by heuristics, similar to Algorithm 14.1, which observe whitespace, punctuation, and capitalization. A linguistic unit roughly equivalent to a sentence is created by the system for the subsequent stages of processing.

Tokenization is performed by system or user-supplied routines. The basic function is to recognize potentially speakable items and to strip irrelevant whitespace or other nonspeakable text features. Note that some punctuation is retained as a feature on its nearest word.

Text normalization is implemented by token-to-word rules, which return a standard orthographic form that can, in turn, be input to the phonetic analysis module. The token-to-word rules have to deal with text normalization issues similar to those presented in Section 14.4. As part of this process, token-type-specific rule sets may be applied to disambiguate tokens whose pronunciations are highly context dependent. For example, a disambiguation routine may be required to examine context for deciding whether *St.* should be realized as *Saint* or *street*. For general English-language phenomena, such as numbers and various

[7] SABLE and other TTS markup systems are discussed further in Chapter 15.

symbols, a standard token-to-word routine is provided. One interesting feature of the Festival system is a utility for helping to automatically construct decision trees to serve text normalization rules, when system integrators can gather some labeled training data.

The linguistic analysis module for the Festival system is mainly a POS analyzer. An *n*-gram based trainable POS tagger is used to predict the likelihoods of POS tags from a limited set given an input sentence. The system uses both a priori probabilities of tags given a word and *n*-grams for sequences of tags. The basic underlying technology is similar to the work in [6] and is described in Section 14.5. When lexical lookup occurs, the predicted most likely POS tag for a given word is input with the word orthography, as a compound lookup key. Thus, the POS tag acts as a secondary selection mechanism for the several hundred words whose pronunciation may differ by POS categories.

14.10.3. Phonetic Analysis

The homograph disambiguation is mainly resolved by POS tags. When lexical lookup occurs, the predicted most likely POS tag for a given word is input with the word orthography as a compound lookup key. Thus, the POS tag acts as a secondary selection mechanism for the several hundred words whose pronunciation may differ by POS categories.

If a word fails lexical lookup, LTS rules may be invoked. These rules may be created by hand, formatted as shown below:

(# [c h] C = /k /) // ch at word start, followed by a consonant, is /k/, e.g., Chris

Alternatively, LTS rules may be constructed by automatic statistical methods, much as described in Section 14.8 above, where CART LTS systems were introduced. Utility routines are provided to assist in using a system lexicon as a training database for CART rule construction.

In addition, Festival system employs *post-lexical rules* to handle *context coarticulation*. Context coarticulation occurs when surrounding words and sounds, as well as speech style, affect the final form of pronunciation of a particular phoneme. Examples include reduction of consonants and vowels, phrase final devoicing, and *r*-insertion. Some coarticulation rules are provided for these processes, and users may also write additional rules.

14.11. Historical Perspective and Further Reading

Text-to-speech has a long and rich history. You can hear samples and review almost a century's worth of work at the Smithsonian's Speech Synthesis History Project [19]. A good source for multilingual samples of various TTS engines is [20].

The most influential single published work on TTS has been *From Text to Speech: The MITalk System* [1]. This book describes the MITalk system, from which a large number of research and commercial systems were derived during the 1980s, including the widely used DECTalk system [9]. The best compact overall historical survey is Klatt's *Review of Text-to-Speech Conversion for English* [15]. For deeper coverage of more recent architectures, refer to [7]. For an overview of some of the most promising current approaches and pressing issues in all areas of TTS and synthesis, see [30]. One of the biggest upcoming issues in TTS text processing is the architectural relation of specialized TTS text processing as opposed to general-purpose natural language or document structure analysis. One of the most elaborate and interesting TTS-specific architectures is the multilingual text processing engine described in [27]. This represents a commitment to providing exactly the necessary and sufficient processing that speech synthesis requires, when a general-purpose language processor is unavailable.

However, it is expected that natural language and document analysis technology will become more widespread and important for a variety of other applications. To get an idea of what capabilities the natural language analysis engines of the future may incorporate, refer to [12] or [2]. Such generalized engines would serve a variety of clients, including TTS, speech recognition, information retrieval, machine translation, and other services which may seem exotic and isolated now but will increasingly share core functionality. This convergence of NL services can be seen in a primitive form today in Japanese *input method editors* (IME), which offload many NL analysis tasks from individual applications, such as word processors and spreadsheets, and unify these functions in a single common processor [18].

For letter-to-sound rules, NETalk [25], which describes automatic learning of LTS processes via neural network, was highly influential. Now, however, most systems have converged on decision-tree systems similar to those described in [14].

REFERENCES

[1] Allen, J., M.S. Hunnicutt, and D.H. Klatt, *From Text to Speech: the MITalk System*, 1987, Cambridge, UK, University Press.

[2] Alshawi, H., *The Core Language Engine*, 1992, Cambridge, US, MIT Press.

[3] Black, A.W., P. Taylor, and R. Caley, "The Architecture of the Festival Speech Synthesis System," *3rd ESCA Workshop on Speech Synthesis*, 1998, Jenolan Caves, Australia, University of Edinburgh, pp. 147-151.

[4] Boguraev, B. and E.J. Briscoe, *Computational Lexicography for Natural Language Processing*, 1989, London, Longmans.

[5] Chomsky, N. and M. Halle, *The Sound Patterns of English*, 1968, Cambridge, MIT Press.

[6] Church, K., "A Stochastic Parts Program and Noun Phrase Parser for Unrestricted Text," *Proc. of the Second Conf. on Applied Natural Language Processing*, 1988, Austin, Texas, pp. 136-143.

[7] Dutoit, T., *An Introduction to Text-to-Speech Synthesis*, 1997, Kluwer Academic Publishers.

[8] Francis, W. and H. Kucera, *Frequency Analysis of English Usage*, 1982, New York, N.Y., Houghton Mifflin.

[9] Hallahan, W.I., "DECtalk Software: Text-to-Speech Technology and Implementation," *Digit Technical Journal*, 1995, **7**(4), pp. 5-19.

[10] Higgins, J., *Homographs*, 2000, http://www.stir.ac.uk/celt/staff/higdox/wordlist/homogrph.htm.

[11] Hitzeman, J., *et al.*, "On the Use of Automatically Generated Discourse-Level Information in a Concept-to-Speech Synthesis System," *Proc. of the Int. Conf. on Spoken Language Processing*, 1998, Sydney, Australia, pp. 2763-2766.

[12] Jensen, K., G. Heidorn, and S. Richardson, *Natural Language Processing: the PLNLP Approach*, 1993, Boston, MASS, Kluwer Academic Publishers.

[13] Jiang, L., H.W. Hon, and X. Huang, "Improvements on a Trainable Letter-to-Sound Converter," *Proc. of Eurospeech*, 1997, Rhodes, Greece, pp. 605-608.

[14] Jiang, L., H.W. Hon, and X.D. Huang, "Improvements on a Trainable Letter-to-Sound Converter," *Eurospeech'97*, 1997, Rhodes, Greece.

[15] Klatt, D., "Review of Text-to-Speech Conversion for English," *Journal of Acoustical Society of America*, 1987, **82**, pp. 737-793.

[16] LDC, *Linguistic Data Consortium*, 2000, http://www.ldc.upenn.edu/ldc/noframe.html.

[17] Levine, J., Mason, T., Brown, D., *Lex and Yacc*, 1992, Sebastopol, CA, O'Rielly & Associates.

[18] Lunde, K., *CJKV Information Processing Chinese, Japanese, Korean & Vietnamese Computing*, 1998, O'Reilly.

[19] Maxey, H., *Smithsonian Speech Synthesis History Project*, 2000, http://www.mindspring.com/~dmaxey/ssshp/.

[20] Möhler, G., *Examples of Synthesized Speech*, 1999, http://www.ims.uni-stuttgart.de/phonetik/gregor/synthspeech/.

[21] Nye, P.W., *et al.*, "A Plan for the Field Evaluation of an Automated Reading System for the Blind," *IEEE Trans. on Audio and Electroacoustics*, 1973, **21**, pp. 265-268.

[22] OMF, *CML - Chemical Markup Language*, 1999, http://www.xml-cml.org/.

[23] Richardson, S.D., W.B. Dolan, and L. Vanderwende, "MindNet: Acquiring and Structuring Semantic Information from Text," *ACL'98: 36th Annual Meeting of the Assoc. for Computational Linguistics and 17th Int. Conf. on Computational Linguistics*, 1998, pp. 1098-1102.

[24] Sable, *The Draft Specification for Sable version 0.2*, 1998, http://www.cstr.ed.ac.uk/projects/sable_spec2.html.

[25] Sejnowski, T.J. and C.R. Rosenberg, *NETtalk: A Parallel Network that Learns to Read Aloud*, 1986, Johns Hopkins University.

[26] Sluijter, A.M.C. and J.M.B. Terken, "Beyond Sentence Prosody: Paragraph Intonation in Dutch," *Phonetica*, 1993, **50**, pp. 180-188.

[27] Sproat, R., *Multilingual Text-To-Speech Synthesis: The Bell Labs Approach*, 1998, Dordrecht, Kluwer Academic Publishers.

[28] Sproat, R. and J. Olive, "An Approach to Text-to-Speech Synthesis," in *Speech Coding and Synthesis*, W.B. Kleijn and K.K. Paliwal, eds. 1995, Amsterdam, pp. 611-634, Elsevier Science.

[29] Turing, A.M., "Computing Machinery and Intelligence," *Mind*, 1950, **LIX**(236), pp. 433-460.

[30] van Santen, J., *et al.*, *Progress in Speech Synthesis*, 1997, New York, Springer-Verlag.

[31] van Santen, J., *et al.*, "Report on the Third ESCA TTS Workshop Evaluation Procedure," *Third ESCA Workshop on Speech Synthesis*, 1998, Sydney, Australia.

[32] Vitale, T., "An Algorithm for High Accuracy Name Pronunciation by Parametric Speech Synthesizer," *Computational Linguistics*, 1991, **17**(3), pp. 257-276.

[33] W3C, *Aural Cascading Style Sheets (ACSS)*, 1997, http://www.w3.org/TR/WD-acss-970328.

[34] W3C, *W3C's Math Home Page*, 1998, http://www.w3.org/Math/.

[35] W3C, *Extensible Markup Language (XML)*, 1999, http://www.w3.org/XML/.

[36] Wall, L., Christiansen, T., Schwartz, R., *Programming Perl*, 1996, Sebastopol, CA, O'Rielly & Associates.

CHAPTER 15

Prosody

It isn't **what** you said; it's **how** you said it! Sheridan pointed out the importance of prosody more than 200 years ago [53]:

> *Children are taught to read sentences, which they do not understand; and as it is impossible to lay the emphasis right, without perfectly comprehending the meaning of what one reads, they get a habit either of reading in a monotone, or if they attempt to distinguish one word from the rest, as the emphasis falls at random, the sense is usually perverted, or changed into nonsense.*

Prosody is a complex weave of physical, phonetic effects that is being employed to express attitude, assumptions, and attention as a parallel channel in our daily speech communication. The semantic content of a spoken or written message is referred to as its *denotation*, while the emotional and attentional effects intended by the speaker or inferred by a listener are part of the message's *connotation*. Prosody has an important supporting role in guiding a listener's recovery of the basic messages (denotation) and a starring role in signaling connotation, or the speaker's attitude toward the message, toward the listener(s),

ing connotation, or the speaker's attitude toward the message, toward the listener(s), and toward the whole communication event.

From the listener's point of view, prosody consists of systematic perception and recovery of a speaker's intentions based on:

- **Pauses**: to indicate phrases and to avoid running out of air.
- **Pitch**: rate of vocal-fold cycling (fundamental frequency or F0) as a function of time.
- **Rate/relative duration**: phoneme durations, timing, and rhythm.
- **Loudness**: relative amplitude/volume.

Pitch is the most expressive of the prosodic phenomena. As we speak, we systematically vary our fundamental frequency to express our feelings about what we are saying, or to direct the listener's attention to especially important aspects of our spoken message. If a paragraph is spoken on a constant, uniform pitch with no pauses, or with uniform pauses between words, it sounds highly unnatural.

In some languages, the pitch variation is partly constrained by lexical and syntactic conventions. For example, Japanese words usually exhibit a sharp pitch fall at a certain vowel on a consistent, word-specific basis. In Mandarin Chinese [52], word meaning depends crucially on shape and register distinctions among four highly stylized syllable pitch contour types. This is a grammatical and lexical use of pitch. However, every language, and especially English, allows some range of pitch variation that can be exploited for emotive and attentional purposes. While this chapter concentrates primarily on American English, the use of some prosodic effects to indicate emotion, mood, and attention is probably universal, even in languages that also make use of pitch for signaling word identity, such as Chinese. It is tempting to speculate that speakers of some languages use expressive and affective lexical particles and interjections to express some of the same emotive effects for which American English speakers typically rely on prosody.

We discuss pausing, pitch generation, and duration separately, because it is convenient to separate them when building systems. Bear in mind, however, that all the prosodic qualities are highly correlated in human speech production. The effect of loudness is not nearly as important in synthesizing speech as the effect of the other two factors and thus is not discussed here. In addition, for many concatenative systems this is generally embedded in the speech segment.

15.1. THE ROLE OF UNDERSTANDING

To date, most work on prosody for TTS has focused exclusively on the utterance, which is the literal content of the message. That is, a TTS system learns whatever it can from the isolated, textual representation of a single sentence or phrase to aid in prosodic generation. Typically a TTS system may rely on word identity, word part-of-speech, punctuation, length of a sentence or phrase, and other superficial characteristics. As more sophisticated NLP capabilities are deployed for use by TTS systems, deeper properties of an utterance, including its document or discourse context, can be taken into account.

Good prosody depends on a speaker or reader's understanding of the text's or message's meaning. As noted in [64], the *golden rule* of the Roman orator Quintilian (c. A.D. 90) states [32] "*That to make a man speak well, and pronounce with a right emphasis, he ought thoroughly to understand all that he says, be fully persuaded of it, and bring himself to have, those affections which he desires to infuse in others.*" This is clearly a tall order for today's computers! How important is understanding of the text's meaning, in generation of appropriately engaging prosody? Consider a stanza from Lewis Carroll's nonsense poem "Jabberwocky" [10]:

Twas brillig, and the slithy toves
Did gyre and gimble in the wabe;
All mimsy were the borogoves,
And the mome raths outgrabe.

Here, a full interpretation is not possible, owing primarily to lexical uncertainty (our ignorance of the meaning of words like *brillig*). However, you can recover a great deal of information from this passage that aids prosodic rendition. Foremost is probably the metrical structure of the poetic meter. This imposes a rhythmic constraint on prosodic phrasing (cadence, timing, and pause placement). Second, the function words such as *and*, *the*, *in*, etc. are interpretable and give rich clues about the general type and direction of action being specified. They also give us contextual hints about the part-of-speech of the neighboring nonsense words, which is a first crude step in interpreting those words' meaning. Third, punctuation is also important in this case. Using these three properties, with some analogical guesses about LTS conversions and stress locations in the nonsense words, would allow most speakers of English to render fairly pleasant and appropriate prosody for this poem.

Can a computer do the same? Where will a computer fall short of a human's performance on this task, and why? First, the carrier voice quality of a human reader is generally superior to synthesized voices. The natural human voice is more pleasant to a listener, all else being equal. As for the prosody per se, most TTS systems today use a fairly simple method to derive prosody, based on a distinction between closed-class function words, such as determiners and prepositions, which are thought to receive lesser emphasis, and open-ended sets of content words such as nouns like *wabe*, which are more likely to be accented. For this nonsense poem, that is essentially what most human readers do. Thus, if accurate LTS conversions are supplied, including main stress locations, a TTS system with a good synthetic voice and a reasonable default pitch algorithm of this type could probably render this stanza fairly well. Again, though the computer does not recognize it explicitly, the constrained rhythmic structure of the poem may be assisting.

But listeners to nonsense poems are generally not fully participating in the unconscious interpretive dialog, the attempt on the part of the listener to actively construct useful meaning from prosodic and message-content cues supplied in good faith by the speaker. Therefore, judgments of the prosodic quality of uninterpretable nonsense materials must always be suspect. In ordinary prose, the definition and recovery of *meaning* remains a slippery question. Consider the passage below [56], which is not metrically structured, has few or no true nonsense words, and, yet, was deliberately constructed to be essentially meaningless.

> *In mathematical terms, Derrida's observation relates to the invariance of the Einstein field equation under nonlinear space-time diffeomorphisms (self-mappings of the space-time manifold which are infinitely differentiable but not necessarily analytic). The key point is that this invariance group 'acts transitively': this means that any space-time point, if it exists at all, can be transformed into any other. In this way the infinite-dimensional invariance group erodes the distinction between observer and observed; the pi of Euclid and the G of Newton, formerly thought to be constant and universal, are now perceived in their ineluctable historicity; and the putative observer becomes fatally decentered, disconnected from any epistemic link to a space-time point that can no longer be defined by geometry alone.*

Should the fact that, say, a professional news broadcaster with no prior knowledge of the author's intent could render this supposedly meaningless passage rather well, make us suspicious of any claims regarding the necessity of deep semantic analysis for high-quality prosody? Though perhaps meaningless when taken as a whole, once again, the educated human reader can certainly recover fragments of meaning from this text sufficient to support reasonable prosody. The morphology and syntax is all standard English, which takes us a long way. The quality of the announcer's rendition degrades somewhat under the condition the computer truly faces, which can be simulated by replacing the *content words* of a sentence above with content words randomly chosen from "Jabberwocky":

> *In brillig toves, Derrida's wabe gimbles to the bandersnatch of the Tumtum whiffling raths under frumious slithy diffeomorphisms (borogoves of the mimsy mome which are beamishly vorpal but not frabjously uffish).*

It is likely the human reader can still outperform the computer by reliance on morphological and syntactic cues, such as the parallelism determining the accent placements in the contrastive structure "*...which ARE...but NOT ...*" Nevertheless, the degree of *understanding* of a message's content that is required for convincing prosodic rendition remains a subtle question. Clearly, the more the machine or human reader knows, the better the prosodic rendition, but some of the most important knowledge is surprisingly shallow and accessible.

There is no rigorous specification or definition of meaning. The meaning of the rendition event itself is more significant than the inherent meaning of the text, if any. The meaning of the rendition event is determined primarily by the goals of the speaker and listener(s). While textual attributes such as metrical conventions, syntax, morphology, lexical semantics, topic, etc., contribute to the construction of both kinds of meaning, the meaning of the rendition event incorporates more important pragmatic and contextual elements, such as goals of the communication event, and speaker identity and attitude projection. Thus the concept-to-speech discussed in Chapter 17 has a much better chance of generating good prosody, since the content of the sentence is known by the SLU system.

15.2. Prosody Generation Schematic

Figure 15.1 shows schematically the elements of prosodic generation in TTS, from pragmatic abstraction to phonetic realization. The input of the prosody module in Figure 15.1 is parsed text with a phoneme string, and the output specifies the duration of each phoneme and the pitch contour. One possible output representation of that output prosody is shown in Figure 15.2 for the sentence *The cat sat.* Up to four points per phoneme were included in this example. Often one point per phoneme is more than sufficient, except for words like *john*, where two points are needed for the phoneme /ao/ to achieve a natural prosody.

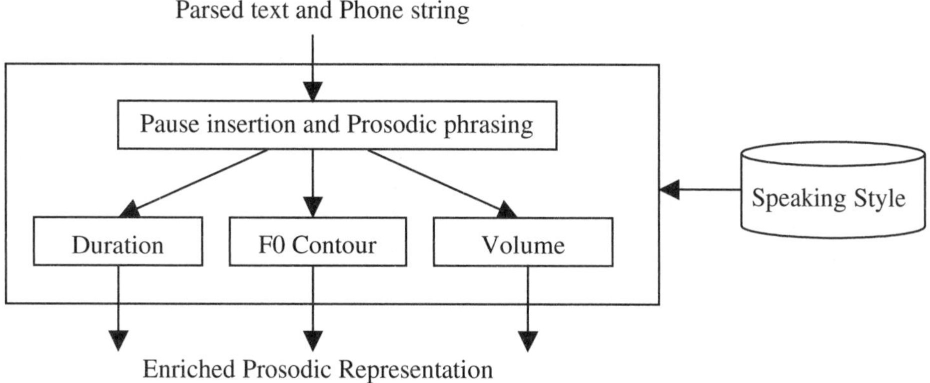

Figure 15.1 Block diagram of a prosody generation system; different prosodic representations are obtained depending on the speaking style we use.

```
DH,  24         (0,178);
AH0, 104        ;
#;
K,   80         (25,178)    (50,184)    (75,201);
AE1, 152        (0,214)     (25,213)    (50,204)    (75,193);
T,   40         (0,175)     (25,175)    (50,174)    (75,172);
#;
S,   104        (0,171)     (25,172)    (50,180)    (75,189);
AE1, 104        (0,198)     (25,196)    (50,168)    (75,137);
T,   112        (0,120)     (100,120);
#;
```

Figure 15.2 Enriched prosody representation, where each line contains one phoneme containing the phoneme identity, the phoneme duration in milliseconds, and a number of prosody points specifying pitch and possibly volume. Each prosody point is determined by a time point, expressed as a percentage of the phoneme's duration, and its corresponding pitch value in Hz. The symbol # is a word delimiter. For example, the fourth line specifies values for phoneme K, which lasts 80 ms and has three prosody points: the first is located at 25% of the phoneme duration, i.e., 20 ms into the phoneme, and has a pitch value of 178 Hz. Pitch in this case is specified in absolute terms in Hz, but it could also be in a logarithmic scale such as quarter-semitones relative to a base pitch.

In the next sections we describe the modules of Figure 15.1: the speaking style, symbolic prosody (including pause insertion), duration assignment, and pitch generation in that order, as usually followed by most TTS systems.

15.3. Speaking Style

Prosody depends not only on the linguistic content of a sentence. Different people generate different prosody for the same sentence. Even the same person generates a different prosody depending on his or her mood. The *speaking style* of the voice in Figure 15.1 can impart an overall tone to a communication. Examples of such global settings include a low register, voice quality (falsetto, creaky, breathy, etc.), narrowed pitch range indicating boredom, depression, or controlled anger, as well as more local effects, such as notable excursion of pitch, higher or lower than surrounding syllables, for a syllable in a word chosen for special emphasis. Another example of a global effect is a very fast speaking rate that might signal excitement, while an example of a local effect would be the typical short, extreme rise in pitch on the last syllable of a *yes-no* question in American English.

15.3.1. Character

Character, as a determining element in prosody, refers primarily to long-term, stable, extralinguistic properties of a speaker, such as membership in a group and individual personality. It also includes sociosyncratic features such as a speaker's region and economic status, to the degree that these influence characteristic speech patterns. In addition, idiosyncratic features such as gender, age, speech defects, etc., affect speech, and physical status may also be a background determiner of prosodic character. Finally, character may sometimes include temporary conditions such as fatigue, inebriation, talking with mouth full, etc. Since many of these elements have implications for both the prosodic and voice quality of speech output, they can be very challenging to model jointly in a TTS system. The current state of the art is insufficient to convincingly render most combinations of the character features listed above.

15.3.2. Emotion

Temporary emotional conditions such as amusement, anger, contempt, grief, sympathy, suspicion, etc. have an effect on prosody. Just as a film director explains the emotional context of a scene to her actors to motivate their most convincing performance, so TTS systems need to provide information on the simulated speaker's state of mind. These are relatively unstable properties, somewhat independent of character as defined above. That is, one could imagine a speaker with any combination of social/dialect/gender/age characteristics being in any of a number of emotional states that have been found to have prosodic correlates, such as anger, grief, happiness, etc. Emotion in speech is actually an important area for future research. A large number of high-level factors go into determining emotional effects in speech. Among these are point of view (can the listener interpret what the speaker is really feeling or expressing?); spontaneous vs. symbolic (e.g., acted emotion vs. real feeling); culture-specific vs. universal; basic emotions and compositional emotions that combine basic feelings and effects; and strength or intensity of emotion. We can draw a few preliminary conclusions from existing research on emotion in speech [34]:

- Speakers vary in their ability to express emotive meaning vocally in controlled situations.
- Listeners vary in their ability to recognize and interpret emotions from recorded speech.
- Some emotions are more readily expressed and identified than others.
- Similar intensity of two emotions can lead to confusing one with the other.

An additional complication in expressing emotion is that the phonetic correlates appear not to be limited to the major prosodic variables (F0, duration, energy) alone. Besides these, phonetic effects in the voice such as jitter (inter-pitch-period microvariation), or the mode of excitation may be important [24]. In a formant synthesizer supported by extremely sophisticated controls [59], and with sufficient data for automatic learning, such voice effects might be simulated. In a typical time-domain synthesizer (see Chapter 16), the lower-level phonetic details are not directly accessible, and only F0, duration, and energy are available.

Some basic emotions that have been studied in speech include:

- **Anger**, though well studied in the literature, may be too broad a category for coherent analysis. One could imagine a threatening kind of anger with a tightly controlled F0, low in the range and near monotone; while a more overtly expressive type of tantrum could be correlated with a wide, raised pitch range.
- **Joy** is generally correlated with increase in pitch and pitch range, with increase in speech rate. Smiling generally raises F0 and formant frequencies and can be well identified by untrained listeners.
- **Sadness** generally has normal or lower than normal pitch realized in a narrow range, with a slow rate and tempo. It may also be characterized by slurred pronunciation and irregular rhythm.
- **Fear** is characterized by high pitch in a wide range, variable rate, precise pronunciation, and irregular voicing (perhaps due to disturbed respiratory pattern).

15.4. SYMBOLIC PROSODY

Abstract or *symbolic prosodic* structure is the link between the infinite multiplicity of pragmatic, semantic, and syntactic features of an utterance and the relatively limited F0, phone durations, energy, and voice quality. The output of the prosody module of Figure 15.2 is a set of real values of F0 over time and real values for phoneme durations. Symbolic prosody deals with:

- Breaking the sentence into prosodic phrases, possibly separated by pauses, and
- Assigning labels, such as emphasis, to different syllables or words within each prosodic phrase.

Words are normally spoken continuously, unless there are specific linguistic reasons to signal a discontinuity. The term *juncture* refers to prosodic phrasing—that is, where do words cohere, and where do prosodic breaks (pauses and/or special pitch movements) occur. Juncture effects, expressing the degree of cohesion or discontinuity between adjacent words, are determined by physiology (running out of breath), phonetics, syntax, semantics, and pragmatics. The primary phonetic means of signaling juncture are:

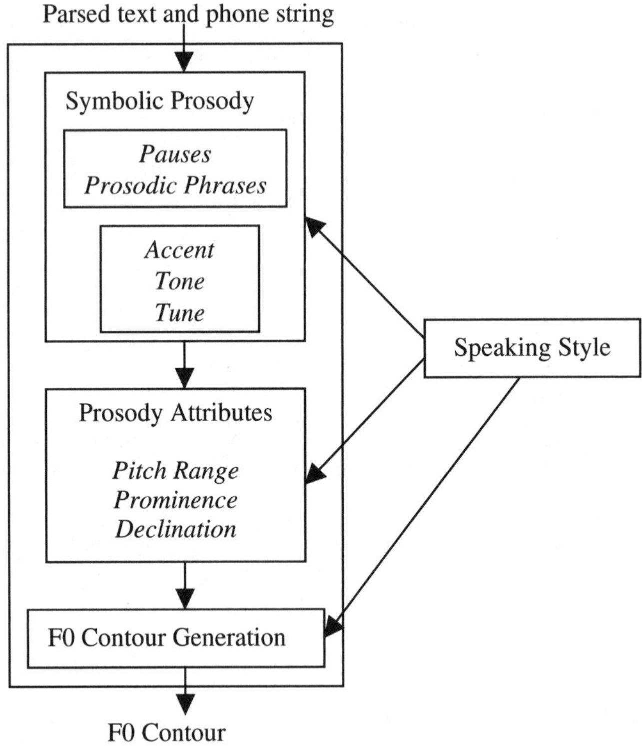

Figure 15.3 Pitch generation decomposed in symbolic and phonetic prosody.

- Silence insertion. This is discussed in Section 15.4.1.
- Characteristic pitch movements in the phrase-final syllable. This is discussed in Section 15.4.4.
- Lengthening of a few phones in the phrase-final syllable. This is discussed in Section 15.5.
- Irregular voice quality such as vocal fry. This is discussed in Chapter 16.

Abstract prosodic structure or annotation typically specifies all the elements in the top block of the pitch-generation schematic in Figure 15.3, including *accents* (corresponding conceptually to *heads* in standard syntactic structure). The accent types are selected from a small inventory of *tones* for American English (e.g., high, low, rising, late-rising, scooped). The sequence of accent and juncture tones in a given prosodic structure may cohere to yield *tune*-like effects that have some holistic semantic interpretation. While we center our description in an abstract representation called ToBI, we also describe other alternate representations in Section 15.4.6. Finally, though in principle the prosody attributes module applies to all prosody variables, it is mostly used for F0 generation in practice, and as such is discussed in Section 15.6.1.

15.4.1. Pauses

In a long sentence, speakers normally and naturally pause a number of times. These pauses have traditionally been thought to correlate with syntactic structure but might more properly be thought of as markers of information structure [58]. They may also be motivated by poorly understood stylistic idiosyncrasies of the speaker, or physical constraints. In spontaneous speech, there is also the possibility that some pauses serve no linguistic function but are merely artifacts of hesitation.

In a typical system, the most reliable indicator of pause location is punctuation. After resolution of abbreviations and special symbols relevant to text normalization (Chapter 14), the remaining punctuation can be reclassified as essentially prosodic in nature. This includes periods, commas, exclamation points, parentheses, ellipsis points, colons, dashes, etc. Each of these can be taken to correspond to a prosodic phrase boundary and can be given a special pitch movement at its end-point.

In predicting pauses, although you have to consider both their occurrence and their duration, the simple presence or absence of a silence (of greater than 30 ms) is the most significant decision, and its exact duration is secondary, based partially on the current rate setting and other extraneous factors.

There are many reasonable places to pause in a long sentence, but a few where it is critical *not* to pause. The goal of a TTS system should be to avoid placing pauses anywhere that might lead to ambiguity, misinterpretation, or complete breakdown of understanding. Fortunately, most decent writing (apart from email) incorporates punctuation according to exactly this metric: no need to punctuate after every word, just where it aids interpretation.

Therefore, by simply following punctuation in many writing styles, the TTS system will not go far wrong.

Consider the opening passage from Edgar Allan Poe's classic story *The Cask of Amontillado* (1846) arranged sentence-by-sentence:

1. *The thousand injuries of Fortunato I had borne as I best could, but when he ventured upon insult, I vowed revenge.*
2. *You, who so well know the nature of my soul, will not suppose, however, that I gave utterance to a threat.*
3. *At length I would be avenged; this was a point definitively settled—but the very definitiveness with which it was resolved precluded the idea of risk.*

If we place prosodic pauses at all and only the punctuation sites, the result is acceptable to most listeners, and no definite mistakes occur. Some stretches seem a bit too long, however. Perhaps the second part of sentence 3 could be broken up as follows:

but the very definitiveness with which it was resolved PAUSE precluded the idea of risk.

While commas are particularly useful in signaling pause breaks, as seen above, pauses may be optional following comma-delimited listed words (*berries, melons, and cheese*), though the special small pitch rise typical of a minor (nonpause) break is often present.

Cases where placing a boundary in certain locations critically affects interpretation include tag questions and verb particle constructions (where the verb must not be separated from its particle), such as:

Why did you hit Joe?

Why did you hit PAUSE Joe?

He distractedly threw out the trash.

(NOT ... threw PAUSE out ...)

He distractedly gazed PAUSE out the window.

(NOT ... out PAUSE the ...)

Supplying junctures at the optimal points sometimes requires deep semantic analysis provided by the module described in Chapter 14. The need for independent methods for pause insertion has motivated some researchers to assume that no independent source of natural language analysis is available. The CART discussed in Chapter 4 can be used for pause assignment [36]. You can use POS categories of words, punctuation, and a few structural measures, such as overall length of a phrase, and length relative to neighboring phrases to construct the classification tree. The decision-tree-based system can have correct prediction of 81% for pauses over test sentences with only 4% false prediction rates. As the algorithm proceeds successively left to right through each pair of words, the following questions can be used:

- *Is this a sentence boundary (marked by punctuation)?*
- *Is the left word a content word and the right word a function word?*
- *What is the function word type of word to the right? (Certain function words are more likely to signal a break)*
- *Is either adjacent word a proper name (capitalized)?*
- *How many content words have occurred since the previous function word (If > 4 or 5 words, a break is more likely)*
- *Is there a comma at this location?*
- *What is the current location in the sentence?*
- *What is the length of current proposed major phrase?*

These questions summarize the relevant knowledge, which could be formulated in expert-system rules, and augmented by high-quality syntactic knowledge if available, or trained statistically from tagged corpora.

15.4.2. Prosodic Phrases

An end-of-sentence period may trigger an extreme lowering of pitch, a comma-terminated prosodic phrase may exhibit a small *continuation rise* at its end, signaling more to come, etc. Rules based on these kinds of simple observations are typically found in commercial TTS systems. Certain pitch-range effects over the entire clause or utterance can also be based on punctuation—for example, the range in a parenthetical restrictive clause is typically narrower than that of surrounding material, while exclamations may have a heightened range, or at least higher accent targets throughout.

Prosodic junctures that are clearly signaled by silence (and usually by characteristic pitch movement as well), also called *intonational phrases*, are required between utterances and usually at punctuation boundaries. Prosodic junctures that are not signaled by silence but rather by characteristic pitch movement only, also called *phonological phrases*, may be harder to place with certainty and to evaluate. In fast speech, the silence demarcating fruits in the sentence 'We have blueberries, raspberries, gooseberries, and blackberries.' may disappear, yet a trace of the continuation rise on each 'berries' typically remains. These locations would then still qualify as minor intonation phrases, or phonological phrases.

In analyzing spontaneous speech, the nature and extent of the signaling pitch movement may vary from speaker to speaker. A further consideration for practical TTS systems is a user's preferred rate setting: blind people who depend on TTS to access information in a computer usually prefer a fast rate, at which most sentence-internal pauses should disappear.

To discuss linguistically significant juncture types and pitch movement, it helps to have a simple standard vocabulary. ToBI (for *Tones and Break Indices*) [4, 55] is a proposed standard for transcribing symbolic intonation of American English utterances, though it can be adapted to other languages as well. The *Tones* part of ToBI is considered in greater detail in Section 15.4.4.

The *Break Indices* part of ToBI specifies an inventory of numbers expressing the strength of a prosodic juncture. The Break Indices are marked for any utterance on their own discrete *break index tier* (or layer of information), with the BI notations aligned in time with a representation of the speech phonetics and pitch track. On the break index tier, the prosodic association of words in an utterance is shown by labeling the end of each word for the subjective strength of its association with the next word, on a scale from 0 (strongest perceived conjoining) to 4 (most disjoint), defined as follows:

- **0** for cases of clear phonetic marks of clitic[1] groups (phrases with appended reduced function words), e.g., the medial affricate in contractions of *did you* or a flap as in *got it.*
- **1** most phrase-medial word boundaries.
- **2** a strong disjuncture marked by a pause or virtual pause, but with no tonal marks, i.e., a well-formed tune continues across the juncture. OR, a disjuncture that is weaker than expected at what is tonally a clear intermediate or full intonation phrase boundary.
- **3** intermediate intonation phrase boundary, i.e., marked by a single phrase tone affecting the region from the last pitch accent to the boundary.
- **4** full intonation phrase boundary, i.e., marked by a final boundary tone after the last phrase tone.

For example, a typical fluent utterance of the following sentence: *Did you want an example?* might have a 0 between *Did* and *you*, indicating palatalization of the /d j/ sequence across the boundary between these words. Similarly, the break index value between *want* and *an* might again be 0, indicating deletion of /t/ and subsequent flapping of /n/. The remaining break index values would probably be 1 between *you* and *want* and between *an* and *example*, indicating the presence of a mere word boundary, and 4 at the end of the utterance, indicating the end of a well-formed intonation phrase. The annotation is thus:

Did-0 you-1 want-0 an-1 example-4?

Without reference to any other knowledge, therefore, a system would place a 1 after every word, except where utterance-final punctuation motivates placement of 4. Perhaps comma boundaries would be marked by 4. Need any more be done? A BI of 0 correlates with any special phone substitution or modification rules for reduction in clitic groups that a TTS system may attempt. By marking the location of clitic (close association) phonetic reduction, such a BI can serve as a trigger for special duration rules that shorten the segments of the cliticized word. Whatever syntactic/semantic processing was done to propose the cliticization can serve to trigger assignment of 1. The 2 mark is generally more useful for

[1] Pronounced as part of another word, as in *ve* in *I've*.

analysis than for speech generation. You may observe that in the literature on intonation, a 3 break is sometimes referred to as an *intermediate phrase* break, or a *minor phrase* break, while a 4 break is sometimes called an *intonational phrase* break or a *major phrase* break.

15.4.3. Accent

We should briefly clarify use of terms such as stress and accent. *Stress* generally refers to an idealized location in an English word that is a potential site for phonetic prominence effects, such as extruded pitch and/or lengthened duration. This information comes from our standard lexicon. Thus, the second syllable of the word *em-**ploy**-er* is said to have the abstract property of lexical stress. In an actual utterance, if the word as a whole is sufficiently important, phonetic highlighting effects are likely to fall on the lexically *stressed* syllable:

*Acme **In**dustries is the **big**gest em**ploy**er in the **ar**ea.*

Accent is the signaling of semantic salience by phonetic means. In American English, accent is typically realized via extruded pitch (higher or lower than the general trend) and possibly extended phone duration. Although lexical stress as noted in dictionaries is strictly an abstract property, these accent-signaling phonetic effects are usually strongest on the lexically stressed syllable of the word that is singled out for accentuation (e.g., *em**ploy**er*).

In the sentence above, the word *employer* is not specially focused or contrasted, but it is an important word in the utterance, so its lexically stressed syllable typically receives a prosodic accent (via pitch/duration phonetic mechanisms), along with the other syllables in boldface. In cases of special emphasis or contrast, the lexically specified preferred location of stress in a word may be overridden in utterance accent placement:

*I didn't say employ**er**, I said employ**ee**.*

It is also possible to override the primary stress of a word with the secondary stress where a neighboring word is accented. While normally we would say *Massa**chu**setts*, we might say ***Ma**ssachusetts **leg**islature* [51].

Let's consider what might make a word accentable in context. A basic rule based on the use of POS category is to decide accentuation by *accenting all and only the content words*. Such rule is used in the baseline F0 generation system of Section 15.6.2. The content words are major open-class categories such as noun, verb, adjective, adverb, and certain strong closed-class words such as negatives and some quantifiers. Thus, the *function words*, made up of closed-class categories such as prepositions, conjunctions, etc., end up on a kind of *stop list* for accentuation, analogous to the stop lists used traditionally in simple document indexing schemes for information retrieval. This works adequately for many short, isolated sentences, such as "*The **cat sat** on the **mat***", where the words selected for accentuation appear in boldface. For more complex sentences, appearing in document or dialog context, such an algorithm will sometimes fail.

How often does the POS class-based stop-list approach fail? Let's consider a slightly more elaborate variant on the theme. A model was created using the Lancaster/IBM Spoken

English Corpus (SEC) [3]. This includes a variety of text types, including news, academic lectures, commentary, and magazine articles. Each word in the corpus has a POS tag automatically assigned by an independent process. The model predicts the probability of a word of POS having accent status. The probability is computed based on POS class of a sequence of words in the history in the similar way as *n*-gram models discussed in Chapter 11. This simple model performed at or above 90% correct predictions for all text types. As for stress predictions that were *incorrect*, we should note that in many cases accents are optional—it is more a game of avoiding plain wrong predictions than it is of finding optimal ones. Clearly, however, there are situations that call for greater power than a simple POS-based model can provide. Even different readings of the exact same text can result in different accents [46].

Consider a simple case where a word or its base form is repeated within a short paragraph. Such words may have the necessary POS to trigger accentuation, but, since they have already been mentioned (perhaps with varying morphological inflection), it can sound strange to highlight them again with accentuation. They are *given* or *old* information the second time around and may be deaccented. For example, the second occurrence of the noun 'switch' below is best not accented:

> *At the **cor**ner of the keyboard are **two switch**es.*
> *The **top** switch is **us**er-defined.*

To achieve this, the TTS system can keep a queue of most recently used words or normalized base forms (if morphological capability is present), and block accentuation when the next word has been used recently. The queue should be reset periodically, perhaps at paragraph boundaries [54].

Of course, the surface form of words, even if reduced to a base form or lemma by morphology, won't always capture the deeper semantic relations that govern accentuation. Consider the following fragment extracted from Roger Rosenblatt's essay:

> ***Kid**s to**day** are being neglected by the **old**er generation. Adults spend **hours** every day on the **Stair**Master, trying to be**come** the youth they should be at**tend**-ing to.*

A simple content-word-based accentuation algorithm accents the word *youth*, because it is a noun. In context, however, it is not optimal to accent *youth*, because it is co-referent with the subject of the fragment, which is *kids today*. Thus it is, by some metrics, old or *given* information, and it had better remain unaccented. The surrounding verbs *become*, *should*, and *attending* may get extra prominence. The degree to which coreference relations, from surface identity to deep anaphora, can be exploited depends on the power of the NL analysis supporting the TTS function.

Other confusions can arise in word accentuation due to English complex nominals, where lack of, or location of, an accent may be a lexical (static) rather than a syntactic or dynamic property. Consider:

> *I in**vit**ed her to my **birth**day party, but she **said** she **can't** at**tend** any parties until her **grades** improve.*

One possible accent structure is indicated in boldface. Here *birthday party* functions as a complex nominal, with lexical stress on *birthday*. The word *party* should not receive stress at all, nor should its later stand-alone form *parties*. Accentuation of *improve* is optional: it is a full content word, yet somehow it also feels predictable from the context, so deaccentuation is possible. Some of the complex nominals, like *birthday party*, are fully fixed and can be entered into the lexicon as such. Others form small families of binary or *n*-ary phrases, which may be detected by local syntactic and lexical analysis. Ambiguous cases such as *moving van* or *hot dog*, which could be either nominals or adjective-noun phrases, may have to be resolved by user markup or text understanding processes.

Dwight Bolinger opined that *Accent is predictable—if you're a mind reader* [6], asserting that accentuation algorithms will never achieve perfect performance, because a writer's exact intentions cannot be inferred from text alone, and understanding is needed. However, work in [20] (similar to [3] but incorporating more sophisticated mechanics for name identification and memory of recent accented items), showed that reasonably straightforward procedures, if applied separately and combined intelligently, can yield adequate results on the accentuation task. This research has also determined that improvement occurs when the system learns that not all 'closed-class' categories are equally likely to be deaccented. For example, *closed accented* items include the negative article, negative modals, negative *do*, most nominal pronouns, most nominative and all reflexive pronouns, pre- and postqualifiers (e.g., quite), prequantifiers (e.g., *all*), postdeterminers (e.g., *next*), nominal adverbials (e.g., *here*), interjections, particles, most wh-words, plus some prepositions (e.g., *despite, unlike*).

One area of current and future development is the introduction of discourse analysis to synthesis of dialog. Discourse analysis algorithms attempt to delimit the time within which a given word/concept can be considered newly introduced, given, old, or reintroduced, and combined with analysis of segments within discourse and their boundary cues (turn-taking, digressions, interruptions, summarization, etc.) can supplement algorithms for accent assignment. This kind of work improves the naturalness of computer responses in human-computer dialog, as well as the accentuation in TTS renditions of pure text, when dialog must be performed (e.g., in reading a novel out loud) [44].

As noted above, user- or application-supplied annotations, based on intimate knowledge of the purpose and content of the speech event, can greatly enhance the quality by offloading the task of automatic accent prediction. The /EMPHASIS/ tag described in Section 15.7, with several levels of strength including *reduced accent* and *no accent*, is ideally suited for this purpose.

15.4.4. Tone

Tones can be understood as labels for perceptually salient levels or movements of F0 on syllables. Pitch levels and movements on accented and phrase-boundary syllables can exhibit a bewildering diversity, based on the speaker's characteristics, the nature of the speech event, and the utterance itself, as discussed above. For modeling purposes, it is useful to

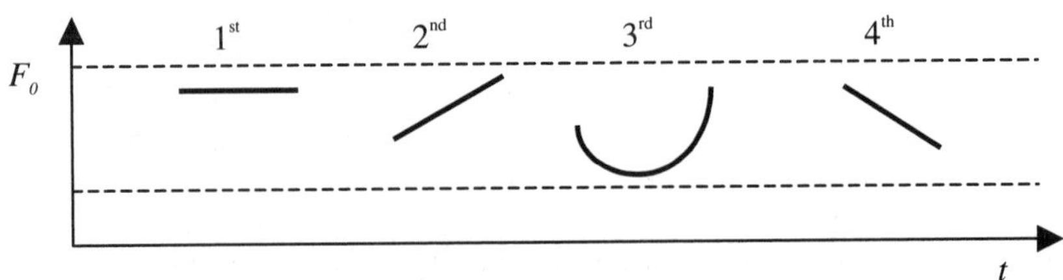

Figure 15.4 The four Chinese tones.

have an inventory of basic, abstract pitch types that could in principle serve as the base inventory for expression of linguistically significant contrasts. Chinese, a lexical tone language, is said to have an inventory of 4 lexical tones (5 if neutral tone is included), as shown in Figure 15.4. Different speakers can realize these tones differently according to their physiology, mood, utterance content, and the speech occasion. But the variance in the tones' shapes, and contrasts with one another, remain fairly predictable, within broad limits.

By analogy, linguists have proposed a relatively small set of tonal primitives for English, which can be used, in isolation or in combination, to specify the gross phonological typology of linguistically relevant contrasts found in theories of English intonational meaning [17, 28, 39]. A basic set of tonal contrasts has been codified for American English as part of the Tones and Break Indices (ToBI) system [4, 55]. These categories can be used for annotation of prosodic training data for machine learning, and also for internal modular control of F0 generation in a TTS system. The set specifies 2 abstract levels, H(igh) and L(ow), indicating a relatively higher or lower point in a speaker's range. The H/L primitive distinctions form the foundations for 2 types of entities: pitch accents, which signal prominence or culmination; and boundary tones, which signal unit completion, or delimitation. The boundary tones are further divided into phrase types and full boundary types, which would mark the ends of intonational phrases or whole utterances.

While useful as a link to syntax/semantics, the term *accent* as defined in Section 15.4.3 is a bit too abstract, even for symbolic prosody. What is required is a way of labeling linguistically significant types of pitch contrast on accented syllables. Such a system could serve as the basis for a theory of intonational meaning. The ToBI standard specifies six types of pitch accents (see Table 15.1) in American English, where the * indicates direct alignment with an accented syllable, two intermediate phrasal tones (see Table 15.2), and five boundary tones [4] (see Table 15.3).

In American English one sometimes hears a string of strictly descending pitch accent levels across a short phrase. When judiciously applied, this *downstep* effect can be pleasantly natural, as in the following sentence:

*"I saw a big-H**

*fat-!H**

pig-!H (L-L%)"*

A basic rule used in the baseline F0 generation system of Section 15.6.2 consists in having all the pitch accents realized as H*, associated with the lexically stressed syllable of accented words. In general, ToBI representations of intonation should be sparse, specifying only what is linguistically significant. So, words lacking accent should not receive ToBI pitch accent annotations, and their pitch must be derived via interpolation over neighbors, or by some other default means. Low excursions can be linguistically significant also, in the crude sense that if you dip very low in your range on a given word, it may be perceived as prominent by listeners. L*+H and L+H* are both F0 rises on the accented syllable, but in the case of L*+H, the association of the starred tone (L*) with the accented syllable may push the realization of H off to the following syllable. !H* can be used for successively lowered high accents, such as might be found on *big red car*, or *tall, dark, and handsome*. A ToBI labeled utterance is shown in Figure 15.5.

A typical boundary tone is the *final lowering*, the marked tendency for the final syllable in all kinds of noninterrogative utterances to be realized on a pitch level close to the absolute bottom of a speaker's range. The final low (L-L%) may 'pull down' the height of some few accents to its left as well [41].

Table 15.1 ToBI pitch accent tones.

ToBI tone	Description	Graph
H*	*peak accent*—a tone target on an accented syllable which is in the upper part of the speaker's pitch range.	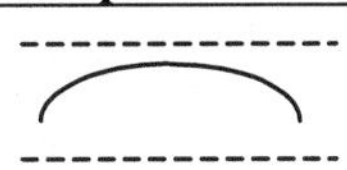
L*	*low accent*—a tone target on an accented syllable which is in the lowest part of the speaker's pitch range	
L*+H	*scooped accent*—a low tone target on an accented syllable which is immediately followed by a relatively sharp rise to a peak in the upper part of the speaker's pitch range.	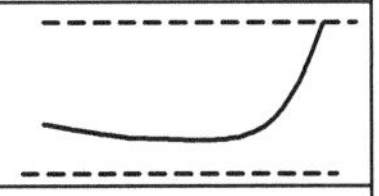
L*+!H	*scooped downstep accent*—a low tone target on an accented syllable which is immediately followed by a relatively flat rise to a downstep peak.	
L+H*	*rising peak accent*—a high peak target on an accented syllable which is immediately preceded by a relatively sharp rise from a valley in the lowest part of the speaker's pitch range.	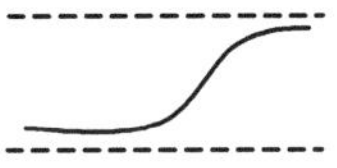
!H*	*downstep high tone*—a clear step down onto an accented syllable from a high pitch which itself cannot be accounted for by an H phrasal tone ending the preceding phrase or by a preceding H pitch accent in the same phrase.	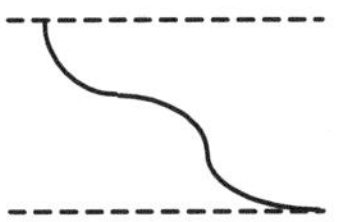

Table 15.2 ToBI intermediate phrasal tones.

ToBI tone	Description
L-	Phrase accent, which occurs at an intermediate phrase boundary (level 3 and above).
H-	Phrase accent, which occurs at an intermediate phrase boundary (level 3 and above).

Ultimately, abstract linguistic categories should correlate with, or provide labels for expressing, contrasts in meaning. While the ToBI pitch accent inventory is useful for generating a variety of English-like F0 effects, the distinction between perceptual contrast, functional contrast, and semantic contrast is particularly unclear in the case of prosody [41]. For example, whether or not the L*, an alternative method of signaling accentual prominence, functions in full linguistic contrast to H* is unclear.

In addition, we have mentioned that junctures are typically marked with perceptible pitch movements that are independent of accent. The ToBI specification also allows for combinations of the H and L primitives that signal phrase, clause, and utterance boundaries. These are called phrasal tones. The ToBI specification further points out that since intonation phrases are composed of one or more intermediate phrases plus a boundary tone, full intonation phrase boundaries have two final tones.

The symbolic ToBI transcription alone is not sufficient to generate a full F0 contour. The remaining components are discussed in Section 15.6.

Table 15.3 ToBI boundary tones.

ToBI tone	Description
L-L%	For a full intonation phrase with an L phrase accent ending its final intermediate phrase and an L% boundary tone falling to a point low in the speaker's range, as in the standard 'declarative' contour of American English.
L-H%	For a full intonation phrase with an L phrase accent closing the last intermediate phrase, followed by an H boundary tone, as in 'continuation rise.'
H-H%	For an intonation phrase with a final intermediate phrase ending in an H phrase accent and a subsequent H boundary tone, as in the canonical 'yes-no question' contour. Note that the H- phrase accent causes 'upstep' on the following boundary tone, so that the H% after H- rises to a very high value.
H- L%	For an intonation phrase in which the H phrase accent of the final intermediate phrase upsteps the L% to a value in the middle of the speaker's range, producing a final level plateau.
%H	High initial boundary tones; marks a phrase that begins relatively high in the speaker's pitch range when not explained by an initial H* or preceding H%.

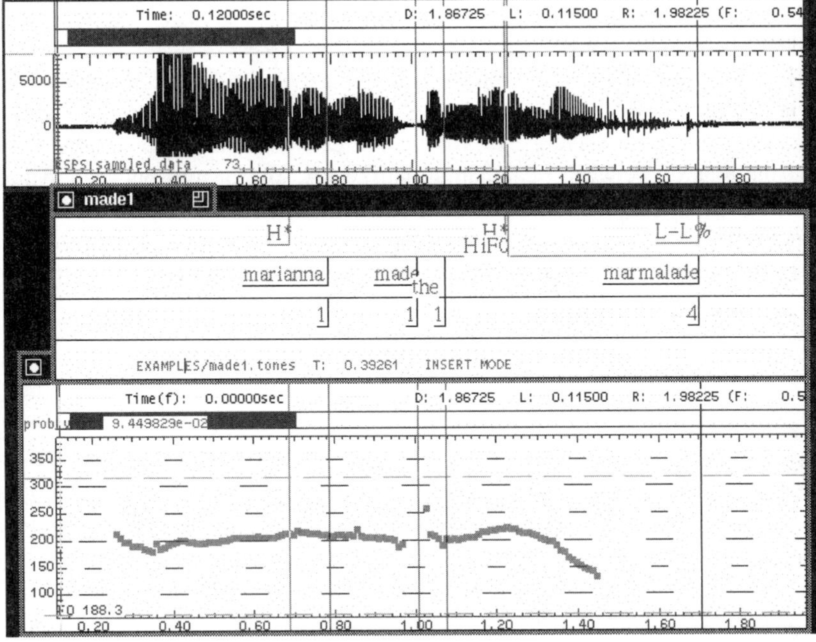

Figure 15.5 "*Marianna made the marmalade*", with an H* accent on *Marianna* and *marmalade*, and a final L-L% marking the characteristic sentence-final pitch drop. Note the use of 1 for the weak inter-word breaks, and 4 for the sentence-final break (after Beckman [4]).

15.4.5. Tune

Nyaah *nuh nyaah* *you* *get*

nyaah *nyaah,* *can't* *me!*

—*Children's chant*

Some pitch contours appear to be immediately recognizable and emotionally interpretable, independent of lexical content, such as the English *children's chant* above [40]. Can this idea of stylized *tunes*, perhaps decomposable into the *tones* we examined above, be applied to the intonation of ordinary speech? In fact, the ideal use of the ToBI pitch accent labels above would be as primitive elements in holistic prosodic contour descriptions, analogous to the role of phonemes in words. Ultimately, a dictionary of meaningful contours, described abstractly by ToBI tone symbols to allow for variable phonetic realization, would constitute a theory of intonational meaning for American English. Ideally, the meanings of such contours could perhaps be derived compositionally from the meanings of their constituent pitch accent and boundary tones, thus allowing us to dispense with the dictionary altogether. Contour stylization approaches describe contours holistically and index them for application on the basis of utterance type, usually based on a naïve syntactic typology, e.g., question, declarative, etc.

The holistic representation of contours can perhaps be defended, but the categorizing of types via syntactic description (usually triggered by punctuation) is questionable. Typically, use of punctuation as a rule trigger for pitch effects is making certain hidden assumptions about the relation between punctuation and syntax, and in turn between syntax and prosody. An obvious example is question intonation. If you find a question mark at the end of a sentence, are you justified in applying a final high rise (which might be denoted as H-H% in ToBI)? First, the intonation of yes-no questions in general differs from that of *wh*-questions. *Wh*-questions usually lack the extreme final upturn of F0 heard in some yes-no questions:

i. *Are you going?*
ii. *Where are you going?*

However, there are cases where an extreme final upturn is acceptable on ii. As [8] puts it, "It has been emphasized repeatedly … that no intonation is an infallible clue to any sentence type: any intonation that can occur with a statement, a command, or an exclamation can also occur with a question."

Admittedly, there is a rough correspondence between syntactic types and speech acts,[2] as shown in Table 15.4. Nevertheless, the correspondence between syntactic types and acts is not deterministic, and prosody in spontaneous speech is definitely mediated via speech acts (the pragmatic context and use of an utterance) rather than syntactic types. Thus, it is difficult to obtain high-quality simulation of spontaneous speech based on linguistic descriptions that do not include speech acts and pragmatics. Likewise, even simulation of prose reading without due consideration of pragmatics and speech acts, and based solely on syntactic types, is difficult because prose that is read may:

- Include acted dialog
- Have limited occurrence of most types other than declarative, lessening variety in practice
- Include long or complex sentences, blunting 'stereotypical' effects based on utterance type
- Lack text cues as to syntactic type, or analysis grammar may be incomplete

Table 15.4 Relationship between syntactic types and speech acts.

Type	Speech Act	Example
interrogative	Questioning	Is it good?
declarative	Stating	It's good.
imperative	Commanding	Be good!
exclamatory	Exclaiming	How good it is!

[2] For a more in-depth coverage of speech acts, consult Chapter 17.

Thus, description of an entire speech event, rather than inferences about text content, is again the ultimate guarantor of quality. This is why the future of automatic prosody lies with *concept-to-speech* systems (see Chapter 17) incorporating explicit pragmatic and semantic context specification to guide message rendition.

For commercial TTS systems that must infer structure from raw text, there are a few characteristic fragmentary pitch patterns that can be taken as tunes and applied to special segments of utterances. These include:

- Phone numbers—downstepping with pauses
- List intonation—downstepping with pauses (melons, pears, and eggplants)
- Tag and quotative tag intonation—low rise on tag (*Never*! he blurted. Come here, Jonathan.)

15.4.6. Prosodic Transcription Systems

ToBI, introduced above, can be used as a notation for transcription of prosodic training data and as a high-level specification for the symbolic phase of prosodic generation. Alternatives to ToBI also exist for these purposes, and some of them are amenable to automated prosody annotation of corpora. Some examples of this type of system are discussed in this section.

PROSPA was developed specially to meet the needs of discourse and conversation analysis, and it has also influenced the Prosody Group in the European ESPRIT 2589 SAM (Multilingual Speech Input/Output Assessment, Methodology and Standardization) project [50]. The system has annotations for general or global trends over long spans shown in Table 15.5, short, accent-lending pitch movements on particular vowels are transcribed in Table 15.6, and the pitch shape after the last accent in a () sequence, or *tail*, is indicated in Table 15.7.

Table 15.5 Annotations for general or global trends over long spans.

( )	extent of a sequence of cohesive accents
F	globally falling intonation
R	globally rising intonation
H	level intonation on high tone level
M	level intonation on middle tone level
L	level intonation on low tone level
H/F	falling intonation on a globally high tone level
...	sequence of weakly accented or unaccented syllables

Table 15.6 Annotations for accent-lending pitch movements on particular vowels.

+	Upward pitch movement
-	Downward pitch movement
=	level pitch accent

Table 15.7 Annotations for pitch shape after the last accent in a () sequence, or *tail.*

	falling tails
/	rising tails
-	level tails
/\	combinations of tails (rising-falling here)

INTSINT is a coding system of intonation described in [22]. It provides a formal encoding of the symbolic or phonologically significant events on a pitch curve. Each such target point of the stylized curve is coded by a symbol, either as an absolute tone, scaled globally with respect to the speakers pitch range, or as a relative tone, defined locally in conjunction with the neighboring target points. Absolute tones in INTSINT are defined according to the speaker's pitch range as shown in Table 15.8. Relative tones are notated in INTSINT with respect to the height of the preceding and following target points, as shown in Table 15.9.

Table 15.8 The definition of absolute tones in INTSINT.

T	top of the speaker's pitch range
M	initial, mid value
B	bottom of the speaker's pitch range

Table 15.9 The definition of relative tones in INTSINT.

H	target higher than both immediate neighbours
L	target lower than both immediate neighbours
S	target not different from preceding target
U	target in a rising sequence
D	target in a falling sequence

In a transcription, numerical values can be retained for all F0 target points. TILT [60] is one of the most interesting models of prosodic annotation. It can represent a curve in both its qualitative (ToBI-like) and quantitative (parametrized) aspects. Generally any 'interesting' movement (potential pitch accent or boundary tone) in a syllable can be described in terms of TILT events, and this allows annotation to be done quickly by humans or machines without specific attention to linguistic/functional considerations, which are paramount for ToBI labeling. The linguistic/functional correlations of TILT events can be linked by subsequent analysis of the pragmatic, semantic, and syntactic properties of utterances.

The automatic parametrization of a pitch event on a syllable is in terms of:

- starting F0 value (Hz)
- duration
- amplitude of rise (A_{rise}, in Hz)
- amplitude of fall (A_{fall}, in Hz)
- starting point, time aligned with the signal and with the vowel onset

The tone shape, mathematically represented by its *tilt*, is a value computed directly from the F0 curve by the following formula:

$$tilt = \frac{|A_{rise}| - |A_{fall}|}{|A_{rise}| + |A_{fall}|} \tag{15.1}$$

A likely syllable for tilt analysis in the contour can be automatically detected based on high energy and relatively extreme F0 values or movements. Human annotators can select syllables for attention and label their qualities according to Table 15.10.

Table 15.10 Label scheme for syllables.

sil	Silence
c	Connection
a	Major pitch accent
fb	Falling boundary
rb	Rising boundary
afb	Accent+falling boundary
arb	Accent+rising boundary
m	Minor accent
mfb	Minor accent+falling boundary
mrb	Minor accent+rising boundary
l	Level accent
lrb	Level accent+rising boundary
lfb	Level accent+falling boundary

15.5. Duration Assignment

Pitch and duration are not entirely independent, and many of the higher-order semantic factors that determine pitch contours may also influence durational effects. The relation between duration and pitch events is a complex and subtle area, in which only initial exploration has been done [63]. Nonetheless, most systems often treat duration and pitch independently because of practical considerations [61].

Numerous factors, including semantics and pragmatic conditions, might ultimately influence phoneme durations. Some factors that are typically neglected include:

- The issue of speech rate relative to speaker intent, mood, and emotion.
- The use of duration and rhythm to possibly signal document structure above the level of phrase or sentence (e.g., paragraph).
- The lack of a consistent and coherent practical definition of the phone such that boundaries can be clearly located for measurement.

15.5.1. Rule-Based Methods

Klatt [1] identified a number of first-order perceptually significant effects that have largely been verified by subsequent research. These effects are summarized in Table 15.11.

Table 15.11. Perceptually significant effects for duration. After Klatt [1].

Lengthening of the final vowel and following consonants in prepausal syllables.
Shortening of all syllabic segments[3] in nonprepausal position.
Shortening of syllabic segments if not in a word final syllable.
Consonants in non-word-initial position are shortened.
Unstressed and secondary stressed phones are shortened.
Emphasized vowels are lengthened.
Vowels may be shortened or lengthened according to phonetic features of their context.
Consonants may be shortened in clusters.

The rule-based duration-modeling mechanism involves table lookup of minimum and inherent durations for every phone type. The minimum duration is rate dependent, so all phones could be globally scaled in their minimum durations for faster or slower rates. The inherent duration is the raw material for the rules above: it may be stretched or contracted by a prespecified percentage attached to each rule type above applied in sequence, then it is finally added back onto the minimum duration to yield a millisecond time for a given phone. The duration of a phone is expressed as

$$d = d_{min} + r(\bar{d} - d_{min}) \tag{15.2}$$

[3] Syllabic segments include vowels and syllabic consonants.

where d_{min} is the minimum duration of the phoneme, $\bar{d}$ is the average duration of the phoneme, and the correction r is given by

$$r = \prod_{i=1}^{N} r_i \tag{15.3}$$

for the case of N rules being applied where each rule has a correction r_i. At the very end, a rule may apply that lengthens vowels when they are preceded by voiceless plosives (/p/, /t/, /k/). This is also the basis for the additive-multiplicative duration model [49] that has been widely used in the field.

15.5.2. CART-Based Durations

A number of generic machine-learning methods have been applied to the duration assignment problem, including CART and linear regression [43, 62]. The voice datasets generally rely on less than the full set of possible joint duration predictors implied in the rule list of Table 15.11. It has been shown that a model restricted to the following features and contexts can compare favorably, in listeners' perceptions, with durations from natural speech [43]:

- Phone identity
- Primary lexical stress (binary feature)
- Left phone context (1 phone)
- Right phone context (1 phone)

In addition, a single rule of vowel and post-vocalic consonant lengthening (rule 1 in Table 15.11) is applied in prepausal syllables. The restriction of phone context to immediate left and right neighbors results in a triphone duration model, congruent with the voice triphone model underlying the basic synthesis in the system [23]. In perceptual testing this simple triphone duration model yielded judgments nearly identical to those elicited by utterances with phone durations from natural speech [43]. From this result, you may conjecture that even the simplified list of *first-order* factors above may be excessive, and that only the handful of factors implicit in the triphones themselves, supplemented by a single-phrase final-syllable coda lengthening rule, is required. This would simplify data collection and analysis for system construction.

15.6. PITCH GENERATION

We now describe the issues involved in generating synthetic pitch contours. Pitch, or F0, is probably the most characteristic of all the prosody dimensions. As discussed in Section 15.8, the quality of a prosody module is dominated by the quality of its pitch-generation component.

Since generating pitch contours is an incredibly complicated problem, pitch generation is often divided into two levels, with the first level computing the so-called symbolic prosody described in Section 15.4 and the second level generating pitch contours from this symbolic prosody. This division is somewhat arbitrary since, as we shall see below, a number of important prosodic phenomena do not fall cleanly on one side or the other but seem to involve aspects of both. Often it is useful to add several other attributes of the pitch contour prior to its generation, which are discussed in Section 15.6.1.

15.6.1. Attributes of Pitch Contours

A pitch contour is characterized not only by its symbolic prosody but also by several other attributes such as pitch range, gradient prominence, declination, and microprosody. Some of these attributes often cross into the realm of symbolic prosody. These attributes are also known in the field as phonetic prosody (termed as an analogy to phonology and phonemics).

15.6.1.1. Pitch Range

Pitch range refers to the high and low limits within which all the accent and boundary tones must be realized: a floor and ceiling, so to speak, which are typically specified in Hz. This may be considered in terms of stable, speaker-specific limits as well as in terms of an utterance or passage. For a TTS system, each voice typically has a characteristic pitch range representing some average of the pitch extremes over test utterances. This speaker-specific range can be set as an initial default for the voice or character. These limits may be changed by an application.

Another sense of pitch range is the actual exploitation of zones within the *hard* limits at any point in time for linguistic purposes, having to do with expression of the content or feeling of the message. Pitch-range variation that is correlated with emotion or other aspects of the speech event is sometimes called *paralinguistic*. This linguistic and paralinguistic use of pitch range includes aspects of both symbolic and phonetic prosody. Since it is quantitative, it certainly is a phonetic property of an utterance's F0 contour. Furthermore, it seems that most linguistic contrasts involving pitch accents, boundary tones, etc. can be realized in any pitch range. These settings can be estimated from natural speech (for research purposes) by calculating F0 mean and variance over an utterance or set of utterances, or by simply adopting the minimum and maximum measurements (perhaps the 5th and 95th percentile to minimize the effect of pitch tracker errors).

But, although pitch range is a phonetic property, it can be systematically manipulated to express states of mind and feeling in ways that other strictly phonetic properties, such as characteristic formant values, rarely are. Pitch range interacts with all the prosodic attributes you have examined above, and certain pitch-range settings may be characteristic of particular styles or utterance events. For example, it is noted [8] that: "*we cannot speak of an into-*

nation of exclamation ... Exclamation draws impartially upon the full repertory of up-down patterns. What characterizes the class is not shape but range: exclamations reach for the extreme—usually higher but sometimes lower." In this sense, then, pitch range cannot be considered an arbitrary or physiological attribute—it is directly manipulated for communicative purposes.

In prosodic research, distinguishing emotive and iconic use of pitch (analogous to gesture) from strictly linguistic (logical, syntactic, and semantic expression, with arbitrary relation between signifier and signified) prosodic phenomena has been difficult. Pitch-range variation seems to straddle emotional, linguistic, and phonetic expression.

A linguistic pitch range may be narrowed or widened, and the zone of current pitch variation may be placed anywhere within a speaker's wider, physically determined range. So, for example, a male speaker might adopt a falsetto speaking style for some purpose, with his pitch range actually narrowed, but with all pitch variation realized in a high portion of his overall range, close to his physical limits.

Pitch range is a gradient property, without categorical bounds. It seems to trade off with other model components: accent, relative prominence, downstep, and declination. For example, if our model of prosody incorporates, say, an accent-strength component, but if we also recognize that pitch range can be manipulated for linguistic purposes, we may have difficulty determining, in analysis, whether a given accent is at partial strength in a wide range or at full strength in a reset, narrower range. This analytic uncertainty may be reflected in the quality of models based on the analysis.

A practical TTS system has to stay within, and make some attempt to maximize the exploitation of, the current system default or user-specified range. Thus, for general TTS purposes, the simplest approach is to use about 90% of the user-set or system default range for general prose reading, most of the time, and use the reserved 10% in special situations, such as the paragraph initial *resets*, exclamations, and emphasized words and phrases.

15.6.1.2. Gradient Prominence

Gradient prominence refers to the relative strength of a given accent position with respect to its neighbors and the current pitch-range setting. The simplest approach, where every accented syllable is realized as a H(igh) tone, at uniform strength, within an invariant range, can sound unnatural. At first glance, the prominence property of accents might appear to be a phonetic detail, in that it is quantitative, and certainly any single symbolic tonal transcription can be realized in a wide variety of relative per-accent prominence settings. However, the relative height of accents can fundamentally alter the information content of a spoken message by determining focus, contrast, and emphasis. You would hope that such linguistic content would be determined by the presence and absence, or perhaps the types (H, L, etc.), of the symbolic accents themselves. But an accented syllable at a low prominence might be perceived as unaccented in some contexts, and there is no guaranteed minimum degree of prominence for accent perception. Furthermore, as noted above, the realization of prominence of an accent is context-sensitive, depending on the current pitch-range setting.

The key knowledge deficit here is a theory of the interpretation of prominence that would allow designers to make sensible decisions. It appears that relative prominence is related to the information status of accent-bearing words and is in that sense linguistic, yet there is no theory of prominence categories that would license any abstraction. For the present, many commercial TTS systems adopt a pseudorandom pattern of alternating stronger/weaker prominence, simply to avoid monotony. If a word is tagged for emphasis, or if its information status can otherwise be inferred, its prominence can be heightened within the local range.

In the absence of information on the relative semantic salience of accented words in the utterance, successive prominence levels are varied in some simple alternating pattern, to avoid monotony. Rather than limiting the system to a single peak F0 value per accented syllable, several points could be specified, which, when connected by interpolation and smoothing, could give varied effects within the syllable, such as rising, falling, and scooped accents.

15.6.1.3. Declination

Related to both pitch range and gradient prominence is the long-term downward trend of accent heights across a typical reading-style, semantically neutral, declarative sentence. This is called *declination*. Although this tendency, if overdone, can simply give the effect of a bored or uncomprehending reader, it is a favorite prosodic effect for TTS systems, because it is simple to implement and licenses some pitch change across a single sentence. If a system uses a 'top line' as a reference for calculating the height of every accent, the slope of that top line can simply be declined across the utterance. Otherwise, each accent's prominence can be realized as a certain percentage of the height of the preceding one. Declination can be reset at utterance boundaries, or within an utterance at the boundaries of certain linguistic structures, such as the beginning of quoted speech. Intrasentence phrase and clause types that typically narrow the pitch range, such as parentheticals and certain relative clauses, can be modeled by suspending the declination, or adopting a new declination line for the temporary narrowed range, then resuming the suspended longer-term trend as the utterance progresses. Needless to say, declination is not a prominent feature of spontaneous speech and in any case should not be overdone.

The minor effect of declination should not be confused with the tendency in all kinds of nonquestioning utterances to end with a very low pitch, close to the bottom of the speaker's range. In prosodic research this is called *final lowering* and is well-attested as a phenomenon that is independent of declination [29]. The ToBI notation used to specify final lowering is the complex boundary tone L-L%. In Figure 15.6 we show the declination line together with the other two *downers of intonation*: downstep and final lowering described in Section 15.4.4.

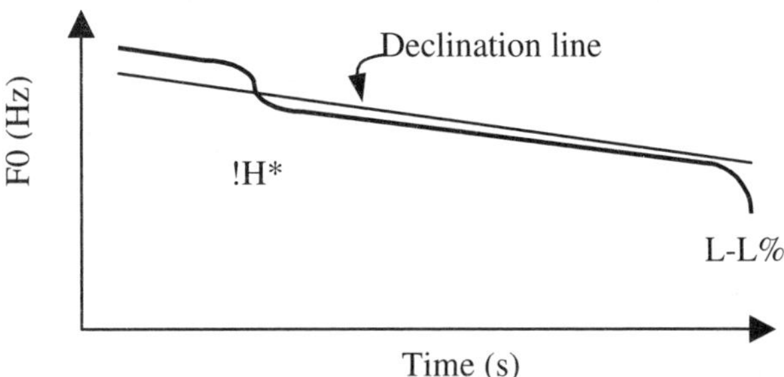

Figure 15.6 The three downers of intonation: the declination line, a downstep (!H*), and the final lowering (L-L%).

15.6.1.4. Phonetic F0—Microprosody

Microprosody refers to those aspects of the pitch contour that are unambiguously phonetic and that often involve some interaction with the speech carrier phones. These may be regarded as *second-order* effects, in the sense that rendering them well cannot compensate for incorrect accentuation or other mistakes at the symbolic level. Conversely, making no attempt to model these but putting a great deal of care into the semantic basis for determining accentuation, contrast, focus, emphasis, phrasing, etc. can result in a system of reasonable quality. Nevertheless, all else being equal, it is advisable to make some attempt to capture the local phonetic properties of natural pitch contours.

If the strength of accents is controlled semantically, by having equal degrees of focus on words of differing phonetic makeup, it has been observed that high vowels described in Chapter 2 carrying H* accents are uniformly higher in the phonetic pitch range than low vowels with the same kinds of accent. The distinction between high and low vowels correlates with the position of the tongue in articulation (high or low in the mouth). The highest English vowels by this metric are */iy/* (as in *bee*) and */uw/* (as in *too*), while the lowest vowel is */aa/* as in *father*. The predictability of F0 under these conditions may relate to the degree of tension placed on the laryngeal mechanisms by the raised tongue position in the high vowels as opposed to the low. In any case, this effect, while probably perceptually important in natural speech, is challenging for a synthesizer. The reason relates again to the issue of gradient prominence, discussed above. Apart from experimental prompts in the lab, there is currently no principled way to assign prominence for accent height realization based on utterance content in general TTS. It may therefore be difficult for a listener to correctly factor pitch accent height that is due to correctly (or incorrectly) assigned gradient prominence from height variation related to the lower-level phonetic effects of vowel height.

Another phonetic effect is the level F0 in the early portion of a vowel that follows a voiced obstruent such as /b/, contrasted with the typical fall in F0 following a voiceless obstruent such as /p/. This phonetic conditioning effect, of both preceding and following consonants, can be observed most clearly when identical underlying accent types are assigned to the carrier vowel, and may persist as long as 50 ms or more into the vowel. The exact contribution of the pre-vocalic consonant, the post-vocalic consonant, and the underlying accent type are difficult to untangle, though [54] is a good survey of all research in this area and adds new experimental results. For commercial synthesizers, this is definitely a second-order effect and is probably more important for rule-based formant synthesizers (see Chapter 16), which need to use every possible cue to enforce distinctions among consonants in phoneme perception, than for strictly intonational synthesis. However, in order to achieve completely natural prosody in the future, this area will have to be addressed.

Last, and perhaps least, *jitter* is a variation of individual cycle lengths in pitch-period measurement, and *shimmer* is variation in energy values of the individual cycles. These are distinct concepts, though somewhat correlated. Obviously, this is an influence of glottal pulse shape and strength on the quality of vowels. Speech with jitter and shimmer over 15% sounds pathological, but complete regularity in the glottal pulse may sound unnatural. For a deeper understanding of how these could be controlled, see Chapter 16.

15.6.2. Baseline F0 Contour Generation

We now examine a simple system that generates F0 contours. Although each stage of an F0 contour algorithm ideally requires a complete natural language and semantic analysis system, in practice a number of rules are often used. The system described here illustrates most of the important features common to the pitch-generation systems of commercial synthesizers.

First, let's consider a natural speech sample and describe what initial information is needed to characterize it, and how an artificial pitch contour can be synthesized based on the input analysis. The chosen sample is the utterance *"Don't hit it to Joey!"*, an exclamation, from the ToBI Labeling Guidelines sample utterance set [4]. The natural waveform, aligned pitch contour, and abstract ToBI labels are shown in Figure 15.7. This utterance is about 1.63 seconds and it has three major ToBI pitch events:

H*	high pitch accent on *Don't*
L*+!H	low pitch accent with following downstepped high on *Joey*
L-L%	low utterance-final boundary tone at the very end of utterance

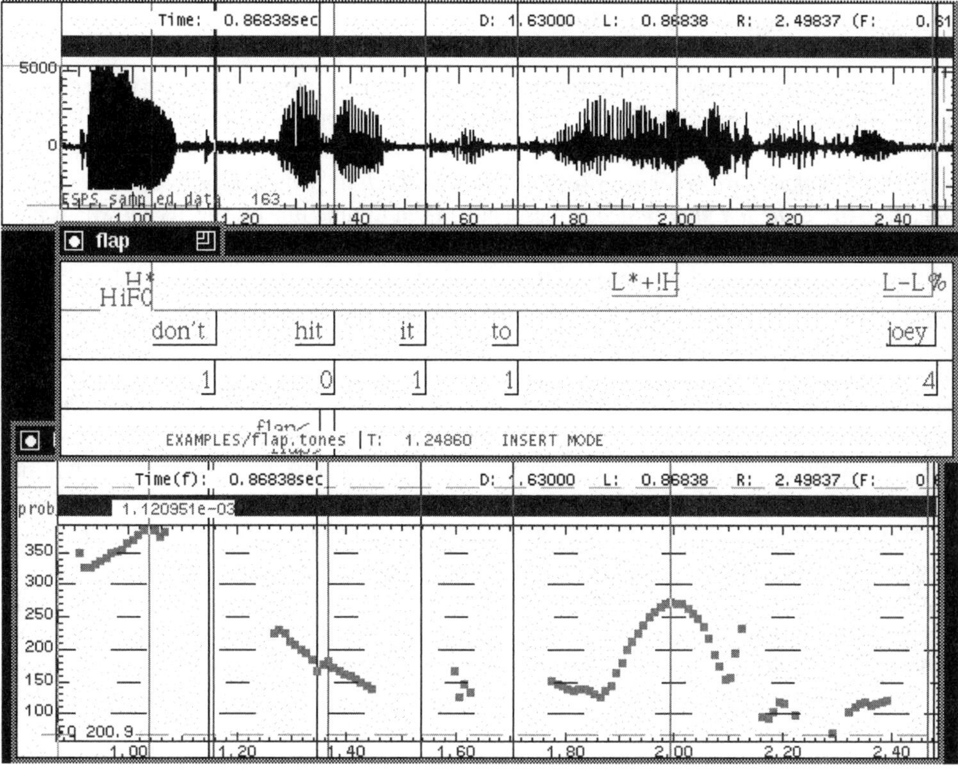

Figure 15.7 Time waveform, segmentation, ToBI marks, and pitch contour for the utterance *"Don't hit it to Joey!"* spoken by a female speaker (after Beckman [4]).

The input to the F0 contour generator includes:

- Word segmentation.
- Phone labels within words.
- Durations for phones, in milliseconds.
- Utterance type and/or punctuation information.
- Relative salience of words as determined by grammatical/semantic analysis.
- Current pitch-range settings for voice.

15.6.2.1. Accent Determination

Although accent determination ideally requires a complete natural language and semantic analysis system (see Section 15.4.3), in practice a number of rules are often used. The first rule is: *Content word categories of noun, verb, adjective, and adverb are to be accented, while the function word categories (everything else, such as pronoun, preposition, conjunction, etc.) are to be left unaccented.* Rules can be used to tune this by specifying which POS is accented or not and in which context.

If we apply that simple metric to the natural sample of Figure 15.7, we see that it does not account for the accentuation of 'hit', which, as a verb, should have been accented. In a real system perhaps we would have accented it, and this might have resulted in the typical overaccented quality of synthetic prosody. For this sample discussion, let's adopt a simplified version of a rule found in some commercial synthesizers: *Monosyllabic common verbs are left unaccented.*

What about "*Don't*"? A simplistic view would state that the POS-based policy has done the right thing, after all "*Don't*" can be regarded as a verbal form. However, usually *do* is considered an auxiliary verb and is not accented. For now we adopt another rule that says: *In a negative imperative exclamation, determined by presence of a second-person negative auxiliary form and a terminal exclamation point, the negative term gets accented*. The adoption of these corollaries to the simple POS-based accentuation rule accounts for our accent placement in the present example, but of course it sometimes fails, as does any rigid policy. So our utterance would now appear (with words selected for accent in upper case) as "*DON'T hit it to JOEY!*"

15.6.2.2. Tone Determination

In the limit, tone determination (see Section 15.4.4) also requires a complete natural language and semantic analysis system, but in practice a number of rules are often used. Generally, in working systems, H* is used for all pitch accent tones, and this is actually very realistic, as H* is the most frequent tone in natural speech.

Sometimes complex tones of the type L*+!H are thrown in for a kind of pseudovariety in TTS. In our sample natural utterance this is the tone that is used, so here we assume that this is the accent type assigned.

We also need to mark punctuation-adjacent and utterance-final phonemes as rise, continue, or fall boundaries. In this case we mark it as L-L%.

15.6.2.3. Pitch Range

To determine the pitch range, we are going to make use of three lines as a frame within which all pitches are calculated. The top line and bottom line would presumably be derived from the current or default pitch-range settings as controlled by an application or user. Here we set them in accord with the limits of our natural sample. Note that while, for this example, the pitch contour is generated within an actual pitch range, it could also be done within an abstract range of, say, 1–100, which the voice-generation module could map to the current actual setting. So we set the top line at T = 375 Hz and the base line at B = 100 Hz.

It is more advantageous to work in a logarithmic scale, because it is more easily ported from males to females, and because this better represents human prosody. There are 24 semitones in an octave; thus a semitone corresponds to a ratio of $a = 2^{1/24}$. The pitch range can be expressed in semitones as

$$n = 24\log_2(T/B) \simeq 80\log_{10}(T/B) \tag{15.4}$$

so that we can express frequencies in semitones as

$$f_0 = 80\log_{10} F_0 \tag{15.5}$$

and its inverse

$$F_0 = 10^{f_0/80} \tag{15.6}$$

Using Eq. (15.5), the top line is t = 206 and the base line is b = 160. The reference line is a kind of midline for the range, used in the accent scaling calculations, and is set halfway between the bottom and top lines, i.e., r = 183, and using Eq. (15.6), R = 194 Hz.

15.6.2.4. Prominence Determination

The relative prominence of the words (see Section 15.6.1.2) allows the pitch module to scale the pitch within any given pitch range. Here we assume (arbitrarily) that N = 5 degrees of abstract relative prominence are sufficient. This means that, e.g., an H* pitch accent with prominence 5 will be at or near the very top of the current pitch-range setting, while an L* with the same prominence will be at or near the very bottom of the range. Smaller prominence numbers indicate less salience, placing their pitch events closer to the middle of the range.

Converting the abstract tone types plus prominence into pitch numbers is more art than science (but see Section 15.6.4 for a discussion of data-based methods for this process). Here we assume a simple linear relationship between the tone's type and relative prominence:

$$f_0[i] = r + (t - r) * p[i] / N \tag{15.7}$$

In the limit, prominence determination also requires a complete natural language and semantic analysis system, but in practice a number of heuristics are often used. One such heuristic is: *In a negative imperative exclamation, the negative term gets the most emphasis*, leading to a relative prominence assignment of 5 on 'don't.' Using Eqs. (15.7) and (15.6), the anchor equals the top range of 375 Hz.

Then, since the L*+!H involves a *downstepped* term, it must by definition be lower than the preceding H* accent, so we arbitrarily assign it a relative prominence of '2'. The L*+!H is more complex, requiring calculation and placement of two separate anchor points. For simplicity we are using a single prominence value for complex tones like L*+!H, but we could also use a value-per-tone approach, at the cost of greater analytical complexity. Using Eq. (15.7), it corresponds to 192 semitones, and with Eq. (15.6), the value of !H is 251 Hz. For L*, we use a prominence of –2 (we use negative values for L tones), which, using Eq. (15.7), results in an anchor of 174, or alternatively 149 Hz.

The L-L% tone is a boundary tone, so it always goes on the final frame of a syllable-final (in this case, utterance-final) phone. The L-L% in most ToBI systems is not treated as a two-part complex tone but rather as a *super low* L% boundary, falling at the very bottom of the speaker's pitch range, i.e., prominence of 5, for a few frames. Thus the F0 value of these anchor point is 100 Hz.

We also need to set anchors for the initial point. The initial anchor is usually set at some arbitrary but high place within the speaker's range (perhaps a rule looking at u tterance type can be used). A prominence of 4 can be used, yielding a value of 329 Hz.

Finally we need to determine where to place the anchors within the accented syllable. Often they are placed in the middle of the vowel. All the anchor points are shown in Figure 15.8.

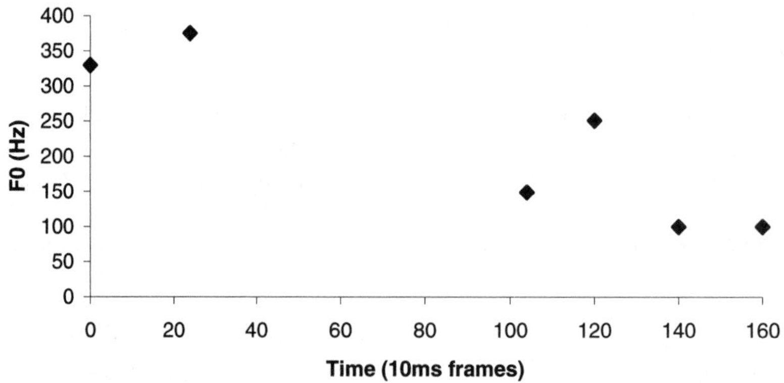

Figure 15.8 Anchor points of the F0 contour.

15.6.2.5. F0 Contour Interpolation

To obtain the full F0 contour we need some kind of interpolation. One way is to interpolate linearly and follow with a multipoint moving-average window over the resulting (angular) contour to smooth it out. Another possibility is a higher-order interpolation polynomial. In this case a cubic interpolation routine is called, which has the advantage of retaining the exact anchor points in a smoothed final contour (as opposed to moving average, which smears the anchor points). In general the choice of interpolation algorithm makes little perceptual difference, as long as no sharp 'corners' remain in the contour. In Figure 15.9 the contour was interpolated fully, without regard to voicing properties of underlying phones. In the graph, the sections corresponding to unvoiced phones have been replaced with zero, for ease of comparison to the sample in Figure 15.7. The interpolation can be done in the linear frequency, as in Figure 15.9, or in the log-frequency.

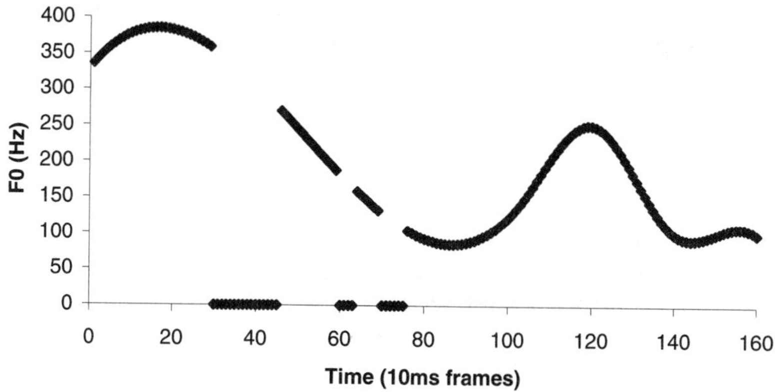

Figure 15.9 F0 contour of Figure 15.8 after cubic interpolation. Sections corresponding to unvoiced phones have been replaced with zero.

In order for the interpolation algorithm to operate properly we need to have phone durations so that the anchor points are appropriately spaced apart. In this baseline algorithm, we followed the algorithm described in Section 15.5.2.

15.6.2.6. Interface to Synthesis Module

Finally, most synthesizers cannot accept an arbitrary number of pitch controls on a given phoneme, nor it this necessary. We can downsample the pitch buffer to allocate a few characteristic points per phoneme record, and, if the synthesizer can interpolate pitch, it may be desirable to skip pitch controls for unvoiced phones altogether. The F0 targets can be placed at default locations (such as the left edge and middle of each phone), or the placements can be indicated by percent values on each target, depending on what the synthesizer supports. This has to be in agreement with the specific interface between the prosody module and the synthesis module as described in Section 15.2.

15.6.2.7. Evaluation and Possible Improvements

In comparing the output contour of Figure 15.9 to the natural one of Figure 15.7, how well have we done? As a first-order approximation, from visual inspection, it is somewhat similar to the original. Of course, we have used hand-coded information for the accent property, accent type, and prominence! However, these choices were reasonable, and could apply them as defaults to many other utterances. At a minimum, almost exactly the code given above would apply without change and give a decent contour for a whole 'family' of similar utterances, such as "*Don't hit the ball to Joey!*" or "*Never give the baseball to Henry!*" A higher-order discourse module would need to determine that *ball* and *baseball* are not accented, however, in order to use the given contour with the same rhetorical effect (presumably *ball* and *baseball* in these cases could be given/understood information).

Something very much like the system described here has been used in most commercially marketed synthesizers throughout the 1990s. This model seems overly simple, even crude, and presumably it could be substantially augmented, or completely replaced by something more sophisticated.

However, many weaknesses are apparent also. For one thing, the contour appears very smooth. The slight *jitter* of real contours can be easily simulated at a final stage of pitch buffer processing by modifying +/- 3 or 4 Hz to the final value of each frame. The degree to which such niceties actually affect listener perceptions depend entirely on the quality of the synthetic speech and the quality of the pitch-modification algorithms in the synthesizer.

The details of peak placement obviously differ between the natural and synthetic contours. This is partly due to the crude uniform durations used, but in practice synthesizers may incorporate large batteries of rules to decide exactly (for example) which frame of a phone the H* definition point should appear in—early, middle, late? Sometimes this decision is based on surrounding phonetic structure, word and syllable structure, and prosodic context. The degree to which this matters in perception depends partly on synthetic speech quality overall.

15.6.3. Parametric F0 Generation

To realize all the prosodic effects discussed above, some systems make almost direct use of a real speaker's measured data, via table lookup methods. Other systems use data indirectly, via parametrized algorithms with generic structure. The simplest systems use an invariant algorithm that has no particular connection to any single speaker's data, such as the algorithm described in the baseline F0 generation system of Section 15.6.2. Each of these approaches has advantages and disadvantages, and none of them has resulted in a system that fully mimics human prosodic performance to the satisfaction of all listeners. As in other areas of TTS, researchers have not converged on any single standard family of approaches. Once we venture beyond the simplest approaches, we find an interesting variety of systems, based on different assumptions, with differing characteristics. We now discuss a few of the more representative approaches.

Even models that make little or no attempt to analyze the internal components of an F0 contour must be indexed somehow. System designers should choose indexing or predictive factors that are derivable from text analysis, are general enough to cover most prosodic situations, and are powerful enough to specify high-quality prosody. In practice, most models' predictive factors have a rough correspondence to, or are an elaboration of, the elements of the baseline algorithm of Section 15.6.2. A typical list might include the following:

- Word structure (stress, phones, syllabification)
- Word class and/or POS
- Punctuation and prosodic phrasing
- Local syntactic structure
- Clause and sentence *type* (declarative, question, exclamation, quote, etc.)
- Externally specified focus and emphasis
- Externally specified speech style, pragmatic style, emotional tone, and speech act goals

These factors jointly determine an output contour's characteristics, as listed below. Ideally, any or all of these may be externally and directly specified, or they may be inferred or implied within the F0 generation model itself:

- Pitch-range setting
- Gradient, relative prominence on each syllable
- Global declination trend, if any
- Local shape of F0 movement
- Timing of F0 events relative to phone (carrier) structure

The combinatorial complexity of these predictive factors, and the size of the resulting models, can be serious issues for practical systems that strive for high coverage of prosodic variability and high-quality output. The possibility of using *externally specified* symbolic markups gives the whole system a degree of modularity, in that prosodic annotation can be specified directly by an authoritative outside source or can be derived automatically by the symbolic prosody prediction process that precedes F0 contour generation.

Parametric models propose an underlying architecture of prosodic production or perception that constrains the set of possible outputs to conform to universal constants of the human speech mechanism. Naturally, these models need settings to distinguish different speakers, different styles, and the specifics of utterances. We describe superposition models and ToBI Realization models.

15.6.3.1. Superposition Models

An influential class of parametric models was initiated by the work [35] for Swedish, which proposed additive superposition of component contours to synthesize a complex final F0 track. In the version refined and elaborated in [14], the component contours, which may all have different strengths and decay characteristics, may correspond to longer-term trends, such as phrase or utterance declination, as well as shorter-time events, such as pitch accents

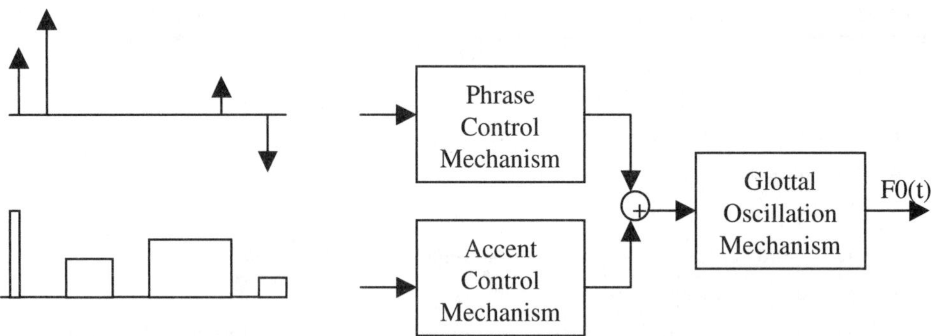

Figure 15.10 Fujisaki pitch model [15]. F0 is a superposition of phrase effects with accent effects. The phrase mechanism controls things like the declination of a declarative sentence or a question, whereas the accent mechanism accounts for accents in individual syllables.

on words. The component contours are modeled as the critically damped responses of second-order systems to impulse functions for the longer-term, slowly decaying phrasal trends, and step or rectangular functions of shorter-term accent events. The components so generated are added and ride a baseline that is speaker specific. The basic ingredients of the system, known as Fujisaki's model [15, 19], are shown in Figure 15.10. The resulting contour is shown in Figure 15.11. Obviously, similar effects can be generated with linear accent shapes as described in the simpler model above, with smoothing. However, there are some plausible claims for the articulatory correlates of the constraints imposed in the second-order damping and superposition effects of this model [33].

Superposition models of this type can, if supplied with accurate parameters in the form of time alignments and strengths of the impulses and steps, generate contours closely mimicking natural examples. In this respect, the remaining quality gap for general application is in the parametric knowledge driving the model, not in the model structure per se. These kinds of models have been particularly successful in replicating the relatively constrained Japanese reading-style. Whether these models can account straightforwardly for the immense variety of a large range of English speakers and text genre, or whether, on the contrary, the parameters proliferate and the settings become increasingly arbitrary, remains to be seen.

Figure 15.11 Composite contour obtained by low-pass filtering the impulses and boxes in the Fujisaki model of Figure 15.10.

15.6.3.2. ToBI Realization Models

One simple parametric model, which in its inherent structure makes only modest claims for principled correspondence to perceptual or articulatory reality, is designed to support prosodic symbols such as the *Tones and Break Indices* (ToBI) system. This model, variants of which are developed in [2, 54], posits two or three control lines, by reference to which ToBI-like prosody symbols can be scaled. This provides for some independence between symbolic and phonetic prosodic subsystems. In the model shown in Figure 15.12, the top line is an upper limit of the pitch range. It can be slanted down to simulate declination. The bottom line represents the bottom of the speaker's range. Pitch accents and boundary tones (as in ToBI) are scaled from a reference line, which is often midway in the range in a logarithmic scale of the pitch range, as described in the baseline algorithm of 15.6.2. You can think of this scaling as operating within a percentage of the current range, rather than absolute values, so a generic method can be applied to any arbitrary pitch-range setting. The quantitative instantiation of accent height is done at the final stage. The accents and boundary tones consist of one or more points, which can be aligned with the carrier phones; then interpolation is applied between points, and smoothing is performed over the resulting contour.

In Figure 15.12, t, r, and b are the top, reference, and baseline pitch values, respectively. They are set from the defaults of the voice character and by user choice. The base b is considered a physiological constraint of voice. P is the prominence of the accent and N is the number of prominence steps. Declination can be modeled by slanting the top and/or reference lines down. The lowered position of the reference in Figure 15.12 reflects the observation that the realization of H(igh) and L(ow) ToBI abstract tones in a given pitch range is asymmetric, with a greater portion available for H, while L saturates more quickly. This is why placing the reference line midway between the top and base lines in a log-frequency scale automatically takes care of this phenomenon. After target points are located and scaled according to their gradient prominence specifications, the (hopefully sparse) targets can be interpolated and the resulting contour smoothed. If the contour is calculated in, say, 10-ms frames, two pitch targets sampled from the contour vector per phone usually suffice to reproduce the intended prosodic effects faithfully.

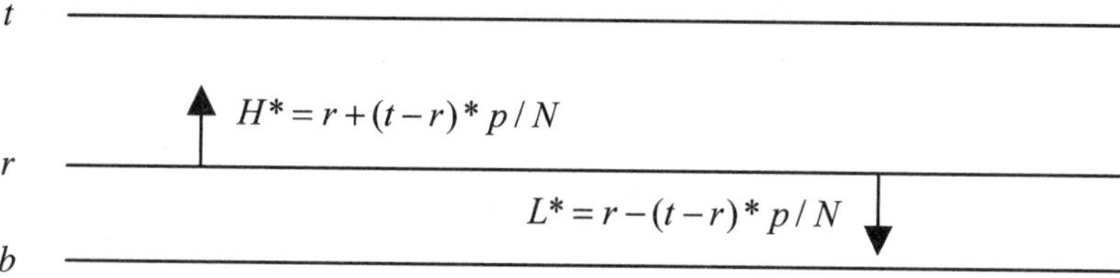

Figure 15.12 A typical model of tone scaling with an abstract pitch range.

If a database of recorded utterances with phone labeling and F0 measurements has been reliably labeled with ToBI pitch annotations, it may be possible to automate the implementation of the ToBI-style parametrized model. This was attempted with some success in [5], where linear regression was used to predict syllable initial, vowel medial, and syllable final F0 based on simple, accurately measurable factors such as:

- ToBI accent type of target and neighbor syllables
- ToBI boundary pitch type of target and neighbor syllables
- Break index on target and neighbor syllables
- Lexical stress of target and neighbor syllables
- Number of syllables in phrase
- Target syllable position phrase
- Number and location of stressed syllable(s)
- Number and location of accented syllable(s)

Models of this sort do not incorporate an explicit mechanism (like the scaling direction from r in Figure 15.12) to distinguish H(igh) from L(ow) tone space, beyond what the data and its annotations imply.

The work in [47] consists of a ToBI realization model in which the 'smoothing' mechanism is built-in as a dynamical system whose parameters are also learnt from data. This work could be viewed as a stochastic realization of Fujisaki's superposition model without the phrase controls and where the accents are given by ToBI labels.

Both the ToBI realization models and the superposition models could, if supplied with sufficiently accurate measurements of an example contour, reproduce it fairly accurately. Both models require much detailed knowledge (global and local pitch range; location, type, and relative strength of accents and boundary tones; degree of declination; etc.) to function at human-level quality for a given utterance. If a system designer is in possession of a completely annotated, high-quality database of fully representative prosodic forms for his/her needs, the question of deployment of the database in a model can be made based on performance tradeoffs, maintenance issues, and other engineering considerations. If, on the other hand, no such database is available for the given application purpose, extremely high prosodic quality, including lively yet principled variation, should not be expected to result simply from choosing the 'mathematically correct' model type.

15.6.4. Corpus-Based F0 Generation

It is possible to have F0 parameters trained from a corpus of natural recordings. The simplest models are the direct models, where an exact match is required. Models that offer more generalization have a library of F0 contours that are indexed either from features from the parse tree or from ToBI labels. Finally, there are F0 generation models from a statistical network such as a neural network or an HMM. In all cases, once the model is set, the parameters are learned automatically from data.

15.6.4.1. Transplanted Prosody

The most direct approach of all is to store a single contour from a real speaker's utterance corresponding to every possible input utterance that one's TTS system will ever face. This seems to limit the ability to freely synthesize any input text. However, this approach can be viable under certain special conditions and limitations. These controls are so detailed that they are tedious to write manually. Fortunately, they can be generated automatically by speech recognition algorithms.

When these controls (*transplanted prosody*), taken from an authentic digitized utterance, are applied to synthetic voice units, the results can be very convincing, sometimes nearly as good as the original digitized samples [43]. A system with this capability can mix predefined utterances having natural-quality prosody, such as greetings, with flexible synthesis capabilities for system response, using a consistent synthetic voice. The transplanted prosody for the *frozen* phrases can be derived either from the original voice data donor used to create the synthetic voice model, or any other speaker, with global adjustment for pitch-range differences. Another use of the transplanted prosody capability is to compress a spoken message into ASCII (phone labels plus the prosodic controls) for playback, preserving much of the quality, if not the full individuality, of the original speaker's recording.

15.6.4.2. F0 Contours Indexed by Parsed Text

In a more generalized variant of the direct approach, once could imagine collecting and indexing a gigantic database of clauses, phrases, words, or syllables, and then annotating all units with their salient prosodic features. If the terms of annotation (word structure, POS, syntactic context, etc.) can be applied to new utterances at runtime, a prosodic description for the closest matching database unit can be recovered and applied to the input utterance [23]. The advantages here are that prosodic quality can be made arbitrarily high, by collecting enough exemplars to cover arbitrarily large quantities of input text, and that detailed analysis of the deeper properties of the prosodic phenomena can be sidestepped. The potential disadvantages are:

- Data-collection time is long (which affects the capability to create new voices).
- A large amount of runtime storage is required (presumably less important as technology progresses).
- Database annotation may have to be manual, or if automated, may be of poor quality.
- The model cannot be easily modified/extended, owing to lack of fundamental understanding.

- Coverage can never be complete, therefore rulelike generalization, fuzzy match capability, or back-off, is needed.
- Consistency control for the prosodic attributes (to prevent unit boundary mismatches) can be difficult.

The first two disadvantages are self-explanatory. The difficulty of annotating the database, to form the basis of the indexing and retrieval scheme, depends on the type and depth of the indexing parameters chosen. Any such scheme requires annotations to identify the component phones of each unit (syllable, word, or phrase) and their durations. This can usually be obtained from speech recognition tools [23], which may be independently required to create a synthetic voice (see Chapter 16). Lexical or word stress attributes can be extracted from an online dictionary or NLP system, though, as we have seen above, lexical stress is neither a necessary nor a sufficient condition for predicting pitch accent placement.

If only a very high level of description is sought, based primarily on the pragmatics of utterance use and some syntactic typology, it may not be necessary to recover a detailed symbolic pitch analysis. An input text can be described in high-level pragmatic/semantic terms, and pitch from the nearest matching word or phrase from the database can be applied with the expectation that its contour is likely correct. For example, such a system might have multiple prosodic versions of a word that can be used in different pragmatic senses, such as *ok*, which could be a question, a statement, an exclamation, a signal of hesitation or uncertainty, etc. The correct version must be selected based on the runtime requirements of the application.

Direct prosody schemes of this type often preserve the original phone carrier material of each instance in order to assure optimal match between prosody and spectrum. However, with DSP techniques enabling arbitrary modifications of waveforms (see Chapter 16), this is not strictly necessary; the prosodic annotations could stand alone, with phone label annotation only. If more detailed prosodic control is required, such as being aware of the type of accent, its pitch range, prominence, and other features, the annotation task is much more difficult.

A straightforward and elegant formulation of the lookup-table direct model approach can be found in [30]. This system, created for Spanish but generally adaptable, is based on a large single-speaker recorded database of a variety of sentence types. The sentences are linguistically analyzed, and prosodic structure is hypothesized based on syllables, accent groups (groups of syllables with one lexical stress), breath groups (groups of accent-groups between pauses regardless of the duration of the pause), and sentences. Note that these structures are hypothesized based on the textual material alone, and the speaker will not always perform accordingly. Pitch (initial, mid, and final F0) and duration data for each spoken syllable is automatically measured and stored. At runtime, the input sentence is analyzed using the same set of structural attributes inferred from the text, and a vector of candidate syllables from the database is constructed with identical, or similar, structural context and attributes for each successive input syllable position.

The best path through the set of such candidates is selected by minimizing the F0 distance and disjuncture across the utterance. This is a clean and simple approach to jointly utilizing both shallow linguistic features and genuine phonetic data (duration and F0), with dynamic programming to smooth and integrate the output F0 contour. However, as with any direct modeling approach, it lacks significant generalization capabilities outside the textual material and speaking style specified during the data collection phase, so a number of separate models may have to be constructed.

The CHATR system of ATR (Japan) [9] takes a similar approach, in that optimal prosody is selected from prerecorded units, rather than synthesized from a more general model. The CHATR system consists of a large database of digitized speech, indexed by the speaker identity, the phoneme sequences of the words, and some pragmatic and semantic attributes. Selection of phonemes proceeds by choosing the minimal-cost path from among the similarly indexed database candidate units available for each phoneme or longer segment of speech to be synthesized. This system achieves high quality by allowing the carrier phones to bear only their original prosody—pitch modification of the contour is minimized or eliminated. Of course, the restriction of DSP modification implies a limitation of the generalizability of the database. This type of approach obtains the prosody implicitly from the database [31], and as such combines both the prosody and speech synthesis modules. This type of minimal-cost search is described in more detail in Chapter 16.

15.6.4.3. F0 Contours Indexed by ToBI

The architecture for a simple and straightforward direct model indexed by ToBI is diagrammed in Figure 15.13. This model combines the two often-conflicting goals: it is empirically (corpus) based, but it permits specification in terms of principled abstract prosodic categories. In this model, an utterance to be synthesized is annotated in terms of its linguistic features—perhaps POS, syntactic structure, word emphasis (based on information structure), etc. The utterance so characterized is matched against a corpus of actual utterances that are annotated with linguistic features and ToBI symbols, Corpus (a). A *fuzzy* matching capability based on edit distance or dynamic programming can be incorporated. If Corpus (a) is sufficiently large and varied, a number of possible ToBI renderings of either the entire utterance or selected parts of it may be recovered. At this level of abstraction, the ToBI labels would not encode relative prominence specifications (strength of pitch extrusions) or pitch range. The set of such abstractly described contours can then be fuzzy matched into Corpus (b), a set of ToBI annotated actual contours, and the best set of matches recovered.

Note that while it is possible that Corpus (a) is the exact same base material as Corpus (b), the model does not enforce an identity, and there may be reasons to desire such flexibility and modularity, depending on the degree and quality of data and annotation at each level. Once a number of likely actual contours have been identified, they can be passed to a voice-unit selection module. The module can select the combination of segmental strings (sometimes called 'long units,' since they may combine more than one phoneme) from the voice database whose original prosody is closest to one of the candidate contours, using root-mean-square-error, correlation, or other statistical tests. Those units are then concatenated (with their prosody unmodified) and sent to the application or played out.

A model of this type has some of the disadvantages of direct models as listed above. It also assumes availability of large and varied databases of both prosodic contours and segmental (phone) long units for concatenation (see Chapter 16). It further requires that these databases be annotated, either by human labelers or automated systems. However, it has certain advantages as well:

- It allows for symbolic, phonological coding of prosody.
- It has high-quality natural contours.
- It has high-quality phonetic units, with unmodified pitch.
- Its modular architecture can work with user-supplied prosodic symbols.

It also allows the immediate, temporary use of data that is collected for deeper analysis, in the hope of eventual construction of smaller, parametrized models. The model of Figure 15.13 is a generalization of the prosody system described in [23].

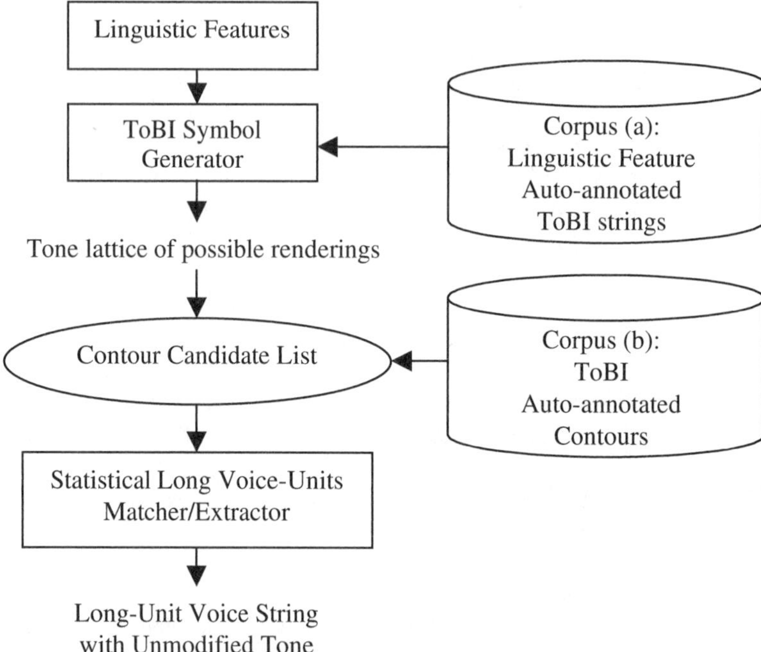

Figure 15.13 A corpus-based prosodic generation model.

15.7. PROSODY MARKUP LANGUAGES

Chapter 14 discussed generalized document markup schemes for text analysis. Most TTS engines provide simple text tags and application programming interface controls that allow at least rudimentary hints to be passed along from an application to a TTS engine. We expect to see more sophisticated speech-specific annotation systems, which eventually incorporate current research on the use of semantically structured inputs to synthesizers, sometimes called concept-to-speech systems. A standard set of prosodic annotation tags would likely include tags for insertion of silence pause, emotion, pitch baseline and range, speed in words-per-minute, and volume. This would be in addition to general tags for specifying the language of origin if not predictable, character of the voice, and text normalization context such as address, date, email, etc.

For prosodic processing, text may be marked with tags that have scope, in the general fashion of XML. Some examples of the form and function of a few common TTS tags for prosodic processing, based loosely on the proposals of [65], are introduced below. Other tags can be added by intermediate subcomponents to indicate variables such as accents and tones. This extension allows for even finer research and prosody models.

- **Pause** or **Break** commands might accept either an absolute duration of silence in milliseconds, or, as in the W3C proposal, a mnemonic describing the relative salience of the pause (Large, Medium, Small, None), or a *prosodic punctuation* symbol from the set ',', '.', '?', '!', '…', etc., which not only indicates a pause insertion but also influences the typical pitch contour of the phone segments entering and leaving the pause area. For example, specifying ',' as the argument of a Pause command might determine the use of a continuation rise on the phones immediately preceding the pause, indicating incompletion or listing intonation.
- **Rate** controls the speed of output. The usual measurement is *words per minute*, which can be a bit vague, since words are of very different durations. However, this metric is familiar to many TTS users and works reasonably well in practice. For non-IndoEuropean languages, different metrics must be contemplated. Some power listeners who use a TTS system routinely can tolerate (in fact, demand) rates of over 300 words per minute, while 150 or fewer might be all that a novice listener could expect to reliably comprehend.
- **Baseline Pitch** specifies the desired average pitch: a level around which, or up from which, pitch is to fluctuate.
- **Pitch Range** specifies within what bounds around the baseline pitch level line the pitch is to fluctuate.
- **Pitch** commands can override the system's default prosody, giving an application or document author greater control. Generally, TTS engines require some freedom to express their typical pitch patterns within the broad limits specified by a Pitch markup.

- **Emphasis** emphasizes or deemphasizes one or more words, signaling their relative importance in an utterance. Its scope could be indicated by XML style. Control over emphasis brings up a number of interesting considerations. For one thing, it may be desirable to have degrees of emphasis. The notion of gradient prominence—the apparent fact that there are no categorical constraints on levels of relative emphasis or accentuation—has been a perpetual thorn in the side for prosodic researchers. This means that in principle any positive real number could be used as an argument to this tag. In practice, most TTS engines would artificially constrain the range of emphasis to a smaller set of integers, or perhaps use semantic labels, such as *strong*, *moderate*, *weak*, *none* for degree of emphasis. Emphasis may be realized with multiple phonetic cues. Thus, if the user or application has, for example, set the pitch range very narrowly, the emphasis effect may be achieved by manipulation of segmental duration or even relative amplitude. The implementation of emphasis by a TTS engine for a given word may involve manipulation (e.g., de-accentuation) of surrounding words as much as it involves heightening the pitch or volume, or stretching the phone durations, of the target word itself. In most cases the main phonetic and perceptual effect of emphasis or accentuation is heard on the lexically main stressed syllable of the word, but this can be violated under special conditions of semantic focus, e.g., "*I didn't say employ**er**, I said employ**ee**.*" This would require a more powerful emphasis specification than is currently provided in most TTS systems, but alternatively it could be specified using phone input commands such as "*The <emp>truth</emp>, the <emp>whole truth</emp>, and nothing <emp> but</emp> the truth.*" For more control, future TTS systems may support degree emphasis: "*... nothing <emp level="strong">but</emp> the truth*" or even deemphasis: "*... nothing <emp level= "reduced">but</emp> the truth*". Emphasis is related to prominence, discussed in Section 15.6.1.2.

15.8. PROSODY EVALUATION

Evaluation of a complete TTS system is discussed in Chapter 16. We limit ourselves here to evaluating the prosody component. We assume that the text analysis module has done a perfect job, and that the synthesis module does a perfect job, which cannot be done in general, so that approximations need to be made.

Evaluation can be done automatically or by using listening tests with human subjects. In both cases it's useful to start with some natural recordings with their associated text. We start by replacing the natural prosody with the system's synthetic prosody. In the case of automatic evaluation, we can compare the enriched prosodic representations described in Section 15.2 for both the natural recording and the synthetic prosody. The reference enriched prosodic representation can be obtained either manually or by using a pitch tracker and a speech recognizer.

Automated testing of prosody involves the following:

- **Duration**. It can be performed by measuring the average squared difference between each phone's actual duration in a real utterance and the duration predicted by the system.
- **Pitch** contours. It can be performed by using standard statistical measures over a system contour and a natural one. When this is done, duration and phoneme identity should be completely controlled. Measures such as root-mean-square error indicate the characteristic divergence between two contours, while correlation indicates the similarity in shape across difference pitch ranges. In general, RMSE scores of 15 Hz or less for male speech over a long sentence, with correlation of .8 or above, indicate quality that may be close to perceptually identical to the natural reference utterance. In general, such exactness of match is useful only during model training and testing and cannot be expected during training on entirely new utterances from random text.

Listening tests can be performed to evaluate a prosody module. This involves subjects listening to the natural recording and the synthetic speech, or to synthetic speech generated with two different prosody modules. This can lead to a more precise evaluation, as humans are the final consumer of this technology. However, such tests are more expensive to carry out. Furthermore, this method results in testing both the prosody module and the synthesis components together. To avoid this, the original waveform can be modified to have the synthetic prosody using the signal processing techniques described in Chapter 16. Since such techniques introduce some distortions, this measuring method is still somewhat biased. In practice, it has been shown that its effect is much smaller than that of the synthetic prosody [43].

It is shown that synthesizing pitch is more difficult than duration [43]. Subjects scored significantly higher utterances that had natural pitch and synthetic duration than utterances with synthetic pitch and natural duration. In fact, using only synthetic duration had a score quite close to that of the original recording. While duration modeling is not a solved problem, this indicates that generation of pitch contours is more difficult.

15.9. Historical Perspective and Further Reading

Prosodic methods have been incorporated within the traditional fields of rhetoric and elocution for centuries. In ancient Greece, at the time of Plato, written documentation in support of claims in legal disputes was rare. To help litigants plead their cases persuasively, systematic programs of rhetorical instruction were established, which included both content and form of verbal argument. This 'prescriptive' tradition of systematic instruction in verbal style uncovered issues that remain central to the descriptively oriented prosodic research of today. A masterful and entertaining discussion of this tradition and its possible relevance to the task of teaching computers to *plead a case* can be found in [64]. The Greeks were particularly concerned about an issue that, as usual, is still important for us today: the separation of rhetorical effectiveness from considerations of truth. If you are interested in this, you cannot do better than to begin with Plato's dialog *Phaedrus* [42].

Modern linguists have also considered a related, but more narrowly formulated question: Should prosody be treated as a logical, categorical analog to phonological and syntactic processes? The best discussion of these issues from a prosodic (as opposed to strictly neurological) point of view is found in [7, 8]. If you are interested in the neurological side, you can begin with [13]. For emotional modeling, before slogging through the scattered and somewhat disjointed papers on emotion in speech that have appeared sporadically for years, the reader would be well advised to get a basic grounding in some of the issues related to emotion in computation, as treated in [38].

Going in the other direction, there are many subtle interactions in the phonetics of prosody: the various muscles, their joint possibilities of operation in phonation and articulation, as well as the acoustics properties of the vocal chambers. For an excellent introduction to the whole field, start with [27].

The most complete and accessible overview of modern prosodic analysis as embedded in mainstream linguistic theory is Ladd's *Intonational Phonology* [26], which covers the precursors, current standard practice, and remaining unsolved issues of the highly influential auto segmental theory of intonational phonology, from which ToBI has arisen. ToBI was devised by speech scientists who wanted a prosodic transcription standard to enable sharing of databases [4]. For most practical purposes, the ToBI definitions are sufficient as a starting point for both research and applications, but for those who prefer alternative annotation systems aligned with the British tradition, conversion guidelines have been attempted [45]. Another major phonological approach to English intonation has been the British school described in [11]. Bridging the two is IViE, a labeling system that is philosophically aligned with ToBI but may be more appropriate for non-U.S. dialects of English [18].

The first prosodic synthesis by rule was developed by Ignatius Mattingly in 1968 in Haskins Laboratories. In 1971, Fujisaki [15] developed his superposition model that has been used for many years. The development of the ToBI in 1992 [55] marked a milestone in automatic prosody generation. The application of statistical techniques, such as CART, for phoneme durations during the 1990s constituted a significant step beyond the rule-based methods. Finally, the development of the CHATR system in the mid-1990s ignited interest in the indexing of massive databases. It is possible to attempt smoothing over both the index space and the resulting prosodic data tracks by means of generalized learning methods, such as neural nets or HMMs. These models have built-in generalization over unseen inputs, and built-in smoothing over the concatenated outputs of unit selection. The network described in [57] codes every syllable in a training database in terms of perceived prominence (human judged), a number from 1 to 31, as well as the syllable's phonemes, rising/falling boundary type for phrase-edge syllables, and distance from preceding and following phrase boundaries, for all syllables. When tested with reasonably simple text material of similar type, these networks yielded high-quality simulations.

A potential research area for future generalizations of this system is to increase the degree and accuracy of automation in labeling the training features of the recordings, such as perceived prominence. Another area is to either expand the inventory of model types, or to determine adequate generalization mechanisms. By training HMMs on accented syllables of differing phonetic structure, some of this fine alignment information can be automatically captured [16]. Another approach consists in generating pitch contours directly from a hidden Markov model, which is run in generation mode [66].

Recently, just as in speech synthesis for voice, there has been a realization that the *direct* and *parametric* prosodic models have a great deal in common. Direct models require huge databases of indexed exemplars for unmediated concatenation and playback of contours, in addition to generalized back-off methods, while parametric models are generalized for any input, but also require phonetic databases of sufficient variety to support statistical learning of parameter settings for high quality. We can, therefore, expect to see increasing numbers of hybrid systems. One such system is described in [47], which could be viewed as a stochastic realization of Fujisaki's superposition model without the phrase controls, where the accents are given by ToBI labels and the smoothing is done by means of a dynamical system.

While this chapter has focused on U.S. English, many similar issues arise in prosodic modeling of other languages. An excellent survey of the prosodic systems of every major European language, as well as Arabic and several major East Asian languages, can be found in [21].

Though not explicitly covered in this chapter, analysis of prosody for speech recognition is a small but growing area of study. Anyone who has digested this chapter should be prepared to approach the more specialized work of [25, 37] and the speech recognition prosody studies collected in [48]. Those with a psycholinguistic bent can begin with [12].

References

[1] Allen, J., M.S. Hunnicutt, and D.H. Klatt, *From Text to Speech: the MITalk System*, 1987, Cambridge, UK, University Press.

[2] Anderson, M.D., J.B. Pierrehumbert, and M.Y. Liberman, "Synthesis by Rule of English Intonation Patterns," *Proc. of Int. Conf. on Acoustics, Speech and Signal Processing*, 1984, pp. 2.8.1-2.8.4.

[3] Arnfield, S., "Word Class Driven Synthesis of Prosodic Annotations," *Proc. of the Int. Conf. on Spoken Language Processing*, 1996, Philadelphia, PA, pp. 1978-1981.

[4] Beckman, M.E. and G.M. Ayers, *Guidelines for ToBI Labelling*, 1994, http://www.ling.ohio-state.edu/phonetics/ToBI/main.html.

[5] Black, A. and A. Hunt, "Generating F0 Contours from ToBI labels using Linear Regression," *Proc. of the Int. Conf. on Spoken Language Processing*, 1996, pp. 1385-1388.

[6] Bolinger, D., "Accent is predictable (if you're a mind-reader)," *Language*, 1972, **48**, pp. 633-44.

[7] Bolinger, D., *Intonation and its parts*, 1986, Stanford, Stanford University Press.

[8] Bolinger, D., *Intonation and its uses*, 1989, Stanford, Stanford University Press.

[9] Campbell, N., "CHATR: A High-Definition Speech Re-sequencing System," *ASA/ASJ Joint Meeting*, 1996, Honolulu, Hawaii, pp. 1223-1228.

[10] Carroll, L., *Alice in Wonderland, Unabridged ed.*, 1997, Penguin USA.

[11] Crystal, D., *Prosodic Systems and Intonation in English*, 1969, Cambridge University Press.

[12] Crystal, D., "Prosody and Parsing," P. Warren, Editor, 1996, Lawrence Erlbaum Associates.

[13] Emmorey, K., "The Neurological Substrates for Prosodic Aspects of Speech," *Brain and Language*, 1987, **30**, pp. 305-320.

[14] Fujisaki, H., "Prosody, Models, and Spontaneous Speech" in *Computing Prosody*, Y. Sagisaka, N. Campbell, N. Higuchi, Editors, 1997, New York, Springer.

[15] Fujisaki, H. and H. Sudo, "A Generative Model of the Prosody of Connected Speech in Japanese," *Annual Report of Eng. Research Institute*, 1971, **30**, pp. 75-80.

[16] Fukada, T., *et al.*, "A Study on Pitch Pattern Generation Using HMM-based Statistical Information," *Proc. Int. Conf. on Spoken Language Processing*, 1994, Yokohama, Japan, pp. 723-726.

[17] Goldsmith, J., "English as a Tone Language" in *Phonology in the 1980's*, D. Goyvaerts, Editor 1980, Ghent, Story-Scientia.

[18] Grabe, E., F. Nolan, and K. Farrar, "IViE - a Comparative Transcription System for Intonational Variation in English," *Proc. of the Int. Conf. on Spoken Language Processing*, 1998, Sydney, Australia.

[19] Hirose, H. and H. Fujisaki, "Analysis and Synthesis of Voice Fundamental Frequency Contours of Spoken Sentences," *IEEE Int. Conf. on Acoustics, Speech and Signal Processing*, 1982, pp. 950-953.

[20] Hirschberg, J., "Pitch Accent in Context: Predicting Intonational Prominence from Text," *Artificial Intelligence*, 1993, **63**, pp. 305-340.

[21] Hirst, D., A.D. Cristo, and A. Cruttenden, *Intonation Systems: A Survey of Twenty Languages*, 1998, Cambridge, U.K., Cambridge University Press.

[22] Hirst, D.J., "The Symbolic Coding of Fundamental Frequency Curves: from Acoustics to Phonology," *Proc. of Int. Symposium on Prosody*, 1994, Yokohama, Japan.

[23] Huang, X., *et al.*, "Whistler: A Trainable Text-to-Speech System," *Int. Conf. on Spoken Language Processing*, 1996, Philadephia, PA, pp. 2387-2390.

[24] Klasmeyer, G. and W.F. Sendlmeier, "The Classification of Different Phonation Types in Emotional and Neutral Speech," *Forensic Linguistics*, 1997, **4**(1), pp. 104-125.

[25] Kompe, R., *Prosody in Speech Understanding Systems*, 1997, Berlin, Springer.

[26] Ladd, R.D., *Intonational Phonology*, Cambridge Studies in Linguistics, 1996, Cambridge, Cambridge University Press.

[27] Ladefoged, P., *A Course in Phonetics*, 1993, Harcourt Brace Jovanovich.

[28] Liberman, M., *The Intonation System of English*, PhD Thesis in Linguistics and Philosophy, 1975, MIT, Cambridge.

[29] Liberman, M. and J. Pierrehumbert, "Intonational Invariance under Changes in Pitch Range and Length" in *Language and Sound Structure*, M. Aronoff, Oerhle, R., ed., 1984, Cambridge, MA, MIT Press, pp. 157-233.

[30] Lopez-Gonzalo, E., and J.M. Rodriguez-Garcia, "Statistical Methods in Data-Driven Modeling of Spanish Prosody for Text to Speech," *in Proc. ICSLP 1996*, 1996, pp. 1373-1376.

[31] Malfrere, F., T. Dutoit, and P. Mertens, "Automatic Prosody Generation Using Supra-Segmental Unit Selection," *Third ESCA/COCOSCA Workshop on Speech Synthesis*, 1998, Jenolan Caves, Australia, pp. 323-328.

[32] Mason, J., *An Essay on Elocution*, 1st ed, 1748, London.

[33] Möbius, B., "Analysis and Synthesis of German F0 Contours by Means of Fujisaki's Model," *Speech Communication*, 1993, **13**(53-61).

[34] Murray, I. and J. Arnott, "Toward the Simulation of Emotion in Synthetic Speech: A Review of the Literature on Human Vocal Emotion," *Journal Acoustical Society of America*, 1993, **93**(2), pp. 1097-1108.

[35] Öhman, S., *Word and Sentence Intonation: A Quantitative Model*, 1967, KTH, pp. 20-54.

[36] Ostendorf, M., and N. Veilleux, "A Hierarchical Stochastic Model for Automatic Prediction of Prosodic Boundary Location," *Computational Linguistics*, 1994, **20**(1), pp. 27-54.

[37] Ostendorf, M., "Linking Speech Recognition and Language Processing Through Prosody," *Journal for the Integrated Study of Artificial Intelligence, Cognitive Science and Applied Epistemology*, 1998, **15**(3), pp. 279-303.

[38] Picard, R.W., *Affective Computing*, 1997, MIT Press.

[39] Pierrehumbert, J., *The Phonology and Phonetics of English Intonation*, PhD Thesis in *Linguistics and Philosophy* 1980, MIT, Cambridge, MA.

[40] Pierrehumbert, J., and M. Beckman, *Japanese Tone Structure*, 1988, Cambridge, MA, MIT Press.

[41] Pierrehumbert, J. and J. Hirschberg, "The Meaning of Intonational Contours in the Interpretation of Discourse" in *Intentions in Communication*, P.R. Cohen, J. Morgan, and M. E. Pollack, ed., 1990, Cambridge, MA, MIT Press.

[42] Plato, *The Symposium and The Phaedrus: Plato's Erotic Dialogues*, 1994, State University of New York Press.

[43] Plumpe, M. and S. Meredith, "Which is More Important in a Concatenative Text-to-Speech System: Pitch, Duration, or Spectral Discontinuity," *Third ESCA/COCOSDA Int. Workshop on Speech Synthesis*, 1998, Jenolan Caves, Australia, pp. 231-235.

[44] Prevost, S. and M. Steedman, "Specifying Intonation from Context for Speech Synthesis," *Speech Communication*, 1994, **15**, pp. 139-153.

[45] Roach, P., "Conversion between Prosodic Transcription Systems: 'Standard British' and ToBI," *Speech Communication*, 1994, **15**, pp. 91-97.

[46] Ross, K. and M. Ostendorf, "Prediction of Abstract Prosodic Labels for Speech Synthesis," *Computer, Speech and Language*, 1996, **10**, pp. 155-185.

[47] Ross, K. and M. Ostendorf, "A Dynamical System Model for Generating Fundamental Frequency for Speech Synthesis," *IEEE Trans. on Speech and Audio Processing*, 1999, **7**(3), pp. 295-309.

[48] Sagisaka, Y., W.N. Campbell, and N. Higuchi, *Computing Prosody*, 1997, Springer-Verlag.

[49] Santen, J.V., "Contextual Effects on Vowel Duration," *Speech Communication*, 1992, **11**(6), pp. 513-546.

[50] Selting, M., *Prosodie im Gespräch*, 1995, Max Niemeyer Verlag.

[51] Shattuck-Hufnagel, S. and M. Ostendorf, "Stress Shift and Early Pitch Accent Placement in Lexical Items in American English," *Journal of Phonetics*, 1994, **22**, pp. 357-388.

[52] Shen, X.-n.S., *The Prosody of Mandarin Chinese*, 1990, Berkeley, University of California Press.

[53] Sheridan, T., *Lectures on the Art of Reading*, 3rd ed, 1787, London, Dodsley.

[54] Silverman, K., *The Structure and Processing of Fundamental Frequency Contours*, Ph.D. Thesis, 1987, University of Cambridge, Cambridge, UK.

[55] Silverman, K., "ToBI: A Standard for Labeling English Prosody," *Int. Conf. on Spoken Language Processing*, 1992, Banff, Canada, pp. 867-870.

[56] Sokal, A.D., "Transgressing the Boundaries: Towards a Transformative Hermeneutics of Quantum Gravity," *Social Text*, 1996, **46/47**, pp. 217-252.

[57] Sonntag, G., T. Portele, and B. Heuft, "Prosody Generation with a Neural Network: Weighing the Importance of Input Parameters," *Proc. Int. Conf. on Acoustics, Speech and Signal Processing*, 1997, pp. 930-934.

[58] Steedman, M., "Information Structure and the Syntax-Phonology Interface," *Linguistic Inquiry*, 2000.

[59] Stevens, K.N., "Control Parameters for Synthesis by Rule," *Proc. of the ESCA Tutorial Day on Speech Synthesis*, 1990, pp. 27-37.

[60] Taylor, P.A., "The Tilt Intonation Model," *Proc. Int. Conf. on Spoken Language Processing*, 1998, Sydney, Australia.

[61] van Santen, J., "Assignment of Segmental Duration in Text-to-Speech Synthesis," *Computer Speech and Language*, 1994, **8**, pp. 95-128.

[62] van Santen, J., "Segmental Duration and Speech Timing" in *Computing Prosody*, Y. Sagisaka, N. Campbell, and N. Higuchi, eds., 1997, New York, Springer, pp. 225-250.

[63] van Santen, J. and J. Hirschberg, "Segmental Effects of Timing and Height of Pitch Contours," *Proc. of the Int. Conf. on Spoken Language Processing*, 1994, pp. 719-722.

[64] Vanderslice, R.L., *Synthetic Elocution: Considerations in Automatic Orthographic-to-Phonetic Conversion of English with Special Reference to Prosody*, PhD Thesis, 1968, UCLA, Los Angeles.

[65] W3C, *Speech Synthesis Markup Requirements for Voice Markup Languages*, 2000, http://www.w3.org/TR/voice-tts-reqs/.

[66] Yoshimura, T., *et al.*, "Simultaneous Modeling of Spectrum, Pitch and Duration in HMM-Based Speech Synthesis," *EuroSpeech*, 1999, Budapest, Hungary, pp. 2347-2350.

CHAPTER 16

Speech Synthesis

The speech synthesis module of a TTS system is the component that generates the waveform. The input of traditional speech synthesis components is a phonetic transcription with its associated prosody. The input can also include the original text with tags, as this may help in producing higher-quality speech.

Speech synthesis can be classified into three types according to the model used in the speech generation. *Articulatory synthesis*, described in Section 16.2.4, uses a physical model of speech production that includes all the articulators described in Chapter 2. *Formant synthesis* uses a source-filter model, where the filter is characterized by slowly varying formant frequencies; it is the subject of Section 16.2. *Concatenative synthesis* generates speech by concatenating speech segments and is described in Section 16.3. To allow more flexibility in concatenative synthesis, a number of prosody modification techniques are described in Sections 16.4 and 16.5. Finally, a guide to evaluating speech synthesis systems is included in Section 16.6.

Speech synthesis can also be classified according to the degree of manual intervention in the system design into *synthesis by rule* and *data-driven synthesis*. In the former, a set of manually derived rules is used to drive a synthesizer, and in the latter the synthesizer's parameters are obtained automatically from real speech data. Concatenative systems are, thus, data driven. Formant synthesizers have traditionally used synthesis by rule, since the evolution of formants in a formant synthesizer has been done with hand-derived rules. Nonetheless, formant transitions can also be trained from data, as we show in Section 16.2.3.

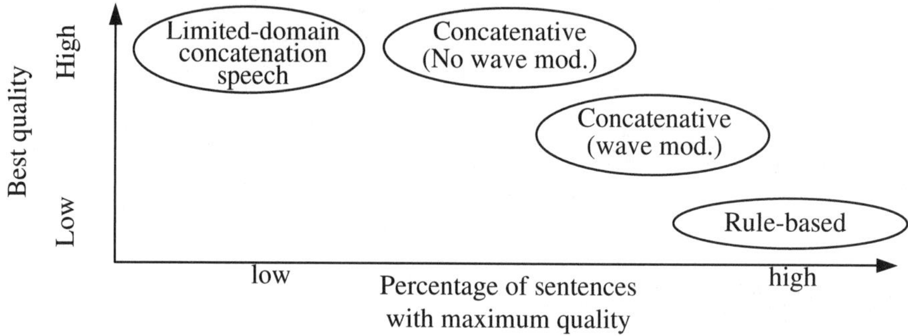

Figure 16.1 Quality and task-independence in speech synthesis approaches.

16.1. ATTRIBUTES OF SPEECH SYNTHESIS

The most important attribute of a speech synthesis system is the quality of its output speech. It is often the case that a single system can sound beautiful on one sentence and terrible on the next. For that reason we need to consider the quality of the best sentences and the percentage of sentences for which such quality is achieved. This tradeoff is illustrated in Figure 16.1, where we compare four different families of speech generation approaches:

- *Limited-domain waveform concatenation.* For a given limited domain, this approach can generate very high quality speech with only a small number of recorded segments. Such an approach, used in most interactive voice response systems, cannot synthesize arbitrary text. Many concept-to-speech systems, described in Chapter 17, use this approach.
- *Concatenative synthesis with no waveform modification.* Unlike the previous approach, these systems can synthesize speech from arbitrary text. They can achieve good quality on a large set of sentences, but the quality can be mediocre for many other sentences where poor concatenations take place.

- *Concatenative systems with waveform modification.* These systems have more flexibility in selecting the speech segments to concatenate because the waveforms can be modified to allow for a better prosody match. This means that the number of sentences with mediocre quality is lower than in the case where no prosody modification is allowed. On the other hand, replacing natural with synthetic prosody can hurt the overall quality. In addition, the prosody modification process also degrades the overall quality.
- *Rule-based systems.* Such systems tend to sound uniformly across different sentences, albeit with quality lower than the best quality obtained in the systems above.

Best-quality and quality variability are possibly two of the most important attributes of a speech synthesis system, but not the only ones. Measuring quality, difficult to do in an objective way, is the main subject of Section 16.6. Other attributes of a speech synthesis system include:

- *Delay.* The time it takes for the synthesizer to start speaking is important for interactive applications and should be less than 200 ms. This delay is composed of the algorithmic delays of the front end and of the speech synthesis module, as well as the computation involved.
- *Memory resources.* Rule-based synthesizers require, on the average, less than 200 KB, so they are a widely used option whenever memory is at a premium. However, required RAM can be an issue for concatenative systems, some of which may require over 100 MB of storage.
- *CPU resources.* With current CPUs, processing time is typically not an issue, unless many channels need to run in the same CPU. Nonetheless, some concatenative synthesizers may require a large amount of computation when searching for the optimal sequence.
- *Variable speed.* Some applications may require the speech synthesis module to generate variable speed, particularly fast speech. This is widely used by blind people who need TTS systems to obtain their information and can accept fast speech because of the increased throughput. Fast speech is also useful when skimming material. Concatenative systems that do not modify the waveform cannot achieve variable speed control, unless a large number of segments are recorded at different speeds.
- *Pitch control.* Some spoken language systems require the output speech to have a specific pitch. This is the case if you want to generate voice for a song. Again, concatenative systems that do not modify the waveform cannot do this, unless a large number of speech segments are recorded at different pitch.

- *Voice characteristics.* Other spoken language systems require specific voices, such as that of a robot, that cannot be recorded naturally, or some, such as monotones, that are tedious to record. Since rule-based systems are so flexible, they are able to do many such modifications.

The approaches described in this chapter assume as input a phonetic string, durations, a pitch contour, and possibly volume. Pauses are signaled by the default phoneme SIL with its corresponding duration. If the parsed text is available, it is possible to do even better in a concatenative system by conducting a matching with all the available information.

16.2. Formant Speech Synthesis

As discussed in Chapter 6, we can synthesize a stationary vowel by passing a glottal periodic waveform through a filter with the formant frequencies of the vocal tract. For the case of unvoiced speech we can use white random noise as the source instead. In practice, speech signals are not stationary, and we thus need to change the pitch of the glottal source and the formant frequencies over time. The so-called *synthesis-by-rule* refers to a set of rules on how to modify the pitch, formant frequencies, and other parameters from one sound to another while maintaining the continuity present in physical systems like the human production system. Such a system is described in the block diagram of Figure 16.2.

In Section 16.2.1 we describe the second block of Figure 16.2, the formant synthesizer that generates a waveform from a set of parameters. In Section 16.2.2 we describe the first block of Figure 16.2, the set of rules that can generate such parameters. This approach was the one followed by Dennis Klatt and his colleagues [4, 30]. A data-driven approach to this first block is studied in Section 16.2.3. Finally, articulatory synthesis is the topic of Section 16.2.4.

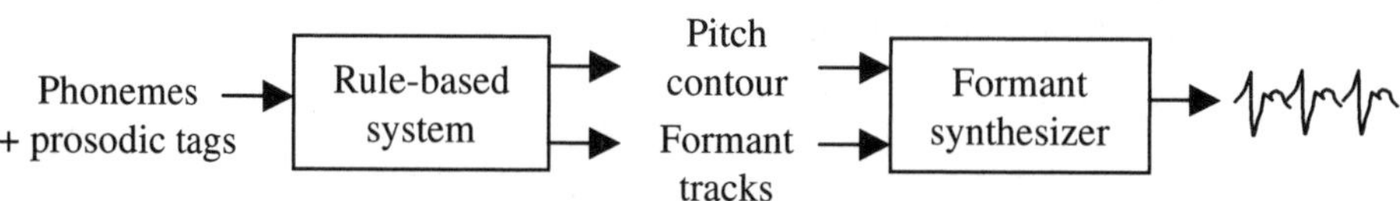

Figure 16.2 Block diagram of a synthesis-by-rule system. Pitch and formants are listed as the only parameters of the synthesizer for convenience. In practice, such system has about 40 parameters.

16.2.1. Waveform Generation from Formant Values

To be able to synthesize speech by rule, a simple model for the filter is needed. Most rule-based synthesizers use the so-called *formant synthesis*, which is derived from models of speech production. The model explicitly represents a number of formant resonances (from 2 to 6). A formant resonance can be implemented (see Chapter 6) with a second-order IIR filter

$$H_i(z) = \frac{1}{1 - 2e^{-\pi b_i} \cos(2\pi f_i) z^{-1} + e^{-2\pi b_i} z^{-2}} \tag{16.1}$$

with $f_i = F_i / F_s$ and $b_i = B_i / F_s$, where F_i, B_i, and F_s are the formant's center frequency, formant's bandwidth, and sampling frequency, respectively, all in Hz. A filter with several resonances can be constructed by cascading several such second-order sections (*cascade model*) or by adding several such sections together (*parallel model*). Formant synthesizers typically use the parallel model to synthesize fricatives and stops and the cascade model for all voiced sounds.

Unlike the cascade model, the parallel model requires gains to be specified for each second-order section, which often are chosen proportional to the formant's frequency and inversely proportional to the formant's bandwidth. The cascade model results in an all-pole filter, whereas the parallel model has zeros in addition to poles. Such a combination is shown in Figure 16.3, where R1 through R6 are the resonances 1 to 6 and each one represents a second-order IIR filter like that in Eq. (16.1). RNP represents the nasal resonance, and RNZ is an FIR filter with the nasal zero. A1 through AB are the gains for each filter, used for the parallel model. Switch SW controls whether the cascade model or parallel model is used.

For voiced sounds the excitation model consists of an impulse train driving a low-pass filter RGP and then a bandpass filter created by the parallel of RGZ and RGS. For unvoiced sounds the excitation consists of white noise driving a low-pass filter LPF. The excitation for voiced fricatives is a combination of the two sources. In practice, this mixed excitation is used for all voiced sounds to add some breathiness. Klatt [30] showed that this model could reproduce quite faithfully a natural recording if the parameters had been manually selected.

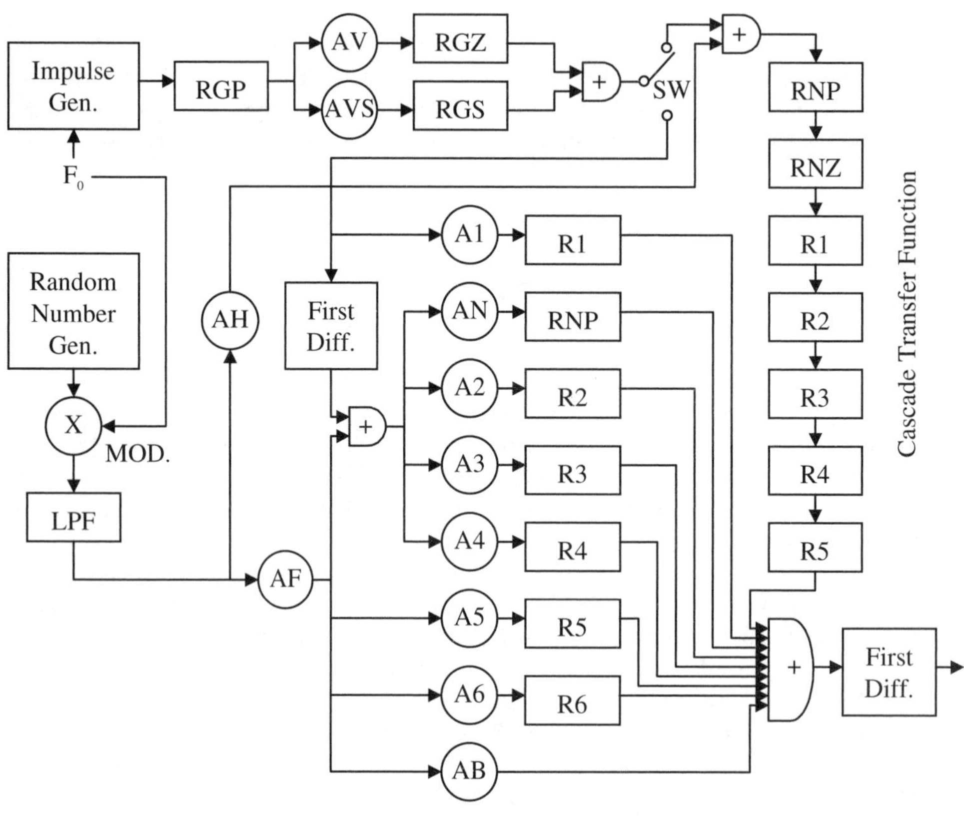

Figure 16.3 Block diagram of the Klatt formant synthesizer (after Allen [4]).

The parameter names and their minimum and maximum values are listed in Table 16.1, where the switch SW can be in voiced (V) or consonant (C) mode. For example, in Figure 16.3, R2 is the resonator corresponding to the second formant, whose center frequency F2 and bandwidth B2 are given in Table 16.1. In addition to the six resonances associated to the six formants, there are other resonances: RGP, RGZ, RGS, RNP, and RNZ. Other source models are also possible [43].

Table 16.1 Parameter values for Klatt's cascade/parallel formant synthesizer with the parameter symbol, full name, minimum, maximum, and typical values (after Allen [4]).

N	Symbol	Name	Min	Max	Typ
1	AV	Amplitude of voicing (dB)	0	80	0
2	AF	Amplitude of frication (dB)	0	80	0
3	AH	Amplitude of aspiration (dB)	0	80	0
4	AVS	Amplitude of sinusoidal voicing (dB)	0	80	0
5	F0	Fundamental frequency (Hz)	0	500	0
6	F1	First formant frequency (Hz)	150	900	500
7	F2	Second formant frequency (Hz)	500	2500	1500
8	F3	Third formant frequency (Hz)	1300	3500	2500
9	F4	Fourth formant frequency (Hz)	2500	4500	3500
10	FNZ	Nasal zero frequency (Hz)	200	700	250
11	AN	Nasal formant amplitude (Hz)	0	80	0
12	A1	First formant amplitude (Hz)	0	80	0
13	A2	Second formant amplitude (Hz)	0	0	0
14	A3	Third formant amplitude (Hz)	0	80	0
15	A4	Fourth formant amplitude (Hz)	0	80	0
16	A5	Fifth formant amplitude (Hz)	0	80	0
17	A6	Sixth formant amplitude (Hz)	0	80	0
18	AB	Bypass path amplitude (Hz)	0	80	0
19	B1	First formant bandwidth (Hz)	40	500	50
20	B2	Second formant bandwidth (Hz)	40	500	70
21	B3	Third formant bandwidth (Hz)	40	500	110
22	SW	Cascade/parallel switch	0	1	0
23	FGP	Glottal resonator 1 frequency (Hz)	0	600	0
24	BGP	Glottal resonator 1 bandwidth (Hz)	100	2000	100
25	FGZ	Glottal zero frequency (Hz)	0	500	1500
26	BGZ	Glottal zero bandwidth (Hz)	100	9000	6000
27	B4	Fourth formant bandwidth (Hz)	100	500	250
28	F5	Fifth formant frequency (Hz)	3500	4900	3850
29	B5	Fifth formant bandwidth (Hz)	150	700	200
30	F6	Sixth formant frequency (Hz)	4000	4999	4900
31	B6	Sixth formant bandwidth (Hz)	200	2000	100
32	FNP	Nasal pole frequency (Hz)	200	500	250
33	BNP	Nasal pole bandwidth (Hz)	50	500	100
34	BNZ	Nasal zero bandwidth (Hz)	50	500	100
35	BGS	Glottal resonator 2 bandwidth (Hz)	100	1000	200
36	SR	Sampling rate (Hz)	500	20000	10000
37	NWS	Number of waveform samples per chunk	1	200	50
38	G0	Overall gain control (dB)	0	80	48
39	NFC	Number of cascaded formants	4	6	5

16.2.2. Formant Generation by Rule

As described in Chapter 2, formants are one of the main features of vowels. Because of the physical limitations of the vocal tract, formants do not change abruptly with time. Rule-based formant synthesizers enforce this by generating continuous values for $f_i[n]$ and $b_i[n]$, typically every 5–10 milliseconds. Continuous values can be implemented through the above structures if a lattice filter is used, because it allows its reflection coefficients to vary at every sample (see Chapter 6). In practice, the values can be fixed within a frame as long as frames are smaller than 5 ms.

Rules on how to generate formant trajectories from a phonetic string are based on the *locus theory* of speech production. The locus theory specifies that formant frequencies within a phoneme tend to reach a stationary value called the *target*. Targets for formant frequencies and bandwidths for a male speaker are shown in Table 16.2 (nonvocalic segments) and Table 16.3 (vocalic segments). This target is reached if either the phoneme is sufficiently long or the previous phoneme's target is close to the current phoneme's target. The maximum slope at which the formants move is dominated by the speed of the articulators, determined by physical constraints. Since each formant is primarily caused by the position of a given articulator, formants caused by the body of the tongue do not vary as rapidly as formants caused by the tip of the tongue or the lips. Thus, rule-based systems store targets for each phoneme as well as maximum allowable slopes and transition times.

For example, a transition between a vowel and a sonorant can follow the rule shown in Figure 16.4 with a_1 being the target of unit 1 and a_2 the target of unit 2. The values of T_{cb} and T_{cf} are 40 and 80 ms, respectively, and $a_b = a_2 + 0.75(a_1 - a_2)$. The time $T_{cb} + T_{cf}$ specifies how rapid the transition is. While linear interpolation could be used, a_b and the ratio T_{cb} / T_{cf} are sometimes used to further refine the shape of the formant transition.

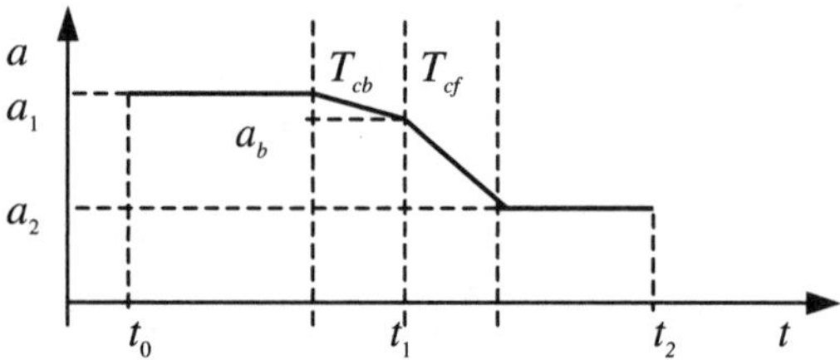

Figure 16.4 Transition between two vowels in a formant synthesizer.

Other rules can allow a discontinuity, for example, when a transition out of an unvoiced segment takes place. To improve naturalness, all these parameters can be made dependent on the immediate phonetic context.

Table 16.2 Targets used in the Klatt synthesizer: formant frequencies and bandwidths for non-vocalic segments of a male speaker. Note that in addition to the phoneme set used in Chapter 2, there are several additional phonetic segments here such as *axp, dx, el, em, en, gp, hx, kp, lx, qq, rx, tq, wh*, that allow more control (after Allen [4]).

	F1	F2	F3	B1	B2	B3
axp	430	1500	2500	120	60	120
b	200	900	2100	65	90	125
ch	300	1700	2400	200	110	270
d	200	1400	2700	70	115	180
dh	300	1150	2700	60	95	185
dx	200	1600	2700	120	140	250
el	450	800	2850	65	60	80
em	200	900	2100	120	60	70
en	200	1600	2700	120	70	110
ff	400	1130	2100	225	120	175
g	250	1600	1900	70	145	190
gp	200	1950	2800	120	140	250
h	450	1450	2450	300	160	300
hx	450	1450	2450	200	120	200
j	200	1700	2400	50	110	270
k	350	1600	1900	280	220	250
kp	300	1950	2800	150	140	250
l	330	1050	2800	50	100	280
lx	450	800	2850	65	60	80
m	480	1050	2100	40	175	120
ng	480	1600	2050	160	150	100
n	480	1400	2700	40	300	260
p	300	900	2100	300	190	185
qq	400	1400	2450	120	140	250
r	330	1060	1380	70	100	120
rx	460	1260	1560	60	60	70
sh	400	1650	2400	200	110	280
sil	400	1400	2400	120	140	250
s	400	1400	2700	200	95	220
th	400	1150	2700	225	95	200
tq	200	1400	2700	120	140	250
t	300	1400	2700	300	180	220
v	300	1130	2100	55	95	125
wh	330	600	2100	150	60	60
w	285	610	2150	50	80	60
y	240	2070	3020	40	250	500
zh	300	1650	2400	220	140	250
z	300	1400	2700	70	85	190

Table 16.3 Targets used in the Klatt synthesizer: formant frequencies and bandwidths for vocalic segments of a male speaker. Note that in addition to the phoneme set used in Chapter 2, there are several additional phonetic segments here such as *axr, exr, ix, ixr, oxr, uxr, yu* that allow more control (after Allen [4]).

	F1	F2	F3	B1	B2	B3
aa	700	1220	2600	130	70	160
ae	620	1660	2430	70	130	300
ah	620	1220	2550	80	50	140
ao	600	990	2570	90	100	80
aw	640	1230	2550	80	70	110
ax	550	1260	2470	80	50	140
axr	680	1170	2380	60	60	110
ay	660	1200	2550	100	120	200
eh	530	1680	2500	60	90	200
er	470	1270	1540	100	60	110
exr	460	1650	2400	60	80	140
ey	480	1720	2520	70	100	200
ih	400	1800	2670	50	100	140
ix	420	1680	2520	50	100	140
ixr	320	1900	2900	70	80	120
iy	310	2200	2960	50	200	400
ow	540	1100	2300	80	70	70
oxr	550	820	2200	60	60	60
oy	550	960	2400	80	120	160
uh	450	1100	2350	80	100	80
uw	350	1250	2200	65	110	140
uxr	360	800	2000	60	60	80
yu	290	1900	2600	70	160	220

Klatt showed that for a given natural utterance, he could manually obtain a sequence of formant tracks $f_i[n]$ and $b_i[n]$, such that the synthesized utterance not only had good quality but also sounded very similar to the original. This shows that the formant synthesizer of Section 16.2.1 appears to be sufficient for generation. On the other hand, when the formant tracks are obtained automatically through rules such as that of Figure 16.4 and Table 16.2 and Table 16.3, the output speech does not sound that natural, and the voice does not resemble the voice of the original recording.

Formant synthesis is very flexible because it can generate intelligible speech with relatively few parameters (about 40). The use of context-dependent rules can improve the quality of the synthesizer at the expense of a great deal of manual tuning. The synthesized speech is, by design, smooth, although it may not resemble any given speaker and may not sound very natural.

Because of their flexibility, formant synthesizers can often generate many different voices and effects. While not as flexible, voice effects can also be produced in concatenative speech systems (see Section 16.5.3).

16.2.3. Data-Driven Formant Generation

While, in general, formant synthesis assumes the formant model of Section 16.2.1 driven by parameter values generated by rules, as in Section 16.2.2, data-driven methods to generate the formant values have also been proposed [3]. An HMM running in generation mode emits three formant frequencies and their bandwidths every 10 ms, and these values are used in a cascade formant synthesizer similar to that described in Section 16.2.1. Like the speech recognition counterparts, this HMM has many decision-tree context-dependent triphones and three states per triphone. A Gaussian distribution per state is used in this work. The baseline system uses a six-dimensional vector that includes the first three formant frequencies and their bandwidths. Initially it is assumed that the input to the synthesizer includes, in addition to the duration of each phoneme, the duration of each state. In this case, the maximum likelihood formant track is a sequence of the state means and, therefore, is discontinuous at state boundaries.

The key to obtaining a smooth formant track is to augment the feature vector with the corresponding delta formants and bandwidths (the difference between the feature at time t and that feature at time $t-1$) to complete a twelve-dimensional vector. The maximum likelihood solution now entails solving a tridiagonal set of linear equations (see the discussion on statistical formant tracking in Chapter 6). The resulting formant track is smooth, as it balances formant values that are close to the state means with delta values that are also within the state means. In addition, the synthesized speech resembles that of the donor speaker. More details on the analysis and model training can be found in the formant tracking section of Chapter 6.

16.2.4. Articulatory Synthesis

Articulatory synthesis is another model that has been used to synthesize speech by rule, by using parameters that model the mechanical motions of the articulators and the resulting distributions of volume velocity and sound pressure in the lungs, larynx, and vocal and nasal tracts. Because the human speech production articulators do not have that many degrees of freedom, articulatory models often use as few as 15 parameters to drive a formant synthesizer.

The relationship between articulators and acoustics is many-to-one [17]. For example, a ventriloquist can produce speech sounds with very different articulator positions than those of normal speech. The *speech inversion* problem is therefore ill posed. By using the assumption that the articulators do not change rapidly over time, it is possible, however, to estimate the vocal-tract area from formant frequencies [37]. In [8] the model uses five articulatory parameters: area of lip opening, constriction formed by the tongue blade, opening to the nasal cavities, average glottal area, and rate of active expansion or contraction of the vocal tract volume behind a constriction. These five parameters are augmented with the first four formant frequencies and F0.

Those area parameters can be obtained from real speech through X-rays and magnetic resonance imaging (MRI), though positioning such sensors in the vocal tract alters the way speech is produced (such as the sensors in the lips) and impedes completely natural sounds. The relationship between articulatory parameters and acoustic values has typically been done using a nonlinear mapping such as a neural network or a codebook.

While one day this may be the best way to synthesize speech, the state-of-the-art in articulatory synthesis does not generate speech with quality comparable to that of formant or concatenative systems.

16.3. CONCATENATIVE SPEECH SYNTHESIS

While state-of-the-art synthesis by rule is quite intelligible, it sounds unnatural, because it is very difficult to capture all the nuances of natural speech in a small set of manually derived rules. In *concatenative synthesis*, a speech segment is synthesized by simply playing back a waveform with matching phoneme string. An utterance is synthesized by concatenating together several speech fragments. The beauty of this approach is that unlike synthesis-by-rule, it requires neither rules nor manual tuning. Moreover, each segment is completely natural, so we should expect very natural output.

Unfortunately, this is equivalent to assembling an automobile with parts of different colors: each part is very good yet there is a color discontinuity from part to part that makes the whole automobile unacceptable. Speech segments are greatly affected by coarticulation [42], so if we concatenate two speech segments that were not adjacent to each other, there can be spectral or prosodic discontinuities. Spectral discontinuities occur when the formants at the concatenation point do not match. Prosodic discontinuities occur when the pitch at the concatenation point does not match. A listener rates as poor any synthetic speech that contains large discontinuities, even if each segment is very natural.

Thus, when designing a concatenative speech synthesis system we need to address the following issues:

1. What type of speech segment to use? We can use diphones, syllables, phonemes, words, phrases, etc.
2. How to design the acoustic inventory, or set of speech segments, from a set of recordings? This includes excising the speech segments from the set of recordings as well as deciding how many are necessary. This is similar to the training problem in speech recognition.
3. How to select the best string of speech segments from a given library of segments, and given a phonetic string and its prosody? There may be several

strings of speech segments that produce the same phonetic string and prosody. This is similar to the search problem in speech recognition.

4. How to alter the prosody of a speech segment to best match the desired output prosody. This is the topic of Section 16.4.

Generally, these concatenative systems suffer from great variability in quality: often they can offer excellent quality in one sentence and terrible quality in the next one. If enough good units are available, a given test utterance can sound almost as good as a recorded utterance. However, if several discontinuities occur, the synthesized utterance can have very poor quality. While synthesizing arbitrary text is still a challenge with these techniques, for restrictive domains this approach can yield excellent quality. We examine all these issues in the following sections.

We define *unit* as an abstract representation of a speech segment, such as its phonetic label, whereas we use *instance* as a speech segment from an utterance that belongs to the same unit. Thus, a system can keep several instances of a given unit to select among them to better reduce the discontinuities at the boundaries. This abstract representation consists of the unit's phonetic transcription at the minimum, in such a way that the concatenation of a set of units matches the target phonetic string. In addition to the phonetic string, this representation can often include prosodic information.

16.3.1. Choice of Unit

A number of units have been used in the field, including context-independent phonemes, diphones, context-dependent phonemes, subphonetic units, syllables, words, and phrases. A compilation of unit types for English is shown in Table 16.4 with their coverage in Figure 16.5.

Table 16.4 Unit types in English assuming a phone set of 42 phonemes. Longer units produce higher quality at the expense of more storage. The number of units is generally below the absolute maximum in theory, i.e., out of the $42^3 = 74{,}088$ possible triphones, only about 30,000 occur in practice.

Unit length	Unit type	#Units	Quality
Short	Phoneme	42	Low
↓	Diphone	~1500	↓
	Triphone	~30K	
	Demisyllable	~2000	
	Syllable	~15K	
	Word	100K–1.5M	
	Phrase	∞	
Long	Sentence	∞	High

The issues in choosing appropriate units for synthesis are similar to those in choosing units for speech recognition (described in Chapter 9):

- The unit should lead to *low concatenation distortion*. A simple way of minimizing this distortion is to have fewer concatenations and thus use long units such as words, phrases or even sentences. But since some concatenations are unavoidable, we also want to use units that naturally lead to "small" discontinuities at the concatenation points. For example, it has been observed that spectral discontinuities at vowels are much more noticeable than at fricatives, or that a discontinuity within a syllable is more perceptually noticeable than a discontinuity across syllable boundaries, and similarly for within-word and across-word discontinuities [55]. Having several instances per unit is an alternative to long units that allows the choice of instances with low concatenation distortion.
- The unit should lead to *low prosodic distortion*. While it is not crucial to have units with the same prosody as the desired target, replacing a unit with a rising pitch with another with a falling pitch may result in an unnatural sentence. Altering the pitch and/or duration of a unit is possible (see Section 16.4) at the expense of additional distortion.
- The unit should be *generalizable*, if unrestricted text-to-speech is required. If we choose words or phrases as our units, we cannot synthesize arbitrary speech from text, because it's almost guaranteed that the text will contain words not in our inventory. As an example, the use of arbitrarily long units in such a way that no concatenation between voiced sounds occurs by cutting at obstruents results in low concatenation distortion but it is shown [47] that over 180,000 such units would be needed to cover 75% of a random corpus. The longer the speech segments are, the more of them we need to be able to synthesize speech from arbitrary text. This generalization property is not needed if closed-domain synthesis is desired.
- The unit should be *trainable*. Our training data should be sufficient to estimate all our units. Since the training data is usually limited, having fewer units leads to better trainability in general. So the use of words, phrases, or sentences may be prohibitive other than for closed-domain synthesis. The other units in Table 16.4 can be considered trainable depending on the limitations on the size of our acoustic inventory.

A practical challenge is how to balance these selection criteria. In this section we compare a number of units and point out their strengths and weaknesses.

16.3.1.1. Context-Independent Phonemes

The most straightforward unit is the phoneme. Having one instance of each phoneme, independent of the neighboring phonetic context, is very generalizable, since it allows us to generate every word/sentence. It is also very trainable and we could have a system that is very compact. For a language with N phonemes, say $N = 42$ for English, we would need only N unit instances. The problem is that using *context-independent* phones results in many audible discontinuities. Such a system is not intelligible.

16.3.1.2. Diphones

A type of subword unit that has been extensively used is the so-called *dyad* or *diphone* [41]. A diphone `s-ih` includes from the middle of the `s` phoneme to the middle of the `ih` phoneme, so diphones are, on the average, one phoneme long. The word `hello /hh ax l ow/` can be mapped into the diphone sequence: `/sil-hh/`, `/hh-ax/`, `/ax-l/`, `/l-ow/`, `/ow-sil/`. If our language has N phonemes, there are potentially N^2 diphones. In practice, many such diphones never occur in the language, so that a smaller number is sufficient. For example, the phonetic alphabet of Chapter 2 has 42 phonemes for English, and only about 1300 diphones are needed. Diphone units were among the first type of unit used in concatenative systems because they yield fairly good quality. While diphones retain the transitional information, there can be large distortions due to the difference in spectra between the stationary parts of two units obtained from different contexts. For example, there is no guarantee that the spectra of `/ax-l/` and `/l-ow/` will match at the junction point, since the instance `/ax-l/` could have been excised from a very different right context than `/ow/` or the instance `/l-ow/` could have been excised from a very different left context than `/ax/`.

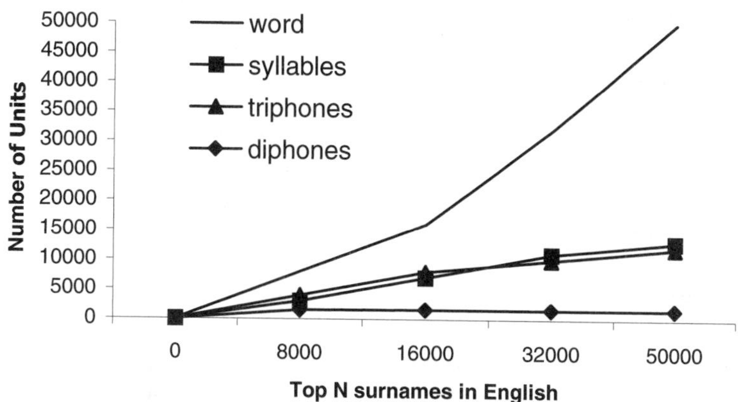

Figure 16.5 Coverage with different number of units displays the number of units of different types required to generate the top N surnames in the United States [34].

For this reason, many practical diphone systems are not purely diphone based: they do not store transitions between fricatives, or between fricatives and stops, while they store longer units that have a high level of coarticulation [48]. If only one representative of a dyad is available, there are pitch discontinuities. Prosody modification techniques, such as those described in Section 16.4, can be applied to correct this problem. Otherwise many instances of each diphone are needed for good prosodic coverage. Diphones are trainable, generalizable and offer better quality than context-independent phones.

16.3.1.3. Context-Dependent Phoneme

Another subword unit used in the literature [24] is the *context-dependent* phoneme. If the context is limited to the immediate left and right phonemes, the unit is known as *triphone*. As in speech recognition, not all N^3 need to be stored, because not all combinations will occur in practice. For English, typically there can be in excess of 25,000 triphones: 12,000 within-word triphones and another 12,000 across-word triphones. Because of the increased number of units, more contextual variations can be accommodated this way. Drawing from experience in speech recognition, we know that many different contexts have a similar effect on the phoneme; thus, several triphones can be clustered together into a smaller number of *generalized triphones*, typically between 500 and 3000. All the clustering procedures described in Chapter 9 can be used here as well. In particular, decision-tree clustered phones have been successfully used. Because a larger number of units can be used, discontinuities can be smaller than in the case of diphones while making the best use of the available data. In addition to only considering the immediate left and right phonetic context, we could also add stress for the current phoneme and its left and right context, word-dependent phones (where phones in particular words are considered distinct context-dependent phones), quinphones (where two immediate left and right phones are used), and different prosodic patterns (pitch ranges and/or durations). As in speech recognition, clustered context-dependent triphones are trainable and generalizable.

Traversing the tree for a given phoneme is equivalent to following the answers for the branching nodes from root to leaves, which determines the clusters for similar context-dependent phones. The decision trees are generated automatically from the analysis database to obtain minimum within-unit distortion (or entropy) for each split. Therefore, one must be able to acquire a large inventory of context-dependent phone HMMs with a decent coverage of the contexts one wishes to model. All the context-dependent phone units can be well replaced by any other units within the same cluster. This method generalizes to contexts not seen in the training data, because the decision tree uses questions involving broad phonetic categories of neighboring contexts, yet provides detailed models for contexts that are represented in the database. Given the assumption that these clustering decision trees should be consistent across different speakers, the use of ample speaker-independent databases instead of limited speaker-dependent databases allows us to model more contexts as well as deeper trees to generate a high-quality TTS voice. These techniques also facilitate the creation of acoustic inventories with a scalable number of units that trade off size with quality. Thus, we can use questions (about the immediate left/right phonetic contexts, stress, pitch, duration, word, etc.) in the decision-tree clustering methods of Chapter 4 to reduce all the possible combinations to a manageable number.

16.3.1.4. Subphonetic Unit

Subphonetic units, or senones, have also been used with some success [13]. Typically, each phoneme can be divided into three states, which are determined by running a speech recognition system in forced alignment mode. These states can also be context dependent and can also be clustered using decision trees like the context-dependent phonemes. The HMM state has proved to be more effective than the context-dependent phone in speech recognition, also trainable and generalizable, but for synthesis it means having more concatenations and thus possibly more discontinuities. If multiple instances per subphonetic unit are used, higher quality can be obtained.

A *half phone* goes either from the middle of a phone to the boundary between phones or from the boundary between phones to the middle of the phone. This unit offers more flexibility than a phone and a diphone and has been shown useful in systems that use multiple instances of the unit [7].

16.3.1.5. Syllable

It has been observed that discontinuities across syllables stand out more than discontinuities within syllables [55], so syllables are natural units. There are more than 10,000 syllables in English, depending on the exact definition of syllable, so even a context-independent syllable system needs to store at least as many if one instance per syllable is needed for full generalizability. There will still be spectral discontinuities, though hopefully not too noticeable. More than one instance per unit may be needed to account for varying acoustic contexts or varying prosodic patterns, particularly if no waveform modification is to be used.

16.3.1.6. Word and Phrase

The unit can be as large as a word or even a phrase. While using these long units can increase naturalness significantly, generalizability and trainability are poor, so that it is difficult to have all the instances desired to synthesize an output utterance. One advantage of using a word or longer unit over its decomposition in phonemes, as in the above units, is that there is no dependence on a phonetically transcribed dictionary. It is possible that the phoneme string associated to a word by our dictionary is not correct, or not fluent enough, so that using a whole-word model can solve this problem. Of course, the system may have a combination of all units: a set of the most frequent words, sentences, or phrases for best quality some percentage of the time, and some smaller units for full generalizability and trainability.

16.3.2. Optimal Unit String: The Decoding Process

The goal of the decoding process is to choose the optimal string of units for a given phonetic string that best matches the desired prosody. Sometimes there is only one possible string, so that this process is trivial, but in general there are several strings of units that result in the same phonetic string yet some of them sound better than others. The goal is to come up with an objective function that approximates this sound quality that allows us to select the best string. The quality of a unit string is typically dominated by spectral and pitch discontinuities at unit boundaries. Discontinuities can occur because of:

1. *Differences in phonetic contexts.* A speech unit was obtained from a different phonetic context than that of the target unit.
2. *Incorrect segmentation.* Such segmentation errors can cause spectral discontinuities even if they had the same phonetic context.
3. *Acoustic variability.* Units can have the same phonetic context and be properly segmented, but variability from one repetition to the next can cause small discontinuities. A unit spoken in fast speech is generally different from another in slow or normal speech. Different recording conditions (amplitude, microphone, sound card) can also cause spectral discontinuities.
4. *Different prosody.* Pitch discontinuity across unit boundaries is also a cause for degradation.

The severity of such discontinuities generally decreases as the number of units increases. More importantly, the prosody of the concatenation has, in general, no resemblance with the prosody specified by the TTS front-end unless we have several instances of each unit, each with a different prosody, or use a prosody modification algorithm (see Section 16.4).

16.3.2.1. Objective Function

Our goal is to come up with a numeric measurement for a concatenation of speech segments that correlates well with how well they sound. To do that we define unit cost and transition cost between two units.

Let θ be a speech segment with phonetic transcription $p = p(\theta)$. Let $\Theta = \{\theta_1, \theta_2, \cdots, \theta_N\}$ be a concatenation of N speech segments whose combined phonetic transcription is $P = \{p_1, p_2, \cdots, p_N\}$. P is a string of M phonemes, and since each segment has at least one phoneme, it holds that $M \ge N$.

For example, the phonetic string P = "*hh ax l ow*" corresponding to the word *hello* has M = 4 phonemes and can be decomposed in N = 4 segments $\Theta_1 = \{\theta_1, \theta_2, \theta_3, \theta_4\}$, where $p(\theta_1) = /hh/$, $p(\theta_2) = /ax/$, $p(\theta_3) = /l/$, $p(\theta_4) = /ow/$, each segment being a phoneme. Or it can be decomposed into N = 2 segments $\Theta_2 = \{\theta_5, \theta_6\}$, where $p(\theta_5) = /hh\ ax/$,

$p(\theta_6) = / l\ ow /$, so that each segment has two phonemes. There are 8 possible such decompositions for this example (in general there are 2^{M-1} possible decompositions[1]).

The distortion or cost function between the segment concatenation Θ and the target T can be expressed as a sum of the corresponding unit costs and transition costs [27, 46] as follows:

$$d(\Theta, T) = \sum_{j=1}^{N} d_u(\theta_j, T) + \sum_{j=1}^{N-1} d_t(\theta_j, \theta_{j+1}) \quad (16.2)$$

where $d_u(\theta_j, T)$ is the *unit cost* of using speech segment θ_j within target T and $d_t(\theta_j, \theta_{j+1})$ is the transition cost in concatenating speech segments θ_j and θ_{j+1}. The optimal speech segment sequence of units $\hat{\Theta}$ can be found as the one that minimizes the overall cost

$$\hat{\Theta} = \arg\min_{\Theta} d(\Theta, T) \quad (16.3)$$

over sequences with all possible numbers of units. Transition and unit costs are described in Sections 16.3.2.2 through 16.3.2.5.

Let's analyze the second term in the sum of Eq. (16.2), also shown in Figure 16.6. If all transition costs were identical, the word string with fewest units would have lowest distortion. In practice transition costs are different and, thus, the string with fewest units is not necessarily the best, though there is clearly a positive correlation.

When a large number of speech segments are available, finding the segment sequence with lowest cost is a search problem like those analyzed in Chapter 12. Often a Viterbi algorithm is needed to make this efficient.

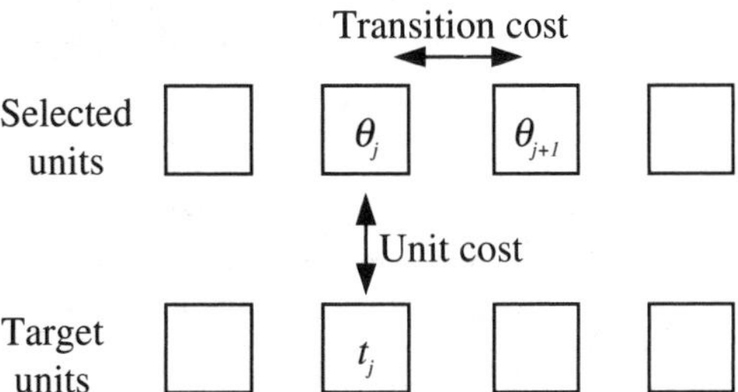

Figure 16.6 Tradeoff between unit and transition costs.

[1] This assumes that one instance per unit is available. If there are several instances per unit, the number of decompositions grows exponentially.

The art in this procedure is in the exact definition of both transition and unit costs, for which no standard has been defined that works best to date. In Sections 16.3.2.2 and 16.3.2.3 we present an approach for which both transition and unit costs are empirically set after perceptual experiments. Such a system is easy to build, and study of those costs gives insight into the perceptual effects.

Costs obtained using a data-driven criterion are described in Sections 16.3.2.4 and 16.3.2.5. While more complicated than that of empirical costs, this method addresses the shortcomings of the previous method. Finally, we need to estimate some weights to combine the different costs for spectrum and prosody, which can be done empirically or by regression [26].

16.3.2.2. Empirical Transition Cost

If spoken in succession, two speech segments have a zero transition cost. But, when they are excised from separate utterances, their concatenation can have varying degrees of naturalness. The transition cost incorporates two types of continuity measures: coarticulatory and prosodic.

An approximation to the prosodic continuity measure is to make it proportional to the absolute difference of the F0 or log F0 at the boundaries, if the boundary is voiced for both units. If we use the prosody modification techniques of Section 16.4, this cost could be set to a small value to reflect the fact that prosody modification is not a perfect process. More sophisticated cost functions can be used to account for prosody mismatches [10].

Regarding the coarticulatory effect, it has been empirically observed that a concatenation within a syllable is more perceptible than when the concatenation is at the syllable boundary. Yi [55] proposed an empirical cost matrix for the concatenation of two speech segments when that concatenation occurs within a syllable (Table 16.5) or at a syllable boundary (Table 16.6). Phonemes are grouped by manner of articulation: vowel/semivowels, fricatives, stops, and nasals. The rows represent the left side of the transition and the columns represent the right side, and NA represents a case that does not occur. These costs reflect perceptual ratings by human listeners to unit concatenations between different phonemes. Values of 10, 2000, 5000, 7500, and 10,000 were used to indicate different degrees of goodness from very good to very bad concatenations.

Table 16.5 Cost matrix for intrasyllable concatenations (after Yi [55]). The rows represent the left side of the transition and the columns represent the right side, and NA represents a case that does not occur.

	vowel	semivowel	nasal	obstruent	/h/
vowel	10,000	10,000	7500	10	NA
semivowel	10,000	7500	7500	10	NA
nasal	5000	10	NA	10	NA
/h/	5000	NA	NA	NA	NA
obstruent	10	10	10	10,000	NA

Table 16.6 Cost matrix for intersyllable concatenations (after Yi [55]). The rows represent the left side of the transition and the columns represent the right side, and NA represents a case that does not occur.

	vowel	semivowel	nasal	obstruent	/h/	silence
vowel	NA	7500	5000	10	5000	10
semivowel	7500	7500	2000	10	10	10
nasal	2000	10	10	10	10	10
obstruent	10	10	10	5000	10	10
/h/	NA	NA	NA	NA	NA	NA
silence	10	10	10	10	10	10

16.3.2.3. Empirical Unit Cost

The unit cost is generally a combination of the coarticulation cost and the prosodic cost. Prosodic mismatches can be made proportional to the F0 difference between the candidate unit and the target unit or set to a fixed low value if the prosody modification techniques of Section 16.4 are used.

A way of determining the cost associated with replacing a phonetic context with another was proposed by Yi [55], who empirically set cost matrices for phone classes by listening to concatenations where such contexts were replaced. These ad hoc values also bring some sense of where the coarticulation problems are. Replacing a vowel or semivowel by another with a context that has a different place of articulation or nasalization results in audible discontinuities. The rows represent the context class of the target phoneme and the columns represent the context class of the proposed unit. Table 16.7, Table 16.8, Table 16.9, Table 16.10, Table 16.11, and Table 16.12 include an empirical set of costs for such mismatches between the target's context and a candidate unit's context for the case of vowel/semivowels, fricatives, stops, and nasals. These costs reflect human listeners' perceptual ratings of speech units with an incorrect phonetic context. Values of 10, 100, 500, and 1000 were used to indicate very good, good, bad, and very bad units. These values are chosen to match the values for transition costs of Section 16.3.2.2.

Table 16.7 Unit coarticulation cost matrix (after Yi [55]) for left and right context replacements for vowels and semivowels.

	labial	alv/den/pal	velar	m	n	ng	front	back	none
labial	10	1000	1000	1000	1000	1000	1000	1000	1000
alv/den/pal	1000	10	1000	1000	1000	1000	1000	1000	1000
velar	1000	1000	10	1000	1000	1000	1000	1000	1000
m	1000	1000	1000	10	1000	1000	1000	1000	1000
n	1000	1000	1000	1000	10	1000	1000	1000	1000
ng	1000	1000	1000	1000	1000	10	1000	1000	1000
front	1000	1000	1000	1000	1000	1000	10	1000	1000
back	1000	1000	1000	1000	1000	1000	1000	10	1000
none	1000	1000	1000	1000	1000	1000	1000	1000	10

Table 16.8 Unit coarticulation cost matrix (after Yi [55]) for left and right context replacements for fricatives.

	retroflex	round	sonorant	other
retroflex	10	100	100	100
round	100	10	100	100
sonorant	100	100	10	100
other	100	100	100	10

Table 16.9 Unit coarticulation cost matrix (after Yi [55]) for left context replacements for stops.

	front	back	retroflex	round	other
front	10	10	10	10	10
back	10	10	10	10	10
retroflex	10	10	10	10	10
round	10	10	10	10	10
other	500	500	500	500	10

Table 16.10 Unit coarticulation cost matrix (after Yi [55]) for right context replacements for stops.

	front	back	retroflex	round	schwa	other
front	10	100	100	100	500	100
back	100	10	100	100	500	100
retroflex	100	100	10	100	500	100
round	100	100	100	10	500	100
schwa	500	500	500	500	10	500
other	100	100	100	100	500	10

Table 16.11 Unit coarticulation cost matrix (after Yi [55]) for left context replacements for nasals.

	obstruent	sonorant
obstruent	10	1000
sonorant	1000	10

Table 16.12 Unit coarticulation cost matrix (after Yi [55]) for right context replacements for nasals.

	voiced	unvoiced	sonorant
voiced	10	100	1000
unvoiced	100	10	1000
sonorant	1000	1000	10

16.3.2.4. Data-Driven Transition Cost

The empirical transition costs of Section 16.3.2.2 do not necessarily mean that a spectral discontinuity will take place, only that one is likely, and that if it occurs within a syllable it will have a larger perceptual effect than if it occurs across syllable boundaries. While that method can result in a good system, the cost is done independently of whether there is a true spectral discontinuity or not. Thus, it has been also proposed to use a measurement of the spectral discontinuity directly. This is often estimated as:

$$d_t(\theta_i,\theta_j) = \left|\mathbf{x}_i(l(\theta_i)-1) - \mathbf{x}_j(0)\right|^2 \tag{16.4}$$

the magnitude squared of the difference between the cepstrum at the last frame of θ_i and the first frame of θ_j. The quantity $l(\theta_i)$ denotes the number of frames of speech segment θ_i, and $\mathbf{x}_i(k)$ the cepstrum of segment θ_i at frame *k*.

This technique can effectively measure a spectral discontinuity in a region with slowly varying spectrum, but it can fail when one of the segments is a nasal, for example, for which a sharp spectral transition is expected and desired. A better way of measuring this discontinuity is shown in Figure 16.7, in which we measure the cepstral distance in an overlap region:[2] the last frame of segment 1 and the first frame before the beginning of segment 2:

$$d_t(\theta_i,\theta_j) = \left|\mathbf{x}_i(l(\theta_i)-1) - \mathbf{x}_j(-1)\right|^2 \tag{16.5}$$

When many speech segments are considered, a large number of cepstral distances need to be computed, which in turn may result in a slow process. To speed it up an approximation can be made where all possible cepstral vectors at the boundaries are vector quantized first, so that the distances between all codebook entries can be precomputed and stored in a table.

[2] This means extra frames need to be stored.

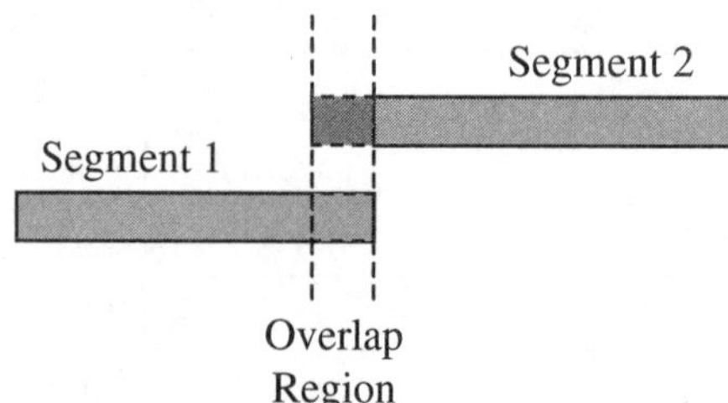

Figure 16.7 Measurement of the spectral discontinuity in the overlap region. The dark gray area is the speech region that precedes segment 2 and does not form part of segment 2. This area should match the last part of the segment 1.

A spectral discontinuity across, say, fricatives is perceptually not as important as if it happens across vowels [48]. For this reason, the cepstral distance described above does not correlate well with perceptual distances. To solve this problem, it is possible to combine both methods, for example by weighting the spectral/cepstral distance by different values.

Even if no spectral discontinuity is present, a phase discontinuity may take place. The pitch periodicity may be lost at the boundary. This can be generally solved by fine adjustment of the boundary using a correlation approach as described in Section 16.4. You need to keep in mind that such methods are not perfect.

16.3.2.5. Data-Driven Unit Cost

Spectral discontinuities across concatenations are often the result of using a speech segment with a different phonetic context than the target. One possibility is, then, to consider only speech segments where the phonetic contexts to the left and right match exactly. For example, if we use a phoneme as the basic speech segment, a perfect match would require on the order of at least 25,000 different segments. In this case, the coarticulation unit cost is zero if the target and candidate segment have the same phonetic context and infinite otherwise. When longer segments are desired, this number explodes exponentially. The problem with this approach is that it severely reduces the number of potential speech segments that can be used.

Generalized triphones, as described in Section 16.3.1.3, are used in [24]. In this approach, if the speech segments have the same generalized triphone contexts as the target utterance, the unit cost is zero, otherwise the cost is infinite. The technique allows us to use many more possible speech segments than the case above, yet it eliminates those speech segments that presumably have context mismatches that in turn lead to unnatural concatenations. When using a large training database, it was found that bringing the number of triphones from 25,000 down to about 2000 did not adversely impact the quality, whereas some degradation was perceived when using only 500 phoneme-length segments. Thus, this technique allows us to reduce the size of the speech segment inventory without severely degrading the voice quality.

If we set the number of decision-tree clustered context-dependent phonemes to be large, there will be fewer choices of long speech segments that match. For instance, in a system with 2000 generalized triphones, the phonetic context of the last phoneme of a long segment and the context of the target phoneme may be clustered together, whereas in a 3000-generalized-triphone system, both contexts may not be clustered together, so that the long segment cannot be used. This would be one example where using a larger number of generalized triphones hurts speech naturalness because the database of speech segments is limited. This problem could have been avoided if we didn't have to match generalized triphones and instead allowed context substitutions, yet penalized them with a corresponding cost. In the framework of decision-tree clustered context-dependent phonemes, this cost can be computed as the increase in entropy when those contexts are merged, using the methods described in Chapter 9. The larger the increase in entropy, the larger the penalty is when doing that context substitution between the candidate segment and the target segment. This approach gives more flexibility in the number of speech segments to be considered. In this case, there is a nonzero unit coarticulation cost associated with replacing one phonetic context with another.

Speech segments that have low HMM probability can be discarded, as they are probably not representative enough for that unit. Moreover, we can eliminate outliers: those units that have parameters too far away from the mean. Eliminating pitch outliers helps if prosody modification is to be done, as modifying pitch by more than a factor of 2 typically yields a decrease of quality, and by keeping units with average pitch, this is less likely to occur. Eliminating duration or amplitude outliers may signal an incorrect segmentation or a bad transcription [13].

16.3.3. Unit Inventory Design

The minimal procedure to obtain an acoustic inventory for a concatenative speech synthesizer consists of simply recording a number of utterances from a single speaker and labeling them with the corresponding text.

Since recording is often done in several sessions, it is important to maintain the recording conditions constant to avoid spectral or amplitude discontinuities caused by changes in recording conditions. The same microphone, room, and sound card should be used throughout all sessions [49].

Not all donor speakers are created equal. The choice of donor speaker can have a significant effect in voice quality (up to 0.3 MOS points on a 5-MOS scale) [7, 51, 52].

We can obtain higher-quality concatenative synthesis if the text read by the target speaker is representative of the text to appear in our application. This way we will be able to use longer units, and few concatenations will be needed.

Then the waveforms have to be segmented into phonemes, which is generally done with a speech recognition system operating in forced-alignment mode. Phonetic transcription, including alternate pronunciations, is generated automatically from text by the phonetic analysis module of Chapter 14. A large part of the inventory preparation includes checking correspondence between the text and corresponding waveform. Possible transcription errors may be flagged by phonemes whose durations are too far away from the mean (outliers) [13, 24].

Once we have the segmented and labeled recordings, we can use them as our inventory, or create smaller inventories as subsets that trade off memory size with quality [21, 25]. A database with a large number of utterances is generally required to obtain high-quality synthesis. It is noteworthy to analyze whether we can reduce the size of our database while obtaining similar synthesis quality on a given set of utterances. To do this, we can measure the cost incurred when we use a subset of the units in the database to synthesize our training database. A greedy algorithm can be used that at each stage eliminates the speech unit that increases the total distortion the least, repeating the approach until the desired size is achieved. This is an iterative analysis-by-synthesis algorithm.

The above procedure can also be used to find the set of units that have lowest cost in synthesizing a given text. For efficiency, instead of a large training text, we could use representative information from such text corpus, like the word trigrams with their corresponding counts, as an approximation.

In concatenative systems, you need to store a large number of speech segments, which could be compressed using any of the speech coding techniques described in Chapter 7. Since many such coders encode a frame of speech based on the previous one, you need to store this context for every segment you want to encode if you are to use such systems.

16.4. Prosodic Modification of Speech

One problem of segment concatenation is that it doesn't generalize well to contexts not included in the training process, partly because prosodic variability is very large. There are techniques that allow us to modify the prosody of a unit to match the target prosody. These prosody-modification techniques degrade the quality of the synthetic speech, though the benefits are often greater than the distortion introduced by using them because of the added flexibility.

The objective of prosodic modification is to change the amplitude, duration, and pitch of a speech segment. Amplitude modification can be easily accomplished by direct multiplication, but duration and pitch changes are not so straightforward.

We first present OLA and SOLA, two algorithms to change the duration of a speech segment. Then we introduce PSOLA, a variant of the above that allows for pitch modification as well.

16.4.1. Synchronous Overlap and Add (SOLA)

Time-scale modification of speech is very useful, particularly voice compression, as it allows a user to listen to a voice mail or taped lecture in a fraction of the time taken by the original segment user to listen to information The overlap-and-add (OLA) technique [12] shown in Figure 16.8 shows the analysis and synthesis windows used in the time compression. Given a Hanning window of length $2N$ and a compression factor of f, the analysis windows are spaced fN. Each analysis window multiplies the analysis signal, and at synthesis time they are overlapped and added together. The synthesis windows are spaced N samples apart. The use of windows such as Hanning allows perfect reconstruction when f equals 1.

In Figure 16.8, some of the signal appearance has been lost; note particularly some irregular pitch periods. To solve this problem, the synchronous overlap-and-add (SOLA) [45] allows for a flexible positioning of the analysis window by searching the location of the analysis window i around fNi in such a way that the overlap region had maximum correlation. The SOLA algorithm produces high-quality time compression. A mathematical formulation of PSOLA, an extension of both OLA and SOLA, is presented in Section 16.4.2.

While typically compression algorithms operate at a uniform rate, they have also been used in a nonuniform rate to take into account human perception, so that rapid transitions are compressed only slightly, steady sounds are compressed more, and pauses are compressed the most. It's reported in [11], that while uniform time compression can achieve a factor of 2.5 at most without degradation in intelligibility, nonuniform compression allows up to an average compression factor of 4.

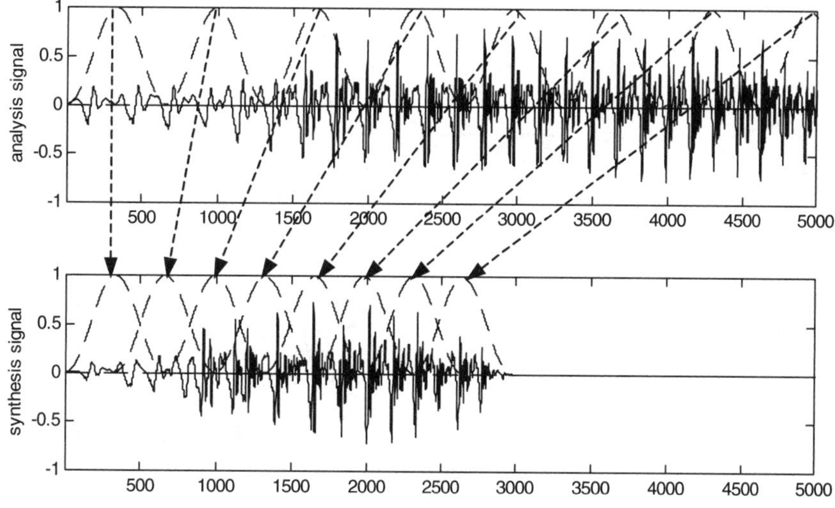

Figure 16.8 Overlap-and-add (OLA) method for time compression. Hanning windows of length $2N$, $N = 330$, are used to multiply the analysis signal, and resulting windowed signals are added. The analysis windows, spaced $2N$ samples, and the analysis signal $x[n]$ are shown on the top. The synthesis windows, spaced N samples apart, and the synthesis signal $y[n]$ are shown below. Time compression is uniform with a factor of 2. Pitch periodicity is somewhat lost, particularly around the fourth window.

16.4.2. Pitch Synchronous Overlap and Add (PSOLA)

Both OLA and SOLA do duration modification but cannot do pitch modification. On the other hand, they operate without knowledge of the signal's pitch. The most widely used method to do pitch modification is called pitch synchronous overlap and add (PSOLA) [38, 39], though to do so it requires knowledge of the signal's pitch. This process is illustrated in Figure 16.9.

Let's assume that our input signal $x[n]$ is voiced, so that it can be expressed as a function of pitch cycles $x_i[n]$

$$x[n] = \sum_{i=-\infty}^{\infty} x_i[n - t_a[i]] \tag{16.6}$$

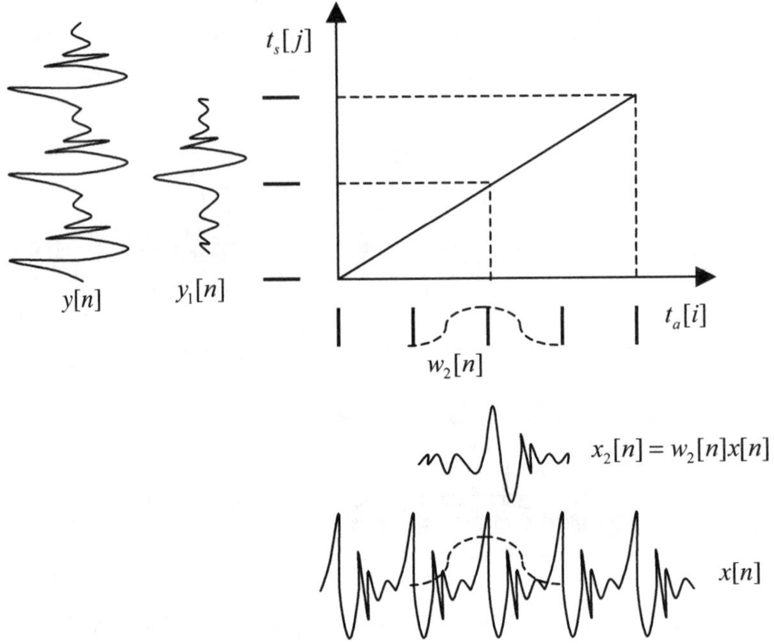

Figure 16.9 Mapping between five analysis epochs $t_a[i]$ and three synthesis epochs $t_s[j]$. Duration has been shortened by 40% and pitch period increased by 60%. Pitch cycle $x_2[n]$ is the product of the analysis window $w_2[n]$, in dotted line, with the analysis signal $x[n]$, which is aligned with analysis epochs $t_a[i]$. In this case, synthesis pitch cycle $y_1[n]$ equals $x_2[n]$ and also $y_0[n] = x_0[n]$ and $y_2[n] = x_5[n]$. Pitch is constant over time in this case.

where $t_a[i]$ are the epochs of the signal, so that the difference between adjacent epochs $P_a[i] = t_a[i] - t_a[i-1]$ is the pitch period at time $t_a[i]$ in samples. The pitch cycle is a windowed version of the input

$$x_i[n] = w_i[n]x[n] \tag{16.7}$$

which requires the windows $w_i[n]$ to meet the following condition:

$$\sum_{i=-\infty}^{\infty} w_i[n - t_a[i]] = 1 \tag{16.8}$$

which can be accomplished with a Hanning window, or a trapezoidal window that spans two pitch periods.

Our goal is to synthesize a signal $y[n]$, which has the same spectral characteristics as $x[n]$ but with a different pitch and/or duration. To do this, we replace the analysis epoch sequence $t_a[i]$ with the synthesis epochs $t_s[j]$, and the analysis pitch cycles $x_i[n]$ with the synthesis pitch cycles $y_j[n]$:

$$y[n] = \sum_{j=-\infty}^{\infty} y_j[n - t_s[j]] \tag{16.9}$$

The synthesis epochs are computed so as to meet a specified duration and pitch contour, as shown in Figure 16.9. This is equivalent to an impulse train with variable spacing driving a time-varying filter $x_t[n]$ which is known for $t = t_a[i]$, as shown in Figure 16.10. The synthesis pitch cycle $y_j[n]$ is obtained via a mapping from the closest corresponding analysis pitch cycle $x_i[n]$. In the following sections we detail how to calculate the synthesis epochs and the synthesis pitch-cycle waveforms.

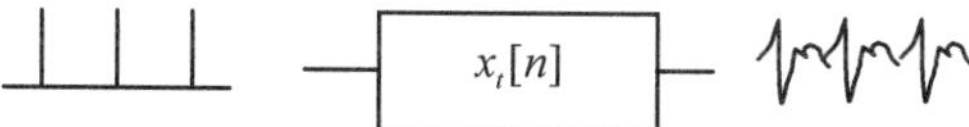

Figure 16.10 PSOLA technique as an impulse train driving a time-varying filter.

The term overlap-and-add derives from the fact that we use overlapping windows that we add together. The pitch-synchronous aspect comes from the fact that the windows are spaced a pitch period apart and are two pitch periods long. As you can see from Figure 16.9, the synthesis waveform has a larger pitch period than the analysis waveform and is shorter in duration.

For unvoiced speech, a set of epochs that are uniformly spaced works well in practice, as long as the spacing is smaller than 10 ms. If the segment needs to be stretched in such a way that these characteristic waveforms are repeated, an artificial periodicity would appear. To avoid this, the characteristic waveform that was to be repeated is flipped in time [38].

This approach is remarkably simple, yet it leads to high-quality prosody modification, as long as the voiced/unvoiced decision is correct and the epoch sequence is accurate.

To do prosody modification, the PSOLA approach requires keeping the waveform of the speech segment and its corresponding set of epochs, or time marks. As you can see from Eq. (16.6), if no prosody modification is done, the original signal is recovered exactly.

16.4.3. Spectral Behavior of PSOLA

Let's analyze why this simple technique works and how. To do that let's consider the case of a speech signal $x[n]$ that is exactly periodic with period T_0 and can be created by passing an impulse train through a filter with impulse response $s[n]$:

$$x[n] = s[n] * \sum_{i=-\infty}^{\infty} \delta[n - iT_0] = \sum_{i=-\infty}^{\infty} s[n - iT_0] \tag{16.10}$$

If we know the impulse response $s[n]$, then we could change the pitch by changing T_0. The problem is how to estimate it from $x[n]$. Let's assume we want to build an estimate $\tilde{s}[n]$ by multiplying $x[n]$ by a window $w[n]$:

$$\tilde{s}[n] = w[n]x[n] \tag{16.11}$$

The Fourier transform of $x[n]$ in Eq. (16.10) is given by

$$X(\omega) = \frac{2\pi}{T_0} S(\omega) \sum_{k=0}^{T_0-1} \delta(\omega - k\omega_0) = \frac{2\pi}{T_0} \sum_{k=0}^{T_0-1} S(k\omega_0)\delta(\omega - k\omega_0) \tag{16.12}$$

where $\omega_0 = 2\pi / T_0$. The Fourier transform of $\tilde{s}[n]$ can be obtained using Eqs. (16.11) and (16.12):

$$\tilde{S}(\omega) = \frac{1}{2\pi} W(\omega) * X(\omega) = \sum_{k=0}^{T_0-1} S(k\omega_0) \frac{W(\omega - k\omega_0)}{T_0} \tag{16.13}$$

If the window $w[n]$ is pitch synchronous, a rectangular window with length T_0 or a Hanning window with length $2T_0$, for example, then the above estimate is exact at the harmonics, i.e., $\tilde{S}(k\omega_0) = S(k\omega_0)$, because the window leakage terms are zero at the harmonics. In-between harmonics, $\tilde{S}(\omega)$ is an interpolation using $W(\omega)$, the transfer function of the window. If we use a rectangular window, the values of $S(\omega)$ in between $S(k\omega_0)$ and $S((k+1)\omega_0)$ are not determined only by those two harmonics, because the leakage from the other harmonics is not negligible. The use of a Hanning window drastically attenuates this leakage, so the estimate of the spectral envelope is better. This is what PSOLA is doing: getting an estimate of the spectral envelope by using a pitch-synchronous window.

Since it is mathematically impossible to recover $S(\omega)$ for a periodic signal, it is reasonable to fill in the remaining values by interpolation with the main lobes of the transform of the window. This approach works particularly well if the harmonics form a dense sampling of the spectral envelope, which is the case for male speakers. For female speakers, where the harmonics may be spaced far apart, the spectral envelope estimated by interpolating across harmonics could be far different from the real envelope.

16.4.4. Synthesis Epoch Calculation

In practice, we want to generate a set of synthesis epochs $t_s[j]$ given a target pitch period $P_s(t)$. If the desired pitch period $P_s(t) = P$ is constant, then the synthesis epochs are given by $t_s[j] = jP$.

In general the desired pitch period $P_s(t)$ is a function of time. Intuitively, we could compute $t_s[j+1]$ in terms of the previous epoch $t_s[j]$ and the pitch period at that time:

$$t_s[j+1] - t_s[j] = P_s(t_s[j]) \tag{16.14}$$

though this is an approximation, which happens to work well if $P_s(t)$ changes slowly over time.

Now we derive an exact equation, which also can help us understand pitch-scale and time-scale modifications of the next few sections. Epoch $t_s[j+1]$ can be computed so that the distance between adjacent epochs $t_s[j+1] - t_s[j]$ equals the average pitch period in the region $t_s[j] \le t < t_s[j+1]$ between them (see Figure 16.11). This can be done by the following expression

$$t_s[j+1] - t_s[j] = \frac{1}{t_s[j+1] - t_s[j]} \int_{t_s[j]}^{t_s[j+1]} P_s(t)dt \tag{16.15}$$

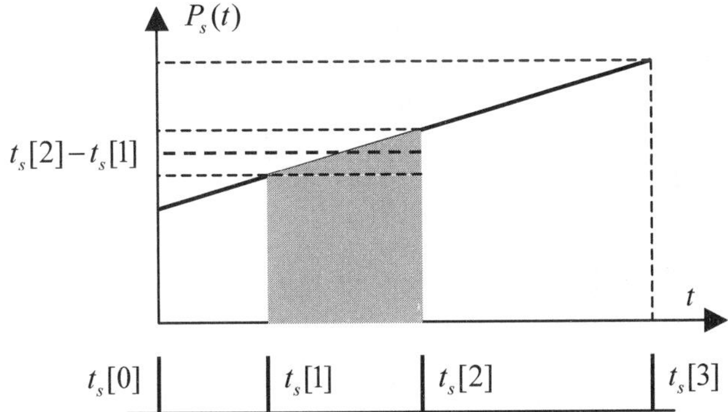

Figure 16.11 The desired pitch period $P_s(t)$ is a linearly increasing function of time such that the pitch period is doubled by the end of the segment. The four synthesis epochs $t_s[j]$ are computed to satisfy Eq. (16.15). In particular, $t_s[2]$ is computed such that $t_s[2] - t_s[1]$ equals the average pitch period in that region. Note that the growing spacing between epochs indicates that pitch is growing over time.

It is useful to consider the case of $P_s(t)$ being linear with t in that interval:

$$P_s(t) = P_s\left(t_s[j]\right) + b\left(t - t_s[j]\right) \tag{16.16}$$

so that the integral in Eq. (16.15) is given by

$$\int_{t_s[j]}^{t_s[j+1]} P_s(t)dt = \delta_j \left[P(t_s[i]) + b\frac{\delta_j}{2} \right] \tag{16.17}$$

where we have defined δ_j as

$$\delta_j = t_s[j+1] - t_s[j] \tag{16.18}$$

Inserting Eqs. (16.17) and (16.18) into Eq. (16.15), we obtain

$$\delta_j = P(t_s[i]) + b\frac{\delta_j}{2} \tag{16.19}$$

which, using Eq. (16.18), gives a solution for epoch $t_s[j+1]$ as

$$t_s[j+1] - t_s[j] = \delta_j = \frac{P_s(t_s[j])}{\left(1 - b/2\right)} \tag{16.20}$$

from the previous epoch $t_s[j]$, the target pitch at that epoch $P_s(t_s[j])$, and the slope b. We see that Eq. (16.14) is a good approximation to Eq. (16.20) if the slope b is small.

Evaluating Eq. (16.16) for $t_s[j+1]$ results in an expression for $P_s(t_s[j+1])$

$$P_s(t_s[j+1]) = P_s\left(t_s[j]\right) + b\left(t_s[j+1] - t_s[j]\right) \tag{16.21}$$

Equations (16.20) and (16.21) can be used iteratively. It is important to note that Eq. (16.20) requires $b < 2$ in order to obtain meaningful results, which fortunately is always the case in practice.

When synthesizing excitations for speech synthesis, it is convenient to specify the synthesis pitch period $P_s(t)$ as a piecewise linear function of time. In this case, Eq. (16.20) is still valid as long as $t_s[j+1]$ falls within the same linear segment. Otherwise, the integral in Eq. (16.17) has two components, and a second-order equation needs to be solved to obtain $t_s[j+1]$.

16.4.5. Pitch-Scale Modification Epoch Calculation

Sometimes, instead of generating an epoch sequence given by a function $P_s(t)$, we want to modify the epoch sequence of an analysis signal with epochs $t_a[i]$ by changing its pitch while maintaining its duration intact. This is called *pitch-scale* modification. To obtain the corresponding synthesis epochs, let's assume that the pitch period $P_a(t)$ of the analysis waveform at time t is constant and equals the difference between both epochs

$$P_a(t) = t_a[i+1] - t_a[i] \tag{16.22}$$

as seen in Figure 16.12.

The pitch period of the synthesis waveform $P_s(t)$ at the same time t now falls in between epochs j and $j + 1$

$$t_s[j] \le t < t_s[j+1] \tag{16.23}$$

with $t_s[j]$ being the time instant of the j epoch of the synthesis waveform. Now, let's define a relationship between analysis and synthesis pitch periods

$$P_s(t) = \beta(t) P_a(t) \tag{16.24}$$

where $\beta(t)$ reflects the pitch-scale modification factor, which, in general, is a function of time. Following the derivation in Section 16.4.4, we compute the synthesis epoch $t_s[j+1]$ so that

$$t_s[j+1] - t_s[j] = \frac{1}{t_s[j+1] - t_s[j]} \int_{t_s[j]}^{t_s[j+1]} \beta(t) P_a(t) dt \tag{16.25}$$

which reflects the fact that the synthesis pitch period at time t is the average pitch period of the analysis waveform times the pitch-scale modification factor. Since $\beta(t)P_a(t)$ is piecewise linear, we can use the results of Section 16.4.4 to solve for $t_s[j+1]$. In general, it needs to be solved recursively, which results in a second-order equation if $\beta(t)$ is a constant or a linear function of t.

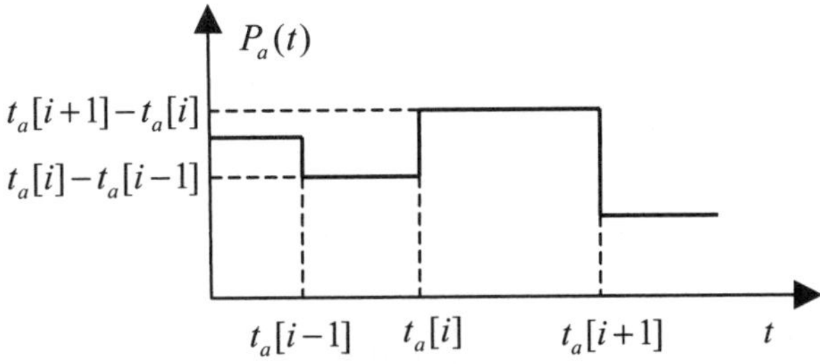

Figure 16.12 Pitch period of the analysis waveform as a function of time. It is a piecewise constant function of time.

16.4.6. Time-Scale Modification Epoch Calculation

Time-scale modification of speech involves changing the duration of a speech segment while maintaining its pitch intact. This can be realized by defining a mapping $t_s = D(t_a)$, a time-warping function, between the original signal and the modified signal. It is useful to define the duration modification rate $\alpha(t)$ from which the mapping function can be derived:

$$D(t) = \int_0^t \alpha(\tau) d\tau \tag{16.26}$$

Let's now assume that the duration modification rate $\alpha(t) = \alpha$ is constant, so that the mapping $D(t)$ in Eq. (16.26) is linear. If $\alpha > 1$, we are slowing down the speech, whereas if $\alpha < 1$, we are speeding it up. Let's consider time t in between epochs i and $i + 1$ so that $t_a[i] \le t < t_a[i+1]$:

$$\begin{aligned} &D(t_a[0]) = 0 \\ &D(t) = D(t_a[i]) + \alpha(t - t_a[i]) \end{aligned} \tag{16.27}$$

So that the relationship between analysis and synthesis pitch periods is given by

$$P_s(D(t)) = P_a(t) \tag{16.28}$$

To solve this it is useful to define a stream of virtual time instants $t'_a[j]$ in the analysis signal related to the synthesis time instants by

$$t_s[j] = D(t'_a[j]) = \alpha t'_a[j] \tag{16.29}$$

as shown in Figure 16.13.

Now we try to determine $t_s[j+1]$ such that $t_s[j+1] - t_s[j]$ is equal to the average pitch period in the original time signal between $t'_a[j]$ and $t'_a[j+1]$:

$$t_s[j+1] - t_s[j] = \frac{1}{t'_a[j+1] - t'_a[j]} \int_{t'_a[j]}^{t'_a[j+1]} P_a(t) dt \tag{16.30}$$

which, using Eq. (16.29), results in

$$t_s[j+1] - t_s[j] = \frac{\alpha}{t_s[j+1] - t_s[j]} \int_{t_s[j]/\alpha}^{t_s[j+1]/\alpha} P_a(t) dt \tag{16.31}$$

which again results in a second-order equation if $P_a(t)$ is piecewise constant or linear in t.

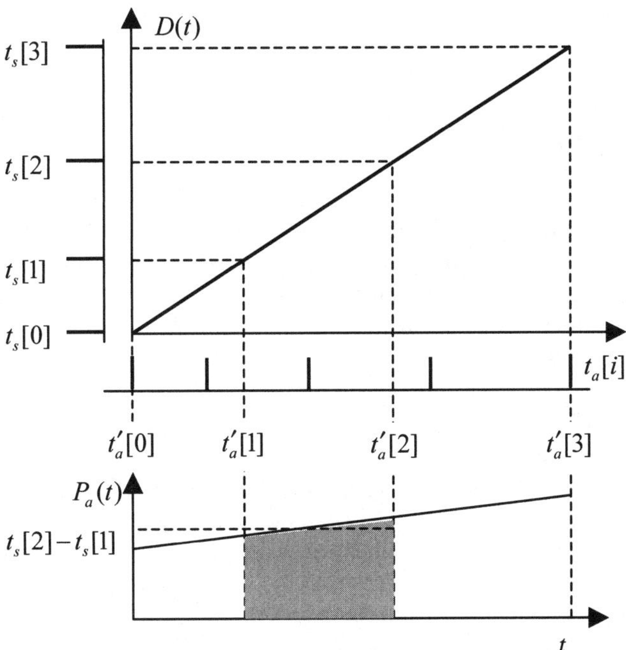

Figure 16.13 Time-scale modification of speech. The five analysis epochs $t_a[i]$ are shown in the x-axis and the four synthesis epochs $t_s[i]$ in the ordinate. Duration is shortened by 25% while maintaining the same pitch period. The corresponding virtual analysis epochs $t'_a[i]$ are obtained through the mapping $D(t)$, a linear transformation with $\alpha = 0.75$.

16.4.7. Pitch-Scale Time-Scale Epoch Calculation

The case of both pitch-scale and time-scale modification results in a combination of Eqs. (16.25) and (16.31):

$$t_s[j+1]-t_s[j] = \frac{\alpha}{t_s[j+1]-t_s[j]} \int_{t_s[j]/\alpha}^{t_s[j+1]/\alpha} \beta(t)P_a(t)dt \tag{16.32}$$

which again results in a second-order equation if $\beta(t)P_a(t)$ is piecewise constant or linear in t.

16.4.8. Waveform Mapping

The synthesis pitch waveforms can be computed through linear interpolation. Suppose that $t_a[i] \le t'_a[j] < t_a[i+1]$, then $y_j[n]$ is given by

$$y_j[n] = (1-\gamma_j)x_i[n] + \gamma_j x_{i+1}[n] \tag{16.33}$$

where γ_j is given by

$$\gamma_j = \frac{t'_a[j] - t_a[i]}{t_a[i+1] - t_a[i]} \tag{16.34}$$

Using this interpolation for voiced sounds results in smooth speech. For unvoiced speech, this interpolation results in a decrease of the amount of aspiration. Since smoothness is not a problem in those cases, the interpolation formula above is not used for unvoiced frames. A simplification of this linear interpolation consists of rounding γ_j to 0 or 1 and, thus, selecting the closest frame.

16.4.9. Epoch Detection

In the PSOLA approach, the analysis epochs $t_a[i]$ were assumed known. In practice this is not the case and we need to estimate them from the speech signal. There can be errors if the pitch period is not correctly estimated, which results in a rough, noisy voice quality. But estimating the epochs is not a trivial task, and this is the most sensitive part of achieving prosody modification in PSOLA.

Most pitch trackers attempt to determine F0 and not the epochs. From the $t_a[i]$ sequence it is easy to determine the pitch, since $P(t) = t_a[i+1] - t_a[i]$ for $t_a[i] < t < t_a[i+1]$. But from the pitch $P(t)$ the epoch placement is not uniquely determined, since the time origin is unspecified.

Common pitch tracking errors, such as pitch doubling, pitch halving, or errors in voiced/unvoiced decisions, result in rough speech. While manual pitch marking can result in accurate pitch marks, it is time consuming and error prone as well, so automatic methods have received a great deal of attention.

A method that attains very high accuracy has been proposed through the use of an *electroglottograph* (EGG) [32]. It consists of a pair of electrodes strapped around the neck at both sides of the larynx that measures the impedance of the larynx. Such a device, also called *laryngograph*, delivers a periodic signal when the vocal cords are vibrating and no signal otherwise. The pitch shape of a laryngograph signal is fairly stationary, which makes it relatively easy to determine the epochs from it (see Figure 16.14).

High-quality epoch extraction can be achieved by performing peak picking on the derivative of the laryngograph signal. Often, the derivative operation is accomplished by a first-order preemphasis filter $H[z] = 1 - \alpha z^{-1}$, with α being close to 1 (0.95 is a good choice).

In practice, the signal is preprocessed to filter out the low frequencies (lower than 100 Hz) and high frequencies (higher than 4 kHz). This can be done with rectangular window filters that are quite efficient and easy to implement. There is a significant amount of energy outside this band that does not contribute to epoch detection, yet it can complicate the process, as can be seen in Figure 16.14, so this bandpass filtering is quite important.

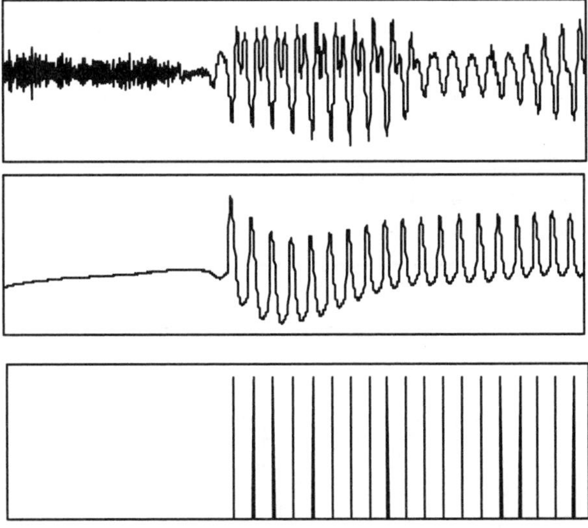

Figure 16.14 Speech signal, laryngograph signal, and its corresponding epochs.

The preemphasized signal exhibits peaks that are found by thresholding. The quality of this epoch detector has been evaluated on recordings from two female and four male speakers, and the voiced/unvoiced decision errors are lower than 1%. This is definitely acceptable for our prosody-modification algorithms. The quality of prosody modification with the epochs computed by this method can vastly exceed the quality achieved when standard pitch trackers (as described in Chapter 6) are used on the original speech signal [2].

16.4.10. Problems with PSOLA

The PSOLA approach is very effective in changing the pitch and duration of a speech segment if the epochs are determined accurately. Even assuming there are no pitch tracking errors, there can be problems when concatenating different segments:

- *Phase mismatches.* Even if the pitch period is accurately estimated, mismatches in the positioning of the epochs in the analysis signal can cause glitches in the output, as can be seen in Figure 16.15. The MBROLA [15] technique, an attempt to overcome phase mismatches, uses the time-domain PSOLA method for prosody modification, but the pitch cycles have been preprocessed so that they have a fixed phase. The advantage is that the spectral smoothing can be done by directly interpolating the pitch cycles in the time domain without adding any extra complexity. Since MBROLA sets the phase to a constant, the algorithm is more robust to phase errors in the epoch detection. Unfortunately, setting the phases constant incurs the added perceived noise described before.

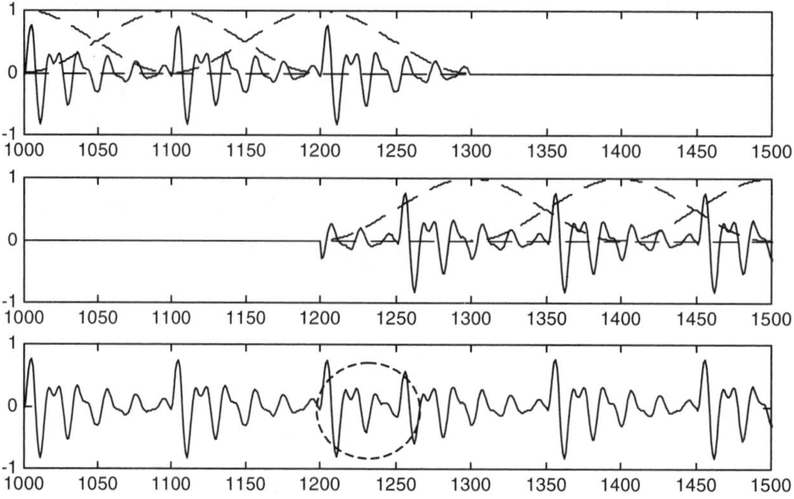

Figure 16.15 Phase mismatches in unit concatenation. Waveforms are identical, but windows are not centered on the same relative positions within periods.

- *Pitch mismatches.* These occur even if there are no pitch or phase errors during the analysis phase. As shown in Section 16.4.3, if two speech segments have the same spectral envelope but different pitch, the estimated spectral envelopes are not the same, and, thus, a discontinuity occurs (see Figure 16.16). In addition, pitch and timbre are not independent. Even when producing the same sound in the same phonetic context, a vastly different pitch will likely result in a different spectral envelope. This effect is particularly accentuated in the case of opera singers, who move their formants around somewhat so that the harmonics fall near the formant values and thus produce higher output.
- *Amplitude mismatch.* A mismatch in amplitude across different units can be corrected with an appropriate amplification, but it is not straightforward to compute such a factor. More importantly, the timbre of the sound will likely change with different levels of loudness.
- *Buzzy voiced fricatives.* The PSOLA approach doesn't handle well voiced fricatives that are stretched considerably because of added buzziness (repeating frames induces periodicity at the high frequency that wasn't present in the original signal) or attenuation of the aspirated component (if frames are interpolated).

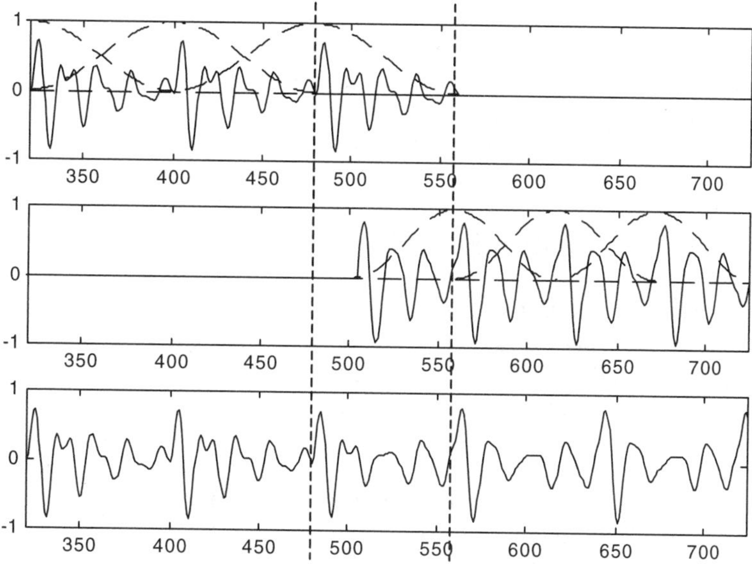

Figure 16.16 Pitch mismatches in unit concatenation. Two synthetic vowels were generated with a pitch of 138 Hz (top) and 197 Hz (middle) and exactly the same transfer function. There is no pitch tracking error, and windows are positioned coherently (no phase mismatch). The pitch of the second wave is changed through PSOLA to match the pitch of the first wave. There is a discontinuity in the resulting waveform and its spectrum (see Section 16.4.3), which is an artifact of the way the PSOLA approach estimates the spectral envelope.

16.5. SOURCE-FILTER MODELS FOR PROSODY MODIFICATION

The largest problem in concatenative synthesis occurs because of spectral discontinuities at unit boundaries. The methods described in Section 16.3 significantly reduce this problem but do not eliminate it. While PSOLA can do high-quality prosody modification on speech segments, it doesn't address these spectral discontinuities occurring at unit boundaries. It would be useful to come up with a technique that allows us to smooth these spectral discontinuities. In addition, PSOLA introduces buzziness for overstretched voiced fricatives. In the following sections we describe a number of techniques that have been proposed to cope with these problems and that are based on source-filter models.

The use of source-filter models allow us to modify the source and filter separately and thus maintain more control over the resulting synthesized signal. In Section 16.5.1 we study an extension of PSOLA that allows filter modification as well for smoothing purposes. Section 16.5.2 describes mixed excitation models that also allow for improved voiced fricatives. Finally, Section 16.5.3 studies a number of voice effects that can be achieved with a source-filter model.

16.5.1. Prosody Modification of the LPC Residual

A method known as LP-PSOLA that has been proposed to allow smoothing in the spectral domain is to do PSOLA on LPC residual. This approach, thus, implicitly uses the LPC spectrum as the spectral envelope instead of the spectral envelope interpolated from the harmonics (see Section 16.4.3) when doing F0 modification. If the LPC spectrum is a better fit to the spectral envelope, this approach should reduce the spectral discontinuities due to different pitch values at the unit boundaries. LP-PSOLA reduces the *bandwidth widening*. In practice, however, this hasn't proven to offer a significant improvement in quality, possibly because the spectral discontinuities due to coarticulation dominate the overall quality.

The main advantage of this approach is that it allows us to smooth the LPC parameters around a unit boundary and thus obtain smoother speech. Since smoothing the LPC parameters directly may lead to unstable frames, other equivalent representations, such as line spectral frequencies, reflection coefficients, log-area ratios, or autocorrelation coefficients, are used instead. The use of a long window for smoothing may blur sharp spectral changes that occur in natural speech. In practice, a window of 20–50 ms centered around the boundary has been proven useful.

While time-domain PSOLA has low computational complexity, its use in a concatenative speech synthesizer generally requires a large acoustic inventory. In some applications this is unacceptable, and it needs to be compressed using any of the coding techniques described in Chapter 7. You need to keep in mind that to use such encoders you need to store the coder's memory so that the first frame of the unit can be accurately encoded. The combined decompression and prosody modification is not as computationally efficient as time-domain PSOLA alone, so that the LP-PSOLA approach may offer an effective tradeoff, given that many speech coders encode the LPC parameters anyway.

16.5.2. Mixed Excitation Models

The block diagram of PSOLA shown in Figure 16.10 for voiced sounds also works for unvoiced sounds by choosing arbitrary epochs. The time-varying filter of Figure 16.10 can be kept in its time-domain form or in the frequency domain $X_t[k]$ by taking the FFT of $x_t[n]$.

It has been empirically shown that for unvoiced frames, the phase of $X_t[k]$ is unimportant as long as it is random. Thus, we can pass white noise through a filter with magnitude response $|X_t[k]|$ and obtain perceptually indistinguishable results. This reduced representation is shown in Figure 16.17. Moreover, it has been shown that the magnitude spectrum does not need to be encoded accurately, because it doesn't affect the synthesized speech. The only potential problem with this model occurs when voiced frames are incorrectly classified as unvoiced.

Maintaining the phase of $X_t[k]$ is perceptually important for voiced sounds. If it is set to 0, two audible distortions appear: the reconstructed speech exhibits a noisy quality, and voiced fricatives sound buzzy.

The perceived noise may come from the fact that a listener who hears a formant, because of its amplitude spectrum, also expects the 180° phase shift associated with a complex pole. In fact, it is not the absolute phase, but the fact that if the formant frequency/bandwidth changes with time, there is a phase difference over time. If such a phase is not present, scene analysis done in the auditory system may match this to noise. This effect can be greatly attenuated if the phase of the residual in LP-PSOLA is set to 0, possibly because the LPC coefficients carry most of the needed phase information.

The buzziness in voiced fricatives is the result of setting phase coherence not only at low frequencies but also at high frequencies, where the aspiration component dominates. This is the result of treating the signal as voiced, when it has both a voiced and an unvoiced component. In fact, most voiced sounds contain some aperiodic component, particularly at high frequencies. The amount of aspiration present in a sound is called *breathiness*. Female speech tends to be more breathy than male speech [29]. Mixed-excitation models, such as those in Figure 16.18, are then proposed to more accurately represent speech.

Such a model is very similar to the waveform-interpolation coding approach of Chapter 7, and, hence, we can leverage much of what was described there regarding the estimation of $x_t^v[n]$ and $x_t^u[n]$. This approach allows us to integrate compression with prosody modification.

The harmonic-plus-noise [50] model decomposes the speech signal $s(t)$ as a sum of a random component $s_r(t)$ and a harmonic component $s_p(t)$

$$s(t) = s_r(t) + s_p(t) \tag{16.35}$$

where $s_p(t)$ uses the sinusoidal model described in Chapter 7:

$$s_p(t) = \sum_{k=1}^{K(t)} A_k(t)\cos(k\theta(t) + \phi_k(t)) \tag{16.36}$$

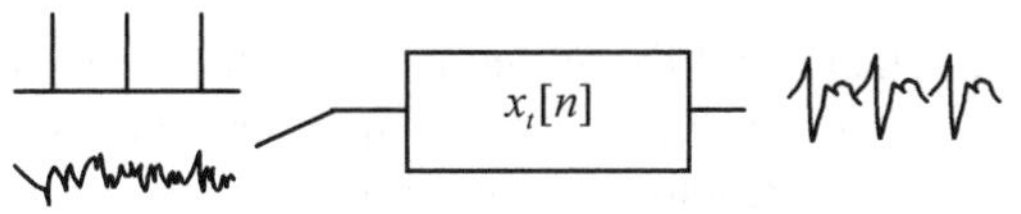

$x_t[n]$

Figure 16.17 Speech synthesis model with white noise or an impulse train driving a time-varying filter.

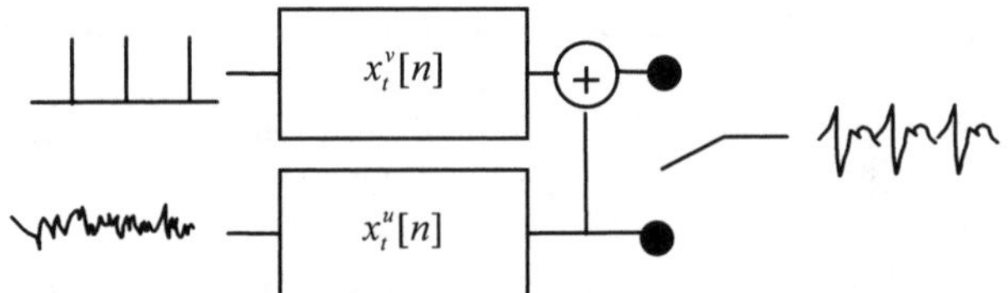

Figure 16.18 Mixed excitation model for speech synthesis.

where $A_k(t)$ and $\phi_k(t)$ are the amplitude and phase at time t of the kth harmonic, and $\theta(t)$ is given by

$$\theta(t) = \int_{-\infty}^{t} \omega_0(l)dl \tag{16.37}$$

16.5.3. Voice Effects

One advantage of using a spectral representation like those described in this section is that several voice effects can be achieved relatively easily, such as whisper, voice conversion, and echo/reverberation.

A whispering effect can be achieved by replacing the voiced component by random noise. Since the power spectrum of the voiced signal is a combination of the vocal tract and the glottal pulse, we would need to remove the spectral roll-off of the glottal pulse. This means that the power spectrum of the noise has to be high-pass in nature. Using white noise results in unnatural speech.

Voice conversion can be accomplished by altering the power spectrum [6, 28]. A warping transformation of the frequency scale can be achieved by shifting the LSF or the LPC roots, if using an LPC approach, or a warping curve if using an FFT representation.

Adding a controlled number of delayed and attenuated echoes can enhance an otherwise *dry* signal. If the delay is longer, it can simulate the room acoustics of a large hall.

16.6. Evaluation of TTS Systems

How do we determine whether one TTS system is better than another? Being able to evaluate TTS systems allows a customer to select the best system for his or her application. TTS evaluation is also important for developers of such systems to set some numerical goals in their design. As in any evaluation, we need to define a metric, which generally is dependent on the particular application for which the customer wants the TTS system. Such a metric consists of one or several variables of a system that are measured. Gibbon et al. [19] present a good summary of techniques used in evaluation of TTS systems.

Here we present a taxonomy of a TTS evaluation:

- *Glass-box vs. black-box evaluation.* There are two types of evaluation of TTS systems according to whether we evaluate the whole system or just one of its components: black-box and glass-box. A black-box evaluation treats the TTS system as a black box and evaluates the system in the context of a real-world application. Thus, a system may do very well on a flight reservation application but poorly on an e-mail reading application. In a glass-box evaluation, we attempt to obtain diagnostics by evaluating the different components that make up a TTS system.
- *Laboratory vs. field.* We can also conduct the study in a laboratory or in the field. While the former is generally easier to do, the latter is generally more accurate.
- *Symbolic vs. acoustic level.* In general, TTS evaluation is normally done by analyzing the output waveform, the so-called acoustic level. Glass-box evaluation at the symbolic level is useful for the text analysis and phonetic module, for example.
- *Human vs. automated.* There are two fundamentally distinct ways of evaluating speech synthesizers, according to how a given attribute of the system is estimated. One is to use human subjects; the other to automate the evaluation process. Both types have some issues in common and a number of dimensions of systematic variation. But the fundamental distinction is one of cost. In system development, and particularly in research on high-quality systems, it can be prohibitively expensive to run continuously a collection of human assessments of every algorithmic change or idea. Though human-subject checkpoints are needed throughout the development process, human testing is of greatest importance for the integrated, functionally complete system in the target field setting. At all earlier stages of development, automated testing should be substituted for human-subject testing wherever possible. The hope is that someday TTS research can be conducted as ASR research is today: algorithms are checked for accuracy or performance improvements automatically in the lab, while human subjects are mainly used when the final integrated system is deployed for field testing. This allows for rapid progress in the basic algorithms contributing to accuracy on any given dimension.
- *Judgment vs. functional testing.* Judgment tests are those that measure the TTS system in the context of the application where it is used, such as what percentage of the time users hang up an IVR system. System A may be more appropriate than system B for a banking application where most of the speech consists of numerical values, and system B may be better than system A for reading e-mail over the phone. Nonetheless, it is useful to use functional tests that measure task-independent variables of a TTS system, since such tests allow an easier comparison among different systems, albeit a nonoptimal one.

Since a human listener is the consumer of a TTS system, tests have been designed to determine the following characteristics of synthesized speech: intelligibility, overall quality, naturalness, suitability for a given task, and pleasantness. In addition, testing has been used for ranking and comparing a number of competing speech synthesizers, and for comparing synthetic with natural speech.

- *Global vs. analytic assessment.* The tests can measure such global aspects as overall quality, naturalness, and acceptability. Analytic tests can measure the rate, the clarity of vowels and consonants, the appropriateness of stresses and accents, pleasantness of voice quality, and tempo. Functional tests have been designed to test the intelligibility of individual sounds (phoneme monitoring), of combinations of sounds (syllable monitoring), and of whole words (word monitoring) in isolation as well as in various types of context.

It should be noted that all the above tests focus on segments, words, and sentences. This is a historical artifact, and as the field evolves, we should see an emphasis on testing of higher-order units. The diagnostic categories mentioned above can be used as a basis for developing tests of systems that take other structure into account. Such systems might include document-to-speech, concept-to-speech, and simulated conversation or dialog. A good system will reflect document and paragraph structure in the pausing and rhythm. Concept-to-speech systems claim to bring fuller knowledge of the intended use of information to bear in message generation and synthesis. Simulated dialog systems, or human-computer dialog systems, have to mimic a more spontaneous style, which is a subtle quality to evaluate. The tricky issue with higher-order units is the difficulty of simple choice or transcription-oriented measures. To develop tests of higher-order synthesizers, the word and sentence metrics can be applied to components and the overall system until reasonable intelligibility can be verified. Then tests of the special issues raised by higher-order systems can be conducted. Appropriate measures might be MOS overall ratings, preference between systems, summarization/gist transcription with subjective scoring, and task-based measures such as following directions. With task-based testing of higher-order units, both the correctness of direction-following and the time to completion, an indirect measure of intelligibility, pleasantness, and fatigue, can be recorded.

Furthermore, speech perception is not simply auditory. As discussed in Chapter 2, the McGurk effect [36] shows that perception of a speech sound is heavily influenced by visual cues. Synthetic speech is thus perceived with higher quality when a talking head is added as a visual cue [9, 18].

Glass-box evaluation of the text analysis and phonetic analysis modules, requiring evaluation at the *symbolic* level, is done in Chapter 14. A glass-box evaluation of the prosody module is presented in Chapter 15. In this section we include glass-box evaluation of the synthesis module, as well as a black-box evaluation of the whole system.

16.6.1. Intelligibility Tests

A critical measurement of a TTS system is whether or not human listeners can understand the text read by the system. Tests that measure this are called *intelligibility tests*. In this section we describe the Diagnostic Rhyme Test, the Modified Rhyme Test, the Phonetically Balanced word list test, the Haskins Syntactic Sentence Test, and the Semantically Unpredictable Sentence Test. The first three were described in a procedure approved by the American National Standards Institute [5].

Among the best known and most mature of these tests is the Diagnostic Rhyme Test (DRT) proposed by Voiers [54], which provides for diagnostic and comparative evaluation of the intelligibility of single initial consonants. The test runs twice through the list of 96 rhyming pairs shown in Table 16.13. The test consists of identification choice between two alternative English (or target-language) words, differing by a single phonetic feature in the initial consonant. For English the test includes contrasts among easily confusable paired consonant sounds such as *veal/feel, meat/beat, fence/pence, cheep/keep, weed/reed*, and *hit/fit*. In the test, both *veal* and *feel* are presented with the response alternatives *veal* and *feel*. Six contrasts are represented, namely voicing, nasality, sustention, sibilation, graveness, and compactness. Each contrast is included 32 times in the test, combined with 8 different vowels. The percentage of right answers is used as an indicator of speech synthesizer intelligibility. The tests use a minimum of five talkers and five listeners; larger subject groups reduce the margin of error. Even for high-quality speech coders, 100% *correct* responses are rarely achieved, so synthesizer results should be interpreted generously.

Table 16.13 The 192 stimulus words of the Diagnostic Rhyme Test (DRT).

Voicing		Nasality		Sustenation		Sibilation		Graveness		Compactness	
veal	feel	meat	beat	vee	bee	zee	thee	weed	reed	yield	wield
bean	peen	need	deed	sheet	cheat	cheep	keep	peak	teak	key	tea
gin	chin	mitt	bit	vill	bill	jilt	gilt	bid	did	hit	fit
dint	tint	nip	dip	thick	tick	sing	thing	fin	thin	gill	dill
zoo	sue	moot	boot	foo	pooh	juice	goose	moon	noon	coop	poop
dune	tune	news	dues	shoes	choose	chew	coo	pool	tool	you	rue
vole	foal	moan	bone	those	doze	joe	go	bowl	dole	ghost	boast
goat	coat	note	dote	though	dough	sole	thole	fore	thor	show	so
zed	said	mend	bend	then	den	jest	guest	met	net	keg	peg
dense	tense	neck	deck	fence	pence	chair	care	pent	tent	yen	wren
vast	fast	mad	bad	than	dan	jab	gab	bank	dank	gat	bat
gaff	calf	nab	dab	shad	chad	sank	thank	fad	thad	shag	sag
vault	fault	moss	boss	thong	tong	jaws	gauze	fought	thought	yawl	wall
daunt	taunt	gnaw	daw	shaw	chaw	saw	thaw	bong	dong	caught	thought
jock	chock	mom	bomb	von	bon	jot	got	wad	rod	hop	fop
bond	pond	knock	dock	vox	box	chop	cop	pot	tot	got	dot

A variant of this is the Modified Rhyme Test (MRT) proposed by House [22], which also uses a 300-entry word list for subjective intelligibility testing. The modified Rhyme Test (shown in Table 16.14) uses 50 six-word lists of rhyming or similar-sounding monosyllabic English words, e.g., *went, sent, bent, dent, tent, rent.* Each word is basically Consonant-Vowel-Consonant (with a few consonant clusters), and the six words in each list differ only in the initial or final consonant sound(s). Listeners are asked to identify which of the words was spoken by the synthesizer (closed response), or in some cases to enter any word they thought they heard (open response). A carrier sentence, such as "Would you write <test word> now," is usually used for greater naturalness in stimulus presentation. Listener responses can be scored as the number of words heard correctly; or the frequency of confusions of particular consonant sounds. This can be viewed as *intelligibility* of the synthesizer.

Though this is a nice isolation of one property, and as such is particularly appropriate for diagnostic use, it is not intended to substitute for fuller evaluation under more realistic listening conditions involving whole sentences. Segmental intelligibility is somewhat overestimated in these tests, because all the alternatives are real words and the subjects can adjust their perception to match the closest word. A typical human voice gives an MRT score of about 99%, with that of TTS systems generally ranging from 70% to 95%.

Table 16.14 The 300 stimulus words of the Modified Rhyme Test (MRT).

went	sent	bent	dent	tent	rent	same	name	game	tame	came	fame
hold	cold	told	fold	sold	gold	peel	reel	feel	eel	keel	heel
pat	pad	pan	path	pack	pass	hark	dark	mark	bark	park	lark
lane	lay	late	lake	lace	lame	heave	hear	heat	heal	heap	heath
kit	bit	fit	hit	wit	sit	cup	cut	cud	cuff	cuss	cud
must	bust	gust	rust	dust	just	thaw	law	raw	paw	jaw	saw
teak	team	teal	teach	tear	tease	pen	hen	men	then	den	ten
din	dill	dim	dig	dip	did	puff	puck	pub	pus	pup	pun
bed	led	fed	red	wed	shed	bean	beach	beat	beak	bead	beam
pin	sin	tin	fin	din	win	heat	neat	feat	seat	meat	beat
dug	dung	duck	dud	dub	dun	dip	sip	hip	tip	lip	rip
sum	sun	sung	sup	sub	sud	kill	kin	kit	kick	king	kid
seep	seen	seethe	seek	seem	seed	hang	sang	bang	rang	fang	gang
not	tot	got	pot	hot	lot	took	cook	look	hook	shook	book
vest	test	rest	best	west	nest	mass	math	map	mat	man	mad
pig	pill	pin	pip	pit	pick	ray	raze	rate	rave	rake	race
back	bath	bad	bass	bat	ban	save	same	sale	sane	sake	safe
way	may	say	pay	day	gay	fill	kill	will	hill	till	bill
pig	big	dig	wig	rig	fig	sill	sick	sip	sing	sit	sin
pale	pace	page	pane	pay	pave	bale	gale	sale	tale	pale	male
cane	case	cape	cake	came	cave	wick	sick	kick	lick	pick	tick
shop	mop	cop	top	hop	pop	peace	peas	peak	peach	peat	peal
coil	oil	soil	toil	boil	foil	bun	bus	but	bug	buck	buff
tan	tang	tap	tack	tam	tab	sag	sat	sass	sack	sad	sap
fit	fib	fizz	fill	fig	fin	fun	sun	bun	gun	run	nun

The set of twenty phonetically balanced (PB) word lists was developed during World War II and has been used very widely since then in subjective intelligibility testing. In Table 16.15 we include the first four PB word lists [20]. The words in each list are presented in a new, random order each time the list is used, each spoken in the same carrier sentence. The PB intelligibility test requires more training of listeners and talkers than other subjective tests and is particularly sensitive to SNR: a relatively small change causes a large change in the intelligibility score.

Tests using the *Haskins Syntactic Sentences* [40] go somewhat farther toward more realistic and holistic stimuli. This test set consists of 100 semantically unpredictable sentences of the form *The <Adjective> <Noun1> <Verb> the <Noun2>*, such as *"The old farm cost the blood,"* using high-frequency words. Compared with the rhyme tests, contextual predictability based on meaning is largely lacking, the longer speech streams are more realistic, and more coarticulation is present. Intelligibility is indicated by percentage of words correct.

Another test minimizing predictability is *Semantically Unpredictable Sentences* [23], with test sets for Dutch, English, French, German, Italian, and Swedish. A short template of syntactic categories provides a frame, into which words are randomly slotted from the lexicon. For example, the template *<Subject> <Verb> <Adverbial>* might appear as *"The chair ate through the green honesty."* Fifty sentences (10 per syntactic template) are considered adequate to test a synthesizer. Open transcription is requested, and *sentences correct* is used to score a synthesizer's intelligibility. Other such tests exist, and some include systematic variation of prosody on particular words or phrases as well.

The Harvard Psychoacoustic Sentences [16] is a set of 100 meaningful, syntactically varied, phonetically balanced sentences, such as *"Add salt before you fry the egg,"* requiring an *open response identification,* instead of a multiple-choice test.

Table 16.15 Phonetically balanced word lists.

List 1	are, bad, bar, bask, box, cane, cleanse, clove, crash, creed, death, deed, dike, dish, end, feast, fern, folk, ford, fraud, fuss, grove, heap, hid, hive, hunt, is, mange, no, nook, not, pan, pants, pest, pile, plush, rag, rat, ride, rise, rub, slip, smile, strife, such, then, there, toe, use, wheat
List 2	awe, bait, bean, blush, bought, bounce, bud, charge, cloud, corpse, dab, earl, else, fate, five, frog, gill, gloss, hire, hit, hock, job, log, moose, mute, nab, need, niece, nut, our, perk, pick, pit, quart, rap, rib, scythe, shoe, sludge, snuff, start, suck, tan, tang, them, trash, vamp, vast, ways, wish
List 3	ache, air, bald, barb, bead, cape, cast, check, class, crave, crime, deck, dig, dill, drop, fame, far, fig, flush, gnaw, hurl, jam, law, leave, lush, muck, neck, nest, oak, path, please, pulse, rate, rouse, shout, sit, size, sob, sped, stag, take, thrash, toil, trip, turf, vow, wedge, wharf, who, why
List 4	bath, beast, bee, blonde, budge, bus, bush, cloak, course, court, dodge, dupe, earn, eel, fin, float, frown, hatch, heed, hiss, hot, how, kite, merge, lush, neat, new, oils, or, peck, pert, pinch, pod, race, rack, rave, raw, rut, sage, scab, shed, shin, sketch, slap, sour, starve, strap, test, tick, touch

16.6.2. Overall Quality Tests

While a TTS system has to be intelligible, this does not guarantee user acceptance, because its quality may be far from that of a human speaker. In this section we describe the Mean Opinion Score and the Absolute Category Ratings.

Human-subject judgment testing for TTS can adapt methods from speech-coding evaluation (see Chapter 7). With speech coders, *Mean Opinion Score* (MOS) is administered by asking 10 to 30 listeners to rate several sentences of coded speech on a scale of 1 to 5 (1 = Bad, 2 = Poor, 3 = Fair, 4 = Good, 5 = Excellent). The scores are averaged, resulting in an overall MOS rating for the coder. This kind of methodology can be applied to speech synthesizers as well. Of course, as with any human subject test, it is essential to carefully design the listening situation and carefully select the subject population, controlling for education, experience, physiological disorders, dialect, etc. As with any statistically interpreted test, the standard analyses of score distributions, standard deviation, and confidence intervals must be performed. The range of quality in coder evaluations by MOS are shown in Table 16.16.

Since we are making the analogy to coders, certain ironies can be noted. Note the lowest-range descriptor for coder evaluation: *synthetic*. In using MOS for synthesis testing, output is being evaluated by implicit reference to real human speech, and the upper range in the coder MOS interpretations above (3.5–4.5) is probably not applicable to the output of most TTS systems. Even a *good* TTS system might fare poorly on such a coder MOS evaluation. Therefore, the MOS interpretive scale, when applied to synthesis, cannot be absolute as the above coding-based interpretive table would imply. Furthermore, subjects participating in MOS-like tests of synthesizers should be made aware of the special nature of the speech (synthetic) and adjust their expectations accordingly. Finally, no matter how carefully the test is designed and administered, it is difficult to correlate, compare, and scale such measures. Nevertheless, MOS tests are perhaps suited to relative ranking of various synthesizers. The 1-to-5 scale is categorical, but similar judgment tests can be run in *magnitude* mode, with the strength of the quality judgment being indicated along a continuous scale, such as a moving slider bar.

Table 16.16 Mean opinion score (MOS) ratings and typical interpretations.

MOS Scores	Quality	Comments
4.0–4.5	Toll/Network	Near-transparent, "in-person" quality
3.5–4.0	Communications	Natural, highly intelligible, adequate for telecommunications, changes and degradation of quality very noticeable
2.5–3.5	Synthetic	Usually intelligible, can be unnatural, loss of speaker recognizability, inadequate levels of naturalness

Table 16.17 Listening Quality Scale.

Quality of the Speech	**Score**
Excellent	5
Good	4
Fair	3
Poor	2
Bad	1

The International Telecommunication Union (ITU) has attempted to specify some standards for assessing synthetic speech, including spliced digitized words and phrases, typically with the expectation of delivery over the phone. The *Absolute Category Rating* (ACR) system recommended by ITU P.800 offers instructions to be given to subjects for making category judgments in MOS-style tests of the type discussed here. The first is the *Listening Quality Scale*, shown in Table 16.17, and the second the *Listening Effort Scale* shown in Table 16.18.

It is sometimes possible to get subjects to pay particular attention to various particular features of the utterance, which may be called *analytic* as opposed to *global* listening. The desired features generally have to be described somehow, and these descriptions can be a bit vague. Thus, standard measures of reliability and validity, as well as result normalization, must be applied. Typical descriptors for important factors in analytic listening might be: *smoothness*, *naturalness*, *pleasantness*, *clarity*, *appropriateness*, etc., each tied to a particular underlying target quality identified by the system designers. For example, smoothness might be a descriptor used when new algorithms for segment concatenation and blending are being evaluated in a concatenative system. Naturalness might be the quality descriptor when a formant-based system has been made more natural by incorporation of a richer glottal source function. Some elements of the speech can be more directly identified to the subject in familiar terms. For example, pleasantness might be a way of targeting the pitch contours for attention, or the subject could be specifically asked to rank the pitch contours per se, in terms of naturalness, pleasantness, etc. Appropriateness might be a way of getting at judgments of accentuation: e.g., a stimulus that was accented as "… *birthday PARTY*" might be judged less appropriate, in a neutral semantic context, than one that was perceived as "… *BIRTHDAY party*." But no matter how the attributes are described, in human-subject MOS-style testing there cannot be a clear and consistent separation of effects.

Table 16.18 Listening Effort Scale.

Effort Required to Understand the Meanings of Sentences	**Score**
Complete relaxation possible; no effort required	5
Attention necessary; no appreciable effort required	4
Moderate effort required	3
Considerable effort required	2
No meaning understood with any feasible effort	1

16.6.3. Preference Tests

Normalized MOS scores for different TTS systems can be obtained without any direct preference judgments. If direct comparisons are desired, especially for systems that are informally judged to be fairly close in quality, another ITU recommendation, the *Comparison Category Rating* (CCR) method, may be used. In this method, listeners are presented with a pair of speech samples on each trial. The order of the *system A system B* samples is chosen at random for each trial. On half of the trials, the *system A* sample is followed by the *system B* sample. On the remaining trials, the order is reversed. Listeners use the instructions in Table 16.19 to judge the quality of the second sample relative to that of the first. Sometimes the granularity can be reduced as much as simply "*prefer A/prefer B*."

Assuming (A,B) is the reference presentation order, scores for the (B,A) presentations may be normalized by reversing their signs (e.g., –1 in B,A order becomes 1, etc.). Subsequently, standard statistical summarizations may be performed, like the one described in Chapter 3.

Table 16.19 Preference ratings between two systems. The quality of the second utterance is compared to the quality of the first by means of 7 categories. Sometimes only better, same, or worse are used.

3	Much Better
2	Better
1	Slightly Better
0	About the Same
-1	Slightly Worse
-2	Worse
-3	Much Worse

16.6.4. Functional Tests

Functional testing places the human subject in the position of carrying out some task related to, or triggered by, the speech. This can simulate a full field deployment, with a usercentric task, or can be more of a laboratory situation, with a testcentric task. In the laboratory situation, various kinds of tasks have been proposed. In *analytic* mode, functional testing can enforce isolation of the features to be attended to in the structure of the test stimuli themselves. This can lead to a more precise form of result than the MOS judgment approach. There have been a wide variety of proposals and experiments of this type.

One of the well-known facts in TTS evaluation is that the quality of a system is dominated by the quality of its worst component. While it may be argued that it is impossible to separate the effects of the front-end analysis and back-end synthesis, it is convenient to do so to gain a better understanding of each component. An attempt to study the quality of the speech synthesis module has been done via the use of natural instead of synthetic prosody. This way, it is presumed that the prosody module is doing the best possible job, and that any problem is then due to a deficient speech synthesis. The natural pitch contour can be obtained with a pitch tracker (or using a laryngograph signal), and the natural durations can be obtained either through manual segmentation or through the use of a speech recognition system used in forced-alignment mode. Plumpe and Meredith [44] conducted a preference test between original recordings and waveforms created when one of the modules of a concatenative TTS system used synthetically generated values instead of the natural values. The results indicated that using synthetic pitch instead of natural pitch was the cause of largest degradation according to listeners, and, thus, that pitch generation was the largest bottleneck in the system. The pitch-generation module was followed by the spectral discontinuities at the concatenation points, with duration being the least damaging.

Some functional tests are much more creative than simple transcription, however. They could, in theory, border on related areas, such as memory testing, involving summarizing passages, or following synthesized directions, such as a route on a map. The ultimate test of synthesis, in conjunction with all other language interface components, is said to be the Turing test [53]. In this amusing scenario, a human being is placed into conversation with a computational agent, represented vocally for our purposes, perhaps over the telephone. As Turing put it: "It is proposed that a machine may be deemed intelligent, if it can act in such a manner that a human cannot distinguish the machine from another human merely by asking questions via a mechanical link." Turing predicted that in the future "an average interrogator will not have more than a 70 percent chance of making the right identification, human or computer on the other end, after five minutes of questioning" in this game. A little reflection might raise objections to this procedure as a check on speech output quality per se, since some highly intelligent people have speech disabilities, but the basic idea should be clear, and it remains a Holy Grail for the artificial intelligence field generally. Of course, no automated or laboratory test can substitute for a real-world trial with paying customers.

16.6.5. Automated Tests

The tests described above always involved the use of human subjects and are the best tests that can be used to evaluate a TTS system. Unfortunately, they are time consuming and expensive to conduct. This limits their application to an infrequent use, which can hardly have any diagnostic value. Automated *objective* tests usually involve establishing a test corpus of correctly tagged examples of the tested phenomena, which can be automatically checked. This style of testing is particularly appropriate when working with isolated components of the TTS system, for diagnosis or regression testing (glass-box testing). It is not particularly productive to discuss such testing in the abstract, as the test features must closely track each system's design and implementation. Nevertheless, a few typical areas for testing can be noted. In general, tests are simultaneously testing the linguistic model and content as well as the software implementation of a system, so whenever a discrepancy arises, both possible sources of error must be considered.

Several automated tests for text analysis and letter-to-sound conversion are presented in Chapter 14. A number of automated tests for prosody are discussed in Chapter 15. Here we touch on automated tests for the synthesis module.

The ITU has created the P.861 proposal for estimating perceptual scores using automated signal-based measurements. The P.861 specifies a particular technique known as *Perceptual Speech Quality Measurement* (PSQM). In this method, for each analysis frame, various quantified measures based on the time signal, the power spectrum, the Bark power spectrum, the excitation spectrum, the compressed loudness spectrum, etc. of both the reference and the test signal can be computed. In some cases the PSQM score can be converted to an estimated MOS score, with interpretations similar to those of Table 16.16. At present such methods are limited primarily to analysis of telephone-quality speech (300–3400 Hz bandwidth), to be compared with closely related reference utterances. This method could perhaps be adapted to stand in for human judgments during system development of new versions of modules, say glottal source functions in a formant synthesizer, comparing the resulting synthetic speech to a standard reference system's output on a given test sample.

16.7. Historical Perspective and Further Reading[3]

In 1779 in St. Petersburg, Russian Professor Christian Kratzenstein explained physiological differences between five long vowels (*/a/*, */e/*, */i/*, */o/*, and */u/*) and made apparatus to produce them artificially. He constructed acoustic resonators similar to the human vocal tract and activated the resonators with vibrating reeds as in music instruments. Von Kempelen (1734–1804) proposed in 1791 in Vienna a mechanical speaking machine that could produce not just vowels but whole words and sentences (see Figure 16.19). While working with his speaking machine, he demonstrated a speaking chess-playing machine. Unfortunately, the main mechanism of the machine was a concealed, legless chess-player expert. Therefore, his real speaking machine was not taken as seriously as it should have been. In 1846, Joseph Faber developed a synthesizer, called speech organ, that had more control of pitch to the extent it could sing *God Save the Queen* in a performance in London.

The first electrical synthesis device was introduced by Stewart in 1922 [4]. The device had a buzzer as excitation and two resonant circuits to model the acoustic resonances of the vocal tract and was able to generate single static vowel sounds with the first two formants. In 1932 Japanese researchers Obata and Teshima added a third formant for more intelligible vowels.

Homer Dudley of Bell Laboratories demonstrated at the 1939 New York World's Fair the *Voder*, the first electrical speech synthesizer, which was human-controlled. The operator worked at a keyboard, with a wrist bar to control the voicing parameter and a pedal for pitch control (see Figure 16.20 and Figure 16.21), and it was able to synthesize continuous speech. The *Pattern Playback* is an early talking machine that was built by Franklin S. Cooper and his colleagues at Haskins Laboratories in the late 1940s. This device synthesized sound by passing light through spectrograms that in turn modulated an oscillator with a fixed F0 of 120 Hz and 50 harmonics.

---

[3] Chapter 6 includes a historical perspective on representation of speech signals that is intimately tied to speech synthesis.

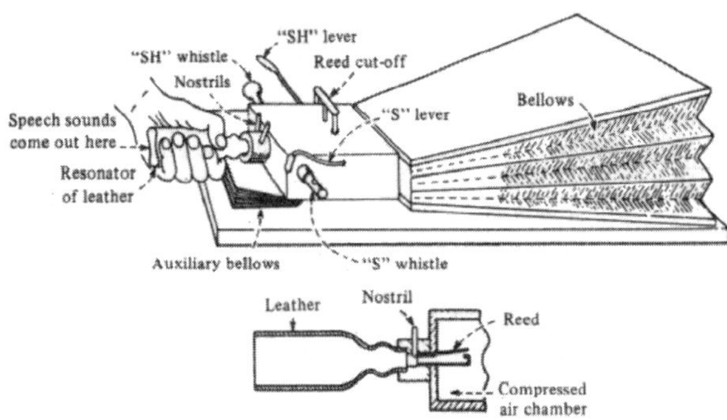

Figure 16.19 Wheatstone's reconstruction of von Kempelen's speaking machine [14] (after Flanagan [17]).

The first analog parallel formant synthesizer, the *Parametric Artificial Talker* (PAT), was developed in 1953 by Walter Lawrence of the Signals Research and Development Establishment of the British Government. Gunnar Fant of the KTH in Sweden developed an analog cascade formant synthesizer, the OVE II. Both Lawrence and Fant showed in 1962 that by manually tuning the parameters, a natural sentence could be reproduced reasonably faithfully. Acoustic analog synthesizers were also known as terminal analogs, resonance-synthesizers. John Holmes tuned by hand the parameters of his formant synthesizer so well that the average listener could not tell the difference between the synthesized sentence "*I enjoy the simple life*" and the natural one [31].

Figure 16.20 The Voder developed by Homer Dudley of Bell Labs at the 1939 World's Fair in New York. The operator worked at a keyboard, with a wrist bar to control the voicing parameter and a pedal for pitch control.

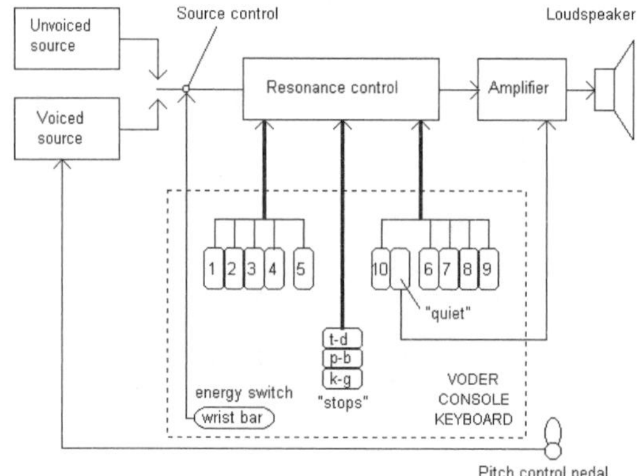

Figure 16.21 Block diagram of the Voder by Homer Dudley, 1939 (after Flanagan [17]).

The first articulatory synthesizer, the DAVO, was developed in 1958 by George Rosen at M.I.T. Cecil Coker designed rules to control a low-dimensionality articulatory model in 1968. Paul Mermelstein and James Flanagan from Bell Labs also used articulatory synthesis in 1976. Articulatory synthesis, however, never took off, because formant synthesis was better understood at the time.

The advent of the digital computer prompted John Kelly and Louis Gerstman to create in 1961 the first phonemic-synthesis-by-rule program. John Holmes and his colleagues Ignatius Mattingly and John Shearme developed a rule program for a formant synthesizer at JSRU in England. The first full text-to-speech system was developed by Noriko Umeda in 1968 at the Electrotechnical Laboratory of Japan. It was based on an articulatory model and included a syntactic analysis module with sophisticated heuristics. The speech was quite intelligible, but monotonous and far from the quality of present systems.

In 1976, Raymond Kurzweil developed a unique reading machine for the blind, a computer-based device that read printed pages aloud. It was an 80-pound device that shot a beam of light across each printed page, converting the reflected light into digital data that was transformed by a computer into synthetic speech. It made reading of all printed material possible for blind people, whose reading has previously been limited to material translated into Braille. The work of Dennis Klatt of MIT had a large influence in the field. In 1979 together with Jonathan Allen and Sheri Hunnicut he developed the MITalk system. Two years later Klatt introduced his famous Klattalk system, which used a new sophisticated voicing source.

The early 1980s marked the beginning of commercial TTS systems. The Klattalk system was the basis of Telesensory Systems' Prose-2000 commercial TTS system in 1982. It also formed the basis for Digital Equipment Corporation's DECtalk commercial system in 1983, probably the most widely used TTS system of the twentieth century. The Infovox TTS system, the first multilanguage formant synthesizer, was developed in Sweden by Rolf Carlson, Bjorn Granstrom, and Sheri Hunnicutt in 1982, and it was a descendant of Gunnar Fant's OVE system. The first integrated circuit for speech synthesis was probably the Votrax chip, which consisted of cascade formant synthesizer and simple low-pass smoothing circuits. In 1978 Richard Gagnon introduced an inexpensive Votrax-based Type-n-Talk system. The first work in concatenative speech synthesis was done in 1968 by Rex Dixon and David Maxey, where diphones were parametrized with formant frequencies and then concatenated. In 1977, Joe Olive and his colleagues at Bell Labs [41] concatenated linear-prediction diphones. In 1982 Street Electronics introduced the Echo system, a diphone concatenation synthesizer which was based on a newer version of the same chip as in the Speak-n-Spell toy introduced by Texas Instruments in 1980.

Concatenative systems started to gain momentum in 1985 with the development of the PSOLA prosody modification technique by France Telecom's Charpentier and Moulines. PSOLA increased the text coverage of concatenative systems by allowing diphones to have their prosody modified. The hybrid Harmonic/Stochastic (H/S) model of Abrantes [1] has also been successfully used for prosody modification. The foundation of corpus-based concatenative systems was developed by a team of researchers at ATR in Japan in the early 1990s [10, 27]. The use of a large database of long units was also pioneered by researchers at AcuVoice Inc. Other corpus-based systems have made use of HMMs to automatically segment speech databases, as well as to serve as units in concatenative synthesis [13, 24]. Microsoft integrated a concatenative TTS [24] in Windows 2000.

For more detailed description of speech synthesis development and history see, for example, [31] and [17] and references in these. A number of audio clips are available in Klatt [31] showing the progress through the early years. You can hear samples at the Smithsonian's Speech Synthesis History Project [35]. A Web site with comparison of recent TTS systems can be found at [33].

REFERENCES

[1] Abrantes, A.J., J.S. Marques, and I.M. Trancoso, "Hybrid Sinusoidal Modeling of Speech without Voicing Decision," *Proc. Eurospeech*, 1991, Genoa, Italy, pp. 231-234.

[2] Acero, A., "Source-Filter Models for Time-Scale Pitch-Scale Modification of Speech," *Int. Conf. on Acoustics, Speech and Signal Processing*, 1998, Seattle, WA, pp. 881-884.

[3] Acero, A., "Formant Analysis and Synthesis Using Hidden Markov Models," *Eurospeech*, 1999, Budapest, pp. 1047-1050.

[4] Allen, J., M.S. Hunnicutt, and D.H. Klatt, *From Text to Speech: The MITalk System*, 1987, Cambridge, UK, University Press.

[5] ANSI, *Method for Measuring the Intelligibility of Speech Over Communication Systems*, 1989, American National Standards Institute.

[6] Arslan, L.M. and D. Talkin, "Speaker Transformation Using Sentence HMM Based Alignments and Detailed Prosody Modification," *Int. Conf. on Acoustics, Speech and Signal Processing*, 1998, Seattle, pp. 289-292.

[7] Beutnagel, M., *et al.*, "The AT&T Next-Gen TTS System," *Joint Meeting of ASA*, 1999, Berlin, pp. 15-19.

[8] Bickley, C.A., K.N. Stevens, and D.R. Williams, "A Framework for Synthesis of Segments Based on Pseudoarticulatory Parameters" in *Progress in Speech Synthesis,* 1997, New York, Springer-Verlag, pp. 211-220.

[9] Bregler, C., M. Covell, and M. Slaney, "Video Rewrite: Driving Visual Speech with Audio," *ACM Siggraph*, 1997, Los Angeles, pp. 353-360.

[10] Campbell, W.N. and A.W. Black, "Prosody and the Selection of Source Units for Concatenative Synthesis" in *Progress in Speech Synthesis*, J.V. Santen, *et al.*, eds., 1996, pp. 279-292, Springer Verlag.

[11] Covell, M., M. Withgott, and M. Slaney, "Mach1: Nouniform Time-Scale Modification of Speech," *Proc. of IEEE Int. Conf. on Acoustics, Speech and Signal Processing*, 1998, Seattle, WA, pp. 349-352.

[12] Crochiere, R., "A Weighted Overlap-Add Method of Short Time Fourier Analysis/Synthesis," *IEEE Trans. on Acoustics, Speech and Signal Processing*, 1980, **28**(2), pp. 99-102.

[13] Donovan, R. and P. Woodland, "Improvements in an HMM-based Speech Synthesizer," *Proc. of the EuroSpeech Conf.*, 1995, Madrid, pp. 573-576.

[14] Dudley, H. and T.H. Tarnoczy, "The Speaking Machine of Wolfgang von Kempelen," *Journal of the Acoustical Society of America*, 1950, **22**, pp. 151-166.

[15] Dutoit, T., *An Introduction to Text-to-Speech Synthesis*, 1997, Kluwer Academic Publishers.

[16] Egan, J., "Articulation Testing Methods," *Laryngoscope*, 1948, **58**, pp. 955-991.

[17] Flanagan, J., *Speech Analysis, Synthesis and Perception*, 1972, New York, Springer-Verlag.

[18] Galanes, F.G., *et al.*, "Generation of Lip-Synched Synthetic Faces from Phonetically Clustered Face Movement Data," *AVSP*, 1998, Terrigal, Australia.

[19] Gibbon, D., R. Moore, and R. Winski, *Handbook of Standards and Resources for Spoken Language Systems*, 1997, Berlin & New York, Walter de Gruyter Publishers.

[20] Goldstein, M., "Classification of Methods Used for Assesment of Text-to-Speech Systems According to the Demands Placed on the Listener," *Speech Communication*, 1995, **16**, pp. 225-244.

[21] Hon, H., *et al.*, "Automatic Generation of Synthesis Units for Trainable Text-to-Speech Systems," *Int. Conf. on Acoustics, Signal and Speech Processing*, 1998, Seattle, WA, pp. 293-296.

[22] House, A., *et al.*, "Articulation Testing Methods: Consonantal Differentiation with a Closed Response Set," *Journal of the Acoustical Society of America*, 1965, **37**, pp. 158-166.

[23] Howard-Jones, P., *SOAP, Speech Output Assessment Package*, 1992, ESPRIT SAM-UCL-042.

[24] Huang, X., *et al.*, "Whistler: A Trainable Text-to-Speech System," *Int. Conf. on Spoken Language Processing*, 1996, Philadephia, pp. 2387-2390.

[25] Huang, X., *et al.*, "Recent Improvements on Microsoft's Trainable Text-To-Speech System - Whistler," *Int. Conf. on Acoustics, Signal and Speech Processing*, 1997, Munich, Germany, pp. 959-962.

[26] Hunt, A.J. and A.W. Black, "Unit Selection in a Concatenative Speech Synthesis System Using a Large Speech Database," *Int. Conf. on Acoustics, Speech and Signal Processing*, 1996, Atlanta, pp. 373-376.

[27] Iwahashi, N., N. Kaiki, and Y. Sagisaka, "Concatenation Speech Synthesis by Minimum Distortion Criteria," *IEEE Int. Conf. on Acoustics, Speech and Signal Processing*, 1992, San Francisco, pp. 65-68.

[28] Kain, A. and M. Macon, "Text-to-Speech Voice Adaptation from Sparse Training Data," *Int. Conf. on Spoken Language Systems*, 1998, Sydney, Australia, pp. 2847-2850.

[29] Klatt, D. and L. Klatt, "Analysis, Synthesis and Perception of Voice Quality Variations among Female and Male Talkers," *Journal of the Acoustical Society of America*, 1990, **87**, pp. 737-793.

[30] Klatt, D.H., "Software for a Cascade/Parallel Formant Synthesizer," *Journal of Acoustical Society of America*, 1980, **67**, pp. 971-995.

[31] Klatt, D.H., "Review of Text to Speech Conversion for English," *Journal of the Acoustical Society of America*, 1987, **82**, pp. 737-793.

[32] Krishnamurthy, A.K. and D.G. Childers, "Two Channel Speech Analysis," *IEEE Trans. on Acoustics, Speech and Signal Processing*, 1986, **34**, pp. 730-743.

[33] LDC, *Interactive Speech Synthesizer Comparison Site*, 2000, http://morph.ldc.upenn.edu.

[34] Maachi, M., "Coverage of Names," *Journal of the Acoustical Society of America*, 1993, **94**(3), pp. 1842.

[35] Maxey, H., *Smithsonian Speech Synthesis History Project*, 2000, http://www.mindspring.com/~dmaxey/ssshp/.

[36] McGurk, H. and J. MacDonald, "Hearing Lips and Seeing Voices," *Nature*, 1976, **264**, pp. 746-748.

[37] Mermelstein, P. and M.R. Schroeder, "Determination of Smoothed Cross-Sectional Area Functions of the Vocal Tract from Formant Frequencies," *Fifth Int. Congress on Acoustics*, 1965.

[38] Moulines, E. and F. Charpentier, "Pitch-Synchronous Waveform Processing Techniques for Text-to-Speech Synthesis Using Diphones," *Speech Communication*, 1990, **9**(5), pp. 453-467.

[39] Moulines, E. and W. Verhelst, "Prosodic Modifications of Speech" in *Speech Coding and Synthesis*, W.B. Kleijn and K.K. Paliwal, eds. 1995, Elsevier, pp. 519-555.

[40] Nye, P. and J. Gaitenby, *The Intelligibility of Synthetic Monosyllabic Words in Short, Syntactically Normal Sentences*, 1974, Haskins Laboratories.

[41] Olive, J., "Rule Synthesis of Speech from Dyadic Units," *Int. Conf. on Acoustics, Speech and Signal Processing*, 1977, Hartford, CT, pp. 568-570.

[42] Olive, J.P., A. Greenwood, and J.S. Coleman, *Acoustics of American English Speech: a Dynamic Approach*, 1993, New York, Springer-Verlag.

[43] Oliveira, L., "Estimation of Source Parameters by Frequency Analysis," *Proc. of the Eurospeech Conf.*, 1993, Berlin, pp. 99-102.

[44] Plumpe, M. and S. Meredith, "Which Is More Important in a Concatenative Text-to-Speech System: Pitch, Duration, or Spectral Discontinuity," *Third ESCA/COCOSDA Int. Workshop on Speech Synthesis*, 1998, Jenolan Caves, Australia, pp. 231-235.

[45] Roucos, S. and A. Wilgus, "High Quality Time-Scale Modification of Speech," *Int. Conf. on Acoustics, Speech and Signal Processing*, 1985, pp. 493-496.

[46] Sagisaka, Y. and N. Iwahashi, "Objective Optimization in Algorithms for Text-to-Speech Synthesis" in *Speech Coding and Synthesis*, W.B. Kleijn and K.K. Paliwal, eds., 1995, pp. 685-706, Elsevier.

[47] Santen, J.V., "Combinatorial Issue in Text-to-Speech Synthesis," *Proc. of the Eurospeech Conf.*, 1997, Rhodes, Greece, pp. 2511-2514.

[48] Sproat, R., *Multilingual Text-To-Speech Synthesis: The Bell Labs Approach*, 1998, Dordrecht, Kluwer Academic Publishers.

[49] Stylianou, Y., "Assessment and Correction of Voice Quality Variabilities in Large Speech Databases for Concatenative Speech Synthesis," *Int. Conf. on Acoustics, Speech and Signal Processing*, 1999, Phoenix, AZ, pp. 377-380.

[50] Stylianou, Y., J. Laroche, and E. Moulines, "High Quality Speech Modification Based on a Harmonic + Noise Model," *Proc. Eurospeech*, 1995, Madrid.

[51] Syrdal, A., A. Conkie, and Y. Stylianou, "Exploration of Acoustic Correlates in Speaker Selection for Concatenative Synthesis," *Int. Conf. on Spoken Language Processing*, 1998, Sydney, Australia, pp. 2743-2746.

[52] Syrdal, A., *et al.*, "Voice Selection for Speech Synthesis," *Journal of the Acoustical Society of America*, 1997, **102**, pp. 3191.

[53] Turing, A.M., "Computing Machinery and Intelligence," *Mind*, 1950, **LIX**(236), pp. 433-460.

[54] Voiers, W., A. Sharpley, and C. Hehmsoth, *Research on Diagnostic Evaluation of Speech Intelligibility*, 1975, Air Force Cambridge Research Laboratories, Bedford, MA.

[55] Yi, J., *Natural Sounding Speech Synthesis Using Variable-Length Units*, Masters' Thesis in EECS Dept. 1998, MIT, Cambridge, MA.

PART V

SPOKEN LANGUAGE SYSTEMS

CHAPTER 17

Spoken Language Understanding

Formal methods for describing sentences are discussed in Chapter 11. While the context-free grammars and *n*-gram models have mathematically well-understood formulations and bounded processing complexity, they are only partial aids in interpreting semantic meaning of the sentences. Suppose a recognizer correctly transcribes a series of spoken words into the written form—the system still has no idea what to do, because there is often no direct mapping between a sequence of words (or the syntactic structure of the sentence) and the functions that the system provides. The problem can also be approached from the opposite direction, i.e., solving the recognition problem itself may require semantic analysis, or domain and language knowledge for perplexity reduction.

What is meant by *meaning* or *understanding*? We could define it operationally: understanding is when a computer we interact with understands our desires and delivers the goods. Or we could define it propositionally: the computer has an accurate and unambiguous representation of *who did what to whom* corresponding to a real-world situation. In practice, the concept of understanding is situation dependent, and both conceptions above have their places. Meaning is often a constellation that emerges from a conversational environment.

There are four main interacting areas in *spoken language understanding* (SLU) systems from which meaning arises:

- **Intent**: goals of listener and speaker in the interaction
- **Context**: the pressures, opportunities, interruptions, etc. of the interaction scene and communication media
- **Content**: the propositional or literal content of each utterance and the discourse as a whole
- **Assumptions**: what each participant can assume about other participants' mental state, abilities, limitations, etc.

In this chapter we take a functional view of SLU systems, where the basic principle is to link linguistic expressions to concrete real-world entities. Currently, only with systems that are restricted to limited domains can understanding be attempted in practice. The domain restrictions allow the creation of specific, highly restricted language models and fully interpretable semantic descriptions that enable high accuracy and usability. Such systems are in contrast to speech recognition approaches that use large dictionaries, but make relatively *loose* or probabilistic predictions of word sequences for general dictation/transcription.

The need for spoken language understanding is double-edged. We generally want more than a string of word choices as a system's output. Instead, we want some interpretation of the word string that helps in accomplishing complex tasks. At the same time, being able to determine *what makes sense in context*, what is more or less likely as a speaker's input, could make a major contribution toward improving speech recognition word accuracy and search efficiency. SLU systems that combine the semantic precision of grammars with the probabilistic coverage of statistical language models can guide recognition and simultaneously control interpretation.

Figure 1.4 in Chapter 1 illustrates a basic SLU system architecture. The SLU problem can be broadly viewed as yet another pattern recognition problem. Namely, given a speech input **X**, the objective of the system is to arrive at actions **A** (including dialog messages and necessary operations) so that the cost of choosing **A** is minimized. Assuming uniform cost, the optimal solution, known as the maximum a posteriori (MAP) decision, can be expressed as

$$\begin{aligned} \mathbf{A}^* &= \arg\max_{\mathbf{A}} P(\mathbf{A} \mid \mathbf{X}, S_{n-1}) \\ &\approx \arg\max_{\mathbf{A}, S_n} P(\mathbf{A} \mid S_n) \sum_{\mathbf{F}} P(S_n \mid \mathbf{F}, S_{n-1}) P(\mathbf{F} \mid \mathbf{X}, S_{n-1}) \end{aligned} \tag{17.1}$$

where **F** denotes semantic interpretation of **X** and S_n, the discourse semantics for the nth dialog turn.

Based on the formulation in Eq. (17.1), a dialog system consists of three pattern recognition components:

- **Semantic parser**—use semantic model $P(\mathbf{F} \mid \mathbf{X}, S_{n-1})$ to convert **X** into a collection of semantic objects **F.** This component is often further decomposed into *speech recognition* module (converting speech signal **X** into textual sen-

tence **W**) and *sentence interpretation* module (parsing sentence **W** into semantic objects **F**). Since the collection of semantic objects **F** is in the linguistic level, it is often referred to as surface semantics.

- **Discourse analysis**—use discourse model $P(S_n \mid \mathbf{F}, S_{n-1})$ to derive new dialog context S_n based on the per-turn semantic parse **F** and the previous context S_{n-1}.
- **Dialog manager**—iterate through the possible actions and pick the most suitable one. The quantitative measures governing operations for dialog manager is called the *behavior model,* $P(\mathbf{A} \mid S_n)$.

The pattern recognition framework can be generalized to multimodal systems as well. For input other than speech signal, you only need to replace the input **X** in the semantic parser with input from an associated modality, e.g., **X** could be input from keyboard typing, mouse clicking, pen input, video, etc. As long as the new semantic parser (replacing speech recognizer and sentence interpretation modules in Figure 1.4) can convert it into appropriate semantic representation, the rest of the system can be identical. Similarly, for different output modality, you just need to replace *message generation* and *text-to-speech* modules with a new rendering mechanism.

In this chapter we first describe the characteristics of spoken languages in comparison with written languages. The structure of dialog is discussed in Section 17.2. Understanding is the most fundamental issue in the field of *artificial intelligence*. The kernel of understanding lies on the representation of semantics (knowledge). Several state-of-the-art semantic representation schemes are discussed in Section 17.3. Based on the architecture of SLU systems illustrated in Chapter 1 (Figure 1.4), major modules are discussed in detail, with the Dr. Who SLU system serving as an example to illustrate important issues.

17.1. WRITTEN VS. SPOKEN LANGUAGES

To construct SLU systems, we need to understand the characteristics of spoken languages. It is worth thinking about possible differences between spoken and written use of language that could be relevant to developing spoken language systems. The following is a typical example of two-agent, task-oriented dialog in action:

Sys: Flight reservation service, how can I help you?
User: One ticket to Honolulu, please
Sys: Anchorage to Honolulu, when would you like to leave?
User: Next Thursday
Sys: Next Tuesday, the 30th of November; and at what time?
User: No, Thursday, December 2nd, late in the evening, and make it first class.
Sys: OK, December 2nd United flight 291, first class. Will you need a car or hotel?
User: No.

17.1.1. Style

In both spoken and written forms, a communicative setting is established. Both forms involve participants. In the case of written language, we normally expect passivity on the part of the addressee(s), though with e-mail bulletin boards, Web chat rooms, and the like, this assumption can be challenged. The communicative event emerges from personal characteristics of the participants—their mood, goals, and interests. Communication depends both on the actual world knowledge and shared knowledge of the participants and on their *beliefs* about one another's knowledge. Communication can be influenced by the setting in which it takes place, whether in spoken or written mode. Also, different subchannels of supportive communication, such as visual aids, gesture, etc., may be available.

A number of grammatical and stylistic attributes have been found to distinguish conversational from written forms. Biber's analysis [8] distinguishes not only a dimension of modality, but also formality; for example a panel discussion is a relatively formal, yet spoken, modality. Some typical features for which distinctions can be measured include the number of passives, the number of pronouns, the use of contractions, and the use of nominalized forms.[1] An example of the grammatical and stylistic difference continuum that Biber uses is illustrated in Figure 17.1. The variation can be measured along multiple orthogonal scales for different genres. In the SLU case, style can be orthogonal to the modality (dialog or dictation, spoken or written). A crossover case is speech dictation used to create a written document that may never be orally rendered again.

Fortunately, much of the disjuncture between spoken and written forms in grammatical style and lexical choice can be handled by training task-specific and modality-specific language models for the recognizer. For this, only the data need vary, not necessarily the modeling methods. In Figure 17.1, the right-hand side is toward the spoken style, while the

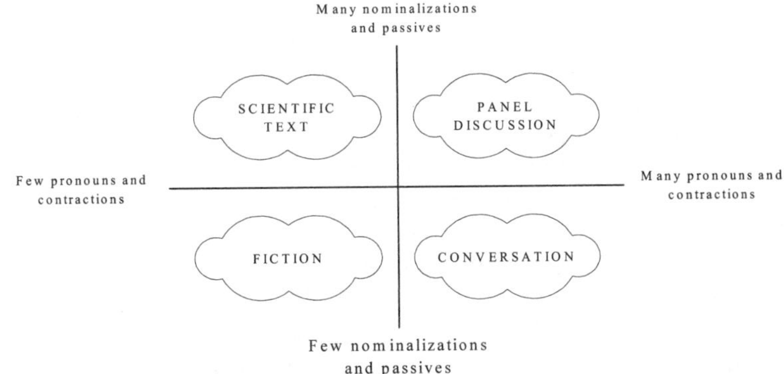

Figure 17.1 Dimensions of written vs. spoken language variation.

[1] Nominalization is a stylistic device whereby a main verb is converted to a noun. For example, *The dean rejected the application unexpectedly* may become: *The rejection of the application by the dean was unexpected.*

left-hand side is toward the written one. The difference in styles is best illustrated by the fact that the statistical *n*-gram trained from newspaper text exhibits a very high perplexity when evaluated on conversational Air Travel Information Service (ATIS) texts.

17.1.2. Disfluency

Another issue for spoken language processing is *disfluency*. Spoken dialogs show a large set of problems such as interruptions, corrections, filled pauses, ungrammatical sentences, ellipses, and unconnected phrases. These challenges are unique to spontaneous spoken input and represent a possible further degradation of speech recognizer performance, as current systems often rely on acoustic models trained from read speech, and language models trained on written text corpora. When speech input is used as dictation for document creation, of course, the models would presumably be most appropriate.

There are a number of types of disfluencies in human-human dialog, and, possibly to a lesser extent, in human-computer dialog as well. The more common types of nonlinguistic disfluencies are listed below:

- Filled pauses: *um*
- Repetitions: *the – the*
- Repairs: *on Thursday – on Friday*
- False Starts: I like – what I always get is …

Early work in discourse led to the determination that discourses are divided into *discourse segments,* much as sentences are divided into phrases [18]. In the experiments of [46], CART methods (see Chapter 4) were used to predict occurrence and location of each of the above types of disfluency. A tree was trained from labeled corpora for each type, and the resulting system classified each interword boundary as having no disfluency or one or more of the above types. The feature types used to derive the classification questions included duration of vocalic regions and pauses, fundamental frequency and its derivatives, signal-to-noise ratios, and distance of the boundary from silence pauses. The basic classification task consisted in selecting each of the four disfluency types listed above (D), given the list of prosodic features ($\mathbf{X}$), by computing the maximum of $P(D \mid \mathbf{X})$. When decision trees were used to supplement the language-model scoring of hypothesis word strings, performance improved.

A number of intriguing regularities were also observed in this work. For example, it was noted that the marked (less common) pronunciation of *the* - */dh iy/* was often used just prior to a production problem, e.g., a disfluent silent pause. Also, it has been noted that the leftmost word of a major phrase or clause is likely to be repeated, as in their example, "*I'll I'll do what I can.*" Continued research on disfluencies may contribute an important secondary knowledge source to supplement text-based language models and *read speech* acoustic models in the future.

17.1.3. Communicative Prosody

Prosodic attributes of utterances, such as fundamental frequency and timing (cf. Chapter 15), are crucial cues for detecting disfluency. However, prosody can be deliberately manipulated by speakers for deep communicative purposes as well. The speaker may intentionally or subconsciously manipulate the fundamental frequency, timing, and other aspects of voice quality to communicate attitude and emotion. If a conversational interface is equipped to recognize and interpret prosodic effects, these can be taken into account for understanding.

In addition to serving as a disfluency detector, as described above, prosodic analysis modules could aid recognition of:

- Utterance type—declarative, *yes-no* questions, *wh*-question, etc.
- Speech act type—directive, commissive, expressive, representative, declarative, etc. Different speech acts will be described in Section 17.2.2.
- Speaker's attentional state.
- Speaker's attitude toward his/her utterance(s).
- Speaker's attitude to system presentations.
- Speaker's mood or emotional state.

Consider the simple utterance *OK*. This may be used along a range of attitudes and meanings, from *bored contempt*, to *enthusiastic agreement*, to *questioning* and *uncertainty*. The interpretation will depend on both the dialog state context of expectations-to-date and the prosody. Generally, a higher relative F0 in a wider range correlates with submission, involvement, questioning, and uncertainty, while a lower relative F0 in a narrower range correlates with dominance, detachment, assertion, and certainty. Even though acknowledgement words such as *yeah* and *ok* are potentially ambiguous among: true agreement; intention of the listener to initiate a new turn; and simple passive encouragement from listener to speaker, the system may rely on a longer duration and greater pitch excursion of a lexical item such as *yeah* or *ok* to hypothesize genuine agreement with a speaker statement, as opposed to mere acknowledgement.

In addition to correlating with speech acts, F0 and timing can be used to demarcate utterance and turn segments. For example, certain boundary pitch movements and phonemic lengthening systematically signal termination of clauses. In general, a fall to the very bottom of a speaker's range, in a prepausal location, coincides with a clause or sentence boundary. A sharp upturn preceding a significant silence gives an impression of incompletion, perhaps signaling a yes-no question, or may signal an intention by the speaker to carry on with further information, as in the case of list intonation.

The disfluent and prosodic characteristics of the conversational speech are in general very distinct from those of read speech. Thus, we often refer conversational speech as *spontaneous* speech.

17.2. DIALOG STRUCTURE

The analysis methods discussed in Chapter 11 are focused on single sentences. They are steps along the way, helping to map vague and ambiguous natural language constructions into precise *logical forms* of propositions. In reality, however, the communicative function of language is not a simple, uncomplicated assembly of discrete logical propositions derived from sentences in a one-to-one fashion. In *discourse*, each sentence or utterance contributes to a larger abstract information structure that the user is attempting to construct. Sometimes feedback is directly available to the user or can be inferred. These considerations take us beyond the process of mapping of isolated utterances into logically structured propositions (with simple truth-values).

A set of principles, known collectively as the cooperative principle, is introduced by Grice [9]. It consists of a set of conversational maxims, the violation of which may lead to a breakdown in communication.

GRICE'S MAXIMS

Quantity: speaker tries to be as informative as possible, and gives only as much information as needed
Quality: speaker tries to be truthful, and does not give information that is false or that is not supported by evidence
Relevance: speaker tries to be relevant, and says things that are pertinent to the discussion
Manner: speaker tries to be as clear, as brief, and as orderly as possible, and avoids obscurity and ambiguity

In general, there are five main domains of operation that must be modeled for intelligent conversation systems, although all these areas are linked:

- **Linguistic forms**: all the knowledge a human-computer dialog system requires to perform semantic and syntactic analysis and generation of actual utterances.
- **Intentional state**: goals related to both the task (*Show me all flights …)*, and the dialog process itself (*Please repeat …*) of the users.
- **Attentional state**: the set of entities at any point in time that can be felicitously discussed and referred to, i.e., the main topic of any stage of interaction.
- **World knowledge**: common sense knowledge and inference. Examples include temporal and spatial concepts and the relation of these to linguistic forms.
- **Task knowledge**: all information relevant to achieving the user's goal in a complete, correct, and efficient fashion.

Human-computer dialog is multiagent communication. Each agent has to form a notion of the other's beliefs, desires, and knowledge, all of which underlie their intentions, plans, and actions. In a limited application, deep inference may not be possible, and the system may have more or less *hardwired* assumptions about the user, the interaction, and the flow of action. An interaction may be controlled by the system's own rigid schedule of information acquisition. In the research community, such a dialog system—always leading the interaction flow control and not allowing the user to digress—is called *system initiative*. On the other hand, a dialog system is called *user initiative* if it always lets the user decide what to do next. It is often more natural, however, to allow for *mixed initiative* systems, where interaction starts with a user's query or command and the system attempts to derive, via inference or further questioning of the user, all information needed to understand and process a complete transaction. When the user knows clearly what he wants and the system has no trouble catching up, the user is in the driver's seat. However, when the system detects that the user is in a state of confusion, or when it has trouble getting user's intention, the machine will offer guidance or negotiate with the user to steer the dialog back on track.

Whether it is system-initiative, user-initiative, or mixed-initiative, however, the fundamental structure of dialog consists of initiative-response pairs as indicated in Figure 17.2. The *Initiatives* (I) are often issued by users while the *Responses* (R) are issued by the system. As shown in Section 17.2.2, there are many types of Initiatives and Responses and there may also be higher-order structure subsuming a number of I/R pairs in a dialog.

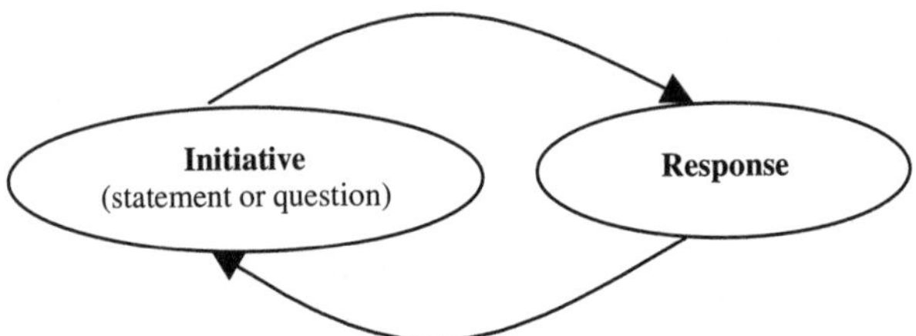

Figure 17.2 The fundamental structure of dialog: initiative and response.

17.2.1. Units of Dialog

The words uttered in a dialog are the surface manifestation of a complex underlying layer of participants' shared interaction knowledge and desires, even when one participant is a computer simulation. It is natural to assume that the *sentence* is a clear and simple chunking unit for dialog, by analogy with written communication. However, since sentences are artificially delimited in written text, researchers in dialog communication usually speak of the *utterance* as the basic unit. An Initiative or Response could consist of one or more utterances. The utterance, however, is not necessarily trivial to define.

It is tempting to posit an equivalence of the notion utterance with *turn*, i.e., an uninterrupted stream of speech from one participant in a dialog. This formulation makes it easy to segment dialog data into utterance units—they are just each speaker's turns. The downside is that this kind of *utterance* possibly spans grammatical units that really do have some rough correspondence to traditional sentences (predicate-argument structures), and to which much of the hard-won gains in natural language processing would apply fairly directly. Thus, the use of *turn* as synonymous with utterance unit is probably too broad, though the *turn* may be independently useful for higher-level segmentation.

Turns are building blocks for constructing a common task-oriented understanding among participants. This process is called *grounding*, a set of discourse strategies by which dialog actors (humans in most current research) attempt to achieve a common understanding, and come to feel confident of the other participants' understanding. In other words, conversational partners are finding or establishing *common ground.*

Turns may have their own typology. For example, a *speaking turn* conveys new information, while a *back-channel turn* is limited to acknowledgement or encouragement, such as *OK, really?*, etc. The turns themselves consist of linguistic substructures, such as sentences, clauses, and phrases. If we assume that turns can be segmented, by grammatical and/or prosodic criteria, into utterances, we can then begin to explore distinct types of utterances, their properties, and their communicative functions.

Finally, dialogs are not flat streams of unrelated turns or utterances. The utterances that make up a dialog have higher-order affiliations with one another. A discourse segment would thus consist of groups of related utterances organized around a common dialog subtask, perhaps spanning turns.

17.2.2. Dialog (Speech) Acts

In simpler applications, the amount and sophistication of world knowledge can be kept to a minimum, and attentional state can be modeled simply as the complete set of task-specific entities. A layer of structure has therefore been sought to link linguistic forms with task knowledge or operations in a theoretically appropriate fashion, which also yields an implicit understanding of intentional state. This is necessary because the function of utterances in discourse cannot be predicted strictly on the basis of their surface grammatical form. The layer of structure that can abstract away from linguistic details and can map well to formulation of goals is called *dialog acts* [42]. Dialog acts are also often referred to as *speech acts* that group infinite families of surface utterances into abstract functional classes. They are traditionally classified into five broad categories:

- **Directive**: The speaker wants the listener to do something.
- **Commissive**: The speaker indicates that he or she will do something in the future.
- **Expressive**: The speaker expresses his or her feelings or emotional response.
- **Representative**: The speaker expresses his or her belief about the truth of a proposition.
- **Declarative**: Speaker's utterance causes a change in external, nonlinguistic situation.

Table 17.1 A simple dialog analyzed with dialog acts.

Utterance	**Form**	**Function**
Do you have the butter?	Y/N-question	REQUEST-ACT
Sure. (passes butter)	statement	COMMIT-TO-ACTION-ACT

While this analysis is somewhat coarse, speech act theory has influenced all current work on human-computer dialog, except the very simplest and most rigid systems. Because dialog functions can be realized with a bewildering variety of linguistic forms, researchers have posited systems of functional abstractions. Speech acts are functional abstractions over variation in utterance form and content. Declare, request, accept, contradict, withdraw, acknowledge, confirm, and assert are all examples of speech acts—things we are attempting to do with speech. An example of dialog acts and their relation to syntactic form is shown in the two-turn dialog in Table 17.1.

The relation between speech acts and linguistic forms (utterances) is a many-to-many mapping. That is, a single linguistic form, such as *OK*, could realize a large number of speech acts, such as *request for acknowledgment* or *confirm*, etc. Likewise, a single speech act, such as *agreement*, could be realized by a variety of linguistic forms, such as *ok, yes, you bet*, etc. In a particular application, special task-specific speech acts may be used to supplement the universal inventory.

Tagging of dialog utterance data with speech-act labels can add useful information for training models. There are a number of ways that dialog-act analysis could be useful:

- *Speech recognition*: Given a history, we can predict the most likely dialog act type for the *next* utterance, so that specialized language models may be applied.
- *Spoken language understanding*: Given a history, and a transcription/parse of the current utterance, we can identify the user's intentions, so that the system can respond appropriately.
- *Semantic authoring*: It is tedious for each team designing or customizing a new application area for SLU to have to wrack their brains for all the ways a given generic function, such as *request* or *confirm*, might be realized linguistically. Libraries of speech acts (form-to-function mappings) may reduce the work in new-domain adaptation of systems.

An example of a practical dialog tagging system that could be the foundation of speech-act analysis is the *Dialog Act Markup in Several Layers* (DAMSL) system [14], which has been used and adapted for a variety of projects. This is a system for annotating dialog transcriptions with speech-act labels and corresponding structural elements. The structuring is based on a loose hierarchy of: discourse segment, turn, utterance, and speech act. The tags applied to utterances fall into three basic categories:

- **Communicative Status**: records whether the utterance is intelligible and whether it was successfully completed. It is mainly used to flag problematic utterances that should be used for data modeling only with caution—Uninterpretable, Abandoned, or Self-talk. *Uninterpretable* is self-explanatory. *Abandoned* marks utterances that were broken off without, crucially, adding any information to the dialog. *Self-talk* is a note that, while an utterance may contain useful information, it did not appear to be intentionally communicated. Self-talk can be considered reliable only when the annotator is working from speech data.
- **Information Level**: a characterization of the semantic content of the utterance. This is used to specify the kind of information the utterance mainly conveys. It includes *Task* (Doing the task), *Task-management* (Talking about the task), *Communication-management* (Maintaining the communication), and *Other-level*. Task utterances relate directly to the business of the transaction and move it toward completion. Task-management utterances ask or tell about the task, explain it perhaps, but do not materially move it forward. Communication-management utterances are about the dialog process and capabilities. The Other level is for unclear cases.
- **The Forward/Backward Looking Function**: how the current utterance constrains the future/previous beliefs and actions of the participants and affects the discourse. Forward Looking functions introduce new information or otherwise move the dialog or task completion forward, while Backward Looking Functions are tied to an antecedent, a prior utterance which they respond to or complete. This distinction is the DAMSL reflection of the common observation that dialogs have a tendency to consist of Initiation/Response pairs. The core of the system is the set of particular act types. The core Forward/backward Looking tags are listed in Table 17.2 and Table 17.3.

Table 17.2 Forward looking tags.

Forward Looking Tags	Example
assert	I always fly first class.
reassert	No, as I said, I always fly first class.
action-directive	Book me a flight to Chicago.
open-option	There's a red-eye flight tonight …
info-request	What time is it?, Tell me the time.
offer	I can meet at 3 if you're free.
commit	I'll come to your party.
conventional opening	May I help you?
conventional closing	Goodbye.
explicit-performative	Thank you, I apologize.
exclamation	Ouch! Darn!

Table 17.3 Backward looking tags.

Backward Looking Tags	Example
accept	(Will you come?) Yes. [and/or, I'll be there at 10.]
accept-part	(Will you come with your wife?) I'll come, she may be busy.
reject	(Will you come?) No.
reject-part	(Want fries and a shake with that burger?) Just the hamburger and fries, please.
maybe	Maybe.
signal-nonunderstanding	What did you say?
acknowledgment	OK.
answer	(Can I fly nonstop from Anchorage to Kabul?) No.

Multiple tags may appear on any given utterance. In the example shown in Figure 17.3, B's utterance is coded as opening the option of buying (from B), asserting the existence of the sofas, and functioning as an offer or solicitation.

Action-directive	`A: Let's buy the living room furniture first.`
Open-option/Assert/Offer	`B: OK, I have a red sofa for $150 or a blue one for $200`

Figure 17.3 A tagged dialog fragment.

The DAMSL system is actually more complex than the example demonstrated above, since subsets of the tags are grouped into mutually exclusive options for a given general speech function. For example, there is a general *Agreement* function, under which the *accept*, *accept-part*, *reject*, and *reject-part* tags are grouped as mutually exclusive options. Above the level of those groupings, however, a single utterance can receive multiple nonexclusive tags. For example, as illustrated in Figure 17.4, the assistant may respond with a countersuggestion (a kind of action-directive) that rejects part of the original command.

Action-directive	`utt1 oper: Take the train to Avon via Bath`
Action-directive/Reject-part(utt1)	`utt2 asst: Go via Corning instead.`

Figure 17.4 A tagged dialog fragment, showing that utterances can be tagged with multiple nonexclusive tags.

The prototypical dialog turn unit in simple applications would be the I/R pair *info-request/answer*, as in the interaction shown in Figure 17.5 between an operator (planner) and an assistant regarding railroad transport scheduling [1].

The example in Figure 17.5 illustrates a dialog for a railway-scheduling task. The turns are numbered T1–T4, the utterances within turns are also numbered sequentially, and the speaker identity alternates between oper: and asst:. The tagging is incomplete, because, for example, within the *ans*| sequence, each utterance is performing a function, asserting, acknowledging, etc. The example in Figure 17.6 is a more completely annotated fragment, omitting turn numbers.

```
info-req    T1 utt1 oper:   where are the engines?
ans|        T2 utt2 asst:   there's an engine at Avon
 |          T3 utt3 oper:   okay
 |          T4 utt4 asst:   and we need
               utt5 asst:   I mean there's another in Corning
```

Figure 17.5 A tagged dialog fragment in railroad transport scheduling.

```
info-req/assert      utt1 oper:  and it's gonna take us also an
                                 hour to load boxcars right
ans/accept(utt1)     utt2 asst:  right
assert               utt3 oper:  and it's gonna take us also an
                                 hour to load boxcars
accept(utt1)         utt4 asst:  right
```

Figure 17.6 A tagged dialog fragment, showing backward-looking utterances.

The example in Figure 17.6 shows backward-looking utterances, where the relevant antecedent in the dialog is shown (in parentheses) as part of the dialog coding.

More elaborate variants of DAMSL have been developed that extend the basic system presented here. Consider, for example, the SWITCHBOARD Shallow-Discourse-Function Annotation SWBD-DAMSL [27]. This project used a shallow discourse tag set of 42 basic tags (frequent composed tags from the large set of possible multitags) to tag 1155 5-minute conversations, comprising 205,000 utterances and 1.4 million words, from the SWITCHBOARD corpus of telephone conversations. Distributed by the Linguistic Data Consortium[2] [28], SWITCHBOARD is a corpus of spontaneous conversations that addresses the growing need for large multispeaker databases of telephone bandwidth speech. The corpus contains 2430 conversations averaging 6 minutes in length—in other words, over 240 hours of recorded speech, and about 3 million words of text, spoken by over 500 speakers of both genders from every major dialect of American English.

More detailed tags are added to DAMSL to create SWBD-DAMSL, most of which are elaborations of existing DAMSL broad categories. For example, where DAMSL has the simple category *answer*, SWBD-DAMSL has: *yes answer*, *no answer*, *affirmative non-yes*

[2] http://www.ldc.upenn.edu

answer, *negative non-no answers*, *other answers*, *no plus expansion*, *yes plus expansion*, *statement expanding y/n answer*, *expansions of y/n answers*, and *dispreferred answer*. SWBD-DAMSL is intended for the annotation and learning of structure in human-human dialog, and could be considered overkill as a basis for describing or constructing grammars for most limited-domain human-computer interactions of today. But the more sophisticated agent-based services of the future will need to assume ever-greater linguistic sophistication along these lines.

One fact noted by the SWBD-DAMSL researchers, which may not apply directly to task-directed human-computer interactions but which casts interesting light on human communication patterns, is that out of 1115 conversations studied, simple nonopinion statements and brief acknowledgements together constituted 55% of the conversational material! If statements of opinion (including simple stuff like *I think it's great!*), expressions of agreement (*That's right!*), turn breakoffs and no-content utterances (*So…*), and appreciative acknowledgements (*I can imagine.*) are added to this base, 80% of the utterances are accounted for. This relative poverty of types may bode well for future attempts to annotate and predict utterance function automatically. The DAMSL scheme is challenging to apply automatically, because it relies on complex linguistic and pragmatic judgments of the trained annotators.

17.2.3. Dialog Control

The system's view of how the dialog should proceed is embodied in its management strategy. Strategy is closely connected to the concept of *initiative* in dialog, meaning basically who is controlling the interaction. Different dialog initiatives are defined in Section 17.2. Initiative can be seen as a continuum from system controlled to user controlled. As background for the dialog management discussion, some important steps along this continuum can be identified:

- *System directs*—The system retains complete dialog control throughout the interaction. The system chooses the content and sequence of all subgoals and initiates any dialog necessary to obtain completion of information from the user for each transaction. This style is often referred as *system initiative*.
- *System guides*—The system may initiate dialog and may maintain a general plan, but the sequence of information acquisition from the user may be flexible, and system subgoals and plans may be modified in response to the user's input. This style is often referred as *mixed initiative.*
- *System inform*—The user directs the dialog and the system responds as helpfully as possible, which may include presentation of relevant data not specifically requested by the user but which the system believes could be helpful. This style also belongs to *mixed initiative*, though users control most of the dialog flows.
- *System accepts*—This is the typical human-computer interaction in traditional systems (whether it is a GUI-, command-line-, or natural language-based sys-

tem). The system interprets each command without any attempted inference of a deeper user plan, or recommendation of any suitable course of action. This style is referred as *user initiative.*

17.3. Semantic Representation

Most SLU systems require an internal representation of *meaning* that lends itself to computer processing. In other words, we need a way of representing semantic entities, which are used at every possible step. In general, an SLU system needs to deal with two types of semantic entities. The first type is *physical entities*, which correspond to the real-world entities. Such representation is often referred as knowledge representation in the field of artificial intelligence. The second type is *functional entities*, which correspond to a way of unambiguously representing the meaning or structure of situations, events, and concepts that can be expressed in natural language. Such representations are often similar to the *logical form* introduced in Chapter 2. Processing may include operations such as determining similarity or identity of events or entities, inference from a state of affairs to its logical consequences, and so on. Here, we briefly review some general properties of the common semantic representation frameworks.

17.3.1. Semantic Frames

Semantic objects are used to represent real world entities. Here, we assume that the domain knowledge conforms to a relational or objected-oriented database, of which the schema is clearly defined. We use the term *entity* to refer to a data item in the domain (a row in a database table), or a function (command or query) that can be fulfilled in the domain. A column in the database table is called an entity attribute, and each database table is given an entity type. Through a small subset of its attributes, an entity can be realized linguistically in many fashions. We call each of them a *semantic class*. For example, a person can be referred to in terms of her full name (*Angela*), a pronoun anaphora (*her*), or her relationships to others (*Christina's manager*). In this case, one can derive three semantic classes for the entity type.

Semantic classes can be viewed as a type definition to denote the objects and describe the relations that hold among them. One of the most popular representations for semantic classes is the *semantic frame* [31]—a type of representation in which a semantic class (concept) is defined by an entity and relations represented by a number of attributes (or slots) with certain values (the attributes are filled in for each instance). Thus, frames are also known as *slot-and-filler* structures.

We could, for example, define a generalized frame for the concept *dog*, with attributes that must be filled in for each particular instance of a particular dog. A type definition for the concept dog appears in Figure 17.7. Many different notational systems have been used for frames [51]. For these introductory examples, we use a simple declarative notation that should be fairly intuitive.

```
[DOG:]-
  [SUPERTYPE]->[mammal]
  [NAME]->()
  [BREED]->()
  [LOC]->()
  [Color]->()
```

Figure 17.7 A semantic frame representation for *dog.*

When we need to describe a particular dog, say *Lassie*, we create an *instance definition*, as shown in Figure 17.8. The knowledge base supporting a typical dialog system consists of a set of type definitions, perhaps arranged in an inheritance hierarchy, and a set of instances.

```
[DOG:]-
  [NAME]->(Lassie)
  [BREED]->(Collie)
  [LOC]->()
  [Color]-()
```

Figure 17.8 An instance of semantic frame *dog.*

Fillers in semantic frames can be attained by attachment of inheritance, procedures or default. Attributes in frame can typically be inherited, as the Lassie instance inherits mammalian properties from the DOG type definition. In some cases, properties of a particular dog may be dynamic. Sometimes *attached procedures* are used to fill dynamic slots. For example, the location of a dog may be variable, and if the dog has a *Global Positioning System* (GPS) chip in its collar, the LOC property could be continually updated by reference to the GPS calculations. Furthermore, procedures of the type `when-needed` or `when-filled` can also be attached to slots. Finally, some default value could provide a typical value for a slot when the information for that slot is not yet available. For example, it might be appropriate to set the default color for *dog* frame as white when such information is not available. For frames without a default-value slot, it is natural to define *mandatory* slots (slots' values must be filled) and optional slots (slots could have null value). For the *dog* frame, it is reasonable to assume the NAME slot should be mandatory while the COLOR slot can be optional.

Often *descriptions* can be attached to slots to establish constraints within or between frames. Description may have connectives, co-referential (description attached to a slot are attached to another) and declarative conditions. For example, the *return-date* slot of a *round-trip* itinerary frame must be no earlier than the *departure-date* slot, and this constraint can be specified by descriptions in both slots. Descriptions can also be inherited and are often implemented by a special procedure (different from the slot-filling procedure) attached to the slot.

The main motivation for having multiple semantic classes for each entity type is to better encapsulate the language, semantic, and behavior models based on the domain knowledge. While the entity relationships capture the domain knowledge, the semantic class hierarchy represents how knowledge can be expressed in the semantics of a language and thus can cover linguistic variation. The concept of semantic objects/classes is similar to that of objects/classes in modern *object-oriented programming*. The semantic classes in Dr. Who [59, 60] are a good illustration of borrowing some important concepts from object-oriented programming to enhance the effectiveness and efficiency of using semantic objects/classes to represent domain knowledge and linguistic expressions.

The semantic grammar used in the Dr. Who Project [58] contains the definitions of semantic classes that refer to real-world or functional entities. A semantic class is defined as a semantic frame containing a set of slots that need to be filled with terminal (verbatim) words or with recursive semantic class objects. For example, *ByRel* is a semantic class that has the type PERSON. The semantic grammar specifies that it has two slots—one has to be filled with an object of a semantic class having the type PERSON, and the other has to be filled with an object of a semantic class having the type P_RELATION. On the other hand, the syntax grammar for this semantic class is specified by the <cfg> tags. Within <cfg> tags, several production rules can be specified to provide linguistic constraints (orders) of possible expressions for this semantic class. The syntactic aspect of semantic classes will be described further in Section 17.4.1.

17.3.1.1. Type Abstraction

As described above, a physical *entity* is an element in the real world that an application has to deal with and wishes to expose to the user via natural language. Since a physical entity can be referred to in many different ways, different semantic classes may have the same type. In Figure 17.9, a person can be referred to in terms of his name (*Peter*) or his relation to another person (*Peter's manager*); therefore, both semantic classes ByName and ByRel can share the same type, PERSON.

Semantic classes are designed to separate the essential attributes of a semantic object from its physical realizations. A semantic class may refer to an entity, and the entity is called the type of the semantic class. The attributes of a semantic class can, in turn, be semantic classes themselves. The concept behind semantic classes is identical to the mechanism known as type abstraction commonly employed in software engineering using a strongly typed programming language. Semantic class can be recursive, as demonstrated in Figure 17.9; a ByRel semantic class of type PERSON contains an attribute of PERSON type. Since the entities can be *nested*, i.e., a database column can in turn refer to another table, an attribute in the semantic class can also be an entity type. From an understanding point of view, a semantic class is an abstraction of the collection of semantic objects that have the same attributes and usually can be expressed, and hence be understood, in similar manners. Under this view, a semantic object is just an instantiation.

Another argument for type abstraction is that the multitude of semantic objects is usually a result of the numerous ways and perspectives that can be used to describe a physical entity. Quite often in an understanding system it is more important to correctly identify the entity of interest than to capture the mechanism that describes it. For instance, one may refer to a person by his name, job function, relations to others, or, in a multimodal environment, by pointing to his photo on a display. All these references lead to semantic objects that are apparently distinct yet should be associated with the same physical entity. Accordingly, it is often useful to segregate the conceptual manifestation and its realizations into different layers of abstraction so that the semantic objects can be better organized and managed. Type abstraction allows the discourse sentence interpretation module to perform robust parsing, since sentence fragments can be parsed into its semantic class type that can be filled into slots with the same correspondent semantic type, as discussed in Section 17.4.1.

```
<!-- semantic class definition for ByRel that has type PER-
SON -->
<class type="PERSON" name="ByRel">
    <slot type="PERSON" name="person"/>
    <slot type="P_RELATION" name="p_relation"/>
    <cfg>
          <prod> [person] [p_relation] </prod>
          <prod> [p_relation] of [person] </prod>
    </cfg>
</class>
<!-- semantic class definition for ByName that has type PERSON too -->
<class type="PERSON" name="ByName">
    <slot type="FIRSTNAME" name="firstname"/>
    <slot type="LASTNAME" name="lastname"/>
    <cfg>
          <prod> [firstname] [lastname] </prod>
          <prod> [firstname] </prod>
          <prod> [lastname] </prod>
    </cfg>
</class>
<!-- semantic class definition for FIRSTNAME and LASTNAME -->
<verbatim type="FIRSTNAME"
    <cfg>
          <prod> john | john's | peter … </prod>
    </cfg>
</verbatim >
<verbatim type="FIRSTNAME"
    <cfg>
          <prod> smith | smith's | shaw … </prod>
    </cfg>
</verbatim >
<!-- semantic class definition for P_RELATION -->
<verbatim type="P_RELATION"
    <cfg>
          <prod> manager | father | mother | … </prod>
    </cfg>
</verbatim >
```

Figure 17.9 The semantic classes of type PERSON as implemented in Dr. Who.

Finally, type abstraction provides a unified framework for resolving *relative expressions* in the discourse analysis module. Type matching often serves to impose strong constraints between real-world entities and relative expressions. The resolution of relative expressions is discussed in Section 17.5.

17.3.1.2. Property Inheritance

Introducing inheritance into the semantic class hierarchy further augments the multilayer abstraction mentioned above. Class *A* is said to inherit or be derived from class *B* if class *A* possesses all the attributes of class *B*. In this case, class *A* is called the derived class and class *B* the base class. Inheritance is a mechanism to propagate knowledge and properties through the structural relationships of semantic classes. It is crucial for many types of intelligent behavior, such as deducing presumed facts from general knowledge and assuming default values in lieu of explicit and specific facts.

Perhaps the strongest motivation to employ inheritance is to facilitate the multilayer abstraction mentioned above. Very often, a base class is constructed with the general properties of a type of semantic objects, and a collection of more specific classes are derived from the base class to support the various embodiments of the underlying type of the semantic objects. For example, a semantic class hierarchy for the reference to a person can have the methods (e.g., by name, job function) and the media (e.g., speech, handwriting) of reference as the first layer of derived classes. One can then cross-match the viable means (e.g., by name via speech, by name via handwriting) and develop the second layer of derived classes for use in the real applications.

17.3.1.3. Functionality Encapsulation

The goal of abstraction is to reduce the complexity in describing the world—in this case, the semantic objects and their relations. One can inspect the quality of abstraction by examining the extent to which the constructs, i.e., semantic classes, are self-contained, and how proliferating they have to become in order to account for novel scenarios. Studies in data structure and software engineering propose the notion of encapsulation, which suggest that individual attributes have local rather than global impacts. This principle also serves as a guideline in designing the semantic class.

Semantic class encapsulation can be elaborated in two aspects: syntactic and semantic. The syntactic encapsulation refers to the constraint that each attribute in a semantic class can only have relations to others from the same class. The collection for these relations is called the *semantic grammar,* which specifies how a semantic object of this type can be identified. In Figure 17.9, the tag <CFG> specifies how the semantic class can be referred to syntactically via a context-free grammar (CFG). For the class ByRel, the specified syntax indicates that expressions like *Peter's manager* and *manager of Peter* are legal references to semantic class ByName. The semantic encapsulation, on the other hand, dictates the actions and the discourse context under which they may be taken by a semantic class. This is discussed further in Section 17.5.

As described in 17.2.2, it is a nontrivial task to determine the types of speech acts. The semantic frame is an abstraction of the speech acts, the domain knowledge, and sometimes even the application logic. Once we have this rich semantic representation, how to parse spoken utterances into the semantic frames becomes the critical task. Nonetheless, the combination of semantic frames and the semantic parser alleviates the need for a dedicated module for determining speech acts.

Semantic frames and associated robust parsing (described in Section 17.4.1) have been widely used in spoken language understanding. For detailed description of semantic classes and frames, you can refer to [58, 65].

17.3.2. Conceptual Graphs

The semantic-representation requirement has led to development of a proposal to standardize the logical form that may form the basis of the internal semantics and semantic interchange of natural language systems, including dialog processing, information retrieval, and machine translation. The proposal is based on conceptual graphs derived from Charles Sanders Peirce [38] and the various types of semantic networks used in artificial intelligence research.

The *conceptual graph* (CG) proposal [53] specifies the syntax and semantics of conceptual graphs as well as formats for graphical and character-based representation and machine-based exchange. In the terms of the proposed standard, a conceptual graph (CG) is an abstract representation for logic with nodes called *concepts* and *conceptual relations*, linked together by arcs. In the graphical representation of a CG, concepts are represented by rectangles, and conceptual relations are represented by circles or ovals. The ordinary phrasing for the association of relations (circles) to concepts (rectangles) is *has a(n)* for arrows pointing toward the circle and *is a(n)* for arrows pointing away.

Figure 17.11 illustrates a conceptual graph for the sentence *Eric is flying to Boston by airplane*. The mnemonic meaning of the arrows is: *Fly* has an agent who is a person, *Eric*, and a destination *Boston*. The proposal also specifies a linear form, as shown in Figure 17.10. In the form, concepts are in square brackets and conceptual relations are in parentheses. The hyphen means that relations of a given concept continue on subsequent lines, as shown in Figure 17.11.

```
[Fly] -
     (Agent)->[Person: Eric]
      (Dest)->[City: Boston]
      (Inst)->[Airplane]
```

Figure 17.10 A linear form representation of *Fly* has an agent who is a person, *Eric*, and a destination *Boston*.

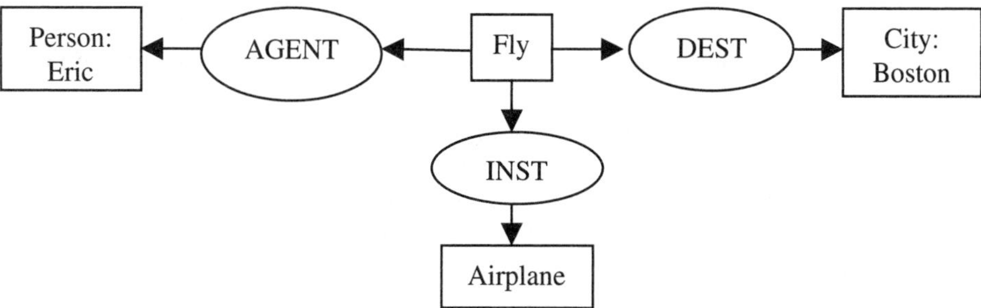

Figure 17.11 CG display form for *Eric is flying to Boston by airplane* [53].

Each concept has a type and a (possibly empty) referent. An empty referent means that at least one, unspecified example of the type is assumed to exist somewhere (an existential quantifier). So, in Figure 17.10, the type is present, but the referent is left unspecified. In an application, the referent can be completed by referring to a train-schedule database and inserting a particular instance of a scheduled train departure time, location, and number. The *valence* of a relation is the number of required concepts that it links. For example, as shown in Figure 17.12, the relation *between* would be a conceptual relation of valence 3, because typically (something/somebody) is *between* one (something/somebody) and another (something/somebody), as in the familiar English idiom "*somebody is between a rock and a hard place*" (meaning, *to be in great difficulty*). This corresponds to the linear form, as shown in Figure 17.13.

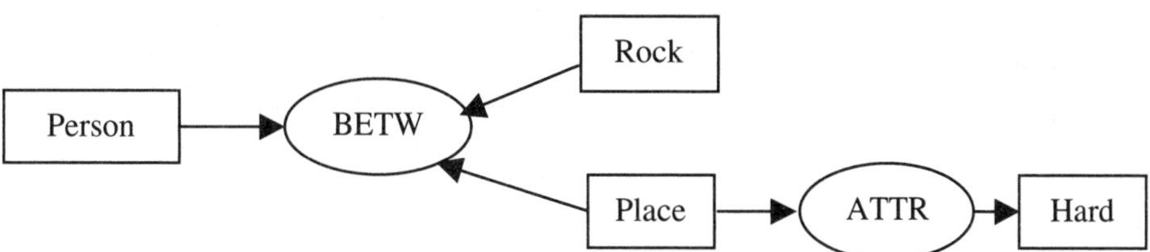

Figure 17.12 CG display form for *a person is between a rock and a hard place* [53].

```
[Person]<-(Betw) -
        <-1-[Rock]
        <-2-[Place]->(Attr)->[Hard]
```

Figure 17.13 A linear form representation of *A person is between a rock and a hard place.*

17.4. SENTENCE INTERPRETATION

We follow the convention of most modern SLU systems—treating semantic parser as a two-step pattern recognition problem (speech recognition followed by sentence interpretation). This convention has the advantage of modular design of SLU systems. Thus, the same SLU system can be used for text input. However, a unified semantic parser [50, 62] may achieve better accuracy, because no hard decision needs to be made before picking the optimal semantic representation.

The heart and soul of the *sentence interpretation* module is how to convert (translate) a user's query (sentence) into the semantic representation. In other words, one has to fill the semantic slots with information derived from the content (words) in the sentence. In this section we describe two popular approaches to accomplish this task. Although they can be perceived as pattern matching methods, they differ in the matching mechanism.

17.4.1. Robust Parsing

Due to the nested nature of semantic classes, the semantic representation **F** in Eq. (17.1) can itself be a tree of semantic objects. A user's utterance may consist of disjoint fragments that may make sense at the discourse level. For instance, in the context of setting up a meeting, the utterance "*Peter Duke at a quarter to two*" can be parsed into two semantic objects: a person and the meeting time. Therefore, the sentence interpretation module must deal with sentence fragments.

The analysis of spoken language is a more challenging task than the analysis of written text. The major issues that come to play in parsing spontaneous speech are speech disfluencies, the looser notion of grammaticality that is characteristic of spoken language, and the lack of clearly marked sentence boundaries. The contamination of the input with errors of a speech recognizer can further exacerbate these problems. Most natural language parsing algorithms are designed to analyze grammatical input. These algorithms are designed to detect ungrammatical input at the earliest possible opportunity and to reject any input that is found to be ungrammatical in even the slightest way. This property, which requires the parser to make a complete and absolute distinction between grammatical and ungrammatical input, makes such formal parsers fragile for spontaneous speech, where completely grammatical input is the exception more than the rule. This is why a robust parser is needed.

In Chapter 11, context-free grammars (CFG) can be written to analyze the structure of entire sentences. It is natural to extend CFG as a pattern matching vehicle. For example, a question such as "*Where would you like to go*?" might be used to solicit a response from a user, who might respond, "*I would like to fly to Boston.*" The following grammar might be used to parse the response:

```
S → NP VP
NP → N
VP → VCluster PP
VCluster → would like to V
V → go | fly
PP → prep NP
N → Boston | I
Prep → to
```

The resulting phrase structure, characterizing the entire sentence, would be:

```
[S [NP [N I ] ] [VP [VCluster would like to [V fly ] ] [PP
[prep to ] [NP [N Boston]]]]]]
```

This structure in turn can provide the foundation for subsequent semantic analysis. Thus, the grammar is adequate for the example response and can be easily extended to cover more city names by expanding the **N** → rule, i.e., by enlarging the lexicon. It has some deficiencies, however. Some of the problems are purely formal or logical in nature, such as the fact that "*Boston would like to go to I* " can be equally parsed. These flaws can be addressed with formal fixes (e.g., a more refined category system), but, in any case, they are not crucial for the practical system designer, because pathological examples are rare in real life. The deeper problem is how to deal with legitimate, natural variations.

The user might respond with any of the following:

```
To Boston
I'm going to Boston.
Well, I want to start in New York and get to Boston by the
day after tomorrow.
I'm in a big hurry; I've got a meeting in Boston.
OK, um, wait a second… OK, I think I've gotta head for Bos-
ton.
```

The above sentences incorporate different kinds of variation for which a *sentence coverage* grammar typically has trouble accounting. For this reason, dialog system designers have gravitated to the idea of *robust parsing*. Robust parsing is the idea of extracting all and only the usable chunks of simple meaning from an utterance, ignoring the rest or treating it as noise or filler. Small grammars can be written that scan a word lattice (see Chapter 13) or a word sequence for just those particular items in which they specialize. For example, a *Destination* grammar, not intended to span an entire utterance, can skim each of the complex utterances above and find the Destination in each case:

```
Destination → Preposition CityName
Preposition → to | for | in
CityName → Boston | …
```

The noise or *filler* elements might include nonspeech noise (cough, laugh, breath, hesitation), elements of *phatic* communication (greetings, polite constructions), irrelevant comments, unnecessary detail, etc. As a user becomes accustomed to the limited yet practical domain of a system's operations, it is expected that variant phrasings would diminish, since they take longer to utter and contribute very little, though disfluencies would always be an issue.

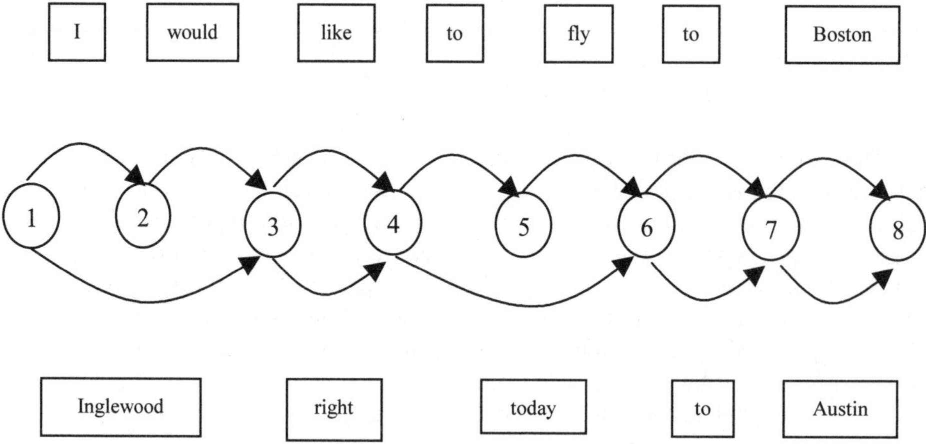

Figure 17.14 Word graph for hypotheses [61].

The original word graph or lattice from the speech recognizer might consist of nodes, representing points in time, and edges representing word hypotheses and acoustic scores for a given span in the utterance. Figure 17.14 illustrates a sample of word graph for the example "*I would like to fly to Boston*" with competing hypotheses. Using the *Destination* grammar on the word graph in Figure 17.14 will skip the earlier parts of the possible sentence hypotheses and identify the short fragment from node 6 to node 8 as a destination. If only the *Destination* grammar were active, a new view of the word graph would result in Figure 17.15.

This example shows that potential and legitimate ambiguities can persist even with flexible grammars of this type, but the key potential meanings have been identified. A robust parser that is capable of handling the example needs to solve the following three problems:

- *Chunking*: appropriate segmentation of text into syntactically meaningful units;
- *Disambiguation:* selecting the unique semantically and pragmatically correct analysis from the potentially large number of syntactically legitimate ones returned; and
- *Undergeneration:* dealing with cases of input outside the system's lexical or syntactic coverage.

Grammars developed for spontaneous speech should concentrate on describing the structure of the meaningful clauses and sentences that are embedded in the spoken utterance. The goal of the parser is to facilitate the extraction of these meaningful clauses from the utterance, while disregarding the surrounding disfluencies. We use the semantic grammar in the Dr. Who SLU engine [61] to illustrate how this works.

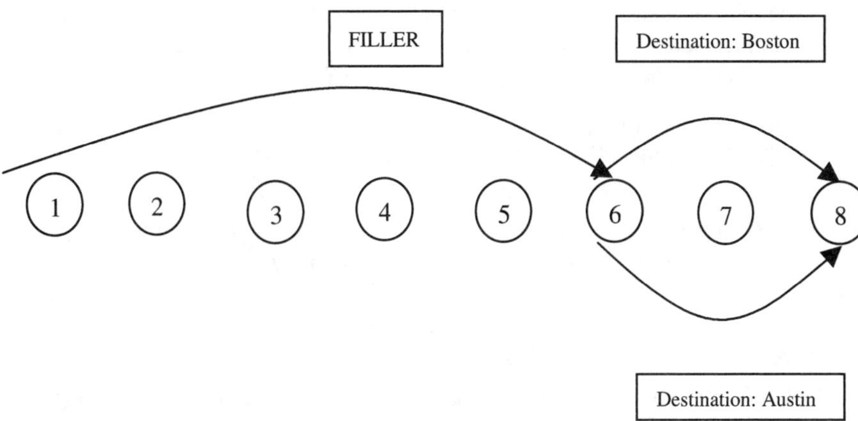

Figure 17.15 Word graph for hypotheses if only the *Destination* grammar is active [61].

As shown in Figure 17.9, a Dr. Who semantic class mostly contains a set of slots that need to be filled with terminal words (verbatim) or with recursive nonterminal semantic classes. Strictly speaking, this semantic class grammar can hardly be called a grammar, since it is primarily used to define the conceptual relations among Dr. Who entities rather than the language expressions that are used to refer to the entities. The syntactic expression is specified by optional CFGs associated with each semantic class. In general, the syntactic grammars need to deal with three kinds of variation in surface linguistic form:

1. *Variation within a slot*—When CFG is missing in the definition of semantic classes, the grammar could allow flexible assembly of an expression. For example, if the <cfg> tags in Figure 17.9 are omitted, any sequence that contains a word of a P_RELATION typed class and a word of a PERSON typed class can be an expression referring to a semantic object of *ByRel* such as *John's father, father of John,* or even *John loves his father.* Thus, CFGs are often specified within the semantic slot to provide linguistic constraints without over-generating.
2. *Variation in the order of frame presentation*—Many systems [64, 66] employ an island-driven robust parsing strategy where the slots in the semantic frames are filled by language fragments from parsing. Parsing of the slots is order independent. Thus utterances "*Schedule a meeting with <u>John</u> at <u>3 PM</u>*" and "*Schedule a meeting at <u>3 PM</u> with <u>John</u>*" can be processed without problems.
3. *Disfluencies and irrelevancies*—Disfluencies and irrelevancies are unavoidable for spoken language input. The system has to deal with real utterances such as "*I'd really like to know whether a meeting by <u>3 PM</u> would be at all possible for <u>John</u>.*"

To cope with these requirements, the robust parsing algorithm [61] is typically implemented as an extension of the bottom-up chart-parsing algorithm discussed in Chapter 11. There are a number of additional requirements for robust parsing:

- The requirement that a hypothesis and a partial parse have to cover adjacent words in the input is relaxed here. This effectively skips the words and enables the parser to omit unwanted words in input sentences.
- The combination of a hypothesis with a new partial parse taken from *agenda* results in multiple new hypotheses. Those hypotheses may have different critical position number. In other words, they are expecting different partial parses. This effectively skips the symbols in a rule, so the parser can continue its operation even if something expected by the grammar does not exist.
- The sequential order in which the partial parses are taken out from the agenda is crucial here. A partial parse that has the minimum span and highest score and that covers the word closest to the sentence start position (in that order) has the highest priority.

In a robust parser, if there is already a parse g that has the same symbol and span as the new parse h, we need to compare their scores so we only keep the better one. The parse scoring can be the likelihood of the parse with respect to a heuristic CFG enhanced with a mechanism of assigning probability for insertions and deletions. It can also be based on heuristics when no training data is available. The typical heuristic values may include the number of words covered by a parse; the number of rule symbols skipped in the parse tree; the number of nodes in the parse tree; the depth of the parse tree; and the leftmost position of the word covered by the parse.

17.4.2. Statistical Pattern Matching

The use of CFGs to capture the semantic meaning of an utterance can be augmented with probabilistic CFGs or the unified language model described in Chapter 11. In the statistical parser, the application developers first define semantic nonterminal and preterminal nodes. A large number of sentences are then collected and annotated with these semantic nodes. The statistical training methods are used to build the parser to extract semantic meaning from an utterance.

For example, a statistical parsing algorithm [15, 26] takes one step further toward automatic discovery of complex CFG rules. Instead of relying on hand-written CFG rules, it builds a statistical parser based on the *tree-banked* data where sentences are labeled with parsing-tree structure. It identifies simple named classes like *Date, Amount, Fund*, or *Percent* and only handles simple classes using the local context. Words that are not part of a class are tagged as *word*, indicating that the word is passed on to the subsequent parser. The subsequent statistical parser takes a classed sentence. It generates the most likely semantic parse in a bottom-up leftmost derivation order. At each step in the derivation, the parsers use

CART (see Chapter 4) to assign probabilities to primitive parser actions such as assigning a tag to a word or deciding when to begin a new constituent. A beam search is used to find the parse with highest probability. The two-step parsing for the sentence "*Please transfer one hundred dollars from voyager fund to fidelity fund*" is illustrated in Figure 17.16.

The *hidden understanding model* (HUM) [29, 30] is another statistical pattern matching techniques. Let **W** denote the sequence of words and S denote the meaning of the utterance. According to Bayes' rule, we have the following equation:

$$P(S \mid \mathbf{W}) = \frac{P(\mathbf{W} \mid S)P(S)}{P(\mathbf{W})} \tag{17.2}$$

The task of sentence interpretation can then be translated into finding the meaning representation $\hat{S}$, such that

$$\hat{S} = \arg\max_{S} P(\mathbf{W} \mid S)P(S) \tag{17.3}$$

$P(S)$ is the *semantic language model* that specifies the prior statistical distribution of meaning expressions. The semantic language model is based on a tree-structured meaning representation where concepts are represented as nodes in a semantic tree with subconcepts represented as child nodes. Figure 17.17 illustrates such a tree-structured meaning representation for the sentence "*United flight 203 from Dallas to Atlanta.*" The Flight concept has Airline, Flight_Ind, Flt_Num, Origin, and Destination subconcepts. Origin and Destination subconcepts have terminal nodes Origin_Ind and City and Dest_Ind and City, respectively. Each terminal node (like City) could be composed of a word or of a sequence of words.

Semantic language model $P(S)$ is modeled as $P(S_i \mid S_{i-1}, concept)$, where *concept* is the parent concept for S_i and S_{i-1}. Based on this definition, the probability $P(\text{Destination} \mid \text{Origin}, \text{Flight})$ is bigger than $P(\text{Origin} \mid \text{Destination}, \text{Flight})$, since users often omit the origin for a flight in an airline reservation system.

$P(\mathbf{W} \mid S)$ is called a *lexical realization* model, which is basically a word bigram model augmented with the context of the parent concept:

$$P(\mathbf{W} \mid S) = \prod P(w_i \mid w_{i-1}, concept) \tag{17.4}$$

Both the semantic language model and lexical realization model are estimated from a labeled corpus. Viterbi search is applied to find the best path of meaning representation $\hat{S}$ according to Eq. (17.3).

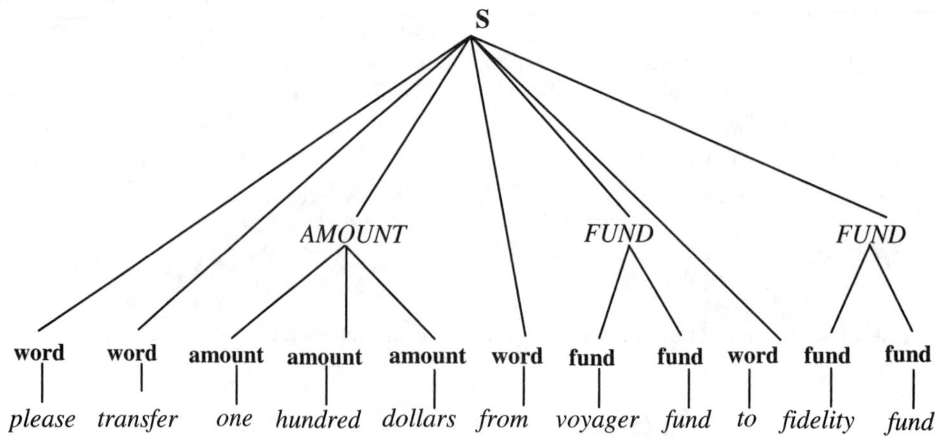

(a)

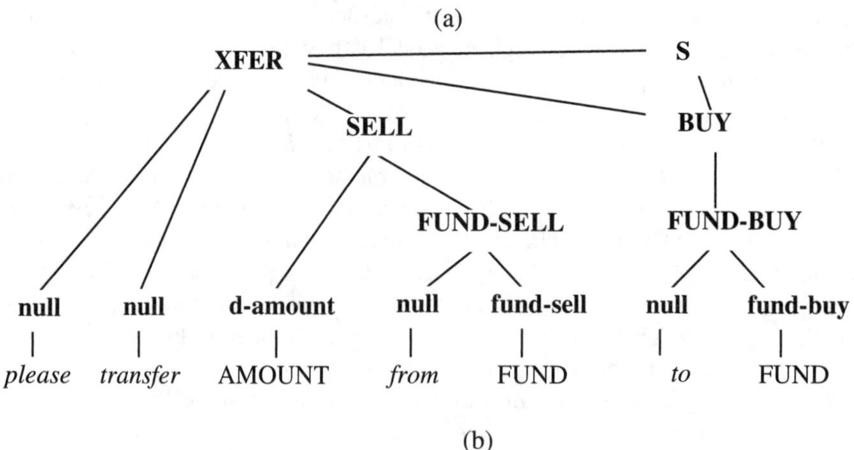

(b)

Figure 17.16 An example class tree in IBM's statistical class parser. (a) The sentence is classified into semantic classes. (b) The classed sentence is parsed into the semantic tree based on CART [15].

```
FLIGHT  [AIRLINE[United]
          FLIGHT_IND[flight]
          FLIGHT_NUM[203]
          ORIGIN[ORIGIN_IND[from] CITY[DALLAS]]
          DESTINATION[DEST_IND[to] CITY[Atlanta]]]
```

Figure 17.17 A tree-structured meaning representation for *United flight 203 from Dallas to Atlanta* in BBN's HUM system [29].

17.5. Discourse Analysis

The sentence interpretation module only attempts to interpret each sentence without knowledge about the current dialog status or discourse. As we mentioned in Section 17.2, sometimes it is impossible to get the right interpretation without discourse knowledge. For example, in the sentence "*Show me the morning flight*" one must have the knowledge what *the morning flight* refers to in order to derive the real-world entitiy, even though the sentence interpretation module comprehends perfectly what *morning flight* means.

Discourse information formed by dialog history is necessary not only for semantic inference but also for *inconsistency detection*. Inconsistency detection is important in a dialog system, since the *dialog management* module (described in Section 17.6) needs such information to disambiguate the dialog flow when needed. For example, in an airline reservation system, the returning date should not proceed the departure date, which may be conveyed in the previous dialog turns. The discourse analysis module needs to maintain a stack of discourse trees so that the semantic representation remains the same whether the information is obtained through several dialog turns or a single one.

The goal of the *discourse analysis* module is to collapse the discourse tree by resolving the semantic objects into the domain entities. This process is also called *semantic evaluation*. When the resolution is successful, the physical semantic object is officially bound to the domain entities. The last process is often called semantic binding. Because an entity can be identified by partial information (e.g., last name of a person), binding is necessary for the system to grasp the whole attributes of the objects the dialog is concerned with. Semantic binding is also critical for intelligent behaviors such as setting the discourse context for reference resolution. The semantic evaluation and binding are the basics for driving the dialog flow. The communication mechanism between discourse analysis and dialog manager is typically event driven. Events that can be passed to the dialog manager are *evaluation succeeded*, *evaluation failed*, *invalid information*, and *value to be determined*. The discourse analysis module often needs to tap into the knowledge base with the semantic object attributes and entity memory for semantic evaluation. The semantic evaluation usually proceeds from the leaves up toward the root of the discourse tree. The process ends when the root node is converted, which indicates the dialog goal has been achieved. The functions of *Discourse analysis* module are the following:

- Converting the *relative expressions* (like *tomorrow*, *next week*, *he*, *it*, *the morning flight*, etc.) in the semantic slots into real-world objects or concepts (such as 1/5/2000, *the week of 2/7/2000*, *John*, *John's dog*, etc.).
- Automatic inference—Based on dialog history, the module may decide some missing information for certain slots. For example, an airline reservation system could infer the destination city for the origin of the return flight even though it is not specified.
- Inconsistency/ambiguity detection—Since the discourse analysis module can perform automatic inference for some slots, it can perform consistency checking when it is explicitly specified during the current dialog turn.

17.5.1. Resolution of Relative Expression

There are two types of relative expression. The first type is the *reference*, relating linguistic expressions to real-world entities. This may involve disambiguation, by inference or direct user query. When a user says, "*Give me Eric's phone number,*" many people with first name *Eric* may exist in the database. The second type of relative expression is the *co-reference.* Co-reference occurs when different names or referring expressions are used to signify the same real-world entity. For example, in the sentence "*Nelson Mandela* has a long history of leadership within the African National Congress, but *he* is aging and nobody was surprised yesterday when *Mandela* announced his successor" the terms *Nelson Mandela* and *Mandela* refer to the same person.

In linguistics, there are three different types of co-references. The example above is an *ellipsis*, where the omitted word(s) can be understood from the context. The other type is *deixis*. A deixis refers to the use of a word such as that, now, tomorrow, or here, whose full meaning depends on the extralinguistic context in which it is used. Location deictic co-references are very common for multimodal applications where pointing devices (modalities) like pens can be used to indicate the real locations. The most common type of co-reference is *anaphora*, which is a special type of co-reference, where a word or phrase has an indirect, dependent meaning, standing for another word or concept previously introduced. The pronoun *he* in the sentence above is an anaphor referring to *Nelson Mandela* too.

Time deictic co-references like *tomorrow*, *next week, the week of 2/ 7/ 2000*, etc., are among the easiest category for resolution (requiring only simple domain knowledge). The resolution of other relative expressions usually requires deep natural language processing. We focus our discussion on anaphora resolution, since it represents the most challenge one among others and approaches of solving this problem are typical of the kind of methods appropriate for resolving a variety of other relative expressions.

17.5.1.1. Priority Entity Memory

We introduce a simple resolution method [60] that is based on semantic class type abstraction and priority entity memory. This method is straightforward and is very powerful to handle most cases even without complex natural language processing.

Whenever a conversion of a relative expression occurs, the consequent entity is added to the entity memory. The entity memory consists of *turn* and *discourse* memories. Either type of memory consists of a number of priority queues that are delineated by entity types. An entity can only be remembered into the queue of compatible types (e.g., through inheritance). When referred to, the memory item increases its priority in the queue. This treatment resembles the *cache language* model described in Chapter 11.

The turn memory is a cache for holding entities in each turn. There are two types of turn memories. The *explicit* memory holds the entities that are resolved directly from semantic objects. In contrast, the *implicit* memory is for entities that are deduced from relative expressions. In accessing the memory, the explicit turn memory takes precedent over the discourse memory, which in turn has a higher priority than the implicit. At the end of the system's turn, all the turn memory items are moved and sorted into the discourse memory.

The distinctions between the three kinds of memories and the rules to operate them are designed as a simple mechanism for most common but not all possible scenarios. It is worth noting that the design has a bias toward *direct* and *backward* reference. For example, in the expression "*Forward this mail to John, his manager, and his assistant,*" the second *his* will be evaluated as referring to *John*, not to *his manager*. The implicit memory, however, provides a back-off for expressions like "*Send email to John, his manager, and her assistant*" in which the pronoun *her* should be taken as indicating John's manager is a female and resolved accordingly. However, since we store only the entities and not the semantic objects into the memory, the mechanism is not suitable for forward or pleonastic references, as in the examples like "*Since his promotion last May, John has been working very hard*" or "*It being so nice, John moved the meeting outside.*" Fortunately, these natural language phenomena are rare in a spoken dialog environment.

It is sensible to confirm[3] the resolved entities with users due to possible resolution errors. In cases where many entities in the entity memory can be matched with a semantic object, a decision of not performing any resolution and directly inquiring the user for disambiguation may be a better solution. In general, name references can be resolved by a sequence of simple rules. In the example of "*Give me Eric's phone number*" the SLU system may just generate the query message "*What is Eric's last name*?" when many people in the entity memory have the same name *Eric*.

17.5.1.2. Resolution by NLP

Extensive understanding is crucial for perfect resolution for relative expressions (in particular, anaphora). Though morphology, lexical semantics, and syntax can be helpful for disambiguation, ultimately it is a problem of inference using real-world knowledge and dialog state or context. In a discourse model of focus, it is assumed that speakers usually center their attention on a single main topic called the *focus*. Some utterances introduce or reintroduce a focus; others elaborate on it. Focus elements typically change (by being suspended, interrupted, resumed, etc.) over the course of a dialog. Once a focus element has been introduced, anaphora is usually used to represent it, making dialog more efficient.

Anaphora resolution specifies the referent of a pronoun or other anaphoric expression. This association should be supported by inference about properties and probabilities in the real world. Anaphora resolution can be done with a simple *entity focus* principle. For example, in the very common *schedule a meeting* type of dialog application, an exchange such as that shown in Figure 17.18 is centered on the initial focus element—the proposed meeting—and anaphora are likely to relate to that central topic, at least early on in the exchange. The

[3] One might decide which confirmation strategies (explicit or implicit confirmation) to use based on the confidence of the resolutions. The details of confirmation strategies are described in Section 17.6.

```
(1) I'd like to schedule a [meeting]i with [Christoph]j.
(2) [It]i can be anytime after 4.
(3) Tell [him]j [he]j can [grab a cab over here]k.
(4) [That]k should be only if he's running late.
```

Figure 17.18 A *schedule a meeting* dialog example showing different anaphora usage.

subscript indicates the co-reference to the same entity. The focus is the *meeting* proposed in (1). The pronoun *it* in (2), by the very simple mechanism discussed here, can be interpreted as referring to the meeting. Some grammatical knowledge and the semantic class type should help the system to resolve *him* in (3) as *Jim* rather than the *meeting*. In sentence (3) the focus has shifted to the action of *taking a cab*, to which *that* refers in sentence (4). The locative *here* in (3) must also be resolved to the speaker's location.

Most formal models of anaphora resolution originated from research into discourse and human-human dialog. They tend to be overpowered, in making elaborate provision for greater topic and reference variation than exists in typical computer speech dialog applications of the present time. On the other hand, while they can provide resolution for some complicated situations, they tend to be underpowered, in failing to deal robustly with the realities of imperfect speech recognition and parsing.

Some of the work on anaphora resolution in dialog relies on elaborate focus-tracking mechanisms [47]. These tend to be somewhat circular in nature, in that anaphoric reference resolution is required for the focus-tracking algorithms to operate, while the anaphoric resolution itself relies on the currently identified focus structure of the dialog or discourse. Rather than elaborate on these possibilities, we instead present a number of relatively straightforward heuristics for anaphora resolution, some of which have been developed based on textual studies, but which may be relevant to increasingly complex human-computer dialog in the future. The discussion here is limited to the resolution of intersentential and intrasentential pronominal anaphora. Full noun-phrase anaphora, where one synonymous noun phrase is co-referent with another, requires even more powerful grammatical and semantic resources.

Syntactic conditions can be tested when a parse tree showing syntactic constituency is available. The most obvious syntactic filter for disallowing co-reference is simple grammatical feature agreement. For example, the following proposed co-indexed relation is not semantically possible in ordinary discourse, and the restriction is explicitly provided through the lexical morphology and syntax of the language:

```
The [girl]i thought [he]i was frightening.
```

Though the theoretical details can be complex [37], the basic intuition of syntax-based anaphoric resolution is that nonreflexive pronouns that are syntactically too close to a candidate co-referential NP (antecedent) are disfavored. For example, in a sentence such as:

```
[Bill's]i photo of [him]i is offensive.
```

the coindexing of *Bill* with *him* is disallowed. By disallowed, we mean that your innate sense of proper English grammar and interpretation will balk at the proposed relation. The language provides a mechanism to override some proximity restrictions, as in the following repaired version:

`[Bill's]`$_i$ `photo of [himself]`$_i$ `is offensive.`

So, when is a pronoun *too close* to a possible antecedent? The most important syntactic concepts for determining anaphoric relations rely on structural attributes of parse trees. In fact, treatment of this problem represents a very large and highly argumentative subfield within theoretical linguistics. Nevertheless, any treatment of anaphora resolution on purely syntactic grounds is very likely to end with a list of conditions that can mostly be subsumed under some form of *x-bar* theory [25], as it is called in the theoretical linguistics.

17.5.2. Automatic Inference and Inconsistency Detection

Automatic inference can be carried out through the same framework of priority entity memory described in Section 17.5.1.1. During semantic evaluation, a partially filled semantic object is first compared with the entities in the memory based on the type compatibility. If a candidate is found, the discourse analysis module then computes a goodness-of-fit score by consulting the knowledge base and considering the position of the entity in the memory list. The semantic object is converted immediately to the entity from the memory if the score exceeds the threshold. In the process, all the actions implied by the entities are carried out following the order in which the corresponding semantic objects are converted.

In general, automatic inference can be implemented as description procedures attached to semantic slots as described in Section 17.3.1. In the example of an airline reservation system, a procedure or rule can be attached to automatically infer the destination city for the returning flight. The other powerful strategy for automatic inference is *slot inheritance*. When changing dialog turn for different semantic objects under the same service, the system may allow such slot inheritance to free users from repeating the same attributes. For example, after a user asks "*What is Peter Hon's office number*?" he may abbreviate his next query to "*How about Derek Acero's*?" Slot inheritance will allow the second semantic object regarding *Derek Acero* to inherit the *office number* slot even though it is not explicitly specified.

Inconsistency checking is crucial to initiate necessary events for *dialog repair*. A dialog may be diverted away from the ideal flow for various reasons (e.g., misrecognition, out-of-domain reference, conflicting information), many of which require domain- and application-specific knowledge to guide the dialog back to the desired course. This process is called dialog repair. Similar to automatic inference, inconsistency checking can be implemented as description procedures attached to semantic slots. In addition, inconsistency checking can also be triggered when semantic binding for a partially filled semantic object fails (e.g., indicated by a failed database lookup). The discourse analysis module is responsible only for sending the dialog repair events to the dialog manager, and it leaves the realization of the repair strategy to the corresponding event handler in the dialog manager.

For example, consider a query: "*Find me the cheapest flight from Seattle to Memphis on Sunday.*" The semantic binding fails because there is actually no flight available on Sunday from Seattle to Memphis based on the flight database. Thus, the discourse manager passes such event to the dialog manager, and the dialog manager will generate an appropriate message to let the users be aware of this fact.

17.6. Dialog Management

For most applications, it is highly unlikely that a user can access or retrieve the desired information with just a single query. The query might be incomplete, imprecise, and sometimed inconsistent with respect to the discourse history. Even if the query is unambiguous, the speech recognition and sentence interpretation modules in a SLU system may make mistakes. Thus the SLU system needs to provide an interactive mechanism to perform clarification, completion, confirmation, and negotiation dialogs with users. By default, the objective of such a dialog is to help users accomplish the required tasks more efficiently. Being user-friendly is also one of the major objectives for dialog systems as discussed in Chapter 18. Since the goal of a SLU system is to provide a natural conversation interface for users, the ultimate SLU system should act like a real human, yet still possessing perfect memory and superfast computation. Based on these criteria, it is not hard to see why mix-initiative systems are preferred over system-initiative systems.

The dialog manager controls the interactive strategy and flow once the semantic meaning of the query is extracted and stored in the system's representation (discourse trees). The architecture of SLU dialog systems resembles the one used in event-driven GUI systems. In the same way that GUI events are assigned to graphical objects, the dialog events are assigned to semantic objects that encapsulate the knowledge for handling events under various discourse contexts. As mentioned in Section 17.5, the discourse tree with domain entity binding is passed along with necessary dialog events generated from the discourse analysis module to the dialog manager. The dialog manager acts as an intelligent domain knowledge handler that uses the semantic meaning of the query to check against domain-specific knowledge (including domain database and application logic) and generates the desired answer for the query or produces other necessary dialog strategy.

In this sense, the dialog manager functions as a GUI application that contains an event handler. The event handler handles dialog events passed from the discourse analysis module and generates appropriate responses to engage users to solve the problems. In addition, the dialog manager needs to implement the application logic to generate appropriate actions (e.g., make real airline and hotel reservation). In this section we discuss two modeling techniques for implementing application logic, and different dialog behaviors related to event handling.

17.6.1. Dialog Grammars

Dialog grammars use constrained, well-understood formalisms such as *finite state* machines to express sequencing regularities in dialogs, termed *adjacency pairs*. The rules state sequential and hierarchical constraints on acceptable dialogs, just as syntactic grammar rules state constraints on grammatically acceptable strings. For example, an answer or a request for clarification is likely to follow a question, just as a finite state grammar might provide for a noun or an adjective, but not a verb, to follow a determiner such as *the*. In most dialog grammar systems, dialog-act types (*explain*, *complain*, *request*, etc. cf. Section 17.2.2) are categorized ,and the categories are used as terminals in the dialog grammar. This approach has the advantage that the formalism is simple and tractable. At every stage of processing the system has a basis for setting expectations, which may correspond to activating state-dependent language models, and for setting thresholds for rejection and requests for clarification.

In its essence, the dialog grammar model is exemplified by a rigid flowchart diagramming system control of the type and sequence of interaction. Figure 17.19 shows a finite state dialog grammar for an airline reservation SLU system. In this simple example, dialog-act categorization is omitted, and the interactions are controlled based on bare information items. This grammar makes simple claims: the interaction is basically question-answer; the topic queries are answered on-topic if possible, and presumably with a confirmation statement to catch the existence of a problem.

This system is easily programmed. The challenge lies in providing tools to application authors to ease the tedium and minimize the errors in the construction of grammars, and to allow for more flexibility and spontaneous deviations from the expected transitions in the grammar. Such deviations may be important for novice users, who may more naturally tend to give their information (origin, destination, time) in one single utterance or in a different order.

In general, the dialog grammar approach has the following potential disadvantages

- The interaction may be experienced by a user as brittle, inflexible, and unforgiving, since it is difficult to support mix-initiative systems.
- Dialog grammars have difficulty with nonliteral language (indirection, irony, etc.).
- A speech act might be expressed by several utterances, complicating the grammar.
- A single utterance might express several speech acts, complicating the grammar.

To address these issues, more sophisticated approaches to enhance hand-built finite state dialog grammars have been attempted. For example, once can add statistical knowledge based on realistic data to dialog grammars. The statistical learning methods, like come CART, *n*-grams, or neural networks [3] can be used to learn the association between utterances and states in the training data.

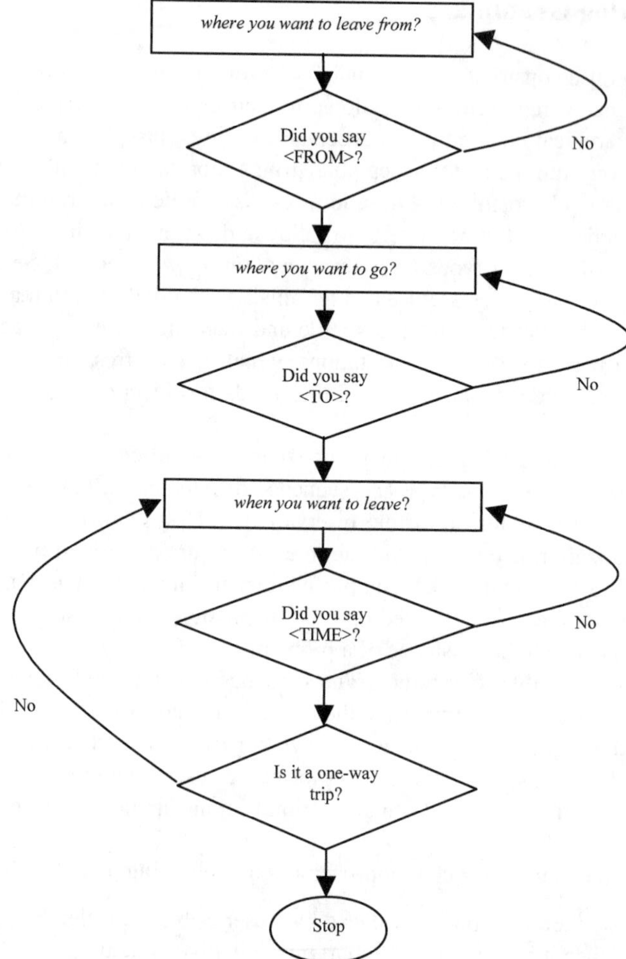

Figure 17.19 A finite state dialog grammar for airline reservation (after [19]).

17.6.2. Plan-Based Systems

Plan-based approaches [2, 41] seek to overcome the rigidity and shallowness of dialog grammars and templates. They are based on the observation that humans *plan* their actions to achieve various *goals*. Thus, plans and goals are in some degree of correspondence. A system operating under these assumptions needs to *infer goals, construct and activate plans*. A user may have a preconceived plan for achieving his/her goals or may need to rely on the system to supplement or construct appropriate plans.

Plan-based systems are well studied in artificial intelligence (AI) [32, 65]. The mathematical foundation of the plan-based approach is inference. The behaviors of the system and the knowledge of the domain are programmed as a set of logical rules and axioms. The system interacts with the user to gather facts, which consequently trigger rules and generate more facts as the interaction progresses. As illustrated in Eq. (17.1), the goal of the dialog manager is to derive the action **A** based on discourse semantic S_n. Taking this view, the dialog manager is a natural outgrow of the semantic evaluation process. It is the step where the system's intent is computed. The outcome of the dialog manager is a message (via different rendering) the system conveys to the user.

In essence, a plan-based system is an embodiment of a state machine for which different discourse semantics are regarded as states. The difference, however, is that the *states* for the plan-based system are generated dynamically and not limited to a predetermined finite set. This capability of handling an unbounded number of states is a key strength of plan-based systems in terms of scalability.

Even a simple interaction can involve a variety of complex subgoals and pragmatic inferences. A partial plan for the airline reservation example in Section 17.6.1 is illustrated in Figure 17.20. One wants to know if a flight itinerary (F12) is an available one. The relationships among the goals and actions that compose a plan can be represented as a directed graph, with *goals*, *preconditions*, *actions,* and *effects* as nodes and relationships among these as arcs. These graphs illustrate the compositional nature of plans, which always include nested subplans, down to an almost infinite level of detail. The appropriate level of planning specification is thus a judgment call and must be application dependent.

The arcs are labeled with the relationship that holds between any two nodes. *SUB* shows that the child arc is the beginning of a subplan for the parent. At some point appropriate to the domain of the planning application, the SUBs will be suspended and represented as a single subsuming node. In Figure 17.20, ENABLE indicates a precondition on a goal or action. EFFECT indicates the result of an action. ENABLE indicates an enabling relationship between parent and child nodes.

Plan-based approaches incorporate a rich and deep model of rational behavior and, thus, in theory, permit a more flexible mode of interaction than do dialog grammar approaches. However, they can be complex to construct and operate in practice, due to reliance on logical and pragmatic inference, and due to the fact that no fully understood theoretical underpinning exists for their specification. The complexity of the domain of modeling often requires significant efforts from human experts to author the logical rules and axioms.

In plan-based theories of agent interaction, each dialog participant needs to construct and maintain a model of all participants' goals, commitments, and beliefs. Plans are, thus, a relatively abstract notion, leading to the hope that plans could be designed in an application-independent fashion, which would permit the development of *plan libraries*. Such libraries could be easily adapted to a variety of domains; just as specific entity models are derived from generic classes via inheritance in object-oriented programming.

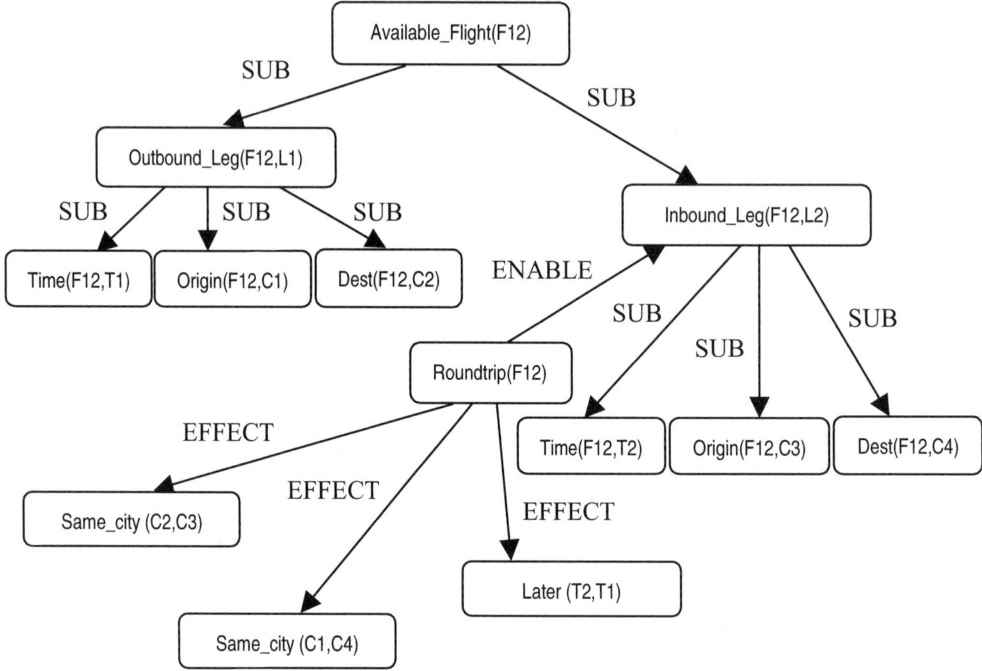

Figure 17.20 A partial plan for the airline reservation example in Figure 17.19 represented as a graph.

The following operational cycle exemplifies the plan approach, describing interaction of two agents, X (the helpful assisting agent) and Y (the client). Interaction is stated from X's point of view [10].

- Observe Y's act(s)
- Infer Y's plan (using X's model of Y's beliefs and goals)
- Debug Y's plan, finding obstacles to success of plan, based on X's beliefs
- Adopt the negation of the obstacles as X's goal
- Plan to achieve those goals and execute the plan

A flight itinerary that at least contains an *Outbound_Leg* subgoal and another possible *Inbound_Leg* subgoal is a round trip. Let's assume F12 is a round trip itinerary. At the *Inbound_Leg* node, the interesting question is how much of the underlying goal (*Time*(F12,T2), *Origin*(F12,C3) and *Dest*(F12,C4)) can be inferred by the information provided by the system from the dialog so far, or from other known conditions. For example, the destination of the *Inbound_Leg* can be inferred from the origin in the outbound leg. The origin city can be inferred similarly. Going one step further, you can also infer that the departure time for the inbound leg must occur after the departure time of the outbound leg (T1 < T2). Those three inferences are shown in the *Effect* arcs in Figure 17.20.

The goal inference could be a cooperative process, with the system making the minimal queries needed to verify and choose among alternative hypotheses. Or, it could be based on pure inference, with perhaps a confirmation step. Inference modeling can get very complicated. The technologies of inference are complex models of the beliefs, desires, and intentions of agents, making use of generic logical systems, which operate over the propositions corresponding to the nodes in a plan structure such as shown in Figure 17.20. Both user and system are assumed to be operating from partially shared world and discourse models consisting of beliefs about all relevant entities and their relationships. If utterances and speech acts are not in conflict with the constraints implied by the world models, communication and action can proceed. Otherwise, either the utterance itself must be further interpreted, supplemented, or clarified, or the world models need to be changed.

The natural expression of rational behavior, communication, and cooperation is some form of first-order logic. We define axioms and inference rules for *Belief* and *Intention.* If the modal operator for belief is *B*, axioms and inference rules for an agent *i* with respect to proposition schemata ϕ or ψ could be formalized in the following logical expression.

$$\begin{aligned}
&(B_i(\phi) \wedge B_i(\phi \Rightarrow \psi)) \Rightarrow B_i(\psi) \\
&B_i(\phi) \Rightarrow \neg B_i \neg \phi \\
&B_i(\phi) \Rightarrow B_i(B_i(\phi)) \\
&\neg B_i(\phi) \Rightarrow B_i(\neg B_i(\phi)) \\
&\neg B_i(\phi) \Rightarrow \neg B_i(B_i(\phi)) \\
&\forall x B_i(\phi) \Rightarrow B_i(\forall x \phi)
\end{aligned} \tag{17.5}$$

These describe appropriate conditions on beliefs of rational agents, such as entailment and consistency. Intentions, in turn, are formalized with respect to beliefs. For example, if an agent is to form an intention to bring about a state of affairs, it is reasonable that s/he believes this state of affairs is not currently in force:

$$I_i(\phi) \Rightarrow B_i(\neg \phi) \tag{17.6}$$

Other such axioms formalize related constraints on intentions, e.g., having an intention entails a commitment to achieving any preconditions, and belief in the possibility of doing so. Many more axioms involving all aspects of rational behavior, and formalizing, to some extent, the Gricean Maxims can be devised. For example, a kind of conversational cooperation occurs when a participant *i* is willing to come to believe what *i* believes his/her conversational partner *j* is attempting to communicate (at least for the limited operational domains in question!), unless *i* holds beliefs to the contrary:

$$B_i(I_j(B_i(\phi(j)))) \wedge \neg B_i(\neg \phi(j)) \Rightarrow B_i(\phi(j)) \tag{17.7}$$

When beliefs and intentions are modeled in this fashion, it may be possible to directly construct the core of a dialog engine based on rational principles as a theorem prover. Such a treatment is, however, beyond the scope of this discussion.

A few desirable system behaviors that would naturally follow from limited inference and goal tracking can be briefly examined. Unlike the dialog grammar approach, a plan-based system allows digression, since the user's intention model has been built into the plan. When a system is confused about a user's input, a cooperative system could begin to perform the critical pragmatic steps that uniquely distinguish the conversational interface. A chain of inferring the user's goal, based on the system's axioms, dialog history, and current knowledge, would be triggered.

It is essential for a system to track the *dialog focus*, or temporary centers of attention, in order to understand things that are unspoken but assumed to be salient across utterances. In this case, the user's input is ambiguous—*June 22* is for outbound or inbound flight? If the dialog architecture provides a method of tracking focus, it may be simple to resolve the legs from an earlier query.

Focus is a useful concept in dialog understanding. The basic idea is similar to the entity memory tracking in anaphora resolution (see Section 17.5.1.1)—at any given point in a conversational exchange, a few items are at the center of attention and are given preference in disambiguation. Other items are in the background but may be revitalized as centers of attention at some later point. A static area can be used to contain items that are assumed background knowledge throughout the exchange. The main goal of conversation can initialize the stack. As subgoals are elaborated, new focus sets are pushed on the stack, and when these subgoals are exhausted, the corresponding focus object is popped from the stack and earlier, presumably broader topics are resumed. Focus shifts that are not naturally characterized as refinements of a broader current topic may be modeled by initiating a new independent focus stack. Focus shifts may be cued by characteristic linguistic signals, such as cue words and phrases (*well now, ok!, by the way, wait!, hey*, etc.). In many cases, focus structure tracks the recursively embedded plan structures, such as that shown in Figure 17.20.

17.6.3. Dialog Behavior

Even though the behavior of the dialog manager is highly dependent on the domain knowledge and the applications, some general styles of dialog behavior are worth investigating. The first important dialog behavior is the *dialog initiative* strategies. System initiative systems have the advantage of narrowing the possible inputs from users, while paying the price for extreme inflexibility. Although user initiative strategy is often adopted for GUI-based systems, it is seldom implemented for SLU systems, since total flexibility is translated into high perplexity (resulting low system performance). For many applications, a flexible mixed initiative style is preferred over a rigidly controlled one. Although it is possible to implement a mixed initiative system using either dialog grammars or plan-based approach, the latter is more flexible because it can handling an unbounded number of states.

Most often, the response generated by the dialog manager is either a *confirmation* or a *negotiation*. Confirmation is important due to possible SLU errors. There are two major confirmation strategies—explicit or implicit confirmations. An *explicit* confirmation is a response solely for confirmation of what the system has heard. On the other hand, an *implicit* confirmation is a response containing new input query and embedded confirmation with the hope that the user can catch and correct the errors if the embedded confirmation is wrong. The examples in Figure 17.21 illustrate both confirmation strategies.

SLU systems usually use a confidence measure as to when to use explicit and implicit confirmation. Obviously, explicit confirmation is used for low-confidence semantic objects while implicit confirmation is for high-confidence ones.

A negotiation response can arise whether a semantic object is fully filled or not. In the case of underspecification, there are some attributes of the semantic objects that cannot be inferred by the discourse manager. Possible actions range from simply pursuing the unfilled attributes in a predefined order, to gathering the entities in the knowledge base sorted by various keys. For cases of ill specification, an entity that matches the semantic object attributes does not exist. The planner can simply report such fact, or suggest removal or replacement of certain attributes, depending on how much domain knowledge is to be included in the planning process.

Often in the design process, we find it desirable to segregate a dialog into several self-contained sessions, each of which can employ specialized language, semantic, and even behavior models to further improve the system performance. Basically, these sessions are subgoals of the dialog, which usually manifest themselves as *trunk* nodes on the discourse tree. We implement a tree stack in which each trunk node is treated as the root for a discourse tree. The stack is managed in a first-in last-out fashion, as currently no digression is allowed from one subdialog to another. So far, the no-digression rule is considered to be a reasonable trade-off for dynamic model swapping.

Consider the example domain of travel itinerary planning [13]. At the top level is the *scenario*, which is the intended output of the interaction. The scenario is the entire itinerary, consisting of reservations for flights, hotels, rental cars, etc., all booked for the user at workable, coordinated times and acceptable prices and quality levels. A scenario might be: a flight out of the user's home city of Boston, from Logan airport, on April 2, at 4:00 PM on a particular flight, connecting in Dallas-Ft. Worth to another flight to a regional airport, an overnight hotel stay, a meeting the next day in the morning, a drive to a second local afternoon meeting, a flight from the regional airport in the evening to LA for a late meeting, another overnight stay in LA, a morning meeting at the hotel, and a return flight back to Boston later that same morning.

```
I:  I would like to fly to Boston.
R1: Do you want to fly to Boston?              (explicit confirmation)
R2: When do you want to fly to Boston?         (implicit confirmation)
```

Figure 17.21 With the input *I would like to fly to Boston*, explicit confirmation response R1 *Do you want to fly to Boston*? only allows the user to confirm the destination, while implicit confirmation response R2 *When do you want to fly to Boston*? allows the user to provide departure-time information and have a chance to confirm the destination as well.

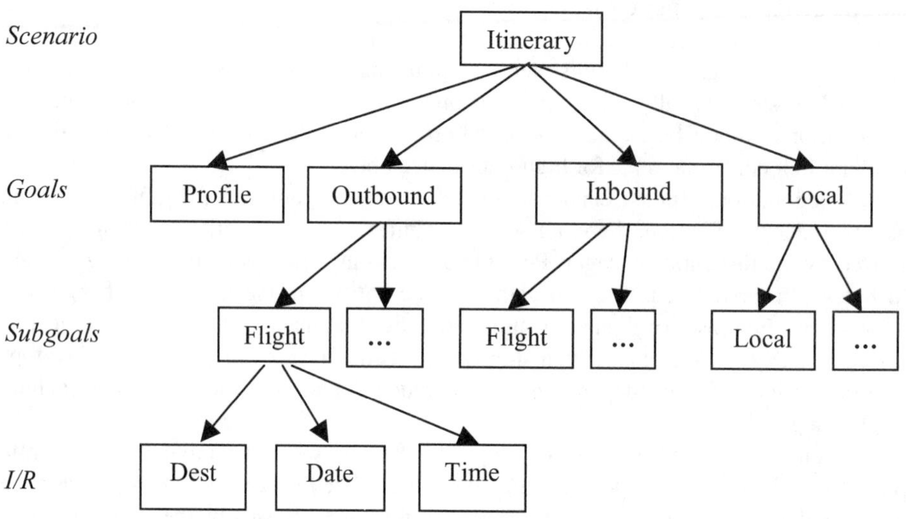

Figure 17.22 A dialog structure hierarchy for travel.

Creating the finished itinerary for this scenario involves *goals*, generated by the system or the user. Goals might include: access user travel profile, book outbound and inbound flights, and make local arrangement (hotel reservation and car rental). Goals in turn may subsume *subgoals*. Subgoals are concerned with the details of planning. These would include establishing particular desired cities and airports for the flights, price investigation, queries about hotel location and quality, etc. The subgoals in their turn are generally realized via speech acts forming I/R pairs. A simplified schematic of the structure of the itinerary structure described above might appear as shown in Figure 17.22.

This structure lends itself to a variety of control mechanisms, including system-led and mixed initiative. For example, the system may ask guiding questions such as "*Where would you like to go*?" followed by "*What day would you like to leave?*" or the system could begin processing from the user's point of view by accepting an utterance like *I want to go from Boston to LA,* corresponding to the *Dest* node of a flight on the outbound flight, and responding with a query about the next needed item, e.g., *What day would you like to leave*? This system can also accommodate a user who may wish to talk about his or her hotel reservation immediately after making the outbound flight reservation, before arranging the inbound flight.

17.7. Response Generation and Rendition

Response generation, also known as the *message generation,* is the process in which the message is physically presented to the user. This is the stage that significantly involves human-factor issues, as discussed in Chapter 18. It is more susceptible to application-specific or user interface considerations. For example, to handle a message requesting the user to select a sizable list of alternatives, a system with a suitable visual display might choose to present the whole list, while a speech-only system might require a more clever way. In this section we mainly focus on speech output modality and provide some thoughts on other popular output modalities.

A conversational interactive system requires a speech-output capability. The speech output may be comprised of system requests for clarification, disambiguation or repeat of garbled input; confirmation; prompting for missing information; statements of system capabilities or expectations; and presentation of results. At the lowest level, this is done via a text-to-speech engine, as discussed in Part III (Chapters 14, 15, and 16) and shown as a component in Figure 1.4. However, most text-to-speech engines have been designed for a read speech style. Moreover, such systems typically perform only shallow syntactic and semantic analysis of their input texts to recover some text features that may have prosodic correlates. Because the topic space of a task-oriented dialog system is narrower, there are opportunities to tune prosodic and other attributes of the speech output for better quality.

There are two major concerns in voice-response rendering. First is the creation or selection of the content to be spoken, and second is the rendition of it, which may include special prosodic markups as guidance to a TTS engine.

17.7.1. Response Content Generation

The response content can be explicitly tied to the semantic representation of the domain task and objects. The semantic class could incorporate custom prompts for specific slots or even for whole semantic classes. Whenever the dialog manager finds that specification for a particular slot is missing from a semantic object, it can check if it contains prompts. If prompts are present, one could be selected at random for presentation to the user.

Response prompts can be embedded in semantic representation. Prompts are usually provided for each slot to provide direction for users to fill the slot in the next dialog term. For example, the semantic class `ByName` defined in Figure 17.9 can be enhanced with the prompts in Figure 17.23.

Prompts could be associated with conditions. For example, in a flight information system, a conditional prompt can be inserted into the semantic class definition to inform users of the flight arrival time based on whether the flight has landed or not, as shown in Figure 17.24.

Other systems may include some categorization of prompts for different functions. For example, at the task level of an airline reservation system, the categorized message list might appear as shown in Figure 17.25. The grammar format makes provision for convenient authoring of messages that can be specified and accessed by functional type at runtime. The `BEMsg` is a special type of message. In this particular architecture, communication with the database engine (cf. the boxes application and database in Figure 17.2) is controlled by messages that are authored in the task specification. The URL attribute indicates a database

```
<!-- semantic class definition for ByName that has type
PERSON too -->
<class type="PERSON" name="ByName">
    <slot type="FIRSTNAME" name="firstname"
          prompt="Please specify the last name for [firstname]/>
    <slot type="LASTNAME" name="lastname"
          prompt="Please specify the first name for [lastname]/>   />
    <cfg>
          ..............
    </cfg>
</class>
```

Figure 17.23 Semantic class ByName in Figure 17.9 is enhanced with prompts specified for the case of missing a particular slot information.

access. The </rclist> is the set of possible return codes from the back-end application (as it attempts to perform the specified command from the message). Again, every return condition is associated with a message by the task specification author. Those shown here include a simple confirmation of a successful completion, as well as a warning for *flight sold out* and a generic failure of transaction message.

```
<class type="FLIGHT" name="Flight">
    <slot type="FLIGHTNO" name="flight_no">
    <slot type="TIME" name="sch_time">
    <slot type="TIME" name="actu_time">
    <slot type="CITY" name="dep_city">
    <slot type="CITY" name="arr_city">
    <slot type="AIRLINE" name="airline">
    <prompt condition= "$SYS_TIME > [actu_time]">
          Flight [flight_no] is landed at [actu_time]
    </prompt>
    <prompt condition= "default">
          Flight [flight_no] is scheduled to land at [sch_time]
    </prompt>
    <cfg>
          .........................
    </cfg>
</class>
```

Figure 17.24 A semantic class Flight contains a conditional prompt to inform users when invalid [depart_time] is detected.

```
<messages>
  <msg id="Help"> Please specify the flight time, origin and destination </msg>
  <msg id="Cancel"> Canceling itinerary... </msg>
  <msg id="Confirm"> Buying ticket from [origin] to [dest] on [time]? </msg>
  <msg id="BEMsg" url="http://server/...?op=buy&time=[time]&flight=[flight].." >
    <rclist>
       <rc id="OK"> Complete buying </rc>
       <rc id="SO"> The flight is sold out </rc>
       <rc id="ERROR"> Cannot complete transaction </rc>
    </rclist>
  </msg>
</message>
```

Figure 17.25 An example of categorization of prompts for an airline reservation SLU system.

Such systems can incorporate other kinds of categorization as well. For example, a system might provide a battery of responses to a given task or subtask situation, varying depending on a speech recognition confidence metric. Thus a set of utterances ordered by decreasing confidence might appear as:

```
You want to fly to Boston?
Did you say Boston?
Could you repeat that, please?
Please state a flight reservation.
```

Systems of this type are sometimes referred to as *template systems* for response generation. They have the advantages of direct authoring and simplicity of implementation and may provide very high quality if the message templates of the application can be played with matching digitized speech utterances or carrier phrases in the synthesizer.

The specificity and application-dependent qualities of template-based systems are sometimes perceived as weaknesses that could potentially be overcome by more general, flexible, and intelligent systems. In these systems the *message generation* box could subsume discrete modules, as shown in Figure 17.26. The semantic representation would typically be akin to logical forms (see Chapter 2) expressed via semantic frames or conceptual graphs. The representation would include abstract expression of content as well as speech-act type and other information to guide the tactical or low-level aspects of utterance generation, such as word choice, sentence type choice, grammatical arrangement, etc.

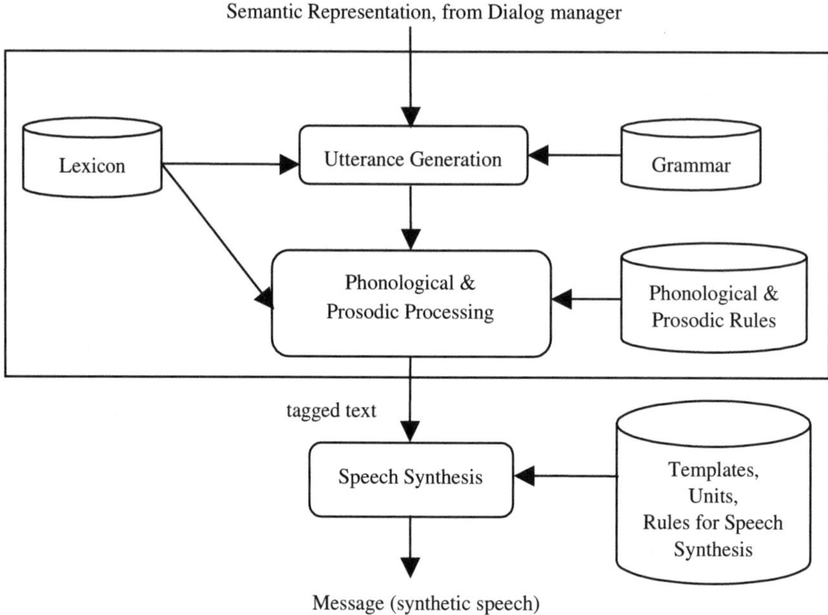

Figure 17.26 Natural language generation and rendition modules.

Natural language generation from abstract semantic input is a deep and complex field. Let us briefly consider a slightly more abstract form of template-selection mechanism that could gracefully either accommodate a simple set of static, authored response utterances or, alternatively, serve as a form of semantic input to a generalized, NLP-based utterance generation module. Imagine that instead of simply providing lists of prompt strings with embedded slot identifiers, a system of parameterization can be used [24]. The parameters could be at varying levels of abstraction and would function as descriptors of static content when preauthored prompts were being used, or would serve as a kind of input semantic representation when a general natural language was used. The set of parameters might include attributes of utterances such as the following:

- *Utterance type*: mood of the sentence, i.e., declarative, *wh*-question, yes/no question, or imperative.
- *Dialog or speech act*: confirmation, suggestion, request, command, warning, etc.
- *Body*: some characteristic lexical content for the utterance, apart from any situation-dependent words and concepts. This could serve as a hint to a generator. In many cases this would be the main verb of a sentence and might

also include characteristic cue words, especially for functional transitions, e.g., *however*, *now*, etc.

- *Given*: information that is understood from the discourse history. This is usually represented as pronouns or other anaphora in the generated utterance.
- *New*: anything that is in the informational foreground, due to lack of prior mention, but may not be precisely the purpose of the prompt, per se. New material typically receives some kind of prosodic prominence in speech.

Examples of these parameter indices for templates from a theater ticket-reservation domain might appear as in Table 17.4. The basic idea of the parametric approach is that such a level of medium abstraction allows for flexibility in the choice of deployment tactics. If a full set of static prompts and response utterances is available for all cases, then this approach reduces to a template system, though it does provide the potential for separation of grammars and prompt files. If, however, a natural language generation component is available for dynamic message generation, a parameter set like that above can serve as input.

Table 17.4 Sentence generation indices for an airline reservation SLU system.

Act	Type	Body	Given	New	Example
Meta [sorry]	Decl	*no*	-	-	*No, sorry.*
Verify	Y/N-Q			Boston	*Boston?*
Request-info	WH-Q	*fly*	you	thing	*When do you want to fly?*
Request-info	WH-Q	*want*	you	airline tomorrow	*Which airline would you like to fly tomorrow?*
Stmt[sorry]	Decl	*sold out*	it	-	*Sorry it is sold out.*
Stmt	Decl	*sold out*	USAir	-	*Sorry, USAir is sold out.*

17.7.2. Concept-to-Speech Rendition

Once the response content is generated, the SLU system needs to render it into a waveform to play to the users. The task is naturally assigned to a text-to-speech component. However, the response generated in the previous session is more than text message. It contains the underlying semantic information, because it is usually embedded in the semantic representation as shown in Figure 17.23 and Figure 17.24. This is why the speech rendition is often done through a *concept-to-speech* module. A concept-to-speech system can be considered as a text-to-speech system with input text enhanced with domain knowledge tags. With these extra tags, a concept-to-speech system should be able to generate tailored speech output to better convey the system intention.

Chapter 15 discussed the role of prosody in human perception. When messages are generated, it is expected that they are supplemented with hints as to their information structure. At a minimum, the message generation component can identify which parts of the utterance constitute the *theme*, which is material understood, previously mentioned, or somehow extending a longer thread of coherence in the dialog, from the *rheme*, which is the unique contribution of the present utterance to the discourse [36]. If such a distinction is marked on the generated utterances, or templates, it can be associated with characteristic pitch contour, prosodic phrasing, and other effects (see Chapter 15).

For example, in the question-answer pair shown in Figure 17.27 (from ordinary human conversation), the theme and rheme components are bracketed. The theme of the answer consists of a mention of Mary, and the act of driving, both carried forward from the question. The theme consists of new information, the answer to the question, embedded in a kind of placeholder noun phrase. Clearly, the input to the message generation component requires some indication of which entities of the input semantic representation are linked to discourse history.

Q: `Which car did Mary drive?`
A: `(Mary drove)`$_{th}$ `(the RED car.)`$_{rh}$

Figure 17.27 A question-answer pair with theme and rheme components marked.

Prosodic rules are triggered by information structure. In general, a theme in the early part of a statement may be realized with a rise-fall-rise pitch contour, often with turning points in the contour aligned with lexically stressed or other salient syllables of the words in the theme. Rheme marking by pitch contour is also essential for naturalness, and a common rheme tune in English declaratives is a slight rise up to the final lexically stressed syllable, followed by a fall to the bottom of the speaker's pitch range. The actual alignment of pitch extrema will depend on the position of focus, or maximum contrast and information value, within either the theme or the rheme.

In Figure 17.27, the word *RED* is in focus within the rheme. If the question had implied a contrast between Mary's car and other people's cars, it would be acceptable to esta b-lish a focus on *Mary* in the theme as well, marked by a pitch accent (see Chapter 15). Sometimes the portion of either theme or rheme that is not in focus (e.g., *drove* or *car*) is called the *ground* [54, 55].

The response generator could add such rheme-theme information that may be used to trigger more specialized prosodic rules. For example, one experimental system is based on a message generator that dynamically creates concise descriptions of individual museum objects during a tour, while attempting to maximize correlations to objects a museum visitor has already seen [21]. During the response generation phase, simple entities and factual statements are combined, first into a semantic graph and then into a text, in which the rhetorical functions of utterances and clauses, and their relations to one another, are known. This information can be passed along to a synthesizer in the form of markup tags within the text. A synthesizer can then select appropriately interesting pitch contours that indirectly reflect rhetorical functions.

In a dialog system, other attributes beyond rheme-theme kinds of information structure, such as speech-act type, may have characteristic intonation patterns. This might include a regretful-sounding contour (perhaps sampled from real speaker data) applied when apologizing (*Sorry, that flight is sold out*) or a cheerful-sounding greeting. Although the concept-to-speech module can be implemented as just a text-to-speech system that take the advantage of the extra semantic knowledge to generate appropriate prosody, the most natural speech rendition is still to play back a prestored waveform for the entire message. This is why the concept-to-speech module usually relies heavily on playback of template waveform. However, it is obvious that we can't record every possible message like "*Flight [flight_no] is schedule to land at [sch_time]*" in Figure 17.24. Instead, a carrier sentence can be recorded and the slots can then be replaced with real information. The slot can be synthesized with an adapted TTS, which essentially eliminates the need for a front end in the TTS system.

One problem of this approach is that the same prosody is used for a word regardless of where it appears, which results in lower naturalness, because prosodic context is important for natural speech. Enhanced quality can be achieved by having different instances of those slot words, depending their contexts. For example, we can have different *one* recordings depending on whether it is the first digit on a flight number, the second, or the last. Determining the number of different contexts where a slot needs to be recorded is typically done much like the context-dependent acoustic modeling discussed in Chapter 9. This technique increases the naturalness, at the expense of increasing the number of necessary recordings.

17.7.3. Other Renditions

So far, we have assumed that a dialog system may be used only in a speech-only modality. Although such systems have found many applications, multi+modal interaction may be more compelling, as discussed in Chapter 18. In fact, voice output might not be the best information carrier in such an environment. For example, the latest wireless phones are equipped with an LCD screen that allows for e-mail and Web access. If a high-resolution screen is available, the renditions mechanism will likely be visually oriented.

When renditions become visually oriented, the message generation component needs to be replaced by a graphic display component. Since GUI has been the dominant platform for deploying major computer applications today, the behavior and technique of such a display component is well studied and documented [17]. The SLU system needs only to pass the semantic representation from the dialog management module to a GUI rendering module. Of course, the GUI rendering module should also be equipped with domain knowledge to generate best rendering to convey the dialog message. MiPad [22] is such an example and is discussed in Chapter 18.

17.8. EVALUATION

How do we define a quantitative measure for understanding? Evaluation of understanding and dialog is a research topic on its own. We review a number of research techniques being pursued.

17.8.1. Evaluation in the ATIS Task

An application used for development, testing, and demonstration of a wide variety of dialog systems is the Air Travel Information Service (ATIS) task, sponsored by the DARPA Spoken Language Systems program [20]. In this task, users ask about flight information and make travel arrangements. To enable consistent evaluation of progress across systems, a corpus of data for this task has been collected and shared among research sites.

The application database contains information about flights, fares, airlines, cities, airports, and ground services, organized in a relational schema. Most user queries, though they may require some system interaction in order to specify fully, can be answered with a single relational query. The ATIS data collection is done using the wizard-of-oz framework.[4] A user interacts with the system as though working with a fully automated travel planner. Hidden human *wizards* were used in the data-collection process to provide efficient and correct responses to the subjects. A typical scenario presented as a task for a subject to accomplish by means of the automated assistant is as follows:

> Plan the travel arrangements for a small family reunion:
> First pick a city where the get-together will be held. From three different cities (of your choice), find travel arrangements that are suitable for the family members who typify the *economy*, *high class*, and *adventurous* life styles.

After data collection, each query was classified as context dependent or context independent. A context-dependent query relies partially on past queries for specification, such as "*Is that a non-stop flight*?" Many of the system tests based on ATIS require not only accuracy of speech recognition (the user's spoken query), but also semantic interpretation sufficient to construct an SQL query to the database and correctly complete the desired transaction. Evaluation of ATIS was based on three benchmarks: SPREC (speech recognition performance), NL (natural language understanding for text transcription of spoken utterances), and SLU (spoken language understanding). For SLU systems we are interested only in the last two benchmarks.

With the help of constrained domain of ATIS, correct understanding can be translated into correct database access. Since database access is usually done via SQL database query, the evaluation of understanding can be performed in the domain of generated SQL queries. However, it is still ambiguous when someone would like to query flights around 11:00 a.m. For the purpose of understanding, how wide a time frame is *around* considered to be?

Many examples of queries contain some ambiguities. For instance, when querying about the flights between city X and Y, should the system display only the flights from X to Y; or flights in both directions. To alleviate the ambiguity, each release of ATIS training corpus was accompanied by a *Principles of Interpretation* document that has standard definitions of the meaning of such terms like *around* (means within a 15-minute window) and *between* (means only *from*).

Once the correct understanding is represented as an SQL query, ATIS can be easily evaluated by comparing the SQL queries generated by SLU systems against the standard labeled SQL queries. The utterances in ATIS are classified into three types:

[4] The wizard-of-oz data collection framework is described in Chapter 18.

- A—semantically independent of earlier utterances, so per-turn semantic interpretation can uniquely identify the semantic intent.
- D—semantically dependent upon earlier utterances, so discourse knowledge is required to provide full interpretation.
- X—unevaluatable, so a response such as *No answer* or *I don't understand you, could you repeat yourself* is considered a right answer.

The other debatable item is whether a *No answer* output for type A and D utterances should be treated equally as a false SQL query. In the original 1991 ATIS evaluation, a false SQL query for type A and D utterances is penalized twice as heavily as a *No answer* output for type A and D utterances. However, the decision was dropped for the 1993 ATIS evaluation. ATIS decided not to evaluate dialog component for three reasons. First, dialog alters users' behavior during data collection. Users' utterances are highly contingent on the performance of the wizard-of-oz system, so the data collected has little use for systematic training and testing. Second, the SLU systems would likely have to be tested by real subjects. Third, the evaluation of dialog behavior is highly subjective, since effectiveness and user friendliness are generally vaguely defined.

17.8.2. PARADISE Framework

The evaluation of a dialog system is subjective in nature and is typically done in an end-to-end fashion. In such a framework, objective criteria like number of dialog turns and system throughput, and subjective measures like user satisfaction, are typically used.

One of the most sophisticated systems for evaluating dialog systems ever developed is the PARAdigm for Dialog System Evaluation (PARADISE) [57]. The designers of this framework took a comprehensive view of the many potential factors affecting dialog evaluation, in particular the distinction between measuring success of transaction (quality) and cost of the dialog, both in human and system terms. A decision-theoretic method, as shown in Figure 17.28, is used to explicitly weight these various disparate factors to achieve a unified measure. In addition, the PARADISE metrics can derive discrete scores for subdialogs, which is useful for diagnosis, comparison across systems, and tuning.

A simple measure for task success can be the following question: "*Was all the needed information exchanged, in the correct direction (user to system, system to user) at each step*?" PARADISE provides a framework for defining, for any interaction in a limited domain, a simplified representation of the minimal required information and its directionality. In PARADISE terms, this is an *attribute-value matrix* (AVM) showing the names and instantiations of required elements at dialog completion. This could be derived from reference frames for each required concept in a dialog exchange, with mandatory slots marked for legal completions. Once such reference frames or matrices are available, different dialog strategies that address the same function can be compared over many instantiations (test dialog sessions), using statistical measures that assess confusability and length.

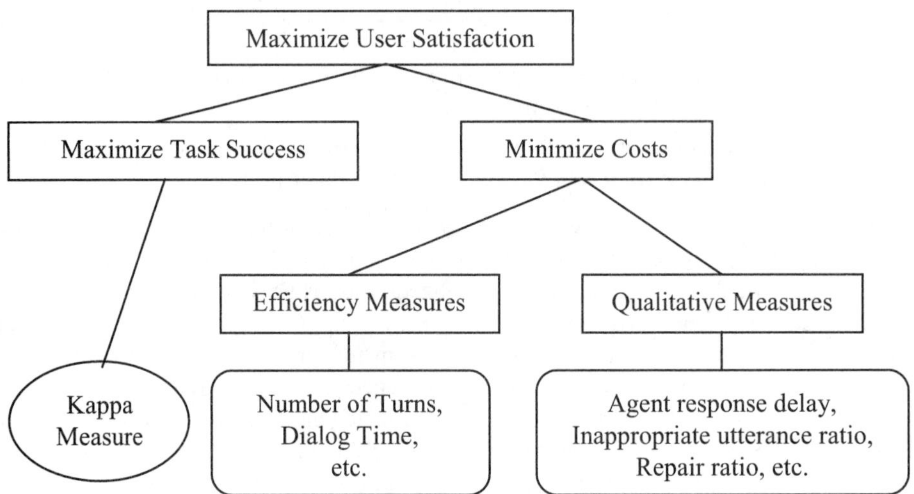

Figure 17.28 PARADISE's structure of objectives for spoken dialog performance [57].

For example, imagine an ATIS-like application that had the following information attributes, with the possible values listed in Table 17.5. An utterance such as "*I want to go from Torin to Milan*" communicates legal DC and AC attribute values from user to system. This is a limited-domain system by assumption, so confusions are assumed to occur within the possible values of the application. For example, if the system instantiates the *Depart-City* (DC) slot with *Trento* instead of *Torin* after processing the given sample utterance, it is a confusion that can be recorded in a confusability matrix over all dialog test sessions. A subsection of such a possible confusability matrix, covering only the DC and AC attributes, is shown in Table 17.6, which shows only confusion within an attribute type that covers a consistent vocabulary (city names, instantiating the DC and AC attributes). In practice, however, the full matrix might show confusions across attribute types, such as *morning* for *Milan*, etc.

Given a confusability matrix M over all possible attributes in the application, we can apply the Kappa coefficient [48] to measure the quality characterizing the task's success at meeting the information requirements of the application:

Table 17.5 Attribute-value table [57].

Attribute	**Possible Values**
Depart-City (DC)	Milan, Rome, Torin, Trento
Arrival-City (AC)	Milan, Rome, Torin, Trento
Depart-Range (DR)	Morning, evening
Depart-Time (DT)	6am, 8am, 6pm, 8pm

Table 17.6 Confusability matrix for city identification [57].

	Depart-City				**Arrival-City**			
Data	Milan	Rome	Torin	Trento	Milan	Rome	Torin	Trento
Milan (depart)	**22**		1		3			
Rome (depart)		**29**						
Torin (depart)	4		**16**	4			1	
Trento (depart)	1	1	5	**11**			1	
Milan (arrive)	3				**20**			
Rome (arrive)						**22**		
Torin (arrive)			2		1	1	**20**	5
Trento (arrive)			1		1	2	8	**15**
sum	30	30	25	15	25	25	30	20

$$\kappa = \frac{P(A) - P(E)}{1 - P(E)} \tag{17.8}$$

where $P(A)$ is the proportion of times that the AVMs for the actual set of dialog agree with the AVMs for the interpreted results, and $P(E)$ is the proportion of times that AVMs for the dialog and interpreted results are expected to agree by chance. $P(E)$ can be estimated by $P(E) = \sum_{i=1}^{n} \left({}^{t_i}\!/_{T} \right)^2$, where t_i is the sum of the frequencies in column i of M and T is total frequencies $(t_1 + \cdots + t_n)$ in M. The measure of $P(A)$ (how well or poorly the application did in information extraction) is calculated simply by examining how much of the total count occurs on the diagonal: $P(A) = \sum_{i=1}^{n} M(i,i)/T$.

In addition to task success, system performance is also a function of several cost measures. Cost measures include efficiency measures, such as the number of dialog turns or task completion time; as well as qualitative measures, such as style of dialog or how good the repair mechanism is. If a set of test dialogs is available, with experimentally measured user satisfaction (the predicted categories), the kappa measure, and quantitative measures of cost (denoted as c_i, such as counts of repetitions, repairs etc.), linear regression can be used, over the z-score normalization of these predictor terms, to identify and weight the most important predictors of satisfaction for a given system. Thus, the performance can be defined as:

$$\text{Performance} = \alpha * \mathfrak{A}(\kappa) - \sum_{i=1}^{n} w_i * \mathfrak{A}(c_i) \tag{17.9}$$

where $\mathfrak{A}$ is the z-score normalization function $\mathfrak{A}(x) = \dfrac{x - \overline{x}}{\sigma_x}$.

Evaluating a dialog system involves having a group of users perform tasks with ideal outcomes. Then the cost measures and task success kappa measure are estimated. These measures are used to derive the regression weights in Eq. (17.9). Once the regression weights are attained, one could possibly predict the user satisfaction when a subpart of the dialog system is improved.

17.9. CASE STUDY—DR. WHO

Dr. Who is a project at Microsoft Research on its multimodal dialog system development. It incorporates many of the dialog technologies described in this chapter. We use Dr. Who's SLU engine as an example to illustrate how to effectively create practical systems [22, 58–61]. It follows the mathematical framework illustrated in Eq. (17.1). The system architecture is shown in Figure 17.29. Since it intends to serve as a general architecture for multimodal dialog systems, it makes some simple assumptions at the architecture level. First, it replaces the *speech recognizer* and *sentence interpretation* modules with a *semantic parser* for each modality. The *response rendering* is merged into *dialog manager* with different XSL style sheets for each media output.

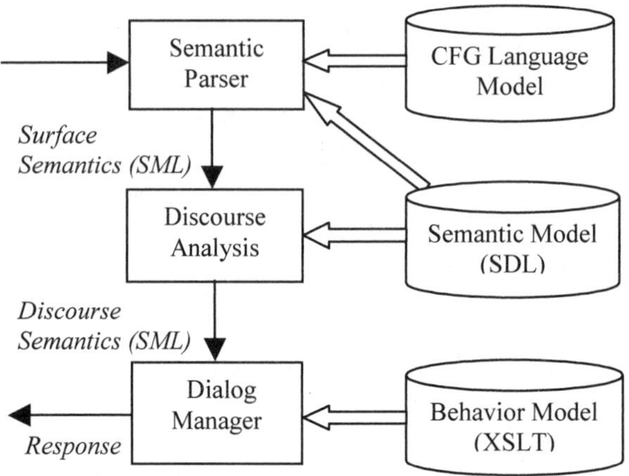

Figure 17.29 The Dr. Who system architecture [60].

17.9.1. Semantic Representation

Semantic representation is a critical part in Dr. Who's SLU engine design. Essentially, the semantic objects are an abstraction of the speech acts, the domain knowledge, and the application logic. They are designed to encapsulate the respective language models and dialog actions that govern their creation and behaviors. The system components communicate with one another through events surrounding the semantic objects. In this view, the dialog (including logic inferences) is an integral part of the discourse semantic evaluation process.

There are two types of semantic objects in Dr. Who. The first type is the *functional* semantic object that is used to represent linguistic expressions in the user's utterance. The second type is the *physical* semantic object that is used to represent real-world entities related to the application domain. Both types of semantic objects are represented by semantic frames and specified in the semantic markup language (SML), which is an extension of XML. Following the principles of the XML schema, Dr. Who defines the schema of SML in another XML called semantic definition language (SDL). SDL is designed to support many discourse and dialog features. In addition, SDL is suited to represent the domain knowledge via the application schema, the hierarchy of the semantic objects, and the semantic inference rules.

The format of various semantic classes follows SDL representations in Dr. Who. The terminal and nonterminal nodes on the parse are denoted in SDL with tags `<verbatim>` and `<class>`, respectively. These tags refer to the semantic objects and have the *`name`* and *`type`* attributes. The `type` attribute corresponds to the entity type the semantic object eventually would be converted to; it plays a key role in inheritance and polymorphism, as described in Section 17.3.1. When a semantic object is unique in its type, SDL can automatically assume its type as the name. In addition, SDL defines a `<cfg>` tag for the language model that governs the instantiation of a semantic object ,and the language model could be stored in another file. An `<expert>` tag can be defined for the system resource to physically convert a semantic object to a domain entity. Finally, the tag `<slot>` in SDL defines the descendant for a nonterminal node.

Take the semantic class for Microsoft employee directory as an example. The simple application answers queries on an employee's data such as office location, phone number, hiring date, etc. An item that can be asked is a semantic terminal `DirectoryItem` as defined in Figure 17.30. To allow users to ask more than one directory item at one dialog turn, a multiple semantic class `DirectoryItems` is also defined recursively, as shown in Figure 17.30.

```
<verbatim type="DirectoryItem" ...>
   <prod name="office"/>
   <prod name="phone"/>
   <prod name="hiring date"/>
   ...
</verbatim>
<class type="DirectoryItems" ...>
   <slot type="DirectoryItem"/>
   <slot type="DirectoryItems"/>
   <cfg ref="DirectoryItems.cfg"/>
</class>
```

Figure 17.30 The terminal semantic class `DirectoryItem` and nonterminal semantic class defined in Dr. Who using SDL. Note that the definition of `DirectoryItems` contains a recursive style, which can accommodate more than one `DirectoryItem` [59].

The <prod> tags inside a terminal semantic object indicate that the terminal is of an enumeration type, and all the possible values are text normalized to the string values of the *name* attribute. The main speech act, the query, is modeled by the functional semantic class DirectoryQuery, as shown in Figure 17.31.

```
<class type="DirectoryQuery" …>
    <slot type="Person"/>
    <slot type="DirectoryItems"/>
    <expert clsid="…"/>
    <cfg ref="Directory.cfg"/>
</class>
<include ref="PeopleGrammar.sdl"/>
```

Figure 17.31 The main semantic class DirectoryQuery defined in Dr. Who using SDL [59].

The semantic object can be instantiated following the language model in "Directory.cfg" and, once instantiated, is handled by a system object identified by its class id (clsid). The system object then formulates the query language that retrieves the data from the database. It is also possible to embed the XML version of the query language (e.g., XQL) within the <expert> tag. Semantic models can be nested and reused, as shown in the <include> tag in the above example, where the semantic model for people is referred.

17.9.2. Semantic Parser (Sentence Interpretation)

For speech modality, Dr. Who employs a speech recognizer with unified language models [62] that take advantage of both rule-based and data-driven approaches, as discussed in Chapter 11. Once we have text transcription of user's utterances, a robust chart parser [61] similar to the one described in Section 17.4.1 is used for sentence interpretation.

The emphasis of sentence interpretation is to annotate the user's utterance in a meaningful way to generate functional semantic entities. Essentially, the surface SML represents a semantic parse. Thus, after a successful parse, the corresponding surface semantic objects are instantiated based on the semantic classes whose CFG grammars are fired. While in SDL we use static tags such as <class> and <verbatim> for the semantic classes, the instances of a semantic object use the object name as the tag in SML. For example, the surface SML for an utterance "*What is the phone number for Kuansan*" is shown in Figure 17.32.

```
<DirectoryQuery …>
   <PersonByName type="Person" parse="kuansan">
       Kuansan
   </PersonByName>
   <DirectoryItem type="DirectoryItem" parse="phone num-
ber">
       phone
   </DirectoryItem>
</DirectoryQuery>
```

Figure 17.32 The surface semantic object DirectoryQuery represented in SML after a successful parse [59].

17.9.3. Discourse Analysis

As mentioned in Section 17.5, the goal of discourse analysis is to resolve surface semantic objects to discourse semantic objects. For the surface semantic object in Figure 17.32, the discourse engine binds the three semantic objects (i.e., the person, the directory item, and the directory query itself) to real-word and functional entities represented in the SML example, as shown in Figure 17.33.

```
<DirectoryQuery ...>
   <Person id="kuansanw" parse="kuansan">
      <First>Kuansan</First>
      <Last>Wang</Last>
      ...
   </Person>
   <DirectoryItem parse="phone number">
      <phone>+1(425)703-8377</phone>
   </DirectoryItem>
</DirectoryQuery>
```

Figure 17.33 The discourse semantic objects for the surface semantic object illustrated in Figure 17.32 [59].

Note that the parse string from the user's original utterance is kept so that the rendering engine can choose to rephrase the response using the user's wording.

When an error occurs, the semantic engine inserts an <error> tag in the offending semantic objects with a code indicating the error condition. For example, if the query is for a person named *Derek*, the discourse SML might appear as shown in Figure 17.34.

```
<DirectoryQuery status="TBD" focus="Person" ...>
   <PersonByName type="Person" parse="Derek" status="TBD"...>
        <error scode="1" count="27"/>
        <Person id="derekba">
           <First>Derek</First>
           <Last>Baines</Last>
           ...
        </Person>
        <Person id="dbevan">
           <First>Derek</First>
           <Last>Bevan</Last>
           ...
        </Person>
           ...
    </PersonByName>
           ...
</DirectoryQuery>
```

Figure 17.34 A discourse semantic object in Dr. Who contains an <error> tag indicating the error condition [59].

In Figure 17.34, semantic objects that cannot be converted (e.g., `DirectoryQuery` and `PersonByName`) are flagged with a status "`TBD`". Discourse SML also marks the dialog focus, as in the `DirectoryQuery`, that indicates the places where the semantic evaluation process fails to continue. These two cues assist the behavior model in deciding the appropriate error-repair responses.

Dr. Who uses three priority types of entity memory (discourse memory, explicit, and implicit turn memory) to resolve relative expressions. Anaphora and deixis are treated as common semantic classes, so they can be resolved according to the algorithm described in Section 17.5.1.1. Ellipsis is treated as an automatic inference. Unless marked as NO _INFER in the semantic class definition, every slot in a semantic class can be automatically inferred. The strategy to automatically resolve partially specified entities is as follows.

During the evaluation stage, a partially filled semantic object is first compared with the entities in the three-entity memory based on the type compatibility. If a candidate is found, the discourse analysis module then computes a goodness-of-fit score by consulting the knowledge base and considering the position of the entity in the memory list. The semantic object is converted immediately to the entity from the memory if the score exceeds the threshold. In the process, all the actions implied by the entities are carried out following the order in which the corresponding semantic objects are converted. For example, the second user's query in the dialog illustrated in Figure 17.35 contains an ellipsis reference to `DirectoryItem` *`office`*, which can be resolved using the discourse entity memory.

U: Where is his office?
S: The office is in building 31, room 1362.
U: How about Kuansan's?
S: The office is in building 31, room 1363.

Figure 17.35 A dialog example in the Dr. Who system. The second user's query contains an ellipsis reference to `DirectoryItem` *`office`* [59].

17.9.4. Dialog Manager

To support mixed-initiative multimodal dialogs, Dr. Who employs a plan-based approach instead of dialog grammars. The dialog manager that handles dialog events surrounding semantic objects is very similar to a GUI program that handles GUI events surrounding graphical objects. These events can be handled synchronously or asynchronously based on various implementation considerations. In addition, the design enables a seamlessly integrated GUI and speech interface for multimodal applications to embrace the same human-computer interaction model.

Dr. Who SLU engine can use XSL-transformations (XSLT) [59] for specifying the behavior of a plan-based dialog system. XSLT, a recent World Wide Web Consortium (W3C) standard, is a specialized XML intended for describing the rules of how a structured document in XML can be transformed into another, say in a text-to-speech markup language for speech rendering or the hypertext markup language (HTML) for visual rendering. Its core construct is a collection of predicate-action pairs: each predicate specifies a textual pattern in the source document, and the corresponding action will produce a text segment in the output whenever the pattern specified by the predicate is seen in the source document. The output segment is specified through a programmable, context-sensitive template. XSLT defines a rich set of logical controls for composing the templates. The basic programming paradigm bears close resemblance to a logical programming language, such as Prolog, which facilitates logic inference in plan-based systems. As a result, XSLT possesses sufficient expressive power for implementing crucial dialog components, ranging from defining dialog plans, realizing dialog strategies, and generating natural language, to manipulating prosodic markup for text-to-speech synthesis and creating dynamic HTML pages for multimodal applications.

Assuming TTS output, the planning rules that render the discourse SML of Figure 17.33 in text can be expressed in XSLT as shown in Figure 17.36.

```
<xsl:template match="DirectoryQuery[@not(status)]">
    For <xsl:apply-templates select="Person"/>, the
    <xsl:apply-templates select="DirectoryItem"/>.
</xsl:template>
<xsl:template match="Person">
    <xsl:value-of select="First"/>
    <xsl:value-of select="Last"/>
</xsl:template>
<xsl:template match="DirectoryItem">
    <xsl:apply-templates/>
</xsl:template>
<xsl:template match="phone">
    phone number is <xsl:value-of/>
</xsl:template>
```

Figure 17.36 A TTS response-rendering rule for discourse SML of Figure 17.33. This rule generates a text message "*For Kuansan Wang, the phone number is +1(425)703-8377*" [59].

This rule leads to a response *For Kuansan Wang, the phone number is +1(425)703-8377*. Elaborated functions, such as prosodic manipulations in text to speech markup, can be included accordingly. To change the output to Web presentation, the above XSLT style sheet can be slightly modified for rendering in HTML as a table, as shown in Figure 17.37.

The Dr. Who SLU engine has a concept called *logical container* as a dialog property to be encapsulated in a semantic class. Three types of logical containers can be accessed in the definition of semantic classes. A semantic class is an AND type container if all its attributes must be evaluated successfully. If this requirement is not met, the evaluation of the AND type semantic object is considered failed, which will prompt the system to post a dialog-repair event. An OR type container requires at least one attribute to be successfully evaluated. Similarly, for an exclusive or (XOR) type container, one and only one attribute must be successfully evaluated.

```
<xsl:template match="DirectoryQuery[@not(status)]">
<TABLE border="1">
    <THEAD><TR>
          <TH>Properties</TH>
          <TH><xsl:apply-templates select="Person"/> </TH>
    </TR></THEAD>
    <TBODY><xsl:apply-templates select="DirectoryItem"/>
    </TBODY>
</TABLE>
</xsl:template>
<xsl:template match="phone">
    <TR> <TD>phone</TD> <TD> <xsl:value-of /> </TD> </TR>
</xsl:template>
```

Figure 17.37 An HTML response-rendering rule for discourse SML of Figure 17.33. It generates a visual table representation rather than a text message [59].

Figure 17.38 shows a semantic class hierarchy corresponding to the partial plan shown in Figure 17.20. The dialog goal—to gather information for booking a flight—corresponds to the highest-level semantic class *Book Flight*. Evaluating this semantic class drives the dialog system to traverse down the semantic class structure, eventually fulfilling all the steps necessary to achieve the dialog goal. This is achieved by recursively evaluating the attributes, instantiating semantic objects actively if necessary. The logical relation of each semantic class determines the rules of instantiation and dialog repair. For instance, if the user specifies the trip to be one way only, the evaluation of the *One Way Flag* semantic class becomes successful. As the *Inbound Trip* semantic class is an XOR container, the dialog system bypasses the evaluation of the *Itinerary* attribute in the *Inbound Trip* semantic class.

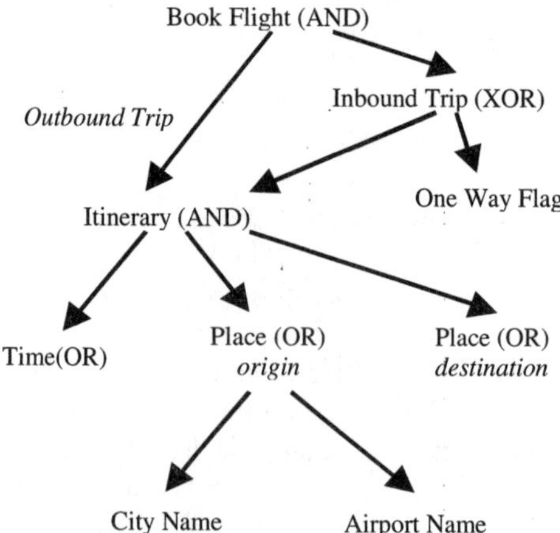

Figure 17.38 A semantic tree hierarchy corresponding to the partial plan shown in Figure 17.20 in an airline reservation application [58].

The *Itinerary* semantic class encapsulates the basic elements to specify a one-way trip. Since it is designated as an AND type container, the dialog manager tries to acquire any missing information by actively instantiating the corresponding semantic classes it contains. The active instantiation event handlers for these classes solicit information from the user by implementing certain prompting strategy. On the other hand, the *Place* semantic class, which is used to denote both the origin and the destination, is implemented as an OR container. The user may specify the location by either the city name or the airport name.

17.10. Historical Perspective and Further Reading

Traditional natural language research has its roots in symbolic systems. Motivated by the desire to—understand cognitive processes, the underlying theories tend to be from linguistics and psychology. As a result, coverage of phenomena of theoretical interest (usually a rare occurrence) has traditionally been more important than developing systems with a broad coverage.

On the other hand, speech recognition research is driven to produce practical usable applications. Techniques motivated by knowledge of human processes have been less important than techniques that can be used for real applications. In recent decades, interest has grown in the use of engineering techniques in computational language processing, although the use of linguistic knowledge and techniques in engineering has lagged somewhat. The ATIS program sponsored by DARPA had a very significant influence upon the SLU research community [34]. For the first time, the research community started seriously evaluating SLU systems on a quantitative basis, which revealed that many traditional NL techniques designed for written language failed to deal with spoken language in practice.

For limited-domain SLU applications, vocabularies are typically about 2000 words. CMU's Phoenix SLU system [63] set the benchmark for domain-specific spoken language understanding in the DARPA ATIS programs. It is based on an island-driven semantic parsing approach. After years of engineering, the speech understanding error rate ranges from 6% to 41%. Since conversational repairs in human-human dialog can often be in the same range for these systems, the determining factor in these domain-specific SLU applications may not be the error rates but instead the ability of the system to manage and recover from errors. Many of these were described in detail in the *Proceedings of the DARPA Spoken Language Systems Technology Workshop* published by Morgan Kaufmann from 1991 to 1995. The special issue *of Speech Communication on Spoken Dialog* [45] also includes several state-of-the-art system descriptions.

Allen's *Natural Language Understanding* [1] is a good book on natural language understanding with a comprehensive coverage of syntactic processing, semantic processing, discourse analysis, and dialog agent. Knowledge and semantic representation comprise the most import fundamental issue for symbolic artificial intelligence. Several AI textbooks [33, 56, 65] contain comprehensive description of knowledge representation. The use of semantic frames can be traced back to case frames or structures proposed by Fillmore [16]. SAM [44] is among the first systems using semantic frames and template matcher for natural language processing. The description of semantic classes and frames in this book mostly follows the systematic treatment of semantic classes in the Dr. Who system [58-60].

Speech-act (sometimes called dialog-act) theory was first proposed by Austin [4] and further developed by Searle [42]. It is an important concept in dialog systems. You can acquire more information about speech-act theory and its application to dialog systems from [12, 40, 43]. Cohen [10] provides a good comparison of different approaches for dialog modeling, including dialog grammar (finite state), plan-based and agent-based (dialog as teamwork). We treat agent-based dialog modeling as an extension of plan-based dialog modeling, as described in Section 17.6.2. Agent-based approach is a very popular framework for multimodal user interface, and interested readers can refer to [11]. Hudson and Newell [23] incorporate probability into finite state dialog management to handle uncertainty in input modalities, such as pen-based interface, gesture recognition, and speech recognition. J. Allen's book [1] has a systematic description of plan-based dialog systems. De Mori's *Spoken Dialogs with Computers* [39] is another excellent book that contains dialog systems and related technologies.

Much of the content in this chapter follows the architecture and implementation of semantic frame based approaches. In particular, we use plenty of descriptions and examples of the Dr. Who SLU engine developed at Microsoft Research [22, 58-60]. The description of plan-based systems is based on semantic frame representation and pattern matching. There is no need for explicit dialog-act analysis and logic reasoning, since these important knowledge sources are encapsulated in the semantic frames.

In addition to the semantic frame-based approach, there other approaches that rely on formal NL parsing, logic form representation, speech acts, and logic inference [2, 41]. Message generation for telephone application is well studied and reported in [5, 6, 49], which provide experimental results for various prompting strategies. Most evaluation schemes for the SLU systems focus on the end-to-end system. Human factors are important in overall evaluation [7, 35, 52].

REFERENCES

[1] Allen, J., *Natural Language Understanding*, 2nd ed., 1995, Menlo Park CA, The Benjamin/Cummings Publishing Company.

[2] Allen, J.F., *et al.*, "Trains as an Embodied Natural Language System," *AAAI-95 Symposium on Embodied Language and Action*, 1995.

[3] Andernach, T., M. Poel, and E. Salomons, "Finding Classes of Dialogue Utterances with Kohonen Networks," *Proc. of the NLP Workshop of the European Conf. on Machine Learning (ECML)*, 1997, Prague, Czech Republic.

[4] Austin, J.L., *How to Do Things with Words*, 1962, Cambridge, MA, Harvard University Press.

[5] Basson, S., "Integrating Speech Recognition and Speech Synthesis in the Telephone Network," *Proc. of the Human Factors Society 36th Annual Meeting*, 1992.

[6] Basson, S., "Prompting The User in ASR Applications," *Proc. of COST232 (European Cooperation in Science and Technology) Workshop*, 1992.

[7] Basson, S., *et al.*, "User Participation and Compliance in Speech Automated Telecommunications Applications," *Proc. of the Int. Conf. on Spoken Language Processing*, 1996, pp. 1680-1683.
[8] Biber, D., *Variation Across Speech and Writing*, 1988, Cambridge University Press.
[9] Clark, H.H. and S.E. Haviland, "Comprehension and the Given-New Contract" in *Discourse production and comprehension*, R.O. Freedle, Editor 1977, Norwood, NJ, Ablex Publishing Corporation, pp. 1-38.
[10] Cohen, P., "Models of Dialogue," *Proc. of the Fourth NEC Research Symposium*, 1994, SIAM Press.
[11] Cohen, P.R., "The Role of Natural Language in a Multimodal Interface," *Proc. of the ACM Symposium on User Interface Software and Technology*, 1992, pp. 143-149.
[12] Cohen, P.R. and C.R. Perrault, "Elements of a Plan-Based Theory of Speech Acts," *Cognitive Science*, 1979, **3**(3), pp. 177-212.
[13] Constantinides, P.H., S. Tchou, C. Rudnicky, "A Schema Based Approach to Dialog Control," *Proc. of the Int. Conf. on Spoken Language Processing*, 1998, pp. 409-412.
[14] Core, M. and J. Allen, "Coding Dialogs with the DAMSL Annotation Scheme," *Proc. AAAI Fall Symposium on Communicative Action in Humans and Machines*, 1997.
[15] Davies, K., *et al.*, "The IBM Conversational Telephony System For Financial Applications," *EuroSpeech'99*, 1999, Budapest, Hungary, pp. 275-278.
[16] Fillmore, C.J., "The Case for Case" in *Universals in Linguistic Theory*, E. Bach and R. Harms, eds. 1968, New York, NY, Holt, Rinehart and Winston.
[17] Galitz, W.O., *The Essential Guide to User Interface Design: An Introduction to Gui Design Principles and Techniques*, 1996, John Wiley & Sons.
[18] Grosz, B., M. Pollack, and C. Sidner, eds. *Discourse*, in Foundations of Cognitive Science, ed. M. Posner, 1989, MIT Press.
[19] Heeman, P.A., *et al.*, "Beyond Structured Dialogues: Factoring Out Grounding," *Proc. of the Int. Conf. on Spoken Language Processing*, 1998, Sydney, Australia.
[20] Hemphill, C.T., J.J. Godfrey, and G.R. Doddington, "The ATIS Spoken Language Systems Pilot Corpus," *Proc. of the Speech and Natural Language Workshop*, 1990 pp. 96-101.
[21] Hitzeman, J., *et al.*, "On the Use of Automatically Generated Discourse-Level Information in a Concept-to-Speech Synthesis System," *Proc. of the Int. Conf. on Spoken Language Processing*, 1998, Sydney, Australia, pp. 2763-2766.
[22] Huang, X., *et al.*, "MIPAD: A Next Generation PDA Prototype," *Int. Conf. on Spoken Language Processing*, 2000, Beijing, China.
[23] Hudson, S.E. and G.L. Newell, "Probabilistic State Machines: Dialog Management for Inputs with Uncertainty," *Proc. of the ACM Symposium on User Interface Software and Technology*, 1992, pp. 199-208.

[24] Hulstijn, J. and A.V. Hessen, "Utterance Generation for Trans. Dialogues," *Int. Conf. on Spoken Language Processing*, 1998, Sydney, Australia.

[25] Jackendoff, R.S., *X' Syntax: A Study of Phrase Structure*, 1977, Cambridge, MA, MIT Press.

[26] Jelinek, F., *et al.*, "Decision Tree Parsing using Hidden Derivational Model," *Proc. of the ARPA Human Language Technology Workshop*, 1994, pp. 272-277.

[27] Jurafsky, D., L. Shriberg, and D. Biasca, *Switchboard SWBD-DAMSL Shallow-Discourse-Function Annotation Coders Manual, Draft 13*, 1997, http://www.colorado.edu/linguistics/jurafsky/manual.august1.html.

[28] LDC, *Linguistic Data Consortium*, 2000, http://www.ldc.upenn.edu/ldc/noframe.html.

[29] Miller, S., *et al.*, "Recent Progress in Hidden Understanding Models," *Proc. of the ARPA Spoken Language Systems Technology Workshop*, 1995, Austin, Texas, Morgan Kaufmann, Los Altos, CA, pp. 22-25.

[30] Miller, S. and R. Bobrow, "Statistical Language Processing Using Hidden Understanding Models," *Proc. of the Spoken Language Technology Workshop*, 1994, Plainsboro, New Jersey, Morgan Kaufmann, Los Altos, CA, pp. 48-52.

[31] Minsky, M., "A Framework for Representing Knowledge" in *The Psychology for Computer Vision*, P.H. Winston, Editor 1975, New York, NY, McGraw-Hill.

[32] Nilsson, N.J., *Principles of Artificial Intelligence*, 1982, Berlin, Germany, Springer-Verlag.

[33] Nilsson, N.J., *Artificial Intelligence: A New Synthesis*, 1998, Academic Press/Morgan Kaufmann.

[34] Pallett, D.S., *et al.*, "1994 Benchmark Tests for the ARPA Spoken Language Program," *Proc. of the 1995 ARPA Human Language Technology Workshop*, 1995, pp. 5-36.

[35] Polifroni, J., *et al.*, "Evaluation Methodology for a Telephone-Based Conversational System," *The First Int. Conf. on Language Resources and Evaluation*, 1998, Granada, Spain, pp. 42-50.

[36] Prevost, S. and M. Steedman, "Specifying Intonation from Context for Speech Synthesis," *Speech Communication*, 1994, **15**, pp. 139-153.

[37] Reinhart, T., *Anaphora and Semantic Interpretation*, Croom Helm Linguistics Series, 1983, University of Chicago Press.

[38] Roberts, D., *The Existential Graphs of Charles S. Peirce*, 1973, Mouton and Co.

[39] Sadek, D. and R. De Mori, "Dialogue Systems" in *Spoken Dialogues with Computers*, R. De Mori, Editor 1998, London, UK, pp. 523-561, Academic Press.

[40] Sadek, M.D., "Dialogue Acts are Rational Plans," *Proc. of the ESCA/ETRW Workshop on the Structure of Multimodal Dialogue*, 1991, Maratea, Italy, pp. 1-29.

[41] Sadek, M.D., *et al.*, "Effective Human-Computer Cooperative Spoken Dialogue: The AGS Demonstrator," *Proc. of the Int. Conf. on Spoken Language Processing*, 1996, Philadelphia, Pennsylvania, pp. 546-549.

[42] Searle, J.R., *Speech Acts: An Essay in the Philosophy of Language*, 1969, UK, Cambridge University Press.

[43] Searle, J.R. and D. Vanderveken, *Foundations of Illocutionary Logic*, 1985, Cambridge University Press.

[44] Shank, R., *Conceptual Information Processing*, 1975, North Holland, Amsterdam, The Netherlands.

[45] Shirai, K. and S. Furui, "Special Issue on Spoken Dialogue," *Speech Communication*, 1994, **15**.

[46] Shriberg, E.E., R. Bates, and A. Stolcke, "A Prosody-Only Decision-tree Model for Disfluency Detection," *Proc. Eurospeech*, 1997, Rhodes, Greece, pp. 2383-2386.

[47] Sidner, C., "Focusing in the Comprehension of Definite Anaphora" in *Computational Model of Discourse*, M. Brady, Berwick, R., eds., 1983, Cambridge, MA, pp. 267-330, The MIT Press.

[48] Sidney, S. and N.J. Castellan, *Nonparametric Statistics for the Behavioral Sciences*, 1988, McGraw Hill.

[49] Sorin, C. and R.D. Mori, "Sentence Generation" in *Spoken Dialogues with Computers*, R.D. Mori, Editor 1998, London, UK, Academic Press, pp. 563-582.

[50] Souvignier, B., *et al.*, "The Thoughtful Elephant: Strategies for Spoken Dialog Systems," *IEEE Trans. on Speech and Audio Processing*, 2000, **8**(1), pp. 51-62.

[51] Sowa, J.F., *Knowledge Representation: Logical, Philosophical, and Computational Foundations*, 1999, Brooks Cole Publishing Co.

[52] Springer, S., S. Basson, and J. Spitz, "Identification of Principal Ergonomic Requirements for Interactive Spoken Language Systems," *Int. Conf. on Spoken Language Processing*, 1992, pp. 1395-1398.

[53] Standards, N.C.f.I.T., *Conceptual Graph Standard Information*, 1999, http://www.bestweb.net/~sowa/cg/cgdpansw.htm.

[54] Steedman, M., ed. *Parsing Spoken Language Using Combinatory Grammars*, in Current Issues in Parsing Technology, ed. M. Tomita, 1991, Kluwer Academic Publishers.

[55] Steedman, M., "Information Structure and the Syntax-Phonology Interface," *Linguistic Inquiry*, 2000.

[56] Tanimoto, S.L., *The Elements of Artificial Intelligence: An Introduction Using Lisp*, 1987, Computer Science Press, Inc.

[57] Walker, M., *et al.*, "PARADISE: A Framework for Evaluating Spoken Dialogue Agents.," *Proc. of the 35th Annual Meeting of the Association for Computational Linguistics (ACL-97)*, 1997, pp. 271-280.

[58] Wang, K., "An Event Driven Model for Dialogue Systems," *Int. Conf. on Spoken Language Processing*, 1998, Sydney, Australia pp. 393-396.

[59] Wang, K., "Implementation of Dr. Who Dialog System Using Extended Markup Languages," *Int. Conf. on Spoken Language Processing*, 2000, Beijing, China.

[60] Wang, K., "A Plan-Based Dialog System With Probabilistic Inferences," *ICSLP*, 2000, Beijing, China.

[61] Wang, Y., "A Robust Parser For Spoken Language Understanding," *Eurospeech*, 1999, Budapest, Hungary, pp. 2055-2058.

[62] Wang, Y., M. Mahajan, and X. Huang, "A Unified Context-Free Grammar and N-Gram Model for Spoken Language Processing," *Int. Conf. on Acoustics, Speech and Signal Processing*, 2000, Istanbul, Turkey, pp. 1639-1642.

[63] Ward, W., "Understanding Spontaneous Speech: The Phoenix System," *Proc. Int. Conf. on Acoustics, Speech and Signal Processing*, 1991, Toronto, Canada, pp. 365-367.

[64] Ward, W. and S. Issar, "The CMU ATIS System," *Proc. of the ARPA Spoken Language Systems Technology Workshop*, 1995, Austin, Texas, Morgan Kaufmann, Palo Alto, CA.

[65] Winston, P.H., *Artificial Intelligence*, 3rd ed, 1992, Reading, MA, Addison-Wesley.

[66] Young, S.R., *et al.*, "High Level Knowledge Sources in Usable Speech Recognition Systems," *Communications of the Association for Computing Machines*, 1989, **32**(2), pp. 183-194.

CHAPTER 18

Applications and User Interfaces

The ultimate impact of spoken language technologies depends on whether you can fully integrate the enabling technologies with applications so that users find it easy to communicate with computers. How to effectively integrate speech into applications often depends on the nature of the user interface and application. This is why we group user interface and application together in this chapter. In discussing some general principles and guidelines in developing spoken language applications, we must look closely at designing the user interface.

A well-designed user interface entails carefully considering the particular user group of the application and delivering an application that works effectively and efficiently. As a general guideline, you need to make sure that the interface matches the way users want to accomplish a task. You also need to use the most appropriate modality at the appropriate time to assist users to achieve their goals. One unique challenge in spoken language applications is that neither speech recognition nor understanding is perfect. In addition, the spoken command can be ambiguous, so the dialog strategy described in Chapter 17 is necessary to clarify the goal of the speaker. There are always mistakes you have to deal with. It is critical that applications employ necessary interactive error-handling techniques to minimize the impact of these errors. Application developers should therefore fully understand the strengths and weaknesses of the underlying speech technologies and identify the appropriate place to use the spoken language technology effectively.

This chapter mirrors Chapter 2, in the sense that you need to incorporate all the needed components of speech communication to make a spoken language system work well. It is important also to have your applications developed based on some standard application programming interfaces (API), which ensures that multiple applications work well with a wide range of speech components provided by different speech technology providers.

18.1. APPLICATION ARCHITECTURE

A typical spoken language application has three key components. It needs an engine that can be either a speech recognizer or a spoken language understanding system. An *application programming interface* (API) is often used to facilitate the communication between the engine and application, as illustrated in Figure 18.1. Multiple applications can interact with a shared speech engine via the speech API. The speech engine may be a CSR engine, a TTS engine, or an SLU engine. The interface between the application and the engine can be distributed. For example, you can have a client-server model in which the engine is running remotely on the server.

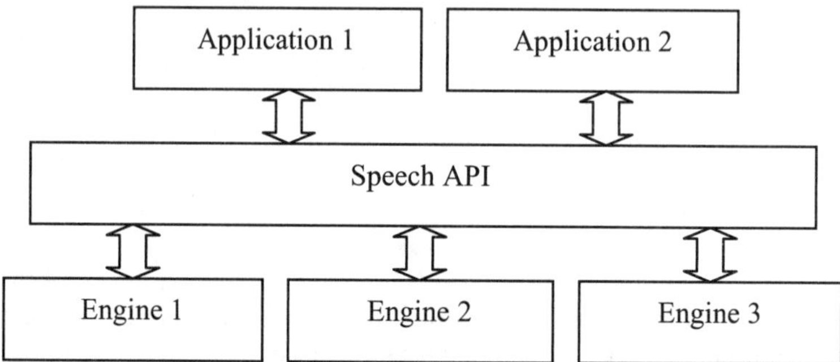

Figure 18.1 In a typical spoken language application architecture, multiple applications can interact with a shared speech engine via the speech API. The speech engine may be a speech recognizer, a TTS converter, or an SLU engine.

For a given API, there is typically an associated toolkit that provides a good development environment and the tools you need in order to build speech applications. You don't need to understand the underlying speech technologies to fully take advantage of state-of-the-art speech engines. Industry-standard based applications can draw upon support from many different speech engine vendors, thus significantly minimizing the cost of your applications development. For the widely used Microsoft Windows®, Microsoft's speech API

(SAPI) brings both engine and application developers together.[1] Alternative standards are available, such as VoiceXML[2] and JSAPI[3].

18.2. TYPICAL APPLICATIONS

There are three broad classes of applications that require different UI design:

- *Office:* This includes the widely used desktop applications such as Microsoft Windows and Office.
- *Home*: TV and kitchen are the centers for home applications. Since home appliances and TV don't have a keyboard or mouse, the traditional GUI application can't be directly extended for this category.
- *Mobile*: Cell phone and car are the two most important mobile scenarios. Because of the physical size and the hands-busy and eyes-busy constraints, the traditional GUI application interaction model requires significant modification.

This section provides descriptions of typical spoken language applications in these three broad classes. Spoken language has the potential to provide a consistent and unified interaction model across these three classes, albeit for these different application scenarios you still need to apply different user interface design principles.

18.2.1. Computer Command and Control

One of the earliest prototypes for speech recognition is *command and control*, which is mainly used to navigate through operating system interfaces and applications running under them. For example, Microsoft Agent is a set of software services that supports the presentation of software agents as interactive personalities within the Microsoft Windows or the Internet Explorer interface. Its command-and-control speech interface is an extension and enhancement of the existing interactive modalities of the Windows interface. It has a character called Peedy, shown in Figure 18.2, which recognizes your speech and talks back to you using a Microsoft SAPI compliant command-and-control speech recognizer and text-to-speech synthesizer.

The speech recognizer used in these command-and-control systems is typically based on a context-free grammar (CFG) decoder. Either developers or users can define these grammars. Associated with each legal path in the grammar is a corresponding executable event that can map a user's command into appropriate control actions the user may want. They possess a built-in vocabulary for the menus and other components. The vocabulary can also be dynamically provided by the application. Command-and-control speech recognition allows the user to speak a word, phrase, or sentence from a list of phrases that the computer is expecting to hear. The number of different commands a user might speak at any time can

---

[1] http://www.microsoft.com/speech

[2] http://www.voicexml.org/

[3] http://java.sun.com/products/java-media/speech/

be in the hundreds. Furthermore, the commands are not just limited to fixed ones but can also contain other open fields, such as "*Send mail to <Name>*" or "*Call <digits>*". With all of the possibilities, the user is able to speak thousands of different commands. As discussed in Chapter 17, a CFG-based recognizer is often very rigid, since it may reject the input utterance that contains a sentence slightly different from what the CFG defines, leading to an unfriendly user experience.

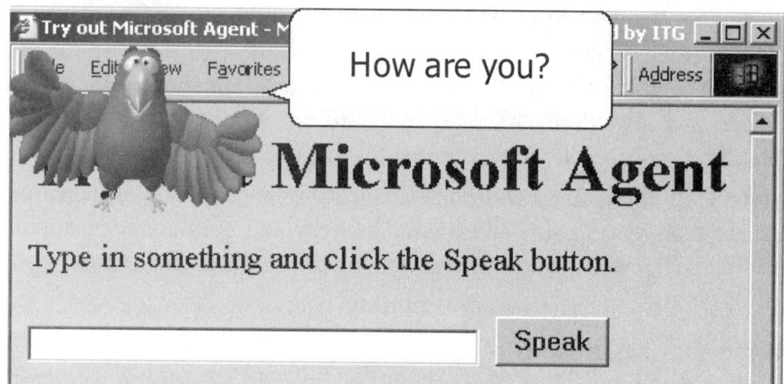

Figure 18.2 A talking character *Peedy*[4] as used in Microsoft Agent. Reprinted with permission from Microsoft Corporation.

Command-and-control recognition might be useful in some of the following situations:

- *Answering questions*. An application can easily be designed to accept voice responses to message boxes and wizard screens. Most speech recognition engines can easily identify *Yes*, *No*, and a few other short responses.
- *Accessing large lists*. In general, it's faster for a user to speak one of the names on a list, such as "*Start running calculator,*" than to scroll through the list to find it. It assumes that the user knows what is in the list. Laurila and Haavisto [23] summarized their usability study of inexperienced users on name dialing. Although the study is based on the telephone handset, it has a similar implication for computer desktop applications.
- *Activating macros*. Speech recognition lets a user speak a more natural word or phrase to activate a macro. For example, "*Spell check the second paragraph*" is easier for most users to remember than the CTRL+F5 key combination after selecting the second paragraph. But again, the user must know the command. This is where most simple speech applications fail. The competition is not CTRL+F5 itself, it is the memory of most users.
- *Facilitating dialog between the user and the computer*. As discussed in Chapter 17, speech recognition works well in situations where the computer essen-

[4] *Peedy* © 1993-1998 Microsoft Corporation.

tially asks the user: "*What do you want to do*?" and branches according to the reply (somewhat like a wizard). For example, the user might reply, "*I want to book a flight from New York to Boston.*" After the computer analyzes the reply, it clarifies any ambiguous words (*Did you say New York*?). Finally, the computer asks for any information that the user did not supply, such as "*At what day and time do you want to leave*?"

- *Providing hands-free computing*. Speech recognition is an essential component of any application that requires hands-free operation; it also can provide an alternative to the keyboard for users who are unable to or prefer not to use one. Users with repetitive-stress injuries or those who cannot type may use speech recognition as the sole means of controlling the computer. As discussed in later sections, hands-free computing is important for accessibility and mobility.
- *Humanizing the computer*. Speech recognition can make the computer seem more like a person—that is, like someone whom the user talks to and who speaks back. This capability can make games more realistic and make educational or entertainment applications friendlier.

The specific use of command and control depends on the application. Here are some sample ideas and their uses:

- *Games and entertainment:* Software games are some of the early adopters of command-and-control speech recognition. One of the most compelling uses of speech recognition technology is in interactive verbal exchanges and conversation with the computer. With games such as flight simulators, for example, traditional computer-based characters can now evolve into characters the user can actually talk to. While speech recognition enhances the realism and fun in many computer games, it also provides a useful alternative to game control. Voice commands provide new freedom for the user.
- *Document editing*: Command and control is useful for document editing when you wish to keep your hands on the keyboard to type, or on the mouse to drag and select. This is especially true when you have to do a lot of editing that requires you to move to menus frequently. You can simultaneously speak commands for manipulating the data that you are working on. A word processor might provide commands like "*bold, italic*" and "*change font.*" A paint package might have "*select eraser*" or "*choose a wider brush.*" Of course, there are users who won't prefer speaking a command to using keyboard equivalents, as they have been using the latter for so long that the combinations have become for them a routine part of program control. But for many people, keyboard equivalents are a lot of hard-to-remember shortcuts. Voice commands provide these users with the means to execute a command directly.

For most of the existing applications, before an application starts a command-and-control recognizer, it must first give the recognizer a *list* of commands to listen for. The list might include commands like "*minimize window*," "*make the font bold*," "*call extension <digit> <digit> <digit>*," or "*send mail to <name>*." If the user speaks the command as it is designed, he/she typically gets very good accuracy. However, if the user speaks the command differently, the system typically either does not recognize anything or erroneously recognizes something completely different. Applications can work around this problem by:

- Making sure the command names are intuitive to users. For many operations like minimizing a window, nine out of ten users will say *minimize* window without prompting.
- Showing the command on the screen. Sometimes an application displays a list of commands on the screen. Users naturally speak the same text they see.
- Using word spotting as discussed in Chapter 9. Many speech recognizers can be told to just listen for one keyword, like *mail*. This way the user can speak, "*Send mail,*" or "*Mail a letter,*" and the recognizer will get it. Of course, the user might say, "*I don't want to send any mail,*" and the computer will still end up sending mail.
- Employing spoken language understanding components as discussed in Chapter 17.
- Employing user studies to collect data on frequently spoken variations on commands so that the coverage is enhanced.

18.2.2. Telephony Applications

Speech is the only available modality for telephony applications besides the awkward-to-use DTMF interface. The earliest uses of speech technology in business were *interactive voice response* (IVR) systems. These systems include *infoline* services in the ad-supported local newspapers, offering everything from world news to school homework assignments at the touch of a few buttons. So what's the big deal with a speech telephony application? It offers greater breadth, ease of use, and interactivity. Navigating by voice rather than by keypad offers more options and quicker navigation. It also works better while you're driving.

To make a successful IVR application, you need to have speech input, output, and related dialog control. People have used IVR systems over the telephone to navigate the application based on the menu option to provide digit strings, such as the credit card numbers, to the application. Such system typically has a small to medium vocabulary. Today, you can use IVR to get stock quotes, people's telephone number, and other directory-related information. For example, you can call AT&T universal card services and the application asks you to speak your 16-digit card number. Most of these IVR systems use recorded messages instead of synthetic speech because the quality of TTS is still far from humanlike. Since speech output is a slow method to present information, it is important to be as brief as possible. Reducing the presentation of repetitive data can shorten the speech output significantly.

Voice portals that let you talk your way to Web-based information from any phone are one class of compelling telephony applications. Linked to specially formatted Web sites and databases, the portals deliver what amounts to customized real-time news radio. You can tailor voice portals much as you do Web portals like *Yahoo!*®, AOL®, or MSN®. But surfing is restricted to the very limited subsets of information the portals choose to offer. These services typically avoid using synthesized speech. For options like news updates they rely on sound bites recorded by announcers. There are a number of free voice portals available, including TellMe, BeVocal, HeyAnita, Quack.com. Table 18.1 illustrates some of their features.

Table 18.1 Some free voice portal features. These portals are being developed and will roll out more features.

Category	Audiopoint[5]	Tell Me[6]
Traffic	Yes	Yes
Weather	U.S. and world cities	U.S.
News	Yes	Yes
Financial	Yes	Yes
Sports	Yes	Yes
Airline info	No	Yes
Restaurants	No	U.S.
Entertainment	Yes	Yes
Personalization	Yes	Yes

Digital wireless telephony applications could make full use of a client-server architecture because of limited computing resources of the client. The server performs most of the needed processing. The client can either send the speech waveform (as used in standard telephone) or the spectral parameters such as MFCC coefficients. Using a quantized MFCC (see Chapter 6) at 4.5 kbps, no loss of accuracy can be achieved [18]. The Aurora project tries to standardize the client server communication protocol based on the quantized MFCC coefficients[7] [6].

When people are engaged in a conversation, even if they have the graphical interface in front of them, they seldom use the vocabulary from the interface (unless prompted by TTS or the speaker). This has an important implication for the UI design. The use of a discourse segment pop cue such as "*What now?*" or "*Do you want to check messages?*" could reorient users (especially after a subdialog) and help them figure out what to say next. Wildfire[8] uses such pop cues extensively. The right feedback is essential, because speech recognition is not perfect. Designers should verify only those commands that might destroy data or trigger future events. People become frustrated very quickly if the error feedback is re-

[5] http://www.myaudiopoint.com or call 1 -888-38-AUDIO.

[6] http://www.tellme.com or call 1-800-555-TELL.

[7] http://www.etsi.org/stq

[8] http://www.wildfire.com/

petitive. It would be nice to design the error messages with different prompts according to recognition confidence measures and the dialog context.

Most current telephony systems do not have much intelligent dialog management beyond a simple state-based dialog model. Advanced dialog management as discussed in Chapter 17 is critical for error repairs, anaphora resolution, and context management. There are a number of institutions pursuing advanced telephony spoken language interfaces. You can find some of these excellent programs at Carnegie Mellon University[9] and Massachusetts Institute of Technology.[10]

18.2.3. Dictation

Dictation is attractive to many people, since speaking is generally faster than typing, especially in East Asian languages such as Chinese. There are a number of general-purpose dictation software products such as IBM's ViaVoice®, Dragon's NaturallySpeaking®, Lernout&Hauspie's Voice Xpress®, as well as the dictation capability built into Microsoft Office 2001. There are also vertical markets for dictation applications. Radiologists and lawyers are two examples of such niche markets. They are generally pressed for time. Radiologists are often hands- and eyes-busy when at work. Lawyers' work is often language intensive and they need to write long documents and reports. Both radiologists and lawyers have been using dictation devices, and they have a strong need to have customized language models for their specialized vocabulary. In fact, because of the constraints of the language they use, the perplexity is often much smaller than in the general business articles, leading to improved performance for both the medical and legal segments.

In dictation applications, it is important to convert speech into the text you would like to see in its written form. This implies that punctuation should be added automatically, dates and times should be converted into the form that is conventionally used, and appropriate capitalization and homonym disambiguation should all be seamless to users. Dictation is typically associated with word processing applications, which provide for a nice separation of input modes based upon the division of the primary task into subactivities:

- Text entry
- Command execution, such as *cross-out* to delete the previous word, or *capitalize-that*
- Direct manipulation activities such as cursor positioning and text selection

It is likely that separate single input modes for each activity can increase efficiency. In a study conducted by Morrison et al. [28], subjects switched modality from speech input to typed input while issuing text-editing commands and found the switch of modality *disruptive*. Karat et al. [19] compared the state-of-the-art dictation product with keyboard input.

---

[9] http://www.speech.cs.cmu.edu/speech/

[10] http://www.sls.lcs.mit.edu/sls/

They reported that several of the initial-use subjects commented that keyboard entry seemed *much more natural*. This reflects the degree to which some people have become used to using keyboards and the fact that speech recognition still makes too many mistakes. Experienced users can enter transcription text at an average rate of 107 uncorrected words per minute. However, correction took them over three times as long as entry time on average, leading to an average input rate nearly twice as slow as keyboard (including correction). Thus, unless the accuracy of speech recognition and its error correction can be significantly improved for the mass market, text entry is best performed by keyboard.

The mouse is generally accepted as being well suited to direct manipulation activities. Further, based on human-factors studies, full typed input of commands, single keyboard presses, and accelerator keys is not as efficient as speech-activated commands [26, 33]. The case for the utility of speech input in word-processing applications relies upon its superiority over the mouse with respect to the activation of commands. Word-processing and text-entry applications are also naturally *hands-busy, eyes-busy* applications. It is inconvenient and time consuming for the user to have to interrupt the typing of text or to move his/her eyes from the work in order to execute word-processing commands. This is not necessary with speech activation of commands.

To improve dictation throughput, you should make it as easy as possible for users to correct mistakes. As illustrated in Figure 18.3, you can provide easy access to the *Correction Window* so the user can correct mistakes that the recognizer made. Thus the quality of *n*-best or word-lattice decoding discussed in Chapter 13 is critical. We use examples drawn from Microsoft Dictation[11] to illustrate these concepts.

You should also allow the user to train the system so it can adapt to different accent and environment conditions. A typical training wizard, illustrated in Figure 18.4, asks users to read a number of sentences. It is important to use a confidence measure, as discussed in Chapter 9, to filter out unwanted training data so that you can eliminate possible divergences of training the acoustic model. In Figure 18.4, the user is prompted with the sentence shown in the window. As the user reads the sentence, the black progress bar covers the corresponding word. If the user misreads the word, the progress bar typically stays where it is.

Another practical problem for most dictation products is that they have about 60,000 to 100,000 words in the vocabulary, which contains all the inflected forms, so *work* and *works* are considered as two different words. As the vocabulary is limited (for detailed discussion on vocabulary selection, see Chapter 11), you should provide a capability to dynamically add new words to the system. This capability is illustrated in Figure 18.5. You can listen to the new word you added to see if the computer guessed the pronunciation of the word correctly or not. If you want to pronounce the word differently, you can either type in the words that sound the way you want it to or type in the phonetic sequence.

---

[11] Microsoft Dictation was developed by the authors at Microsoft Research, and was included for free in the SDK4.0.

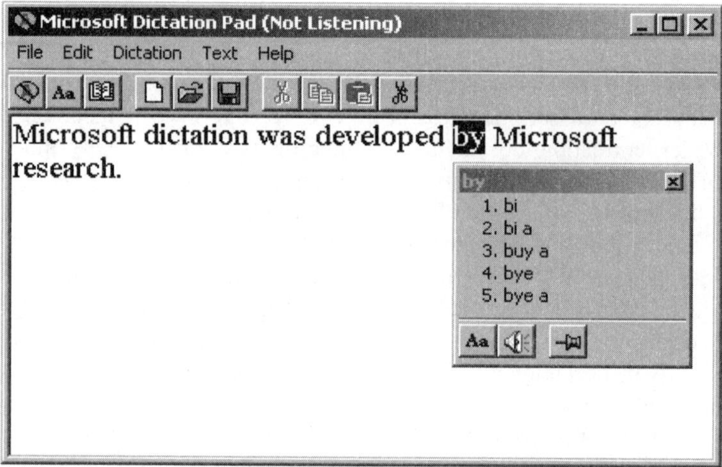

Figure 18.3 *Correction window* in Microsoft Dictation. By clicking the dictated word, the alternative words are listed in the correction window. If the dictated word is wrong, you can click the word in the correction window to easily replace the misrecognized word. Reprinted with permission from Microsoft Corporation.

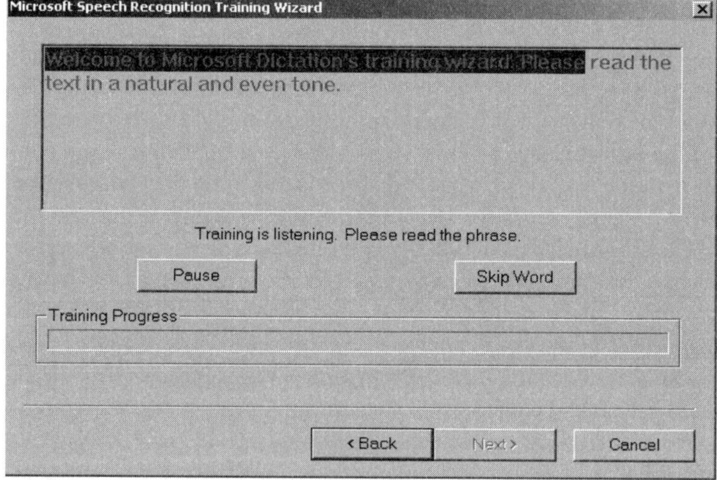

Figure 18.4 Training wizard in Microsoft Dictation. Reprinted with permission from Microsoft Corporation.

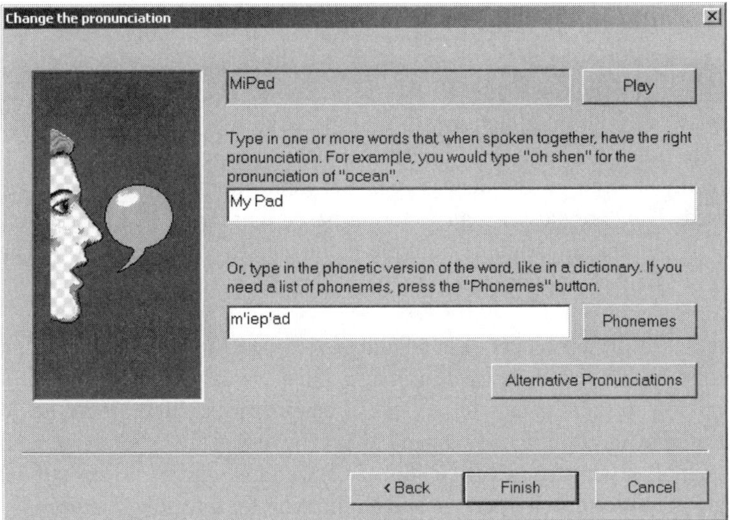

Figure 18.5 New word addition in Microsoft Dictation. Reprinted with permission from Microsoft Corporation.

18.2.4. Accessibility

Approximately 10–15 percent of the population has an impairment that affects their ability to use computers efficiently. Accessibility is about designing applications that everyone can use. The GUI actually makes accessibility a more critical issue. People with disabilities face more usability issues than those without. There are a number of Web sites dedicated to accessibility.[12]

People who cannot effectively use a keyboard or a mouse can use speech commands to control the computer and use the dictation software to input text. For example, Dragon's dictation software lets you control everything by voice, making it extremely useful for people with repetitive stress injury (RSI). A screen reader is a program that uses TTS to help people who have reading problems or are vision impaired. Windows® 2000 includes one such screen reader, called Narrator, which reads what is displayed on the screen: the contents of the active window, menu options, and text that the user types. Users can customize the way these screen elements are read. They can also configure Narrator to make the mouse pointer follow the active item on the screen and can adjust the speed, volume, and pitch of the narrator voice.

[12] http://www.el.net/CAT/index.html (The Center for Accessible Technology).
http://www.lighthouse.org/print_leg.htm (Lighthouse International).
http://www.w3c.org/wai (World Wide Web Consortium's Accessibility Page).
http://www.microsoft.com/enable (Microsoft Accessibility Home Page).

18.2.5. Handheld Devices

For handheld devices such as *smart phones* or *Personal Digital Assistants* (PDA), speech, pen, and display can be used effectively to improve usability. The small form-factor implies that most of the functions have to be hidden. Speech is particularly suitable to access information that can't be seen on the display. Current input methods are generally pen- or soft keyboard-based. They are slower than dictation-based text input. For form filling, speech is also particularly suitable to combine multiple parameters in a single phrase. The increasing popularity of cell phones is likely to be the major thrust for using handheld devices to access information.

Potential usability problems of handheld devices are the lack of privacy and the problem of environmental noise. It is hard to imagine that we can have 100 people in a conference room talking to their handheld devices simultaneously. Since these devices are also often used in a noisy environment, you need to have a very robust speech recognition system to deal with signals around 15 dB. In Section 18.5 we illustrate the process of designing and developing an effective speech interface and application for a handheld device.

18.2.6. Automobile Applications

People already use cell phones extensively when driving. Safety is an important issue whenever a driver takes his/her eyes off the road or his/her hands off the steering wheel. In many countries it is illegal to use a cellular phone while driving. Speech-based interaction is particularly suitable for this eyes-busy and hands-busy environment. However, it is a challenge to use speech recognition in the car due to the relative high noise environment. Environment robustness algorithms, discussed in Chapter 10, are necessary to make automobile applications compelling.

Figure 18.6 Clarion in-dash AutoPC.

One such system is Clarion's AutoPC, shown in Figure 18.6. It enables drivers to access information by speech while driving. You can tell your car what to do by speech. AutoPC has a vocabulary of more than 1200 words. It understands commands to change radio stations, provide turn-by-turn directions based on its Global Positioning System (GPS) navigation unit, or read e-mail subject lines using TTS. Running on Microsoft's Windows CE operating system, AutoPC's control module fits in a single dashboard slot, with the computer and six-disc CD changer stored in the trunk.

18.2.7. Speaker Recognition

The HMM-based speech recognition algorithms discussed in earlier chapters can be modified for speaker recognition, where a speaker model is needed for each speaker to be recognized. The term speaker recognition is an umbrella that includes speaker identification, speaker detection, and speaker verification. Speaker identification consists in deciding who the person is, what group the person is a member of, or that the person is unknown. Speaker detection consists in detecting if there is a speaker change. This is often needed in speech recognition, as a better speaker model can be used for improved accuracy. Speaker verification consists in verifying a person's claimed identify from his/her voice, which is also called speaker authentication, talker verification, or voice authentication.

In text-dependent approaches, the system knows the phrase, which can be either fixed or prompted to the user. The speaker speaks the required phrase to be identified or verified. The speaker typically has to enroll in the system by presenting a small number of speech samples to build the corresponding speaker model. Since the amount of training data is small, the parameter tying techniques discussed in Chapter 9 are very important.

Speaker recognition is a performance biometric, i.e., you perform a task to be recognized. It can be made fairly robust against noise and channel variations, ordinary human changes, and mimicry of humans and tape recorders. Campbell [3] and Furui [9] provided two excellent tutorial papers on recent advances in speaker recognition.

In general, accuracy is much higher for speaker recognition than for speech recognition. The best speaker recognition systems typically use the same algorithms, such as the ways to deal with environment noises and hidden Markov modeling, found in speech recognition systems. Applicable speaker recognition services include voice dialing, banking over a telephone network, telephone shopping, database access services, information and reservation services, forensic application, and voice login. There are a number of commercial speaker verification systems, including Apple's voice login for its iMac® and Sprint's Voice FONCARD®.

18.3. Speech Interface Design

No single unified interaction theory has emerged so far to guide user interface design. Study of the interaction has been largely associated with developing practical applications, having its roots in human factors and user studies. Practitioners have postulated general principles that we discuss in this section.

18.3.1. General Principles

There are a number of books on graphical user interface (GUI) design [10, 11, 17, 29, 41]. Usability research also generated a large body of data that is useful to UI designers. The Human Factors and Ergonomic Society (HFES) is developing the HFES-200 standard—

Ergonomics of Software User Interface, which has many insightful implications for interface design. The most important criteria, in order of importance, are:

- Effectiveness or usefulness
- Efficiency or usability
- User satisfaction or desirability

These three general points mean that, first and foremost, a user interface must work, so the user can get the job done. After that, it is important that the user interface accomplishes its task as productively as possible. Last, but not least, the user interface can be further improved by focusing on user satisfaction as measured by means of questionnaires and user feedback.

The major difference between GUI and spoken language interface is that speech recognition or understanding can never be perfect. Imagine you are designing for a GUI and your input method is a keyboard and a mouse where the error rate is about 5% for the key or mouse presses. These 5% errors make it much more difficult to have an effective user interface. In GUI applications, when you mistype or misclick, you regard these errors as your own mistakes. In spoken language applications, all the misrecognition errors are regarded as system errors. This psychological effort is a very important factor in the UI design. Thus it is not stretching to say that most of the design work for spoken language applications should be on how to effectively manage what to do when something goes wrong. This should be the most important design principle to remember.

We also want to stress the importance of iteratively testing the UI design. In practice, users, their tasks, and the work environment are often too varied to be able to have a good design without going through a detailed and thorough test program, especially for new areas like spoken language applications.

18.3.1.1. Human Limitations

Applications make constant demands on human capabilities. When we design the user interface, we must understand that humans have cognitive, auditory, visual, tactile, and motor limits. As a general principle, we should make the user interface easy to use, and not overload any of these limits.

For example, if the font is too small, the speech volume is too low, or the application requires people to constantly switch between a mouse and a keyboard, we have a poor interface, leading to lowered productivity. People have trouble remembering long complex messages spoken to them in a continuous fashion. If these messages are delivered with synthesized speech, the problem is worsened, because of the greater attention required to understand the synthesized speech. Effective user interface design requires a balance of easy access to information and an awareness of human limitations in processing this information. In general, people tend to handle well 7 (+ or – 2), pieces of information at a time.

18.3.1.2. User Accommodation

The interface and application should accommodate users as much as possible. This means that the design needs to match the way users want to do their work, not the other way around. People always have, or quickly create, a mental model of how the interface operates. You need to design and communicate a conceptual model that works well with the users' mental model. Their mental model should match the actual interface and the task, which is very important for the designer to consider consciously.

To accommodate users' needs, you must ask whether the interface you are designing helps people get their work done efficiently. Will it be easy to learn and use? Will it have an attractive and appropriate design? Will the interface fit individual tasks with a suitable modality? Will the interface allow the user to perceive that they are in control? Will the interface be flexible enough to allow the user to adjust the design for custom use? Does the interface have an appropriate tempo? Notice that speech is temporal. Does the interface contain sufficient user support if the user needs or requests it? Does the interface respond to the use in a timely fashion? Does the interface provide appropriate feedback to the user about the results of the actions and application status?

18.3.1.3. Modes of Interaction

The interface is unimodal if only one mode is used as both input and output modality. By contrast, multimodal systems use additional modalities in exchanging information with their users. When we interact with each other, we often rely on multimodal interaction, which is normal for us.

When speech is combined with other modalities of information presentation and exchange, it is generally more effective than a unimodal interface, especially for dealing with errors from speech recognition and understanding.

As an input modality, speech is generally not as precise as mouse or pen to perform position-related operations. Speech interaction can also be adversely affected by the ambient noise. When privacy is of concern, speech is also disadvantageous, since others can overhear the conversation. Despite these disadvantages, speech communication is not only natural but also provides a powerful complementary modality to enhance the existing keyboard- and mouse-based graphical user interface. As an input-output modality, speech has a number of unique characteristics in comparison with the traditional modalities:

- Speech does not require any screen real estate in the interface. For ubiquitous computing, this is particularly suitable for devices that are as small as our watches.
- Speech can be used at a distance, which makes it ideal for hands-busy and eyes-busy situations such as driving.

- Speech is expressive and very suitable to bring information that is hidden to the forefront. This is particularly suitable in the GUI environment or with handheld devices as the monitor space is always limited.
- Speech is omnidirectional and can communicate information to multiple users easily. However, this also has implications relating to privacy.
- Speech is temporal and sequential. Once uttered, auditory information is no longer available. This can place extra memory burden on the user and severely limit the ability to scan, review, and cross-reference information.
- For social interfaces, speech plays a very important role. For example, if we need to pose a question, make a remark, and delegate a task, there is no more natural way than speech itself.

Because of these unique features, we need to leverage the strengths and overcome the technology limitations that are associated with the speech modality. The conventional GUI relies on the visual display of objects of interest, the selection by pointing, and continuous feedback. Direct manipulation is generally inadequate for supporting fundamental transactions in applications such as word processing and database queries, as there are only limited means to identify objects, not to mention that ambiguity in the meanings of icons and limitations in screen display space further complicate the usability of the interface.

Generally, multimodality can dramatically enhance the usability of speech, because GUI and speech have complementary strengths and weaknesses. The strengths of one can be used to offset the weaknesses of others. For example, pen and speech can be complementary and they can be used very effectively for handheld devices. You can use *Tap and Talk* as discussed in Section 18.4 to activate microphone and select appropriate context for speech recognition. The respective strengths and weaknesses are summarized in Table 18.2. The advantage of pen is typically the weakness of speech and vice versa. This implies that user interface performance and acceptance could increase by combining both. Thus, visible, limited, and simple actions can be enhanced by nonvisible, unlimited, and complex actions. The multimodal interface is particularly suitable to people for completing contrastive functions or when task attributes are perceived as separable. For example, speech and pen can be used in different ways to designate a shift in context or functionality. Although most people may not prefer to use speech and pen simultaneously, they like to use speech to enter data and pen for corrections. Other contrastive tasks may include data versus commands and digits versus text. When inputs are perceived as integral, unimodal interfaces are efficient and work best.

Table 18.2 Complementary strengths of pen and speech for multimodal user interface.

Pen	Speech
Direct manipulation, requires hands and eyes	Hands/eyes-free manipulation
Good at performing simple actions	Good at performing complex actions
Visual feedback	Visual and location independent references
No reference ambiguity on the display	Multiple ways to refer to objects

18.3.1.4. Technology Considerations

Speech recognition applications can be divided into many classes, as illustrated in Figure 18.7. The easiest category is the system that can only deal with isolated speech with limited vocabulary for a command-and-control application, where typically a context-free grammar is used to constrain the search space, resulting in a limited number of ways users can issue a command. The hardest is the one that can support channel- and speaker-independent continuous speech recognition with both dictation and robust application-specific understanding capability. This is also the one that can really add value to a wide range of speech applications. In Figure 18.7, a *command-and-control* application implies application-specific restrictive understanding capability—typically with less than 1000-word vocabulary and with a lower perplexity context-free grammar. A *dictation* application implies general-purpose dictation capability—typically with more than 60,000 words in the lexicon and a stochastic *n*-gram language model. A *dictation and SLU* application implies both general-purpose dictation and less restrictive application-specific spoken language understanding capability—typically with 60,000 words in the lexicon with both *n*-gram and semantic grammar language models for both recognition and understanding.

Although a number of parameters can be used to characterize the capability of speech recognition systems, user-centered design often dictates that you need to use the most challenging technology solutions. The speaking style (from isolated speech to continuous speech) can dramatically change the performance of a speech recognition system. Natural, spontaneous, continuous speech remains the most difficult to recognize; but it is more user friendly. Small vocabulary (less than 50 words) is generally more accurate than large vocabulary (greater than 20,000 words); but the small vocabulary limits the options the user has on the task. The lower-perplexity language model generally has a great recognition accuracy; but it imposes a considerable constraint on how the user can issue a command. Most systems may work in the high-SNR (>30 dB) environment; but it is desirable that the system should still work when the SNR is below 10 dB.

Command & Control	Dictation only	Dictation & SLU	
Easiest			Isolated Speech
			+Continuous speech
			+Environment-independence
		Hardest	+Speaker-independence

Figure 18.7 Illustrative degrees of difficulty for different classes of speech recognition (gray level from easy to hard).

Associated with the input modality is the speech output modality for many speech applications. Speech as an output medium taxes user memory and may even cause loss of context and distraction from the task if it is too verbose. Most of the text-to-speech systems have a problem generating a prosody pattern that is dynamic and accurate enough to convey the underlying semantic message. Recorded speech still has the best quality; but it is generally inflexible in comparison to TTS. There are many ways to provide additional information to the TTS so that the final synthesized quality can be improved dramatically. Examples include the use of recorded speech for constant messages, transplanted prosody, and concept-to-speech synthesis as discussed in Chapters 16 and 17.

Improved quality and ease of use often need expensive authoring. This is true for speech recognition, synthesis, and understanding. For example, to support channel independence, you need not only additional channel normalization algorithms to deal with both additive and convolutional noises but also realistic training data to estimate model parameters. To support robust understanding, you need to author semantic grammars that have broad coverage. This often requires tedious authoring and labor-intensive data collection and annotation to augment the grammar. Ultimately, you need a modular and reusable system that can warrant the development costs. To make effective use of dialog and error-repair strategies, you need to adopt tools for preventing dialog design problems during the early stages, ones that guide the choice of words in dialog and feedback and ensure usability and correctness in the context, etc. In addition, as discussed in Chapter 17, whether you have a system-initiative dialog, user-initiative dialog, or mixed-initiative dialog will have significant ramifications on the authoring cost.

A spoken language application requires certain hardware and software on the user's computer in order to run.[13] Most new sound cards work well for speech recognition and text-to-speech, including Sound Blaster™. Some old sound cards are only *half duplex* (as opposed to *full duplex*). If a sound card is half duplex, it cannot record and play audio at the same time. Thus it cannot be used for barge-in and echo cancellation. The user can choose among different kinds of microphones: a close-talk headset microphone that is held close to the mouth, a medium-distance microphone, or a microphone array device that rests on the computer 30 to 60 centimeters away from the speaker. A headset microphone is often needed for noisy environments, although microphone array or blind separation techniques have the potential to close the gap in the future (see Chapter 10). Most speech products also need to calibrate the microphone gain by speaking one or two utterances, as illustrated in Figure 18.8. It adjusts the gain of the amplifier to make sure there is an appropriate gain without clipping the signal. This can be done in the background without distracting the user.

---

[13] Speech recognition engines currently on the market typically require a Pentium 200 or faster processor. Speech recognition for command and control consumes 1 to 4 MB RAM, and dictation requires an additional 16-32 MB. Text-to-speech engines use about 1-3 MB, but high-quality TTS can require more than 64 MB.

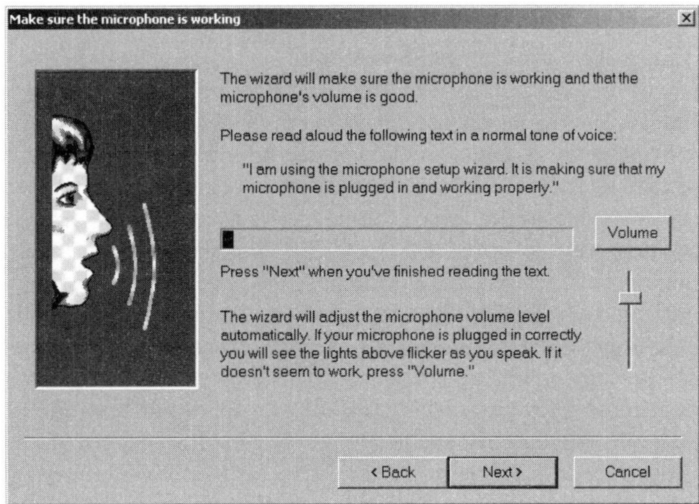

Figure 18.8 Microphone setup wizard used in Microsoft Speech SDK 4.0. Reprinted with permission from Microsoft Corporation.

18.3.2. Handling Errors

Speech recognition and understanding can never be perfect. Current spoken language processing work is very much a matter of minimizing errors. Together with the underlying speech-engine technologies, the role of the interface is to help reduce the errors or minimize their severity and consequences. As discussed in earlier chapters, there are errors associated with word deletion, word insertion, and word substitution. The recognizer may be unable to reject noises and classify them as regular words. The speaker may also have new words that are not in the lexicon, or may have less clear articulation due to accent, stress, sickness, and excitement.

To repair errors in the system, we need to take each individual application into consideration, since the strategy is often application-dependent. To have a compelling application, we must design the application with both the limitation of the engine and the interaction of the engine and UI in mind. For example, multimodal applications are generally more compelling and useful than speech-only solutions as far as error handling is concerned. A multimodal application can overcome many limitations of today's SLU technologies, but it may be inappropriate for hands-busy or eyes-busy scenarios.

18.3.2.1. Error Detection and Correction

We need to detect errors before we can repair them. This is a particularly difficult problem that often requires high-level context and discourse information to decipher whether there are errors. We can certainly use confidence measures, as discussed in Chapter 9, to detect potential errors, although confidence measures alone are not particularly reliable.

There are many ways to correct the errors. The choice of which one to use depends on the types of the applications you are developing. We want to emphasize that correction based on another modality is often desirable, since it is unusual that two independent modalities make the same mistake. For example, speech recognition errors are often different from handwriting recognition errors. In addition to the complementary nature of these different errors, we can also leverage other modalities for improved error-correction efficacy. For example, a pen is particularly suitable for pointing, and we can use that to locate misrecognized words. A camera can be used to track whether the speaker is speaking in order to eliminate background noise generated from other sound sources.

Let's consider a concrete example. If you have a handheld device, you may be able to see the errors and respeak the desired commands or use a different modality such as pen to help task completion. When you want to create an email using speech, and a dictation error occurs, you can click the misrecognized word or phrases to display a list of n-best alternatives. You can select the correct word or phrase from the n-best list with the pen. Alternatively, you can respeak the highlighted word or phrase. Since the correct answer should not be the previous misrecognized word or phrase, you can exclude them in the second trial. If it is an isolated word in the correction stage, you can also use the correct silence context triphone models for the word boundary as the correct acoustic model. You can also use the correct n-gram language model from the left context with the assumption that the left words are all correct. With these correct acoustic and language models, it is expected that the second trial would have a better chance to get the right answer.

If you have telephony-based applications, you can count on keywords such as *no*, *I meant* to identify possible recognition or understanding errors. When errors occur, you should provide specific error-handling messages so that the user knows what response the system wants. An adequate error message should tell the user what is wrong, how to correct it, and where to get help if needed. Let's consider an example of the following directory assistant application:

Computer: Please speak the name

User: Mike

Computer: Please speak the name

User: Mike

Clearly, the user does not know what responses the computer wants. An improved dialog can be:

> **Computer: Please speak the first and last name**
>
> **User: Mike Miller**
>
> **Computer: Which division is Mike Miller working?**
>
> **User: Research group**

You can use alternative guesses intelligently for telephony applications, too. For example, if the user says "*Mike Miller*" and the computer is unsure of what was said, it can prompt the user to repeat with a constrained grammar. In the example above, if the computer is unsure whether the user said Mike or Mark, it may want to respond with "*Did you say Mike Miller or Mark Miller*?" before it proceeds to clarify which division Miller is working in.

When an error is detected, the system should be adapted to learn from the error. The best error-repair strategy is to use an adaptive acoustic and language model so that the system can learn from the errors and not repeat the same mistake again. This is why adaptive acoustic and language modeling, as discussed in Chapters 9 and 11, is critical for real acceptance of speech applications. When an error occurs and the computer understands the error via user correction, you can use the corrected words or phrases to adapt related components in the system.

18.3.2.2. Feedback and Confirmation

Users need feedback from the system so that they understand the status of the interaction. When a command is issued, the user expects the system to acknowledge that the command is heard. It is also important to tell the users if the system is busy so that they know when to wait. Otherwise, the users may interpret the lack of response or feedback as errors of recognition or understanding.

If the user has requested that an action be taken, the best response is to carry out the action as requested, but if the action cannot be carried out, the user must be notified. Sometimes, the user may need this feedback to know whether it is her turn to speak again. If the interface is multimodal, you can use visual feedback to indicate the status of the interaction. For example, in Microsoft's MiPad research prototype, when the microphone is activated with a pen pointing to a specific field, the microphone levels are indicated in the feedback rectangle in strip-chart fashion as shown in Figure 18.9 for the corresponding field. Initially this wipes from left to right, but when the right edge is reached, the contents scroll to the left as each new sample is added. The height of each sample indicates the audio level; there is some indication given for the redline level for a possible saturation. It also shows the total number of frames as well as the number of remaining frames to be processed. Then, as the

Figure 18.9 MiPad's dynamic volume meter. Reprinted with permission from Microsoft Corporation.

recognizer processes, the black bar wipes from left to right to indicate its progress. When the recognizer completes (or is canceled by the user), the dynamic volume meter is removed. The volume meter feedback described addresses a number of issues that are critical to most users: if the computer is listening or not; if the volume is too high or not, and when the computer finishes processing the speech data.

If the interface has no visual component, you can use auditory cues or speech feedback. If there is more than a 3-second delay after the user issues a command and the system responds, an auditory icon during the delay, such as music, may be used. A delay of less than 2-3 seconds is acceptable without using any feedback.

Since there are errors in spoken language systems, you may have to use confirmation questions to assure that the computer heard the correct message. You can associate risks with different commands, so you can have a different level of strategy for confirmation. For example, *format disk* should have a higher threshold or may not be allowed by voice at all. You can have an explicit confirmation before executing the command. You can also have implicit confirmation based on the level of the risks in the corresponding confidence scores from the recognizer. For example, if the user's response has more than one possible meaning (X and Y) or the computer is not confident about the recognized result (X or Y), the computer could ask, "*Do you want to do X or Y?*" You want to be specific about what the system needs. A prompt like "*Do you mean X or Y?*" is always better than *Please repeat.* You have to balance the cost of making an error with the extra time and annoyance in requiring the user to confirm a number of statements.

In a telephony-based application, you need to tell the users specifically that they are talking to a computer, not a real person. When AT&T used a long greeting for its telephone customer service, their prompts were as follows:

"AT&T Automated Customer Service. This service listens to your speech and sends your call to the appropriate operator. How may I help you?"

The longer version resulted in shorter utterances from the people calling in. The short automated version did not help shorten the utterances, because people did not seem to catch the automated connotation.

You can also design the prompt in such a way that you can constrain how people respond or use their words. Let's consider an example of the following directory assistant application:

Computer: Welcome to Directory Assistant. We look forward to serving you. How can I help you?

User: I am interested in knowing if you can tell me the office number of Mr. Derek Smith?

To have the user follow the lead of the computer, an improved dialog can be:

Computer: Welcome to Directory Assistant. Please say the full name of the person.

User: Derek Smith

When the system expects the user to respond with both the first name and last name, you should still design the grammar so that the system can recognize a partial name with lower probabilities. This makes the system far more robust—a good example of combining engine technology and interface seamlessly for improved user satisfaction.

18.3.3. Other Considerations

When speech is used as a modality, you should remember the following general principles:

18.3.3.1. Don't Use Speech as an Add-On Feature

It's poor design to just bolt speech recognition onto an application that is designed for a mouse and keyboard. Such applications get little benefit from speech recognition. Speech recognition is not a replacement for the keyboard and mouse, but in some circumstances it is a better input device than both. Speech recognition makes a terrible pointing device, just as the mouse makes a terrible text entry device, or the keyboard is bad for drawing. When speech recognition systems were first bolted onto the PC, it was thought that speaking menu names would be really useful. As it turns out, very few users use speech recognition to access a window menu.

18.3.3.2. Give Users Options

Every feature in an application should be accessible from all input devices: keyboard, mouse, and speech recognition. Users naturally use whichever input mechanism provides them the quickest or easiest access to the feature. The ideal input device for a given feature may vary from user to user.

18.3.3.3. Respect Technology Limitations

There are a number of examples of poor use of speech recognition. Having a user spell out words is a bad idea for most recognizers, because they are too inaccurate unless you constrain the recognizer for spelling. An engine generally has a hard time detecting multiple speakers talking over each other in the same digital-audio stream. This means that a dictation system used to transcribe a meeting will not perform accurately during times when two or more people are talking at once. An engine cannot hear a new word and guess its spelling unless the word is specified in the vocabulary. Speakers with accents or those speaking in nonstandard dialects can expect more errors until they train the engine to recognize their speech. Even then, the engine accuracy will not be as high as it would be for someone with the expected accent or dialect. An engine can be designed to recognize different accents or dialects, but this requires almost as much effort as porting the engine to a new language.

18.3.3.4. Manage User Expectations

When you design a speech recognition application, it is important to communicate to the user that your application is speech aware and to provide him or her with the commands it understands. It is also important to provide command sets that are consistent and complete. When users hear that they can speak to their computers, they instantly think of *Star Trek* and *2001: A Space Odyssey*, expecting that the computer will correctly transcribe every word that they speak, understand it, and then act upon it in an intelligent manner. You should also convey as clearly as possible exactly what an application can and cannot do and emphasize that the user should speak clearly, using words the application understands.

18.3.4. Dialog Flow

Dialog flow is as important to speech interface as screen design is to GUI. A robust dialog system is a prerequisite to the success of the speech interface. The dialog flow design and the prompting strategy can dramatically improve user experience and reduce the task complexity of the spoken language system.

18.3.4.1. Spoken Menus

Many speech-only systems rely on tree-structure spoken menus as the main navigation vehicle when no sophisticated dialog system is used. The design of these menus is, therefore, critical to the usability of the system. If there are only two options, you can have "*Say 1 for yes, or 2 for no.*" However, if a menu has more than two options, you should describe the result before you specify the menu option. For example, you want to have the system say "*For email say 1. For address book say 2…*" instead of "*Say 1 for email. Say 2 for address book….*" This is because the user may forget which choice he wants before hearing all the menu options.

You should not use more than five options at each level. If there are more than five, you should consider submenus. It is important to place the most frequently used options at the top of the menu hierarchy.

You should also design the menu so that commands sound different. As a rule of thumb, the more different phonemes you have between two commands, the more different they sound to the computer. The two commands, *go* and *no,* only differ by one phoneme, so when the user says, "*Go*" the computer is likely to recognize "*No.*" However, if the commands were "*Go there*" and "*No way*" instead, recognition would be much easier.

18.3.4.2. Prompting Strategy

When you design the prompt, use as few words as possible. You should also avoid using a personal pronoun when asking the user to respond to a question. This gives the user less to remember, as speech output is slow and one-dimensional. Typical users are able to remember three to four menu or prompt options at a time. You should also allow users to interrupt the computer with *barge-in* techniques, as discussed in Chapter 10.

It is often a good idea to use small steps to query the user at each step if application has a speech-only interface progressively. You can start with short high-level prompts such as "*How can I help you?*" If the user does not respond appropriately, the system can provide a more detailed prompt such as "*You can check your email, calendar, address book, and your home page.*" If the user still does not respond correctly, you can give a more detailed response to tell the user how to speak the appropriate commands that the computer can understand.

If the application contains users' personal information, you may want to treat novice users and experienced users differently. Since the novice users are unfamiliar with the system and dialog flow, you can use more detailed instructions and then reduce the number of words used the next time when the user is going through the same scenario.

18.3.4.3. Prosody

For TTS, prosodic features, including pitch, volume, speed, and pause, are very important for the overall user experience. You can vary pitch and volume to introduce new topics and emphasize important sections. Increased pitch dynamic range also makes the system sound lively.

Users tend to mimic the speed of the system. If the system is reacting and speaking slowly, the user may tend to slow down. If the computer is speaking quickly, the user may speed up. An appropriate speech rate for natural English speech is about 150 to 170 words per minute.

Pauses are important in conversation, but they can be ambiguous, too. We generally use pauses to suggest that it is the user's turn to act. On the other hand, if the system has requested a response from the user and the user has not responded, you should use the time-out period (typically 10 seconds) to take appropriate actions, which may be a repeat of a more detailed prompt or may be a connection to an operator. People are generally uncomfortable with long pauses in conversation. If there are long pauses in the dialog flow, people tend to talk more and use less meaningful words, resulting in more potential errors.

18.4. INTERNATIONALIZATION

To meet the demands of international markets you need to support a number of languages. For example, Microsoft Windows 2000 supports more than 40 languages. Software internationalization typically involves both globalization and localization phases. Globalization is the process of defining and developing a product such that its core features and code design do not make assumptions about a single language. To adapt the application for a specific market, you need to modify the interface, resize the dialog boxes, and translate text to a specific market. This process is called localization. The globalization phase of software internationalization typically focuses on the design and development of the product. The product designers plan for locale-neutral features to support multicultural conventions such as address formats and monetary units. Development of the source code is typically based on a single code base to have the capability to turn locale-specific features on and off without modifying the source code, and the ability to correctly process different character encoding methods. Products developed with the single worldwide binary model typically include complete global functionality, such as support for Input Method Editors used in Asian languages and bidirectional support for Arabic and Hebrew languages.

In addition to the most noticeable change in the translated UI, spoken language engines such as speech recognition or text-to-speech synthesis require significant undertakings. You need to have not only the lexicon defined for the specific language but also a large amount of speech and text data to build your acoustic and language model. There are many exceptional rules for different languages. Let's take dictation application as an example. When you have inverse text normalization, whole numbers may be grouped differently from country to country, as shown in Table 18.3. In the United States, we group three digits separated by a comma. However, many European locales use a period as the number separator, and Chinese conventions typically do not group numbers at all.

Table 18.3 Examples of how numbers are grouped by different locales for inverse text normalization.

123,456,789.0	United States and most locales
123456789.0	Chinese
12,34,56,789,0	Hindi

For the speech recognition engine localization, there are a number of excellent studies on both Asian [5, 15, 16, 21, 24, 32, 34, 37] and European languages [7, 8, 20, 22, 27, 31, 40]. Many spoken language processing components, such as the speech signal processing module, the speech decoder, and the parser, are language independent. Major changes are in the *content* of the engine, such as the lexicon and its associated processing algorithm, the grammar, and the acoustic model or language model parameter.

In general, it is easier to localize speech recognition than either text-to-speech or spoken language understanding. This is because a speech recognition engine can be mostly automatically derived from a large amount of speech and text data, as discussed in Chapters 9 through 13. For Chinese, you need to have special processing for tones, and the signal-processing component needs to be modified accordingly [4]. For both Chinese and Japanese the lexicon also needs to be carefully selected, since there is no space between words in the lexicon [12, 46]. As discussed in Chapter 17, the parser can be used without much modification for SLU systems. Most of localization efforts would be on the semantic/syntactic grammar and dialog design, which is language specific by nature.

As discussed in Chapters 14 and 15, the TTS front end and symbolic prosody components are much harder to localize than the TTS back end. The TTS back end can be easily localized if you use the trainable approach discussed in Chapter 16 on the assumption that the speech recognizer for the language is available. The internationalization process mainly consists of creating tools that can generate training phrase lists from target language texts and dictionaries, and ensuring that any reasonable-sized set of symbolic distinctive language sound inventories (phone sets) can be accepted by the system. The human vocal tract is largely invariant across language communities, and this relative invariance is reflected in the universality of the speech synthesis tools and methods.

When the lexical, syntactic, and semantic content is localized, you should be aware that the incorrect use of sensitive terms can lead to outright product bans in some countries, while other misuses are offensive to the local culture. For example, Turkey forbids the use of the term *Kurdistan* in text and any association of Kurdistan with Turkey. Knowing proper terminology is critical to the success of your application internationalization.

18.5. Case Study—MiPad

An effective speech interface design requires a rigorous process that typically consists of three steps with a number of iterations based on the user evaluation:

- Specifying the application
- Rapid prototyping
- Evaluation

We use Microsoft's research prototype MiPad [18] as an example to illustrate how you can use the process to create an effective speech application. As a wireless Personal Digital Assistant (PDA), MiPad fully integrates continuous speech recognition (CSR) and spoken language understanding (SLU) to enable users to accomplish many common tasks using a multimodal interface and wireless technologies. It tries to solve the problem of pecking with tiny styluses or typing on minuscule keyboards in today's PDAs or smart phones. It also avoids the problem of being a cellular telephone that depends on speech-only interaction. MiPad incorporates a built-in microphone that activates whenever a field is selected. As a user taps the screen or uses a built-in roller to navigate, the tapping action narrows the number of possible instructions for spoken language processing. MiPad runs on a Windows CE Pocket PC with a Windows 2000 Server where speech recognition is performed. The CSR engine has a 64,000 word vocabulary with a unified context-free grammar and n-gram language model as discussed in Chapter 11. The SLU engine is based on a robust chart parser and a plan-based dialog manager discussed in Chapter 17.

18.5.1. Specifying the Application

The first step to develop an application is to identify key features that can help users. You need to understand what work has been done, and how that work can be used or modified. You need to ask yourself why you need to use speech for the applications and review project goals and time frames. The first deliverable will be the spec of the application that includes a documented interface design and usability engineering project plan. Some of the key components should include:

- Define usability goal, not technology
- Identify the user groups and characteristics, interview potential users, and conduct focus-group studies
- Create system technology feasibility reports
- Specify system requirements and application architecture
- Develop development plan, QA test plan, and marketing plan

You should develop task scenarios that adopt a user's point of view. Scenarios should provide detail for each user task and be in a prose or dialog format. You may also want to include frequency information as a percentage to illustrate whether a particular case is happening more than 20% or 80% of the time, which helps you focus on the most common features first.

To create a conceptual interaction model, not the software architecture, we need to have a mental model for the users. Users have to manipulate interface objects either literally or figuratively in the navigation. Interfaces without clear and consistent user objects are difficult to use. Generally, you need a script flow diagram if speech is used. This describes how the user moves through dialogs with the system. To create a script flow diagram from your scenarios, you can start at the beginning of a scenario and derive a set of possible dialogs the user needs to have with the system to complete the scenario. You can draw a box representing a dialog with arrows that shows the order and relationship among the dialogs. If your interface involves graphic display, you need to include window flow diagrams as well, which organize objects and views and define how the user moves from one object to another. Of course, these two diagrams may be combined for effective multimodal interaction.

When the initial spec is in place, you can use the conceptual model to create actual paper windows and dialogs for the interface. You must iterate the design several times and get feedback early. Paper prototyping allows you to do these tasks most efficiently. It is useful to conduct walkthroughs with a number of people as though they were performing the task. You may want to start usability testing to collect real data on real users before the software is implemented. Even a system that is well designed and follows the best proactive approach may reveal some usability problems during testing. Usability testing has established protocols, procedures, and methodologies to get reliable data for design feedback [36, 42]. Generally you need to decide what to test, create usability specifications for the tasks, select the right user groups, specify exact testing scenarios, conduct the actual tests with participants, analyze the data, and report critical findings.

How did we do this for MiPad? Since MiPad is a small handheld device, the present pen-based methods for getting text into a PDA (Graffiti, Jot, soft keyboard) are barriers to broad market acceptance. As an input modality, speech is generally not as precise as mouse or pen to perform position-related operations. Speech interaction can be adversely affected by the ambient noise. When privacy is of concern, speech is also disadvantageous, since others can overhear the conversation. Despite these disadvantages, speech communication is not only natural but also provides a powerful complementary modality to enhance the pen-based interface. Because of these unique features, we need to leverage the strengths and overcome the technology limitations that are associated with the speech modality. Pen and speech can be complementary and they can be used very effectively for handheld devices. You can tap to activate a microphone and select appropriate context for speech recognition. The advantage of pen is typically the weakness of speech and vice versa. This implies that user interface performance and acceptance could increase by combining both. Thus, visible, limited, and simple actions can be enhanced by nonvisible, unlimited, and complex actions.

Table 18.4 Benefits of having speech and pen for MiPad.

Action	Benefit
Ed uses MiPad to read an email, which reminds him to schedule a meeting. Ed taps to activate microphone and says *Meet with Peter on Friday.*	Using speech, information can be accessed directly, even if not visible. Tap and Talk also provides increased reliability for speech detection.
The screen shows a new appointment to meet with Peter at 10:00 on Friday for an hour.	An action and multiple parameters can be specified in only a few words.
Ed taps Time field and says *Noon to one thirty.*	Field values can be easily changed using field-specific language and semantic models.
Ed taps Subject field dictates and corrects a couple of sentences explaining the purpose of the meeting.	Bulk text can be entered easily and faster.

People tend to like to use speech to enter data and pen for corrections and pointing. As illustrated in Table 18.4, MiPad's Tap and Talk interface offers a number of benefits. MiPad has a Tap and Talk field that is always present on the screen, as illustrated in MiPad's start page in Figure 18.10 (a) (the bottom gray window is always on the screen).

The user can give spontaneous commands by tapping the Tap and Talk field and talking to it. The system recognizes and parses the command, such as showing a new appointment form. The appointment form shown on MiPad's display is similar to the underlying semantic objects. The user can have conversation by *tapping and talking* to any subfield as well. By tapping to the attendees field in the calendar card shown in Figure 18.10 (b), for example, the semantic information related to potential attendees is used to constrain both CSR and SLU, leading to a significantly reduced error rate and dramatically improved throughput. This is because the perplexity is much smaller for each subfield-dependent language and semantic model. General text fields, such as the title or body of an email, call for general dictation. Dictation is handled by the same mechanism as other speech entries: The user dictates while tapping the pen where the words should go. To correct a misrecognized word, a tap on the word invokes a correction window, which contains the n-best list and a soft keyboard. The user may hold the pen on the misrecognized word to dictate its replacement.

From the start page, you can reach most application states. There are also a number of short-cut states that can be reached via the hardware buttons. The design is refined incrementally in a number of iterations.

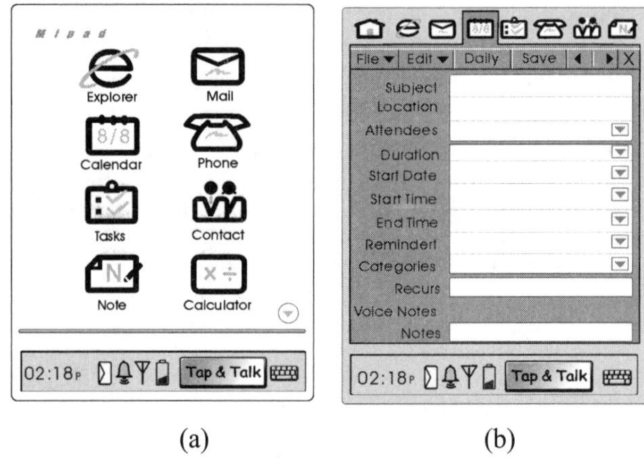

(a) (b)

Figure 18.10 Concept design for (a) MiPad's main card and (b) MiPad's calendar card.

18.5.2. Rapid Prototyping

When you have the spec for the application, you need to communicate the design to the development team so they can rapidly develop a prototype for further iterations. In practice, handing off an interface to a development team without involvement in its implementation seldom works. This is because the developers always have questions about the design document. You need to be available to them to interpret your design. Technical and practical issues may also come up during development, requiring changes to the interface design.

For speech applications, there are a number of key components that have a significant implication for the user interface design. As your design iterations progress, it is useful to track the changes in these components and the effect that these changes have on the user experience of the prototype.

- **Semantic classes**: Each semantic class, as discussed in Chapter 17, is representing real-world entities or associated with the action that applications can take. We need to abstract the application to a level such that we can represent all the actions that the application can take with a number of semantic classes.

Each semantic class has needed slots that require a context-free grammar with a decent coverage. You can start with example sentences and reuse some of the predefined task-independent CFGs, such as date and time. The interface design can help to constrain users to say the sentences that are within the domain.

- **Dialog control**: When appropriate semantic objects are selected, you need to decide the flow of these semantic objects with an appropriate dialog strategy, either speech-based or GUI-based, which should include both inter- and intra-frame control and error-repair strategy. In addition, appropriate use of confidence measures is also critical.
- **Language models**: You need to use appropriate language models for the speech recognizer. Context-free grammars used for the semantic classes can be used for the recognizer. As coverage is generally limited, it is desirable to use the n-gram as an alternative. You can use a unified context-free grammar and n-gram for improved robustness, as introduced in Chapter 11.
- **Acoustic models**: In comparison to the semantic classes and language models, subphonetic acoustic models can be used relatively effectively for task-independent prototyping. Nevertheless, mismatches such as microphones, environment, vocabulary, and speakers are critical to the overall performance. You may want to consider speaker-dependent or environment-dependent prototyping initially, as this minimizes the impact of various mismatches.
- **TTS models**: If the speech output is well defined, you can adapt the general-purpose TTS systems with the correct prosody patterns. This can be achieved with the transplanted prosody in which variables can be reserved for TTS to deal only with proper nouns or phrases that cannot be predecided. You can also adapt acoustic units such that the concatenation errors can be minimized.

The work needed for rapid prototyping is significant, but less than it might at first appear to be. Earlier tasks can be repeated in later iterations when more data become available in order to further improve the usability of the overall system.

18.5.3. Evaluation

It is vital to have a rigorous evaluation and testing program to measure the usability of the prototype. It is a good practice to evaluate not only the entire system but also each individual component. The most important measure should be whether users are satisfied with the entire system. You can ask questions such as: *Is the task completion time much better? Is it easier to get the job done? Did we solve the major problem existing in the previous design*? Most of the time, we may find that the overall system can't meet your goal. This is why you need to further analyze your design and evaluate each individual component from UI to the underlying engines.

Before the prototype becomes fully functioning, you may want to use Wizard-of-Oz (WOZ) experimentation to study the impact of the prototype. That is, you have a real person (the wizard) to control some of the functions that you have not implemented yet. The person can carry out spoken interactions with users, who are made to believe that they are interacting with a real computer system. The goal of WOZ is to study the behavior of human-computer interactions. By producing data on the interaction between a simulated system and its users, WOZ provides the basis for early tests of the system and its feasibility prior to implementation. It can also simulate the error patterns in the existing spoken language system.

WOZ is still expensive, as you have the cost of training a wizard in addition to the usual costs of selecting experimental subjects, transcribing the session, and analyzing overall results. The use of WOZ has so far been justified in terms of relatively higher cost of having to revise an already implemented system that turned out to be flawed in the usability study.

In practice, WOZ may be replaced by the implementation-test-revise approach based on available tools. Whether or not WOZ is preferable depends on several factors, such as the methods and tools available, the complexity of the application to be designed, and the risk and cost of implementation failure. Low complexity speaks in favor of directly implementing the interface and application without interposing a simulation phase. On the other hand, high complexity may advocate iterative simulations before a full-fledged implementation.

To evaluate each component, you often need to select users that typify a wide range of potential users. You can collect representative data from these typical users based on the prototype we have. The most important measures include language-model perplexity, speech recognition accuracy and speed, parser accuracy and speed (on text and speech recognition output), dialog control performance, prompting strategy and quality, error-repair efficacy, impact of adaptive modeling, and interactions among all these components.

Effective evaluation of a user interface depends on a mix of objective and subjective measures. One of the most important characteristics is repeatability of your results. If you can repeat them, then any difference is likely due to the changes in the design that you are testing. You will want to establish reference tasks for your application and then measure the time that it takes subjects to complete each of those tasks. The consistent selection of subjects is important; as is the use of enough subjects to account for individual differences in the way they complete the tasks.

How do we evaluate MiPad? For our preliminary user study, we did a benchmark study to assess the performance in terms of task-completion time, text throughput (WPM), and user satisfaction. The focal question of this study is whether the MiPad interface can provide added value to the existing PDA interface.

Is the task-completion time much better? We had 20 computer-savvy users test the partially implemented MiPad prototype. These people had no experience with PDA or speech recognition software. The tasks we evaluated include creating a new e-mail, checking a calendar, and creating a new appointment. Task order was randomized. We alternated tasks to different user groups using either pen-only or Tap and Talk interfaces. The text throughput was calculated during email paragraph transcription tasks.[14] Compared to using the pen-only iPaq interface, we observed that the Tap and Talk interface is about twice as fast transcribing email.[15] For the overall command-and-control operations such as scheduling appointments, the Tap and Talk interface was about 33% faster than the existing pen-only interface.[16] Error correction for the Tap and Talk interface was one of the most unsatisfied features. In our user study, calendar access time using the Tap and Talk methods was about the same as pen-only methods, which suggests that simple actions are very suitable for pen-based interaction.

Is it easier to get the job done? Most users we tested stated that they preferred using the Tap and Talk interface. The preferences were consistent with the task completion times. Indeed, most users' comments concerning preference were based on ease of use and time to complete the task, as shown in Figure 18.11.

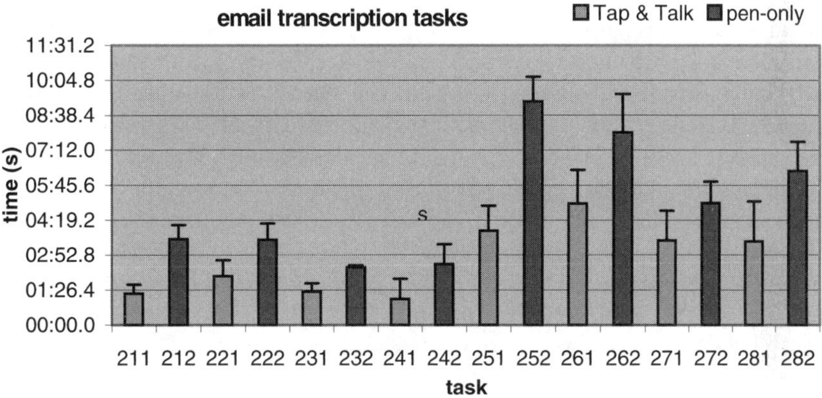

Figure 18.11 Task completion time of email transcription between the pen-only interface and Tap and Talk interface.

18.5.4. Iterations

With rigorous evaluations, you can revisit the components that are on the critical path. In practice, you need to run several iterations to improve user interface design, semantic classes, CFGs, *n*-gram, dialog control, and acoustic models. Iterative experiments are particularly important for deployment, as the coverage of the grammars can be dramatically improved only after we accumulate enough realistic data from end users.

[14] Transcription may not be a realistic task and gives an advantage to speech, because speech avoids a lot of attention shifting that is involved with pen input.

[15] The corresponding speech recognition error rate for the tasks is about 14%, which is based on using a close-talk microphone and a speaker-dependent acoustic model trained from about 20 minutes of speech.

[16] The SLU (at card level) error rate for the task is about 4%.

For MiPad, the performance improvement from using real data has been more dramatic than any other new technologies we can think of. When we collected domain-specific data, MiPad's SLU error rate was reduced by more than a factor of two. The major improvement came from broadening the semantic grammars from real data. We believe that data collection and iterative improvement should be an integral part of the process of developing interactive systems.

Despite unquestionable progress in the last decade, there are many unsolved problems in the procedures, concepts, theory, methods, and tools. Designing an effective speech application and interface remains as much of an art as it is an exact science with established procedures of good engineering practice. The quest for an effective speech application has been the goal for many speech and interface researchers. A successful speech application needs to significantly improve people's productivity. We believe that speech will be one of the most important technology components to enable people access information more effectively in the near future.

MiPad is a work in progress to develop a consistent interaction model and engine technologies for three broad classes of applications. A number of discussed features are yet to be fully implemented and iteratively tested. Our currently tested features include PIM functions only. Despite our incomplete implementation, we observed in our preliminary user study that speech and pen have the potential to significantly improve user experience. Thanks to the multimodal interaction, MiPad also offers a far more compelling user experience than standard telephony interaction. The success of MiPad depends on spoken language technology and always-on wireless connection. With upcoming 3G wireless deployments in sight,[17] the critical challenge for MiPad remains the accuracy and efficiency of our spoken language systems, since MiPad may be used in noisy environments without a close-talk microphone, and the server also needs to support a large number of MiPad clients.

18.6. HISTORICAL PERSPECTIVE AND FURTHER READING

User interface design is particularly resistant to rigid methodologies that attempt to define it simply as another branch of engineering. This is because design deals with human perception and human performance. Issues affecting the user interface design typically include a wide range of topics such as technology capability and limitations, product productivity (speed, ease of use, range of functions, flexibility), esthetics (look and feel, user familiarity, user impression), ergonomics (cognitive load, memory), and user education [2, 13, 14, 30, 35, 38, 44, 45]. The spoken language interface is a unique and specialized human interface medium that differs considerably from the traditional graphical user interface, and there are a number of excellent books discuss how to build an effective speech interface [1, 25, 39, 43].

When we communicate with speech, we must hear and comprehend all information sequentially. This very serial or sequential nature makes speech less effective as an information presentation medium than the graphics. As an input medium, speech must be supplied

[17] http://www.wirelessweek.com/issues/3G

one command or word at a time, requiring the user to remember and order information in a way that is more taxing than it is with visual and mechanical interfaces. The powerful and parallel presentation of information in today's GUI in the form of menus, icons, and windows makes GUI one of the most effective media for people to communicate. GUI is not only parallel but also persistent. That is, once presented, information remains on the screen until replaced as desired. Speech has no such luxury, which places the burden of remembering machine output onto the user. It is no wonder that *a picture is worth a thousand words*. We believe this is one of the key reasons why the spoken language interface has not taken off in comparison to the graphical user interface.

Although speech is natural, as humans already know how to talk effortlessly, this common shared experience of humans may not necessarily translate to human-machine interaction. This is because machines do not share a common cultural heritage with humans and they do not posses certain assumptions about the reality of the world in which we live. We have no precedent for effective speech interaction with nonsentient devices that are self-aware. Our experience with natural speech interaction is based on the assumption that when we talk to another person, the other person has some stake in the outcome of the interaction. The result of this expectation is that structured and goal-oriented conventions are necessary to steer people away from social speech behaviors toward task-oriented protocols, which essentially eliminate our expectation of naturalness.

The fact that speech interactions are one dimensional and time consuming has significant ramifications for interface design. The major challenge is to help the user accomplish as much as possible in as little time as possible. The designer thus must encourage the user to construct a mental model of the task and the interface that serves it, thereby creating an illusion of forward movement that exploits time by managing and responding to the user goals. Many telephony applications have carefully considered these constraints and provided a viable alternative to GUI-based applications and interfaces.

Despite the challenges, a number of excellent applications set milestones for the spoken language industry. Widely used examples include DECTalk (TTS), AT&T universal card services, Dragon's NaturallySpeaking dictation software, and Microsoft Windows 2000 TTS accessibility services. With the wireless smart devices such as smart phones, we are confident that MiPad-like devices will find their way to empower people to access information anywhere and anytime, leading to dramatically improved productivity.

There are a number of companies offering a variety of new speech products. Their Web sites are the best reference points for further reading. The following Web sites contain relevant spoken language applications and user interface technologies:

- AOL (http://www.aol.com)
- Apple (http://www.apple.com/macos/speech)
- AT&T (http://www.att.com/aspg)
- BBN (http://www.bbn.com)
- Dragon (http://www.dragonsystems.com)
- IBM (http://www.ibm.com/software/speech)

- Lernout & Hauspie (http://www.lhs.com)
- Lucent Bell Labs (http://www.lucent.com/speech)
- Microsoft (http://www.microsoft.com/speech)
- Nuance (http://www.nuance.com)
- NTT (http://www.ntt.com)
- Philips (http://www.speech.be.philips.com)
- Speechworks (http://www.speechworks.com)
- Tellme (http://www.tellme.com)
- Wildfire (http://www.wildfire.com)

REFERENCES

[1] Balentine, B. and D. Morgan, *How to Build a Speech Recognition Application*, 1999, Enterprise Integration Group.

[2] Beyer, H. and K. Holtzblatt, *Contextual Design: Defining Customer-Centered Systems*, 1998, San Francisco, Morgan Kaufmann Publishers.

[3] Campbell, J., "Speaker Recognition: A Tutorial," *Proc. of the IEEE*, 1997, **85**(9), pp. 1437-1462.

[4] Chang, E., *et al.*, "Large Vocabulary Mandarin Speech Recognition with Different Approaches in Modeling Tones," *Int. Conf. on Spoken Language Processing*, 2000, Beijing, China.

[5] Chien, L.F., K.J. Chen, and L.S. Lee, "A Best-First Language Processing Model Integrating the Unification Grammar and Markov Language Model for Speech Recognition Applications," *IEEE Trans. on Speech and Audio Processing*, 1993, **1**(2), pp. 221-240.

[6] Dobler, S., "Speech Control In The Mobile Communications Environment," *IEEE ASRU Workshop*, 1999, Keystone, CO.

[7] Dugast, C., X. Aubert, and R. Kneser, "The Philips Large-Vocabulary Recognition System for American English, French and German," *Proc. of the European Conf. on Speech Communication and Technology*, 1995, Madrid, pp. 197-200.

[8] Eichner, M. and M. Wolff, "Data-Driven Generation of Pronunciation Dictionaries In The German Verbmobil Project - Discussion of Experimental Results," *IEEE Int. Conf. on Acoustics, Speech and Signal Processing*, 2000, Istanbul, pp. 1687-1690.

[9] Furui, S., "Recent Advances in Speaker Recognition," *Pattern Recognition Letters*, 1997, **18**, pp. 859-872.

[10] Galitz, W.O., *Handbook of Screen Format Design*, 1985, Wellesley, MA, Q. E. D. Information Sciences Inc.

[11] Galitz, W.O., *User-Interface Screen Design*, 1993, Wellesley, MA, Q. E. D. Information Sciences Inc.

[12] Gao, J., *et al.*, "A Unified Approach to Statistical Language Modeling for Chinese," *Int. Conf. on Acoustics, Speech and Signal Processing*, 2000, Istanbul, pp. 1703-1706.

[13] Hackos, J. and J. Redish, *User and Task Anaysis for User Interface Design*, 1998, New York, John Wiley.

[14] Helander, M., *Handbook of Human-Computer Interaction*, 1997, Amsterdam, North-Holland.

[15] Hon, H.W., *et al.*, "Towards Large Vocabulary Mandarin Chinese Speech Recognition," *Int. Conf. on Acoustics, Signal and Speech Processing*, 1994, Adelaide, Australia, pp. 545-548.

[16] Hon, H.-W., Y. Ju, and K. Otani, "Japanese Large-Vocabulary Continuous Speech Recognition System Based on Microsoft Whisper," *The 5th Int. Conf. on Spoken Language Processing*, 1998, Sydney, Australia.

[17] Horton, W., *The Icon Book*, 1994, New York, John Wiley.

[18] Huang, X., *et al.*, "MIPAD: A Next Generation PDA Prototype," *Int. Conf. on Spoken Language Processing*, 2000, Beijing, China.

[19] Karat, C., *et al.*, "Patterns of Entry and Correction in Large Vocabulary Continuous Speech Recognition Systems," *CHI'99*, 1999, Pittsburgh, pp. 15-20.

[20] Karat, J., "Int. Perspectives: Some Thoughts on Differences between North American and European HCI," *ACM SIGCHI Bulletin*, 1991, **23**(4), pp. 9-10.

[21] Kumamoto, T. and A. Ito, "Structural Analysis of Spoken Japanese Language and its Application to Communicative Intention Recognition," *Proc. of the Fifth Int. Conf. on Human-Computer Interaction—Poster Sessions: Abridged Proc.*, 1993, pp. 284.

[22] Lamel, L. and R.D. Mori, "Speech Recognition of European Languages" in *Proc. IEEE Automatic Speech Recognition Workshop* 1995, Snowbird, UT, pp. 51-54.

[23] Laurila, K. and P. Haavisto, "Name Dialing—How Useful Is It?," *IEEE Int. Conf. on Acoustics, Speech and Signal Processing*, 2000, Istanbul, pp. 3731-3734.

[24] Lyu, R.Y., *et al.*, "Golden Mandarin (III)—A User-Adaptive Prosodic Segment-Based Mandarin Dictation Machine for Chinese Language with Very Large Vocabulary," *Proc. of the IEEE Int. Conf. on Acoustics, Speech and Signal Processing*, 1995, Detroit, pp. 57-60.

[25] Markowitz, J., *Using Speech Recognition*, 1996, Upper Saddle River, Prentice Hall.

[26] Martin, G.L., "The Utility of Speech Input in User-Computer Interfaces," *Int. Journal of Man-Machine Studies*, 1989, **30**(4), pp. 355-375.

[27] Möbius, B., "Analysis and Synthesis of German F0 Contours by Means of Fujisaki's Model," *Speech Communication*, 1993, **13**(53-61).

[28] Morrison, D.L., *et al.*, "Speech-Controlled Text-Editing: Effects of Input Modality and of Command Structure," *Int. Journal of Man-Machine Studies*, 1984, **21**(1), pp. 49-63.

[29] Nielsen, J., *Usability Engineering*, 1993, Boston, MA, Academic Press.

[30] Norman, D.A., *Things That Make Us Smart: Defending Human Attributes in the Age of the Machine*, 1993, Reading, MA, Addison-Wesley.

[31] O'Shaughnessy, D., "Specifying Accent Marks in French Text for Teletext and Speech Synthesis," *Int. Journal of Man-Machine Studies*, 1989, **31**(4), pp. 405-414.

[32] Pierrehumbert, J., and M. Beckman, *Japanese Tone Structure*, 1988, Cambridge, MA, MIT Press.

[33] Poock, G.K., "Voice Recognition Boosts Command Terminal Throughput," *Speech Technology*, 1982, **1**, pp. 36-39.

[34] Rao, P.V.S., "VOICE: An Integrated Speech Recognition Synthesis System for the Hindi Language," *Speech Communication*, 1993, **13**, pp. 197-205.

[35] Royer, T., "Using Scenario-Based Designs to Review User Interface Changes and Enhancements," *Proc. of DIS'95 Symposium on Designing Interactive Systems: Processes, Practices, Methods, & Techniques*, 1995, pp. 237-246.

[36] Rubin, J., *Handbook of Usability Testing: How to Plan, Design, and Conduct Effective Tests*, 1994, New York, John Wiley.

[37] Sagisaka, Y. and L.S. Lee, "Speech Recognition of Asian Languages" in *Proc. IEEE Automatic Speech Recognition Workshop* 1995, Snowbird, UT, pp. 55-57.

[38] Salvendy, G., *Handbook of Human Factors and Ergonomics*, 1997, New York, John Wiley.

[39] Schmandt, C., *Voice Communication with Computers*, 1994, New York, Van Nostrand Reinhold.

[40] Schmidt, M., *et al.*, "Phonetic Transcription Standards for European Names (ONOMASTICA)," *Proc. of the European Conf. on Speech Communication and Technology*, 1993, Berlin, pp. 279-283.

[41] Shneiderman, B., *Designing the User Interface: Strategies for Effective Human-Computer Interaction*, 1997, Reading, MA, Addison-Wesley.

[42] Spencer, R.H., *Computer Usability Testing and Evaluation*, 1985, Englewood Cliffs, NJ, Prentice-Hall.

[43] Weinschenk, S. and D. Barker, *Designing Effective Speech Interfaces*, 2000, New York, John Wiley.

[44] Wixon, D. and J. Ramey, *Field Methods Casebook for Software Design*, 1996, New York, John Wiley.

[45] Wood, L., *User Interface Design: Bridging the Gap from User Requirements to Design*, 1997, CRC Press.

[46] Yang, K.C., *et al.*, "Statistics-Based Segment Pattern Lexicon—A New Direction for Chinese Language Modeling," *Int. Conf. on Acoustics, Speech and Signal Processing*, 1998, pp. 169-172.

INDEX

A

A* search, 603, 606, 626-39
  admissible heuristics, 630-31
  extending new words, 631-34
  fast match, 634-38
  multistack search, 639
  stack decoders, 592
  stack pruning, 638-39
Abbreviations, 709-12
  disambiguation, 711
  expansion, 711-12
Absolute Category Rating (ACR), 841
Absolute discounting, 568
Absolute threshold of hearing, 23, 36
Abstract prosody, 745-61
  accent, 751-53
  pauses, 747-49
  prosodic phrases, 749-51
  prosodic transcription systems, 759-61
    INTSINT, 760
    PROSPA, 759
    TILT, 760
  tone, 753-57
  tune, 757-59
Accent, 751-53
Accessibility, 929
Acoustical environment, 477, 478-86
  additive noise, 478-80
  babble noise, 479
  cocktail party effect, 479
  Lombard effect, 480
  model of the environment, 482-86
  pink noise, 478
  reverberation, 480-82
  white noise, 478-79
Acoustical model of speech production, 283-90
  glottal excitation, 284
  lossless tube concatenation, 284-88
  mixed excitation model, 289
  source-filter models, 288-90
Acoustical transducers, 486-97
  active microphones, 496
  bidirectional microphones, 490-94
  carbon button microphones, 497
  condenser microphone, 486-89
  directionality patterns, 489-96
  dynamic microphones, 497
  electromagnetic microphones, 497
  electrostatic microphones, 497
  passive microphones, 496
  piezoelectric microphones, 497
  piezoresistive microphones, 497
  pressure gradient microphones, 496
  pressure microphones, 496
  ribbon microphones, 497
  unidirectional microphones, 494-96
Acoustic modeling, 415-75
  adaptive techniques, 444-53
    clustered models, 452-53
    maximum likelihood linear regression (MLLR), 447-50
    maximum a posteriori (MAP), 445-47
    MLLR vs. MAP, 450-51
  confidence models, 453-57
    filler models, 453-54
    transformation models, 454-55
  historical perspective, 465-68
  neural networks, 457-59
    integrating with HMMs, 458-59
    parametric trajectory models, 460-62
    recurrent, 457-58
    time delay neural network (TDNN), 458
  scoring acoustic features, 439-43
    HMM output distributions, 439-41
    isolated vs. continuous speech training, 441-43
  speech signals:
    context variability, 417
    environment variability, 419
    speaker variability, 418-19
    style variability, 418
  word recognition errors, types of, 420
Acoustic pattern matching, 545
Acronyms, 711-12
Active arcs, 550
Active constituents, 550-51
Active microphones, 496
Adaptive codebook, 356-57
Adaptive delta modulation (ADM), 347
Adaptive echo cancellation (AEC), 497-504
  LMS algorithm, 499-500
  normalized LMS algorithm (NLMS), 501-2
  RLS algorithm, 503-4
  transform-domain LMS algorithm, 502-3
Adaptive language models, 575-78
  cache language models, 574-75
  maximum entropy models, 576-78
  topic-adaptive models, 575-76
Adaptive PCM, 344-45
Adaptive spectral entropy coding (ASPEC), 350-51
Additive noise, 478-80
Adjectives, 54
ADPCM, 348
Adverbs, 60
AEC, *See* Adaptive echo cancellation (AEC)
Affricates, 44
Agent-based dialog modeling, 914
Air Travel Information Service (ATIS) task, 901-3
A-law compander, 343

Algebraic code books, 358-59
Algorithmic delay, 339
Aliasing, 245
Allophones, 47-49
Alphabet, 121
Alternative hypothesis, 114
Alternatives question, 62
Alveolar consonants, 46
American Institute of Electrical Engineers (AIEE), 272
American National Standards Institute (ANSI), 360
Amplitude modulator, 208
Analog signals, 202
  analog-to-digital conversion, 245-46
  digital processing of, 242-48
  digital-to-analog conversion, 246-48
  Fourier transform of, 243
  sampling theorem, 243-45
Analysis by synthesis, 353-56
Analysis frames, 276
Anaphora, 882
Anaphora resolution, 883-84
Anechoic chamber, 480
Anger, and speech, 745
Anti-causal system, 211
Anti-jam (AJ), 361
Application programming interface (API), 920-21
Applications:
  accessibility, 929
  automobile, 930
  classes of, 921-31
  computer command and control, 921-24
  dictation, 926-29
  handheld devices, 930
  hands-busy, eyes-busy, 927
  speaker recognition, 931
  telephony applications, 924-26
Approximants, 42
Articulation, of English consonants, 42, 45
Articulators, 24-25
Articulatory speech synthesis, 793, 803-4, 846
Artificial intelligence (AI), 855, 889
Attentional state, 859
Audio coding, 338
Aurora project, 925
Autocorrelation method, 324-27
Automobile applications, 930
AutoPC (Clarion), 930
Auto-regressive (AR) modeling, 290

B

Babble noise, 479
Back-channel turn, 861
Backoff models, 563-64
Backoff paths, 618-19
Backoff smoothing, 565-70
Back propagation algorithm, 163
Backtracking, 598
Backus-Nauer form (BNF), 69
Backward prediction error, 297
Backward reasoning, 595
Bandlimited interpolation, 245
Bandpass filter, 242
Bark scale, 32-33
Bark scale functions, 35
Basic search algorithms, 591-643
  blind graph, 597-601
    breadth-first, 600-601
    depth-first, 598-99
    heuristic graph, 601-8
  complexity, 592
  general graph searching procedures, 593-97
  graph-search algorithm, 597
  historical perspective, 640
  search algorithms for speech recognition, 608-12
  stack decoding (A* search), 626-39
    admissible heuristics, 630-31
    extending new words, 631-34
    fast match, 634-38
    multistack search, 639
    stack pruning, 638-39
  time-synchronous Viterbi beam search, 622-26
    use of beam, 624-25
    Viterbi beam search, 625-26
Baum-Welch algorithm, 389-93
Bayes' classifiers:
  comparing, 148-49
  representation, 138-39
Bayes' decision theory, 133, 134-49, 159
  curse of dimensionality, 144-46
  discriminant functions, 138-41
  error rate, estimating, 146-48
  Gaussian classifiers, 142-44
  minimum-error-rate decision rules, 135-38
Bayes' estimator, 99
Bayesian estimation, 107-13
  general, 109-10
  prior and posterior estimation, 108-9
Bayes' risk, 136
Bayes' rule, 75-78
Bayes' theorem, 128
Beamforming, 507
Beam search, 606-8
Behavior model, 855
Best-first search, 602-6
Bidirectional microphones, 490-94
Bidirectional search, 595-96
Bigrams, 559, 563
  search space with, 617-18
Bilinear transforms, 315-16
Binaural listening, 31
Binomial distributions, 86
Bit, 121
Bit reversal, 224
Blind graph search algorithms, 597-601

Blind source separation (BSS), 510-15
Block coding, 125
Bolzmann's constant, 488
Bottom-up chart parsing, 550-53
Bound stress, 431
Branch-and-bound algorithm, 604, 640
Branching factor, 593-95
Breadth-first search, 600-601, 624
Breathy-voiced speech, 330
Bridging, 488
Brute-force search, 601

C

Cache language models, 574-75, 882
Carbon button microphones, 497
Cardioid mikes, 495
CART, *See* Classification and regression trees (CART)
CART-based durations, 763
Cascade/parallel formant synthesizer, 797-802
  parameter values, 799
  targets, 801-2
Case relations, 63, 66
Cauchy-Schwarz inequality, 263
CCITT (Comite Consultatif International Telephonique et Telegraphique), 343
CELP, *See* Code excited linear prediction (CELP)
Central Limit Theorem, 93
Centroid, 165
Cepstral mean normalization (CMN), 522-24, 540
Cepstrum, 306-15
  cepstrum vector, 309
  complex, 307-8
  LPC-cepstrum, 309-11
  of periodic signals, 311-12
  of pole-zero filters, 308-98
  real, 307-8
  source-filter separation via, 314-15
  of speech signals, 312-13
Chain rule, 75, 77-78, 124
Channel coding, 126-28
Character, and prosody, 744
Chart, 550
Chart parsing for context-free grammars, 549-53
  bottom-up, 550-53
  top down vs. bottom up, 549-50
CHATR system of ATR, 781
Chebychev polynomials, 237
Chemical formulae, 719
Chemical Markup Language (CML), 719
Chomsky hierarchy, 547-48
Chunking, 876
Chunks, 691
Circular convolution, 227
Class-conditional pdfs, 133, 140, 144
Classification and regression trees (CART), 134, 175-89, 191, 729-30, 748, 879
  CART algorithm, 189
  choice of question set, 177-78
  complex questions, 182-84
  growing the tree, 181
  missing values and conflict resolution, 182
  right-sized tree, 184-89
    cross-validation, 188-89
    independent test sample estimation, 187-88
    minimum cost-complexity pruning, 185-87
  splitting criteria, 178-81
Class inclusion, 66
Class *n*-grams, 570-74
  data-driven classes, 572-73
  rule-based classes, 571-72
Clauses, 61-62
  relative, 61
Clear /l/, 48
Cleft sentence, 62
Closed-loop estimation, 356
Closed-phase analysis, 319
Closed POS categories, 54
Cluster, 572
Clustered acoustic-phonetic units, 432-36
Clustered models, 452-53
CMU Pronunciation Dictionary, 436
Coarticulation, 47, 49-51
Cochlea, 30
Cocke-Younger-Kasmi (CYK) algorithm, 584
Cocktail party effect, 479
Codebook, 164-65
Code Division Multiple Access (CDMA), 360-61
Code excited linear prediction (CELP), 353-61
  adaptive codebook, 356-57
  analysis by synthesis, 353-56
  LPC vocoder, 353
  parameter quantization, 358-59
  perceptual weighting/postfiltering, 357-58
  pitch prediction, 356-57
  standards, 359-61
Coder delay, 339
Codeword, 164-65
Colored noise, 270
Combination models, 456-57
COMLEX dictionary (LDC), 436-37
Command and control speech recognition, 921-24
  application ideas/uses, 923
  situations for, 922-23
Commissives, 861
Communicative prosody, 858
Compact Disc-Digital Audio (CD-DA), 342
Compander, 342
Comparison Category Rating (CCR) method, 842
Complements, 58, 59
Complex cepstrum, 307-8
Complexity parameter, 185
Compressions, 21
Computational delay, 339

Concatenative speech synthesis, 793-94
  choice of unit, 805-8
  context-dependent phonemes, 808-9
  context-independent phonemes, 806-7
  diphones, 807-8
  with no waveform modification, 794
  optimal unit string, 810-17
    data-driven transition cost, 815-16
    data-driven unit cost, 816-17
    empirical transition cost, 812-13
    empirical unit cost, 813-15
    objective function, 810-12
  subphonetic units (senones), 809
  syllables, 809
  unit inventory design, 817-18
  with waveform modification, 795
  word and phrase, 809
Concept-to-speech rendition, 899-901
Conceptual graphs, 872-73
Condenser microphone, 486-89
Conditional entropy, 123-24
Conditional expectation, 81
Conditional likelihood, 151
Conditional maximum likelihood estimator (CMLE), 151
Conditional probability, 75-76
Conditional risks, 136
Conditioning, 320
Conference of European Posts and Telegraphs (CEPT), 360
Confidence models, 453-57
  combination models, 456-57
  filler models, 453-54
  transformation models, 454-55
Conflict resolution procedure, 182
Conjugate quadrature filters, 251-54
Conjunctions, 54
Connotation, message, 739
Consonants, 24, 42-46
  affricates, 44
  alveolar, 46
  dental, 46
  fricatives, 42, 44
  labial, 46
  labio-dental, 46
  nasal, 43-44
  obstruent, 43
  palatal, 46
  plosive, 42-43
  stop, 43
  velar, 46
Consumer audio, 351-52
Content words, 54
Context coarticulation, 735
Context dependency, 430-31
Context-dependent phonemes, 808-9
Context-dependent units and inter-word triphones, 658-59
Context-free grammar (CFG), 465, 547, 921
  vs. *n*-gram models, 580-84
  search space, 613-16
Context-free grammars (CFGs), search architecture, 676-77
Context-independent phonemes, 806-7
Context variability, 417
Continuation rise, 749
Continuous distribution, 78
Continuous-frequency transforms, 209-16
  Fourier transforms, 208-10
  z-transforms, 211-15
Continuously listening model, 422
Continuously variable slope delta modulation (CVSDM), 347
Continuous mixture density HMMs, 394-96
Continuous random variable, 78
Continuous speech recognition (CSR), 591, 611-12, 945
Continuous speech training, vs. isolated speech training, 441-43
Continuous-time stochastic processes, 260
Contrastive stress, 431
Contrasts, 66
Convolution operator, 207
Co-references, 882
Corpora, 545-46
Corpus-based F0 generation, 779-82
  F0 contours indexed by parsed text, 779-81
  F0 contours indexed by ToBI, 781-82
  transplanted prosody, 779
Corrective training, 158
Correlation, 82-83
Correlation coefficient, 82-83
Covariance, 82-83
Covariance matrix, 84
Critical region, 114
Cross-validation, 188-89
Cumulative distribution function, 79
Currency, 716
Curse of dimensionality, 144-46

D

DAMSL system, *See* Dialog Act Markup in Several Layers (DAMSL)
Dark /*l*/, 48
DARPA, 11
DARPA ATIS programs, 913
Data-directed search, 549
Data-driven parallel model combination (DPMC), 533
Data-driven speech synthesis, 794, 803
Data flow, 694-97
DAVO, 846
DCT, *See* Discrete Cosine Transform (DCT)
Decibels (dB), 22

Decimation-in-frequency, 223
Decimation-in-time, 223
Declaratives, 861
Declarative sentence, 62
Declination, 766-67
Decoder, 5
Decoder basics, 609
Decoder delay, 339
DECTalk system, 736, 846
Degree of displacement, 21-22
Deixis, 882
Deleted interpolation smoothing, 564-65
Deletion errors, 420
Delta modulation (DM), 346
Denotation, message, 739
Dental consonants, 46
Depth-first search, 598-99
Derivational morphology, 56-57
Derivational prefixes, 57
Derivational suffixes, 57
Descrambling, 224
Deterministic signals, 260
Development set, 419-20
Diagnostic Rhyme Test (DRT), 837
Dialects, 725
Dialog Act Markup in Several Layers (DAMSL), 862-66
Dialog-act theory, 914
Dialog control, 866-67, 949
Dialog flow, 942-43
  prompting strategy, 943
  prosody, 943
  spoken menus, 942
Dialog grammars, 887-88
Dialog management, 886-94
  dialog grammars, 887-88
  plan-based systems, 888-92
Dialog management module, 881
Dialog Manager, 7-8, 855
Dialog repair, 885
Dialog (speech) acts, 861-66
Dialog structure, 859-67
  attentional state, 859
  Dialog Act Markup in Several Layers (DAMSL), 862-66
  intentional state, 859
  linguistic forms, 859
  task knowledge, 859
  units of dialog, 860-61
  world knowledge, 859
Dialog system, 854-55
  dialog manager, 855
  discourse analysis, 855
  semantic parser, 854-55
Dialog turns, 705-6
Dictation, 926-29, 935, 948
Dielectric, 486
Differential pulse code modulation (DPCM), 345-48
Differential quantization, 345-48
Digital Audio Broadcasting (DAB), 352
Digital filters and windows, 229-42
  generalized Hamming window, 231-32
  ideal low-pass filter, 229-30
  rectangular window, 230-31
  window functions, 230-32
Digital signal processing, 201-73
  of analog signals, 242-48
    analog-to-digital conversion, 245-46
    digital-to-analog conversion, 246-48
    Fourier transform of, 243
    sampling theorem, 243-45
  circular convolution, 227
  continuous-frequency transforms, 209-16
    of elementary functions, 212-15
    Fourier transform, 208-10
    properties, 215-16
    z-transforms, 211-12
  digital filters and windows, 229-42
    generalized Hamming window, 231-32
    ideal low-pass filter, 229-30
    rectangular window, 230-31
    window functions, 230-32
  digital signals/systems, 202-8
  Discrete Cosine Transform (DCT), 228-29
  discrete-frequency transforms, 216-29
    discrete Fourier transform (DFT), 218-19
    Fourier transforms of periodic signals, 219-22
  Fast Fourier Transforms (FFT), 222-27
    FFT subroutines, 224-27
    prime-factor algorithm, 224
    radix-2 FFT, 222-23
    radix-4 algorithm, 223
    radix-6 algorithm, 223
    radix-8 algorithm, 223
    split-radix algorithm, 223
  filterbanks, 251-60
    DFTs as, 255-58
    multiresolution, 254-55
    two-band conjugate quadrature filters, 251-54
  FIR filters, 229, 232-38
    first-order, 234-35
    linear-phase, 233-34
    Parks McClellan algorithm, 236-38
    window design FIR lowpass fiters, 235-36
  IIR filters, 238-42
    first-order, 239-41
    second-order, 241-42
  multirate signal processing, 248-51
    decimation, 248-49
    interpolation, 249-50
    resampling, 250-51
  stochastic processes, 260-70
    continuous-time, 260
    discrete-time, 260

Digital signal processing, stochastic processes *(cont.)*
LTI systems with stochastic inputs, 267
noise, 269-70
power spectral density, 268-69
stationary processes, 264-67
statistics of, 261-64
Digital Signal Processing (DSP), 202-3, 339
Digital signals/systems, 202-8
digital systems, 206-8
linear time-invariant (LTI) systems, 207
linear time-varying systems, 208
nonlinear systems, 208
other digital signals, 206
sinusoidal systems, 203-5
Digital systems, defined, 202
Digital-to-analog conversion, 246-48
Digital wireless telephony applications, 925
Diphones, 807-8
Diphthongs, 40
Directionality patterns, 489-96
Directives, 861
Disambiguation, 876
Discourse analysis, 7, 753, 855, 881-86
resolution, 882-85
Discourse memories, 882
Discourse segments, 857
Discrete Cosine Transform (DCT), 228-29
Discrete distribution, 77
Discrete-frequency transforms, 216-29
discrete Fourier transform (DFT), 218-19
Fourier transforms of periodic signals, 219-22
Discrete joint distribution, 83-84
Discrete random variables, 77
Discrete-time Fourier transform, 209, 210
Discrete-time stochastic processes, 260
Discriminative training, 150-63
gradient descent, 153-55
maximum mutual information estimation (MMIE), 150-52
minimum-error-rate estimation, 156-58
multi-layer perceptrons, 160-63
neural networks, 158
single-layer perceptrons, 159-60
Disfluency, 857
Distortion measures, 164-66
Distribution function, 79
Document structure detection, 692, 699-706
chapter and section headers, 700-701
dialog turns and speech acts, 705-6
email, 704-5
lists, 701-2
paragraphs, 702
sentences, 702-4
Web pages, 705
Dolby Digital, 351
Domain knowledge, 2
Domain-specific tags, 718-20
chemical formulae, 719
mathematical expressions, 718-19
miscellaneous formats, 719-20
Dragon NaturallySpeaking, 926
Dr. Who case study, 906-13
dialog manager, 910-13
discourse analysis, 909-10
semantic parser (sentence interpolation), 908
semantic representation, 906-8
Dr. Who Project, 869, 876-77
DTS, 351-52
Duration assignment, 761-63
CART-based durations, 763
rule-based methods, 762-63
Dynamic microphones, 497
Dynamic time warping (DTW), 383-85

E

Ear:
cochlea, 30
eardrum, 29
middle ear, 29
outer ear, 29
oval window, 29
physiology of, 29-32
sensitivity of, 30
Eardrum, 29
Earley algorithm, 584
Eigensignals, 209
Eigenvalue, 209
Electret microphones, 487
Electroglottograph (EGG), 828
signals, 319-20
Electromagnetic microphones, 497
Electronic Industries Alliance (EIA), 360
Electrostatic microphones, 497
Ellipsis, 882
EM algorithm, 134, 170-72
Embedded ADPCM, 348
Emotion, and prosody, 744-45
Emphatic stress, 431
End-point detection, 422-24
Entropy, 120-22
conditional, 123-24
Entropy coding, 350-51
Environmental model adaptation, 528-38
model adaptation, 530-31
parallel model combination, 531-34
retraining on compensated features, 537-38
retraining on corrupted speech, 528-39
vector Taylor series, 535-37
Environmental robustness, 477-544
acoustical environment, 477, 478-86
additive noise, 478-80
babble noise, 479
cocktail party effect, 479

Lombard effect, 480
model of the environment, 482-86
pink noise, 478
reverberation, 480-82
white noise, 478-79
acoustical transducers, 486-97
active microphones, 496
bidirectional microphones, 490-94
carbon button microphones, 497
condenser microphone, 486-89
directionality patterns, 489-96
dynamic microphones, 497
electromagnetic microphones, 497
electrostatic microphones, 497
passive microphones, 496
piezoelectric microphones, 497
piezoresistive microphones, 497
pressure gradient microphones, 496
pressure microphones, 496
ribbon microphones, 497
unidirectional microphones, 494-96
adaptive echo cancellation (AEC), 497-504
convergence properties of the LMS algorithm, 500-501
LMS algorithm, 499-500
normalized LMS algorithm (NLMS), 501-2
RLS algorithm, 503-4
transform-domain LMS algorithm, 502-3
environmental model adaptation, 528-38
model adaptation, 530-31
parallel model combination, 531-34
retraining on compensated features, 537-38
retraining on corrupted speech, 528-39
vector Taylor series, 535-37
environment compensation preprocessing, 515-28
cepstral mean normalization (CMN), 522-24
frequency-domain MMSE from stereo data, 519-20
real-time cepstral normalization, 525
spectral subtraction, 516-19
use of Gaussian mixture models, 525-28
Weiner filtering, 520-22
multimicrophone speech enhancement, 504-15
blind source separation (BSS), 510-15
microphone arrays, 505-10
nonstationary noise, modeling, 538-39
Environment compensation preprocessing, 515-28
cepstral mean normalization (CMN), 522-24
frequency-domain MMSE from stereo data, 519-20
real-time cepstral normalization, 525
spectral subtraction, 516-19
use of Gaussian mixture models, 525-28
Wiener filtering, 520-22
Environment variability, 419
Epoch detection, 828-29
Equal-loudness curves, 31
Ergodic processes, 265-67
Ergonomics of Software User Interface, 932
Error handling, 937-41
error detection and correction, 938-39
feedback and confirmation, 939-41
Estimation, 98-99
Estimation theory, 98-113
Bayesian estimation, 107-13
general, 109-10
prior and posterior estimation, 108-9
least squared error (LSE) estimation, 99-100
for constant functions, 100
for linear functions, 101-2
for nonlinear functions, 102-4
MAP estimation, 111-13
maximum likelihood estimation (MLE), 104-7
minimum mean squared error (MMSE), 99-104
for constant functions, 100
for linear functions, 101-2
for nonlinear functions, 102-4
Euclidean distortion measure, 165-66
Eureka 147 DAB specification, 352
European Telecommunication Standards Institute (ETSI), 360
Evaluation of understanding and dialog, 901-3
and ATIS task, 901-3
PARADISE framework, 903-6
Exact *n*-best algorithm, 666-67
Exception list, 697, 728
Excitation signal, 301
Exclamative sentence, 62
Exhaustive search, 597
Expectation (mean) vector, 84
Expectation of a random variable, 79
Exponential distribution, 98

F

F0 contour interpolation, 772-73
F0 jumps, 330
F1/F2 targets, 39
Factored language probabilities, 650-53
Factored lexical trees, 652-53
Fast Fourier Transforms (FFT), 222-27
FFT subroutines, 224-27
prime-factor algorithm, 224
radix-2 FFT, 222, 223
radix-4 algorithm, 223
radix-6 algorithm, 223
radix-8 algorithm, 223
split-radix algorithm, 223
Fast match, 634-38, 661-62
look-ahead strategy, 661-62
Rich-Get-Richer (RGR) strategy, 662
Fear, and speech, 745
Feedback, 490
Feedforward adaptation, 345
Fenones, 467
Festival, 732-35
FFT, *See* Fast Fourier Transforms (FFT)

Filler models, 453-54
Filterbanks, 251-60
  DFTs as, 255-58
  multiresolution, 254-55
  two-band conjugate quadrature filters, 251-54
Filters, 210
Finite-impulse response (FIR), *See* FIR filters
Finite-state automaton, 547
Finite-state grammar, 613-16
Finite-state machines (FSM), 548
Finite state network, 654
FIR filters, 229, 232-38
  first-order, 234-35
  linear-phase, 233-34
  Parks McClellan algorithm, 236-38
  window design, 235-36
First coding theorem, 124
First-order FIR filters, 234-35
First-order IIR filters, 239-41
First-order moment, 261
Focus, 883
Focus shifts, cueing, 892
Formal language modeling, 546-53
  chart parsing for context-free grammars, 549-53
    bottom-up, 550-53
    top down vs. bottom up, 549-50
  Chomsky hierarchy, 547-48
Formant frequencies, 319-23
  statistical formant tracking, 320-23
Formants, 27-28
Formant speech synthesis, 793, 796-804
  cascade model, 797
  formant generation by rule, 800-803
  Klatt's cascade/parallel formant synthesizer, 797-99
  locus theory of speech production, 800
  parallel model, 797
  waveform generation from formant values, 797-99
Formant targets, 39
Forward algorithm, 385-87
Forward-backward algorithm, 389-93, 442-43, 557
Forward-backward search algorithm, 670-73
Forward error correction (FEC), 352
Forward prediction error, 297
Forward reasoning, 595
Fourier series expansion, 218
Fourier transforms, 208-10
  Fast Fourier Transforms (FFT), 222-27
    FFT subroutines, 224-27
    prime-factor algorithm, 224
    radix-2 FFT, 222-23
    radix-4 algorithm, 223
    radix-6 algorithm, 223
    radix-8 algorithm, 223
    split-radix algorithm, 223
  properties of, 215-17
Fourier transforms of periodic signals, 219-22
  complex exponential, 219-20
  general periodic signals, 221-22
  impulse train, 221
Frames, 339
Free stress, 431
Frequency analysis, 32-34
Frequency domain, advantages of, 348-49
Frequency-domain MMSE from stereo data, 519-20
Frequency masking, *See* Masking
Frequency response, 210
Fricatives, 42, 44
Fujisaki's model, 776
Full duplex sound cards, 936
Functionality encapsulation, 871-72
Functional tests, 842-43
Function words, 54
Fundamental frequency, 25

G

G.711 standard, 343-44, 348, 359, 371
G.722 standard, 348, 359
G.723.1 standard, 359
G.727 standard, 348
G.728 standard, 359
G.729 standard, 359
Game search, 594
Gamma distributions, 90-91, 95
Gaussian distributions, 92-98
  Central Limit Theorem, 93
  lognormal distribution, 97-98
  multivariate mixture Gaussian distributions, 93-95
  standard, 92-93
  χ^2 distributions, 95-96
Gaussian mixture models, 525-28
Gaussian processes, 264-65
General graph searching procedures, 593-97
Generality, of grammar, 546
Generalized Hamming window, 231-32
Generalized Lloyd algorithm, 168
Generalized triphones, 808
General Packet Radio Service (GPRS), 361
Geometric distributions, 86-87
Gibbs phenomenon, 235
Gini index of diversity, 181
Glides, 42
Global Positioning System (GPS), 868, 930
Glottal cycle, 26
Glottal excitation, 284
Glottal stop, 43
Glottis, 25, 288
Glyphs, 36
Goal-directed search, 549
Goodness-of-fit test, 116-18
Good-Turing estimates and Katz smoothing, 565-67
Gradient descent, 153-55, 190
Gradient prominence, 765-66
Graham-Harison-Ruzzo algorithm, 584
Grammar, 545

Granular noise, 346
Grapheme-to-phoneme conversion, 692-93
Graphical user interface (GUI), 1
Graph-search algorithms, 591, 597
Greedy symbols, 558
Ground, 900
Grounding, 861

H

H.323, 359
Half duplex sound cards, 936
Half phone, 809
Hamming window, 232, 258, 278-80, 283
  generalized, 231-32
Handheld devices, 930
Hands-busy, eyes-busy applications, 927
Hanning window, *See* Hamming window
Hard palate, 25
Harmonic coding, 363-67
  parameter estimation, 364-65
  parameter quantization, 366-67
  phase modeling, 365-66
Harmonic errors, 330
Harmonic sinusoids, 218
Harmonic/Stochastic (H/S) model, 847
Harvard Psychoacoustic Sentences, 839
Has-a relations, 65
Haskins Syntactic Sentence Test, 839
Head-noun, 59
Head of a phrase, 59
Headset microphone, 936
Hearing sensitivity, 30
Hermitian function, 265
Hertz (Hz), 21
Hessian of the least-squares function, 504
Hessian matrix, 154
Heuristic graph search, 601-8
  beam search, 606-8
  best-first (A* search), 602-6
Heuristic information, 601-2
Heuristic search methods, 601
Hidden Markov models (HMM), 56, 134, 170, 377-413, 416, 547, 931
  Baum-Welch algorithm, 389-93
  continuous mixture density, 394-96
  decoding, 387-89
  definition of, 380-93
  deleted interpolation, 401-3
  dynamic programming, 384-85
    advantage of, 384
    algorithm, 385
  dynamic time warping (DTW), 383-85
  estimating parameters, 389-93
  evaluating, 385-87
  forward algorithm, 385-87
  forward-backward algorithm, 389-93
  historical perspective, 409-10
  initial estimates, 398-99
  limitations of, 405-9
    conditional independence assumption, 409
    duration modeling, 406-8
    first-order assumption, 408
  Markov chain, 378-80
    Markov assumption for, 382
    output-independence assumption, 382
  model topology, 399-401
  observable Markov model, 379-80
  parameter smoothing, 403-4
  practical issues, 398-405
  probability representations, 404-5
  semicontinuous, 396-98
  training criteria, 401
  Viterbi algorithm, 387-89
Hidden understanding model (HUM), 879-80
High-frequency sounds, 31
High-pass filters, 235
Hill-climbing style of guidance, 601
H method, 147
HMM, *See* Hidden Markov models (HMM)
Holdout method, 147
Home applications, 921
Homograph disambiguation, 693, 723, 724-25
Homographs, 721
Homomorphic transformation, 306, 312
Huffman coding, 125-26
Human Factors and Ergonomic Society (HFES), 931-32
Human-machine interaction, 1

I

Ideal low-pass filter, 229-30
IIR filters, 238-42
  first-order, 239-41
  second-order, 241-42
Imperative sentence, 62
Implicit confirmation, 892-93
Implicit memory, 882-83
Impulse response, 207
Inconsistency checking, 885-86
Inconsistency detection, 881
Independent component analysis (ICA), 510
Independent identically distributed (iid), 82
Independent processes, 264
Independent test sample estimation, 187-88
Indistinguishable states, 654
Infinite-impulse response (IIR) filters, *See* IIR filters
Inflectional morphology, 56
Inflectional suffix, 57
Infomax rule, 512-13
Information theory, 73-131
  channel coding, 126-28
  conditional entropy, 123-24
  entropy, 120-22
  mutual information, 126-28

Information theory *(cont.)*
  origin of, 74
  source coding theorem, 124-26
Informed search, 604
Infovox TTS system, 846-47
Inner ear, 29
Input method editors (IME), 736
Insertion errors, 420
Insertion penalty, 610
Inside constituent probability, 555
Inside-outside algorithm, 555
Instance definition, 868
Instantaneous coding, 125
Instantaneous mixing, 510
Instantaneous mutual information, 151
Institute of Electrical and Electronic Engineers (IEEE), 272
Institute of Radio Engineers (IRE), 272
Intelligibility tests, 837-39
Intentional state, 859
Interactive voice response (IVR) systems, 924
Intermediate phrase break, 751
International Conference on Acoustic, Speech and Signal Processing (ICASSP), 272
Internationalization, 943-45
International Telecommunication Union (ITU), 343
International Telecommunication Union-Radiocommunication (ITU-R), 352
Interpolated models, 564-65
Interword-context-dependent phones, 430
Inter-word triphones, 658-59
Intonational phrase break, 751
Intonational phrases, 53, 749
INTSINT, 760
Inverse filter, 290
Inverse-square-law effect, 494
Inverse z-transform, 212
  of rational functions, 213-15
Is-a taxonomies, 64-66
Isolated vs. continuous speech training, 441-43
Isolated word recognition, 610-11

J

Japanese vowels, 46-47
Jensen's inequality, 122
Jitter, 768
Joint distribution function, 84
Jointly strict-sense stationary, 264
Joint probability, 74
Joy, and speech, 745
JSAPI, 921
Juncture, 746-47
Just noticeable distortion (JND), 35

K

Kalman filter, 522
Karhunen-Loeve transform, 426
Katz' backoff mechanism, 618
Katz smoothing, 565-67
Klattalk system, 846
Klatt's cascade/parallel formant synthesizer, 797-802
  parameter values for, 799
  targets used in, 801-2
K-means algorithm, 166-69
Kneser-Ney smoothing, 568-70, 573
Knowledge sources (KSs), 646, 663, 673-74
Kolmogorov-Smirnov test, 118
Kronecker delta, 220
Kth moment, 80
Kullback-Leibler (KL) distance, 122, 581

L

Labial consonants, 46
Labio-dental consonants, 46
Lancaster/IBM Spoken English Corpus, 751-52
Language modeling, 545-90
  adaptive, 575-78
    cache language models, 574-75
    maximum entropy models, 576-78
    topic-adaptive models, 575-76
  CFG vs. n-gram models, 580-84
  complexity measure of, 560-62
  formal, 546-53
    chart parsing for context-free grammars, 549-53
    Chomsky hierarchy, 547-48
  historical perspective, 584
  n-gram pruning, 580-81
  n-gram smoothing, 562-74
    backoff smoothing, 565-70
    class n-grams, 570-74
    deleted interpolation smoothing, 564-65
    performance of, 573-74
  stochastic language models, 554-60
    n-gram language models, 558-60
    probabilistic context-free grammars, 554-58
  vocabulary selection, 578-80
Language model probability, 610
Language models, 4, 949
Language model states, 613-22
  backoff paths, 618-19
  search space:
    with bigrams, 617-18
    with FSM and CFG, 613-16
    with trigrams, 619-20
    with the unigram, 616-17
  silences between words, 621-22
Lapped Orthogonal Transform (LOT), 260
Large-vocabulary search algorithms, 645-85
  context-dependent units and inter-word triphones, 658-59
  exact n-best algorithm, 666-67
  factored language probabilities, 650-53
  factored lexical trees, 652-53
  finite state network, 654

forward-backward search algorithm, 670-73
historical perspective, 681-82
HMM:
different layers of beams, 660-61
fast match, 661-62
lexical successor trees, 652
lexical trees, 646-48
handling multiple linguistic contexts in, 657-58
linear tail in, 655
optimization of, 653
lexical tree search, 648
Microsoft Whisper, 676-81
CFG search architecture, 676-77
n-gram search architecture, 677-81
n-best and multipass search strategies, 663-74
one-pass *n*-best and word-lattice algorithm, 669-70
one-pass vs. multipass search, 673-74
polymorphic linguistic context assignment, 656-57
prefix trees, 647
pronunciation trees, multiple copies of, 648-50
search-algorithm evaluation, 674-76
sharing tails, 655-56
single-word subpath, 655
subtree dominance, 656
subtree isomorphism, 654
subtree polymorphism, 656-58
tree lexicon, 646-59
word-dependent *n*-best and word-lattice algorithm, 667-70
word-lattice generation, 672-73
Larnyx, 25
vocal fold cycling at, 26
Laryngograph, 828
Lateralization, 31
Lateral liquid, 42
Law of large numbers, 82
LBG algorithm, 169-70
Least squared error (LSE) estimation, 99-100
for constant functions, 100
for linear functions, 101-2
for nonlinear functions, 102-4
Least squared regression methods, 180
Least square error (LSE), 160
Leave-one-out method, 147
Left-recursive grammar, 550
Lempel-Ziv coding, 126
Lernout&Hauspie's Voice Xpress, 926
Letter-to-sound (LTS) conversion, 437, 693, 728-30
Letter-to-sound (LTS) rules, 697
Level of significance, 114-15
Levinson-Durbin recursion, 297-98, 333
Lexical baseforms, 436-39
Lexical knowledge, 545
Lexical part-of-speech (POS), 53-56
Lexical successor trees, 652
Lexical trees, 646-48
handling multiple linguistic contexts in, 657-58
linear tail in, 655
optimization of, 653
Lexicon, 697-98
Light /l/, 48
Likelihood function, 104
Likelihood ratio, 139
Limited-domain waveform concatenation, 794
Linear bounded automaton, 548
Linear Discriminate Analysis (LDA), 427
Linear-phase FIR filters, 233-34
Linear predictive coding (LPC), 290-306
autocorrelation method, 295-96
covariance method, 293-94
equivalent representations, 303-6
lattice formulation, 297-300
line spectral frequencies (LSF), 303-5
log-area ratios, 305-6
orthogonality principle, 291-92
prediction error, 301-3
reflection coefficients, 305
roots of the polynomial, 306
solution of the LPC equations, 292-300
spectral analysis via, 300-301
Linear pulse code modulation (PCM), 340-42
Linear time-invariant (LTI) systems, 207
eigensignals of, 209
with stochastic inputs, 267
Linear time-varying systems, 208
Line spectral frequencies (LSF), 303-5
Linguistic analysis, 692, 720-23
homograph disambiguation, 723, 724-25
noun phrase (NP) and clause detection, 723
POS tagging, 722-23
sentence tagging, 722
sentence type identification, 723
shallow parse, 723
Linguistic Data Consortium (LDC), 467
Linguistic forms, 859
Linguistics, co-references in, 882
Lips, 25
Liquid group, 42
Listening Effort Scale, 841
Listening Quality Scale, 841
LMS algorithm, 540
Load loss of signal level, 488
Localization issues, 696-97
Locus theory of speech production, 800
Log-area ratios, 305-6
Logical form, 67-68
Lognormal distribution, 97-98
Lombard effect, 480
Long-term prediction, 353
Look-ahead strategy, 661-62
Lossless compression, 338
Lossless tube concatenation, 284-88
Lossy compression, 338
Loudness, 740

Low-bit rate speech coders, 361-70
  harmonic coding, 363-67
    parameter estimation, 364-65
    parameter quantization, 366-67
    phase modeling, 365-66
  mixed-excitation LPC vocoder, 362
  waveform interpolation, 367-70
Lower bound of probability, 74
Low-frequency sounds, lateralization of, 31
Low-pass filter, bandwidth of, 240
Low-pass filters, digital, 229-30, 235
Low probability of intercept (LPI), 361
LPC, *See* Linear predictive coding (LPC)
LPC-cepstrum, 309-11
LPC vocoder, 353
LP-PSOLA, 832-33
LSE estimation, *See* Least squared error (LSE) estimation
LTI systems, *See* Linear time-invariant (LTI) systems
LTS conversion, 437, 728-30
LTS rules, 697
Lungs, 25

M

McGurk effect, 69
Machine-learning methods, 56
McNemar's test, 148-49, 190
Magnitude-difference test, 119-20
Magnitude subtraction rule, 519
Mahalanobis distance, 166, 168
MAP, *See* Maximum a posteriori (MAP)
MAP estimation, 111-13
Marginal probability, 76, 77-78
Markov chain, 378-80
Masking, 30-31, 34-36, 349-50
  Bark scale functions, 35
  just noticeable distortion (JND), 35
  spread-of-masking function, 35-36
  temporal masking, 35-36
  tone-masking noise, 35
Matched pairs test, 118-20, 148
Mathematical expressions, 718-19
MathML, 718
Maximal projection, 58
Maximum entropy models, 576-78
Maximum likelihood estimation (MLE), 73, 104-7, 134, 141, 168-69
Maximum likelihood estimator, 99
Maximum likelihood linear regression (MLLR), 447-50
  vs. MAP, 450-51
Maximum mutual information estimation (MMIE), 134, 150-52, 156
  defined, 151
Maximum phase signals, 309
Maximum a posteriori (MAP), 73, 111, 141, 331, 445-47, 854
Maximum substring matching problem, 420
MBROLA technique, 829
Mean, 79-81
Mean-ergodic process, 266
Mean opinion score (MOS), 338-39, 840
Mean squared error (MSE), 99
Mean vector, 84
Median, 81
Median smoother of order, 208
Mel-frequency cepstral coefficients (MFCC), 424-26
Mel frequency scale, 34
Message generation, 894-901
  See also Response generation
Message generation box, 897
Metaunits, 658-59
Microphone, 936
Microphone arrays, 505-10
  delay-and-sum beamformer, 505-6
  goals of, 505
  steering, 505
Microprosody, 767-68
Microsoft Dictation, 928-29
Microsoft Speech SDK 4.0, 937
Microsoft's speech API (SAPI), 921
Microsoft Whisper case study, 676-81
  CFG search architecture, 676-77
  n-gram search architecture, 677-81
Middle ear, 29
Mid-riser quantizer, 340
Mid-tread quantizer, 340
Minimum-classification-error (MCE), 156
Minimum cost-complexity pruning, 185-87
Minimum-error-rate decision rules, 135-38
Minimum-error-rate estimation, 134, 156-58
Minimum mean squared error (MMSE), 73, 99-104
  for constant functions, 100
  for linear functions, 101-2
  for nonlinear functions, 102-4
Minimum mean square estimator, 99
Minimum phase signals, 309
Minimum squared error (MSE) estimation, 100
Minor phrase break, 751
MiPad case study, 945-52
  evaluation, 949-51
  iterations, 951-52
  rapid prototyping, 948-49
  specifying the application, 946-48
MITalk System, 735-36, 846
Mixed-excitation LPC vocoder, 362
Mixed excitation model, 289
Mixed initiative systems, 860
Mixture density estimation, 172
MMIE, *See* Maximum mutual information estimation (MMIE)
MMSE, *See* Minimum mean squared error (MMSE)
Mobile applications, 921
Mode, 81

Modified Discrete Cosine Transform (MDCT), 259
Modified Rhyme Test (MRT), 838
Modifiers, 61
Modular (component) testing, 731
Modulated Lapped Transform (MLT), 259
Money and currency, 716
Monolithic whole-system evaluation, 731
Morphological analysis, 693, 725-27
  algorithm, 727
  suffix and prefix stripping, 726-27
Morphological attributes, 55
Morphology, 56-57
  derivational, 56-57
  inflectional, 56
Move generator, 594
MP3, 371
MPEG, 351-52, 371
Multi-layer perceptrons, 160-63
Multimicrophone speech enhancement, 504-15
  blind source separation (BSS), 510-15
  microphone arrays, 505-10
    delay-and-sum beamformer, 505-6
    steering, 505
Multinomial distributions, 87-89
Multipass search, 663-74, 682
  n-best lists and word lattices, 664-66
  n-best search paradigm, 663
Multipass search vs. one-pass search, 673-74
Multiple tree combination, 730
Multiplexing delay, 339
Multirate signal processing, 248-51
  decimation, 248-49
  interpolation, 249-50
  resampling, 250-51
Multiresolution filterbanks, 254-254
Multistack search, 639
Multistyle training, 419
Multivariate distributions, 83-85
Multivariate Gaussian mixture density estimation, 172-75
Multivariate mixture Gaussian distributions, 93-95
Musical noise, 517
Musical pitch scales, and prosodic research, 32
MUSICAM, 352
Mutual information, 126-28

N

Narrow-band filtering, 330
Narrow-band spectrograms, 282
Nasal, 42
Nasal cavity, 25
Nasal consonants, 43-44
Natural gradient, 513
Natural language, linguistic analysis of, 720-23
Natural language generation from abstract semantic input, 898
Natural language process (NLP) systems, 693-94
N-best lists, 664-66
N-best search paradigm, 663
Near-miss list, 158
Negative correlation, 83
Negotiation, 892
NETALK, 729
Neural networks, 134, 158, 457-59
  integrating with HMMs, 458-59
  recurrent, 457-58
  time delay neural network (TDNN), 458
Neural transduction process, 20
Neural units, 158
Neuromuscular signals, 20
Newton's algorithm, 155
N-gram language models, 558-60
N-gram pruning, 580-81
N-grams, search architecture, 677-81
N-gram smoothing, 562-74
  backoff smoothing, 565-70
    alternative backoff models, 568-70
    Good-Turing estimates and Katz smoothing, 565-67
  class *n*-grams, 570-74
    data-driven classes, 572-73
    rule-based classes, 571-72
  deleted interpolation smoothing, 564-65
  performance of, 573-74
Noise-canceling microphone, 490
Noiseless channels, 127
Noisy conditions, 330
Nonbranching hierarchies, 65
Noncausal Wiener filter, 522
Non-hierarchical relations, 65
Non-informative prior, 112
Nonlinear systems, 208
Nonstationary noise, modeling, 538-39
Normalized cross-correlation method, 327-29
Normalized LMS algorithm (NLMS), 501-2
Noun phrases (NPs), 58-59
Nouns, 54
NP-hard problem, 593
N-queens problem, 593, 598
Nucleus, 52
Number formats, 712-20
  account numbers, 716
  cardinal numbers, 717-18
  dates, 714-15
  money and currency, 716
  ordinal numbers, 717
  phone numbers, 712-14
  times, 715
Nyquist frequency, 243, 245

O

Object-oriented programming, 869
Observable Markov model, 379-80
Obstruent, 43

Octaves, 32
Office applications, 921
Omnidirectional condenser microphones, 489-90
One-pass n-best and word-lattice algorithm, 669-70
One-pass vs. multipass search, 673-74
One-place predicates, 67
On-glides, 42
Onset, 52
Open-loop estimation, 356
Open POS categories, 54
Operations research problems, 604
Oral cavity, 25
Ordinal numbers, 717
Orthogonality principle, 291-92
Orthogonal processes, 263
Orthogonal variables, 83
Outer ear, 29
Out-Of-Vocabulary (OOV) word rate, 578
Outside probability, 556
Oval window, ear, 29
Overall quality tests, 840-41
  Absolute Category Rating (ACR), 841
  Listening Effort Scale, 841
  Listening Quality Scale, 841
  Mean Opinion Score (MOS), 840
Overlap-and-add (OLA) technique, 818-19
Overlapped evaluation scheme, 463
Oversampling, 246
Oversubtraction, 519

P

Paired observations test, 114
Palatal consonants, 46
Palate, 46
Paradigmatic properties, 53
PARADISE framework, 903-6
Paragraphs, 702
Paralinguistic, use of term, 764
Parameter space, 98
Parametric Artificial Talker (PAT), 845
Parks McClellan algorithm, 236-38
Parsers, 721
Parse tree representations, 62-63
Parseval's theorem, 216
  for random processes, 268
Parsing algorithm, 545
Partial correlation coefficients (PARCOR), 299
Partition, 74
Part-whole, 66
Passive microphones, 496
Passive sentence, 62
Pattern recognition, 133-97
Pauses, 747-49
Pausing, 740
Penn Treebank project, 55
Perceived loudness, 30
Perceived pitch, 30
Perceptron training algorithm, 159
Perceptual attributes, sounds, 30
Perceptual Audio Coder (PAC), 351, 371
Perceptual linear prediction (PLP), 318-19
Perceptually-based distortion measures, 166
Perceptually motivated representations, 315-19
  bilinear transforms, 315-16
  mel-frequency cepstrum coefficients (MFCC), 316-18
  perceptual linear prediction (PLP), 318-19
Perceptual Speech Quality Measurement (PSQM), 844
Perceptual weighting, 357-58
Periodic lobe, 26
Periodic signals, 203
  cepstrum of, 311-12
Perplexity, 122, 560-62, 579
Personal Digital Assistants (PDAs), 930, 945
Phantom power, 488
Phantom trajectories, 463
Pharyngeal cavity, 25
Pharynx, 288
Phonemes, 20, 24, 36-38, 611
Phoneme trigram rescoring, 730
Phone numbers, 712-14
Phonetically balanced word list test, 839
Phonetic F0 (microprosody), 767-68
Phonetic languages, 692-93
Phonetic modeling, 428-39
  clustered acoustic-phonetic units, 432-36
  comparison of different units, 429-30
  context dependency, 430-31
  lexical baseforms, 436-39
Phonetics, 36-50
  allophones, 47-49
  clauses, 61-62
  coarticulation, 49-51
  consonants, 42-46
  lexical part-of-speech (POS), 53-56
  lexical semantics, 64-66
  logical form, 67-68
  morphology, 56-57
  parse tree representations, 62-63
  phonemes, 36-38
  phonetic typology, 46-47
  phrase schemata, 58-61
  semantic roles, 63-64
  semantics, 58
  sentences, 61-62
  speech rate, 49-51
  syllables, 51-52
  syntactic constituents, 58
  syntax, defined, 58
  vowels, 39-42
  word classes, 57
  words, 53-57
Phonetic typology, 46-47
Phonological phrases, 749

Phonology, 36-50
Phrase schemata, 58-61
Phrase-structure diagram, 63
Physical vs. perceptual attributes of sounds, 30-32
Physiology of the ear, 29-32
Pickup pattern, microphone, 489
Piezoelectric microphones, 497
Piezoresistive microphones, 497
Pink noise, 270, 478
Pitch, 25, 30, 47, 740
  autocorrelation method, 324-27
  normalized cross-correlation method, 327-29
  role of, 324-32
  signal conditioning, 329-30
  tracking, 330-32
Pitch generation, 763-82
  accent termination, 770
  attributes of pitch contours, 764-68
  baseline F0 contour generation, 768-69
  corpus-based F0 generation, 779-82
    F0 contours indexed by parsed text, 779-81
    F0 contours indexed by ToBI, 781-82
    transplanted prosody, 779
  declination, 766-67
  evaluations/improvements, 773-74
  F0 contour interpolation, 772-73
  gradient prominence, 765-66
  interface to synthesis module, 773
  parametric F0 generation, 774-75
  phonetic F0 (microprosody), 767-68
  pitch range, 764-65, 770-71
  prominence determination, 771-72
  superposition models, 775-76
  ToBI realization models, 777-78
  tone determination, 770
Pitch prediction, 356-57
Pitch range, 764-65, 770-71
Pitch-scale modification epoch calculation, 825
Pitch-scale time-scale epoch calculation, 827
Pitch synchronous analysis, 283, 302-3
Pitch synchronous overlap and add (PSOLA), 820-23, 831, 847
  problems with, 829-31
    amplitude mismatch, 830
    buzzy voiced fricatives, 830
    phase mismatches, 829
    pitch mismatches, 830-31
  spectral behavior of, 822-23
Pitch tracking, 330-32
Pitch tracking errors, 828
Plan-based dialog modeling, 914
Plan-based systems, 888-92
Plan libraries, 889
Plosive, 42
Plosive consonant, 42-43
Poisson distributions, 89
Poles, 213
Pole-zero filters, cepstrum of, 308-98
Polymorphic linguistic context assignment, 656-57
Polysemy, 65
Positive correlation, 83
Positive-definite function, 262
POS tagging, 56, 722-23
Posterior probability, 135, 142, 156
Postfiltering, 357-58
Post-lexical rules, 735
Postmodifiers, 58-61
Power function, 114
Power spectral subtraction rule, 519
Power spectrum, 216
Predicate, 61, 67
Predicate logic, 68
Pre-emphasis filtering, 235, 320
Preference tests, 842
Prefix nodes, 658
Prefix trees, 647
Premodifiers, 58-59
Prepositions, 54, 60
Pressure gradient microphones, 496
Pressure microphones, 496
Prime-factor algorithm, 224
Principal-component analysis (PCA), 426
Priority entity memory, 882-83
Prior probability, 133, 135, 140
Probabilistic CFGs (PCFGs), 554
Probabilistic context-free grammars, 554-58
Probability density function (pdf), 78, 261
Probability function (pf), 77
Probability mass function (pmf), 77
Probability theory, 73-131
  Bayes' rule, 75-78
  binomial distributions, 86
  chain rule, 75, 77-78
  conditional probability, 75-76
  correlation, 82-83
  covariance, 82-83
  gamma distributions, 90-91, 95
  Gaussian distributions, 92-98
  geometric distributions, 86-87
  law of large numbers, 82
  marginal probability, 76, 77-78
  mean, 79-81
  multinomial distributions, 87-89
  multivariate distributions, 83-85
  Poisson distributions, 89
  probability density function (pdf), 78
  random variables, 77-79
  random vectors, 83-85
  uniform distributions, 85
  variance, 79-81
Prominence determination, 771-72
Prompting strategy, 943
Pronouns, 54
Pronunciation trees, 648-50

Proper noun, 53-54
Property inheritance, 871
Propositional phrases (PPs), 59-60
Prosodic analysis module, 7
Prosodic modification of speech, 818-31
  epoch detection, 828-29
  evaluation of TTS systems, 834-44
    automated tests, 843-44
    Diagnostic Rhyme Test (DRT), 837
    functional tests, 842-43
    glass-box vs. black-box evaluation, 835-36
    global vs. analytic assessment, 836
    Haskins Syntactic Sentence Test, 839
    human vs. automated, 835
    intelligibility tests, 837-39
    judgment vs. functional testing, 835-36
    laboratory vs. field, 835
    Modified Rhyme Test (MRT), 838
    overall quality tests, 840-41
    phonetically balanced word list test, 839
    preference tests, 842
    Semantically Unpredictable Sentence Test, 837
    symbolic vs. acoustic level, 835
  pitch-scale modification epoch calculation, 825
  pitch-scale time-scale epoch calculation, 827
  pitch synchronous overlap and add (PSOLA), 820-23, 831, 847
    problems with, 829-31
    spectral behavior of, 822-23
  source-filter models for prosody modification, 831-34
    LP-PSOLA, 832-33
    mixed excitation models, 832-34
    prosody modification of the LPC residual, 832
    voice effects, 834
  synchronous overlap and add (SOLA), 818-19
  synthesis epoch calculation, 823-24
  time-scale modification epoch calculation, 826-27
  waveform mapping, 827-28
Prosodic phrases, 749-51
Prosodic transcription systems, 759-61
Prosody, 739-91, 943
  and character, 744
  duration assignment, 761-63
    CART-based durations, 763
    rule-based methods, 762-63
  generation, 721-22
  generation schematic, 743-44
  loudness, 740
  pausing, 740
  pitch, 740
  pitch generation, 763-82
    accent termination, 770
    attributes the pitch contours, 764-68
    baseline F0 contour generation, 768-69
    corpus-based F0 generation, 779-82
    declination, 766-67
    evaluations/improvements, 773-74
    F0 contour interpolation, 772-73
    gradient prominence, 765-66
    interface to synthesis module, 773
    parametric F0 generation, 774-75
    phonetic F0 (microprosody), 767-68
    pitch range, 764-65, 770-71
    prominence determination, 771-72
    superposition models, 775-76
    ToBI realization models, 777-78
    tone determination, 770
  prosody markup languages, 783-85
  rate/relative duration, 740
  role of understanding, 740-44
  speaking style, 744-45
    character, 744
    emotion, 744-45
  symbolic, 745-61
    accent, 751-53
    pauses, 747-49
    prosodic phrases, 749-51
    prosodic transcription systems, 759-61
    tone, 753-57
    tune, 757-59
Prosody markup languages, 783-85
PROSPA, 759
Pruning, 609
Pruning error, 675
PSOLA, *See* Pitch synchronous overlap and add (PSOLA)
Psychoacoustics, 30
Pulse code modulation (PCM), 271, 340-42
Pure tones, 31
Push-down automation, 548
Push-to-talk model, 422-23
P-value, 115-16

Q

Quantization noise, 246
Questioned noun phrase, 61

R

Radix-2 FFT, 222, 223
Randomness, 73
Random noise, 276
Random variables, 77-79
  expectation of, 79
Random vectors, 83-85
Rapidly evolving waveforms (REW), 368-70
Rapid prototyping, 948-49
Rate/relative duration, 740
Read speech acoustic models, 857
Real cepstrum, 307-8
Real-time cepstral normalization, 525
Recognition problem, 554
Rectangular window, 230-31
Recurrent neural networks, 457-58

Recursive least squares (RLS) algorithm, 504
Recursive transition network (RTN), 548, 613-14
Reflection coefficients, 296, 299, 305
Region of convergence (ROC), 211-12
Regular Pulse Excited-Linear Predictive Coder (RPE-LPC), 360
Relative clauses, 61
Relative expressions, 871
Relative frequency, 74
Relative spectral processing (RASTA), 525
Renditions, 899-901
Repetitive stress injury (RSI), 929
Residual signal, 301
Resonances of vocal tract, excitation of, 27
Response generation, 894-901
  message generation box, 897
  natural language generation from abstract semantic input, 898
  response content generation, 895-99
  template systems for, 897
Retraining on compensated features, 537-38
Retraining on corrupted speech, 528-39
Retroflex liquid, 42
Reverberation, 480-82
Ribbon microphones, 497
Rich-Get-Richer (RGR) strategy, 662
Right-sized tree, 184-89
  cross-validation, 188-89
  independent test sample estimation, 187-88
  minimum cost-complexity pruning, 185-87
RLS algorithm, 503-4
Robust parsing, 874-78
Roll-off, 281
Rule-based duration-modeling methods, 56
Rule-based speech synthesis systems, 795-96
  CPU resources, 795
  delay, 795
  memory resources, 795
  pitch control, 795
  variable speed, 795
  voice characteristics, 796

S

Sadness, and speech, 745
Sample mean, 82
Sample variance, 82
Sampling theorem, 243-45
SAM system, 913-14
Scalable coders, 371
Scalar frequency domain coders, 348-52
  consumer audio, 351-52
  Digital Audio Broadcasting (DAB), 352
  frequency domain, advantages of, 348-49
  masking, 349-50
  transform coders, 350-51
Scalar waveform coders, 340-48
  adaptive PCM, 344-45
  differential quantization, 345-48
  linear pulse code modulation (PCM), 340-42
  μ-law and A-law PCM, 342-44, 348
Screen reader, 929
Search, defined, 592
Search-algorithm evaluation, 674-76
Search algorithms:
  beam, 606-8
  best-first, 602-6
  blind graph, 597-601
  breadth-first, 600-601
  depth-first, 598-99
  forward-backward, 670-73
  large vocabulary, 645-85
  speech-recognition, 608-12
    combining acoustic and language models, 610
    continuous speech recognition, 611-12
    decoder basics, 609
    isolated word recognition, 610-11
  tree-trellis forward-backward, 671
Search error, 675
Second-order IIR filters, 241-42
Second-order resonators, 242
Segment models, 459-60
Segment-model weight, 462
Selectivity, of grammar, 546
Semantically Unpredictable Sentence Test, 837
Semantic authoring, 862
Semantic classes, 948-49
Semantic grammars, 585
Semantic language model, 879
Semantic parser, 854-55
Semantic representation, 867-73
  conceptual graphs, 872-73
  functionality encapsulation, 871-72
  property inheritance, 871
  semantic frames, 867-69
  type abstraction, 869-71
Semantic roles, 63-64
Semantics:
  defined, 58
  language, 545
  lexical, 64-66
Semicontinuous HMMs, 396-98
Semi-tones, 32
Semivowels, 42
Senones, 433-36, 467, 809
Senone sequence, 658
Sentence interpolation, 873-80
  robust parsing, 874-78
    defined, 875
  statistical pattern matching, 878-80
  syntactic grammars, 877
Sentence interpretation, 7
Sentence interpretation module, 855
Sentence-level stress, 431

Sentences, 61-62, 702-4
  diagramming in parse trees, 63
Sentence tagging, 722
Sentence type identification, 723
Shades of meaning, 67
Shallow parse, 723
Shannon's channel coding theorem, 127
Shannon's source coding theorem, 124-25, 128
Sharing, 609
Sharing tails, 655-56
Shimmer, 768
Short-term prediction, 353
Short-time Fourier analysis, 276-83
  pitch-synchronous analysis, 283
  spectrograms, 281-83
Sigma-delta A/D, 246
Sigma-delta modulation, 346
Signal acquisition, 422
Signal conditioning, 329-30
Signal processing module, 421-28
  end-point detection, 422-24
  feature transformation, 426-28
  mel-frequency cepstral coefficients (MFCC), 424-26
  signal acquisition, 422
Signals, 201
Signal-to-noise ratio (SNR), 339, 486, 489
Significance testing, 98, 113-20
  goodness-of-fit test, 116-18
  level of significance, 114-15
  magnitude-difference test, 119-20
  matched pairs test, 118-20
  normal test, 115-16
  sign test, 119
  Z test, 115-16
Sign test, 119
Silences between words, handling, 621-22
Similars, 65-66
Simple questions, 177
Sinc function, 229-30
Single-layer perceptrons, 159-60
Singleton questions, 177
Single-word subpath, 655
Sinusoidal coding, 371
Sinusoidal systems, 203-5
Slope overload distortion, 346
Slot inheritance, 885
Slowly evolving waveform (SEW), 367-70
SLU, *See* Spoken language understanding (SLU) systems
Smart phones, 930
SNR, *See* Signal-to-noise ratio (SNR)
Soft palate, 25
Sound, 21-23
Sound Blaster, 936
Sound pressure level (SPL), 23
Source coding theorem, 124-26
Source-filter models for prosody modification, 831-34
Source-filter models of speech production, 288-90
Source-filter separation, via the cepstrum, 314-15
Speak & Spell, 271
Speaker-adaptive training techniques, 419
Speaker-dependent speech recognition, 418-19
Speaker-independent speech recognition, 418
Speaker recognition, 931
Speaker variability, 418-19
Speaking style, 744-45
  character, 744
  emotion, 744-45
Speaking turn, 861
Specificity ordering conflict resolution strategy, 182
Specifier position, 61
Spectral analysis via linear predictive coding (LPC), 300-301
Spectral leakage, 279
Spectral subtraction, 516-19
Spectrograms, 27-28, 276, 281-83
Speech:
  defined, 283
  interfacing with computers, 1
  prosodic modification of, 818-31
  supplemented by information streams, 2
  using as an add-on feature, 941
Speech acts, 705-6
Speech-act theory, 914
Speech coding, 337-74
  code excited linear prediction (CELP), 353-61
    adaptive codebook, 356-57
    analysis by synthesis, 353-56
    LPC vocoder, 353
    parameter quantization, 358-59
    perceptual weighting/postfiltering, 357-58
    pitch prediction, 356-57
    standards, 359-61
  coder delay, 339
  low-bit rate speech coders, 361-70
    harmonic coding, 363-67
    mixed-excitation LPC vocoder, 362
    waveform interpolation, 367-70
  scalar frequency domain coders, 348-52
    consumer audio, 351-52
    Digital Audio Broadcasting (DAB), 352
    masking, 349-50
    transform coders, 350-51
  scalar waveform coders, 340-48
    adaptive PCM, 344-45
    differential quantization, 345-48
    linear pulse code modulation (PCM), 340-42
    μ-law and A-law PCM, 342-44, 348
  speech coder attributes, 338-39
Speech communication, history of, 1
Speech end-point detector, 423
Speech interaction, modes of, 933-34

Speech interface design, 931-43
  general principles of, 931-37
    human limitations, 932
    modes of interaction, 933-34
    technological considerations, 935-36
    user accommodation, 933
  handling errors, 937-41
    error detection and correction, 938-39
    feedback and confirmation, 939-41
  internationalization, 943-45
Speech inversion problem, 803-4
Speech perception, 29-36
Speech processing:
  digital signal processing, 201-73
  speech coding, 337-74
  speech signal representations, 275-336
Speech production, 24-28
  acoustical model of, 283-90
  articulators, 24-25
  formants, 27-28
  frequency analysis, 32-34
  masking, 34-36
  physical vs. perceptual attributes of sounds, 30-32
  physiology of the ear, 29-32
  spectrograms, 27-28
  speech perception, 29-36
  voicing mechanism, 25-27
  See also Acoustical model of speech production
Speech production process, start of, 19
Speech rate, 49-51
Speech recognition, 2, 3, 375-685, 862
  acoustic modeling, 415-75
    context variability, 417
    environment variability, 419
    scoring acoustic features, 439-43
    speaker variability, 418-19
    speech recognition errors, 419-21
    style variability, 418
    variability in speech signals, 416-19
  hidden Markov models (HMM), 377-413, 416
    Baum-Welch algorithm, 389-93
    continuous mixture density, 394-96
    decoding, 387-89
    definition of, 380-93
    deleted interpolation, 401-3
    dynamic programming, 384-85
    dynamic time warping (DTW), 383-85
    estimating parameters, 389-93
    evaluating, 385-87
    forward algorithm, 385-87
    forward-backward algorithm, 389-93
    initial estimates, 398-99
    limitations of, 405-9
    Markov chain, 378-80
    model topology, 399-401
    observable Markov model, 379-80
    parameter smoothing, 403-4
    practical issues concerning, 398-405
    probability representations, 404-5
    semicontinuous, 396-98
    training criteria, 401
    Viterbi algorithm, 387-89
  phonetic modeling, 428-39
    clustered acoustic-phonetic units, 432-36
    comparison of different units, 429-30
    context dependency, 430-31
    lexical baseforms, 436-39
  signal processing module, 421-28
    end-point detection, 422-24
    feature transformation, 426-28
    mel-frequency cepstral coefficients (MFCC), 424-26
    signal acquisition, 422
  speech recognition errors, 419-21
    word error rate, 420
    word recognition errors, types of, 420
Speech recognition search algorithms, 608-12
  combining acoustic and language models, 610
  continuous speech recognition, 611-12
  decoder basics, 609
  isolated word recognition, 610-11
Speech recognition system, 4-5
  basic system architecture of, 5
  components of, 4
  source-channel model for, 5
  vocabulary, 58
Speech signal representations, 275-336
  acoustical model of speech production, 283-90
    glottal excitation, 284
    lossless tube concatenation, 284-88
    mixed excitation model, 289
    source-filter models of speech production, 288-90
  cepstrum, 306-15
    cepstrum vector, 309
    complex, 307-8
    LPC-cepstrum, 309-11
    of periodic signals, 311-12
    of pole-zero filters, 308-98
    real, 307-8
    source-filter separation via, 314-15
    of speech signals, 312-13
  formant frequencies, 319-23
    statistical formant tracking, 320-23
  linear predictive coding (LPC), 290-306
    autocorrelation method, 295-96
    covariance method, 293-94
    equivalent representations, 303-6
    lattice formulation, 297-300
    line spectral frequencies (LSF), 303-5
    log-area ratios, 305-6
    orthogonality principle, 291-92
    prediction error, 301-3
    reflection coefficients, 305
    roots of the polynomial, 306

Speech signal representations, linear predictive coding *(cont.)*
solution of the LPC equations, 292-300
spectral analysis via, 300-301
perceptually motivated representations, 315-19
bilinear transforms, 315-16
mel-frequency cepstrum coefficients (MFCC), 316-18
perceptual linear prediction (PLP), 318-19
pitch:
autocorrelation method, 324-27
normalized cross-correlation method, 327-29
pitch tracking, 330-32
role of, 324-32
signal conditioning, 329-30
short-time Fourier analysis, 276-83
pitch-synchronous analysis, 283
spectrograms, 281-83
Speech signals, 5, 20
cepstrum of, 312-13
context variability, 417
environment variability, 419
speaker variability, 418-19
style variability, 418
variability in, 416-19
Speech synthesis, 6, 793-852
articulatory speech synthesis, 793, 803-4
attributes of, 794-96
concatenative synthesis with no waveform modification, 794
concatenative synthesis with waveform modification, 795
limited-domain waveform concatenation, 794
rule-based systems, 795-96
concatenative speech synthesis, 793-94
choice of unit, 805-8
context-dependent phonemes, 808-9
context-independent phonemes, 806-7
diphones, 807-8
optimal unit string, 810-17
subphonetic units (senones), 809
syllables, 809
unit inventory design, 817-18
word and phrase, 809
data-driven synthesis, 794, 803
formant speech synthesis, 793, 796-804
cascade model, 797
formant generation by rule, 800-803
Klatt's cascade/parallel formant synthesizer, 797-802
locus theory of speech production, 800
parallel model, 797
waveform generation from formant values, 797-99
prosodic modification of speech, 818-31
epoch detection, 828-29
pitch-scale modification epoch calculation, 825
pitch-scale time-scale epoch calculation, 827
pitch synchronous overlap and add (PSOLA), 820-23, 847
synchronous overlap and add (SOLA), 818-819
synthesis epoch calculation, 823-24
time-scale modification epoch calculation, 826-27
waveform mapping, 827-28
synthesis by rule, 794
Speech-to-speech translation, 3
Split-radix algorithm, 223
Splits, 182
Spoken language, 19
Spoken language interface, 2-3
Spoken language processing, 4, 133
Spoken language structure, 19-72
Spoken language system, 2
Spoken language system architecture, 4-8
automatic speech recognition, 4-6
spoken language understanding, 7-8
text-to-speech conversion, 6-7
Spoken language understanding, 7-8
basic system architecture of, 8
Spoken language understanding (SLU) systems, 853-918, 945
assumptions, 854
content, 854
context, 854
dialog management, 886-94
dialog grammars, 887-88
plan-based systems, 888-92
dialog structure, 859-67
attentional state, 859
dialog (speech) acts, 861-66
intentional state, 859
linguistic forms, 859
task knowledge, 859
units of dialog, 860-61
world knowledge, 859
dialog system, 854-55
dialog manager, 855
discourse analysis, 855
semantic parser, 854-55
discourse analysis, 881-86
resolution by NLP, 883-85
resolution of relative expression, 882-85
Dr. Who case study, 906-13
evaluation, 901-6
in the ATIS task, 901-3
PARADISE framework, 903-6
historical perspective, 913-14
intent, 854
rendition, 899-901
response generation, 894-901
concept-to-speech rendition, 899-901
natural language generation from abstract semantic input, 898
response content generation, 895-99
semantic representation, 867-73

conceptual graphs, 872-73
functionality encapsulation, 871-72
property inheritance, 871
semantic frames, 867-69
type abstraction, 869-71
sentence interpolation, 873-80
robust parsing, 874-78
statistical pattern matching, 878-80
syntactic grammars, surface linguistic variations in, 877
written vs. spoken languages, 855-58
communicative prosody, 858
disfluency, 857
style, 856-57
Spoken menus, 942
Spread-of-masking function, 35-36
Spread spectrum, 360-61
Stable LTI system, 211
Stack decoding, 592
advantage of, 628
defined, 627
formulating in a tree search framework, 629
Stack decoding (A* search), 626-39
admissible heuristics for remaining path, 630-31
extending new words, 631-34
fast match, 634-38
multistack search, 639
stack pruning, 638-39
Stack pruning, 638-39
Standard deviation, 80
Standard Gaussian distributions, 92-93
State-space search paradigm, 592
Stationary processes, 264-67
ergodic processes, 265-67
Stationary signal, 276
Statistical formant tracking, 320-23
Statistical inference, 98, 113
Statistical language models, 583
Statistical pattern matching, 878-80
Statistical pattern recognition, 190
Statistics, 73-131
Stochastic language models, 546, 554-60
n-gram language models, 558-60
probabilistic context-free grammars, 554-58
Stochastic processes, 260-70
continuous-time, 260
discrete-time, 260
LTI systems with stochastic inputs, 267
noise, 269-70
power spectral density, 268-69
stationary processes, 264-67
statistics of, 261-64
Stop, 43
Stress, 751
Stressed vowels, 430-31
Strict-sense stationary (SSS), 264
Style, 856-57
Style variability, 418
Subgoals, 894
Sub-harmonic errors, 330
Subject, sentence, 61
Subphonetic units (senones), 809
Subscripting, 67
Substitution errors, 420
Subtree dominance, 656
Subtree isomorphism, 654
Subtree polymorphism, exploiting, 656-58
Successor operator, 594
Sum-of-squared-error (SSE), 99, 160
Superposition models, 775-76
Supervised learning, 134, 141
Surrogate questions, 182
SWITCHBOARD Shallow-Discourse-Function Annotation SWBD-DAMSL, 865-66
Syllable parse tree, 52
Syllables, 20, 51-52, 430, 809
Syllables centers, 52
Symbolic prosody, 745-61
accent, 751-53
pauses, 747-49
prosodic phrases, 749-51
prosodic transcription systems, 759-61
tone, 753-57
tune, 757-59
Symmetrical loss function, 136
Symmetric channel, 127
Synchronous overlap and add (SOLA), 818-19
Syntactic constituents, 58
Syntactic theory, 69
Syntagmatic properties, 53
Syntax:
defined, 58
language, 545
Synthesis-by-rule, 794, 796
Synthesis epoch calculation, 823-24
System initiative, 860

T

Tag question, 62
Tags, 7
Tail area, 115
Tap and Talk interface, 934, 947-48, 951
Task knowledge, 859
TDMA Interim Standard 54, 360
Telecommunication Industry Association (TIA), 360
Telephone speech, 338
Telephony applications, 924-26
Temporal masking, 35-36
Testing set, 141
Test procedure, 114
Text analysis phase, 7
Text normalization (TN), 692, 706-20
abbreviations, 709-12
acronyms, 711-12

Text normalization (TN) *(cont.)*
  domain-specific tags, 718-20
    chemical formulae, 719
    mathematical expressions, 718-19
    miscellaneous formats, 719-20
  evaluation, 730-32
  Festival case study, 732-35
  historical perspective, 735-36
  letter-to-sound (LTS) conversion, 728-30
  linguistic analysis, 720-23
    and closed-class function words, 722
    homograph disambiguation, 723, 724-25
    noun phrase (NP) and clause detection, 723
    POS tagging, 722-23
    sentence tagging, 722
    sentence type identification, 723
    shallow parse, 723
  morphological analysis, 725-27
    algorithm, 727
    suffix and prefix stripping, 726-27
  number formats, 712-20
    account numbers, 716
    cardinal numbers, 717-18
    dates, 714-15
    money and currency, 716
    ordinal numbers, 717
    phone numbers, 712-14
    times, 715
Text and phonetic analysis, 689-738
  American-English vocabulary relevant to, 698
  data flow, 694-97
    skeleton, 695
  defined, 692
  document structure detection, 692, 699-706
  grapheme-to-phoneme conversion, 692-93
  homograph disambiguation, 693
  letter-to-sound (LTS) conversion, 693
  lexicon, 697-98
  linguistic analysis, 692
  localization issues, 696-97
  modules, 692-94
  morphological analysis, 693
  natural language process (NLP) systems, 693-94
  phonetic languages, 692-93
  text normalization (TN), 692, 706-20
Text-to-speech (TTS) conversion, 6-7
Text-to-speech (TTS) system, 687-850
  basic system architecture of, 6
  goals of, 689-90
  phonetic analysis component, 7
  prosody, 739-91
  speech synthesis, 793-850
  speech synthesis component, 7
  tags, 7
  text analysis component, 6-7
  text and phonetic analysis, 689-738
Text-to-speech (TTS) system evaluation, 834-44
  automated tests, 843-44
  Diagnostic Rhyme Test (DRT), 837
  functional tests, 842-43
  glass-box vs. black-box evaluation, 835-36
  global vs. analytic assessment, 836
  Haskins Syntactic Sentence Test, 839
  historical perspective, 844-47
  human vs. automated, 835
  intelligibility tests, 837-39
  judgment vs. functional testing, 835-36
  laboratory vs. field, 835
  Modified Rhyme Test (MRT), 838
  overall quality tests, 840-41
    Absolute Category Rating (ACR), 841
    Listening Effort Scale, 841
    Listening Quality Scale, 841
    Mean Opinion Score (MOS), 840
  phonetically balanced word list test, 839
  preference tests, 842
  Semantically Unpredictable Sentence Test, 837
  symbolic vs. acoustic level, 835
TFIDF information retrieval measure, 576
Third generation (3G) systems, 361
Threshold of hearing (TOH), 22
Threshold value, likelihood ratio, 139
Throat, 25
TIA/EIA/IS54, 360
TIA/EIA/IS-127-2, 361
TIA/EIA/IS-733-1, 361
TILT, 760
Timbre, 25, 32
Time delay neural network (TDNN), 458
Time Division Multiple Access (TDMA), 360
Time-scale modification epoch calculation, 826-27
Time-synchronous Viterbi beam search, 622-26
  algorithm, 627
  use of beam, 624-25
Time-synchronous Viterbi search, 666-67
TN, *See* Text normalization (TN)
ToBI realization models, 777-78
ToBI (Tones and Break Indices) system, 749-50, 754, 777
  boundary tolerance, 756
  intermediate phrasal tones, 756
  pitch accent tones, 755
Toeplitz matrix, 296
Toll quality, 344
Tone, 753-57
Tone determination, 770
Tone-masking noise, 35
Tongue, 25
Top-down chart parsing, 549-51
  top-down vs., 549-50
Topic-adaptive models, 575-76
Trachea, 25
Trainability, 145

Training corpus, 559
Training problem, 554
Training set, 141, 419-20
Transducers, acoustical, 486-97
Transfer function, 210
Transformation models, 454-55
Transform coders, 350-51, 371
Transform-domain LMS algorithm, 502-3
Transition network, 548
Transmission delay, 339
Transparent quality, 358
Tree-banked data, 878
Tree lexicon, efficient manipulation of, 646-59
Tree structure, 646
Tree-trellis forward-backward search algorithms, 671
Triangular windows, 280
Trigram grammar, 465
Trigrams, 559, 583
  search space with, 619-20
Trilled *r* sound, 47
Triphone model, 430
Triphones, 808
TTS models, 949
TTS system, *See* Text-to-speech (TTS) system
Tune, 757-59
Turing machine, 548
Turing test, 3, 843
Turn memories, 882
Twiddle factors, 224
Two-band conjugate quadrature filters, 251-54
Twoing rule, 181
Two-place predicates, 67
Two-tailed test, 115-16
Type abstraction, 869-71
Type-I filter, 233
Type-n-Talk system, 847

U

U method, 147
Uncertainty, 73, 121
Uncorrelated orthogonal processes, 263
Undergeneration, 876
Understandability, of grammar, 546
Unicode, 36
Unidirectional microphones, 494-96
Unification grammar, 584
Unified frame- and segment-based models, 462-64
Uniform distributions, 85
Uniform prior, 112
Uniform quantization, 340
Uniform search, 597
Unigram, 559
  search space with, 616-17
Unimodal distribution, 95
Uniquely decipherable coding, 125
United States Public Switched Telephone Network (PSTN), 371
Units, choice of, in concatenative speech synthesis, 805-8
Units of dialog, 860-61
Universal encoding scheme, 126
Universal Mobile Telecommunications System (UMTS), 361
Unknown word, defined, 563
Unstressed vowels, 430
Unsupervised estimation methods, 163-75
  EM algorithm, 170-72
  multivariate Gaussian mixture density estimation, 172-75
  vector quantization (VQ), 164-70
Unsupervised learning, 141
Upper bound of probability, 74
U.S. Defense Advanced Research Projects Agency (DARPA), 467
User expectations, managing, 942
User initiative, 860, 867
Utterance unit, 861

V

Variance, 79-81
Vector quantization (VQ), 164-70, 191
  distortion measures, 164-66
  EM algorithm, 170-72
  K-means algorithm, 166-69
  LBG algorithm, 169-70
Vector Taylor series, 535-37
Velar consonants, 46
Velum, 25
Verbs, 54
Verbs phrases (VPs), 59-61
V-fold cross-delegation, 188-89
V-fold cross validation, 147
ViaVoice (IBM), 926
Viterbi algorithm, 387-89, 409, 609
Viterbi approximation, 623
Viterbi beam search, 625-26
Viterbi decoder, 592
Viterbi forced alignment, 630
Viterbi stack decoder, 592
Viterbi trellis, 624
Vocabulary independence, 433
Vocabulary selection, 578-80
Vocal cords, 25
Vocal fold cycling at the larnyx, 26
Vocal fry, 330
Vocal tract normalization (VTN), 427
Voder, 6
Voice conversion, 834
Voice effects, 834
Voice FONCARD (Sprint), 931
Voiceless plosive consonants, 43
Voice over Internet protocol (Voice over IP), 359
Voice portals, 925
VoiceXML, 921

Voice Xpress, 926
Voicing mechanism, 25-27
Vowels, 24, 39-42
  Japanese, 46-47
VQ, *See* Vector quantization (VQ)

W

Wall Street Journal (WSJ) Dictation Task, 11-12
Waveform-approximating coders, 361
Waveform interpolation, 367-70, 371
Waveform mapping, 827-28
Waveforms, fundamental frequency, 27
Web pages, 705
Whisper case study, 464-65
Whispering effect, 834
White noise, 269-70, 478-79
Whole-word models, difficulty in building, 428
Wh-question, 62
Wide-band spectrograms, 282
Wideband speech, 338
Wide-sense stationary (WSS), 265
Wiener filtering, 520-22, 540
  noncausal, 522
Wiener-Hopf equation, 521
Wiener-Khinchin theorem, 269
Window design filter, 235-36
Window design FIR lowpass filters, 235-36
Window function, 255, 277-78
Window functions, 230-32
  generalized Hamming window, 231-32
  rectangular window, 230-31
Wizard-of-Oz (WOZ) experimentation, 950
Word classes, 57
Word-dependent *n*-best and word-lattice algorithm, 667-70
Word error rate, 420-21
  algorithm to measure, 421
Word error rate comparisons, humans vs. machines, 12
Word-final unit, 659
Word graphs, 664-66
Word-initial unit, 658
Word-lattice algorithm:
  one-pass *n*-best and, 669-70
  word-dependent *n*-best and, 667-70
Word-lattice generation, 672-73
Word lattices, 664-66
Word-level stress, 431
Word recognition errors, types of, 420
Words, 20, 53-57
  natural affinities/disaffinities, 65
Word-spotting applications, 454
World knowledge, 859
Written vs. spoken languages, 855-58
  disfluency, 857
  style, 856-57

X

χ^2 distributions, 95-96
X-bar theory, 885
XML, 699-700
X-template, 58

Y

Yes-no question, 62
Yule-Walker equations, 291-92, 299

Z

Zero-mean process, 262
Zero-one loss function, 136
Zero padding, 227, 280
Zeros, 213
Z test, 115-16
Z-transforms, 211-12
  of elementary functions, 212-15
    inverse *z*-transform of rational functions, 213-15
    left-sided complex exponentials, 213
    right-sided complex exponentials, 212-13
  properties of, 215-17
    convolution property, 215
    power spectrum and Parseval's theorem, 216

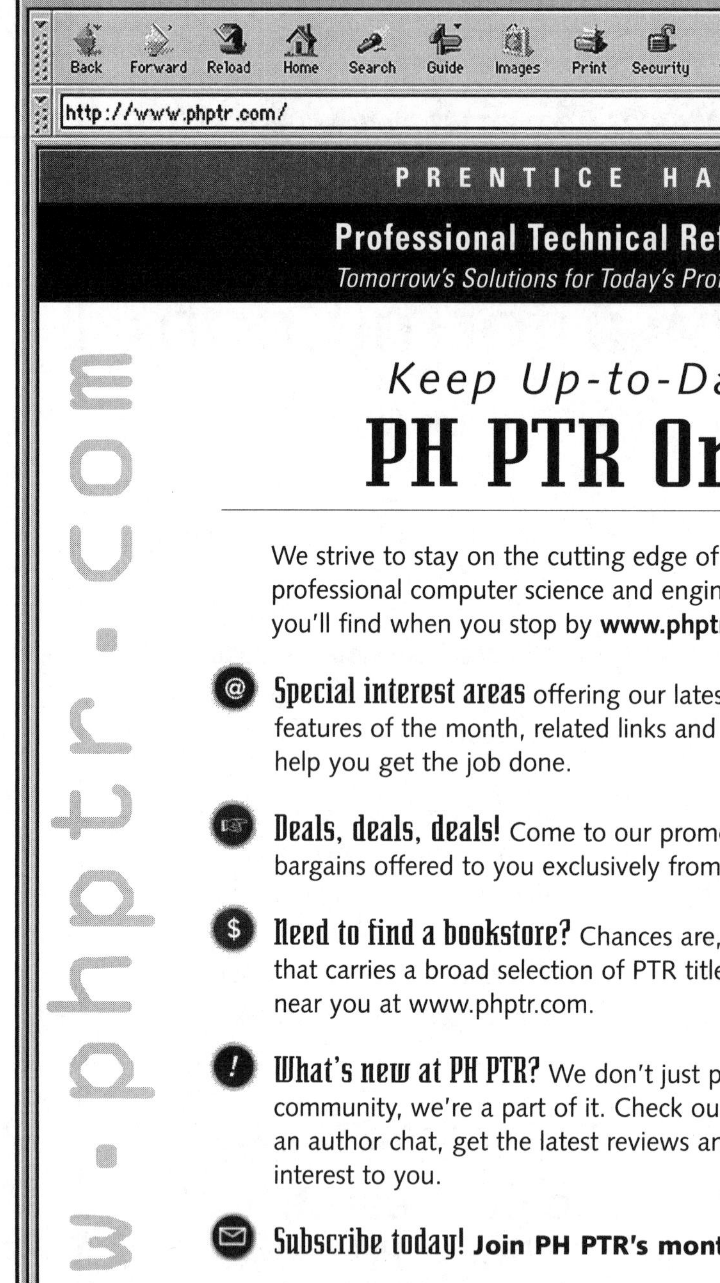

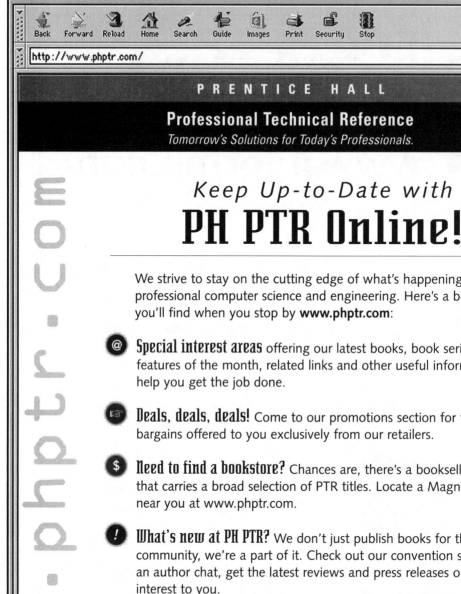